1989
2nd EUROPEAN CONFERENCE ON ARCHITECTURE

Commission of the European Communities

1989 2nd EUROPEAN CONFERENCE ON ARCHITECTURE

Science and Technology at the Service of Architecture

Proceedings of an International Conference held at Paris, France, 4-8 December 1989

Edited by: Theo C. Steemers and Wolfgang Palz,
Commission of the European Communities

Kluwer Academic Publishers

ISBN 0–7923–0732–1

Publication arrangements by
Commission of the European Communities
Directorate-General Telecommunications, Information Industries and Innovation,
Scientific and Technical Communications Unit, Luxembourg

EUR 12730

Published by Kluwer Academic Publishers,
P.O. Box 17, 3300 AA Dordrecht, The Netherlands.

Kluwer Academic Publishers incorporates the publishing programmes of
D. Reidel, Martinus Nijhoff, Dr W. Junk and MTP Press.

Sold and distributed in the U.S.A. and Canada
by Kluwer Academic Publishers,
101 Philip Drive, Norwell, MA 02061, U.S.A.

In all other countries, sold and distributed
by Kluwer Academic Publishers Group,
P.O. Box 322, 3300 AH Dordrecht, The Netherlands.

Printed on acid-free paper

Printed in the Netherlands

CONTENTS

Alphabetical list of authors
at the end of the book

The papers which were not submitted for
publication are not listed hereunder

ARCHITECTURE, ENERGY + EDUCATION

PREFACE

After the first successful conference in Munich in 1987, the Commission of the European Communities has been organising its 2nd Conference on Architecture in Paris. It is part of the Community's development programme on solar energy.

The use of solar energy in dwellings and buildings in general was actually the starting point of this conference and it still is a major part of it. But the development of solar energy has not only led to a better energy utilisation and a new type of comfort in our buildings; it just showed an example of the immense technological potential which still remains unexploited in modern architecture.

The approach needed is multi-disciplinary: these conferences are meant as the forum where architects, scientists, engineers and industrialists get together. I express my hope that they can lead the way to an architecture stimulated by technology and a new start for technology developments at the service of that architecture.

Wolfgang Palz
Commission of the European Communities
Brussels

Suite à la première conférence réussie de Munich en 1987, la Commission des Communautés Européennes a organisé sa 2e Conférence de l'Architecture à Paris. Elle fait partie du programme communautaire de développement de l'énergie solaire.

L'utilisation de l'énergie solaire dans l'habitat et la construction en général est effectivement à l'origine de la conférence et en reste un élément majeur. Mais le développement du solaire n'a pas seulement conduit à une meilleure utilisation de l'énergie et à un nouveau type de confort dans nos bâtiments; il a donné l'exemple de l'immense potentiel technologique qu'il reste à mobiliser dans le développement architectural d'aujourd'hui.

L'approche nécessaire est multidisciplinaire: ces conférences se veulent forum d'échanges entre architectes, scientifiques, ingénieurs et industriels. Mon voeu est qu'elles soient aussi le révélateur d'une architecture enrichie par la technique et un nouveau départ pour les développements technologiques appropriés au service de cette architecture.

Wolfgang Palz
Commission des Communautés Européennes
Bruxelles

The auditorium of the UNESCO in Paris where the 2nd European Conference on Architecture was held. 463 participants from 26 countries were registered for this event.

Introduction

The R&D-programme on building energy of the Commission of the European Communities, which was part of the Solar Energy R&D-programme and the R&D-programme on the Rational Use of Energy, has seen the last decade a gradual but persistent shift in emphasis away from active modular energy system development. Typical for these modules were that they could be added or taken out of a building without consequences for the design of the building and in particular without drawbacks for the building envelope.
The reason for abandoning active modular energy system development was twofold:

-these systems and in particular active solar space heating systems where in general not cost-effective and became less so with decreasing oil prices in the eighties, and

-a growing interest in bioclimatic architecture and at the same time a successful introduction of cost-effective passive solar technology offered an excellent alternative for primary energy conservation in buildings.

It should, however, be said that there are signs that the decline of interest in active solar heating and cooling systems is, as could be expected, only temporarily.

The renewed interest in bioclimatic architecture marked the start of the end of an era where energy considerations in building construction only determined the size of the cooling and heating plant. The environmental impact of this type of squandermania was also gradually realized.

A similar development can be observed for lighting in buildings. At a time when artificial lighting was extremely inadequate, it was particular important to get daylight design right. The introduction of (cost-)effective artificial lighting equipment and the availability of cheap energy tended to overlay the earlier European tradition of daylight design.

This all led to the development of styles of architecture that were neither environmentally conscious nor energy conscious. And one must admit that in the architectural profession at large the energy and environmental issues still don't get the priority they deserve. Spinosa already said: "We are part of universal nature, thus we obey its laws". This is exactly the quintessence of bio- climatic architecture: design with the climate (and topography), and not against it. If we accept this principle than the design of an office for Copenhagen will in the future look differently from that of one designed for Napels. One often refers with a certain melancholy and admiration to vernacular architecture, but one forgets to draw the conclusions from this type of architecture, which contains very often a wealth of concentrated experiences.

One should avoid giving the impression that the architectural profession as a whole is to be blamed for the waste of energy in the building sector. A critical examination of the requirements for the introduction of a new technology in industry makes clear that the state of the art of bioclimatic architecture in general and passive solar technology in particular does not satisfy yet all the necessary demands for a smooth adoption in the building sector.
The three essential requirements concern:

1) the development of the technology,
2) the technology transfer,
3) incentives for the application of the technology.

The degree of **development of the technology** is in general satisfactory for passive solar heating, less so for daylighting and little progress has been made in passive cooling.
Software (design tools) is emerging in a user-friendly form mainly as relatively simplified models only. The more sofisticated models developed sofar require still quite some expertise to handle. In general, building performance modelling is still addressed in a rather sectorial manner, i.e. dealing with e.g. thermal, visual and comfort performance of a building separately. Also passive solar urban planning has been addressed as a separate issue. A first effort towards the development of a European Integrated Building Design System is being made in project COMBINE.

R&D on the performance assessment of passive solar building components is likely to result in common European test procedures by 1992. The main thrust will come from project PASSYS. The development of innovative passive solar building components, including daylighting components and control systems is receiving currently the attention it deserves.

However, in spite of all this progress a designer has to be very determined in order to gain access to the know-how and tools which would allow him to incorporate passive solar features in his building design. It is in general essential for the introduction of this technology that all actors in the building sector have easy access to passive solar know-how. The importance of **technology transfer** has, therefore, to be emphasized, nationally as well as internationally. The Commission is playing a modest, but exemplary, role in this field with the project SOLINFO.
The forthcoming publication of the " European Passive Solar Handbook ", and the " European Daylighting Reference Book ", will certainly be of great help in this process of technology transfer.

The **incentives for the application bio-climatic architecture** should come from:

* more stringent building regulations with respect to energy consumption in buildings, and
* the market pull.

The advocates of bio-climatic architecture are convinced that the market pull will make headway because of the greater visual and thermal comfort of bio-climatic buildings together with their energy economy as soon as sufficient examples (demonstration projects) are in easy reach of the European citizen.

Theo C. Steemers,

Commission of the
European Communities.

SPONSORED BY

Commission of the European Communities

CO-SPONSORED BY

Ministère de l'Equipement, du Logement, des Transports et de la Mer, Direction Architecture et Art urbain, France

Federal Ministry for Research and Technology, FRG

Eurosolar, FRG

Liaison Committee of the Architects of United Europe, Belgium

Union Internationale des Architectes, France

European Association for Architectural Education, Belgium

The Royal Institute of the Architects of Ireland

ORGANIZED BY WIP MUNICH

SECA CONFERENCE COMMITTEE

General Chairman

Architect Dipl. Ing. Alex Lohr, M.S. Arch.
Büro für energiegerechtes Bauen, Cologne, FRG

Conference Director

Dr. Wolfgang Palz, CEC, Brussels, Belgium

Conference Secretary

Theo C. Steemers, CEC, Brussels, Belgium

Local Coordinator

Michel Retbi, Ministère de l'Equipement, du Logement, des Transports et de la Mer, Direction Architecture et Art urbain, France

International Organization, Administration, Exhibitions

Peter Helm, WIP, Munich, FRG
Huguette Riegel, WIP, Munich, FRG

Architectural Excursions

Alex Lohr, Cologne, FRG

Publications

Theo C. Steemers, CEC, Brussels, Belgium
Alex Lohr, Cologne, FRG

Posters

Gabriele Willbold-Lohr, Cologne, FRG
Catherine Sako Atayi, Cologne, FRG

Public Relations

J. Owen Lewis, Univ. College Dublin, Ireland
Shane O'Toole, Univ. College Dublin, Ireland

SECA STEERING COMMITTEE

Mr. Della Porta	AFME, Valbonne, France
Mr. J.P. Marie	Plan Construction, Paris, France
Dr. W. Palz	Commission of the European Communities
Mr. M. Retbi	Ministère de l'Equipement, MELTM, Paris, France
Mr. T.C. Steemers	Commission of the European Communities
Mrs. M. Tilmont	Plan Construction, Paris, France

SCIENTIFIC COMMITTEE

Comfort

Prof. P.O. Fanger	Technical Univ., Lyngby, Denmark
E. Mayer	Fraunhofer Inst. für Bauphysik, Holzkirchen, FRG

Daylighting

Prof. J. Krochmann	PRC Krochmann GmbH, Berlin, FRG
Prof. D. O'Connell	Univ. College Dublin, Ireland
Prof. A. Fanchiotti	I.C.I.E., Rome, Italy

Acoustics

M. Val	Paris, France
Prof. F. Rostock	Cologne, FRG

Passive Solar Systems

Prof. A. de Herde	Univ. Catholique de Louvain, Belgium
Dr. M. Antinucci	Conphoebus, Catania, Italy
M. Gerber	Treilles, France
Prof. W.L. Dutré	Katholieke Univ. Leuven, Belgium
A. Rabl	Ecole Nat. Supérieure des Mines de Paris, France
Prof. E. Maldonado	Univ. of Porto, Portugal

Thermal Building Performance

G. Olive	BEGO, Paris, France
Dr. M. Santamouris	Protechna Ltd., Athens, Greece
Dr. N. Baker	The Martin Centre for Architectural and Urban Studies, Cambridge, UK

Urban Planning

Prof. A. Dupagne	Univ. de Liège, Belgium
K. Duyvestein	Delft Univ. of Technology, The Netherlands
Prof. A. Enguita	Univ. of Madrid, Spain

Realisations of Buildings and Urban Plans

M. Retbi	Ministère de l'Equipement, MELTM, Paris, France
A.N. Tombazis	Athens, Greece
Prof. D. O'Connell	Univ. College Dublin, Ireland

WELCOME ADDRESS TO SECA

Theo C. Steemers
Secretary of the Conference
Commission of the European Communities, DG XII
Rue de la Loi, 200
B-1049 BRUSSELS, Belgium

On behalf of the Commission of the European Communities I would like to welcome you here in Paris to the Second European Conference on Architecture.

It has gradually become an important element in CEC-Work to organise conferences in all the fields covered by the Commission's research, development and demonstration programmes. The separate programmes cover such diverse fields as energy, information technology and biology. Furthermore these areas, combined in the so-called framework programme, are increasing both in scope and funding. For example for the period 1990 to 1994 the budget for the Commission's framework programme will, subject to approval of the European Council of Ministers, amount to 8.8 billion ECU, which is about 62 billion French francs. It is difficult to assess how much exactly goes to building related science and technology, but it is no doubt a modest amount compared to the size and importance of the building industry.

An enquiry amongst contractors in the Commission's R.D. and D programmes revealed, by the way, the great importance given to the transnationality of the collaboration in the CEC-programmes.

The underlying theme of this conference is energy conservation in buildings, which serves not only those aspects of European policy, which strive for a greater European independence of imported oil, but more importantly, those which aim to reduce the burden of pollution on our environment.

European architecture has a tradition of energy consciousness, but this requires constant updating to meet changing requirements and living standards. Besides, new technologies should assist the building designer to improve on the current records of energy conscious buildings.

40% of Europe's primary energy is used for heating, cooling, lighting and for appliances in buildings, this makes the potential for energy conservation in buildings impressive. However, the effects on Europe's primary energy consumption of the introduction of passive solar and other energy saving technologies in the building sector are slow mainly due to two factors:
Firstly the replacement rate and the renovation rate of the existing building stock is only 1 to 2% per year and secondly the building industry is a rather conservative and fragmented industry slow in picking up new technologies.
On the other hand the efficiency of investment in energy conservation in buildings is high; it has been stated that it is on average 7-times higher than in nuclear power generation.

Having said this, I would at the same time warn against exaggeration. There are at least three issues, leaving aside aesthetics, which in building design take precedence over the energy issue and which will even be of increasing importance in the future; these are health, comfort and safety.

In practical terms this means that there is for instance a limit to the reduction of the air infiltration rate or the air renewal rate because it can create a moisture problem and the chance of increased exposure to radiation due to Radon decay.

In practical terms it means also that the South facade can not be opened up indiscriminately because of the risk of discomfort due to glare and overheating.

With increasing prosperity people are less prepared to accept uncomfortable living conditions. Let me give some examples:
In 1988 for the first time ever the peak electricity load in Italy occurred in the summer due to the use of air conditioning systems.
The second example is from Greece where in the last three years the installed airconditioning power has increased 6-fold.

Quite a considerable part of the housing stock in Portugal has no heating system; one does not need to be a clairvoyant to see that this situation will not last for long.

The increased comfort requirements can be satisfied using passive solar measures. This conference will show you how far Europe already has progressed towards passive solar architecture. It will show you the state of the art and it will identify the way ahead.

I wish you a fruitful meeting.

WELCOME ADDRESS TO SECA

Dr. Hermann Scheer
Member of the German Parliament
President of EUROSOLAR
Postfach 12 06 18
D-5300 BONN, FRG

Since the Report of the Club of Rome in the early Seventies, the Report "Global 2000" to the American President in 1980, the United Nations Environment Report in 1982 and the Brundtlandt-Report in 1987 we hear the argument - with increasing frequency - that we must not leave behind a destroyed environment for future generations.

The Toronto Conference THE CHANGING ATMOSPHERE, held in June 1988, states in the first sentence of its final report that mankind is currently conducting an experiment on the atmosphere, the potential consequences of which can only be compared to those of a nuclear war.

The greenhouse effect is of concern for all and it is caused by all. While in the industrial countries the forests are dying, in developing countries tropical forests are being cut down to meet their current energy requirements or for export as raw materials while at the same time gaining more land for farming.

Some 1 billion people in the industrialised countries, who make up only 20 percent of the world's population, use 70 percent of the annual primary energy stock of approximately 10 billion tons of coal equivalents, of which about 90 percent are based on fossil raw materials. The industrialised countries are thus the main cause of carbon dioxide being released into the atmosphere. Western Europe produces 15 percent of global CO_2 emissions and the USA, on its own, 25 percent.

An effort unprecedented in human history is already needed. The reason for this is that we are facing an ecological threat of a kind never before known in the history of human civilisation. We will only be able to achieve this effort if we have a clear perception of <u>what</u> political objectives are needed - and how we <u>can</u> meet the new priorities.

We need to introduce more economical forms of energy use. But this approach is not sufficient in itself. This is clearly demonstrated in a study ENERGY FOR A SUSTAINABLE WORLD put out by the World Resources Institute in Washington, in which ambitious energy-saving potentials are projected for the next 30 years. The prediction is that, even if the most feasible energy savings are achieved, by the year 2020 primary energy use will be approximately the same as it is today because of the growth in world population and the inevitable increase in energy consumption in the developing countries to satisfy their basic needs.
Thus, if the systematic implementation of an energy saving scenario on a global scale cannot lower the current level of primary energy consumption, a climatic disaster is apparently an ineluctable trend.

To avoid this disaster there is, parallel to energy-saving, preservation of forests and reforestation, an urgent need to convert the energy-supply sector as rapidly as possible to the use of direct and indirect solar energy - solar power, wind power, water power and biomass. Instead of assuming that there is still time for this, or that the developing state of solar energy technolgies will not yet permit it on a broad scale, we need to work actively towards bringing about a technological revolution in this area.

What is involved in this context is not just making use of solar technologies for the production of electricity, but also the use of solar energy to replace the main energy sources used for transport systems, industrial production and heating. Solar Energy must replace engine and heating fuels.

That is the mission and the importance of this Second European Conference on Architecture. The position can be stated in three sentences:

- Our previous sources of energy must be replaced by solar energy.
- In the meantime, we have to use energy more efficiently.
- Solar methods of construction in harmony with the environment are the most comprehensive and fastest means of achieving this aim.

Almost half of all energy consumption in the industrial societies of the northern hemisphere takes place in buildings. This means that architecture and construction technology represent a key factor in using solar energy and thus preventing global environmental disaster. So responsibility lies in these fields to a degree that cannot be overestimated, because there are three obstacles to the possibilities inherent in solar construction methods:

- There is widespread ignorance even among architects and building engineers as to the state of research and development in the field of solar design and construction methods;

- Present political frameworks and conditions in building laws and regulations make it difficult - or even impossible - to apply solar construction methods;

- Conventional building aesthetics which have developed over centuries.

There is still a widespread view among town planners and architects that solar buildings are inconsistent with aesthetic quality. In past years solar research has focussed on the problems of proving the applicability of solar technology and improving its efficiency. It was not possible to meet all aesthetic requirements as well. In this we are almost at the beginning. It is a task to be undertaken by architects, with the objective of creating aesthetically integrated solar technology.

But aesthetics are also a matter of perception. <u>What</u> we perceive and <u>how</u> we perceive it is subject <u>to</u> constant cultural <u>change</u>. And the latter is determined by what we can afford economically and by the political conditions we live in.

The architecture of rich countries is different from the architecture of poor countries, as is the case with architecture in monarchies, dictatorships and democracies. Therefore, architecture which faces up to its responsibility for the environment will be and must be different from what it was in those times during which the environment was neglected. Thinking in terms of the environment changes our aesthetic perception.

Buildings, airplanes and cars that use up much conventional energy and thus contribute to environmental destruction may still be considered perfect in an aesthetic sense. But the more we know about the environmental risks arising from them, the earlier our aesthetic perception will change. The more we appreciate the negative implications for the environment for any object, the quicker it may happen that the same object will acquire a negative aesthetic perception. And the more people are aware that a solar building contributes to preserving our environment, the more positive will be their aesthetic perception of it.

Solar design meets in principle the two crucial requirements of our time - environmental responsibility and democracy - and with this, it meets the necessary cultural change:

- Solar design means transparent construction, and transparency is a basic requirement for further developing democracies.

- Solar technology encourages individual autonomy rather than the supply by others.

For these reasons - and for saving the environment - solar design will also meet the aesthetic needs of the future.

Solar architecture is, in a comprehensive sense, the architecture of the future - a future that will reconcile individual autonomy and social responsibility.

In philosophical terms solar energy technologies make a synthesis of individual freedom and common interest which has not been the case with any major technology before. Up to now, individual freedom and the idea of community have tended to be in opposition to each other. Either an enhancement of individual freedom was achieved at the expense of common interests (including nature as the source of all human life), as in the case of cars and the energy consumption in buildings, or common efforts have been made at the expense of individual freedom, as is illustrated by the increasing dependence of the individual on large-scale technologies and large companies.

Solar technology has got to change people's behaviour if it is to be used sensibly and to an optimum degree. The use of solar power requires the structure of an open society. It reserves the trend of humans becoming the slaves of their technical achievements and it gives us the chance to use our technological potential for genuine human progress.

I am speaking to you as President of the international political Association for Solar Energy, EUROSOLAR, which was founded in 1988 and which co-sponsors this Conference. It is our aim to promote solar energy politically, economically and in the cultural field in a comprehensive way. EUROSOLAR is a non-governmental, non-profit organisation.

We are working out policy frameworks for actions on the basis of scientific and technological progress, making them popular with the public, and introducing them into the political decision-making process.

I wish this Conference success - so that by your results, your help, arguments for the potential of solar energy in the field of building construction can be substantiated and stated more precisely.

We need to stimulate our practical imagination on the way from an "energy economy of death" to an "energy economy of survival" on our planet.

EUROSOLAR wishes to contribute to making your results so widely known that no one can ignore them anymore. To this end, we need your expertise.

One of the biggest problems we have is the tendency to close our eyes in the face of these problems.

It is our common task to open people's eyes to the problems and to the new potentials alike.

DESIGN TOOLS - FRIEND OR ENEMY

PROFESSOR P.E. O'SULLIVAN
BARTLETT SCHOOL OF ARCHITECTURE AND PLANNING
UNIVERSITY COLLEGE LONDON
22, GORDON STREET
LONDON WC1H 0QB

The job of the designer can be thought of as one of combining disparate building components into a cost effective, energy efficient, aesthetically pleasing solution, and this in order to resolve a real problem. In performing this service, the designer acts primarily on his own, as he functions in a predominately component based industry, which has little or no knowledge of a systems approach to producing a building solution.

The purpose of a design tool is presumably to aid the design professional in carrying out his function, namely the designing and making of a building. In this process the designer has to make choices between things that are known, and judgements between and about things that are not. Now I would argue that the role for the professional only truly exists when the body of knowledge on which he depends is incomplete, for the primary professional role is to make judgements from the base of an incomplete knowledge base. If the body of knowledge is complete, then the professional role is not necessary and that of the technician suffices. For example, whereas the maintenance of a motor car requires a technician, the design of a motor car requires a professional. In real life of course, it is all a matter of degree.

From those to whom society gives the responsibility of making design judgements, it requires two things.
a) That the judgements be made on the best information and knowledge possible, and not on myth, magic or intuition. This means that the design professional must be fully conversant with his professions body of knowledge.
b) That the design professionals, must in making their judgements, act so as to increase the body of knowledge, and so reduce the risk of failure in the future.
Design tools, therefore, in my view are aids to help the designer make better judgements (and better choices) using the incomplete professional knowledge base in the production of his design. As such, they are professionally supportive, in that they support and improve the designer, rather than professionally replacive.
They also form part of the designers' mid-career education process, in that by arriving at a judgement with the help of a design tool (which contains more knowledge than that of the designer), the designer is improving his own knowledge base.

Buildings proceed through three separate stages, namely:
a) A building is first designed by a designer whose end point is a set of working drawings that act as instructions and/or directions to the contractor. The working drawings tend to specify of what the building will comprise, rather than how it should be made.
b) The building is then constructed by a builder, whose skills are in the making of the building, and in so doing, interpreting the instructions of the designer. He realises the designers intentions.
c) Finally, the building is occupied and put into operation.

Now the building is designed according to current design knowledge, is constructed according to the current standards of the construction industry (availability of personnel, materials, etc.) and is operated according to current management processes. My point is that these are three largley independent free-standing elements, linked if at all in complex as opposed to simple ways. So that the conventional idea that one or more design aids/tools can be the linkage between these elements, for example between operation and design is likely to be unfounded.

The design tool, therefore, is more likely to be purely an aid to the designer. It will not necessarily aid the constructor or the operator, and may not have an effect on the construction process or the mode of operation of the building. It will hopefully help the designer to make better judgements.
How then does the designer improve his knowledge base. I suspect it can happen in each of two ways:
a) When the designer has a current project on hand, and is using his own knowledge base to make the judgements he needs to make. In this situation, if he reaches a point when he recognises that his knowledge base is not sufficient, my experience is that he asks a professional colleague that he trusts to help him in making the judgement. He does not at this point, turn to any other sort of aid or tool. By asking the additional person to help, the designer is in effect improving his knowledge base, by adding to it that of his colleagues. This may well include for example, computational and predictive processes, but the judgement is made by the combination of the two persons. The second one acting as the design aid/tool to the first.

b) When the designer has no specific project on hand. During this period he reads, looks, listens, and learns. In other words, generally improves his body of knowledge. The rate of improvement depends on the level of his "question asking ability". The "good" designers seem to have a higher question asking ability than the "bad". It is for this reason that the "good" get better and the "bad" do not. Indeed, once the designer's question asking ability is at a high level, then the rest follows. For this reason, a lot of mid-career education is geared to improving the level of question asking ability.

Now design aids/tools are normally thought of as being employed only when the designer is working on a project.
They appear to exist in two forms:

i) When the ideas/basis for the design are being formulated. It is at this time that the person to person (professional to professional) discussions take place that result in the better judgements.
ii) When the design that results from the ideas developed above is being refined and tested.

Taking these in order:
i) Design aids/tools that inform the professional when he is formulating his ideas, and help him to make better judgements. These design aids exist in three types:
a) Aids/Tools that help the designer to think about the problem under consideration. For example, the theory of Climatic Modification, the Lack of Discomfort theory, Balance Point theory, the variety of spatial theories, e.g. compartmentalisation.
b) In the form of simple rules, for example, the number, size and shape of the windows in a room.
c) In the form of partial solutions (solution models). What a typical operating theatre suite usually looks like.
The purpose of the first is to stimulate ideas, and of the second to make judgements, and therefore design progress possible. These latter are often called "Variety reducing Constraints", and the way in which they are used, the Art of Coarse Design.

ii) Those aids/tools against which the designer tests his developing design. These often take the form of statutory safeguards for society. They would include for example, the Building regulations, the Fire regulations, and the Energy regulations. The curious thing is that once again they are often made manifest by the people whose responsibility it is to apply, interpret, and explain them to the designer. Namely the building inspector and the fire officer.

More interestingly, it is these very interpreters who do provide the links between the three building stages mentioned above. For example between design and construction via the mechanism of the idea of "buildability".
Are then design aids/tools friends or enemies? Do they help or not?

My view is that the most helpful design aid/tool is another competent professional, whose skills are complementary. That such a tool(person), is probably of most use at the stage when the designer is formulating his design ideas, and making his judgements. His skill(range) should spread across theories, rules, and partial solutions. He is not a common commodity.

The least helpful and perhaps the most common is the "Regulator", who rigidly requires compliance with his Regulations, in a component based way.

Perhaps more time should be spent than at present on developing the former.

DAYLIGHTING IN THE U. S.: TRENDS IN TECHNOLOGY AND DESIGN

C. C. Benton

Associate Professor
Department of Architecture
University of California, Berkeley

Faculty Research Associate
Applied Science Division
Lawrence Berkeley Laboratory
Berkeley, CA 94720

ABSTRACT. Engineering advancements at mid-century promised the ability to produce an ideal interior environment. The reality has been more prosaic. In the United States, designers are rediscovering daylight and its abilities to animate architectural space and promote occupant well-being. Our utilities and national laboratories are interested in daylighting's potential for reducing energy consumption and peak power demand. This paper reviews the historical context of daylighting in United States architecture and describes activities over the last decade to revive interest in daylight. Following an assessment of educational efforts in daylighting, the paper provides examples of recent daylighted buildings, a discussion of new programs in support of daylighting practice, and a brief assessment of new technologies.

1. INTRODUCTION

1.1 Historical Context

This paper combines a personal view of daylighting in the United States with a survey of current trends linking daylighting and architecture. Through the first half of this century we unselfconsciously designed buildings to admit daylight. Compared to open-flame and incandescent sources, daylight provided interiors with relatively high levels of illumination as well as strong spatial modeling, excellent color rendering, animation, and a sense of connection with the exterior world. During the tenure of our present generation of architects, however, the relationship between buildings and daylight has been altered. Since the late 1940s, architecture has shared with other human endeavors the heady advances of science, an abandoning of older ways to embrace the potential of a technological frontier. For buildings these advances included an expanded palette of electric lighting sources, mechanical air-conditioning equipment, and solar-control glazing plus unlimited domestic utilities and a fair dose of engineering acumen. Following World War II U. S. architects, after briefly debating the role of technology in design, cast prior architectural tradition aside and marched boldly into a scientific future. Architectural Forum, in a 1948 editorial [1], captured the spirit of the era by noting that "*vast strides have been made by science and industry and, significantly, in design. We have learned more about the kinds of environment which people need in order to carry on their daily lives most effectively. Materials and devices have been developed to produce such environmental conditions. And, since such material and devices must be incorporated into buildings, the buildings themselves have come to be looked on as means of conditioning the working and living spaces which they enclose.*" Explicit in the scientific approach was a belief that engineered environments would be an exact and optimal setting for the activities of building occupants.

We have since witnessed profound change in the physical character of commercial building interiors. The new technological standards called for steady-state interior environments, for lighting and thermal conditions uniform over time and space. Rapidly developing environmental control systems offered convenience and control but not without cost. The craft of coupling the interior and exterior environments, historically a source of challenge and order in architecture, was dismissed with the flip of a switch.

There have been changes in the professional as well as the profession. The architects of today have an appreciably different set of skills than those in practice forty years ago. Wright, Aalto, and Kahn, master architects at mid-century, were clearly talented in the manipulation of natural light. The products of their generation evidence training in the patterns of light and a care in observing light as an architectural element. Graves, Meier and Eisenman, architectural luminaries of our day, are demonstrably less interested, and less skilled, in this aspect of architecture. In fact, it is quite rare to find presentation drawings from a contemporary professional or student that represent daylight in a proposed architectural space. As a design element, daylight was absent for twenty years from the curricula of our schools (especially the studios) and, as a consequence, is now something of a lost art among building professionals. A recent report by the Windows and Daylighting Group at Lawrence Berkeley Laboratory [2] concluded that "*the technical and design skills to optimize daylighted building designs to produce maximum energy and load savings (while maintaining or improving visual quality and comfort) are practically non-existent. The designers and consultants who venture beyond the simplest of the proven solutions find themselves in unexplored territory*". Like ruling pens and slide rules, the skills of daylighting disappeared from the domain of architects.

The production of commercial buildings now requires a team of consulting specialists: each generally compartmentalized within the rigors of their discipline, each striving to meet specialized and well-bounded performance standards. While the architect remains the titular head of the design team, it is increasingly difficult for anyone to assess the accumulative effect of engineering decisions on the quality of interior environments. The result in contemporary architectural practice is often a tacit surrender, the acceptance of a default interior. The character of these spaces, and the U. S. produces 25 million m^2 of office per year, is largely determined by conventions in systems engineering: seal the glazing to control air exchange, use low-transmittance glass to reduce heat gain, provide an evenly-spaced grid of direct fluorescent fixtures for uniform interior illuminance, etc. As a result, our office occupants are routinely denied access to daylight, natural ventilation, and contact with the exterior.

1.2 A Reassessment

The bloom is off the mechanistic rose. After four decades of mechanically conditioning interior environments, engineering and technology, while appreciated, are no longer seen as panaceas. In retrospect, uniform steady-state environments are rarely considered optimal by the majority of office occupants, a situation exacerbated by the widespread absence of individual controls. In fact, the luminous and thermal environments are consistently rated as primary sources of dissatisfaction in the workplace [3]. Plentiful and inexpensive energy provided the cornerstone of the technological vision of the 1950s. We now have substantial economic, environmental and political impediments to expanded energy use.

Our building professions are reassessing the standard practices of environmental engineering to address concerns regarding energy efficiency, operating costs and environmental quality. In particular, the current professional and research literature cites daylighting as a strategy that offers lower energy costs, reduced peak demand, and increased employee productivity while providing a more humane quality of light [4]. Renewed interest in daylighting has increased activity in the design of lighting-related architectural features, supplemental electric lighting systems, and lighting control systems that take advantage of daylight as well as our technological prowess. These are relatively new developments, and the techniques involved are often unproven. In consequence, few contemporary buildings feature applied daylighting elements and fewer still can serve as useful exemplars of daylighting practice. Unfortunately, the shortage of suitable exemplars is problematic in an building industry that possesses remarkable inertial tendency toward "business as usual".

The reintroduction of daylighting in the United States is a major undertaking. The most difficult obstacle we face is not a shortage of daylighting means but rather the compartmentalization of decision making within the various building design disciplines. Our immediate challenge is convincing architects, as the designers charged with the whole of a building rather than its parts, to adopt a concern for daylight. As allies in this endeavor the U. S. architecture schools, our national laboratories, and the domestic electrical utilities are actively engaged in the advancement of daylighting practice. The remainder of this paper describes daylighting activity in several categories: architectural education, design practice, case studies of completed buildings, and new building technologies.

2. DAYLIGHTING in ARCHITECTURAL EDUCATION

2.1 The Daylighting Network of North America

Our architecture schools provide a natural setting for the exploration of current issues in daylighting and the shaping of future design values. For several decades many issues of environmental control were noticeably absent from architectural curricula, a situation directly related to the compartmentalization of design responsibilities. As educators Don Prowler and Harrison Fraker noted in 1981 [5]: "*The success of environmental control technology freed the architect of this century from pragmatic concerns of comfort and climate. This has been achieved by the application of increasingly sophisticated and energy-consuming devices which can be located remote from the space that they serve. This circumstance has reinforced the widely-held belief that technics in general, and environmental controls in particular, are independent of form. From this premise, the simple pedagogical result is that technology has become separated from the primary educational experience in schools of architecture today, the design studio.*" In the last ten years many of our schools, motivated by issues of aesthetics, occupant well-being, energy, and ecology, have reexamined the physical environment as an architectural issue. Many have responded to these issues by expanded their faculties, facilities, and course offerings related to building science. While we continue to grapple with relationship between the design studio and physical building performance, there is a growing recognition of the architectural potential inherent in this connection.

Daylighting, an issue immediately tangible to architects, has been at the forefront of the building science curricular expansion. Over twenty of the nation's schools of architecture are members of the Daylighting Network of North America (DNNA). Each of these institutions has developed coursework and facilities related to daylighting (see Table I.) Many of these schools are remarkably active. For example, Rensselaer Polytechnic Institute in New York is assembling a lighting research center staffed by architects, vision scientists, lighting scientists, a psychologist, an energy analyst, and technical support personnel. The DNNA schools now serve as regional centers of expertise, providing a resource that extends beyond the schools to support practicing architects, private-sector research, and government policy making. They are working to return the mastery of light as an innate architectural skill. Until this skill is intuitive, our schools are particularly interested in exploring techniques for the prediction and evaluation of daylighting performance.

2.2 Daylighting Design Tools

Today's daylighting tools vary widely in usefulness and rigor. Analysis methods use graphic, numerical or physical model techniques to predict the quantity of light, its distribution and, in some cases, the visual appearance of a daylighting scheme. The skills of a mature designer are ultimately based on observation of daylight and experience in its manipulation. For the developing student, and today's daylighting-innocent professional, evaluation tools should support the development of daylighting skills so that they become second nature as experience deepens. As this objective has both quantitative and qualitative dimensions, the ideal design tool should serve as an agent of discovery as well as a vehicle for analysis. Important features in this endeavor are the ability to model complex spaces, qualitative visualization of daylighting effects, and an ease of use appropriate for the early stages of design.

TABLE I: MEMBER SCHOOLS OF THE DAYLIGHTING NETWORK OF NORTH AMERICA

The Daylighting Network of North America (DNNA) Regional and Instructional Centers offer information services, seminars, and workshops. Many of the centers have additional resources related to daylighting--including spot meters, color samples, model shops, daylighting libraries, local weather data, and electric lighting analysis equipment. For more information, contact the DNNA centers. All Regional and Instructional Centers are based in the Architecture Programs of their respective universities with the exception of the University of New Hampshire (Electrical Engineering).

Center	Photo-meters	Sky simulator	Shadow simulator	Computer simulation
Arizona State University	HH,R,DA	MB		M,MF
Ball State University	HH,R,DA	MB	H	
Cal Poly, Pomona	HH,R,DA		H,T	M,MF
Idaho, Univ. of	R			M
Iowa State University	HH,R		T,H	M,MF
Kent State University	HH	MB	T	M
Louisiana State University	HH,R		T	M,MF
Miami University	HH,R,DA	MB		M,MF
Michigan, Univ. of	HH,R,DA	MB	T,H	M,MF
Minnesota, Univ. of	HH,R,DA		T	M
New Hampshire, Univ. of	HH,R,DA			M,MF
New School of Architecture	HH			M
Oregon, Univ. of	HH,R,DA	MB		M
Pennsylvania, Univ. of	HH,R,DA		T	M,MF
Rensselaer Polytechnic	HH,R,DA	MB	T,H	M,MF
Texas A&M University	HH,R.DA	HS	T	M,MF
UCLA	HH,R,DA	MB	H	M,MF
UC, Berkeley	HH,R,DA	HS	H,T	M,MF
USC	HH,R		H	M,MF
Virginia, Univ. of	HH			M
Washington, Univ. of	HH,R,DA	MB	T	M,MF
Washington State University	HH	MB		

KEY

Meters:

HH = *hand-held light meters*
R = *remote reading or model light meters*
DA = *data acquisition system*

Sky simulator:

MB = *mirrored box sky*
HS = *hemispherical sky*

Shadow simulator:

T = *tilt table*
H = *movable sun heliodon*

Computer simulation:

M = *microcomputer*
MF = *main frame*

Simplified design tools include calculation procedures formulated for rapid execution in the early stages of the design process. There are literally dozens of calculation techniques in this category based on first order principles, correlation with laboratory measurements and graphic representations of the sky [6]. The Daylight Factor Method and the Lumen Method are well known examples in the U. S. As a category, simplified design tools are well-suited to the formative period of preliminary design. They are widely accessible to building designers and have become easy and quick in their microcomputer implementations. However, after a first impression of reasonable utility, the designer often discovers that design solutions are limited by the base assumptions of the simplified method. Speed and convenience require the compromise of limited applicability and, within these limits, the outcome of simplified analyses quickly seem intuitive. Harvey Bryan [7], an architecture professor involved in the development of simplified tools, lauds the pedagogical value of these simple geometry predictions but also concludes that it is "*increasingly difficult to propagandize the use of design tools that I rarely use in my own design practice*".

A second category of daylight prediction techniques, advanced design tools, is not restricted to simple geometries. Elaborate programs like Lawrence Berkeley Laboratory's SUPERLITE and the daylighting routines in DOE 2.1C are capable of simulating relatively complex spaces under changing sky conditions. Their power exacts the price of complexity. It takes long hours, and a computationally apt mind, to master the features of these programs. In return, they can provide detailed accounts of illuminance distribution on multiple planes, lighting energy consumption and interaction with space-conditioning loads. At this time, few computer programs have the capability of visually depicting the appearance of daylighted spaces. One notable example of this genre is Greg Ward's ray-tracing program RADIANCE [8], an extremely capable program that requires advanced engineering workstation hardware.

With recent developments in computing hardware and increasing interest in daylight prediction, we are beginning to see a third category of computer-based design tool. This category promises to combine an intuitive user interface with computational finesse. An interesting example of the new genre is the Architecture and Engineering Series produced by Skidmore, Owings, and Merrill. Running on an networked engineering workstation, this program provides an integrated CADD environment that includes tools for drawing management, structural analysis, and elaborate visual rendering. The effective visualization of predicted daylight will be practical in design offices within the next ten years. For the present, however, computer-based daylighting tools cannot provide a practical visual prediction of daylighting effects nor can they offer the elusive combination of easy use and computational versatility.

Physical scale models, on the other hand, do support the visual assessment of daylighting quality. Models are a traditional architectural tool and have been used to assess lighting effects since the beginning of the century [9]. "*A fully-detailed scale model of an architectural space is a true photometric analog and permits a precise study to be made of daylighting in its photometric aspects. In almost every respect the model also provides a true subjective simulation of the daylighting conditions in the full-scale room*" [10]. This combination of quantitative and qualitative capability makes modeling a powerful technique indeed. Architects can make relatively quick models of complex spatial configurations using skills that are common in the design studio. The finished product is a powerful analytical tool which, as a bonus, serves as a presentation medium. Model studies can be conducted under all types of real sky conditions and in the controlled environment of a sky simulator. With an impressive list of advantages, daylighting models are gaining popularity in today's education community.

There are challenges associated with using scale models for daylighting analysis. The accuracy of the model is a function of the skill of the model builder and the resources available for the task. In practical application, the level of model detail is often compromised by limits of time and money. In this context, the selection of model features is important. To effectively test models, designers must have access to photometric and photographic equipment, some of which can be quite expensive. Finally, the testing environment can be problematic. Model tests under the real sky provide wonderfully realistic predictions vis-a-vis color, interior luminance, and the relative importance of beam versus diffuse radiation. The down side is that tests are dependent on access to the desired type of sky which, as a function of nature, is beyond the immediate control of the designer. As an alternative to the real sky, sky simulators are becoming common features in our schools of architecture (see Table I).

All things considered, physical model testing offers an attractive method for the analysis of daylighting schemes. When building their first daylighting model, designers are faced with decisions unique to this model type. Common questions during the learning curve for daylighting models relate to material selection, the detailing of fenestration, and planning for model testing procedures. Guidance on daylight modeling is available in several references, including books recently published in the United States [11-13].

3. DAYLIGHTING in ARCHITECTURAL PRACTICE

3.1 Recent Efforts

To qualify as a daylighted building, new construction should include specific features for admitting daylight without excessive glare, an electric lighting system designed to work with daylight, and a lighting control system to coordinate the two sources. By these measures fewer than one new U.S. building in 1,000 can be considered a daylighted building. Yet major daylighting schemes are being built where there were none before and each successful project adds momentum to the case for daylighting. Professional offices, often uncertain about the specifics of daylighting, are turning to the schools, the utilities, and the national laboratories for information on daylighting. The schools provide recent graduates, and faculty consultants, to participate in design development. As a result, the daylighting evaluation techniques used in practice follow trends established in education. Physical models are commonly used for performance evaluation and energy simulation codes, particularly DOE 2.1, furnish annual performance predictions. Table II lists recent North American daylighting projects and the role of physical models in their design.

The virtual absence of daylight from recent mainstream architectural practice has produced something of a pedagogical challenge. Architects and lighting designers in mid-career are seeking a basic conceptual understanding of daylighting principles. In response, the manufacturers of lighting equipment, electrical controls, and architectural fenestration products have produced improved daylighting-related literature. Professional organizations in the building design industry have organized continuing education workshops on daylighting design and these have been popular. Within the last year, several electric utility companies have undertaken the construction of laboratory-like lighting design centers featuring daylighting assistance. Finally, the Windows and Daylighting Group at the Lawrence Berkeley Laboratory (LBL) is assembling an interactive, multimedia daylighting instructional tool. The latter two developments merit further comment.

3.2 Utility-Sponsored Lighting Design Centers

The United States' electric utilities are sensitive to the potential impacts and utility-side benefits that might arise from a wider application of daylighting in commercial structures. This sensitivity, combined with public pressure on ecological and energy conservation issues, is now producing tangible support for daylighting. Electric utilities in Seattle, San Francisco and Los Angeles are developing an innovative series of lighting design centers. Part lighting laboratory and part demonstration facility, the new centers provide building designers with a physical setting for exploring lighting design options. Purposefully designed for a broad constituency, the centers fulfill several functions. They provide classrooms, microcomputer clusters, and libraries in support of professional seminars and short courses on all aspects of architectural lighting. Throughout the centers' spaces, electric lighting systems are installed side-by-side providing designers with visual comparisons of system performance. Exterior rooms are equipped with alternative fixtures and dimming systems to provide reconfigurable demonstrations of control system performance.

The lighting center under development in San Francisco will have a range of features directly related to daylighting. Physi-

TABLE II: EXAMPLE DAYLIGHTED BUILDINGS IN NORTH AMERICA (with notes on physical models)

Ventura Coastal Corporation, Ventura, California Rasmussen and Ellinwood, Architects 1982	An early daylighting design, this scheme combines lightshelves, a clerestory, and atrium glazing in a two-story, 3,000 m^2 building. In the interior, open-plan workstations are a maximum of 10 m from an exterior wall. The architects used a rough model representing two bays of the building at 1:24 to test daylighting design assumptions. The model was tested in several configurations under real sky conditions (on a football field) and measured with simple instrumentation. The resulting office building has won wide acclaim for 'character and visual interest' as well as energy conservation.
Lockheed Building 157, Sunnyvale, California Leo A. Daly, Architects 1983	This 56,000 m^2 office structure houses 3,000 engineering personnel in five floors of open-plan offices. Physical models were used early in the design process to establish overall building massing, then interior models were tested in the Sky Simulator at Lawrence Berkeley Laboratory to fine tune the geometry of interior lightshelves, exterior shading devices, and exterior reflectors. The design process included a full-scale mock-up of a typical office bay (built at the Owens-Corning Research Center in Granville, Ohio).
Johnson Controls Building, Salt Lake City, Utah Johnson Controls Corp. Facilities Group, Architects Donald Watson and William Lam, Daylighting Consultants 1983	This modest, 1,400 m^2 office structure features a series of effective daylighting techniques. Physical model tests, conducted at the Yale architecture school, demonstrated the visual consequences of specular lightshelves and a clerestory designed to wash the opaque northern wall with light. The daylighting model was used to select reflector finishes, light monitor geometry, and interior colors. For instance, the building's ductwork is painted a light color rather than the dark finish originally proposed. The model studies suggest that this produced a twofold increase in illuminance from daylight at the rear of the building.
Pacific Bell San Ramon Administrative Center, San Ramon, California Skidmore, Owings & Merrill, Architects Van der Ryn, Calthorpe & Matthews, Daylighting Consultants 1986	Located in a growing suburban area, this building features eight kilometers of strip windows with clear glazing extending from just above the floor to a 3 meter ceiling height. These windows provide a dramatic view, and daylight, to 7,200 occupants in the 160,000 m^2 building. A 1:24 daylighting model was used to test the effectiveness of a continuous 2.5-m horizontal overhang, evaluate glazing options, and to predict the performance of a four-step, on-off electric lighting control system. Results from these tests were used as input in DOE 2.1A simulations of cooling loads. A second model was used to evaluate skylight options for an atrium space with tests emphasizing trade-offs between solar heat gain and adequate lighting for plant growth.
Olympic Oval, Calgary, Alberta Graham-McCourt, Architects Mojtaba Navvab, Daylighting Consultant 1987	The Olympic Oval is an indoor arena for speed skating, normally an outdoor sport. The vast building, with interior spaces 32 meters from an exterior wall, was designed using lightshelves inspired by Lockheed Building 157. A physical model "built as big as would fit on an airplane" was used to test lightshelf geometry with measurements made under the real sky in Calgary and in the LBL Sky Simulator. Under many conditions the building can be entirely lit by daylight. Data from the model studies were also used in the design of supplementary electric lighting for television cameras.
National Gallery of Canada, Ottawa, Ontario Parkin Safdie, Architects 1987	Museum display spaces are a particularly sensitive case in daylighting design. In addition to meeting the normal criteria of visual comfort and adequate illuminance, daylighting on artwork must meet strict curatorial requirements. Moshe Safdie's design for the National Gallery of Canada provides a rich variety of daylighted spaces that are both comfortable and effective. Scale model studies were used to establish gallery proportions and to design daylighting control systems using operable devices. Some of the gallery spaces are toplit by skylights located two floors, and 16 meters, above the gallery.
Triton Museum of Art, Santa Clara, California Barcelon + Jang Associates, Architects BETA Associates, Daylighting Consultants 1987	For the 3,000 m^2 Triton Museum design, the architects studied over twenty aperture and light-well configurations for gallery space. The final candidates for a top-lighting solution were tested using a 1:16 scale model in the LBL Sky Simulator. The tests examined the effectiveness of two complicated shading and control systems using louvers in a pyramidal-shaped skylight. The final scheme included a manually-controlled, motorized shading baffle.
Hyatt Regency, Greenwich, Connecticut Kohn Pedersen Fox Associates, Architects Mojtaba Navvab, Daylighting Consultant 1987	The atrium of this Connecticut hotel features extensive planting of species native to New England. Design of the atrium glazing system had to balance the light requirements of the plants and cooling loads on the interior space. A 1:48 scale model was tested under real sky conditions and in the LBL Sky Simulator. The tests included evaluation of 15 different interior finish options and three roof fenestration systems. Data from the model provided input for computer simulations using DOE 2.1B and SUPERLITE 1.0. The final design was strongly influenced by results of the model tests.
3M Corporation Research & Development Complex, Austin, Texas CRS Sirrine, Inc., Architects Daylighting Center, Texas A&M, Daylighting Tests 1988	Another large office example, this project will ultimately house 5,000 office workers along a four-story building featuring an atrium 150 meters long. A very detailed model of a portion of the atrium was built at 1:24 for the study of over two dozen roof covering alternatives, including novel systems using Fresnel lenses. The finished model, weighing 725 kilograms and having a 3-m length, 2.1-m width, and 1.4-m height, was considered very effective as a design evaluation tool.

cal model testing will be supported by a roof-top testing platform complete with photometric equipment and daylight availability monitoring. Tests requiring a sky simulator will be conducted in the LBL Sky Simulator located across San Francisco Bay. Sunlight penetration and shadow studies will take place on the center's fixed-sun, moving-earth heliodon. An exterior wall is designed to provide side-by-side demonstrations using scale models of daylighting features. Similar to the side-by-side demonstration of lighting fixtures, this technique allows the visual comparison of glazing alternatives, window placement, wall treatment, and interior finish options. The center will feature a collection of generic daylighting models to support instruction plus a model shop for their construction. Control will be demonstrated in a room featuring variable dimensions, variable window size, and several combinations of photocell placement, control algorithm, and switching/dimming hardware. Data acquisition systems in the room will provide a visual record of control system performance while the room itself offers an opportunity for experiencing the performance of control options. Beyond providing a reference collection of popular and technical works on daylighting, the center's library will lend equipment for photometric measurement and data acquisition off-site. The San Francisco center, to be staffed by a group of seven lighting specialists, represents a substantial commitment to the support of daylighting.

3.3 LBL's 'Advanced Design Tool' Program

With support from the U. S. Department of Energy, Lawrence Berkeley Laboratory is developing an innovative, microcomputer-based design tool oriented toward practicing designers [14]. The LBL 'Advanced Design Tool' applies hypermedia concepts to combine a wide range of daylighting resource materials into a single, versatile teaching tool. In a hypermedia document information is accessed in user-controlled, non-linear sequences providing a uniquely accessible reference system for the non-linear processes of design. Within the 'Advance Design Tool', architects can follow daylighting topics through numerous organizational paths, a feature best explained by example. A user's interest in sidelighting techniques might raise questions concerning lightshelf performance. Clicking a mouse on the keyword 'lightshelf' leads to a lightshelf tutorial including references and animated diagrams of lightshelf theory. From this tutorial the user might select information on case study examples of applied lightshelves. Within the case studies the user selects a specific building, say the Mt. Airy Library, and the system provides videodisc images of the library's lightshelf, architectural drawings of its construction, and diagrams of the lightshelf's performance. Intrigued by the library in general, the user then clicks the mouse on a drawing of the building's entry, an action that produces a still image of that entry on the videodisc monitor. A second click in the drawing's interior produces a series of images representing the path from entry to the selected location. Additional case study information includes a history of the building's development, press reviews, interviews with the building's designers (including audio or video statements), the results of post-occupancy surveys, and monitored performance data. The appeal of this concept is the ability to browse through a large body of daylighting material while selecting the specific topics of interest and an appropriate level of detail. Information included in the database can be extremely broad, ranging from case study documentation to simulation codes, tutorials, reference tables, and detail libraries.
The LBL project is forward looking, and seeks to apply the best of emerging computer technologies to create an advanced design reference environment. The current version of the tool combines an Apple Macintosh II microcomputer with a touchscreen color monitor, an optical disk player (20-cm analog optical disks for image storage), and a second color video monitor. Planned additions to the hardware environment include CD-ROM data storage, digital optical image storage, CD audio, and access to a central computer network.

4 CASE STUDIES

4.1 The Field Study of Daylighting

The emerging stock of daylighted commercial structures offers an empirical opportunity for valuable lessons concerning the character and performance of daylighting schemes. It is an opportunity that we are just beginning to embrace. In a 1985 assessment of daylight as a commercial-sector energy conservation strategy, Lawrence Berkeley Laboratory concluded that "*performance data for daylighted buildings is virtually nonexistent. A review of over 40 'daylighted' buildings described in the architectural and engineering press provided virtually no useful data on the magnitude of daylight savings. These data are necessary, not only to validate computer models that provide guidance to designers, but also to convince hard-nosed decision makers that these approaches are viable and cost effective.*" [15] The field assessment of daylighting system performance is now quite manageable with the advent of portable, microprocessor-based data acquisition systems and an increasingly sophisticated set of diagnostic tools including luminance mapping cameras and laptop-computer-based subjective surveys. The DNNA schools and our national laboratories are actively seeking sponsorship to add documented post-occupancy evaluations to the literature on daylighting. In work now underway, the DNNA is developing a standardized daylighting case study protocol while LBL is adding case study information to the 'Advanced Design Tool' database.

4.2 An Example: Lockheed Building 157

Lockheed Building 157, described in Table II, provides an interesting example of applied case study technique. The building is located in the San Francisco Bay Area and incorporates a coordinated set of daylighting features designed to displace electrical energy consumption for ambient lighting. In this ambitious design the building must provide daylight, without glare and in proper quantity, to all five floors of its 122 m by 73 m open-plan office area. Toward this endeavor, the architects have assembled a system that combines architectural features for the admission and distribution of daylight, a dimmable electric lighting system, and an automatic control system to operate the electric lights in response to available daylight.
Lawrence Berkeley Laboratory has completed two phases of field monitoring to characterize the performance of this building. The first phase, conducted in 1985, concluded that the building's architectural features (lightshelves, atrium, 4.5 m floor-to-floor height) were performing well in admitting daylight [16]. It also identified operational shortcomings in the electric light control system and concluded that without modifications to the control system, the owners would not realize the substantial electric energy savings made possible by the building's architectural features. During Phase II, LBL designed revisions to the electric lighting control system, implemented these changes in a representative zone of the building, and monitored year-round effectiveness of the new control scheme [17]. It was our thesis that low-cost modifications to the existing control system would produce immediate dividends in reduced power demand and lower electrical energy consumption.
The modifications were relatively minor and included relocation of the control photosensors, attenuation of the control signal, change in the response pattern of the photocells, and implementation of a new calibration procedure. Control system revisions required approximately one-half day of labor and $650 in materials for the 2,000 m^2 test zone.
To test the control revision hypothesis, we installed four data-acquisition systems in the test zone. They monitored illuminance data in 15 locations plus lighting energy data and control photocell signals for eight lighting circuits. Data were collected for two-week periods in the summer, equinox, and winter seasons. Analysis of these data indicates that LBL's remedial actions were very successful in establishing proper operation of the electric light dimming system. The revised system accurately maintains target interior illuminance with minimum electric energy consumption. Substantial savings in peak electric demand were also obtained. The monitored data present a convincing portrait of daylighting successfully applied to a major contemporary office building.

5. NEW TECHNOLOGIES

5.1 New resources

As architects reconsider daylighting they find a certain historical intrigue, a sense of connection with architectural techniques and skills long past. There is rich architectural challenge in applying these traditional skills within the context of contemporary technology. The architect's palette now includes advanced glazings, light sources, and logical controls that would have been difficult to imagine fifty years ago. Wisely applied, these new resources extend the daylighting potential of new designs. Lighting controls are becoming more sophisticated with the development of local microprocessors, improved ballasts, and occupancy sensors. One interesting scheme sends control signals to ballasts using a subcarrier frequency on the power line. Control systems, once switching entire quadrants at a time, are now being applied to much smaller zones. The trend is leading toward local occupant control of workstation lighting and HVAC output.

5.2 Variable Property Glazings

Architectural glazings continue to advance at a rapid pace. During the last decade, selective transmitting glazings for thermal energy conservation have developed from speculative projection to a standard product in domestic glass sales. This decade should witness similar developments in the commercial market with glazings optimized for daylighting. In daylighting the designer faces tremendous variability in the environmental forces that impinge on the building exterior. The traditional techniques of exterior shading, interior window treatments, and low-transmittance glass usually produces adverse effects on daylighting performance. Research is in progress in a number of locations internationally to perfect a new generation of coatings for "smart windows". Light-sensitive (photochromic), heat-sensitive (thermochromic), and particularly electrically-activated (electrochromic) coatings offer attractive profiles of physical performance. The electrochromic or optical switch glazings, when coupled with a smart controller, will provide considerable gains in fenestration performance.

6. CONCLUSIONS

Daylighting as a concept is alive and well in the United States. The United States' education, research, and utility sectors are providing advocates for daylighting. Through these advocates, support programs are in place to assist practicing designers. The schools are producing architectural graduates with new daylighting skills, a resource for the mainstream of the profession. The technological capacity for analysis and implementation of daylighting systems is expanding. The issues of wise energy consumption, ecological restraint, and economy combined with concerns about employee well-being and productivity are contributing to a market for daylighting. The future holds a continuation of these trends. The key question remains one of mainstream architectural values. Will architects seize the day and accept the challenges inherent in daylighting? Time will tell.

REFERENCES

1. Wright, H. ed., "Measure," Forum, Vol. 89, No. 5, Nov., 1948, pp. 97-100.
2. Usibelli, A., et al., Commercial-Sector Conservation Technologies, Report LBL-18543, Lawrence Berkeley Laboratory, Berkeley, 1985.
3. Schiller, G., Arens, E., Bauman, F., Benton, C., et al., "A Field Study of Thermal Environments and Comfort in Office Buildings," ASHRAE Transactions, Vol. 94, Part 2, 1988.
4. Selkowitz, S. and Griffith, J., "Effective Daylighting in Buildings - Revisited," Lighting Design and Application, Vol. 16, No. 3, March, 1986, pp. 34-47.
5. Fraker, H. and Prowler, D., Project Journal: Teaching Passive Design in Architecture, University of Pennsylvania, Philadelphia, PA, 1981.
6. Commission Internationale de l'Eclairage, Daylight: International Recommendations for the Calculation of Natural Light, CIE Publication No. 16, E-3.2, CIE, Paris, 1970.
7. Bryan, H., "Daylighting Design Tools: a Personal Review," Lighting Design and Application, Vol. 16 No. 3, March, 1986, pp.49-56.
8. Ward, G., and Rubinstein, F., "A New Technique for Computer Simulation of Illuminated Spaces," Journal of the Illuminating Engineering Society, Vol 17, No. 1, Winter, 1988, pp. 80-91.
9. Hirons, F., "The Use of Scale Models,"Pencil Points, Vol.1, No. 6, Nov., 1920, pp. 5-8.
10. Hopkinson, R., et al., Architectural Physics: Lighting, Her Majesty's Stationery Office, London, 1963.
11. Schiler, M., ed., Simulating Daylight with Architectural Models, Daylighting Network of North America, University of Southern California, 1986.
12. Moore, F., Concepts and Practice of Architectural Daylighting, Van Nostrand - Reinhold, New York, 1985, pp. 167-178.
13. Robbins, C., Daylighting Design and Analysis, Van Nostrand - Reinhold, New York, 1986, pp. 221-234.
14. Schuman, J., et al., "A Daylighting Design Tool Using Hypercard on the Macintosh," Proceedings of the Third National Conference on Microcomputer Applications in Energy, Tuscon, AZ, Nov., 1988.
15. Usibelli, A., et al., Commercial-Sector Conservation Technologies, Report LBL-18543, Lawrence Berkeley Laboratory, Berkeley, 1985.
16. Benton, C., Warren, M., Selkowitz, S., et al., "Lighting System Performance in an Innovative Daylighted Structure," Proceedings of the 2nd International Daylighting Conference, ASHRAE, Atlanta, Georgia, Sept., 1989.
17. Benton, C., "The Lockheed Building 157 Monitoring Project Phase II: The Lighting Control System," Report LBL-18970, Lawrence Berkeley Laboratory, Berkeley, 1989.

ACKNOWLEDGMENTS

Steve Selkowitz and the Windows & Daylighting Group at Lawrence Berkeley Laboratory have generously supported the development of daylighting curricula at the University of California, Berkeley and elsewhere. Berkeley students Alison Kwok and Marc Fountain provided welcome assistance in the production of this paper.

QUALITY - THE CRUCIAL METRIC IN DAYLIGHT DESIGN

C M Parry
The Welsh School of Architecture
University of Wales College of Cardiff
P O Box 25
Cardiff CF1 3XE

ABSTRACT. In the recent past the quantitative aspects of Daylight have been separated from the qualitative to give design recommendations for lighting building interiors. Daylight design practice, particularly within Passive Solar strategies is discussed with regard to the quality of the visual environment. Recommendations for establishing qualitative assessment techniques are proposed in order to reintegrate quality and quantity into design guidance.

Many buildings of the 1960s and '70s were characterised as being overglazed. A concentration on the single design parameter - maximise the use of Daylight as a working illuminant - led to buildings which were thermally unstable often giving glare and overheating problems. Design guidance at the time relating to the visual environment, concentrated on the provision of Daylight quantities for visual performance of the tasks and functions to be housed in building interiors. The studies which gave rise to guidance of this sort were theoretical and empirical in nature and led to somewhat simplistic design recommendations of Daylight illumination provision or the application of esoteric Glare formulae. This could be expressed in absolute terms as an illuminance level in lux or, more usually, as a ratio of the total Daylight simultaneously available out-of-doors from the whole unobstructed overcast sky. This ratio expressed as a percentage is known as the Daylight Factor - DF. Recommendations for DF in an interior normally are expressed as averages or minima to be achieved. Daylight Factors related to overcast skies, are sensibly regarded as being the critical lighting conditions for the cloudy conditions associated with the climates of NW Europe and parts of North America. However Daylight by its nature is dynamic and critical design conditions will not apply for more than a short part of the year. For a small percentage of Daylight hours conditions will be worse than those produced under the base standard 5000 lux CIE Sky. For another smaller percentage 5000 lux Skies will be achieved, but for the majority of time, the critical 5000 lux Sky will be exceeded by much higher levels of illumination from overcast or partially clouded skies.

Daylight studies, most significantly developed under Hopkinson at the Building Research Station in the '50s, /1/ attempted to improve and give a scientific basis to Daylight design guidance which for years had relied on experience and a number of 'rules of thumb', for example:

window areas should be one fifth the floor area of the room, /2/

no office space can be 'let' which is more than twenty seven and a half feet from the window wall, /3/

make windows smaller as you go higher in the building, /4/

Inherent in some of these 'rules' were a complex set of visual environmental characteristics, so that their application would inevitably give rise to interior environments where the inter-relationships of Daylight's qualitative and quantitative aspects were given their proper importance.

The problems with the first set of 'scientific' Daylighting studies were that they produced recommendations that dealt solely with the quantitative aspects of the subject. This was not the fault of the scientists who were working in the area. They were aware of the complexity of real design situations and their writing abounds with provisos and caveats as to the use of the suggested quantitative design tools that they had developed. Unfortunately legislation and Architectural design practice responded positively to the recommendations for Daylight quantity in interiors. 'Daylight as working illuminant' was implemented at the expense of other thermal or acoustic environmental parameters and, possibly the quality of the resultant visual environments.

The limitations of these recommendations which covered explicitly quantitative Daylight illumination and, implicitly perhaps, qualitative visual conditions, were that they obtained only in so far as the provision of Daylight in the interior was produced by normal side lighting through windows. Daylight provision by other means, roof lights or reflected light methods for example, while they produced the quantity of Daylight, changed the quality of the resultant visual environments.

It is instructive to see the way that design approaches to a single building type - the school - present different characteristics with regard to the visual environment in response to the stresses placed on particular environmental design parameters over a period of time.

Immediately post-war the emphasis was placed on light and air and on Daylight specifically as a working illuminant. /5/ This led to legislation in the United Kingdom which still insists on a minimum 2% Daylight Factor being achieved throughout classroom interiors./6/ Normal unilateral side lighting in a classroom would limit its plan depth to two or two and a half times the ceiling height if the glazing were to run continuously from working plane to ceiling for the length of the window wall. Application of this standard produced the typical late 40s - 60s school building characterised by a shallow single-storey plan form with continuous bands of windows around complex perimeter configurations, difficult to cover with anything other than a flat roof, which was punctured frequently with rooflights. Associated with this type of Daylight provision would be those characteristics of the visual environment which could be said to be qualitative viz:

windows with a view connecting the room's occupants with external surroundings and a relationship with the weather,

a strong directional component to the lighting which would give a comprehensible 'modelling' to people and objects viewed in the interior, often represented by physical measurements of scalar,

vector or cylindrical illumination,
a luminance or brightness pattern where walls are lit obliquely and the ceiling is lit by reflected light from the floor inside and outside the room,
sunlight penetration 'stencilling' window openings throughout the interior which is capable of control at certain times of the day or season.

The lighting quantity was generally adequate in these interiors, although often field measurements showed that minimum Daylight Factor values were not achieved throughout all classrooms due to shadowing from partitions, furniture, curtains and display material. /7/

In terms of the visual character of the rooms, the lighting was generally successful, side lighting lent the spaces an intelligibility appropriate in terms of seeing surfaces and objects within the interiors left something to be desired. Many of the classrooms were unilaterally lit with resultant harsh modelling to objects in the room and a pattern of brightnesses within the visual field that gave rise to frequent glare problems. When the plan form was articulated to give lighting from adjacent walls then the environment was improved but except for corner locations this was rarely the case. In deep spaces where the 2% DF was not achieved roof lights supplemented the side lighting giving a different character to the modelling and because of their orientation to the zenith - the brightest part of the overcast sky - they produced pools of high illuminance in remote parts of the horizontal plane very different to the glazing illuminance associated with lighting from the windows. Solar overheating was a common place with little or no control incorporated into the window walls.

The form gave rise to problems - in energy terms it was extremely wasteful, overheating and inadequate ventilation gave difficulties in controlling the thermal environment, the flat roofs and rooflights were a constant source of building failure and maintenance problems. /8/

Schools buildings of the 1970s attempted to address some of these problems based on guidance from the Department of Education and Science which gave broader categories of environmental design advice based on a recognition that environmental parameters were interrelated. The recommendations regarding lighting and the visual environment suggested a percentage of window to external wall area related to the depth of the plan space beyond. In general, the deeper the space the higher the percentage of glazing that was to be incorporated. /9/

Recognition was given to the fact that electric lighting could act in conjunction with Daylight in producing suitable visual environments although Daylight was singled out as being the appropriate, dominant illuminant for school buildings.

The buildings resulting from these recommendations have contained a wide variety of internal environments, mostly the result of energy conserving approaches and in a number of cases, passive solar strategies.

It is worthwhile considering the visual environments produced by these strategies in order to see if there are any generally applicable lessons to be learnt, since it would appear that there is an increase in the use of passive solar methods in designing all types of buildings.

The approach to Passive solar design as currently manifested has been described as one where you "wrap up well and face South". What this has led to are a series of buildings which share certain characteristics viz:

long axis East-West orientation,
highly insulated envelope,
small windows on the North facade and
a heavily glazed South Facade.

The visual environments associated with this approach are:

i. South facing major rooms, mostly unilaterally lit, fairly shallow plan depth normally not over 6 metres. High levels of illumination near the window giving strong to harsh modelling to objects seen in the room. Patterns of brightnesses or luminances within the visual scene will often prove too varied - excessively high near the window and comparatively dark within the room depth - with the result that Glare can be experienced for long periods of time and controls will normally be necessary to obviate its effects. These controls may be expensive to supply if they are to operate automatically and a nuisance to the occupants of the space if they have to manipulate cheaper means of control such as blinds or other shading devices. Experience has shown that occupants use sun controls badly even if the correct methods of application have been pointed out to them. Reports show that blinds are often left in place long after the glare condition which gave rise to their use has passed /10/ and that blinds are operated incorrectly through ignorance or laziness so that unpleasant visual conditions result.

Views through the window may be good but this can work both ways so that problems of privacy can result. The clip-on 'sun space' while offering advantages in thermal terms may also create privacy difficulties if the occupants can see out the passer by can see in, particularly if large sheets of glass are used. Experience with high density passive solar housing schemes with privacy problems have shown that it is worthwhile considering a technique of the past - the use of an increased number of transomes and mullions to break up the glass surface. /11/ The result is that it becomes more difficult to see into the interior directly from the outside, due to the eyes adapting to the higher luminances of the glazing bars rather than the lower luminances of the internal scene beyond. Sun spaces on the ground floor are particularly prone to privacy complaints if used in a high density development. Unlike the bay window, which in some ways they resemble, it is less easy for occupants to retreat back into the room if they feel they are being watched. The use of sun spaces above the ground floor so that occupants can overlook external activity without being seen directly from the street has so far not been used frequently in passive solar design. This 'pattern' as a design approach has been recognised by Alexander /12/ who quotes the use of the Peruvian mirador as an example of a successful upper storey window strategy.

ii. Rooms on the North facade of the building are generally unattractive in terms of the visual environment. Daylight as it is experienced in a Northerly aspect is an illuminant source cold in its colour appearance, conveying a somewhat flat and lifeless character to the interior. In summer there will be some sunlight in the rooms but only for a short time early morning and early evening - times of the day which would be rarely experienced in working interiors. Most of the time the small windows will be facing the Northerly part of the sky vault which will be blue, overcast or partially cloudy without the possibility of sunlight. In many cases of Passive Solar design the rooms on the North facade are underlit in terms of the quantity of illumination. In housing perhaps this is not too important for service rooms such as lavatories and bathrooms but bedrooms and kitchens which are often delegated to the 'back' of the building are often seriously underlit.

Recommendations for Daylight in Dwellings /13/ would suggest that the kitchen is the room in the house which should have the most Daylight and yet in order to balance the number of rooms on the South facade the kitchen is often placed on the North facade with a small window opening.

Consequently it is often underlit, particularly when compared to the major sunlit spaces in the interior. In other building types it is often difficult when adopting a Passive Solar strategy to balance the North and South facades with an equal weighting of the elements of accommodation. Normally there are more demands in the 'brief' for accommodation to face South than to face North.

There have been few studies of the effects of solar strategies on visual environments /15/, more will be needed before solar thinking becomes totally accepted into architects' design approaches. The first findings would seem to indicate that active solar buildings are not much liked by the general public, mainly due to a disaffection for the associated hardware, but also because of the quality of the visual environment in these buildings was unacceptable. Shading of windows, obstruction to view, confined aspect and restrictions on spatial planning are characteristic complaints. Passive solar buildings do not suffer as badly from complaints of this kind, on the contrary they appear to be well liked, yet there are joint problems that all solar buildings have. Certainly the trend towards general acceptance of passive solar strategies will be accelerated only if the positive quality visual characteristics which have merited favourable responses from building users can be maximised while the negative ones are minimised.

Case studies would usefully develop information on users' responses to the following qualitative characteristics:
i. View, its depth and breadth. Feelings of contact between interior and exterior.
ii. Size and type of window opening. Feelings of privacy.
iii. Luminance pattern of typical views in the interior. Feelings of concentration or relaxation produced.
iv. Sunlight penetration into interior. Times of day and seasons of the year. Attitudes towards this and possibilities of control.
v. Glare from sky. Control and acceptability.
vi. Modelling of objects and people in the interior. Acceptable or desirable.
vii. Visual performance. Suitability of lighting to support the designated activities.
The subjective responses of the building occupants should be supported with relevant physical data. This would take the form of luminance and illuminance/Daylight Factor measurements, scalar vector or cylindrical illuminance levels. Contrast Rendering Factor and visual performance tests and time lapse photography to illustrate the dynamic nature of Daylight and the geometric relationships of window and view.
Complex measurements and analyses of the visual environment have occasionally been undertaken in the past /15/ and doubts have been expressed about the possibilities of feeding these research findings into sensible guidance for the designer. Information regarding these types of qualitative factors in the visual environment can now be readily absorbed into the visual design tools which are coming into increased use with CAD. In the new thinking that use of these tools will entail, the reintergration of the quantitative and qualitative aspects of lighting and the visual environment will again take its place in new and better design recommendations.

REFERENCES

/1/ Hopkinson, R.G., Architectural Physics - Lighting, London, H.M.S.O., 1963.
/2/ Hopkinson, R.G. and Kay, J.D., The Lighting of Buildings, London, Faber, 1969.
/3/ Karp, W., The Center: a history and guide to Rockerfeller Center. New York: Van Nostrand Reinhold, 1982.
/4/ Alexander, C., Ishikawa, Silverstein, M., A Pattern Language: Towns, Building Construction, New York, Oxford University Press, 1977.
/5/ Department of Scientific and Industrial Research, Post-War Building Studies No.12 The Lighting of Buildings, London, H.M.S.O., 1944.
/6/ Department of Education and Science, Lighting in Schools, Building Bulleting No.33, London, H.M.S.O., 1967.
/7/ Gloucestershire County Council, Integrated Environmental Design: A feasibility study for School buildings, 1971.
/8/ Hardy, A.C. and O'Sullivan, P.E., Insolation and Fenestration, Newcastle, Oriel, 1967.
/9/ Department of Education and Science, Guidelines for Environmental Design and Fuel Conservation in Educational Buildings, London, H.M.S.O., 1982.
/10/Nelson, G. and Simpson, F.M., The seasonal management of windows, Oxford County Council, 1981.
/11/Alexander, C. et al, A Pattern Language: Towns, Building Construction, New York, Oxford University Press, 1977.
/12/Alexander, C. et al, Ibid.
/13/Code of Functional Requirements of Buildings CP3, Chapter 1(A) Daylight (dwellings and schools), London, H.M.S.O., 1949.
/14/Crisp, V. et al, Daylight as a Passive Solar Energy option, Building Research Establishment, 1988.
/15/Ward, C., British School Buildings:Designs and Appraisal 1964-74, London, Architectural Press, 1976.

THE SCALE OF LIGHT - A NEW CONCEPT AND ITS APPLICATION

S. Frandsen
Lighting Laboratory, School of Architecture
1 Peder Skramsgade, DK-1054 Copenhagen K, Denmark

The paper presents three planning-devices developed in recent years by the Lighting Laboratory in Copenhagen.
1. The Four Shadows, which according to place and size, categorizes the shadows existing in ordinary side-lit rooms, and without any modification is adaptable to conditions of artificial lighting. The extent of the shadows is related to the diameter of the human trunk, two experienced as bigger, and two as smaller. A. The big room shadow. B. The big object shadow. C. The small object shadow. D. The detail shadow.
2. A Scale of Shadows defined by the angular relation between source and object. The scale has 10 subjectively evaluated equal intervals, showing on a sphere the result of the change from parallel to diffuse light sources. Geometrically it is quite simple, being defined by the percentage of the (partially) shaded areas on the sphere (0%, 10%, 20% etc.) and produced by a circular diffuse light source of which the maximum, corresponding sky factor is 0%, 1%, 4% etc. - all the whole squares between 1 and 100.
3. Using The Scale of Shadows as a precise description of each of The Four Shadows, we can derive The Scale of Light which, theoretically at least, not only describes, but also indicates the degree of diffusion that should be found in everyday conditions.

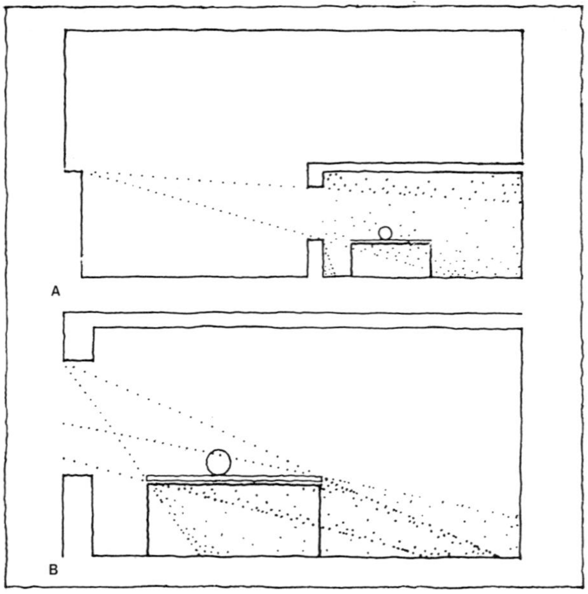

fig. 1

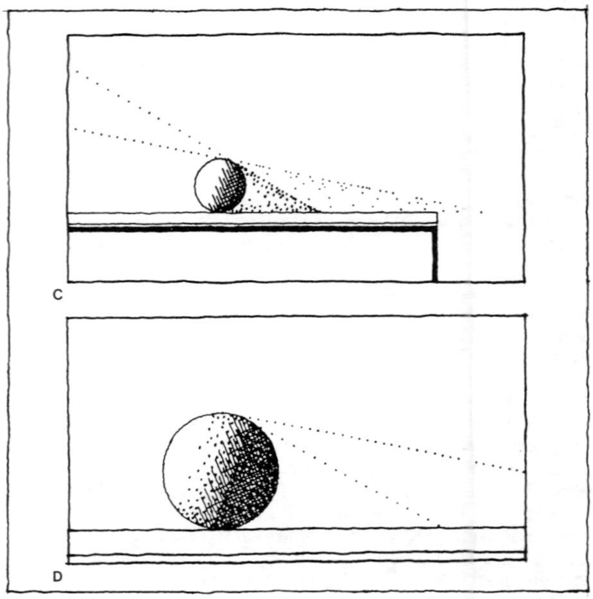

fig. 2

Light has a special quality that very often goes unnoticed - it has scale. In almost the same way that small objects are overlooked and get lost in a larger space, they tend to get lost if the light is 'large', coming from a large area. Attention simply shifts from detail to totality when light sources become larger and, inversely, from totality to details as the sources become smaller.
For centuries artists have known and used these effects - and with a little effort they can be found everywhere in our surroundings. Yet in modern lighting research and engineering, they seem next to unknown. Daylighting is almost exclusively based on the wellknown calculation of daylight factors, on a working plane, on a wall etc. No methods exists that can lead to a balanced use of ambient vision as opposed to focal vision, providing feeling of space versus interest in objects, respectively. It is to handle this conflict between the visual reality and the way in which it is treated, that the new devices of this paper have been developed.

1. THE FOUR SHADOWS

On the basis of experience with an ordinary sidelit room, prevailing shadows can be categorized according to their size as follows: (fig. 1-2).
A. The big room shadow
B. The big object shadow
C. The small object shadow
D. The small detail/texture shadow
The terms big and small are to be seen in relation to the diameter of the human trunk, which seems to be a usefull yardstick of size perception.
In the room the two big shadows, A, and to a certain degree B, occur where the daylight is relatively weak and dominated by the light reflected from room surfaces. The two small shadows, C and D, occur where daylight is relatively strong, dominated by light from the window.
Consequently, the two big shadows are those related to the lighting of the room as such, shadow A in the upper part and shadow B behind the furniture and larger architectural elements.

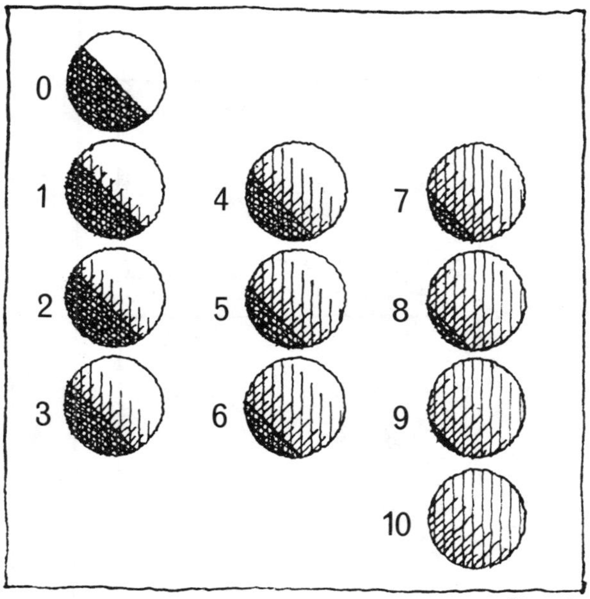

fig. 3

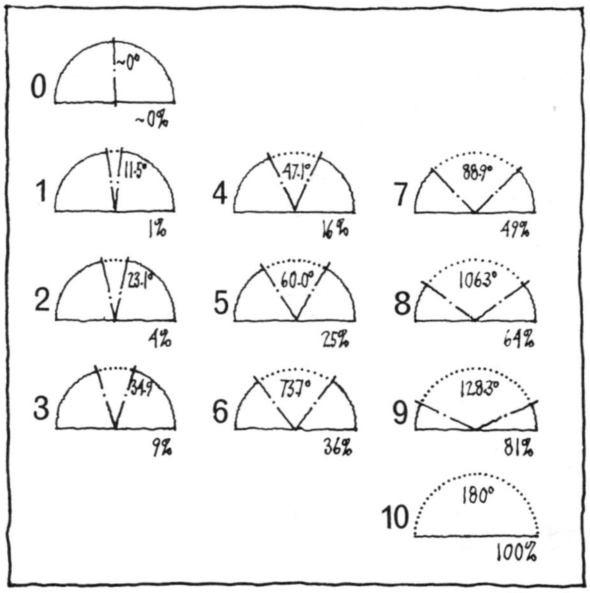

fig. 4

The two small shadows are those associated with the light on small objects; shadow C on all things that we handle and, shadow D in the tiniest visible detail.
In summary, the four categories differ from each other in four ways: by their position in the room; by their size; by their illumination level; and, by their precision. The smaller the shadow the stronger the contrast, since the gradation from bright to dark occurs with a decreasing viewing angle.
The perceptual outcome of The Four Shadows is that the light in the room is perceived as being more or less soft, without the sharp transitions that distract and easily get misinterpreted if they occur in peripheral vision (who has not stumbled over a step that wasn't there, because a shadow on the floor was misinterpreted?).
Conversely, the light on objects is characterised by a distinctness which facilitates the perception of three-dimensional form and detail and, as long as the reflected light is reasonably balanced will the contrast from light to shade be kept within acceptable limits.

2. A SCALE OF SHADOWS

Obviously the handling of shadows asks for more than just two words: 'soft' and 'sharp'. Nearly everything inbetween seems to be without name and explanation. So what'sproposed here is a scale of shadows.
In principle it is constructed as a grey scale, each step being visually equivalent to all other steps. The scale comprises 10 intervals and 11 types of shadows (fig. 3) each characterised by a specific illumination of three separate areas on the sphere.
1) The top segment facing the circular light source receives light from the entire source.
2) The bottom segment, turned away from the light source receives no light from the source, and is therefore in full shadow.
3) The middle zone of the sphere, located between area 1 and area 2, receives a varying amount of light from the source, which is sometimes called semi-shadov.
Of these the middle zone has the decisive role. It is here that the shading occurs from the brighter part to the darker part of the sphere. In completely parallel light there is no shade, only two hemispheres, one with and one without light from the source. In totally diffuse light it is all shading, and the gradation extends over the entire surface
In all intermediate situations, individual shadow-types can then be characterised by the size of the shaded middle zone, which leads us to the following conclusion: it is the size of the middle zone that determines the intervals on the scale. On this basis the shadows can be categorized by type and area of coverage: type 0: zero shading, type 1:1/10 shading, type 2:2/10, type 3:3/10, etc., ending with type 9:9/10, and type 10:10/10 shading.
Assuming that the circular light sources are very large in relation to the sphere, then the solid angles having the centre of the sphere as their apex will follow a simple pattern. Their maximum illumination will be determined by the projected solid angle, the very same method used for daylight calculations and so we get: type 0: (the so-called sky-factor%), type 1:1%, type 2:4%, type 3:9%, etc., ending with type 9:81%, and type 10:100%. (Fig. 4).
It must, however be mentioned that the shadow scale was developed in a black room with no reflections, apart from the small horizontal areas beneath the balls. Against a brighter background, all shadow types will shift 0.5-1 step up to the scale, just as in a grey scale.

fig. 5

fig. 6

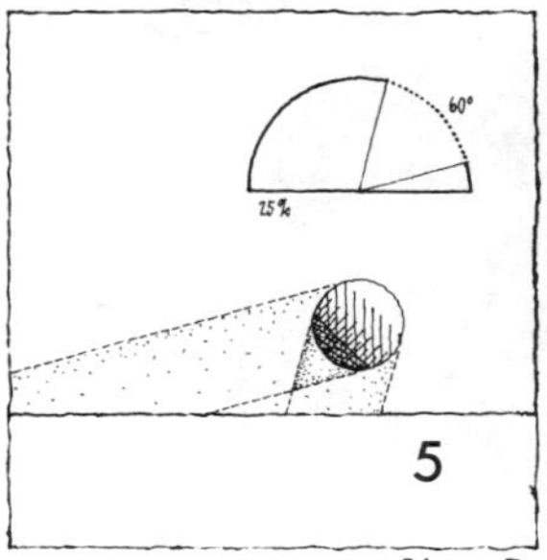

fig. 7

fig. 8

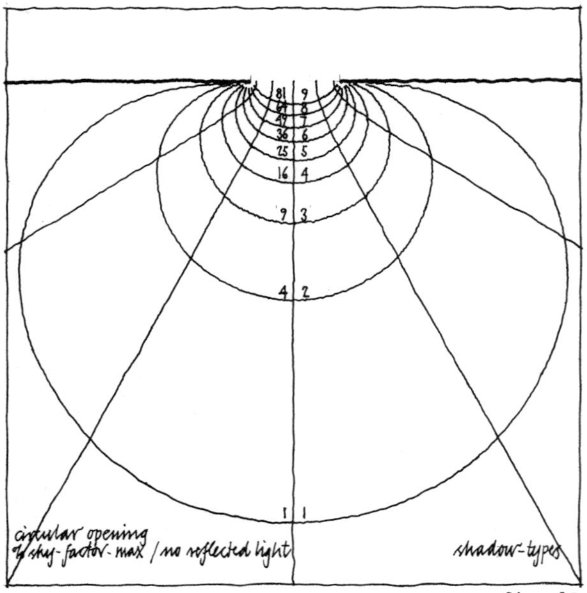

fig. 13

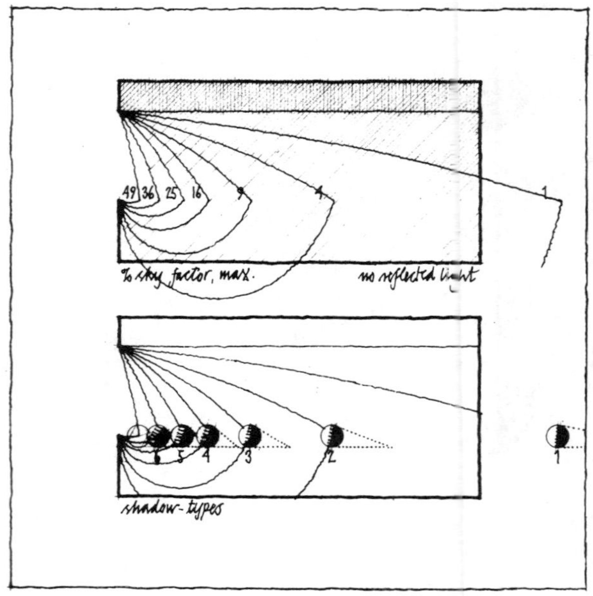

fig. 14

3. SHADOWS, FORM, AND TEXTURE

The visual effects of the eleven steps are obvious. In predominently parallel light (fig. 5 - shadowtype 1) the two balls look as if they were protruding from the picture plane. In diffuse light (fig. 6 - shadowtype 9) they get flat as a full moon. As to the texture of the tennis ball, there is a corresponding change from the obvious articulation of every little hair to the impression of a fur. For practical purposes, however, it is possible to consider the whole shadow-scale as composed of three separate parts. The two extremes dominated by parallel light (no. 0-3), respectively diffuse light (no. 7-10), and the middle one (no. 4-6) combining the characteristics of the two extremes. In the middle one (fig. 7-8 - shadowtype 5) a new quality emerges. Neither dramatically three-dimensional nor flat and dull. Capable of showing the form of the ball as well as the texture of the surface. One could probably say that the middle is for everyday use, stimulating what psychologists call constancy.
The two extremes are then for artistic purposes - or for advertising.
It should be kept in mind, however, that The Scale of Shadows tells us about solid angles and their effect. This means that the same angular relations are valid irrespective of the size of the object. The greater the difference between the physical size of the main form and that of the detail, the more difficult it becomes to simultaneously optimise the lighting on both.
This conflict is essential. Just as details are pushed into the background when a large form receives 'large' lighting, they can also be emphasised by directional lighting, in which case the main form tends to get lost. We cannot expect to have both as a matter of course, so we must make a choice that choice will normally be determined by the need for emphasis. As an example consider the three balls (fig. 9-12). (The changes in arrangement are evident on the silver ball).
In parallel light (fig. 9) the shadows are so sharp and so dense that the spheres almost lose their form, it's like camouflage. Even the small pits in the orange are big enough to create harsh and disturbing shadows. Only the hairs of the ball and the fibres of the paper stand out in a natural way, primarily as a result of gloss.
In very diffuse light (fig. 10) the lack of shadows means lack of threedimensional form. The spheres do not appear spherical and the texture is missing. The tennis ball looks like cotton wool while the orange appears remarkably unappetising.
A combination of parallel and diffuse light (fig. 11) maintains the sharp shadow edges and pronounced texture of fig. 9 while providing a shadow density that permits visual penetration. The spheres become spheres, even though they still may call for some interpretation. The texture is - if not beautiful - at least recognisable.
When parallel light is replaced by a primarily directional one (fig. 12), still in combination with diffuse light from an infinite source, the ball becomes really round, soft, and downy, and ready for touch, just as the orange appears ready for eating. The corresponding shadowtypes are 3½ and 10.

4. SHADOWS AND SPACE

The next question is then: Where in a room are the different shadow-types to be found? Or, since the shadow-type and the sky-factor are linked to each other: Where in the room are the contours of the maximum sky-factor to be found, and what is the local direction of the shadows (fig. 13) gives the

fig. 9

fig. 10

fig. 11

fig. 12

fig. 15

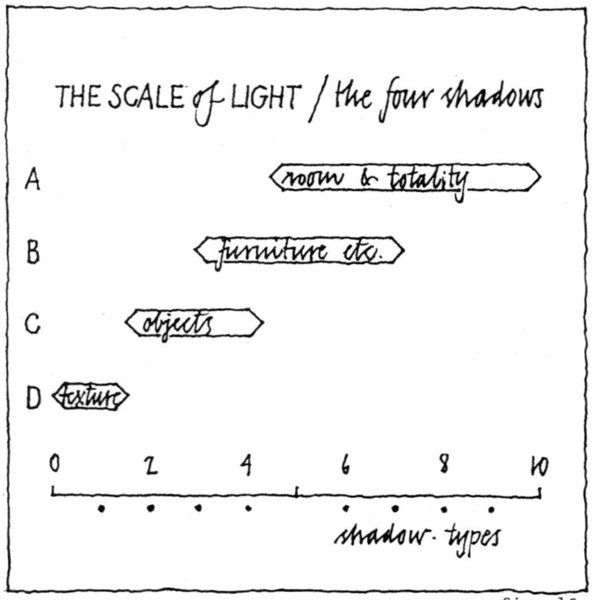

fig. 16

answer for a circular opening. Each of the more or less elliptic contours represents one of the maximum obtainable sky-factors, 1%, 4%, 9%, etc. corresponding to shadow-type no. 1, 2, 3, etc.
The hyperbolas then show the direction of the shadows, and their deformation as well. Perpendicular to the opening all the shadows on the sphere itself and projected from the sphere will be circular. Everywhere else they are more or less elliptic in proportion to the slope of the asymptotes.
Please note that this kind af curves - although they look almost as common iso-daylight contours - are quite different from those. The daylight contours normally describe the distribution over plane surfaces. But these curves represent what could be called the morphology of light. It is a three-dimensional map of the light we are sitting in and moving through, and of course that which gives form to all objects in our surroundings, from the smallest to the biggest.
With the help of such curves it is possible to analyse the effect of lamps and of window-arrangements. The combination of shadow-type 10 with shadow-type 3,5 (fig. 12) in order to create a certain visual result is actually the first step in the design of a lighting fixture. If you look at a Danish PH-lamp, (fig. 15) based on principles that are now more than 60 years old, it is easy to imagine how the four shades, seen from a given position in the room, each produces a different shadow-type. And that means that the fixture as a whole covers a larger interval of texture and form than for example a translucent globe of the same size and light distribution.
Turning to the sections of a sidelit room (fig. 14) the conditions are probably more recognizable - and the addition of approximately 1% diffused reflected light wouldn't change the overall picture.
The relation sky-factor/shadow-type explains itself. (It should be mentioned that the sphere are oversize). It even becomes evident that the shadow-types can be used in the analysis of gloss. Just think of these spheres as red or white billard balls.
Sections of this kind could be the basis of a more thorough understanding of daylight planning than we have today. And in addition it now makes it possible to analyse f.x. the light in the interiors of Vermer van Delft, or on the nudes of Ingres, or the apples of Paul Cezanne.
Up to now it has all been questions of more or less sophisticated lighting techniques. But there is one thing more.
It can't be illustrated here but it is easy to observe if rooms of identical size are influenced by different shadow-types: Directed light - the smaller shadow-types - shortens all distances, whereas diffuse light increases them.
Both can be influenced by brightness, but they exist independently. Exactly the same phenomenon can be seen on the balls of the shadow-scale; the diffusely illuminated no. 8-10 seem to be farther away, no. 0-2 seem to be nearer. On a big scale it can be observed in the mountains. It is the 'aerial perspective' of Leonardo da Vinci.
The conclusion in simple. Distance and background are connected with the blurring of form and lack of texture. Inversely, nearness and foreground are linked to precision of form and emphasis on detail. That is part of the 'figure/ground problem' which has engaged visual psychologists since the beginning of the century. It occurs everywhere and conscious planning of light should try to take it into consideration.

5. THE SCALE OF LIGHT

As I mentioned in the introduction, we see in two different ways, simultaneously using our focal vision and our ambient vision. Outside of the visual axis and its immediate surroundings we are not observers, but rather 'perceivers'. And what we perceive is space. The very same light that centrally placed means attraction and concentration, could easily, placed in the periphery, cause distraction and even insecurity. In everyday life the periphery calls for softness and peace, which again means that it should be dominated by relatively diffuse light. This leads to The Scale of Light in which The Four Shadows are combined with The Scale of Shadows.

A. the big room shadow: shadow-types 4½-10
B. the big object shadow: shadow-types 3-7
C. the small object shadow: shadow-types 1½-4½
D. the small detail/texture shadow: 0-1½

These indications are approximate, The Scale of Light is still a conceptual framework. But as a guide for observation and analysis, in education and in planning it seems to work.

Sophus Frandsen: The Scale of Light, ILR 1987/3.

DAYLIGHTING IN GLAZED ATRIA

Gabriele Willbold-Lohr

RWTH Aachen
Lehrstuhl für Baukonstruktion II
Prof. W. Döring
Schinkelstraße 1, D-5100 Aachen

now:
Büro für energiegerechtes Bauen
Oldenburgerstraße 68, D-5000 Köln 60, FRG

ABSTRACT
Glazed atria have become an interesting international architectural style in recent years for different use. Mainly considerations as - the desire for an attractive environment using lots of greenery which also protects from noisy and polluted surroundings -and the seek for a building type with a better energy balance by providing buffer spaces which protect the occupied parts of the building from the full impact of external temperatures, wind, rain and insolation have contributed to the relatively fast implementation of this building type into the architectural language.
This paper presents results of extensive scale model studies for defining the influencing parameters for the illumination of the atrium and the illumination of the adjacent office spaces for overcast sky conditions.

1. DAYLIGHT IN THE ATRIUM

Daylight in the atrium design is important for two reasons
- for plant growth and
- for contribution to the illumination of the adjacent spaces

1.1 Plants in the atrium

The architectural design of atria can create a positive impression of quality with the use of plants. Depending on the type of plants they require different amounts of daylight and for a longer period of time than office hours, usually more than 10 hours/-day. The minimum illuminance for survival depends on the size and the species of the plants and varies from 250 lux for small plants to 2000 lux for trees and tropical plants.
In daylighting terms, a factor of 5 to 40% will be needed under the 5000 lux standard sky.
Besides the attractive space which can be created with the use of plants they effect the illumination in the atrium and the adjacent spaces by reducing the reflectances of the floor and the walls up to 0.2 (if the plants cover completely the surfaces).

Species	Required Illuminance (lux)
Tropical plants	above 2000
Trees (1.5 - 10 m)	750 - 2000
Floor plants (0,6 - 1,8 m)	250 - 750
Desk plants (below 0,6 m)	250 - 2000

Table I /1/:
Recommended illuminance for acclimatized plants

1.2 Contribution to the illumination of adjacent spaces

The design of atria in office buildings provides a new quality of working spaces facing the atrium. The design of the atrium and the use of plants influence the contribution of natural light to the illumination of the office space to a very high extent.
Following the path of the light travelling through the building until it reaches finally the desk of an office and the proirity of design decisions the following design parameters for a good daylit atrium have to be taken into consideration:
- geometry of the courtyard
- roof construction and glazing
- reflectances of atrium walls and floors
- glazing to the adjacent space
- dimensions and reflectances of the adjacent space

Fig.1: Scale model of the 10-storey square shaped atrium

Fig.2: Fisheye photograph of the 10-storey square shaped atrium (scale model) from the floor

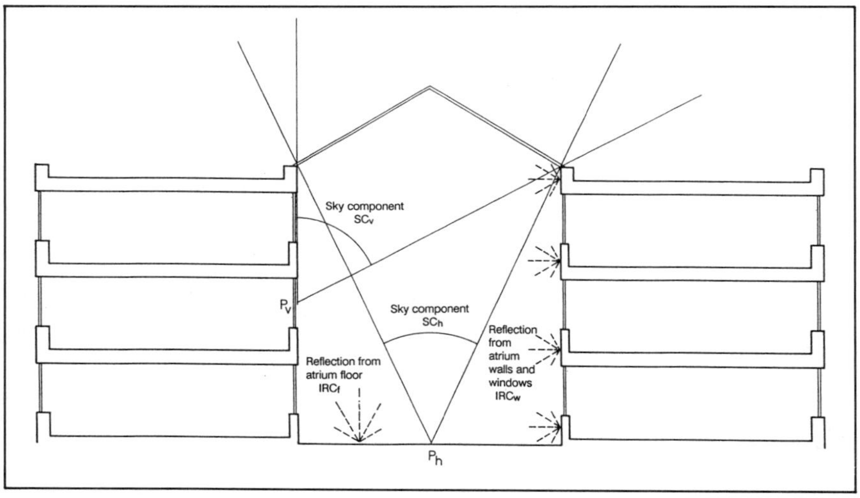

Fig.3: Natural illumination of the atrium and influencing parameters

2. OUTLINE OF THE SCALE MODEL TESTING

The interrelationship of all mentioned design parameters on the illumination of the atrium and the office space is quite complex and therefore analysed with the aid of scale models. All scale models are equipped with numerous photometric sensors and have been tested in an artificial sky of the mirror box type at RWTH Aachen. All measurements are focussing mainly on daylighting under overcast sky conditions. Three types of scale models have been built:

- square shaped atrium
- triangular shaped atrium
- rectangular shaped atrium.

The scale of the models was 1:50; the geometric aspect of the atrium space could be altered by changing the level of the floor of the atrium. Also the reflectances of the surfaces (walls and floors) could be changed easily. All scale models were designed with the same width and height of the atrium but with different length. The size of the square shaped atrium was 20 by 20 meters. Each storey height was 3 meters. Maximum height of all atria was 10 storeys.
Different facade apertures

- completely white facade (reflectance 0.7)
- facade with 50% window/wall ratio (average reflectance 0.4)
- only glazing (average reflectance 0.1)
- completely black facade (reflectance 0.05)

have been tested with glazing material for taking into account the reflection of light off the glazing in an atrium.
The size of the tested office space was 9 by 9 by 2.8 meters with a window opening of 1.5 by 9 meters. The reflectances of the surfaces were chosen as
- walls 0.6, - ceiling 0.7, - floor 0.3.

3. ILLUMINATION OF THE ATRIUM

The illumination in the atrium is defined by two components:

- the direct light reaching the bottom and the walls of the atrium SC (horizontal or vertical)
- and the reflected light off the surrounding surfaces of the atrium IRC (IRC_{floor} and IRC_{walls}).

3.1 Horizontal illumination at atrium floor level

The horizontal illumination in the center of the atrium floor (presented as DF-curve in fig. 4) depends mainly on the geometrical aspects of the atrium space given as aspect ratio of the atrium at the x-axis. A square shaped atrium with the proportions 1x1x1 has an aspect ratio of 1, with the proportions 1x1x0.5 an aspect ratio of 4.

The main contribution of light is the direct light coming from the sky with little variations due to the shape of the light admitting area. The influence of the shape of the atrium is evident in the performance of the internally reflected component: due to the fact that the surrounding surfaces of the square shaped floor plan are minimised compared to the rectangular and triangular floor plan the number of interreflections in the atrium is smaller. Therefore the square shaped atrium provides about 10% more daylight than the rectangular shaped atrium.
The total illumination varies from 60 to 80% for wider atria than 1x1x1; for higher atria the total illumination decreases rapidly as both components are decreasing. For an atrium with 1.5 times the width only a Daylight Factor of 30% can be reached. All these values given are based on a complete white atrium and indicate the maximum contribution of daylight.
However the design of the facades may influence the illumination due to altered reflectances. A facade aperture with 50% window openings will reduce the contribution of the IRC by half and having only glazed walls as separation between the office and the atrium the IRC is reduced to 1/3 of the white walls, and almost reduced to the contribution of the skylight alone.

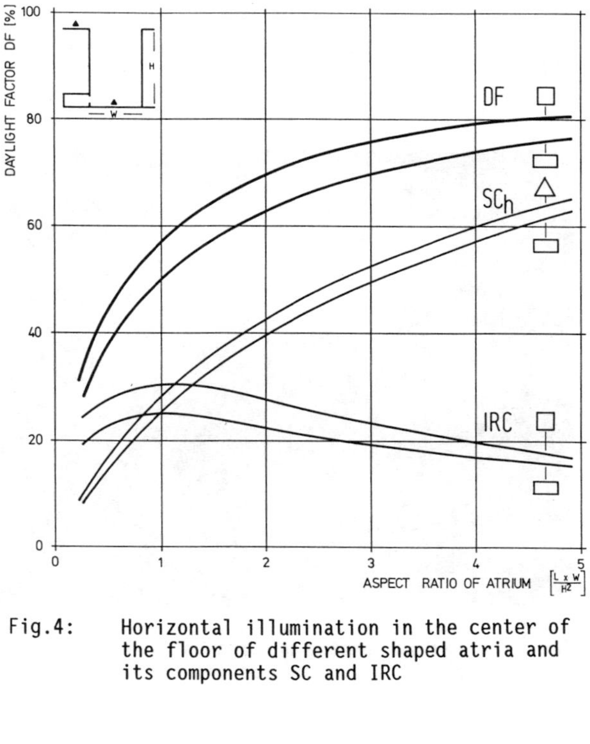

Fig.4: Horizontal illumination in the center of the floor of different shaped atria and its components SC and IRC

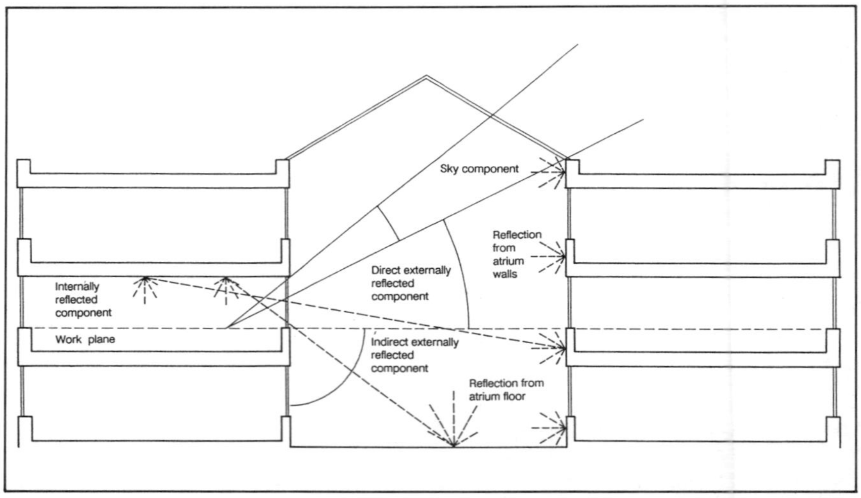

Fig.5: The atrium as a source of daylight for adjacent spaces

Much more influence on the illumination in the atrium than the shape of the light admitting area however is the reduction of the light by the roofing of the atrium.
The construction of the roofing has to be minimised but still the most advanced roofing construction would reduce the light admitting area by at least 10%. The glazing material used again will reduce the amount of light entering the atrium by its transmission porperties. The minimum reduction using single pane glazing therefore still would reduce the daylight entering the space by 20%. This means instead of having a horizontal illumination of 60% (1x1x1) in an open courtyard there would be an contribution of daylight of 45%. Using double pane glazing would reduce the daylight at floor level to 40%.
Especially for climates with mainly overcast sky there is a necessity to reduce any obstruction in the light admitting area to a minimum. This also concerns fixed shadings and fixed light directing elements in the roofing area which can be seen as heavy obstructions in the light admitting area and therefore have a negative impact on the illumination for most of the time of the year. Using sunlighting systems to redirect sunlight in the atrium therefore have to be thought carefully and checked concerning their efficiency both on daylighting and sunlighting.
Also movable shading is recommended instead of fixed shading devices.

3.2 Vertical illumination at atrium walls

The vertical illuminance at the atrium walls defines due to its composition the amount of light and the way it is entering the adjacent space.
Fisheye photographs from the window of an office space towards the atrium demonstrate the different parameters influencing the vertical illumination at the atrium facade:

- direct light coming from the sky reduced by the roofing of the atrium
- reflected light from the walls facing the office space and from the floor of the atrium.

4. ILLUMINATION OF THE SPACE FACING THE ATRIUM

Both the Sky Component and the Internally Reflected Component of the atrium can be considered as external light sources for the illumination in the office space which have to pass through the thermal and visual separation between the office and the atrium. Again the amount of light will be reduced due to the framing of the window and the transmission of the glazing. In order to provide as much daylight in the office as possible the optimization of framings and transmission properties of glazing materials is necessary.
The illumination in the tested space which represents a medium sized office space of 9x9x2.8m with a window opening of 1.5x9m and a sill height of .85m

Fig.6: Fisheye photograph from a window of an adjacent space towards an atrium

Fig.7: Fisheye photograph of an office space facing the atrium

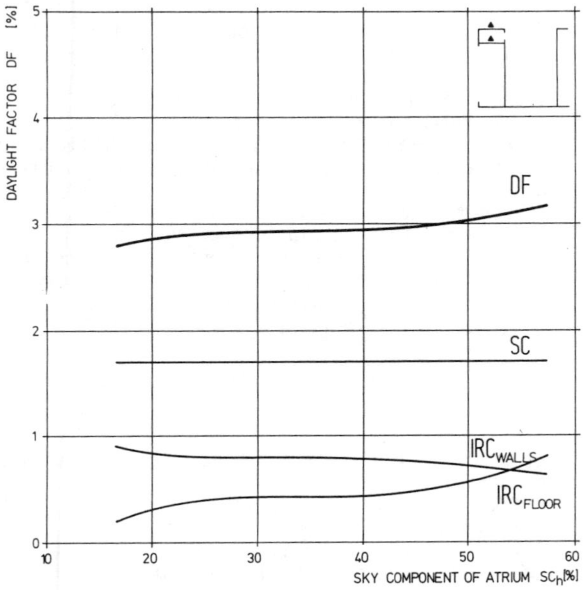

Fig.8: Illumination at workplane height at a distance of 4.5m from the window and its components due to the illumination of the atrium, room located next to the roofing of the atrium, overcast sky

is illustrated in fig.10:
- for a 5 and 9 storey square shaped atrium building with an atrium wall of 50% glazing and 50% white wall elements
- and for a 5 and 9 storey square shaped atrium building with completely white atrium walls.

The dottet line in the graphs indicates the daylight distribution in the same space facing the opposite direction of the atrium (to the exterior).

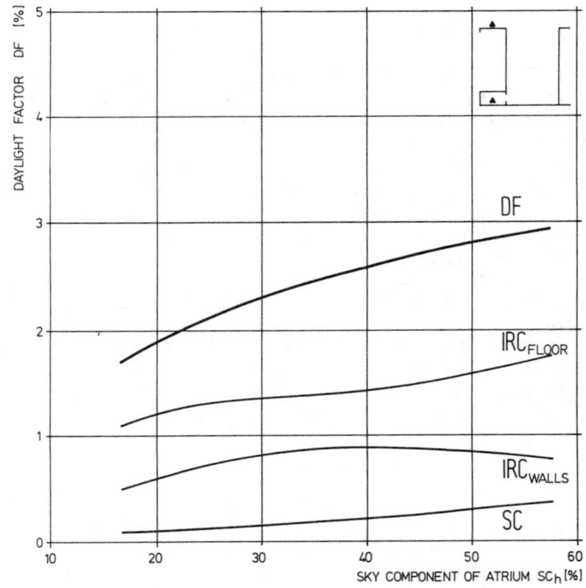

Fig.9: Illumination at workplane height at a distance of 4.5m from the window and its components due to the illumination of the atrium, room located next to the floor of the atrium, overcast sky

Curve A illustrates the daylight distribution in the room near the roofing of the atrium and Curve C gives the daylight distribution for the lowest room. The contribution of daylight to the illumination varies with the location of the room within the height of the building. As the illumination of the room located at floor level is mainly determined by the light reflected from the floor, the room located near the roofing receives most light directly from the sky.

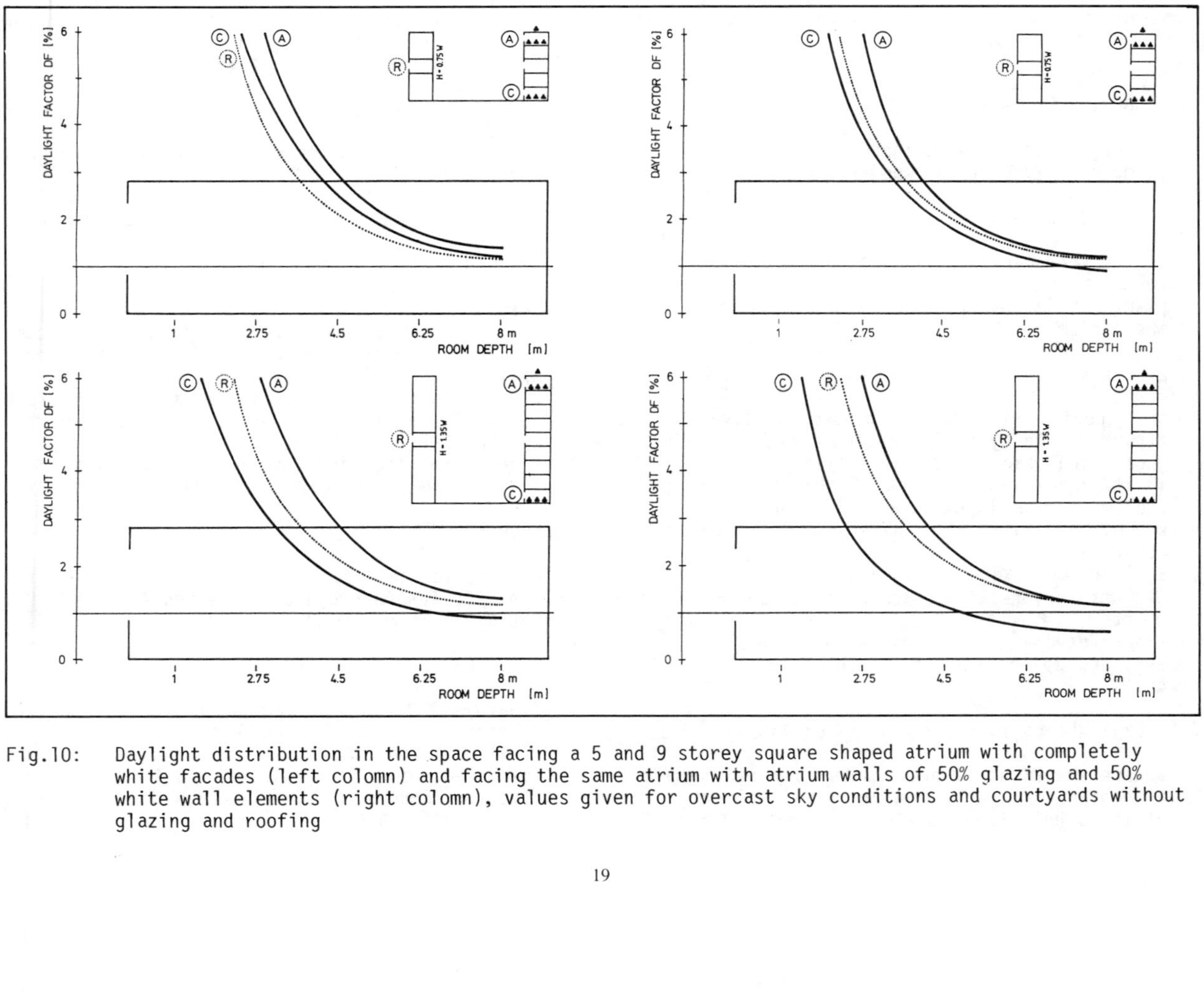

Fig.10: Daylight distribution in the space facing a 5 and 9 storey square shaped atrium with completely white facades (left colomn) and facing the same atrium with atrium walls of 50% glazing and 50% white wall elements (right colomn), values given for overcast sky conditions and courtyards without glazing and roofing

All rooms depending mainly on reflected light as main daylight source require a high amount of reflected light for acceptable illuminance levels; this means that the reflectances of the atrium surfaces have to be as high as possible to reflect down as much light as possible.
In the presented series of measurements only atria (square shaped) with a height smaller than 0.75 times the width contribute in a positive way to the illumination of the adjacent spaces by increasing the quantity of light in the rooms. The quantitative improvement of natural illumination is based on the contribution of the reflected light off the surrounding atrium walls and off the atrium floor.
Although the quantity of light is decreased in higher atria than 0.75 of the width, the quality of illumination is improved as the uniformity in these spaces illuminated mainly by reflected light is better and glare problems in these rooms are reduced.
The use of plants on the atrium floor affects the contribution of reflected light IRC_{floor} to the illumination of adjacent spaces: landscaping design in the atrium using dark coloured materials and plants reduces the reflectance of the atrium floor to 0.2. The natural illumination of all rooms facing the atrium and depending very much on reflected light from the atrium floor (almost all rooms except the rooms located near the roofing) will be very poor as the quantity of light is decreased.
However the quantity of light can be improved by using light directing elements in the facade apertures which direct zenithal light onto the ceiling. Additional studies /2/ show that this measure only helps improving the quantity of daylight in the space if the the reflectances of atrium walls and floor are very low; using these elements in combination with white atrium facades only the quality of illumination in these spaces can be improved. The additional studies also show that the quantity of light in the spaces due to white atrium facades (high contribution of reflecetd light) is higher compared with the quantity of light having a low contribution due to reflected light and using light directing elements.

5. DESIGN RECOMMENDATIONS

for closed atria designed for locations with mainly overcast sky conditions

5.1 Purpose and use of the atrium

- in the very early design stage the decision has to be made whether the atrium is used for growing plants or improving illumination in spaces facing the atrium

5.2 Plant growth in the atrium

- the species of plants intended to live in the atrium define by their requirements for light the geometrical aspects of the atrium
- if the geometrical aspect of the atrium is given and also the light available for survival is known, the species of plants have to be chosen accordingly
- another important selection criteria for plants is whether they require sunshine for growing/blooming or not
- plants need a careful maintenance for surviving in enclosed spaces such as cleaning the leaves from dust

5.3 Daylighting in atria and spaces facing atria

- square shaped atria provide the highest contribution of natural illumination
- atria higher than the width are not suitable for daylighting as the contribution of direct light coming from the sky is rapidly decreasing with the height
- facades with high reflective materials result in the highest natural illumination
- specular reflective materials perform better than diffuse, but there is a potential of discomfort due to glare
- sloped facades of atria opening the atrium to the sky improve the contribution of the skylight
- atrium facades with high reflectivity in the upper storeys reflect down as much light aspossible
- the size of the windows should be adjusted according the available light level

Roofing of atria

- minimization of the reduction of the light admitting area is strongly recommended
- glazing materials should have high transmission properties
- diffusing glazing materials reduce the light transmittance for diffuse skylight
- fixed shading devices reduce the light admitting area and therefore are not recommended (movable shadings)
- the construction for the atrium roof should sit on top of the adjacent building or be fixed in front of the atrium walls with acertain distance so that light can wash the walls

Shading

- movable shading is recommended (fabrics)
- shading can also be provided by designing large high reflective vertical baffles in the roofing area excluding the sun with minimized reduction of the light admitting area (problem: weight)
- diffuse glazing materials can cause discomfort to the occupants due to the bright sky

Illumination of spaces facing atria

- the size of the windows should be adjusted according the available light level
- transmission properties of the separating glazing should be as high as possible, if building codes and fire regulations allow single pane glazing would suit best
- minimization of the framing in the window area is necessary
- rooms with a high contribution of reflected light from the atrium floor have to have high reflective ceilings
- light directing elements only increase the quantity of light in atria with dark surrouncing surfaces
- for improving the quality of light light directing elements also work in atria with high reflectiv walls
- closed square shaped atria with a height <0.75 of the width result in a quantitative improvement of the natural illumination
- light directing elements in the roofing area for sunlighting do not suit for northern climates

ACKNOWLEDGEMENT

These studies have been carried out within the frame of the CEC funded project: Concerted Action on Daylighting, Program PASTOR.

REFERENCES

/1/ Navvab,M:Daylight Design for Plant Growth, Draft, CIE TC.3.11, June 1987
/2/ Willbold-Lohr, G: Verbesserung der Tageslichtbelichtung des Hinterhauses Oranienstraße 9 in Berlin, Studie im Rahmen des CEC Projektes Building 2000, 1989

SIMULATION OF LIGHT PROPAGATION APPLIED TO DESIGN OF DAYLIT BUILDINGS

M. Fontoynont and B. Paule
Laboratoire Sciences de l'Habitat
Ecole Nationale des Travaux Publics de l'Etat
F 69120 - Vaulx en Velin, Lyon.

ABSTRACT. The accurate prediction of the behaviour of a building with respect to daylight is a technological challenge. It involves the understanding of the various phenomena associated with the process of light propagation. The photons generation technique which is presented appears to be a very attractive approach. It is clear however that such a capability is not sufficient. This paper presents some constraints in the use of simulation programs as well as its limitations.

1.INTRODUCTION:

There has been a strong interest in these recent years in the development of computer programs for daylighting design. However, their real use has often be poorly explained. The consequence is often an overestimation of their capabilities. The programs are not created to solve daylighting problems, but to inform the user of the manner in which a given configuration will match his expectations. This means that the major difficulty in a simulation program is not the precision of the simulation itself, but the clear expression of the criteria which will be selected to assess a daylighting system.

2. COMPARISON OF SIMULATION TOOLS.

The question which is most often raised regarding these tools stands in their validity. The designer must have some hints regarding the level of confidence he can have with the results.

Light penetration in a real building depends on:

1 - type of sky : overcast, clear, w/o sun, ...
2 - exterior environment (vegetation, buildings, obstructions)
3 - material properties (indoor and outdoor surface finishes)
4 - building geometry
5 - furniture

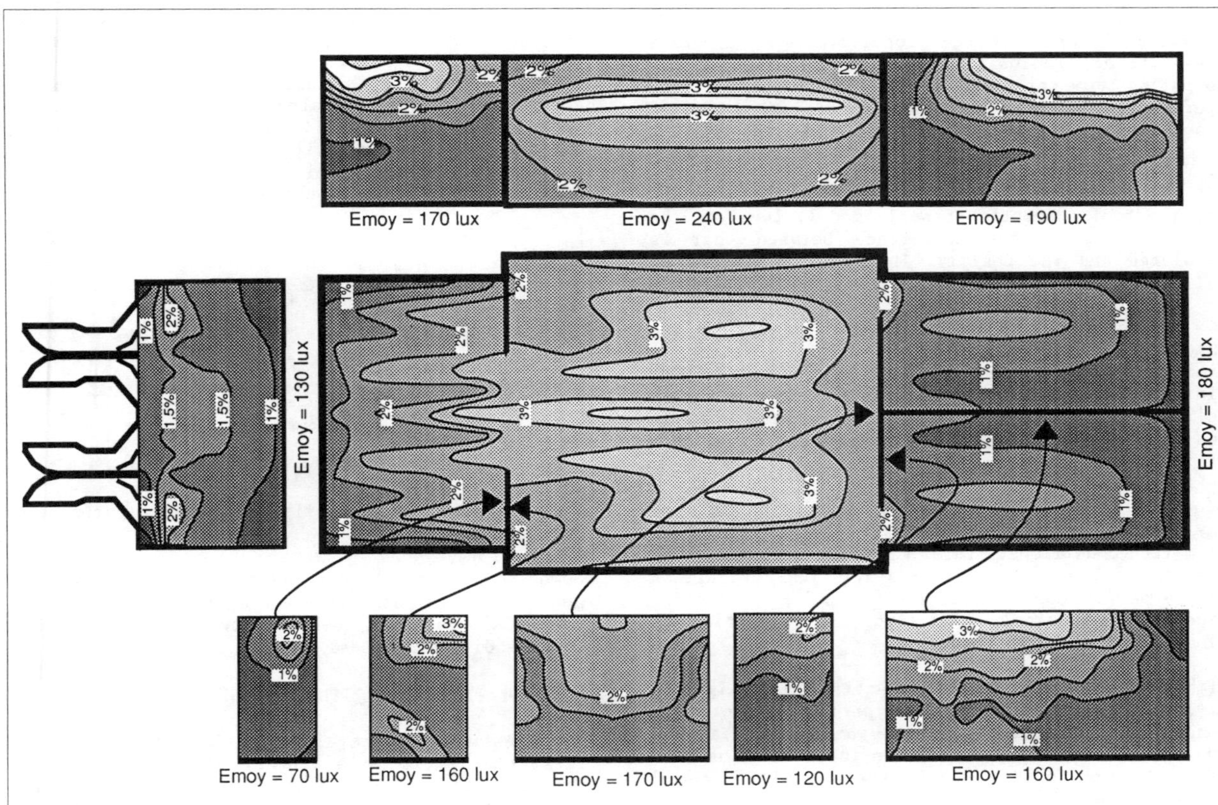

Figure 1. Examples of results showing illuminance distributions for an overcast sky (uniform model) on indoor surfaces, including partitions (Musée d'Intérêt National in Grenoble, Groupe 6, Felix-Faure and Macary architects)

Using a tool such as a scale model or a computer program means that some choices have to be made.

For scale models, problem N°1 (skies) requires either an artificial sky (reference overcast sky) or well identified real skies. To identify a real sky, one needs to evaluate the luminance distribution which involves precise equipment and leads to a huge number of parameters of little use. One can also use the half sky technique (Fontoynont, [1]) which permits a quick correlation of Daylight Factors with parameters which describe the sky "non-uniformity". Problem 2 is difficult to simulate since it involves the simulation of the entire environment. Problems 3 and 4 are the easiest on scale models: real building materials can be used and scale model can have the exact complex shape of the real building. For furniture simulation (5), it is just a matter of finding mock-up of furniture at the right scale.

In case of computer tools, the analytic solving of the problem of the light penetration process requires a selection of numerical models to simulate the light source (problem N°1), the environment (2), materials (3), building geometry (4) and furniture (5). Each of the tools which have been presented suppose a well defined and ideal type of sky (1), usually a CIE standard overcast or clear sky. Simulation of the environment is usually very simplified: simulation of the obstructions, sometimes facade brightness, rarely outdoor multiple reflexions. Material photometry is supposed to be perfectly diffusing except for ray tracing-programs, as well as photons generation and some radiosity algorithms which simulate specular surfaces. Intermediate surfaces (glossy, smooth) are simulated by the most powerful programs, but they require the knowledge of the photometric properties of the material which is being simulated. Problems 4 and 5 are approached in a simplified manner. With the large majority of programs, rooms shape must be rather simple and borrowed light systems cannot be simulated. With more sophisticated programs (ray-tracing, radiosity, photons generation), only computing time as well as the time spend by the operator is the real limit to the complexity of item 4 and 5.

3. VALIDATION

Using a tool (computer or scale model) involves the understanding of the difference between what is simulated and the reality. They are two levels of validation:

A - one is aimed at checking if illuminances are the same than in a real building where everything is similar to what has been simulated. Due to the limitations of tools, such a building must be a test cell under an artificial sky. For a real sky, we would need to know the exact sky luminance distribution.

B - the second concerns the prediction of the daylighting performance of a real buiding for real climatic conditions. This requires adjustment of results of simulation and a statistical definition of the performances. Here, the role of a daylighting expert is essential. This type of validation is almost impossible to conduct.

Within the European Concerted Action on Daylighting [2], a validation task of A-type was conducted to compare design tools. The room was rectangular and two apertures were employed: one in the roof and the other in the facade. Results were very close and allow the identification of the influence of surface finishes and construction details.

A user or a client of a design tool should remember that only A-type validations have been conducted on design tools. Therefore, analysis of results of these tools requires the input of observations and comments based on the designer experience.

4. PRINCIPLE OF THE PHOTONS GENERATION

Light propagation is a physical phenomenon. It involves a source of light, a medium, and surfaces which transmit, reflect or absorb the associated energy.

To simulate natural light penetration in an enclosure, the difficulty stands in the process in which light interacts with matter, and the geometric difficulty of handling multiple reflexions and transmissions in a volume with a particularly complex shape.

The interraction of light and matter is complex, and the restriction to perfectly diffusing surfaces as well as mirrors cannot be judged satisfactory. For this reason, we developed a technique very similar to the ray-tracing technique, but which follows closely the physics of light propagation.

The technique is called **"photons generation"**. The principle is to generate rays of elementary energy (packs of photons) from the source (the sun, sky or lamps) and to follow them during the multiple reflexions until final absorption. Dispersion and diffusion of light are simulated using a Monte-Carlo method. The energy accounting is followed during this process in order to monitor the various optical phenomena and bring the understanding of the energy exchanges. After complete absorption, the program produces illuminance values on elements of each surfaces of the building.

5. SIMULATION OF PHOTOMETRY

The dispersion of light can be varied from specular to perfectly diffusing, by adjusting the dispersion angle. This angle is the angle for which the intensity is half the maximum intensity. Figure 2 shows the reflected intensity graph for various types of photometries. When combining a perfectly diffusing photometry with a specific dispersion photometry, it is possible to approach the behaviour of most indoor surface coatings.

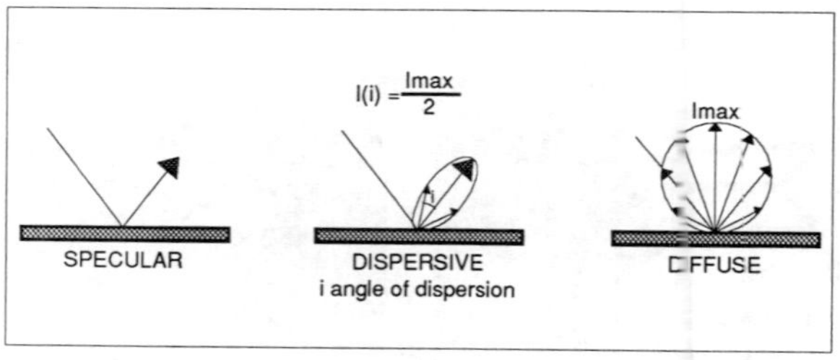

Figure 2. Various types light reflection patterns on surfaces.

Some polycarbonate components transmit light in generating two peaks of specularity. Such components can be simulated in combining two dispersion models and one diffusing model (Figure 3).

6. SIMULATION OF MULTIPLE COMPONENTS

Complex materials are simulated in two manners. The first one refers to a library of photometries. It includes, for each incidence angle, a 2 dimensionnal intensity curve for transmission and another for reflexion. This fits photometries with a symetry with respect to the perpendicular to the surface. But it requires experimental measurements for complex photometries.

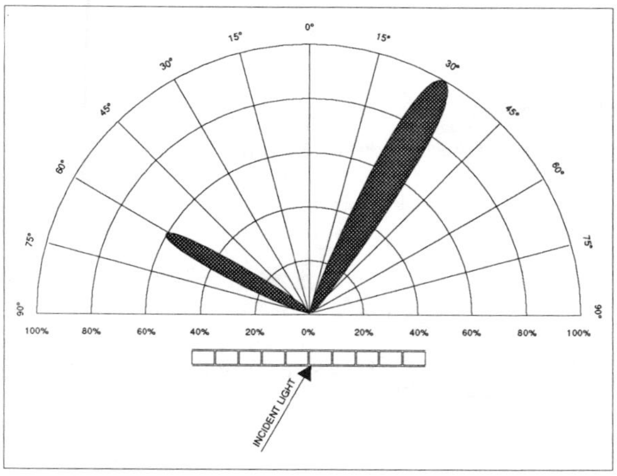

Figure 3. Light transmission through a double-layer polycarbonate panel (doc LASH-ENTPE).

The second technique consists of using basics of optics to simultate multiple layer components. In this case, the program does not refer to a library of materials, but simulates "in real time" the component during the simulation of light propagation in the building. Such a technique was used to develop a reflective blind component adapted to the climate of south of France (Figure 4).

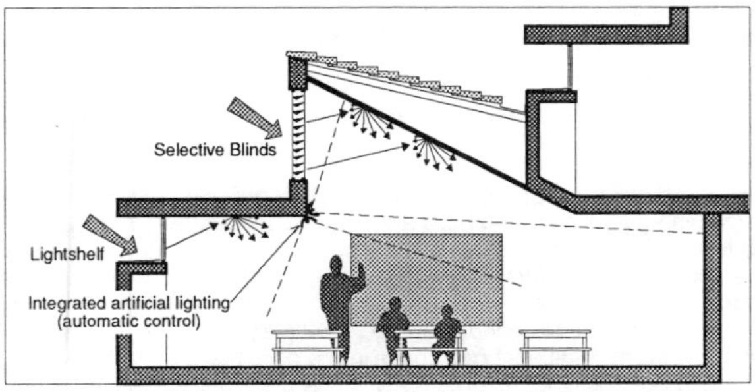

Figure 4. Reflective blind system developed for the school of Collioure, France (M. Gerber & P. Pous architects), [3].

7. CONCLUSION

The tool which has been presented brings a very significant amount of information to the daylighting expert or the designer. But it should be remembered that it does not produce a solution. It should be used by specialists who have a very clear opinion of what they aim at. The performance of a Daylighting system cannot be judged only on the base of the distribution of the Daylight Factors in the interior. Not only the efficacy of the Daylighting strategy needs to be assessed, but also the quality of the indoor space: comfort, performance and agreeableness.

The missing link in the chain of analysis of Daylighting systems is therefore a reference to a precise set of specifications.

8. REFERENCES

1. Marc Fontoynont, Prise en compte du rayonnement solaire dans l'éclairage naturel de locaux: méthode et perspectives, Thèse de doctorat, Ecole Nationale Supérieure des Mines de Paris, 1987.

2. European Concerted Action on Daylighting, Commission of the European Communities, Directorate General XII, 1987-1989.

3. Marc Fontoynont and Bernard Paule, Sunlighting techniques developed for the school of Collioure, France. Conference proceedings, ASES'88 Cambridge USA, juin 1988.

9. ACKNOWLEDGMENTS

R. Mitanchey provided a significant help for the computer graphics, particularly for the oral presentation. Jean Marc Duport helped us for the management of the data for the light simulations. The city of Grenoble provided the necessary funds for the luminous analysis of the future Musée d'Intérêt National. The daylighting analysis for the school of Collioure was made possible with a grant from Agence Francaise pour la Maîtrise de l'Energie.

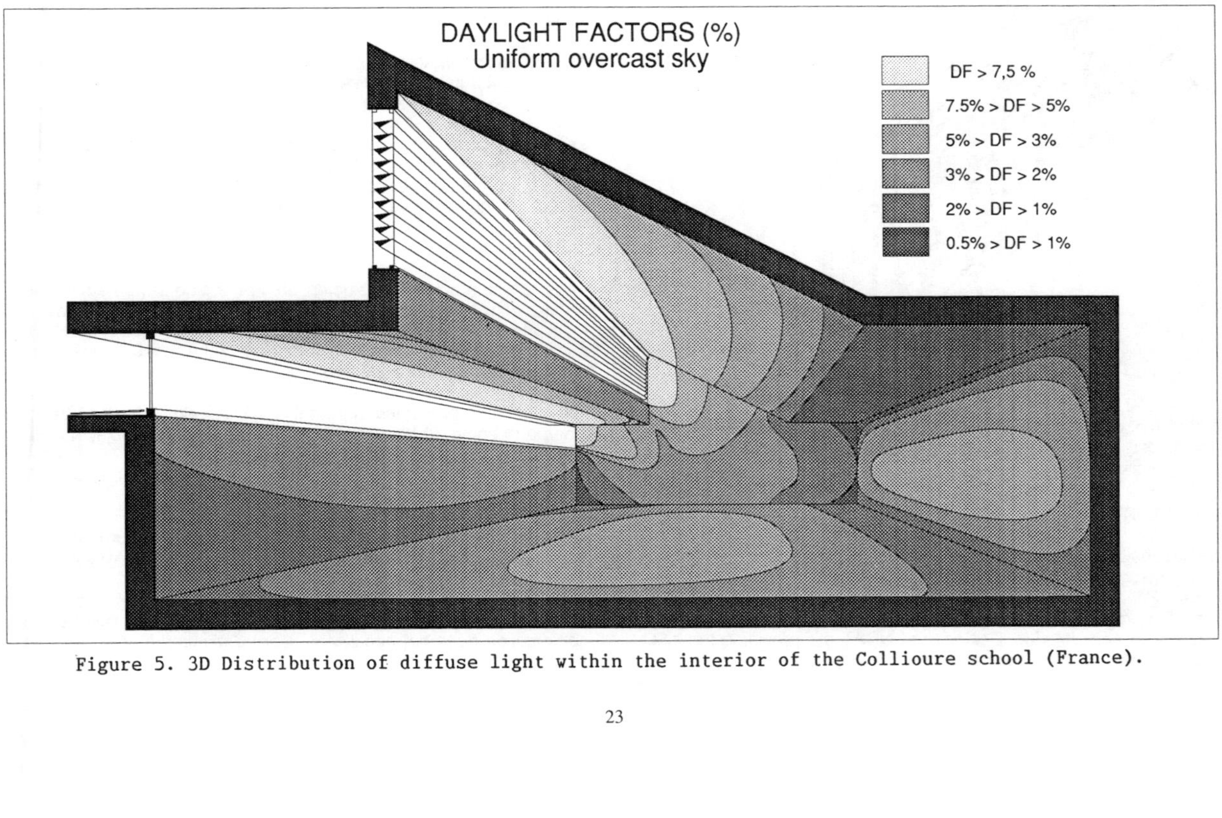

Figure 5. 3D Distribution of diffuse light within the interior of the Collioure school (France).

SOLAR ARCHITECTURE AS AN INITIATOR FOR UNCONVENTIAL PLANNING AND DAYLIGHTING TECHNIQUES

Y.Dean, M.Wilson, J.Walker.

Low Energy Architecture Research Unit, Polytechnic of North London, Holloway Road, London N7 8DB United Kingdom

This is a joint project between the Low Energy Architecture Research Unit, of the Polytechnic of North London,and the National Centre for Alternative Technology, Wales. The project is situated in a north western maritime climate which is not suitable for long term mass wall storage systems. The building is part of the CEC Building 2000 programme.

early studies decided aginst the cluster plans typical of residential study centre and consequently an east/west linear plan arrangement developed showing a southerly orientation. Areas with the same function ie study bedrooms, or teaching areas had consistent orientations. Early decisions included the buffering of the building from extrem temperature variations by externalising the main circulations spaces. A large proportion of ventilation air will consequently be pre-heated. This also saved on the lighting of internal circulation spaces. Splitting the building in two halves allowed for side daylighting penetration via the centrally glazed area.

Daytime occupancy is on the first floor and in teaching areas, a fast response timber framed construction which from early morning onward, benefits from dierct solar gain through vertical glazing and rooflights. Evening occupancy is on the ground floor in study bedrooms which have a greater thermal mass and benefit from the accumulated gain.

Studies were carried out to optimise daylighting quality using 1:25 models in ana artificial sky, and computing daylighting analysis made using "Daylight," a programme developed by Essex IHE UK. The ability to toplight the first floor which has greater levels of illuminance, and occupancy modes co-inciding with daylighting availibility ensures minimum lighting demand. This contrasts with general layouts study/conference centres where accommodation is often planned over teaching facilities. Designing with passive solar strategies in mind has produced a building that has a more successful planning arrangement showing also structural benefits by the placement of large span teaching spaces over short span study bedrooms.

Results from the 17 zone Seri-Res model which has now been used to optimise the design shows the performance as:

27.5KwHR/sq m annum. (ie under 1 ecu per sq m per annum)
12.8 Whr/sq m degree day per (degree days =2342)or 87% energy saving over UK Building Regulations Standards
47% of the electric lighting load is displaced by daylighting.

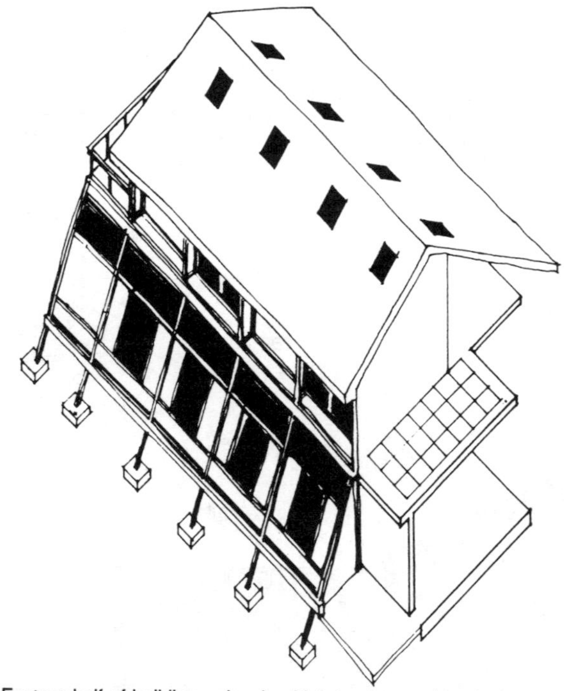

Eastern half of building, showing high level openable glazing to the ground floor, and on the upper floor the solar chimneys alternating with opening balcony/external doors.

INTRODUCTION

The National Centre for Alternative Technology and the Low Energy Unit at the Polytechnic of North London are in partnership to provide a study centre. PNL have taken the role of architects and energy consultants using the project as an educational tool within the Department of Architecture and Interior Design, and NCAT are providing the site, practical help, fundraising, and will be the end users.

This new building will provide teaching and accommodation for courses on low energy building, self build and organic agriculture The accommodation will consist of 8 study bedrooms for 16 people with associated washing and eating facilities, one large teaching space study areas, a library/reading room and a public demonstration and research room. The bedrooms are designed primarily for single occupancy but are high spaces with an additional bed space in a 'crog loft', (a mezzanine deck above storage and utilities). The study areas are either side of a central conservatory/atria space. The building will use heating from an existing hot water district supply ring main on site.

Floor area heated : 400 sq m
Volume heated 1325 cu m
Total Area : 560 sq m
Total volume 1703 cu m

The building orientation is offset 10 degrees to the west of south with a linear layout from east to west.

The building is located in the valley of the River Dulas, a tributary of the River Dovey near the west coast of Wales, famous for its giant sea trout. The centre has developed over a disused slate quarry and has open aspect to the south west but is bounded by hills to the south east.

Site Microclimate

The site is characterized as being a temperate north western maritime climate with prevailing south westerlies and high rainfall. Being in a hilly location and some 15 km inland some wind sheltering is provided and sunshine availability is reduced compared to the coast. This gives the area a low temperature climate in summer, but milder in winter.

Weather data presented is from the CEC Test Reference Year for Aberporth some 70 km to the south west, which is subject to 2342 degree days to an 18.3 deg C base.

CLIMATE, CONSTRUCTION AND PLANNING

On this particular site the climatic conditions are unfavourable for large scale solar gains. There is a mountainous landscape which is quite small scale, with a good deal of overshading. The site has many trees and bushes which are quite dense, and so the reliance on trees losing their leaves for winter solar gain is not reliable. In this climate heavy construction does not necessarily lead to mimimum auxillary energy usage.

Thermal mass sensitivity studies showed that optimum performance could be achieved by differing screed thicknesses in the North and South zones. However, as underfloor heating was to be used, the optimum thickness of 50mm had to be increased to 75mm. Decreasing the mass increased the temperatures by 1.3 degrees C on average.

On the hottest day, temperatures exceeding 30 degrees were predicted in the buffer zones but the air change rate could be substantially increased in these zones to minimise the overheating. Of the sensitive areas the temperature was predicted to rise to 34 degrees in the library and 30 degrees in the teaching space. Both these arease have ample openable windows and rooflights to considerably increase ventilation rates beyond those used in computer simulation.

Unconventional planning arising from pragmatic decision making in solar architecture.

Many of the planning decisions undertaken arose from a combination of difficult site conditions that forced a particular configuration, but this gave natural advantages. The site forced a linear solution being the only space available between a set of buildings and an old water course which initiated the East West layout. It was logical to use linear planning for circulation but it was decided quite early on that this could not be buried element that needed artificial lighting, but should be on the outside of the building giving access to our study bedrooms. This corridor element was then moved to the South side of the building to optimise daylighting and immediately became a prime solar feature. Not only did it become a useful buffer space but provided some pre-heating, and contributed amenity value, giving unusual access to spaces which gave the feeling of being less institutional.

Floor Planning

Areas subject to daytime occupancy would be most sensitive to the provision of daylight. Due to quite heavy tree growth and shading from neighbouring hills, restricting daylight from vertical glazing, an early decision was taken to raise the teaching level to the upper floor, exposing this level to the total sky vault by flexibility in the use of rooflighting and control of light quality. If this decision had not been made early on we would have had a poorer quality of lighting on the ground floor for these functions, and a greater reliance on more energy wasting electric lighting. This has proved to be by far the best strategy for large teaching spaces as it also makes sense structurally with the larger spaces carried over small span structural load bearing support system of the bedroom units on the ground floor. Disabled access to a public building of this nature is an important factor and will still have to be achieved for both levels.

Construction, planning and daylighting

The building consists of a lightweight timber framed structure for the large teaching spaces on the upper floor, supported by the monolithic block-work cross wall system which encloses the study bedrooms, separating them from other teaching and utility spaces on the ground o73 floor. This solid internal core is wrapped by buffer spaces which are timber framed, and provide the main points of entry. The ground floor uses a 100mm screed for the installation of the underfloor heating system, separated from the ground floor slab by 100mm mineral fibre bats. The 150mm cavity in the timber frame walls is completely filled with mineral fibre bats. The roof has a total of 190mm of mineral fibre in the form of a cross lattice of 140mm and 50mm. All windows are double glazed.

U values: Walls .26W/m2
Roof .18W/m2
Floor .32W/m2

Construction type is responsive to the timed use of the building. The upper floor is a fast response system rapidly benefiting from direct solar gain in the mornings, and pre-heated warm air entering from the south facing conservatory. The ground floor construction which has the greater mass, accumulates the gain during the day and is warm for evening use when the study bedrooms will be occupied. The basis of the passive solar design concentrates on extensive thermal buffering for indirect gain and ventilation pre-heat. The buffer spaces are functional, provididng circulation areas and therefore do not represent a solar add-cost and may provide a cost reduction.

As the study centre adjoins existing buildings, rooflights have been introduced along the whole length of the ground floor northern rooms, which simultaneously light these rooms and upper teaching spaces. A glazed stairwell penetrates the core of the building, introducing sidelighting for the study areas in the northern rooms on the ground floor, and leads to a south facing conservatory also providing circulation space. Together with a southern highly glazed corridor, which links all bedrooms, the kitchen, and bathrooms, all circulation areas are 'maximum daylit' spaces. The teaching area and library on the first floor receive daylighting from four sides and from rooflights located to provide an even distribution, avoiding daylight and sunlight glare. Low length vertical windows with overhangs on the Southern aspect together with external shading from the hills minimise the area that could be affected by winter sun.

There are good precedents for large scale spaces in teaching spaces being located on top of smaller scale service and administrative areas such as in Middlex Polytechnic (Faculty of Art and Design) by Howell Killick, Partridge and Amis or the School of Architecture and Design in Newcastle (Newcastle City Architects Department). However, the orientation of these buildings does not make use of the passive solar possibilities . Daylighting design is not used to counteract the energy load of electric lighting needed for some deep plan spaces in the Newcastle project although it does provide daylighting for circulation spaces.

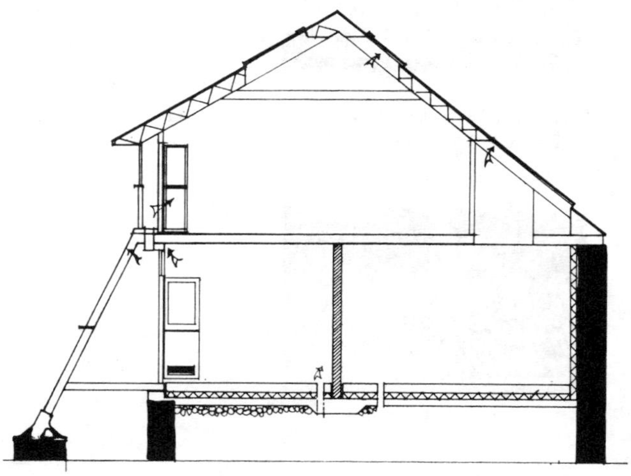

Section through building showing the large teaching spaces on the upper floor and the bedrooms opening on to the sloping, south facing sunspace, with rear study areas behind, that are chiefly top-lit.

VALIDATING THE PLANNING ARRANGEMENTS

Thermal and daylight modelling of the study centre was carried out using the thermal simulation model SERI-RES Version 1.2, (1,2), which is public domain in the UK, and interactively with the Essex IHE Daylight programme. (3,4)

Summary of conclusions from Simulation results

Lighting energy use was far greater when switching on minimum daylight factor than average daylight factor but sensitivity to fenestration changes was found to be similar. Increased fenestration was found to reduce primary energy demand up to approximately an 8% average daylight factor for plain double glazing in areas having morning and early evening use. Lower average daylight factors of 3% to 4% were optimum for areas which were not used after 6pm or which were used very late at night.

When low-E coating replaced plain units total energy consumption was little altered because lower transmissivity led to increased lighting load offsetting the improved thermal properties. The sensitivity to increased fenestration was better than for plain double glazing leading to high average daylight factors (12% or more) being optimal. Additional annual savings of approximately 2600kWh possible with low-E units were not felt to justify the additional cost of 5500 ecu over plain double glazing.

A manual switching system will not realise the full potential lighting savings of greater fenestration predicted for an automatic system although thermal losses are likely to be as predicted. Taking into account the low sensitivity to increased fenestration, base case window sizes, as originally developed in the model studies, have been retained. A full discussion of these points is available. (5,6)

Northern aspect showing lighting arrangement.

Part of model showing structure and solar chimneys used in summer to ventilate corridor sunspace.

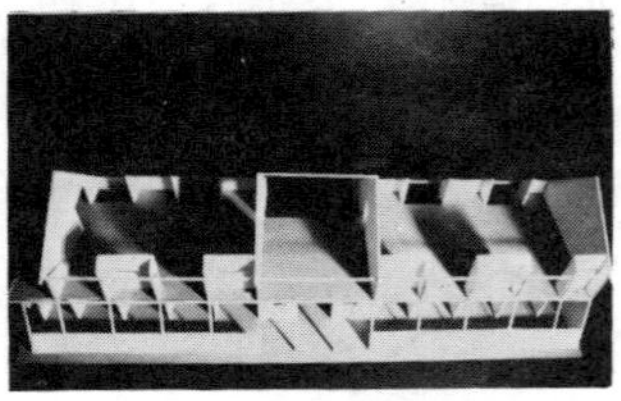

Model of the upper floors showing large teaching spaces and the method of lighting the ground floor rooms.

Model of lower floor showing study **bedroom spaces** on the south side opening onto the sunspace/corridor.

Example of daylight study on the **upper floor**.

Planning Conclusions

From an early design stage the planning of the building aimed to minimise space heating and lighting energy use. Design features found to be useful, with potential for replication include:

*The use of a fast response timber frame structure on the first floor, but a slow response mass wall structure on the ground floor, relates directly to the timed use of the building.
*Corridors and the circulation has been moved to the perimeter of the building to avoid daytime internal lighting.
*Multiple use of glazed spaces for solar gain circulation and ventilation pre-heat.
*Deep plan spaces have a mixture of daylighting sources (eg rooflights and atria) to achieve better light quality and distribution.

Modelling conclusions

Planning design features tested through lighting simulation and dynamic thermal modelling produced the following conclusions.
*Most daylighting benefit is obtained reaching daylight factors of 2%-4%.
*Large savings result from avoiding high lighting standards where not needed.
*Summer evening daylighting makes an important contribution in non-office accommodation.
*Ventilation pre-heat from buffer spaces provides large energy savings in one of the remaining areas of major heat loss for a well insulated building.
*Low-E coated double glazing was found to reduce space heating demand but taking into account the reduced light transmission was not economically viable overall.
*Mass sensitivity studies working to +20% or -20% of the base case produced increases in energy consumption.
*In the modelling study overheating became of prime concern and natural cooling mechanisms an important area to develop.
*Heating Load = 27.5KwHR/sq m annum, 12.8 Whr/sq m degree day per annum.
The building shows an 87% saving over UK Building Regulations Standards. 47% of the electric lighting load is displaced by daylighting. A technical report is available describing the modelling study in full.

REFERENCES

1 I.Frame, S.Burch, Manual for 'Daylight' Program, Essex Institute of Higher Education UK, July 1988

2 L.Palmiter, T.Wheeling, SERI-RES Solar Energy Research Institute Residential Energy Simulator Version 1.0, Solar Energy Research Institute USA.

3. P.Haves, J.Littler, Refinements to SERI-RES, UK Energy Technology Support Unit Report S-1130, 1987.

4 P.Haves, J.Littlefair, Daylight in Dynamic Thermal Modelling Programs: A Lighting Facility for SERI-RES, Polytechnic of Central London and Building Research Establishment UK, 1988

5 D.R.G.Hunt, The use of Artificial Lighting in Relation to Daylight Levels and Occupancy, Building and Environment 14, 1979

6 Daylight Evaulation in a Total Energy Context, Wilson M, Walker J, Dean Y CIBSE National Lighting Conference 199[illegible].

7 Principia Mechanica, Building Performance Analysis - Swanley Downsview Primary Infants School, 1989

APPLICABILITY OF SHORT REFERENCE YEARS FOR PASSIVE SOLAR BUILDING SIMULATION

P.A. Waide & B. Norton

School of Mechanical Engineering
Cranfield Institute of Technology
Bedford MK43 OAL
England

Department of Building
University of Ulster
Newtownabbey BT37 OQB
N. Ireland

ABSTRACT. Ten "short reference years" tested against three full year weather sets. Using the SERI-RES building thermal analysis programme comparative simulations were performed on range of buildings .

1. INTRODUCTION.

With the increasing application of computers in the building services industry it has become common practise to use full year weather sets for the simulation or prediction of building thermal performance. These full years (1,2,3,4,5,6,7,8) are comprised of 365 days, i.e. 8760 hourly values for each weather parameter while additionally several Short Reference Years (SRY) have been produced (5,6,9,10,11) which also contain collections of hourly data for a specific location but use less than 365 days to represent a whole year. SRYs have been developed for the dual purposes of reduced computer time and more rapid real-time of simulation. The aim of this study is to assess the areas of validity whereby SRYs might be used in preference to full years. A range of building types were simulated with the SERI-RES building thermal analysis programme (12). The results from using the SRYs were compared with the mean values from the full year weather sets. The latter being the data sets currently in use, represent the "truth" to which the SRYs aspire.

In consultation with other participants of the UK Department of Energy Passive Solar Programme (PSP) it was decided to limit the study to Kew weather years. Three full year data sets are commonly available for Kew, these are the CIBSE Example Year (1,2,3), the CEC Typical Reference Year (TRY) (5,6,7,8), and the TRY (4). Ten SRYs have been produced for Kew, which are 9 "Loxsom" SRYs (11) of either 3,5, or 7 days per month with one of three different statistical methods used in their selection, and 1 "CEC" SRY (5,6,10) which comprises 16 days per season. The characteristics of these weather files are summarised in Table 1. The varying weather parameters and data formats were converted into a form suitable for use with SERI-RES (12).

SERI-RES (12,15) is one of three detailed passive solar building thermal simulation programmes employed in the U.K.-PSP, the others being HTB-2 (16) and ESP (17). It was chosen for use in this study, because it is a well established and tested program, and has the fastest running speed of the three, (running speed ratios are SERI-RES = 1, ESP = 2, and HTB-2 = 11). The program uses hourly weather data.

Version 1.2 of SERI-RES (15), has the facility to perform simulations using SRY's providing they are of a consistent number of days per month. This is true of the Loxsom SRYs but not the CEC SRY. The CEC SRY was designed to have 14 days per season to allow week/weekend sequences of occupancy to be modelled. The annual total of 56 days corresponded to monthly values from January to December of 6,4,4,6,4,4,6,4,4,6,4,4 days per month respectively. A special modified version of SERI-RES was created which permitted this sequence of days per month and enabled the CEC SRY to be used for simulations. To ensure that the solar angles corresponded correctly to each "monthly" data sequence, the hour numbers for each sequence were set for the middle most days of each month. It is worth noting however, that if the SRYs, as produced, are to gain acceptance then the programmes which designers use should be capable of running them without user modification.

Table 1. The Weather Years Examined

	Alternative name	Total No. of days	No. of days per month
CIBSE Example Year	CIB	365	28,30,31
CEC Typical Reference Year	ETY	365	28,30,31
Loxsom Typical Reference Year	LTY	365	28,30,31
CEC Short Reference Year	ESY	56	4,6
Loxsom Short Reference Years			
SF3	LF3	36	3
SF5	LF5	60	5
SF7	LF7	84	7
SR3	LR3	36	3
SR5	LR5	60	5
SR7	LR7	84	7
ST3	LT3	36	3
ST5	LT5	60	5
ST7	LT7	84	7

2. THE BUILDING DESCRIPTION FILES (BDF s)

Ideally a very diverse set of building types would be used for a wide range of simulations using both SRYs and full weather years and the differences in performance noted. Unfortunately the constraints of time and computing costs require both the number and complexity of the building types to be limited. The decision was taken to confine the first appraisal to a selection of BDFs already generated within the U.K.-PSP (18, 19) These BDFs are single zone, direct-gain rooms designed to represent various typical south-facing rooms in British houses. Their simplicity allows a great many simulations to be performed for the equivalent cost of simulating a single whole building, but the effects of interzone heat transfer are not modelled (instead the partition walls are made of dummy materials designed to mimic the behaviour of adjoining buffer zones). Nonetheless, a sufficient diversity of types was available to permit testing of the sequences of the key weather parameters; i) the dry bulb temperature, ii) the direct normal solar radiation, iii) the diffuse solar radiation and iv) the correlations between the above parameters. A set of 9 was selected, which would be suitable for testing responses of the key weather parameters. To these were added a further set of 8 building density variants for analysis of time constant behaviour.

The BDFs chosen are identical to the base case building, detailed in Table 2., in all but one aspect each. The final set, included variants with:-

i) 1.0 and 0.25 air-changes/hour ventilation rate, which because this addresses a direct coupling between the internal and external air temperatures, allows dry bulb temperature responses to be studied.

ii) Window orientation due south (base case) and due west, to permit testing of the diurnal solar distribution.

iii) Window areas of 20%, 33%, 45%(base case) and 55% of the floor area, to test the solar gain responses as well as the solar load ratio behaviour.

iv) Four construction methods corresponding to traditional heavy weight, modern heavy weight, and modern light weight, with the base case being advanced heavy weight.

In addition to those buildings mentioned, two lots of four buildings, with one, two, three, or four times the base case density were generated for simulations to observe the SRY sensitivity to varying building time constant. The first set of four, had 24 hour operation of the heating & ventilating system with a low thermostat set point of 28°C, and a high set point of 21°C. The second set of four buildings were free running with no set points.

Table 2. Base Case Building Description

GENERAL
An living room with no internal gains, unfurnished but carpeted.
FORM
Shape: Rectangular box, 2.83m deep, 5.66 wide, 2.5m high. Floor area: 16 sq m Window: orientation; South, area; 7.19 sq m, type; double-glazed.
HEATING
Solar Gains: re-reflected to ambient; 0.08 coupled to air-node; 0.1 falling on floor; 0.82 falling on walls; 0.0 Air-change rate: 1.0 air change per hourHeating capacity: 100 kW Thermostat set points (on 24 hours) low: 21 degrees Celcius high: none
FABRIC
General: Advanced heavyweight, post 1985 U.K. building regulations. Walls: South & East; (Cavity wall) Outer brick leaf, insulation, inner block, plaster West; (Pseudo buffer zone wall) Dummy material YYY, plaster, block, plaster North; (Pseudo buffer zone wall) Dummy material XXX, plaster, block, plaster Floor: Earth, hard core, polystyrene, concrete slab, screed, R - 0.18. Roof: R - 0.083, chipboard, R - 0.083, chipboard, R - 0.18, plaster board.

3. SELECTING AN SRY FOR SIMULATIONS

There are diverse criteria by which a potential user might select an SRY, dependent on the type of building undergoing simulation, the desired output, and the competition between accuracy and speed. None of the SRY's were found to be clearly superior in all aspects to the others, thus rather than make presumptions concerning a user's requirements, the SRY results are presented in a accessible visual form from which can be chosen readily the most appropriate SRY for a specific task. Figures 1 to 4 show the monthly comparative energy index versus the comparative indices of monthly solar gain, useful loss, mean temperature and maximum temperature respectively. The comparative indices are correlation-weighted measures of the divergence between each SRY result and the mean of those for the full years. A linear relationship can be assumed between the comparative indices and the ability of the SRY results to mimic those of the full years. Thus the SRY with the lowest user-weighted sum of indices for the desired combination of parameters should be selected as the weather set for the simulation.

For example, suppose that a user is simulating the performance of a well insulated building with low glazing area and seeks output for the energy requirement and the monthly mean temperatures. It is evident from Fig. 2 that the SRY producing the lowest energy error is ESY, and the SRY producing the lowest useful loss error is LR5. Knowing that the building is well insulated, but with a low solar gain, the user may decide that the "energy error" will be more influenced by the fabric loss than average, and decide that LR5 is more appropriate (although ESY is almost as good at predicting useful loss). However, the user requires output for monthly mean temperature, and inspection of Fig. 3 shows that the seven day SRY's are the best at predicting this parameter. In the end, the user decides to select an SRY by giving equal weighting to the errors in energy consumption, useful loss, and mean temperature, so LR5 = 1.4 + 1.4 + 2.1 = 4.9, LT7 = 3.0 + 2.7 + 0.25 = 5.75, and ESY = 0.75 + 1.5 + 3.5 = 5.75. Thus, LR5 has the lowest combined error, and is the most suitable for the users needs. It should be stressed that the error magnitudes only indicate the relative predictive performance of the SRY's for a given parameter.

4. CONCLUSIONS

The SRY predictions always underestimate the size of the annual parameters, but to varying degrees. This is a reflection of damping arising from the deficiency of extreme values within the SRY data distributions.

In general the annual energy consumption, useful loss, and south facing solar gain were reasonably well predicted using the SRY's, showing underestimates for the range of buildings examined of between 0.3% and

Fig 1. Plot of Comparative SRY Indices for Monthly Heat Consumption against Monthly Solar Gain

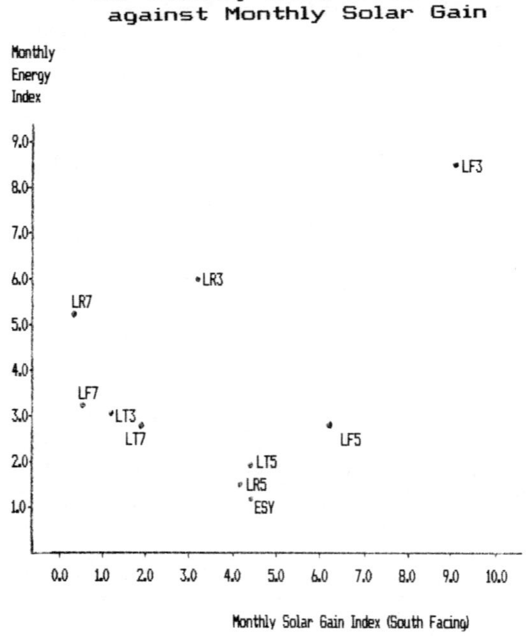

Fig 2. Plot of Comparative SRY Indices for Monthly Heat Consumption against Monthly "Useful Loss"

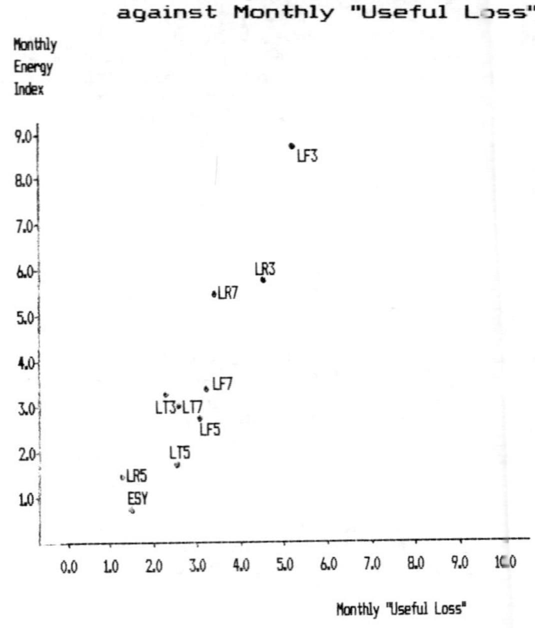

Fig 3. Plot of Comparative SRY Indices for Monthly Heat Consumption against Monthly Mean Temperature

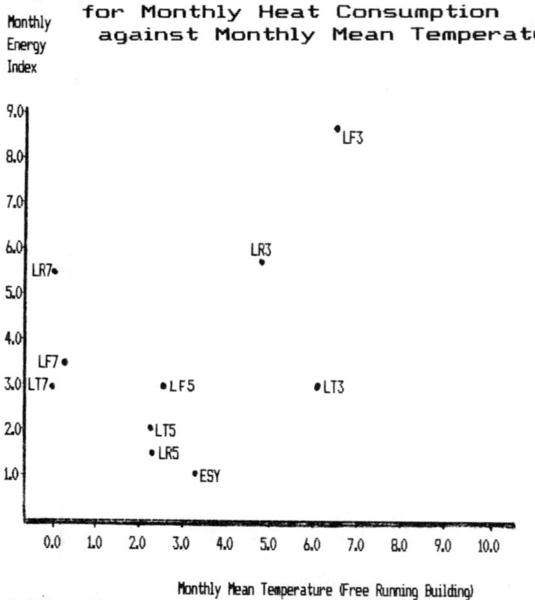

Fig 4. Plot of Comparative SRY Indices for Monthly Heat Consumption against Monthly Maximum Temperature

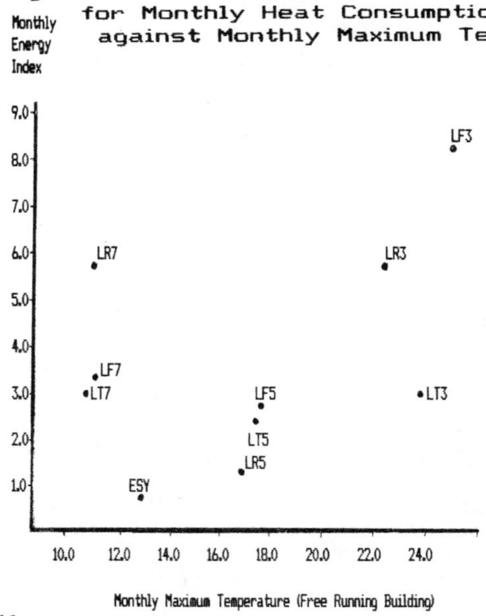

7.0%. The seven day sets performed particulary well for the solar gain estimates, but were less impressive for the annual energy and useful loss. It seems that while the longer SRY's are more likely to contain representative extremes of monthly values than the shorter SRY's, there are less possible sequences available to be selected from the ten year data base, thus the shorter SRY's may contain sequences more representative of the data base's long term mean values and variances. In general the three-day SRYs perform badly; but the five-day SRYs are often the best all round years: dependent though on the weighting ascribed to the various parameters.

ACKNOWLEDGEMENTS

This work form part of a study supported financially by the Energy Technology Support Unit, (ETSU), Harwell, Oxfordshire, for the UK Dept. of Energy(20).

REFERENCES

1. CIBS Guide, Section A2.

2. Holmes, M.J. and Hitchen, E.R., "An 'Example Year' for the calculation of energy demands in buildings", Build. Serv. Eng., 45(9), 186 (January 1978).

3. Holmes, M.J., Hitchen, E.R., et al, "The CIBS Example Weather Year", Build. Serv. Eng. Res. & Tech., Vol.4, No.3, 1983.

4. Loxsom, F.M., "Selection and testing of a Test Reference Year for Kew", as part of "Meteorological Data for the Passive Solar Programme", Polytechnic of Central London, Research in Building Unit, Collection of reports for ETSU Model Refinement Study (1985-86).

5. Lund, H., "Short Reference Years - Test Reference Years", Technical University of Denmark, Thermal Insulation Laboratory, Final Report, contract ESF-011-DK, Report EUR 9402.

6. Lund, H., "Short Reference Years and Test Reference Years for EEC Countries", Technical University of Denmark, Thermal Insulation Laboratory, Final Report, Contract ESF-029-DK., 1985, Report EUR 10208 EN.

7. Lund, H., "Test Reference Years TRY's: Weather Data Sets for Computer Simulations of Solar Energy Systems and Energy Consumption in Buildings", Report EUR 9765. Commission of the European Communities, 1985.

8. Lund, H., Eidorff, S., "Selection Methods for the Production of Test Reference Years", Final Report, short version. Report EUR 7306 N. 1982.

9. Liem, S.H., Paassen, A.H.C., "Development of a Basic Statistical Weather Model Valid in Europe for Design and Energy Calculations", Final Report on CEC Contract ESF-128-NL (CEC Report Number EUR 8968 EN), Delft University of Technology, Netherlands, 1984.

10. Lund, H., "E.C. Work on Short Reference Years (SRY)", Proc. UK-ISES Conference: Advances in European Solar Radiation Climatology, London, January 1986.

11. Loxsom, F.M., "Selection and Testing of Short Reference Years for Kew", as part of "Meteorological Data for the Passive Solar Programme", Polytechnic of Central London, Research in Building Unit, Collection of Reports for ETSU Model Refinement Study, (1985-86).

12. Palmiter, L., Wheeling, T., "SERI-RES, Residential Energy Simulator", Version 1.0, Ecotope Group, 2328 East Madison, Seattle, Washington 98112, under contract to The Solar Energy Research Institute, 1617 Cole Boulevard, Golden, Colorado 80401, USA.

13. Principia Mechanica Ltd., 233 High Holborn, London WC1V 7DJ, Tel: 01-831-6144, Telex: 267152 CAPLON G, Fax: 01-242-5968.

14. The Met. Office Advisory Services, Room 228, London Road, Bracknell, Berkshire RG12 2SZ, Tel: 0344-420242 Ext.2296.

15. Haves. P., Littler. J., "Refinements to SERI-RES", Polytechnic of Central London, Research in Building Group, Final Report, ETSU S-1130, (1987).

16. Lewis, P.T., Alexander, D.K., "HTB2 - A Model for the Thermal Environment of Buildings in Operation", Welsh School of Architecture, User Manual Release 1, Rev. 0, 155-pp, Technical Reference Manual, Release 1., Rev. 0, Vol 1, 378-pp, Vol. 2 434-pp (1985)

17. Clarke, J.A., ESP documentation set, Arch. and Bldg., Aids Computer Unit Strathclyde, Univ. of Strathclyde, 17 Sections, (1982).

18. Anon., Paper Relating to Future R&D Studies on Performance Analysis, ETSU Passive Solar Design Programme, (1986).

19. Lomas, K.J., Bowman, N.T., "Applicability Study 1. Phase II - Programme Proposals", Leicester Polytechnic, June 1988.

20. Waide, P.A. , Norton, B., "Applicability of Short Reference Years for Passive Solar Building Simulation - Phase 1 Report", Energy Technology Support Unit, Harwell, Oxfordshire, U.K. Sept 1989.

MADRID'S MICROCLIMATES

A. González Barrios
Inst. c.c. Eduardo Torroja
c/ Serrano Galvache s/n
28.033 Madrid - España
Fax 2.02.07.00

R.-I. González Barrios
Translator and Interpreter
c/ Santa Engracia, 126
28.003 Madrid - España
Tf. 4.42.13.26

ABSTRACT, The most recent studies about Madrid's urba climate have been carried out by an investigation team of the Institute of Applied Economy and Geography from the C.S.I.C., in cooperation with Madrid Autonomous University, led by Mr. A. López Gómez and published recently with the title "Madrid's Urban Climate: Te Heat Island" This work studies more in depth the work on Madrid's urban climate, observing the microclimates existing within the city.

1. MADRID'S MICROLIMATES

This subject came out at the beginning of 1988, when the information started to go be sollected in order to be able to check for Madrid environment the reliability of an air heating and cooling computer driven calculation method carried out by D.A. Alamán Simón in I.C.C.E.T. /1/, which belongs to C.S.I.C. /2/.
The climatological data obtained for Madrid, were those from Barajas, Retiro and University National Observatories and those from the Municipal weather stations in Plaza Castilla, Plaza de Roma and Plaza de España, all of them located in places where the city influence was the least possible, and it was checked that the most advanced methods on climatology in Madrid, were beeing carried out by an investigation team from the Inst. of Economy and Geography applicated belonging to C.S.I.C. /2/, in cooperation with U.A.M. /3/, leaded by D. A. López Gómez and that have been published at the end of 1988 with the title "Madrid's Urban Climate. The Heat Island".
In this situation and because of the need of obtaining temperature data more accurate to the Urban ones, a daily observation was made in some fixed of the city and of the real circunstances of the momente, such as trafic density, cludiness, rain, win, outstanding events, etc.
This work, is a foretaste in the knowledgment of the climate phenomena within Madrid town, since in the previously quoted study it was proved that in Madrid as well as in general in towns, there is the so called "The Heat Island", and in this work we have been looking for the cooler and warmer zones than the mean of the town in order to observe the existing Madrid Microclimates.
By lack of media to carry out measurements in a continuous form, we have observed daily 200 fixed points during a year and a hald, in variable itineraries, in such a way that we have been able to see the temperature difference between every two points of the above mentioned, comparing them at the same time with the temperature obtained by the Automatic weather station located in Retiro Observatory, belonging to the National Institute of Metheorology which takes temperature every 10 minutes.
Below we present a series of appreciations that have not been able to be studied in depth, but they are pointed out as plans for later feasible studies.
It has been observed how zones closer to parks have always a lower temperature than the others. An example of this are zones next to Retiro Park, where temperature is at least 1 or 2 centigrate degrees in respect to the rest. This phenomenon is observed up of all if these zones are located just behind the park, according to the main wind's direction.
A clear example of it is Doctor Esquerdo Street, between Plaza de Roma or Manuel Becerra and Goya Street, since it is behind Maria Eva Duarte de Perón Park, in the dominant wind's direction.
Another example of this phenomenon is observed in Plaza de la República Argentina, Wich even if it is not next to a park, but it is behind El Viso, which is a zone of villas and green areas surrounding them.
The Urban design has an influence also in the mecroclimate, since an hypodamic design like in the expansion areas, allows two directions of the wind, but an irregular design like the one in the centre prevents the wind circulation, and this makes temperature to be maintained.
In figure 1 you can see the compass with wind's speeds and directions corresponding to Barajas, obtained from a publication of ATECYR: "Condiciones de diseño. Datos climáticos". on climatological conditions in Madrid, among whose authors is D. J.M. de las Casas Ayala.
Figure 2 corresponds to Madrid's street-plan with the street directions, NS-EW, NE-SW and NW-SE.
It is noted that the Madrid's north and east zones, besides beeing designed mainly with villas or blocks of flasts of low density of ocupation, they are with the street orientation along with the outstanding direction of the wind, which allow its passage, but in the Expansion zone (Ensanche), though there is a layout with wide streets it opposes to the wind passage. In the centre of the town we can see still much more impossibility for the passage of the air, due to the more irregular layout of streets and it aids to form the Heat Island in Madrid.
The height of the building affects also temperatures, since it produces shadows proportional to their height. This phenomenon has been checked by Dr.Milo Hoffman in Israel using his calculation method CTTC "Cluster thermal Time Constant", for the urban design, which predicts temperature for a given urban layout.
It has been observed as well the effect of the temperature lag of the buildings and in days in which no heating is put on, in the neighbourhood of the buildings, the temperature rises with respect to global environment.
This has been observed in open building zones, where the cause of this change of temperature is obvious.
The effect of temperature lag is observed veri neatly in the city, since a fenomenon which usually occurs is that the more opened zones, such as it occurs with Juan Zorrilla Square, next to two open zones and the green zone from Canal de Isabel II, respect to Rios Rosas or Santa Engracia Streets, next to them and more populated, is that temperatures are more extreme in the square or open space than in the streets. this phenomenontakes place in wintes as well as in summer, since in winter the extremes are cooler and in summer warmer due to their exposure.
It has also been observed the effect of population, that is, how temperature rises in precise zones whe_

re there are agglomeration because of public acts, such as public demonstrations, even if no big crowd is observed, specially in the Holly Week, etc.. Likewise it could be said that contamination produced by trafic, the energy given off by heated buildings, the various activities carried out in the city, etc., are the causing part of the microclimates.
Figure 3 whows the daily mean intensity of vehicules of Madrid's streets, handed out by the Counciliery of trafic and transports of Madrid, which to give us an idea of the variation of intensity by different sqares we can see here a brief list with figures corresponding to 1.984 by ascending order.

Fiqure 4 qives us a light idea of temperature differences existing between distinct zones of Madrid. In it we want to make clearer which zones are usually with lower or with higher temperatures. They are represented in a plan together with the bus lines so that it is easier to see how zones with more trafic and more activity are those with the highest temperature.

/1/ I.C.C.E.T. Instituto de Ciencias de la Construccion Eduardo Torroja.
/2/ C.S.I.C Consejo Superior de Investigaciones Científicas.
/3/ U.A.M. Universidad Autonoma de Madrid.

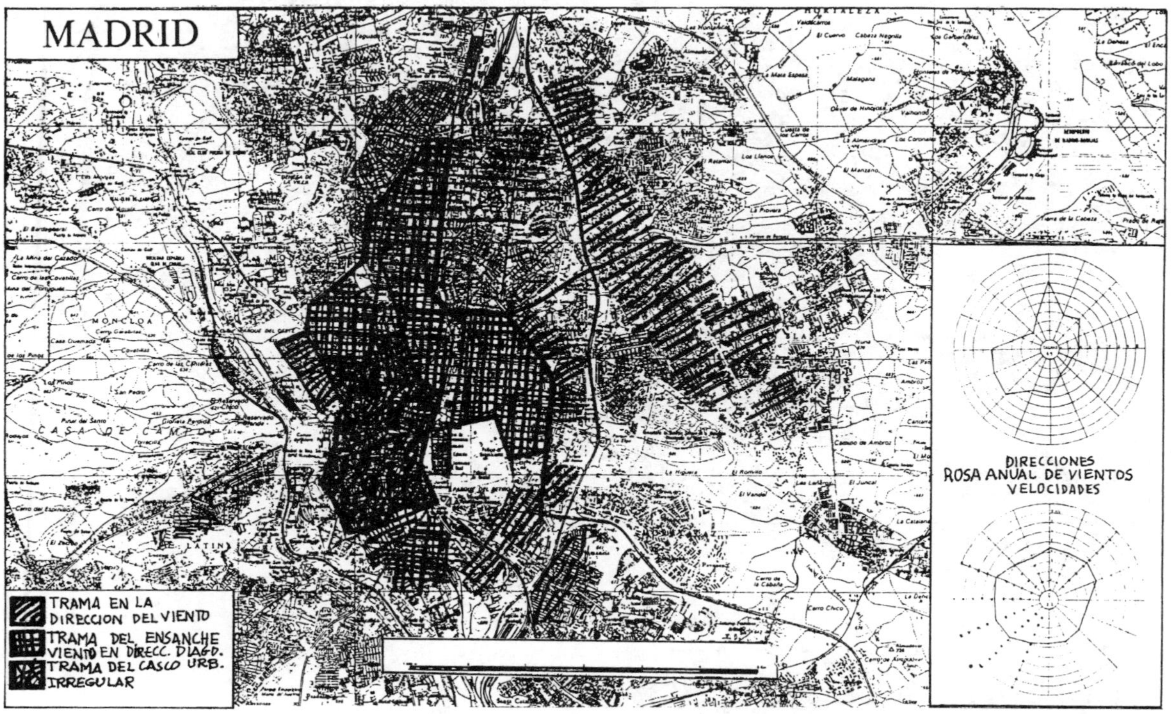

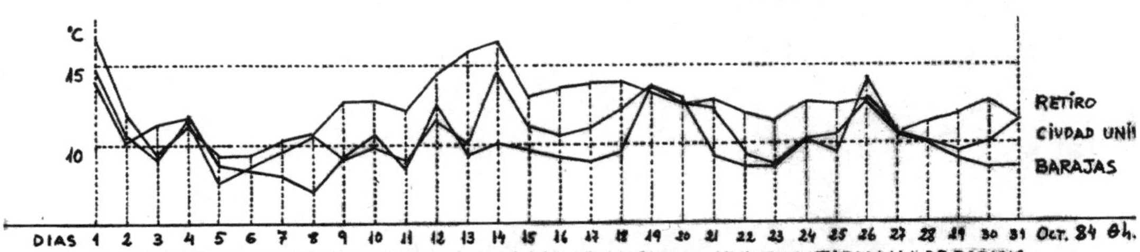

LA TEMPERATURA DEL RETIRO ES SUPERIOR A LAS DE LA CIUDAD UNIVERSITARIA Y LA DE BARAJAS

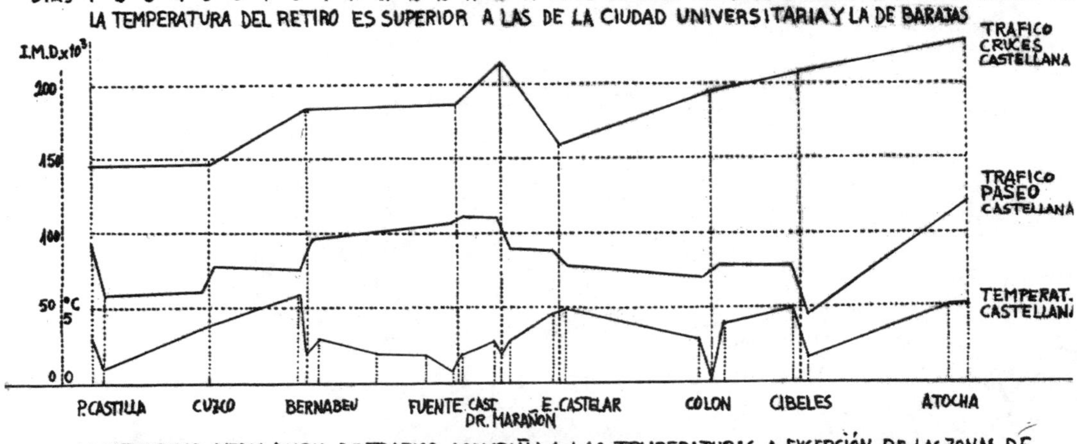

LA INTENSIDAD MEDIA DIARIA DE TRAFICO ACOMPAÑA A LAS TEMPERATURAS A EXCEPCIÓN DE LAS ZONAS DE BERNABEU, FUENTE CASTELLANA, DR.MARAÑON, COLON Y PASEO DEL PRADO POR SER ESPACIOS ABIERTOS Y ZONAS VERDES QUE BAJAN SU TEMPERATURA Y EMILIO CASTELAR QUE LA SUBE POR LA EDIFICACIÓN Y TOPOGRAFÍA.

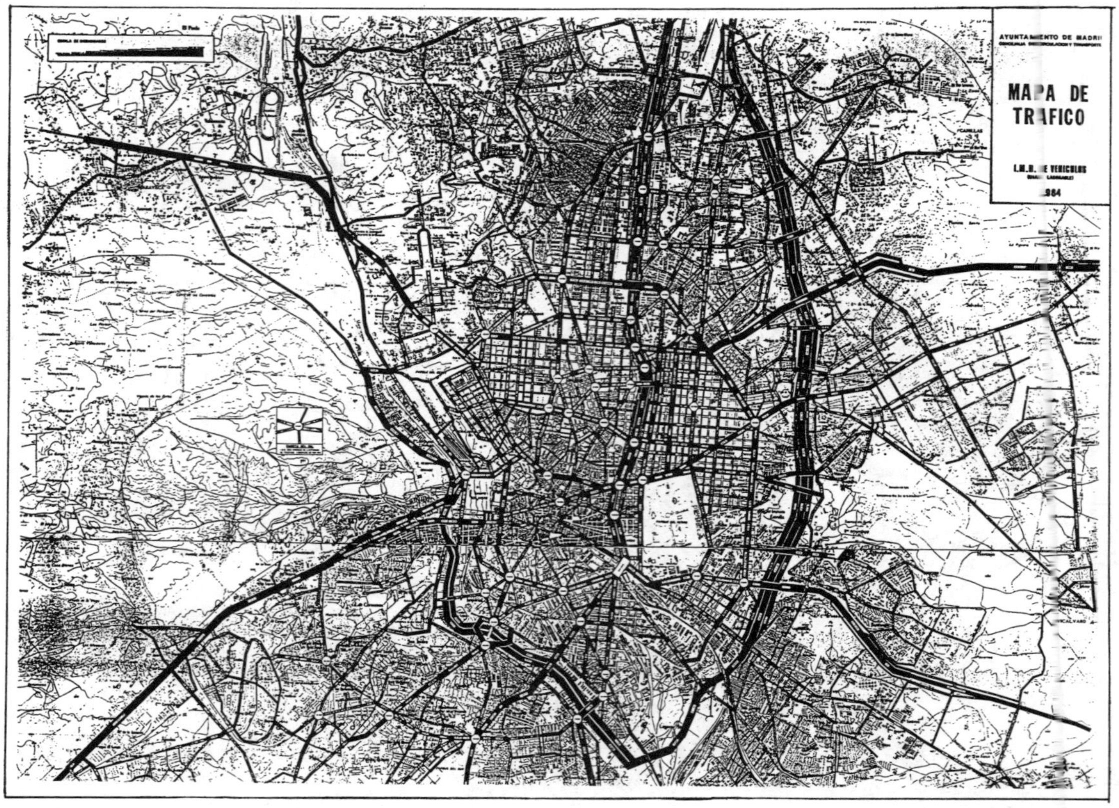
AYUNTAMIENTO DE MADRID
MA A DE
TR FICO

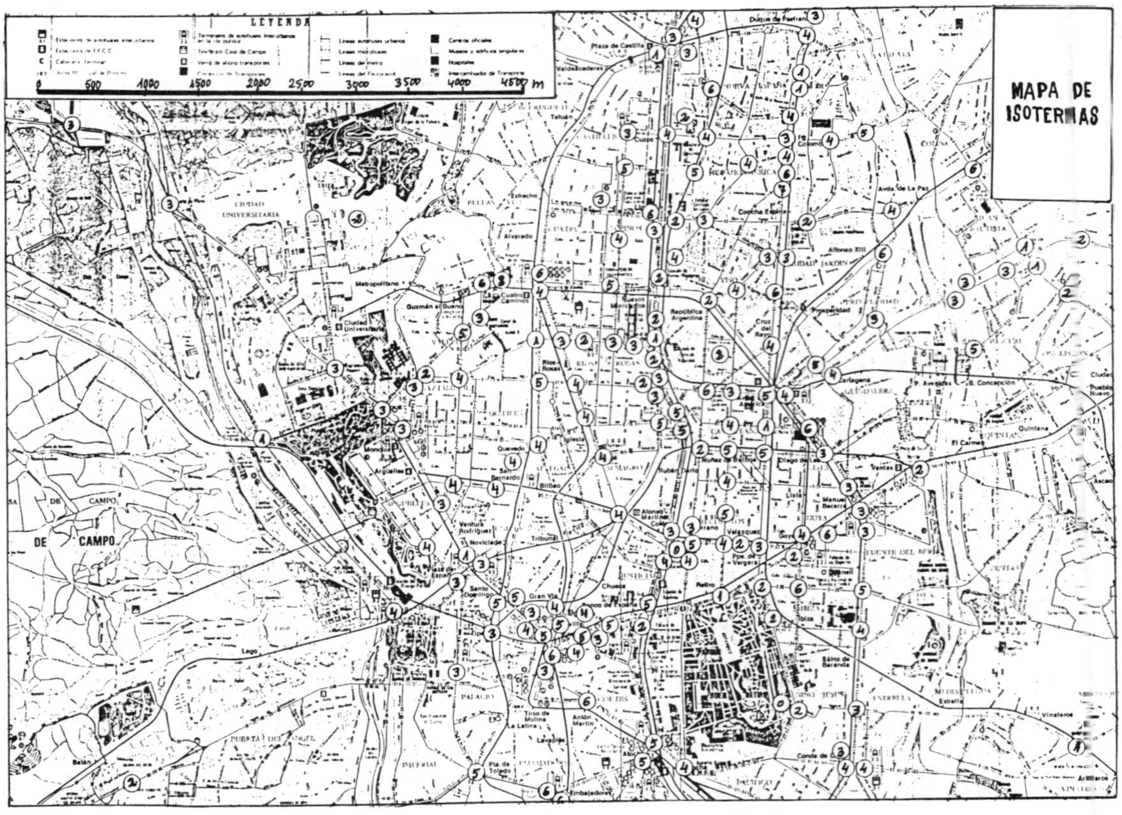
LEYENDA
MAPA DE
ISOTER AS

SOUND TOWN PLANNING.

B. DELAGE - Architect
ESPACES NOUVEAUX - 56 Bd Davout 75020 Paris

Listening to a city's soundscape.

A city's din is directly caused by its activity. This "soundscape" is so familiar that we often overlook it. However, as soon as it is unfamiliar, then we become aware of it very suddenly and, sometimes, very violently.
There are countless sources of sound which intermix and drown each other out : voices, footsteps, alarms, bells, roaring motors, sirens, music, fountains or chimes. Stimulating or suffocating, these noises become clear as soon as we lend them an ear so that, in fact, listening to the din suffices to decipher it.

Recognizing forms in space.
Unfamiliar as we may be with a city, we are familiar with (or, at least, are hardly surprised by), many of its sources of sound. Because of the world-wide distribution of industrial products such as vehicles, construction machines and variety music, cities'sound backgrounds are composed of the same elements everywhere.
More specific sound signs (language, the sounds of local activities, etc) are discernible against this background but differences lie mainly in the urban and architectural forms. It is these forms which compose a city's soundscape and acoustic identity.

Identifying the materials.
Form supports urban materials and the materials which clothe form meld with the emotional content of the soundscape. Glass and steel facades give off sparkling or distinct echoes. On the other hand, parks and public gardens seize on sounds lost in their thickness. The buildings give the sounds thickness and texture. In marble-paved squares, sounds interweave and intermingle. Courtyards thick with balconies produce an astonishing stratification of sounds : the low notes float far above the high ones and voices bounce comfortably from balcony to balcony. Under the open arcades, sounds wear haloes to keep them from sudden outbursts.

Accepting the city's energy.
The city is a complex of urban forms all interwoven into a single fabric.
These forms, static elements, make the flow of dynamic elements possible : pedestrians, cars, motorcycles, subways, and helicopters. Movement reveals the city and makes it ring with life. This flow of movement is the energy which produces soundscapes. A city is dead without it. Therefore, we must not only accept this energy but control, manage and use it so as not to be controlled by it :
This means adapting a city's development (expansion or renovation) to its changing lifestyles and policies (industrial zones unblocked, free circulation of vehicles and pedestrians either separately or together, elevated gardens, public transportation and so forth).

Managing this energy means using particular edifices and technology : including anti-noise walls and slopes, screen buildings, hooding fixed sources, lining facades and windows. But we can also use (not to the exclusion of control or management) the modern city's fantastic energy. Rather than rejecting the whole, why not salvage some of the parts ? Why not fight against noise with new sounds ?

Rediscovering Sounds.
These new sounds can be old ones, for long hidden by the background noise and now coming to the forefront when the background is less powerful. Or they can be created, or provoked, so as to fit the acoustic properties of a urban environment :
- natural sounds such as fountains, chimes, bamboo leaves under the wind, footsteps on a wooden floor.
- artifical sounds such as delivered by equipped traffic lights, warnings related to infra red-sensors, musical sculptures, musical designs.

Methods and tools.
The use of acoustic materials (reflective, absorbant, diffusing), the influence of the shapes of the facades (concentrating, echoing, splittering), the role of the volumes created by the buildings along the street allow "sound town planners" to modify, combine, orchestrate the sounds of the city. They can strengthen or weaken some, harden or soften other, etc.
The organization of these many elements, in time and space, must be done on audio-models, so as to test the projects before realisation. Audio-simulation is therefore an important step. It is the only way of appreciating the improvement of the global environment allowed by the accordance of the sounds, the city and the citizens. We work on such audio-simulations in a specifically equipped recording studio, just as others do sketches, photographies or video to test the visual integration of new buildings, new highways, new colorations.
The realization of what has been simulated, and accepted, can be done through acoustic, or electro-acoustic, or computer controlled devices, depending on the particular site : there is no general rule. There is only a goal to reach : install a better balance between the city and the citizens, "tune" them considering their shared past and future.

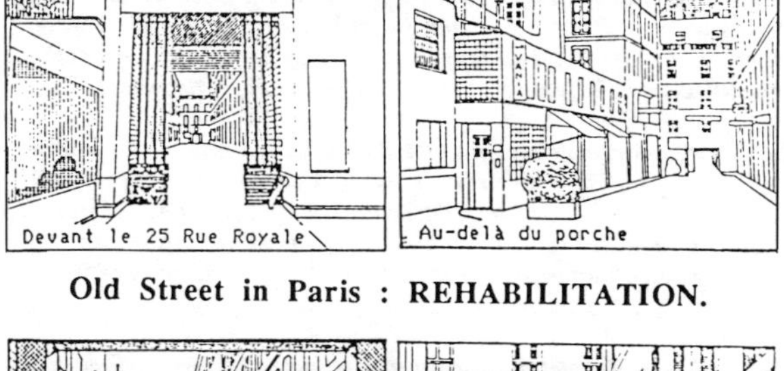

Old Street in Paris : REHABILITATION.

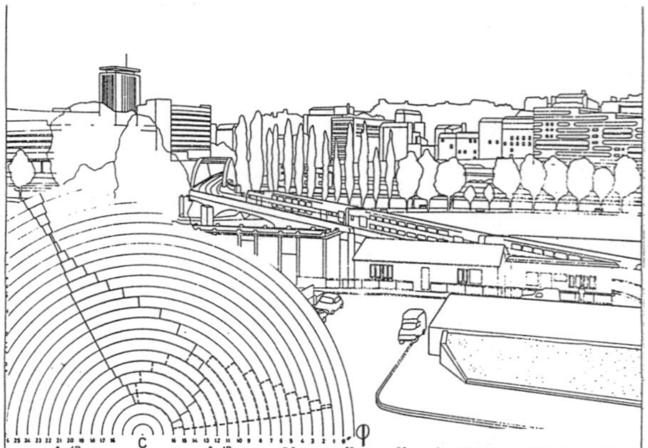

New railway line in Paris : INSERTION.

Experiments and results.

In year 1988, we realized two experiments in Hong Kong : one (electro-acoustics) in a garden at the base of Norman Foster's Hong Kong and Shangaï Bank, the other (computer controlled) in a footbridge linking the Star Ferry's pier to Connaught Road.

The Garden of Sounds : a sound architecture.

Sound architecture is the art of creating space and sound.

The ear interprets sounds in relation to their source and gives us an idea of the place where we are : room or cathedral, building or house. Better yet, the ear provides information about places where we are not. From the fifth-storey, one's ear can tell whether the street below is a boulevard or an alley and whether the staircase is bare or carpeted, made of stone or wood.

In fact, the structuring of space by sounds is one of the keys to understanding the environment we live in. In general, our sense of hearing is so under-stimulated and so badly educated that we are unaware of the sounds that make up our environment. We use our ears without being conscious of it and with only a hundredth of its capacity.

We will use geometric vocabulary - point, line, surface and volume - to describe sound architecture. More specifically, this means that the sound of a car honking is a point, a sparrow spiralling through the sky a line, the wet asphalt hissing under tyres a surface, while the din of passing trucks, cars and buses represents volume.

Like art, sound architecture has at least the three following objectives : to structure or restructure an existing space, to create a space other than the existing one (an "espace nouveau" - a "new space") and to create an independent space with no links to the visible universe.

To achieve this, sound architecture uses acoustic or electroacoustic resources, sounds produced "naturally" in the space to be constructed or artifical sounds. Fiction has its rules. On the other hand, work carried out in a created space must take on constructive principles, if only to rechannel, contradict or amplify them. "Potential" sound architecture of sound and "real" materials must work together.

More than by its degree of materiality, sound architecture is distinguished by its position in time.
Earthly architecture exists in linea time whereas sound architecture is beyond time. A building stays where it was built until it is destroyed. But sound architecture, even when part of a territory, appears and disappears and changes form. It does not know when it's birthday is and at every second carries the seeds of its own destruction.

Sound architecture does not decay : it either exists or not. This is one of the paradoxes of sound, which exists for a time it is free of whereas stone is unaware of the time it is a victim of.

Statue Square as the place of an experience.

Statue Square has particularly discernible architectural and sound characteristics.

It is a square esplanade, bordered on the north and south by two very buzy streets and on three sides by two buildings : Prince's building, the famous Hongkong and Shanghai Bank, designed by the architect Norman Foster and the long Legislative Council building. Statue Square has four entrances facing each other on median axes.

Statue Square, with its sheets of water, fountains and vegetation, is meant to be a rest area. The square occupies an important position in the Central quarter, having underground stations and access to the Star Ferry underground passageway in Chater Road and tram stations in Des Voeux Road.

Except on rare occasions, the sounds in the square are extremely confused. Traffic noises dominate the soundspace, which is as formless as the visual space is geometric : the high-pitched screeching of bus brakes, and, at present, even the noises from the construction site of the Bank of China, designed by the architect I.M. Pei, can hardly be distinguished against the continuous background of traffic noises. One has to walkway back into the square to be able to hear children's voices or jets of water spluttering against the fountain wall. The soundscape has no character and the territory, victim to "noises of prey", is nebulous.

But the social importance of Statue Square is undeniable.Filipino women chose it for their weekend meetings (during which the territory is thick with the high-pitched sound of thousands of voices). Special events, like the Lantern Festival also take place at Statue Square. Statue Square is an important meeting place but it needs a responsive, structured soundspace.

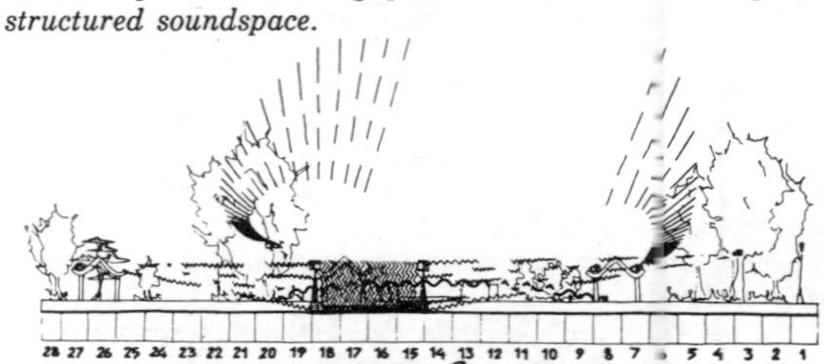

south façade

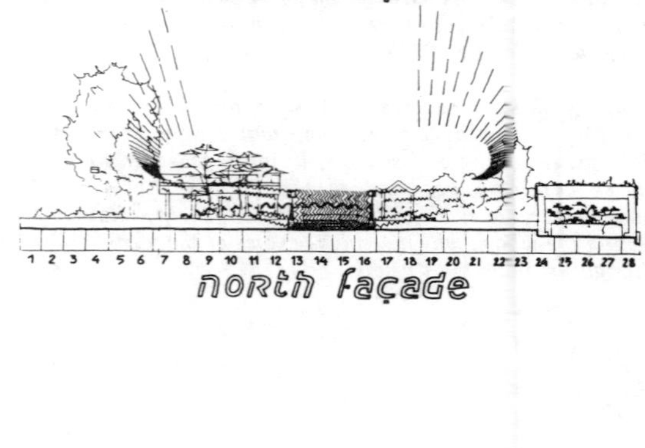

north façade

Putting space to music.

Urbasonic designed an experimental sound structure in Statue Square called the Garden of Sounds. Its objectives are the following :
- to protect the space as much as possible from outside noise.
- to improve the quality of the space.
- to clearly determine the territory and give it more character.

These objectives have to be translated into musical terms because music, or art, will combine these sounds in space. Sound construction consists of the following :
- analyzing the site's sounds. Statue Square was recorded and a sound map was drawn up.
- composing the sound architecture on the basis of acoustic constraints and an artistic project. A simulation was carried out at the Espaces Nouveaux workshop, in order to evaluate the results of combining new sound architecture with existing soundscapes.

The Garden of Sounds consists of the following :

- The four sides of the esplanade are delineated by sounds which can be heard in the sound background (this background of sound cannot be entirely masked, but our ears can be attracted elsewhere). The entrances are thresholds, or barriers, that must be crossed to enter the square and which serve as acoustic break. The sounds move very rapidly and in a high register, like the cicadas on top of the Peak.
- A calm zone around a small bridge in the middle of the square provides sounds for the sheets of water. One of the characteristics of water sounds is that they mask medium and medium bass sounds like car engines.
- At each corner, four larger airier sounds ressembling the phoenix's cry which echo and rise slowly in the sky.
The sound materials are both acoustic and synthetic, coming from the pipa, a Chinese instrument and modulated frequency synthesizers. The use of a Chinese instrument, part of Hong Kong's culture, is very important : it gives the sound architecture its unity and character. Because the city mixes modern forms and Chinese traditions, the Garden of Sounds attempts to preserve this mixture.

The Elevated Path : A moving sound design.

Hong Kong has an extensive network of pedestrian walkways. By putting pedestrians above and motor vehicules below, these walkways mean less traffic, free circulation and better safety.

But besides being useful, why shouldn't these walkways be pleasant places ? Far from the hustle and bustle of the street, the pedestrian should be able to use these passageways between sky and earth without the slightest disturbance. There is not much to see up there and even less to hear. On the sidewalk, he perceives sounds, movements and details.
From the air, all these elements are blurry.

The "Elevated Path" on Connaught Place walkway has two objectives : to regulate the pedestrian flow and to give the place an aesthetic value. To this end, sound trajectories are designed on the surface of the walls.

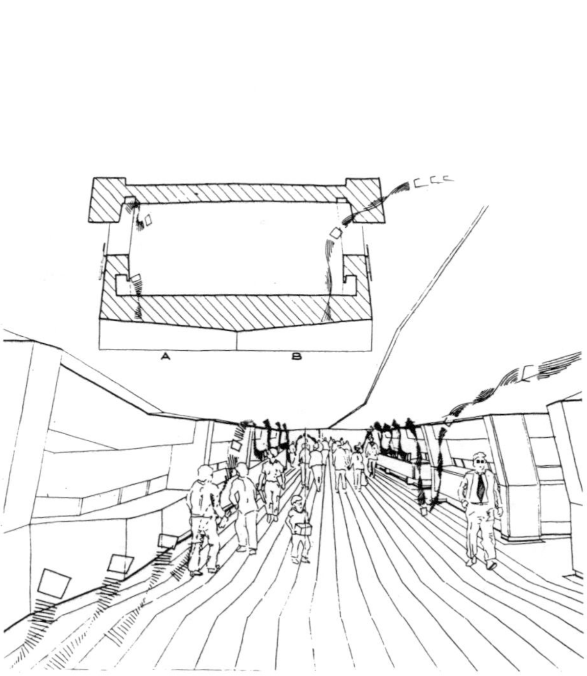

This walkway is an interesting experiment. Being covered and partially enclosed, it provides pedestrians with protection and comfort, reduces sound nuisances and has a pleasant interior design, all conditions favorable to smooth sound management.

In all truth, the Elevated Path is not sound architecture but rather a "moving sound design". More specifically, as the pedestrian enters the walkway, he hears sounds which either move towards or away from him, depending on where he is. In fact, on the walls to his left and right is a wiring system (Swift) which broadcasts sounds the lenght of the walkway. On the left wall, the sounds move away from him, inviting him to follow. On the right wall, the sounds move towards him, making him move away and rejoin the "natural" flow of circulation to the left.
Pedestrians naturally fall into two ascending and descending lines, but without being forced. Sound signals are hosts not dictators.

In addition to regulating, theses moving sounds give the walkway added character. Because of moving in the sound forefront, they attract the pedestrian's ear, Because of being heard in the sound background, they diminish the perception of it. The walkway becomes an identifiable and original sound territory.

What are these sounds ? They are made up of different sequences and based on a model consisting of the following sounds in order :
- points, or short sounds which travel along the floor.
- short segments, like points extending in time, which rise and move at ear level.
- a line which abruptly descends and rises to the ceiling and even beyond, a smooth sound which drops before rising.

The model has numerous varied forms, which are recognizable but not uniform. Like those at Statue Square, the sounds have both acoustic and synthetic sources which come from the guqin and a modulated frequency synthesis system.
In addition to its usefulness, the sound composition of the Elevated Path is meant to give the walkway a dynamic and airy atmosphere and to give the pedestrian such a sensation of lightness that he feels as though he were really walking between heaven and earth.

In year 1989, we realize an other experiment in Osaka, Japan, on an outdoor plazza.By the end of year 1989, other projects for Osaka will have been designed, and computered images and sounds assembled for simulation.
The results of such experiments can be easily appreciated, by comparing recordings of the sound environment without and with our installations. But what is still to be evaluated is how these experiments will react on architecture and urban modern art. We are just at the beginning of a time where the quality of the sounds in the street will be as important as the quality of the music in concert halls actually is.

INSOLATION IN THE BUILT ENVIRONMENT: CRITERIA FOR ITS NORMALIZATION AND REGULATION

F.O. Ruttkay Pereira
J.A.C. Mincache
Dept. of Building Science, University of Sheffield, Sheffield S10 2TN - UK
Depto. de Arq. e Urb., Univ. Federal. Sta. Cat., 88049-Fpolis/SC, BRASIL

ABSTRACT. The main purpose of this paper is to present data which will help improve the quality of the built environment based on the rational use of solar radiation, with the intention of advancing insolation and sun control utilization, in order to accommodate solar heat gains or cooling benefits. A real situation of insolation regulation, based on "property lines", is analysed, and other possible regulation values are suggested by applying the concept of desirability and undesirability of solar radiation.

1. INTRODUCTION

Insolation, at the present time, is considered to be one of the most important design parameters in the built environment, and one that assumes a large role in external and internal spatial-composition resolutions. However, it is not a new concept. It is enough to call to mind Gropius with his studies of the distance and height of rectangular blocks of houses /1/, Le Corbusier with the Tower of Shadows in Chandigar, India /2/, and many other historical examples of the correct use of the geometrical relation existing between Sun and Earth.
The first time in recorded history when regulations were set down establishing sun access was in Rome, sixth century A.D., with the Justinian Code of Law: "If any object is so placed as to take away the sunshine from a heliocaminus, it must be affirmed that this object creates a shadow in a place where sunshine is an absolute necessity. Thus it is in violation of the heliocaminus' right to the sun" /3/.
Nevertheless, the achievement of an adequate and fulfilling normalization of insolation is not an easy task; many researchers have demonstrated that it is a multi-criteria problem, which involves climatic, biological, socio-architectural and technico-economic parameters.
Some relevant research, which was carried out by the Scientific Research Institute for Construction Physics, Moscow, has demonstrated that the most effective system of insolation normalization must be based on the normalization of the annual amount of effective solar radiation that is incident on the buildings /4/. However, one of the most significant barriers to the introduction of this concept in practice, the correct computation of cloudiness condition, is not very well understood as yet.
As far as the variant of indirect normalization is concerned, normative spatial indicators, which can be checked using spatial-geometric graphics to investigate the contextual shading of sites, have been used, with good results, to allow the right balance of shadow effect.
The best known indicators are those based:
i) on the insolation hours of a given facade;
ii) on the sectoral angles which are positioned with their vertex on an important design point that must have a view free from site obstructions.
The use of these spatial indicators is very simple and practical, and can provide good comprehension of the site contextual shielding. However, most of these indicators are based on the duration of insolation, and since an effective normalization must deal with multi-criteria solutions (as stated above), it has been agreed that the time of insolation alone, except in particular situations, cannot be an adequate normative indicator /4/. To explore the possibilities of the development of one of these spatial indicators, the emphasis of this paper is on analysing a real situation of regulation of insolation, which has been applied in Florianopolis, SC, Brazil, by the local government. Further, with the help of a simple instrument (On-Site Shielding Analyser) and by applying the Bioclimatic-Chart /5/, which presents a temperature limit that determines whether the solar radiation is to be accepted or not, other possible regulation values obtained through the application of the concept of desirability and undesirability of solar radiation are given.
Meanwhile, is not our intention to get deeper in the related parameters to solar radiation, but just to stimulate the discussion with respect to the sun control techniques in urban planning and how we can achieve the best results from its use.

2. PRESENT SITUATION

The Prefecture of Florianopolis, SC, Brazil, has applied a regulation (Law N. 1851) that regulates about zoning, use and occupation of the insular urban area of the city /6/. This legislation states that: "to ensure adequate insolation and ventilation to the public area, the height of buildings could determine the requirement of frontal removing and in any case the building are not able to surpass the projection line of a 70 (seventy) degrees angle measured from the street's longitudinal axis until the most elevated point of the facade".
The main objections that can be made to this regulation are:
i) the standard value (70 degrees) was obtained by analysing only North orientation;
ii) it takes into account only the time of insolation criteria (2 hours), dependent of the time of year.
These objections become more significant with the fact that the present regulation was conceived without any consideration of the actual climatic conditions of Florianopolis and ambient comfort requirements /7/.

3. PROPOSED APPROACH

The Southern region of Brazil, where Florianopolis (lat. 27.5° S) is located, is characterized by a climate that can be defined as Sub-Tropical, composed of three main well-defined seasons : hot summer, spring-autumn and mild winter. Using any method of bio-climatic analysis it is possible to point out the overheating period (5.5 months), the comfort period (4 months) and the underheating period (2.5 months) /8/.
In this way, overlaying the local climatic data over the Bioclimatic-Chart it is possible to obtain the climatic needs and, consequently, the period throughout the year when insolation is desirable or

not. Then, this interval in which shading is required (with temperatures above 22°C) can be transferred to a sun-path diagram. This other diagram will provide the basis for the evaluation of the shielding of buildings in a site context with respect to shading exigencies /5/. (Fig. 1)
The present proposal seeks to integrate the diagram of Figure 1 with the Shading Mask Protactor inputing the concept of Desirability and Undesirability Index (DI and UI), where:

$$DI = \frac{\text{hours of desirable sunshine}}{\text{total hours of sunshine (profile angle = 0°)}}$$

$$UI = \frac{\text{hours of undesirable sunshine}}{\text{total hours of sunshine (profile angle = 0°)}}$$

These numbers of hours were obtained overlaying a Shading Mask Protactor on the diagram of Fig. 1, according to:
i) the vertical obstruction angles lined from the street axis varying from 0° (horizon line) to 80°, assuming built site with no lateral gaps considering the sheltering effect, as leaves of a tree, of the constructions in the urban area;
ii) the direction of the street (N - S and W - E);
iii) the period of the day/year:
SM - summer morning (Oct-Mar, sunrise to 12:00 a.m.) presenting both desirable and undesirable insolation periods.
SA - summer afternoon (Oct-Mar, 12:00 a.m. to sun-set) shading required almost all the time.
WM - winter morning (Apr-Sept, sunrise to 12:00 a.m.) insolation required for heating purpose almost all the time.
WA - Winter afternoon (Apr-Sept, 12:00 a.m. to sun-set) the same needs as SM.
With the values obtained from the above procedures, it was possible to draw graphics showing the relationships between the variables (Fig. 2 and 3).

4. RESULT ANALYSIS

Initially, analysing the graphic results one can confirm the different insolation requirements in each period of day/year. In succession, without diverging from a daily average time of insolation between 1.50 and 3.00 hours all the year (approximately the values that has been claimed by many authors as an adequate normative indicator), and trying to keep DI as higher as possible and UI as lower as possible, we do the following considerations:
i) with the street orientated North-South there are two distinct situations, the side of the street facing West and the other side facing East. In the former situation the 70° angle can be reduced to 65°-67° in order to allow better reception of desirable sunlight at WM (DI=0.2), without introducing much losses at the SM period. The sun in this period will assume high altitudes and as has been reported by Lee /9/, the shadows produced are short and move quickly providing little shading benefit, particularly around noon when the most effective way to create shadow casting in the facades is properly designed overhangs. In SA and WA periods the limitation of UI=<0.2 and DI>=0.15 (minimum desirable insolation) led to an angle range of 70°-72° for the street side facing East (Fig. 4).
ii) in the case of the street orientated East-West the discrepancies are greater than those in the other direction. Now one street side is facing North and the other side is facing South, presenting very much different insolation conditions. For the SM and SA periods the 70° angle provides excessive solar radiation incident, especially in SA when almost all the radiation is undesirable by thermal aspects (UI=0.6). Acceptable insolation levels should be achieved with angles around 82°-85° (UI=0.2-0.3 in SA). Considering the WM and WA periods the 70° angle does not provide any insolation on the North side of the street. In this way, attempting to provide a minimum quantity of desirable sunlight the angle should be around 53°-55° (Fig. 5) with DI=0.15 in WM.
As can be seen from the above considerations, the concept of desirability and undesirability of solar radiation has led to different values for limiting projection line ("property line") angles which can be used at the same mode to minimize and/or maximize the sun's impact on the urban area. The main difference in relation to the "70 degrees angle law" is that at the periods in which desirable insolation is available, the DI Index (DI>=0.15) provides part of this insolation. On the other hand, taking in account the UI Index (UI=<0.2) one can avoid excessive undesirable insolation which is the worst problem on overheated periods and the most common error in applying only the time of insolation normalization to regions that presents opposite needs along the year.
Transferring the angles from Fig. 4 and 5 to the edges and corners of a site, the result will be a set of geometric limits derived from the expected balance between DI and UI Indices. The resulting three-dimensional form in which the surfaces represents the boundaries of the site has been defined by Knowles /10/ as the solar envelope - the largest volume that a building can occupy whilst allowing solar access to adjacent buildings.
One of the main restrictions to the application of this concept is the resulting low density. In the present study, however, this consequence has not been so significant because the rules used to generate the property line angles has not been as restricting as those developed by Knowles /10/ and /11/. According to a simple verification, taking into account a regular block in the centre of Florianopolis, the solar envelope obtained with the angles from Fig. 4 and 5 is only 2 - 3% shorter than the solar envelope defined by the "70 degrees law", and, in addition, the former situation provides more balanced insolation conditions with respect to thermal requirements.

5. CONCLUSIONS

Insolation in the built environment depends mainly on decisions concerning "design parameters" made by urban planners and architects. Therefore, they must be able to manage with the involved physical parameters.
The authors seek to introduce into the discussion one of these important parameters - thermal desirability and undesirability of solar radiation, to help urban planners and architects to avoid poor urban and architectural configurations in relation to sun access which cannot be easily reversed later.
In the present work, a simplified analysis was made about how the built environment spatial configuration can perform in order to provide good insolation conditions in the urban area. The approach used was to check the possibility of introducing restrictions as property lines which, by extension, confirm the solar envelope as defined above.
The main parameter chosen to control the study as a target to be balanced, instead of the operating schedules applied by Knowles /10/ and /11/ and Robbins /12/, was the accessibility (heating) and intercepting (cooling) of solar radiation requirements through the year, translated into the concept of desirability and undesirability of solar radiation.
The results above have confirmed that the time of insolation alone (which has led to the "70 degrees law" /7/) cannot provide a good normative indicator and define the real conditions of insolation in the built environment, and that the conscious application of property lines regulations can

provide a reasonable method for defining setbacks, easements and building height limitations.
Further research is needed to verify the situation for more orientations and to vary particular site parameters (size, shape, orientation and use of the site, topography and surrounding urban context). The case of lateral gaps between constructions, and interactions with others no less important design parameters such as daylighting, air flow around the buildings, urban acoustic, must also be examined. Attempts should be made to assess the effective solar radiation incident on the usually complex configuration or urban areas.
Finally, the On-Site Shielding Analyser has shown itself to be a very useful instrument to carry out a full investigation about insolation conditions in the built environment.

6. REFERENCES

/1/ Cornoldi, A. and Los, S. (1985): Habitat y Energia. Gustavo Gilli Ed., Madrid.
/2/ ENARCH'83 (1983): Architettura Bioclimatica. De Luca Editore, Italia.
/3/ Butti, K. and Perlin, J (1980): A Golden Thread. Marion Boyars Publishers Ltd., London.
/4/ Obolensky, N.V. and Korzin, O.A. (1982): Insolation and Sun Control in the Field of Construction: The Progressive Ways of Their Normalization and Regulation. CIB Proceedings of Symposium Building Climatology, Moscow.
/5/ Olgyay, V. (1973): Design with Climate. Princeton University Press, Princeton.
/6/ PMF (1982): Dispoe sobre o zoneamento, o uso e a ocupacao do solo em setor urbano da area insular de Florianopolis. Lei N. 1851 (Regulations), Prefeitura Municipal de Florianopolis.
/7/ Souza, A.V. (1982): Parametros de afastamento entre edificacoes em funcao da insolacao minima na regiao de Florianopolis. Internal Report, IPUF, Florianopolis.
/8/ Pereira, F.O.R., de Abreu, R.C. and Schmidt, A.F.C. (1982): Edificacao para o Clima de Florianopolis Internal Report, Dept. de Arq. e Urb., UFSC, Florianopolis.
/9/ Lee, K. (1985): Angular Velocity of Shadows - Site Impact. Proceedings of International Symposium on Thermal Application of Solar Energy, Hakone, Japan.
/10/ Knowles, R.L. and Berry, R.D. (1980): Solar Envelope Concepts: Moderate Density Building Applications. Solar Energy Res. Inst., SERI/SP-98155-1. Golden, Colorado.
/11/ Knowles, R.L. (1981): Sun Rhythm Form. Cambridge, Massachussets, MIT Press, London.
/12/ Robbins, C.L. (1986): Daylighting: Design and Analysis. Van Nostrand Reinhold Company, New York.

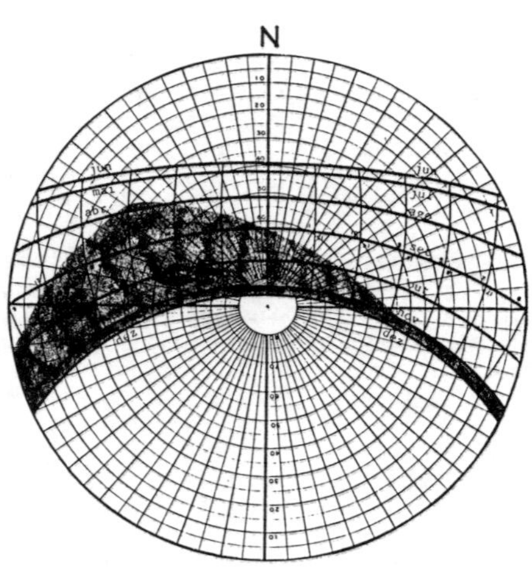

Figure 1: Shading requirement period for Florianopolis (Lat. 27.5°S) /8/.

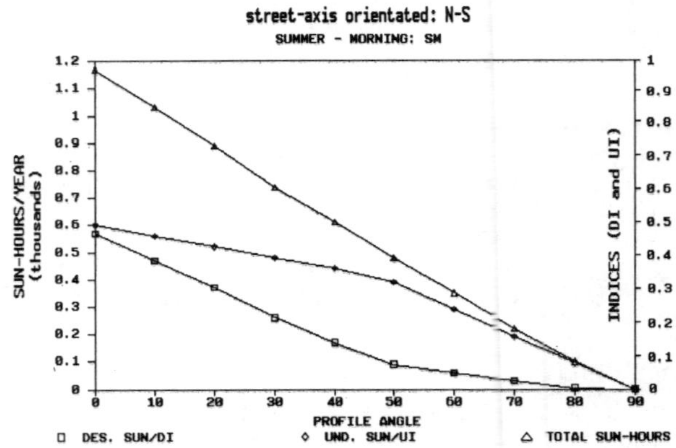

Figure 2: Sun-hours X Profile Angle (NS-SM).

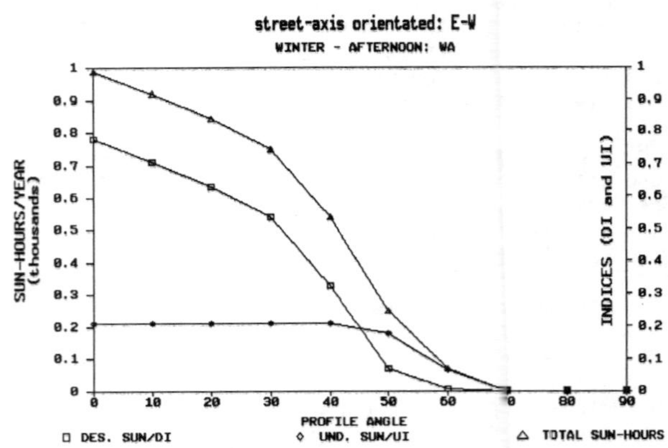

Figure 3: Sun-hours X profile Angle (EW-WA).

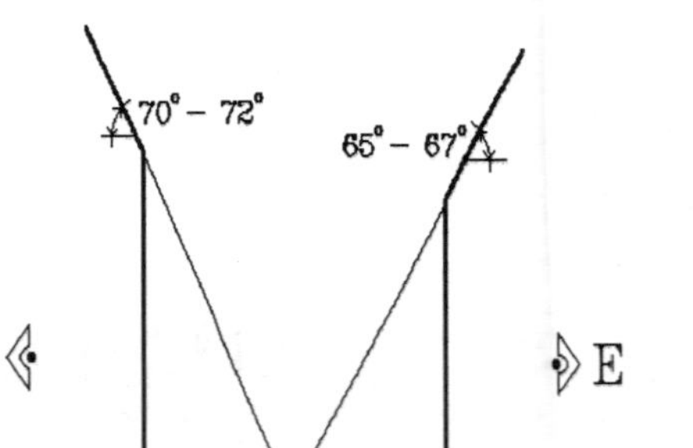

Figure 4: Angles proposed/street orientated N-S

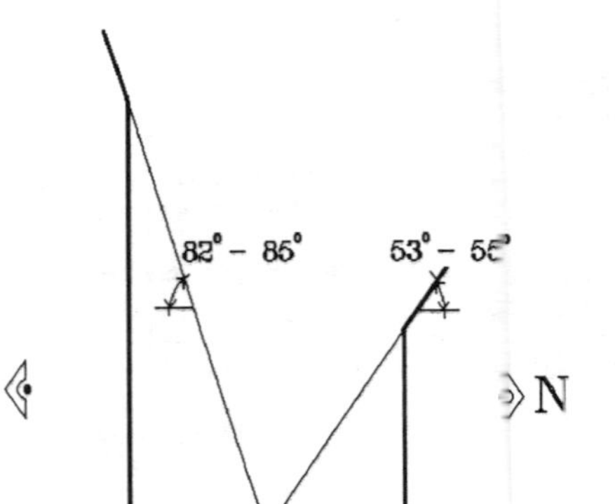

Figure 5: Angles proposed/street orientated E-W.

LES ECONOMIES D'ENERGIE DANS LE CENTRE HISTORIQUE DE PALERME

Guli A.,Bellanca L.,Bertolino S.,Cammalleri A.,Crimi L.,
Lo Iacono M.,Pino M.L.,Schilleci F.
Département Cité et Territoire de l'Université de Palerme
323 rue Resuttana Colli 90146 - Palerme Italie

ABSTRACT. The research on the vegetation in Palermo old centre is based on a comparison between the present state and the plan aimed at the saving of energy and improvement of the city environment.We can get appro - priate informations on the choice of plants and above all on the arrangment of trees considering the local climatic conditions and the following parameters : leaves thickness,foliation period,size and shape,growth speed and duration.Through the reclamation of the present vegetation and the creation of new spaces,the plan underlines the importance of greenery as an important element in climatic regulation : variation in tempe- rature,humidity,air speed,transmitted sun radiation.

1. L'ETAT ACTUEL

1.1 L'histoire du paysage urbain

La ville ancienne de Palerme,entourée de remparts espagnols,se divise en quatre arrondissements:Tribunaux,Palais Royal,Mont-de-Piété,Château à la mer. Sa morphologie dérive d'une sequence d'interventions qui commencent dans le XVIe siècle et qui brisent le schéma fondé sur un axe:le "Cassaro".C'est le schéma de la ville dès l'époque Punique et le Moyen-Age,jusq'à la Rainaissance.L'architecture , souvent aulique,a saturé les espaces de la ville rappelés dans les descriptions des voyageurs arabes.Après le XVIIIe siècle,avec l'expansion hors les remparts,la ville ancienne devient la partie d'une structure urbaine qui,en deux siècles,a envahi le territoire de la Conque d'Or.En constituant une charnière,dedans et hors le centre historique, les parties restantes des remparts gagnent une particulière valeur dans le paysage urbain.

1.2 Analyse de la végétation

Dans le centre historique il y a des espèces typiques du maquis méditerranéen (chêne,olivier,cyprès, laurier,romarin,myrte,etc.);pendant l'invasion musulmane on a introduit des arbres à fruits /1/. Après le XVIe siècle on a importé de nouvelles espèces exotiques (provenantes des pays d'outre-mer)qui s'acclimatent bien dans notre région.Les jardins historiques sur les remparts,les cloîtres plantés de arbres des grands complexes religieux,les cours des palais nobiliaires,les jardins du XIXe siècle le long du "Cassaro",constituent aujourd'hui un patrimoine végétal sur lequel intervenir,pour en sauvegarder les valeurs historiques et en exploiter les potentialités à fin de créer un nouvel équilibre de l'ambiance.

1.3 Les jardins sur les remparts

Les murs d'enceinte espagnols,nés pour des raisons défensives,sont otés au domaine de l'Etat sur les remparts et,à la fin du XVIIIe siècle,on a dressé, à leur place des jardins parmi lesquels en 1789 le Jardins des plantes sur le rempart "Aragona" à "Porta Carini".Ce dernier et bien d'autres remparts seront démolis à partir de la seconde moitié du XIXe siècle,même pour des mesures d'hygiène.Le chateau à la mer sera démoli en 1922.Il ne reste de lui qu'un donjon (peut-etre de l'époque arabe) et la Porte Aragonaise.Les jardins sur les remparts restants méritent,donc,une particulière attention,parce que leur récupération et leur utilisation paraissent intéressantes pour en rappeler le mémoire historique. Il s'agit d'espaces à l'abandon comme celui du "Spasimo" et des "Mura delle Cattive" ou bien d'espaces rendus privés comme les murs de "S. Vito"et le rempart de "Porta Cuccia",mais en réalité ils gardent encore des témoignages architectoniques et une remarquable quantité de végétation à récupérer et à intensifier.D'autre part ils ont une position singulière entre la cité et la ville du XIXe siècle et ils pourraient acquérir,une fois valorisés,des roles certainement significatifs.

1.4 Les jardins publics

Le jardin "Garibaldi",la villa "Bonanno" et la place "S. Spirito",situés le long de l'axe de la cité, l'avenue "Vittorio Emanuele" (ancien Cassaro),avec leur masse végétale par rapport aux architectures auxquelles ils se rapportent,effacent la désolation des vides et créent des épisodes significatifs d'architecture arborescente.L'autenticité de la valeur sociale du jardin du XIXe siècle a gravé sur les courtines,édifiées en des époques différentes, une unité et un équilibre pour le rapport entre les constructions et l'espace libre.Le jardin "Garibaldi" dessiné en 1863 par G.B. F. Basile,repropose urbanisement le modèle du "square" anglais.Il se trouve là où il y avait jadis un espace libre et accidenté et il est caractérisé par d'exceptionnelles masses arborescentes constituées par des palmiers et des ficus extraordinaires.Au-delà de la végétation,ce qui définit architectoniquement le jardin, ce sont la grille en fer forgé,les pavillons et les autres nombreuses décorations urbaines.Il est relié à l'avenue "Vittorio Emanuele" par une charnière formée de la place et la fontaine du "Carraffo", place ombragée par de nombreux ficus.Même Basile avait projeté en 1861 la place "S.Spirito" là ou se dressait,avant,paroisse de St.Nicolas à la "Kalsa". C'est un espace à la marche dynamique qui trouve un rythme cadencé dans la présence des palmiers alternés avec des buissons de hibiscus et le caractère classique dans la fontaine centrale du Cheval marin. Il faudra attendre le début du siècle successif pour voir se dresser une autre villa publique:il s'agit de la plaine en face du Palais Royal;c'est là qu'en 1922 le maire Bonanno a délibéré la création d'un jardin au centre duquel il y a une grande masse de palmiers,tandis que la rangée des platanes périmétraux sert à rythmer et à protéger l'oasis intérieure du trafic routier.Juste en ce lieu,dans la seconde moitié de ce siècle,dans le secteur sud-est, on a retrouvé quelquels pièces correspondantes à la paleopolis,desquelles seulement les mosaiques romaines ont été exposées et protégées par un pavillon/2/.

1.5 Les plans existants

Le Plan Régulateur Général de 1959 reflète la technique urbaniste liée aux demolitions et aux restructurations;la verdure répond à la logique des standards.Le Plan Programme de 1983 se relie à la poétique urbaniste du respect absolu de l'ambiance ancienne,où la végétation est au service de la collectivité.Le nouveau Plan Détaillé,rédigé par les architectes L. Benevolo,L.Cervellati et I. Insolera, n'est pas encore évaluable,parce qu'on est dans la attente de son approbation définitive.Le Plan a été présenté à la ville le 15 Novembre 1989;dans ce plan la verdure a pris une fonction de grande importance pour la récupération de la cité.Des 6 hectares d'aujourd'hui on devrait arriver,même à travers la démolition des parties baties,à 30 hectares environ de surfaces à verdure.

2. LA VEGETATION COMME FACTEUR DE REGULATION CLIMATIQUE

2.1 Le climat

La Sicile a un climat méditerranéen,Palerme en particulier méditerranéen chaud sub-tropical.La ville confine au Nord-Est avec la mer et elle est renfermée à l'Ouest et au Sud par une enceinte de reliefs montagneux qui vont de 700 à 1000 mètres;elle en est séparée par un amphithéatre naturel appelé "Conca d'Oro" qui,jusqu'aux années quarante a été entèrement le "jardin" de la ville.Son emplacement géographique aux centre de la mer Méditerranée (mer relativement fermée à basse amplitude) la morphologie du territoire,la protection des collines environnantes et d'autres facteurs particuliers comme la disposition spatiale des édifices et l'expansion urbaine des derniers décennies vers le Nord,ont entrainé à un progressif "réchauffement" du centre urbain /3/.La tache verte dans laquelle la ville était plongée du Parc de la "Favorita" à la vallée de l'Oreto s'est progressivement amenuisée et découpée;la disparition de plusiers terrains verts à l'exterieur et à l'intérieur des espaces bâtis a entrainé à une diminution de l'humidité relative et à un manque de protection à l'égard des vents chauds qui viennent du Sud et du Sud-Est,avec des excès thermiques de 35-40°C pendant la période Juillet-Aout.

2.2 Les effets de la végétation sur le climat

Dans ces conditions climatique le besoin d'énergie est plus grand pour le conditionnement de l'été que pour le réchauffage de l'hiver,le choix,donc,des espèces végétales a comme objectif premier le refroidissement de l'air en été en produisant,en meme temps,un effet bénéfique sur la régulation de la temperature de l'hiver aussi.Dans les zones vertes suffisamment vastes on peut relever aussi de sensibles reductions de la température,dues surtout aux procédés d'évapotranspiration et à l'effet de l'ombre. Une couverture végétale avec des cultivations arborescentes même à une évapotranspiration d'environ le 50% de plus au regard d'une superficie cultivée à pré.De toute facon les variations climatiques causées par la végétation se ràvèlent surtout sur les édifices directement contigus à elle.Les effets les plus évidents sont donnés par la reduction de l'intensité du rayonnement solaire et de la vitesse du vent.Le calcul de la reduction du rayonnement solaire dépend de la densité de la chevelure,de sa forme et de sa dimension (des paramètres étroitement liés à la vitesse de la croissance et à la durée de chaque espèce),en plus,de la disposition et du coefficient de transmission des feuilles et donc de l'inclinaison des rayons solaires.La densité de la chevelure varie d'une espèce à l'autre avec le changement de la forme et de la dimension de la chevelure meme,mais dans la meme espèce aussi,au cours de son développement ou bien d'une saison à l'autre relativement à la persistance ou à la caducité du feuillage.Les espèces sempervirens ne modifient fondamentalement pas la densité de leur chevelure en hiver,tandis que les espèces caduques absorbent le 50 % en moins,par rapport à l'été,du rayonnement solaire avec une économie d'énergie pour réchauffage du 7% /4/.

2.3 Le relief de la végétation

La restauration des jardins historiques existants "devra etre entrepris seulement après une étude approfondie qui va du percement à la récolte de toute la documentation pour ce qui concerne le jardin et les jardins analogues ..."(art. 15 de la Carte de la restauration des jardins historiques).Il est fondamental toutefois la connaisance du patrimoine végétal dans sa consistance quantitative et qualitative.Le relief de l'état actuel indique la distinction des espèces botaniques présentes dans la cité de Palerme,il laisse de coté les espèces herbacées annuelles et vivaces et fait une première subdivision en arbres et arbustes,à l'intérieur de laquelle,d'autres spécifications morphologiques, physiologiques et génétiques,ont conduit à une distinction entre: latifoliés,arbres à feuilles aciculaires,palmiers,cycadophites et xérophites.On a groupé à l'intérieur des latifoliés toutes les espèces à feuilles large indépendamment de leurs hauteurs et de leur provenence:du platane à l'eucalyptus,au chêne vert,au citronnier parmi les arbres, de la troene,à l'hibisque à l'oléandre parmi les arbustes;à l'intérieur des arbres à feuilles aciculaires toutes les conifères:du pin au cyprès,à l'auracaria;à l'intérieur des palmiers aussi bien le palmier des Canaries que le dattier et la Washingtonia parmi les arbres,tandis que parmi les arbustes aussi bien la Chamadorea que la Phoenix roboelini

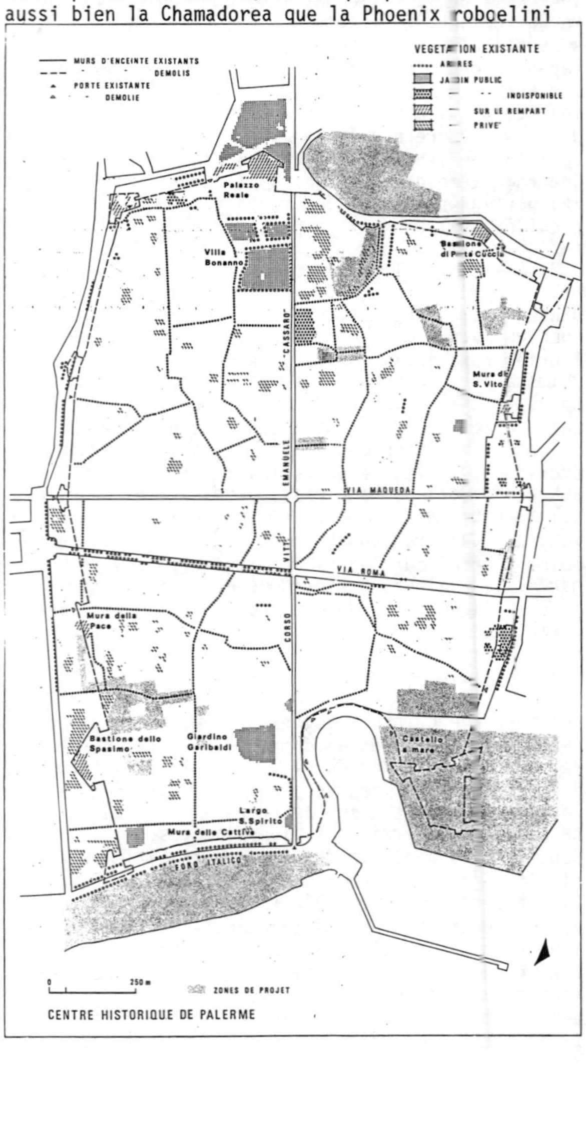

CENTRE HISTORIQUE DE PALERME

et la Chamaerops humilis (palmier nain,le seul genre spontané de la flore méditerranéenne).On peut faire un discours à part pour les xérophites,c'est à dire ces plantes qui présentent une résistance très remarquable aux fortes insolations et à la carence d'eau qui se repercute sur la structure et sur la forme de l'espèce qui présente un aspect unique et en caracterise les différents types: à feuilles réduites,ou épaisses,à feuilles transformées en épines,à tiges très réduites ou presque absentes, avec des feuilles à rosette ou bien à futs grossis à la bse,dracénas,dolines,cordylines parmi les arbres,euphorbes,aloes,oponces parmi les arbustes.

3. LE PROJET

3.1 Les jardins existants et les nouvelles espaces

La recherche conduite dans le centre historique de Palerme se fonde sur une comparaison entre l'état actuel et le projet,qui veut obtenir une meilleure utilisation des surfaces libres et la spécification des espèces d'arbres en fonction des besoins climatique locaux.La thématique de la restauration dans le centre historique se présente sous un double aspect:la sauvegarde des monuments et la qualification de la verdure,en considérant sa potentialité comme facteur de régulation climatique.Le projet, partant,est adressé à l'économie d'énergie et à la remise en valeur de l'ambiance urbaine,par la restauration des jardins historiques existants et la aménagement de nouvelles espaces."L'occasion de créer des surfaces vertes" comme dit la Charte d'Athènes (1933) est donnée par "la destruction de taudis à l'entour des monuments historiques",ces taudis qui à Palerme sont un témoignage,encore aujourd'hui,de la dernière guerre mondiale et du conséquent abandon du centre historique.A travers la transformation en espaces vertes de toutes les surfaces dégradées,le projet se propose une récupération immédiate et un reclassement de l'habitat urbain.Un long parcours pour piétons planté d'arbres relie les espaces localisées existantes et à l'étude en redonnant une continuité aux restes du mur d'enceinte et en proposant une image nouvelle pour la cité qui,tout en étant désormais une partie difficilement identifiable de la ville,peut devenir le "Parc" de la ville même /5/.

3.2 La végétation

Parmi les plantes aptes aux différentes fonctions on a donné la préférence aux semper virens natives de la Sicile,représentées en communauté de maquis et déjà adaptées aux conditions de sub-aridité(Mirtus communis,Pistacia lentiscus,Viburnum tinus,Phillyrea angustifolia) en tant que l'avantage de l'économie énergétique pour le réchauffage avec l'usage d'espèces caduques,plus délicates,est anéanti par la difficulté liée à la carence chronique d'eau dont ces dernières plantes ont plus besoin;toutefois on n'a pas écarté des espèces caduques (Fraxinus ornus,Punica granatum ...) ou éxotiques (Citrus sinensis,Citrus limon,Citrus deliciose,Ligustrum lucidum,Nolina recurvata,Yucca aloifolia,Aloe barbadensis,Agave americana,Euphorbia splendens ...) déjà présentes et longuement expérimentâes dans le territoire palermitain.Le choix a tenu compte non seulement de leur adaptation à la sécheresse,mais de leur résistence à la salinité et à la pollution et on a préféré des espèces qui germent précocement et qui,après une longue saison végétative,perdent tard les feuilles et qui sont d'une pousse rapide (Celtis australis,Jacaranda mimosaefolia parmi les arbres et Hibiscus rosa-sinensis,Lagerstroemia indica parmi les arbustes) afin que la capacité d'ombrager se révèle en des temps courts,à implanter melés à des espèces à pousse lente (Ceratonia siliqua,Quercus ilex ...) dont la durée est bien longue,de facon à pouvoir les substituer quand les autres auront atteint les dimensions désirées.La vitesse du vent peut etre attenuée surtout à la présence de barrières végétales réalisées par des rangées d'arbres (Erythrina viarum) entremelées avec de bas arbustes (Tamarix gallica,Nerium oleander, Chamaerops humilis,Laurus nobilis,Rosmarinus officinalis,Lippia citriodara,Lavandula spica,...),qui forment des barriâres assez perméables à réduire les infiltrations d'air sans toutefois en empecher excessivement le passage avec la formation consécutive sous-vent de zones de turbulence comme il pourrait en créer par exemple,une rangée compacte d'arbres à la forme de colonnes.On obtient la plus grande réduction par les distances entre la barrière et le batiment de deux-quatre fois la hauteur de la barrière même;la protection augmente,en augmentant la hauteur des arbres.

BIBLIOGRAPHIE

/1/ H. BRESC,Les jardins de Palerme (1290-1460),en Mélanges de l'école française de Rome,tome 84-1972.

/2/ E.BALDI - A.GULI,I verde pubblico a Palermo negli ultimi cento anni,Palermo 1975;A. GULI,Il problema del verde a Palermo attraverso i piani urbanistici,en Quaderno n. 7 dell'Istituto di Urbanistica della Facoltà di Architettura di Palermo, Palermo 1980.

/3/ A. GULLOTTI - C. SPANO',Il problema delle zone verdi con particolare riguardo alla città di Palermo,Palermo 1965.

/4/ S. ALESSANDRO - G. BARBERA - G. SILVESTRINI, Stato dell'arte delle ricerche concernenti l'interazione energetica tra vegetazione e ambiente costruito ,en Quaderno n. 13 del CNR,Istituto per l'edilizia ed il risparmio energetico,Palermo 1987.

/5/ A. CAMMALLERI - F. PALMA - F. SCHILLECI ,Il recupero delle porte e dei giardini nella città antica ,tesi di laurea,Università di Palermo,anno accademico 1986-87.

PROPOSAL FOR THE PLANIFICATION AND REVITALITATION OF RURAL AREAS.

LUXAN, Margarita; TENDERO, Ricardo; VILLOTA, Isidro; y el
SEMINARIO DE ARQUITECTURA INTEGRADA EN SU MEDIO AMBIENTE

Escuela Técnica Superior de Arquitectura, Universidad Politécnica, Madrid.
Avenida Juan de Herrera s.n. Ciudad Universitaria. 28040. Madrid.

The realization of intervention studies in Rural Areas is of particular interest because of their actuality and their importance as a contribution of datas in order to interrumpt the present process of degradation that meny zones are presently undergoing.

The Urbanistic Norms, in theri evolution, have been developed fron the problems of the nucleus of large populations with powerful economic and management resources. Posteriorly, they have been applied to smaller and smaller urban units without taking into acount the change of scale, tis implies a reduction of the possibilities of control and increases the enviromental influence. It has provoked an imbalanced condition betwen the habitual planning propositions and the reality, making te former impractical.

We intend to offer a possibility of taking into acount the bioclimatic analysis as criteriums for deciding on the buildings recuperation designs and the fitness of urban units.

We can understand this by studyng small rural village conditions. Our antecesors did not really have a scientific way of making their settlement, but the trial and error system obteined particular results, conditioning their realization on natural conditions, with appreciable degrees of confort. Now, we can verify that, taking mesures analyzing and reinterpretating with use of the present tecnical knowlege.

The enviromental studies have been made on a discontinuos and extreme scale. It is very easy to find projets of a large scale that plan analysis, models and propositions for Territorial Zoning; it is also possible to encounter cases of building designed by making use of natural energies; however, it is very dificult to find small settlement wich have been analyzed and designed using enviromental techniques and directives.

We propose a method wich is based on the differenciated analysis of the diverse physical scopes, beginig with the most general analysis to the most specific of the buildings and their parts, studyng for each of the spaces all factors that intervene in their beaviour.

So, we have a mixed methodology of approach, that takes elements and analysis methods from:

- The Territorial Planning patterns.
- The habitual analysis of the arcuitectural form and their units.
- The factors derived from the ecological analysis.
- The analysis of stetic and dynamic human behaviuor.

This method and analysis are studied in three scales:

- On the level of the intrinsic qualities of the zone.
- On the level of the urban unit.
- On the level of the buildings.

The following objectives are used as the guideline of the project.

We present the application of this methodology in two places, each of them with climatic, ecological, and social conditions, almost opposite.
El Gaucín, Málaga: mediterranean climate (dry and soft); regression in the population and less explotation of its own economical, and ecological resources.
Patones de Arriba, Madrid: mountain climate with ecological enviroment practicaly wasted and, in fact, saved from desertization by its special characteristics, landscape, scientific interest.

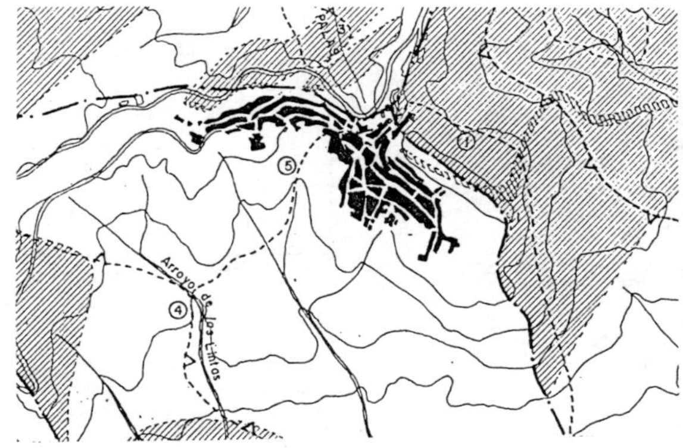

EL GAUCIN -MALAGA-

DIAGNOSIS.

The village is in economic and demographic crisis.

They depend basically of the agricultural sector.

Working population is about 24%.

Return of the emigrate population.

Depopulation of the old quarters.

Need of housing for emigrate population who come back.

External pression for secon residence houses.

Need of educational and sporting sites.

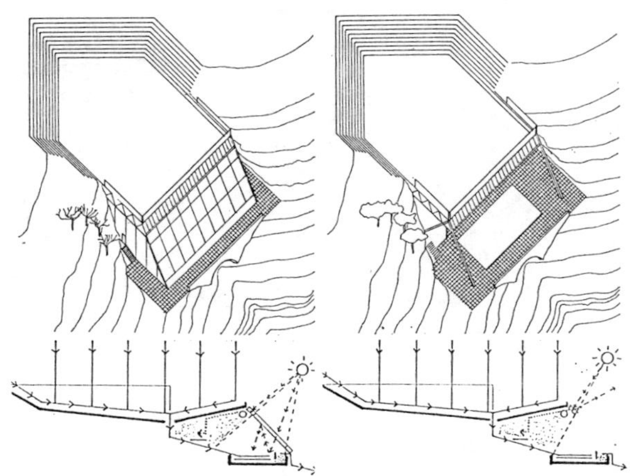

PROPOSAL.

Rehabilitation of old housing keeping the traditional systems of bioclimatic adaptation.

Adequation of belvedere on landscape interesting points.

Reutilization of a ousing square as apartamenthotel use.

Creation of new housing quarters with a communitary bioclimatic over-roof.

Consolidation of castle structures and desing of public garden and etnogrphic museum.

Promotion of a cork-cooperative with a temporal market wich its tilt-roof provides shades and plays like advertising.

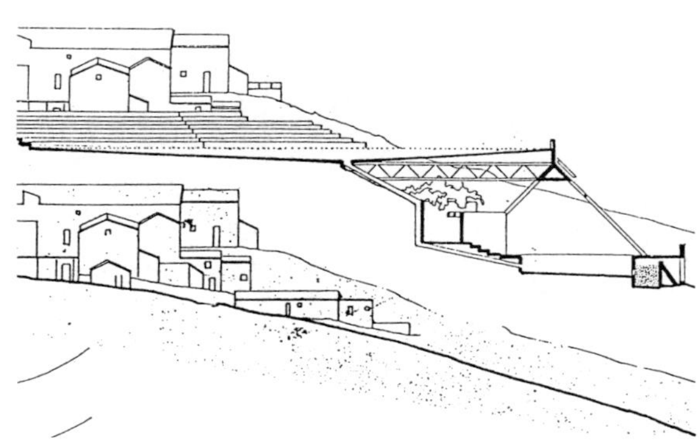

Multi-sports space taking advantage of the natural capacities of the natural capacities of the waterway adding a rain water collecting system and reutilization of the water used in swimming pool for irrigation, also use of the solar energy.

Agriculture school with vivarium , communitary multi-use sites and experimental eolic energy.

Pork stabling with to obtein methane.

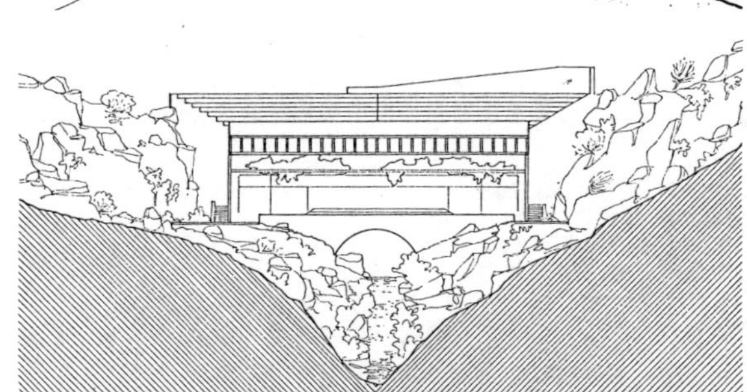

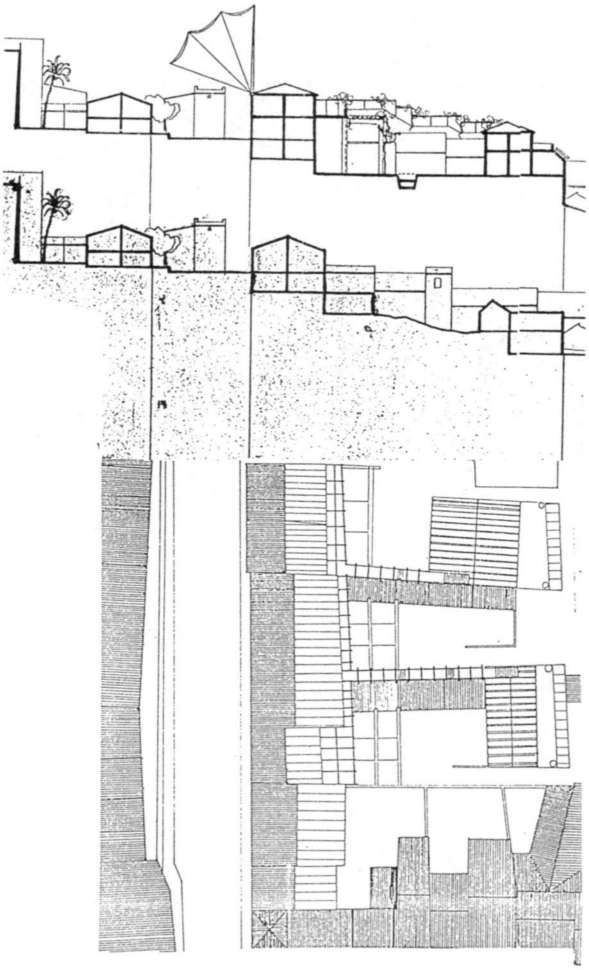

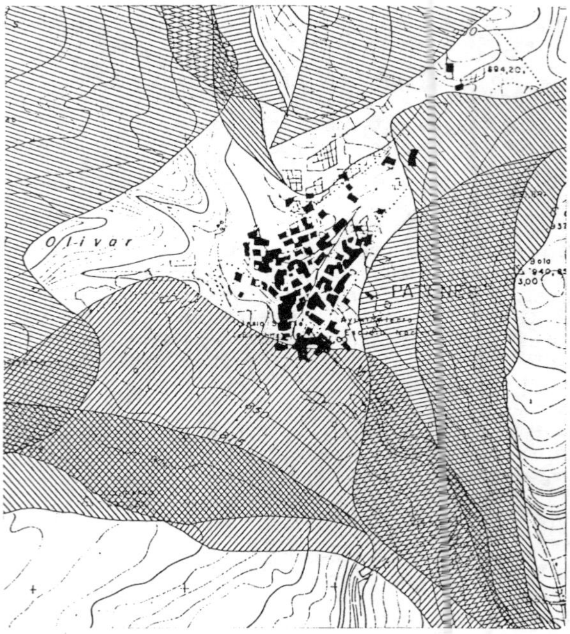

WINTER SUMMER

PATONES DE ARRIBA -MADRID-

Population with limited capacity of 450 inhabitants.

It is geomorfologyc attraction and it is good integration in the landscape in wich the roof system constitutes the main facade.

We suggest didactic and ludic uses.

The economical system is so deteriorated that a revitalitation demands a new social structure in the agricultural and cattle-raising the explotation are not possible, because of the enviromental lack of production conditions.
We foresee te comunitary-temporal type of resiaency.

The ampliation of the urban limits are conditioned by thesunlighting necessity (six hours a day) to reach the confort conditions.

Location of protected view zones for ocasional parking.

Determination of interesting "wandern".

Revitalization of the communitary free spaces with specific solutions for its manteinance and conservation, proposing ludical and working uses in relation with the proposed social structure.

Specific suggestions about the restauration criteria of the traditional housing arquitecture.

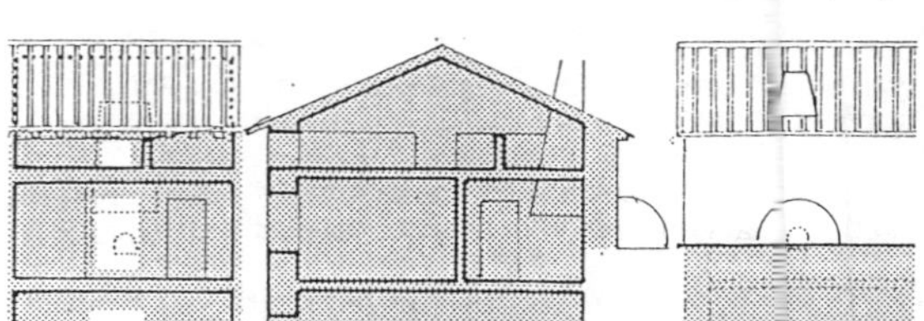

SUR ZONE C2 ▷ NORTE

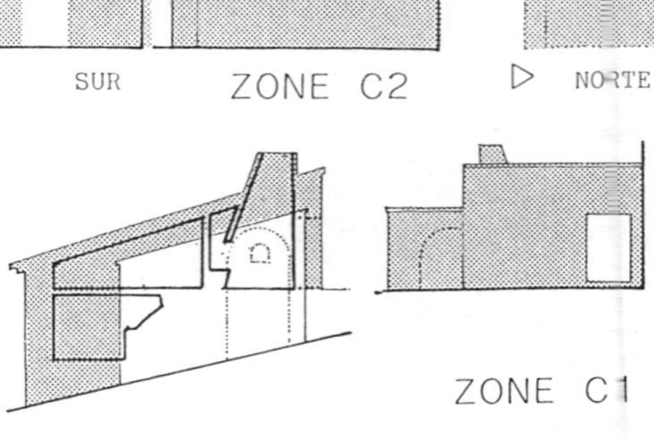

ZONE C1

THE COMPUTER PROGRAMME "GOSOL", AN INSTRUMENT TO INCREASE PASSIVE SOLAR ENERGY USE IN URBAN PLANNING

Peter Goretzki

Universität Stuttgart - Institut für Bauökonomie
Zinsholzstraße 11, D-7000 Stuttgart 75

Windows are solar systems which can easily reduce the auxiliary heating load of buildings by about 30%. But this can only happen if urban planning sets the preconditions. To help the urban planner to do this we developed GOSOL.
GOSOL is a DOS Personal-Computer programme for quantitative energy analysis of passive solar energy use in urban planning which includes the capabilities of a CAAD Programme. It is used as a design utility to set the structural preconditions for the passive solar energy use through windows. With GOSOL it is possible to create a three-dimensional computer model of the planning area in a short time, consisting of topography and environment, trees and buildings. This model can be displayed in various isometric views. It is also possible to show the shading on the area at any moment. The solar gains can be calculated for each window, each building and the total planning area. An automatic comparison mapping allows quick analysis of the energetic results. A graphic editor enables comfortable redesigning to improve the solar-energetic conditions.

1. DEVELOPMENT

Talking about passive solar energy use, means mostly special "Solar-Houses", but each normal building has a very efficient solar system - the windows. Today the passive solar fraction of new buildings in the FRG is up to 30% . Disadvantageous urban planning can reduce the possible solar gains by about half. This stresses the importance of including thoughts about reduction of energy consumption in the urban planning. Unfortunately, there was a lack of instruments for such practical use.

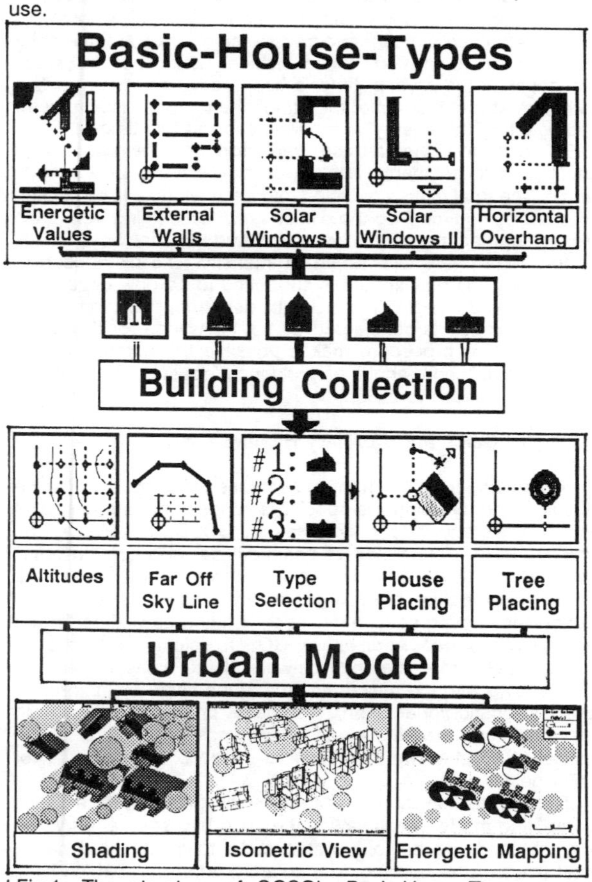

Fig.1 The structure of GOSOL. Basic-House-Types are defined by energy data as well as size and location of external walls, windows and horizontal overhangs and then placed several times into the urban model consisting of the terrain, (far skyline being optional), the Basic-House-Type selection table and location of buildings and trees.

As early as 1986 and 1987 the first pilot projects to incorporate passive solar energy use in new settlements by urban planning were launched in the communities of LAHR /1/ and STERNENFELS /2/ by using GOSOL I, developed in 1984. During these projects the necessity of a comfortable and simple to use computer programme for the urban planner became obvious, as the disconnectedness between urban planners and solar specialists resulted in delayed planning and prevention of optimal results. This provided the impulse to develop GOSOL II. GOSOL II enables the urban planner to analyze and correct the qualification of an urban planning for the potential use of solar energy in buildings without additional specialists or prolonged duration of the normal design process.

Nevertheless, insolation of rooms is more than an energetic factor. It means living quality, especially in colder climates with less sunshine during the winter, which has an effect on human well-being. Therefore, besides energetic evaluation during in the process of urban design, the duration of insolation in living quarters should be controlled, especially on high density projects.

2. THE COMPUTER MODEL:

GOSOL was developed as a design instrument with CAD similar surface which handles the evaluation of space quality as well as energetic computing. GOSOL is based on the placing of buildings and vegetation in a topographic computer model. The energy flows of all buildings into this computer model are successively calculated. Simultaneously, the buildings represent the shading environment for other buildings. The three-dimensional computer model can be displayed isometricly from any point of view in order to check up on the design quality .

2.1 The House Model

Starting point is the modelling of as many **Basic-House-Types** as you like which can be placed into the planning area in a second step. Each of these Basic-House-Types includes attributes necessary for the description of the building, related to topology such as external walls and horizontal overhangs, location/orientation and size of windows as well as energetic attributes like apparent absorptance efficiency of the rooms, the solar transmittance of glazing materials, the average temperature of the rooms, the internal gains, the specific heat loss and the relation between insolation and heat storage describing Solar-System-Factor. The specific heat loss can be calculated automatically, by defining the infiltration rate, using these building parameters either for any U-values (Windows, external walls, roof and ground) or for both methods of the German "Wärmeschutzverordnung 1982". These Basic-House-Types represent a collection, from which any types needed for the planning area can be selected. If any Basic-House-Type is not available during the urban modelling, the house-modelling module can be selected from the urban planning module.

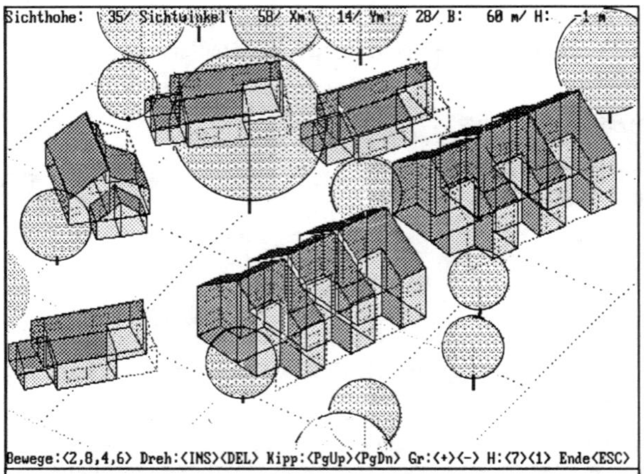

Fig.2 The three dimensional computer model of the area consisting of buildings, trees and terrain can be displayed from any point of view.

2.2 The Urban Model

The **Basic-House-Types** selected and marked by number in a table are now copied as often as needed, by using a graphics tablet (input of coordinates and orientation angle is also possible), into the topographic computer model of the planning area. The **altitude of the buildings** above the terrain is automatically calculated, yet can be changed at any time. To place the **buildings** only three inputs are needed:

- the number of the Basic-House-Type,
- its location point and
- a direction point to define the orientation.

Trees, which can be defined in a table by using the height and the diameter, are placed by a location point and the number of the tree in the table. Free input of the height and diameter of the tree is also possible. The altitude above the terrain is calculated automatically.

The computer model has its limit at about 500 trees plus about 500 buildings (with a total maximum of 3200 solar elements and 5300 external walls). This is sufficient even for a greater planning area.

A comfortable **graphic editor** allows the moving, rotating, deletion or insertion of buildings and trees on the computer screen with the graphics tablet or the arrow keys from the numeric keypad. Also, each Basic-House-Type can be automatically replaced by another.

The **topographic computer model** is defined by the input of the altitudes of the terrain at the section points of a grid over the planning area (which can be freely defined). That topographic model can be optionally supplemented by the input of the sky line forming lines outside the planning area such as, for example, far off hills (Fig.3).

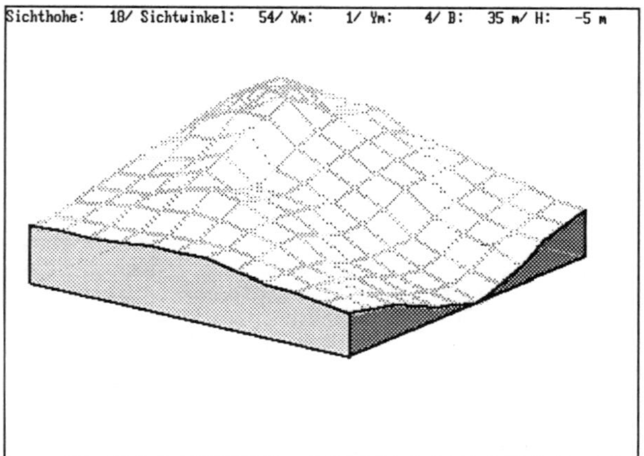

Fig.3 Computer model of the [terrain] topography. The size of the grid can be chosen.

3. COMPUTING METHODS

3.1 Solar Irradiation

Using the **projection of the shade** on the windows, the shading of the windows through beam irradiation is calculated hourly, considering the shading by opaque objects like neighbour buildings, the building itself (external walls, horizontal overhangs,...) and topography, as well as translucent objects, like trees. The translucence of vegetation for beam and diffuse irradiation depending on foliage, as well as the ground reflectance, can be defined for each month in the weather data file. The portions of the window areas exposed to sunshine are printed out in a table for the 15th day of each month (Tab. I). This table makes possible the analysis of the lighting quality of rooms,

BESONNUNG IN % Fenster Nr. 1 Orientierung 18.79 Neigung 90 NEUSTADT.7

Zeit	!Juli	Aug.	Sept.	Okt.	Nov.	Dez.	Jan.	Feb.	März	April	Mai	Juni
4- 5	.	.	.	.	.	.	.	.	.	.	.	/*
5- 6	/*	/*	.	.	.	.	.	.	.	.	/*	/*
6- 7	/*	/*	/*	.	.	.	.	.	.	/*	/*	/*
7- 8	/*	/*	/*	2	.	.	.	.	/*	/*	/*	/*
8- 9	/*	/*	39	27	69	66	47	63	76	/*	/*	/*
9-10	100	78	63	38	73	69	69	72	84	92	94	100
10-11	100	100	100	100	100	100	100	100	95	98	100	100
11-12	100	100	100	100	100	100	100	100	100	100	100	100
12-13	100	100	100	97	87	82	82	95	100	100	100	100
13-14	100	100	100	100	87	79	83	91	100	100	100	100
14-15	100	100	100	100	100	100	100	100	100	100	100	100
15-16	100	100	100	100	100	100	100	100	100	100	100	100
16-17	100	100	87	20	69	.	100	81	93	100	100	100
17-18	100	100	100	.	.	.	.	69	74	100	100	100
18-19	44	100	.	.	.	.	.	.	.	100	37	52
19-20	/*	/*	.	.	.	.	.	.	.	.	/*	/*

Tab.I Insolated area of a window in percent during time of day for the heating season.

The **monthly beam radiation** is calculated depending on the shading for each window using the monthly sunshine hours per hour interval, an atmospheric transmission factor and the solar constant. In doing so, the sunshine hours per hour are representing the probability for a beam radiation from a certain direction at a certain time, the atmospheric transmission factor representing the energy of the irradiation at this time.

To calculate the **diffuse radiation** the hemisphere is partitioned in segments with an certain azimuth- and altitude angle. The total diffuse radiation into each window is calculated monthly by summing up the diffuse radiation coming from each visible segment between skyline, fins and horizontal overhangs of the hemisphere with a certain intensity. The distribution of radiation at the hemisphere is supposed to be only dependent on the altitude angle. The azimuth dependent fraction of the diffuse radiation is considered as the allowance for the beam radiation. This procedure makes possible the computation of the "shading" and angle of incidence dependent on transmission of diffuse radiation.

3.2 Solar Gains and Auxiliary Heating Load

Proceeding from the monthly heat loss of the building (which is calculated by using the specific heat loss, the internal gains and the average indoor temperature) GOSOL calculates the solar radiation (separated in beam-diffuse- and ground reflex radiation for each window), the monthly degree days, the utilized solar gains, the auxiliary heating and the solar fraction during the heating season, for each building as well as for the total planning area. Additionally, GOSOL calculates on a comparative basis the difference between these measurements of the building with the measurements of the same building in an optimal location/orientation. This helps finding buildings with less insolation than the mean of buildings and control of the efficiency of planning variations to increase the potential for solar gains.

The calculation of the utilized solar gains uses a static computing model which values the absorbed solar energy, the net heat load and the solar-system-factor to imitate the dynamic heat flows of the building. The solar-system-factor which itself is adapted dynamically to the local irradiation of the building during the calculation, is a parameter specific to a building.

3.3 Mapping

To eliminate planners' evaluation of tables and to offer fast survey, GOSOL creates an automatic mapping of all energetic

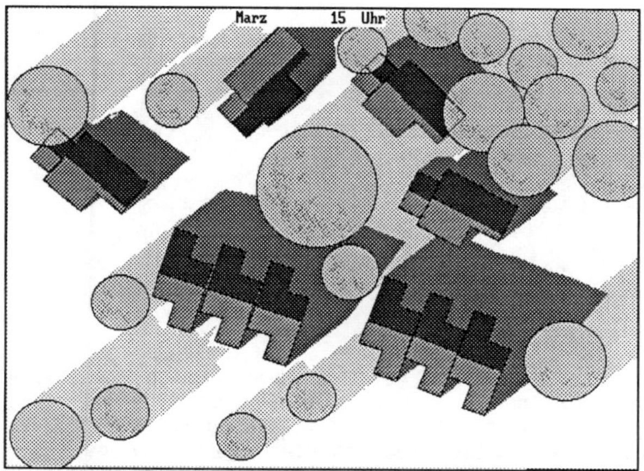

Fig.4 The shading of the area, its respective source differentiated by different colours, can be displayed for any moment (here at March, 15th, 3:00 pm). The scale or the cut can be chosen.

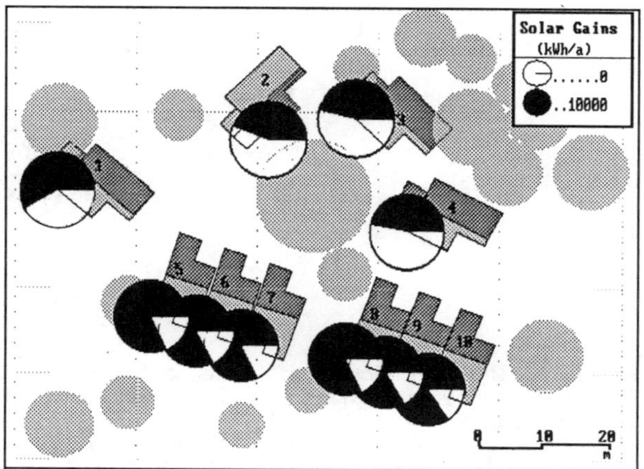

Fig.5 Several energetic results (here: Solar Gains) can be mapped and displayed either for a selected cut of the area without scale or for the total area in a fixed scale (several screens).

results (Solar gain, Auxiliary heating, Solar Fraction,...) by circle diagrams. To offer visible impression of insolation, GOSOL shows, besides tables of the sunshine hours for each window, the mapping of the shading of the area at any moment. The presentation distinguishes between shading originating from buildings, trees and terrain by using different colours.

All mappings can be printed either as a cut (selected part) of the area without scale (one page) or in form of the total area in a fixed scale (with matrix printers mostly in several pages).

4. APPLICATION

GOSOL was developed to analyze the reciprocal energetic effects of buildings in urban context. The focus therefore is not the description of thermal effects within the building (there is no sufficient database with urban planning).

The applications reach from drawing up empirical design rules for early planning stages which consider local climates, through to building elements.

4.1 Design Rules

With GOSOL it is possible to create design standard rules by empirical analysis of effects under the condition of a certain climate. Helpful, for example, are rules about distance between buildings (e.g. south distance/height > 2.2) and trees (e.g. south distance/height > 1.5), orientation,... . These rules /3/ help design in an early stage without computer.

4.2 Early Urban Planning

GOSOL is used in early urban planning to analyze the qualification of areas for the use of solar energy (The LAHR /1/ example shows that hills may reduce the potential utilisation of solar gains of an area during the heating period up to 30%). Therefore, buildings without walls and with only solar openings, are distributed within a fixed distance over the planning area; the calculated insolation is mapped in form of lines with equal insolation energy density. This map informs about the possible energy irradiation during the heating period for any part of the area.

4.3 Urban planning

In final stages of urban planning GOSOL is used, alongside evaluation of space quality, to evaluate energetic and daylighting quality of the urban design and to modify, if necessary, the orientation, location and distance of buildings and trees to increase useful solar gains. The result of redesigning can be evaluated in a few minutes, as buildings can be selected for calculation.

4.4 Building planning

During building design stages, GOSOL is used to analyze and design the location, the orientation and the size of windows as well as horizontal overhangs (roofs, balconies, fixed shading devices,...) and vertical fins (walls,...). GOSOL offers an energetic optimizing of the windows as well as the evaluation of thermal heat storage yet not thermal comfort.

5. WEATHER DATA

The weather data file for GOSOL is based on the monthly means of global irradiation, diffuse irradiation, outdoor temperature, irradiation transmittance of trees and ground reflectance. This database is supplemented by the monthly mean of sunshine hours per hour. A special module calibrates the mean beam radiation according to time of day, using the sunshine hours and the monthly mean of beam radiation, by calculating an atmospheric transmissions factor. The profit of this database is its easy availability for almost any location.

6. PRECISION OF CALCULATION

GOSOL was developed to compare buildings, but it also can be utilized to calculate energy consumption of a planning area.

The hourly divergence between the calculated insolation and the observed insolation to the horizontal is less then 3% of the daily insolation. Regarding the calibration of beam radiation the daily insolation remains unchanged. The divergence of the utilized solar gains is less than +/- 5% between GOSOL and a dynamic simulation programme.

7. TIME EXPOSURE

The consumed time for data input depends, like the simulations duration, on the numbers of the buildings and trees. In any individual case data input requires approximately:

- 10-20 minutes for each Basic-House-Type,
- 1 hour for topography including preparation of data,
- 5-10 minutes per acre for building and trees.

The duration of the simulation depends on the total number of windows as well as the number of external walls of the buildings and increases with the number of buildings in the planning area. Using a 8088 Processor with 8087 Coprocessor (4.77 MHz) you need approximately:

- 5 minutes for 10 buildings
- 1 hour for 50 buildings
- 5 hours for 150 buildings
- 20 hours for 300 buildings

(Each building in this example has 4 windows and 7 external walls. This is sufficient for simple buildings)

/1/ Peter Goretzki: Energetische Bewertung Lahr-Hosenmatten, Gutachten, Universität Stuttgart, 1987.

/2/ H. Küsgen, P. Goretzki. In: Dorfentwicklung. (Hrsg.: Ministerium für Ländlichen Raum, Landwirtschaft und Forsten Baden-Württemberg), MLR 2/1987.

/3/ Peter Goretzki: Energiebewußte Bauleitplanung, Diplomarbeit an der Universität Stuttgart, 1984.

TOWN DEVELOPMENT BY PLANNING AND ADVICE ON THE SPOT

Reinhold Hytrek
Planergruppe Hytrek, Thomas, Weyell und Weyell
Freie Architekten und Städtebauarchitekten
Rollingergasse 1
D 6093 Flörsheim am Main
Tel 06145/2066 Fax 06145/53275

ABSTRACT. The development of towns and cities regarded as a whole can no more rely only on formalized land use= and zoning plans. Especially areas of Urban Renewal need not only a legal frame for decisions concerning buildings and public space. An integrated approach must be able to stimulate public and private initiative and harmonize their efforts to make neighbourhoods function and livable, even loved by their inhabitants and users. The method consists of an "Urban Frame Plan", the improvement of urban public space, public and private buildings at the same time. Private projects are furthered by offering expert advice. It is our attempt to harmonize public and private aims through the planning process. This includes the discovery, description and eventual solving of conflicts.

1. THE URBAN "FRAME PLAN"

1.1 Evaluation

Any planning for cities requires the evaluation of what is there. Qualities and defects of complex structures need description by word and drawing to be usable in public political discussion. "Graveyards" of data with no design result on short or long sight are to be avoided. Conflicts between different interests and factors need to be described ans named directly to be decided upon. Social situation, traffic, technical quality of buildings, historical value, spatial quality, the view of the city are subjects to description by the expert and the public.

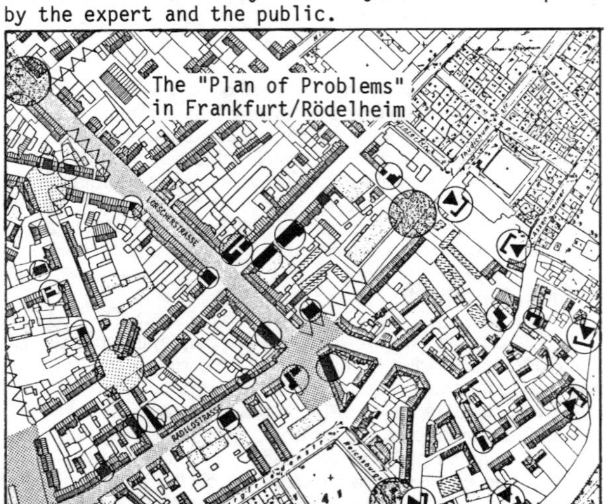

The "Plan of Problems" in Frankfurt/Rödelheim

1.2 Public participation

The opinion of the "(wo)man in the street" about his/her city's problems and values are vitally important for the success of any design action. The interview with the key=persons who can offer general or special inside=views on the city during the process of evaluation. We like to make people draw the beauties and problems of their town.
For the planner it seems even more important to learn to know the problems of a town through the eyes of her inhabitants and users. His design proposals will be more effective.
Certainly, the design proposals have to be introduced to the public and voted upon by the council.

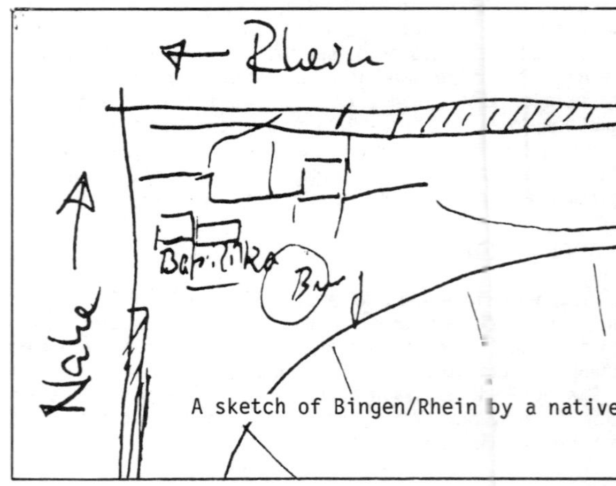

A sketch of Bingen/Rhein by a native

1.3 Design levels

The urban "frame plan" contains of different levels of planning and design proposals. The overall structures like landscape, greenery, traffic or land use need description and alteration if necessary. The overall view is important to set the frame for all specific design proposals.

Situation in Ortenberg
present and design proposal

These design proposals should contain the important key projects, but also show a good selection for the solution of normal and everyday design problems. All cities cosist of a few landmarks but the large majority of ordinary houses make her view and participate to her character. Therefore it is valuable to develope typical designs for the average house, for example by designing a picture book for a specific town.

From the picture book of Geisenheim

1.4 Low scale of abstraction

Planning is a political process. It is necessary to use an everyday language and use techniques to visualize the design attempt with as little abstraction as possible. It is not the technical drawing, but scale models, free=hand scetches and isometric drawings which offer the necessary and comprehendable information ans impression to everyone.

The isometric view of Hochheim

2. PUBLIC PROJEKTS

2.1 Public space

Streets and all public spaces are like the veins of the city. Their use is defined by their surface. Quarters of urban renewal have a need of integrated improvement of the design quality of the streets. Not only the market square is subject to design efforts. It seems important to have the "picture book" for the normal streets as well to avoid the senseless use of yet another shape of pavement. Street furniture also has the tendency to become chaotic. Design quality for the public streets sets a standard for its houses. It should of simplicity and good quality.

2.2 Public buildings

The way the community deals with her buildings sets standards and gives good or bad examples for the private owners. Houses owned by the community can be used as "pilot projects". They can be published during the design and building process to improve the information about building technicques and good design. This is especially useful in areas with "do it yourself" activities. The high quality renovation of a normal house owned by the community is more convincing to other house owners than public appeals by politicians.

The pilot project in Hochheim

3. ADVICE ON THE SPOT

3.1 How to harmonize public and private aims

Public planning often is a one way street. The private owner finds himself restricted,often without need. Design can help to relieve the tension between public and privat aims. To achieve this, planners must be willingto design for the individual owner on his kitchen table. this also modfies the idea of the town's development layed out in the "frame plan". The growing insight of the planner into the small scale problems will also gradually make the frame plan more concrete and eventually alter aims and means of the town's developement.

3.2 Advice improves design quality

Advisory action not only helps to wake private initiative, it also helps to improve design quality. It can clarify on a conceptional scale all important factors of building technique, architectural design, social acceptance,finance, energy consumption,historic value. This is especially valuable for all building projects which are never handled by architects.

The advisory team in action

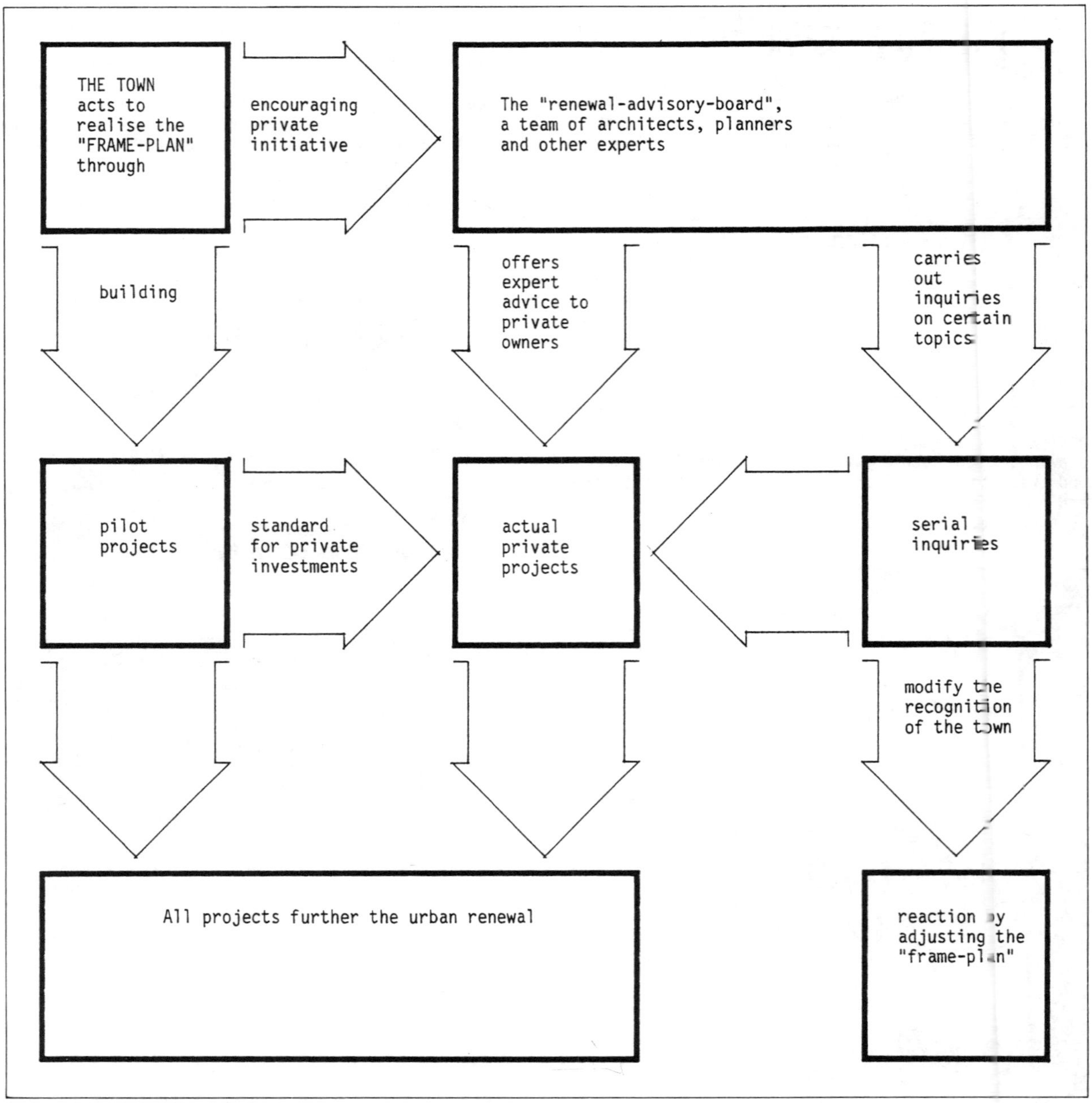

The scheme of town development by planning and advice on the spot

3.3 Availability of information

The quality of the advice depends on the complexity of the advisor's knowledge. He needs to have access to all informations needed to make them parts of a concept which will last into the future. The growing complexity of the factors involved in urban developement and in the renovation of individual houses needs a computer aided management system. The results of the CEC=Project "Computer Aided Management in Urban Renewal" (CAMUR) offer usefull tools to improve the complexity and quality of design and action in zones of urban developement. The projekt needs to continue however to reach the stage of direct applicability by the adviser on the spot.

LOW ENERGY ECOLOGICAL SETTLEMENTS (LEES)

J. Behnsen
University of Oldenburg Postbox 2503 D-2900 Oldenburg FRG
N. Baker
University of Cambridge 6 Chaucer Road Cambridge UK
J. Rosemann
Technical University of Delft Postbus 5043 NL-2600 GA Delft

ABSTRACT: During the last 10 years great efforts have been made in the Community to reduce the energy needs of buildings, i.e. to reduce the needs of fossile fuels for heating demands by improving thermal conservation and the active and passive use of solar energy, with emphasis on the single building. In order to reach more extensive energetical goals, it will be necessary to include other energy needs concerning buildings - e.g. electricity and energy for transportation - and to consider the environmental burdens of the energy use. Therefore this project stresses not only the reduction of all energy demands concerning buildings (heating, electricity, transportation) but also the reduction of environmental pollution caused by energy use, at a larger scale on an urban level. The goals of this project are twofold: First, to develop a European Performance Index as a Standard of Energy Use (incl. all energy demands of housing) and, - energy related - Environmental Pollution (incl. SO2, NOx, CH, CO2) for existing and new urban areas, for all types of urban settlements (like rural settlements, suburban and inner city areas), similar to the Energy Cost Index, currently being used as a planning control in Milton Keynes Energy Park, UK. This shall be done in 2 steps: In the beginning all kinds of energy use and energy related pollution will be modelled and evaluated for all types of agglomeration for the northern region of Europe. In a second step this model shall cover the climatically defined southern region of Europe as well, cooperating with scientists from Spain, Portugal and Italy. Second, the project will identify three possible Demonstration Projects, one in each of the member countries of the European Communities, and, with the client body, develop a technical brief. These demonstrations will serve to test the Performance Index, and to demonstrate the realization of low energy, low pollution planning policy in an economically and ecologically sustainable way. It is anticipated that the Demonstration Projects will each have one of three different target levels. 1) To reduce non-renewable energy use and pollution levels to 30% of existing norms. (This is already being attained for energy alone in several European low energy schemes). 2) To reduce energy use and pollution to 15% of existing norms. 3) the "zero" option - i.e. a target of zero use of non-renewable energy and a zero environmental impact. The Demonstration Projects will be chosen from differing contexts, e.g. urban renewal, suburban new-build, rural new-build, and this will influence the choice of target. For the zero option obviously new-build is required, and in this case also it is anticipated that advanced renewable energy technologies such as biomass, photovoltaics, wind power and gruop solar thermal, will be appropriate. For the initial project, cooperation has already been agreed by the University of Cambridge, UK, the University of Oldenburg, FRG, and the Technical University of Delft, NL. The first stage will run from Jan to Dec 1990.

1. INTRODUCTION / GOALS

During the last 10 years great efforts have been made in the European Communities to reduce the energy needs of buildings, i.e. to reduce the needs of fossile fuels for heating demands by improving thermal conservation, and the active and passive use of solar energy, with emphasis on the single buildings.

But if the energetical goals of the CEC for 1995 shall be reached, it will be necessary to include other energy needs concerning buildings - e.g. electricity and energy for transportation - and to consider the environmental burdens of the energy use.

Therefore this project stresses not only the reduction of all energy demands concerning buildings (heating, electricity, transportation) but also the reduction of environmental pollution caused by energy use, at a more economic and more efficient larger scale on an urban level.

Energetical goals of the CEC for 1995 concerned by this project are:

The longterm resources for energy supply shall be secured, the imports of fuels shall be reduced by diversification of the sources and improvement of the efficiency of energy use: in this project by supply from local, renewable energy sources such as sun, wind, biomass.

The environmenal pollution caused by energy use shall be reduced: in this project by reducing the energy needs (thermal, electrciy, transportation) and using renewable energy sources with less or no pollution of dust, SO2, NOx, CmHn, CO and CO2.

The European Market of 1992 concerning energy supply and energy use shall be strengthened: in this project by developing European Standards of energy use and energy related pollution on an urban level and developing urban areas with low energy use and low energy related pollution stimulating the competition capability of the European industry (buildings, electrical equipment, transportation systems).

Solutions for the energy supply of settlements will be developed that will be worthful to solve the problems of energy supply of the underdeveloped countries: in this project by systems that will be developed with reduced energy needs and energy supply systems using renewable energy sources.

This project stresses the following targets:

To strenthen the efficient energy conservation and energy use in the sectors of casual use developing strategies to integrate the different technologies to reduce the energy needs by thermal conservation, passive solar energy use, improving daylighting conditions, less energy needs of electrical equipment and transportation systems.
To use renewable energies like solar energy, wind energy, photovoltaics, biomass convertion using a network of different energy sources compensating the meteorological caused fluctuations of energy supply relating to the fluctuating energy demands.

PROJECT SCEME

STEP 1: European Energy and Environmental Performance Index (EEEPI)

We want to develop and evaluate a European Performance Index of Energy Use (including all energy demands of housing) and energy related Environmental Pollution (including dust, SO_2, NO_x, C_nH_m, CO, CO_2) for existing and new urban areas of all types of urban settlements (like rural settlements, suburban and inner city areas). The Index will provide a planning tool, and will be used to establish targets for the energetic and environmental performance of settlements.

STEP 2: Design Studies

The project will elaborate design studies to show how settlements of different types with low energy supply and energy related pollution targets can be realized.
Three types of settlements - representating the total spectrum of urban areas - shall be identified, a rural settlement, a suburban settlement and an inner city area.
Three reduction rates shall be identified:
1) a reduction to 30% of the now normally existing energy demands - especially concerning fossile fuels - and energy related pollution;
2) a reduction to 15% of the normally existing energy demands and pollution rates;
3) a "zero" option with a target of zero use of non-renewable energy and a zero environmental impact.

STEP 3: Demonstration Projects

The described design studies shall provide the realisation of three demonstration projects to evaluate the mentioned tools, techniques and strategies.
Three types of settlements (rural, suburban, inner city), three types of realisation (new-build, renewal, improvement) and three levels of reduction (30%, 15%. 0%) shall be realized (each with about 100 units).
First contacts and agreements to local authorities and developers have to be made.
One shall be realized in Amsterdam (NL)(inner city type, mix of renewal and improvement).
The second shall be realized in Hannover (FRG)(suburban type, new-build).
The third shall be realized in Cambridge (UK)(rural type, new-build).

EUROPEAN ENERGY AND ENVIRONMENTAL PERFORMANCE INDEX (EEEPI)

An essential part of this project is the setting of targets for new settlements and existing settlements as well undergoing rehabilitation and renewal. Different reductions of energy consumption of the existing norm shall be realized: 30%, 15%, 0% have been identified. This is based on experience of monitored low energy schemes, which already show that a reduction to 30% is possible.
However, due to particular constraints, not all new settlements, and certainly not all rehabilitations, will be able to adopt all measures. Futhermore, the uptake of measures will clearly be depentend upon the rate at which new settlements and rehabilitations and renewal projects take place.
In order to assess the impact at a National and European level, of the proposed energy and environmental standards, we propose to set up a computer based model. The model, using spreadsheet/data base software will respond to housing development statistics, and to factors which will represent the degree to which the targets, are likely to be met,
gree to which the targets, are likely to be met,
due to physical and other constraints.
The model will be useful to predict the likely impact of the LEES strategy on overall energy demand, energy demand by type, and environmental impact. It would be of interest to planners, politicans, energy specialists, environmental groups, and ecologists.
This model will be used to develop a European Performance Index of Energy Use and energy related Environmental impact (EEEPI).
Performance Indices based on energy performance alone exist already (like the Energy Cost Index (ECI) of the Milton Keynes Development Corporation).
The ECI integrates different energy sources only by money cost. It does not in any way evaluate the environmental costs, e.g. the production of CO_2 or SO_2. It follows that designs with the same ECI do not necessarily cause the same production of pollution.
Our project is concerned with developing an index which as well as evaluating energy cost, also evaluates the environmental cost associated with energy use. The index will also extend its application to larger groups of housing units, permitting the interaction between units and the synergetic effects of scale. Planning will also influence the energy used for tranport. Existing planning models take this into account but these have not generally been integrated with building energy models.
In the beginning all kinds of energy use and energy related pollution will be modelled and evaluated for all types of agglomeration for the northern region of Europe. In a second step this modell shall cover the climatically defined southern region of Europe as well, cooperating with scientists from Spain, Portugal and Italy.

DESIGN STUDIES / AGGLOMERATION TYPES

The Design Studies are aimed at making a contribution to strategies of reducing energy comsumption and pollution output on an urban level. Different types of urban settlements have to be identified which are representative for the urban development in the three countries.
For the selection of representative cases the following criteria will be used:

- the site (inner city areas, suburban areas, rural areas),
- the building structure (old housing stock and new building construction,
  high rised buildings and terrace houses),
- the type of intervention (improvement, reconstruction, new-build),
- different levels of constraints for the reducing of energy comsumption and pollution output.

The Design Studies do not aim at developing new components and techniques but rather to apply proven but isolated technologies at an urban level.
An integrated approach will be developed, focusing the interaction between different technologies and design systems.
The synergetic effect of this integration, involving broader environmental issues, may lead to a higher quality and a higher performance.

The project will be concerned with all aspects of energy use related to housing, i.e. space and water heating, lighting and electrical appliances, transport, building materials and maintenance.
As well as the reduction in energy consumption, alternative sources of energy such as photovoltaic, solar thermal, wind, and biomass conversion will be considered.
Initially the project focuses upon housing, of sufficient number and density, for there to be inter-

active effects on energy consumption. For example these effects may be negative, e.g. overshading of daylighting or sunlight, or positive, making group heating, or combined heat and power systems viable. Later, the principles should be applied to mixed urban development, where more complicated energy interactions are possible. Within the field of housing, the project is not only concerned with new development but also with the improvement to the existing urban fabric.
Beside of the technical approach the Design Studies will also include an investigation of the constraints coming from the building and urban planning laws and regulations and other administrative conditions, including financial systems to realize the strategies of low energy ecological settlements in the different countries.

COOPERATION

This interdisciplinary research project takes place by cooperation between working groups from the following universities:

University of Cambridge, UK
responsible: Dir.BSc PhD Nick Baker, Environmental Design

University of Oldenburg, FRG
responsible: Prof.Dr.Ing. Jörn Behnsen, Technical Infrastructure Planning

Technical University of Delft, NL
responsible: prof.dipl.ing. Jürgen Rosemann - RIW - Research Institute for
Housing and Urban Renewal; other members of the research group:
ir. C. Duivestijn, Ecological Building; ir. H. de Haan, Housing Improvement and Housing Construction; prof.dr. T. de Jong, Environmental Planning

The coordinator of this project is Prof.Dr.Ing. Jörn Behnsen.

The interdisciplinary and international cooperation, this project is based on, shall manage and develop

- the integration and synergetic grouth of knowledge and specialisation as developed in the different Member Countries,
- the integration of different types of settlements and agglomerations, and the types of infrastruc ture technology as developed in the different Member Countries,
- the incorporation of the different Building and Urban Planning Regulations and other administra tive conditions of the different Member Coun tries aiming to regulations on a European level.

Though the discussed strategies are highly influenced by climate, the project is initially concerned with the Northern European Region only.
It is anticipated to whiden the project to the Southern European Region as well following the same methodological approach. Ultimately this could lead to a broadly unified European Standard.
Contacts to scientists from Spain, Portugal and Italy are made and will be intensified during the run of this project stage (1990).

Though a slight division of work is intended (Uni Cambridge: modeling and simulation, Uni Oldenburg: Technical Infrastructure Planning, TU Delft: Housing and Urban Planning) the three universities want to work together in one integrated group. Monthly meetings with changing location at each of the cooperating universities will take place. Realizing this aim, will rise the costs for communication and traveling but, to our conviction, to get in intensified contract with each other is the only way to initiate a real European cooperation.

ADRESSEE

First, the results of this project shall support the authorities of the CEC and the National Governments that develop further recommendations to realize the energetical goals of the Community for 1995.

Second, the project shall support local authorities to develop strategies to reduce energy needs and energy related pollution on a local level, especially the local authorities of the three cities where the demonstration projects will take place.

Third, recommendations shall be given to architects, urban planners, civil engineers, and other experts in the field of environmental planning, urban planning, and housing and building construction.

TIMESCALE

The work shall last from Jan 90 to Dec 90.
In the first six month the Performance Index will be developed (intermediate report to the CEC at July 90).
In the second six month the Design Studies for the proposal of the three demonstration projects will be worked out (final report to the CEC at Jan 91).

REFERENCES

Turrent D, Baker N:
Solar Thermal Energy in Europe (an assessment study).
CEC DG XII, 1984

Milton Keynes Development Corporation, UK:
Milton Keynes ENERGY COST INDEX.
Cambridge 1985

J. Behnsen:
Energy Autonomous Laboratory of Oldenburg University
in: Bauwelt 25/1982
in: 4th Int. Solarforum, Berlin 1982

Jörn Behnsen:
Planning Strategies of Energy Autonomous Housing Settlements. Oldenburg 1982

Rosemann J, e.a.:
Urban Renewal. An International Comparison.
Amsterdam 1988

T. de Jong:
Introduction to Enviromental Planning.
Delft 1988

Duivestijn, De Haan, Rosemann, e.a.:
Integrated Urban Renewal. Problems and Perspectives.
Delft 1988

Duivestijn, C.:
Ecological Building.
in: W. Riedijk (ed): Appropriate Technology in Industrialized Countries.
Delft 1989

THE BUILDING, THE URBAN SPACE, THE SITE AND THE ENVIRONMENT IN THE AREA OF CERGY-PONTOISE: A GLOBAL CONCEPTION METHOD

Jean Bouillot
architecte
3 rue de l'Agent Bailly
F 75009 PARIS

1. GENERAL PRINCIPLE

That global spatial method proceeds by four successive scales, from environment to architecture, by the way of site and urban space, and following four basic elements: the solid one, the spatial one, the liquid and the solar ones.
The bioclimatic dimensions of the living space are developped at each scale, and for each element, starting from sixteen components which form the basic matarials to be identified, in relation with sixteen other components which form the basic programme to be materialized.

2. MANIPULATIONS

The global method allows, in the same time, the processing of the whole spatial problem at a particular scale (horizontal manipulations), the processing of a particular spatial component at different scales (vertical manipulations), and the processing of the whole spatial problem at all the scales (horizontal + vertical manipulations).
The entrance into the method can be done either by the materials or by the programmes.
The central matrixes and charts constitute the obliged way for identification of materials and programmes

SCALES		MATERIALS					PROGRAMMES			
		LITHOSPHERE	ATMOSPHERE	HYDROSPHERE	HELIOSPHERE		PARTICULAR	GENERAL	WATER	ENERGY
S	ENVIRONMENT A	GEOMORPHOLOGY	CLIMATOLOGY	HYDROGRAPHY	HELIOLOGY	ENVIRONMENT MATRIX	HUMAN	NATURAL	RURAL	INDUSTRIAL
C A	SITE B	TOPOGRAPHY	MICROCLIMATO.	HYDRO-VEGET.	SUNNINESS	SITE MATRIX	REPOSE	ACTIVITY	HYDRAUL. SYST.	ENERGY SYST.
L E	URBAN SPACE C	BUILT AREA	FREE SPACE	GREEN SPACE	SOLAR SPACE	BIOCLIMATIC CHART	FUNCTIONAL	GLOBAL	FLUIDS	RADIATIVES
S	ARCHITECTURE D	HEAVY OPAQUE UNIT	LIGHT TRANSP. UNIT	WATER ELEMENT	FIRE ELEMENT	BIOCLIMATIC CHART	PRIVATE	PUBLIC	SERVICE	CONDITIONNING

3. ENTRANCE BY THE MATERIALS

The different possible manipulations are as follow:

1. Materialization: it consists to fix the physical context of the component(s); horizontally.
2. Situation: the component(s) is determined in relation to the other three scales; vertically.
3. Bioclimatic principle: it is elaborated with the support of the multicritera matrix, or of the bioclimatic chart; horizontally.
4. Identification: the achieved principle is transposed in the programmes board so as to fix its vocation(s); horizontally.
5. Information: situation of that vocation(s) at different scales within the programmes; vertically.
6. Combination: the component(s) is carried over and combined in its initial context; horizontally.

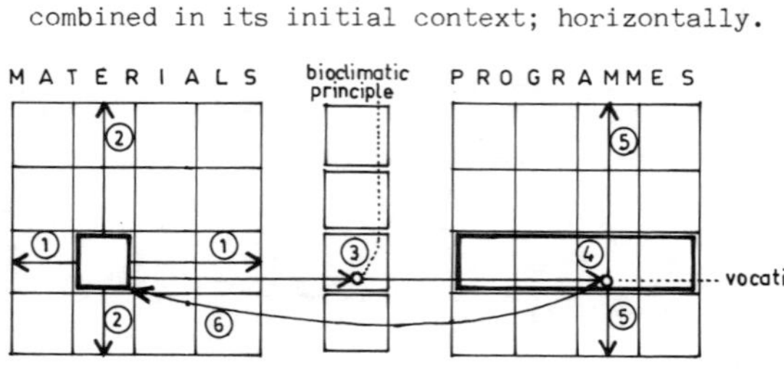

4. ENTRANCE BY THE PROGRAMMES

The different possible manipulations are as follow:

1. Programmation: formulation of the programme at its scale, through its functions and dimensions.
2. Description: with regard to the three other scales; informative and vertically.
3. Setting up: the component(s) is put in situation in the physical context at the same scale.
4. Information: complement of the physical data in the three other scales; vertically.
5. Bioclimatic principle: with the help of the multicritera matrix or of the bioclimatic chart; horizontally.
6. Combination: the component(s) is carried over and combined in its initial context, starting from the bioclimatic principle; horizontally.

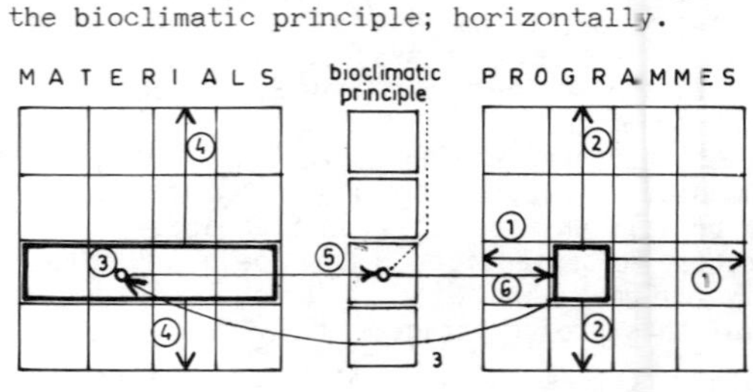

5. CASE STUDY: THE NEW TOWN OF CERGY-PONTOISE.

The new town of Cergy-Pontoise, north-west of Paris, has been choiced as case study for testing that method. After that experience, it appeared that the two kinds of entrance into the method answer to the two large scale groups: architecture and urban space by the materials, site and environment by the programmes. Indeed, for the two first ones, the programme is the first data and the problem of the architect and urban planner consists in giving a form to it, and the materials are the elements of the design.
In the same way, but inversly, for the two twice ones the materials are the first data, and the problem of the planner and town-planner consists in finding the vocations of the site and of the environment, and the programmes are the elements of the design.

This study was realized with the financial support of the Agence Française pour la Maitrise de l'Energie.

6. ENVIRONMENT

The environment space is considered as the combination of the four following elementary components: geomorphology, climatology, hydrography and heliology (sun rythm); they can be identified with the four land programmes, urban, natural, rural and industrial.

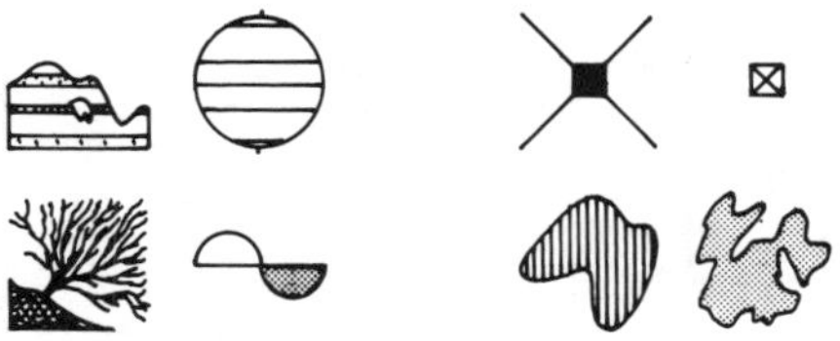

The method is entered by the materials; the map below represents the last step (6), the combination.

MATERIALS		PROGRAMMES: HUMAN	NATURAL	RURAL	INDUSTRIAL
GEOMORPH	TOPOGRAPHY	- LIMIT SLOPE - ACCESSIBILITY SLOPE BASIN	+ NATURAL SITES -	- ACCESSIBILITY - HOLLOWING OUT	- SLOPE BASIN
	PEDOLOGY			+ FERTILE AREAS	
	GEOLOGY	- UNBUILTABLE - SISMIC ZONE			+ MINERAL RES.
HYDRO	HYDROGRAPHY	+ WATER SUPPLY - ACCESSIBILITY	+ HYDRAULIC CYCLE + NATURAL NETW.	+ WATER SUPPLY	+ HYDRO ENERGY + ACCESSIBILITY
	HYDROGEOLOGY	+ WATER SUPPLY		- POLLUTION	+ GEOTHERMIC
	VEGETATION	- POLLUTION	+ CO 2 CYCLE + NATURAL UNITIES	+ WATER SUPPLY	
CLIMAT	WIND	∓ W. EXPOSURE		- W. EXPOSURE	- AIR POLLUTION + W. ENERGY
	PRECIPITATIONS		+ HYDRAUL. CYCLE		
HEL	SUN RYTHM	± SUN EXPOSURE	SUN RYTHM	+ SUN EXPOSURE	SOLAR ENERGY

MULTICRITERA MATRIX OF THE ENVIRONMENT

POTENTIAL SCHEME OF THE ENVIRONMENT - CERGY-PONTOISE - 1989

7. SITE

The site space is considered as the combination of the four following natural components: topography, microclimate, hydro-vegetation and sunniness; they can be identified with the repose and activities areas, and with the hydraulic and energizing systems.

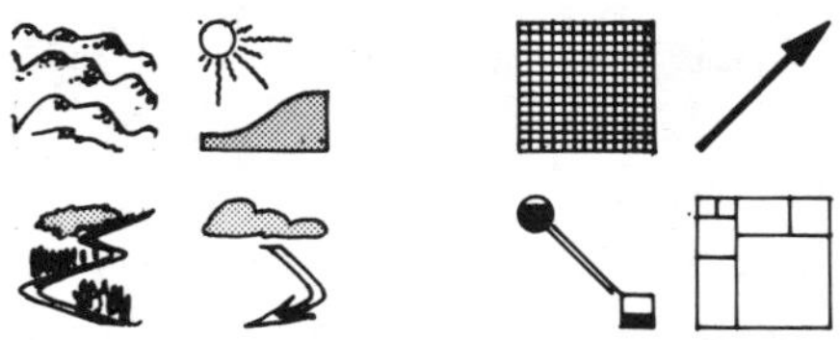

The method is also entered by the materials; the matrix helps to fix the programmes into the urban areas of the environment (step 3) as shown on the map (6).

MATERIALS		PROGRAMMES: REPOSE	ACTIVITIES	HYDRAUL. SYST.	ENERGY SYST.
TOPO	ACCESSIBILITY	0	++	-	++
	VIEWS	++	+	--	--
	LIVING SPACE	+	++	-	-
	DENSITY	+	++	--	--
HYDR	WATER AREA	+	++	++	+
	WATER TREAT.	++	++	++	0
	WATER SUPPL.	++	++	++	0
	VEGETAL AREA	+	+	-	-
MCL	MICROCLIMATE	++	++	--	--
	AIR QUALITY	++	+	+	0
	ACOUSTIC	++	+	-	-
SUN	THERMICAL	++	+	0	++
	LIGHTING	+	++	0	++
	LOCOMOTION	0	++	0	+

MULTICRITERA MATRIX OF THE SITE

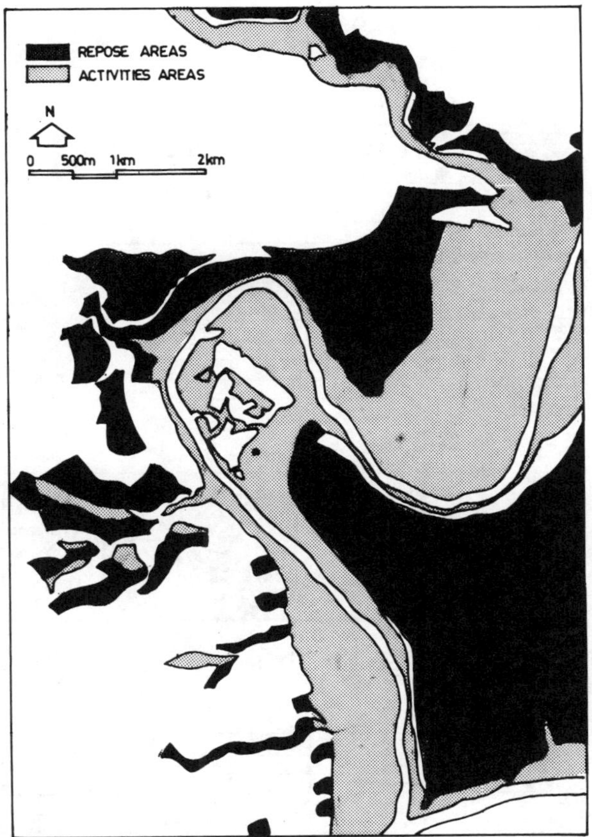

POTENTIAL SCHEME OF THE SITE - CERGY-PONTOISE - 1989

8. URBAN SPACE

The urban space is considered as the combination of the four following space components: the built areas, the free space, the green space (vegetation + water) and the solar space (portion of the free space enlightened by the sun); they can be identified with the functional and global spaces, and with the fluids and radiatives (heating + lighting) systems.

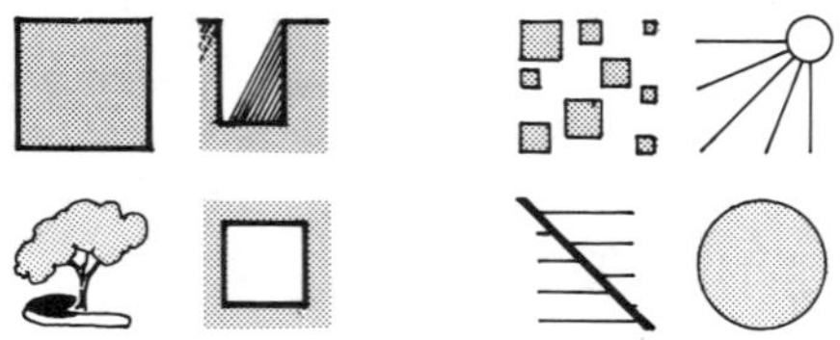

The method is entered by the programmes; the step hereafter described (step 5 of the method) concerns the elaboration of the bioclimatic principle of the urban space, just after the climatological data collection (step 4). The four components are combined so as to offer comfortable outdoors areas in every season, as shown in the appropriations board, below.

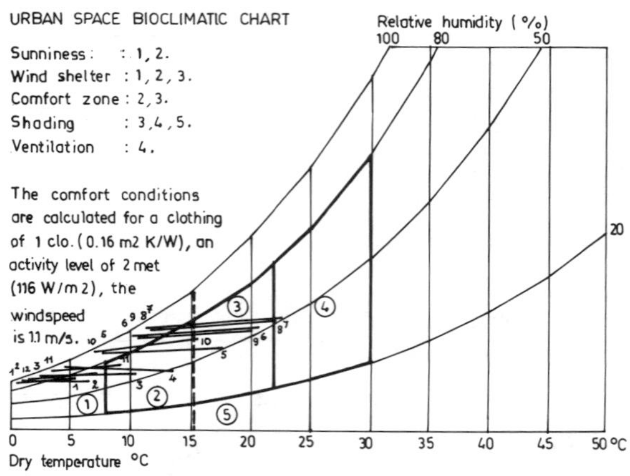

	1	2	3	4	5	6	7	8	9	10	11	12
Wind shelter	●	●	•	•	•	·	·	·	·	•	●	●
Rain shelter	•	•	•	•	•	·	·	·	·	·	•	•
Sunniness	●	●	•	•	•	·	·	·	·	•	●	●
Shading					·	•	●	●	•	·		
Comfort			·	·	•	●	●	●	●	•	·	
Ventilation							·	·				

ANNUAL URBAN COMFORT CONDITIONS

	BUILT SP.	FREE SP.	GREEN SP.	SOLAR SP.
Wind shelter	●		●	
Rain shelter	●			
Sunniness		●		●
Shading	•		•	
Ventilation		·		

IDENTIFICATION

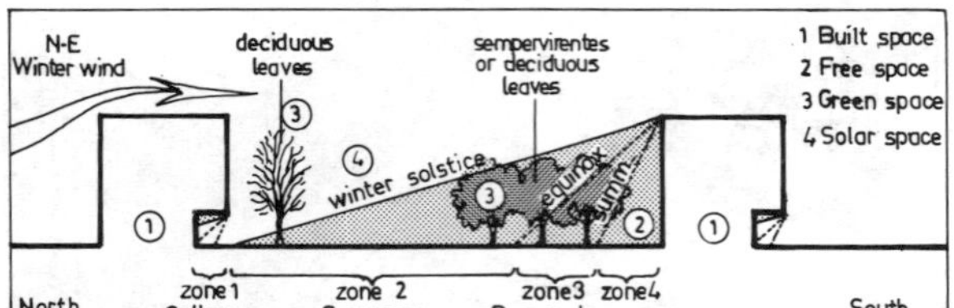

BIOCLIMATIC PRINCIPLE OF URBAN SPACE CERGY PONTOISE

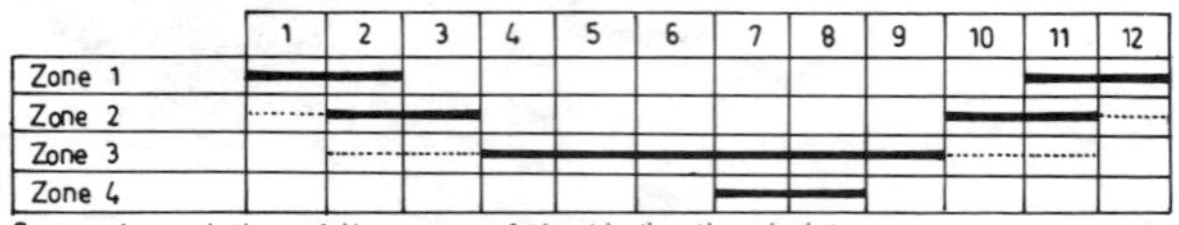

Seasonal appriation of the zones of the bioclimatic principle

9. ARCHITECTURE

The architectural space is considered as the combination of the four following spatial and constructive components: a massive opaque unit, a lightweight transparent one, a water element and a fire one; otherwise, these four components can be identified with the private and public spaces, and with the servant and conditioning devices.

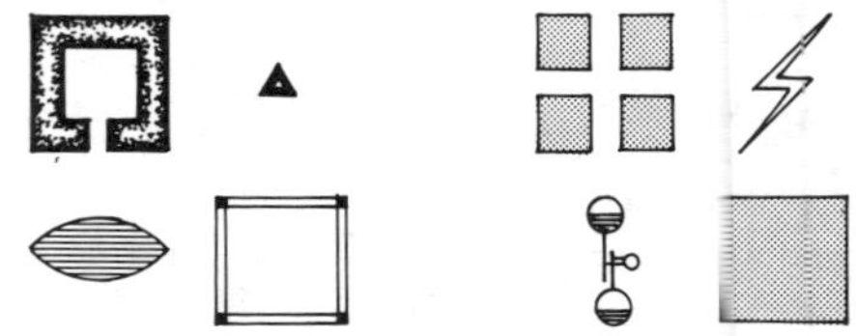

The method is also entered by the programmes; the step 5, hereafter described, concerns the elaboration of the bioclimatic principle of the day-night use building (housing). The combination of the components (here three, the water element don't play any bioclimatic part) gives rise to an intermediate space which plays a part of balance of the internal atmospheres.

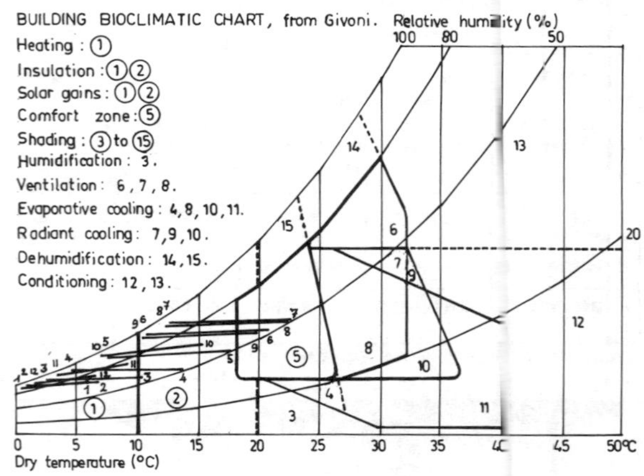

	1	2	3	4	5	6	7	8	9	10	11	12
Heating	●	●	•	•	·					·	•	●
Insulation	●	●	•	•	·	·	·	·	·	·	•	●
Solar gains	●	●	●	●	•	·	·	·	·	•	●	●
Wind shelter	●	●	•	•	·					·	•	●
Rain shelter	•	•	•	•	•	·	·	·	·	·	•	•
COMFORT						•	●	●	•			
Shading						·	•	•	·			

YEARLY COMFORT CONDITIONS OF THE BUILDING

	HEAVY	LIGHT W.	WATER	FIRE
Heating	•			•
Insulation	•			
Solar gains		●		
Rain shelter	●	●		
Wind shelter	●	●		
Shading	●	●		

IDENTIFICATION OF THE 4 COMPONENTS

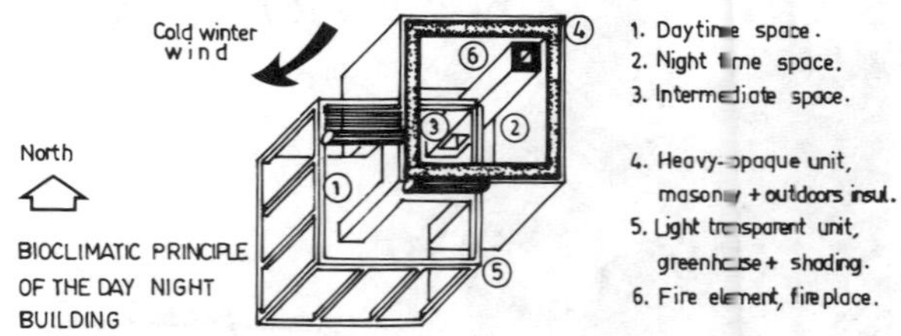

BIOCLIMATIC PRINCIPLE OF THE DAY NIGHT BUILDING

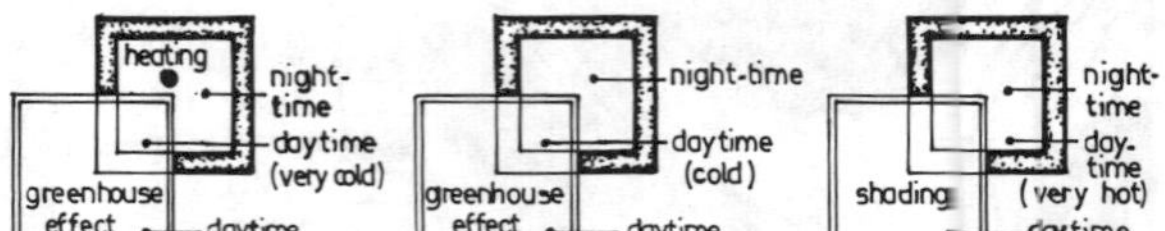

OPTIMISATION OF URBAN LAYOUT WITH RESPECT TO SOLAR IRRADIATION

R.J.Peckham
Commission of the European Communities,
Joint Research Centre, Ispra 21020, Varese, Italy.

E.J.Keeble
Building Research Establishment,
Garston, Watford WD2 7JR, U.K.

ABSTRACT. The paper explores and illustrates the use of a C.A.D. type design aid for assessing the effects of obstructions on solar radiation received on building surfaces. Parametric studies are presented for simple rectilinear structures in order to illustrate the potential of this type of design aid, and to present results of general applicability for urban layout design. Aspects of layout design addressed include the effect of varying building separation to height ratio, sloping ground, staggering successive rows, and orientation.

1. INTRODUCTION

The admission of adequate sunlight into buildings and open spaces is a recognised criterion in urban planning, although the importance attached to it has varied. In the post-war rebuilding boom from the late 1940s, minimum standards for winter sunshine penetrating into living rooms were set for housing in the UK and elsewhere /1/. Later, the value attributed to sunlight seems to have declined, and influences such as higher densities due to increasing land values, greater dependence on artificial light, and the prestige attached to building high, often led to greater obstruction of sunlight.
However, in recent years the value of using solar heat in enegy-efficient buildings has refocussed attention on ways of predicting the effects of obstruction on the radiation reaching solar collectors, windows and facades. In the 1970s local legislation affecting overshadowing and a building occupant's "right to solar energy" was introduced in parts of the USA and Canada /2/, and has since been considered in Australia /3/. For ease of application, the criteria have been very simple, such as specifying a minimum period of insolation on a particular date, for example, the winter solstice.
The design tools previously available, such as sunpath diagrams and related graphical methods of calculation /4/, are generally time consuming to use. This makes it impractical to use them to consider several surface points or design alternatives. Also, many such methods give information only on the availability of direct sunlight, not irradiation.
For energy applications, there is a need for design tools providing easy assessment and comparison of the amount and distribution of solar radiation received on building surfaces or on the ground, accumulated over different periods.
This paper presents data on the effects of obstruction of solar radiation, calculated using SHADOWPACK, a computer aid suitable for use at the early stages of building design /5/. SHADOWPACK allows simple geometric models of possible site layouts to be created, and then automatically calculates the cumulative solar irradiation on any surface, taking account of obstructions. The user selects the latitude and period of the year over which the radiation is accumulated. SHADOWPACK is an example of technology simplifying the assessment, and illustrating important consequences, of architectural decisions about built form and site layout.
As well as illustrating the use of SHADOWPACK, the object of the paper is to analyse the effects of urban obstructions at a parametric level, to show the sensitivity of solar irradiation to elementary differences in built form and spacings.

2. PARAMETRIC STUDIES

In order to explore the possibilities for optimising urban layout with respect to solar irradiation a very simple model consisting of regularly spaced cubic blocks, aligned in E-W rows, was first considered.
While such a model may be less interesting, from an architectural point of view, than more complex structures, in order to arrive at general conclusions it is essential to start with a geometrically simple model. This model has the advantage that it can be characterised by only one parameter, the separation of the blocks, or two parameters if sloping ground is allowed. In more complex models, with more variable parameters it becomes more difficult to arrive at general conclusions which can be usefully applied to layout decisions.
Starting with this simple model, computer simulations were performed in order to answer the following questions regarding the effects of the layout and spacing of the blocks:

i) How does the spacing affect the fraction of energy received on the S and E/W faces?
ii) What is the effect on the energy received of sloping the ground, towards the north and towards the south?
iii) How much does staggering successive rows (so that S faces look onto the gaps between the blocks in front) affect the fraction of energy

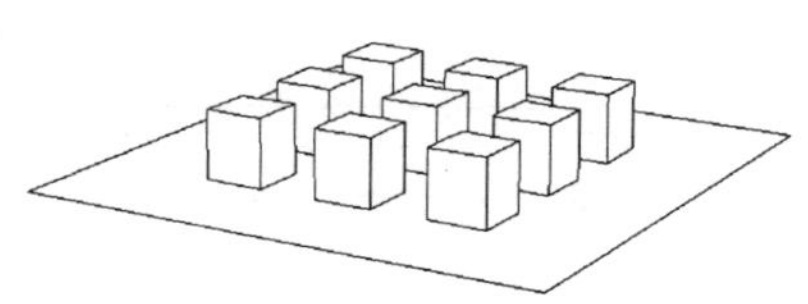

Figure 1. Example of one of the nine cube models (Separation = 1.0).

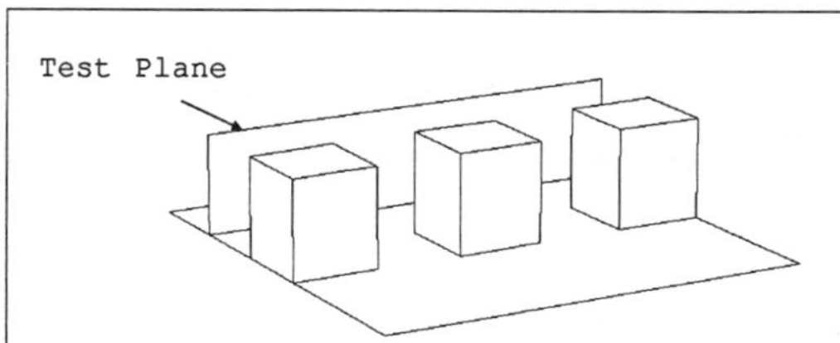

Figure 2. Three cube model with vertical plane to north for examining insolation distribution.

received?

The model used consisted of a central cube and its eight nearest surrounding neighbours in a regular grid (Fig. 1). Taking the blocks to be cubes of unit side three models were created with values for the separation of 0.5, 1.0 and 2.0. The value of 0.5 is exceptionally close for an urban situation but is useful in order to evaluate how great the impact of obstruction can be in an extreme situation. Results were generated for the model on level ground, and for north sloping, and south sloping ground, with a slope of 1/10.

Table (i) reports the fraction of direct energy received (compared with the unobstructed face) on S and E/W faces in the month of December for latitude 51°. We find that in the extreme case of separation = 0.5, on north sloping ground, only 6% of the direct energy is received on the south face, while in the "best" case, separation = 2.0 and south sloping, 92% of the direct energy is received. The remaining values in the table could be useful as design guidelines for the December shading effects, when planning layouts.

Table (ii) gives the results averaged over the winter six months, from October to March inclusive. In this case the closest separation on north sloping ground reduces the direct energy received on the south face to 14% of that for the unobstructed face, while in the best case the reduction is barely significant with 97% of the direct energy being received.

In Tables (iii) and (iv) similar results are reported for the effects of the obstructions on the global (direct plus diffuse) energy received on the faces. The model used takes into account the shape factors seen by each face for other model faces, sky and ground, but does not take into account multiple reflections between adjacent blocks.

A value of 0.2 was assumed for the albedo of both the model faces and the ground so the effect of multiple reflections should be small.

Values for the incident radiation were taken from a Test Reference Year for KEW, latitude 51 deg. For the global radiation we find the "worst" case reduction in December is to 26% of that received by the unobstructed face, and the "best" case is 92%. The equivalent figures for the winter six months are 29% and 97% respectively.

For the separation = 1.0 examples, which are typical of many urban situations, the fraction of global energy received over the winter six month period ranges from 55% to 81%.

It is worthwhile noting that, while the fractional results reported here are calculated for the whole S or E/W faces, they can also be applied to glazing on those faces, provided that the glazing is distributed evenly over the faces.

In order to address question (iii) above regarding the staggering of the successive rows models were created with a vertical test plane in the appropriate position behind a single row of blocks (see Fig. 2), and the distribution of direct energy received on this test plane was found using the program ISOSUN. These distributions are shown in Figs. 3.1, 3.2 and 3.3.

It is clear from Figs. 3.1 and 3.2, which both show pronounced minima, that for separations of 0.5 and 1.0 moving successive rows to face the gaps between the rows in front will make a significant difference to the energy received on the S faces. For the separation = 2.0 case however the contour lines reveal little benefit to be gained from staggering. As a further check on this, the global energy received on the south and E/W faces for the staggered cases was calculated, and compared with the unstaggered cases. The results for the period October to March inclusive are reported in Table (v) as percentage "improvement" (increase) over the unstaggered case. The increase for the south face in the separation = 0.5 case, is very significant - 27%, and staggering is still worthwhile at separation = 1.0. The benefit of staggering the buildings at separation = 2.0 is not significant.

Contour plots of the type shown in Figure 3 can also be created for examining the distribution of solar radiation received on the ground plane. This can be useful in layout decisions regarding for example the location of recreation areas or swimming pools.

3. OPTIMISATION WITH RESPECT TO ORIENTATION

For a given design or structure, the radiation received on surfaces will vary with orientation of the structure within the chosen site.

This effect can also be investigated quantitatively using SHADOWPACK.

Once a model has been created the variation of energy received on surfaces with orientation can be investigated by performing repeated calculations

Table (i)

Values for the fraction of energy received, compared with the unobstructed face, in December. Computed for direct radiation and clear skies, lat. 51°.

Face:	N.Slope S	N.Slope E/W	Level Gnd. S	Level Gnd. E/W	S.Slope S	S.Slope E/W
Sepn.						
0.5	0.06	0.21	0.15	0.32	0.30	0.43
1.0	0.28	0.42	0.43	0.55	0.59	0.69
2.0	0.65	0.65	0.78	0.77	0.92	0.90

Table (ii)

Values for the fraction of energy received, compared with the unobstructed face, averaged over October-March incl. Computed for direct radiation and clear skies, lat. 51°.

Face:	N.Slope S	N.Slope E/W	Level Gnd. S	Level Gnd. E/W	S.Slope S	S.Slope E/W
Sepn.						
0.5	0.14	0.29	0.27	0.37	0.41	0.45
1.0	0.51	0.58	0.59	0.69	0.74	0.78
2.0	0.79	0.84	0.90	0.91	0.97	0.96

Table (iii)

Estimates of the fraction of energy received, compared with the unobstructed face, in the month of December.
Computed for global radiation using KEW T.R.Y data.

Face:	N.Slope S	N.Slope E/W	Level Gnd. S	Level Gnd. E/W	S.Slope S	S.Slope E/W
Sepn.						
0.5	0.26	0.47	0.34	0.50	0.47	0.54
1.0	0.47	0.63	0.58	0.68	0.69	0.74
2.0	0.76	0.78	0.82	0.83	0.92	0.89

Table (iv)

Estimates of the fraction of energy received, compared with the unobstructed face, averaged over October-March incl. Computed for global radiation using KEW T.R.Y data.

Face:	N.Slope S	N.Slope E/W	Level Gnd. S	Level Gnd. E/W	S.Slope S	S.Slope E/W
Sepn.						
0.5	0.29	0.50	0.40	0.54	0.53	0.58
1.0	0.55	0.72	0.67	0.77	0.79	0.81
2.0	0.85	0.91	0.91	0.93	0.97	0.95

Sepn.	S.Face	E/W Face
0.5	27%	8%
1.0	10%	7%
2.0	2%	2%

Table (v) Percentage increase in global radiation achieved by staggering successive rows.

Table (vi) Global radiation received in canyon model as percentage of that for unobstructed S. face. (KEW T.R.Y. data)

Sepn.	S Face	SE or SW	E or W Face
0.5	38%	37%	28%
1.0	58%	53%	36%
2.0	84%	71%	43%

varying only a single orientation parameter. In order to illustrate this, and to generate results of general use, a simple "canyon" model was created consisting of two opposed vertical faces of unit height and ten units length. Three values for the separation of the faces were again used: 0.5, 1.0, and 2.0 units. The energy received on one face was calculated for different orientations by rotating the whole model. Results are presented for the face looking East (or West), SE (or SW), and S, calculated for global radiation, integrated over October-March inclusive, using KEW T.R.Y. data. A value of 0.2 was again used for the albedo of the ground and the model faces.
In Table (vi) are reported values for the percentage of energy received compared with the unobstructed south face. We see that in the case of closest spacing considered the E or W faces receive approximately one quarter of the global energy received by an unobstructed S face.

4. CONCLUSIONS

Parametric studies have been used to investigate the use of a C.A.D. type design aid in evaluating and optimising different urban layouts with respect to solar irradiation. Simple models have been used to arrive at general conclusions and quantitative results regarding the variation of energy received on facades with varying separation of buildings, and inclination of the ground. The benefits to be obtained by staggering regularly spaced rows of buildings have been evaluated for different separation/height ratios. The effect of orientation in a simple canyon model has also been presented. The tabulated values presented could be used to obtain initial estimates for the effects of obstructions in regularly spaced layouts, though particular projects, especially those with more complex geometries, would normally require the creation and analysis of an appropriate three dimensional model using a program such as the one used here. This type of analysis is now possible using P.C. based design aids.

REFERENCES

/1/ British Standard Code of Practice 3, Chapter 1 (B); Sunlight (1945).

/2/ Tabb S.E. Solar Energy Planning: A guide to residential settlement; McGraw-Hill, New York (1984).

/3/ Williamson T.J. Solar Access; Environment and Planning Law Journal, 1985 (June), 143-154.

/4/ Petherbridge P. Sunpath diagrams and overlays for solar heat gain calculations; BRE Report; HMSO London (1969).

/5/ Peckham R.J. A P.C.-Based Design Aid for Siting Solar Components. Proc. 6th Int. PLEA Conference, Pergamon Press (1988).

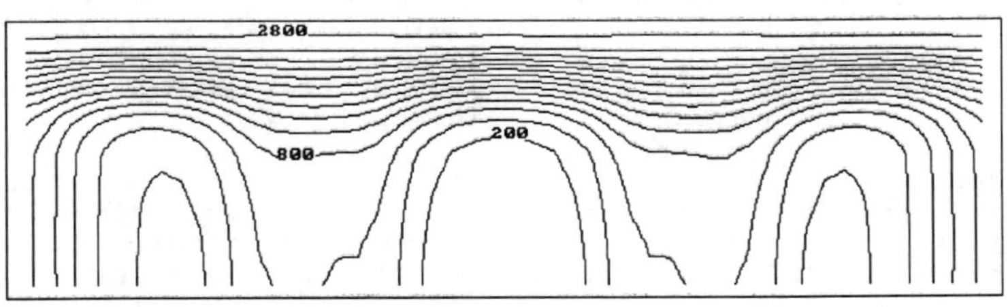

Fig.3.1 Separation = 0.5

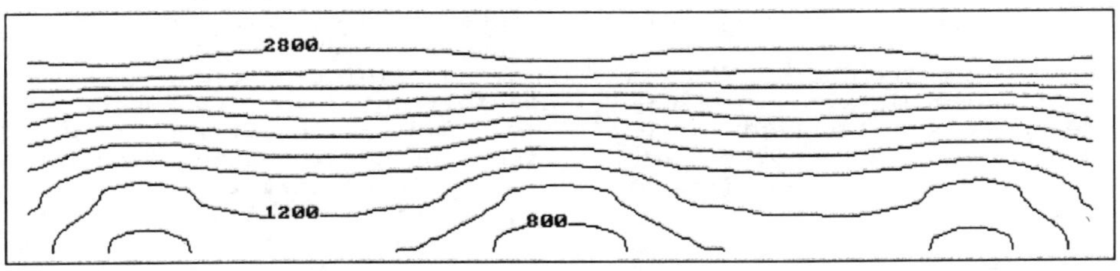

Fig.3.2 Separation = 1.0

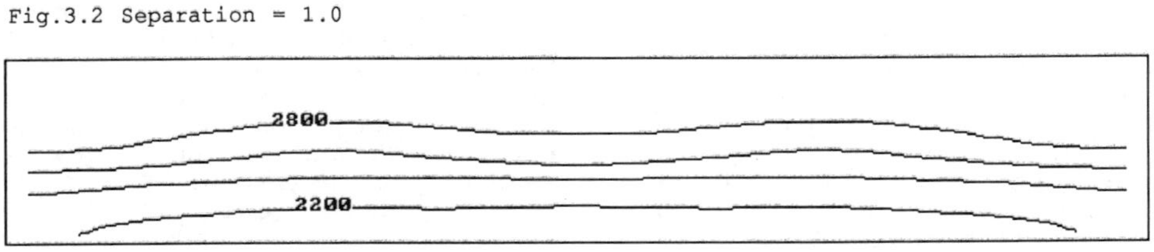

Fig.3.3 Separation = 2.0 (scale reduced with respect to figures above)

Figure 3. Contour plots of the distribution of radiation received on the test planes. Results for direct radiation, with clear skies, integrated over October to March inclusive.

LE LOGICIEL DE GESTION DES ENERGIES
ENERGIE-TERRITORIA

RAHAL.D, MIRAMOND.M
Laboratoire Méthodes, Institut National des Sciences Appliquées de LYON
20 Avenue A.EINSTEIN 69621 VILLEURBANNE

AULAGNE.B (GAZ DE FRANCE)
92080 PARIS LA DEFENSE Cedex n°8

GRILLOT.J (Ingénieurs des Villes de France)
91 rue du Faubourg Saint Honoré 75008 PARIS

ABSTRACT. More than 70% of all energy is consummed in the urban area, wich represents between 5% and 20% of the functioning commune budgets.
The software market for energy management is very small either regards to quantity and the quality of the services that programms can offer.
Therefore, Methods Laboratory of INSA-LYON (National Institute of Applied Sciences), Gaz de France (GDF) (Gas National Supplier Company), and Ingénieurs des Villes de France (IVF) (National Association of Cities Engineers) including over 30 towns, decided to form a research team on this particular aspect.
After a year of research, this collaboration resulted in a new methodological approach of energy management. This success is based on confrontations and analysis of great deal of large towns. The aim was to publish the results in concret form, by realising a software, named ENERGIE-TERRITORIA.
The originality of this project was that the futur users were implicated in this research process.

1. L'ENERGIE DANS LA VILLE

1.1 Les besoins énergétiques au plan communal

Les pays industrialisés sont de gros consommateurs d'énergie principalement dans les villes. Ainsi l'énergie consommée est utilisée à raison de 65% pour l'industrie et de 75% pour les secteurs residentiels et tertiaires.
La moitié de l'énergie totale consacrée aux transports l'est dans les agglomérations. Plus de 70% de l'énergie totale consommée dans la France l'est dans la ville.
L'énergie représente entre 5 et 20% du budget de fonctionnement des communes. C'est un poste budgétaire susceptible de compression sans nuire au confort des usagers et sans investir nécessairement dans les travaux.

1.2 Les objectifs de la gestion des energies

La gestion des énergies tend à :
* Réaliser des économies, en améliorant les rendements aussi bien des installations importantes que des plus petites, car toutes les consommations s'additionnent. Ainsi, une bonne gestion permettra de connaître et de limiter les dépenses en optimisant les installations considérées dans leur ensemble.
* Accroître le confort, en faisant une étude attentive des besoins, afin que les techniques appropriées pour les satisfaire pleinement puissent être mises en oeuvre.
* Assurer le fonctionnement des installations, en ayant un contrôle continuel qui permettra de détecter et de signaler toute anomalie.

1.3 La nécessité d'une méthode et de son informatisation

Il y va de soi que parler de gestion sans évoquer l'aspect méthodologique serait une utopie. En effet, tout repose sur la qualité de la conception et du choix d'une méthode de travail. Celle-ci est d'autant plus efficace, que les données sont plus fines et donc plus nombreuses, ce qui conduit indubitablement au recours aux moyens informatiques. Ces derniers permettront, non seulement, de supprimer les tâches répétitives et de gagner un temps précieux pouvant être consacré à une véritable exploitation des résultats, mais également le stockage d'un grand nombre d'informations utilisables à tout moment pour un usage quelconque. Sans oublier la fiabilité et la rapidité quant à l'obtention des résultats.
Le marché des Logiciels destinés à la gestion des énergies dans les communes est limité aussi bien par leurs nombre (2), que par la qualité des services qu'ils rendent. Il existe biens sûr des outils spécifiques qui ont été développés par certaines communes dont l'inconvénient st d'une part la nécessité des gros moyens financiers, et d'autre part leur grande spécificité pose un problème de transposabilité d'une ville à une autre.

2. LE LOGICIEL ENERGIE-TERRITORIA

2.1 Généralités

Tout cet état de choses, a conduit au montage d'une véritable association, outre l'I.N.S.A, la société GAZ de FRANCE qui ne voudrait pas se contenter d'être un simple fournisseur d'énergie, et l'association des INGENIEURS des VILLES de FRANCE (I.V.F) (groupe de travail énergie qui englobe une trentaine de villes environ) et qui sont en fait les futurs utilisateurs.
Cette collaboration a conduit, après plus d'une année de travail, à la concrétisation d'une méthodologie de gestion, qui est le résultat d'une confrontation de réflexions d'un grand nombre de villes. Parmi ces dernières, quatre ont été choisies pour former un échantillon suffisamment représentatif, et qui auront un rôle de villes-pilotes.
L'idée consistait ensuite à concrétiser, ces résultats par le développement d'un Logiciel, dénommé ENERGIE-TERRITORIA.

2.2 Présentation

Conçu pour renforcer l'efficacité des Services Techniques, ce Logiciel s'affirme comme un "tronc commun" capable de répondre aux problèmes les plus couramment rencontrés dans la pratique quotidienne. Il contient toutes les bases indispensables pour faire une bonne gestion sur le plan énergétique comme sur le plan du suivi budgétaire.

2.3 Les axes principaux de l'outil

Pour les villes ne souhaitant pas utiliser toutes les possibilités offertes par le Logiciel, dés le départ, différents niveaux d'utilisation sont prévus pour simplifier la saisie et les traitements.
* Analyse rapide de la facturation
Le coeur du Logiciel repose sur la saisie des factures et des relevés. Compte tenu du nombre élevé de saisies (plusieurs milliers), les programmes seront optimisés dans le sens de la rapidité d'utilisation et de la fiabilité des données.
Différentes vérification seront effectuées:
- Un contrôle rapide de cohérence lors de la saisie, réalisé à l'aide d'une série de tests, afin de pouvoir corriger immédiatement d'éventuelles fautes de frappe.
- Un contrôle plus approfondi, effectué par comparaison des données introduites avec l'ensemble des données disponibles (valeurs correspondantes pour la même période de l'année précédente, intégration des degrés-jours,etc..).
* Analyse des coûts induits par service rendu
La mise en place du fichier "client" permettra un suivi analytique des consommations d'énergie. L'objectif dans ce cas sera de mieux analyser les consommations et les coûts induits pour chaque service rendu par la municipalité.
Il s'agit là de coûts ne reflètant pas a priori la ventilation comptable, mais permettant plutôt un suivi par nature, se rapprochant ainsi d'une comptabilité analytique.
* Aide à l'ajustement et à la prévision budgétaire
Les factures d'énergie sont ventilées par Chapitre et Sous-Chapitre de la comptabilité des communes (M12 pour les villes de plus de 10.000 habitants et M11 pour les autres), permettant ainsi un suivi comptable par poste budgétaire voté en début d'exercice. L'aide à l'ajustement et à la prévision budgétaire intervient à plusieurs niveaux dont principalement:
- Restitution d'un tableau récapitulatif par poste budgétaire.
- Calcul de l'incidence d'éventuelles modifications du coût des énergies.
- Situation budgétaire (en pourcentage).
- Estimation des factures "à venir" sur l'exercice en cours.
* Analyse technique des consommations
La répartition des consommations par équipement, batiment, ou zone géographique permettra d'établir des ratios nécessaires aux comparaisons entre villes, ou pour différents batiments d'une même ville. Ces ratios seront obtenus à partir des unité suivantes: Kwh, Francs, m2, m3, degrés-jours, unités spécifiques.
* Optimisation tarifaire gaz

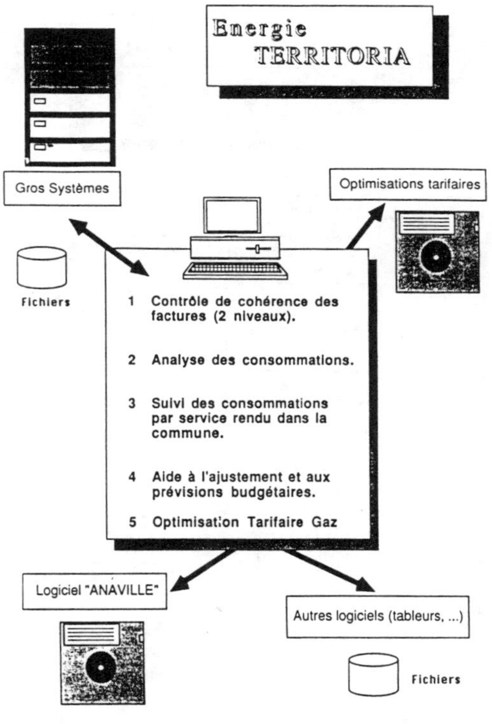

2.4 Organisation de la base de données

ENERGIE-TERRITORIA se compose de deux parties:

- Une partie relative à la répartition géographique et technique (Equipement, Batiment, Zone, ..)
- Une autre partie concernant l'approche budgétaire (Chapitre, Sous chapitre, ..).

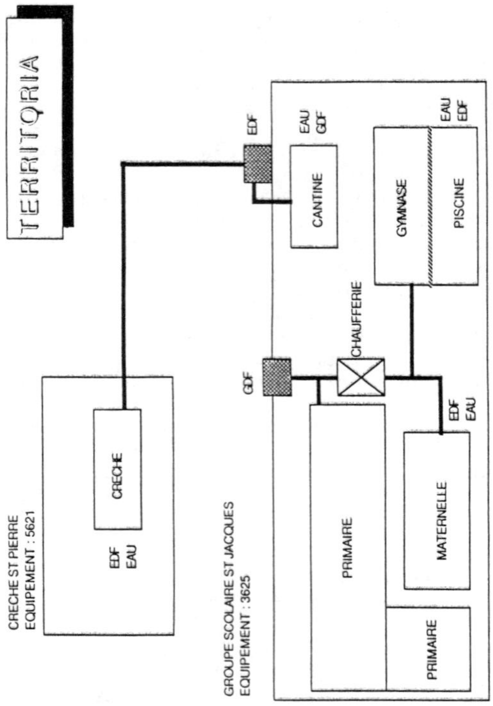

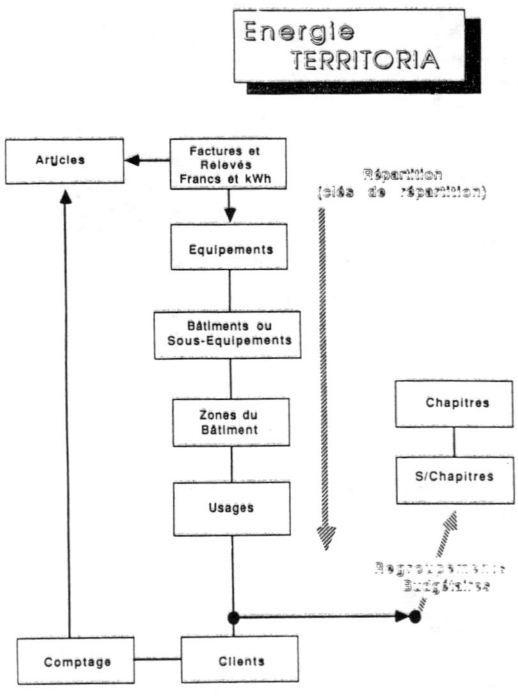

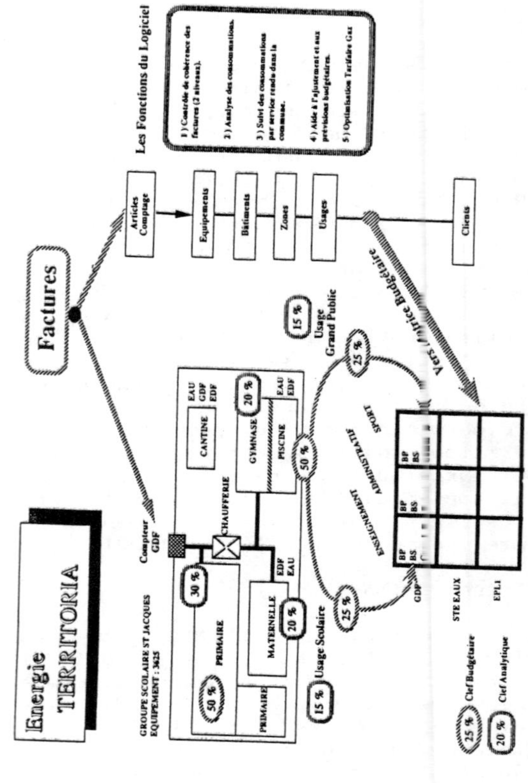

3. CONCLUSION

L'originalité de ce projet, en est que les utilisateurs sont eux mêmes les concepteurs. Ainsi chaque commune exprime ses besoins et ses limites.
Essayer de faire un outil qui puisse s'adapter à la diversité du marché, tel a été l'objectif premier. En effet, ENERGIE-TERRITORIA est un produit qui s'élabore au fur et à mesure des demandes exprimées, mais aussi des remises en cause.
Cet outil équipe actuellement 25 communes.

THE EVALUATION OF URBAN INFRASTRUCTURES IN THE PLANIFICATION OF URBAN NETWORKS

SAUGY B., GROUX R., HADORN J.-C., REVAZ J.-M.

CREM (Centre de Recherche et d'Enseignement en Energie et Techniques Municipales),
Rue des Morasses 5 / CP 48 / CH-1920 MARTIGNY

The main objective of SYSURB - a research project aiming at computer aided management of urban systems - is to undertstand the structure and geographical distribution of energy demand in order to plan the competitive development of power, gas and urban heating networks. Thorough knowledge of this structure allows planification of network extensions, thus satisfying new demands with a minimum of supplementary investments, and anticipating negative consequences on a network of renewal or modernization of users' equipements. The information system has been extended to underground cadaster, since the costs resulting from incomplete knowledge of the network are high in comparison to the costs brought about by the management of infrastructures with the help of micro-informatics. Moreover, the greatest part of the polluting impacts on the environment are caused by processes of energy transformation, such as engines, oil burners, rubbish putrefaction and industrial processes. By reducing the costs of urban networks, computer aided network management brings about increased competitiveness in comparison to decentralized oil combustion and an increase of their share of the market, thus reducing impacts on the environment. Furtehermore, urban heating allows transformation of household refuse and industrial wastes - the CO2 they contain returns to the atmosphere during the decomposition process - into fuel and endogeneous resources which are to be managed as an independant production source at the level of the City.

1. INTRODUCTION

Tout système urbain pour fonctionner importe des ressources sous forme d'énergie d'eau et de matière et les restitue sous forme d'ordures, de déchets et d'eau usée. Les problèmes actuels conduisent à limiter ces flux en recyclant une partie des déchets ou en remplaçant ces importations de ressources par du savoir-faire et des produits industriels d'un haut niveau technologique tels que la micro-informatique *(fig. 1)*. Les travaux entrepris dans le cadre du projet SYSURB entrent dans cet objectif par l'analyse de la demande d'énergie au niveau de sa structure et de sa répartition géographique.

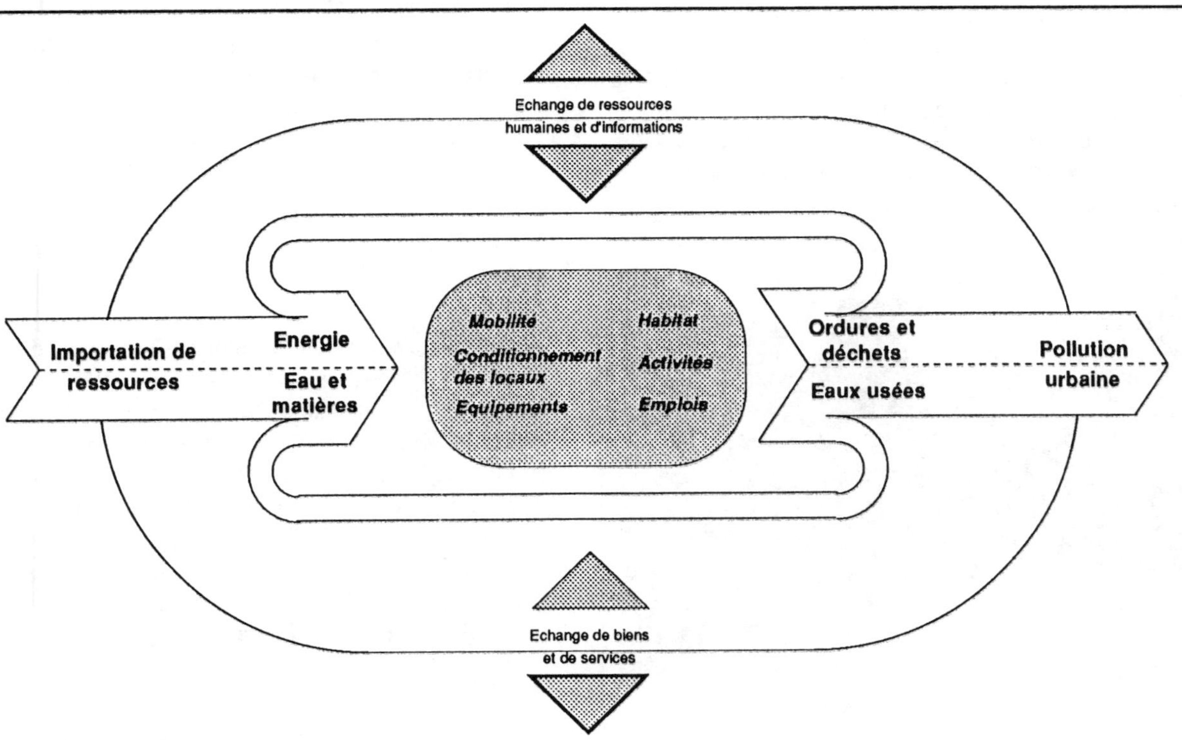

CREM | Fig. N° 1 Le système urbain | N° 114.004 29 mai 1989 Sg / Dr

Une bonne connaissance de cette structure permet une meilleure gestion des infrastructures dans ses trois volets que sont :

- la planification
- le cadastre souterrain
- la conduite en temps réel des réseaux

L'intégration de ces trois phases dans un même système permet une utilisation commune des données. A titre d'exemple, la conduite en temps réel fournit des données de consommation ou de réserves de capacités indispensables à la planification.

2. GESTION DES INFRASTRUCTURES

2.1 Planification des infrastructures

A cette étape, il s'agit d'établir, pour chacun des acteurs du développement urbain, un plan d'extension des infrastructures pour satisfaire à de nouvelles demandes avec un minimum d'investissements supplémentaires et de prévoir les conséquences d'une rénovation ou modernisation des équipements des usagers.

Dans le cadre de travaux de recherche de l'EPFL(Ecole Polytechnique Fédérale, Lausanne), l'analyse énergétique de la commune de Martigny a abouti à l'élaboration d'un plan directeur énergétique spécifiant les vecteurs énergétiques distribués dans les différents secteurs de la ville *(fig. 2)*. Ce résultat est basé sur l'analyse de la demande d'énergie à l'état initial, puis, de différents paramètres régissant l'évolution de cette demande.

- Etude des plans de zone fixant les évolutions possibles de densité de logement et d'emploi des différents secteurs de la ville.
- Etude des améliorations des procédés et évaluation de l'évolution possible et probable de la qualité du parc immobilier.
- Etude des infrastructures existantes avec leurs réserves de capacité actuelles et probables.

Ces divers paramètres permettent d'établir des scénarios de développement qui serviront de base au modèle de distribution. Ce modèle repose sur un découpage en éléments finis des réseaux et du territoire. Cette méthode permet de traiter tous les réseaux et ainsi de déterminer l'agent adapté à chaque portion du territoire.

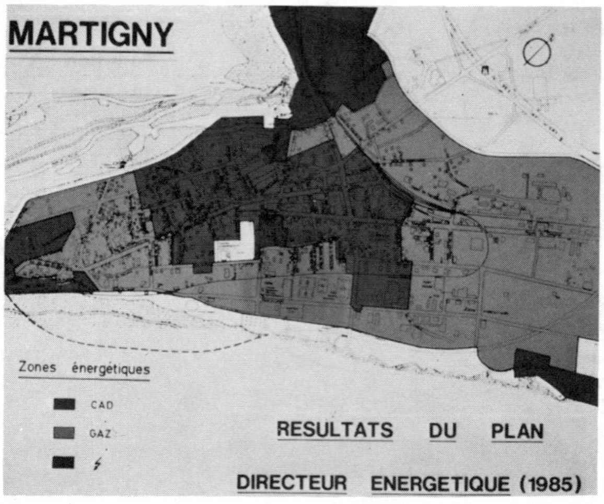

Fig. N° 2 Plan directeur

2.2 Cadastre souterrain

Le système d'information a été étendu à la tenue du cadastre souterrain car les coûts induits par une mauvaise connaissance du réseau sont élevés en comparaison d'une gestion micro-informatique des infrastructures.

Le choix d'une structure de données basée sur la méthode de découpage en éléments finis pour le calcul des flux physiques, pour l'organisation des données ou pour la visualisation ainsi que l'utilisation de fonds de plan non interprétés permettent le traitement de gros réseaux urbains sur micro ordinateurs. Une telle structure réduit le volume d'informations "actives"et facilite ainsi les échanges entre les différents acteurs urbains.

De ces concepts sont issues des maquettes, développées sur PC, PS2 et Macintosh, mises en oeuvre à Martigny comme élément de collaboration inter-services et à Neuchâtel pour la gestion et la planification du réseau de gaz.

2.3 Conduite en temps réel des réseaux

L'essentiel des émissions de polluants dans l'environnement sont le fait de procédés de transformation d'énergie tels que moteurs, brûleurs à mazout, putréfaction des ordures et procédés industriels. La conduite des réseaux permet, en baissant le coût de ceux-ci, d'être plus compétitif par rapport à la combustion décentralisée du mazout et ainsi d'augmenter leur part de marché pour réduire les émissions dans l'environnement.

Un chauffage urbain permet de transformer les ordures et déchets en combustible, dit ressource endogène, qu'il s'agira de gérer comme une production indépendante au niveau de la ville, dans le but de limiter l'importation de ressources.

Dans un objectif d'optimisation des différents agents énergétiques, la gestion des générateurs de secours de certains immeubles permet une diversification des ressources.

Au niveau des consommateurs de Martigny, des tableaux de bord énergétiques sont installés dans les immeubles; ils informent les habitants sur la performance de leurs installations, qui est le rapport entre le confort requis et la consommation *(photo 3)*. Dans un même temps, ces tableaux de bord permettent aux gestionnaires de réseaux d'acquérir les données pour la conduite fine des réseaux ou de proposer des consignes aux habitants. Cet échange d'informations au niveau de la ville donne ainsi naissance au terme d'URBISTIQUE.

3. IMPACTS SUR L'ENVIRONNEMENT

Le fait de limiter les importations de ressources et de recycler une partie des déchets a pour effet de diminuer les émissions dans l'environnement.

Fig. N° 3 Economètre

Pour la ville de Martigny, les conséquences sur l'environnement de quatre scénarios sont décrites à la figure 4.

Ces 4 scénarios sont les suivants :

- Maintien de l'état actuel avec les réseaux existants.
- Développement du chauffage urbain et du réseau de gaz.
- Encouragement de la gestion optimale de l'énergie et des ressources.
- Extension des réseaux et gestion optimale.

Il est impressionnant de constater à quel point l'encouragement d'une gestion optimale correspondant aux techniques informatiques et aux aspirations actuelles offre des solutions simples et efficaces.

Références:

/1/ Dr Bernard Saugy, projet SYSURB "Modélisation des flux énergétiques, gestion des réseaux urbains centrée sur l'usager" - IHE 623.141 - octobre 1988

/2/ Bernard Saugy, Patrick Pralong, Jean-Marc Revaz et Cie, "Dialogue Energétique, bilan intermédiaire de la préparation du Dialogue Energétique" - CREM 108.042 - mars 1989

CREM - 906 - 28.11.89-Gr/dl

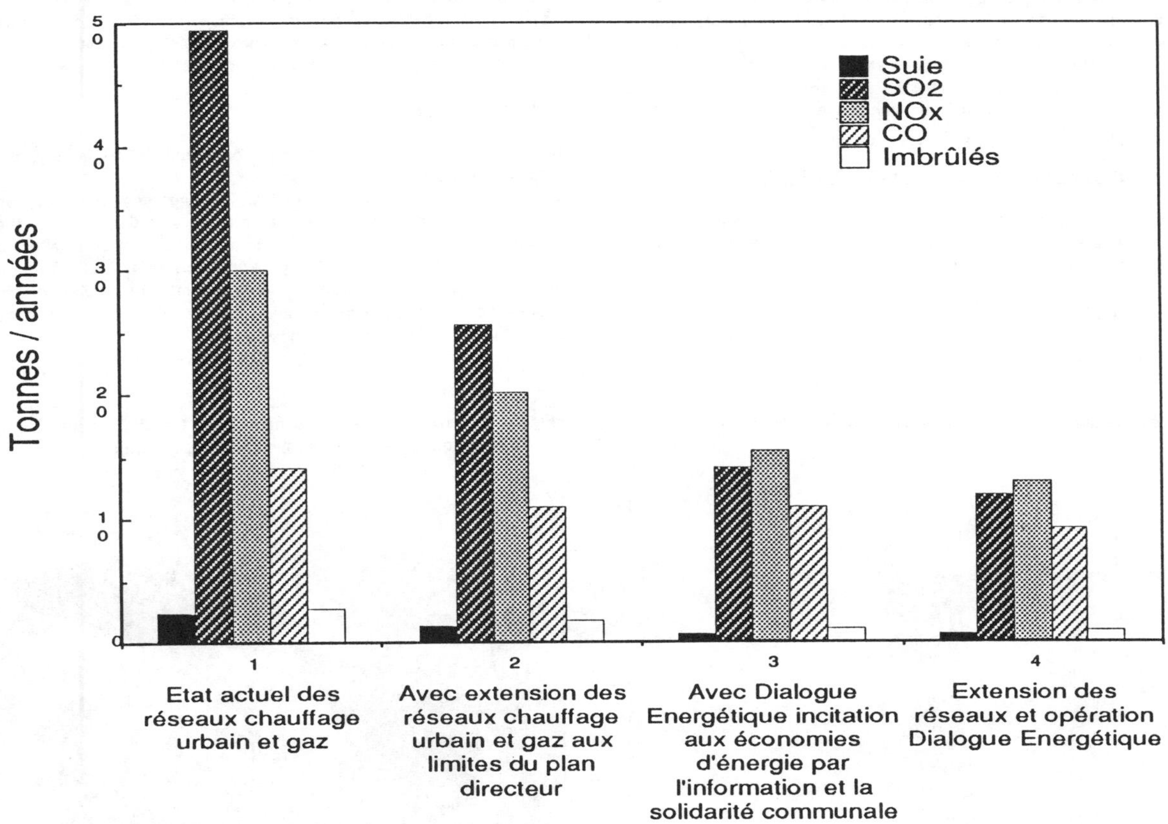

CREM	***Fig. N° 4 Comparaison des émissions selon divers scénarios pour 15'000 habitants et 5'000 emplois***	N° 108.136 20 mars 1989 SI Martigny

A MODELISED MANAGEMENT OF MACHINE TRANSFORMATION

B. Feuillie
"Welcome to the Machine"
Blauzac - 30700 Uzès - France.

Any similarity with already existing sites would be a mere coincidence.

This analogical attitude is directly derived from the 'Ready Made' concept of Marcel Duchamp, created in 1913 with "The Bicycle Wheel" (M.D. "I liked the idea of having a bicycle wheel in my workshop. I took pleasure in looking at it just as I take pleasure in watching the flames dancing in the chimney".)/4/, then defined in 1917 with the "Fountain" (M.D."Whether Mr. Mutt had made the fountain with his own hands or not is of no importance. He had chosen it. He had taken an element of ordinary existance and arranged it so that its utilitarian significance disappeared under its new title and from the new point of view. He had created a new thought for this object.")/4/.
Since Marcel Duchamp's time, Ready Made forms may also be found in waste, but the application of creativity is still moved from the domain of shape towards the domain of the senses, of identity; this new position might be seen as the acknowledgement of a lack: our modern world overproduces objects and forms which have only a rationalist justification for their existance, or which, once transformed into waste have no possible meaning. The machine allows the production of series of identical objects. The Ready Made concept creates a meaning for this multiutude by taking it out of its functional and/or conventional context, that is by diverting it. However in contrast to Art, whose modern function involves the pure creation of identities, the realisation of this concept in Architecture is inevitably linked to other domains, such as the economy, the recycling of (industrial) waste and the search for other life-styles in the larger sense of the term.

1. CULTURE IS THE OBJECT OF ARCHITECTURE

The field of the present analysis concerns the domain of industrial production, with particular interest in the constructions associated with it. Closely linked to technological knowledge, itself derived directly from scientific research, industrial production is a human creation which, at all levels, invests us with an artificial environment, and is thereby a cultural projection.
Thus a landscape is created and may be modelled into the expression of a culture. The means developed in the 19th century allowed increased manipulation at the geographical scale and, with the passage of time, it is sometimes difficult to recognise which part is natural and which results from Human modelling.
Taken as a cultural - that is to say artificial - interference with a previously "natural" environment, the domain of industrial production becomes the domain of architecture, since culture is the field of architecture.

2. THE FIELD OF THE ANALYSIS: INDUSTRIAL PRODUCTION

Since industrial production contitutes a major element of our modern culture, this analysis will attempt to distinguish several different and separate sub-components of it.

2.1 Industrial Form

Industrial form is dictated by the machine, either through its product, or directly by the machinery itself (which is, in general, the product of other machinery). The *raison d'être* of the machine is serial production, so its product is a multitude of identical objects, and a list of such series might be contained in a catalogue or dictionnary of industrial forms associated with their individual definitions - that is, their functions ("The machine declares its function through its name", Jean Baudrillard /2/).

2.2 The Process

The basis of all industrial production, a process is a series of abstract manipulations which organise the flow of primary materials or of products into a complex of operations and functions which may be organised linearly or in parallel.

2.3 The Factory

Hence, while the process is the programme of the factory; the factory is the site of the process.
We will restrict the present analysis to one type of factory as defined by the industrial architect V. Grenier - the 'Factory-Machine' :"It is a machine, a lime kiln or a smelter, a cement works or a glass oven, a pithead gear. Its form is technical, guided by economy of material and technological makeshifts. It is the work of an Engineer, often beautiful in its simplicity. It is, in itself, a landscape" /3/.

Figure1: Idealised section of a vertical cement kiln.

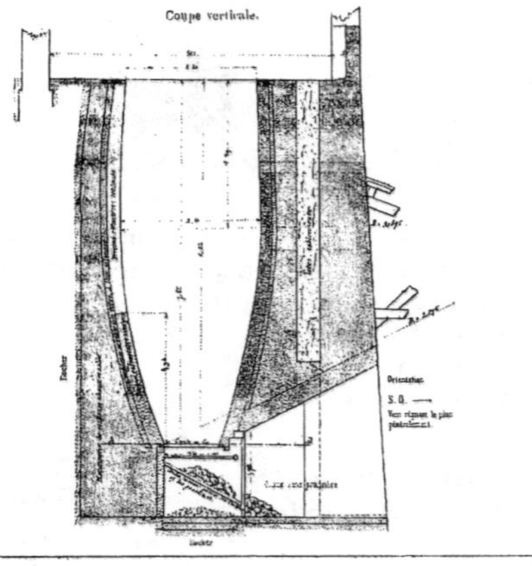

Figure 2: View of old vertical cement kilns: the disused machine as a site or a considerable pre existing shaping of the slop.

2.4 The Machine

Generally, however, the process and the factory are totally dependant on the machinery which justifies their existance: at the same time the machine is a purely functional principle which makes no concessions to values irrelevant to its operation and its composition is minimal with regard to the task it is designed to perform : the relationship between form and function creates an objective, recursive and obligatory duality between the physical constraints of its design and the end result of its construction.
However, in reality, the only construction satisfying this definition is V. Grenier's 'Factory Machine'. On the scale of the building, only this type of factory is itself the machine - the form of the process - and not merely its wrapping.

2.5 The machine as Waste

Once abandoned, such a factory-process is called industrial wasteland. At the scale of the building, it is a disused machine, frozen in a prolonged instant following the end of an activity whose specific nature often renders any alternative use problematic: for, while the machine is an excellent example of a technological object whose production is the direct result of scientific research, its eventual destiny as waste has been taken into account neither by its makers nor by Science.

Figure 3: Project for the reconversion of a site into offices (stepped double block united by an interior covered roadway) including a commercial showroom (ground floor of the upper block and backing on the old kilns).
Photographs of a model of a typical section: to build on an existing functional model implies a new functional model.
Interior view of the showroom: transformed into light wells serving sales alcoves in the showroom, the machine "vertical cement kilns" is an example of management of industrial wasteland.

3. ECOLOGICAL DIMENSIONS OF THE INDUSTRIAL WASTELAND

The present state of affairs is indicated by a qualitative measurement of the areas involved which shows this to be a phenomenon on at least an European scale: in a report by Lacaze /4/, 30,000 Ha of industrial wastelands were identified in the Rhur valley in 1980; Great Britain in 1982 had 45,700 Ha, of which 17,000 were being reclaimed, and in France the stock of industrial wasteland in 1986 was estimated at 20,000 Ha.
Elsewhere the same report addresses the need to change our attitudes to industrial wasteland from an approach to rehabilitation based on clasical urbanisation procedures to an approach through the restitution of the landscape and long-term land management. It specifies that "..re-use practically never consists of (...) a return to the previous industrial installation, (...), but will consist in a new change of use, and therefore of appearance, in a zone which has already suffered other transformations." /4/.
From an ecological point of view, it is clear that an industrial sequence can have no "natural" future, being a fundamentally artificial medium. Industrial wastelands therefore represent the special perspective of a site already linked to Human history, prepared ground, and, sometimes, a complex site structure within an urban environment. The problem with this type of structure is generally concerned with its abandonment, that is, a lack of management in the cycle of its transformations.

4. ARCHETECTURE IS A PROJECT FOR HABITABILITY

4.1 The Machine as a Model

Since the machine is a structure built to satisfy the organic requirements of its function and the Factory-Machine is a process comprising a defined number of functions, the identity of function between two machines allows the existance of similar forms in different localities, with minor variations due to site morphology. Viewed from the standpoint of the building, the machine plays the role of a model in the archetectural sense - that is; it is a representation of the set of concepts, references and techniques needed for the realisation of a project.

4.2 A Modelised management of machine transformation

Therefore, seen as an archetectural model, the machine can find its transformation in a counter-model which seeks a new qualification for it and whose elaboration proceeds from Marcel Duchamp's 'Ready Made' concept: "form follows from function". Such a project has the property of a transportable system in that the machines themselves are in principal international, but the counter-model will only be valid in its specific interactions with the environment at the level of the different systems necessary to render the space inhabitable (Lighting, heating, urbanization etc...).
At the scale of the site, a cement works is a process of industrial production. It contains constructions which have the identity of machines. These constructions either have no internal space, or their internal space is entirely closed off (non-habitable) and usually has the function of a container (of cement, in the various forms imposed on it by the manufacturing process). However, seen as a model, each of these function-machines represents a multitude of analagous forms which may be found on every continent: the appearance of a new and more profitable technique heralds the end of an industrial sequence and is linked to its reappearance as waste, again at multiple sites.
This is the case for the machine "Vertical cement kilns braced into sloping ground" transformed into light wells serving sales alcoves in a commercial showroom (Architectural project for the reconversion of a cement works) ; this is an example of management of industrial wasteland. The project conforms to a multiple 'Ready Made' concept; a new function, which, through its value as a new model, may be repeated as required.

5. PRECESSES BY WHICH INDUSTRIAL FORMS ACQUIRE AN IDENTITY AND A HISTORY

From a cement works to a can, there is an analogy between forms produced by industry; they are cultural forms which modify the original so-called 'natural' environment.
At the scale of the object, the can is an industrial object whose function is to contain a drink. Present in large numbers, and also the visible sign of a cultural imperialism, it may be found on every continent. Emptied and useless, it leaves the shop window to enter the environment. Crushed under a car wheel, it plays an active part in the sedimentology of the town's undersoil.
Although the two forms mentioned above differ radically by their size and by belonging to quite different domains of production: an endproduct on the one hand and an engineering work on the other, we may relate them by the abstraction of their sculptural properties. The analogy may now be found at the level of the geometry of simple forms, of an organic structure guided by the laws of forces and actions, and of a purely operational identity of funtion: filling/storage/emptying. And, in particular, as waste unconsidered by the production process, they both bear witness to the end of their functional state and exhibit a fragment of dead history.
When transformed into an oil lamp (by the artisans of the Comores), the can undergoes mutation and acquires an identity that its industrial destiny did not foresee.
This is the nature of manufacture in recent history: management of the order of transformations where each product is the creation of a cycle, an enrichment of the object with little identity other than its function.

Figure 4: Usually take as a drinking's contain, the can is an industrial product which may be found on every continent. Crushed under a car whell, it plays an active part in the sedimentology of the town's undersoil.

Figure 5: When transformed into an oil lamp (by the artisans of the Comoros islands), the can undergoes mutation and acquires an identity that its industrial destiny did not foresee.

6. CONCLUSIONS AND PERSPECTIVES

As described above, this approach defines the bases of a prospective research which aims to (better) organize the perpetual transformation of the environment which we produce. Based on an example taken from the interior operation of an economy of scarcity emerging from a subsistance economy, and by admitting the analogy that industrial products are involved in a cycle of transmutations which excludes ' left-overs' (particularly 'disposibles'), our economic production system could, at the moment of the conception of a machine or other industrial form, take an analagous attitude and incorporate possible hypotheses for its destiny at the end of its phase of productivity or service, in the form of a catalogue of likely and compatible transformations, taking into account the expected service life of the object.

REFERENCES

/1/ Dixit Marcel Duchamp.
/2/ "Le système des objets" - Jean Baudrillard - © Gallimard -

/3/ "L'archéologie industrielle : quelles usines protéger ?" - V. Grenier - Service de la Recherche Architecturale, Direction de l'Architecture, Ministère de l'Environnement - 1981.
/4/ "Les grandes friches industrielles" - DATAR - Ministère de l'Equipement, du Logement, de l'Aménagement du Territoire et des Transports - La Documentation Française - Paris - 198[illegible].

BIOCLIMATIC URBAN DESIGN IN ANDALUCIA

Jaime López de Asiain
Alberto Luis Ballesteros Rguez
Jose Mª Cabeza Lainez
Jose Pérez de Lama

Seminario de Arquitectura Bioclimática
Escuela Superior de Arquitectura
Reina Mercedes s/n. 41012-Sevilla, Spain
Tno.: 5 - 461 26 00 Ext. 36

ABSTRACT. Urban space conceived as basic space for circulation and communication acquires different forms depending on its importance and dimension, from the small street of the neighbourhood to the big avenue. The street, in its dimension for a neighbourhood, predominantly pedestrian, incorporates a series of design elements very characteristic in the urban furniture such as tress, benches, lamps and an adequate pavement. Shops and some ludics activities mingle with them forming a random tissue with the houses and the tertiary services. The general scheme acquires forms and dimensions which depend on the scale, from the small plazuela for the resting and playground for children, to the Main Square. The urban space as a net for transportation has been studied due to, nowadays, it presents very complex problems, many times in contradiction with the habitability of the city.

1. PRINCIPLES

We understand that the urban structure has to provide a real habitable space where the jerarchy of the different uses should be considered carefully.

there are two main structures:

1.1 Transportation

It supposes to provide by means of transportation in both two ways:

- Public (bus, underground, etc.) and
- Private (car, bycicle, etc.).

It also supposes to provide the necessary space and facilities for private and public parking.

1.2 Habitable Space

The urban structure has to provide real living space as the environment of the housing, commercial or offices zones (parks, pedestrian zones, plazas, etc.), without any contradiction with the communications net, but in relation to it.
We also consider the urban space as an extension of the habitable area of the house, for an use of seasonal (gardens, streets, parks, etc.) or daily (patios, balconies, etc.) character.

2. EXPERIMENT

It matters on the ellaboration of the design and construction of 124 SINGLE FAMILY HOUSE in Osuna, Sevilla, with an average of three bedrooms and 65 m² surface, with some variations: 4 bedrooms, 2 bedrooms, or just 1 bedroom.
The character of the dwelling and its situation is rur-urban, attached to the Osuna urban area at its S.E. border over the Sevilla-Malaga road and it performs a new neighbourhood continuing the existing town.
The bioclimatic design has been realized based on an exhausting analysis of the town and the land considering the climatic, constructive and environmental aspects. The climatic aspects consider the topics of solar radiation, temperature, wind, rain, etc. The constructive aspects consider the local construction systems, materials, forms and elements. The environmental aspects consider the geographic profile, situation and topography, vegetation, noises, pollution, and the historic, antropological, cultural and esthetic factors.

The design process has been developped based on a two stories single house prototype adapted to the norms and dimensions of the Housing Public Promotion Institute.
The bioclimatic design of the urbanization stablish different levels of circulation: general for cars and persons, semiprivate and private, using the concept and design of street-patio space and multifamily meeting plaza (corral de vecinos). The last is a revival of an old and very important andalusian urban pattern.
The houses have all the rooms looking to the South receiving the Sun in winter whilst are protected from it in summer. The stepped organization of the group of houses allows the Sun gets into all of them creating at the same time spaces for living outside, to the landscape, and the Sun shining. Trees, vegetation and urban furniture will defend against external noises.
The basic constructive elements are the tile-roof, and the white-washed wall. The rest of materials are usual and easy to find in the market but carefully selected.
All of them are controlled and expressed in a rationalistic language in order to avoid the possible excesses of a historicistic form.
For this project we select some typologies which are relevant in the country.

3. PROTOTYPE

A prototype house was built, implementing in it a monitoring and evaluation system during 18 months which results were incorporated to the final design.

Figure 1: Prototype; Summer Performance

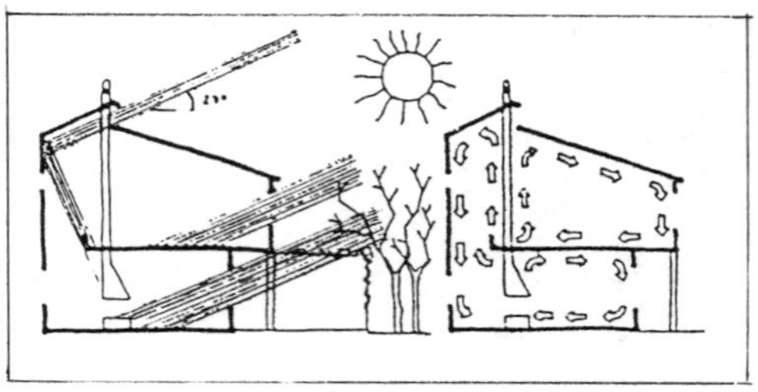

Figure 2: Prototype; Winter Performance

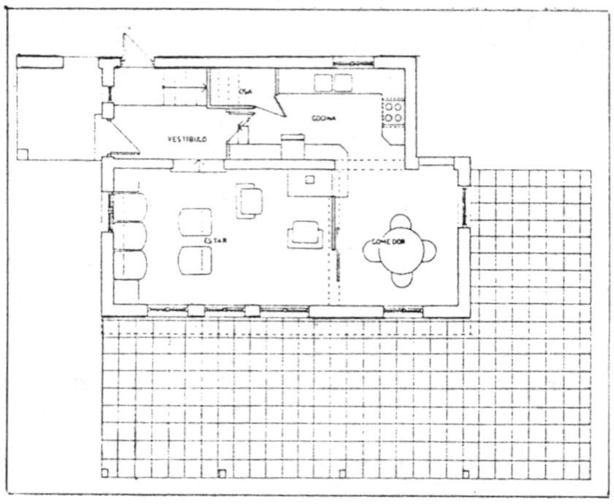

Figure 3: Prototype; Ground floor

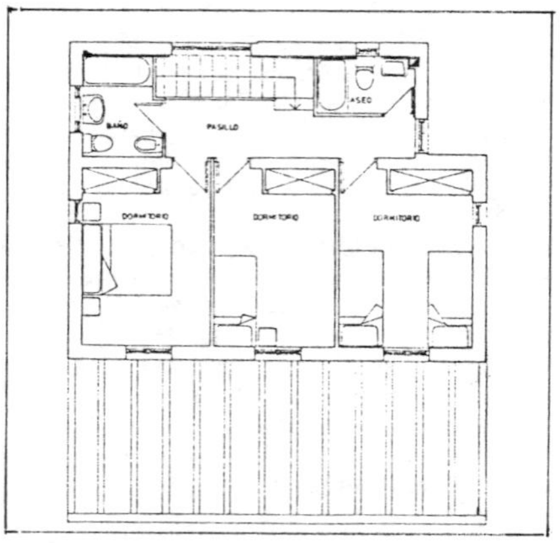

Figure 4: Prototype; Upper floor

4. FINAL PROJECT

Finally, the whole project of the neighbourhood was made and it's right now under construction for the Dirección General de la Vivienda de la Junta de Andalucía.

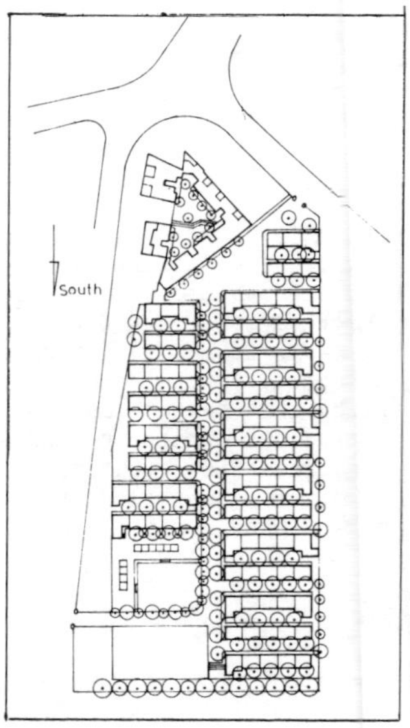

Figure 5: Final Project: General Plan

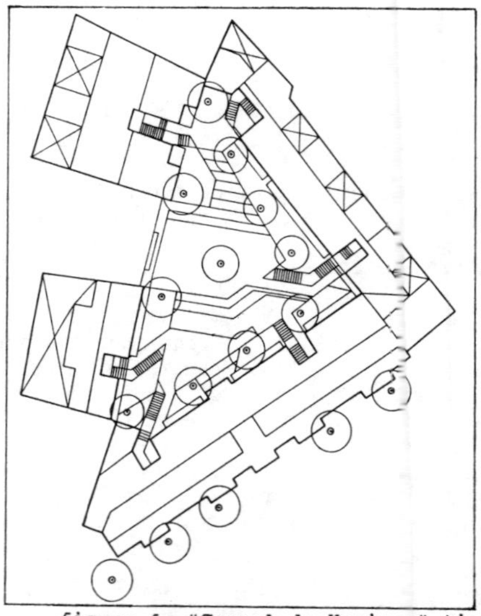

figure 6: "Corral de Vecinos" tipology

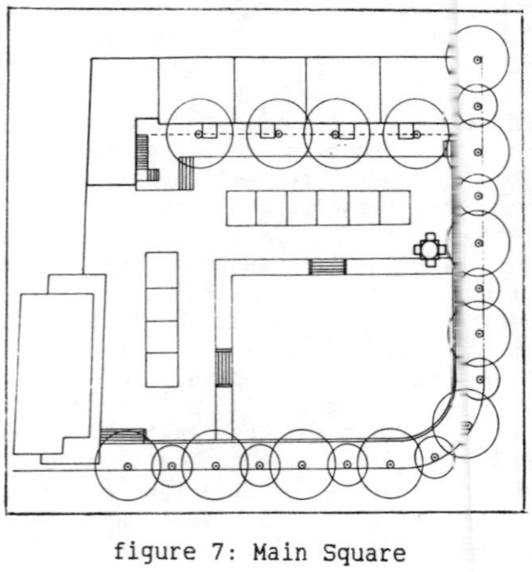

figure 7: Main Square

CLIMATIC STUDY OF THE OLD TOWN IN NICE

Christophe PETITCOLLOT
Architecte D.E.S.A.
Tourrettes S/Loup
F-06140 VENCE
Tél.: 93 59 36 66

Emmanuel RAUZIER
Ingénieur Conseil
203, Chemin de la Gorgue
F-06410 BIOT
Tél. : 93 65 02 56

ABSTRACT. The old town of Nice is really interersting by its architecture, and also by the passive cooling which provides a fresh microclimate in summer. This can be explained by the narrow streets and good thermal inertia of the constructions materials, but above all, by the good ventilation. This ventilation is provided by thermosiphon effect in the yards of buildings covered by sheet glass and widely opened to the stairs well and the adjacent streets. Our goal is to verify this assumptions and we do this by historical and architectural research, mesearements of air temperature, humidity and air velocity. These data should be used in a computer simulation comparing the town as a porous material. This is completed by a sociological investigation in which people describes how they feel thermal comfort and how they manage with their environment. This will help, we hope, architects and designers to use at the best natural elements when they create or renovate urban sites.

1. PRESENTATION OF THE OLD NICE

This old town is a triangle limited by the sea in the South, the river of Paillon on North-West and the castle hill on North-East. The town has been built probably at the 5th century before J.C. but most of buildings are from the 18th century. There is a mixing of tightened buildings with narrow streets and small flats, and rich houses with large yards.
The streets are narrow and the sun reaches the bottom only half an hour per day. This is really appreciate in summer, and the absence of solar gains is balanced by the good thermal intertia of walls. In fact people had no heating before and now the charges of it are quite weak.
The place beside the sea is protected by a double too floors barrier. It breaks the wind that climbs over the roofs.
We classified the buildings in four types :

- Type 1 : the stairs are covered by a closed roof and there is no ventilation. This exists in buildings having gone through modifications. They have no thermal interest but are good witness.

- Type 2 : The most typical of Nice architecture : the stairs wall is widely opened to a small yard. This yard is a shaft for a good ventilation. Over the entrance door, there is a gate, as shown on the photo of figure 1.1. the air flows through it even if the door is closed.

- Type 3 : same as type 2 but the yard is covered by a glass sheet that prevents from rain but greenhouse effect is bad for cooling. Some of them are covered by nearly opaque yellow plastic sheet. People has no light but also is not over-heated. The glass sheet is sustained one half meter over the roof to let the air escape.

- Type 4 : some of stairs well are opened on the adjacent streets. No windows and no frames show us that this has been done on purpose.

The figure 3.1. shows shematically these configurations.

2. THERMAL EXCHANGES IN THE TOWN

Old town of Nice has the following interesting aspects :

- narrow streets provides shadow and prevent from direct insolation
- large wall gives a good thermal inertia
- ventilation is really homogenous and air flow every where inside the yards, the stains wall and the flats.

Figure 1.1. Typical gate over a door which provides air flow from street to inside

The two first aspects are confirmed by measurements shown in figure 2.1. We see that mean air temperature is 1 or 2°C less than outside streets from others districts. The amplitude is about 2,5°C in old Nice and 4,5°C outside.

The third aspect is not directly obvious with measurements compilation and we observed its importance by calculation. For this purpose, we made a computer simulation based on exchanges in a porous material. Porosity of the town is the rate between filled and void. The second important parameter is what we call "mesh size" or thinness. Town is represented by a juxtaposition of regular solid elements whose mean size

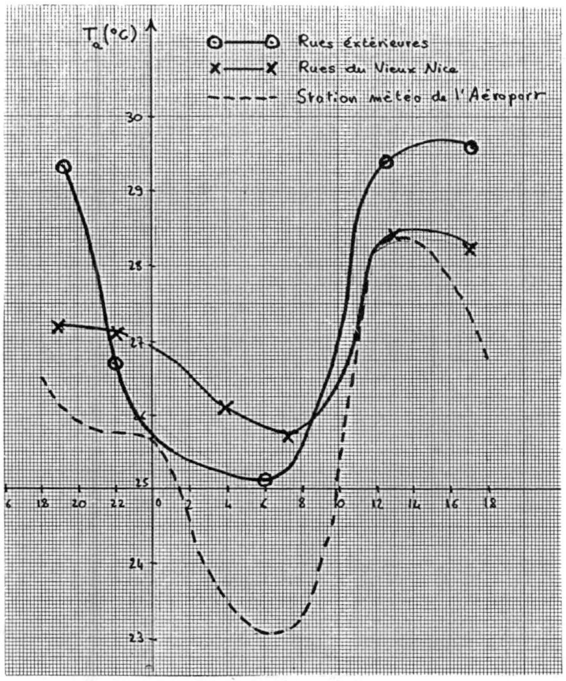

Figure 2.1. Comparison of temperature air cycle during a day
O-O streets of others districts
X-X streets of old town
--- meteo station

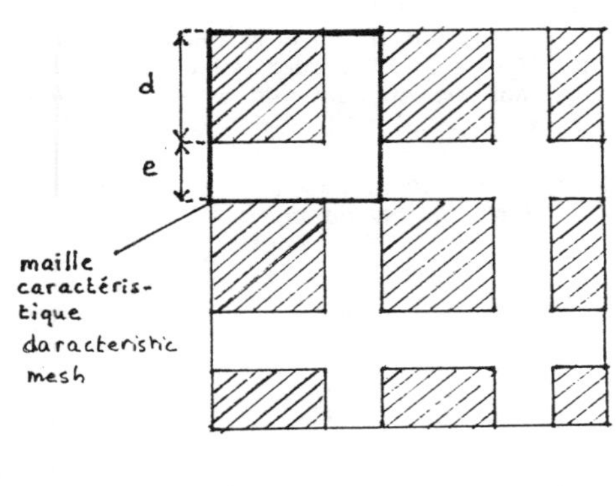

Figure 2.2. Shema of daracteristical mesh of an urban site compared with a porous material

is this "mesh size". "Mesh size" may be the distance between two doors or between a door and an other street. Figure 1.1 shows a typical mesh of our simulation. D is the mean distance between two paths for air and e the mean thickness of this path. So porosity p can be calculate as

$$p = 1 - \left(\frac{D}{Dte}\right)^2$$

The exchange surface is

$$Sm = \frac{4\,Dh}{(Dte)^2}$$

Where h is the tallness of buildings and the exchange coefficient is from ECKERT

$$h_v = \frac{6 h_{cv}(1 - p)}{1 + \frac{D h_{cv}}{20} p} \quad \text{and} \quad h_{cv} = 14{,}3 \left(\frac{1 - P}{p}\right)^{0{,}3} V^{0{,}7}$$

where V is air velocity

A new modern district is a bad porous, a rough material. An old disctrict is "thiner" and exchanges are better.
The effect of thinness has been calculated and corresponds to the comparative measurements between old town and other districts. This is illustrated by the figure 2.3. In a modern district, the wind rushes violently in the avenues and not in the buildings. The "mesh size" is important and buildings not cooled. In the old town, the wind spread softely in the streets and in the yards and in the stairs well. The "mesh size" is small and the contrasts have weak amplitude. The buildings are cooled.

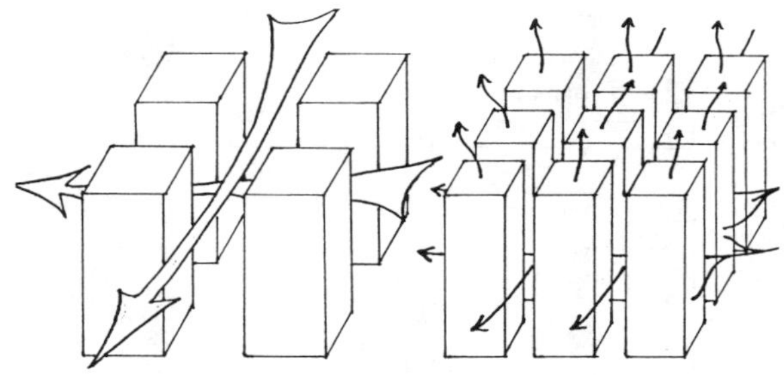

Figure 2.3. Illustrated comparison of a thick porous town (left) corresponding to a modern district and a thin porous town corresponding to an old one

3. THERMAL EXCHANGES IN THE BUILDINGS

Air temperature gradient in the stairs well depends of time in the day or in the night and is a function of air temperature in the building and in the street below. When this gradient is about 0,8°C in the old town, it is between 1,5 to 1,8°C in the buildings cf the adjacent districts. This means that the last one are lees cooled than the first one. During the day, no thermosiphon effect is observed. Cool air lays below because of the stratification. During the night, it depends on configuration as we shall discuss. Forced convection may occurs because the large gate over the doors allows air to spread into the yards and this car explain the weak temperature difference between top and bottom in the old Nice.

Figure 3.1. presents some measurements of air temperature for the four types of buildings exposed in chapter 1 in the day and in the night. These measurements are written comparatively to the air temperature in the bottom of the street.

Let's call ΔT_y, the difference of temperature between top and bottom in the yard and ΔT_w in the stairs well.

In type 1, ΔT_w is quite important. There is absolutly no wind and during the night, the upper flats could not be cooled.
Type 2 is the most interesting and in fact the most representative of Nice architecture. ΔT_y is quite low during the day and negative in the night. This allows a thermosiphon ventilation effect that is a good cooling. Upper walls are fresh from the night while the lower one are keeping heat. The consequence of this is that the temperature inside are all less than street air temperature during the day.

Covering the yard by a glass sheet has not a good effect as shown in Type 3. It is good against rain but ΔT_y increases. In the night ΔT_y is too important to become negative.

Type 4 has good performances but not as efficient as type 2.

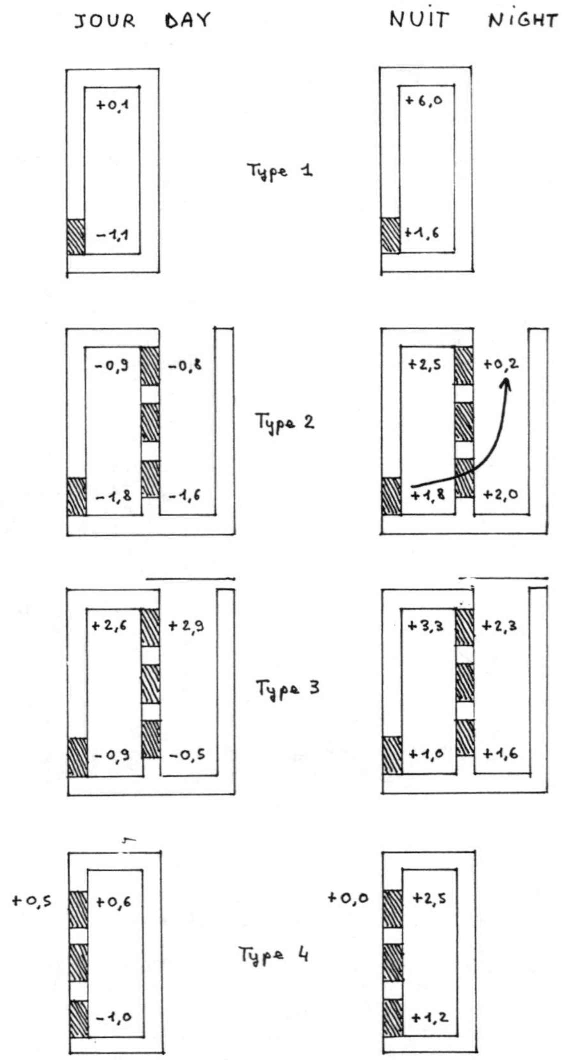

Figure 3.1. Air measurements in the four types of building in old Nice. Temperatures are related by their difference with the street air temperature

4. SOCIOLOGICAL INVESTIGATION

All results with scientifical instruments are not of great interest if they are not related to what feel and live people. Observing that comfort is finaly a psychological concept, we decide to ask a pattern of population with a sociological investigation.

The goal of this inquiry is to anower to the following questions:

- what people feel realy ? Hot, warm, too shady, humid, etc...
- how do they appreciate ? What is most important ? the desagreement of not sun in winter or the agreement of not sun in summer !
- why do they live here ? To the thermal comfort a reason and how they compare their district to the other point of the town ?
- how do people manage with the doors, the windows, the shutters, etc... ?
- Can they give witness of transformations : the glass cover sheet, the streets wavering, etc...?

5. CONCLUSION

This study is interesting because it gives a good example of a cool town. But we want to do more and use the teachings of our study for others towns, to concept or to restore them. In fact, it will be not realistic to copy the model of old Nice adapted to a particular climate, and also because it will be artificial and not adapted to build such narrow streets.

But some elements can be used for other towns : proximity of fresh areas ; the sea, lakes, wood, but especially porosity and thinness. The most imporotant teaching our a study is this one : in comfort, it is good to have low temperature good ventilation and not too much sun, but it is better to reduce contrasts and amplitudes. Human being can support warm ambiance if he is used. What is very unpleasant is to walk from hot to fresh sensations. This is not the case when the wind can go everywhere, not too strongly and when the streets are regulary shaded.

We shall go on and want to elaborate a simple calculation and recommandations by comparing differents distincts and different towns.

ENERGY USE AND URBAN DEVELOPMENT
Urban Block Analysis

Koen Steemers, Nick Baker and Dean Hawkes
The Martin Centre for Architectural and Urban Studies
University of Cambridge Department of Architecture
6 Chaucer Road, Cambridge, UK. Tel.: 0223 332981

1 INTRODUCTION

One of the most important recent developments in the theory and practice of architectural design has been the formal recognition that any building is a complex system of interrelated elements and processes. In the field of environmental design and energy conservation these inter-relationships have been explicitly represented, in both conceptual models and in the computer simulation programs which have become important tools for research and design.

Very few buildings exist in isolation, and in central urban locations the relation of a building to its immediate neighbours can be shown to have a significant effect upon the environment within it and, hence, upon its patterns of energy use.

The work presented in this paper should be read in conjunction with the introductory paper entitled *Parametric Energy Modelling For Urban Situations: Detailed Analysis* by Steemers, Baker and Hawkes (1) in these Proceedings. Both papers represent the first stage of a research project, funded by the Science and Engineering Research Council, which links together a computer model of building energy use (developed originally by Baker and Hoch (2)) and a representation of the form and variables of the characteristic urban block of the British city.

Using the computer model, which allows an integrated investigation of detailed urban and building parameters, the energy balance of a variety of generic urban block forms have been studied. Although the parametric studies of the first paper reveal the degree of significance of *urban* variables, complete blocks have been assessed in this paper to examine the implications of the wide range of parametric variations that are combined in any urban block. For example, the lack of access to daylight at street level due to obstructions may be counterbalanced by a greater availability of natural light to a top floor; it is therefore necessary to assess and integrate all the situations present in a block.

The total primary energy use for a range of typical generic blocks, and for a limited number of variations in *building* parameters (e.g. an energy conscious design, another designed to building regulations standards, and a third for poor or old buildings) is presented. This is to examine and highlight the interrelationship of both *urban* and *building* variables.

2 METHODOLOGY

The sensitivity of the primary energy balance to each input parameter has been tested so that the required definition of these data can be estimated, and to extract those factors related to urban design that are most important. This definition enables an easier approach to the collection of real data from case studies and the concentration of efforts to gain a greater understanding of the fundamental parameters relevant to energy use of buildings in the urban context.

The main feature of this computer program is that it integrates the lighting (natural and artificial) and thermal (heat losses, solar and casual gains) consequences of glazing on the energy balance. LT also allows the input of a simple urban situation, so that the consequences of both building characteristics and urban context can be compared.

In the paper entitled *Parametric Energy Modelling in Urban Situation: Detailed Analysis* (1) a series of parametric studies were carried out using LT for a theoretical room or 'cell' varying a range of characteristics relating to both building and context.

In this paper an array of cells representing the various conditions present in each form, needs to be carried out. The integrated effects of urban parameters over a whole building are tested on a series of generic urban block types.

3 GENERIC URBAN BLOCK FORMS

The urban situation under consideration is a typical UK city centre commercial block. It is assumed that the subject building and its immediate neighbours are on average four storeys high. This is not an unreasonable starting point as UK planning policies often restrict building heights to integrate with the surrounding urban context. The accommodation assumed at this stage consists largely of office type uses.

A series of ten generic block types, as shown below (Figure 1), are modelled for a range of orientations using the LT Program. This range represents only a small sample of possible solutions to assess the significance of basic form strategies. Further fine-tuning of the form may include stepped facades, storey height, building height, glazed courts and other significant manipulations of the built form.

Fundamentally the blocks represent two strategies. One assumes deep plan uninterrupted *blocks*, the other represents a limited plan depth arranged around *courts* or atria.

4 INPUTS AND ASSUMPTIONS

Each urban block is broken down into a series of cells for the purpose of modelling. The cells or rooms in this case are all of a standardised specification to allow comparative analysis of the results, with the possibility of varying any one or more of the factors. Below are listed some of the standard LT inputs indicating those which have been left as variables in this study.

BUILDING CHARACTERISTICS

Location: site	Cambridge	
latitude	52.22	(deg)
orientation	*variable*	
Room size: width	3.6	(m)
depth	6.0	(m)
height	3.0	(m)
Window: width	3.4	(m)
height	1.3	(m)
glazing ratio	40	(%)
tilt	90	(deg)
glazed area (% of W x H)	85	(%)
distance to light reference pt.1	1.5	(m)
distance to light reference pt.2	4.5	(m)
ht. of sill	0.8	(m)
Reflectance: walls	0.50	
ceiling	0.70	
floor	0.30	
window	0.08	
average indoor reflectance	0.48	
U-values: interior walls	0.00	(W/sqmC)
ceiling	*variable*	
floor	*variable*	
window (day)	*variable*	
(night)	*variable*	
exterior wall	*variable*	
Air changes: ventilation	1.0	(ach)
infiltration	0.5	(ach)
Occupancy:		
hours occupied/day	9	(hrs)
incident. gains w/o lights	15	(W/sqm)
datum lux level	400	(lux)

Luminous efficacy of sky	115 (lm/W)
Installed illuminance	15 (W/sqm)
Plant efficiency: heating	*variable*
cooling	*variable*
Energy costs: electricity	3.70
heating	1.00
URBAN CHARACTERISTICS	
Urban Horizon Angle	*variable*
Street Width	15 (m)
Obstruction Sky Angle	*variable*
Refl. opposite wall	0.20
Refl. ground	0.20
Refl. window wall	0.20

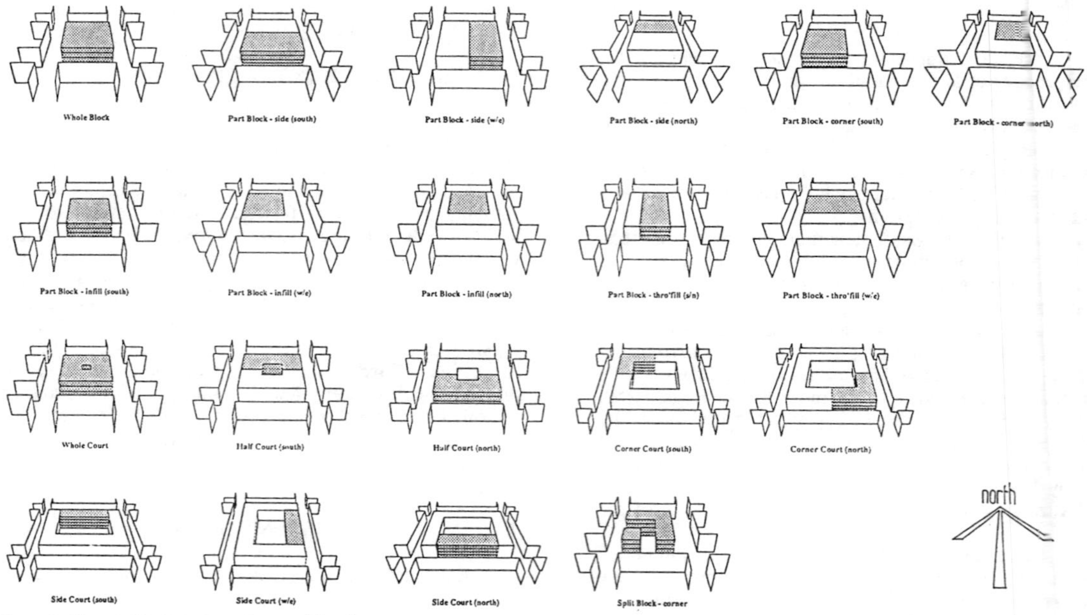

Figure 1: Generic block types under consideration.

For each cell, LT can produce results on a monthly basis broken down into primary energy used for heating, lighting and cooling. This study specifically assesses the annual energy use per square meter.

A cell representative of every floor and each orientation requires individual modelling. The resultant energy consumption rate (GJ/sqm) is then assumed for that area of the building related to the specific conditions.

The same procedure is used for internal spaces, which are defined as those which are greater than six meters from the building envelope. This definition assumes that natural lighting and ventilation are only potentially effective for a maximum of six meters deep, which has been reinforced by the parametric studies (1). Deeper plans will probably require both mechanical ventilation and artificial lighting. A plan depth of six meters may benefit from natural lighting according to the context. The degree of artificial lighting is calculated by LT. There is thus a notion of a potentially *passive zone* (2) in relation to the building envelope.

The energy balance for this zone does however depend on external obstructions in the form of the urban context. In a highly obstructed situation, the potential of the passive zone is diminished in terms of natural lighting, and this is reflected in the LT model.

5 RESULTS

The results show three energy use figures for each of the urban block types. These indicate poor, average and good standards of energy design in terms of the building fabric (U-values of opaque and glazed elements) and engineering details (heating, cooling and lighting systems efficiencies):

	U-values				Systems Efficiencies		
	wall	floor	roof	glass	heat.	cool.	light.
Poor:	0.6	0.45	0.45	5.7	0.55	1.88	20 W/sqm
Ave.:	0.45	0.35	0.35	2.85	0.65	2.0	15 W/sqm
Good:	0.3	0.25	0.25	1.85	0.75	2.12	10 W/sqm

5.1 Energy Conscious Engineering

As can be seen from the figure, the savings, by unilaterally improving all aspects of the building's environmental systems and fabric, can be up to 50%. Systems efficiencies are the main contributory cause, with insulation as secondary. An energy conscious approach to the design and specification of these *engineering parameters* thus proves to be very effective.

5.2 Energy Efficient Urban Block Forms

It is noticeable that alternative block forms also significantly affect the energy use. The drop in consumption

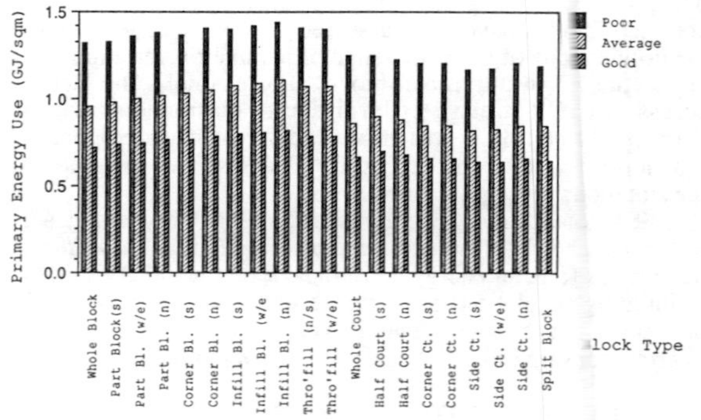

Figure 2: Energy use for three energy design standards.

in the middle of the chart represents a fundamental difference in approach to the design of the generic urban blocks. To the left the blocks are deep plan, whereas to the right they are not. The difference in primary energy use between blocks with the same engineering specifications but alternative forms, can be as great as 30%. The largest difference occurring between a north facing infill block, and a south facing side of a courtyard type. This seems to indicate that, for office type accommodation, it is beneficial to have a block with a large surface area, to allow all the spaces to be in close proximity to the external envelope of the building.

5.3 Orientation

This factor is of limited consequence, as it is only of significance for south facing unobstructed cells in a block. When all the variations are taken into account that are present to make up an urban block, the impact of orientation on energy use is minimised due to the obstructions. This conclusion is reinforced by the chart below, where only insulation and glazing ratios are variables. The adoption of slightly larger areas of glazing (50%) on south facades than for other orientations (40%), (see columns labelled 'EC-solar'), sometimes has a small impact.

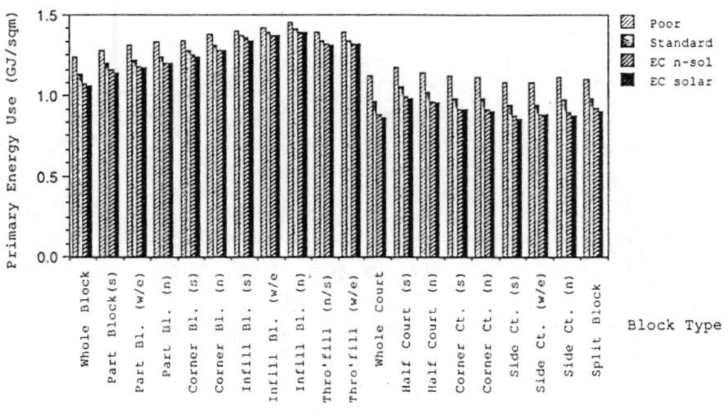

Figure 3: Energy use for three insulation standards.

5.4 Insulation and Form

The chart also indicates, as expected, that compact and protected blocks (infill and thro'fill) gain little from improved insulation, but much from better system performances. Blocks with a greater area of external walls generally perform better and also benefit more significantly from increased insulation standards. The primary energy use of court type blocks can be reduced by about 20% when comparing U-values of 0.3 for walls and high performance double glazing, with 0.6 for walls and single glazing.

6 CONCLUSIONS

At the most basic level the results show that the *urban* factors are as significant in terms of primary energy use as the *building* variables. Furthermore, the urban form of buildings themselves contribute to the energy balance.

For the assumptions made in this paper, such as the climate (UK), location (Cambridge), urban context (four storey city centre), and building type (office use), some conclusions on urban design can be made. Primarily, the importance of lighting in offices, in terms of the energy balance, determines the preference for shallow plan blocks. The improvement is due to better natural lighting, without significant solar gains because of the surrounding urban density. This is reinforced by the potential use of natural ventilation, which is not possible in deep plan buildings. There are obvious design consequences stemming from narrow plan, naturally ventilated buildings. These include a reduced plot density and, noise and pollution problems. However, the provision of an internal court, particularly if it is glazed over to form an atrium, brings greater benefits (3). These may include a useable pleasant intermediate environment, a natural light source, reinforced natural ventilation at a higher temperature than ambient, and less pollution and noise than a busy street.

The blocks modelled here all assumed the same glazing ratio for all floors. However, the results of this paper and *Parametric Energy Modelling in Urban Situations* (1)show that, in the urban context, orientation is likely to be less significant than optimsing glazing strategies. The contribution of natural light is directly associated with the obstructions, which suggests that the strategy should alter progressively from the base to the top of a building. Similarly the disrtibution of various uses within a building which require differing light levels should relate to the optimum situation vertically.

REFERENCES

(1) Steemers, K., Baker, N. and Hawkes, D., *Parametric Energy Modelling in Urban Situations: Detailed Analysis.* The Proceedings of the Second European Conference on Architecture, Paris, 1989.

(2) Baker, N. and Hoch, D., *Experience with a design tool which combines he energy value of daylight and the thermal value of solar gain.* Daylighting Buildings Conference Proceedings, UK-ISES, London, 1989.

(3) Hawkes, D. and Baker, N., *Glazed Coutyards: an Element of the Low-Energy City,* in Energy and Urban Built Form, Hawkes *et al* (eds.), Butterworths, London, 1987.

CONFORT IN URBAN SPACE OF SOUTHERN EUROPE

Jaime López de Asiain

Seminario de Arquitectura Bioclimática

Escuela Superior de Arquitectura
Reina Mercedes s/n 41012. Sevilla, Spain.
Tno.: 5 - 461 26 00 Ext. 36

When we talk about comfort in urban space is obvius that concept acquires some connotations that we must precise. On one side activities that take place "in the street" are very different to those occurring in the buildings and also our living attitudes, clothing habits and the time we spent there, are very different. We should say that urban space constitutes a peculiar space and is in habited in a peculiar way as well. For that reason it is important, beafore studing the matter, to clarify some basic aspects: The urban space is normally used to go from one site to another, what supposes a certain fluidity and easiness of circulation, and besides this, is an space where we go to spend some time, to enjoy open air, to go shopping or just to take a drink. This means that the requirements of comfort we ask for, are referred mainly to the temperature factors and to the protection against weather inclemencies. From the temperature point of view the admisible gradient in heat or cool is much wider than in buildings due to the clothing used, heavy coats in winter or light shirts in summer. However, the factor of protection against the inclemencies is permanent. The rain, the snow, the wind in winter or the excesive solar radiation in summer are elements against which the urban space should protect and that generates diverse design patterns and architectural issues such as walkways, galleries, promenades, boulevards, etc. On the other side, due to the confluence between the resting and circulation activities, often a conflict arouses. We should not forget about the problem of acoustic comfort, that in a secondary level, presents also some exigences. Air cleanness and pollution problems take also a preferent place in our consideration.

1. THE CLIMATIC GRADIENT IN SOUTHERN EUROPE

If we apply all these general ideas to the climate and the city of southern of Europe we will see how these factors will concrete and profile. To take into account the climate of southern Europe means to distinguish among very different situation.
First, the maritime climate, where winter and summers are temperate and temperature rarely reachs excesive levels for the human body.
Besides that, dominant winds normally blow from the seashore to the sea, or from the sea to inland producing milder temperatures and seldom acquiring uncomfortable speed.
The rain aspect, mainly torrential, does not reach arelevant frequency or duration, except in very precise spots.
Second, the continental climate of the hinterland which corresponds usually to arid regions where winters may be colder and in summer temperatures go up to levels difficult to bear and actually uncomfortable. Here protection against solar radiation is very important.
In these zones the existence of prevailing winds, cold in winter and hot in summer becomes an added problem.
Third, the montainous climate, in the mediterranean very often at a short distance of the coast, with cold winters, rain and snow, and on the contrary, fresh summers.
In all of them, as a differential characteristic with the countries of Northern and Central Europe, we should consider the two halves of the year, spring and summer, amd autumn and winter, so we have to study two periods of about six months with opposed conditions.
This means that in urban design is necessary to consider both situations with the same relative importance and solve problems that some times are compatible and some times opposite.
However, keeping in mind all these considerations, we can conclude that is easier for man to bear the cold inclemencies than solar radiation and high temperatures.

Therefore, historically, meridional cities have been planned and designed to be protected against the heat in summer, that is to say, problems of cooling have prevailed, though this is not uncompatible, through an appropiate orientation with sufficient solar gains in winter. Consequently, our first target will be to define the human comfort requirements in the outside space.
In summer, outdoor exposure often causes heat discomfort and stress. In such cases the possibilities of improving the comfort conditions by shading the area and, if possible, lowering its air radiant temperature, should be considered.
When an outdoor space is not shaded the people staying there are exposed first to direct solar beams and second to diffused solar radiation from the sky. Reflected solar radiation from the surrounding ground, may add significant heat load.
During daytime, ground surface temperature, specially in arid regions, is significantly higher than the ambient air temperature. This temperature elevation imposes an extra longwave radiant load on the person on which it impinges, in addition to the direct, diffused and reflected solar radiation.
The physiological severity of the solar and longwave radiant loads depends up-on the ambient wind speed, temperature and humidity, as well as up-on the clothing worn by the people. The higher the temperature and the humidity and the lower the wind speed, the greater is the phisiological heat stress imposed by the radiant load.
The direct solar radiation impinging on people in an outdoor space can be reduced by shading. The efectiveness of a given shading device depends on geometrical factors and on the partial transparency of the shade material.
Any cooling scheme of an outdoor space should start, therefore, by minimizing the solar and the longwave radiant loads although such treatments do not comprise a "positive cooling" of the space, there are equivalent from the comfort aspect to actual lowering of the space air temperature.

COMFORT LEVEL

How comfort temperature differs from one place to another, depending on seasons and clothing habits, is a known fact. In summer and in Seville, we can expect a comfortable situation for temperatures varying between 22 and 30ºC.

Taking the graph presented by Szokolay, from Olgyay, Arens et al. (PLEA 84), the Seminar of Bioclimatic Architecture (S.A.B.) has found that is is possible to expand the comfort zone for pedestrian areas (up to 30ºC is tolerable) and to adopt an intermediate value for waiting and sitting areas. The same applies to the lower limit.

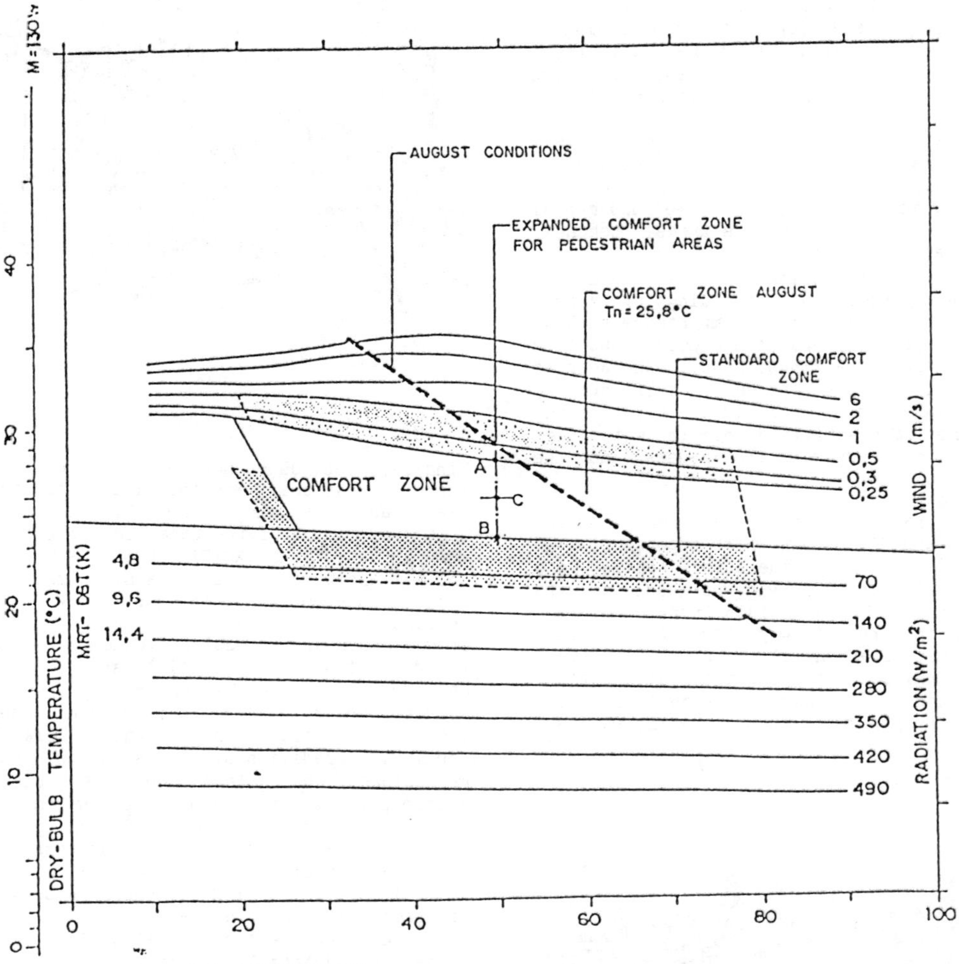

Fig. 7. Proposed diagram of comfort conditions for the Isla de la Cartuja. EXPO 92. S.A.B.

2. THE HISTORICAL, CULTURAL AND AESTHETIC FACTORS

Always when we talk about architecture and urban space we may consider, besides the phisiological aspects, those referring historical, cultural and aesthetic factors. If we accept that people and civilation have built their habitat very much conditioned by climate and geographical situation, this is much more evident in the mediterranean area, southern Europe, source of the western civilization and culture.

Then, one of the most important contribution to the urban space is the Agora or Forum produced by the greek-roman culture and assimilated and formalized in different shapes by the european city; first medieval, then renaissance or baroque and later modern. Together with the agora and the forum appears also, from another culture and civilization of paramount importance for Iberian peninsula and the whole mediterranean, the arabian one, the "souk" or market, that many times is placed in the urban space as an independient feature, but, most of the times, is associated to the agora, constituting the Main Square, the Main Street, etc. The temple and the religious space, mainly the christian, assimilates these fundamentals and generally is associated with the urban space mentioned, reinforcing its importance and significancy. The parish, the Main Church or, some times, the cathedral, take place in that space, complementing its structure.

3. THE URBAN SPACE AS A LIVING SPACE

Let us consider now the diverse aspects and forms of the urban space as a living space, as it is nowadays in southern Europe countries.
First, the street.
Conceived as a basic space for circulation and communication acquires different forms depending on its importance and dimension, from the small street of a neighbourhood to the big avenue or the motorway. The street, in its dimension for a neighbourhood, predominantly pedestrian, incorporates a series of design elements very characteristic such as porticos, galleries, pedestrian passage, etc., all of them protected against the sun and the rain, and some fundamental elements of urban furniture such as trees, benches, lamps and an adequate pavement.
The avenue, of bigger dimension, that structures the city and always links singular buildings or zones, requires design elements of major importance such as boulevards or alamedas, separating pedestrians from urban traffic, linear parkings, etc., and adds to the urban furniture the presence of fountains, monuments, asking for a wider visual perspective. Shops and some ludic activities mingle with them forming a random tissue with the houses and the tertiary services.
Second, the square.
Inheritor of the forum or the medieval market, also acquires forms and dimensions which depends of the scale, from the small plazuela for resting and playground for children in a neighbourhood, to the Main Square representing the idiosincrasy of the city. In them, the protecting porticos or the urban furniture are developed in all their possibilities. The pedestrian dominion grows and the resting areas, bars, coffee shops, etc., occupy a significant part of its space. The monument takes place there, as a key element.

And third, the park.
This space that has a more recent origin, as a consequence of the baroque garden or from the XIXth century city, and that occupies a preferent role in the southern Europe cities, beeing not only a lung or landscape but a real place for resting and staying to be used during the whole year and that, due to the increasing problems produced by the invasion of the cars and the massification of the housing, contitutes today the living urban space by excellence.

4. THE URBAN SPACE AS A TRANSPORT NET

Though this is not the theme of our study now, we should mention the urban space as a net for transportation, that today presents very complex problems, many times in contradiction with the habitability of the city and, occasionally, like in the city of Rome, reach levels of institutional and political problematic.

5. BIOCLIMATIC CONTROL FOR THE OUTDOOR SPACES

Finally, let us enumerate the elements and design factors that deal with the bioclimatic treatment of the urban spaces such as we described them before.
On the occassion of the redaction of the MASTER PLAN FOR THE UNIVERSAL EXHIBITION of 1992, in Seville, the Seminario de Arquitectura Bioclimática developed a research about these factors that we may resume as follows:
It was necessary to design avenues, walkways, free-standing zones and resting areas, parkings and entrance facilities which really lower and make milder the extreme temperatures and conditions of the hot sevillian summer and which, related with the andalusian tradition of cities and gardens, such as a Medina Azahar, la Alhambra de Granada, etc., incorporate at the same time the most advanced techniques and components of the passive solar technologies: micronizing water systems, evaporative humidity, "cool towers", "wet walls", cybernetic fountains, water-cooled pavements, underground ducts for fresh air producing, etc.
We have used the traditional devices of shading, trees and vegetation, cooling by running water, ponds and fountains, large cloths covering large areas, together with the most sophisticated systems of artificial fog, production of fresh air computer control of temperatures and relative humidity, hydroponic cultivation for the construction of large surfaces of pergolas covered by vegetation, etc.
First of all, we identified some typologies that were relevant for the treatment of open spaces.

Then, the first application took place on the Isla de la Cartuja (site of the EXPO) itself in an office area belonging to the State Company for EXPO 92; this was the PILOT EXPERIMENT.

They constructed the first phase of the experiment that was monitorized and evaluated during the summer of 1988 by the department of thermotechniques of the E.T.S.I.I.S. of Seville with the consulting of prof. Givoni from U.C.L.A.

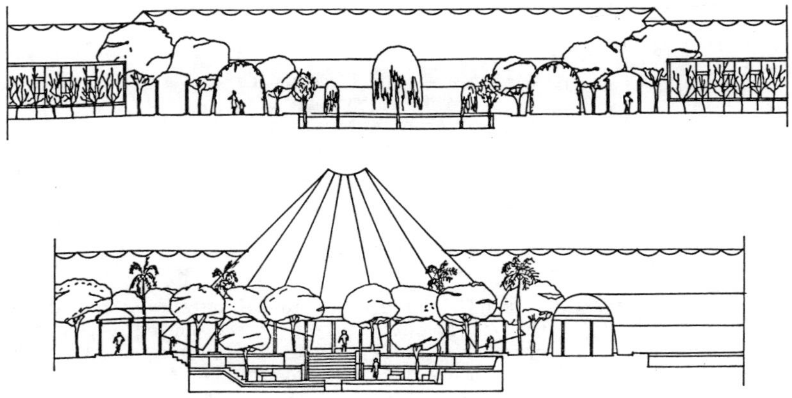

BIOCLIMATIC ROTUNDA. Sections.

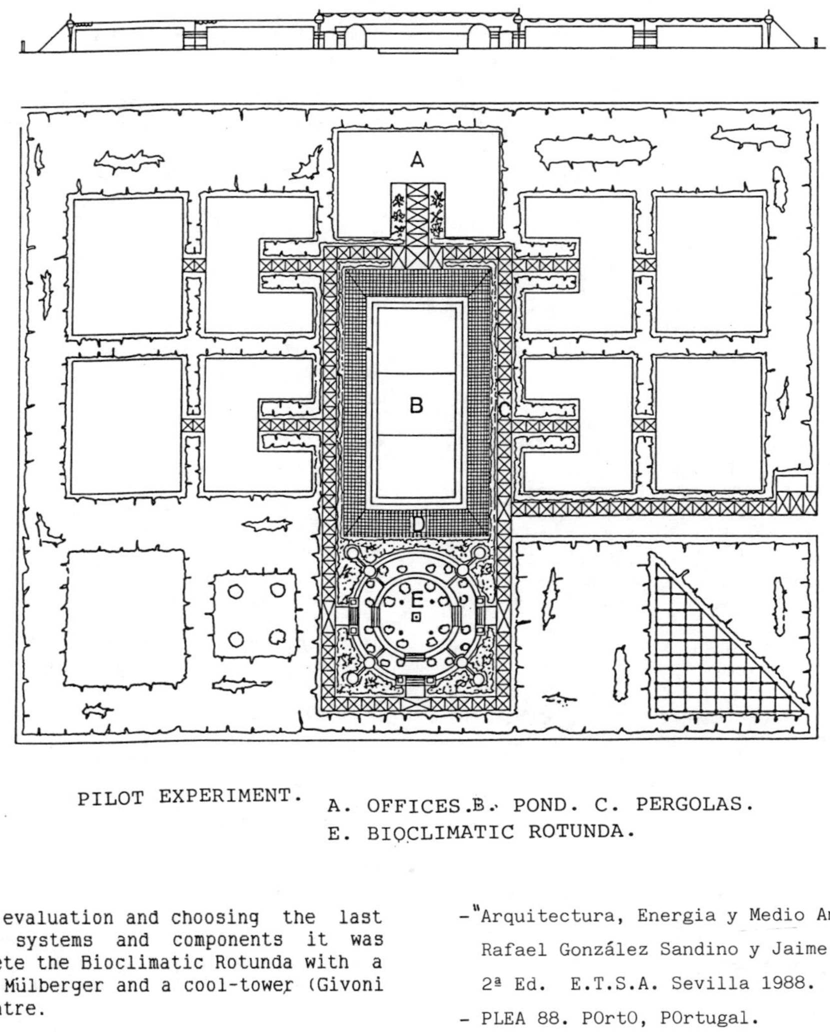

PILOT EXPERIMENT. A. OFFICES. B. POND. C. PERGOLAS. E. BIOCLIMATIC ROTUNDA.

After the first evaluation and choosing the last results of some systems and components it was decided to complete the Bioclimatic Rotunda with a covering built by Mülberger and a cool-tower (Givoni system) in the centre.

It has been monitorized and meassured during the summer of 1989 and now is on the final evaluation and report.

REFERENCES

- "Cuaderno de Trabajo nº 5." "Estudio Bioclimatico." Ed. EXPO 92. Sevilla 1987.
- "Arquitectura, Energia y Medio Ambiente." Rafael González Sandino y Jaime López de Asiain. 2ª Ed. E.T.S.A. Sevilla 1988.
- PLEA 88. POrtO, POrtugal. Ed. Pergamon Press 1988.
- PLEA 89. Nara, Japan.
- I.E.A. Task XI. Cooling Section Source Book.1989
- COOLING OF OUTDOOR SPACES. Baruch Givoni. Paper presented at the "Workshop on Interaction between Physics and Architecture in Environmental Concious Design". I.C.T.P., UNESCO. Trieste. Italy. Sept. 1989.

2ND EUROPEAN CONFERENCE ON ARCHITECTURE 4–8 December 1989 PARIS, FRANCE Sponsored by Commission of the European Communities

CLIMATIC AND CULTURAL PREFERENCES IN WINDOW DESIGN

K. A. Tabet and S. Sharples
Building Science Unit
School of Architectural Studies
University of Sheffield
Sheffield
S10 2TN
UK

ABSTRACT. A scaled model of an office was used for the experimental determination of the minimum acceptable and the optimum window size according to the visual satisfaction of the subject. The attitude towards window size and shape of two groups of observers from different climatic and cultural backgrounds has been contrasted. The dimensions of the selected openings were basically of a horizontal shape and mainly determined by the elevation of the skyline and the type of view. A range of ratios of glazing to wall area from 12 to 17% and 21 to 30% respectively for the minimum acceptable and optimum size of window were obtained.

1. INTRODUCTION

The first step in designing a window should be to establish what purpose it is to serve. Markus /1/ defined the traditional criteria for window design as: 1. Daylight, 2. Sunlight, 3. Ventilation, 4. View and privacy, 5. Contribution to the general visual character. The successful design of a window depends on the weighting of each function against the other in a trial of optimization. The nature and use of the building, the climate, the economic situation, the traditions and the social and cultural environment in which the building will exist influence these criteria. A windowless building might seem to be the best design to overcome these conflicts, but people prefer buildings with windows /2/3/4/. The main benefit from a window becomes the "view" it provides. Windows provide a sense of orientation and allow us to assess the time, the weather and avoid a feeling of isolation and claustrophobia /5/.

However, little is known about the design of windows which would satisfy the visual requirements in relation to the view. Window area, shape and location are all likely to influence people's appreciation of their view. According to Markus /1/ the information content and the dynamic properties of a view are also critical parameters. To maximize the content, when reduced fenestration is required, he suggested the use of narrow vertical windows, which would comprise the three essential layers of the striated view: sky, the cityscape and the ground. Ne'eman /6/ found that the view content was an important factor in determining the minimum window width, which itself was directly proportional to a subject's distance from the window. Wider windows were preferred for close views than for distant views. In the case of multiple apertures visible to the observer within a 60° field of view the width was summated to the width of a single aperture. Keighley /7/, using a scale model of an open plan office, investigated shape preferences for a window area limited to 20% of the window wall. Preferences were found for wide horizontal apertures. A window occupying 10% or less of the external wall was found extremely unsatisfactory, whereas satisfaction increased with 20% window to wall area and was greatest for windows 30% or larger. Ludlow /8/, using a similar technique, found that the preferred size of window is between 50 and 80% of the whole of the window wall.

These results indicate that for view purposes there are windows which are too small to be acceptable, but some part of the experimental settings have to be subject to review as in most of these studies the subject was required to alter a preset combination of variables (limited window area, variation of width only). It is not known what would be the case when none of these limitations exist. The investigations have also left open some important issues, such as the attitude towards window and the view out dependent of people's background and the experience of their usual environment.

2. THE PURPOSE OF THE STUDY AND METHODS

2.1 Scope

This experimental investigation seeks to find out the minimum acceptable window size as well as the preferred (optimum) size and shape of a window, according to the view out, in a medium size office for two different groups of subjects. The two groups comprise of people from temperate climates corresponding to the north of England and people from hot climatic zones (Algeria and Iraq). The purpose of considering these two groups in the experiment was to find out if there are any differences in the approach and attitude towards windows between people from different climatic and cultural backgrounds. The experiment allowed also a comparison to be made of the importance of the view in, (related to a sense of privacy), and the view out between the two groups.

2.2 Experimental Method

For this purpose a one fifth scale model was built in an existing artificial sky; it was furnished to give a realistic impression of a medium size office 8.5m deep, 4.7m wide and 3m from floor to ceiling. The side walls were painted in a matt light grey (reflectance 42%). Permanent artificial lighting was provided by fluorescent lamps. They were covered by a clear polystyrene sheet simulating recessed luminaires. The illuminance at desk level was set at 600 lux.

The window assembly of the model comprised of two sets of pairs of rectangular sliding panels. The horizontal panels moved symmetrically and therefore did not allow the window to be offset from the centre, whereas the upper and lower panels movements were independent. The panels were connected to a computer, via a stepper motor, for a direct data recording. By means of a joy-stick the observer could adjust the aperture to any size and configuration. There were no limiting sizes or steps of different sizes within the limits of the room bounderies.

Three views on slides were back projected beyond the window on a translucent screen. The views represented scenes from ground floor and progressively higher floors. The content of the views was a mixture of natural and man made environment.

Interior of Model

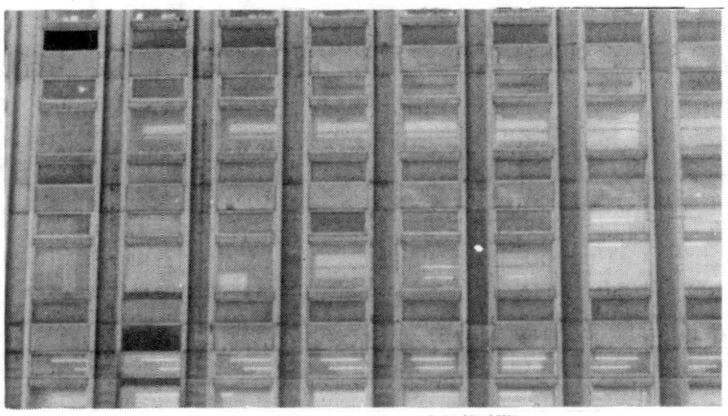

View 1

View 2

View 3

Figure 1: The Model and the three views

Before viewing the model the observer had to adjust his eye level to the corresponding one of a seated person in the scaled model. The observer having his head inside the model viewed it from the far end of the room.

2.3 Procedure

Fifty observers in each group have taken part in the experiment. They were drawn from students and members of staff at the University. To avoid prejudicing the observers two forms were prepared. Before viewing the model the subject read the first form which comprised of a general explanation about the window function in a building and stressed the purpose of the experiment. After adjusting his eye level he was invited to view the model and imagine that he was working in a similar environment. The second form, which included instructions on how to execute the settings, was then read to him by the operator. According to the purpose of the experiment the observer was asked to adjust the window aperture to either the minimum acceptable or the preferred window openings. He was encouraged to comment on any part and had no time constraints to make his assessments. The change of view was operated unseen by the observers when the panels were brought to their initial closed positions. When the subject had achieved his settings the displacements of the panels were recorded on the connected computer. Finally, a questionnaire probed what most influenced his/her settings. The testing of the minimum acceptable and the optimum window sizes were carried out in separate sessions.

3. RESULTS

For the convenience of comparison the overall mean values, obtained from the experiment for respectively the minimum acceptable and the optimum window size, are presented in Tables I and II.

Table I : Mean Dimensions of Minimum Acceptable Apertures for each View. Values Relate to Full Scale (m).
Gr I : Subjects from Hot Climate, Gr II: Subjects from Temperate Climate

Views	N.Observat	Sill Height	Head Height	Window Height	Window Width	Glazing Ratio	Height/ Width
VIEW 1	Gr I : 34	0.89	1.86	0.98	1.58	12 %	0.67
	Gr II : 38	0.84	1.97	1.13	1.70	15 %	0.70
VIEW 2	Gr I : 34	0.81*	1.88	1.07*	1.66	14 %	0.70
	Gr II : 40	0.62*	1.93	1.31*	1.76	17 %	0.84
VIEW 3	Gr I : 33	0.56	1.57	1.01	1.72	13 %	0.65
	Gr II : 38	0.58	1.66	1.08	1.76	15 %	0.69

Table II : Mean Dimensions of Optimum Apertures for each View. Values Relate to Full Scale (m).
Gr I : Subjects from Hot Climate Gr II: Subjects from Temperate Climate

Views	N.Observat	Sill Height	Head Height	Window Height	Window Width	Glazing Ratio	Height/ Width
VIEW 1	Gr I : 14	0.72*	1.98	1.25	2.40	23 %	0.56
	Gr II : 17	0.60*	1.98	1.38	2.10	21 %	0.72
VIEW 2	Gr I : 14	0.77*	2.06	1.29*	2.45	24 %	0.56
	Gr II : 17	0.66*	2.17	1.51*	2.26	26 %	0.70
VIEW 3	Gr I : 14	0.54*	1.86	1.32	2.75	27 %	0.49
	Gr II : 17	0.39*	1.82	1.44	2.72	29 %	0.57

*** Denotes significance at the 5% level (two sided test) between the two groups.**

The mean values obtained indicate in the first place the general horizontal shape of the chosen apertures. The preferred or optimum opening are wider (Height/Width: 0.4 to 0.7) than the minimum window size (H/W: 0.65 to 0.84). Fig. 3 shows the predominant choice of horizontal versus non horizontal shapes. There are more vertical shape choices for both groups for the minimum acceptable window sizes. The width in this case seems to have been compromised in an attempt to include, whenever visible, the three basic layers of a view: ground, silouette and sky. View 2 is typical of this case with a higher height to width ratio. These findings are broadly in agreement with the experimental work of Ne'eman and Hopkinson /6/, Keighley /7/ and Ludlow /8/ but conflict with Markus's /1/ suggestion.

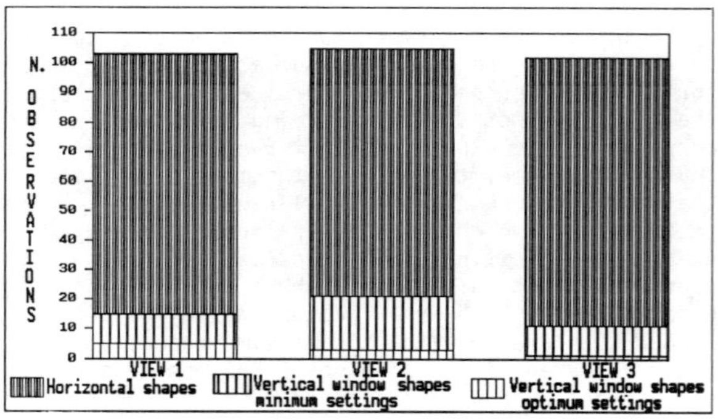

Figure 2: Proportion of Horizontal and Non-Horizontal Window Shapes

4. DISCUSSION

4.1 Window width and distance of viewed objects

The three views (1, 2, 3) respectively presented close to distant objects. Fig. 3 shows a significant difference on the window width for the preferred settings between close and distant views. The more complex and richer in information content the view was then the wider the window width as it stimulated more interest. It is not possible to assert with certainty, due to the limited number of close views used, that for close objects but more diversified than view 1, which was featureless and all repetitive, near objects would not attract more attention and require as wide opening to be wholly seen, as it was found by Ne'eman /6/. Nevertheless, the width is affected by the meaningfulness and the content of view and there is a clear tendency for preferring wider windows for the distant views.

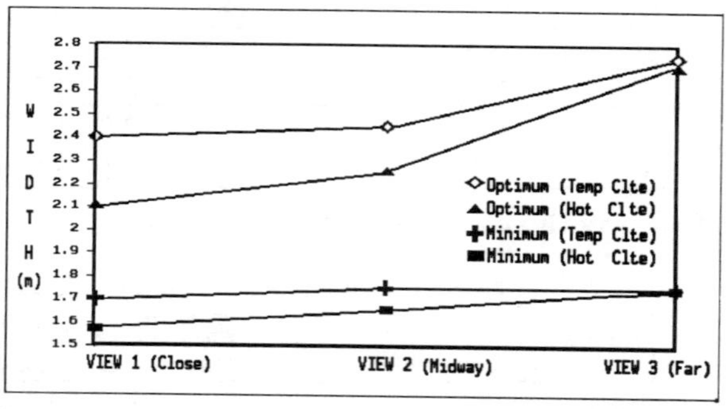

Figure 3: Variation of Window Width with Distance of Viewed Objects

4.2 Variation of window head and sill height

The skyline seems to have a decisive effect on the settings of window head and sill height. Fig. 4,5 reflect their variations with the mean skyline height. For the views having a visible skyline an increase of the skyline height caused the window head height to be set high enough to include it, but the settings followed a different pattern for the fully obstructed view (skyline above the 3m of the ceiling as the view 1). An agreement was found between observers to lower the window head height and set it at about 1.95m.

Some conflicting requirements were found to influence the settings of the window sill height. For high floors the desire was for a very low sill so as to include as much of the cityscape, the part of the view offering most information compared to the sky, but it conflicted with a sense of insecurity and a feeling of vertigo. It resulted in sills ranging from 0.40 to 0.60m. For ground level, although the greenery and the continuous movements of people and cars were attractive features, lowering the sill up to a point was considered by the observers as a nuisance. The view in by passers by was regarded as a form of visual intrusion into their privacy as well as being a disturbance from work. Therefore higher sills were selected ranging from 0.65 to 0.80m.

For the fully obstructed view, in the absence of strong cues (sky, ground), there was less consistancy in the result. The observers found it harder to decide upon the size and shape of the window. Some of them tend to dimension the window in relation to some features of their internal space i.e. desk level. Others, set it according to the view outside. It was set high enough to avoid being overlooked by people in the opposite building. It resulted in sills ranging from 0.60 to 0.90m.

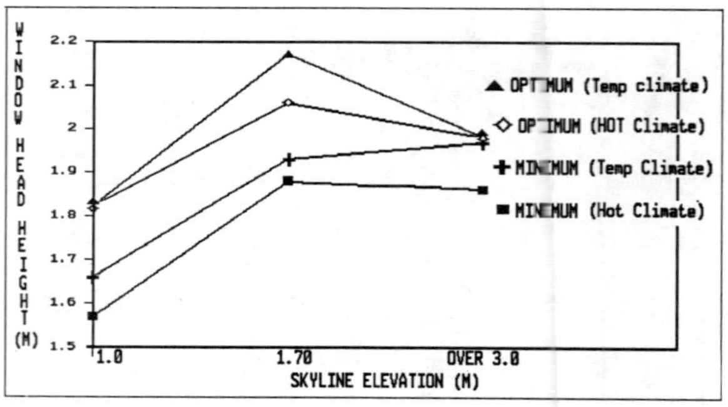

Figure 4: Variation of Window Head Height with Skyline Height

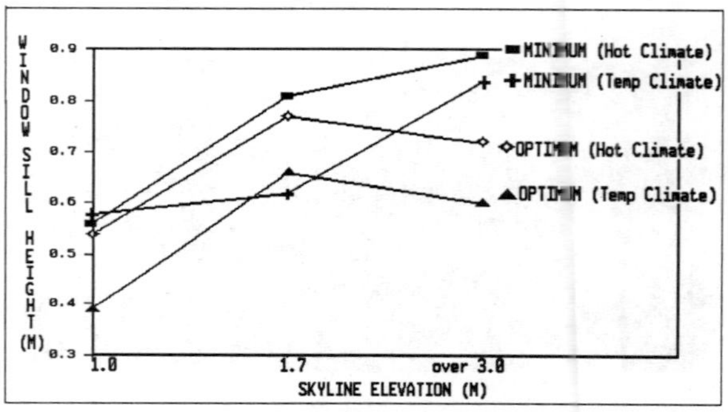

Figure 5: Variation of Window Sill Height with Skyline Height

5. COMPARISON

5.1 Minimum acceptable and the preferred settings

The student test comparison between minimum and optimum mean settings of the window dimensions were, except for window head height, all statistically significant at the 5% level. Two points can be drawn from these findings. The first point confirms the validity of the experiment and the technique used as the observers did adapt to the simulated environment. They agreed upon a minimum acceptable window size which was markedly different from the preferred one.

The second point is a significant finding. A ratio of window head height (H) to eye level (hl) ranged consistently between 1.5 and 2 for both groups and settings (minimum and optimum). Fig. 6 views with low skyline, corresponding to high floors, called for lower window head height and had ratios of 1.49 to 1.69, whereas, views of a higher skyline, as seen from lower floors, resulted in ratios equal to 1.7 to 2. The complexity of the skyline is of a particular importance and had to be included in the view, but the sky itself does not yield much information and therefore did not require an increase in height. This explains the minor variation in the setting of the window head height between the minimum and optimum apertures.

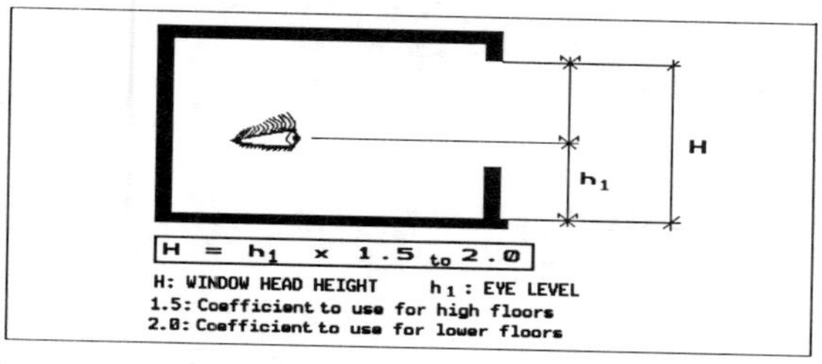

Figure 6: Settings of Window Head Height

5.2 Comparison of settings between the two groups

The consistent difference of settings between the two groups was found for the sill height of the windows. Fig. 7 shows the differences on window sills settings. People from a hot climate tend to set the sill at a higher level. It resulted on smaller window height and therefore wider apertures. Expressive explanations for this difference came from the comments and discussions with the observers. For the minimum acceptable window size, only view 2 where the foreground was visible, had a significantly higher sill. This view accounted for most of the spontaneous comments on visual privacy. The possibility of being overlooked was most disliked. It resulted in a conscious choice of higher sills. Surprisingly however the preferred sills for all the views were unconsciously (as there was no restriction for a minimum) set higher than the selected sills from people of the temperate climate. A conscious desire to see more of the sky was often cited. Here, in the absence of a diligent analysis of the reasons, one can only speculate about them. The plausible one is the fact that they are accoustomed to higher sills as these are often designed in a hot climate to avoid glare reflection from the ground. One can go further and associate their choices with the traditional inward looking architecture, open to the sky vault. It is impossible to deny the effect of some features of the view (skyline, ground) on people's settings in this experiment but taking further the former reasoning would lead to the suggestion that people would generate forms, sizes and shapes (for the windows at least) from their previous experience and their customary environment.

In the absence of a further investigation of the reasons of these choices it will be merely wishful thinking to assert the suggested reasons of these differences, but it is also presumptuous and midleading to consider one type of design as the only valid one for different climatic and cultural regions and people. The difference in settings cannot be ignored but a more thorough investigation is necessarily required to confirm the suggestions before they can be of any usefulness.

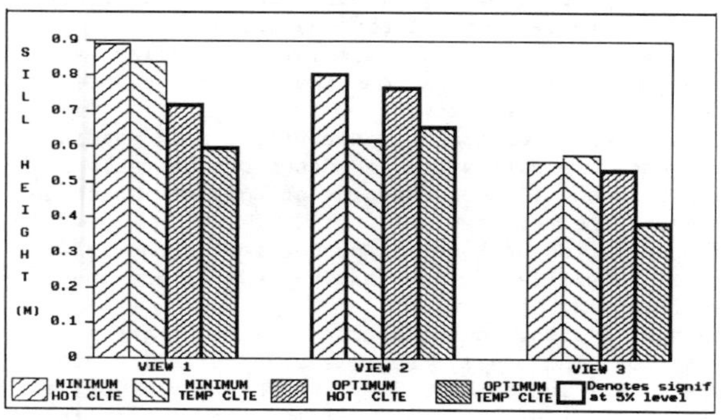

Figure 7: Comparison of Sill Height Setting between the Two Groups.

6. CONCLUSIONS

It must be emphasized that the result obtained in this investigation, although in agreement with most previous work, is just another step forward into the subjective appraisal of windows where more research is required. This experiment does not provide a single solution to window design as a view provided but it indicates the preferred arrangements. It will be the designer's task to reconcile these requirements with the other parameters of window design. The summary of the findings are as follows:

1. The content of the view has a determinant effect on the window size and shape and consequently on people satisfaction of their view.
2. The visual requirements were found to be best met by horizontal apertures, the width of which would occupy a minimum of 35 to 40% of the wall width and an optimum of 45 to 60%. Wider apertures were preferred for the distant views.
3. Window head and sill height are most important for window design purposes. They tend to be lowered the higher it goes in a building, resulting in low horizontal apertures (which would resolve the problem of glare from the sky).
4. For view purposes, the window head height can confidently be set at 1.5 to 2 times the eye level height respectively for high and lower floors.
5. Window ratio of glazing to wall area ranged between 12 and 17% for the minimum acceptable window size and 21 to 30% for the optimum window size. The smaller ratios were obtained for the close view with repetitive features, whereas, views with complex and richer information content called for larger windows.
6. The only difference accounted for, between the two groups, was the setting of a higher window sill from people originally from hot climatic zones. It was associated with the preservation of their visual privacy and a desire of seeing more of the sky.

REFERENCES

/1/ Markus, T.A. (1967),'The Functions of Windows: A Reappraisal', Building Sci., 2, 97-121.
/2/ Ruys, T., (1970), 'Windowless Offices', Master of Arch, University of Washington.
/3/ Collins, B., (1975) 'Windows and People: A Literature Survey', National Bur Stand, Washington D.C. Gov Printing Office.
/4/ Markus, T.A. (1967), 'The Significance of Sunshine for Office Workers', Sunlight in Buildings, Proc. CIE Conference, Rotterdam, 59-93.
/5/ Heerwagen, J.H. (1986), 'Windowscapes: The Role of Nature in the View from the Window', Int. Daylighting Conference, Proc. II, Long Beach California.
/6/ Ne'eman, E., Hopkinson, R.G. (1970), 'Critical Minimum Acceptable Window Size, a Study of Window Design and a Provision of a View', Lighting Research and Technology, 2, 17-27.
/7/ Keighley, E.C. (1973), 'Visual Requirements and Reduced Fenestration in Offices - A Study of Window Shape', J. Building Sci., 8, 311-320.
/8/ Ludlow, A.M. (1976), 'The Functions of Windows in Buildings, Ltg. Res. and Technol., 8, 2, 57-68.

THE GLAZING: A CHALLENGE FOR THERMAL AND VISUAL COMFORT

C.Aghemo, S.Biava, M.Filippi
Dipartimento di Energetica - Politecnico di Torino
c.so Duca degli Abruzzi 24 - 10129 Torino Italia

Abstract. The glazed components of the building envelope are "weak" components as regards the thermal and visual comfort and are "dynamic" components as regards the control of the conduction heat gain, the solar heat gain and the daylighting. The architects must have a knowledge of the available glazing materials and indoor/outdoor shading devices, must understand the phenomena that characterize their thermal and luminous performances and must be well informed about the progress in the scientific and industrial research. A classification of the glazed components is proposed with reference to the used materials and to the components complexity:

- glazing materials (glass or plastic)
- shading devices (internal or external)
- glazing-shading systems
- multi functions integrated systems

and their thermal and luminous behaviour is described by means of some specific parameters such as heat transmission coefficient, shading coefficient, luminous transmission factor, colour temperature

1. THE MATTER IN THE CHALLENGE

The environment for man should be wealthy, appropriate to its purposes and pleasant to live in.
In it human activities, working or not, must be performed without any physical or mental trouble, in favor of the health, of the comfort and of the user's productivity.
The outdoor climate variability makes man create a protected indoor environment in which he may have ideal environmental conditions.
Therefore the function of the building envelope is to be the control surface of the energy and mass fluxes exchanged between outdoor and indoor environments.
In order to insure a proper control surface function the building envelope must be able to change its own performances in time.
The glazed components of the building envelope, which are "weak" components as regards the thermal and visual comfort, present versatile technological solutions. They permit the modulation of the energy and mass fluxes in time, owing to the change of outdoor environmental conditions and of the ones wished indoor: therefore they are "dynamic" components.
The main problem of the glazed components design is both to reduce their "weakness" and to increase their "dynamicity".
In order to obtain that, the architect must understand the glazed components physical phenomena and to be able to define their behaviour through the knowledge of the thermal and luminous performances data of the system and of its parts (glazing materials and shading devices).

2. THE LUMINOUS PHENOMENA

It is known that the incident luminous flux on the glazed components is partly reflected, partly absorbed and partly transmitted (figure 1).
The transmitted ratio is important as regards the indoor visual comfort, since determine the glazed components luminance value, while the reflected ratio concerns expecially the building external look.
From the point of photometric view, the ligth reflection and transmission may be of various types: specular, diffuse, spread wide or narrow,

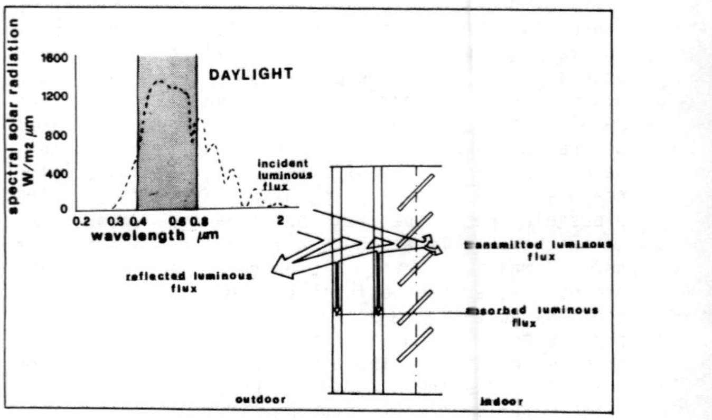

Figure 1: Luminous phenomena

diffuse and specular, complex (prismatic), according to a suggested classification /1/.
The system behaviour depends on the photometric properties of the glazing materials and the shading devices forming it and, when these properties are known, only a physical-mathematical simulation model or an experimentation permit to express its global behaviour.
Also a "diligent" architect meets big difficulties because the glazing materials and shading devices producers do not usually give the photometric properties of their products and, if asked, sometimes cannot give useful data.
However the available simulation models and experimentation data on the global systems behaviour are very few.
A hypotesis of glazing materials and shading devices classification is described in table I and in table II some examples of glazing-shading systems are reported.
The specified photometric properties of the glazing materials and the shading devices are shown in figure 2.
In figure 3 the luminous transmission factor ranges for the mostly used glass and plastic materials are indicated. They are the results of a wide research made on the italian market and the international literature /3/,/4/,/5/,/6/,/7/,/8/, /9/,/10/,/11/,/12/,/13/,/14/,/15/,/16/,/17/,/18/.

Table I Glazing materials and shading devices classification

GLAZING MATERIALS	
GLASS	**PLASTIC**
with fixed performances i.e. - clear glass - heat-absorbing glass - solar reflecting glass - low-emittance coating glass	with fixed performances i.e. - acrylic - polycarbonate
with variable performances i.e. - electrochromic glass - thermocromic glass - photocromic glass	with variable performances ?
SHADING DEVICES	
INTERNAL SHADING DEVICES	**EXTERNAL SHADING DEVICES**
with fixed performances i.e. - draper - roller shade - curtain	with fixed performances i.e. - horizontal projection and eaves - fixed vertical louvre - fixed vertical screen - fixed louvre or canopy - awning and sun blind - vertical roller blind - fixed horizontal louvre - fixed vertical woven mesh
with variable performances i.e. - venetian blind - vertical louvre blind	with variable performances i.e. - retractable louvre - pivoted horizontal or vertical louvre system

Table II Some examples of glazing - shading systems

Double or triple glass/plastic	double	triple
Double or triple glass/plastic with internal shading devices	fixed performances	variable performances
Double or triple glass/plastic with external devices	fixed performances	variable performances
Double or triple glass/plastic with shading in cavity	fixed performances	variable performances
Double glazed envelope		
Double glazed envelope with shading in cavity	fixed performances	variable performances

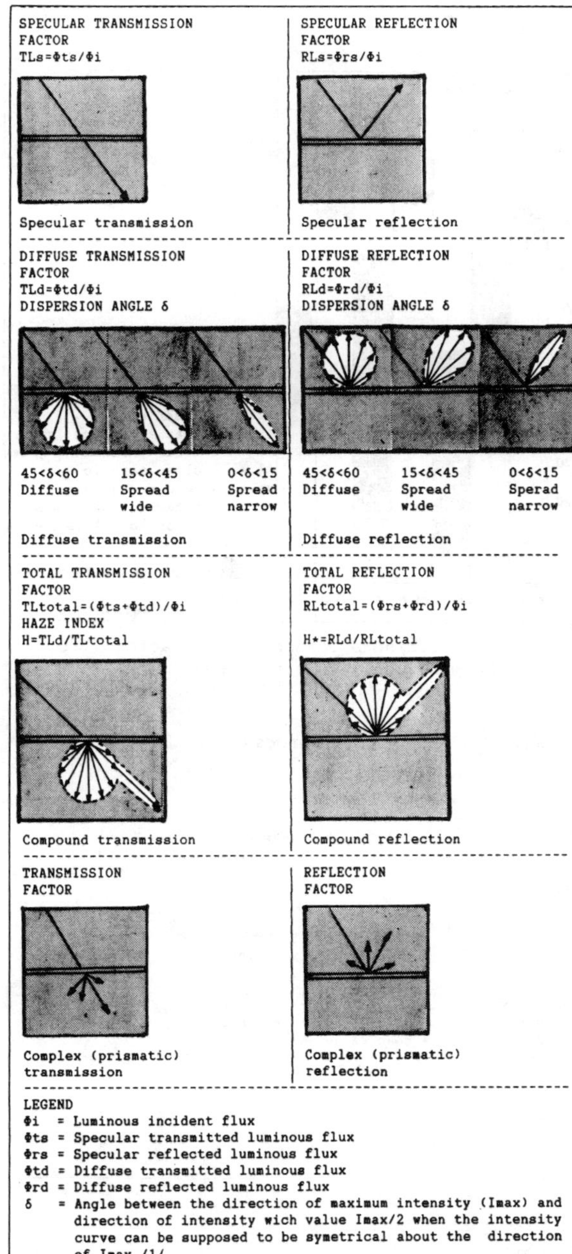

Figure 2: Photometric properties of glazing materials and shading devices

Table III Correlated colour temperature for some glazing materials (K)

GLAZING MATERIALS	COLOUR OF TINT	CORRELATED COLOUR TEMPERATURE (1)
Clear glass	clear	6535
Heat-absorbing glass	gray bronze green	6535 5586 7642
Solar reflecting glass	bronze	4975
Acrylic	opalescent white	5115 (2)
Polycarbonate	opalescent white	5000 (2)

(1) Correlated colour temperature of the light transmitted when exposed to illuminant D65 (daylight 6500 K) /20/
(2) Correlated colour temperature of the light transmitted when exposed to 5250 K source /1/

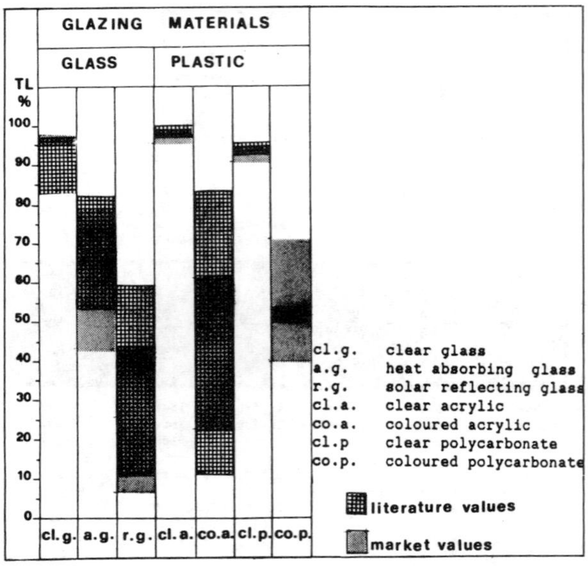

Figure 3: Luminous transmission factor (TL total) - Values ranges

Table IV Identified procedures for TL (total) evaluation

RADIATION	• Solar Direct Radiation - Moon Model /22/ - Thekaekara Model /21/ - CIE Model /14/ • Illuminant D65 Direct Radiation /14/
RADIATION INCIDENCE ANGLE	• Normal Radiation Incidence /7/, /9/, /14/, /18/ • 30° Radiation Incidence /17/ • Mean Value /4/

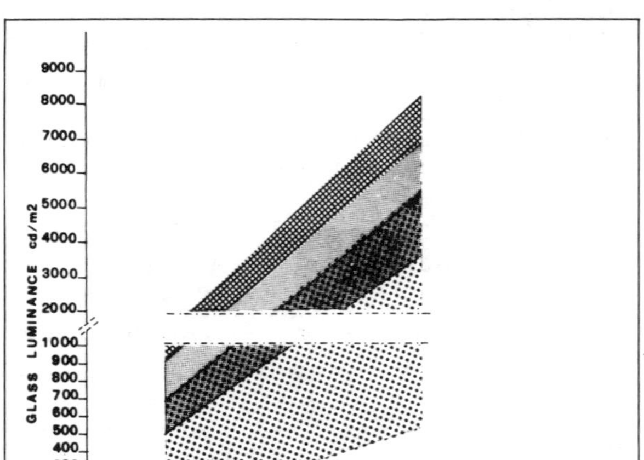

Figure 4: Glass luminance areas versus sky luminance values (with reference to the TL total values)

In figure 4 the glass luminance areas versus the sky luminance values for all the identified glass materials are shown.
In table III are reported the colour temperature data referring to a conventional incident solar radiation /14/.
We must remember that other parameters, as the CIE Colour Rendering Factor /28/ and the Colour Gamut Area /8/ may be used to identify the transmitted light colour.
We must point out that, besides the difficulty of finding data, there is a considerable difficulty in understanding the data themselves because the properties evaluation procedures are different from an Author to other and also from a producer to another even if of the same country. The wide values range is often not only due to the different production, but also to the various data evaluation procedures (see table IV).

3. THE THERMAL PHENOMENA

The thermal phenomena comprehends, as it is known, two different terms: the former is due to the temperature difference between indoor and outdoor environments without solar radiation (conduction heat gain), the latter due to the solar radiation without temperature difference (solar heat gain).
As regards the solar heat gain, similarly to the luminous phenomena, the heat flow is partly reflected, partly absorbed and partly transmitted (figure 5).
The temperature of the inside surface interests the thermal comfort: this temperature depends both on the thermal conductance of the glazed component and on the solar radiation absorbed by the component.
Besides for the evaluation of the total solar radiation transmitted to the indoor environment by the glazed components (thermal behaviour) all the heat flow ratios (reflected, transmitted and absorbed) are important. The quantitative analysis of the thermal behaviour is referred to the value of the overall coefficient of heat transmission (U value) and to a characteristic solar heat gain coefficient of the glazing-shading systems, named shading coefficient (SC).
Particularly in order to characterize the glazed components it is necessary to know separately the various contributions to the total heat flow : they are the ratio directly transmitted as radiation (visible and near infrared) and the ratios exchanged as convection and as radiation (far infrared) /6/. These last data are rarely reported in the literature and in the technical documentation, whilest informations about the SC and U values are usually reported.
However many difficulties of the interpretation and comparison among the various sources are still present because the data evaluation procedures, like for the luminous properties, are different.
The SC and U values ranges referring to the above considered glass and plastic materials are indicated in figures 6 and 7 and the data evaluation procedures are described in table V.
For the glazing-shading systems there are exhaustive studies about the global thermal behaviour /4/,/6/,/7/ and there are sufficient referring data for design, but it is necessary to have more suitable simulation models tested by the experimentation.
With reference to some mean values of the various contributions to the total heat flow it is possible to evaluate the inside surface temperature of some glazing-shading systems; the results are shown in figure 8.

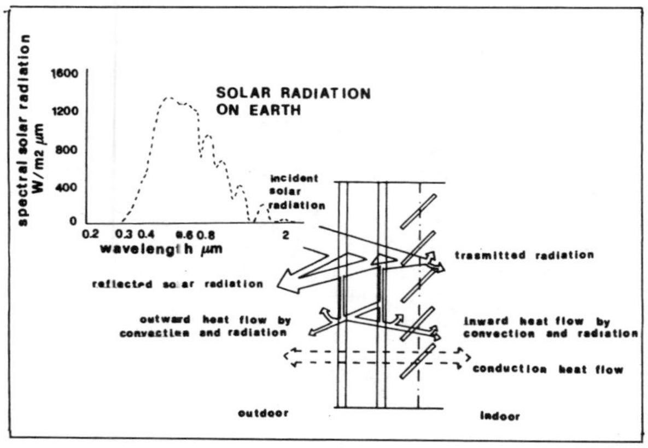

Figure 5: Thermal phenomena

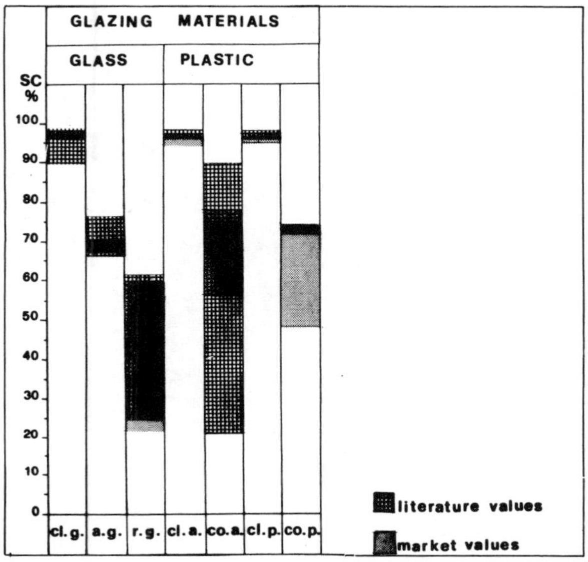

Figure 6: Shading coefficient (SC) - Values ranges

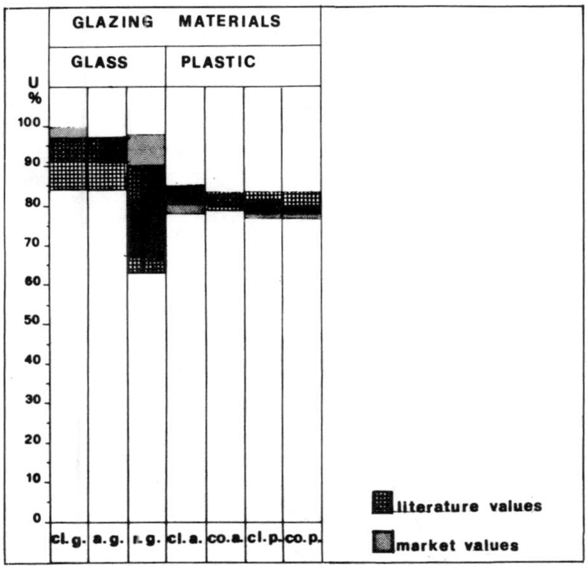

Figure 7: Overall heat transmission coefficient (U) - Values ranges

Table V Identified procedures for SC and U evaluation

SC		
RADIATION	* Solar Direct Radiation	- Moon Model /21/ - Thekaekara Model /22/ - CIE Model /14/
RADIATION INCIDENCE ANGLE	* Normal	
U		
PROJECT DATA	* Summer Data /7/, /18/ * Winter Data /7/, /8/, /10/, /17/ * Yearly Mean Data /4/	

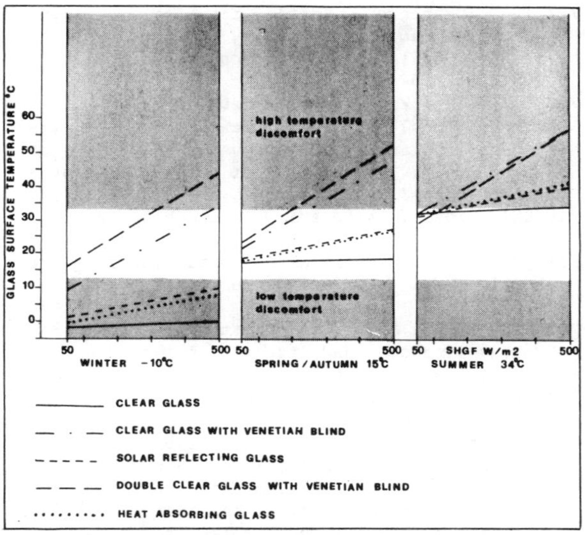

Figure 8: Inside surface temperature versus solar heat gain factor and outdoor temperature (with reference to the mean values)

4. A LOOK TO THE FUTURE

The glazing-shading systems described in table II are certainly a forward step to obtain a lower "weakness" and a greater "dynamicity".
The double/triple glass with indoor/outdoor or in cavity shading devices permit the luminous flux control and the heat flowes (as conduction, as convection and as radiation) control.
However in the last years new and more complex glazing-shading systems have been designed.
These systems integrate functions proper to the building envelope (passive functions) and functions proper to the climatization and lighting installations (active functions); they are characterized by step by step control operated by building automation systems.
Therefore these complex systems may be introduced in the proposed classification as " multi functions integrated systems ".

5. SELECTED REFERENCES

/1/ Fontoynont M., Paule B.
Progress report on materials, European Reference Book on Daylighting, Draft, Meeting Venice, March 1989
/2/ Illuminating Engineering Society
Lighting Handbook,1984 Reference Volume, IES, New York, 1984
/3/ Consiglio Nazionale delle Ricerche
Repertorio delle caratteristiche termofisiche dei componenti edili opachi e trasparenti - Appendice 2 alla Guida al controllo energetico della progettazione, CNR PFE, Roma, 1982

/4/ Brunello P.F.
Aspetti prestazionali delle superfici vetrate nell'involucro edilizio, CNR PFE Seminario Informativo 6, Roma, 1988

/5/ Brunello P.F., Zecchin R.
Analysis of solar properties of windows, Proceeding Windows in building design and maintenance, Goteborg, 1984

/6/ International Energy Agency
Thermal and solar properties of windows - Expert guide, IEA, Delft, December 1987

/7/ American Society of Heating Refrigeration and Air Conditioning Engineer
1989 ASHRAE Handbook - Fundamentals, Chapter 27 Fenestration, ASHRAE, Atlanta, 1989

/8/ The Chartered Institution of Building Services Engineers
Window Design - Application Manual, CIBSE, London, 1987

/9/ Furler R., Williams P., Kneubuhl F.K.
Experimental and theoretical studies on the energy balance of windows, NEFF Report Project No. 177.1, Institute of Quantum Electronics, Zurich, 1988

/10/UNI 7144 Vetri piani - Isolamento termico, 1977

/11/UNI 7885 Prove sul vetro - Determinazione dei fattori di trasmissione dell'energia solare, 1978

/12/DIN 67507 Light transmittance, radiation transmittance and total solar energy transmission coefficient of glazing, 1980

/13/Centre Scientifique et Technique du Batiment Regles Th-K77 - Regles de calcul des caracteristiques thermiques utiles des parois de construction, cahier 2096, 1986

/14/ISO DIS 9050 Glass in building - Determination of light transmittance, direct solar transmittance, total solar energy transmittance and ultraviolet transmittance and related glazing factor, 1988

/15/DTU Calcul du facteur solaires des parois vitrees, cahier 2255, 1988

/16/Aydinli S., Krochmann E., Krochmann J., Lindemuth F.
Characteristics and measurament of luminosus, radiant and heat transfer properties of window materials, CIE 21st Session, pagg.126-129, Venice, 1987

/17/Saint Gobain, Fabbrica Pisana
Manuale tecnico del vetro, 1989

/18/Societa' Italiana Vetro
Documentazione Tecnica, 1989

/19/UNI 8941 Superfici colorate - Colorimetria principi, 1987

/20/Horie C.V.
Solar control films for reducing light levels in building with daylight, Congress Conservation within historic building, pagg. 49-54, Vienna, 1980

/21/Moon P.
Proposed standard solar-radiation curves for engineering use, Journal of the Franklin Institute, pagg. 583-617, November 1940

/22/Thekaekara M.P.
The solar constant and spectral distribution of solar radiant flux, Solar Energy, vol. 9, No. 1, pagg. 7-20, 1965

/23/Commission Internationale de l'Eclairage
Daylighting - International recomandations for the calculation of natural daylighting, CIE Pubblication No. 16, Paris, 1970

/24/International Energy Agency
The state of the art in existing windows and new windows design - A survey of eight countries, IEA, Delft, September 1986

/25/Brunello P.F.
Thermal and optical performance of windows with venetian blinds, Proceeding Solar World Congress, Amburg, 1987

/26/Robbins C.L.
Daylighting - Design and analysis, Van Nostrand Reinhold Company, New York, 1986

/27/Wright J.L., Sullivan H.F.
VISION 2 - A computer program to evaluate the thermal performance of innovative glazing systems, University of Waterloo, Canada, March 1989

/28/Commission Internationale de l' Eclairage
Method of measuring and specifying colour rendering of light sources, CIE Pubblication No. 13.2, Paris, 1988

COMFORT STUDIES BY SPATIAL THERMAL MODELLING OF A DIRECT GAIN ROOM

N.V.Baker, G.R. Newsham
The Martin Centre for Architectural and Urban Studies,
University of Cambridge,
6, Chaucer Road,
Cambridge, UK, CB2 2EB.
Telephone: (0223) 332981.

ABSTRACT. The need to incorporate the comfort of occupants in an assessment of a building's performance is now recognised. The authors have developed a computer based thermal model, FENESTRA, to predict air and fabric temperatures spatially and temporally within a direct gain room. From these temperatures spatially varying comfort indices can be calculated and comfort contours can be mapped. The initial results of this mapping comparing a vertical and horizontal slot window for a variety of daily climate files are presented in this paper. The case of the vertical slot window is examined in more detail with comfort contours in time series.

1. INTRODUCTION

Thermal modelling of buildings, in order to predict the interior air and surface temperatures with the aim of lowering energy consumption, is well established. However, although there is a considerable body of work on comfort, comfort has rarely been used /1, 2/ as an indicator of a building's performance. Comfort should be the most important indicator; energy should not be saved if the comfort of the building's occupants is compromised, as the actions taken by the occupants to improve their comfort in such situations will be, in most cases, energy costly.

However, comfort considerations do not always signal a reduction in energy savings. In domestic situations occupants will tend to sit in the most comfortable parts of a room, it may not be necessary to heat the whole room if some parts of the room are comfortable (for example, a bay window area on a sunny winter's day). To examine such possibilities a thermal model must have spatial sensitivity.

2. THE THERMAL MODEL FENESTRA

2.1 Brief Model Description

The computer based thermal model FENESTRA has been developed by the authors to evaluate temperatures, lighting levels and comfort levels within a single, south facing, rectangular plan, direct gain room /3/. The model has 144 fixed nodes and calculates with a timestep of 15 minutes. Air temperature and radiant temperature, and hence comfort levels, can show extreme spatial variation and FENESTRA seeks to resolve these. Each interior wall surface is divided into 9 (3 x 3) nodes, the air within the room being similarly divided into 27 (3 x 3 x 3) equal volume cells. Sunpatch tracking is incorporated, as is a simplified procedure for determining air temperature gradients, for example;

$$DT_{strat} = c(Q_{upper} - Q_{lower})$$

$$T_{upper} = T_{air}^* + DT_{strat}$$

$$T_{lower} = T_{air}^* - DT_{strat}$$

T_{upper} = air temperature at upper part of room (°C)
T_{lower} = air temperature at lower part of room (°C)
T_{air}^* = mean air temperature (°C)
Q_{upper} = heat input to upper part of room (J)
Q_{lower} = heat input to lower part of room (J)
c = empirically derived constant (°C/J)
(a similar process is followed for horizontal gradients)

thus, both predicted radiant and air temperatures have spatial sensitivity. Daylighting is calculated by the daylight factor method but with an additional term that attempts to quantify the contribution from any beam sunlight. This facility is used to calculate artificial lighting energy, and hence lighting gains, and may also be incorporated in a visual comfort index, though the latter is not related to the subject of this paper. The comfort parameter employed, which is purely thermal, is the predicted percentage dissatisfied (PPD) calculated by the method below. Figure 1 is a simplified flow diagram for FENESTRA.

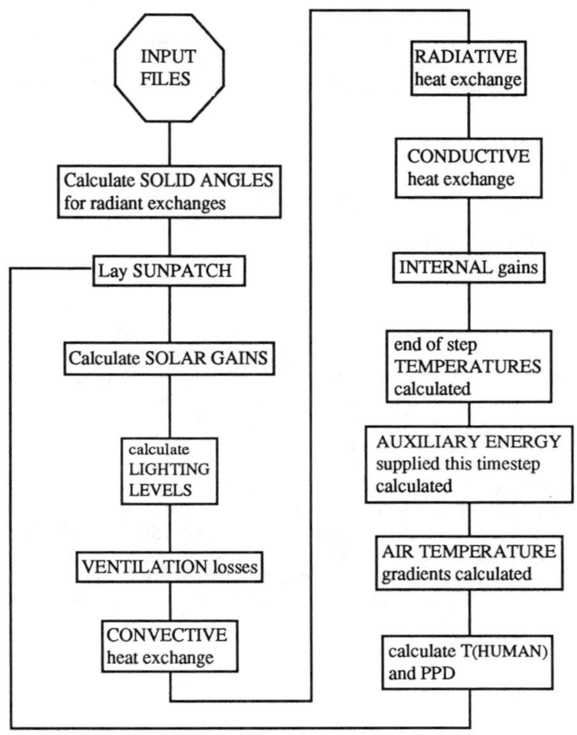

Figure 1. A simplified flow diagram for the thermal model FENESTRA

2.2 Comfort Calculation

The human body loses heat both radiatively and convectively by a number of different processes and therefore if one seeks to represent the human thermal sensation by a single temperature it must accurately reflect the importance of these processes in a weighted mean of the temperatures which govern the rate of each heat loss. Humphreys /4/ suggests

$$T_h = 0.54T_{air} + 0.46T_{mrt}$$

T_{mrt} = mean radiant temperature (°C)
T_h = temperature appropriate to heat loss from body (°C)

It is then necessary to calculate at what value of T_h the subjects will be most comfortable. Fanger's equation /5/ based on the heat balance on the body can predict a T_h, however, the equations are complicated and require inputs (humidity, air speed, subject's clothing and metabolic rate) which may be difficult to compile hourly for the purposes of thermal modelling. Humphreys reports /6/ that 78% of the variation of preferred temperature can be accounted for by the change in the monthly mean ambient air temperature. A reasonable simplification is to relate thermal comfort to air and radiant temperatures only, interior air velocity and humidity are rarely troublesome /7/ in the UK. Carroll /8/ adopts a similar approach,

$$T_{pref} = PT_{base} + k(T_h - PT_{base})$$

T_{pref} ='ideal' value of T_h for that timestep (°C)
PT_{base} = mean T_{pref} (°C)

Since T_h is adopted as the variable this form is well suited to 15 minute timestep modelling. Using Carroll's guidelines and UK preferences the equation becomes

$$T_{pref} = 20.5 + 0.1T_h$$

To relate $|T_h\text{-}T_{pref}|$ to the percentage of people dissatisfied with T_h Fanger's PPD curve in quadratic form is used. Following Carroll, an additional term can be added to the PPD for discomfort due to variations in temperature. It has been demonstrated /9/ that temperature fluctuations cause an uncomfortable sensation and Carroll suggests a further 'bother' effect due to the need to adjust clothing, thermostats, ventilation rates, sunshades etc. The weighting of this term is, at present, arbitrary; fluctuations over periods of 16 minutes (~1 timestep) are found to be particularly uncomfortable. The form of the PPD equation is therefore

$$PPD = 2(T_h - T_{pref})^2 + 5(\partial T)^2 + 5$$

$$\partial T \quad = T_h|^{\text{this timestep}} - T_h|^{\text{last timestep}} \quad (^\circ C)$$

3. INITIAL RESULTS

3.1 The Studied Room

All simulations were performed on a single, south facing, direct gain room, 6m x 4m x 2.4m. Only the south wall is external and there is no net heat exchange with adjoining rooms, however, the floor is in thermal contact with the ground. U-values are, south wall = 0.32 $W/m^2.K$, floor = 0.27 $W/m^2.K$. A panel radiator is positioned at the lower east corner of the south wall, the thermostat being set for an air temperature of 20°C for an occupancy period 0800h - 2300h. Internal gains during occupancy are 150 W plus thermal gains from lighting by incandescent filament to maintain a lighting level of 150 lux. 33% of the south façade is double glazed in two distributions; a vertical slot 2m wide and 2.4m high or a horizontal slot 6m wide and 0.8m high. The air change rate is fixed at 0.5 air changes/hour.

3.2 The Simulations

3.2.1 Air Temperature, T_h and PPD

Figure 2 shows in section the distribution, along the central north-south axis of the vertical slot window, of air temperature, T_h and PPD for day 118, a sunny spring day, at 1300h. The numbers quoted are as would be experienced at that position.
Most of the heat input to the room at this time is from the sunpatch at floor level. This results in good fluid mixing and a close to uniform air temperature. The effect of the sunpatch near the window and of sunlight directly incident on occupant on the radiant temperature is apparent when T_h is mapped. The close relation between T_h and PPD is obvious if the PPD contours and the T_h contours are compared.

3.2.2 Glazing distribution

Figures 3,4 and 5 are plan views at a height of 1.2m comparing PPD for the two glazing distributions, vertical slot and horizontal slot, at 1230h on three very different days; day 5 is a cold winter's day with little sun, day 118 is a sunny spring day and day 239 is a hot, sunny summer day.
The contours for day 5 show that both glazing distributions result in rooms that are cooler than desired with the chosen thermostat. The horizontal slot glazing distribution is more comfortable at most points in the room, however, near the window the horizontal distribution appears considerably less comfortable due to the cold glazed surface and the resulting cool downdraft.
At 1230h on day 118 the horizontal glazing distribution results in a room of comfortable temperature at all points. The vertical distribution results in a room which is comfortable at all points except those near the glazing where the proximity of a sunpatch and sunlight directly incident on the occupant leads to overheating.
During day 239 the rooms with both glazing distributions overheat. In both cases the west side of the room is slightly warmer than the east side, this is due to the morning sunshine that was incident on the side of the room. The vertical glazing distribution shows signs of serious overheat near the glazing.

3.2.3 Comfort Contours in Time Series

Figure 6 shows hourly comfort contours for the vertical slot glazing distribution for the hours 0700h - 1500h on day 118, both in section and in plan. Day 118 was chosen because the room experiences a transition from underheat to overheat.
A gradual warming from 0700h to 1000h is observed due to the switching on of the radiator and some early morning sun. The most uncomfortably cold positions are those near the cold window surface. The west of the room is generally more comfortable due to the incident morning sunlight. At 1100h a change to overheat occurs as the solar gain increases. Overheating increases until 1400h with the positions close to the glazing being particularly uncomfortable due to the proximity of a sunpatch and sunlight directly incident on the occupant combining to produce high radiant temperatures, the back of the room, which receives little direct sunlight and has a poor view of the hottest surfaces, remains comfortable throughout the general overheating period. At 1500h the room begins to cool as the solar gain decreases.

3.2.4 Overall Performance of each Glazing Distribution

Table 1
A summary of the overall performance of each of the two glazing distributions for the three days studied. 'Heat' is the total daily useful energy consumption, 'ppd' is the mean of the mean ppd for all timesteps in each occupant position in the horizontal plane 1.2m from the floor.

	day 5		day 118		day 239	
	vert.	horiz.	vert.	horiz.	vert.	horiz.
heat /kWh	5.9	5.7	0.0	0.0	0.0	0.0
ppd	22.7	23.1	5.7	5.5	9.4	8.8

Table 1 shows that on average there is little to choose between the two fenestration strategies for these conditions. The horizontal slot window performs slightly better and lacks the extremes in the 1.2m plane exhibited at the positions closest to the window; Table 2 details these extremes.

Table 2
The maximum ppd at any timestep at which the room is considered occupied at an occupant position 1.2m from the floor and nearest to the middle of the window wall. (+) indicates discomfort due to overheat and (-) discomfort due to underheat.

	P.P.D. window configuration	
day	vertical	horizontal
239	54.1 (+)	24.4 (+)
118	27.5 (-)	23.1 (-)
	27.0 (+)	9.4 (+)
5	98.3 (-)	84.0 (-)

An interesting concept is that of the 'flexible occupant', a person who is willing to move around the room as the comfort conditions change. If this concept is applied to both glazing options on day 118 for the positions in the horizontal plane 1.2m from the floor, then the vertical slot window is found to be slightly more comfortable as the positions near the window are avoided. For the hours 0700h - 1500h the vertical slot window produces an average lowest ppd of 9%, the corresponding average ppd for the horizontal slot window is 10%. The occupant would not have to move much to find the most comfortable position, this is near the west wall for most hours. An explanation for this is that during the cold morning the sun falls on the west side of the room, in the afternoon, when the room is warmest the sun falls on the east side of the room.

3.2.5 The Influence of the Thermostat

In all the above calculations the thermostat that controlled the panel radiator responded to the mean air temperature. This strategy gave good definition of comfort contours. FENESTRA allows for the use of a thermostat that responds to human comfort with a set point equal to T_{pref} at the observation point.

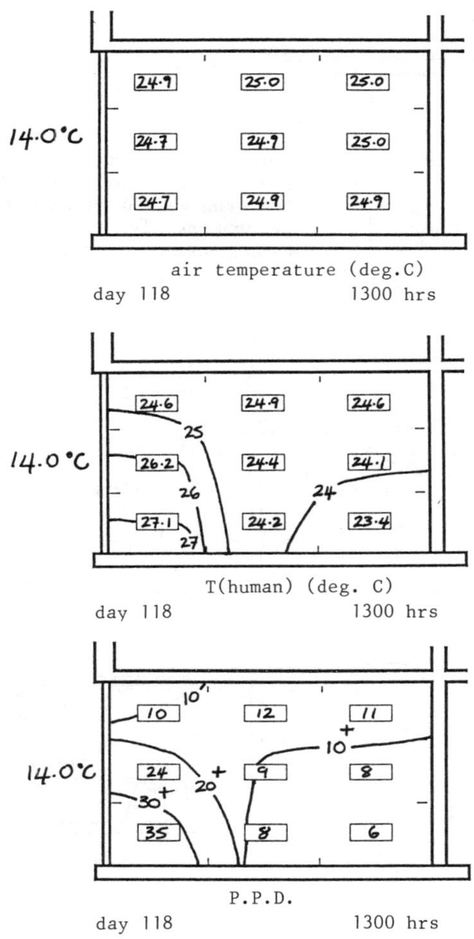

Figure 2. Air temperature, T_h and ppd ('+' indicates overheat, '-' indicates underheat) for the room with a vertical slot window at 1300h on day 118. Spatial variation is shown in section. The number outside the room is ambient air temperature.

If a thermostat of this kind is used the comfort level at all points in the room save for those near the radiator are very similar for all timesteps, however, the auxiliary energy consumption to maintain these comfort levels is very different. This is illustrated in Figure 7.
As one might expect, the positions closest to the radiator require less heat to maintain comfort whereas those nearest the window require more.

4. CONCLUSION

It is demonstrated that an occupant's position in a room can have a significant influence on that occupant's thermal comfort since air temperature and radiant temperature vary spatially; this may be beneficial in terms of energy savings in a domestic situation but in an office or school where seating positions are imposed, resentment and lowering of efficiency could result.
A conclusion from these initial simulations is that where seating positions are fixed and uniformly distributed, a horizontal window is generally better than a vertical window of the same size in providing a comfortable room environment under the above conditions, though if seating position is variable then the vertical window is best. However, nothing more than a demonstration of the technique is intended, simulations of specific cases will follow in the future.

REFERENCES

1. Heerwagen, D.R. et al 1979. "The use of a computer routine to assess human thermal comfort for passive-hybrid building designs." Third National Passive Solar Conference (San Jose), pp. 134-145.
2. Alder, K. et al 1984. "Energy balance and thermal comfort in passive solar housing." 1st E.C. Conferenece on Solar Heating (Amsterdam), pp. 386-390.
3. Newsham, G.R. and Baker, N.V. 1989. "A model for the direct gain passive solar room facilitating performance analysis of fenestration strategies." Thermal Performance of the Exterior Envelopes of Buildings Conference IV (Orlando).
4. Humphreys, M.A. 1974. "Environmental temperature and thermal comfort." BRE CP 80/74. Garston: Building Research Establishment.
5. Sherman, M. 1985. "A simplified model of thermal comfort." Energy and Buildings, 8, pp 37-50.
6. Humphreys, M.A. 1978. "Outdoor temperatures and comfort indoors." BRE CP 53/78. Garston: Building Research Establishment.
7. Hannay, J. et al. "Thermal comfort and energy consumption in winter conditions - a new experimental approach." Institute de Thermodynamique, Université de Liège, Belgium.
8. Carroll, J.A. 1980. "An index to quantify thermal comfort in homes." Fifth National Passive Solar Conference (Amherst), pp. 1210-1214.
9. McIntyre, D.A. 1980. Indoor Climate. London: Applied Science Publishers.

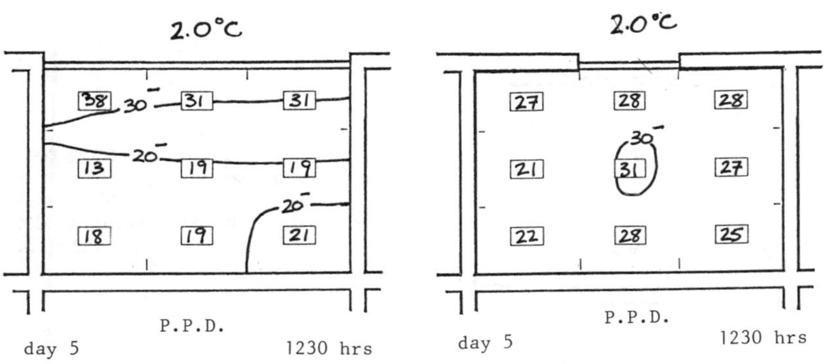

Figure 3. Comparison of ppd ('+' indicates overheat, '-' indicates underheat) in plan at 1230h on day 5 for rooms with horizontal slot and vertical slot window distributions. The number outside the room is ambient air temperature.

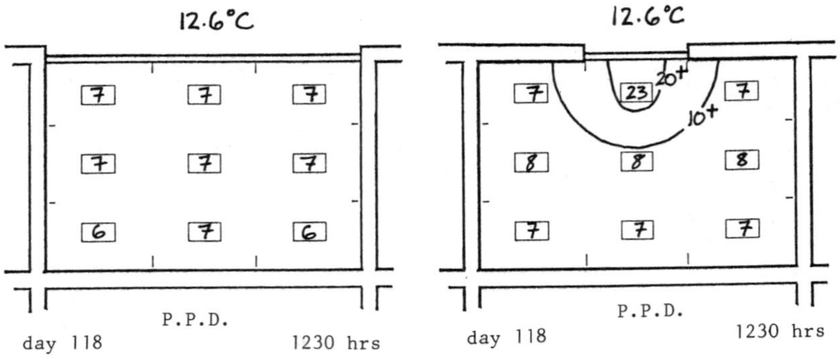

Figure 4. Comparison of ppd ('+' indicates overheat, '-' indicates underheat) in plan at 1230h on day 118 for rooms with horizontal slot and vertical slot window distributions. The number outside the room is ambient air temperature.

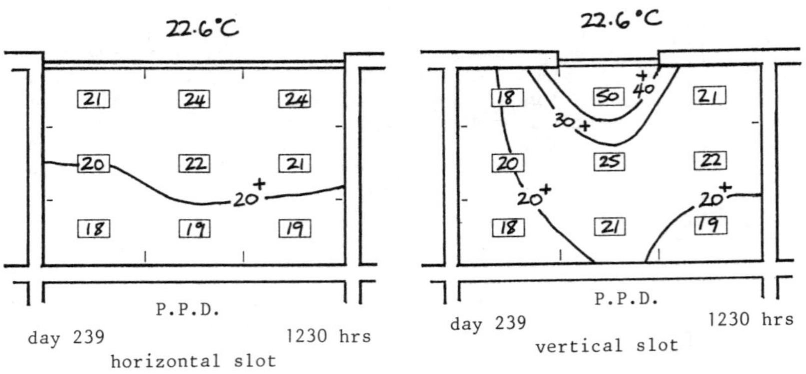

Figure 5. Comparison of ppd ('+' indicates overheat, '-' indicates underheat) in plan at 1230h on day 239 for rooms with horizontal slot and vertical slot window distributions. The number outside the room is ambient air temperature.

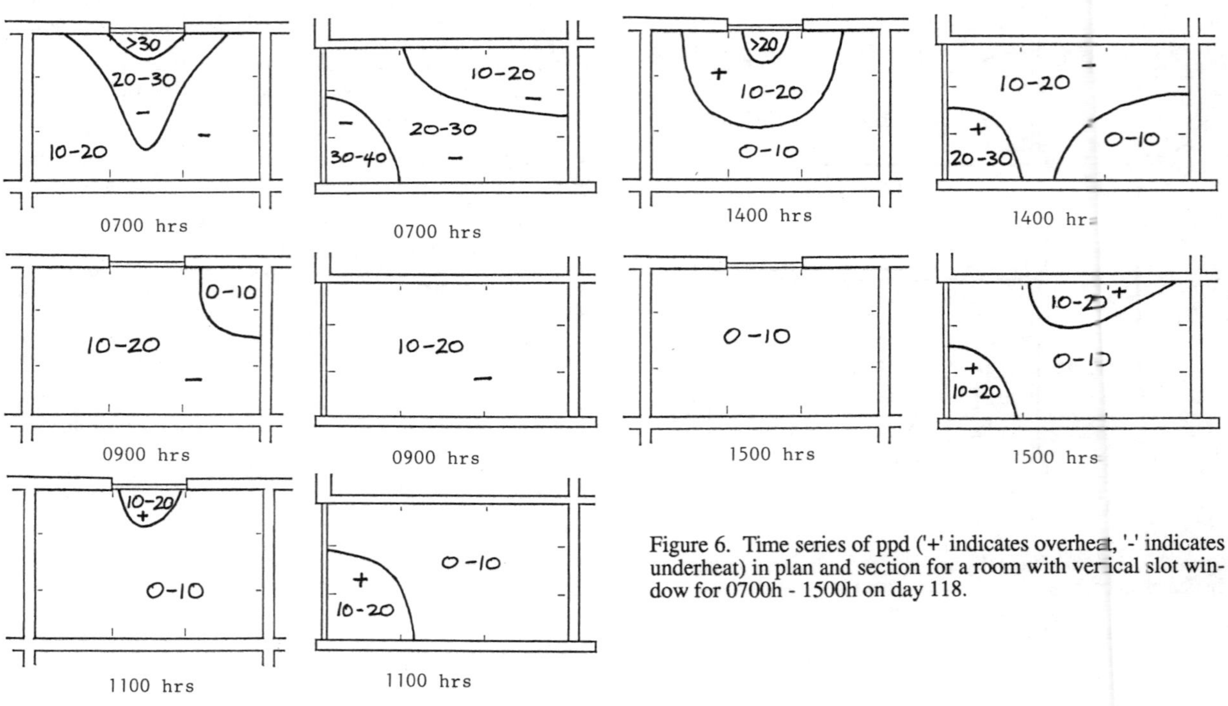

Figure 6. Time series of ppd ('+' indicates overheat, '-' indicates underheat) in plan and section for a room with vertical slot window for 0700h - 1500h on day 118.

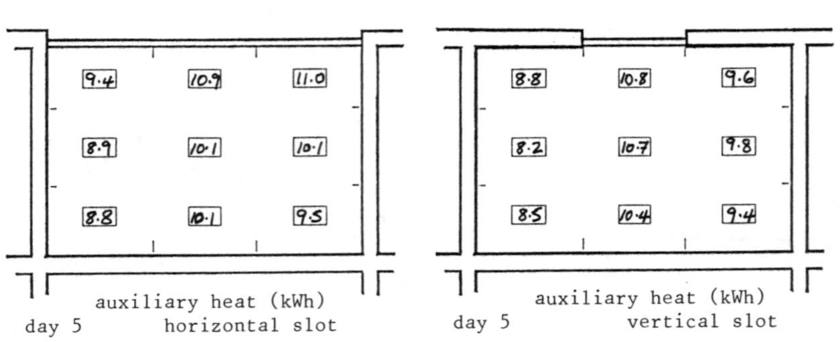

Figure 7. Auxiliary useful energy consumed during day 5 using a thermostat responding to preferred human temperature. Rooms with both horizontal slot and vertical slot windows are shown in plan.

ANALYSIS OF COMFORT LEVEL IN ITALIAN BIOCLIMATIC BUILDINGS

P. Oliaro, V. Corrado

Politecnico di Torino
Dipartimento di Energetica
Corso Duca degli Abruzzi, 24
10129 Torino - Italy

ABSTRACT. In Italy quite a lot of bioclimatic buildings have been constructed, being mainly residences and schools. A sample of those buildings has been analyzed under the comfort point of view. Simplest systems have proved to work best since they require less interventions of occupants. Screens are very important in order to avoid overheatings. Glazings shall assure a visual contact with outside, but they shall avoid a lack of privacy. Finally, good information of users is of the greatest importance.

1. INTRODUCTION

In Italy quite a lot of bioclimatic buildings have been constructed since the last 70's. With a study performed in 1985 ENEA (the main Italian energy agency) collected informations about 116 of such buildings, and found out the existance of some other ones /1/. Moreover, other passive (or ibrid) solar buildings have been realized after 1985, so nowadays we can exstimate that the Italian bioclimatic building stock consists of 200 elements at least.
A wide spectrum of bioclimatic solutions has been tried. Direct gain, sunspaces, Trombe walls, Barra Costantini systems, different kinds of solar air collectors and storage strategies have been employed /2/; together with passive solar features, almost all buildings have been equipped with active solar systems .
As far as the typology distribution is concerned, we have a strong prevalence of residential buildings (70% - 75% of the built volume) and a big number of schools (20% - 25% of the built volume). Other typologies (commercial buildings, etc) don't reach 10% of the built volume.
Anyway, the peculiarity of Italian experience is the kind of bioclimatic residential buildings. While in Europe such buildings are mainly medium-little ones, in Italy almost all of them are large, multi-family ones, containing up to 50 residential units. On one hand this is due to the global prevalence of large buildings in the Italian building stock; on the other hand, it is due to the fact that many bioclimatic buildings have been realized in the frame of agreements between public boards and private companies; in these agreements the interest of both the "contracting parties" was mainly for large buildings rather than for little ones.
Together with the aforementioned realizations, a wide research activity has been performed in the last years on bioclimatic architecture in Italy.
The largest part of the work has been done in the frame of PFE (National Energy Project), a project jointly sponsored by ENEA (National Committee for Nuclear and Renewable Energy Sources) and CNR (National Research Council) with about 30 million ECU per year. Among others, PFE covers the topic of energy saving in buildings: bioclimatic architecture is seen as an energy saving strategy, needing studies both on energy consumptions and on comfort conditions.
More than fifty passive solar buildings have been deeply analyzed and most of them have been monitored in order to evaluate their energergy performance /3/, /4/, /5/. Among these buildings, a sample has been chosen for assessing comfort levels, by means of an interview campaign of occupants.

2. THE STUDY AND THE RESULTS

2.1 The choice of the buildings and the parameters of analysis

In order to give an idea of the Italian situation, a sample of buildings has been chosen according to the typology distribution of the bioclimatic building stock. This led to investigate a majority of residential buildings and a consistant minority of schools /6/, /7/. In particular, in the following paragraphs informations on ten residences and seven schools will be reported. With such sample of cases, whose main characteristics are reported in table I, all the main passive solar systems have been investigated.
As far as the parameters of analysis are concerned, not only the building passive solar features have been taken into account. Obviously they are the main parameters to refer to, but also the "education level" (in the sense of information) of occupants is important for reaching good comfort levels: let's think to the importance of a correct management of the screens, for example. Finally, the auxiliary plant has been analysed too, since the comfort level depends on it as well.

N°	Typology	Site	Size of the building	Notes
1	Residence	Forte dei Marmi	4 flats	Direct gain
2	Residence	Certaldo	18 flats	Direct gain
3	Residence	S. Vincenzo	12 flats	Indirect gain
4	Residence	S. Vincenzo	12 flats	Indirect gain
5	Residence	Salviano	24 flats	Indirect gain
6	Residence	Pisa	45 flats	Dir/indir gain
7	Residence	Borgo Val Taro	16 flats	Dir/indir gain
8	Residence	Piazzola sul Brenta	40 flats	Direct gain
9	Residence	Caorle	24 flats	Indirect gain

10	Residence	Orbassano	40 flats	Dir/indir gain
11	Kinder-garden	Marostica	30 pupils	Dir/indir gain
12	School	Bologna	24 classes	Direct gain
13	School	Finale Emilia	18 classes	Direct gain
14	Kinder-garden	Tredozio	40 pupils	Dir/indir gain
15	School	Rimini	24 classes	Indirect gain
16	Kinder-garden	Ascoli Piceno	25 classe	Indirect gain
17	Kinder-garden	Roma	6 classes	Direct gain

Table I

2.2 Residential buildings

2.2.1 Residence at Forte dei Marmi (figure 1)

The building faces South and has two floors with two flats per floor. It is very well insulated and the windows facing not to South are small and double glazed. The South facade is completely glazed with external movable screens. Each flat is equipped with a loggia operating as a sunspace (direct gain), with some lateral fixed glazed surfaces, inside which heavy concrete elements facing SW have been realized for thermal storage. On the roof, water solar collectors have been installed.
A training action has been organized for occupants, in the form of written instructions about the management of solar components (glazings and screens). The attention of the occupants for the energetic behaviour of the building was find to be excellent.
The comfort level is rather good, but complains were reported about the noise coming from fans and, expecially, about the large glazings at the lower floors that cause a lack of privacy. Besides that, due to aestetic considerations, the heavy concrete elements have been painted in white, instead of being painted in black. Finally, the management of the screens was find rather complicated.

2.2.2 Residence at Certaldo (figure 2)

The building is similar to the one at Forte dei Marmi, except for the heavy concrete elements. It is characterized by direct gain thorough the South facade, which is glazed for 2/3 of its surface. The structures of internal walls and floors are heavy, in order to realize a great thermal inertia.
The occupants have been supplied with instructions about the energy management of their flats, but a rather complicated way of dividing heating expenses makes them only partially aware of the energy savings.
About comfort, the occupants complain about the difficulty of cleaning the external surface of the glazed areas; besides that, glazings facing North are too small and placed too high, so preventing a vision of outside.

2.2.3 Residences at S. Vincenzo (figure 3)

The two passive solar buildings are part of a group of three buildings, one of which is traditionally built, for comparison purposes. Both two solar buildings include a Trombe wall covering about 70% of the South facade; the left 30% of the South facade is glazed, allowing a contribution of direct gain. A high conservation level is planned for both buildings.
The occupants show a good knowledge about solar features of their flats.
A good level was realized; problems arose for cleaning external surfaces of glazings and, partially, due to the complex management of dampers, screens, etc.

2.2.4 Residence at Salviano

Similar indications to the residences of S. Vincenzo sprang out. Furthermore some occupants complain with the architectural distribution, expecially with the single space kitchen-living room that would have been preferred not to be a single room. The management of screens and dampers is felt to be rather hard, expecially for lack of a good information. Finally, the inlets for the air coming from Trombe wall had sometimes to be closed because they didn't allow a convenient arrangement of furniture.

2.2.5 Residence in Pisa

The main solar feature is a greenhouse covering the whole South facade.
A good information about solar systems have been given to occupants, but the buiding energetic management has been rather bad, due to the medium-low cultural level of occupants.
About comfort level, problems have been identified in cleaning the external surfaces of glazings and in an initial lack of internal movable screens, producing overheatings during summer (these overheatings were noted even with internal screens). Moreover, the large glazings caused a lack of privacy at the lower storeys. Finally, the active solar system is not able to supply enough hot sanitary water.

2.2.6 Residence at Borgo Val di Taro (figure 4)

The main solar feature is still a greenhouse covering the whole South facade. It has been realized with high quality components and the glazings are movable in order to allow easy cleaning operations.
Occupants behaviour was not found to be correct, despite of an information campaign, due to their medium-low cultural level.
About comfort level, some complains were noted only for difficulty of cleaning the fixed glazings and for the disposal of windows and doors on the North facade.

2.2.7 Residences at Piazzola sul Brenta

The buildings are two storeys high, with loggias facing South and provided with an horizontal fixed screen; other movable vertical screens are planned for the glazings. Tha passive solar strategy is direct gain.
A good comfort level has been reached, both in winter and in summer, expecially thanks to well designed screens.

2.2.8 Residence at Caorle (figure 5)

The building is provided with triangular loggias; the rooms face SE in order to improve solar collection during winter mornings. Each loggia is provided with an air solar collector.
Some problems concerned with comfort have been detected. Mainly, the air solar collector system is quite noisy and doesn't seem to give appreciable energy savings. The cleaning of the external surface of glazings is difficult. Finally, some overheating problems sprang out during summer, also due to the fact that the plastic movable screens of the

glazings don't allow lighting transmission and so they cannot be adequately managed.

2.2.9 Residences at Orbassano (figure 6)

The building has a triangular plan, it is 10 storeys high and allows for each flat a SW or a SE exposure. The facades are courtani walls, inside which solar spaces are realized.
A good comfort level has been reached.

2.2.10 Kindergarden at Marostica

A wide porch-sunspace facing South constitutes the main passive solar component. Solar air collectors have been installed too.
Some comfort problems have been encountered, mainly for overheating during summer (due to the lack of screens) and noise levels; other problems come from lack of maintenance.

2.2.11 Secondary school in Bologna

The building is provided with a glazed surface, 60° tilted and facing South for direct gain. Movable (rotable) screens have been installed inside the glazings.
Good comfort levels have been realized, also as a consequence of an information campaign that made the users aware of the managing of passive solar systems.

2.2.12 Primary school at Finale Emilia

The building has an articulated plan. Solar radiation is collected through direct gain; concrete storage elements have been realized too.
Some problems have benn detected with reference to overheating during spring/autumn and with reference to glare during sunny days. Moreover, users complain with difficulties in cleaning the glazings.

2.2.13 Kindergarden at Tredozio

Direct gain and Trombe wall have been utilized, together with water solar collectors.
Minor complains are noted for overheating during spring and autumn. Besides that, a lack of maintenance generates a good, but not excellent comfort level.

2.2.14 Secondary school in Rimini

The main passive solar component is a sunspace on the South facade. An air circulation between the sunspace (where vegetation should grow) and the classrooms was planned, with a supply of oxigen by the vegetation. In practice this doesn't work and the comfort level is rather poor, also due to overheating problems.

2.2.15 Kindergarden/Primary school in Ascoli Piceno

Wide sunspaces facing South have been realized. The comfort level is not good, due to both overheating during hot seasons and lack of heating during winter. It should be kept in mind that this building is one of the first realized in Italy.

2.2.16 Kindergarden in Rome

The building is a classic bioclimatic one, totally blind on the North facade and with large glazings facing South for direct solar gain. Because of the simple operating of the systems, no big troubles were detected, except for a bad general level of building maintenence. Comfort conditions are rather good.

3. CONCLUSIONS

3.1 Residential buildings

The disposals of the rooms typical of passive solar architecture is almost always accepted, except sometimes for the kitchen-living room beińg one only space.
The large glazed surfaces are generally appreciated; problems spring out for cleaning their external side if they are not movable, or if they are wasistas. Besides that, on the lower floors, their wideness should be limited in order to avoid lack of privacy, or in alternative selective glasses should be utilized:
As far as windows facing North are concerned, they should not be too small and to high, in order to allow a vision of outside. Furthermore, there should be planned with movable screens, both inside and outside; their management is to be carefully studied, in order to minimize the possibilities of incorrect operations by the occupants.
In the same way, Trombe walls should be designed keeping in mind the necessity of easy management and avoiding conflicts between the positions of air inlets/outlets and the position of the furniture.
Finally, a good "education" of occupants is felt to be necessary, but it is not always enough; the solar systems should be easy to manage.

3.2 Schools

The comfort levels in schools seemed to be a bit worse than in residential buildings: overheatings were noted more often. Apart from error in design, problems generally come from lack of users information, from a too complicate management of passive solar systems (e.g. screens) and, expecially, from lack of maintenance.

3.3 Final remarks

From the aforementioned data, some indications could be given in order to realize a good comfort level in passive solar buildings in Italy.
Firts of all, passive solar systems should be as simple as possible and their management should not be a trouble for occupants (if the case, automatic controls could be adopted). Then, good attention should be payed in designing the screens, in order to avoide overheating. Then, attention should be payed also in designing the glazed surfaces; they should allow a "visual contact" with outdoor even through the North facades and, on the other hand, they should not cause lack of privacy (South facades). Finally, a good information of occupants is felt to be of great importance.
Moreover, for schools, passive solar systems should be designed requiring the minimun maintenace because it is often completely neglected, due to economic and "burocratic" problems of the public boards.

4. REFERENCES

/1/ G. Funaro, A. Fanchiotti, E. D'Errico - "116 edifici solari passivi in Italia" - ENEA, 1985

/2/ CEC - "European Passive Solar Handbook" - 1986

/3/ CNR/ENEA - "L'edilizia bioclimatica in Italia. Situazione attuale e prospettive future" - PFE, 1987

/4/ CNR/ENEA - "Stato dell'arte ed innovazione nell'uso dell'energia in edilizia" - PFE, 1988

/5/ CNR/ENEA - "Solare passivo" - PFE, 1989

/6/ AGIP Petroli - "Analisi tecnica-economica di interventi già realizzati di edilizia bioclimatica" - 1986

/7/ CEC - "Project Monitor Brochures" - 1987/1989

/8/ CNR/ENEA - "Architettura bioclimatica, risparmio energetico e qualità ambientale. Prospettive e applicazioni nell'edilizia pubblica e

nell'edilizia scolastica" - Proceedings of ISES Conference, Rome 9-10/11/1988, PFE, 1989
/9/ G. Contadini, G. De Giorgi, S. Fumagalli, G. Rizzi, A Ferrario, G. Rossi, G. Silvestrini - "Monitoring of a Solar Chimney and different Storage/Distribution Systems in an Office Building located in Northern Italy" - 14th National Passive Solar Conference, Denver 19-23/6/1989.

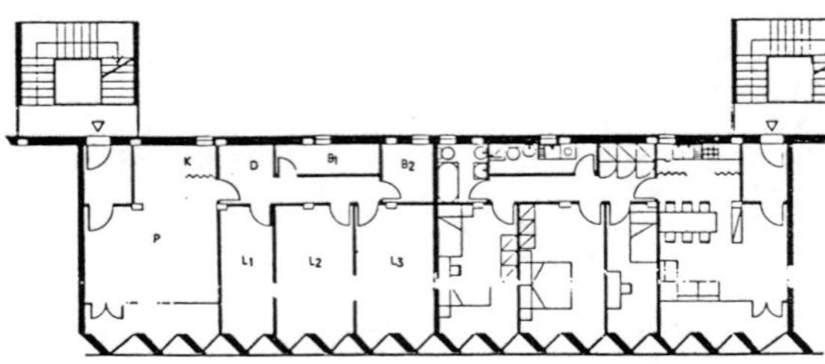

Figure 1 - Forte dei Marmi Residence Plan

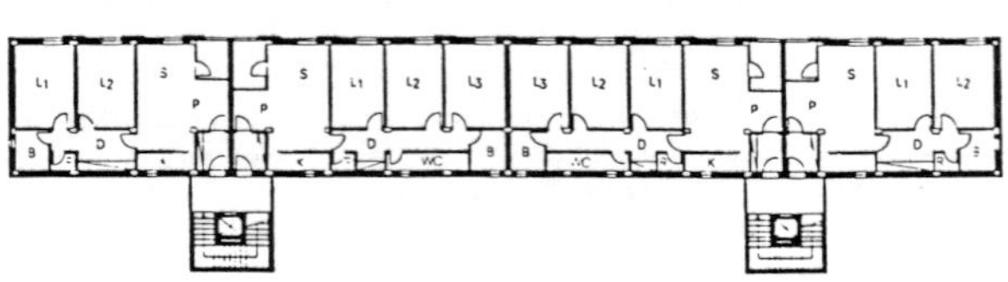

Figure 2 - Certaldo Residence Plan

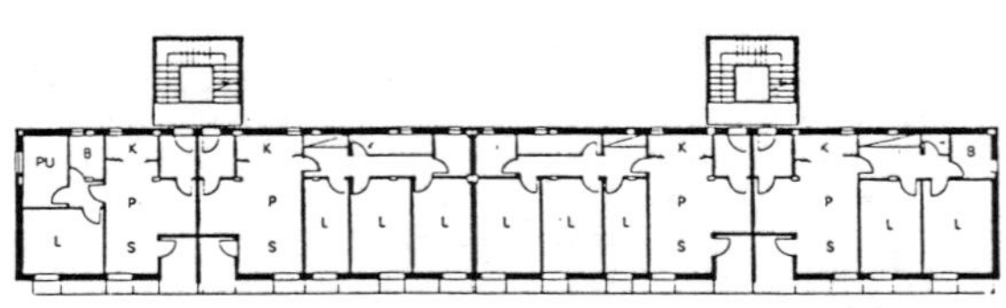

Figure 3 - San Vincenzo Residence Plan

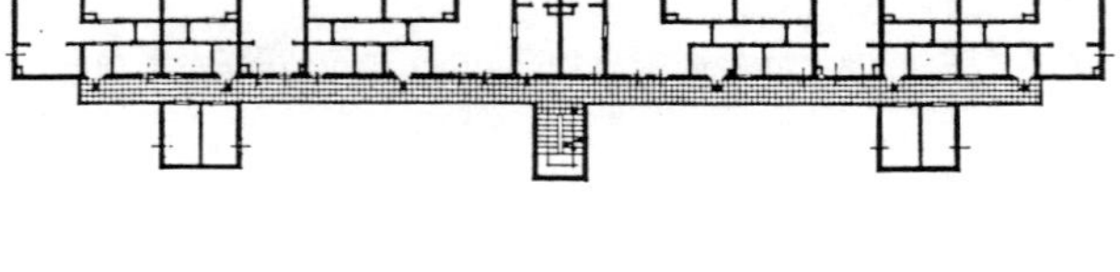

Figure 4 - Borgo Val di Taro Residence Plan

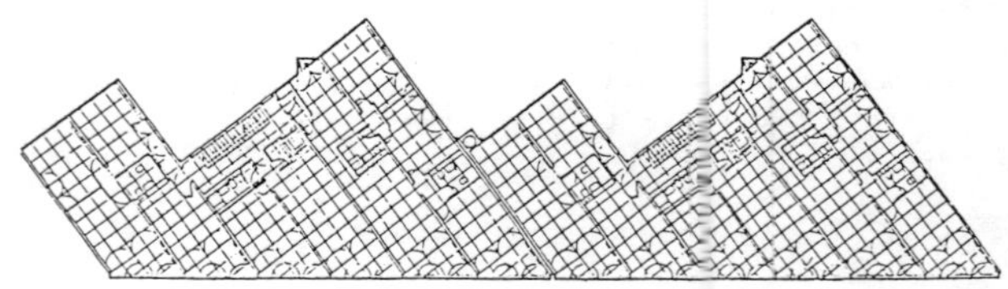

Figure 5 - Caorle Residence Plan

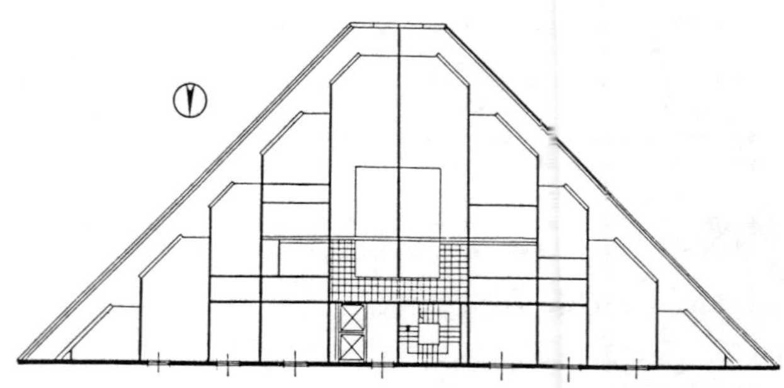

Figure 6 - Orbassano Residence Plan

FIELD STUDIES OF THERMAL COMFORT IN PASSIVE SOLAR BUILDINGS

I D Griffiths
Department of Psychology
University of Surrey
Guildford Surrey GU2 5XH
Great Britain

ABSTRACT The paper reports the results of a study of the occupants of homes and office buildings in France, the Federal Republic of Germany and Great Britain. The research focussed on the occupants' subjective reports of thermal comfort and their relation to simultaneous measures of physical thermal variables carried out in situ and estimates of activity levels and clothing insulation. Evaluation of the results shows that the general assumption that people wish to feel thermally neutral is not supported by these results, that comfort standards do not appear to be different for passive solar buildings as compared with other buildings, that desired temperatures for neutrality are generally lower in both homes and offices than those usually specified, and that there are significant differences in the desired temperatures for homes when compared with offices. A field experiment investigating office workers' responses to temporal and spatial non-uniformities in temperature shows that such effects are predictable from steady-state data.

1. THERMAL COMFORT IN OFFICES AND HOUSES WITH PASSIVE SOLAR FEATURES

1.1 Introduction

1.1.1 The Sample of Buildings

The office buildings which were investigated were, with one exception, in the United Kingdom. They were:
The administration building of the Badische Gemeinde-Versicherungs-Verband (Karlsruhe), an open-plan office with glass-roofed inner courtyards, predominantly natural lighting and ventilation (with the artificial lighting controlled in accordance with natural illuminance levels) and solar collectors for domestic hot water. 40 occupants took part. Building Research Station Building 10, Hertfordshire, a conventional cellular office block, intended as a control building for the low energy office described below. 26 people took part.
Building Research Station Low Energy Office, Hertfordshire designed to make optimum use of solar energy through heat gain and the use of day-light and natural ventilation. 24 of the occupants took part in the study.
Briarcliff House, Hampshire, a five-storey building with a glass solar wall on its south elevation, exterior to a conventional curtain wall. 32 of its occupants were investigated.
The Civic Centre, Chester-le-Street, Tyne and Wear, two burolandschaft offices connected by a glazed mall. The sample was composed of 37 people.
Frank B Hall House, London, a conventional office building occupied by the same organization as Briarcliff House and included as a control for that building. 35 members of staff helped with the study.
Gateway 2, Hampshire, a speculatively built office with an atrium, and naturally ventilated. 42 took part in the study.
International House, London, part of a larger post-modernist scheme and entered via an atrium. 45 staff took part.
Lloyds of London, the headquarters of the insurance organization, with a full height atrium. 53 office workers helped with the study.

The houses studied were situated in France, the Federal Republic of Germany, and Great Britain. In particular, they were:
In France, 24 houses and flats in the departments of Aude, Britanny, Doubs, Haute Garonne, Loire, Normandy, Pyrenees Atlantiques, Pyrenees Orientales and Tarn. The majority of the dwellings were two-story detached houses, but about a third were retrofitted apartments. 18 had sunspaces, ten had direct gain systems and five Trombe walls; five had mixed active and passive systems and nineteen had significant thermal mass. 29 people helped with the study.
In Germany, 19 people living in houses at Am Lindenwaeldle (Freiburg) took part in the study. The houses are described in Project Monitor Issue 6. 10 respondents lived in a joint Swedish/German development at Ingoldstadt, where there are 10 terraced houses with south-facing orientation, glazed southern elevation, heavy concrete structure, and conservatory, porch and sunscreens to the south. A further four respondents came from two individually designed passive solar houses, one near Cologne and one near Freiburg.
In Great Britain 18 people reported on passive solar houses in the new city of Milton Keynes, at Giffard Park (fully described in the first issue of Project Monitor). 19 people from a housing development in Lancashire (Willow Park) took part in the study: their houses are fully described in Project Monitor Issue 44.

1.1.2 Research Methods

Participants in the study completed an extensive questionnaire while the researcher measured air temperature, globe temperature, air velocity and relative humidity. Key elements in the questionnaire were a table of items of clothing which enabled the computation of total clothing insulation, and a table of activities which allowed the assessment of time-weighted activity levels over a period of thirty minutes before interview. From these data theoretical subjective temperature requirements were calculated, following the approach of McIntyre /1/, which produces close approximations to the results of the Fanger equations /2/. Respondents reported their thermal sensations on a nine-point expansion of the seven-point ASHRAE scale. The deviation from the centre point of this scale was used, in conjunction with the physical measurements of the indoor climate, to assess the appropriate temperature for thermal neutrality.

1.2 Results

1.2.1 The Office Buildings

In general the occupants of the offices, whether with passive solar features or conventional in design, were warmer than the theoretical optimum sensation of warmth, feeling neither cool nor warm. The discrepancy was of the order of 1.6 K operative temperature. However, in general people did not wish to feel the proposed optimum sensation but to feel a little warmer than neutrality, equivalent to having room temperature 0.7 K warmer than neutral.

The biophysical theory of thermal comfort, having regard to the activity and clothing insulation levels of the occupants of the offices in the study, produced optimum operative temperatures in the range 22.9°C to 23.9°C, with a mean of 23.2°C. In all nine cases the empirical optimum temperature to provide thermal neutrality (ie feeling neither cool nor warm) was significantly lower than the theoretical value, with an average desired temperature of 20.67°C, range 20.0°C to 21.4°C. The mean difference between theoretical and empirical temperatures is thus 2.5 K, but if allowance is made for the finding that people wish to be slightly warmer than neutral then the optimal temperature would appear to be of the order of 21.4°C. There is no apparent difference in requirement between buildings with passive solar features and those without.

The degree of disagreement with the proposition that the office usually provides comfortable temperatures shows a great deal of variation from office to office: while on the average 42% of the office samples disagreed, the maximum level of disagreement was 70% in the Civic Offices (Chester-le-Street, UK), and the lowest 7.5% (in the BGV office, Karlsruhe, FRG). Both of these offices had passive features, but there is some possibility of association of this feeling with passive solar features, since the two totally conventional offices had below average levels of dissatisfation in this respect.

In terms of the detectability of temperature changes during the working day there was not a very large amount of variation between offices (such changes were noticed by an average of 77% of occupants, with a range from 60% to 93%). Of the four offices below average in this respect, two are of conventional design, but all of those above average have passive solar features. The two offices with significantly lower than average levels of noticeability are Briarcliff House and Frank B Hall House, one of which is conventional and one not. The two offices significantly above average in this respect are both passive solar. No firm conclusion can be drawn from these data but there is some possibility that passive solar offices are subject to more noticeable changes in temperature. There is less variation in the number of people who find such temperature changes unpleasant (average 72%, maximum 86% and minimum 55%) and it would seem unlikely that there is a real problem for passive solar design. Noticeability of differences in temperature between different parts of the building is roughly equal in incidence to the noticeability of changes over time (average 82% maximum 93%, minimum 63%) although the two lowest percentage noticeability rates (which are significantly below average) are found in the two conventional buildings. This difference closely approaches statistical significance. There is no sign of even this degree of pattern in the ratings of unpleasantness. There is probably a need for caution about the possibility that passive solar design can lead to problems with non-uniformity of temperature within the office space.

Investigation of the pattern of physical symptoms which are often related to the sick-building syndrome shows that the office with the lowest incidence of symptoms has passive solar features (the Building Research Establishment's Low Energy Office) as has the office with the highest incidence of such symptoms (International House). Of the two conventional offices one is better than the average and one worse. The office which was considered by its occupants to be the most unhealthy (Civic Offices) has solar features as has the most healthy building (BGV, Karlsruhe). The two conventional buildings came out well on this index, but there is no evidence for a pattern in this set of results. Thus, it cannot be concluded from these data that the passive solar approach necessarily helps reduce the incidence of unpleasant physical symptoms.

In rating the importance of a variety of environmental factors there is considerable consensus and no visible differences between the conventional and passive solar offices. The highest importance rating overall was given to a comfortable room temperature (8.8 out of 10), followed by suitable artificial lighting (8.1) and then fresh air (8.09). The lowest rating was given to the presence of sunlight in the office not shining on the occupant (an average of 6.4 out of 10) The next lowest importance was attributed to freedom from temperature changes during the day (6.68 out of 10) which followed freedom from dry air (6.71). The highest level of dissatisfaction was with air freshness (the average proportion dissatisfied was 49%) followed by the amount of change in room temperature during the day (46% dissatisfied, on average) and the room temperature itself (45% dissatisfied). There was the least dissatisfaction with the amount of direct sunlight shining upon occupants (24% dissatisfied), the level of glare in the office (26%) and the suitability of the artificial lighting (27% dissatisfaction). There are no signs of differences between the two conventional offices and the group with passive features in the levels of dissatisfaction with any of the features investigated.

The two conventional offices were seen as more pleasant places to be in than all the other offices, and this difference approached statistical significance. In all of the other remaining attitudinal variables (internal and external appearance of the office, general satisfaction, office efficiency, pride in the office, desire to spend time out of doors, lighting, odours, ventilation and stuffiness, pleasantness, health and general evaluation of the office design) there was great variation from office to office, but no sign of differences between the conventional offices and those with solar features.

1.2.2 The Houses

In all seven cases there is a significant difference between the theoretical operative temperature required to secure thermal neutrality and that which can be deduced on empirical grounds from the subjective warmth estimates given by participants in the surveys and the physical conditions in which those assessments were made. In all cases the empirical temperature is lower than the theoretical and the difference between the average of each type of temperature is more than four degrees.

All houses provide average conditions which exceed thermal neutrality in terms of the warmth assessments given by the respondents. The average

desired warmth is in the category 'slightly warm' rather than 'neither cool nor warm' and in all seven cases desired warmth is greater than the numerical value associated with sensations of 'neither cool nor warm': statistically this indicates that the discrepancy between theoretical expectation of a desire for thermal neutrality and the empirical findings is significant.
In terms of the importance which the occupants place upon a variety of environmental factors there is a clear ordering which varies little from subsample to subsample. Comfortable room temperatures is, overall, the most important consideration (securing an overall average of nearly nine out of ten, with relatively little variation from subsample to subsample). Second overall comes the quantity of daylight, averaging 8.5, again with little variation. The third rank among the variables of possible relevance to passive solar features is held by freedom from draughts, with an average of 8. In descending order, the following have averaged scores between 7 and 8: air freshness, suitable artificial lighting and the presence of indirect sunlight. The absence of glare is of the next importance, with an average of 6.3. The least important variables are: dryness of the air (5.8), freedom from changes in temperature during the day (5.5) and the absence of direct sunlight on the occupant (5.1).
Considering the same variables from the point of view of the degree of dissatisfaction which they cause on the average, then it is clear that the most significant is the presence of draughts. More than 20% express dissatisfaction in this regard, although there is a great deal of variation from subsample to subsample.
Temperature changes during the day is next, with about 16% average and the most common level of dissatisfaction is slight. Glare has a very similar average dissatisfaction level but is subject to greater variation between subsamples. Room temperatures give rise to dissatisfaction in about 14% on the average, and this is once again subject to a great deal of variation.
The suitability of artificial lighting produces dissatisfaction to about 13% on the average and the presence or absence of indirect sunlight does so to about 11%. Dryness of the air causes problems to about 11% on the average, and the other variables (daylight, freshness of the air and sunlight directly on the occupant) cause less than 10% dissatisfaction. About 13% on the average found their control over temperature not to be effective and 16% found the same for lighting. 20% found their control of ventilation to be ineffective. All of these were subject to a great deal of variation between the subsamples.
Nearly 90% of each subsample, on average, agreed that they were generally very satisfied with their home, and slightly lower proportions felt that their homes helped them to do the things that they wanted to do and that they were proud to show them to others.

2. FIELD STUDIES OF THERMAL TRANSIENTS

2.1 Introduction

It has often been claimed that passive solar buildings may well cause discomfort to their occupants because of changes in temperature. These would arise either because of differences in temperature between the solar and nonsolar aspects of the building and experienced as the occupants move from one part to another, or they could arise as temporal transients caused by, for example, the sun going behind clouds.

2.1.1 The Sample

The study of temporal transients was carried out in the Senate House building of the University of Surrey and at the offices of the Midland Montague merchant bank in the City of London. The first building is a 9-storey highly glazed office block, the second a 13-storey office development where the office accommodation is grouped around an atrium.

2.1.2 Research Methods

At four times during the day each of 33 subjects was asked to rate their warmth using a continuous version of the ASHRAE 7-point warmth scale (ie one in which they were not constrained to the use of integer values), and to provide details of the breakdown of activities within the 30 minutes prior to the warmth assessment, as well as their clothing at the time of assessment. Measurements of air temperature and speed, globe temperature and relative humidity were made simultaneously with the subjective assessment of warmth.
The study of spatial transients was carried out in the AD academic building of the University of Surrey, which houses the departments of Linguistics and International Studies, Psychology, Sociology and Economics. Subjects made assessments of their warmth in offices in different parts of the building, at intervals of 12 minutes, each assessment being made according to the same methodology as that used in the spatial study. The hypothesis to be tested was that the repetition assessments would not follow the same quantitative relationship with the physical warmth variables, and activity and clothing levels as the first assessment made. Since in a field experiment it is not possible to ensure that experimental subjects all experience exactly the same physical conditions their subjective assessments of the environmental conditions they have been exposed to, their assessment on the ASHRAE 7-point scale has been used to correct the actual subjective temperature experienced into an ideal temperature (ie one corresponding to the sensation 'neither cool nor warm'). This is done quite simply: the subjective temperature experienced by each individual is corrected to thermal neutrality by adding or subtracting their rating's deviation from zero at 3 degrees per scale interval. A subsidiary hypothesis was that theoretical assessments of required temperatures for each individual (ie assessments made on the basis of known clothing insulation and activity levels) would differ from empirical assessments (where objective warmth and warmth vote would allow the prediction of an optimum subjective temperature).
The major hypotheses for both spatial and temporal transients were tested by repeated-measures analysis of variance, and the minor hypothesis by matched pairs t-tests. For the purposes of the latter analyses, metabolic rate was assessed both as a level corresponding to the actual activity level of each subject at the time of assessment (ie filling in a questionnaire) and as the time-weighted mean of the activities engaged in during the half-hour previous to the assessment.

2.2 Results

2.2.1 Temporal Transients

There are no statistically significant differences between the optimal temperatures required at each of the four times of observation, which can therefore be adequately represented by an overall mean subjective temperature of 21.61°C. Since there are no differences between observations made

at different times it is clear that temperature changes of the magnitudes experienced do not cause deviations from steady-state expectations. Empirically determined optimal temperatures have been calculated for each individual subject and for each set of observations a repeated-measures t-test has been carried out to test the hypothesis that there is a significant difference between the theoretical and empirical values. In all four cases the observed differences are significant and in the same direction (the empirical values average 2.23° less than the theoretical). There is thus clear indication that theoretical requirements of temperatures for these particular groups of office workers are one or two degrees higher than seem to be warranted.
The general conclusions to be drawn are that temperature changes over time do not follow rules which are different from those obtaining in the steady-state, and that there is evidence that theoretical determinations of the subjective temperature may lead to higher temperatures than necessary.

2.2.2 Spatial Transients

There is no significant effect of the change in location on temperature requirement. Thus, once again, it can be concluded that within the range of conditions experience by these subjects there is no specific effect of transients. In both the case in which the relevant metabolic rate is assumed to be that representative of the activity of completing the research questionnaire and that in which a time-weighted mean is calculated, the overall mean difference is small (about three-quarters of a degree) and none of the t-tests is significant. Thus, for this small number of subjects (9) the theoretical determinations of neutral temperatures seem to be adequate.

3. References

McIntyre D A (1973) A guide to thermal comfort Applied Ergonomics 4 (2) 66-72

Fanger P O (1970) Thermal Comfort Danish Technical Press Copenhagen

Table 1: Average Optimal Temperatures for Offices

	Mean	S D
Theoretical Subjective Temperature	23.2	0.29
Empirically Derived Subjective Temperatures	20.6	0.46

Table 2: Average Optimal Temperatures for Houses

	Mean	S D
Theoretical Subjective Temperatures	23.8	0.99
Empirically Derived Subjective Temperature	19.0	1.73

Table 3: Relationship Between Theoretical and Empirical Subjective Temperatures for all Samples and Mean Warmth

	Mean Warmth	
	<5	>5
Theoretical < T Sub	0	2
Theoretical > T Sub	1	13

(n = Buildings)

PASSIVE COOLING - STATE OF THE ART

M.ANTINUCCI[1], B.FLEURY[2], J.LOPEZ D' ASIAIN[3], E.MALDONADO[4], M.SANTAMOURIS[5], A.TOMBAZIS[6], S.YANNAS[7]

1: CONPHDEBUS, Italy, 2: ENTPE, France, 3: Univers.Seville, Spain, 4: Univ. Porto, Portugal, 5: PROTECHNA, Greece, 6: MELETITIKI, Greece, 7: Arch. Assos., UK.

ABSTRACT. This paper summarizes the results of an horizontal study on Passive and Hybrid Cooling for buildings, carried out by the authors, on behalf of the EEC Research Program "Building 2000" /1/. The state of the art of the strategies for the prevention of thermal gain is presented. Also the existing situation on the design tools and the computational methods regarding the Passive and Hybrid cooling techniques and components is analysed. Finally the data of several designed and constructed buildings in Europe using Passive and Hybrid cooling techniques and components are given and analysed.

1. INTRODUCTION

The major objective of the cooling strategy in a building is to provide comfortable internal conditions at a minimum cost. The overal strategy is based on the prevention of overheating as well as on the promotion of heat losses. The approach is illustrated in figure 1.

Equally, as with passive solar heating and passive solar design the study and application of passive cooling is a multilayered and multidisciplinary process. It is especially important to treat the subject in conjuction with other aspects of architectural design strategy. From such viewpoint passive cooling processes should be considered not as isolated phencmena but in close relationship to building types, occupancy patterns and the sources of heat gain to buildings in the climatic conditions of Europe.

Therefore the main objective of this paper is the critical reconsideration of the purposes and functions of the known cooling strategies in the light of present knowledge of building physics and environmental design, and not the presentation of novel techniques.

2. THE CLIMATIC ZONES

The Mediterranean climatic conditions are characterized by moderate high and low temperatures during the annual cycle. The design dry bulb temperatures for nine Mediterranean cities are given in figure 2 /2/. These climatic conditions require a proper balance of the design strategies on passive heating natural cooling and daylighting in buildings. Therefore passive strategies in the Mediterranean should not be approached singly, but in the context of an overall environmental design strategy.

3. PREVENTION AND MODULATION OF HEAT GAINS

The prevention and modulation of the heat gains involves consideration of the following design measures :

- Thermal insulation
- Storage mass
- Solar control and shading of building surface
- Landscaping and the use of outdoor and semi-outdoor spaces
- Building form, layout and external finishings.

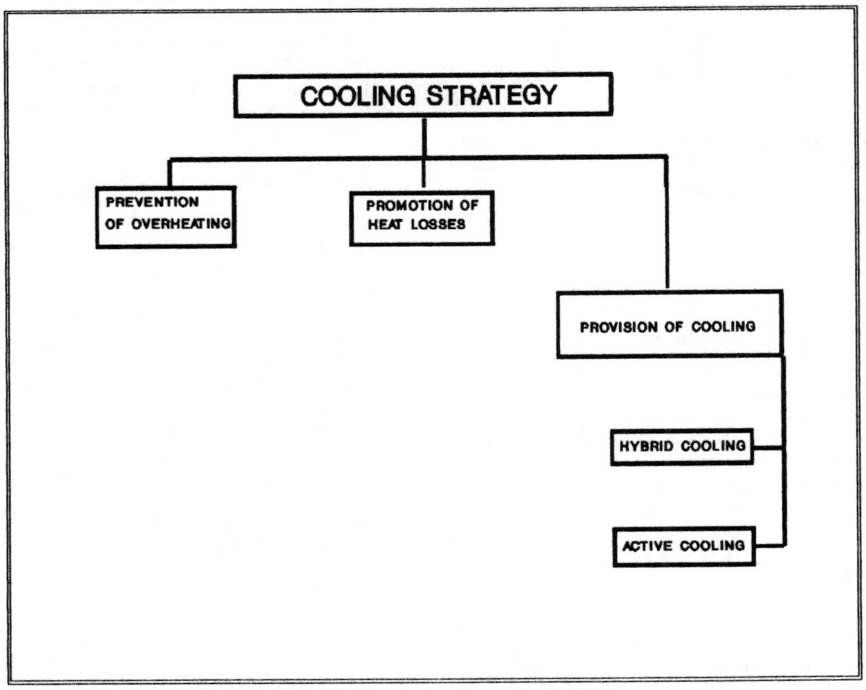

Figure 1 : The Cooling strategy in buildings.

Such measures tend to be generally applicable under all types of climate and site conditions and are relevant to all building types.

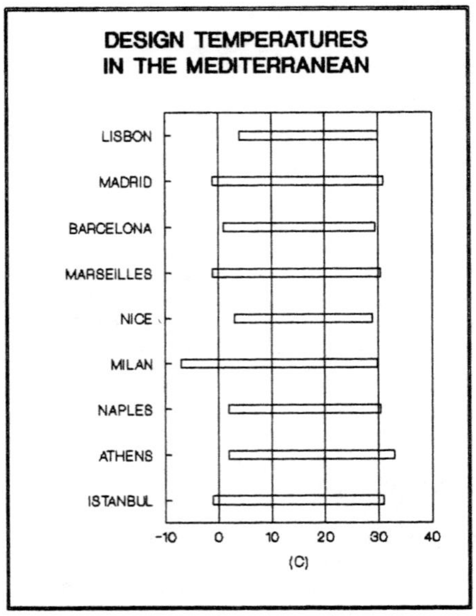

Figure 2: Design Summer and Winter Temperatures for Nine Mediterranean cities.

3.1. The role of the thermal mass

The capacity for heat storage in the building structure offers a "daily strategy" providing attenuation of peaks in cooling load and modulation of internal temperature with heat discharge at a later time.The larger the swings in outdoor temperature, the more important the effects of such storage capacity. However, when auxillary heating and cooling systems are used, an increase of the mass over a threshold value can increase the load. Therefore the determination of the optimum thermal mass taking into account the winter and summer performance of the building is necessary. The influence of the thermal mass on the energy performance, of the building during winter and summer in Israel is discussed in /3/.
The question if it is better to have the thermal mass as a part of the external walls or as an internal partition has many architectural, implications and influences highly the performance of the building. The results of such an investigation are given in /3/.
However it should be pointed out that there are not standarts on the optimum depth and the placement which take into account the Summer and Winter operation of the building.

3.2. The role of shading

Minimizing heat gains is the first design rule in any passive cooling application. Increase of the shading levels decrease directly the cooling load of the buildings but in the same time the lighting level is increased. Overshading reduces also the amount of solar energy penetrating through the window in winter increasing so the winter heating load. Therefore the problem of optimal window shading devices, taken into account heating, cooling and daylighting, arises in the Southern European Countries. An interesting method to define the optimum shading levels for a building in Israel, using intergral criteria, is proposed in /4/.
However, it should be pointed out that there is no criteria for optimal shading levels which take into account the heating and cooling performance of the building and the lighting requirements as well.

3.3. The role of the building form and shape

The architectural configuration of the building enclosure can be a major factor in preventing heat gains and encourages losses.
An important discussion on the building form, in order to achieve the best proportion for heat gains and heat losses is given in /5/.
However questions like :
- What is the best volume to surface ratio
- What is the optimum shape and
- What is the optimum spacing

are not systematically answered and further research on this topic is required.

4. THE PROMOTION OF HEAT LOSSES

The promotion for disposal of excess heat by natural means depend : a) On the availability of an appropriate environmental heat sink for the heat that is to be rejected and b) On the appropriate thermal coupling and sufficient temperature differences for the transfer of heat from indoor spaces to sink.The main processes of passive cooling are identified in the Table below.

Process	Heat Sink	Heat Transfer node
Radiative Cooling	Sky	Radiation
Evaporative Cooling	Water	Evaporation
Convective Cooling	Air	Convection
Earth contact Cooling	Ground	Conduction

The use of the environmental heat sinks for heat dissipation is possible over some climatic threshold conditions while in some cases the heat transfer to sink may be assisted by mechanical devices on a mean of overcoming low temperature differences or other limitations.

4.1. Passive Ventilation

Ventilation provides cooling by using moving air to carry out away from the building or/and from the human body itself. Air movement created by natural forces is due either to wind or to stack effect. Nightime ventilation may offer also a high potential for building cooling because of the lower outside temperature. Experiments using night time ventilation described in /6/, show an important decrease of the daytime indoor temperature.
In designing window system for passive ventilation cooling, four essential steps should be followed /7/. First it is important to have a clear picture of the directional range of the wind on the site involved through both diurnal and annual cycles. Second an assessment on the ventilative cooling needs for thermal comfort should be done. Third the shielding or "wind shadow" effects of the neighboring structures and of the topography should be evaluated. Also it is vital to choose a window type and size which corresponds to both the ventilation and the solar heating, cooling and daylighting requirements of the building.
However, regarding the four above steps it should be pointed out that winds are extremely variable and weather data, are not available at most sites. Also the answer on the question : "what is the optimum size of the building openings", is that there is no standarts which take into account the balance of heating, cooling and daylighting performance of the building, together with the ventilation requirements.

4.2. Radiative, Evaporative and Ground Cooling

The use of the sky, water and ground as a heat sink for dissipation of the buildings excess heat has gained an increasing importance during the last years. An overview of the theoretical background and a description of possible uses is given in /8/. However the use of these type of cooling is associated with various problems like :

- The definition of the threshold values allowing the application of the sink.
- The definition of the selection and sizing procedure.
- The calculation of the overal building efficiency and
- The luck of information on construction details.

Unfortunately there is an important lack on appropriate climatological data while there is no available design tools for designers as well as information material, handbooks, etc.

5. THE DESIGN TOOLS ON COOLING

In order to identify the design tools available for cooling purposes, an extensive survey of the existing design tools for passive solar applications has been performed.

The results of the survey are given in figure 3. Hundred twenty eight programs having capabilities to calculate the cooling load of a building have been found. From them 107 programs come from U.S. However between these programs only 54 can calculate the cooling load and the variation of the space temperature as well. Also it is found that 67 programs can simulate the effects of mass, while 65 programs take into account and simulate the effects of shading devices. However there is only 49 programs which can estimate shading and mass effects together.

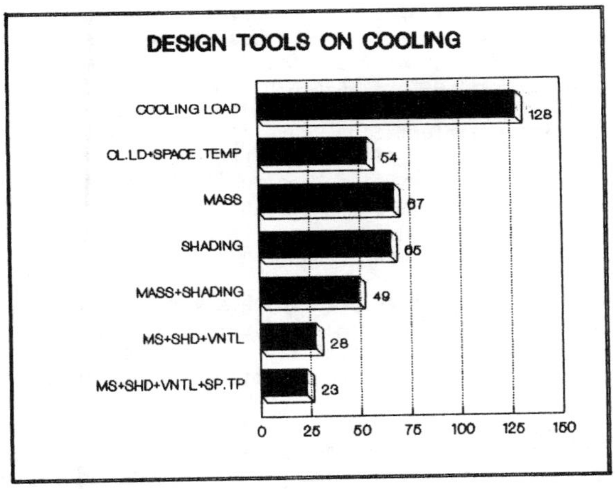

Figure 3: Design Tools on cooling Classification regarding the calculation capabilities.

The simulation of the various natural ventilation strategies is performed by a very small number of tools. It is found that there are 28 programs which simulates the natural ventilation strategies and in the same time, take into account the shading and mass effects.

Finally there are 23 programs which can considered as "complete programs", which calculate the cooling load and the space temperature variation,while take into account the mass and shading effects and the natural ventilation systems as well.

From the above 23 "complete programs", 8 between them are designed for mainframe computers and are used especially for research purposes. Four of these programs come from the US, two from Switzerland and one from Italy and UK. Also two of the "complete programs" are designed for programable calculators, which come from the US.

Finally nine programs are designed for PC's. Seven of these programs come from the US and two from Greece. Also there are four more programs from the US designed for various type of microcomputers.All the collected information are placed in a data base. An extract of the data base offering important information on the identified design tools is given in figures 3-5.

As it can be seen from these figures there are 113 programs suitable for architects, 123 programs for engineers, 55 programs for techniciens, 28 programs for researchers and 42 programs for builders. The majority of the programs are written for micro's and PC's except the research programs which are mainly written for mainframes.

Also the most of the programs can be used for residential and small commercial buildings. Fifty five of these programs are suitable for the predesign phase, also fifty five for the schematic design phase, fifty two for the design development, 21 for Post design services and 16 tools can be used for research purposes. For the large commercial buildings there are 26 programs suitable for the predesign phase, 39 for the schematic design phase, 47 for the design development, 16 for post design services and 11 for research purposes. Regarding the design tools available to simulate the use of environmental, sinks, three tools managing the evaporative cooling techniques, have been identified. Also there are 42 programs capable to simulate convective cooling techniques, 2 for radiative cooling and 19 for ground cooling.

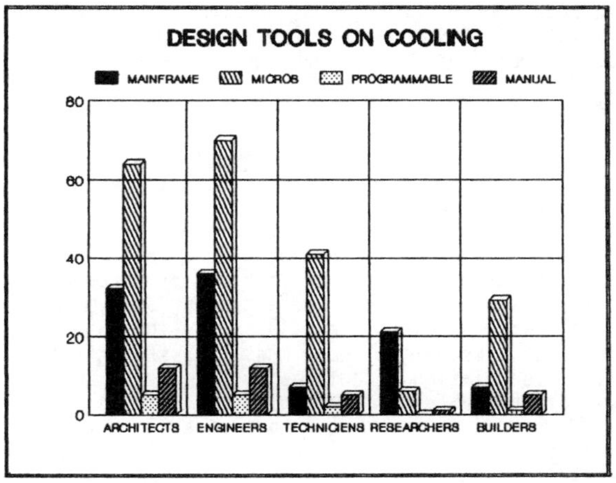

Figure 4: Design Tools on Cooling Classification regarding the computer type and the users.

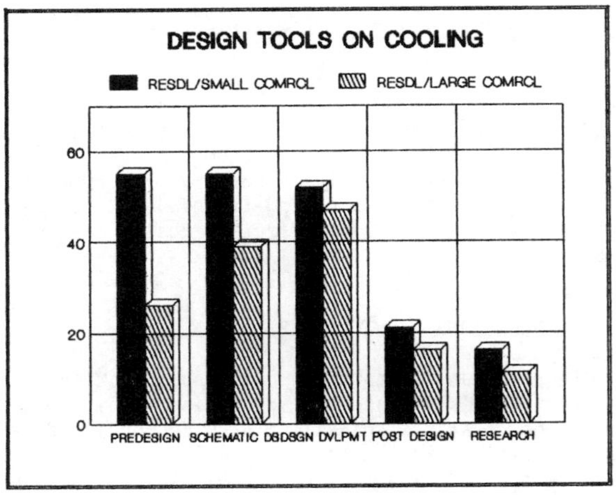

Figure 5: Design Tools on Cooling Classification regarding the design phase and the building's type.

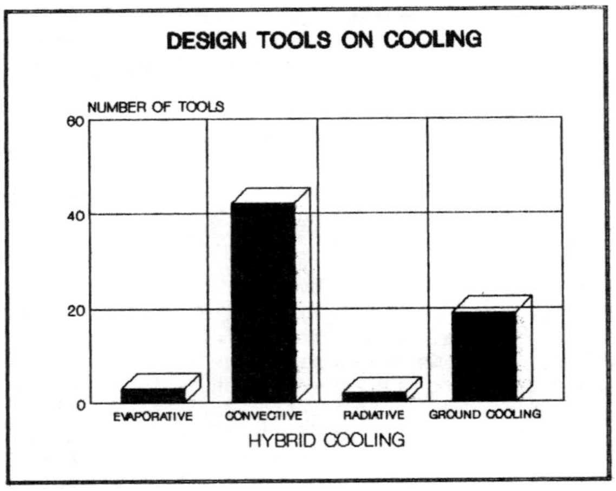

Figure 6: Design Tools on Cooling Classification regarding the heat sink.

6. EXAMPLES OF PROJECTS USING PASSIVE COOLING TECHNIQUES AND COMPONENTS

In the frame of the horizontal study on Passive and Hybrid cooling, data from European Buildings which use such cooling techniques and components have been identified. Seven of them are situated in Greece, two in Spain and three in Portugal. Three of the buildings are educational centers, five are hotels and the rest four are residential and commercial buildings.
Six of the reffered buildings are already constructed and six are under construction. All the studied buildings are designed taken into account all the basic principles of bioclimatic design for Summer protection. Therefore the buildings are firstly well insulated and are characterized by increased mass levels and very efficient shading systems.
For all the buildings, ventilation is the more important heat dissipation technique. Cross and stack effect ventilation techniques are used. Also the most of the buildings use cooling techniques based on the use of environmental heat sinks,like the sky, (radiative), the water, (evaporative),and the ground (earth contact).
A layout of the studied buildings as well as the used systems and the cover percentage of the cooling needs is given in figure 7. As it is shown from the above figure, the passive and hybrid cooling systems offer a very high percentage of the cooling needs, which varies from 20 per cent to 100 per cent.
However only two of the studied building are monitored systematically while for the rest of the buildings the cover percentage is the theoretical estimation. For each building a six page brochure has been created, where all the necessary information, like design strategies cooling techniques, energy efficiency, climatological data and cost can be found.

7. CONCLUSIONS - FUTURE RESEARCH ACTIONS

In the previous chapters the existing situation in the area of Passive and Hybrid cooling has been described. Various scientific gaps concerning different areas of knowledge have been identified since the new technologies require further theoretical and experimental investigation, while the more traditional ones often need better design tools and performance evaluation criteria. Moreover,several techniques lack practical guidelines to facilitate their use by architects.
Therefore three main areas where scientific research should be developed in order to improve the knowledge and possibility of application of Passive cooling technologies are so identified
a) Solar Control in mild climates
b) Passive heat removal techniques
c) Interaction between building and Environment

7.1. Solar Control in Mild climates

Several solar gain techniques have been developed in cold climates, but the application of these techniques in Southern regions requires lot of care, and still large uncertainties do exist in the theoretical and experimental knowledge necessary for the current applications. Among the solar gain techniques and products we consider of interest those that specially challenge the architects skill in southern climates :
- Thermochronic and electrochronic glasses
- Transparent insulation and
- Solar buffering

Therefore a research action in this field has to cover a wide range of aspects, spanning from component and material development to software application for control and design, performance evaluation both in laboratories and field conditions, detailed simulation and components and building behaviour.

7.2. Passive heat removal techniques

A considerable amount of research has been devoted so far to the passive heat removal techniques but still several dark areas remain in this area. The theoretical background is quite well known for eg.
- Passive ventilation (wind and stack induced)
- Radiative, Evaporative and buried pipes cooling.

The main obstacle to a larger use of these techniques is the absence of practical design methods supported by well established experimental evidences. To fill this gap a research activity should gather the existing information, develop significant experiences, collect applications, run detailed simulations comparing results and experiments and finally supply design tools and guidelines.

7.3. Interaction between building and environment

The influence of the surrounding environment on the building thermal performance has long be studied. The aim of a research activity should be :
- To reduce the energy consumption and the thermal discomfort
- To improve amenity and to attenuate noise.

Specific studies to be considered are :
- Selection of suitable autochtonous vegetable species, capable of resisting local climatic conditions
- Study of single tree shapes in different periods of the year, identification of leave front and fall time, evaluation of shading factors
- Study of wind screening effect of natural barriers (earth, banks, vegetation, screens)
- Evaluation of possible landscaping solutions
- Alternative uses of water
- Measurement campaign on existing landscaping applications
- Production of catalogues, handbook, performance indexes.

8. REFERENCES

1. M. Antinucci, B. Fleury, J. Lopez d' Asiain, E. Maldonado, M. Santamouris, A. Tombazis, S.Yannas: "Passive Cooling - State of the art" J.2000 Research Project, M.Santamouris (ed), 1989.
2. ASHRAE : Handbook of Fundamentals, 1981.
3. E. SHAVIV : "On the determination of the Optimum Thermal mass in the Mediterranean climate". PLEA 88, E. de Oliveira Fernandes and Simos Yannas

(eds), p. 385, 1988.
4. E. SHAVIV : "On the optimum design of Shading : Devices for windows", Ibid 3, p. 279, 1988.
5. J. Cook : "Cooling as the absence of heat. Strategies for the prevention of thermal gain".Inter. Passive and Hybrid Conference, A. Bowen and E. Clark, K. Labs (eds), p. 612, 1981.
6. M.E. Hoffman and M. Gideon : "Window design practical directions for passive heating and cooling in heavy and light buildings". Proc. Conference "Windows in building design and Maintenance", p. 277, 1984.
7. H. Sobin : "Window design for passive ventilative cooling. An experimental model scale study". Ibid 5, p. 191, 1981.
8. J. Yellot : "Passive and Hybrid Cooling research". Advances in Solar Energy, p. 241,1983.

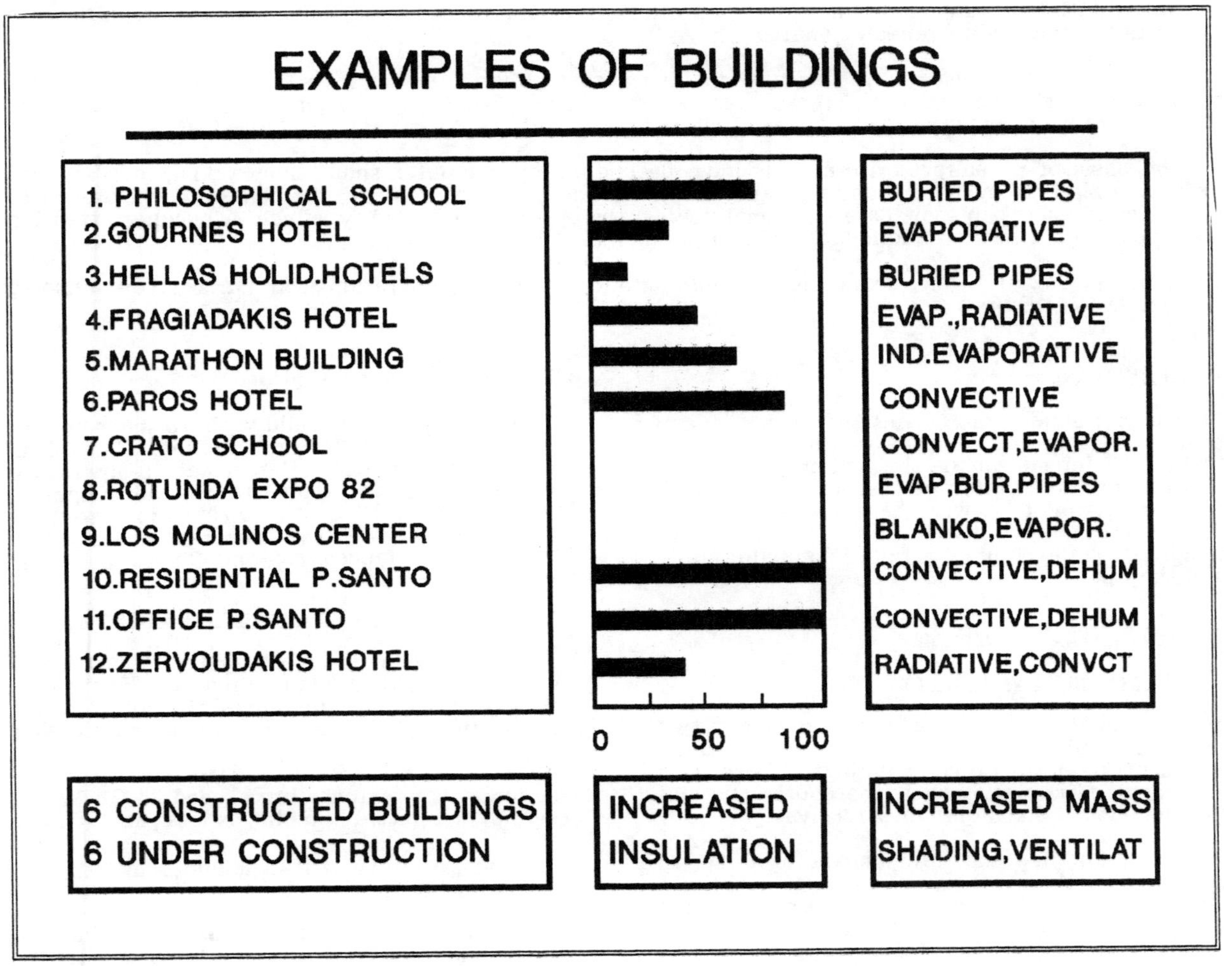

Figure 7: Data of the buildings using passive and hybrid cooling techniques and components.

EVAPORATIVE COOLING FOR SUMMER COMFORT IN OFFICE BUILDINGS

JR. MILLET*, E. HUTTER*, P. PICARD**

* Centre Scientifique et Technique du Bâtiment, 4 av. du Recteur Poincaré 75782 PARIS Cédex 16

** Gaz de France, 361 av. du Président Wilson, B.P. 33 - 93211 LA PLAINE-SAINT-DENIS Cédex

1 - ABSTRACT : In France, air conditioning of offices is often required, especially in areas with high noise levels and in the Mediterranean climates.

Evaporative cooling systems appear able to give a cost effective solution to the growing demand for summer comfort : hence, the CSTB and Gaz de France are currently conducting a joint research on the efficiency of such systems.

On the basis of a comfort criteria, this have been done by computer simulations taking into account differents kinds of systems, buildings, and climatic areas. As a general rule, the comfort level proved to be sufficient except in some cases, especially when the outdoor air humidity is high. A peculiar attention must yet be paid to the building design (thermal inertia, solar protection).

Tests in the Gaz de France experimental building and a full scale operation near Paris have confirmed the simulation results.

2 - APPROACH.

The performances, in terms of comfort and energy consumption, of the systems included in this study depend on the thermal characteristics and geographic location of the building in which they are installed. It was therefore necessary to develop detailed models of typical office buildings, of the cooling installations under study and their control systems.

3 - THE SYSTEMS STUDIED.

3.1 - Representative buildings.

A typological survey was made to select two reference buildings representative of a large share of modern office buildings. These two buildings differ only by their thermal inertia and the amount of solar input they receive.

The first building has low inertia (100 kg/m^2 according to the Th-B rules from the CSTB) and receives little solar input (the ratio of glazed areas to the floor surface is 4 %).

The second has moderate inertia (250 kg/m^2) and receives moderate amounts of solar input (10 %).

All the other parameters are identical. They include : 5000 m^3 volume, 1500 m^2 office area, North-South orientation, Coefficient of volume losses through walls G1 = 0,5 W/m^3. °C, Controlled mechanical ventilation.

3.2 - Different cooling systems.

3.2.1 - The reference.

In order to determine the performances of different systems, a reference case was established : non-cooled building in which the occupants open the window as soon as the inside temperature exceeds both 24°C and the outside temperature. The air change rate is in this case 7 volumes per hour. The windows remain closed during the innocupied periods.

3.2.2. - Night-time ventilation.

One way to cool the building is to make use of its inertia and night-time cooling. Control is simple : as soon as the inside temperature exceeds the outside temperature, the maximum flow rate (4, 7, 12 or 20 vol/h were tested) is triggered. Otherwise, the flow rate is nul when the building is inoccupied and 0,8 volumes per hour when occupied.

3.2.3 - Simple flux with direct humidification.

A first improvement of the system 2 is to cool the fresh air by direct humidification. However it is necessary to limit this humidification to avoid condensation problems. The air humidification acts if :

- the indoor air temperature exceed 24°C,
- the indoor relative humidity is less than 70 %.

3.2.4 - Double flux with humidification of exhaust air.

When a system of warm air heating with double flux ventilation is used, a heat exchanger between supply air and exhaust air must be installed. This equipment can be used during the summer if a humidifier is fitted on the exhaust air before the exchanger.

The above-mentioned method of night-time ventilation is, of course, conserved.

A simple control method, based on temperature thresholds, was developed. It is described in figure 1.

3.2.5 - Double flux with humidification of exhaust air and fresh air.

In order to fully exploit the air network, it was deemed worthwhile to test additional fresh air humidification.

This additional stage is triggered if the inside temperature exceeds 24°C, providing that relative humidity in the offices is not greater than 70 %.

3.2.6 - Location stations.

Three types of contrasting climate were selected according to hygrometry and temperature differences between day and night. The stations chosen are TRAPPES, AGEN and NICE.

non occupation

Indoor temperature (Ti °C)	humidifier	heat exchanger	Flow rate
	A	A	0
19	A	A	0 if $T_s > T_i$ 1 else
20	M	A if Taeh>Te	id
21	M	id	0 if $T_s > T_i$ 2 else

occupation

	A	A	1
22	M	A if Taeh>Te M else	1
23	M	id	1 if $T_s > T_i$ 2 else

A : off Ts : supply air temp 0 = nul
M : on Te : outdoor air temp 1 = 0,8 vol/h
Taeh : air temp behind humidifier 2 = maximum

Figure 1 : cooling system control.

4 - CRITERION FOR COMFORT ASSESSMENT.

A simple calculation of the number of hours during which the threshold of 26°C is exceeded gave an initial evaluation of systems. This method takes account of the duration of discomfort but not its degree.

For this reason, a criterion taking account of both parameters (duration and degree of discomfort) based on the work conducted at the JB Pierce Foundation (P. GAGGE) and the Technical University of Denmark (P.O. FANGER), was chosen. At any moment, the effective temperature, the PMV (Predicted Mean Vote) and the PPD (Predicted Percentage of Dissatisfied) are determined on the basis of operative temperature and humidity.

A mean discomfort duration is calculated over the entire hot season by integrating the difference (PPD - 0,10) when PPD is greater than 0,10 (the comfort zone corresponding to a PPD value less than 0,10). By separing warm and cold mean discomfort durations(WMDD or CMDD), it is then possible to make a more accurate classification of the systems.

5 - SIMULATION RESULTS.

The tool used for this study is the ASTEC 3 software package, an algebrodifferential system solver developed initially for the description and simulation of electric circuits. The models are thus described in the form of electric circuits using the electric thermal analogy.

The simulation results have been studied using the mean discomfort duration (MDD) criterion. Energy and water needs have also been evaluated. As an exemple, table 1 indicates the WMDD value for Trappes site according to the building and system type and the orientation (South or North facing).

inertia	orientation	1	2		3		4		5	
		7 vol/h	4	7	4	7	4	7	4	7
med.	N	23	17	11	8	2	9	3	5	1
	S	30	25	16	14	4	14	5	10	2
low.	N	29	30	21	16	6	18	8	11	4
	S	38	43	28	24	10	28	12	18	6

Warm mean discomfort duration (hours) for different cooling systems, airflow rate and buildings at Trappes

Table 1

For the three weather stations, the system 5 which is the most complete is also the most efficient. Systems 3 and 4 give close results. The system 2 is not efficient and don't give sensibly better results than the reference case and will not be further considered. Detail results are the following :

Trappes : For the low inertia building and an efficient solar protection, systems 3, 4 and 5 need 7 vol/h air flow.
For a medium inertia building systems 3, 4 and 5 can be used with a 4 vol/h air flow if the solar protection is good.

Agen : The possibilities are the same as above if the solar protection is very efficient, and if internal heat gain are low.

If not, low inertia building need a 12 vol/h air flow and medium inertia building a 7 vol/h air flow.

Nice : A 7 vol/h air flow can be used only for the medium inertia building with very efficient solar protection and low internal heat gain. For others buildings, a 12 vol/h or 20 vol/h air flow is needed.

6 - TESTS IN THE GAZ DE FRANCE EXPERIMENTAL BUILDING.

The main aim of the tests was to verify the gain in comfort obtained by indirect evaporative cooling. On three floors of the building, the air network was used to distribute air from a central unit equippedwith a double flux system with exhaust air humidification at a rate of 4 office volume/h (approx. 6000 m^3/h). The top and bottom floors were not cooled. The same occupation scenario as for the numerical simulations was adopted for all floors. A control similar to that of the simulations was applied.

During occupation, the temperature drops by about 3°C and the relative humidity remains at around 50 % for the cooled floor.

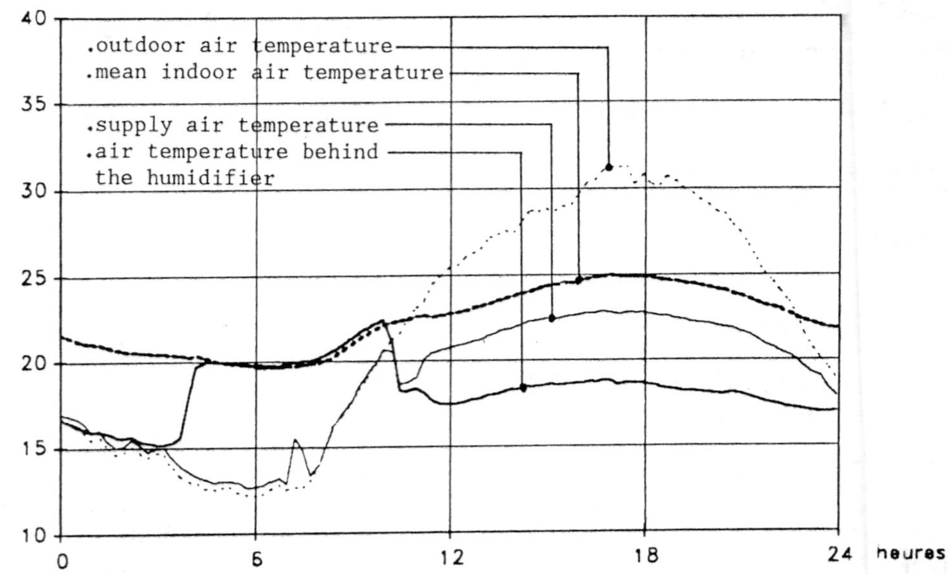

Figure 2 : air temperatures for a warm day in the Senlis pilot building.

7 - PILOT OPERATION.

A pilot operation carried out at Senlis near Paris has been followed up during the summer of 1989. The three level building has a floor area of 3000 m^2. The floors are free of inside walls, except for some rooms situated in the corners of the building. Two indirect evaporative systems have been settled up respectively for the north and the south part of the building.

Air temperature and humidity were mesured in the climatisation units and in some points inside the building.

Though the outdoor air temperatures have been sensibly warmer than for the mean summer, indoor air temperature remain generally between 20°C and 26°C, with a daily variation less than 4°C. (fig. 2 shows the different air temperatures for a warm day). These values are in good agreement with the simulation results.

Air temperatures are higher in the South and West oriented rooms which are situated in the corners of the building and have large window area.

A good window solar protection is needed for these rooms to obtain the same comfort level than for the rest of the building.

Besides, we have pointed out the necessity of verifying precisely the regulation set point temperatures : the cooling effect being limited, it is important that the different running stages perform as intented. The electricity and water consumptions have been of about 13 kWh and 40 l per floor m^2 for the summer period.

8 - CONCLUSION.

These studies have shown that it is possible to obtain satisfactory comfort in summer by using evaporative cooling system . The efficiency will yet depend on the thermal characteristics of the building (solar window protection, thermal inertia) and of the site. This imply that for these systems the building and the cooling system have to be designed at the same time.

For the Paris district and in a building with moderate inertia, the flow rate required (4 office vol/h) is perfectly compatible with an air heating system, and the slight additional cost is mainly due to the humidifier.

For hotter regions such as Agen, the air flow rate must be higher (up to 7 vol/h) but remains acceptable, whereas in Nice (Mediterannean coast) a network with a flow rate of 12 vol/h or more would be difficult to implement.

Experimental studies have confirmed the simulation results both for obtained comfort level and electricity and water consumptions. Further researches are now conducted : Firstly on the use of dessicant wheels to improve comfort and lower required air flow rates and secondly on others techniques such as cooled water circulation in floors or ceilings.

9 - BIBLIOGRAPHY.

WATT J.R., "Evaporative Air Conditioning Handbook", Second edition, Chapman and Hall, 1986.

ESKRA N., "Indirect/Direct Evaporative Cooling System", ASHRAE Journal, May 1980.

SUPPLE R.G., Evaporating cooling for comfort", ASHRAE Journal, August 1982.

ASHRAE Systems Handbook, "Evaporative air cooling", 1984.

CURT C.W., "Control concepts for evaporative cooling systems", ASHRAE TRANSACTIONS, 1985.

MILLET J.R., HUTTER E., "Etude sur la climatisation sans groupe frigorifique des immeubles de bureaux", rapport CSTB GEC n° 89-4733, 1989.

BOHLER A., MILLET J.R, VILLENAVE J.G., "Climatisation sans groupe frigorifique : Suivi de l'opération pilote de Senlis", rapport CSTB GEC n° 89-4938, 1989.

CONTROLLED VENTILATION AS A COOLING AGENT IN SCHOOL DESIGN FOR ARID TROPICAL AREAS

KHAFAJI, DR. O. M. A., and MURTA, PROFESSOR K. H.
KING ABDULAZIZ UNIVERSITY
JEDDAH, SAUDI ARABIA
UNIVERSITY OF SHEFFIELD
WESTERN BANK
SHEFFIELD S10 2TN

ABSTRACT. This paper studies the utilization of natural ventilation in school buildings for hot arid regions. It attempts to provide natural ventilation by the controlled movement of cool air. This is to be assisted by special use of stack effect system in buildings. To achieve such goals, two typical problems have to be resolved; firstly, external temperatures should be reduced to the minimum; secondly, pressure differences between inside and outside should be increased to the maximum. The outcome of this study indicated that, typical layouts (central hall, double banked corridors, etc...), should be avoided. The solution should incorporate functional and what it may call climatical layouts. The conventional methods of applying the stack effect system are not effective in these circumstances. An original method has been developed to utilise the stack effect system to use convection currents to draw cool air through the building.

1. CONCEPT

This paper describes an experimental and innovative design for school building in Jeddah, Saudi Arabia. It is based on the functional requirements for educational provision in that country, which proscribes particular planning form and layout. Beyond that functional aspect there were conceptual aspects which were as follows:

1. The design would respect the cultural tradition of the region.
2. That where possible mechanical air-conditioning would be avoided, and in particular class rooms and teaching areas would be cooled by natural means.

A great deal of investigation and experimental work was undertaken by Dr. Khafaji. This lead to the following principles being adopted for the prototype design.

1. That natural air movement, mainly convectional, which had been used traditionally in houses at Jeddah, could be used in school design. In particular, the use of cross ventilation during early morning and night to cool the building fabric. (Fig. 1)
2. The use of suitable construction and materials to restrict internal radiation to those periods when the building was not normally in use.
3. The recognition that in Jeddah, sun angles are such that careful consideration must be given to the formation of self shading buildings by suitable planning and appropriate planting as well as functional shading devices.
4. The development of shaded areas to form cool air 'pools' which could be utilised for cooling purposes.
5. To utilise the propriety of heated air to rise and create convection currents throughout the building.
6. The need for flexibility in the building operation to take advantage of the diurnal rhythm and changing thermal conditions.

Design using these principles is often described as 'passive' to reflect the lack of mechanical aids. The authors would instead regard this as 'active' solar design as thermal gain is used in solar 'chimneys' to create air movements which can then be directed for cooling and/or provision of comfortable conditions.

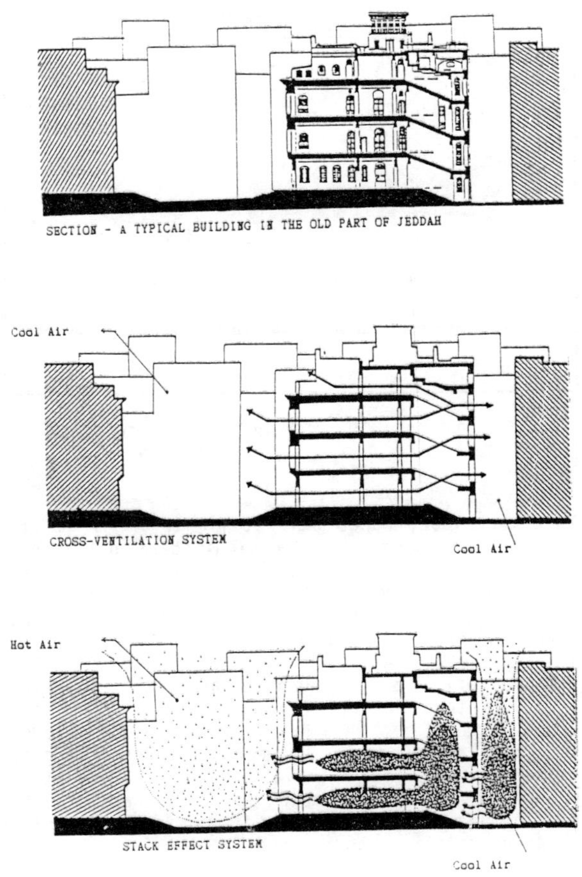

Fig. 1

2. DESIGN POLICIES AND PROPOSALS

The starting point for the environmental part of the work was an analysis of traditional buildings in Jeddah. This showed that the self shading principle was soundly based and led to the rejection of some standard layouts for school buildings which were popular in other climatic zones. A clear example was the finger plan layout orientated east/west which is often adopted in tropical areas.

The double banking of teaching spaces around a central circulation route, so economical in temperate zones was also rejected. A surprising high number of such layouts were found in the survey of existing educational buildings in the area. Fig. 2 represents this quite usual situation showing three temperature zones. The creation of cluster groups has been devised to improve this situation. These form 'climatical' layouts which enable the strategy of control of air movement, seen as the driving force for natural cooling to take place. Fig. 3 shows how the three temperature zones have been increased to five, giving greater flexibility and a hierarchy of temperature related areas.

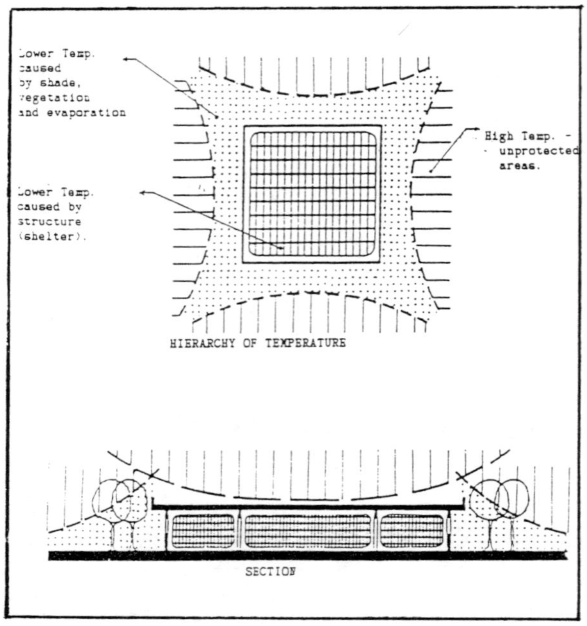

Fig. 2

This technique, based on traditional forms, is to create areas which have a significant difference between external air temperature and that of the shaded areas. The resultant pressure differences being used to create air movement which will be directed across body levels.

The theoretical stance adapted was that air passing into the teaching areas would pass through four zones (the last one being the teaching space itself), the boundaries between the zones acting as filters, and the travel paths deliberately elongated by the planning process, assist the cooling process. Where conditions are suitable, cross ventilation can be used.

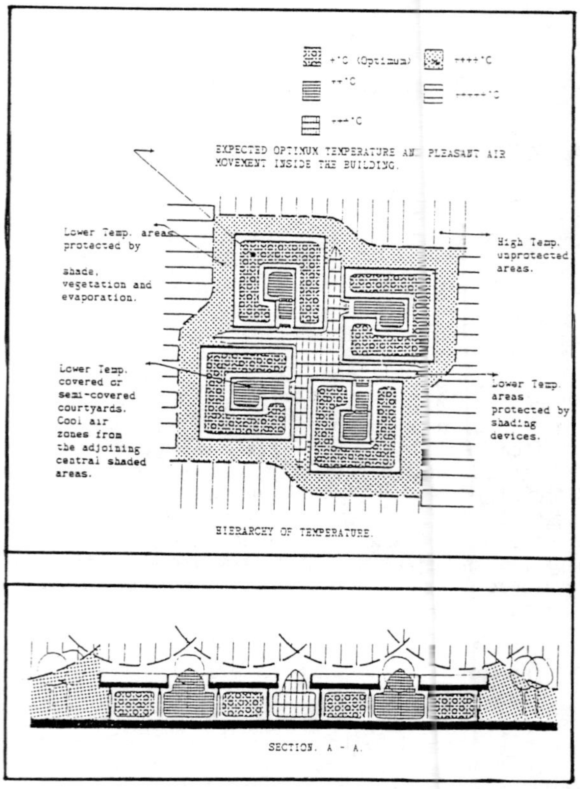

Figs. 3

The first zone, the shaded courtyard would employ water, either in the form of removable pools or other evaporative devices, to reduce air temperature. Normally, ventilation in arid tropical buildings is reliant upon cross winds and natural convection. An innovative design for forcing the flow of air to body levels in the teaching areas is by placing solar 'chimneys' on the exterior and drawing air from the inner zone by convection. The flues having a suitable cross section to give some control by use of dampers or louvres. The higher part of the 'chimney' is of glass and steel to increase temperature to a high level. The principle is illustrated in the sections. (Fig. 4) The stack is expected to be of use in the later part of the day as well as when the cross ventilation system is closed down in mid-morning.

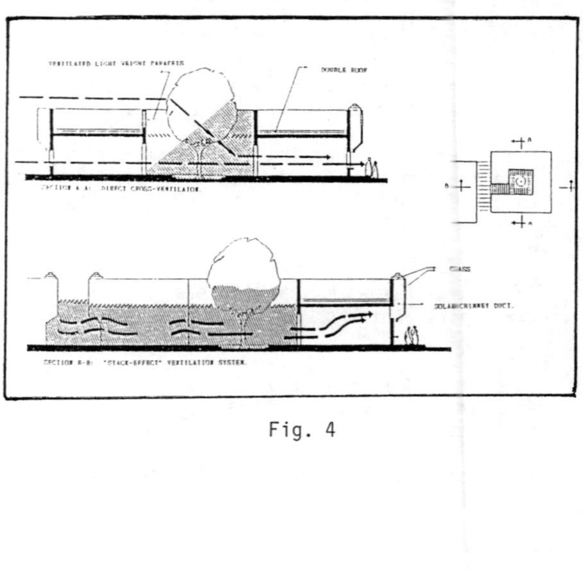

Fig. 4

The detail of the 'chimney' is shown in Fig. 5 while Fig. 6 shows the model made for wind tunnel testing. Fig. 7 shows the wind tunnel test in operation.

Fig. 6

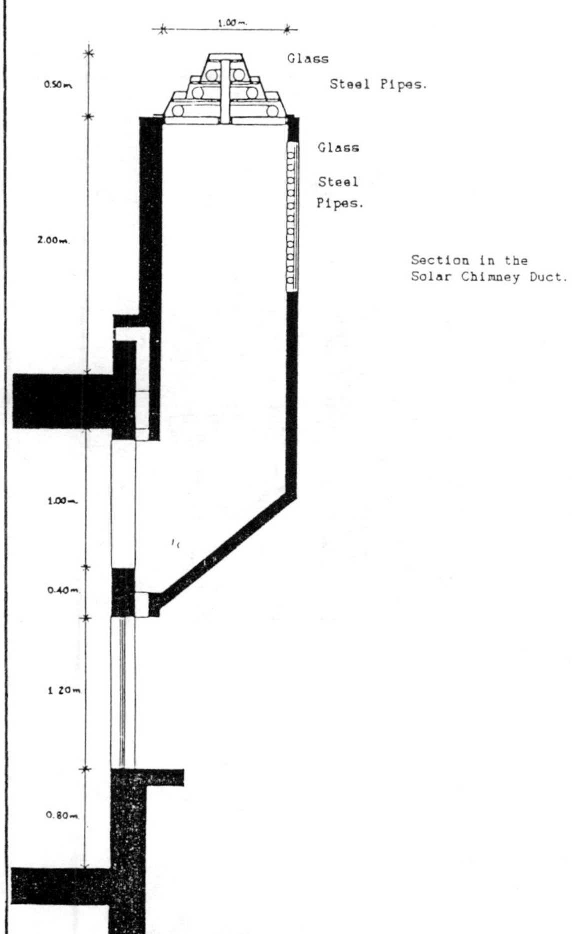

Fig. 5

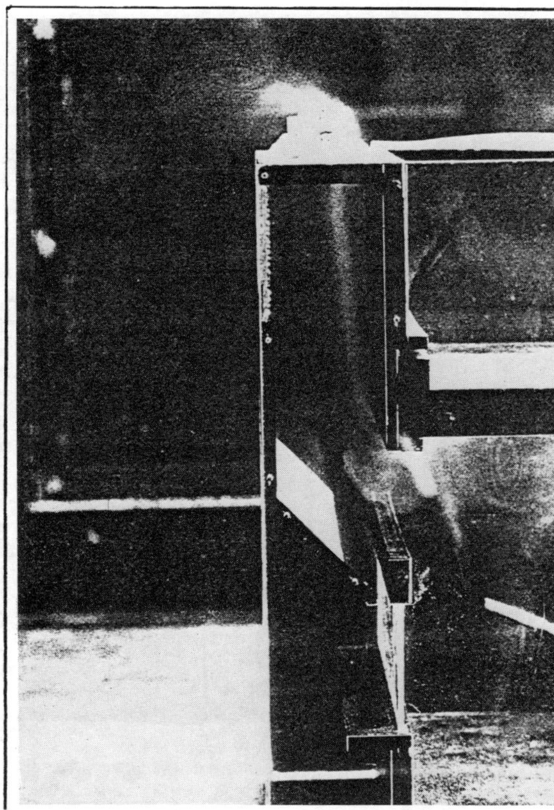

A model of the chimney duct has been made to examine the air flow by using the wind tunnel. This picture shows the flow of the smoke in and out of the chimney.

Fig. 7

3. CONCLUSIONS

The testing and theoretical studies indicate that a combination of cross ventilation, and active cooling from the technique described could provide architectural solutions within the original conceptual framework. It also became clear that the design of buildings in hot arid climates demanded attention and selection to materials which provide a thermal time lag so that radiation from internal surfaces took place after school hours. In addition, the design of roofs needed spatial attention to prevent radiation occurring quickly in teaching areas and thus distorting the convectional flow. There is still detailed work to be done. Lighting levels, which if provided by using natural daylight, means heat has to be carefully thought out. Again, the plan allows for carefully shaded openings and the use of light from courtyards etc. Dust penetration is a hazard which has to be prevented by filters at appropriate places before the ventilation air reaches sensitive areas.

There are, of course, may other factors to be considered in school design. The plans, Figs. 8, 9 and 10 show the application of the climatic cluster technique to two functional areas with a variety of teaching functions, thus demonstrating realistic and practical application of the system in a positive manner.

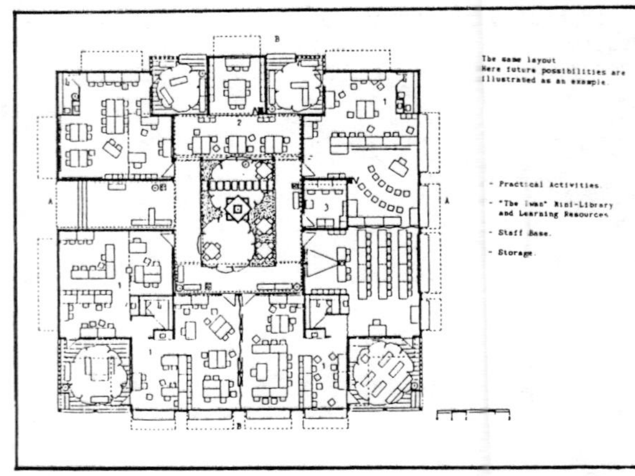

Fig. 9

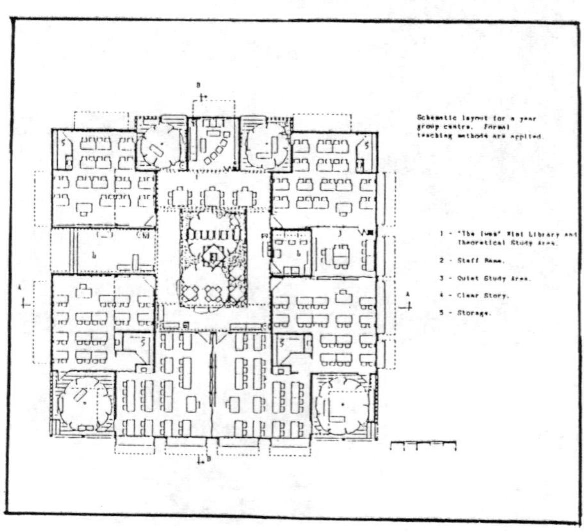

Fig. 8

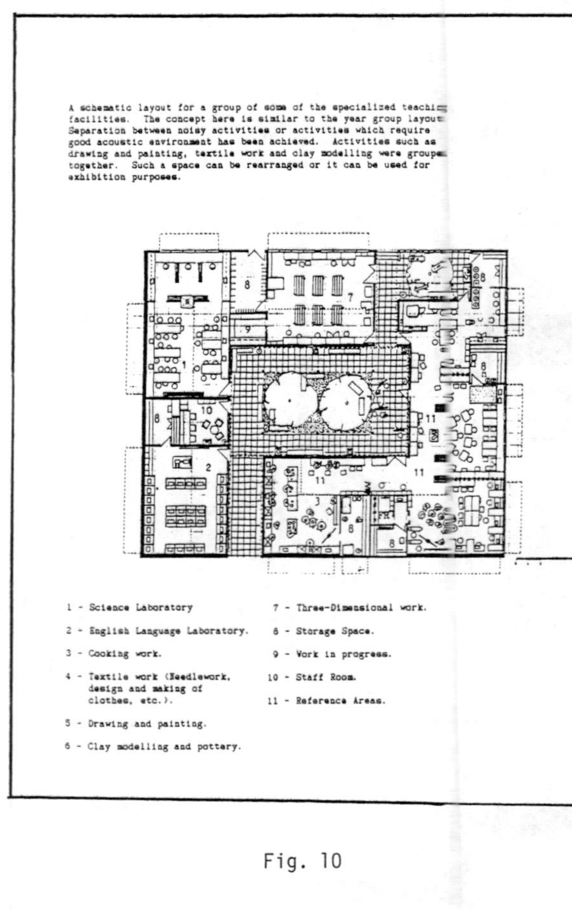

Fig. 10

OVERHEATING IN BIOCLIMATIC ARCHITECTURE IN MEDITERRANEAN CLIMATES

J.A.Cusidó(1), J.J.Escobar(2), J.Esteve(2), J.Jorge(1)
A.Mitjà(2), J.Puigdomènech(1), C.Torra(2)

(1) Departament de Meteorologia i Ciència del Sòl
P.O.Box 508, 08220 Terrassa (Spain)

(2) Direcció General d'Energia, Generalitat de
Catalunya, Diagonal 514, 2n. floor, 08006 Barcelona

ABSTRACT. For the last years Thermal Behaviour Analysis research on Buildings has been focused more on heating requirements than on the aspects related to their natural refri--geration. We present our experience on tracking buildings placed in Mediterranean area and we discuss the problems related to evaluate the energy and confort performances.

1. INTRODUCTION

The Mediterranean climate region covers the coastal areas and most of the islands of the Mediterranean Sea, and some conti--nental lands, between paral.lels 35 N and 45 (Fig. 1). It presents a seasonal vari--ation between overheating and underheat--ing with no severe periods.
The range of average outdoor temperatures is from 4ºC to 13ºC in winter and from 21ºC to 27ºC in summer, except the moun--tainous zones in the region. In some ca--ses, temperatures above 37ºC are not rare.
The annual degree-days for a base tempe--rature of 18.3ºC in this region area about 1300-2100.
The mean sunshine duration per day is more than 6.5 hours for 7 month, and the annual average global irradiation on an horizon--tal plane is above 4Kwh/m^2 day (1.5 in winter, 6.5 in summer).
The precipitation, between 38 and 76 cm, falls in spring and autumn, with a pre--dominant drought in summer.
There is an important relative humidity near the sea that makes water evaporation difficult and may reduce solar irradiance, which is moved by the breezes coming from the sea.
Winds are of local nature, important for local evaporation studies. "Pirineos", "Alps" and "Balcans" are a barrier for the cold polar air.
In Mediterranean climates the cooling energètic needs can exceed appreciably the heating ones. In warmer zones (i.e. Athens) the cooling nees can be three times higher than the heating ones, but in colder zones these quantities can be similar or slightly inferior in a small percentage (-25% to 50%) /1/.

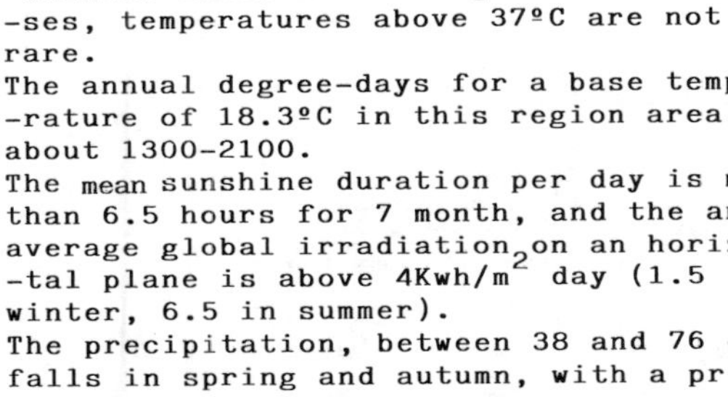

Figure 1: European Mediterranean Climate Map.

The Mediterranean zone has had a delayed industrialization process compared to the rest of Europe. For this reason, there is an acceptance of confort levels that are under the average of those in Europe, as well as the adoption of special charac--teristics in the behavior of excess ener--getic periods. We can foresee that forth--coming generations will no longer accept this low confort levels and then the use of HVAC systems, that is the energetic needs will increase, by diminishing the long summer periods of the Mediterranean area, with a general lack of labour acti--vities due to the climate.
Research on natural climatization systems of both summer and winter in mild climates has a long way to develop, specially in the experimental field and prototypes rea--lization, and can be of great influence in the energy savings and the energetic diversification policies, where we are on time to incise. In this way we need to deepen more than to enunciate simple reco--mendations like white walls and light colored roofs (radiative cooling), high thermal inertia (block or earth thicker than 50 cm), ventilation (low during the day, high at night), outside insulation of walls and roof, evaporative cooling in some cases, etc. and achieve simple methods to quantify and evaluate biocli--matic constructive solutions for natural cooling (Fig.2).

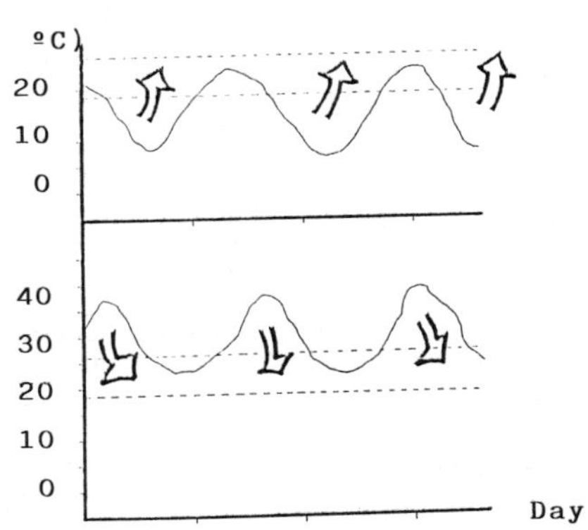

Figure 2. Very different strategies for natural climatization during lack period (winter) and excess period (summer).

2. THE EXPERIENCE OF PROJECT MONITOR IN MEDITERRANEAN SUMMER BUILDINGS

If in itself the definition and the reali-zation of a full scale monitoring is difficult, it increases when we try to carry it out in exces periods. Nevertheless if we analyze the results shown on the brochures published in the Project Monitor /2/ and select those that belong to the Mediterranean zone (shown in Table I), we can extract the following conclusions:

a) Although Global Irradiation on horizon-tal plane can achieve a 50% surplus in Mediterranean locations against others with higher latitudes, Solar Contribu-tion to Net Space Heating Load depends on the suitable bioclimatic design, whe-re the priority to avoid overheating problems often causes its moderate si-zing and consequently the building performs similarly to the buildings placed in colder climates.

b) The substractive method for data analy-sis is not accurate when applied dur-ing summer conditions in Mediterranean climates and it is not clearly applica-ble for some typologies either (i.e. sheltered buildings, earth-air tunnels, evaporative cooling, etc.).

c) The 31% at least summer tracking has not been carried out.

d) Most of the buildings monitorized pre-sent overheating problems during sum-mer (70%), when the site is clearly Mediterranean, otherwise, the auxili-ry energy consumtion is high.

e) No summer passive systems are used in the selected buildings. They only ex-ist, practically, for the heating seasons. More specific summer natural climatizations can be developed and integrated in Mediterranean Architec-ture Tradition.

f) On the other hand, in orther to select or continue building monitoring a re-markable energetic response must be required as well as its formal quali-ty to prevent the decrease of poten-tials users amount.
We must also promote the non-residenti-al buildings, like public buildings, industrial, collective dwellings, etc., where we hope it will be easier to reach a future development of Bioclima-tic Architecture.

TABLE I. Selected Solar Passive Buildings in Mediterranean area from published Project Monitor Schedules.

Brochure Number	Building	Summer Behaviour	Comparison with conventional design		Contribution to total space heating requeriment		
			Actual Load	Saving	Aux. Heat.	Int. Gain	Pass. Solar Gain
3	Polysportive Esterri (Spain)	No problems in summer (Climate Alpine)	27%	73%	4%	26%	70%
5	Liceo Scientifico Casalpalocco (I)	Blinds and shutters insulate in winter and in summer pre-vent overheating and allow control over the intensity of daylighting. No summer tracking.	24%	76%	37%	20%	11.5%
9	Casa Termicamente Optimizada (P)	Overhangs, shutters, cross ventilation and large ther-mal mass prevent overheating in summer.	27%	73%	38%	17.5%	44.5%
10	La Salut. Barce-lona (Spain)	Overheating in the sunspace in summer. Stratification temperature in building.	13%	87%	6%	27%	67%
14	Marostica Marostica (I)	Barra Constantini is adopted and may be operated as a sol-lar chimney for summer cool-ing of dwelling. No summer tracking.	59%	41%	60%	14%	26%
19	Barbaricina Pisa (I)	Westerly winds 2.1 m/s in summer are considered. No summer tracking.	49%	51%	49%	25%	24%
20	Lou Souleu Avignon (F)	Remarkably good thermal per-formance is achieved in win-ter, the solar contribution to the heat load reaches 80%, but at the expense of summer overheating.	33%	67%	18%	12%	70%
24	Los Molinos Crevillent (S)	Overheating in summer in different parts of the buil-ding.	-	-	-	20%	80%
26	Orbasano Torino (I)	Overheating in sunspace and rooms, in summer.	49%	51%	48%	16%	30%
33	Collegno Torino (I)	The shading and ventilation measures have been adequate in controlling overheating. (65% auxiliar energy)	66%	34%	65%	13%	16%
34	Hospital Vielha Vielha (S)	No overheating in summer. (Climate Alpine)	38%	62%	44%	23%	33%
37	Llavaneres Semi-buried House Llavaneres (S)	Good summer confort without the need for auxiliary ener-gy.	38%	62%	58%	7%	35%
38	Istituto Tecnico Montefiascone (I)	Monitoring data indicate sum-mer overheating in the top floor classrooms. No summer tracking.	49%	51%	31%	41%	28%

3. PROBLEMS IN FULL-SCALE COOLING MONITO--RIZATION

3.1. Degree-day based energy building evaluation.

Despite the popularity of degree--day-based techniques, beyond theo--retical considerations about their accuracy in practice, these methods are often used in the energetic effi--ciency evaluation of a building in full-scale perpective and in a prac--tical point of view in the architec--tural development. Some other exis--ting methods of more accurate theo--retical foreseign are very costly, in part, because they require measurement of all influences on building usage and weather.

The most widespread method user consi--ders the following steps:

- -Calculate the Gross space heating load (G).
- -Calculate the Internal heat gains(I).
- -Calculate the Net heat load (N= G-I).
- -Substract Auxiliary (A) to obtain useful Solar (S= N-A).

Obviously, in this case cooling de--gree-days for weather normalization should be considered:

$$\text{Cooling DD} = \sum_1^N |\text{base temp} - \text{average daily temp}|$$

where:

Average daily temperature=

(max day temp + min day temp)/2

and when:

(base temp - aver day temp) < 0

The definition of the base temperature are usually 18.3ºC but the physical significance can be thought of as the outdour temperature at which internal plus solar and other gains offset heat losses /3/. This difficulty increases when cooling base temperature is re--quired.

In a steady-state model heat loss e--quation is:

$$T_{out} = T_{in} - (Q_{gain} / G.V)$$

T_{out} is known as the "balance point" temperature since lower outside temperatures mean that heating energy will be required to maintain T_{in} at a constant level. In practice it is very to calculate, given the large amounts of data required. This is also the reason which a correct definition for cooling degree-days should be given.

In this sense, it seems more suitable to use alternative methods, derived from a regressive analysis, in energy evaluation. With this method we take into account that the two most impor--tant factors afecting auxiliary ener--gy consumption in a building are the difference between indoor and outdoor temperatures and the received irradia--tion. Both variables and auxiliary energy can be measured with good accu--racy in daily or weekly periods.

Every data set give a surface whose mathematical fitting gives us thermal losses coefficient, useful solar con--tribution and net internal gains. However in hot seasons there are some incertainity in the losses evaluation (or gains) due to the poor response of auxiliary systems and ventilation.

A variation of the last method is to consider /3/:

Heating energy use= $\alpha + \beta \cdot DD$

where:

α = intercept non weather-sensiti--ve component (MJ)

β = heating or cooling slope or weather sensitive-component (MJ/DD)

DD = heating or cooling degree-days to base temperature

In this technique, the base temperatu--re corresponds to the best regression of energy use on degree-days, as mea--sured by R^2 values.

3.2. Summer ventilation.

In a design point of view, the seaso--nal energy savings in the refrigera--tion by ventilation concept is impor--tant.

The evaluation of structural refrige--ration is a complex objective, becau--se dynamical effects of thermal mass and the mass-ventilation air coupling can be considered. It is obvious that we cannot use tools to design structu--ral refrigeration but heavy and power--ful simulation methods /4/.

We cannot forget, that in high tempe--ratures and high humidity levels, very common in the Mediterranean area, ven--tilation is the most efficient cooling system.

When we use the susbtractive method, the determination of the renovation air index, ACH, usually 0.5 to 1.5 in winter, can become from 13 to 30, and can reach 50 depending on internal gains. In this case, ventilation beco--me the most important contribution to the Gross space heating load (G).

3.3. Convective loop in buildings placed in mediterranean climates.

In climates with a high solar irradia--tion, an suitable driving force is present which can be able to generate convective loops with heat distribution in the building; this influences the system efficiency in winter and the refrigeration potential in summer. In this case the building surface with a close contact with the air is a cri--tical parameter of the summer passive system that cannot be dimensioned in easy way.

3.4. Difficulties of internal gains deter--mination and the definition of con--fort parameter for summer in Medite--rranean climates.

It is necessary to carry out an speci--fic research to enunciate a confort parameter, together with a thermal evaluation. One índex can be related to:

$$I_{cnf} = 1/4 \int_{8am}^{12pm} (T_{air} - 26)^{+} \, dt$$

The increse of complexity takes into account latent heat dissipation in accordance with tipycal behaviours in natural climatization in the Mediterra--nean area.

4. SOME CASES

To illustrate the discussion, we show some exemples of summer behaviour where simple methods of analysis cannot be user.

a) Llavaneres Semi-buried house.
Arch. F.Xavier Barba
External and internal wave temperature and positive or negative heat flux in sheltered building placed in Medite--rranean climates (Catalonian, Spain, Issue 37 from Project Monitor).
In this case, the substractive method lack on calculate the Gross Space Heating, G.

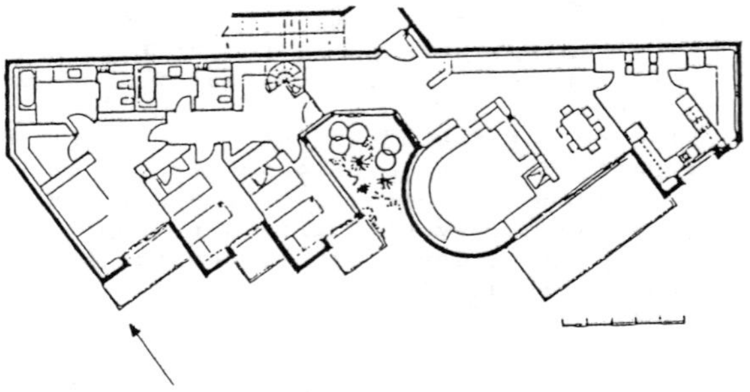

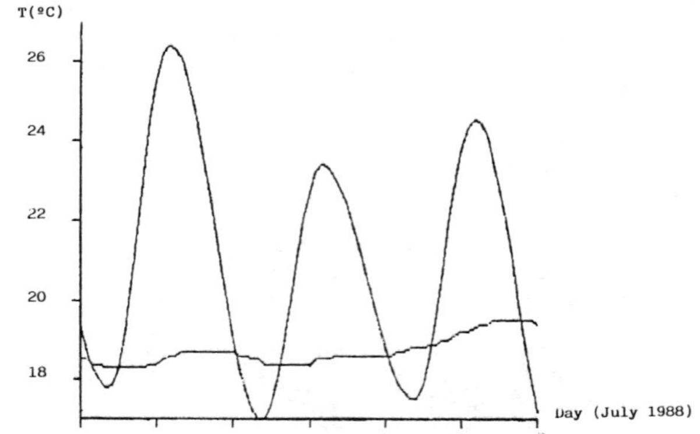

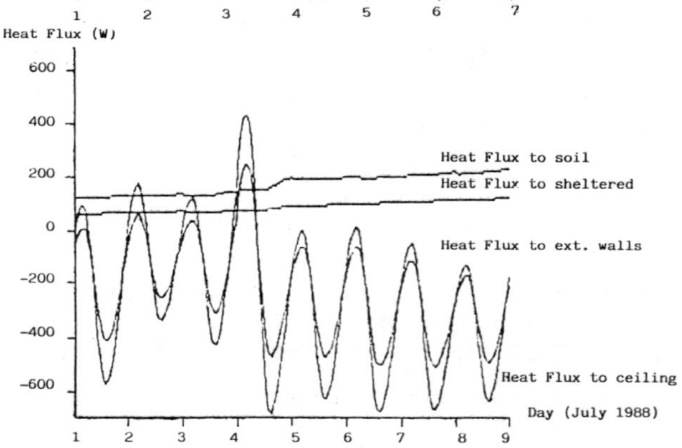

b) Double wall Sapé hôuse.
Arch. J.I.Llorens and A.Soldevila
In this case, the high thermal mass induced by the masonry produces a negative effect during summer. Temperature stratification and dissi--pation problems are observed.

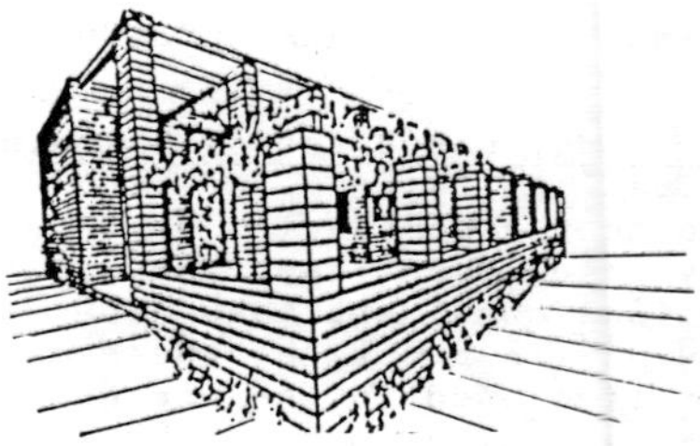

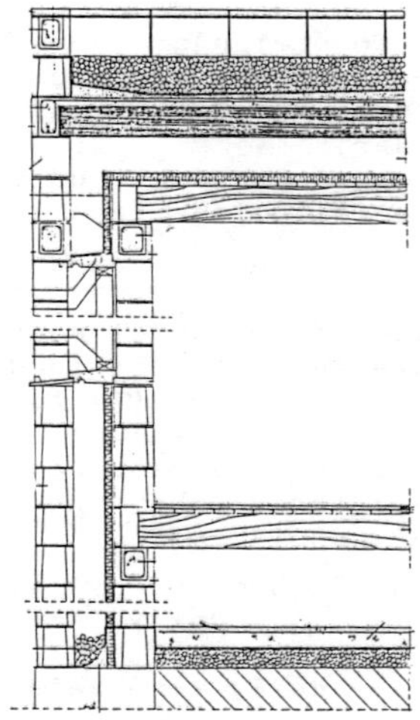

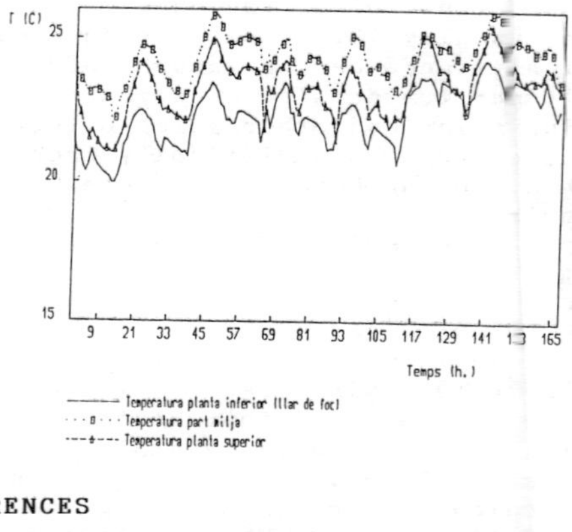

REFERENCES

/1/ EUROPEAN PASSIVE SOLAR HANDBOOK, Edited by P.Achard and G.Gicquel. Published for the Commission of the European Com--munities (1986).

/2/ PROJECT MONITOR BROCHURES, CEC, J.Owen Lewis, University College, Dublin Rich--view, Clonskeagh, Dublin 14 (Ireland).

/3/ ON USING DEGREE-DAYS TO ACCOUNT FOR THE EFFECTS OF WEATHER ON ANNUAL ENERGY USE IN OFFICE BUILDINGS, J.H.Eto, Energy and Buildings, 12, 113-127 (1988).

/4/ FEASIBILITY STUDY FOR MEDITERRANEAN TEST CELLS, CEC Contract no. EN3S-0152--I(A), M.Antinucci, Conphoebus s.c.r.l., Via Leopardi, 148, 95127 Catania (Italy).

NEW VENTILATION CONCEPTS WITH RESPECT TO INDOOR AIR QUALITY AND ENERGY CONSERVATION

P. Wouters
Belgian Building Research Institute
avenue Pierre Holoffe 21
1342 LIMELETTE - Belgium

ABSTRACT. The strong increase in energy prices during the seventies has focused most of the efforts on energy conservation. The fact that many European countries with a moderate to cold climate (e.g. Belgium, United Kingdom, West Germany, Switserland) haven't yet appropriate ventilation requirements for dwellings reflects the priority given to energy conservation measures. The eighties are more characterised by an increased awareness that a healthy indoor and outdoor environment is not evident. This growing awareness is a result of mould growth problems, the increased number of allergic reactions, the findings with respect to dangerous gasses (radon, ...) and particles (asbestos, ...). It becomes more and more clear that a good indoor air quality has its price.

As a result of it, there is at present in the field of ventilation a clear trend to optimize the combination of a good indoor air quality with a reasonable energy consumption. Many research projects in the eighties were focused on a better understanding of the indoor air quality. At present, the various pollutant sources are better known, the allowable concentration levels in occupied zones are better defined and new ventilation concepts with respect to the distribution system as well as with respect to the control strategy (e.g. demand controlled ventilation), are becoming operational.

A second clear trend is the strong increase in knowledge with respect to ventilation. Most significant is the evolution in computer modelling and measurement techniques.

However, there is still a lot of work to be done with respect to the development of simplified design rules for designers as well as with respect to user friendly software codes for more detailed calculations.

1. INTRODUCTION

To realise a good indoor air quality is less evident than many people thought 5 - 10 years ago. Various types of pollution are possible. The same is true for the outdoor environment. In its first part this paper is focused on the indoor air quality. The important role of ventilation becomes clear from this limited analysis. In the 2nd part more attention is paid to ventilation related research activities and some new trends in building practice.

2. THE SIXTIES AND SEVENTIES : FIRST MORE, THEN LESS

The technology available in the sixties, the strong growth in economy and the increase in income have stimulated new construction types, new heating systems and also new ventilation systems. The main aim of these changes was to increase comfort. The application of mechanical ventilation systems, especially in office buildings and appartments, was increased significantly. Due to the low energy prices, high ventilation rates per person were no real problem. An energy conscious design was in these circumstances less crucial.

The strong increase in energy prices due to the energy crises in the seventies significantly changed this situation. This resulted in :

* increased thermal insulation of the building envelope to reduce heat losses;
* use of appliances with a higher efficiency;
* lower indoor temperatures;
* efforts to reduce ventilation losses, e.g. by draught-stripping and low standard values for ventilation. The 1981 version of ASHRAE-standard 62 /1/ requires only 9 m^3/h person (ASHRAE = American Society for Heating, Refrigerating and Air Conditioning Engineers). The effect of these changes on the specific energy consumption is clearly shown in figure 1.

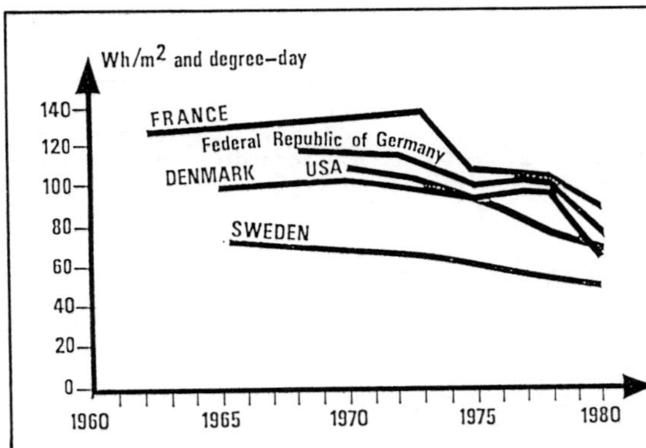

Figure 1 : Specific energy consumption per m^2 of heated floor surface (corrected for degree-days)/2/

Research activities in the second half of the seventies and the beginning of the eighties were primarely focused on strategies to reduce the energy consumption. In most countries the concern for a good indoor air quality was a marginal part in the research activities.

The Programme on Energy Conservation in Buildings and Community Systems (ECBCS), operating within the International Energy Agency (I.E.A.) started various research activities /3/. The Air Infiltration Center (A.I.C.) was created as an international organisation for coordination of research and technology transfer /4/.

One can summarise these 2 decades by stating that on the one hand the sixties have led to a strong increase in energy demand (and ventilation) due to <u>comfort reasons</u>. On the other hand the second half of the seventies and the beginning of the eighties have led to a strong need to reduce the energy demands and ventilation losses due to <u>budgetary reasons</u>. This is illustrated in figure 2.

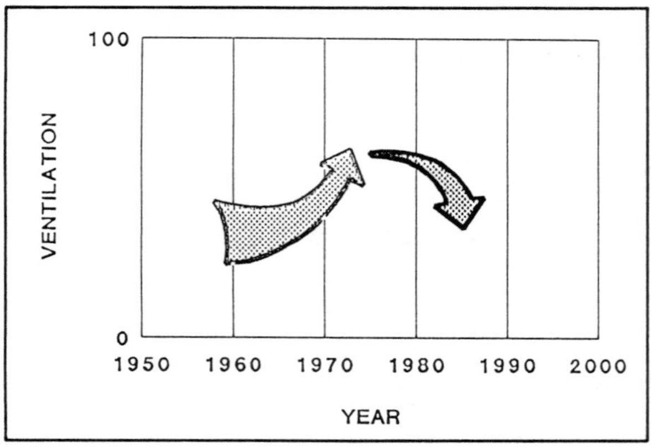

Figure 2 : General trends in energy consumption and ventilation losses

3. THE LATE EIGHTIES : MORE AND LESS AT THE SAME TIME

Health issues related to the quality of the indoor environment have received in the late eighties. large interest by the public and the media :

- radon-related problems;
- asbestos and other particles;
- allergic reactions due to a bad air quality;
- problems in the field of the so-called "sick-building syndrome";
- moisture and mould growth.

Increased ventilation seems to be an appropriate solution to improve the air quality.
The strong reduction in energy prices during the last 5 years has given many people the impression that energy efficiency has become less crucial.
Figure 3 shows the evolution of the imported crude oil price in US $ and ECU according to information and estimations by CEC-DG XVII /5/.

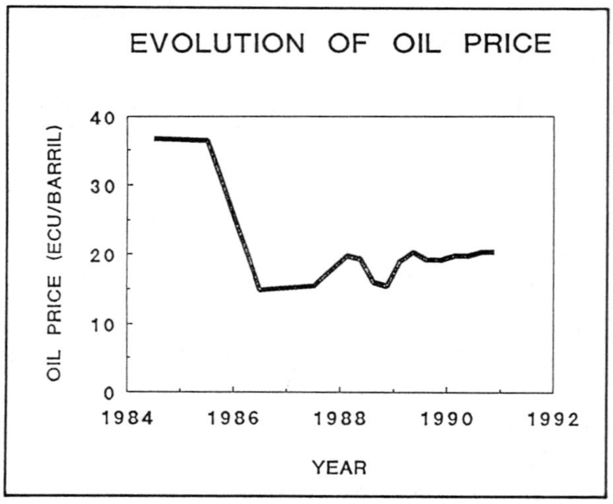

Figure 3 : Estimated evolution of oil price at short term /5/

The reduction in the energy prices has in any case a negative impact on the financial benefits of investments in energy efficiency improvements. The advantage being higher ventilation rates at a reasonable cost.

However, environmental issues have become more and more important towards the end of the eighties :

- a number of important industrial and nuclear accidents have shown how seriously we are at present capable of damaging large parts of the outside environment;
- the problems of acid rain, deterioration of the ozon layer, warming up of the atmosphere due to CO_2-increase are seriously threating our future.

One of the main effects of the growing interest by the public in these environmental problems is the very recent political statements that energy efficiency is an important issue, NOT only for budgetary reasons BUT also for environmental reasons. Important budgets will be reserved in the near future to improve energy efficiency.
Mr. J. Millhone, Director Buildings and Community Systems at the U.S. Department of Energy declared recently at the European Conference on Energy Efficiency in Buildings in Lille (F) /6/, that energy efficiency is again an important issue for the U.S. government for 6 reasons :

. economic efficiency;
. economic competitivity (with other countries);
. global warming (CO_2);
. acid rain;
. ozon depletion (limit emission of CFC);
. national security (limit dependency of energy import).

Figure 4 gives a graphical representation.

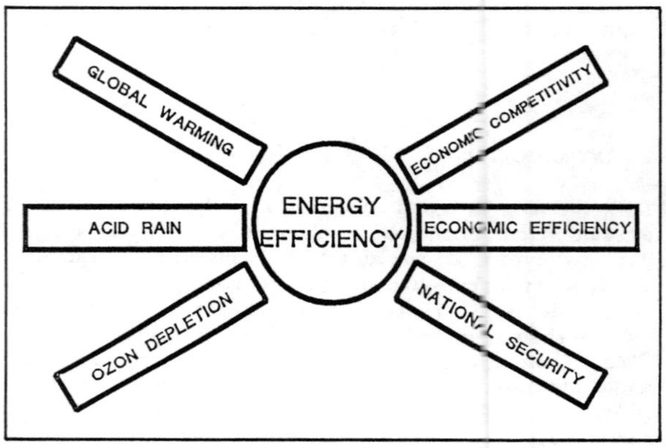

Figure 4 : Main reasons of U.S. Department of Energy for policy of energy efficiency /6/

One can summarize the situation with respect to the ventilation related topics, by stating that 2 types of requirements should be met in the near future :

- on the one hand, better indoor air quality is needed in most building, implicating probably more fresh air;
- on the other hand, the need to reduce to pollution of the outdoor air requires lower energy consumption and less ventilation losses.

This 'more and less'-requirement is shown in figure 5.

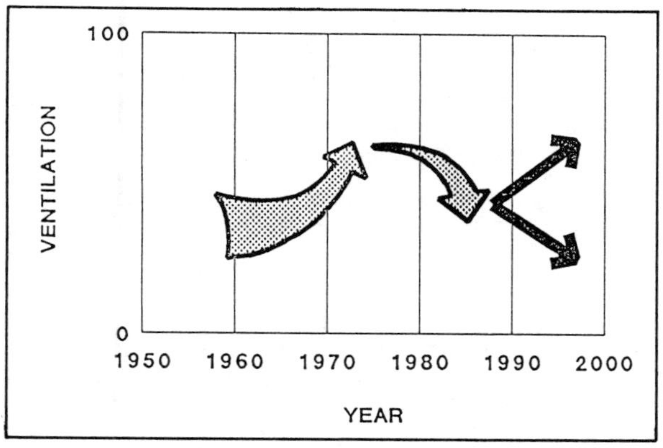

Figure 5 : The 'more and less' trends in the late eighties

One part of this trend is already reflected in the 1989 draft proposal of the ASHRAE- standard 62 which requires 7.5 l/s/person instead of 2.5 l/s/person in the 1981 version.

4. THE ROLE OF VENTILATION WITH RESPECT TO HEALTHY BUILDINGS AND THE SO-CALLED SICK BUILDING SYNDROME (SBS)

Many reports published during the last decade in Scandinavia, North America and Western Europe indicate high incidence of sickness among people in certain office buildings. This phenomenon is often described as the sick building syndrome. It is without any doubt a rather controversal issue, partly due to the commercial misuse by certain firms who want to sell products, houses,... based on doubtful arguments related to the 'building health'.
However it is sure that a number of buildings suffer serious problems.
The most common symptomes in such SBS-buildings are /7/ :
- eye, nose and throat irritation;
- sensation of dry mucous membranes and skin;
- erythema (skin rash);
- mental fatigue;
- headaches, high frequency of airway infections and cough;
- hoarseness, wheezing, itching, and unspecified hypersensitivity (allergic reactions);
- nausea, dizziness.

The WHO (World Health Organisation) identifies a number of features common to sick buildings, /7/, /8/ :
- they often have 'forced' ventilation (WHO does not make specific reference to air conditioning although air conditioning will fall into this general category);
- they are often of lightweight construction;
- indoor surfaces are often covered in textiles (carpets, furnishing fabrics,..);
- they are often energy efficient, kept relatively warm, and have a homogeneous thermal environment;
- they are often airtight, i.e. windows etc. cannot be opened.

According to Sykes in a review paper on the Sick Building Syndrome /7/, many causes have been suggested for sick building syndrome, including :
- airborne pollutants :
  . chemical pollutants from occupants, fabric, furnishing, ...;
  . dusts and fibres;
  . microbiological contaminant from carpets,furniture, occupant, HVAC-system;
- odours;
- lack of negatively charged small air ions;
- inadequate ventilation/fresh air supply;
- low relative humidity;
- poor working environment/discomfort due to
  . high temperatures;
  . inadequate air movement/stuffiness;
  . poor lighting;
- general dissatisfaction or psychosomatic curves.

It is clear that the energy crisis has contributed "to better conditions" for SBS-problems :
- unappropriate thermal insulation techniques have led to mould growth problems;
- financial problems resulted in lower or no maintenance of HVAC systems;
- lower ventilation rates;
- lower indoor temperatures.

Many questions with respect to SBS and healthy buildings still need a precise answer. However, the important role of good ventilation design as an element in a strategy to evacuate the pollutants as well as to reduce air pollution by the system itself is generally accepted.
The second part of the paper is more focused on ventilation related research activities and the impact of these activities on the building practice in the future. The activities are for practical reasons split up in 3 parts :

5. Activities related to a better description of the actual problems and the required performances
6. Trends in detailed research activities on ventilation
7. New trends in ventilation design

5. ACTIVITIES TO EVALUATE ACTUAL PROBLEMS AND TO SET UP AIR QUALITY REQUIREMENT

Many research projects in the eighties aimed to get a better understanding of various aspects of air pollution. A few projects are described below.

5.1. IEA-ECBCS Annex 9 : "Minimum ventilation"

An international project including participants of 12 countries was set up in 1982 to clarify the problem of air pollution and the role of ventilation. Table 1 summarizes the types of pollutant which were investigated.

Indoor Pollutant	Minimum ventilation rate
Tobacco smoke	5 - 120 m^3 per cigaret
Particles	17,5 l/s person (based on average smoking habits)
Body odour	8 l/s person
Humidity	0,5 - 1 vol/h
Radon	no general rules
Combustion products	no general rules
Selected organic compounds	no general rules

Table 1 : Types of pollutant investigated in IEA-ECBCS Annex 9.

This project finished in 1987 and a report is available free of charge for people in the participating countries /9/.

5.2. Research on air quality perception

In order to come at long term to a global evaluation of the air quality as perceived by the occupants, a research project led by O. Fanger /10/, /11/ was set up. As a result of it, 2 new units were defined : the OLF and the DECIPOL /10/. An important part of the research activities concerned 20 non-domestic buildings in the region of Copenhagen. One very important finding was the determination of the various pollution sources. Figures 6 and 7 show clearly that the occupants by their body odours are on the average, for the 15 office buildings, only responsible for 13 % of the perceived air pollution. The ventilation system and the building materials are responsible for 42 % and 20 %.

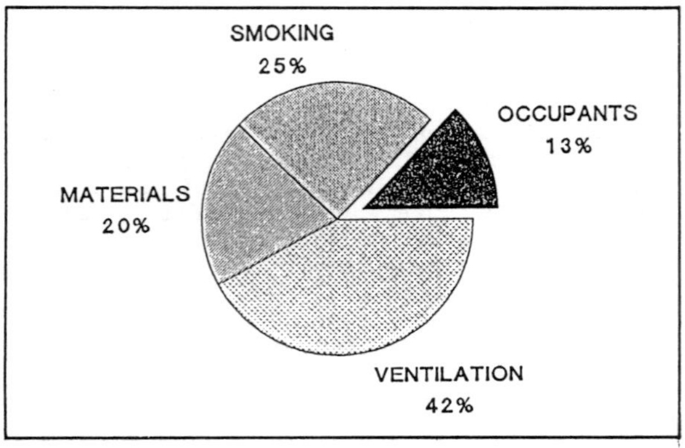

Figure 6 : Average air pollution by various sources in 15 office buildings in Copenhagen /11/.

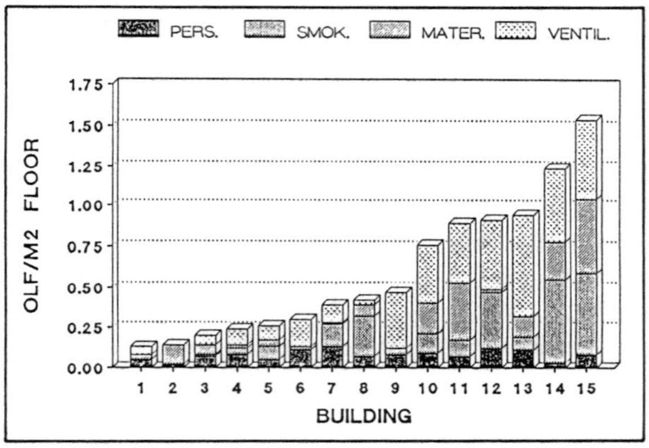

Figure 7 : Air pollution by various sources for 15 office buildings in Copenhagen /11/

5.3. IEA-ECBS Annex 14 "Energy and condensation" /12/

The energy crises of the seventies have directly resulted in a strong increase in thermal insulation of buildings. Mainly due to poor design and/or bad execution of the work, a strong increase in condensation and mould growth problems was found afterwards. Several studies mention figures of 5 to 20 % of social houses with moisture problems. The IEA project "Energy and condensation" focuses on a better understanding of the causes and the remedial strategies.
Some preliminary conclusions relating to ventilation :

- a continuous even limited ventilation of occupied rooms is very important
- the hygric inertia of building finishing and furnishing may be very important.
  Intensive ventilation of short duration (5 .... 10 min) is therefore not an appropriate strategy with respect to moisture.

It also seems to be proven that a significant part (figures up to 20 %) of occupants living in buildings with severe mould growth problems, have allergic reactions after 2 years.

5.4. Radon

Radon migration into buildings may create health risks for people living in these buildings. Surveys in many countries indicate that the problems largely depend on the ground below the building, the airtightness of the building envelope, the location of these leakages and the ventilation strategy. Ventilation can play an important role :

- it can reduce the radon levels by dillution;
- it can also increase the problems especially in the case of a mechanical extraction system bringing the building in underpressure and sucking the radon into the building.

A project involving several CEC-countries is now set-up by CEC DG XII to study radon source models and counter measures.

5.5. Better understanding of draught conditions

Draught problem is one of the most common complaints in mechanically ventilated buildings with mechanical air supply. A research project, led by O. Fanger /13/ aimed to have a more accurate formula for estimating the draught risk. It includes the air temperature and the turbulency. Figure 8 shows the formula proposed /13/.

MODEL OF DRAFT RISK

$$PD = (34 - t_a)(\bar{v} - 0.05)^{0.62}(0.37\,\bar{v}\,Tu + 3.14)$$

for $\bar{v} < 0.05$ m/s insert $\bar{v} = 0.05$ m/s, for PD > 100% use PD = 100%

where PD = Percentage Dissatisfied people due to draught (%)
t_a = air temperature (°C)
$\bar{v}$ = mean air velocity (m/s)
Tu = turbulence intensity (%)

Figure 8 : Formala for estimating draught risk /13/

6. TRENDS IN DETAILED RESEARCH ACTIVITIES ON VENTILATION

The limited knowledge on how ventilation really happens in buildings at the beginning of the eighties, the increased importance of ventilation losses in the energy balance and the general feeling that ventilation is an important element in the health issue, has stimulated research on ventilation related topics. They can be subdivided in multi-zone studies and single-zone studies.

6.1. Multi-zone studies

The study of air flows between zones is an important topic, experimental and simulation activities are going on.

6.1.1. Experimental studies

. Tracer gas systems are used for more than a decade in ventilation research. Most emphasis at present goes to the development of new measurement techniques allowing to measure not only the fresh air flow rates from the outside but also the interzonal flow rates. This leads in most cases to the use of several tracer gasses.
. Another important activity is the experimental determination of the leakage paths in buildings. This involves the use of advanced pressurisation techniques. The knowledge obtained by these activities will allow to better evaluate which ventilation systems are appropriate in certain building types. The results of the leakage distribution will be usefull as input data for simulation.

6.1.2. Theoretical studies

The prediction of air flows between zones is an important aid for evaluating the functioning of various ventilation strategies. The COMIS project /14/ (Conjunction of Multizone Infiltration Specialists) was set-up in 1988 to allow the development of a powerfull and well validated programme for predicting air flows in multi-zone buildings.
A market study of available programmes during the preparation phase of COMIS resulted in the data of table 2 /15/.

Table 2 Review of Multizone Infiltration Network Models		
Program Language:	FORTRAN	33
	BASIC	6
	PASCAL	1
	C	1
	HPL	3
Computer Type:	Main Frame Computer	23
	Personal Computer	18
Solver:	Newton	22
	others	8
Input Features:	interactive input	10
	CAD-input	1
	weather data from weather files	18
	3-D building description	8
	schedules (e.g. occupants)	14
Output Features:	file of arrays used by the model	23
	graphical output	7
	statistical functions	
Miscellaneous:	combined with thermal model	12
	combined with pollution model	8
Program Available?	yes	7
	no	11
	yes, but	11

Table 2 : Overview of features of multizonal simulation programmes /15/

Some 10 researchers worked together during 1 year in Berkeley (U.S.) and at present a first version of the programme is available.
It is clear that there are still large uncertainties with respect to e.g. the air flow through large openings, the pressure distribution around buildings, occupant behaviour. These topics as well as others are studied in the framework of various national programmes (e.g. Belgium, France, Switserland) as well as in IEA-ECBS Annex 20 "Air flows in buildings".
A very important challenge for the future is the validation of these programmes. It seems realistic to expect a lot of results during the next 5 years.

6.2. Single-zone studies

The study of flow patterns within one zone is a topic to which a lot of attention is paid. This is on the one hand due to the need to know more about the pollution distribution within a room, new ventilation systems, thermal comfort studies, ...) and on the other hand due to new measurement and simulation tools allowing for detailed studies.

6.2.1. Experimental studies

. The need to describe the efficiency of ventilation systems and of the evacuation of pollutants has led to new measurement techniques allowing to determine various kinds of efficiency. A detailed description is found in /17/ and /18/. A number of these methods are already standardised in Scandinavie. These methods allow to evaluate the performance of classical ventilation systems as well as of new ventilation concepts, e.g. displacement ventilation /19/.
. The need to measure the flow patterns within one zone has resulted in research on new measurement techniques. Laser Doppler Anometry (LDA) which allows velocity measurements 'at distance' seems to be a very promising technique.

Many measurements of air flows and temperature distribution with more classical methods are at present done. The purpose of these measurements is on the one hand to evaluate the performances of various ventilation systems and on the other hand to collect data-sets for validating predictions by computer programms (see 6.2.2.).

6.2.2. Theoretical studies

According to P. Nielsen /20/ the cost of full-scale experiments is in the range of 3 000 to 20 000 US $ or even more. The accuracy of the results can be very good. One alternative is flow simulation. The use of flow simulations is a question of the price level and of the quality of the predictions. At present the price of a full-scale experiment and of a flow simulation are of similar magnitude. Airflow simulations will increase in importance in the coming years due to the economics. Chapman /20/ /21/ shows how the relative computation cost is decreasing every year and this trend will continue in the coming years. Figure 9 shows that the cost of performing a given calculation has been reduced by a factor of 10 every 8 years. There is also the trend from mainframe computers in computer centers to work stations located close to the user. This will increase the use of simulations. There are promising results with respect to the future of flow simulations. Figure 10 is an example of results found by Hestad and Nielsen /20/. There is a fairly good agreement between measurements and calculations.

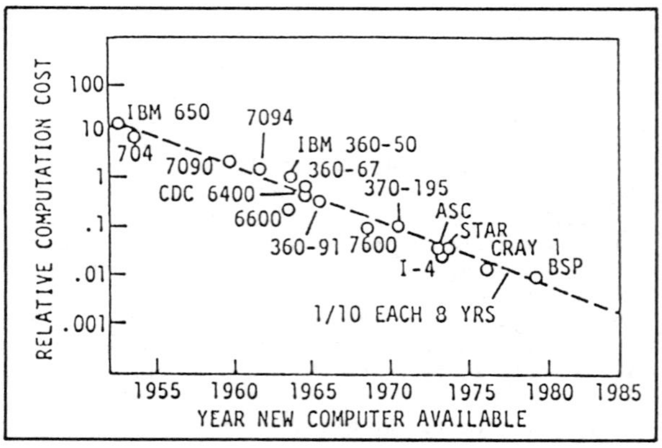

Figure 9 : Trend of relative computation cost for a given flow and algorithm /20/ /21/

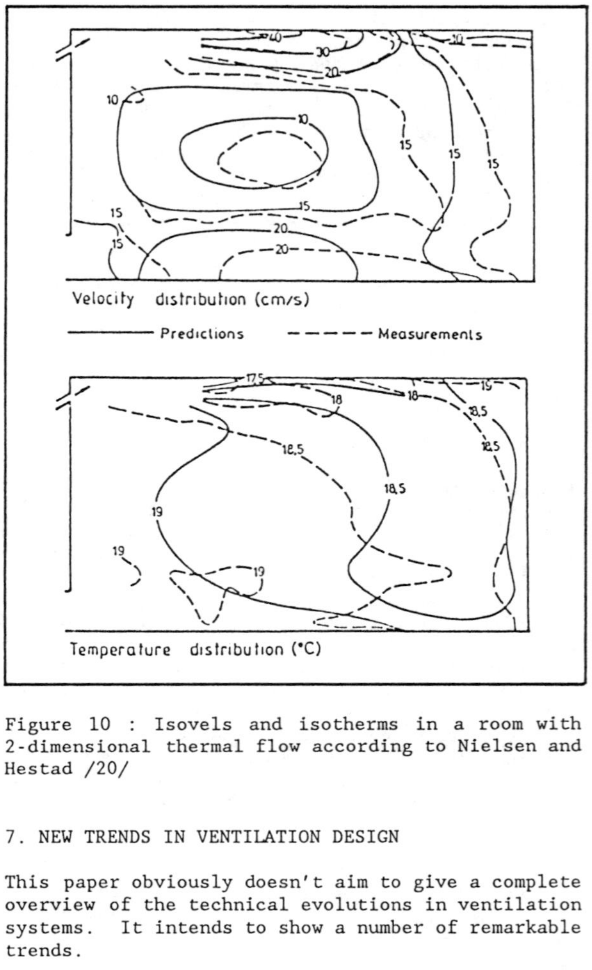

Figure 10 : Isovels and isotherms in a room with 2-dimensional thermal flow according to Nielsen and Hestad /20/

7. NEW TRENDS IN VENTILATION DESIGN

This paper obviously doesn't aim to give a complete overview of the technical evolutions in ventilation systems. It intends to show a number of remarkable trends.

7.1. Installation of ventilation previsions in the domestic sector

In several European countries (France, The Netherlands, Scandinavian countries) the systematical installation of ventilation previsions is more or less the rule. In other countries (e.g. Belgium, Switserland, UK, ...) this is not so at all. The growing interest in air quality issues and the standardisation intentions in a number of these countries indicate the large probability of significant changes in these countries. This trend - not involving new technological findings - is probably one of the most important evolutions with respect to ventilation.

7.2. Demand controlled ventilation

The need to guarantee a good air quality at a low energy cost and the fact that there can be large variations in occupancy and pollution profiles requires optimisation and control strategies. Building Energy Management Systems (BEMS) are at present already used and can control the whole HVAC and technical appliances systems. BEMS systems are studied in the framework of IEA-ECBCS Annex 16 (hardware) and Annex 17 (software) /3/.
The optimal control of the ventilation is at present possible by using e.g. variable air volume (VAV) systems and by installing air quality sensors in various parts of the building. This so-called demand controlled ventilation (DCV) is thoroughly studied in the framework of IEA-ECBCS Annex 18 "Demand Controlled Ventilation" /22/. At present, a list of commercial sensors is available and detailed performance testing of a number of these sensors is started by the Swedish participant. The types of sensors which are under analysis are :

- humidity sensors
- CO_2-sensors
- CO-sensors
- mixed gas sensors

7.3. Integrated components

Various reasons (energy recuperation, no draught problems, higher surface temperature) have led to the development of building components which combine various functions, one of them being ventilation. Figure 11 and 12 show some examples.

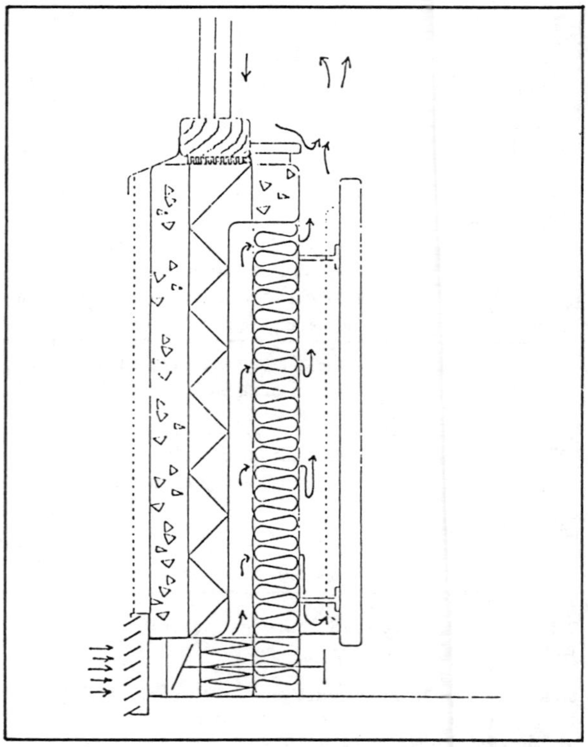

Figure 11 : Wall components also functioning as air inlet aiming to eliminate draught problems /23/

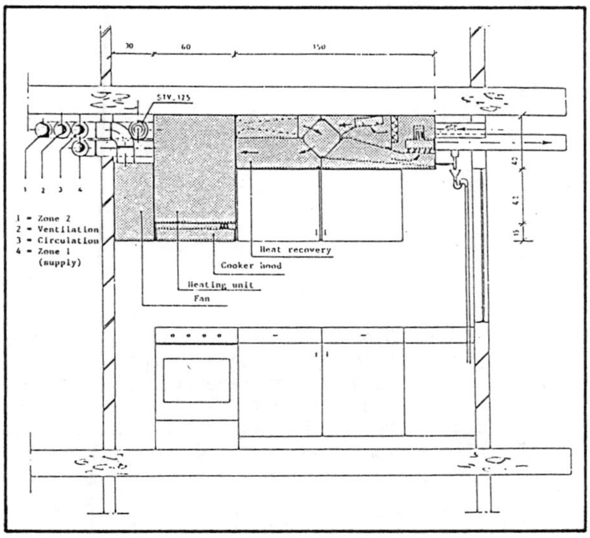

Figure 12 : Example of unit integration of various functions (heating, ventilation, hot water, cooking hood) /24/

7.4. Maintenance procedure HVAC-system

Several studies have highlighted the bad quality of the air entering the room by the HVAC system. The main reasons are : no or poor maintenance and/or no replacement of filters,... It is expected that in the future more attention will be paid (and more budget) to the keeping of the HVAC-system in a good condition.
Several firms are at present specialised in cleaning these systems.

8. CONCLUSIONS

The concern to have a cleaner indoor air and to protect the outdoor environment is a very remarkable fact of the last few years. Ventilation design and use is of paramount importance to reach both goals.

9. ACKNOWLEDGMENTS

The author wishes to thank the organizations who have sponsored various research projects allowing to make this paper : the Belgian Ministry for Science Policy (D.P.W.B./S.P.P.S), the Institute for Encouragement of Scientific Research in Industry and Agriculture (I.W.O.N.L./I.R.S.I.A.), the Ministry for Economic Affairs. The paper itself is the result of the efforts of the author, L. Vandaele, R. Bossicard, E. Bruyninckx, D. L'Heureux and P. Voordecker. Mrs. A. De Keyser, B. Deffense and A.F.Stalport took care of the typing and the lay-out.

10. REFERENCES

/1/ ASHRAE 62-1981, Ventilation for acceptable indoor air quality, ASHRAE Atlanta (US), 1981.
/2/ Swedish Council for Building Research, Energy 1985, Energy use in the built environment.
/3/ International Energy Agency, Energy Conservation in Buildings and Community Systems Programme, March 1985.
/4/ Air Infiltration Review, Publication by Air Infiltration and Ventilation Center, quarterly, AIVC, Coventry.
/5/ Energy in Europe, Short term energy outlook for the European Community, CEC DG XVII, Brussels, May 1989.
/6/ Millhone, J., Energy efficiency in buildings. Conference "Energy Efficiency in buildings", Lille, 28-30 November, 1989.
/7/ Sykes, J.M., Sick Building Syndrome, Building Serv. Eng. Res. Technol, 10 (1) 1-11, 1989.
/8/ World Health Organisation, Indoor Air Pollutants : Exposure and Health Effects, Report Norlingen, 8-11 June 1982, ISSBN 92 890 12 447.
/9/ Haberda, F., Trepte, L., IEA Annex 9, Minimum ventilation rates and measures for controlling indoor air quality, Technical Note 26, AIVC, Coventry, 1989.
/10/ Fanger, P.O., The Olf and Decipol, ASHRAE Journal, October 1988.
/11/ ASHRAE, HVAC Handbook, Chapter 27, Environmental control for survival, ASHRAE 1987.
/12/ IEA Annex 14, Contact person : Prof. Hens, H., Operating Agent, K.U.Leuven, Kasteel Arenberg, 3030 Heverlee, Belgium
/13/ Fanger, P.O., Melikov, A.K., Hanzawa, H., Ring, J., Turbulence and draught, ASHRAE Journal, April 1989.
/14/ The Comis Infiltration Model, The Comis group, 10th A.I.V.C. Conference, Dipoli, Finland, 25-28 September 1989.
/15/ Feustel, H., Mathematical modelling of Infiltration and Ventilation, 10th A.I.V.C. Conference, Dipoli, Finland, 25-28 September 1989.
/16/ Moser, A., Flow Flash Newsletter, ETH Zurich, 1989.
/17/ Liddament, M., A review and bibliography of ventilation effectiveness - definitions, measurement, design and calculation, Technical Note 21, AIVC, Coventry, 1987.
/18/ A new, more detailed technical note will be published by A.I.V.C. in 1990.
/19/ Nielsen, P., Hoff, L., Pedersen, L.G., Displacement ventilation by different types of diffusers, 9th A.I.V.C. Conference, Gent, Belgium, 12-15 September 1988.
/20/ Nielsen, P., Airflow simulation techniques, progress and trends, 10th A.I.V.C. Conference, Dipoli, Finland, 25-28 September 1989.
/21/ Chapman, D.R., Computational Aerodynamics Development and Outlook, AIVC, J., vol. 17, pp. 1293-1313, 1979.
/22/ Demand Controlled Ventilation Systems : State-of-the-art-Review, IEA-ECBCS, Annex 18, 1989.
/23/ Werner, G., Outlet air inlet without draught problem, 10th A.I.V.C. Conference, Dipoli, Finland, 25-28 September 1989.
/24/ Stork Ventilatoren, b.v. Zwolle, Nederland.

THERMAL PERFORMANCE OF NON-AIR-CONDITIONED OFFICES IN THE U.K.

D.K.Alexander and P.J.Jones
Welsh School of Architecture
University of Wales College of Cardiff
PO BOX 25, Cardiff,
U.K.
CF1 3XE.

ABSTRACT. The acceptability of summertime office environments, in the absence of air-conditioning systems, was investigated using a dynamic thermal building model. A realistic combination of solar shading, thermal mass, and night ventilation purging was found to allow reasonable conditions (average temperatures approximately 26°C), effectively dealing with internal gains of up to 40 W/sqm. Higher gain levels, of up to 60 W/sqm can also be controlled, but at the expense of higher mean temperatures. The options tested also lead to a significant reduction in cooling loads, had air-conditioning systems been supplied.

1. INTRODUCTION

Many office buildings in the U.K. are air-conditioned to produce what is considered to be an acceptable internal environment regardless of the outside conditions.

It has emerged that, far from producing a beneficial environment, such buildings can have a distinctly negative effect on the occupants. The causes of these effects can often be identified with the use of, or the mis-use of, air-conditioning plant. /1/

There is a movement towards the reduction of, and possible avoidance of, air conditioning systems in new office buildings. This has arisen from many concerns :

- the desire to reduce capital and maintenance costs,
- the desire to reduce energy usage and costs,
- the desire to reduce the usage of CFC's, and
- the desire to avoid the phenomena of "sick building syndrome".

Recent design endeavours have sought to alleviate such problems, for instance atria are provided for enhanced daylight and ventilation, and control over lighting and window opening is often returned to the occupants.

On the other hand, internal gains can be seen to be increasing (due to the spreading use of IT office electronics), as are fabric insulation levels.

Results of the monitoring of a naturally ventilated, thermally massive office building /2/ showed that at internal gain levels of approximately 30 W/sqm, overheating in summer was not a problem.

The question arises, to what level of internal gains, can active cooling be avoided through the use of architectural/engineering options?

This paper describes the use of a dynamic thermal model (HTB2) to investigate the reductions in cooling requirement, and the allied environmental consequences, caused by combinations of such design options as:

- the provision of controllable ventilation,
- the use of external solar shading devices,
- the use of high internal thermal mass to absorb internal heat gains, and
- the use night time purging to flush out those stored heat gains.

The investigation is based around an office design, and around internal heat gains, typical of modern practice.

The current study has evolved from a project option for third year architectural students, wherein they are encouraged to subject their own office designs to appraisal by simulation, and to explore the consequences of shading devices, glazing areas and types, ventilation options and varying incidental gains.

2. SUMMERTIME OVERHEATING AND COMFORT CRITERIA

There are in the U.K. guide-lines and procedures available to the designer wishing to provide an acceptable internal environment, such as those published by CIBSE /3/ and BRE /4/. These often lay out a set of criteria for acceptable conditions.

Six indicators of the summertime environmental performance of offices, and their recommended maxima, were selected as test criteria for this work. They are summarized in Table I.

Parameter	Criteria
Peak Air Temperature	< 26 °C
Peak Resultant Temperature	< 26 °C
MRT - Air Temperature Difference	< 4 °C
Average Temperature in Working Day	< 24 °C
Temperature Range in Working Day	< 4 °C
Psychometic limits	< 27 °C at 20%rh, < 23 °C at 80%rh

Table I. Comfort Criteria for Summertime Conditions in U.K. Offices.

Realistically, these criteria cannot be absolute maxima; it has been suggested that they not be exceeded more than 30 times in 10 years.

3. THERMAL MODELLING

The model used in this work was HTB2 /5/. HTB2 is capable of simulating, to fine time detail, the interactions of fabric, ventilation, plant and incidental energy systems, and the climate. This model was developed at the Welsh School of Architecture (R&D) as a general research and teaching tool for the investigation of energy and environmental performance of buildings.

HTB2 has been subject to external appraisal, and has been included in the recent inter-model comparisons of IEA Task VIII. It is in use in a number of higher education and commercial organisations in the U.K.

4. TEST CASE DESCRIPTIONS

4.1 Location and Time

The simulations were carried out for 30 days from mid-July through mid-August. The building was assumed to be on an unobstructed site near London. Kew 1967 meteorological data was used to provide data for external air temperature, humidity, and solar gains.

4.1 Layout

Although generic in nature, the office area and layout for the case study has been developed from a live (and fully air-conditioned) example. It contained a mixture of open plan and cellular accommodation as in Figure 1. The office areas under test were assumed to be situated on an intermediate floor of a multi-floor block. The gross floor area of the test offices was 900 sqm.

4.2 Structure

Two basic construction types for the office structure were considered, representing light-weight and heavy-weight structural options, as in Table II. The heavy-weight option included an exposed concrete ceiling; raised flooring was provided to allow provision for services. A third option further increased the thermal mass by assuming a waffled ceiling structure.
All opaque walls had U-values of 0.45 W/sqm/°C, and windows were double glazed.

Type -	Lightweight	Heavyweight
External Walls	40mm concrete 60mm insulation 200mm cavity 12mm plasterboard	105mm brick 55mm foamboard 105mm heavy block 12mm heavy plaster
Internal Partitions	12mm plasterboard 50mm stud/cavity 12mm plasterboard	12mm heavy plaster 105mm heavy block 12mm heavy plaster
Floor/ Ceiling	25mm flooring 65mm concrete 500mm cavity 12mm plasterboard	25mm flooring 500mm cavity 65mm concrete

Average Admittances (W/sqm):

	Light	Heavy	Very Heavy
Open Plan areas	6	9	13
Cellular Offices	6	12	17

Table II Construction Type Options and Equivalent Admittance Values (Very Heavy as Heavy with a waffled ceiling structure).

It was noted that the use of these structures did not achieve the admittance range suggested in the manuals, due to the relatively large open spaces in the test case.

4.3 Glazing areas and Shading

The amount of external glazing was fixed for all simulations and was based on 50% of the external elevation for all orientations (apart from the blank north elevation).
In cases where external shading was applied, the equivalent of a projecting horizontal awning of 2.5m depth was assumed on all facades.

4.4 Ventilation and Air Movement

A minimum ventilation rate of 1.5 air changes per hour during occupancy was chosen, based on the recommended fresh air supply for the assumed occupancy rates. Overnight infiltration was set to be 0.25 ac/h. Initially the ventilation rate was fixed under all conditions.
Ventilation options tested were the provision of boosted ventilation at 6 ac/h during high temperatures (>26 °C) in occupancy periods, and the provision of night-time purging at 6 and 12 ac/h. It was considered that the latter high rate could be achieved in practice as at night there would be no acoustic penalties. Mechanical ventilation was assumed in all cases.

The open plan areas were zoned to perimeter and deep areas, as in figure 1, and were connected thermally by the equivalent of a mixing air flow of 0.1 m/s.

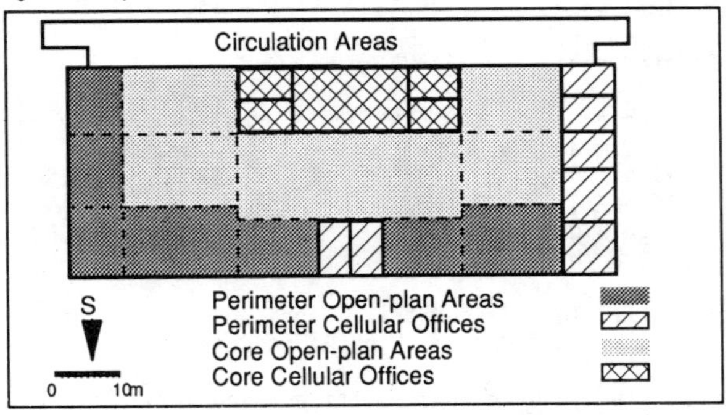

Figure 1. Test Office Floor Plan, Showing Zoning.

4.5 Internal Gains

Three levels of internal gains were tested, summarized in table II. These gain level do not necessarily reflect the total range found in U.K. offices, but were arrived at through reasonable assumptions on occupancy, lighting, and electronic equipment provision.
The office was assumed to be in use between the hours of 9:00 am and 5:00 pm; comfort criteria were assessed between those times. The internal gains, including lighting, were assumed to be on full over that entire period, and there were no weekend breaks in the building use.

4.5.1 Occupancy and small power gains

Two occupancy rates were assumed for the open plan areas; 15 and 10 sqm/person. Cellular offices were assumed to contain only one person. Each occupant was considered to contribute 100 W heat, and 40 g/h water-vapour to the space.
Small power gains were determined from the occupancy rates. Two scenarios of the provision of electronic office equipment were assumed. For the High and Medium gain cases, each "person" had a desktop computer (or electrical load equivalent) and each group of 5 people had a shared laser-printer. In the Low gain case, PCs' were assumed to be shared between staff, and there were no local printers.

4.5.2 Lighting

The lighting system installed load was taken to be 22 W/sqm for the High and Medium gain cases, and 17 W/sqm for the Low gain case.

Case -	LOW	MEDIUM	HIGH
Open Plan Office Areas	32W/sqm	44W/sqm	59W/sqm
Cellular Office Areas	30W/sqm	39W/sqm	42W/sqm
Mean Overall	33W/sqm	44W/sqm	53W/sqm

Table III Incidental Gains Cases, Breakdown By Office Type

5. RESULTS AND DISCUSSION

5.1 Tests

The test cases progressed incrementally from the lightweight, unshaded and minimally ventilated case, through the addition of design features; controllable ventilation boost, shading, high and very high internal mass. The two options of night purging, at 6 and 12 ac/h, were applied. The cases tested are summarised and coded in table IV.

Option - ǀ	Thermal Mass Light	Heavy	V.Heavy
Minimal vent. (1.5 ac/h)	"Base"	-	-
Boosted vent. (1.5-6 ac/h)	"V"	-	-
+ Shading (2.5m awning)	"VS"	"HVS"	"SVS"
+ Night Purge (6 ac/h)	"VSP"	"HVSP"	"SVSP"
+ High Night Purge (12 ac/h)	-	"HVSP2"	"SVSP2"

Table IV Test Case Matrix, repeated for each of Low, Medium, High Gains Cases.

Open plan and cellular offices were very different in their results, but there was little average difference between perimeter areas and deep areas. Perimeter areas had higher solar gains, and a higher afternoon temperature, whilst deep areas, with their lower fabric heat losses, had higher morning temperatures but lower increases over the day. Discussions on results will concentrate on perimeter areas only.

5.2 Environmental Criteria

Table V shows , for the open and cellular offices (in brackets), the ability of the various test cases to meet the comfort criteria listed in table I. The percentage of time exceeding the criteria are listed. Also listed are the mean daily, and mean peak, air and resultant temperatures, and the mean humidity in the spaces. All data is restricted to the working day.

In general those criteria based on environmental temperature were stricter than those based on air temperature. This was due to the extra radiant gains from solar and casual sources being included in the determination of MRT. The mean resultant temperature in the working day was the most severe criterion used in these tests.

In the open plan areas, the options tested had little effect on that particular criterion; for the best result >70% of working days averaged above 24°C, even for the low gain case.

The results for the criteria on peak temperatures fared better; by the best result <15% of working hours were below 26°C for the low gains case.

The results for the psychometric criteria were similar to those based on environmental temperature. Humidity levels were not a cause for concern, being in the range 40 - 60%rh for the best high gains case.

Due to the relatively higher thermal mass, and lower gain levels, in the cellular offices, conditions in the those spaces were often more satisfactory.

5.3 Environmental Conditions

Figures 2,3 indicate the average (and 1 standard deviation range) daily temperature variations for open plan and cellular offices, respectively. Mean and peak temperatures are listed in Table V.

The addition of shading reduced the overall temperatures, and also acted to reduce the difference between radiant and air temperatures.

The options of increased mass had little significant effect until night purging was introduced. Adding mass alone often increased problems, as heat gains built-up, increasing the average working day temperature. Night purging allowed some of that build-up to be removed. Test cases on night purging without added mass reduced the overall temperatures, but increased the temperature range through day.

The high gain cases could not achieve the comfort criteria laid out, and so this must mark an upper limit of feasibility. The environment produced was

Case	Percentage Time in working day Criteria not met (cf table I) A	B	C	D	E	F	Resultant temperature Mean Daily °C	Mean Peak °C	Humidity Mean RH %
HIGH GAINS									
"Base"	100(100)	100(100)	34(2)	100(100)	100(100)	65(66)	50.6(47.4)	53.5(50.4)	14(16)
"V"	100(99)	100(100)	85(41)	100(100)	100(100)	89(35)	36.4(32.7)	42.5(37.8)	32(36)
"VS"	100(90)	100(99)	1(0)	100(97)	100(100)	46(20)	31.5(28.8)	35.7(32.5)	36(41)
"VSP"	83(83)	77(69)	3(1)	73(66)	100(100)	74(41)	27.7(25.8)	31.6(29.2)	42(47)
"HVS"	100(93)	100(100)	0(0)	100(98)	100(100)	17(3)	31.1(28.7)	35.1(31.9)	37(41)
"HVSP"	75(22)	66(24)	2(0)	64(29)	100(75)	77(3)	27.0(24.7)	30.6(27.7)	43(50)
"HVSP2"	41(12)	51(19)	1(0)	41(20)	90(70)	20(2)	26.2(24.5)	29.4(27.3)	46(50)
"SVS"	100(93)	100(100)	0(0)	100(98)	100(100)	17(3)	30.7(28.5)	34.4(31.5)	37(41)
"SVSP"	65(17)	55(17)	1(0)	55(22)	98(71)	39(3)	26.3(24.4)	29.5(27.3)	45(51)
"SVSP2"	33(10)	39(13)	1(0)	41(20)	90(70)	20(2)	25.6(24.3)	28.6(27.0)	47(51)
MEDIUM GAINS									
"Base"	100(100)	100(100)	30(2)	100(100)	100(100)	82(67)	50.6(47.4)	53.5(50.4)	14(16)
"V"	99(99)	100(100)	88(39)	100(100)	100(100)	84(32)	36.4(32.7)	42.5(37.8)	32(36)
"VS"	94(84)	100(99)	2(0)	99(96)	100(100)	39(16)	31.5(28.8)	35.7(32.5)	36(41)
"VSP"	53(53)	63(49)	2(1)	62(48)	100(97)	47(17)	27.7(25.8)	31.6(29.2)	42(47)
"HVS"	94(87)	100(100)	1(0)	99(97)	100(100)	23(3)	31.1(28.7)	35.1(31.9)	37(41)
"HVSP"	46(15)	53(19)	1(0)	54(24)	89(70)	47(4)	27.0(24.7)	30.6(27.7)	43(50)
"HVSP2"	20(10)	44(16)	1(0)	47(23)	87(69)	35(4)	26.2(24.5)	29.4(27.3)	46(50)
"SVS"	95(87)	100(100)	0(0)	99(96)	100(100)	13(3)	30.7(28.5)	34.4(31.5)	37(41)
"SVSP"	36(12)	40(12)	1(0)	40(19)	80(63)	35(3)	26.3(24.4)	29.5(27.3)	45(51)
"SVSP2"	16(8)	23(11)	1(0)	33(18)	80(60)	30(2)	25.6(24.3)	28.6(27.0)	47(51)
LOW gains									
"Base"	100(100)	100(100)	21(1)	100(100)	100(100)	93(60)	49.3(45.7)	52.6(49.3)	15(18)
"V"	95(98)	100(100)	85(30)	100(100)	100(100)	80(27)	33.1(31.3)	38.8(36.1)	39(38)
"VS"	77(74)	99(96)	1(0)	95(90)	100(100)	27(9)	28.6(27.7)	32.1(30.9)	43(43)
"VSP"	33(33)	48(33)	1(0)	49(35)	85(81)	46(25)	25.7(24.9)	29.1(28.3)	48(50)
"HVS"	77(77)	99(98)	1(0)	95(90)	100(100)	18(3)	28.4(27.4)	31.6(30.2)	43(44)
"HVSP"	23(9)	36(10)	1(0)	39(18)	79(50)	48(4)	25.1(23.7)	28.4(26.9)	49(53)
"HVSP2"	11(7)	31(9)	1(0)	36(17)	79(48)	42(4)	24.9(23.7)	28.0(26.8)	50(53)
"SVS"	79(77)	97(97)	0(0)	90(86)	100(100)	11(3)	28.0(27.4)	31.0(29.8)	43(44)
"SVSP"	15(6)	20(6)	0(0)	26(14)	73(37)	23(0)	24.4(23.4)	27.6(26.5)	51(54)
"SVSP2"	9(5)	16(6)	0(0)	23(14)	71(37)	22(0)	24.3(23.4)	27.3(26.4)	51(54)

Table V Criteria and Environmental Results for All Test Cases; Open Plan (and Cellular) offices.

however thought marginally acceptable. On average, the internal temperature of the open areas peak, at the end of the day, at 27°C. There may therefore be expected some loss of comfort, but note that with local air movement as might be produced by desk fans, the comfort temperature can exceed 30°C.
The cellular offices fared much better, peaking below 25°C on average.

5.4 Equivalent cooling loads

Air-conditioning is sometimes provided for reasons beyond environmental comfort, i.e. for prestige, competition, or increased rentability of speculative buildings.
Some of the high gain tests were repeated with an active cooling system in operation, so that the reduction in cooling requirement, due to these measures, could be assessed. A cooling set point of 23 °C was assumed.
The best case was the same as that for the non-cooled cases, that is with the provision of high mass, shading, and night purging.
The best reduction on average cooling requirement for the whole office area was some 50%, while the peak cooling loads were reduced by 33%. The results are summarized, for the total office area, in table VI.
It was noted that in the initial test cases, the air-conditioned environment fared little better in meeting some comfort criteria, than the non-air-conditioned cases. Though air temperature was completely controlled, the presence of radiant gains meant that other criteria failed.
The addition of shading, thermal mass, and purging options improved the internal environment considerably, in addition to reducing cooling energy demand.

	A/C Load		Resultant Temperature	
	Average	Peak	Mean	Peak
Case	W/sqm		°C	°C
"V"	59	74	26.3	27.3
"VS"	48	60	25.0	25.5
"HVS"	48	59	24.9	25.4
"HVSP"	32	51	23.3	24.5
"SVSP"	29	51	23.0	24.1

Table VI Cooling Loads and Temperatures for Air-Conditioned Test Cases.

CONCLUSIONS

The options investigated have been shown to offer potential for improved comfort and reduced energy costs in offices.
The use of external shading devices are seen to be effective in both reducing cooling loads, in reducing the radiant temperature levels in a space, thus increasing comfort levels.
Through a combination of responsive day time ventilation, night time purging, solar shading and thermal mass, a significant reduction in cooling requirement can be achieved. At moderate levels of internal gains (approximately 40 W/sqm), reasonable internal environments were predicted, without the recourse to active cooling. At higher gain levels (<60 W/sqm), active cooling could also be avoided, although some relaxation of expectation or provision of local air movement would be required.
Even though ventilation rates greater than 6 ac/h were not used during occupancy periods, the effect has been as if much higher rates had been applied throughout the day.
Even in the air-conditioned test cases, some comfort criteria were not always met. This in particular was due to the presence of radiant gains (from solar or internal sources) to the spaces, increasing the resultant temperature. The options could also significantly improve internal conditions, and reduce cooling loads, where air-conditioning had been provided.

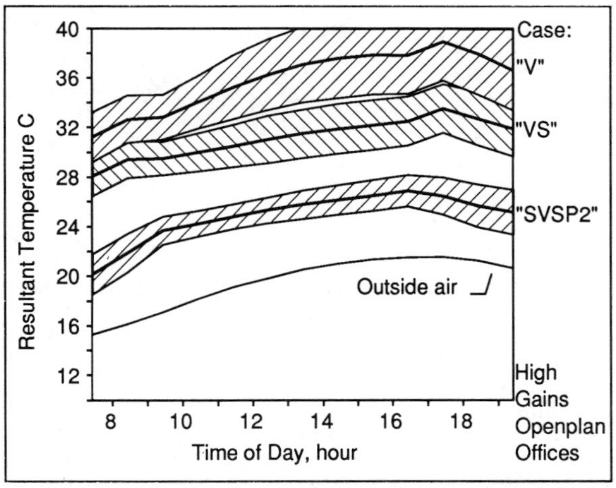

Figure 2. Reduction in Mean Daily Temperature Profiles In Open Plan Offices (shaded area indicates 1 s.d range).

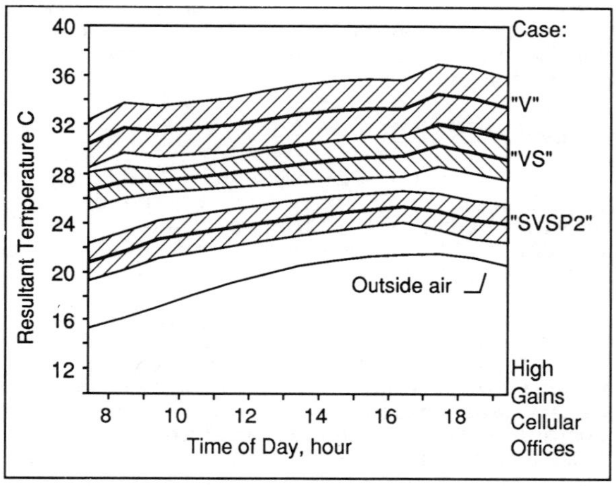

Figure 3. Reduction in Mean Daily Temperature Profiles In Cellular Offices (shaded area indicates 1 s.d range).

REFERENCES

1 Wilson,S., O'Sullivan,P., Jones,P., Hedge,A., "Sick Building Syndrome & Environmental Conditions: Case Studies of 9 Buildings", Building Use Studies Ltd, London, 1987.

2 Databuild Ltd., "South Staffordshire Water Corporation Headquarters: EPA Technical Report", ETSU, 1989 (draft).

3 CIBSE, "A8 - Summertime Temperatures in Buildings", CIBSE, London, 1980.

4 Petherbridge,P., et al, "Environmental Design Manual, Summer Conditions in Naturally Ventilated Office", Building Research Establishment, Garston, Watford, 1988.

5 Lewis,P.T., Alexander,D.K., "HTB2: A Flexible Model for Dynamic Building Simulation", accepted for publication in Building and Environment, 1990.

DESIGN OF VENTILATION SYSTEMS IN RESIDENTIAL BUILDINGS

Dominique BIENFAIT
Centre Scientifique et Technique du Bâtiment
84 avenue Jean Jaurès, BP 02
77421 MARNE LA VALLEE CEDEX 2

ABSTRACT : In most residential buildings, ventilation is naturally achieved by air permeability of the building shell which, in old building, is often associated with chimmey stacks. The architectural evolutions (chimmey stacks become less frequent), as well as energy conservation concern (air leakage reduction) and new requirements in matter of olfactive comfort or acoutic annoyance, makes it desirable to develop specifically designed ventilation systems. Moreover, these systems may be designed in order to achieve ventilation of service rooms, such as bathrooms, with no window, which leads to much more flexibility in the architectural design : Building regulation prevailing in France has made it compulsory, as early as 1969, to use specific ventilation systems in new residential buildings. Different kind of ventilation systems (mechanically powered, temperature driven, hybrid systems,..) have been developped.

This article reviews these systems and outlines their advantages and drawbacks in single family or collective buildings, with respect to architectural flexibility and comfort or safety requirements.In a second part, the article introduces and discuss the scientific grounds (for instance : wind effect) which govern the ventilation rates. Results of theoretical work, with special emphasis given to stack effect ventilation, is given, and design methods in use in France, with their influence on architecture conception are featured.

I INTRODUCTION

Ventilation systems in residential buildings are primarily intended to remove air pollutants : moisture, combustion products, odours,...) and prevent building damage due to condensations and mould growth.

This was previously achieved by natural means : window opening and also air change when fireplaces are working. In the sixties, new architectural trends have appeared : water-closets and bathrooms are often in a central position with no window, and fireplaces have almost disappeared in multifamily houses. On the other hand, greater concern has been paid to energy conservation as well as comfort matters (cold draught avoidance, protection against outdoor noise). These reasons lead to the spreading out of specific ventilation systems such as natural ventilation using vertical shafts, mechanical systems or new techniques such as humidity controlled ventilation : the usual air-flow pattern, using such systems, is shown in fig 1 :

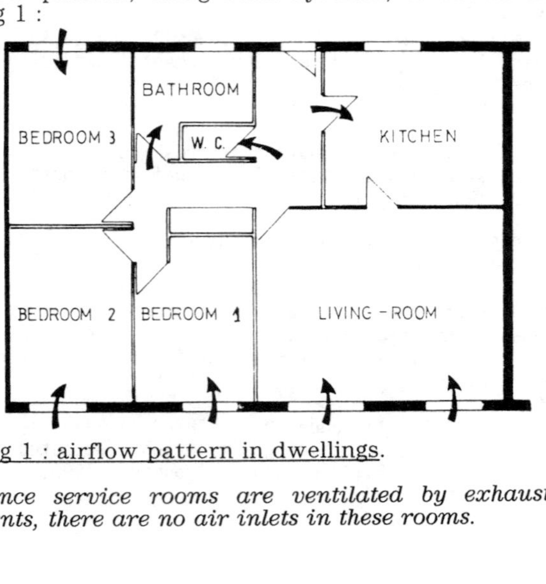

Fig 1 : airflow pattern in dwellings.

Since service rooms are ventilated by exhaust vents, there are no air inlets in these rooms.

These systems must comply with additional requirements :

. limit heat losses to a reasonable value, whatever the climatic conditions (wind, outdoor temperature) are,
. prevent acoustic annoyance :
  In noisy environment, acoustic attenuation of air inlets must be sufficient.
  Mechanically powered system must not lead to excessive pressure, which would cause exhaust vents to generate noise.
  In multifamily houses with common ventilation network, acoustic attenuation between air vents must be sufficient to prevent noise transmission from one flat to another,
. prevent reverse flow through a common ventilation network,
. design and install the air inlets in order to avoid cold draughts,
. avoid drift of ventilation components characteristics due to fouling or aging,
. allow maintenance operations,
. safety in case of vented or unvented combustion appliances.

II VENTILATION SYSTEMS IN USE IN FRANCE

2.1 Mechanical exhaust systems

Mechanical exhaust systems which appeared in France in the sixties are now the most commonly used systems. As opposed to natural ventilation, they provide a much greater flowrate steadiness, whatever the outdoor climatic conditions : Mechanical systems have been improved in recent years : new systems with flowrates depending on the prevailing indoor climatic conditions, enable energy conservation and indoor air quality to be further improved.

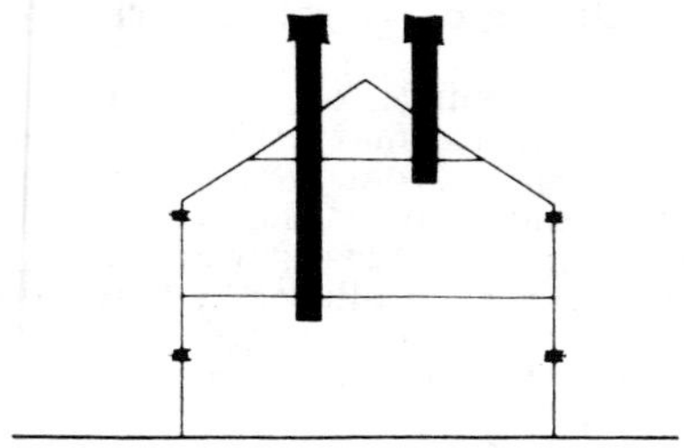

Fig 2 : *Scheme of a natural ventilation system in single family dewlling.*

2.2 Natural exhaust systems

In natural exhaust systems (fig.2) the flowrate is generated by buoyancy and wind effects. Accordingly, flowrates are varying a lot.

2.3 Other systems

Other systems (mechanical balanced systems,...) are also in use and many new systems (humidity controlled vents, hybrid systems conbining natural and mechanical effect,...) have been recently developped.

III VENTILATION SYSTEMS DESIGN AND DIMENSIONING.

Research work relevant to ventilation network calculation has been undertaken in order to improve the design methods :

3.1 Dimensioning of air inlets.

Air inlets dimensioning must lead to a good balance between air quality and energy conservation requirements. As a matter of fact, air inlets are intended to ensure as equal as possible fresh air distribution amidst the different rooms. Therefore their dimensioning must be sufficient with regard to flow disturbance due to building shell permeability : fig 3 depicts the influence of air leakage increase on fresh air flow in a room.

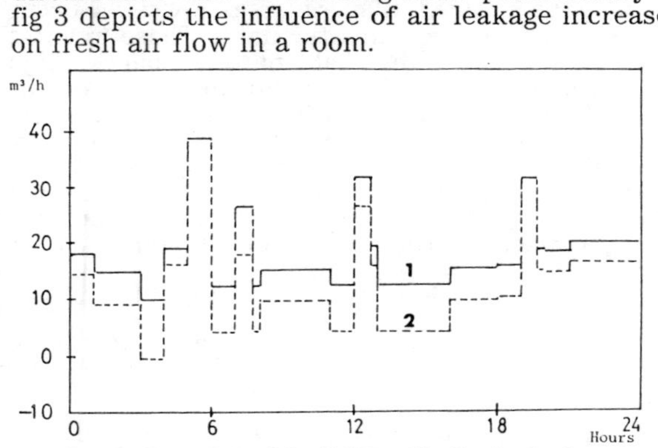

Fig 3 : Influence of building shell air leakage on airflow pattern :

The curves represent the calculated values of fresh air entering a bedroom (cf fig. 1) according two different assumptions :

Curve 1 : *external walls of the kitchen are assumed to be perfectly airtight*

Curve 2 : *air leakage value of the kitchen external walls is assumed to be equal to 20 m^3/h under 1 Pa pressure difference*

On the other hand, air inlets dimensioning must be as low as possible in order to reduce heat losses due to cross ventilation in case of wind : calculations (ref. 6) has shown that cross ventilation heat losses are decreasing when the negative pressure inside the dwelling, caused by the operation of the ventilation system is increasing. More precisely, it was found that the following equation :

$$Q_S = P.\frac{e}{1 + d.\left(\frac{Q}{P}\right)^2}$$

where :

- Q_S (m^3/h) flowrate corresponding to cross ventilation
- Q (m^3/h) exhaust flowrate due to ventilation system operation
- P (m^3/h) flowrate through the building shell air leakage and air inlets when the pressure difference is 1Pa
- e, d numerical coefficients, the values of which depend on building wind exposure .

Taking into account the mean air leakage values of the building shell in new buildings, the above considerations lead to propose that the dimensioning of air inlets should be equal to 30 cm^2 in each room when exhaust flowrate is achieved by a mechanical systems. When a natural exhaust system is used, the air inlet dimensioning must be higher.

3.2 Dimensioning of mechanical exhaust systems.

Fig. 4 depicts an example of airflow calculation in an exhaust network : Airflows are mainly depending on the pressure-flowrate curves of the fan, exhaust vents and network itself.

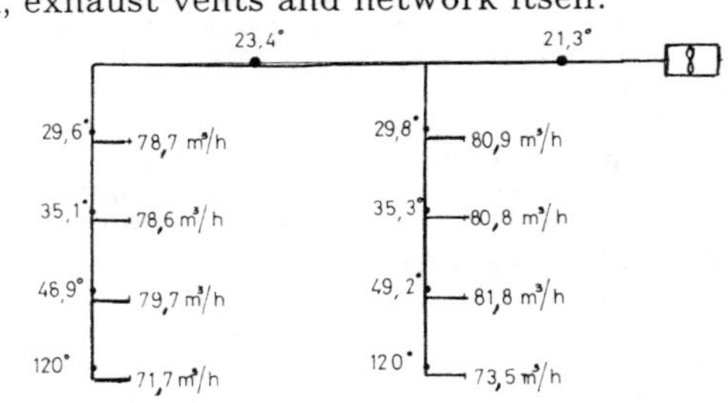

Fig. 4 : Mechanical exhaust network

This figure depicts flowrates and temperatures calculation in an exhaust network using the code GAINE (ref 1). The exhaust vents of the first floor are linked to vented gaz appliances, the flue temperature of which is assumed to be equal to 120°C.

However the negative pressure in the network, (hence the flowrates at each exhaust vent), may vary quite in an important way : as a matter of fact, thermal or wind effect may increase substantially the flowrates. Moreover, according the french regulation, it is allowed that exhaust vents aperture area may be set by the occupant either to a low value, or to a peak value : when most of the exhaust vents are at their peak value, pressure drop in the ducts increase and exhaust flowrates in other vents are lower.

The pressure variation inside the network may be, as illustrated on fig. 5, quite important and cause, either noise production when the pressure is high, or unsufficient ventilation when this pressure is low. The allowed values of noise production (acoustic pressure level) in kitchen is 35 dB(A), which can be usually achieved provided the negative pressure in the network does not exceed a value, which usually is in the range of 120 to 140 Pa.

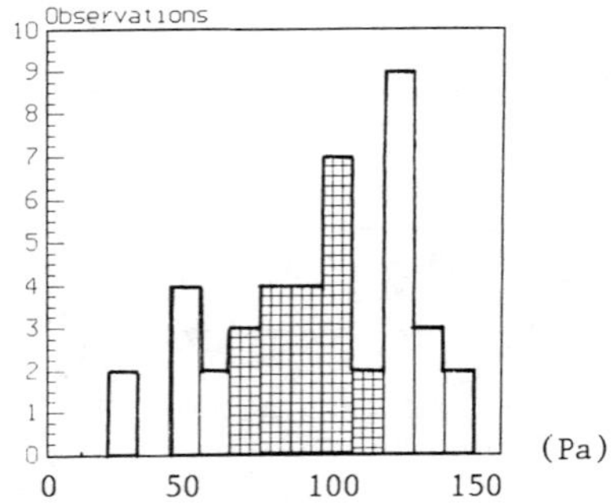

Fig. 5 : pressure variation in a mechanical exhaust network.

This histogramme shows the distribution of negative pressure which were measured in an exhaust mechanical system. Crossed lines indicate the range of pressure for which aeraulic and acoustic requirements of the exhaust vent are met: a great deal of measured values are not in the authorized range ; the dimensioning of the exhaust network must therefore be improved.

On the other hand, the pressure must be important enough in order to limit flowrate variations due to wind or thermal effect. Moreover the exhaust vent aperture area must not be too important in order to reduce noise transmission from one flat to another one, and the pressure must be important enough in order to limit flowrate variations due to wind or thermal effect.

For all these reasons, a typical set of pressure range is 70 - 120 Pa : Most of the exhaust vents in use in France are designed in order to meet the flowrate steadiness and noise requirements in this range of pressure.

The design principle of mechanical exhaust systems in multifamily buildings is therefore to dimension the network in such a way that the pressure drop of exhaust vents remains in the authorized range of value, regardless of the climatic data and the occupant behaviour. This is generally achieved either in using purpose designed fans or, preferably, in using ducts with sufficiently large diameters, which allow to accommodate more easily the pressure drop variation due to human bahaviour (i.e : variation of required exhaust flowrate). A design method (ref. 2) based on this principle is under preparation and will hopefully help designers to improve exhaust network sizing.

3.3 – Dimensioning of natural ventilation systems.

Proper dimensioning of natural ventilation system is a delicate matter because the airflow is depending to a great extent on the outdoor temperature and also on the wind effect, which still needs a lot of investigations. The need for more research can be illustrated by considering the example of shunt ducts which were in common use in high rise buildings in the sixties :

As it appears on fig. 6, the shunt duct is a double duct composed of a small duct used for individual air exhaust of each building and a larger one used to collect individual airflows up to the top, where a cowl is installed in order to transform wind velocity in pressure head. From an architectural point of view, the shunt duct makes it possible, compared with individual ducts, to save space and decrease the building cost. On the other hand, it may lead to higher likelihood of reverse flow. The fig. 6 shows dependance of flowrates according to wind pressure variations.

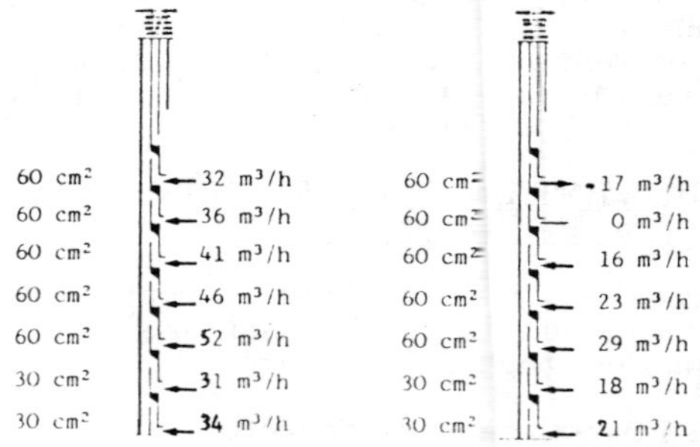

Fig.6 : Shunt ducts : influence of wind velocity and outdoor temperature on exhaust flowrates.

first columm :	*outdoor temperature : 0°C ; wind velocity : 0 m/s*
second columm :	*outdoor temperature : 10°C ; wind pressure : -3 Pa*

It may be noticed that reverse flowrates occur under unfavorable climatic date. These reverse flowrates may be alleviated using high performance cowls. A practical difficulty is that, for a given cowl, the negative pressure caused by the wind is a complex function of wind velocity and air flowrate, thus making design calculation difficult to handle. Gonzalez (ref. 3) investigated that point, and showed that a relationship was existing between the two following quantities :

$$X = \frac{DP}{1/2.\rho e.V^2} \qquad Y = \frac{1/2.\rho i.U^2}{1/2.\rho i.V^2}$$

where :

DP (Pa)	: difference between total pressure at the cowl inlet and outdoor static pressure
U (m/s)	: air velocity in the duct
V (m/s)	: wind velocity in the cowl neighbourhood
$\rho i, \rho e$ (kg/m^3)	: volumic mass of indoor and outdoor air

The simplest way of formulating the relationship between X and Y is to assume that suction effect due to wind and pressure drop due to airflow are independent, which leads to :

$$X = \frac{Y}{A} + C \qquad (3)$$

It may be observed, fig. 7, that the agreement between this formula and experimental points is far to be good. Thus a new relationship (which still has to be confirmed by supplementary experiments),

$$X = \frac{Y}{A} + C + \sqrt{\frac{-C.Y}{2.A}} \qquad (2)$$

was derived. This relation leads to a better agreement and is quite easy to handle because the coefficients,A and C may be readily mesured, according standardized methods :

A is related to the cowl pressure drop when there is a no wind effect,

C is related to the section head due to wind velocity when there is no flowrate.

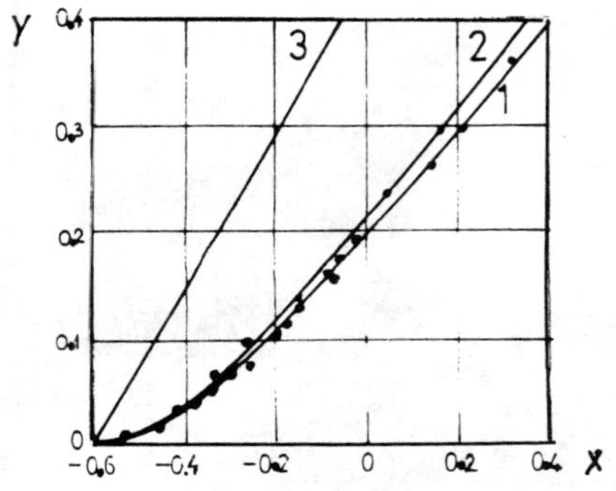

Fig. 7 : Cowls : agreement between experimental data and different formulas.

1 best fit
2 fit using equation (2)
3 fit using linear equation (3)

A last difficulty arises from wind turbulence : A computer code (ref.1) was used in order to calculate the flowrate variation due to wind turbulence in a shunt duct : It can be noticed, fig 8, that exhaust flowrates variations cannot be neglected : taking into account wind turbulence makes it possible to explain reverse flow at the upper floor.

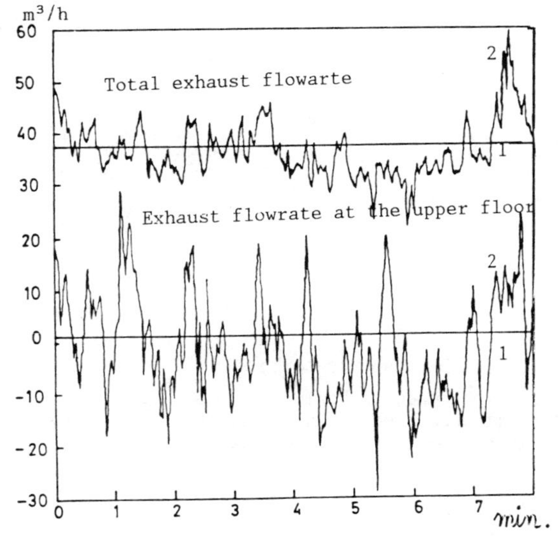

Fig 8 : computed flowrates in a shunt duct.

assumption 1 : flowrates are calculated using time averaged wind pressure

assumption 2 : flowrates are calculated using actual wind pressure records.

IV CONCLUSION

Design methods based on computer codes makes it possible to improve the design of ventilation systems. More research work is still needed in order to ascertain results relevant to natural ventilation.

REFERENCES

1. R.MOUNAJED, fascicule de documentation du code GAINE, CSTB avril 1989

2. projet de DTU 68.1 : Installations de ventilation mécanique. Règles de conception et de dimensionnement, CSTB 1989

3. M.A. GONZALEZ "On the aerodynamics of natural ventilators" Building and environment vol 19 n°3 PP 179-179, 1984

4. J.RIBERON "La ventilation naturelle" AFME, Sophia-Antipolis sept 1989.

5. J.RIBERON "Turbulence du vent et ventilation" AFME, Sophia-Antipolis, sept 1989.

6. R.MOUNAJED, fascicule de documentation du code SIREN 2, CSTB avril 1989.

ACKNOWLEDGEMENTS

The author wish to express his thanks to R.MOUNAJED who developped computer models, J.GANDEMER who provided wind data and J. RIBERON for general assistance. Participation of AFME, France, to research support is gratefully acknowledged

MAPPING INDOOR AIR MOVEMENTS WITH A TRACER-GAS TECHNIQUE

Clito F.A. Afonso and Eduardo A.B. Maldonado
Dept. of Mechanical Engr., University of Porto
Rua dos Bragas, 4099 Porto Codex, Portugal

ABSTRACT. This paper describes how the tracer decay method was used to map all indoor air flows in a passive solar residence in Porto. A specific analysis of the influence of the fireplace chimney was made, showing that it accounted for a 50% increase in air exchange rate, and that indoor air movements were strongly affected in both magnitudes and directions.

Fig.1 - The C.T.O.

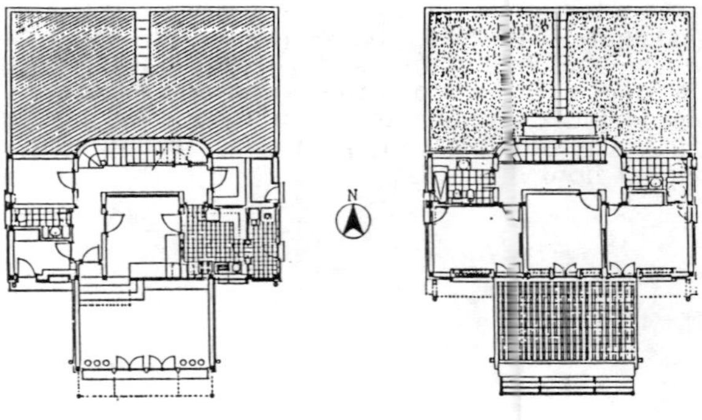

Fig.2. - C.T.O. Plans

1. INTRODUCTION

One of the major difficulties in the characterization of the performance of a passive solar building lies in the knowledge of air movement patterns inside the building. Passive solar buildings are typically multizone spaces at different temperature levels which cause convective currents that combine with outdoor air infiltration for complex circulation patterns.

Direct measurement of these air patterns is almost impossible because of the very low prevelant air velocities, /1/. Indirect forms of measurement are thus required, and the tracer-gas methods have been proven as a viable alternative, /2/. This method can be implemented in different ways, with the most common being the decay,/3,4/, and the steady-state concentration,/5/,techniques. While this latter technique requires a more sophisticated control apparatus, the former has no such requirements but, on the other hand,to obtain meaningful results, the air volumes in each zone must be know. The volume of air in a zone is a difficult quantity to get by direct measurement, because the furniture and finishing inside each zone, even those with well--established boundaries - which is not always the case -, make it all but impossible to accurately account for every detail.

A tracer decay methodologie that enables the simultaneous evaluation of the airflows and the airvolumes in buildings was however developed and validated in a laboratory under controlled conditions, /6/, and in a real house, under real climatic conditions, where airflows fluctuated as indoor and outdoor temperatures and outdoor wind conditions chanted,/7/.

This paper describes how such a technique was used to map all indoor airflows in a passive solar house in Porto and the results that were obtained. The objective was to quantify indoor airflows under several operating modes (e.g., opened or closed room doors, consequences of a fireplace and glazing protections, etc.) to allow for better insight into thermal energy flows, (e.g.,the relative weight of radiative and convective energy transfer) and indoor air quality (effective air exchange rates) in ordinary passive solar buildings.

2. PASSIVE SOLAR HOUSE

The tests were carried out in a two-story passive solar house, the C.T.O., /8/, shown in Figs.1 and 2. The building, a three-bedroom single-family house with 150 m^2, is wholly oriented to the south. There are only small windows on the north, east and west sides, to allow cross ventilation in summer. The main rooms face south and the service areas are on the north side, providing a buffer zone around the main rooms.

The CTO has an all-electric auxiliary heating system that can thermostatically control individual rooms to the desired temperature level, but it also has a fireplace in the center of the living-room, in the lower-floor. Although great care was put during construction to ensure the tightness of the chimney and the duct that brings outdoor air for combustion when the fireplace is in operation, the results described in a later section show that the influence of the fireplace upon air infiltration and indoor air movements is quite important.

The attic above the bedrooms was completely sealed from the rest of the building to minimize its influence upon air infiltration and indoor air move-

ments. As such, the house could be considered as a two-zone space, as long as all the internal doors were kept open all the time, each zone corresponding to each of the two floors.
Tests other than those reported in this paper,/9/, have shown that, when a door is closed, the respective room behaves as a separate zone with little interaction with the rest of the house.

3. TEST METHODOLOGY

The tests were performed with CH_4 as the tracer-gas. For each test, a pulse of CH_4 was released in one of the zones and the ensuying varying concentrations were measured in both zones. The procedure was repeated with the CH_4 pulse released in the other zone. Fig.3 shows the concentration of tracer gas in both zones, in the two experiments of one of the tests carried out.
The concentrations were measured by a single calibrated infrared detector, which received air from all sampling points through a multiplexer programmed in an appropriate sequence.
To enhance uniformity of indoor air properties, small fans were placed in each zone to improve mixing within them. Furthermore, the outlets of the tracer-gas injection lines were placed downstream of the fans in order to spread the gas more evently.

4. RESULTS

Indoor air movements inside the CTO change of course, with the prevailing outdoor weather, i.e., wind speed, wind direction and air temperature, which strongly affect air infiltration. A thorough analysis of air infiltration in the CTO, /10/, carried out by standard tracer-gas techniques, /11/, has shown that the air infiltration could be correlated with wind speed, wind direction and stack effect according to the equation

$$I = 0.512 + 6.184E\text{-}3\ \Delta T + 0.014V^2_{south} + 0.008E\text{-}3V^2 \quad (1)$$

where:
ΔT - Temperature difference
V_{south} - Wind component acting on the south façade
V - Wind component acting on the other façades

The average air infiltration for the mean seasonal climatic conditions is 0.75 RPH.
These tests showed that the south wind plays a fundamental role in the air infiltration: as can be seen in equation (1), the regression coefficient for the south wind component has a significantly higher value when compared with the remainder weather coefficients, namely other wind directions and stack effect. Physicaly, this is related to the large glazed area of the south façade, and small wind variations in that direction have significant effects upon the subsequent air infiltration.
In order to verify the influence of outdoor weather conditions upon the indoor air movements in the CTO several tests were carried out following the methodology described in the previous section. Fig.4 shows the airflows inside the CTO under widely varying outdoor weather conditions, with winds blowing from several directions.
The main conclusions obtained from these results are the following:
- Each floor of this two-story residence exchanges air both with outdoors and with the other floor, through the staircase. There is a downward flow of cooled air along the north wall of the building and an upward flow of warmer air through the main shaft of the staircase.
- There is a net downward flow under most circumstances, which is related with the induced stack effect of the fireplace chimney.

As can be seen, the changes in indoor air movement with outdoor weather are more pronounced in magnitude, as the overall directions of the airflows remain relatively constant.
In order to verify the influence of the fireplace upon air infiltration in the CTO, the fireplace was completely sealed from the rest of the house, which therefore behaved as if the fireplace did not exist. Tracer-gas tests were then performed following the same methodology used in the tests just described. In this new situation, the air infiltration is related with wind speed, wind direction and stack effect according the following equation, /10/:

$$I = 0.294 + 0.014\Delta T + 0.019V^2_{south} + 0.008E\text{-}3V^2 \quad (2)$$

The average air infiltration for the mean seasonal climatic conditions is I = 0.47 RPH.
Once again, the south wind plays a fundamental role upon air infiltration, but there is a clear decrease of the magnitude of air infiltration - almost 50% - when compared with the usual running mode, i.e., with the fireplace not sealed.
Several two zone tracer-gas tests were also carried out when the fireplace was sealed, as shown in Fig.5. The main conclusions obtained from these tests are the following:
- The net airflow rate between the two stories of the CTO is now from the ground floor towards the first floor, i.e., it is the opposite of the direction identified in the previous case. This is related to the absence of the stack effect induced by the fireplace chimney.
- There is a decrease in the global air infiltration of the CTO when compared with the situation of open fireplace. This is in agreement with the single-zone

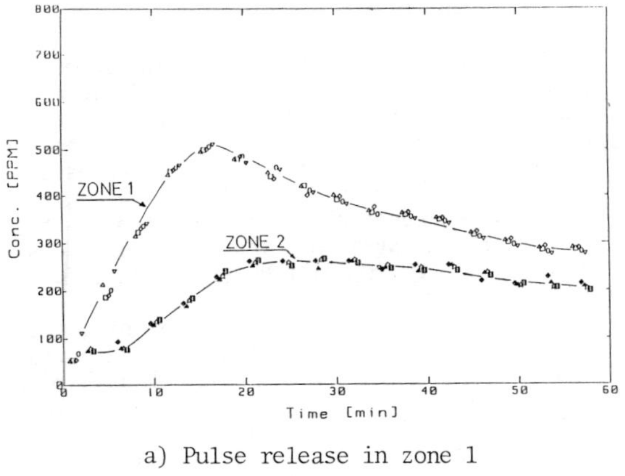

a) Pulse release in zone 1

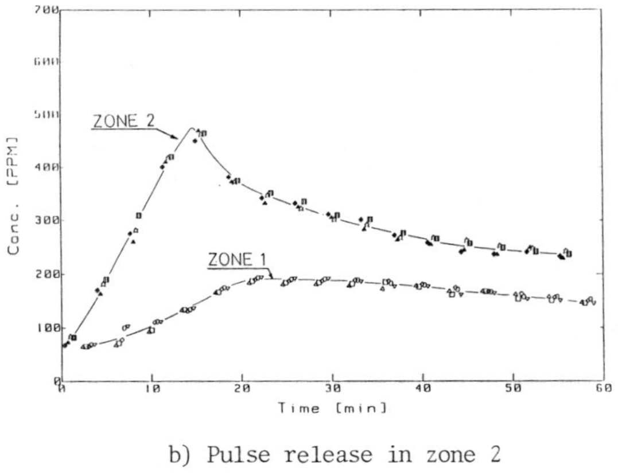

b) Pulse release in zone 2

Fig.3 - Tracer-gas concentration profiles

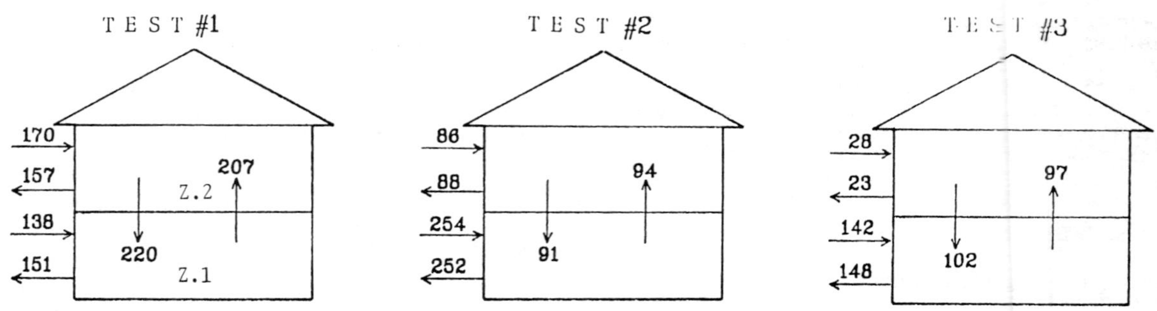

Test #	Climatic Conditions				
	Wind		Temperature [°C]		
	Dir [°]*	Int [Km/h]	Ext	Z.1	Z.2
1	188	26,6	15,6	20,7	21,5
2	315	28,5	13,8	22,0	22,4
3	254	12,6	14,4	20,4	21,4

*Azimute Angle: 0°=North; 90°=East; 180°=South; 270°=West

Test #	Infiltration [RPH] Global	Multizone
1	1,0	0,98
2	1,0	1,0
3	0,52	0,54

Fig.4 - Measured airflow rates in the C.T.O.

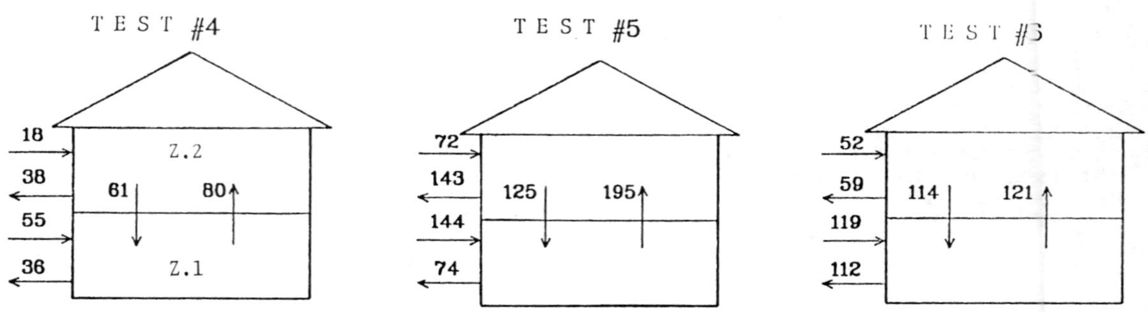

Test #	Climatic conditions				
	Wind		Temperature [°C]		
	Dir [°]*	Int [Km/h]	Ext	Z.1	Z.2
4	108	9,6	13,8	21,0	21,9
5	90	29,2	9,3	18,9	21,4
6	256	18,2	14,7	21,7	22,2

*Azimute Angle: 0°=North; 90°=East; 180°=South; 270°=West

Test #	Infiltration [RPH] Global	Multizone
4	0,22	0,23
5	0,72	0,70
6	0,54	0,56

Fig.5 - Measured airflow rates in the C.T.O. with the fireplace sealed.

whole-house infiltration results obtained earlier (eqn.2).
Once again the sole effect of the outdoor weather upon the indoor air movements is in the magnitude of the airflows. The relative directions remained constant in most of the tests.

5. CONCLUSIONS

The tracer gas methodology used in this work has proven to be reliable for the measurement of indoor air movements in building under real conditions. The tests performed in the CTO showed that:
- When the interior doors are open, each floor of the CTO behaves as an individual zone.
- Under normal conditions, both floors exchange air with outdoors and with each other.
If the doors are closed, the rooms with closed doors behave as additional zones with little interaction with the rest of the building . As the number of interior closed doors increases, the magnitude of the air infiltration rate decreases due to the higher resistance to airflow.
- Indoor air movements change in magnitude as the outdoor weather changes, but the mean directions of the airflows remain relatively constant, for most weather conditions.
- The fireplace is responsible for a significant increase in the global air infiltration of buildings. So, to reduce air infiltration when there is a fireplace, it is necessary to improve the tightness of the chimney and the window frames.
- A fireplace plays a fundamental role in the overall pattern of indoor air movement inside a house. In this particular case, an inversion of the net

airflow rate between both floors was observed: While there was a net downwards airflow when the fireplace was not sealed, there was a net upwards airflow after it was sealed.

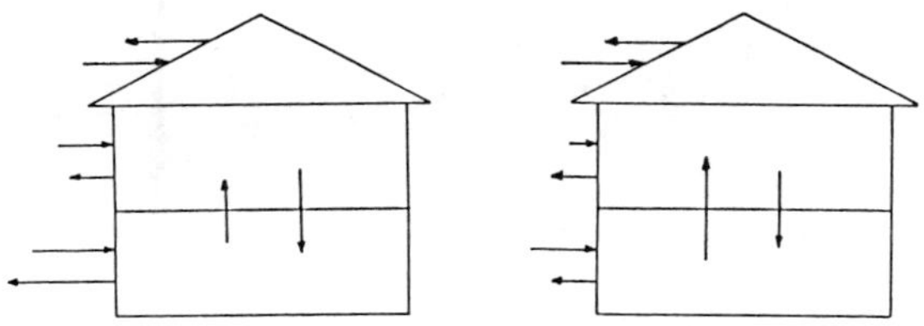

Fireplace open Fireplace sealed

6. REFERENCES

1 - Hitchin, E.R.; Wilson, C.B. "A review of Experimental Techniques for the Investigation of the Natural Ventilation in Buildings". Building and Science, vol.2, pp.53-62. 1967.

2 - Harrje, D.T.; Dutt, G.S.; Bohar, D.L.; Gadsby, K.J. "Documenting Air Movements and Infiltration in Multicell Buildings Using Various Tracer-Gas Techniques". ASHRAE Transactions, vol.91, Part 2B, pp.2012-2026. 1985.

3 - Afonso, C.F.A.; Maldonado, E.A.B.; Skaret, E. "A Single Tracer-Gas Method to Characterize Multi--Room Air Exchanges". Energy and Buildings, vol.9, pp.273-280. 1986.

4 - Afonso, C.F.A.; Maldonado, E.A.B.; Skaret, E. "Measuring Indoor Air Movement: An Essential Task to Characterize Indoor Air Quality". Proc. of Indoor Air Quality Conference. pp.592-601. ASHRAE, Atlanta, USA. 1986.

5 - Dietz, R.N.; Goodrich, R.W.; Cote, E.A.; Wieser, R.F. "Detailled Description and Performance of a Passive Perfluorcarbon Tracer System for Building Ventilation and Air Exchange Measurements". Measured Air Lackage of Building. ASTM Special Technique Publication 94. 1986.

6 - Afonso, C.F.A.; Maldonado, E.A.B. "A Tracer Gas Procedure for the Simultaneous Evaluation of Effective Volumes and Multizone Air Flows". Proc. of the International Congress on Building Energy Management, vol.3, pp.194-201. Lausanne. Switzerland. 1987.

7 - Afonso, C.F.A.; Maldonado, E.A.B. "Characterization of Indoor Air Movement in Passive Solar Buildings Using a Tracer-Gas Technique". Proc. of the 2nd World Congress on Heating, Refrigerating and Air Conditioning - CLIMA 2000. vol.IV, pp.1 - 6 Sarajevo. Jugoslavia. 1989

8 - "Casa Térmicamente Optimizada". Project Monitor - Commission of the European Communities. Issue 9, Dec. 1987.

9 - Afonso, C.F.A. "A Methodology for the Characterization of Multizone Air Movement in Buildings" (In Portuguese).Ph. D. Thesis. Porto. Portugal.1989.

10 - Afonso, C.F.A.; Maldonado, E.A.B. "The Behaviour of the C.T.O. to the Air Infiltration". (In Portuguese). Proc. of the 4th Iberian Congress on Solar Energy, pp.405-412. Porto, Portugal. 1988.

11 - ASTM Standard E741-80- "Standard Practice for Measuring Air Leackage Rate by the Tracer Dilution Method". Philadelphia, Pennsylvania: ASTM, 1980

AKNOWLEDEGMENT

This work was partially supported by L.N.E.T.I. (Laboratório Nacional de Engenharia e Tecnologia Industrial-Portugal) and by INIC-CEEM (Instituto Nacional de Investigação Científica - Centro de Estudos de Engenharia Mecânica-Portugal).

SIMULATION OF NATURAL VENTILATION IN BUILDINGS BY MEANS OF FLUID FLOWS APPLIED TO THE CASE OF CROWN AND COUNTY COURTS

P.F. Linden & G.F. Lane-Serff
Department of Applied Mathematics & Theoretical Physics
University of Cambridge, Silver Street, Cambridge, CB3 9EW, U.K.

and F. Penz
The Martin Centre of Architecture and Urban Studies
University of Cambridge, 6 Chaucer Road, Cambridge, CB2 2EB, U.K.

ABSTRACT. The use of natural ventilation to provide air changes within a building is investigated. The driving forces for the flows considered are buoyancy forces produced by temperature differences within a building and between the interior and the exterior air - the 'stack effect'. Wind-driven ventilation, which usually produces increased ventilation rates is not considered here. The use of small-scale models is described in which water is used as the working fluid and variations in salt concentrations represent different air temperatures in the full scale building. A description of the use of laboratory models to investigate a building with internal heat gains is given and the scaling laws necessary to apply the results at full scale are derived. An outline of a mathematical model is also given and some general features of buoyancy driven natural ventilation are described. These modelling techniques are applied to a proposed design of a Crown Court.

1. INTRODUCTION

Natural ventilation is the use of naturally occuring pressure differences to drive a ventilation flow. These pressure differences may be caused by temperature differences between the internal and the external air - the 'stack effect' - or by the wind. We restrict attention here to the effects of temperature differences. Recently, there has been renewed interest in natural ventilation, especially for the ventilation of large spaces /1/. This interest has arisen partly because of considerations of energy conservation and cost, and partly from dissatisfaction with the performance of mechanical systems. Since these naturally driven flows are less easy to control, it is desirable to accurately determine flow patterns and ventilation rates and to determine how effective control may be established.

The laboratory modelling of these flows is described in section 2 and the outline of a mathematical model (described fully in /2/) is given in section 3. The application of these techniques is the Crown Court design which is described in section 4 and the concluding section 5 contains a discussion of the general features of natural ventilation and the use of modelling in the design process.

2. LABORATORY MODELLING OF VENTILATION FLOWS

2.1 Scaling Laws

For the ventilation flows considered here the effects of viscosity and heat diffusion are small, as is evident from the large values of the Reynolds and Peclet numbers (10^3 and greater). These numbers are given by UL/ν and UL/k respectively, where U is a typical velocity, L is a typical length, ν is viscosity and k is the diffusivity of heat. At such large values the flow will be independent of these numbers, except at the smallest scales. To simulate these flows in the laboratory it is necessary to ensure that the values of the Reynolds and Peclet numbers in the experiment are sufficiently high. This can be achieved in the laboratory in small scale models using water as the working fluid and adding salt to produce density differences /3/. Thus different salt concentrations represent different temperatures. Note that since heat is modelled by salt it is the diffusivity of salt in water that gives the relevant Peclet number.

The driving force in these flows is the buoyancy force caused by density differences between different parts of the fluid. A parcel of fluid of density $\rho+\Delta\rho$ surrounded by fluid of density ρ will experience an acceleration of approximately $g\Delta\rho/\rho$, if $\Delta\rho/\rho$ is small. This acceleration is known as the reduced gravity, and it will be denoted by g'. For air, it is useful to note that $\Delta\rho/\rho$ is approximately equal to $-\Delta T/T$, where T is the temperature measured in Kelvin. It is not necessary to use the same density difference in the model as in the real situation, rather it is chosen to give high enough Reynolds numbers. The relation between the experimental results and the real situation is found by considering the appropriate scalings. The subscripts M and F will be used to denote the scales in the model and real (full-scale) cases. Length-scales will be denoted by L, velocities by U, times by t and buoyancy fluxes by B. These scales can all be constructed from the length-scale and g' as shown in table 1. (The buoyancy flux is the flux of g'.)

scale:		model	full-scale
	times	$(L_M/g'_M)^{1/2}$	$(L_F/g'_F)^{1/2}$
	velocities	$(L_M g'_M)^{1/2}$	$(L_F g'_F)^{1/2}$
	buoyancy fluxes	$L_M^{5/2}\ g'^{3/2}_M$	$L_F^{5/2}\ g'^{3/2}_F$

Table 1. Scales for other variables in terms of the length-scale and the reduced gravity ($g' \equiv g\Delta\rho/\rho$).

Thus, for example, the ratio of velocities in an experiment to those at full scale will be $(L_M g'_M)^{1/2}$: $(L_F g'_F)^{1/2}$. It is useful to note that the buoyancy flux due to a heat source of strength W (in kilowatts) in air is $B = 0.0281W$, where B is measured in m^4s^{-3}. Further discussion of scalings will be given below with reference to the study of a particular building.

2.2 Experimental Techniques

The laboratory models are constructed from clear perspex, which enables the visualisation of the flows. The concentration of salt solutions may be measured either by taking samples or by measurement in situ using conductivity probes, the latter technique makes use of the change in conductivity of salt solutions with changing concentration. Note that since salt water is denser than fresh water, higher salt concentrations represent lower

temperatures. In many cases, including the case described below, it is convenient to turn the model upside down so that higher salt concentrations represent higher temperatures and heat sources can be represented by sources of salty water.
The flows are visualised by shining light through the apparatus onto a translucent screen forming a shadowgraph. Since different concentrations of salt solution have a different refractive index, changes in salt concentration are shown by this technique. Dyes are also used to mark different parts of the fluid as they are advected by the flow. The flow is recorded using video cameras and by taking still photos, and can be examined in detail using image processing techniques.

3. MATHEMATICAL MODEL OF NATURAL VENTILATION

In this section the main results of mathematical models of ventilation flows will be given. Two main kinds of ventilation flow occur: mixing ventilation and displacement ventilation and we shall treat each case in turn.

3.1 Mixing Ventilation

Mixing ventilation occurs where a heated space is connected to cooler ambient air by openings near the top of the space. Warm air leaves through these openings and cool air enters, often mixing throughout the space as it descends giving a fairly uniform temperature. A steady situation will develop, with the heat lost through the openings being balanced by the heat sources within the space. If there is a single opening, of height d and area A, connecting a space at temperature $T+\Delta T$ with ambient air at temperature T, then the air flow through the opening will be

$$F=kA(dg\Delta T/T)^{1/2}, \qquad (1)$$

where k is a constant approximately equal to 0.25 for a window /4/ and 0.05 for a skylight in a horizontal roof /5/. If the space contains a heat source of W kilowatts, and lengths are measured in metres, the steady temperature is given by $T+\Delta T_M$, with

$$\Delta T_M=2.8(W/kAd^{1/2})^{2/3}. \qquad (2)$$

3.2 Displacement Ventilation

If a heated space has openings near the floor and openings near the ceiling connecting the space to cooler ambient air then warm air will leave through the upper openings and cooler air will enter through the lower openings. The air in the space will be highly stratified with cool air near the floor and warm air near the ceiling. We consider here the simple case of a single heat source on the floor of a space with an opening near the ceiling of area a_1 and an opening near the floor of area a_2. A turbulent plume will rise above the heat source, entraining ambient air and becoming cooler as it rises. A layer of warm air will build up near the ceiling and drive a flow through the openings due to the stack effect. Eventually a steady state will develop, with a stationary interface between warm air in the upper layer and cool air in the lower layer as shown in figure 1. The only flow from the lower to upper layer will be within the rising plume so the volume flux in the plume at the interface level must match the volume flux out through the upper opening, which in turn must be equal to the inflow through the lower opening. In addition, the heat flux into the upper layer, due to the plume, must equal the heat flux out through the upper opening. Thus the temperature in the rising plume at the interface level must equal the temperature of the air leaving the space through the upper opening. Consequently, the temperature of the upper layer must be uniform throughout the layer, and equal to the temperature in the plume at the interface level.

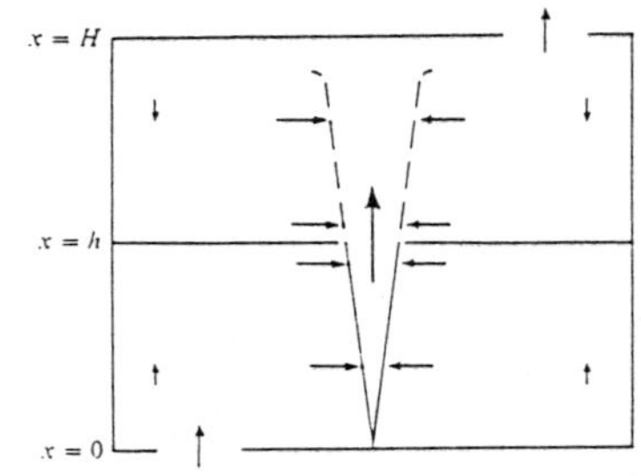

Figure 1. Steady displacement flow in a box with an internal source of buoyancy. The rising plume entrains fluid both above and below the interface. Outside the plume and below the interface the vertical component of velocity is upward: outside the plume and above the interface the vertical component of velocity is downward. Buoyant fluid leaves the space through the upper opening and ambient fluid enters through the lower opening.

Writing H for the height of the space and h for the height of the interface, the steady state balances described above lead to a relation between the area of the openings and the interface height.

$$A/H^2=0.04(h/H)^{5/2}(1-h/H)^{-1/2}, \qquad (3)$$

where A is the effective area, given by

$$A=(a_1 a_2)/(1/2(a_1^2/c+a_2^2))^{1/2}. \qquad (4)$$

In equation (4) c is a constant lying between one half, for a sharp expansion at the inlet, to unity, for a perfectly smooth expansion. The effective area is dominated by the smaller of the upper and lower openings. Note that the strength of the heat source does not appear in equation (3): the interface height is a geometric quantity determined by the geometric relation between the area of the openings and the height of the space. The temperature difference between the upper and lower layer, which will be denoted by ΔT_D, does depend on the heat source, and is given by

$$\Delta T_D=24W^{2/3}h^{-5/3}, \qquad (5)$$

where h is measured in metres, W in kilowatts and ΔT_D in degrees Celsius or Kelvin.
The results for both mixing and displacement flows are shown in graphical form on figures 2 and 3, together with results from laboratory experiments conducted in a simple rectangular box. The results given in this section apply to a single point source but the analysis can be extended to multiple sources and to line and area sources /2/.

4. CASE STUDY: CROWN COURT BUILDING

4.1 Design Considerations

A courtroom has a number of restrictions to the possible ventilation openings. In the majority of cases all windows are above head height. There is also separate access to the main courtroom for the judge, jury, defendants and public. As a result high level vents were placed in the walls and ceiling of a raised roof section as shown in figure 4, to simulate the effects of windows and skylights. Two possible roof heights were examined viz 6m and 8m. Low level vents in both the walls and the floor were included (see figure 5) so that a range of opening configurations could be tested.

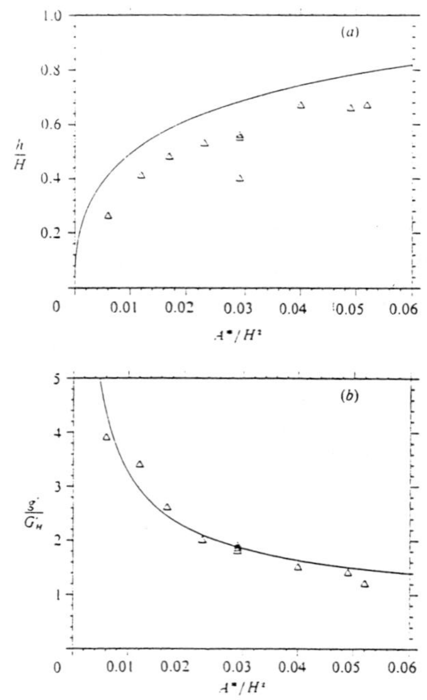

Figure 2. Theoretical and experimental results for displacement flow (a) interface height (b) buoyancy of heated layer.

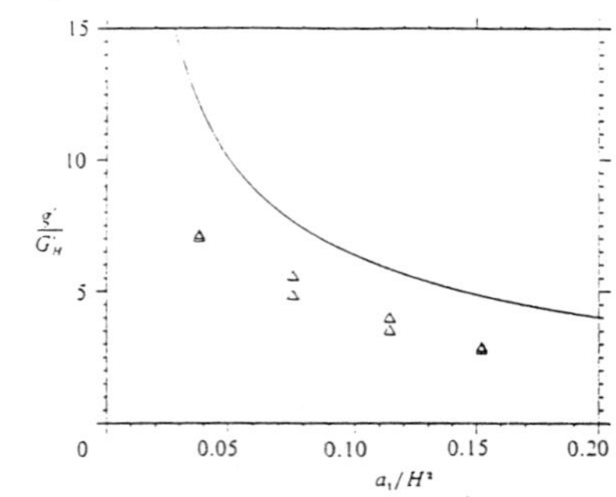

Figure 3. Steady-state buoyancy of space in mixing ventilation: comparison of theory and experiments.

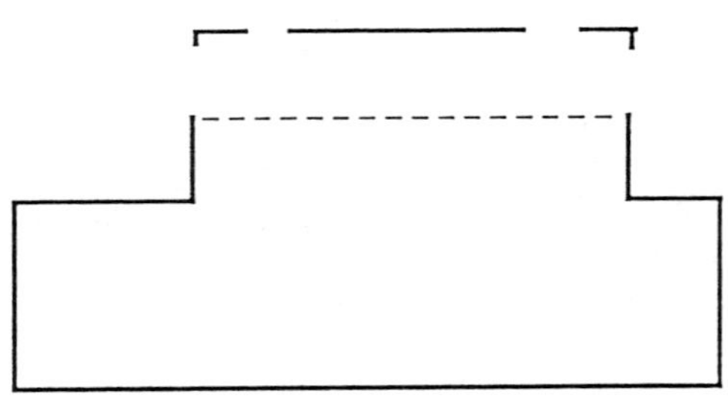

Figure 4. Courtroom model: section.

4.2 Modelling

A 1:30 scale model of a medium size courtroom (see figure 6) was constructed from perspex. Solid perspex blocks were used to represent the interior furniture. The model was completely transparent so that the flow within it could be visualised. Openings in the walls and floor were in the form of 18mm diameter circular holes. In the roof there were 4 square roof lights measuring 25mm x 25mm and windows were constructed in the ends of the roof above the judge and the defendents. These windows measured 50mm x 160mm and were constructed in two parts so they could be either fully or half open. All openings were sealed by plastic plugs or perspex covers and could be opened as desired. The model was suspended upside down in a large tank of water, illuminated by shadowgraph and filmed with an inverted camera. To avoid confusion the terms up, down, floor etc will be used with reference to the actual building.

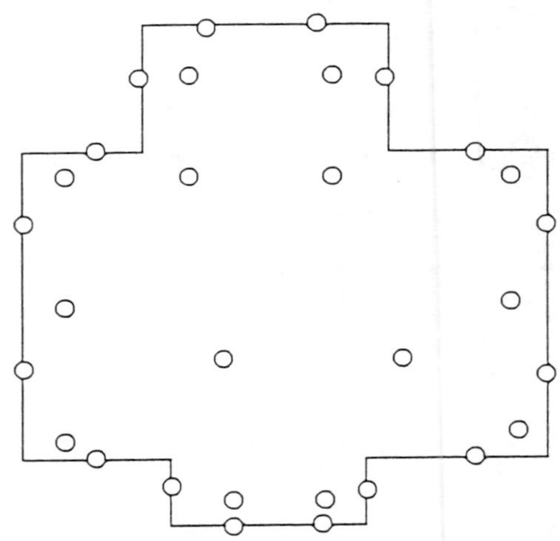

Figure 5. Courtroom model: plan.

Two main series of tests were carried out. In the first, the model was filled with salt solution and the transient flows that developed when one or more vents were opened were observed. The initial density difference was 26kg m^{-3}, and since L_F/L_M=30, the velocities at fullscale are 6 times faster than in the model and the real timescales are 5 times longer when the initial temperature differences are 10 degrees Celcius. In the second series, the effects of occupants were investigated. Three equal sources were placed on the floor to represent the judge and jury, the defendents and counsel, and the public. The strengths of these sources were equivalent to a total of 8 kilowatts (80 people each producing 100W) with the same ratio of velocities and timescales as in the first series.

4.3 Results

4.3.1 Openings in roof only

When windows were opened an exchange flow was established with warm air leaving the building through the upper part of the windows and cooler air entered through the lower half of the window. The cool air descended as a turbulent plume and mixed the interior. This was a mixing ventilation case with very weak stratification within the courtroom. When roof lights were opened the exchange flow was much less organized with interleaving and mixing of the incoming and outgoing air as it passed through the opening. The combination of roof lights and high level windows resulted in outflow from all openings, but inflow only through the windows. Measurements of temperature were in agreement with the predictions of (1) which implies that ΔT decreases as $1/t^2$ as shown in figure 7.

4.3.2 Low level openings only

In this case cool air enters the building with very little mixing and spreads out over the floor. It rises to the top of the highest open vent, and then further ventilation ceases. The details of the flow through the vents depends on the configurations of the openings.

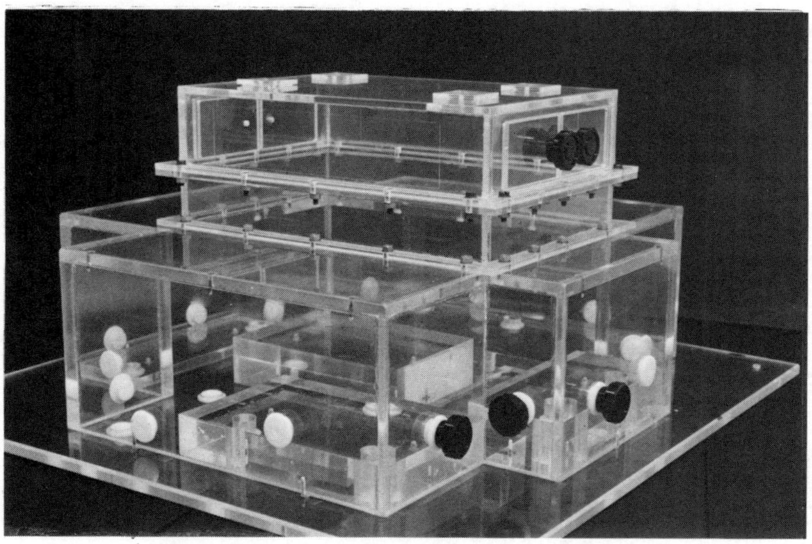

Figure 6. 1:30 scale model of courtroom showing the vents and the interior furniture .

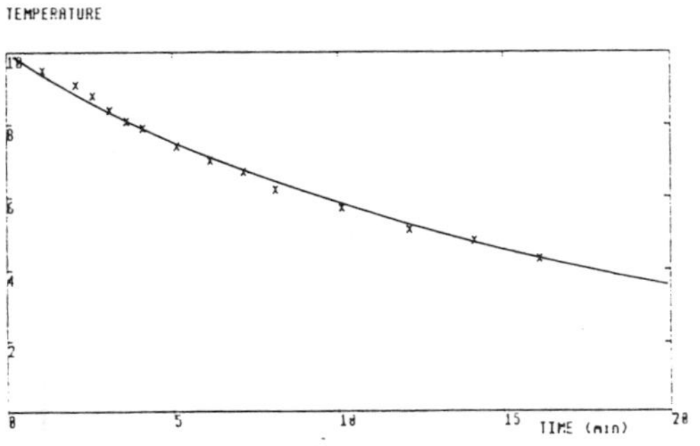

Figure 7. Decay of temperature in model.

4.3.3 High and low level openings

In these circumstances displacement ventilation is established with cold air entering at low level and warm air leaving the courtroom through the roof openings.

As predicted by (4) the rate of ventilation is determined by the vents with the smaller total area. The flow is a two-layer stratification with relatively little mixing between the incoming and interior air. If the total area of the upper level vents exceeds the low level vents by a factor of approximately 5 there is also some inflow through the high level windows. In this case the ventilation is a combination of the displacement and the mixing modes.

4.3.4 Effects of occupants

These tests were carried out with both high and low level openings. A displacement flow was established and a steady-state achieved as described in section 3. As successive low level vents were opened the level of the interface between the incoming air and the warm layer was raised. It was possible to control the level using the openings, so that the judge, at the highest point in the courtroom, remains in the cool air. As predicted by (3) the depth of the cool layer depends of the geometrical arrangement and location of sources and not on the heat flux.

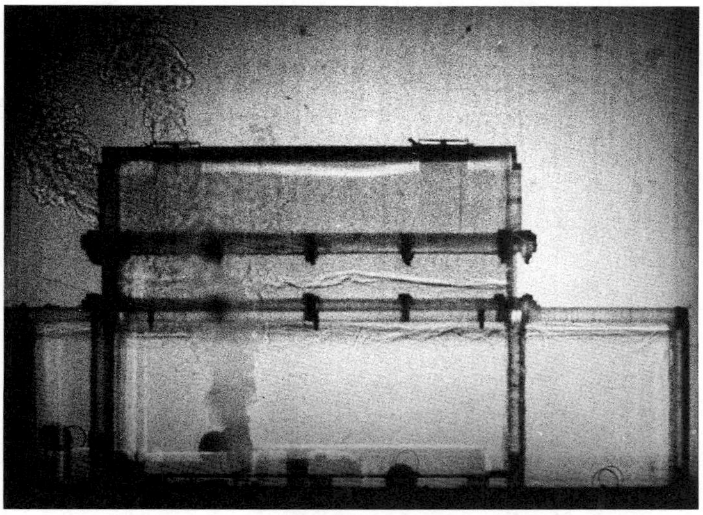

Figure 8. Photograph of displacement ventilation. Several vents near the floor, a window near the ceiling and a skylight are open.

5. CONCLUSIONS

This study has shown it is possible to naturally ventilate a courtroom, and that laboratory and mathematical modelling provides insights and quantitative information on the ventilation rates. Displacement ventilation is a controllable mode and works efficiently provided the ambient air is at a comfortable temperature. On cooler days, some preheating of the incoming air can be achieved by using high level inflows. Laboratory modelling allows the ventilation regimes to be delineated and the mathematical model is confirmed. It is, therefore, possible to predict the effects of other design changes and to decide on other configurations to test. The scaling laws which enable the quantitative results to be extrapolated to full-scale are also given.

ACKNOWLEDGEMENTS

This work was funded by the Property Services Agency (PSA Projects) and G.F. Lane-Serff was supported by a British Gas Research Scholarship. The model was constructed by D. Cheesley.

REFERENCES

1. Penz F.A. 1983 Passive solar heating in existing dwellings. ETSU publication ETSU-5-1056a.

2. Linden P.F., Lane-Serff G.F. & Smeed D.A. 1990 Emptying filling boxes: the fluid mechanics of natural ventilation. J. Fluid Mech. (in press).

3. Linden P.F. & Simpson J.E. 1985 Buoyancy-driven flow through an open doorway. Air Infil. Rev.5

4. Shaw B.H. & Whyte W. 1974 Air movement through doorways - the influence of temperature and its control by forced airflow. J. Inst. Heating and Ventilation Eng. 42: 210-218.

5. Epstein M. 1987 Buoyancy-driven exchange flow through openings in horizontal partitions. Infil Conf. on Cloud Vapor Modelling. Nov. 1987, Cambridge MA.

AN INVESTIGATION INTO THE VENTILATION AND THERMAL PERFORMANCE OF CONTEMPORARY HOUSING IN NORTH AFRICA

Mohamed B. Gadi and Ian C. Ward
Building Science Unit
School of Architectural Studies
University of Sheffield
Sheffield
S10 2TN
UK

ABSTRACT. New housing developments in North Africa often show very little regard for local culture, climate and geographical resources. Evidence has indicated that these new houses consume large amounts of energy to maintain indoor thermal comfort or as is often the case have excessively high internal air temperatures.
In order to be able to design better housing a deeper understanding of the thermal performance is required. One way of doing so is to carry out computer simulations based on a validated computer programme of the proposed design.
This paper outlines a method of establishing the likely indoor air temperatures of a proposed design using local climatic data and a new approach to establishing the likely air change rates resulting from an understanding of the occupancy patterns within the house.
The ventilation aspect of this paper is perhaps the most important as it involved the testing of the proposed design for the house in a boundary layer wind tunnel to obtain the actual variation in internal static pressures between the different rooms.
This variation was then converted to air change rates by using the actual external pressures and the air transfer areas.
Having established the variation in air change rates throughout the day a computer simulation was carried out using a standard thermal admittance programme and local climatic data.
The results were then checked against independently published work on the indoor air temperatures found in housing in the City of Ghadames, Lybia and good agreement was obtained.

1. INTRODUCTION

Most residents throughout the North African region, particularly the older generation, still hold the view that the traditional house is thermally more satisfactory than the contemporary house.
One of the reasons for this dissatisfaction is that the traditional house was constructed of local materials such as rammed earth, puddled mud, lime, straw and palm leaves which gave the house a high thermal mass which prevented a rapid build up of temperature in summer and kept the house warm in winter. The contemporary house on the other hand is constructed of lightweight concrete or burnt clay blocks which have little thermal mass thus allowing heat to pass through readily. One other factor which affects the resultant internal air temperature of such housing is the ventilation rate. In such climates during summer it can be too hot to open windows during part of the day as the resultant ventilation rate would heat the house further. In traditional housing which was designed on an inward looking concept a courtyard was provided in which there was a fountain, grass and shaded by a veranda. This architectural form produced a cooled enclosure in summer and as most of the windows opened onto this courtyard natural cooling of the air was produced. In winter when the air was cooler, the high mass of the veranda floor (which was constructed of stones held in place by lime morter) was warmed by the sun which produced a warming effect to the air before it passed into the house.
On the other hand the contemporary house is designed on a more outwardlylooking aspect with more windows exposed to the external environment. This has the effect of making the rooms on the perimeter of the house very hot under certain summer conditions and the result is that the only cool area in the house is the centre core. By adopting such a plan form the centre core becomes the main living space in summer and in winter when ventilation is needed cross ventilation is promoted by the plan form, Fig. 1.

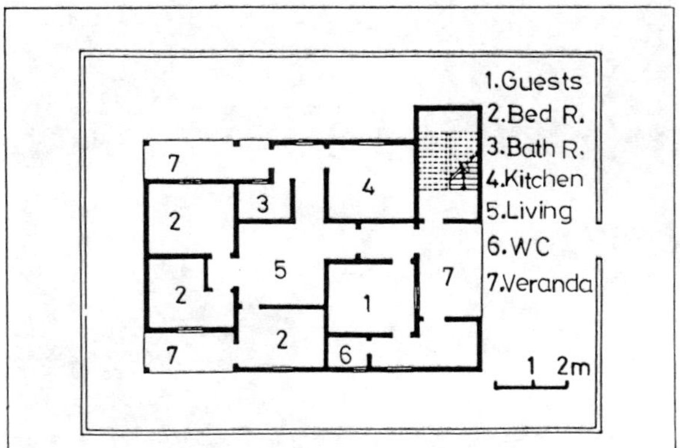

Figure 1: Contemporary North African house plan.

The traditional house could not offer the scope for housing the domestic appliances which go hand in hand with modern life, such as refrigerators, furniture etc. Thus people accepted the new house type despite it's poor thermal performance.
Due to this situation, it has become the feeling among the local population that an alternative house, designed and built with respect to local climate, culture and modern life styles is needed. This is not an easy task because of the lack of local design criteria and local design codes. Most of the design codes which are available are derived from foreign sources without particular reference being made to local conditions. In order to overcome these obstacles and to arrive at relevant solutions designers have two alternatives, either:-
(a) To develop a new version of the old courtyard house.
(b) To investigate the thermal characteristics of the current house in order to improve its performance.
One method of investigating the performance of a new house design is to carry out a series of computer simulations. However in order to do so it is necessary to have available appropriate climatic

data as well as a knowledge of the thermal properties of the building materials. Perhaps the most difficult data to obtain are the wind forces acting on the proposed building. As these are used to generate the pressure differences across the building and hence the natural air change rate it is very important that the correct values are used. There are several techniques for predicting ventilation rates through naturally ventilated buildings, some of which have been discussed by Gadi /1/. However, for the purpose of the present study, it was concluded that scale model techniques were more convenient than other techniques such as computer simulations or full scale techniques.

2. SCALE MODEL STUDIES

2.1 Experimental Set Up

In this technique, a prototype scale model of the proposed building was constructed and mounted in a boundary layer wind tunnel where the appropriate external wind environment characteristic of a particular type of terrain was simulated in order to investigate the resulting dynamic and static pressures of wind outside and inside the building.
The tests were carried out in the University of Sheffield 1.2 x 1.2m Boundary Layer Wind Tunnel. The simulation arrangement of the rural boundary layer and the details of the wind tunnel are given by Lee /2/. A general view of the flow simulation arrangement and the layout of the building model are shown in Plate 1.

Plate 1: The rural Boundary Layer arrangement and the building model in the Wind Tunnel.

The mean wind velocity profile for the rural simulation measured at the turntable centre shows a maximum free stream velocity of 9.3m/s which was attained at a height of 800mm, the physical thickness of the boundary layer. The mean velocity profile plotted in its power law form, shows that the best fit line through the data points has an exponent of $\alpha = 0.19$, /2/. Accepted values for rural terrain lie in the range of 0.14 to 0.17 however a value of $\alpha = 0.19$ is given by ESDU /3/ for a similar type of terrain. Applying these exponents within the power law for the standard meteorological height of 10m:-

$$V_{10} = V_{300}(10/300)^{0.19} = 9.3(10/300)^{0.19} = 4.9 \simeq 5\text{m/s}$$

$\alpha = 0.14$, $V_{10} = 5.8 \simeq 6$m/s, 300m is the gradient height for rural terrain. These two values agree well with the measured mean monthly wind speeds recorded at airport sites for three locations in Libya, Tripoli, Bengazi and Sabha, Table I.

Table I: Frequency of wind speed intervals.

m/s	1-1.5	2-3	3.5-5	6-8	9-1[illegible]	11-17	calm
Tripoli	4.2	14	35.4	30.6	10.[illegible]	0.8	5.0
Bengazi	5.9	13.8	34.9	29.3	9.[illegible]	0.1	6.3
Sabha	9.9	20.2	26.6	22.0	12.0	1.0	8.3

Once the rural boundary layer has been simulated inside the wind tunnel, the next step was to model the building to fit in the tunnel's working section. In a paper published by Aynsley /4/ he discusses the different modelling parameters for boundary layer wind tunnel studies of natural ventilation, and highlighted the importance of the blockage ratio.
The wind tunnel blockage is the ratio of the cross-sectional area of the model to that of the working section. Actually there is no fixed criteria for this ratio and the main requirement in this aspect is to keep the model as small as possible. However it is generally recommended to attain the blockage ratio below 5%, Lawson /5/. In the present work, the maximum blockage ratio for the model employed was 2.4%. This was due to the suitable scale, 1:50 of the model which allowed for the important architectural details of the building to be modelled. This scale was also convenient for the technique used in the pressure measurements, Plate 2.

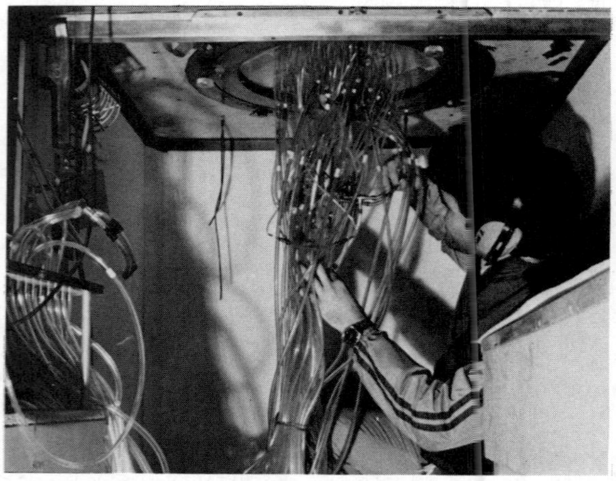

Plate 2: Plastic tubes connected between individual pressure tapping and measuring devices.

It has also kept the resulting Reynolds number within the acceptable range, 3.6×10^6, which is at the upper limits of the critical value (10^5, 3.6×10^6) given by Melaragno /6/, above which the air flow is expected to be turbulent. Generally, it is well known that buildings with sharp edges can be modelled and tested in the wind tunnel as they will give the same flow patterns for a range of air speeds.
The model was made of two materials, aluminium plate and perspex. The choice of the aluminium was due to the need for a sharp edged model with smooth surfaces in order to define the flow separation points at the corners. As it was useful to provide a visual access through the model, both the floor and the roof were made of perspex which is a transparent material.
One limitation associated with this material is that it needs extra care in order to avoid any cracks or scratches on its surface.

2.2 Tests Carried Out

Having modelled the building and its wind environment, the following stage was to prepare the

model for measuring wind pressure distribution on its envelope and through its plan for eight wind directions and according to different configurations of doors and windows. The main purpose for conducting these tests was:-
To predict the air change rates under various patterns of using doors and windows by the local people.
The model employed had eleven cells laid in a generally rectangular plan, Fig. 1. It is surrounded by a solid fence of 2 cm high and at different distances from its sides. All external and internal doors were made openable from thinner aluminium plates and the joints around them were filled with a coat of vaseline in order to control their movement and to avoid overscaling of the joints. Windows were used either fully opened or closed with aluminium plates. External pressure measurements were taken through a linear grid of pressure tapping fitted at the window centre line. Later on, another grid was constructed on the floor of each cell for the measurement of internal pressure, Plate 3.

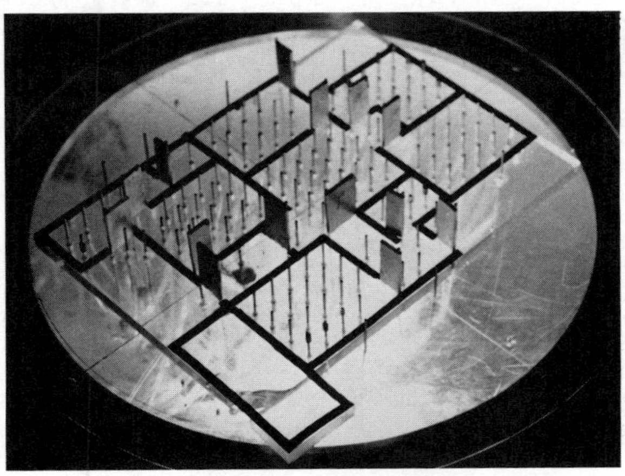

Plate 3: Plan of the building model and the pressure tapping distribution.

In order to ensure maximum accuracy in the pressure measurements and to prevent any leakage of air from one cell to another through joints between walls, floor and roof, all these joints were sealed with vaseline.
After measuring the external pressure, the following series of pressure tests were carried out internally but according to the daily patterns in which doors and windows are usually used by the local people of North Africa influenced by their social and climatic environment. Therefore the following cases were considered:-
- A summer day or a winter night. All external and most internal openings were closed.
- A summer night (bedtime) or a winter day. All external and most internal doors were closed while all windows were opened.
- A summer night,(before bedtime). All internal doors and windows were opened while external doors were closed.
- A hypothetical case in which all openings were opened.

Finally all pressure readings were obtained from a digital micromanometer linked to a 20 channel scanning box. The reference velocity pressure was checked by a pitot tube mounted inside the wind tunnel at a height of 800 mm (the gradient height).

3. PREDICTION OF VENTILATION RATES

After knowing the pressure distribution inside and outside the building, it became possible to identify the different paths and directions in which air will flow through the building bearing in mind that it flows from higher to lower pressure regions.
The consequent step was then to calculate the rates of air entering and leaving the building through and around its openings in order to estimate the resulting air change rates. By studying the plan of the house in question, it can be seen that it is made up of two adjoining apartments; the family zone and the guests zone, between which the only access is through the entrance corridor which is controlled by a swinging door in the middle. When the rates of air flow through each zone were calculated separately, it was found that significant differences exist for all cases and under most wind directions. The mean minimum and maximum air flow rates through the family zone were respectively 0.07 and $1.5m^3/S$, for the first and fourth cases of occupancy paterns mentioned previously. The corresponding values for the guests zone were 0.03 and $1.15m^3/S$.
When the first case was taken as the "Reference Case" in which the minimum rate of air flow ($0.07m^3/S$) was associated with a minimum "Leakage Area", ($0.3m^2$) then all the rates of air flow and their corresponding leakage areas were related to the reference case, a clear relationship was obtained.

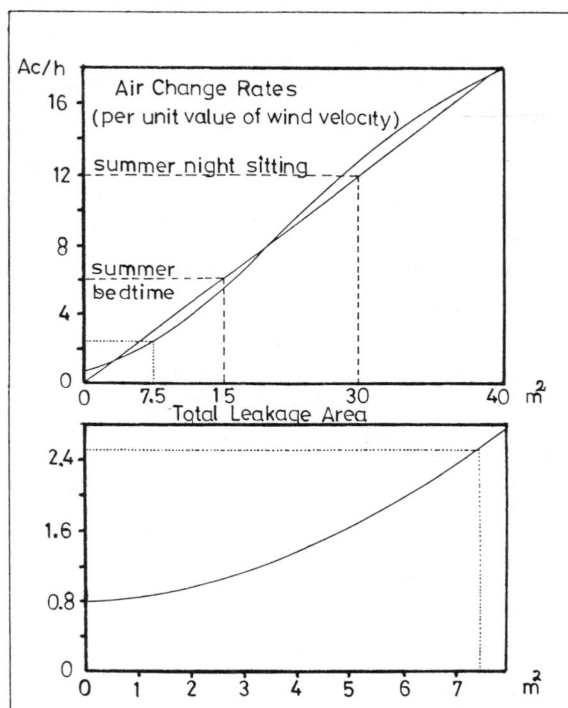

Figure 2: Ventilation Rate versus Leakage Area.

On the other hand, when the resulting rate of air change for the present house (in which the family zone has a space volume of $300m^3$) was plotted against the total leakage area for the four cases, almost a linear relationship was produced , Fig. 2 . If the total leakage area together with the average on-site wind speed are known at any period of occupancy, the resulting rate of air change can be obtained by multiplying the average wind speed by the rate of air change (per unit value of wind velocity) corresponding to the total leakage area presented in Fig. 2. This graph was used to predict the air change rates for various periods of occupancy related to the typical patterns of domestic life in most parts of the region of North Africa, Table II.

Table II: Environmental data (Sabha)

Time	Air Temp (°C)		S.W.Rad. (W/m²)		Wind (m/s)		Vent. (Ac/h)		Casual Gains (W)
	Jun	Jan	Jun	Jan	Jun	Jan	Jun	Jan	
2	24.9	5.7			2.2	2.0	12	2	350
4	23.5	4.6			1.2	1.9	12	2	350
6	24.0	4.0	122		2.4	3.7	12	2	700
8	28.2	6.4	491	130	3.1	3.0	3	8	700
10	34.7	13.9	841	353	3.6	2.6	3	8	140
12	37.6	17.8	986	453	3.8	3.5	3	8	140
14	39.0	19.5	841	353	3.7	4.1	3	8	2380
16	37.6	18.7	491	130	4.4	3.5	3	8	575
18	32.7	14.1	122		3.9	3.2	3	8	575
20	29.7	10.7			2.5	2.7	20	4	1190
22	27.8	8.6			1.5	2.3	20	4	1075
24	26.5	7.0			1.9	2.9	12	2	350

4. CALCULATED AND MEASURED INDOOR AIR TEMPERATURES

The predicted ventilation rates together with assumed casual heat gains and measured local climatic data, Table II, averaged over fifty years, were used within a validated computer programme based on a standard thermal admittance procedure, /7/, to calculate air temperatures inside the contemporary house employed in this study, Fig. 3.

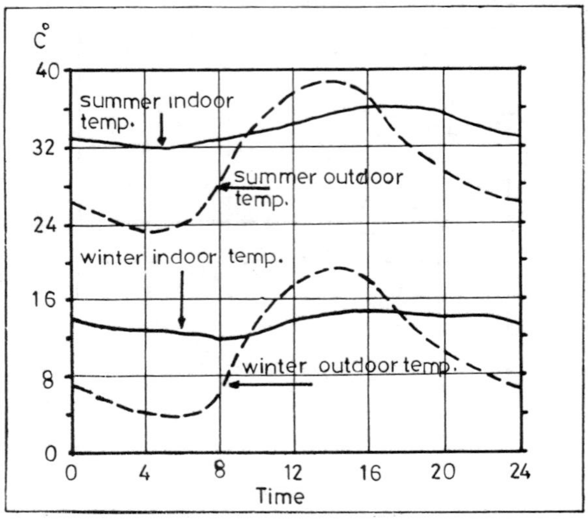

Figure 3: Indoor and outdoor air temperatures for January and June in Sabha.

The climatic data was recorded at a meteorological station in Sabha, (27°00 N, 14°:26E). A comparison with similar data for four other locations in the region, Ghadames, Aswan, El-Cufra and In-Salah, /1/, has shown slight deviations from one another due to differences in altitude and distance from sea.

In a paper published in 1985, Ahmed et al /8/ presented the results of full scale thermal simulation of typical traditional and contemporary houses in the city of Ghadames (30°00 N, 9°:50E). The new houses of Ghadames are of single and double storey detached from each other and built according to the same design and construction concepts of the house tested in the present work.

In the results presented in Ahmed's paper, the summer ambient air temperature in Ghadames varies from 22 to 40°C with an average of about 31°C. On the other hand, the average indoor air temperature in the new house is 35°C and ranges between 34 and 39°C for July-August. The new house reduces the ambient air swing of 18°C to 5°C.

In the present study, the estimated average indoor air temperature for June is 34°C ranging from 32 to 36 in Sabha. Thus the swing in air temperature is reduced from 15 to 4°C. The winter ambient air temperature in Ghadames varies from 4 to 23°C whilst that of the new house ranges between 12 and 14°C in December. In the present work the estimated indoor air temperature for the new house ranges between 13 and 15°C in January.

5. CONCLUSIONS

This investigation has provided building designers with a simple design tool they need to produce the actual daily ventilation rates which are likely to occur inside contemporary houses in North Africa. The provision of such data would make it possible to carry out realistic thermal simulation of any house design for any location in the region. Preliminary simulation of the expected indoor thermal conditions of a particular design is a necessary and important step towards improving the building's thermal performance. The comparison with the results of the full scale experimental observations in Ghadames has emphasized the validity of the information produced in the present study. However, both investigations have illustrated, in terms of air temperature, the poor thermal performance of the new house in comparison with the well known better thermal performance of the old traditional house. An apparent limitation of the present work is that it does not cover a wide range of house designs. This actually is not a significant factor as most house plans in North Africa are based on similar design concepts influenced by almost the same cultural, climatic and geographical environment.

6. REFERENCES

/1/ Mohamed B. Gadi, "An Investigation into the Ventilation and Thermal Performance of Contemporary Housing in North Africa with Special Reference to Libya", University of Sheffield, Building Science Department, M.Phil Thesis, 1987.

/2/ B.E. Lee, 'The Simulation of Atmospheric Boundary Layer in the Sheffield University 1.2 x 1.2m Boundary Layer wind Tunnel', University of Sheffield, Department of Building Science, Internal Report No. 38, July 1977.

/3/ ESDU, 'Fluid Forces, Pressure and Moments on Rectangular Blocks', engineering Societies Data Unit, Data Item 71017, 1979.

/4/ R.M. Aynsley, 'Modelling Parameters for Boundary Layer wind Tunnel Studies of Natural Ventilation', ASHRAE TRANS., 1985, Vol. 91, Part 2B.

/5/ T.V. Lawson, 'Wind Effects on Buildings', Vol. 1, Design Application, Applied Science Publishers Ltd., London, 1980.

/6/ M. Melaragno, 'Wind in Architectural and Environmental Design', Van Nostrand Reinhold Company, New york, 1982.

/7/ M.A. Sattler, 'A Computer Program for the Thermal Design of Unconditioned Buildings', University of Sheffield, Department of Building Science, Internal Report No. 81, 1986.

/8/ I. Ahmed, E. Khetrish and S.M. Abughres, 'Thermal Analysis of the Architecture of Old and New Houses at Gadames', Building and Environment, Vol. 20, No. 1, Pergamon Press, 1985.

DYNAMIC CHARACTERIZATION OF DAYLIGHTED SPACES

L. CANTARINO DE CARVALHO
Laboratório Nacional de Engenharia Civil
Av. do Brasil, 101, 1799 LISBOA. PORTUGAL

ABSTRACT. The daylight factor is the most usual concept in Europe for characterization of daylighting conditions. CIE Publication 16 presents some curves that allow to use this concept, in most latitudes, to calculate the percentage of hours in which illuminance levels will be available or exceeded in a given period of the day. However the method is only applicable if the sky could be assumed to follow the CIE standard overcast sky, because it is for those conditions that the daylight factor is usually calculated.
In European Southern countries sunlight represents a very important component of total daylight and its effects cannot be ignored. In Lisbon, for instance, the sun shines, in average, 53% of the day in January,84% in July and more than 3000 hours all the year round. Often daylighting systems are only conceived for sunshine conditions. A dynamic characterization of daylighted spaces including sunlight effects is needed.
The paper presents some results obtained from a computer and experimental daylighting analysis of several classrooms of a Portuguese passive solar school. These results are discussed and it is considered that they must be complemented by other kind of information able to represent in a simple and short way what happens in average in the course of the year and therefore to allow to compare easily different daylighting solutions and to lay down recommendations. A proposal about the contents of that information is presented and exemplified. A calculation method is also presented.

1. INTRODUCTION

National Laboratory of Civil Engineering (LNEC), in Lisbon, is a Department of the Ministry of Housing and Public Works. In the scope of daylighting studies carried out in this Laboratory characterization of daylighted spaces is often necessary, namely for schools. There is a need also to lay down recommendations for the design of buildings, and requirements for portuguese building regulations.
For the portuguese climatic conditions a simple daylight factor characterization (with overcast skies) is not enough. Sunlight effects are relevant and orientation influence cannot be ignored. In the next paragraph some results of a case study are presented and discussed. Afterwards in paragraph 3 a method of representing dynamic characteristics of daylighted spaces is proposed.

2. EXAMPLE RESULTS FROM A DAYLIGHTING ANALYSIS OF CLASSROOMS

2.1 The Results

Results in figure 2 correspond to a classroom (figure 1) of Crato passive solar school, built in Alentejo, in the south of Portugal. Five more classrooms of the same school have been characterized in the scope of the study carried out/1/.
These results are static representations of daylight conditions. They need to be complemented by other kind of information able to represent in a simple and short way what happens in average in the course of the year. That information must allow easily to compare different daylighting solutions and to lay down recommendations or requirements for Portuguese building regulations.
Next paragraph presents a contribution for a dynamic characterization of daylighted spaces.

3. DYNAMIC CHARACTERIZATION OF DAYLIGHTED SPACES

3.1. Daylight hours of the utilization period

The first item proposed for a dynamic characterization of daylighting conditions is the representation over a solar chart of the daylight hours of the utilization period as well as the quantification of the corresponding annual value (figure 3).

3.2. H indicators

Illuminances at a reference point can be characterized through a curve representing the mean annual number of hours (H) for which reference levels E_{ref} are available or exceeded. However, usually it is only necessary to know particular points of this curve, for instance, for E_{ref}=150 or 300 or 700 lux. These points will be represented by H150, H300, H700 and called H indicators.
CIE Publication 16 /2/ presents some curves that allow to calculate these indicators, for most latitudes. However the method is only applicable if the sky can be assumed to conform to the CIE standard overcast sky. More recent results of research on luminous climates may provide more general methods to calculate H indicators. However, as far as we know, such methods and corresponding climatic information are not yet generally available for the different luminous climates in Europe.
A method using current available climatic information and able to consider sunlight effects is proposed herein.
First we must define the points and planes where illuminance value. For the classroom that is represented in figure 1 two points of the work plane seem to be enough to represent two zones - the half part closer to the windows and the other half part closer to the rear wall. A reference illuminance value of E_{ref}=300 lux is appropriate as a base to distinguish the periods in which daylight is enough from those in which it is not.
H300 indicator is computed separatelly for the periods in which the sun is overcast (H_{oc}) and for those in which the sun is clear (H_{cl}). Average frequencies of those periods are assumed to be equal to 100-S and to S being S the sunshine percent value. Usually monthly sunshine percent values are easily available.
For overcast sun periods an overcast sky is assumed. For these periods climatic information is being prepared at LNEC in collaboration with the Portuguese Meteorological Office. Cumulate frequences of occurrence of horizontal diffuse illuminances when the sun is obscured have been computed. These frequences are referred to the duration of that periods with

the sun obscured. The work is being carried out with records from Lisbon, but it is expected that the same frequencies could be used for other sites in the country with different 1-S values. Then it will be possible to calculate H_{oc} from the external horizontal reference illuminance value (equal to E_{ref}/J, where J is the daylight factor), from the cumulate frequences and from the 1-S values.Whereas this information is not available CIE curves can be used as a first approach, though CIE curves are not applicable for high values of horizontal reference illuminances.
For clear sun periods a clear sky is assumed. The above presented solar chart with daylight hours of the utilization period is now necessary. Using a computer program or a scale model the line that limits the periods when clear sky illuminance levels at the reference point are equal or higher than E_{ref} must be represented over the solar chart. With a computer program this line can be obtained by interpolation, calculating illuminance levels for a grid of positions of the sun over the solar chart. Using a scale model that line can be obtained as follows. Choose a clear sky day and put the scale model outside. For each height of the sun different azimuthal positions can be reproduced just by rotating the model. Measuring illuminances at reference points is possible to mark on the solar charts - one for each point - the positions of the sun where E_{ref} illuminances are obtained. In the course of a half day enough positions will be marked allowing to draw the line. Usually E_{ref} will be exceeded for all azimuthal positions of the sun higher than a relative low value. Therefore it will be possible in most cases to draw the whole line even if measurements are carried out in winter. However if that is not possible the model can be inclined allowing for higher relative positions of the sun.
The H_{cl} indicator can now be easily quantified. Using the solar chart count hours of the periods corresponding to illuminances equal or higher than E_{ref} for each month. H_{cl} is obtained adding the products of these numbers of hours by the monthly percent sunshine values.
Lastly H indicator is computed by adding $H_{oc} + H_{cl}$.

3.3. H indicators with allowance to sun glare and to the use of sunlight movable control devices

In the classroom represented in figure 1 many of the possible positions of the sun may originate glare situations.
Solar chart in figure 4 represents different situations of sunlight penetration conditions.
The first situation represents periods in which sun light is able to penetrate deeply in the classroom through the upper and lower glazings. In this situation particularly severe conditions may occur if sunlight falls over the eyes of the occupants. Some kind of control device is necessary, if possible without elimination of sunlight contribution to day lighting. In the designed solution an internal shelf can be rotated and be fixed in 45 degrees position and diffusing curtains are provided in windows.
The second situation represents periods in which sunlight penetrates through the lower glazing (windows). Sun patches at the work plane level may appear up to 1.5 m far from the windows. There will be some free space between the windows wall and the desks closer to the wall. Therefore it is assumed that occupants most of the time will enjoy sunlight and will keep the curtains opened.
At last the third situation represents the periods in which sunlight does not penetrate inside the classroom and sunlight control devices are assumed to be opened as before.
A corrected H_{cl} value can now be obtained considering, when computing or measuring illuminance values with clear skies, the above mentioned use of movable sunlight devices for each position of the sun.
As earlier, adding H_{cl} and H_{oc} H_{ref} can be obtained. Computed H indicators are represented in figure 6.

3.4. Other information about dynamic characteristics of daylight and future research

For a complete characterization of daylight further information is required.
Daylight stability conditions are different in the course of time. With clear skies illuminance values are usually very stable.
Earlier we assumed that H indicators refer to hourly average illuminances. We think that there is need to consider other indicator called V indicator (V from variability) whose value will be for each hour equal to $(1-E_{min}/E_{ave})*100\%$, where E_{min} represents the minimum instantaneous value and E_{ave} the hourly average value.
With ECC support, in the scope of JOULE program and of a transnational research project, LNEC intends to study H and V indicators recording illuminances inside and outside scale models under real skies.
Another important dynamic characteristic of daylighted spaces refers to the relative internal luminances distribution. In most cases sunlight represents a positive contribution (from the point of view of visual comfort) to this distribution of luminances, because it increases substantially the proportion of external reflected daylight coming from down and falling on the ceiling and upper parts of the walls. This aspect must also be investigated and LNEC will try to obtain resources for that.

3.5. Discussion

The adopted modelling of luminous climates considering only overcasts and clear skies is of course, a simplified approach to the problem It represents, however, a readily available method for H indicator calculations.
The consideration of intermediate skies from a sunlight point of view is very important, because they represent a relevant third case different from the overcast skies (no sunlight contribution) and from clear skies (stable sunlight contribution). Unlike these two cases with intermediate skies there is an unstable sunlight contribution to daylighting. It is expected that the above referred research projet could enable to take account of intermediate skies in the futur.

REFERENCES

1 - LNEC - Lisboa. Iluminação natural e insolação da escola do Crato. Lisboa. LNEC. 1988.
2 - CIE - Paris. Daylight. International recommendations for the calculation of natural daylight. Paris. CIE. 1970. Publication CIE N.16.

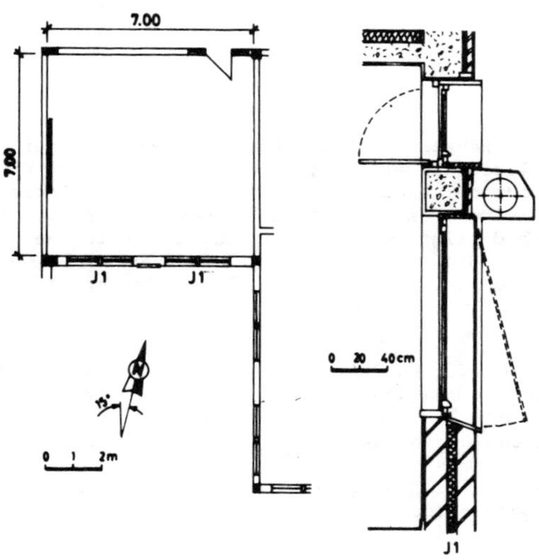

Fig. 1 - Plan and vertical section of the classroom

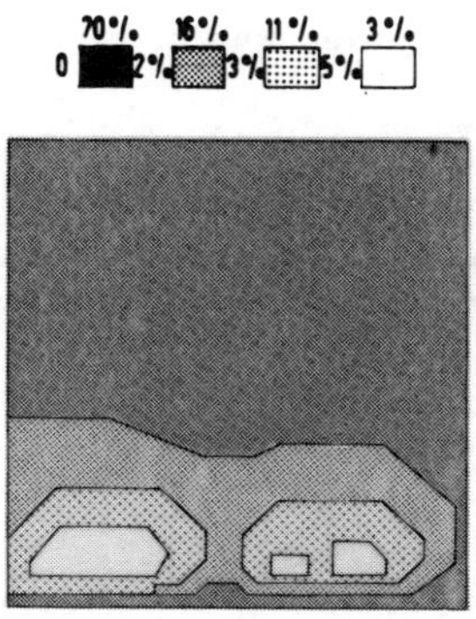

.Daylight factors. Overcast sky

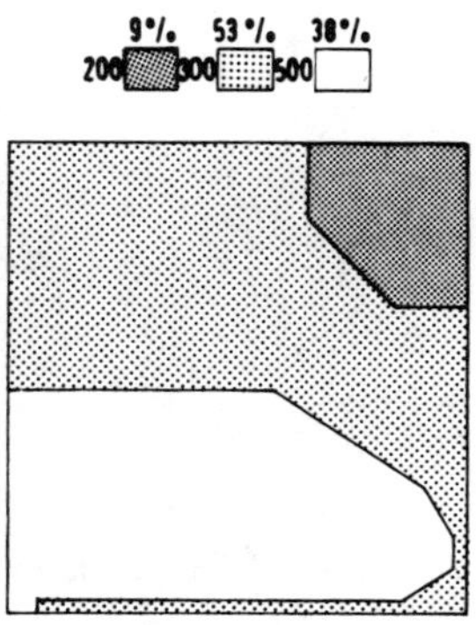

.Illuminances. Clear sky at 18 p.m(summer solstice

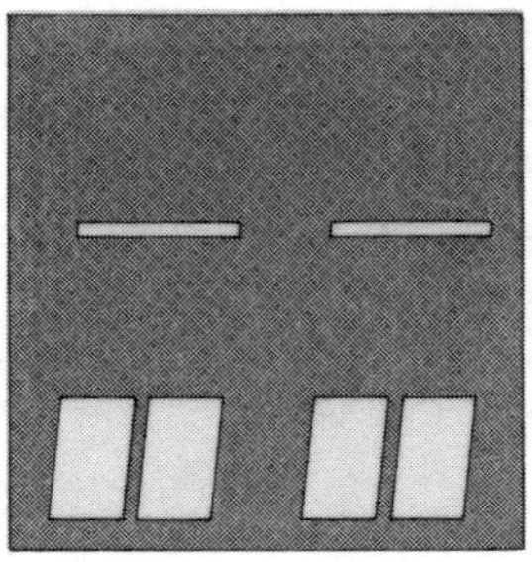

.Sunlight patches

Fig. 2 - Representation of daylight conditions on the work plane

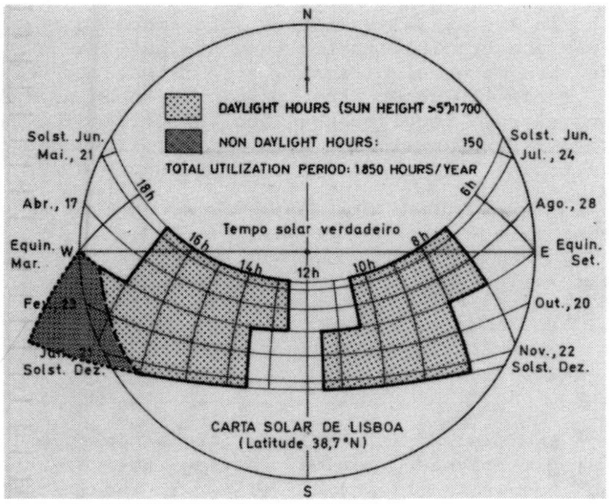

Fig. 3 - Daylight hours of the utilization period

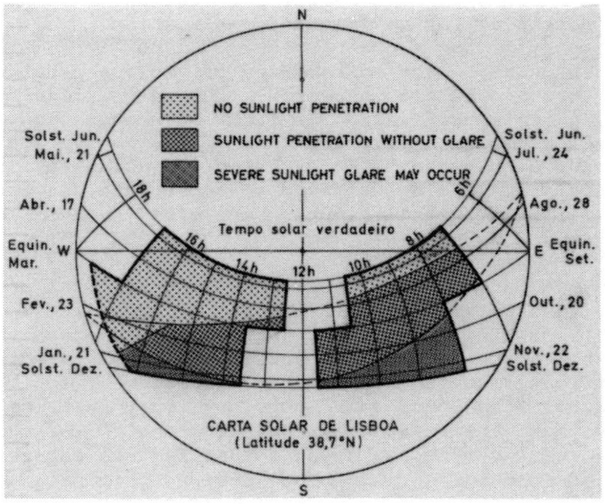

Fig. 4 - Sunlight glare

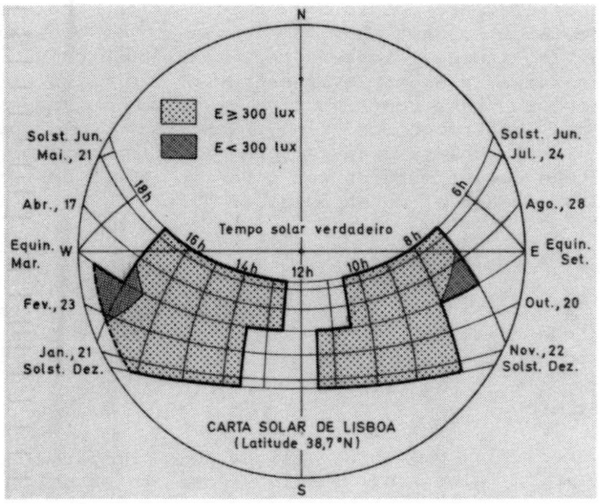

Fig. 5 - Positions of the sun corresponding to illuminances exceeding 300 lux at the reference point, with clear sky

H300 = 1100 H150 = 1350 H75 = 1500
H300 = 1350 H150 = 1500 H75 = 1600

FACADE

Fig. 6 - H indicators

THE ROLE OF DAYLIGHT IN INDUSTRIAL BUILDINGS

P A Ruyssevelt[1], W B Gillett[1], R R Cohen[1] and I McCubbin[2]

1. Halcrow Gilbert Associates Ltd, Burderop Park, Swindon
2. Energy Technology Support Unit, Harwell Laboratory, Oxfordshire

The role of daylight as a passive solar design technique in light industrial buildings is currently being investigated through a group of design studies coordinated by Halcrow Gilbert Associates. Results of the final design assessments indicate that the use of daylight produces significant savings in artificial lighting energy as well as a pleasant working environment. Savings can only be achieved if artificial lighting controls can respond to the availability of daylight, and daylight will only be useful if it can be admitted without the attendant visual comfort problems of glare and disuniformity. The Group A designs display a range of features, including shading devices and light shelves, which endeavour to eliminate glare and give rise to uniform daylight distribution patterns.

1 DESIGN STUDIES

A series of design studies is being sponsored by the Department of Energy to investigate the practical application of passive solar technology in non-domestic buildings. The first set of eight studies is complete /1/, and two further sets are under way /2/.

The twelve non-domestic design studies being coordinated by Halcrow Gilbert Associates have been divided into four groups of three. Each group of design studies concentrates on a particular building sector. The building sectors were selected with reference to a previous study /3/ on the basis of their potential for savings and replication.

In each study, the suitability of various passive solar techniques is explored for a particular building type by asking a firm of architects to propose a design that uses them effectively. The brief for each study is prepared in consultation with a client who has offered to cooperate in the study. These "quasi-clients" are also involved in the evaluation process.

The key factors used to evaluate the proposal are energy performance, environmental performance and capital cost. In order to judge these factors in context, a 'typical' reference building of a more conventional design is also assessed. To be successful the proposal must not only use less energy than the reference, but the value of the energy saving must be seen to justify the capital cost of providing it. The design proposal is developed in two stages with performance and cost assessment at each.

2 INDUSTRIAL BUILDINGS

Industrial buildings commonly incorporate three types of accommodation. These are:

- Offices, including other staff facilities eg canteens.
- Workshops for producing or assembling products.
- Warehouses for storing goods coming in or going out.

Each type of accommodation will have distinct requirements for temperature, lighting levels and other environmental criteria. These criteria may also vary within the type of accommodation according to its use. The architects undertaking the design studies were asked to consider carefully the environmental requirements for each space before embarking on a sketch design.

2.1 Potential for Savings from Daylighting

Recent studies have shown that fairly simple energy conservation measures can reduce the space heating requirement of industrial buildings by up to 40% /4/. These measures include reducing fabric U values to ~0.4W/m²K and incorporating high performance loading bay doors. After applying these energy conservation measures to industrial buildings there is only limited potential for further reducing energy consumption by displacing space heating with passive solar gains. This is particularly true of buildings which have a high level of internal heat gains from machinery and occupants.

There is however, considerable scope for saving energy by displacing artificial lighting with daylight in industrial buildings, and since the fuel source is electricity, rather than say gas, the cost savings are more significant. To illustrate this point, Table I shows the predicted heating and lighting energy requirements and associated costs for two reference buildings used in the first group of industrial building design studies (Group A).

Table I Reference Building Annual Energy Performance and Cost

Reference Building	Floor Area m²	Capital Cost £/m²	Lighting Energy kWh/m²	Lighting Cost £/m²	Space Heating Energy kWh/m²	Space Heating Cost £/m²	Total Cost £/m²	% Year Over-Heating
A2	1234	416	33.0	1.61	35.7	.62	2.23	25%
A3	2140	332	33.8	1.74	36.3	.64	2.38	22%

Based on the figures in Table I, and assuming a 10 year simple payback period, saving 50% of the heating requirement would allow an additional capital expenditure of £2.50/m², whereas saving 50% of the lighting requirement would allow an additional capital expenditure of £10.50/m².

Both the reference buildings reported in Table 1 have rooflights, but they do not result in savings in lighting energy, instead they contribute only to the excessive level of overheating. The size and location of these rooflights give rise to uneven daylight distributions in which it would be difficult to achieve satisfactory lighting control, and little consideration was given in their design to the avoidance of unwanted solar gain.

3 GROUP A DESIGN STUDIES

Design studies in the first group, Group A, feature three industrial buildings, of which two are owner occupied and the third is a speculative development.

3.1 Briefs

The briefs for the three designs were prepared in consultation with the quasi-clients to establish functional requirements and to ensure comparability with the reference buildings. Table II summarises the requirements of each brief in respect of the accommodation to be provided and the relevant environmental criteria.

Table II Summary of Design Study Briefs

Design reference :		A1	A2	A3
Client	:	SunarHauserman	Graphex	English Estates
Designer	:	Jestico+Whiles	Ryder Nicklin	ECD Partnership
Type	:	Owner occupied	Owner occupied	Speculative
Product	:	Furniture	Signs & stands	Unknown
Total floor area	m²	4690	2640	2200
Offices				
area	m²	1532	1200	566
heating set pt.	^{o}C	20	20	20
lighting level	lux	500	500 & 750	500
Workshops				
area	m²	1026	1108	1634
heating set pt.	^{o}C	16	16	16
lighting level	lux	500	500	500
Warehouse				
area	m²	2132	332	---
heating set pt.	^{o}C	15	15	---
lighting level	lux	300	300	---

3.2 Final Designs

The following sections present a description of the designs at the final stage of development concentrating on the role played by daylight and the means of making use of it in the buildings.

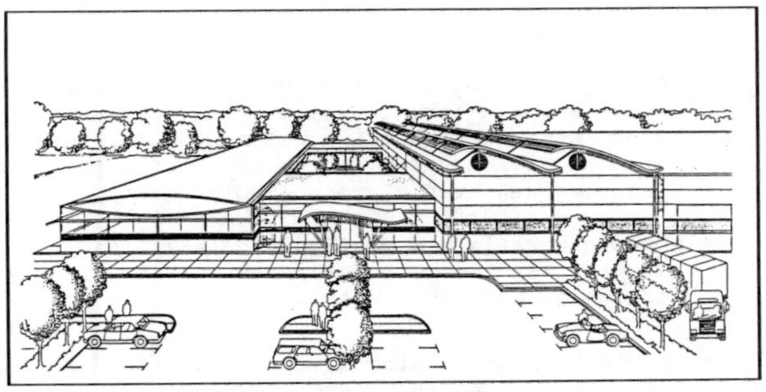

Figure 1 A1 Final Design Viewed from East

3.2.1 A1 Design

In this design for a furniture manufacturer, the offices, assembly and warehouse were each treated as distinct elements in the building form as illustrated in Figure 1. Each element has its own solution to the provision of heating and lighting. Internal heat gains were shown to be low and hence there was scope for passive solar heating. The architects also recognised the benefits of admitting daylight from the brighter southern sky, provided that the associated problems of glare and overheating could be avoided.

A careful analysis of the lighting and heating requirements indicated that daylight and passive solar gain would not be appropriate in the warehouse.

Daylight and solar gain are admitted to the assembly space through two continuous south facing rooflights. Direct solar radiation incident above 35° to the horizontal is rejected by external shading which admits some daylight from the zenith sky as illustrated in Figure 2. The shading takes the form of louvres which extend the curvilinear form of the roof and present an interesting feature to the passer-by. Careful consideration has been given to the location of luminaires and the dimming control strategy to be adopted.

The offices are located to the south of the assembly space from which they are partially separated by a courtyard which is a source of daylight and natural ventilation as well as a pleasant space to look out onto. On the southern elevation, horizontal shading extends through the fenestration to form an internal light shelf. The performance of light shelves is not yet fully understood, but the design intention here is to improve the distribution of daylight by reducing daylight levels close to the window and hopefully increasing them slightly, further away.

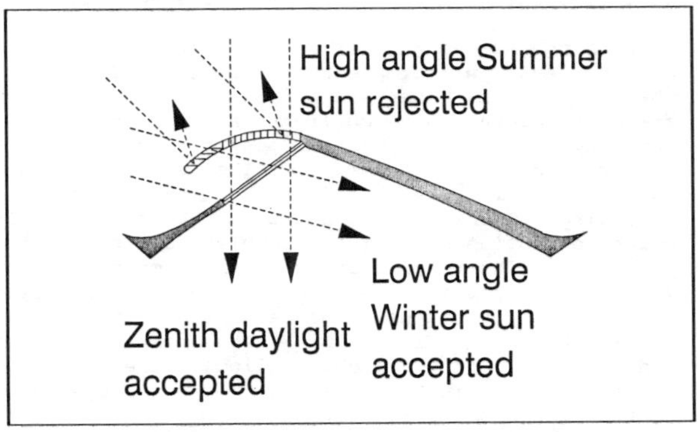

Figure 2 Shading to Assembly Area Rooflights in A1

3.2.2 A2 Design

The design for Graphex Industrial Art houses offices, workshops and stores within a simple rectangular building form as illustrated in Figure 3. Daylight and solar radiation are admitted to the majority of the spaces through six triangular rooflights formed by glazing the external roof structure which runs across the width of the building. The sloping faces of the rooflights face north and south and a responsive internal shading device forms an integral part of the rooflight design as illustrated in Figure 4. The shading device rejects unwanted solar radiation and reduces heat loss.

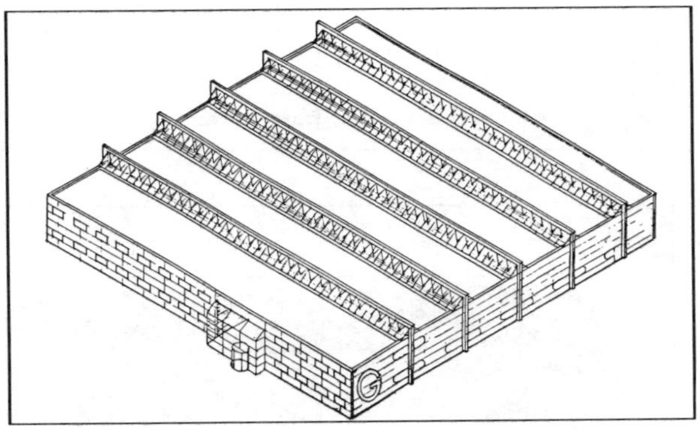

Figure 3 A2 Final Design Viewed from North-West

Offices are located on ground and first floors at the north of the building and are daylit by north facing windows as well as one row of rooflights on the first floor. This row of rooflights has a horizontal leylight at its base acting as a daylight diffuser and also reducing cold down draughts from the otherwise single glazed rooflights.

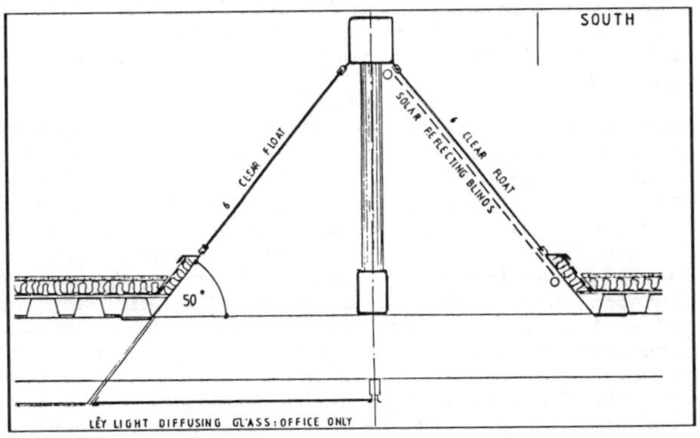

Figure 4 Responsive Shading to Rooflights in A2

3.2.3 **A3 Design**

English Estates require all their speculative industrial developments to have rooflight areas equivalent to 10% of the floor area. In this design solution, illustrated in Figure 5, the architects optimised the area and location of the rooflights.

As with the A2 design, offices, workshops and stores are enclosed in a simple rectangular building form which in this case has an adapted double portal frame structure. GRP rooflights are incorporated in continuous strips as far as possible on the north facing slopes of the roof. The exact size and locations of the rooflights were determined following studies involving physical scale models which established daylight levels and distribution patterns. The results of the scale model studies showed good agreement with the daylight distribution predicted by the performance assessment computer model shown in Figure 6. The resulting even distribution of daylight allows the proposed artificial lighting system for the workshops to be controlled on one circuit.

Offices in this design are located on two floors at the north of the building and are daylit by windows facing north. The artificial lighting system has a two step controller to respond to the availability of daylight.

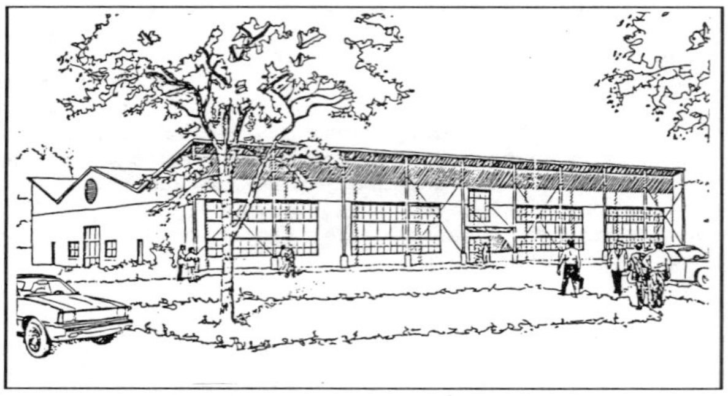

Figure 5 A3 Design Viewed from North

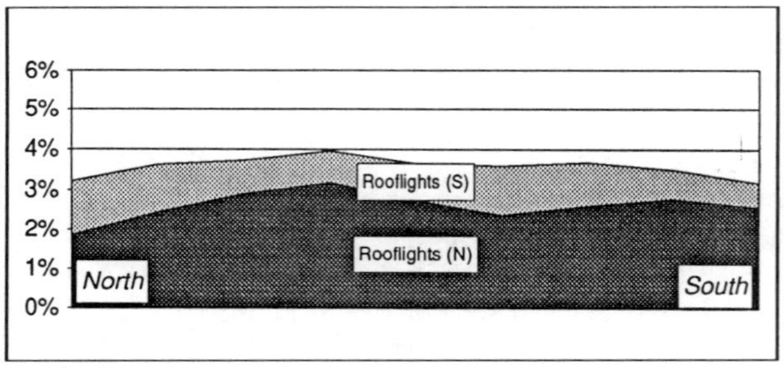

Figure 6 Daylight Factor Distribution in A3 Workshop

3.3 **Results of Performance and Cost Assessments of Final Designs**

The results of the performance and cost assessment of the final design are summarised in Table III. The figures presented for the A1 design exclude the warehouse, thus allowing a comparison to be made with the reference buildings which do not have large warehouses.

At the bottom of Table III the mean differences between the designs and the reference buildings are shown, together with the percentage change. Although lighting energy requirements have been reduced, heating energy requirements have increased marginally. A large portion of the increased heating requirement is due to the reduction in internal heat gains attributable to the displacement of artificial lighting by daylight. Space heating requirements are to some extent offset by additional useful solar gains through the larger glazing areas in the final designs. Table III also shows that lighting energy requirements have been reduced by 56% to very low levels.

The final designs show considerably reduced levels of overheating compared to the reference buildings resulting from the thoughtful design of the rooflights and windows, and the incorporation of shading devices.

The estimated capital cost of the A1 final design is within the client specified cost target, and the capital cost of designs A2 and A3 compare favourably with their respective reference buildings.

Table III Reference Building Annual Energy Performance and Cost

Final Design	Floor Area m²	Capital Cost £/m²	Lighting Energy kWh/m²	Lighting Cost £/m²	Space Heating Energy kWh/m²	Space Heating Cost £/m²	Total Cost £/m²	% Year Over-Heating
A1	2560	450	14.5	.80	53.7	.94	1.74	1%
A2	2640	412	14.3	.91	29.1	.51	1.42	3%
A3	2200	332	15.7	.93	42.3	.74	1.67	7%
Comparison of final designs against reference building								
Mean differenc		-2	-18.57	-.79	5.70	.10	-.70	-20%
% change from		-1%	-56%	-47%	16%	16%	-30%	-84%

4 **CONCLUSIONS**

This group of design studies has so far revealed the following:

- There is considerable potential for saving energy through the exploitation of daylight in industrial buildings.
- Daylight may be admitted through the use of windows which face in all orientations, and rooflights facing both north and south, but shading devices play an important role in reducing unwanted glare and solar gain.
- The integration of daylight and artificial lighting systems and their controls is vitally if energy savings are to achieved.
- A clear understanding of the potential for running cost savings, environmental improvements and other benefits of a passive solar measure is necessary before the associated additional capital expenditure can be justified.

5 **ACKNOWLEDGEMENTS**

The work reported here is fully funded by the Energy Technology Support Unit on behalf of the Department of Energy. The performance assessments were undertaken by YARD and cost assessments were performed by Davis Langdon & Everest. Specialist advice was provided by D Hawkes of Cambridge University, P Owens of Pilkingtons, N Vaughan of UWIST and A Richens of Ferguson & Partners.

6 **REFERENCES**

1. Campbell, J and White, K (Ove Arup R&D). PASSIVE SOLAR DESIGN STUDIES FOR NON-DOMESTIC BUILDINGS (PHASE 1). ETSU report S-1157 (1987).

2. Non-Domestic Design Studies Phase 2, coordinated by Building Design Partnership and Halcrow Gilbert Associates.

3. Duncan, I P and Hawkes, D. PASSIVE SOLAR DESIGN IN NON-DOMESTIC BUILDINGS. ETSU report S-1110 (1983).

4. Hughes, D. LOW ENERGY FACTORY BUILDINGS. Building Services, The CIBSE Journal (February 1989).

5. Fedeski, M. DESIGN LESSONS FROM PASSIVE SOLAR NON-DOMESTIC DESIGN STUDIES. UK-ISES Daylighting Buildings Conference Imperial College, London, 20-21 April 1989.

FLUORESCENT MATERIAL - TECHNICAL IMPROVEMENTS FOR A NEW MEASURING DEVICE

Dr.-Ing. F.W. Grimme
KFA/PBE Jülich
Postbox 1913
D 5170 Jülich 1

Prof. Dipl.-Phys. F. Rostock
Im langen Bruch 28
D 5000 Köln 91

Dr.-rer.nat. V. Wittwer
FhG-ISE
Oltmannsstr. 22
D 7800 Freiburg

ABSTRACT. With the existing fluorescent materials, which concentrate direct and diffuse radiation, a new type of measurement device was developed. The first prototypes were investigated at the Fraunhofer Institute for Solar Energy Systems, Freiburg and tested by Rostock. First examples are being used in the Greek-German research solar demonstration project "Lykovrissi" to analyse the user reaction to climatic conditions by measuring how far and how frequently windows and shading devices are operated. By using two modified fluorescent collectors, connected via a special electronic device, the complete measurement device gives a signal which is proportional to the window or shading under natural or artificial light.

OBJECTIVES

When measuring the radiation flow through regulated openings, e.g. windows with external venitian blinds, it is necessary to calculate the integral of the radiation intensity on a surface which is not shaded uniformly. Conventional measuring sensors, such as pyranometers or photocells, allow only point measurements. Further information can be obtained in principle from the mechanical regulation of the shade and the knowledge of the specific position of the sun. This requires mechanical devices and a sufficiently precise simulation programme for the specific position of the sun and the corresponding external shade. As an alternative, a point sensor with an online integrator scanning the opening fast enough can also be used which leads on the one hand, to an obstruction of the opening and requires, on the other hand, quite a complicated mechanism. In order to make things easier to handle, it would therefore be desirable to have a measuring method which functions without mechanical components and does not have to rely on computer-aided processes.

Thus, at present it is not possible to make integral measurements over the whole surface due to the lack of suitable surface sensors and the difficulties imposed by the changing position of shading devices or illuminated areas. Since most of the employed shading elements have a lamellar structure, it is sufficient to use a linear integrating measuring sensor. A new type of measuring device was developed, using the existing fluorescent materials, which concentrate direct and diffuse radiation. The operation principle of fluorescent planar concentrators (FPC) has been described in literature /e.g. 1, 2/. A sheet of highly transparent material with surfaces of optical quality (e.g. acrylic glass), a few mm thick and with an area of up to a few m² is doped with fluorescent dyes (e.g. organic dyes).

The dye molecules absorb part of the solar radiation incident on the large area and re-emit fluorescent radiation, which is slightly shifted to longer wave lengths, nearly isotropically. About 75 % of this fluorescent radiation is trapped inside the plate and guided to the edges by multiple internal reflections at the surfaces. The concentrated light leaves the sheet at the edges and can be converted to electrical energy by photovoltaic cells.

In order to compensate the length-dependent alteration of the intensity a special sensor geometry is required. Since the fluorescence dye and the installed solar cell, both have different characteristic features, it is necessary for some objectives to record a correlation graph by means of a solarimeter.

The maximum sensor length of two meters results from the delivery conditions concerning the acrylic glass. From the technical viewpoint it is possible to line up several sensors in a row.

APPLICATIONS ACHIEVED

- Integral Surface Measurement

The simple type of sensor as described above is suitable for determining the availability of radiation on a line or a surface. In this case, a calibration is nesessary because absolute values are measured.

- Determination of the Overall Energy Flow

When using two identical calibrated sensors, one installed in front of a light opening, the other one behind it, the overall energy passing through the opening can be determined by means of the quotient formed by the division of the value measured at the interior by the value measured at the exterior of the light opening.

- Determination of the Shading Degree

The approach toward this objective corresponds to that of the overall energy flow determination. By means of such an arrangement, the position of, e.g., a sliding shutter or roller blind can be determined. In this fashion the radiation flow through a partly open follow-up or regulated lamellar shading construction can be determined continuously and due to the integrating function on real time basis.

- Contactless Path Measurement

By means of the same sensor arrangement, a contactless path sensor can be constructed. If the secondary light source, which is nesessary in most cases, is sufficiently constant, only one sensor is required. If the sensor features the corresponding design, contactless angular measuring sensors can be achieved as well.

When using two sensors, the required quotient during signal processing can be formed digitally as well as by analog processes. Compared to the digital procedure, the analog one has several advantages, such as:

- analog to digital converters no longer necessary,
- only one pair of signal lines necessary,
- calibration and therefore adapttaion to local geometry of measuring points electrically possible.

FIRST APPLICATION

The first prototypes using this collector principle were investigated at the Fraunhofer-Institute for Solar Energy Systems, Freiburg and extensively tested by Rostock, Cologne. First examples are being used in the Greek-German research solar demonstration project "Lykovrissi" in the measurement and evaluation programme for active and passive systems and the user behavior to analyse the user reaction to climatic conditions by measuring how far and how frequently windows and shutters are operated. By using the specially designed fluorescent collectors, which are connected with an electronic circuit developed by Rostock, the complete measuring unit gives a signal which is proportional to the window or shutter opening under natural or artificial lighting. At the moment, the prototypes work with light intensity > 50 W/m² in the 2 % error range with a 24 V supply. In further research and development work, not only will the error range improve but the general applicability of such devices will be increased so that they cannot only be used in connection with simple sliding windows but also with more complicated casement windows. An approach is being tried in which an intergrated light source will make an external light source (sun or strong lamp) superfluous.

A second new sensor, suggested and tested by Rostock, is able to distinguish between daylight and 50 Hz artificial light. In the case of artificial light, the sensor signal is inverted without changing the absolute value. All the values are recorded on a logarithmic scale, which avoids switching between different ranges. Simple installation is possible, as only one sensor is necessary for measuring diffuse radiation and detection of artificial light.

Acknowledgement: This work was part of the Concerted Programme "Energetic Development Possibilities of Solar Aperture" sponsored by the German Ministry for Research and Technology (0335003A "System Developments)

/1/ Goetzberger, A. Wittwer, V., Solar Cells 4, 3 (1981)

/2/ Stahl, W., Zastrow, A., Physik in unserer Zeit 16, 167 (1985)

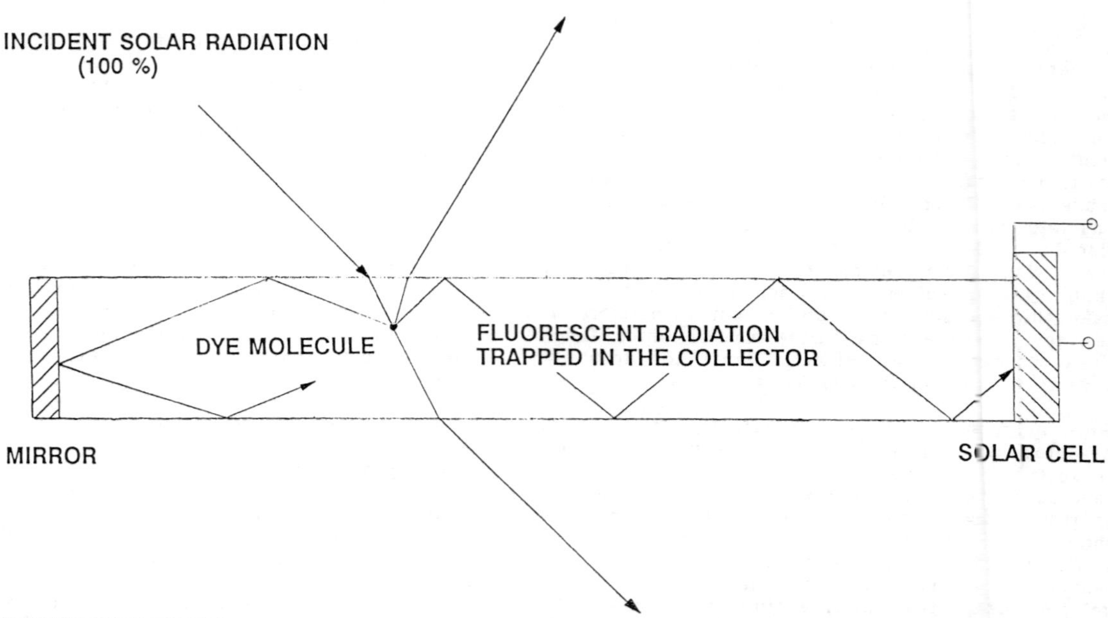

FLUORESCENT COLLECTOR
CROSS SECTION (NOT TO SCALE)

LOW COST INSTRUMENT FOR MEASURING DAYLIGHT FACTORS IN MODELS

Nick Baker
The Martin Centre, Department of Archirecture
University of Cambridge UK

ABSTRACT. The paper presents the design for a low cost instrument for measuring Daylight Factors in scale models. The instrument was developed for participants in the European Daylighting Design Competition "Working in the City"

1 PRINCIPLES

The Daylight Factor (DF) is defined as the following -

$$\frac{\text{(Illumination level in the room at point of interest)}}{\text{(Illumination due to unobsructed sky)}} \times 100\ \%$$

This important parameter is dependent upon room geometry and window design, room finishes, external obstructions, etc. From a known DF, the designer can assess if there is likely to be sufficient daylight for the use of the room by comparing with recommended values, and can check the quality of the daylight from the variation of the DF across th room. Also, if the hourly sky luminance for the site is available, the energy saving of the daylight, by the avoidance of use of artificial light, can be evaluated.

The DF can be calculated for rooms of simple geometry, but for more complex lighting features such as light well, light shelves, atria etc, physical models are often used. These are illuminated under a real sky, or an artificial sky and the illuminance levels measured inside and outside the model with an electrical Luxmeter. The device described in this paper is a low cost alternative to this.

The DF Meter, fig(1), consists in effect of two little white rooms, the *measuring room* and the *sky room* connected by a *dark corridor.* The measuring room is illuminated with light falling onto a *diffusing window* in the top of the measuring room, which in a 1:20 model is in the working plane - 1m above floor level. The sky room is illuminated with light from the sky passing through a *diffusing skylight* and down the *light tower*.

Through a hole in the sky room wall, the *occulus,* you can look down the corridor to the measuring room. However, half way down there is a *mirror* across the lower half of the corridor allowing a view of the sky room also. Look above the mirror and you see into the measuring room; look at the mirror and you see into the sky room. The two room surfaces appear vertically adjacent.

The Daylight Factor in a room is always much less than 100% and so to get a brightness match we have to reduce the light in the sky room. The light tower already reduces this

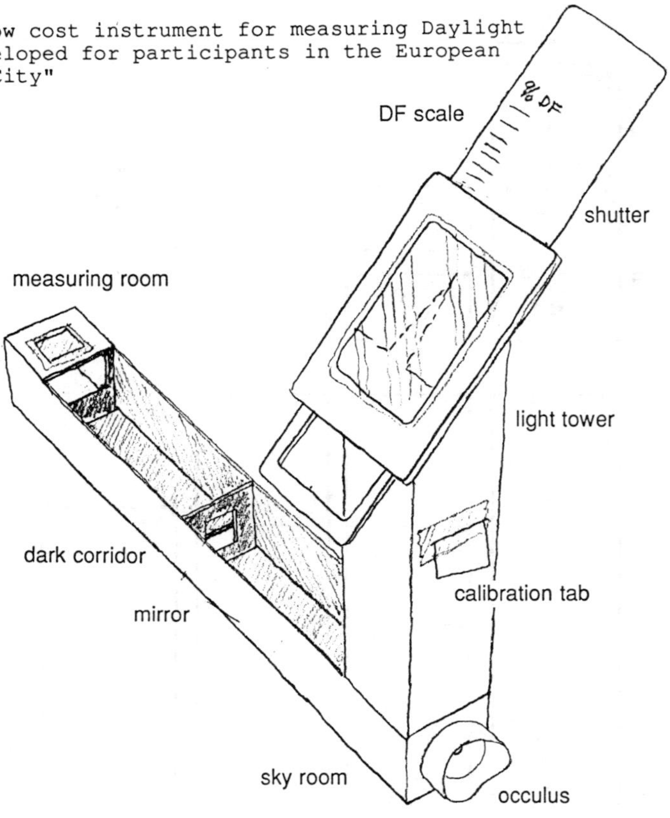

Fig 1. View of DF Meter with top of corridor (9) removed.

to about 25% and the remainder of the reduction is carried out by the *sliding shutter.* The higher the value of DF in the room, the less the shutter will have to be closed. When a brightness match is made, the DF can be read off from the position of the shutter.

2 CONSTRUCTION

2.1 Materials

You will need 1 sheet 4mm mounting board, black one side white other side, or white card and black paper.

Note that the white must be as white as possible without being fluorescent - chose a paper or card which looks noticeably the whitest in tungsten light. You will also need a small mirror cut from a handbag mirror, or failing this a mirror made from aluminised Letrafilm on a substrate cut from a credit card. Make sure that the mirror has one clean cut edge and that this is the upper edge.

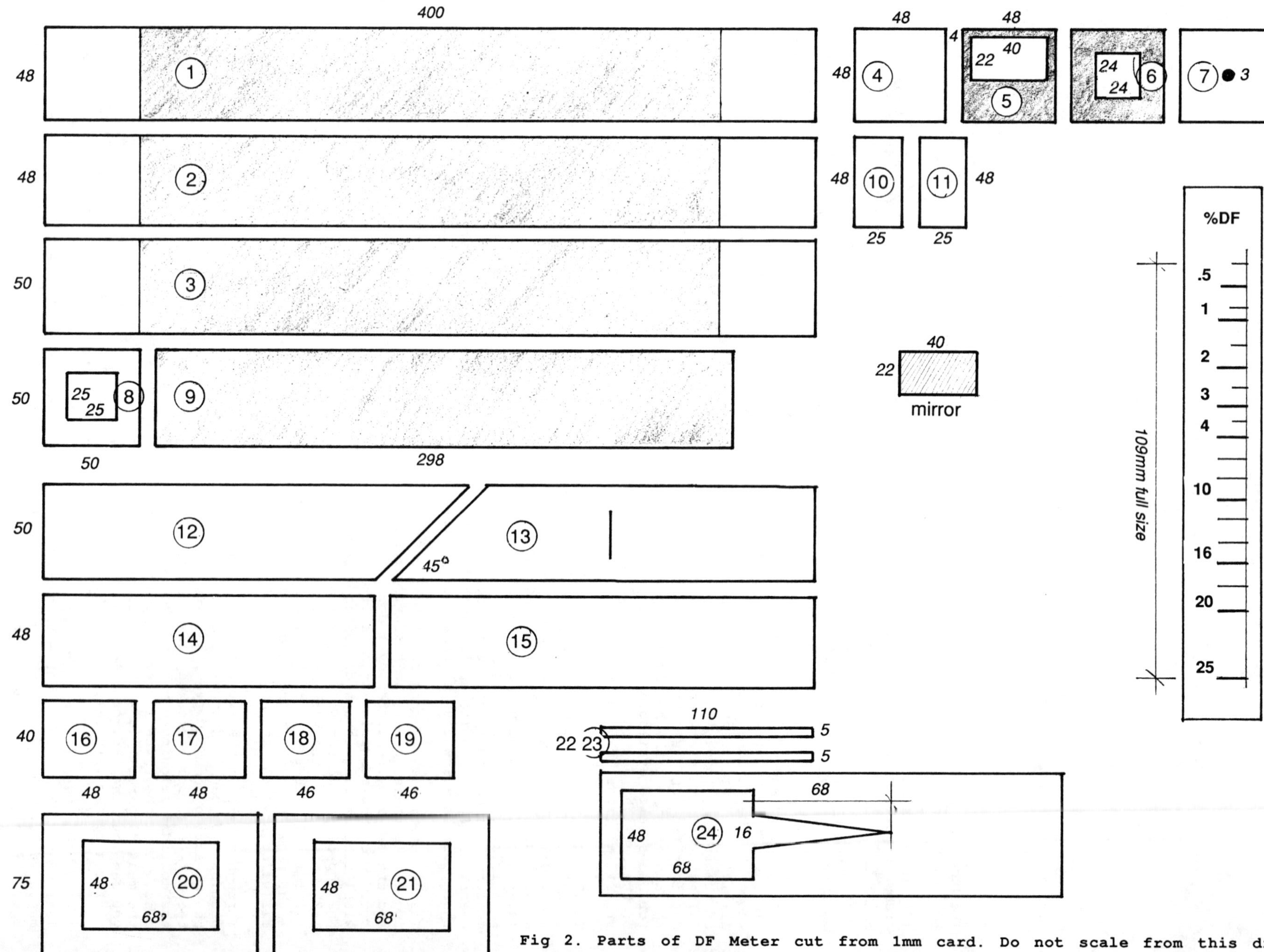

Fig 2. Parts of DF Meter cut from 1mm card. Do not scale from this drawing.

2.2 Cutting out and Assembly

The parts are shown in fig (2). Mount the card on a drawing board and use the rule and set square to mark out, with a knife. Note that many parts are cut from strips 48mm and 50mm wide. Do not separate the strips when you mark out, as in fig (2), i.e. make one cut serve two parts wherever you can. Stick on areas of black or white paper as appropriate.

Cut out the parts with an accuracy of at least 0.5mm and with a clean square edge. Punch or burn the hole in part (7). Save the square, cut from the hole in part (8).

For assembly, refer to fig (1) and the long section fig(3).

Start with the rooms and the corridor. Mark out the base (3) with the positions of parts (4), (5), (6) and (7) and glue them into position making sure they are square in plan and section. PVA glue is best but you may prefer to use a contact adhesive. Part (5) has white side to the measuring room. Glue on side pieces (1) and (2). Complete the measuring room with (8) and add two landing pieces (10) and (11) inside the top of the corridor at each end.

Now position the mirror with the reflective side to the sky room, so that looking through the occulus the hole is just not visible at the edge of the mirror. The top of the mirror showld be horizontal and if the frame (6) is vertical and square, the mirror should fill the lower half of the aperture. The mirror may be stuck onto the frame with glue, double sided tape, or Blue-Tack. The latter will permit easy alignment. Blacken all edges and surfaces other than actually in the rooms.

The top (9) may now be glued on, black side down, and the tower base (16), (17), (18) and (19) glued into the sky room, white side in. Note that these parts should protrude 32mm above the top edge of the room.

The light tower is made from parts (12), (13), (14) and (15). Reinforce the joints at the base with some tape. If everything is accurate, the tower should fit snugly on the base in any of four orientations. You may need to chamfer the corners of the base with a blade or sandpaper. If it is loose, use four map pins.

The skylight is made from parts (20) and (21), and spacers (23) and (24)., The glueing must be clean and the spacers positioned right at the edge. The shutter (24), which must be cut out very accurately, can have sharp edges removed and the sliding surfaces rubbed with a candle.

Two of the top edges of the tower should be beveled to take the sloping skylight . The skylight can now be glued to the tower.

Stick a piece of drawing film over the skylight and the opening in the measuring room. Add a shaped piece of cardboard tube aound the occulus. Check that all joints are light proof. It is useful to draw a scale on the side of the meter in units of 50mm from the centre of the measuring room. This is a metre scale for a 1:20 model. The DF meter is now finished except for calibration.

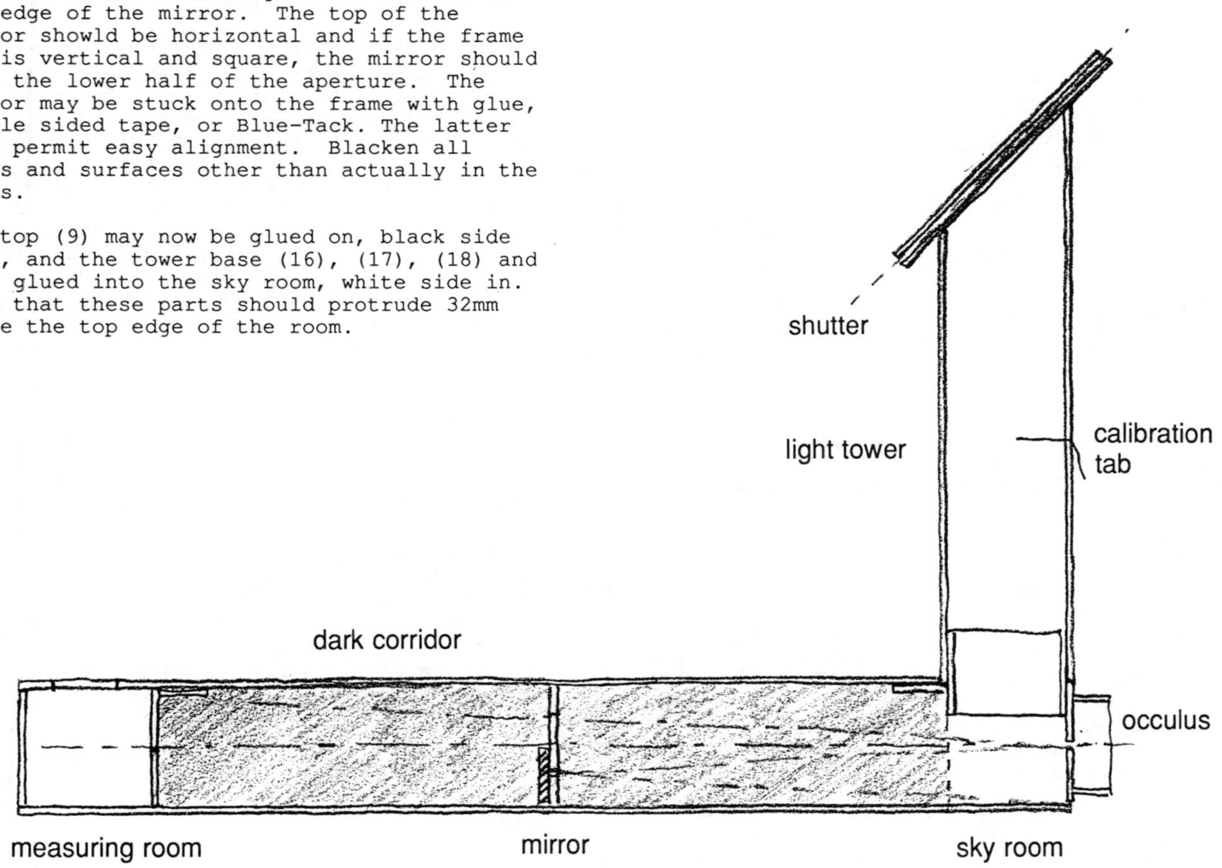

Fig 3. Section of DF Meter (thickness of card exagerated)

3 PRACTICE

Before the final calibration it is a good idea to practice obtaining a brightness match. Outside on an overcast day, insert the meter into a model, as in fig (4) and obtain a match by moving the shutter. You may find that the measuring room appears as a slightly different colour from the sky room, This is due to the internally reflected component from the model. This effect can almost be eliminated by almost closing the eyelids - reducing the amount of light to the eye suppresses colour vision.

Make sure that you wait long enough for your eye to adapt to the low level of light in the model - a black patch over the spare eye may help! You may also find it helps to rock the eye up and down, scanning the edge between the two surfaces. This seems to accentuate any slight diffrence in brightness.

Your performance can now be tested. Fix the positioning of the measuring room in one position in the model, say about 3m back from the window, and adjust the shutter to give a match. Make a pencil mark where the shutter emerges from the skylight. Repeat the operation several times; when your marks are within 1mm of each other, you are ready to calibrate.

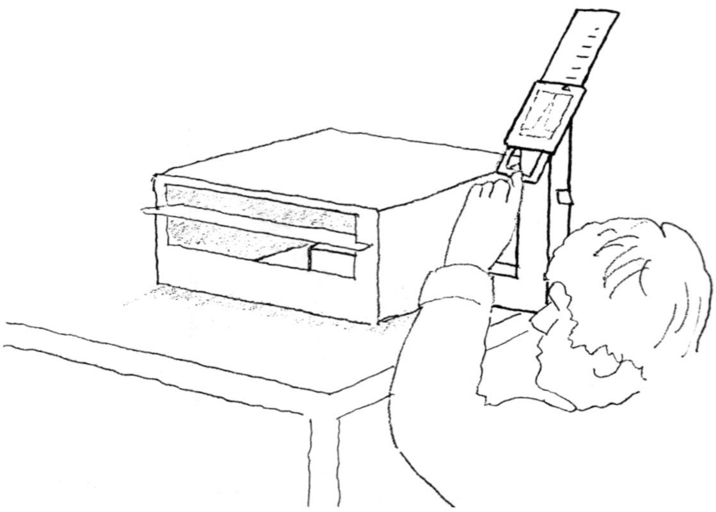

Fig 4. DF Meter in use in model

4 CALIBRATION

With the shutter fully open, the skylight and the tower reduce the light in the sky room to a little over 25% of the illuminance due to the sky. The scale provided in this paper indicates the DF as a function of shutter position, if the DF is exacactly 25% when the shutter is fully open. So before positioning this scale, the meter must be "trimmed" to this condition.

Find the square piece of card that you cut out of part (8) - it is probably in the bin or on the floor - and accurately cut into four pieces across the diagonals. With the model fixed to the table under an overcast sky (not in the model), place the three triangular pieces over the measuring window so that they obscure 75% of the aperture.

Make sure that the shutter is accurately fully open.

It is clear that we have now reduced the illumination of the measuring room to 25%. If all is well, the sky room should appear just a little brighter.

Assuming that it does, now push a strip of stiff white paper through the calibration slot in (15) and use your newly developed optical skill to obtain a really good match. Try moving the meter to face different parts of the sky, and move the triangular pieces around on the window. When you are sure of the match, fold the trimming strip down and secure with tape.

If you find the sky room already dimmer, something has gone a little wrong. If there is nothing obvious, a possible cause is that the card in the light tower is not white enough. There are three possible cures - line the tower with whiter paper, shorten the tower by 10mm or so, or place another layer of drawing film over the measuring room.

Assuming all is now well, stick the scale, which you will have to enlarge to the dimension shown either by a variable magnification photo-copier or by geometry (remember similar triangles), onto the shutter. Obviously the 25% mark lines up with the edge of the skylight.

5 Use of the DF Meter

The main dissadvantage of the DF meter is its size. Clearly the model must be designed with a 50mm x 50mm aperture in the side to allow the meter to be inserted. This should be a good fit to reduce extraneous light. Do not forget that the model should have the realistic reflectances on the inside - colours do not matter but the degree of lightness or darkness must be roughly correct. Also for the final DF values an allowance must be made for the reduction of light due to glazing and glazing bars, not normally present in models. Information on this can usually be found in text books.

It is possible to change the dimensions of the DF Meter. The design for the EC Competition had the corridor and the light tower reduced to 35mm across. If dimensions are changed, it will be necessary to experiment with the height of the tower in order to establish a balance at 25% DF. The rule is to keep the height to width ratio of the tower roughly constant. The length of the corridor makes no difference to light intensity - only the width of the field of view.

A PERSONAL DOSIMETRY DEVICE TO MEASURE LIGHT EXPOSURE.

N.V. BAKER & D.A. CAWTHORNE
THE MARTIN CENTRE FOR ARCHITECTURAL AND URBAN STUDIES
6 CHAUCER ROAD, CAMBRIDGE, UK Tel 0223 332981

Abstract : "Sick Building Syndrome" has been receiving increased research interest recently, particularly in the areas of indoor air quality and thermal comfort. However, a component of the syndrome which appears to have been neglected is the possible influence that daylighting regimes may have upon occupant health and well-being via the light induced circadian cycles which regulate the human body's physiological mechanisms. In order to investigate the daylight modifying properties of buildings which may influence these circadian cycles, a simple non-intrusive device designed to be worn on the lapel has been developed which monitors the total light exposure and its time dependency.

The device is based on an electric watch mechanism. A rotating disk of photographic film is housed within a watch casing which has been made light-tight by painting it matt black both inside and out and sealing all orifices with silicone. Into the casing is incorporated a small transparent window below which lies a slit aperture of approximately 50 microns in width. The slit is made from a piece of finely scored photographic film, a little less in length than the radius of the revolving disk (approximately 7.5mm) and is fixed in a box mounting to give a 90 degree angle of view. Between the scored slit film and the transparent window in the casing is an opalescent acrylic diffuser. The slit film is mounted in such a way as to be gently sprung against the revolving photographic disk. This is to accommodate any unevenness in its rotation and so maintain direct contact between itself and the slit film. Furthermore the disk is positioned with its emulsion side up and the slit with its emulsion side down thus putting slit and emulsion in direct contact. Both of these measures are designed to reduce the amount of divergence as light falls through the slit onto the photographic film.

As light falls through the aperture onto the disk it does so in a narrow band and creates a trace mark which becomes visible when the disk is developed. For the purposes of developing the device a high resolution lithographic film was used to make cutting, loading and developing the disks easier. This is because the particular film used (Kodak ortho 2115 type 3) can be handled in red safelight conditions (filter no.1A). It does have the disadvantage of making the device incapable of registering the red part of the spectrum when in use, but were the device to be used on a wide scale a film corresponding to CIE standard photopic observers curve could be substituted thus giving it full human spectral sensitivity.

The photographic disk is mounted on a revolving base plate and rotates once every twelve hours. By doing so it creates a radially banded sequence of traces on the film which indicate the light intensities that were present during the recording period. By subjecting the device to a fixed exposure of light at the beginning of each recording session (measured with a light meter) a calibration trace can be made on which all other readings may be zeroed. This is necessary to allow for small variations in developing. The device is then attached to the subject for the duration of the recording run.

To measure the density of the traces, the disks are developed and mounted in 35mm clear glass projector slides. They are then "read" by passing a projected beam of light through the disks onto a matt black surface at a standard distance, about 8m is usually adequate. Inscribed on the matt black surface is a circumference divided into 360 degrees. A light meter is placed at the projected calibration trace position (usually the 12 O'clock position) and a reading is taken. This can be equated to the original lightmeter reading taken at the time of exposure for the calibration trace. The lightmeter is then moved to sequential points in increments of one degree of arc around the image circumference, recording the variation in light transmittance. This gives a readable resolution of approximately two minutes over a twelve hour period.

Most films have a non-linear density response which can be described by a simple equation available from the film manufacturers. By incorporating the calibration variable into the linearity equation and applying the resulting equation to the lightmeter readings, true light level readings in lux for the recording session may be obtained. From these a graph of lux against time may be drawn and a statistical analysis of the fractions of the day spent above or below fixed thresholds can be carried out.

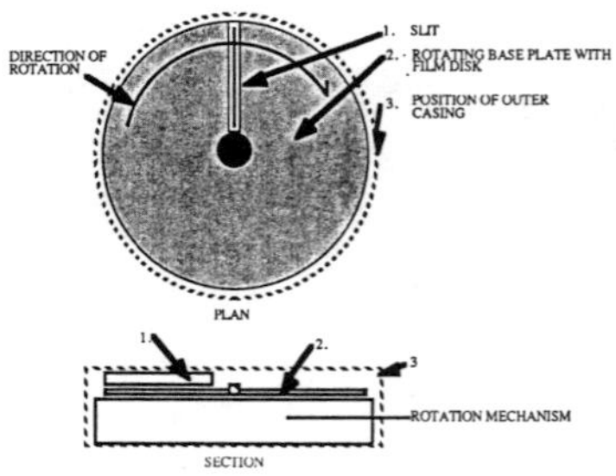

FIGURE 1.

DIAGRAM OF A TYPICAL DISK SHOWING THE RADIAL BANDING OF THE TRACE. N.B. THE BENCHMARK TRACE IN THE 12.00 HOURS POSITION (1).

FIGURE 2.

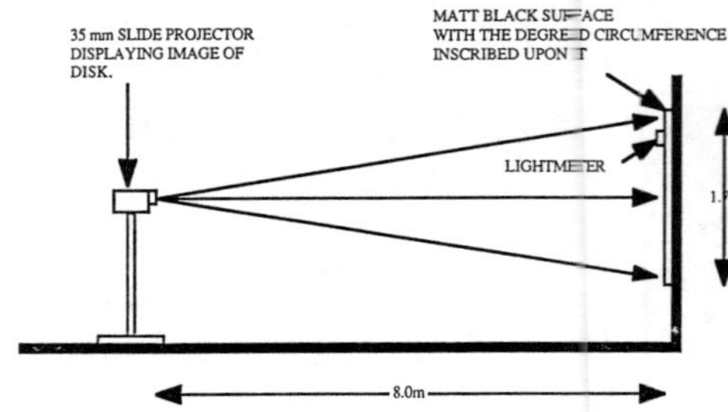

DIAGRAM OF THE PROJECTOR AND SCREEN ARRANGE-MENT FOR READING DISKS

5

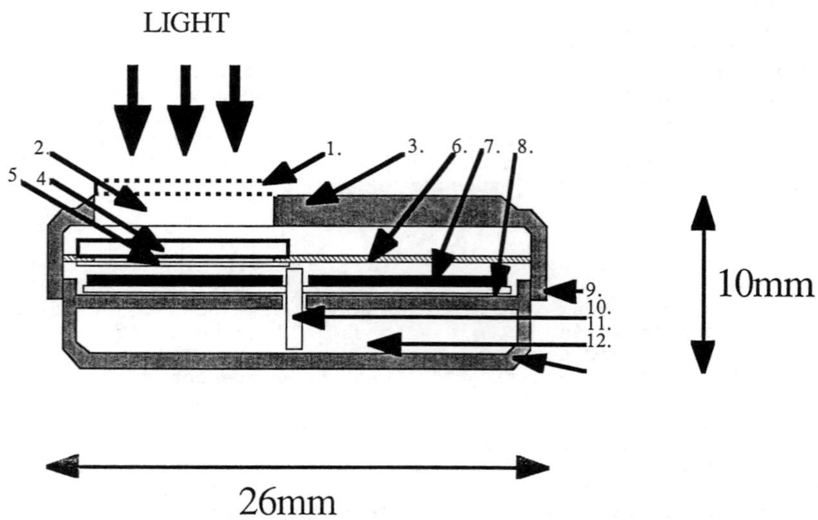

1.OPTIONAL FILTER 2.TRANSPARENT WINDOW IN CASING 3.LIGHT TIGHT CASING 4.DIFFUSER 5.SCORED FILM SLIT	6.RECESSED MOUNTING TO GIVE 90 DEGREE ANGLE OF VIEW 7.PHOTOGRAPHIC DISK 8.ROTATING BASE PLATE 9.O-RING SEAL	10.DRIVE SHAFT 11.WATCH MECHANISM 12.OUTER CASING

SECTION THROUGH DOSE METER

3

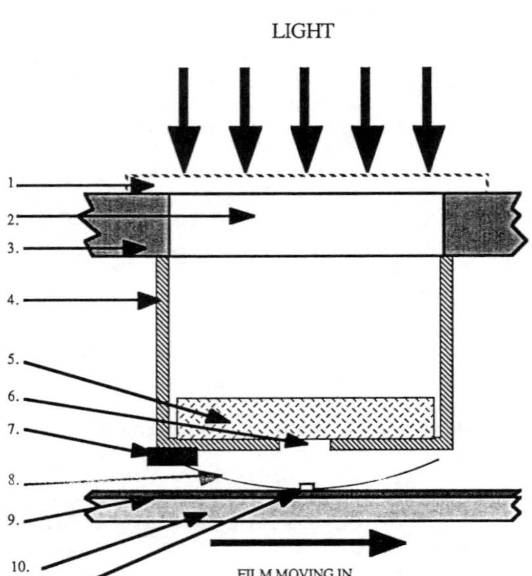

1.OPTIONAL FILTER
2.TRANSPARENT WINDOW IN CASING
3.LIGHT TIGHT CASING
4.RECESSED BOX MOUNTING TO GIVE A 90 DEGREE ANGLE OF VIEW
5.OPALESCENT ACRYLIC DIFFUSER
6.SLIT APETURE IN BOX MOUNTING
7.ADHESIVE FIXING FOR SLIT FILM
8.SLIT FILM
9.PHOTOGRAPHIC FILM DISK
10.REVOLVING BASE PLATE
11.SLIT

NOTE THE METHOD OF FIXING THE SLIT FILM AT ONE EDGE ONLY. THIS UTILISES THE FILMS' NATURAL SPRINGINESS TO ENSURE THAT THE SLIT ALWAYS RESTS GENTLY ON THE MOVING PHOTOGRAPHIC DISK AND ACCOMMODATES ANY UNEVENNESS.

SECTION THROUGH THE APERTURE

4

EXPERIENCES WITH SIMPLIFIED DESIGN TOOLS FOR DAYLIGHTING AND THERMAL DESIGN

N V Baker
The Martin Centre for Architectural and Urban Studies
Department of Architecture, University of cambridge
6 Chaucer Rd Cambridge UK

M Fontoynont
L.A.S.H. Rue de M Audin Lyons FRANCE

ABSTRACT

This paper describes two design aids which were developed for participants in the EC "Working in the City" Architectural competition. The design tools are concerned with the combined effect of glazing and building form on primary energy consumption. Comments on how these models were used in the competition, and their influence on the designs, are given.

1 Introduction

Two simplified calculation methods were developed recently for use in the Third European Passive Architecture Competition/1/. This competition, the results of which were announced in the spring of 1989, was specifically aimed at raising the interest and knowledge in daylighting and other energy related topics in non-domestic buildings.

A requirement of the calculation tools was that they should be simple enough to be used by a non-specialist, using manual methods, and that they should be educative and complimentary to descriptive technical guidelines also issued in the competition package.

The first calculation method, the LT Method, offered a means of estimating the total energy consumption for lighting, heating, cooling and mechanical ventilation for a non-domestic building, in terms of building shape and facade design. This tool was intended to be used early in the design process.

The Daylight Evaluation Proceedure allowed a more detailed appraisal of the energy performance of specific daylighting design, from data which would normally be available later in the design process. The proceedure involved the use of physical models and calculation.

2 THE LT-METHOD

Amongst the considerations made early in the development of a building design, the designer is concerned with two areas in particular:

1) the form of the building, that is its plan depth, and section

2) the design of the facades, that is the area and distributionn of glazing

If energy performance is to be a major factor, then the designer requires a tool which will respond to an input of the design parameters described above, and also to be quick and simple enough to use to allow the designer to explore a number of options. A further requirement is that the model must respond to local climatic conditions.

The LT Method, standing for the Lighting and Thermal energy value of glazing, was developed expressely to for this purpose. A mathematical model, described in more detail elswhere /2/ has been used to predict the annual primary energy consumption as a function of -

1) local climatic conditions
2) orientation of facade
3) fraction of glazing area in facade
4) type of glazing
5) depth of plan
6) incorporation of atria or sunspaces

The LT Method is presented in the form of sets of eight curves, one set for each of five European climate zones, showing -

annual primary energy consumption per sq metre floor area

for three orientations of facade plus one for horizontal glazing (rooflights). Curves are presented for lighting, heating and cooling. Sample curves are shown in fig (1).

The monthly total hours of artificial lighting is calculated using a switching algorithm using inputs including hourly sky luminance, a calculated daylight factor, lighting datum level, and occupancy pattern.

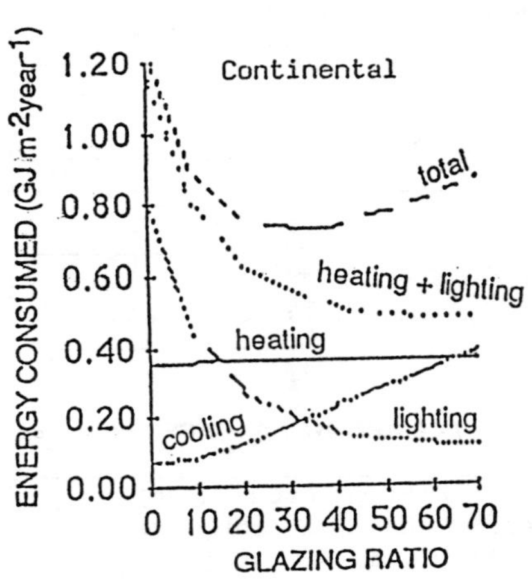

Fig 1. LT-Curves for S-facing double glazing Continental climatic zone

Assuming an internal thermal reference temperature, it then evaluates a monthly thermal balance -

ENERGY DEFICITS	ENERGY GAINS
conduction through glass	solar gain
conduction through opaque fabric	lighting thermal gains
ventilation losses	casual gains

Solar gains are calculated using the correlation curves of van Dijk/3/
The thermal balance (positive or negative) is then used to established delivered thermal energy for heating and cooling, applying a plant efficiency. Thermal energy is then converted to primary energy and added to the primary energy associated with lighting.

3 USE OF THE LT-CURVES

The method uses the concept of the passive zone, as illustrated in fig (2). This makes the assumption that any part of the building within 6m (for normal ceiling heights) of the external envelope, has the potential for daylighting, natural ventilation, and (according to orientation) use of solar gains. Thus the annual energy consumption for each type of passive zone can be read off from the appropriate curves. For areas beneath the roof, all the area is potentially passive zone. For areas outside the passive zones, the energy consumption can be read from the curves at the zero glazing ratio point where a small component has been added for fan power.

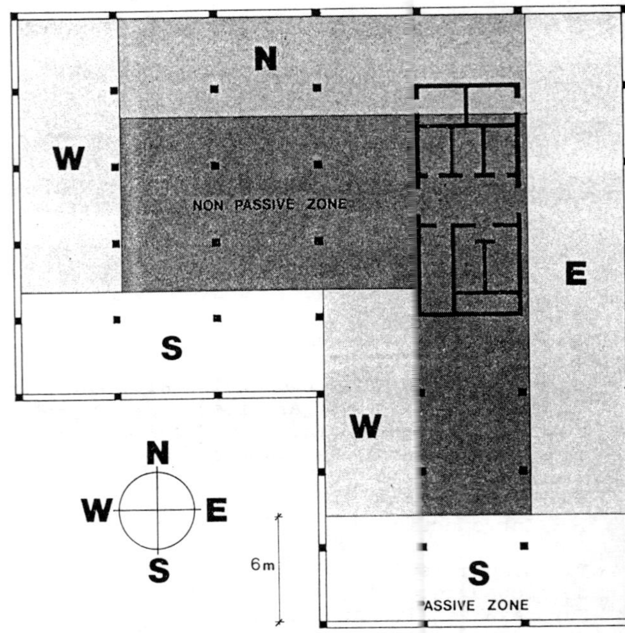

Fig 2. A building plan showing passive and non-passive zones

The affect of an adjacent atrium is dealt with in two stages. Firstly the depth of the passive zone for daylighting is reduced from 6m to 4.5 or 3.0 or zero according to the charcteristics of the atrium. Then a thermal saving is calculated from conductance and "solarity" parameters which is applied to the thermal energy total. These values are presented as a table.

Finally the values are entered on a worksheet, (fig 3), and the separate totals are calculated. A further parameter which the user is invited ti calculated, is the passive area ratio - the ratio of passive zone area to total floor area.

PROJECT **LOCATION**

		PASSIVE ZONES					NON-PASSIVE ZONES	TOTAL	
		north	*east/west*	*south*	*roof*	*buffer adjacent*			
zone areas (sq. m)									sq. m
façade glazing ratio (%)							%		
glazing type							1X or 2X		
energy consumption (per sq. m)	*light*								kWh/sq. m/yr
	heat								
	cool							TOTALS	
energy consumption (total)	*light*								kWh/yr
	heat								
	cool								
TOTAL									

BUFFER SPACE THERMAL SAVINGS (if appropriate)

buffer adjacent area (sq. m)	
'solarity' = ratio south : total glazing	
'protectivity' = ratio sep. wall : ext. glazing	
thermal energy saving (kWh/sq. m/yr)	
total thermal energy saving (kWh/sq. m/yr)	

SUMMARY

Net annual primary energy consumption	*light*		kWh/yr
	heat		
	cool		
Net energy consumption (per sq.m)			kWh/sq. m/yr
Ratio of passive area to total area			%

Fig 3. The LT Worksheet

4 THE DAYLIGHT EVALUATION PROCEEDURE

This consists of two parts. First the use of physical models to evaluate Daylight Factors and characterise the building into zones 0-1%, 1-3%, 3-6%, >6%. The zones are then ascribed lighting datum values and associated lighting power densities e.g.200 Lux 10 W/m^2.

Working back from the datum values and Daylight Factors, a datum sky luminance can be easily calculated. Sky luminance curves published by the Commission Internationale de l'Eclairage, show the percentage of time during the working day that these datum values are exceeded, as a function of latitude.

These values are entered on a worksheet, and the total electrical energy consumption for the building is calculated.

It is important to note that the initial DF values are evaluated from model studies and are thus able to reflect details of the room design and daylighting techniques such as light shelves etc., provided these have been modelled with sufficient accuracy.

The proccedure also included notes on sunlighting analysis, both by use of sunpath diagrams, and by observation using the daylight model.

5 SUBJECTIVE ASSESSMENT

As well as using the model studies as a step in evaluating the detailed lighting energy performance, the evaluation proceedure suggests that the model, illuminated under a real sky or an artificial sky, can be used to make subjective evaluation of the visual appearance of the design, both under diffuse sky and sunlit conditions. Hints are given on how to record the effect photographically, using an easily available wide-angle lense.

6 HOW THE TOOLS WERE USED

Almost all of the entrants completed at least one LT calculation. This was reassuring to the authors - clearly the mechanics of the proceedure was clear and straightforward. However, few participants completed more than one calculation and this goes some way to indicate the intentions of the user. In spite of the fact that the simplicity of the LT Method permitted several calculations without undue effort it was evident that most of the participants felt that it was intended as a once off evaluation, rather than a tool to compare several design options. There were notable exceptions - (fig 5) , but on closer inspection it proved that there was a "sympathetic" or an expert in the design team.

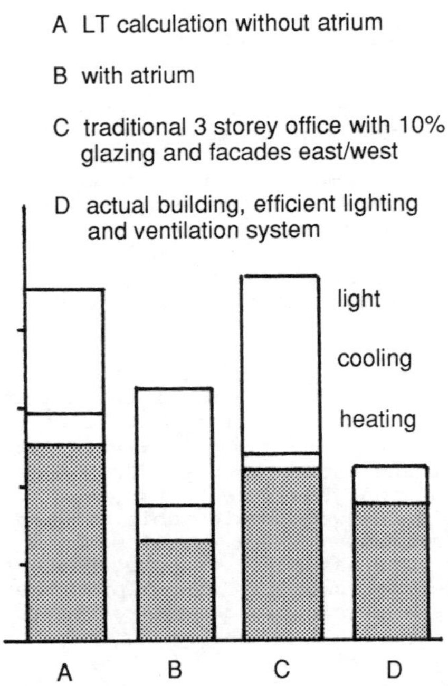

Fig 5. Graph showing comparison between building with and without atrium, by LT-Method.

As authors, following discussions with, and in some cases direct observation of users, we felt that there were two main reasons for this. Firstly the documentation did not stress the design role of the method sufficiently. Secondly, the method, in spite of the explanation of how it worked in terms

Fig 4. Photograph from inside daylight model

of design parameters, the final interface, the graphs and pro-forma, did not look like a familiar design medium. Futhermore, the single number, is not seen by most architects as a contribution to a design descision.

Did the use of the method then, contribute anything to the design quality of the schemes? In their report on the entries, the technical assessors made the following statement :

....."although the calculation excersise did not explicitely contribute to the design quality of the schemes, it contributed to the general aims of the competition in three ways -

1) it introduced the concept of quantitative analysis

2) it demonstrated the complex interaction ofenergy, environment and climate

3) it introduced useful concepts such as the passive zone ..... "

The impression gained about the daylight evaluation proceedure was rather different, and the important factor was the physical modelling and the subjective assessment. Some of the photographs produced were very good and the judges felt that the participants were comfortable working with a familiar medium - i.e. model making. Furthermore, although the proceedure also involved calculation, it followed the model studies and thus had a more realistic basis.

However, it was not generally apparent that designers had explored a number of options, except in a few cases.

Most architects are probably familiar with daylight modelling. However the simple instructions of how to carry it out are not generally available in popular text books. Thus the description of the method filled a gap in the readily available design support material.

7 CONCLUSIONS

So what have we, as authors of the design tools, learned?

It seems that the problem of presenting quantitative tools in such a way that they encourage the iterative mode of use to test proposals, converging upon an optimised solution, has still not been solved. Is it simply a problem of packaging - do we need friendly looking graphics and more explanation? Certainly, the requirement that this was a manual method unable to take advantage of the interactive nature of a computer made our task more dificult.

Or is the problem more fundamental than this? Is it because architects have the notion of designing from precedent, deeply established in their method, and that the use of quantitative analysis is still understood as a "design-folloing, proving" excersise rather than a "design-leading" activity?

If this is the case, then the aim of simplified tools must be changed from an emphasis on a specific quantitative result, to multi-parametric outputs which encourage the designer to consider the quantitative effect of design parameters. If, for example, when the designer tested his proposal, instead of giving a single performance figure, the design tool showed him the position of his proposal on a curve (or surface) drawn by varying other relevant parameters, this surely would be more likely to have the intended effect of a design-leading aid.

Following the experience of the "Working in the City" competition, the simplified design tools described here are being further developed as part of the SOLINFO programme. A manual method, with some improvements particularly in the documentation, will be published shortly. There are also proposals to issue two versions of the LT model as a computer programme, one aimed at the non-specialist designer with a friendly, educative interface, using a multi parameter graphic output as described above. The other will offer the model in a form for more specialist users.

8 REFERENCES

/1/ European Daylighting Design Competition "Working in the City" O'Toole S and Lewis J O, reported at this conference.

/2/ Experience with a Design Tool which combines the Energy Value of Daylight and the Thermal Value of Solar gain., Baker N and Hoch D, UK ISES Conference on Daylighting, London, 1989.

/3/ Windows and spaceheating requirements; parameter studies leading to simplified calculation method. Dijk, Hal and Arkesteijn, The Netherlands National report on activities within Step, Delft, TNO Inst of Applied Physics.

SIMPLIFIED GRAPHICAL-NUMERICAL METHOD FOR DETERMINING DAYLIGHT ILLUMINANCES

A. Fanchiotti
University of Rome "La Sapienza", Via Eudossiana 18, 00184 Rome, Italy
M. Vio
Istituto Universitario di Architettura di Venezia, S. Croce 191, 30125 Venice, Italy

ABSTRACT. The method presented here allows to estimate the illuminance on a horizontal surface due to a portion of any shape of the sky, as a function of sky type, latitude, hour of the day and day of the year. The skies are defined according to the classification recently developed by CSTB: overcast, intermediate overcast, intermediate mean, intermediate blue, and blue. The method requires the use of a set of three diagrams for each sky type: A. a modified Waldram diagram, over which the sky portion has to be outlined; B. a stereographic projection of the sun path diagram for that latitude; C. a dot diagram, showing the effect of the angular distance from the sun.

1. INTRODUCTION

The growing interest for more efficient and conscious ways to use daylight, has produced, over the last few years, a growing demand for design tools, enabling designers to evaluate the consequences of their choices, such as position, shape and size of openings. Such tools can range from sophisticated, mainframe computer programs, to very simple graphical methods. The first ones provide very accurate predictions, but require a very detailed definition of the design and some knowledge of numerical simulation techniques. The second ones are fa less accurate, but require little information, still allowing the study of complex configurations.

European literature on the matter shows several graphical methods for determining the direct and indirect illuminance caused by overcast skies, both with uniform and variable luminance distribution [1]. American literature shows a greater interest for clear blue sky conditions, where the luminance varies with sun position and angular coordinates [2,3].

The use of only overcast sky models can be adequate, at least at the initial design stage, for determining internal illuminances in urban and industrial areas. In sunny and dry climates, it is necessary to use clear sky models, while the use of both is advisable to control glare problems.

The present paper proposes a simplified method, based on a clear sky model, capable of dealing with complex configurations.

2. CLEAR SKY MODELS

As it is known, the luminance of a point in the sky for overcast conditions, increases with the altitude of the point above the horizon, being independent from the position of the sun (see, for instance, the C.I.E. Standard Overcast Sky). With clear skies, such as the Clear Blue Sky, the luminance is strongly dependent on the sun position, with complex but regular distribution, with values of the luminance that vary with the position of the sun, but keeping the same relative distribution.

While the overcast sky luminance can be computed by means of a simple equation (Moon and Spencer [4]), also adopted by C.I.E., after extensive tests against experimental data, the modelization of clear skies appears much more difficult.

The diffuse (sky) component of light can be treated either in terms of illuminance on horizontal and vertical surfaces, or in terms of luminance of each point in the sky.

The first approach (Kimball [6], Chroscicki [7], Elvegard and Sjostedt [8]), usually lead to expressions providing the horizontal illuminance as a function of solar altitude, such as the ones proposed by Krockmann [9] and Gillette [10], based on experiments in Europe and USA, respectively:

$$EH = A + B\,(\sin\alpha)^C$$

Where EH is the horizontal illuminance, α is the solar altitude, A, B and C are coefficients having the values 1.1, 15.5 and 0.5 according to Krockmann, and 0.8, 15.5 and 0.5, according to Gillette.

This type of approach does not allow to estimate the illuminance due to a portion of the sky, as it is the case for internal surfaces.

The second approach is therefore necessary, which requires the integration of the luminance over the portion of the sky seen by the surface.

Kittler [11] proposes an expression to compute the luminance, L_P, of point P as a function of the zenithal luminance, L_Z, of the angle of altitude of the point above the horizon, of the zenithal angle of the sun, and of anle between point and sun.

A new method has been recently proposed by Chauvel and Perraudeau [12], where L_P is expressed as the product of three functions:

$$L_P = f(\gamma,I_N)\,g(z_P,I_N)\,h(z_S,I_N) \quad [\text{kcd/m}^2]$$

where:

- I_N : nuvolosity index, which characterizes the sky type
- z_p : zenithal angle of point P;
- z_0 : zenithal angle of sun;
- γ : angle between sun and point.

The three functions are given by:

$$f(\gamma,I_N) = a_1 + b_1 e^{-k\gamma} + c_1 \cos^2(\gamma/2)$$
$$g(z_P,I_N) = a_2 - b_2 (\cos z_P)^{0.6}$$
$$h(z_S,I_N) = a_3 + b_3 \cos z_S + c_3 \sin z_S$$

where the coefficients a, b and c have been experimentally determined, and depend on the sky type.

3. THE PROPOSED METHOD

It is well known that the illuminance E_P of point P on a horizontal surface can be expressed as:

$$E_P = \int_\delta L \sin\delta \, d\Omega$$

and expressing $d\Omega$ as $d\Omega = \cos\delta \, d\delta \, d\beta$, being δ the altitude angle, $\delta = \pi/2 - z_P$, and β the azimuth angle, defined by the intersection of the solid angle Ω with the unit sphere,

$$E_P = \int_\delta \int_\beta L \sin\delta \cos\delta \, d\delta \, d\beta$$

If the sky is assumed to consist of small areas of uniform luminance:

$$E_P = \Sigma_\beta \, 1/2 \, L \, |\sin^2\delta| \quad \Delta\beta$$

Using the expression (1):

$$E_P = \Sigma_\beta \, f(\gamma) h(z_S) s(\delta) \, \Delta\beta$$

where $s(\delta) = (1/2) g(\delta) \, \Delta \sin^2\delta$

Thus, the illuminance in P is obtained by adding all the contributions $f(\gamma)h(z_S)s(\delta)$, due to the small areas wide $\Delta\beta$. Being the three functions $f(\gamma)$, $h(z_S)$ and $s(\delta)$ independent from each other, they can be evaluated by means of separate graphical tools: a dot diagram for $f(\gamma)$, a sun-path diagram coupled with a graphical function for $h(z_S)$, and an istogram for $s(\delta)$.

Fig. 1 shows a set of such graphs. The curves characterized by different values of the ratio H/D, where H is the is the height and D is the minimum distance from P of lines in the space, allow to rapidly draw the profiles of obstructions and openings (see Fig 2). The curve $h(z_S)$ is the graphical function already mentioned, and the istogram shows the average value of $s(\delta) = (1/2)g(\delta) \, \Delta\sin^2\delta$ in 10° intervals of δ. The sun-path diagram is relative to latitude 42° N. Finally, the dot diagram provides $f(\gamma)$, obtained by dividing the sky vault in 10° azimuth by 10° elevation solid angles, assuming the luminance to be constant in each area, with each dot representing the unit luminance in the baricentre of the area.

The method works in the following way:

a. draw the profiles on the first graph;

b. divide horizontally the portions of sky seen through the openings, according to the istogram;

c. overlay the sun-path diagram for the latitude;

d. find the position of the sun for hour and day being considered, and determine the corresponding value of $h(z_S)$;

e. overlay the dot diagram, with its center on the sun position;

f. count number of dots falling into each horizontal strip marked in b.;

g. multiply this number by the corresponding value of $s(\delta)$ provided by the istogram, and add the results;

h. multiply this figure by the value of $h(z_S)$ found in d.;

i. multiply this result by $\Delta\beta$; for istance, if $\Delta\beta$ is 10°, by $\pi/18 = 0.1745$.

4. CONCLUSIONS

The proposed method allows to rapidly evaluate even complex configurations. The method also allows for various types of sky, as for each of the five types proposed by CSTB it is possible to draw the dot diagram, the istogram and the curve providing, respectively, the functions $f(\gamma)$, $h(z_S)$ and $s(\delta)$.

5. REFERENCES

1. Hopkinson, R.G., Petherbridge, P. and Longmore, J., *Daylighting*, Heinemann, London, 1964.

2. Robbins, C.L., Daylighting, *Design and Analysis*, Van Nostrand Reinhold Co., New York, 1986:

3. Vio, M., "Metodi per la valutazione dell'illuminamento naturale negli ambienti chiusi", *Recuperare*, N, 36, 38, 1988.

4. Moon, P., Spencer, D.E., "Illumination from Non-Uniform Sky", *Illum. Eng.*, N. 37, 1942.

5. C.I.E., "Daylight. Official Recommendations", Paris, 1955.

6. Kimball, H., "Daylight Illumination on Horizontal, Vertical and sloped surfaces", *U.S. Weather Review*, N. 481.

7. Chroscicki, W., "Calculation Methods for Determining the Value of Daylight Intensity on the Ground ...", *Proceedings of the C.I.E. Meeting*, Barcelona, 1971.

8. Elvegard, E., Sjostedt, G., "The Calculation of Illumination from Sun and Sky", *Illum. Eng.*, 333-342, 42, 1940.

9. Krochmann, J., Muller, K., Retzow, U., "The Horizontal Illuminance Intensity and the Zenith Luminance of Clear Skies", *Lichttechnik*, Vol.22, 1970.

10. Gillette, G., Pierpoint, W., Treado, S., "A General Illuminance Model for Daylight Availability", *Journal of I.E.S.*, July 1984.

11. Kittler, R., "Standardization of Outdoor Conditions for the Calculation of Daylight Factor with Clear Sky", *Proceedings C.I.E. Inter-sessional Meeting on Sunlight*, Newcastle-upon-Tyne, 1965.

12. Perraudeau, M., Chauvel, P., "One Year's Measurements of Luminous Climate in Nantes", *Proceedings of the International Daylighting Conference*, Long Beach, CA, 1986.

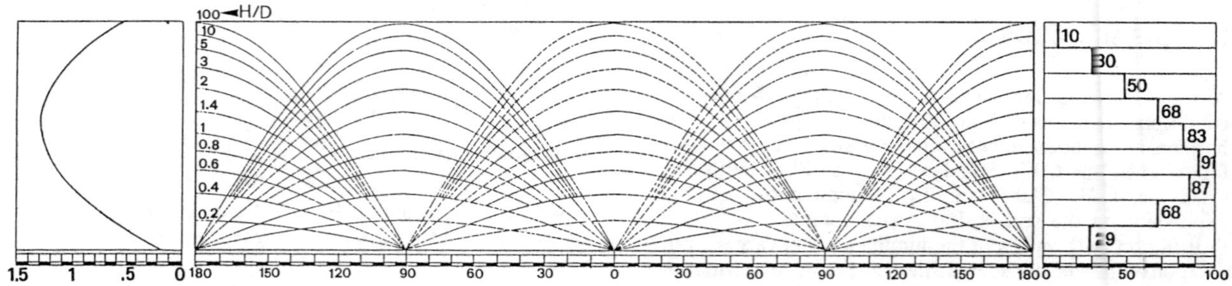

Fig. 1A Sky-vault with H/D lines to plot obstructions profiles; to the left: function $h(z_S)$; to the right: istogram providing values of $s(\delta)$.

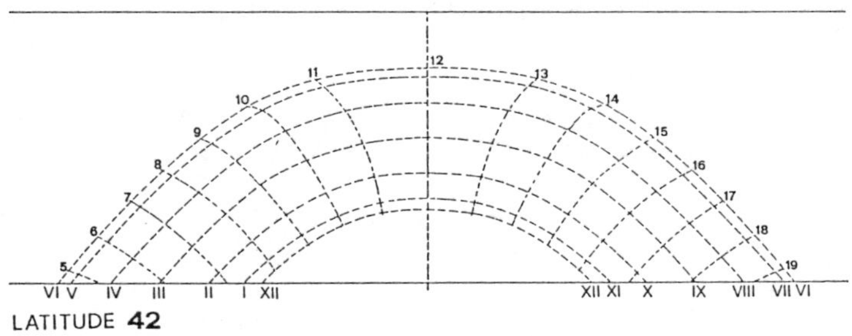

Fig. 1B Sun-path diagram for 42° N latitude

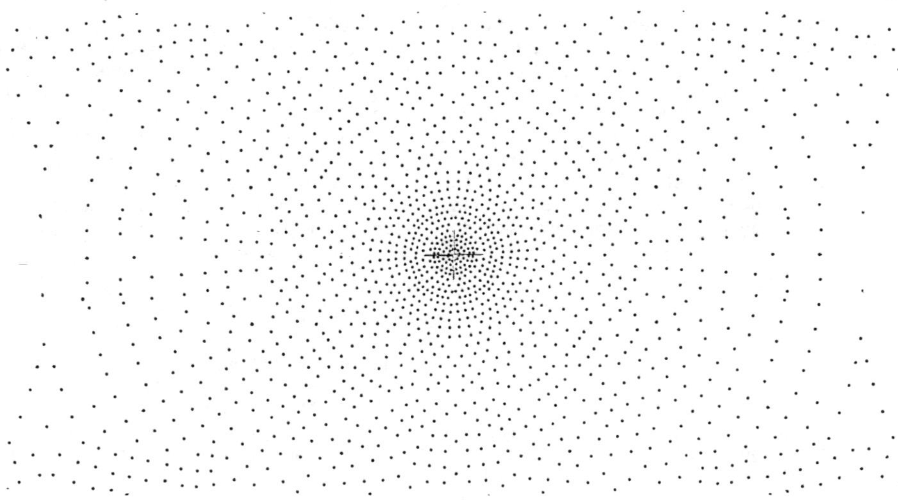

Fig. 1C Dot diagram: function f(γ) for clear sky

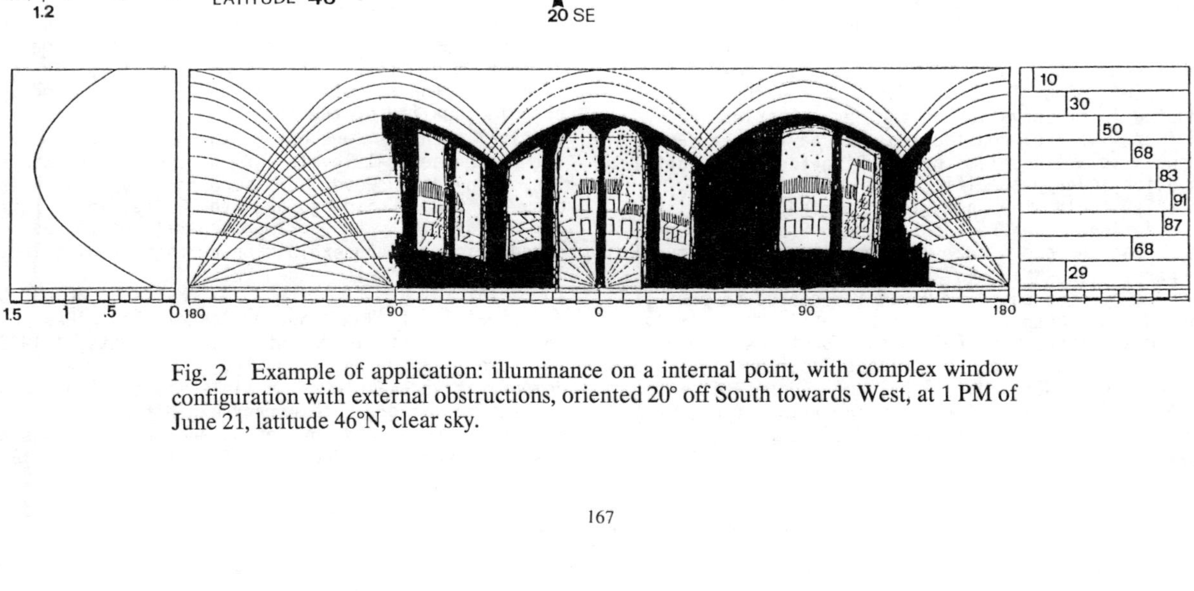

Fig. 2 Example of application: illuminance on a internal point, with complex window configuration with external obstructions, oriented 20° off South towards West, at 1 PM of June 21, latitude 46°N, clear sky.

CLIMATE AND DAYLIGHTING FOR CONSERVING

F. SALVI
Genio Civile of Catania
89, via Lago Nicito 95100 Catania (ITALY)

ABSTRACT. At last, for the increased interest on conservation, monuments are considered not only as a heavy responsability for collectivity, but also as a real estate to conserve and transmit to the future. That it's require a knowledge and setting up of new technics to work in the proper manner in existing buildings by technical operators. Problem becomes more complex to be solved when it involves the introduction of various public functions in historic (and ancient) buildings. The new metodology and practice of conservation carries a different approach of the up to date designer. He has to know ancient materials, historic stratification, historic and structural array, building technics on the course of ages. It's necassary an original cooperation among different knowledges. The modern solutions must be limited as possible to find the most agreable utilisation with respect for everything reached from the past. In particular, it must be considered that heating or conditioning or lighting system have always been practically insignificant in ancient buildings. Now a day, the utilisation in these buildings of technical solutions, just experimented in new buildings, often carries out great risks of being tampered. The aim of this work is to study how to obtain the ambient conditions through which it will be possible to ensure the Comfort and Security of people (users and employeds) and contemporaneously the conservation and Security of buildings, furniture and collections.

1. GENERAL ASPECTS

1.1 Introduction

No architect, archaelogist, museum curator or conservation manager today can ignore the climate and lighting influence on cultural heritage conservation, from a monument to a small piece of antiquity.
The cultural property, passed down from preceeding generations sufficiently intact, enable us to increase our knowledge. But first it must be preserved and restored by carefully studying the climatic conditions and the daylight to which it is exposed.
Secondly, and just as important is the risk run of damage caused because of a different, or wrong, environment.

1.2 Definition of conservation

Conservation is basically an historical discipline. It requires, however, contributions from many other disciplines.
So emphasis must be given to the understanding of cultural heritage (historic town or building or object) through its evolution in time as well as through its technological aspects.
Time by time conscious restaurer will analyse the behaviour of materials in their historical and physical context. Moreover it's important to understand that different solutions coming from different cultural and climatic approaches will be adopted in conservation in various countries.
In fact, we are considering not having ready-made answers to all problems but looking for a respectful solution in each specific case.
In this field the activity carried out by the International Centre for the Study of the Preservation and the Restoration of Cultural Property (ICCROM) is very interesting. It was founded by Unesco in 1959. It is located in part of the ancient "Ospizio di S. Michele" in Rome, Italy. ICCROM exercises the following functions, mainly: collects, studies and circulates documentations and experience concerned with the scientific and technical problems of the preservation of cultural property to raising the standard of restoration work (1).

1.3 Light control

An ancient building, open to the public, or a museum have two contradictory functions. They must protect the treasures of our cultural heritage, and at the same time, show them to all the world to see. This contradiction is based on the climate and light use and control.
Natural light and sunlight are the primary "materials" of architecture, but are also destroyers, when they fall uncontrolled into internal space. The infrared and ultra-violet rays, and the high temperatures generated by sunlight are particularity harmful. Artificial lighting is consequently indispensable in museology, not just for its special protective properties (when well selected), but also as a regulator of the daily and seasonal variations of natural light levels.
One of the simplest ways of assessing how it can best be used, is to analyse the course of the sun, at the latitude of the site during day and year.
From this it is possible to calculate and evaluate the problems of having the wrong light in the wrong place, and in particular unwanted, direct sunlight.
An interesting and stimulative approach is the IMA CENTRE solution in Paris. Each window of this buildings is made as a camera diaphragm having a variable opening used to restrict and control the amount of light.
A museum space can be naturally lit so that exhibits can be placed anywhere and still look the same, or so that a visitor's perception of them is directly related to the object's position in the room and the changing quality of the light.
Vertical openings are often the only solution when there are several levels or in ancient buildings. They provide however an important visual link between the building and the site, most with artificial light. A good approach will be the use of circulation and non-exhibition areas to redistribute light to where it is needed (2).
The exhibition "I Mito' di Miro'" held by Municipality of Rome, Spain Academy and CIDAC was one of the most important cultural events in 1989, in Rome. But, unfortunally, it was also a demonstrative example of possible mistakes, using an ancient building for temporary exhibition.
All of us are agreed in opening ancient buildings to visitors even introducing different functions. But

it's necessary to control climate and expecially lighting for exhibitions. Wrong lighting can damage the building, paintings or visitors apreciation as in this case (3).

1.4 Microclimatic control

In the opinion of specialists, the most ruthless of all destructive agents of artefacts is climate; but contrary to popular belief, it is not so much the variation of temperature as the variation of relative humidity that are the most harmful (4).
It is important to learn how an ancient building or a museum "bahaves" climatically and define different climatic zones and see how climatic variations outside the building are transmitted inside.
This study allows: to assign the most sensitive collections to the zones with the most stable humidity, to improve the passive and active thermal response of the building, to decide where it is necessary to dehumidity and where to humidify, to find a solution to conserve the building and to exhibit collections in the best possible conditions (5).
The Book Sanctuary (Arch. Kiesler and A. Bartos) where the famous Dead Sea Rolls are protected is a part of the Israel Museum in Jerusalem.
It's mainly underground: a corridor divided by trapezoidal arcs introduces visitors under the dome, a circular space through which daylight penetrates.
This manuscript had its origine two towsand years ago and it's very precious. It's exposed in many show-cases, where a thermohigrograph controls temperature and relative humidity (6).
The Riace Bronzes from many years under the sea when descovered, have created a great problem in finding a switable enviroment.
In the future it also be important to avoid too many changes of place, not only because of the danger of mechanical damage, but because changes of humidity and temperature could easily revive the rapid process of deterioration, and thus endanger the two statues which have so splendidly survived the passing of 24 centuries.

2. CASES STUDY

2.1 The Capodimonte Museum Picture Gallery Microclimate

This study was carried out in Italy by ENEA-FARE on the basis of previous experiences. It has concerned mainly the thermohigrometric condition analysis of exhibitioon space. Automatic calculation systems and computer simulation programmes were used to examine experimental data and to select passive and active improving solutions.
The Capodimonte Palace was built by Charles IIIth of Borboun to protect the Farnase Collections and has been used as a museum from 1806.
This building consists of four elevations: ground floor has a porch, first and second floors are exhibition areas and the floor under the roof is partially used as a laboratory. The sheds on the roof guarantee the daylighting of the picture gallery, on the second floor, through a sort of glass and "novopan".
This particular shape causes great problems to the gallery microclimate: the overheating in summer. Moreover the building's west side is subject to the action of damp wind coming from the sea.
The building was monitored for one year.
Temperature and humidity data have showed by grafics. It's quite clear the roof type (sheds) influences the building thermohigrometric behaviour. Practically it acts as a solar collector.
The analysis of collected data indicates the thermohigrometric difficult conditions of exhibition areas. A new climatisation system will be used step by step taking the building conditions slowly to the optimization. But the active conditioning system only cannot control microclimate: the roof of the building must be refitted, because existing conditions are very unacceptable for long periods during the year.
In this situation a climatization system break-down could suddenly change the microclimate situation of exhibition areas, causing irreversible damage to the collections and artefacts. A simple external screen of sheds, made whit movable louvers can decrease the inside temperature with a great advantage.

2.2 The Sistine Chapel

The ambient of the Sistine Chapel has remained through the ages exposed only to natural cycles depending on meteorological and vareing climatic conditions.
Recently this space was open to visitors on mass. So the internal micriclimate has been radically transformed. It's very difficult to evaluate, in advance, the consequences of the increase of dust an ventilation, lighting and thermohigrometric conditions daily and seasonal variations on artefacts.
An interesting study was carried out to analyse the climatic pertubation sources and to remove them in the correct way, or, at least, to reduce their effects.
Monitoring to identify the Chapel's internal microclimate dynamics was considered opportune while the frescoes are under restoring.
The results showed us microclimate changes and dust quantity in the atmosphere of the Chapel, caused by continuos use and heating and lighting systems.
In order to give to future generations the same cultural property image now showed to us, it will be necessary to provide conservation solutions with the technology evolution.
The lunettes are the only part of the ceiling decoration painted on a vertical wall surface and are quite different from the other sections of the Sistine Chapel's cycle, both because of their state of conservation before the recent restoration programme, and because of the technique Michelangelo used in painting them.
Here are fourteen lunettes, large semicircular spaces above the architectural cornice that runs along the top of the series of portraits of popes; the surface of each lunette is interrupted in the middle by the top of the round-arched windows. In these unusually shaped spaces and in the eight spandrels along the side walls, Michelangelo has arranged the members of the Family of David, that is the sixteen families of ancestors of Jesus Christ.
Before the restoration was begun the lunettes appared to be much darker than the rest of the ceiling frescoes, for they were covered by a thick layer of dirt and lampblack; it seems likely, in fact, that the difference in temperature between the vertical walls and the ceiling caused the smoke from candles oil lamps and braziers used in the past to light and heat Chapel, to deposit in greater quantities on the lunettes.
It was difficult to distinguish the overall tonality of the scenes and impossible to make out the individual colours.
The recent cleaning has restored to the frescoes their original pictorial quality, and we can now fully appreciate the extraordinary freshness of execution and the liviliness of the colours, elements that were difficult even to image before the restoration. It is interesting to compare the frescoes, befores and after cleaning. Take for example the Eleazar-Mathan and the Achim-Eliud lunettes as they were before the restoration and in their present condition (7).

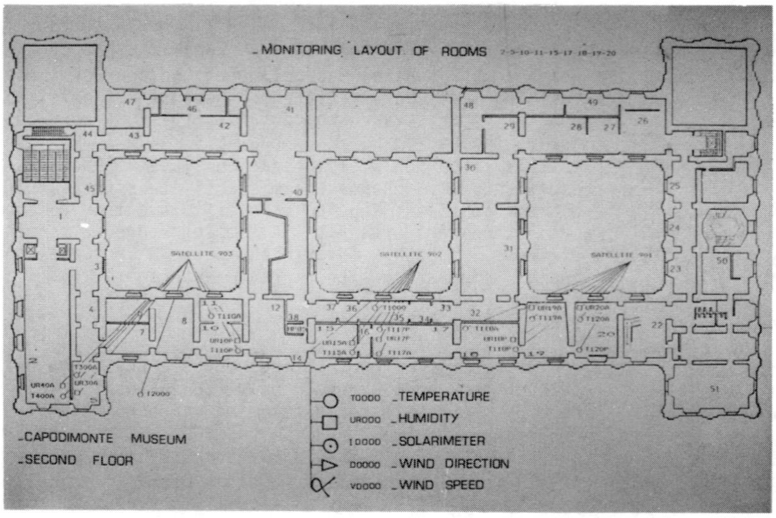

Fig.1 - 2 - CAPODIMONTE MUSEUM - ENEA
Monitoring

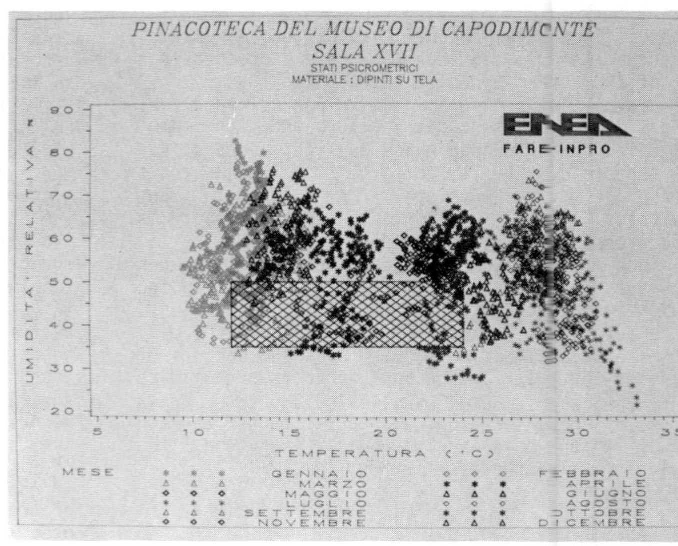

Fig.3 - 4 -IMA Center - Paris

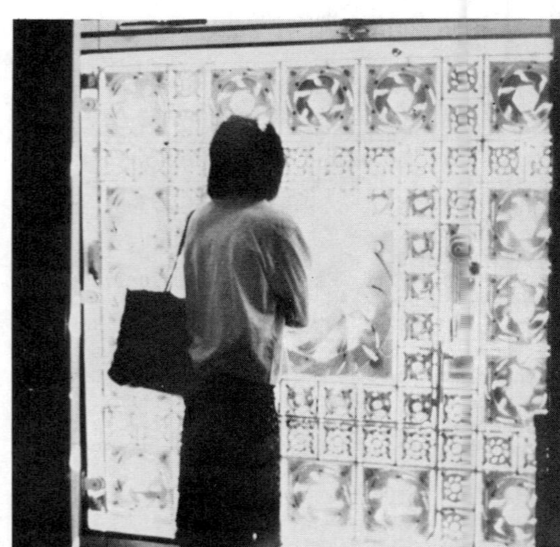

(1) ICCROM - "Architectural Conservation Course" (lessons) - Via di S.Michele 13 I - 00153 Roma
(2) "Techniques & Architecture" N.368-1986
(3) "I Mirò di Mirò" - Newton Compton Ed./ CIDAC - Roma 1989
(4) "Climate in Museums" G.de Guichen-1988
(5) "Il Museo negli anni 80" F.Minissi -Ed Kappa - Roma 1983
(6) "I capolavori del Museo d'Israele"1985
(7) "Michelangelo's cleaned frescoes"Scala Istituto Fotografico Ed. Firenze 1989

EVALUATION OF DAYLIGHTING AND THERMAL PERFORMANCEOF BUILDINGS

Dipl.- Ing. Volker Dingeldein,
Dipl.- Ing. Ingo Lütkemeyer
Technische Universität Berlin - Sekr. Ro 44
Klimagerechtes Bauen, Prof. H. Schreck
Strasse des 17. Juni 135, D -1000 Berlin 12
Tel: 030 / 381 71 59

ABSTRACT: In use of administration buildings relatively high demands determine the quality of artificial and natural illumination, so the spectrum of possible solar energy actions is to be seen in connection with lighting comfort. A synthesis of direct solar radation for passive gains and the use of daylight has to be developed to improve the energy balance and the comfort conditions of buildings. By means of parametric studies daylight and energy relevant aspects are investigated to achieve design guidelines for the architects plannings.

1. INTRODUCTION

At the Technical University of Berlin a research project was established to develop criterions of evaluation for building- and structural concepts, that represent thermal and daylighting demands.
Essentialy two strategies are proceeded:

- Daylighting of deep volumetric spaces in compact buildings by atria
  — Influence of atria to the energy and lighting balance
  — practicability of use and comfort of atria
- Daylighting and solar-energy gains in transparent facade structures
  — limitation of glare, energy and lighting balance

The energy and lighting relevant influences are placed together for a building- and spacetypology to analyse the essential parameters. The analysis of thermal characteristics of buildings and spaces is carried out by dynamic computer simulation. The syntheses of thermal and daylight considerations leads to evaluation criteria, resulting in design guidelines for the architect's plannings.

2. DAYLIGHT SIMULATION

2.1 Artificial Sky

Experiences have shown, that the scaled daylighting model is the most effective instrument in evaluating daylight for detailed or 'difficult to draw' forms. A daylighting-model, built in a scale of 1:50, 1:20 or 1:10, allows measurements on a surface and at a real point of reference.
The analysis of daylight relevant parameters is taken through under controlled conditions in an "artificial sky" at the Technical University of Berlin.
There are significant differences between indoor and outdoor model testing, but the current change of sky conditions outside make it impossible to estimate the influence of single parameters to the daylight illumination. For model testing the artificial sky is the only 'tool' to do parametric studies with comparable results. Artificial sky conditions provide a controllable, constant and reproducable environment for the determination of the influence of varied single parameters . Quantitative and qualitative features are investigated by measurement of illuminance distribution and light ratio - mainly for the overcast sky.
The artificial sky that has been set up to carry out this project consists of a translucent hemisperical (diffuse, white) dome with 16 segments. In a distance of 40 cm behind the diffusor 650 fluorescent lamps are installed in 6 rings that provide a horizontal illuminance of more than 10 klx in the measuring area for uniform sky conditions. The level of light can be reduced by a dimming equipment to 30 % of the

Figure 1: Outside of the artificial sky at TU Berlin

Figure 2: Interior view of the artificial sky

HEMISPHERICAL SKY AT TU BERLIN

Interior diameter:	4.00 m
Exterior diameter:	5.20 m
650 fluoreszent lamps - 18W daylight	
Illuminance in the centre: >10000 lx	
luminance distribution: CIE overcast sky	
Interior envelope: transluzent white acryl glass	
measuring area: 80 cm diameter	

Figure 3: Technical Data of the "sky"

maximum. The division into 32 independant dimmable segments allow an approach to other non uniform sky conditions.
More daylight comfort criterions like contrast and glare are simulated in situations of direct light by an artificial sun and evaluated by photograph and video techniques.
The artificial sun consists of a parabolic mirror (90 cm diameter) with a 1000 W halogen lamp in the focus. So nearly parallel light for sun simulation is achieved.
Different sun heights can be simulated by moving the sun vertically, different azimut angles by rotating the model.

2.2 Scaled Models

In architectural practice and also in teaching architecture the use of physical models is a common method. They illustrate the three dimensional performance of a building more effectivly than any perspective or axonometric drawing.
Also for the prediction of daylight illumination and effects of sunlighting, scaled models are useful.
The figures 4 to 6 show some physical models built in a scale 1:50 for the prediction of daylight illumination and the presentation of the design concept. The design and the models were set up by students of architecture.
The fotos were taken with an ordinary camera with a 28 mm lense under artificial conditions.
The presentation of interior atrium spaces under different daylight conditions within the artificial sky - overcast sky without sun (fig. 4) and with sun simulated by the artificial sun (fig.5) - enable the visualisation of different spatial atmospheres. The control of direct light (avoidance of glare) for working place surroundings an the light reflection to the ceiling for improving the daylight illumination of the depth of a room can be observed by scaled models. (fig. 6)
The strong influence of direct sunlight to the appearance and character of rooms is shown. The model for predetermining studies might be much simpler and easyer to built than the models shown here.

2.3 Parametric Studies

The evaluation of various design or construction details for daylight illumination of rooms and buildings and the influence of different light relevant parameters is investigated by parametric studies by changeable scaled models.
For the evaluation of daylight distribution in sidelighted rooms the facade and the surfaces of the room are removable, different sizes of rooms are to be analysed. The models built in a scale of 1:10 allow the detailed simulation of shading devices for the protection against direct solar radiation (e.g. on working surfaces).
For atrium buildings the daylight illumination in the atrium space and in the adjacent rooms are to be investigated.
The models built in a scale of 1:50 enalble the variation of atrium type (such as core or integrated atria), atrium floor area, atrium height, atrium roof shape and surface reflection. The illuminance inside the atrium is measured by small scale photometer heads (7mm heigh, 4mm light sensor) on horizontal and vertical surfaces.
For the daylight evaluation of rooms linked to an atrium a model room can be attached in different heights and different locations. The surfaceses of this room and the facade (window proportions and size) can be changed. Inside the illuminance distribution is measured on a horizontal plane (working plane).

3. THERMAL SIMULATION

The energy balance of non- residential buildings depends on various influences. Some of the most important parameters and their interdependenceses with daylight relevant parameters are investigated within this project.
Base for the evaluation of the thermal characteristics of Atriumbuildings is a parametric study by dynamic thermal simulation.
For different atriumtypes, mainly Core- and Integrated Atria,

Figure 4: Atrium Model under overcast sky conditions

Figure 5: Atrium model with direct light by the artificial sun

Figure 6: Room with reflecting lightshelfs facing the Atrium

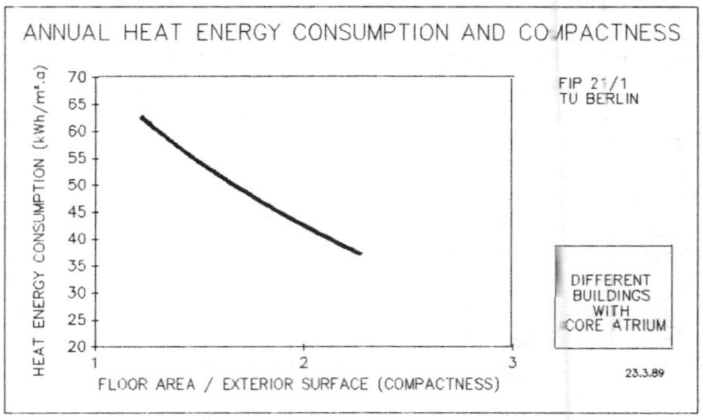

Figure 7: Annual heat energy consumption for different buildings with core atrium for different values of compactness

the influences of geometry, exterior envelope and interior connection between building and atrium are investigated. For the simulation runs a thermal network is built that shows the interchanges between the heated building space, the unheated atriumspace and the ambient situation.

3.1 First Results

The evaluation of building cubes can be done by the grade of compactness. The graphics are reflecting on a core atrium building, where several sizes of atria and buildings have been analysized. The thermal quality of the building envelope has not been changed.
The result of analyses is, that the demand of heat energy is reduced by rising compactness, represented by the proportion of building surface to floor area.
The compactness of the building rises by expanding the atrium floor area or atrium height.
The medium temperatures of atria depend on the proportion of the interior atrium surface to exterior atrium surface. The more compact an atrium is, the smaller is the days and years amplitude of temperatures.

The comparison between different glazings for the roof of a core atrium (atriumsize 18m/18 m, 4 storeys) show the small influence of glasquality to the heat energy demand of this specific building configuration.

4. CONCLUSION

First experiences show that scaled models for the evaluation of daylight illumination of buildings are very useful. The quality and the distribution of daylight can improved by simple means. The positive influence of Atria to the energy and daylighting performance of buildings is known, design guidelines to optimize thermal and visual conditions are to be developed.

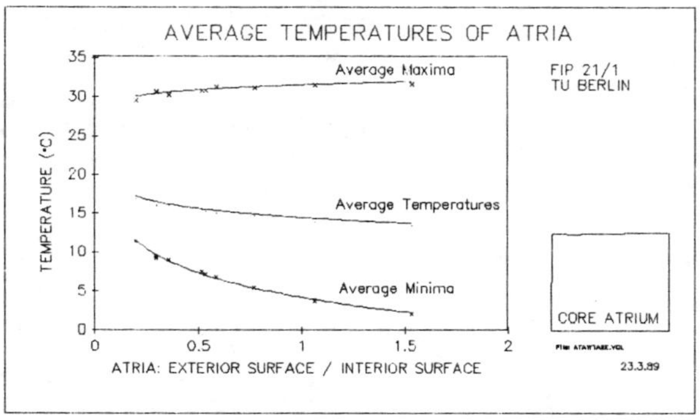

Figure 9: Average temperatures of Atria for core Atria buildings with different compactness

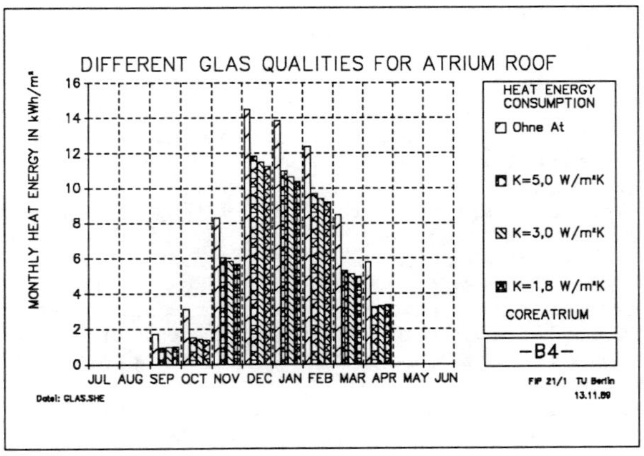

Figure 10: Influence of glass quality of the atrium roof to the monthly heat energy demand of a building (4 storeys) with core atrium (18/18m floor area)

NEW DAYLIGHTING COMPONENTS: A SUN-DUCT IN BUILDING RETROFITING

X. SOLSONA, arch; C. BADINIER, arch; H. COCH, arch.
ESCOLA TECNICA SUPERIOR D'ARQUITECTURA DE BARCELONA (U.P.C.)
Av. Diagonal, 649 BARCELONA 08028 (Spain)

ABSTRACT: The aim of this paper es to show the results of a recently constructed sun-duct in an existing building. A sun-duct is a new daylighting component designed to conduct sun rays into interior parts of the building which have little or insufficient light level. After previous studies carried out by way of simulations on a model and graphic studies, the sun-duct was built. This sun-duct has been calculated to have a great number of hours of direct sun rays in winter and it avoids direct solar radiation in the interior during summer.

This sun-duct is placed in an existing building in Teia, a little village near Barcelona. There we are in a mediterranean climate (hot and humid in summer and cold and humit in winter) as well as theeffects of being near the sea.
When the design of this retrofiting began, one of the most important problems of this house was the low ilumination level of certain spaces especially the kitchen. The architect (X. Solsona) proposed the option of constructing a sun-duct taking advantage of the existing duct of an old chimney. This was approved and this daylighting component was successfully built. It is a duct of 60 x 60 cm. and 4 m. in depth. The construction materials are plastic mirrors supported by a stainless-steel structure. The materials of the mirrors and structure have been selected to avoid the corrosion problems in a sea-side environment.

When designing this duct the sun path had to be taken into account in order to collect direct solar radiation during winter and prevent its entry in summer to avoid overheating. As well as this principal function it can also allow natural ventilation. The external aspect of this duct, especially the top of it has been integrated into the roof of the house using the same materials (arabic tile) as the whole roof. It is situated in the same place where the former chimney was. The final effect can be seen in photograph 1.
It is important to keep in mind that in such a climate there is a greater amount of sun hours in the daylight period (aprox. 2/3 of the total). For ambiental, visual and psycological reasons we intend to have the visual effect of sun rays in the house, especially in winter. This aesthetic effect can be seen in photograph 2.

1.- Exterior view

2.- Interior view

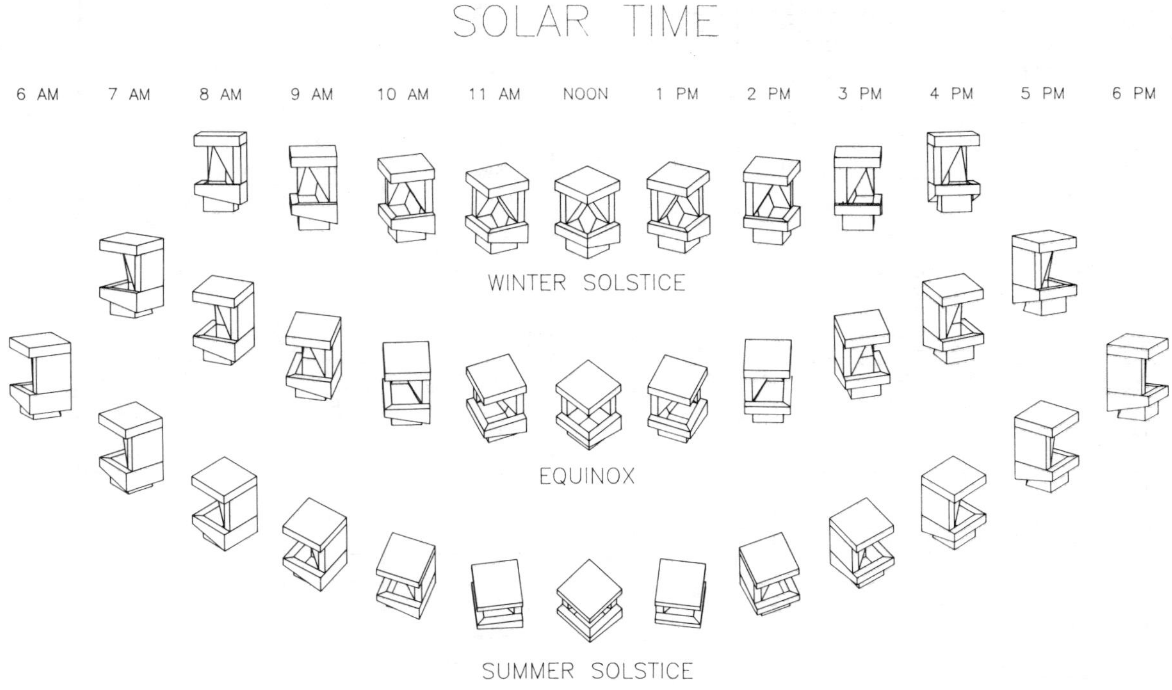

Fig. 1.- Position of the head collector in the different seasons.

The design has been carried out following previous studies and realizations on these kind of daylighting components of Prof. R. Serra, arch; (the earlier ones are from 1967).
The first stage has been the graphic design of the sun-duct with the aim of improving its performance, which for us means having much sun as possible in the interior of the kitchen in winter and at the same time avoiding as much sun as possible during the hot summer days. Its function is at the same time energetic (thermal and luminic) and also aesthetic and psycological.
In figure 1 we can see the graphic representation of the positions of the collector head of the duct in the different seasons (winter & summer solstices and spring & autumn equinoxes) during all daylighting hours. The surface of interior mirror shown in the graph is the real surface that receives solar radiation.
The second stage has been approached by modelling it, and testing the results.

The last stage has been the construction of the sun-duct in the house. After this, measurements have been taken in the spring equinox (March,21) and summer solstice (June,21) and the results have been compared. This kitchen also has a window in façade and a glazed door connected to an illuminated space, but when the measurements and photographs were carried out they were closed. Nevertheless the light level in the interior was not high but there was enough to perform some of the every day activities that take place in a kitchen without the necessity of switching on an electric light.
We can sum up by saying that the daylighting factor is approximately the same in both seasons, so the quantity of luxes are proportional to the exterior luminance level. But the initial desired effect of having a great number of hours with direct sun rays in the interior of the kitchen during winter time has been successfully achieved, while at the same time avoiding direct solar radiation in summer.

SOLAR TIME 8:30	EQUINOX		SUMMER SOLSTICE	
EXTERNAL ILUMINANCE LEVEL	43,000 LUX		45,000 LUX	
	LUX	D.F.	LUX	D.F.
CELL 0	300	0.70	220	0.50
CELL 1	060	0.14	060	0.13
CELL 2	010	0.02	030	0.07
CELL 3	100	0.23	010	0.02
CELL 4	030	0.07	020	0.04
CELL 5	010	0.02	010	0.02
CELL 6	020	0.05	010	0.02
CELL 7	010	0.02	030	0.07

TABLE I.- Measurements 8:30 a.m.

SOLAR TIME 9:00	EQUINOX		SUMMER SOLSTICE	
EXTERNAL ILUMINANCE LEVEL	45,000 LUX		51,000 LUX	
	LUX	D.F.	LUX	D.F.
CELL 0	450	1.00	240	0.47
CELL 1	060	0.13	070	0.14
CELL 2	020	0.04	030	0.06
CELL 3	100	0.22	030	0.06
CELL 4	030	0.07	020	0.04
CELL 5	010	0.02	010	0.02
CELL 6	020	0.04	010	0.02
CELL 7	010	0.02	020	0.04

TABLE II.- Measurements 9:00 a.m.

SOLAR TIME 9:30	EQUINOX		SUMMER SOLSTICE	
EXTERNAL ILUMINANCE LEVEL	40,000 LUX		54,000 LUX	
	LUX	D.F.	LUX	D.F.
CELL 0	700	1.75	290	0.53
CELL 1	060	0.15	240	0.44
CELL 2	020	0.05	030	0.06
CELL 3	020	0.05	020	0.04
CELL 4	030	0.08	030	0.06
CELL 5	010	0.03	010	0.02
CELL 6	020	0.05	020	0.04
CELL 7	010	0.03	040	0.07

TABLE III.- Measurements 9:30 a.m.

SOLAR TIME 10:00	EQUINOX		SUMMER SOLSTICE	
EXTERNAL ILUMINANCE LEVEL	64,000 LUX		[illegible]5,000 LUX	
	LUX	D.F.	LUX	D.F.
CELL 0	370	0.58	[illegible]10	0.56
CELL 1	120	0.19	[illegible]20	0.22
CELL 2	020	0.03	040	0.07
CELL 3	020	0.03	010	0.02
CELL 4	050	0.08	030	0.05
CELL 5	020	0.03	020	0.04
CELL 6	020	0.03	0[illegible]0	0.02
CELL 7	020	0.03	0[illegible]0	0.05

TABLE IV.- Measurements 10:00 a.m.

SOLAR TIME 10:30	EQUINOX		SUMMER SOLSTICE	
EXTERNAL ILUMINANCE LEVEL	70,000 LUX		57,000 LUX	
	LUX	D.F.	LUX	D.F.
CELL 0	350	0.50	310	0.54
CELL 1	070	0.10	080	0.14
CELL 2	020	0.03	040	0.07
CELL 3	030	0.04	020	0.04
CELL 4	090	0.13	030	0.05
CELL 5	030	0.04	020	0.04
CELL 6	020	0.03	020	0.04
CELL 7	020	0.03	030	0.05

TABLE V.- Measurements 10:30 a.m.

SOLAR TIME 11:00	EQUINOX		SUMMER SOLSTICE	
EXTERNAL ILUMINANCE LEVEL	70,000 LUX		6[illegible],000 LUX	
	LUX	D.F.	LUX	D.F.
CELL 0	310	0.44	2[illegible]0	0.44
CELL 1	060	0.09	0[illegible]0	0.08
CELL 2	020	0.03	0[illegible]0	0.05
CELL 3	020	0.03	0[illegible]0	0.05
CELL 4	270	0.39	0[illegible]0	0.05
CELL 5	020	0.03	0[illegible]0	0.00
CELL 6	020	0.03	0[illegible]0	0.02
CELL 7	010	0.01	0[illegible]0	0.02

TABLE VI.- Measurements 11:00 a.m.

SOLAR TIME 11:30	EQUINOX		SUMMER SOLSTICE	
EXTERNAL ILUMINANCE LEVEL	70,000 LUX		62,000 LUX	
	LUX	D.F.	LUX	D.F.
CELL 0	320	0.46	1470	2.37
CELL 1	050	0.07	030	0.05
CELL 2	010	0.01	030	0.05
CELL 3	020	0.03	020	0.03
CELL 4	080	0.11	020	0.03
CELL 5	020	0.03	010	0.02
CELL 6	020	0.03	010	0.02
CELL 7	010	0.01	030	0.05

TABLE VII.- Measurements 11:30 a.m.

SOLAR TIME 12:00	EQUINOX		SUMMER SOLSTICE	
EXTERNAL ILUMINANCE LEVEL	66,000 LUX		63 000 LUX	
	LUX	D.F.	LUX	D.F.
CELL 0	780	1.18	24[illegible]	0.38
CELL 1	050	0.08	03[illegible]	0.05
CELL 2	020	0.03	03[illegible]	0.05
CELL 3	030	0.05	03[illegible]	0.05
CELL 4	060	0.09	01[illegible]	0.02
CELL 5	020	0.03	01[illegible]	0.02
CELL 6	020	0.03	01[illegible]	0.02
CELL 7	020	0.03	01[illegible]	0.02

TABLE VIII.- Measurements 12:00 a.m.

DAYLIGHTING AND THERMAL PERFORMANCE OF SHED COVERINGS FACED SOUTH IN BUILDINGS

Angelo Mingozzi

Istituto di Architettura e Urbanistica
Faculty of Engineering
University of Bologna
Viale Risorgimento, 2 - 40136 Bologna - Italia (*)

ABSTRACT. The general objective of this study, was to reduce the energy required in the heating in both industrial and commercial buildings, while optimising a sense of well-being. To this end, the possibility to act on the covering of the building was examined; since the "shed covering" used in industrial buildings adapts best to this type of operation. In particular, the possibile performance of a commercial type shed was evaluated after an adequate redimensioning it was faced south, rather than north, as they are usually constructed. Thus, a design method was defined, as was a formal solution, optimised under all aspects of quality control. A covering system for three case studies of industrial building, located at different latitudes was designed. The shed system to be chosen must include various features of comfort in the physical environment such as lighting, acoustics, and energy supply using a design which reduce global costs. The research has not been terminated. Currently tests on transparent materials are being conducted. The indicated system (south facing sheds) will be tested on a case study now under construction.

1. INTRODUCTION

In the last ten years the augmented level of comfort demand and the high costs of building management have given rise to the investigation of technologies designed to insure the necessary level of physical comfort while keeping global costs down at the same time. If passive aspects of building design are not optimized without a correct integration with mechanical systems energy waste will be caused.

Keeping this in mind, and given that the building construction field is very slow to accept innovations, an attempt was made to figure out which factors to consider in order to maximize comfort and reduce energy consumption in industrial and coimmercial buildings, in order to demonstrate some of their more successful energy savings features.

The research sumarized in this paper is conducted under the auspices of the Institute of Architecture and Urban Planning and the Institute of Physical Technology of the University of Bologna beginning since 1985 with financing first by the regional government of Emilia Romagna followed by funding from CNR which still continues today.

The research's theme is still quite timely, and the methodological approach has been shown through the years to be correct.

2. OBJECTIVES

The objective of this study is the definition of an operational methodology and a practical solution for the design of a shed system in buildings used for productive activity which will add to comfort conditions, and reduce the management costs.

The typical shed system installed in commercial and industrial buildings could emphasizes their internal enviromental qualities.

The indicated covering system has certain advantages in the areas of energy (direct solar gain and containment of the heated volume, natural lighting (distribution of light and glare control) and acoustics control (due to its shape integrated with the antiglare screen). Figure 1.

The initial specific objectives of the study have already been reached: defining the methodology and project criteria; making an outline of the system function; measuring the performance of some cases of study.

At this stage of the research the object focusses on selecting the most adequate materials and the testing of their performance, in particular the transparent and translucent materials, since the design method indicates that the control of natural lighting and the contribution of solar energy are the primary features of the system.

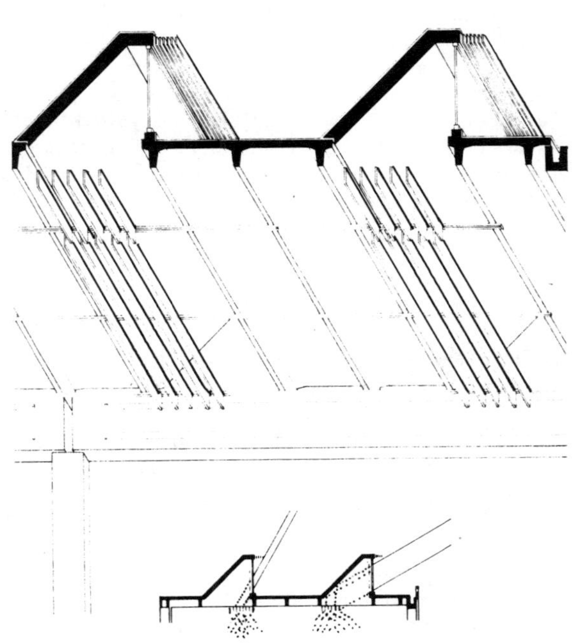

Figure 1: Axonometric scheme of the system.

3. METHODOLOGY OF APPROACH

The Study is divided in five phases. The first three phases are concluded and the construction typology of the system and the design guidelines have been defined.

In the first phase the variables of the problem were defined, thus the project method was outlined and laid out. In the second phase, samples of the covering were tested to maximixe internal and external light control systems, and the theoretic performance of the model (scale 1/50) was tested facing south in various latitudes with the assistance of a "Helidon" /1/. In the third phase, the climatic behaviour of an annual cycle was simulated for three italian climatic conditions, across three types of transparent surfaces /2/.

Now the fourth phase is being carried out. In this phase the study of materials, and in particular experimental tests on honeycomb transparent flat plastic sheets are doing /3/. In the last phase an experimental building will be completed within 1991 using the criteria established by the research (figure 1).

In particular, the possible performance of a commercial type shed was evaluated when, after an opportune redimensioning, it was faced toward south, rather toward north, as they are usually constructed.

The system presents two modular screens, attached to the windows, which control the light and heat gain. The external one is fixed. Its dimensions varies from 0,28 m to 0,32 m according to latitude. The internal one, antiglare and antinoise, is mobile using a simple system. Its size varies according latitude, type of transparent component and specific use of the building.

The transparent surface is vertical and not inclined. This was chosen because of its various advantages. The vertical sheet minimizes the lenght of the protective screen both internally and externally and remains clean longer. This resolves some construction and maintenece problems having to do with the fixtures, especially those which can be opened. Finally, the vertical glass covering cuts down the summer overheat, at the cost of slight reduction in winter gain of solar energy.

The model used for physical and analytical tests measured 20 x 30 x 8 meter long with southern exposure and no windows on the lateral walls butonly in the covering.

4. SIGNIFICANT RESULTS

The Following results were achieved: the operational methodology; the functional outline of the south facing microshed optimizing the control of solar gain, natural light intensity and glare.

The mathematic simulation /4/ has shown that this system could result in a savings of approximately 28% in North Italy; 54% in Central Italy; and 87% in South Italy in comparision which 'traditional' shed covering (table I) and that thermal inertia of the building registers between 2% to 5%.

The results obtained also show that the increased energy gain due to new shape and certain materials tested, can be quickly converted to economic gain thanks to reduced heating costs. By comparing the savings obtained by our study and the costs of the initial transformation, the optimum combination of structure and material can be chosen (table II).

	TRANSPARENT SURFACE	FTR	FAR	RA	RP
Bologna	simple glazing	511,4	367,1	144,3	28,5
	double glazing	467,7	342,4	125,9	26,5
	policarbonate	456,2	326,4	129,8	28,5
Roma	simple glazing	259,1	117,5	141,6	54,5
	double glazing	236,6	123,2	113,4	53,0
	policarbonate	223,3	104,2	119,1	55,0
Messina	simple glazing	85,1	11,0	74,1	87,5
	double glazing	77,4	11,4	66,5	85,5
	policarbonate	76,0	9,8	66,2	87,0

FTR = Fabbisogno Totale di Riscaldamento GJ
FAR = Fabbisogno Ausiliario di Riscaldamento GJ
RA = Risparmio Assoluto GJ
RP = Risparmio Percentuale %

TABLE I: Annual energy balance of the model /4/.

	TRANSPARENT SURFACE	CRER	CREM	SAVING
Bologna	simple glazing	9810	7040	2770
	double glazing	8960	6550	2410
	policarbonate	8750	6260	2490
Roma	simple glazing	4980	2250	2730
	double glazing	4540	2360	2180
	policarbonate	4280	1990	2290
Messina	simple glazing	1630	210	1420
	double glazing	1490	210	1280
	policarbonate	1460	190	1270

CREM = Costo Riscaldamento Edificio di Riferimento £ x 1000
CREM = Costo Riscaldamento Edificio Modello £ x 1000

TABLE II: Thermal energy costs of the model /4/.

One very important feature to point out is that the use of different transparent material means the modication of the anti-glare screen. Comparing the possible savings and the forseeable extra costs it is possible to choose which transparent combination of structure feature according to the type of activity and the climate which will best satisfy expected returns on the investment. It is suggested that for the northern regions it is better to choose material like honeycomb flat plastic sheets. This material, besides the fair savings which it offers, also resolves problems of window safety and permits a return on investment within four to six years.

For the central regions materials of simply diffused or retinate are equivalent to the honeycomb flat plastic sheets, however the return on investment is a bit longer. For the southern regions the model shows its limited applicability in that the extra costs have a return on investment which equals the expected life of the element. Additionally, the problem of summer and even middle season overheating arises. For this climate the geometry of the shed should be adjusted by diminishing the superficial area of the passive solar system.

The use of double glazing in industrial buildings does not seem to be a wise decision in economic terms for any latitude.

Finally, it should be noted that experimental tests on performance of different types of honeycomb flat plastic sheets in relation to light transmission is being analyzed.

(*) Current affiliation of A. Mingozzi: ICIE (Cooperative Institute for Innovation) Via Nomentana, 133 - Roma.

References

1. G.Cuppini, L.Bezzi, A.Mingozzi, S.Piva, "Illuminazione naturale e apporti gratuiti di energia", Quasco n. 87, Ed. Quasco, Bologna 1987.
2. A. Mingozzi, G. Funaro, "Prestazioni termiche ed illuminotecniche delle coperture e shed nell'edilizia industriale", L'Industria delle costruzioni n. 198 - aprile 88. Ed. Edilstampa, Roma, 1988.
3. G. Raffellini, S. Piva, A. Mingozzi, "Sulla determinazione sperimentale del coefficiente di trasmissione della radiazione solare in lastre piane alveolari", Proceedings of 43° Congresso Nazionale dell'Associazione Termotecnica Italiana, Ancona, 1988.
4. P.E.S.P. a computer program of University of Bologna derived from:
   A. Klein e altri, "A design procedure for solar heating system", Solar Energy 18 - 1980;
   M.Raoust e altri, "Méthode 5000", ed. Pyc, 1982;
   CSTB, "Règles Th B 82".

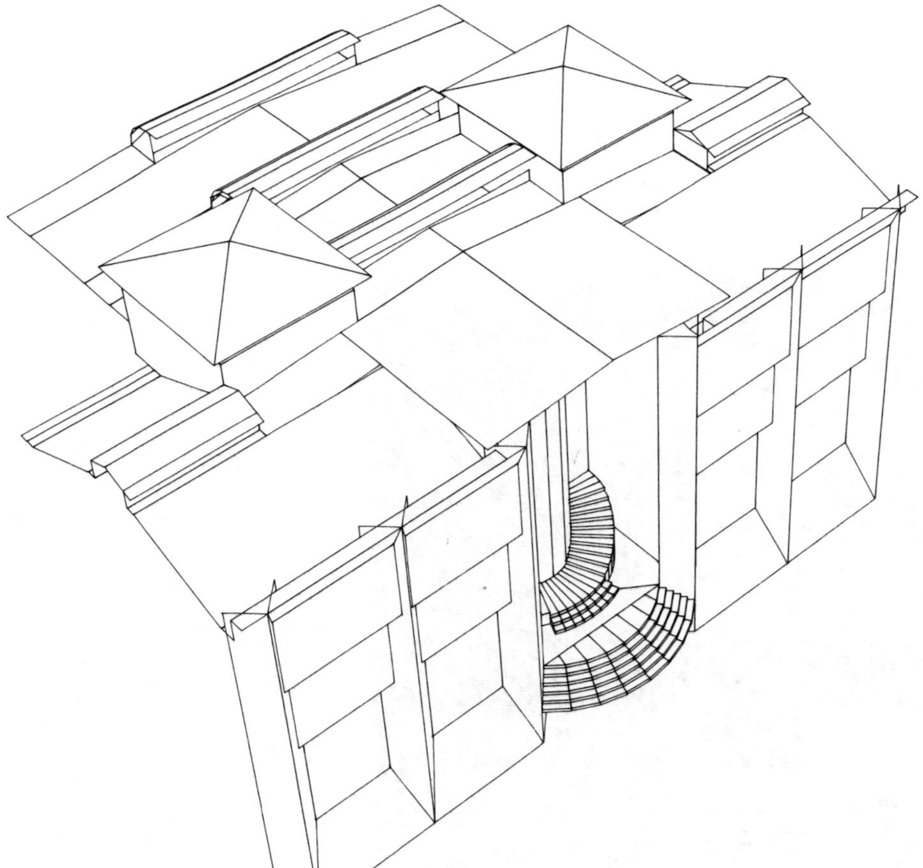

Figure 2. Partial schematic design of the experimental building under construction.

ARCHITECTURE EN FETE

La précision de la poésie, l'enthousiasme de la science, la magie de la fête

ATELIER DU PASSAGE
4 Passage Molière 75OO3 PARIS- Tel.(33)I 48 87 91 84

Jacques DEVAL, Architecte,
Professeur à l'Ecole d'Architecture de Lyon

L'expression des connaissances scientifiques, climatiques et biologiques, est un enjeu aujourd'hui. Le propos est de montrer la force de proposition de la démarche architecturale : en l'occurence, elle s'appuie sur la pérennité d'un travail poétique en quête d'une harmonie avec les élements naturels, le vent, le soleil, l'eau... et sur l'éphémère des architectures de fêtes considérées comme des lieux privilégiées d'invention et de création architecturale.

UN EXEMPLE D'APPLICATION

CITE EN SCENE

SUPERPRODUCTION MONDIALE POUR LA METROPOLE LYONNAISE

Concours "Un avenir pour Démocratie"
Vénissieux, communauté urbaine de Lyon
Projet "Métropoles en scène", pour le quartier des Tours des Minguettes
Lauréat 1ère phase en collaboration avec le GRES.

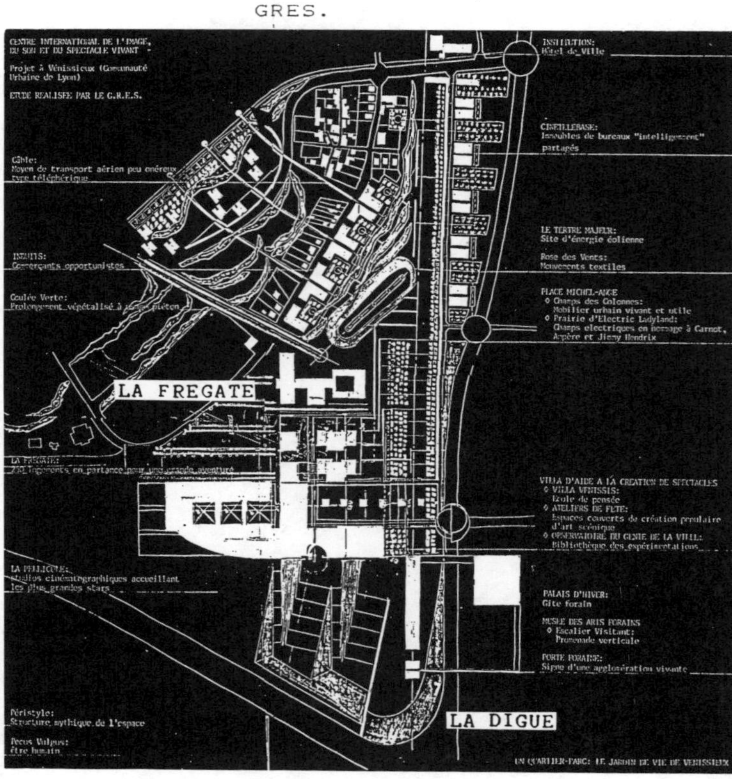

CREATION
UN CENTRE INTERNATIONAL DE L'IMAGE, DU SON ET DU SPECTACLE VIVANT

«Un Avenir pour Démocratie»

par le G.R.E.S.
(Groupe de Recherche et d'Etudes pour le Spectacle)

La précision de la poésie, l'enthousiasme de la science, la magie du spectacle inversent le pathétique de certaines énergies humaines en un merveilleux Jardin de Vie, un lieu populaire d'émotions quotidiennes.

La récherche d'une harmonie avec le cosmos oriente la composition du Jardin de Vie dans une continuité d'espace qui accueille l'ensemble des éléments du programme. Il rassemble les gens du spectacle, les métiers du cinéma, de la publicité, du cirque, des vacances, des parcs à thèmes, et des forains. Ce sont les lieux de la fête d'aujourd'hui, lieux de la dépense sociale où se produisent les savoir-faire les plus avancés de notre époque. C'est une belle occasion de renouveler le paysage de la ville, en créant une diversité de mouvements, de machines aériennes ou nautiques, permis par la haute technologie liée à ces savoir-faire : un moment judicieux pour inventer et construire les nouveaux langages de la construction.

La force de proposition de la démarche architecturale, la puissance de l'imagination transforment les tours des Minguettes à Vénissieux, au sein même de la métropole lyonnaise, en de fabuleux Géants protecteurs. Pour la circonstance, la confection d'habits de fête, de stores de toiles et d'architectures animées par les énergies naturelles et artificielles du site, sensibles au moindre frisson de vent, au premier rayon de soleil, grâce à des motorisations asservies par anémomètres et photopiles, un bel hommage de la Nature à la Ville, pour faire honneur aux extra-terrestres qui nous regardent».

UN EXEMPLE D'APPLICATION:

UN QUARTIER-PARC AUX MINGUETTES, VENISSIEUX METROPOLE LYONNAISE

Ce premier domaine d'application fut choisi en regard de la force émotionnelle d'une population en attente, regard-vidéo sur son histoire récente ou "poésie vidéo" du groupe "TRACTION AVANT"; et de l'énergie sensible du site qui lui a permis d'accueillir, il y a près de 30 ans, une opération d'une telle envergure dont le nom même de "DEMOCRATIE" était à l'époque le reflet d'une magnifique ambition.

La diversité des cultures ethniques présentes sur le site associés au métissage des cultures professionnelles permis par le Monde du spectacle constitue de toute évidence un ancrage essentiel de ce projet pour prendre en charge la mémoire locale, et proposer de beaux métiers d'avenir.

Par la formidable échelle de plan masse de ce grand ensemble, par la volonté première de relation au centre ville et l'atomisation du parcellaire environnant, ce majestueux balcon sur les collines Sud de Lyon, aujourd'hui seuil entre la ville et la campagne, est alors un bel exemple d'application **pour composer, dans l'espace et dans le temps,** la continuité **des mouvements de vie de la ville.**

Le Monde du Spectacle va donc inscrire les tours du quartier Démocratie dans un vaste domaine, suggérant et créant de nouvelles échelles, inventant de nouveaux langages sur les propositions des "Ateliers de fête", les réflexions nées à la "Villa Venissis". Entre l'Hôtel de ville et le Boulevard Urbain Sud une digue structurera l'ensemble de ce domaine, une magnifique promenade aérienne pour aller voir la mer. Un ensemble de logements, "La frégate", verticalité symbolique d'une aventure et une "Porte foraine" ancrent cette composition.

De la naissance du Cinéma à celle de Guignol, de la tradition des métiers du cirque aux arts forains le Monde du Spectacle contribue depuis plusieurs siècles à forger l'identité de la Métropole Lyonnaise.

Nos enquêtes auprès des gens du spectacle concluent, face à la saturation des studios de la Région Parisienne et de la Victorine, sur la nécéssité de créer une nouvelle infrastucture de recherche, de formation et de production autour des métiers du spectacle. La Métropole Lyonnaise, riche de sa mémoire, apparait comme un lieu privilégié pour cette implantation. La diversité de caractères et d'échelles de ses paysages souligne ses potentialités.

Dans ce contexte, notre projet est de définir un domaine au sein de la ville même pour y inscrire le Monde du Spectacle: du cinéma, de la publicité, du cirque, des marionnettes, des fêtes foraines...Réponse noble pour la ville, nous voulons par une démarche poétique que ce domaine accueille et organise le plaisir d'une vie de quartier dans un jardin en harmonie avec les éléments naturels: le vent, le soleil, la terre et l'eau. Imaginer et créer un lieu populaire de bien-être et d'émotions quotidiennes.

Le Monde du Spectacle, fêtes d'aujourd'hui, permettra au travail architectural de renouer avec la convivialité et la force de proposition des lieux de fête désignés de tous temps comme lieu d'expérimentation et de création de la ville.

Le Monde du Spectacle joue alors le rôle d'un véritable organe producteur de sens et de savoir-faire . Deux enjeux de l'évolution actuelle de la ville vont permettre d'illustrer ce propos:

- LE SENS DU MOUVEMENT

La haute technologie des émissions du son et de l'image peuvent donner, au delà de la sécurité, une continuité virtuelle à nos parcours et à nos échanges.

La haute technologie des métiers forains permet aujourd'hui de survoler les carrefours et les fleuves, d'entrer dignement

LE SENS DU MOUVEMENT

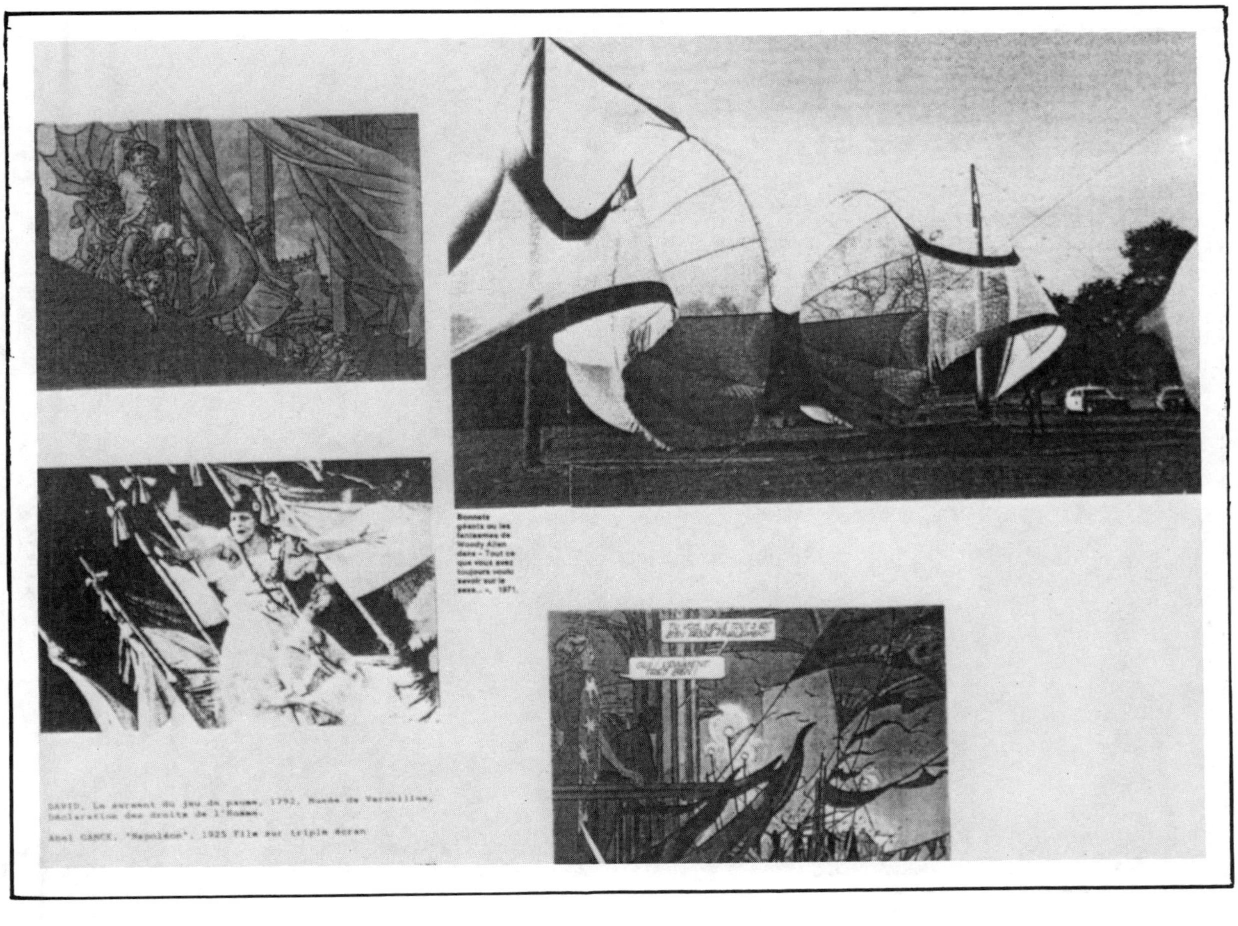

DAVID, Le serment du jeu de paume, 1792, Musée de Versailles, Déclaration des droits de l'Homme.

Abel GANCE, "Napoléon", 1925 Film sur triple écran

par les toits de nos immeubles en prennant place dans des télésièges, des tobogans, des grandes roues... Un "flum-drive" et/ou un "paratower" nous offrira pour un édifice public le plaisir cent fois renouvel é d'un exercice d'incendie...

Une diversité qui ex alte concrètement notre sens du mouvement, qui porte nos projections à l'infini, enrichie nos perceptions sensibles et créent de nouvelles échelles entre Terre et Ciel.

- UN NOUVEAU LANGAGE DE L'ARCHITECTURE

La présence des "métiers" du spectacle permet de rassembler, au sein même de ce Domaine de la ville, les savoir-faire les plus avancés de notre époque.
Une occasion rêvée de dépasser les archaïsmes et/ou héroïsme des chantiers du bâtiment: en interprétant les structures astucieuses des cirques et des manèges; en s'inspirant des prototypes d'espaces, de substances et de textures des ateliers de fabrication de décors; en exprimant la haute technologie de la lumière et du son des lieux de tournage.
Une belle occasion pour ce laboratoire vivant de rechercher les nouveaux langage de l'architecture.

"Cité en scène" veut être une superproduction d'un domaine qui met en scène une dynamique culturelle pour l'expérimentation et la création de la cité d'aujourd'hui. Ainsi ce domaine du spectacle vivant, domaine de haute culture, domaine du sens et des sens, rayonnera sur l'ensemble de la ville, diffusant, dans l'espace et dans le temps, la richesse de son langage renouvel é et la fluidité de ses mouvements de vie, la joie de vivre des expériences poétiques.

QUINT BUCHHOLZ — Einstein meets Monroe, 1986

SOURCES:

. "Architecture de fête" W. OESCHSUN, A. BUSCHOW Ed.MARDAGA,1987
. "Deux essais pour la construction" B.HAMBURGER Ed.MARDAGA,1981
. "Le langage de l'architecture moderne" B.ZEVI Ed.DUNOD,1977
. "Animated architecture" Themepark Architectural Design,1985

De la "sensibilité de pellicule" de Jean VIGO aux émotions nostalgiques de "la Strada de FELLINI"; du sublime des cirques théâtres de la "FURA DELS BAUS" et d'"ARCHAOS" à la poésie pittoresque de "MIR CARAVAN", du carnaval de Jean-Paul GOUDE à la simplicité de l'architecture du cirque "ZINGARO".

UN NOUVEAU LANGAGE DE L'ARCHITECTURE

ECONOMY OF TRANSPARENT INSULATION MATERIALS THROUGH THERMAL COMFORT, DAYLIGHTING PERFORMANCE, AND ENERGY SAVING

W.-S. Wilke

Fraunhofer-Institut für Solare Energiesysteme
Oltmannsstr. 22, D-7800 Freiburg i.Br. (FRG)

ABSTRACT: Transparent insulation material (TIM) combines high translucence with very low heat loss coefficients. This property allows architectural concepts with a high solar heating contribution, extremely comfortable indoor conditions, and substantially better daylighting than conventional facades can ever achieve. Meaningful applications and boundary conditions for an economical use of TIM are pointed out.

1. INTRODUCTION

A new class of transparent insulation material (TIM), combining high translucence with very low heat loss coefficients, allows totally new architectural concepts /1/. The main functions of a building facade are:

1. Protecting the building interior from the exterior climate, aiming at minimum heating and cooling energy consumption,
2. Satisfying thermal comfort requirements by achieving optimal inside surface temperatures,
3. Creating a daylighting atmosphere, and
4. Providing visual contact to the environment.

Using conventional facade elements (opaque walls in combination with window glazing), the points 1. and 2. strongly oppose point 3., whereas point 4. can usually be achieved by relatively small window areas. In addition to these window areas, intelligent TIM elements with shading devices turn out to fulfill the requirements 1. to 3. in an optimal way.

2. TIM FACADE CONCEPTS

In principle, TIM can be used in three different forms.

2.1. TIM daylighting system

Fig.1 shows a typical arrangement for a daylighting facade element with its thermal and optical properties. The material scatters the light, which leads to high penetration depths into the room and avoids reflections on monitor screens, but also causes glare and high brightness contrasts with direct sunlight. Roller or venetian blind systems avoid such unpleasant effects, protect the room against overheating in summer, and improve the U-value during winter nights.
As can be seen from the thermal properties, the daylight gaining area is no longer restricted by thermal comfort requirements ("cold" window glazings).

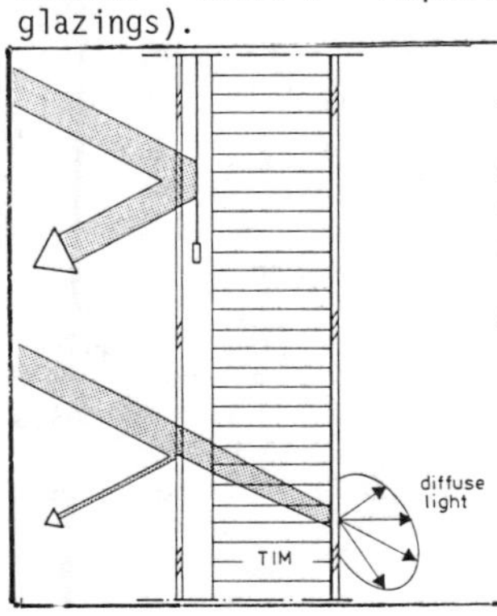

Fig.1: TIM daylighting system: tpical arrangement with 12 cm honeycomb structure and a highly reflective roller blind;
$U_{day/night}$ = 0.7/0.5 W/m^2K
light transmittance: 0.5

2.2. TIM wall heating system

Putting a massive wall behind the TIM facade leads to an attenuated and delayed delivery of the collected solar energy by heat conduction. The time lag can be roughly determined from Fig.2. The radiation from the warm inside wall is perceived by the inhabitants to be pleasant. Oriented to the south, such elements no longer have any heat losses worth mentioning. In combination with directly gaining elements, high solar energy utilization is possible.

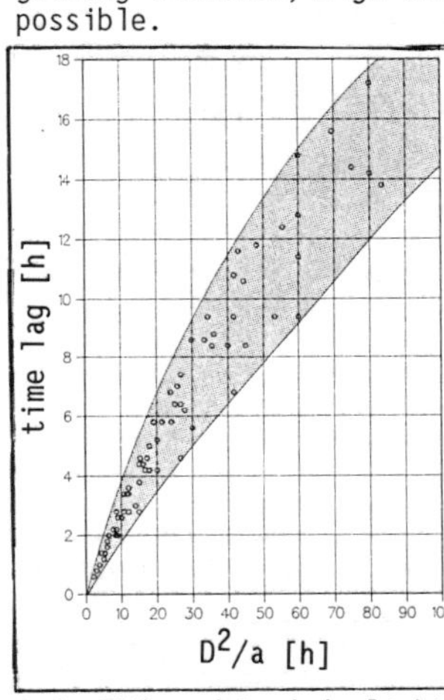

Fig.2: TIM wall heating system: time lag of delivered solar energy, obtained from various simulations (D:thickness, a:thermal diffusivity of the wall)

2.3. TIM combined daylighting and wall heating system

When replacing the opaque wall by a translucent wall (e.g. glass tiles), the advantages of 2.1. and 2.2. can be ideally combined. A high TIM potential exists especially in retrofitting glass tile facades.

3. ECONOMIC ASPECTS

When comparing different concepts for a building construction, two classes of economic considerations, as shown in Fig.3, have to be taken into account.

3.1. "Numeral" related aspects

Assuming some developments for the different influences in Fig.3, this class of arguments can be directly represented in numbers. However, the estimation certainty decreases for the topics from left to right. Every economy computation therefore has a sensitivity range, most dependent on the energy price. One should observe that the other points are also influenced by the energy price. Increasing energy prices lead to general inflation effects and will also encourage laws limiting the energy consumption (to avoid, for example, social conflicts between owners and tenants of apartments). At the moment, it can probably be expected that all the global and environmental problems will force politicians to substantially increase energy prices by taxes within the coming years. Social costs cannot be excluded from future energy price scenarios. That is why "numeral" related economy studies mean, in the first place, energy price predictions.

3.2. "Human" related aspects

Beside these factors, the "human" aspect, influencing economy less obviously but probably much more strongly, has to be considered. The productivity of human beings, which is the largest cost factor in most cases, is coupled to the points listed in Fig.3. At the moment, thermal and optical comfort relations are most thouroughly investigated, whereas the impact of recreation conditions or psycho-social effects on productivity is relatively unknown and can only be assessed subjectively by the individual concerned.

Some examples for scientifically proven facts:
People are more productive
- with lower air temperatures and at the same time higher surface temperatures in the winter, with lower air and surface temperatures than ambient in the summer (comfort temperatures above 24 °C reduce productivities dramatically)
- in rooms which are well illuminated, mainly by daylight
- with pleasant brightness contrasts and no glare and reflection effects,
- with good visual contact to the environment,
- in well ventilated rooms.

These criteria seem to be trivial but cannot be fulfilled by conventional concepts, as can be experienced in many existing buildings. Intelligent TIM facades will help us to achieve more comfortable, and thus more productive, and, finally, more economic buildings.

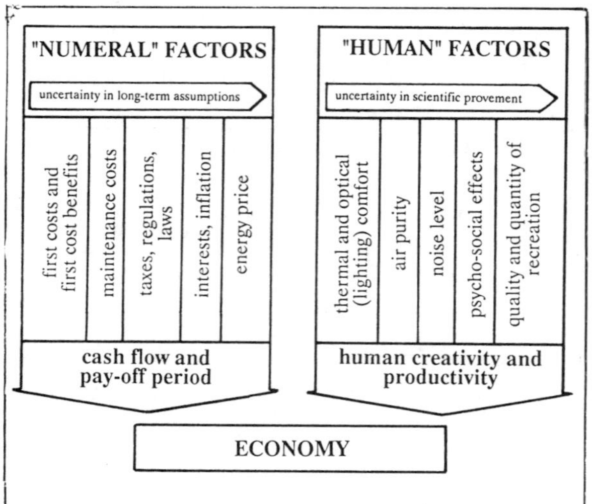

Fig.3: The two different economic factors

4. EXAMPLES

Various simulation programs have been developed at the FhG-ISE /2/ to predict the thermal behaviour of TIM-equipped buildings (WANDSIM, HAUSSIM), comfort temperature and illuminance distribution in rooms (LICOM), and daylight utilization times (DUF). On this basis, some examples are shown in the following sections.

4.1. Office room

Fig.4 shows the floor plan, cross section, and facade view of a 24 m² office room.

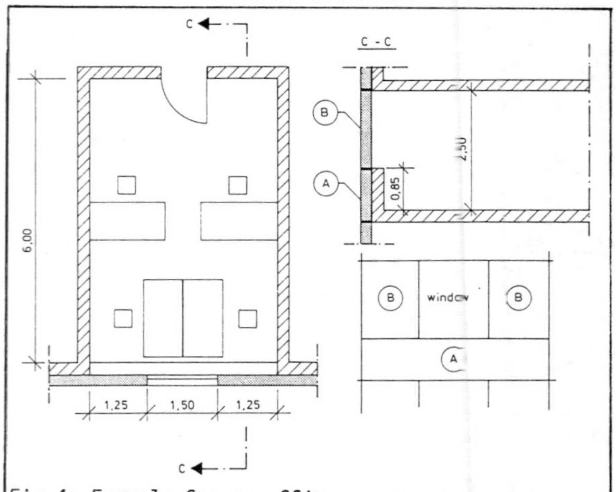

Fig.4: Example for an office room

The window consists of double glazing with a semitransparent roller blind (U=2.8/1.7 W/m²K). Assuming that the window area and placement is fixed by visual contact requirements, the architect can vary the properties of the segments A and B. The following three cases are considered:

I. A and B: opaquely insulated (U=0.5 W/m²K)
II. A: opaquely insulated as above,
B: double glazing
III. A: TIM wall heating system (concrete wall of 18 cm thickness, time lag 3 h),
B: TIM daylighting system as described in 2.1.

Fig.5 shows the radiant temperature and daylight factor for all cases.

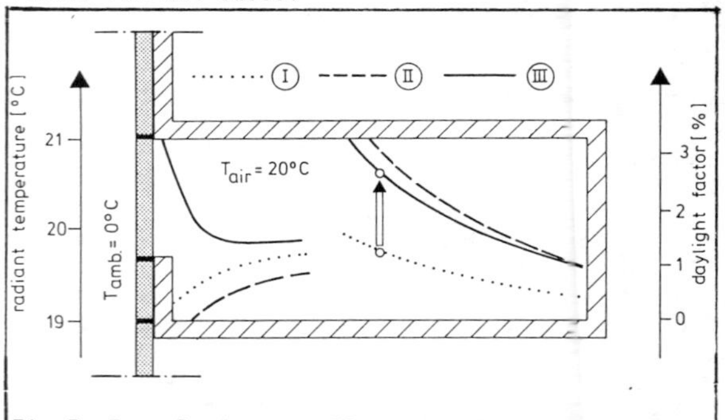

Fig.5: Example for an office room: daylight factor and radiant temperature on an overcast winter day

The incompatibility of I. and II. with human comfort is obvious. In case I. the work places in the back part are poorly illuminated, in case II. the area near the window is uncomfortably cold. TIM (case III.) gives a warm balustrade of 22 °C and scatters the light deeply into the room, so all workplaces are attractive. It is noteworthy, that the TIM daylighting system is able to compensate its lower light transmission (50% in contrast to 72% of the double glazing) by the scattering characteristic.
At a point located at half the room depth, the daylight factor can be increased from 1% to 2.5% (Fig.6). The possible reduction in artificial lighting time is nearly 2000 h/y with a required minimum illuminance of 250 lx. Assuming four fluorescent lamps, each of 18 W (operating costs 0.005 ECU/h each), a cash flow of 10 ECU/(m²TIM y) for the TIM daylighting system can be expected. Demanding a pay-off period of ten years, the additional costs of the TIM system may lie around

100 ECU/m². This value can be reached with the currently available materials. The TIM wall heating system is, in this case, not economic under "numeral" aspects but increases "human" comfort and is relatively easy to achieve in combination with the daylighting elements (uniform facade technology).

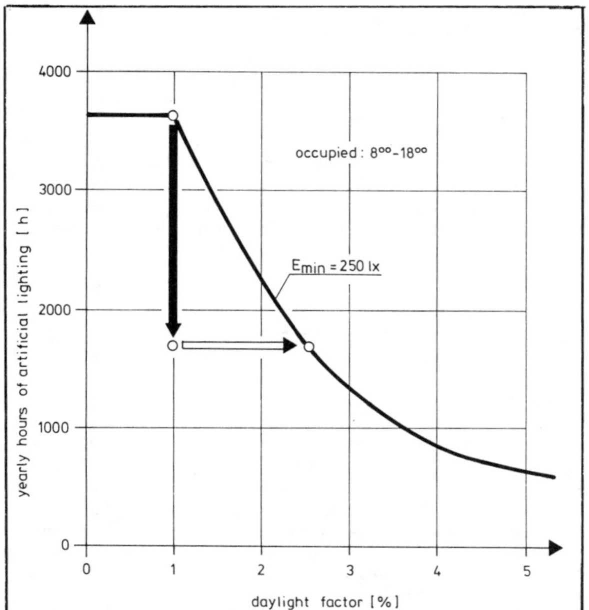

Fig.6: hours of artificial lighting vs daylight factor for a continuously overcast sky

4.2. Apartment

The second example demonstrates the performance of the TIM wall heating system for residential homes. The outline of a living unit in the middle of an apartment complex with the geometric and thermal properties is given in Fig.7.

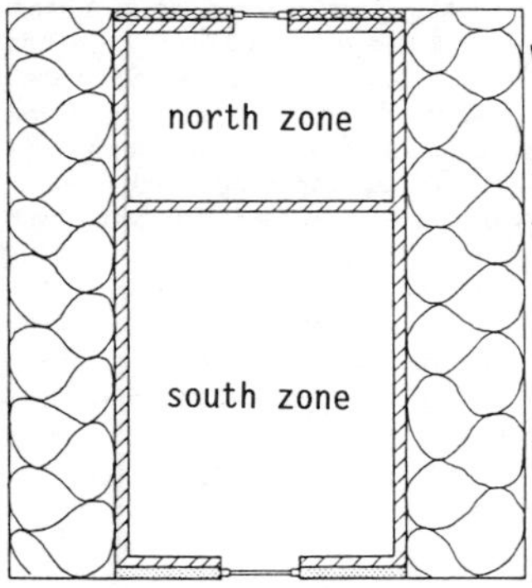

living space: 140 m^2
windows: low-e double glazing with blinds ($U=1.6/1.0$ W/m^2K)
opaque insulation: $U=0.2$ W/m^2K
TIM wall: 10 cm honeycomb with blinds ($U=0.60/0.36$ W/m^2K)
internal gains: 6.3 kWh/d
heat recovery: 70 % efficiency
adiabatic conditions to all sides except south and north

Fig.7: Example for an apartment

The southern brick wall (21 m^2) is in case
I. very well ($U=0.2$ W/m^2K) opaquely insulated,
II. transparently insulated (time lag 8 h).

Looking at the yearly heating energy consumptions of I. 980 kWh, and II. 280 kWh, the cash flow for the TIM investment amounts with an oil price of 0.25 ECU to 0.83 ECU/(m^2TIM y) which is almost negligible. The situation changes completely when the free floating room air temperature without heating is examined (Fig.8). The indoor climate for the TIM case is in such a comfortable range that a heating plant becomes dispensable. For case I. it is still necessary. The economic viability of TIM is achieved here by benefits from the heating plant investment ("numeral" aspect) and the feeling of complete independence from fuel prices and availabilities ("human" aspect).

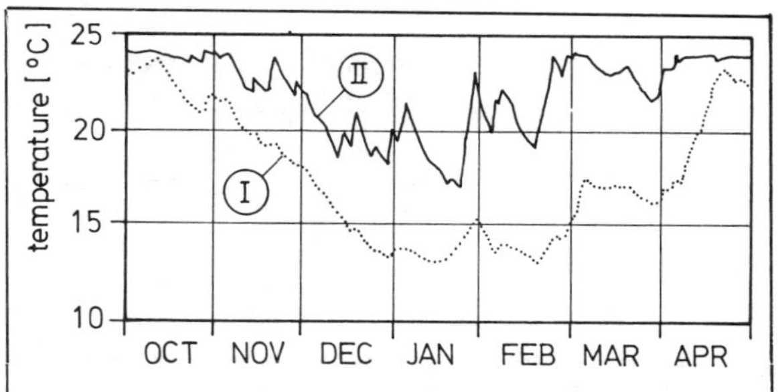

Fig.8: Example for an appartement: free floating room air temperature without any heating

5. CONCLUSIONS

As the examples showed, at the moment there are more weighty arguments for TIM than the energy saving percentages, which will not become important until dramatic energy price increases occur. However there are already economic applications for TIM which are summarized here in the form of recommendations for project planners and architects:

- TIM itself is relatively cheap; integrating TIM in expensive facade elements (glass facades, windows) very often has energetic effects which are already economic.
- TIM daylighting system: main use for office and factory buildings; best suited for skylights and window areas above head level, use below head level only with movable shading for glare protection; scattering most effective in deep rooms and surrounding obstructions.
- TIM wall heating system: main use for residential homes; if heating plant can be replaced by TIM and heat recovery system, then economic; energetic effect in office rooms only with extremely high air exchange rates (e.g. conference rooms) and low internal heat gains; biggest effect on thermal comfort below "cold" windows and in heights up to 1.0 m; often a solution for condensation problems.
- TIM can be used without shading and is therefore more economic in applications requiring high indoor temperature levels (e.g. indoor swimming pools, sauna and fitness centers).
- TIM systems achieve highest utilization factors in extremely cold but sunny climates (high altitudes, e.g. mountain refuges).
- TIM systems provide tepid temperatures with high reliability (condensation protection, freezing protection for water pipes) in irregularly occupied buildings (e.g. holiday apartments).
- TIM systems make rooms more useful which are difficult or uneconomic to connect to a fossil fuel heating system (e.g. sunspaces).

6. REFERENCES

/1/ Schmid J.
New Concepts for Daylighting and Heating of Buildings
Proc. of the 2nd European Conference on Arch., Paris 1989, to be published

/2/ Sick F., Wilke W.S., Kummer J.P.
Simulation Tools for Prediction of the Thermal Behaviour of Transparently Insulated Buildings
Proc. of the 2nd European Conference on Arch., Paris 1989, to be published

A STUDY OF DAYLIGHT AVAILABILITY IN ATHENS, GREECE

Dr G.N. Fatseas
Architect Engineer NTUA, MSc EDE UCL, MTCG MGAA MARCUK
Bartlett School of Arc. & Planning, University College London.
Home address,
24 Evanthias Kairi Street, GR 112 55 Athens, Greece, tel: (01) 2286 545.

This report explores the validity of existing theoretical sky models developed to describe and predict luminance distribution and global horizontal illuminance, for Greece. The validation is based upon measurements carried out over a period of sixteen months, (from 1 April 1987 to 17 July 1988), following a six month pilot study since October 1986. The IESNA RP-21[5] sky model produces the most satisfactory correlation with measured data of global horizontal illuminance for clear sky conditions, 80% of the measured values were within ±10% of the theoretical values.

1. Purpose of work.

This report is part of a PhD research programme[1] carried out at the Bartlett School of Arch. & Planning of University College London, (UCL) and it was partially assisted by the National Technical University of Athens, (NTUA). The research has three major parts. The first part is a study of the daylight availability in Greece[2]. The second part explores the validity of existing theoretical sky models[3-13] developed to describe and predict luminance distribution and global horizontal illuminance, for Greece. Finally the third part develops guidelines for the formulation of a 'Daylighting Code of Practice for Building Design Applications', for Greece. The above research is considered as the necessary base to establish a computer-aided prediction method for the lighting properties of interior spaces in Greece during the building design stage. This environmental parameter prediction process will enable architects to appraise the likely lighting performance in proposed 'physical-systems-buildings' at the early stages of the design process. Thus, a better energy efficiency will be possible by optimising the use of daylight. However, it is of paramount importance that daylighting should be treated as a particular design strategy and not merely as an energy conserving technique. The provision of an energy efficient working environment that ignores user satisfaction, productivity and well being (often related with successful daylighting) is a false economy.

It is **the second part** of the above research that is described here.

2. Methodology of approach.

In Greece, weather factors such as temperature, precipitation, relative humidity, wind speed and direction, cloudiness, and sunshine hours have been recorded in Athens since 1901 by the Meteorological Institute[14]. However, lighting data of the Greek sky vault over a long period have never been recorded. Before the main area of the project could then be explored it was necessary to study the sky conditions in Greece over an extended period. To serve the aims of the research programme and partially comply with CIE TC 3.07[15] recommendations the collection of the following daylight data was decided *:

1 the luminance distribution of the sky vault (cd/m^2).
2 the total horizontal illuminance, sun and sky (lux).

Thus, in order to proceed the development of appropriate instrumentation to gather the above data was essential. The instrumentation was designed and constructed following a critical appraisal of similar approaches on relevant research areas. Known relevant instrumentation developed during 1912-1989 is listed below :

1 F.Herxheimer[18], Berlin W.Germany 1912.
2 J.Longmore[19], UCL London UK 1964.
3 J.Wegner[20], TU Berlin W.Germany 1975.
4 H.Nakamura, M.Oki[21], NIT Nagoya Japan 1975.
5 P.J.Littlefair[22], BRS Garston UK 1980.
6 P.R.Tregenza[23], NU Nottingham UK 1980.
7 S.Treado, G.Gillette[24], NBS Washington DC USA 1981.
8 E.Rowlands, D.L.Loe, N.Brickman[25], UCL London UK 1984.
9 J.B.Murdoch[26,27], UNH Durham USA 1984.
10 SERI[28,29], Colorado USA 1986.
11 M.S.Rea[30,31], NRCC Ottawa Canada 1987.
12 E.Krochmann[32], PRC Gmbh Berlin W.Germany 1988.
13 S.Hayman[33,34],US Sydney Australia 1989.

* Since the commencement of the project in January 1985 the IESNA[17] late in 1985 and CIE[16,40] in 1987 produced interim recommendations of measurement practice for similar studies including a "basic" amount of lighting data to be recorded.

The study of the daylight availability at DMS, Athens (Daylight Measurement Station, (l) 38°00'36"N, (L) 23°44'40"E, Altitude 115.00m above Mean Sea Level, Standard Meridian 30°E, +2hrs GMT), was carried out over a period of sixteen months, (from 1 April 1987 to 17 July 1988), following a six month pilot study since October 1986. The basic data corrected for calibration drift and temperature variations comprise 6,720 measurements of global horizontal illuminance and 349,184 measurements of sky luminance distribution recorded daily on a half-hourly basis from 9:00 to 16:30 Local Clock Time (see Fig 1). The efficiency of the DMS over the above period (percentage of days measured over the total number of days operating) was 88.6% for illuminance readings and 52.3% for luminance readings (ie. 420 and 248 days read out of 474 for illuminance and luminance respectively). The measured values were statistically analysed and compared with predicted ones produced by the following theoretical sky models:

1 CIE, standard skies[3,4].
2 IES RP-21[5,6] (IESNA).
3 Intermediate sky, H.Nakamura, M.Oki[7,8] (NOHI).
4 Homogenous sky, R.Kittler, N.C.Ruck[9,10].
5 Average sky, J.Wegner[11], S.Aydinli, E.Krochmann[12], P.Littlefair[13].

3. Significant results.

The whole system was tested and further improved, on site in Athens, since January 1986. Main problems solved to increase reliability and accuracy, were : (i) Mechanical and electrical ones concerning rotation and positioning accuracy, (ii) electronic ones coming from the microcomputer A/D converter's sensitivity to electrical interference. Its accuracy is increased enormously by replacing the existing reference diodes with a ZN 423 device[35] . (iii) Finally problems with temperature variations which in turn influence the photocell's output. From experimentations Megatron photocells suffer a decrease in their output signal of the order of 0.9% per 5 degrees Centigrade rise. Their response in real terms (the variation of dry temperature in Athens is -2°C to +40°C) for a long time period is not yet known. However, if they are left exposed under high heat levels (above 45°C) for more than a month they cease to function. Considering the various parameters influencing the approximation of the final produced data such as, positioning ±1%, levelling ±0.1%, inherent inaccuracies of the photocells ±7%, A/D conversion ±6.5%, temperature fluctuation +3% to -8%, calibration drift (fatigue) 5% and finally calibration errors 3% it is estimated that the present accuracy of the system as regards the final measurements is of the order of ±11.6% to ±1[illegible].8% (root mean square, RMS). This still needs to be reduced down to ±8% which is certainly adequate for preliminary studies. Furthermore, it may well be acceptable for the ultimate application.

It is not unusual for instantaneous measured values to be more than twice or less than half the mean value given by a daylight availability equation. For 85% of the working year (9:00 to 5:00 LCT five days a week) the measured global horizontal illuminance value exceeded 16 klx under any sky condition. The IESNA RP-21[5] sky model produces the most satisfactory correlation with measured data of global horizontal illuminance for clear sky conditions, 80% of the measured values were within ±10% of the theoretical values (see Fig $2_{a,b}$). The model depends primarily upon the position of the sun in the sky vault. Other factors such as atmospheric turbidity and solar illumination constant are considered fixed values. Thus, the model enables less experienced users to apply the prediction. Furthermore, it has now a known range of accuracy which is not dependent upon assumptions made by the user that may influence the result dramatically. The CIE[4] standard overcast sky luminance distribution model produces relatively acceptable results but overall the homogeneous sky model by Dr R.Kittler[9] describes all sky conditions more accurately (±40%). In between these extremes (clear and overcast skies) the cloudy sky condition occured for 30.7% of the measurement period. 60.4% of the measured global horizontal illuminance values under cloudy sky conditions were within ±40% of the theoretical values produced by the IESNA model. Also the same model was able to describe the overcast sky conditions satisfactorily. Thus, for such sky conditions it is suggested to apply the IESNA partly cloudy sky formula and reduce the obtained result by 40%, ie. <u>multiply the IESNA partly cloudy formula by 0.6</u> . The Kittler model (T_{il}=2.45 for clear sky, T_{il}=6 for cloudy sky and T_{il}=71.6 for overcast sky) is a very promising approach but it is influenced greatly by solar altitude . The recommended correction factors are unavailable in Greece. For both, global sky illuminance and sky luminance distribution 'composite' models (eg. see ref. 3,4,5,6,9,10) performed better than statistical models (eg. see ref. 7,8,11,13) since the latter have been based on measurements made in specific locations with local daylight profile. Also the possibility of the application of a 'CIE Overcast Sky' model is very limited since the assumptions upon which such a model was based are realised very rarely in Athens. In other words the Stratus cloud type and the ratio of zenith to horizon sky luminance 3:1 are not common in the locality (less than 2% of the year). Regarding atmospheric pollution, although only a limited amount of sky data were collected, a reasonable assessment can still be made for the clear sky condition, ie. a reduction by 13% overall.

The mean luminance efficacy value of 118±13 lm/W under all sky types was computed as the average value of instantaneous luminous efficacies.

The most appropriate classification of sky conditions according to the okta scale is as follows;
(1) clear 0-2 oktas,
(2) cloudy 3-6 okta and
(3) 'darker' cloudy (overcast) 7-8 oktas.

In order to establish a more firm opinion about the range of expected accuracy from various models further daylight availability studies, and more collaboration between researchers in relevant fields are necessary. Also the gradual but alerting increase of atmospheric pollution and the over-expansion of the city will ultimately influence dramatically the microclimate of the site and hence will effect the local daylight climate. Thus, any proposed alteration of existing formulae based on short term daylight availability measurements will have a limited application period and a possible long term inaccuracy. Since almost 50% of the year a clear sky is the prevailing sky condition it is essential to investigate further the factors that influence clear sky models. Also if manual techniques are to be recommended for the calculation of daylight availability in interiors the application of simple but reliable formulae suitable for '<u>non experienced users</u>' are of paramount importance.

The CIE designated recently (at the 1987 Venice meeting) 1991 as 'International Daylight Measurement Year'[36,37,38] aiming at this attempt. Finally the research proved that 52% of the working year the sky in Athens is clear providing a horizontal illuminance level range of 6.7 klx to 116.5 klx. The potential of daylight as a light source is significant not only due to its magnitude but also to its inherent quality. Unfortunately the majority of the current building design in Greece continues to neglect daylight[39,1].

REFERENCES

1 G.N.Fatseas, 'A Study of Daylight Availability as an Aid to Natural Lighting Conscious Building Design in Greece', PhD Thesis, submitted June 1989, University College London.

2 G.N.Fatseas, 'Computer-Aided Recording of Sky Luminance and Illuminance Data Over an Extended Period of Time in Greece', Proc. CIBSE Nat. Lighting Conf., Cambridge 1988, pp.316-330.

3 Commission Internationale de l'Eclairage, (CIE), "International Recommendations for the Calculation of Daylight", CIE No.16, 1970.

4 Commission Internationale de l'Eclairage, (CIE), "Standarization of Luminance Distribution on Clear Skies", CIE No.22, 1973.

5 G.Gillette, W.Pierpoint, S.Treado, 'A General Illuminance Model for Daylight Availability', Journal of Illuminating Engineering Society of North America, July 1984, pp.330-340.

6 IES Calculation Procedures Committee, 'Recommended Practice for the Calculation of Daylight Availability', Journal of IES of NA, July 1984, pp.381-392.

7 H.Nakamura, M.Oki, 'Mathematical Description of an Intermediate Sky', Proceedings 21st CIE Session, Venice 1987, pp.230-231.

8 H.Nakamura, M.Oki, "Composition of Mean Sky and its Application to daylight Prediction', Proceedings 20th CIE Session, 1983, pp.D303/1-D/303/4.

9 R.Kittler, 'Luminance Distribution Characteristics of Homogenous Skies', Lighting Research & Technology, Vol.17, No.4, 1985, pp.183-188.

10 R.Kittler, N.C.Ruck, 'Definition of Typical Average Exterior Daylight Conditions in Different Climatic Zones', Energy & Buildings, Vol.6, Nos.2,3&4,pp.253-259.

11 J.Wegner, BRE translation from original German text, "Calculation of the Average Interior Daylight Illuminance on the Basis of the Average Sky Luminance Distribution", his PhD Thesis submitted to the Technical University of Berlin, 1975.

12 S.Aydinli, J.Krochmann, 'Guide on Daylight', Chapter8: Data on Daylight and Solar Radiation, 3rd Draft August 1983, Institut für Lichttechnik TU Berlin.

13 P.Littlefair, 'Designing for Daylight Availability Using the BRE Average Sky', CIBS National Lighting Conference, Warwick 1982, pp.40-62, .(BRE Rep.R2/82).

14 National Observatory of Athens, Meteorological Institute, "Climatological Bulletin Year 1981", pp.173. (37° 58' 18" N, 23° 43' 30"E, Alt. 107m.)

15 CIE Technical Committee 3-07.

16 P.R.Tregenza, 'Guide to Recommended Practice of Daylight Measurement, General Class Stations', CIE research note, suppl. to CIE-Journal, Vol.6, No.2, 1987.

17 U.S. Working Group on Daylight, 'United States Standard Practice for the Measurement and Reporting of Daylight Availability Data', Florida SECenter, 1985.

18 F.Herxheimer, 'Ein Photometer zur Bestimmung der Himmelshelligkeit', Zeitschrift für Instrumentenkunde, XXXII, Feb.1912, pp.55.

19 J.Longmore, 'The Full-Field Camera Photogrammetric Technique in Lighting Research', Photographic Journal, 104,1964, pp.133-141.

20 Op.Cit. 11.

21 H.Nakamura, M.Oki, 'Measurement of Luminance Distribution Under Various Sky Conditions by Orthographic Projection Camera', Proceedings 18th CIE Session, London 1975, P-75-5, pp.493-502.

22 P.J.Littlefair, "The Luminance Distribution of an Average Sky", Lighting Research & Technology Vol.13, No.4, 1980, pp.192-198.

23 P.R.Tregenza, "A Simple Mathematical Model of Illumination from a Cloudy Sky", Lighting Research & Tech. Vol.12, No.3, 1980, pp.121-128.

24 S.Treado, G.Gillette, 'Measurements of Sky Luminance, Sky Illuminance, and Horizontal Solar Radiation', Journal of Illuminating Engineering Society, Vol.12, No.3, April 1983, pp.130-135.

25 E.Rowlands, D.L.Loe, N.T.Brickman, "Instrumentation for Measuring the Luminance Distribution within the Visual Field", National Lighting Conference Proceedings, CIBSE 1984, pp.187-192.

26 J.B.Murdoch, Robert Levin, "A Portable Sky Luminance Meter", Energy & Buildings, Vol. 6, Nos 2,3&4, 1984, pp. 241-246.

27 J.B.Murdoch, W.H.Kelton, 'Sky Luminance Measurement and Analysis', International Daylight Conf. Proc. I, Long Beach USA 1986, pp.9-13.

28 C.Robbins,SERI (Solar Energy Research Institute), 'SERI Pioneers Daylighting Research', Sunworld, Vol.10, No.4, 1986, pp.115-116.

29 N.L.Weaver, C.L.Robbins, T.W.Cannon, K.C.Hunter, 'Development of a Second-Generation, All-Sky, Video-Based Luminance Mapper for Daylighting Research', International Daylight Conference Proceedings I, Long Beach USA 1986, pp.445-448.

30 D.G.Kamblich, M.S.Rea, 'New Canadian Lighting Analysis System', Lighting Magazine, November 1987, pp.10-14.

31 M.S.Rea, 'Lighting Measurement and Calculation: Aview of Future Practice', Architectural Lighting, December 1987, pp.10-12.

32 E.Krochmann, J.Krochmann, 'Daylight Measurement Equipment', Proceedings CIBSE National Lighting Conference, Cambridge 1988, pp.302-315.

33 S.Hayman, 'Daylighting Models and Fibre-Optic Photocells', Lighting in Australia, February 1989, pp.16-21.

34 S.Hayman, 'Sky Scanning Luminance Meters: A proposal Using Fibre-Optic Photocells', to be published in LR&Technology.

35 A.D.Derrick, " A Hardware Guide for the BBC Microcomputer ", Wise Owl Publication, 1983, pp. 107-109.

36 R.McCluney, D.Kendrick, 'An International Daylight Measurement Year', International Daylight Conf. Proc. I, Long Beach USA 1986, pp.3-8.

37 H.W.Bodmann, H.H.Bjørset, Commission Internationale de l'Eclairage President and Vice-President letter to National CIE Committees, 18 Dec 1987.

38 D.Kendrick, 'International Daylight Measurement Year', Proceedings CIBSE Nat. Lighting Conf, Cambridge 1988, pp.299.

39 G.N.Fatseas, 'A Consideration of the Natural Lighting Design of Buildings in Greece', Millar Award Entry, 1988.

40 CIE T.C.307, J.D.Kendrick, 'Draft of Guide to Recommended Practice of Daylight Measurements Parts A (by Dr P.R.Tregenza), B (by J.J.Michalsky and R.Perez), C (by R.Perez, J.J.Michalsky, R.Seals), Draft No.3, Department of Architecture, University of Adelaide, Australia, 1989.

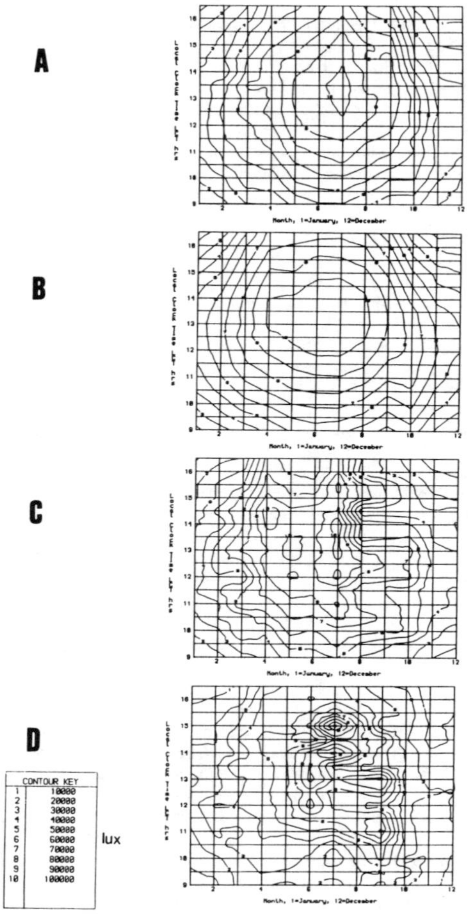

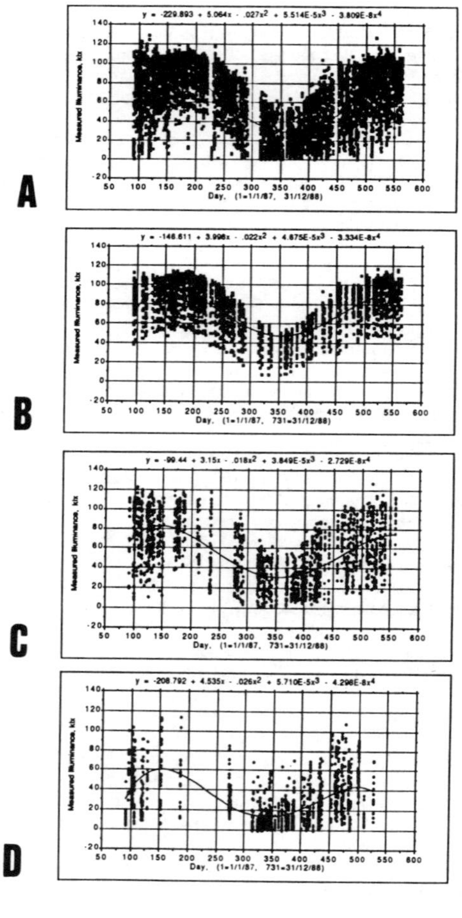

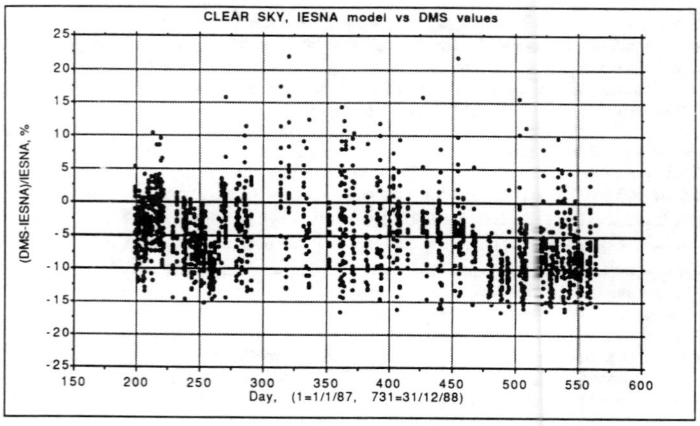

2a Measured percentage difference obtained during the selected one year of the measurement period (18/7/87-17/7/88) at DMS, between the measured values and the predicted values, ie. (DMS-IESNA)/IESNA %, for clear sky condition.

A

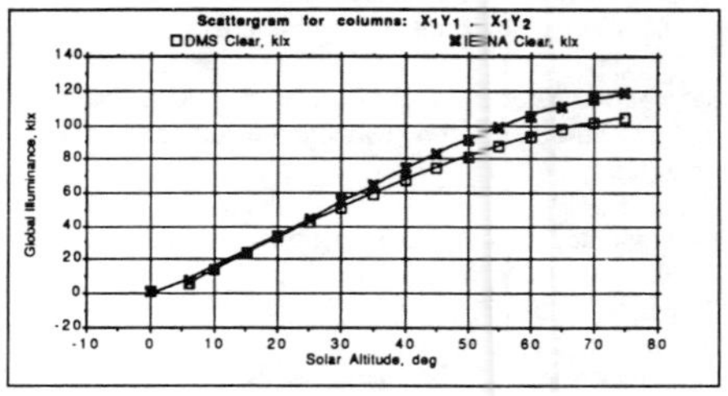

2b The relation of the IESNA model predicted and DMS measured illuminance values with solar altitude for clear sky (0-2 oktas) condition.

B

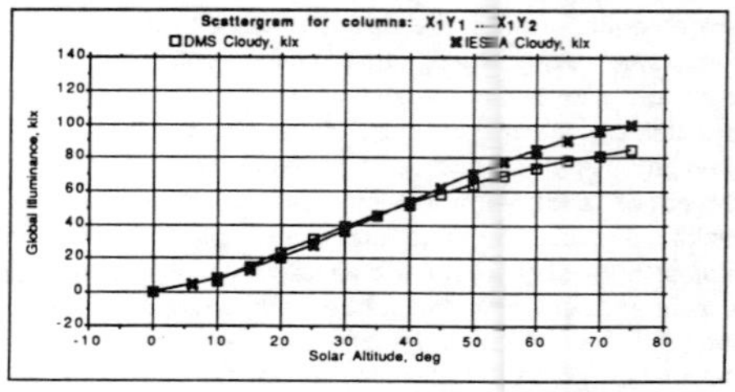

The relation of the IESNA model predicted and DMS measured illuminance values with solar altitude for cloudy sky (3-6 oktas) condition.

C

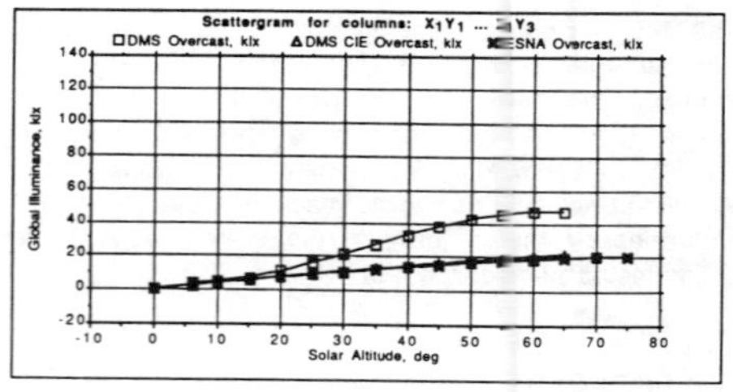

The relation of the IESNA model predicted and DMS measured illuminance values with solar altitude for overcast sky (7-8 oktas) and 'CIE overcast' sky condition.

1 Iso-lux contour plots of measured monthly average illuminance values against local clock time for (a) all sky types, (b) clear sky, (c) cloudy sky and (d) overcast sky conditions.

Range of illuminance values on a daily basis during the period 1/4/87 to 17/7/88 for (a) all sky condition types, (b) clear sky conditions, (c) cloudy sky conditions and (d) overcast sky conditions.

CONDENSATION TARGETER

by

Tadj Oreszczyn Phd
MInstE CEng MCIBSE
Lecturer
Bartlett School
Univ. College London
22 Gordon Street
London WCIH OQB, UK

David Boyd MSc PhD
CPhys MInstP MCIBSE
CEng, Senior Lecturer
School of Architecture
Birmingham Polytechnic
Perry Barr
Birminhgam B42 2SU, UK

Abstract

This paper describes a design tool, 'Condensation Targeter', for assessing condensation risk in dwellings and the effect of remedial measures thereon. The BREDEM energy model is augmented by a moisture model to determine mean internal relative humidity (MIRH). This measure of condensation risk is calculated for two zones in a dwelling from mean internal temperatures, moisture generation and ventilation rates. Primary input data relate to occupancy (fuel expenditure and moisture production) and dwelling characteristics (thermal and ventilation). MIRH results are presented as a function of space heating input for an example dwelling with remedial measures applied (insulation and extract fans). The constraints imposed by household income and the implications for condensation risk are discussed. Comparisons between model predictions and monitored results are discussed as is the sensitivity of predictions to the accuracy of input data.

1 Introduction

Mould growth occurs in 10 to 20% of housing in the UK and several other European countries. Considerable funds are now being allocated to the refurbishment of these buildings. At present some of this money is misdirected, as mistakes are made in the design and installation of anti-condensation measures. One reason for error is the lack of adequate design tools for making decisions on the best means of remedying condensation and mould growth in particular dwellings.

Research at many institutions around the world is now leading to a better understanding of the mechanisms causing condensation and mould growth. However, few comprehensive design tools are available which contain the most up-to-date thinking on remedial measure design. This paper outlines one way of addressing these problems through a combined energy and condensation model. The BREDEM energy model /1/ has been enhanced by incorporating Loudon's condensation model /2/. The final objective of this work has been to produce a computer package for use by building designers when refurbishing housing. This calculates the risk of condensation within dwellings using two main parameter inputs; dwelling characteristics and type of occupant (e.g. pensioners, unemployed families, single persons etc.) as illustrated in Figure 1. Remedial measures (e.g. draught stripping, central heating, etc.) can then be compared for their effect on condensation risk.

Condensation risk is here taken to mean both the risk of surface condensation and the risk of mould growth due to high humidity conditions on a surface without liquid water present.

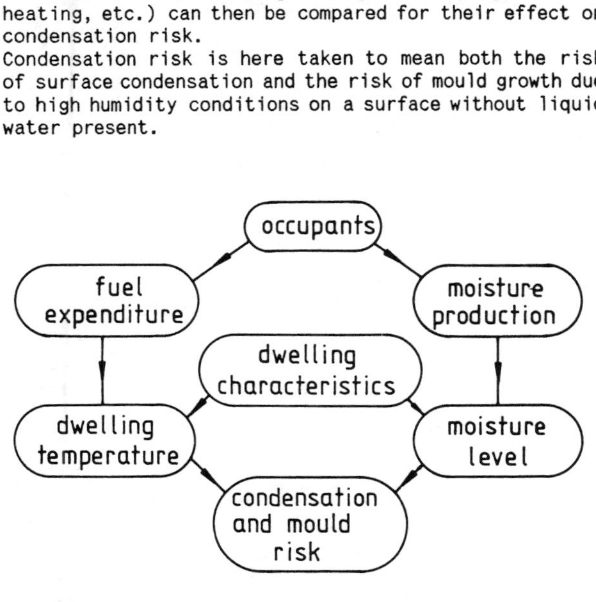

Figure 1 Model for condensation risk prediction

2 Model Selection

A condensation risk prediction model which can analyse the effects of changes to building design, moisture control and fuel expenditure requires components which determine the temperature and moisture content of the air inside the dwelling. In practice two integrated models are required; a thermal model and a moisture model.

Current thermal models which deal with energy consumption and internal temperatures in buildings range in complexity from those based on steady-state heat loss for hand calculation to large dynamic simulation models requiring mainframe computers. The latter are all complex and are unsuitable for use as design tools. However, several simple but sophisticated models have been developed to predict annual energy consumption. The Building Research Establishment (BRE) have designed the core of such a model, called BREDEM /1/, which the present authors considered to be the best vehicle for a condensation risk model. A major advantage is the availability of a very 'user friendly' implementation of version BREDEM-5, which is commercially available as the 'Energy Auditor' computer package/3/. The latter is already widely used as a design tool on refurbishment projects.

BREDEM is a two-zone modified degree-day model which accounts for casual and solar gains, and the efficiency and responsiveness of a variety of heating systems. 'Energy Auditor' was specifically designed for examining the energy savings associated with improvements to the building fabric and heating system of existing buildings. The accuracy of 'Energy Auditor' has been tested and reported by Henderson and Shorrock /4/.

Much less work has been done on moisture and condensation models than on thermal models, since energy has been far more of a national issue than condensation during the past two decades. The most generally accepted model for prediction of surface condensation and mould risk is that due to Loudon /2/. This model uses just two steady-state equations to determine mean internal relative humidity (MIRH); one for energy balance and one for moisture balance in a dwelling or zone. Condensation risk is considered to be unacceptably high if the MIRH in the space exceeds 70%. This work has helped to improve understanding of the interactions between the many factors involved in condensation; it has recently been introduced into a draft revision of BS 5250 /5/ as a simple design tool. The present authors have therefore chosen to incorporate a moisture balance equation into BREDEM.

3 Interaction of thermal and moisture models

The dwelling is treated as two separate zones as set out in BREDEM: the 'living area' taken as Zone 1 and the 'rest of the house' as Zone 2. The kitchen may be allocated to Zone 1 or Zone 2 depending on the layout of the particular dwelling. The thermal model calculates the mean internal temperature (MIT) of each zone, whole house ventilation rate and energy consumption over the heating season. The moisture model calculates mean internal vapour pressure for given moisture generation and ventilation rates in each zone. The division of the dwelling into two zones is an improvement on whole-house assessment since it allows modelling of different modes of heating and moisture production in different parts of the home. The structure of the model is shown schematically in Figure 2. For a detailed description of the model see reference 6.

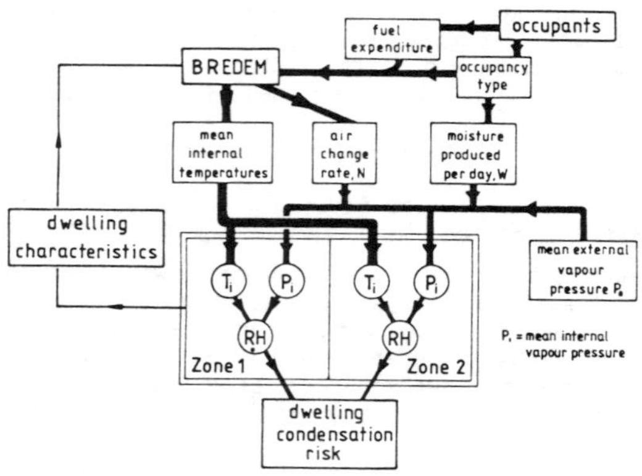

Figure 2 Integration of thermal and moisture models

4 Assessment of condensation risk

4.1 Influence of dwelling characteristics

An assessment of condensation risk must be made from the mean internal relative humidity (MIRH) of the two zones in the dwelling. Within a real occupied dwelling the risk of condensation or mould growth will vary widely due to spatial and temporal effects. BS 5250 asserts that the condensation risk will be unacceptably high when MIRH exceeds 70%. The validity of this limit is open to question, however, it does seek to account for differences between conditions in the bulk air of a room and those on the inside of external surfaces and for the fact that mould may grow where surface RH is less than 100%. Measurements in environmental chambers have shown that mould grows on wall covering materials if the relative humidity is greater than 80% for a period of several days /7/.

In this paper a value of MIRH=70% has been taken as critical with regard to condensation risk. However, this will underestimate the condensation risk in certain dwellings where the surface temperatures are low so that the surface RHs are greater than 80% with a MIRH lower than 70%. For example, external walls with U-values greater than 1.2 $Wm^{-2}K^{-1}$ or thermal bridges in well insulated walls may a have low surface temperatures which will result in mould growth at a MIRH of less than 70% /8/.

4.2 Influence of occupant income

Condensation is critically dependent on dwelling MIT which relates to occupant income through space heating input. For low income groups the amount of money and thus fuel available for space heating is limited. Hence, dwelling MITs decrease with decreasing income as confirmed by a number of studies /9/10/. Field study /10/ data has shown that decreasing income leads to a progressive fall in the temperature of Zone 2, while that in Zone 1 remains relatively constant. This implies that people with limited resources maintain a comfortable temperature in their living room while either reducing the heated area in the rest of the dwelling or reducing the temperature at which the latter is maintained. BREDEM is uniquely structured to model this practice. A number of BREDEM calculations can be performed for different areas of heating in Zone 2 and decreasing demand temperature in Zone 1.

5 Use of model

The integrated thermal and moisture models have been applied to a specific building to illustrate the operation of the condensation prediction method. To facilitate comparison of results with those of earlier studies, a mid-terrace two-storey house with the same construction as the example house used by Loudon /2/ was analysed, see reference 6 for constructional and thermal details. Two generic types of remedial measure are available for combating condensation in dwelling: those which increase mean internal temperature and those designed to reduce internal vapour pressure. An example of insulation (i.e. a measure which increases the MIT) is shown in Figure 3. Taking the original mid-terrace house to be occupied by two adults and three children producing 8.5 kg day^{-1} of moisture, MIRH of 70% in Zone 2 is maintained for a total space heating load of approximately 52 GJ y^{-1}. The space heating required to maintain the same conditions in the insulated house is reduced by more than half to 25 GJ y^{-1}.

Remedial measures which effectively reduce the input of moisture to the dwelling include the installation of kitchen and bathroom extract fans. Figure 4 illustrates the effect of this in the case of the unmodified house occupied by a five-person family. It is assumed that extract fans are fitted in the bathroom and kitchen which remove 80% of the moisture produced by cooking and bathing. The effect of the fan is to reduce the MIRH in Zone 2 by approximately 10% for a given space heating load. The corresponding decrease in the space heating fuel requirement to avoid condensation is 40%.

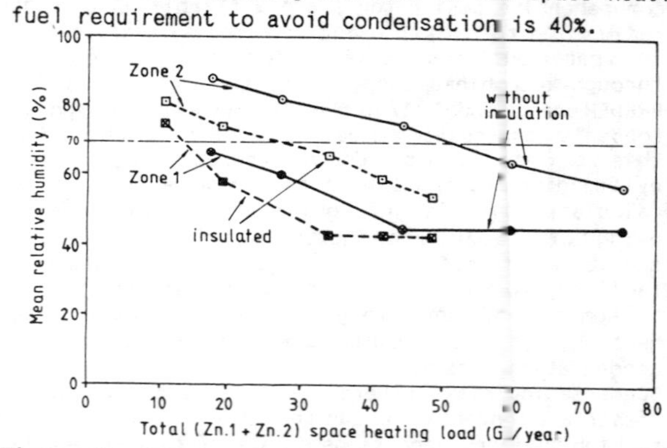

Figure 3 Zone MIRH for standard (U value - wall 1.7 W/m^2K, roof 1.4 W/m^2K) and insulated (U value - wall 0.6 W/m^2K, roof 0.35 W/m^2K) versions of two-storey mid-terrace house (occupied by a five-person family)

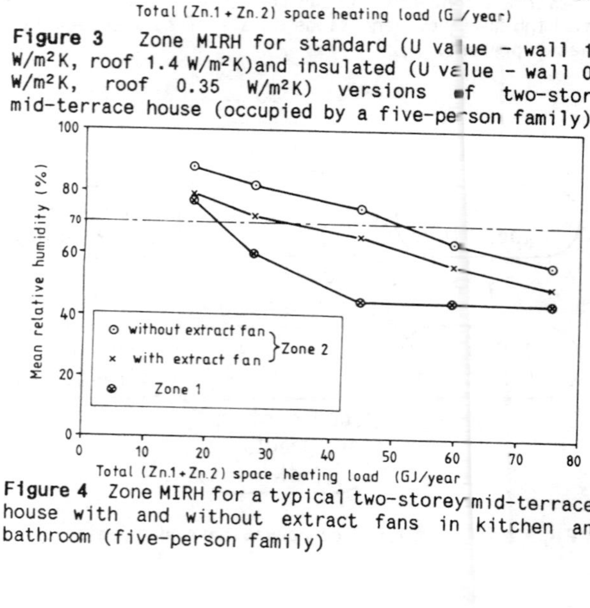

Figure 4 Zone MIRH for a typical two-storey mid-terraced house with and without extract fans in kitchen and bathroom (five-person family)

A more complex measure like draught stripping affects both MIT and vapour pressure. For a relatively leaky house with a mean ACR of 2.0 h^{-1} occupants must supply 52 GJ y^{-1} total space heating to avoid condensation in Zone 2. If draught stripping were applied and the ACR reduced to 0.5 h^{-1}, MITs are significantly increased by about 4°C for the same energy input. However, condensation risk is increased, not reduced, because of the accompanying rise in internal vapour pressure. In fact to avoid the condensation risk in Zone 2 heat input must be increased by about 27%.

It is important that designers are aware of how little households have to spend on fuel and how this relates to condensation risk. Consider a five-person family living in the house described above. If the two adults were unemployed the household income from supplementary benefit payments would only be £113 per week. On average they would only spend £9.29 per week or £483 per year of this on fuel /6/. The total cost of hot water heating, cooking, standing charges, lights and appliances as calculated by 'Energy Auditor' is £380 per year, leaving the hypothetical family with £103 per year to spend on space heating. This is equivalent to a heating load of 28.6 GJ y^{-1} (with gas central heating). Reference to Figure 3 shows that this family could not afford to heat their home sufficiently to avoid condensation without the installation of remedial measures such as wall insulation.

6 Discussion

The tool outlined in this paper enables designers to assess the condensation risk which might result from dwelling refurbishment. It enables quantitative comparison of how this risk may be reduced by various remedial measures, e.g. installation of insulation, efficient heating systems and mechanical ventilation. The approach highlights the necessity for dwellings to be appropriate to the needs of occupants; dwelling insulation and heating systems must allow comfort conditions and space heating fuel bills which the occupants can afford. The result of insufficient heating is that people endure reduced temperatures which may lead to mould growth, condensation and building fabric decay. The use of this model will help to identify failures in design (e.g. inappropriate draught-stripping) and so avoid costly maintenance of remedies.

Developments of this condensation risk model has been led by demand. The imperative of providing an easy-to-use design tool has resulted in a simplified approach which has several shortcomings and unvalidated assumptions. Work is in progress to appraise the assumptions and to validate the model against real data and experience. Two main areas of concern are the use of average MIRH and ventilation analysis.

The assessment of condensation risk by using temperature and moisture generation averaged over time and space within a dwelling may not be entirely appropriate. Time averaged MIRH may be sufficiently low to regard a dwelling as theoretically free from condensation, yet condensation may occur at troughs of temperature and at peaks of moisture generation. In addition, thermal and moisture capacity of the building fabric is neglected in the above model but intermittent heating and occupancy may result in condensation. Similarly, spatial averaging does not account for high condensation risk in specific rooms of a dwelling. Transient affects can not be fully accounted for until the response of mould growth to transient conditions is understood, this work is being carried out the Building Research Establishment.

Partitioning of the dwelling into zones, as effected by BREDEM, may not give sufficient resolution for predicting condensation risk. However, analysis of air and moisture exchange between the two zones and to outside is complex. In the above model it has been assumed that the two zones are decoupled for moisture migration. Allowing for inter-zone moisture transfer can alter the Zone MIRH by up to 5%. Improving this part of the model will depend on the development of an appropriate ventilation model requiring relatively simple dwelling and occupant data. Similarly, the efficiency of extract fans in reducing internal vapour pressure (under both humidistat and occupant control regimes) needs elucidation.

The predictions obtained from any model are only as good as the input data. The above model has therefore been tested to see how sensitive the predictions are to the accuracy of the input data. The model is most sensitive to the fabric loss where a 1% change in fabric loss can result in a 1% change in predicted RH.

Preliminary comparisons between modelled results and measurements in several dwellings has shown good agreement when predicting the average dwelling MIRH (i.e. modelled MIRH within 5% of measured). The Zonal predictions however have been in error by up to 8% suggesting that the model does not adequately account for moisture movement between zones or the allocation of moisture production to each zone.

The condensation risk model discussed in this paper, now called 'Condensation Targeter', has been integrated into a commercial computer package called 'Energy Targeter'/11/. Energy Targeter calculates the total energy cost, and total cost of remedial measures, as well as assessing the risk of surface and interstitial condensation. The integration of condensation assessment into a general refurbishment tool has meant that building designers who would not normally consider the quality of the indoor environment are made aware of condensation risks. Energy Targeter has therefore proved a useful vehicle for carrying research results into practice.

References

1 Anderson B R, Clark A J and Milbank N O *BREDEM-BRE Domestic Energy Model: Background, philosophy and description* Building Research Establishment Report (HMSO)(1985)

2 Loudon A G *The effects of ventilation and building design factors on the risk of condensation and mould growth in dwellings* Building Research Station Current Paper CP31/71 (1971)

3 *Energy Auditor Operating Manual* (Milton Keynes: Energy Advisory Services Ltd) (1983)

4 Henderson G and Shorrock L D BREDEM-The BRE domestic energy model: Testing the predictions of a two-zone model *Building Serv.* Eng. Res. Technol. **7**(2) 87-91 (1986)

5 Draft revision to BS 5250 British Standard Code of Practice: *The control of condensation in buildings* (Revision of BS 5250, 1975)(Milton Keynes: British Standards Institute)(1975)

6 Boyd D, Cooper P and Oreszczyn T Condensation Risk Prediction: Addition of a condensation model to BREDEM *Building Serv. Eng. Res. Technol.* **9**(3) 117-125 (1988)

7 Bravery A F, Grant C and Sanders C H Controlling mould growth in housing *Proc. Unhealthy Housing Conf.* University of Warwick December 1987

8 Oreszczyn T Cold bridging at corners: Surface temperature and condensation risk *Building Serv. Eng. Res. Technol.* **9**(4) 167-175 (1988)

9 *Condensation dampness* Research Project Report (Department of Architecture and Building Science, University of Strathclyde)(1985)

10 Hunt D R G and Gidman M I A national field survey of house temperatures *Building Environ.* **17**(2) 107-1224 (1982)

11 *Energy Targeter Operating Manual* (Milton Keynes: Energy Advisory Services Ltd) (1988)

Acknowledgement

This work was carried out while the authors were employed by the Polytechnic of Central London

COMPUTER MODEL FOR PREDICTING MAN'S THERMAL BEHAVIOUR

L. Imre; A. Bitai; C.D. Horváth and L. Bánhidi
Technical University Budapest
1521.Budapest, Hungary

ABSTRACT. A computer model and the software "HUMTHERM" /HMT/ for IBM PC is offered for the analysis of thermal human comfort problems aiming at the simulation of man's accomodation to changes in the ambient thermal conditions. By the use of the calculating and evaluating programs and of the accompanying data files the time functions of the main parameters correlating with the sensation of comfort /i.e. skin surface temperatures, surface heat flow rates, rate of perspiration/ can be computed.

1. INTRODUCTION

Thermal human comfort is of great importance in various fields of the activity of man influencing his intellectuel and physical performance. That is the reason why great efforts have been devoted to solve thermal comfort problems in housing, in industrial environments, in the city, and in the garment design [2...7]. Solution of comfort problems by mathematical models and numerical solution methods needs satisfactory informations about the human physiological responses to changes in the ambient conditions. The main point is to find a proper concept for the interpretation and description of the thermoregulation of the human body [8...18]. Various analytical and numerical models have been proposed [11,16, 19...23] based on various concepts. Considering the published data referred and the results of recent test measurements, a simulation model has been established considering the thermoregulatory processes of the human body and the simultaneous heat and mass transfer processes through the clothing and between the surface of the clothing and the environment [1]. Based on the model, the software "HUMTHERM" has been elaborated and tested. Herewith the peculiarities of the model and the description of the program is given.

2. MAIN FEATURES OF THE MATHEMATICAL MODEL

The human body is divided into 16 parts /Fig.1/. Body-part models consist of "core" and "shell" but the part 16 /heart and lungs/. The skin is considered as the surface of the shell. Blood flow through the large arteries and veins in the core is considered as adiabatic. Heat transfer to or from the blood is assumed to proceed in the capillaries of the shell. In the model blood is flowing from the heart into each body-part. Body-parts may or may not be covered by clothing. Five clothing layers can be taken into consideration in the model as a maximum. Heat and mass transfer related to the individual body-parts and the corresponding clothing is calculated separately /details are given in [1]/.

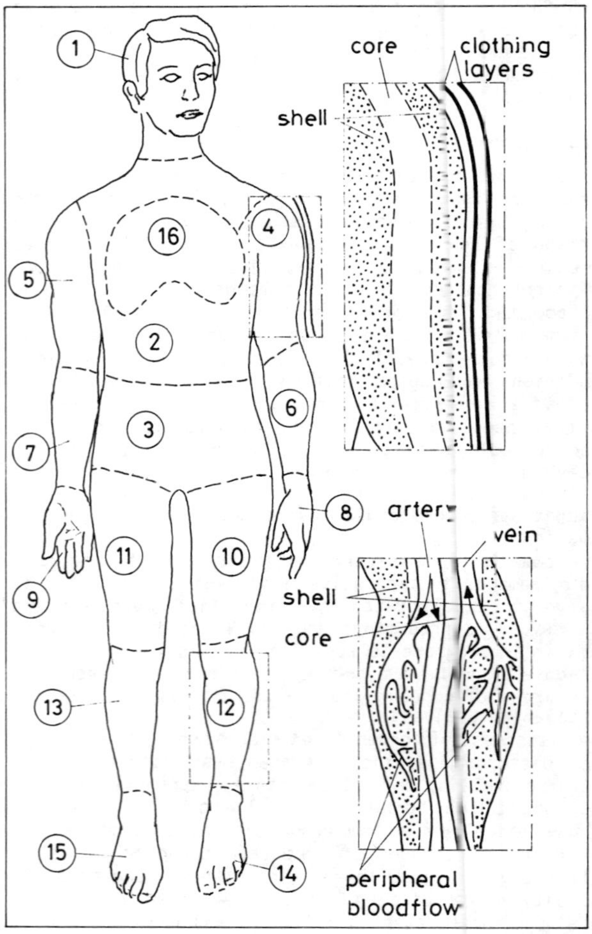

Figure 1: Interpretation and numbering of the body-parts used in "HUMTHERM"

3. INTERPRETATION OF THE MAIN PROCESSES

3.1. Heat and mass transfer

Thermal resistances R_c and R_{sh} are regarded as constants specific for the body parts. Thermal resistances between the clothing layers /including air layers too/ are calculated for the layer number k on the body part number i as follows:

$$R_{cl,i,k} = \delta_{i,k}/[A_i \cdot (\lambda_{i,k}+0.6\ x_{i,k})] \qquad /1/$$

where thicknesses $/\delta/$ and thermal conductivities $/\lambda/$ are input data.
Mass transfer resistances are calculated by considering the mass diffusivity of the clothing and the mass transfer coefficients. For the heat and mass transfer with the environment, convection, radiation, evaporation and the process of respiration are considered.
Convective heat transfer solutions are based on well-known conventional relations of the heat transfer coefficients. The main parameters are: temperatures T_{sf}, T_a; velocity values of air flow around body parts; characteristic diameters of parts; free areas A_f exposed to heat transfer. A_f depends on the position and activity of the body /input data/.
Radiative heat transfer is modelled by assuming that wall temperature T_w and emissivity ε_w may be accounted for as single average values. Emissivity ε is taken as constant for skin and textile. Radiative heat flow rate and thermal resistance for body part number i are calculated as follows:

$$\dot{Q}_{r,i} = \frac{\sigma A_{f,i} \cdot \left(T_{sf}^4 - T_w^4\right)}{1/\varepsilon + (1/\varepsilon_w - 1) \cdot A_{f,i}/A_w} \qquad /2/$$

and

$$R_{rad,i} = \left(T_{sf} - T_w\right)/\dot{Q}_{r,i}\ , \qquad /3/$$

where σ is the Stefan-Boltzmann constant. Evaporative heat losses are calculated by the mass transfer resistances

$$R_{evap,i} = 1/(\beta_i \cdot A_{f,i})\ , \qquad /4/$$

where β_i are the mass transfer coefficients.
Respiration heat loss depends on the overall metabolic heat generation of the body and temperatures T_{16} and T_a.

3.2. Thermoregulation of the human body

3.2.1. Blood flow rate and vasodilatation

Human body is able to accomodate, to some extent, to the thermal effects from the surroundings, by varying the total rate of blood flow in the body and by varying the ratio of the peripheral blood flow rate to that in the internal large blood vessels. Basic values of the two above parameters belonging to thermal comfort are built in the program, as well as the maximum and minimum values of the ratio of vasodilatation for each body part. The actual values of flow rates in the blood vessels indicated in Fig.1 are functions of the hypothalamus temperature T_{hyp}, of the body position and activity, and of the condition that shivering has set in or not.

3.2.2. Perspiration

The source mass flow rate of perspiration for body-part number i is determined as follows:

$$\dot{m}_{x,i} = A_i \cdot f_1\left(T_{hyp}, V_o\right) \sum_{k=0}^{9} \alpha_{i,k} \cdot \left[f_2\left(T_{s,i}\right)\right]^k \qquad /5/$$

where A_i is the surface area of the body-part, $\alpha_{i,k}$ are the coefficients of the approximating function $f_2/T_{s,i}/$ and, V_o is the oxygen intake capacity of the person in unit of lit/m^{2}min. as an input value.

3.2.3. Metabolic heat generation

Generated heat sources in the core and the shell of body-part number i are determined as

$$\dot{Q}_{m,c,i} = \mu_i \dot{Q}_m \quad \text{and} \quad \dot{Q}_{m,sh,i} = \nu_i \dot{Q}_m \qquad /6/$$

respectively, where $\dot{Q}_m$ is the overall metabolic heat flow rate of the body and coefficients μ_i and ν_i are constants specific for the individual body parts. Metabolic heat generation $\dot{Q}_m$ of an adult, at rest, depends only on the position of the body. At higher activities $\dot{Q}_m$ is the function of T_{hyp} and $T_{s,av}$.

4. COMPUTATION

The block-scheme of the computation is represented by Fig.2. The user puts the "macro time step" in as well as the data of

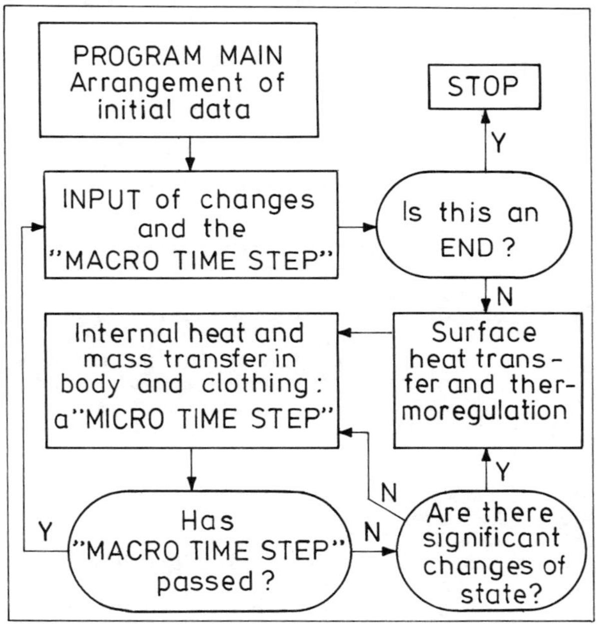

Figure 2: Block-scheme of the computation process

any changes in the state of the body-clothing-environment system. This state is determined by a set of momentary values of properties derived either from previous program runs or a group of standard values of the program HMT. Equations of internal heat and mass transfer are solved in a subroutine by using a finite time-element scheme. Body-to-surrounding heat transfer is recalculated and thermoregulation is assumed to start when a significant change of body state has been occured. During the program run, frequency of thermoregulation activity is indicated on the screen. The program evaluates and stores the history of several time steps.

5. FACILITIES OF THE SOFTWARE

"HUMTHERM" is realized as a batch file of MS-DOS. The main menu offers the following possibilities. There are two variants of the calculating program, one works in an interactive mood while the other lets the user to pre-program the whole task and to leave the computer alone until the end of calculation.
Two other programs serve for numerical and graphic representation of the results.
All important data are continually recorded in groups of data files, in a determined structure.
Previous runnings may be continued and variants of data files derived from different sets of input data produced.

6. SUMMARY

Experiences with HMT show that computed results are in good agreement with the results of the test measurements. The program can be used coupled in series with building simulation programs for IBM PC for solution of indoor comfort problems. In that case the environmental state can be computed by the building simulation program and these state characteristics can be used as input data to HMT.

7. LIST OF SYMBOLS

Letters:

A	Surface, area
C, $\dot{C}$	Thermal capacity and capacity flow rate, respectively
Q	Heat flow rate
R	Heat or mass transfer resistance
T, $\dot{T}$	Temperature and its change per unit time, respectively
X	Humidity content of clothing layer per unit dry mass
c_w	Specific heat capacity of water
Δh_{vap}	Latent heat of evaporation of water
p	Partial pressure of water vapour

Subscripts:

a	Ambient air
av	Average
bl	Blood
c, sh	"Core" and "shell", respectively
cl	Clothing layer
m	Refers to metabolic heat generation
n	The outmost clothing layer
resp	Respiration
s	Skin
sf	On the surface
sat	Saturation
w	Refers to the walls of the room
16	Body part number 16 /heart and lungs/

conv, rad, evap = convective, radiative, evaporative, respectively

8. REFERENCES

1. Imre,L.; Bitai,A.; Horváth,C.D.;Bánhidi, L. and Pammer,Z.: Thermal Analysis of Human Body-Clothing-Environment System, Int.Journal for Num.Meth. in Engng, Vol. 25, 357-371 /1988/.
2. P.O.Fanger, Thermal Comfort, McGraw-Hill, New York, 1970.
3. E.A.McCullogh and D.P.Wyon, 'Insulation characteristics of winter and summer indoor clothing',ASHRAE Trans.89,No.2. /1983/
4. E.A.McCullogh,B.W.Jones and P.J.Zbikowski,'The effect of garment design...', ASHRAE Meeting,Washington,D.C., 1983.
5. E.A.McCullogh, E.J.Arpin,S.A.Konz and F.H.Rohles, 'Heat transfer characteristics of clothing worn in hot industrial environments',ASHRAE Trans.No.1./1982/
6. K.H.Umbach, 'Einfluss der äusseren Luftbewegung...',Hochensteiner Forschungsbericht Bekleidungsphysiologie, Bönningheim, 1982.
7. D.L.Morgan and R.L.Baskett, 'Comfort of man in the city...', Int.J.Biometeorology, 18, 184-198 /1974/.
8. A.R.Atkins and C.H.Wyndham 'A study of temperature regulation in the human body...',Pflügers Arch.307,p.104. /1969/.
9. R.F.Goldman, 'The role of clothing in modyfying the human thermal comfort range', Editions de l'INSERM, 75, 163-176 /1977/.
10. R.C.Seagrave and N.C.Miller, 'A model of human thermoregulation', Proc.24th Annual Conference of Engineering in Medicine and Biology, 1971, p.56.
11. J.A.J.Stolwijk, 'Mathematical model of physiological temperature regulation in man', NASA Report Cr-1855, Washington,D.C., 1971.
12. L.Pedersen, The Thermoregulation of Man, H.C.Oersted Institute, University of Copenhagen, 1973.
13. J.Werner, 'The concept of regulation for human body temperature' J.Therm. Biology, 5, 75-82 /1980/.
14. J.Werner and T.Reents, 'A contribution to the topography of temperature regulation in man' Eur.J.Appl.Physiology, 45, 87-94 /1980/.
15. H.Arkin and A.Shitzer, 'Model of thermoregulation in the human body' I-II--III,EEC-148,ISSN 0333-8048,Haifa,1984.
16. E.Simon, O.Ludwig and E.Vieth, 'Are black-box models of thermoregulatory control obsolete? Yale J.Biology Medicine,59, 349-359 /1986/.
17. E.Simon, F.K.Pierau and D.C.M.Taylor, 'Central and peripheral thermal control of effectors...', Physiological Rev., 66, 235-300 /1986/.
18. J.Timbal, M.Loncle and C.Boutelier, 'Mathematical model of man's tolerance...', Aviation Space Environ. Medicine, 47, 958-964 /1976/.
19. A.P.Gagge, J.A.J.Stolwijk and Y.Niski, 'An effective temperature scale...', ASHRAE Trans. 83, 88-102 /1977/.
20. A.R.Atkins, 'Application of 'Hopscotch' algorithm...', Comp.Biology Medicine,3, 397-405 /1973/.
21. A.F.Emery, R.E.Short, A.W.Guy, K.K. Kranning and J.C.Lin, 'The numerical thermal simulation of the human body...', J.Heat Transfer,98, 284-291 /1976/.
22. E.H.Wissler, 'The use of finite difference techniques...', in J.D.Hardy et al. /eds/, Physiological and Behavioral Temperature Regulation, 1971.pp.367--388.
23. C.E.Huckaba, H.S.Tam, R.C.Darling and J.A.Dewney, 'Prediction of dynamic temperature distribution in the human body', J.AIChE, 71, 1006-1012 /1975/.

A COMPREHENSIVE MODEL OF THE ENVIRONMENTAL PERFORMANCE OF BUILDINGS

A. Giaccone
CNR-IEREN
Viale delle Scienze, 90128 Palermo (Italy)

As it is well known, buildings act as filters placed between men and the outdoor environment in order to delimitate a comfortable thermohygrometric, visual, acoustic and respiratory-olfactive indoor environment for living. A model is here presented, describing outdoor environment and building features (the "source model") as well as indoor environment and human response to it (the "target model"); calculation procedures are also discussed, using "source model" parameters as known input data for analysing energy and mass transfer processes in buildings and so obtaining, as output data, values for the "target model" parameters.

1. THE SOURCE MODEL

Parameters characterizing the building to be examined (together with its interfaces towards the outdoor environment, the plants and the occupants) are first described.
Knowledge of all these parameters, proceeding from design resolutions or, in case of existing buildings, from a building survey, should allow a quite refined analysis of the physical processes determining environmental performance of buildings; if some parameter should not be known, as it is common in initial design stages, or easily measurable, suitable default values could be assumed or simplified calculation models could be used.

1.1 The outdoor environment model

Outdoor climate (intended as including also light and sound climate as well as air quality parameters) should be defined in order to provide input data representing the initial or boundary conditions for the solution of the heat, light, sound, air, moisture and pollutants transfer models of buildings.
Especially for transient physical processes such as heat and moisture transfers, external climatic excitations should be known in form of time-series data, coming from actual measurements or from statistical analysis of historical recordings and relative to long (usually reference years) or short periods, representative of typical and extreme situations.
The requested parameters include:
- dry bulb air temperature
- wet bulb air temperature or relative humidity
- direct and diffuse (or global) solar radiation
- wind speed (or pressure) and direction
- sky temperature
- sky luminance distribution
- sound level
- concentration of pollutants

In some cases, temperature and humidity of the ground, temperature, humidity, sound level and concentration of pollutants in contiguous indoor spaces are also requested.
In absence of some of these parameters, auxiliary models can be used to obtain, for example, unaivalable solar radiation or sky luminance data from cloud cover or sunshine hours data, direct and diffuse radiation data from global radiation data, sky temperature data from cloud cover and dry and wet bulb air temperature data, ground temperature data from air temperature data, sound level or air pollution data from traffic and industrial activities data, etc.
Other models can be used to predict hourly values of temperature, solar radiation, light, noise, etc. from daily overall or average values.
Finally geometrical and physical features of the surrounding scene, as regards its interaction with the examined building, should be considered and models should be provided to deal with microclimatic modifications induced by the building itself or to explore interfacing between the building and its surroundings as, for example: insolation on envelope surfaces of different orientation and tilt; moving of shading patterns on external and internal building surfaces as produced by remote (other buildings, trees and shelter belts, etc.) or facade (overhangs, recesses, etc.) obstructions; wind pressure profile at different heights for windward and leeward facades, as influenced by various microclimatic effects (obstructions, wind hollows, etc.); sound level modifications at the building facades by external sound barriers (vegetations, etc.) and terrain configuration.

1.2 The building model

Together with a suitable geometrical description (size and position of each element), informations have to be provided specifically for every inhabitated spatial unit (i.e. for every inhabitated room) about features of building spaces, opaque or transparent material parts and surfaces influencing energy and mass transfer processes.
In particular, the following parameters should be defined:
- thermal insulation and capacity coefficients relative to heat transient conduction through the building fabric: conducibility, density, specific heat, thickness and relative position of different materials
- natural or forced convection characteristics (surface roughness) for external and internal exposures
- longwave radiation absorptivity/emissivity of external and internal surfaces as well as view factors relating each external surfaces with the three portions (sky, surrounding buildings, ground) of its scene and each surface of the internal ambiences
- shortwave (solar) radiation absorption and transmission (for transparent elements) characteristics, as a function of the angle of incidence
- air permeability of the building enclosures, distribution and magnitude of interzone air leakage paths inside buildings
- daylight transmittance of external glazed surfaces and reflectance of internal surfaces
- sound insulation of the building envelope and absorption of the enclosure surfaces
- vapour permeability of the building envelope
- production of pollutants by the building fabric materials

1.3 The plant interface model

Auxiliary equipment (space heating or cooling, ventilation, illumination plants) is usually added to most buildings in order to provide better comfort conditions to their occupants.
Even though plant operation is not intended to be investigated in combination with building performance analysis, parameters describing the plant/environment interfacing mode have at least to be known, including:
- characteristics of plant final devices, such as heat and cold emitters, artificial light sources, mechanical air distribution units
- location and operative mode of plant control devices

1.4 The occupancy model

In the built environment men rest (or work), so acting as simple users, but also interfere with building performances, just like buildings are not only influenced but also do influence climatic features of their surroundings.
Occupancy parameters are therefore needed, often originating from forecasting or statistical observations of the human behaviour and activities, in form of time schedules of:
- utilization of the building spaces
- indoor production of heat and cold (convectively or radiatively distributed), humidity, noise and pollutants
- manual modification of adjustable building parameters (opening of

windows, operation of solar immission/exclusion devices, etc.)
- utilization of the plants

2. THE TARGET MODEL

The parameters characterizing the quality of a built inhabitated environment under the point of view of thermal, visual, acoustic, respiratory-olfactive comfort are here analysed.
The reference scale to be used for the appraisal of such parameters will be that of the minimum inhabitable unit, i.e. the ambient room.
However, as each parameter will generally change at different locations in the room, values should be known for a sufficient number of points to eventually define, at least for large ambiences, uniformity ratios and iso-line maps.
In a similar way, anisotropy factors should be treated in order to obtain the corresponding polar charts.
Moreover, as parameters will also change according to time- variable sollecitations, periods of investigation should also be selected such as to embrace, if not the complete range of foreseeable situations, the average and the more extreme of them as well as typical time sequences.
Statistical analysis could finally be performed to better assess space and time distributions of such parameters.

2.1. The indoor environment model

As it is well known, thermohygrometric comfort of men wearing a clothing of known thermal resistance and performing an assigned activity depends on air temperature, humidity and velocity, as well as on mean radiant temperature of the surrounding surfaces (walls, glazed surfaces, furniture, etc.) as they are seen by the occupants.
Visual environment is then determined by the illuminance of some reference objects of the vision (workplane, human faces, etc.) as well as, with regards to glare, by the luminance (or reflectance) of the surrounding surfaces (light sources like windows and luminaires, reflecting objects like walls and furniture) according to their position in the visual field of the occupants.
Acoustic comfort, on the other hand, depends on the sound level and the reverberation time, respectively bound to the steady and the transient conditions of the acoustic field in the room.
Last of all, respiratory-olfactive environment can be appraised by the concentration of gases (carbon monoxide and dioxide, nitrogen oxides, radon), odours, smoke, particles and dusts, micro-organisms, formaldehyde and other volatile organic compounds, asbestos and other man-made mineral fibres, aerosol sprays or other pollutants in the ambience.
Thus the following parameters determining the indoor environment can be listed:
- air temperature
- temperature of the enclosure surfaces
- air humidity
- air velocity
- illuminance of the reference surfaces
- luminance of the enclosure surfaces
- sound level
- reverberation time
- concentration rate of pollutants
Special attention should be devoted to the parameters (temperature and illuminance) attaining the enclosure surfaces: their knowledge allows to compute radiant energy fluxes (thermal and luminous) impinging on planes of different location and orientation throughout the room, thus defining the room volume as a vectorial field.
Likewise, air velocity in a room can also be seen as a vectorial field as opposed to the scalar fields related to all the other parameters.
The investigation of all these vectorial quantities is useful to appraise the anisotropic behaviour of the environment as regards thermal (extremely hot or cold surfaces) and visual (relief and shading of objects, glare) sensations.
Local thermal discomfort situations can also be appraised by the consideration of temperature gradients or air draughts.

2.2. The human response model

An estimation of the environmental quality of an ambience can be performed by the comparison of the actual values of the indoor environment parameters described before with some recommended values, according to the space and time layout dictated by the occupancy model of the building spaces.
Otherwise some indexes can be evaluated, directly measuring the mean human physiological or psychological response to that particular indoor environment.
With regards to thermohygrometric environment, for example, "effective temperature" indicates on a thermometric scale the amount of heat exchanged between the human body and the surrounding environment, while the so-called "predicted mean vote" (PMV) and "predicted percentage of dissatisfied" (PPD) are a direct measure of the probable level of satisfaction experienced by an average sample of occupants.
Many indexes have been proposed also to appraise the fitness and the pleasantness of a visual environment: e.g. the "contrast rendering factor" (CRF) and the "equivalent sphere illuminance" (ESI), measuring the visibility of objects in a certain luminous environment; the "glare index" (GI), indicating on a numeric scale the degree of glare problems to be encountered by occupants; the "visual comfort probability" (VCP), that is the probability for an occupant to esteem its visual environment as agreeable, etc.
Indexes can then be assessed evaluating the annoyance produced on occupants by an acoustic environment (the "noise rating" NR, the "perceived noise level" PNL, the "composite annoyance levels" CALs, the "noise pollution level" NPL, the "traffic noise index" TNI, etc.), the capability to understand sequences of spoken words in spite of masking effects due to disturbing noises and sound deformation due to the enclosure (the "speech-to-noise ratio" S/N, the "speech interference level" SIL, the "articulation index" AI, the "modulation transfer function" MTF, the "speech transmission index" STI) or finally the audiological damage risk due to prolonged exposition to a very loud environment (the "exposure level" EXL, the "noise immission level" NIL).
Similar exposure levels can be defined about heath hazard from air pollutants, as well as an index of the air freshness as perceived by people, called "decipol".
Many others parameters, not mentioned here, have been proposed by scientific literature, but all of them can be generally derived by the knowledge of the indoor environment parameters listed in the previous chapter; they are therefore subjected to similar space and time variations.

3. THE CALCULATION MODEL

Calculation models have to be used, dealing with energy (heat, light, sound) and mass (air, moisture, pollutants) transfer occurring in the built environment, in order to assess environmental (thermohygrometric, visual, acoustic, respiratory- olfactive) performance of buildings.
Such models will use the parameters listed in the "source model" as input parameters to obtain as output parameters, through the analysis of these transfer processes, the parameters listed in the "target model".
The discipline known as "building physics" has for long years developed models devoted to analyse the aforementioned physical processes.
Initially, traditional calculative models were based on empirical considerations or simplifying assumptions, but quite recently the larger availability of more powerful computing machines has spread the use of simulation techniques performing a dynamic modelling of the energy flowpaths within buildings and their associated control systems.
Such techniques are more realistically dealing with the complex behaviour of buildings, imagined as composed of several non- homogeneous zones (spaces and material parts) and dynamically responding to variable external sollecitations.
Some of these models are only available for research purposes (generally the most complex and up-to-dated), whereas others (not necessarily the less sophisticated) are already used to investigate alternative design choices on the building and the plants, optimum configuration of particular passive (solar, ventilating, daylighting, etc) building components or impact of different operational strategies.
Many of them, however, devote a major attention to energy-saving, while a different and more demanding approach is generally needed in order to appraise environmental performance of buildings: as a matter of fact, for example, whereas energy consumptions can be computed as a unique determination over a whole building (or large portions of it) as well as over a whole season (or large periods of time), environmental comfort have to be considered as a complex set of local and instant features (the "target model"), thus requiring a more extensive context description (the "source model") and a more detailed definition of the calculation model.
The calculation model here presented lists a quite complete and up-to-dated range of disposable simulation issues; yet, due also to

input data (outdoor environment, building definition) availability, partial or simplified simulations could possibly be performed and some "target" parameters eventually neglected, using "source" parameters for empirically predicting the achievable performance level.

3.1 The heat transfer model

Steady-state or simple and complex dynamic models can be utilized to analyse thermal energy flows through buildings and their thermal behaviour as a response to outdoor climatic variations.
The building is generally regarded as a network of thermal resistances and capacitances linking different regions of the building fabrics or indoor spaces.
In such models nodes will represent portions of the fabric (capacity/insulation layers), air volumes (still as in ambient rooms or moving as in open gaps or ducts) or interface layers between the fabric and indoor or outdoor air volumes, while nodal connections will represent conductive, convective and radiative heat transfer processes.
As steady-state models are unable to deal realistically with many energy flows, often of major importance, these methods play a diminishing role in favour of dynamic models, solving partial differential heat flow equations within the building fabric by analytical (time-domain and frequency-domain response function methods) or by numerical means (finite difference and finite element methods).
Simplified models, on the other hand, can be obtained by neglecting some of the energy flows, by reducing the number of nodes (up to a single-node lumped-parameters model) or by turning, through regression techniques, the results of multiple parametric runs of more detailed models into simple analytic or tabular relationships.
A detailed model should encompass all the following energy flowpaths occurring in a building, dealing with transient conduction of heat through the enclosure envelope with the associated (usually sensible) storage and lag effects; external and internal, forced or natural, surface convection; inter-surface longwave radiation exchange between internal surfaces; external surface longwave radiation exchange with the sky vault and the surroundings; shortwave solar radiation impinging on exposed external surfaces and penetrating through glazings into building enclosure; casual gains from occupants, lights, equipments, etc.; heating and cooling plant interaction.

3.2 The air movement model

Air movement inside buildings is produced by pressure or density (bound to temperature differences) variations between inside and outside or between two adjacent internal volumes of air, communicating through small cracks around windows and doors or through the fabric junctions and materials (infiltration) or through intentional opening of windows and vents (natural ventilation); mechanical ventilation can also move air between or inside building spaces.
Since a mass balance has to be maintained in the whole building and in each of its spaces, any air loss or gain induces a corresponding air movement through all the paths (apertures and leakages) connecting two different air volumes (rooms) as well as inside each volume itself.
Computational models are available ranging through three levels of increasing complexity:
- simplified predictive expressions correlating the air flows in terms of global air change rate to temperature differences and external wind pressure or velocity
- models applying mass (air) balance conditions to nodal networks in which nodes represent discrete volumes of air (zones or portions of them, if movement inside a single zone has to be analysed), characterized by their pressure and temperature, while nodal connections represent leakage occurring through internal and external apertures (doors, windows, vents) or cracks.
- detailed models, particularly dedicated to analyse the distribution of air velocities inside building spaces by numerically solving, often in steady state conditions, the governing partial differential equations relative to conservation of mass, energy and momentum in two- or three-dimensional fluid flow fields.

Since temperature acts as a driving force for air movement in buildings and conversely each air movement between regions of different temperature correspond to an advective (fluid-to-fluid) heat exchange, simultaneous heat and air transfer simulation is desiderable.

3.3 The humidity model

Water vapour is introduced in buildings as latent heat gains from occupants and equipments and is diffused inside building spaces by interzone air movement, passing through the building envelope as the natural content of the building air changes.
Another set of mass balance conditions, relative to water vapour balance, has to be added to heat and air balance conditions for building network nodes in order to deal with phase change processes occurring through the fabric (moisture accumulation, affecting its heat transfer properties), on envelope surfaces (moisture condensation) and inside inhabitated spaces (affecting thermohygrometric comfort of occupants and operation of most air- conditioning plants).
Since phase change processes are strictly bound to temperature, the humidity model cannot be solved separately from the heat transfer model.

3.4 The daylight penetration model

Daylight penetrates buildings through glazed surfaces and is then reflected and diffused in the ambience by its internal surfaces (walls, ceiling, floor or furniture).
Some simplified models, coming from the consideration of a rough scheme of the physical process or from parametric correlations, operates only in reference sky conditions (e.g. overcast sky) and provide results only as an average value or for few specific points of the indoor space.
Only flux transfer models, solving the radiant balance between the glazed and the reflecting surfaces of the enclosure in real daylighting conditions and allowing for beam solar radiation, provide complete information on the visual environment.
Daylighting illumination time-schedules in indoor environments gratly influence operation of the lighting plant, often a major source of casual heat gains in buildings.

3.5 The sound propagation model

As it is well known, sound fields in enclosed room, as produced by generally distributed and multidirectional sound sources (indoor sources or noise from outside), are the sum of a direct and a reverberant field.
Whereas mathematical expressions can be used for a simplified assessment of the characteristics of an acoustic environment, sound spatial propagation and temporal decay inside an enclosure is usually studied through the methods of geometrical acoustics.
Ray-tracing models are used, recording the path of each sound wave in the room and every successive reflection by the enclosure surfaces, or source image models, calculating the sound energy in a single point as the sum of the energy received by the sources (direct field) and by their specular images beyond the enclosure surfaces (reverberant field).

3.6 The pollutants concentration model

Air pollutants can be produced and admitted into buildings by the occupants, the fabric materials, the equipments, the furniture; they, just like humidity, are diffused inside buildings and pass through the building envelope due to air movement.
Models stating the appropriate mass balance conditions for each indoor air volume of the building allow to evaluate, when air flow rates are known, the corresponding concentrations of pollutants.

4. REFERENCES

J.A. Clarke, "Energy simulation in building design", Adam Hilger Ltd., Bristol, 1986
G. Alfano, F.R. d'Ambrosio, F. de' Rossi, "Fondamenti di benessere termoigrometrico", CUEN, Napoli, 1987
C.L. Robbins, "Daylighting", Van Nostrand Reinhold Co., New York, 1986
L.L. Beranek, "Noise and vibration control", McGraw Hill, New York, 1971
U.S. EPA, "The inside story: a guide to indoor air quality", 1988

A STUDY ON THE COMFORT LEVELS IN A ROOM OF AN EXPERIMENTAL BUILDING SITUATED IN A COMPOSITE TROPICAL CLIMATE

M.Srinivasa Reddy S.Krishnamoorthy
Department of Civil Engineering
Indian Institute of Technology
Hauzkhas; NEW DELHI 110 016 INDIA

The objective of the work is to assess the applicability of a specific type of construction with respect to human comfort for a composite climatic place like Delhi. A room in the top floor of an experimental multistorey residential (brick masonry) building was chosen for study. The roof of the room consisted of precast reinforced concrete planks of 60mm thick supported on a joist system unlike a conventional reinforced concrete slab of 100mm thick. Hourly room air temperature and relative humidity were recorded for a continuous period of one year. Two thermal comfort indices viz., Tropical Summer Index and Effective Temperature had been used to evaluate the degree of comfort. It has been observed that the construction more or less keeps the inside environment in the comfort level except for a short period during summer and winter.

1. INTRODUCTION

The main aim of a building is to provide an enclosure which will keep inhabitants at some level of comfort while the external environment is either too hot or too cold. In other words, the building envelope acts as a barrier as well as modifier at the same time. the modifying ability of a fabric depends to some extent on the method of construction also as the different methods of construction or design have been evolved to economise the cost of construction.
The present study, where the cost of reduction in building works had been 13%, deals with an experimental multistorey residential building (4 floors) of brick masonry (230mm thick walls), which has been built with prefabricated floor components unlike the large number of residential quarters are still being constructed with insitu concrete floors. The roof of the room consisted of precast reinforced concrete planks of 60mm thick supported on a joist system unlike a conventional reinforeced concrete slab of 100mm thick.
The main objective of the work is to assess the applicability of this type of construction with respect to human comfort for a composite climatic place like Delhi. The climate of Delhi is characterised by three seasons, viz., hot-dry, warm-humid and cool-dry.

2. EXPERIMENTAL INVESTIGATION

2.1 Description of test room

A room in the fourth storey of a four storeyed residential building in Delhi was selected for the experimental measurements. The test room had an internal dimension of 2.77m x 3.40m x 2.70m high. It had two doors, of 2.0m x 0.85m, one in its north facing wall and the other in its east wall and two windows of 0.9m x 1.2m and 0.5m x 1.2m in the south facing and north facing walls respectively. These windows were fitted with ordinary glass shutters and the one in the south facing wall had a sunshade. In the south facing wall, there was a built in cupboard, which was projected horizontally outwards, covering a vertical area of 3.05m^2 of the wall area. The window in the south facing wall had a horizontal sunshade of 0.55m width and vertical drop of 0.55m on one side. The projected portion of the cupboard acted as a vertical shading device for the window adjacent to it.
All the walls were of common clay brick masonry with unplastered, exposed brick work texture on the external side for the south and west facing walls. The thickness of all walls, except that of north facing wall was 0.23m, whereas the north facing wall was of 0.12m. The walls were plastered internally with a composite cement-lime mortar.
The roof of the room consisted of precast, reinforced concrete planks supported on a joist system. The planks were 0.06m thick and were covered with mudphuska (combination of clay and hay) of 0.1m thick and brick tile of 0.045m thick for thermal insulation.
The floor of the room was also of precast reinforced concrete plank system. However, the flooring consisted of 0.025m thick cement concrete finish, with neat cement finish, over a 0.05m thick base layer of lean cement concrete which, in turn was laid over the planks.
Outside the test room, on the east wall side, there was another room, passage and water-closet and there was a verandah to the north.
Figure 1 shows the plan of the experimental test room.

2.2 Observations

The internal room air temperature and internal relative humidity were recorded continuously for every hour throughout the year by using thermohygrographs /1/. From the values of globe temperature, it is observed that there has been not much difference between the mean radiant temperature of the test room and room air temperature /1/.
Throughout the test period,the test room remained closed except for the purpose of

taking the measurements and hence this room has been considered as one with low air changes per hour. A ventilation rate of 3 air changes per hour had been considered based on the results obtained from the study /1/.

3. ANALYSIS AND RESULTS

For tropical summer conditions, Sharma/2,3/ proposed a Tropical Summer Index (T.S.I) to define the level of thermal comfort. Though this particular index does not specifically take into account the effect of differnet types of clothing ao activity levels, the index has been determined in the Indian context. The index is defined as the air/globe temperature of still air tat 50% relative humidity which produces the same overall thermal senstation as the environment under investigation. The index is given by

$$T.S.I=0.308t_w+0.745t_g-2.06v^{0.5}+0.841$$

where,

t_w = wet-bulb temperature, ^oC
t_g = globe temperature, ^oC
v = air velocity, m/sec
and T.S.I is expressed in ^oC.

Sharma suggested further that the comfort limits are when the T.S.I is in the range of 25°C to 30°C. When the T.S.I is between 19°C to 25°C, it can be considered as slightly cool, when the T.S.I is between 30°C to 34°C, it can be considered as slightly warm.

The T.S.I has been calculated for each hour of each day for the whole year for the air temperatures of the test room. The air velocity has been assumed to be 0.003m/sec. Air temperature is assumed in place of globe temperature.

Koenigsberger et al/4/ have suggested that the comfort limits which are probably valid for most tropical regions would be between 22°C to 27°C when the values are expressed in terms of effective temperature scale with appropriate tropical summer clothing. Sharma has suggested that the comfort zone will be when the values of T.S.I are in the range of 25°C to 30°C/3/. When these limits are converted in to effective temperature scale, the range works out to 22.7°C to 27.5°C. As such it may be noted that the comfort zone suggested by Sharma is almost the same as that suggested by Koenigsberger et al. The limits suggested by Sharma have been used further in the analysis of the data of the test room.

It is observed that 61 days, out of the 365 days of the year, fall into this category of comfort, when, at no time the internal air temperatures exceed the specified limits. It is further considered that the winter season in Delhi is probably more or less similar to the summer season in U.K. in so far as the warm clothing adopted by the inhabitants are concerned. For summer season in U.K. Koenigsberger has suggested that the optimum and maximum limits of effective temperature, for comfort, could be 18°C and 22°C, respectively. Sharma has suggested that a T.S.I range of 19°C to 25°C (17.2°C to 22.7°C in ET scale) as slightly cool. Likewise, Sharma has also suggested that a T.S.I range of 30°C to 34°C (27.5°C to 31.5°C in ET scale) as slightly warm. Usually, in residential buildings in India, the occupants are habituated to use electric fans in summer for the purpose of local cooling of human body though not of the room and as such the slightly warm condition postulated by Sharma can be considered as acceptable for comfort when electric fans are used for the purpose.

Totally, 198 days out of 365 days (including the 61 days mentioned earlier) turn out to be comfortable within the limits of 17.2°C to 31.5°C of ET or 19°C to 34°C of T.S.I with the appropriate proviso of using electric fans during summer and warm clothing in winter. It may be noted that the limits are satisfied at all hours of a day.

The remaining 167 days consisted of 72 transient days and 95 uncomfortable days. During the transient days, the calculated T.S.I for each hour of a day may be within or beyond the comfort limit i.e., 19 to 34°C. The number of transient days during summer and winter seasons are 56 and 16 respectively. Among the 56 transient days of summer period, the duration of discomfort in a day is about 10 to 11 hours, approximately from 17.00 to 6.00 hours. The peak T.S.I during these days is 35.5°C. For the 16 transient days of winter season, it is observed that the period of discomfort occurs at different times of the day and it is difficult to identify a specific time of discomfort. The extreme lower limit of T.S.I observed, however, is 17.5°C, during such transient days.

During the uncomfortable period of the winter season (68 days) excepting for the days of transiency, the T.S.I was below 19°C throughout the 24 hour period of a day. Typically, the T.S.I was around 16 to 16.5°C. On a few occasions, the T.S.I was low at about 13 to 13.5°C. This discomfort due to cold can probably be taken care by sufficient warm clothing unless there are compelling or special reasons to incorporate heating.

During the uncomfortable period in the summer (27 days) excepting for the days of transiency, the T.S.I was above 34°C throughout the 24 hour period in a day. To mitigate this condition, one has to resort to a temporary evaporative or other cooling mechanisms and to shading of the room from direct solar radiation. Cooling of infiltration air entering the room by providing deep verandahs etc. would also be useful.

4. CONCLUSIONS

It has been observed that the construction more or less keeps the inside environment in the comfort level except for a short period during summer and winter. In fact the reaction of the building envelope to the diurnal changes does not keep pace with the changes in the outdoor conditions because of the inertia of its thermal capacity. Thus some discomfort is evident and unavoidable. During the short period of discomfort in summer, it was slightly warmer. But it can be taken care by providing mechanical ventilation which is a normal practice. Similarly, during the short period of discomfort in winter, it was slightly cooler. It can be taken care by sufficient warm clothing so that no heating is necessary.

In so far as the building fabric design is concerned, it appears that the thermal massivity provided by the 23cm thick brick wall, and 20.5cm thick roof seem to be adequate and further increase in thermal massivity, while it may bring down the number of transient days, particularly in summer to a smaller level, the cost effectiveness of such additional thermal massivity is probably marginal and has to be examined.

REFERENCES

1. Srinivasa Reddy,M., "A study on the temperature and heat flow behaviour of a room in top floor of a multistorey residential building located in Delhi", Ph.D.Thesis, Indian Institute of Technology, New Delhi, 1988.

2. "Evaluation of thermal stress - Tropical Summer Index", Building Digest No. 127, Central Building Research Institute, Roorkee, India.

3.Sharma, M.R. and Sharafat Ali (1986), "Tropical Summer Index - a study of thermal comfort of Indian subjects", Building and Environment, Vol. 21, No.1, pp. 11-24.

4.Koenigsberger, O.H., Ingersoll, T.G., Alan Mayhew and Szokolay, S.V. (1975), "Manual of tropical housing and building, part one: Climatic design", Orient Longman, New Delhi.

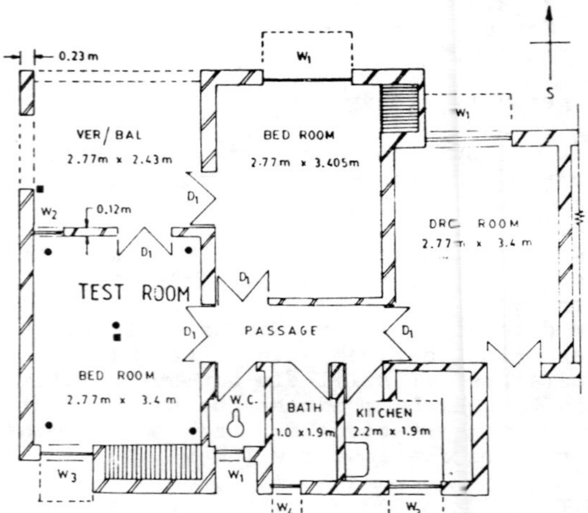

■ The location of thermo-hygrographs

● The location of thermometers

FIG.1. PLAN OF THE TEST APARTMENT

A STUDY INTO CLIMATIC, LUMINIC AND ACOUSTIC COMFORT IN THE ARCHITECTURE

Prof. R. SERRA, arch; H. COCH, arch.
ESCOLA TECNICA SUPERIOR D'ARQUITECTURA DE BARCELONA (U.P.C.)
Av. Diagonal, 649 BARCELONA 08028 (Spain)

ABSTRACT: The following work presents the first stages of the investigation we are starting on the theme of "comfort integrated into architecture". We will be looking at the possibility of establishing a unique system of evaluation that will jointly take into account human response to luminic, acoustic and thermal stimuli. In this draft we propose a primary classification of the elements that influence comfort, distinguishing those that are called "environmental comfort parameters" from those called "comfort factors". Some common aspects are defined to different stimuli that can serve as a base on which develop this study. Lastly, the main lines are examined schematically, which should be an algorithm that will permit a global evaluation to be carried out.

INTRODUCTION

In order to go about this joint analysis of acoustic, thermal and luminic comfort, we must first study the elements that are common to these three types of stimuli. In this way we can find the bases for establishing a joint evaluation system of comfort in buildings.

1.- PRINCIPLES OF THE GLOBAL ANALYSIS

On the one part we must consider that in many cases there is no clear distinction between sensation and perception. Even though the fact of being conscious or not of a concrete stimulus that provokes a reaction does not imply that our comfort has been more or less affected by it, there are reactions that are induced by the "knowledge" of the stimulus that should be taken into account in the evaluation. This type of reactions corrected by knowledge must follow laws of individual behaviour that can be analyzed in parallel in all types of stimuli. As well as this common element, it is proved that a psychological level there are interrelationships within the parameters of different senses, which means a determined stimulus can influence the perception of other stimuli.

This second fact gives us a second field of integrated analysis. If we can define some laws that would evaluate the perception influence of one sense over another (synaesthesia), we will obtain a useful tool with which to connect together the effects of different stimuli. Basing our study on these principles, along with conventional comfort evaluations, we will see straight away which elements should be used in our system.

2.- PARAMETERS AND COMFORT FACTORS

In order to evaluate correctly the elements that intervene in the comfort sensation of a space, we distinghish between the "environmental comfort parameters" and that we call "comfort factors". The parameters are those energetic manifestations of a measureable or not, environment, that sum up its ows characteristics (see table I). Comfort factors are defined as being external to the environmental conditions that have influence over individual appreciation of it.
By analyzing the parameters and factors we can see there is a part of the first ones that are independent to each stimulus (thermal, acoustic and luminic); on the other hand, other parameters and comfort factors can be analyzed globally for the three types.

TABLE I.- PARAMETERS OF ENVIRONMENTAL COMFORT

	ENERGETIC EFFECTS	SPECIFIC PARAMETERS	UNITS	GENERAL PARAMETERS
THERMAL SENSE	$I_{thermal}$ $(I_t = I_{cd/cv} + I_{rad} + I_e)$ $I_{cd/cv} = I_{conduct/convect}$ $I_{rad} = I_{radiation}$ $I_e = I_{evaporation}$	Air temperature.....T_a Radiant effects.....I_r Humidity...........H_e Air velocity.......v	ºC w/m^2 gr/Kg m/s	* Directionality * Temporal dependence * Space dependence
ACOUSTIC SENSE	$I_{acoustics}$	Spectral distribution of intensities.....$I_{(f)}$	$w/m^2/Hz$	
VISUAL SENSE	I_{light}	Energetic illuminance........$E_{(\lambda)}$	$w/m^2/m$	
OLFACTIVE SENSE	Chemical	Air composition	gr/m^3	

TABLE II.- COMFORT FACTORS

SOCIAL	Type of activity, clothing,....
PHYSIOTEMPORAL	Aclimatization (short or long period)
PHYSIOLOGICAL	Type of anatomy and physiology of a given user, food,...
PSICHOLOGICAL	Type of character (nervous, quiet,..), education, perception ability,...

In the study of such parameters and factors of comfort, we discover that many of them do not have a normalized system of evaluation at the moment. In spite of this, we believe they should be taken into account with the same importance that the others have. Because of that, part of our work will be to find such methods of evaluation, even if they are only approximate, for all parameters and factors. We are convinced that by using this system the margin of error commited will be always less than the one provided by ignoring such unmeasureable elements.

3.- EVALUATION SYSTEM

After developing the above mentioned, we try to establish a joint evaluation system, that would include in the first place the three functions corresponding to the thermal, luminic and acoustic sensations; all dependent on parameters and factors These three functions would interrelate themselves in another general function or comfort function, in which the above mentioned common elements would be taken into account. Apart from the former aspects, we must understand that comfort is not a limitative theme nor should it be seen only in a negative light (bothersome). We believe there are positive effects or stimulants that can better the reaction to other negative parameters. In this sense, a global formula should jointly integrate all these aspects, with the corresponding sign from its components. Every parameter should intervene in the formula with a "time" and "space" function that would take into account what we called general parameters in table I. On the other hand, the relations of the exponential kind, between stimulus and sensation, are also introduced in the formulation of each one of them.

$$S_{ther} = f(Ta, Tr, h, v) \qquad (+/-)$$

$$S_{light} = f(L/L, \ldots, I_{\alpha}) + \text{evaluated qualitative factors}$$

$$S_{acous} = f(N, \ldots) + \text{evaluated qualitative factors}$$

$$S_{global} = (S_{ther} + S_{light} + S_{acous})$$

Specific and general parameters determine the physical environment and comfort parameters act as a filter to give a final measurement of comfort.

This can be stated as:

$$S_k = f_{i=k}(\bar{P}_i, \dot{\bar{P}}_i, \nabla\bar{P}_i, F_c) + \sum_{j\neq k} f'_j(\bar{P}_j, \dot{\bar{P}}_j, \nabla\bar{P}_j, F_c)$$

and

$$S_{gl} = \varnothing(S_k)$$

where:

$\bar{P}_i$ or $\bar{P}_j$ = Specific parameters i or j (See table I)

$\dot{\bar{P}}_i$ = Time derivative of specific parameter i

$\nabla\bar{P}_i$ = Space derivative of specific parameter i

F_c = Comfort factors (See table II)

S_k = Comfort measurement (K= thermal, acoustic, visual)

S_{gl} = Global comfort measurement, including synaesthesia effects

$\varnothing$ Represents the function that relates the different kind of sensations

$f_{i=k}$ Represents the influence of a given parameter on the Kth comfort measurement

$f_{j\neq k}$ Represents the influence of the other parameters on the Kth comfort measurement

4.- ACKNOWLEDGEMENT

One of the team members (H.Coch), would like to thank the Ministry of Science and Education for their help in the realization of this work.
We would also like to thank P. Chauvel, J. Roset and F. Daumal for their help in investigating the base information and in the general approach of this work.

IMPROVING THERMAL COMFORT COULD WORSEN VISUAL ENVIRONMENT

INDOOR AIR CONDITION IMPROVEMENT IS NOT ALWAYS FREE

5.- BIBLIOGRAPHY

ANSI/ASHRAE
Standard 55-1981: Thermal environmental conditions for human occupancy. ASHRAE,1981.

Cena, K; Clark, J.A.
Bioengineering thermal physiology and comfort. Elsevier, 1981.

Fanger,P.O.
Thermal Comfort Analysis and Applications in Environmental Engineering. MacGraw-Hill Book Company, New York,1973.

Givoni, B.
Man, Climate and Architecture. Applied Science Publishers, London, 1976.

Humpheys, N.A.
Outdoor temperatures and Comfort indoors. B.R.E., United Kingdom, July 1978.

McIntyre, D.A.
Indoor climate.Applied Science Publ. London,1980

Aubree,A; Chauvel,P.
Eblouissement en eclairage naturel interieur. C.S.T.B., EN-ECL-74-6L. Nantes, 1974.

Chauvel, P.
Les luminances en eclairage interieur. R.G.E., mai, 1974.

Hopkinson, R.G.
Architectural Physics:Lighting. H.M.S.O. 1963.

Lynes, J.A.
The principles of natural lighting. Elsevier,1968.

Harris, C.M.
Manual para el control del ruido. I.E.A.L. Madrid, 1977.

Kurtze, E.
Física y técnica de la lucha contra el ruido. Urmo, Bilbao, 1972.

Purkis, H.J.
Building Physics: Acoustics. Pergamon Press, Oxford, 1966.

How can science help the architect to estimate indoor comfort conditions to be obtained passively within a building at the project stage

P. ACHARD A. AYOOB M. TANTOT

Ecole des Mines de Paris
Centre d'Energétique
Sophia-Antipolis
F - 06560 Valbonne

ABSTRACT. At the project stage most of the parameters conditionning future passive behaviour of a building are definitely fixed : site location, site orientation and slope, climate type, shadow.
One knows that early design choices will strongly define indoor comfort conditions and energy saving capability of a future building. To evaluate his choices the architect has to trust in Rules of thumb which appear to be in most cases unsufficient.
This paper is a contribution trying to enlighten the influence of different sizings of the constitutive elements (glazing area, thermal mass, shading) on the behaviour of a perfectly known building (a monozone test cell equiped with a movable shading device) for a given climate (the Mediterranean one). An experimentally validated simulation code was ran.
For this purpose comfort to be obtained for the different configurations is analysed through three types of diagramms :
- psychrometric charts
- time percentage diagramms of underheating, overheating, comfort conditions expected
- histogramms of indoor temperature

1. Introduction

Modern architecture has the opportunity to take benefit of passive strategies to achieve thermal comfort without increasing energy consumption.

At the project stage most of the parameters conditionning future passive behaviour of a building are definitely fixed : site location, site orientation and slope, climate type, shadow.

The architect has to select at this stage the main orientations and treatment of the façades (glazing area, shading devices), and the type of construction (insulation and constructive mass) which will determine the thermal load, the inertia and the solar aperture of the building.

One knows that these characteristics strongly define indoor comfort conditions and energy saving capability of a building. If regulation calculation methods are available to calculate the thermal load, a great lack faces the characterization of the inertia and solar aperture. The architect has to trust in Rules of thumb which appear to be in most cases unsufficient.

General approach

Accurate simulation tools are today widly available. They generally permit to get for a given climate the evolution of indoor temperature in a building, as well as energy balances. They were developped first to take in account monozone buildings and permit today to treat multizone ones. Initially developped as heavy detailed codes, numerous methods (technics of model reduction...) allow now to simplify them in order to reduce the simulation time, the computer size and memory capacity. Meteorological data are available for a significative number of sites for a given country on an hourly basis (global solar radiation on horizontal or vertical plan, outdoor temperature, relative or absolute humidity).

For the present work we used three tools :

- the simulation code ESP developped by ABACUS (Sthrathclyde University Glasgow) whithin the CEC solar energy program.

- A reduced state model, issued from the Minerve code developped by Centre d'Energétique (Paris) of Ecole des Mines, and which parameters have been identified through the computer processing of a set of experimental data recorded in Sophia Antipolis on our test cell.

- This monozone test cell

To enlighten the influence of the design of a building on the indoor comfort conditions we performed simulations of the different configurations of the cell in free evolution for the Mediterranean climate : heavy and low inertia, 5 m^2 and 1.25 m^2 glazing area.

The results are plotted on 3 types of diagramms :

- psychrometric charts
- time percentage diagramms of underheating, overheating, comfort conditions expected
- histogramms of indoor temperature.

The second one is deduced from the first one by calculating the time percentage for which the indoor temperature is in a comfort zone, an overheating zone or an underheating zone using the formula :

$$T = Tmean + \frac{Tmax - T\ min}{2} \sin \frac{\Pi}{12} (t - 8)$$

These kinds of diagramms will be obtained automatically from the computer when comfort to be obtained in a future building will be a priority for the architect or the designer at the project stage. With such tools, he will be able to compare at the design stage the influence of the different choices he has to do when making the global composition.

Case study

The mediterranean climate is characterized by hot summer and mild winter. Maximum temperatures in summer time are between 30°C and 40°C, therefore some cooling is required. In winter heating is needed.

Thermal characteristics of building in this area must be such as to prevent from solar gain in summer and cold in winter. Since the outdoor average daily temperature swings in winter and summer are rather high, passive systems can improve the comfort effectively.

Traditionally, heavy building structures and shading devices are widely used in this region to maintain indoor thermal condition in agreement with human comfort requirements.

Thermal mass is mainly located in walls and floors of buildings. It prevents the building of overheatings by absorbing direct solar gains and restitutes the so accumulated heat the evening by radiation and convection.

Direct gain solutions are adopted in most buildings using large glazing area for obtaining some heating in the cold winter. The problem is in the arrangement of such window's orientation and dimensions. Here the main problem of overheating in summer is arisen.

Evaporative is an interesting method to cool. Evaporative cooling systems can be used to improve the summer comfort in building in this region. Also night ventilation can be considered as effective and essential for heat rejection from building with such design if there is a good heat exchange between the thermal mass of the building and the air stream. Buildings operate often in free floating conditions. This makes the design task of the architects to create a good thermal comfort in building much harder.

Hereunder figures 1 to 12 :

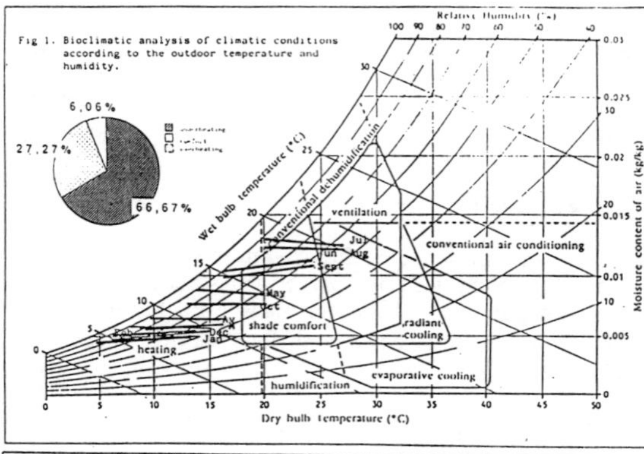

Fig 1. Bioclimatic analysis of climatic conditions according to the outdoor temperature and humidity.

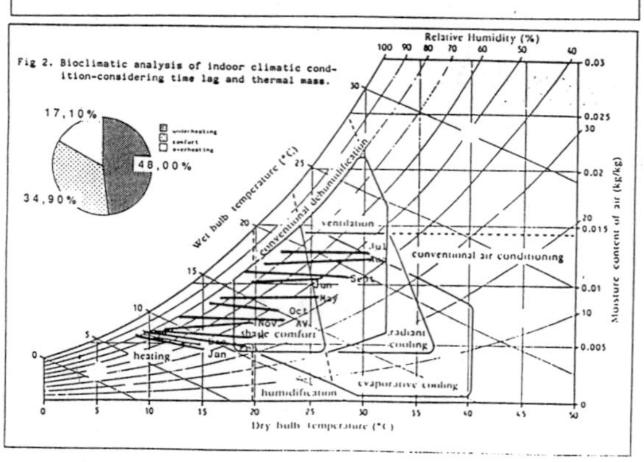

Fig 2. Bioclimatic analysis of indoor climatic condition-considering time lag and thermal mass.

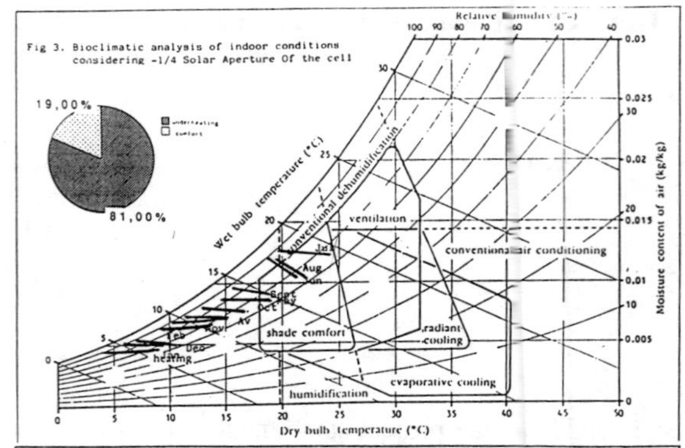

Fig 3. Bioclimatic analysis of indoor conditions considering -1/4 Solar Aperture Of the cell

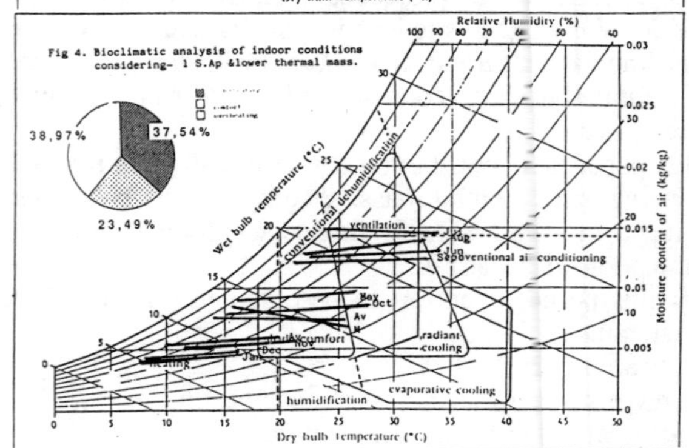

Fig 4. Bioclimatic analysis of indoor conditions considering- 1 S.Ap &lower thermal mass.

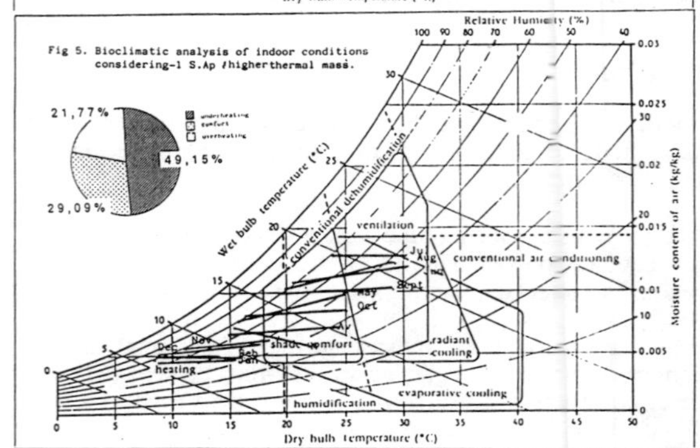

Fig 5. Bioclimatic analysis of indoor conditions considering-1 S.Ap /higherthermal mass.

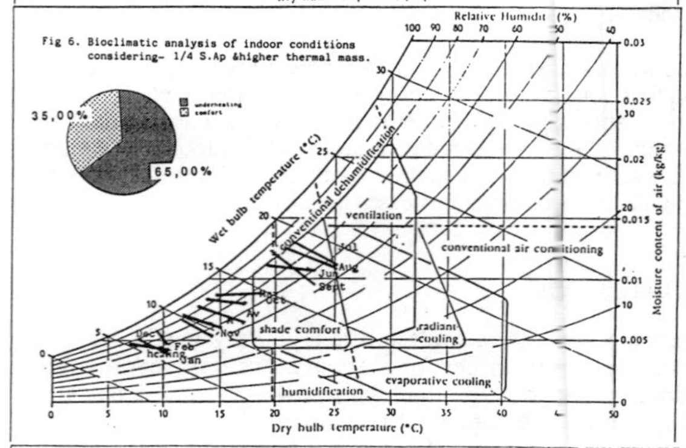

Fig 6. Bioclimatic analysis of indoor conditions considering- 1/4 S.Ap &higher thermal mass.

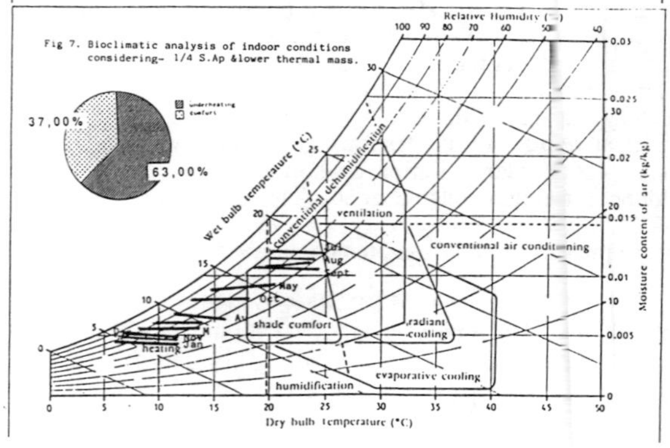

Fig 7. Bioclimatic analysis of indoor conditions considering- 1/4 S.Ap &lower thermal mass.

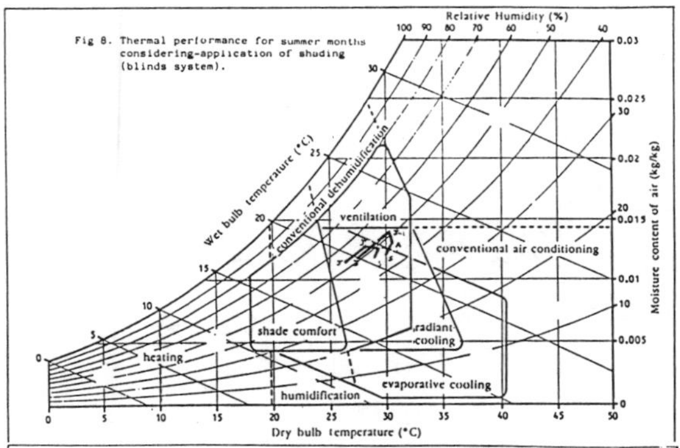

Fig 8. Thermal performance for summer months considering-application of shading (blinds system).

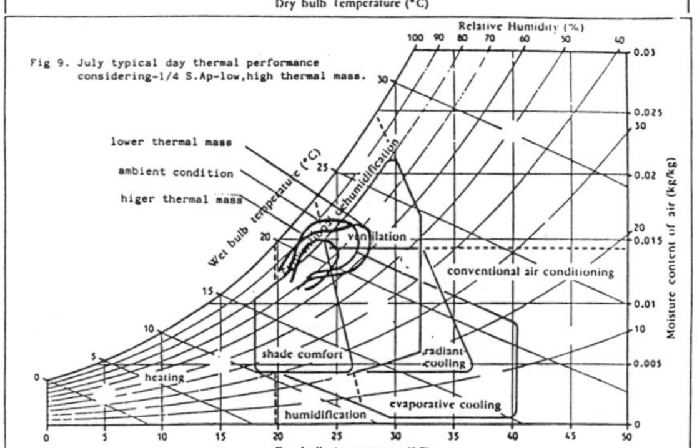

Fig 9. July typical day thermal performance considering-1/4 S.Ap-low,high thermal mass.

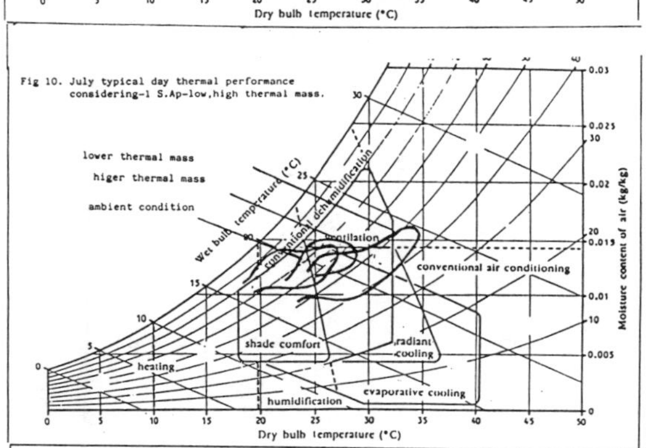

Fig 10. July typical day thermal performance considering-1 S.Ap-low,high thermal mass.

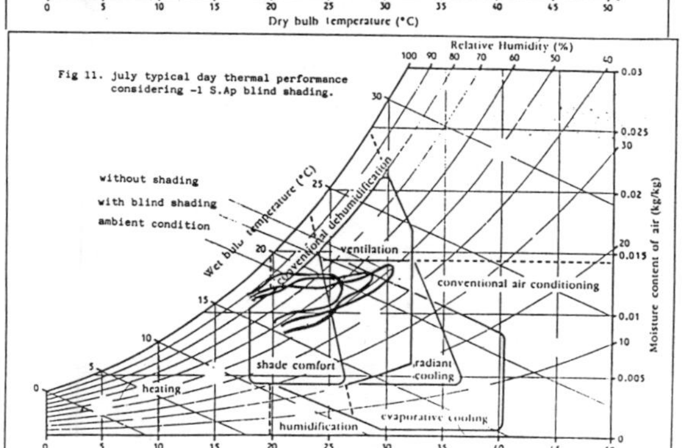

Fig 11. July typical day thermal performance considering -1 S.Ap blind shading.

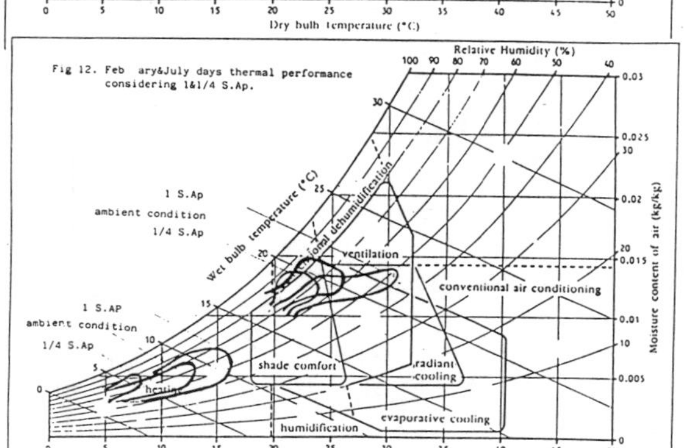

Fig 12. Feb ary&July days thermal performance considering 1&1/4 S.Ap.

Seasonaly sensitivity analysis

In order to clarify the results shown on a yearly basis by the time percentage diagramm deduced form the pychrometric chart, the same type of diagramm has been established but on seasonaly basis.

Figure 13 exhibits the differences corresponding to figures 2 and 3 : one can see that first configuration (1 glazing area, reference thermal mass) compared to 2nd one (1/4 glazing area, reference thermal mass) induces more comfort in mid seasons but generates a high percentage of overheating in summer.

Vice-versa, the 2nd one is responsible of a great percentage of underheating in all seasons except in summer where it generates a large fraction of comfort conditions and no overheating.

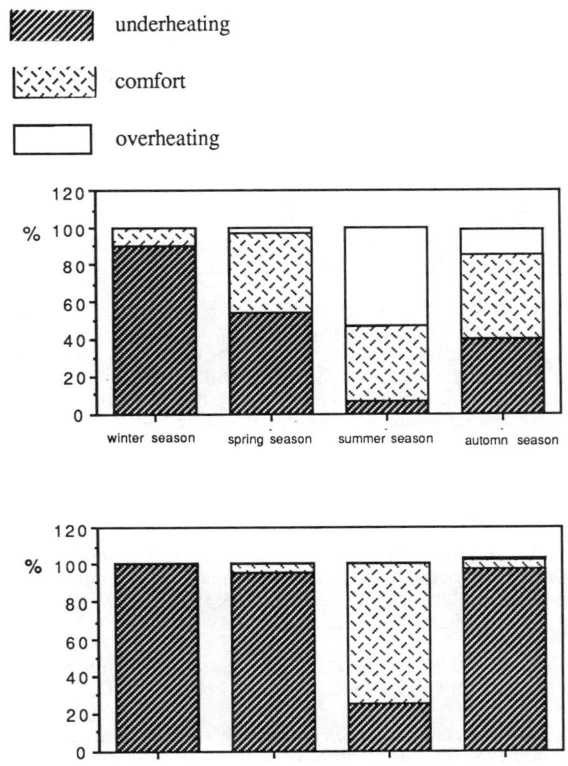

Figure 13 :Time percentage of respectively underheating, comfort and overheating within the four seasons for the two configurations :
up -> the reference test cell (1 glazing area, reference thermal mass)
down -> test cell with 1/4 glazing area and reference thermal mass

Histogramms of monthly temperature distribution

This kind of diagramm permits to compare the influence of the choices of the configuration of a building on the comfort to be obtained by another qualitative way.

Figure 14 shows the influence of a permanent shading. Here the comparison concerns the reference test cell with a permanent shading and without any shading. One can see that with shading, the frequency of indoor conditions between 16°C and 22°C is higher than in the case without shading and that reversely, without shading, the frequency of indoor conditons between 24°C and 32°C is higher than in the case with shading. In particular one can notice a high frequency around 30°C.

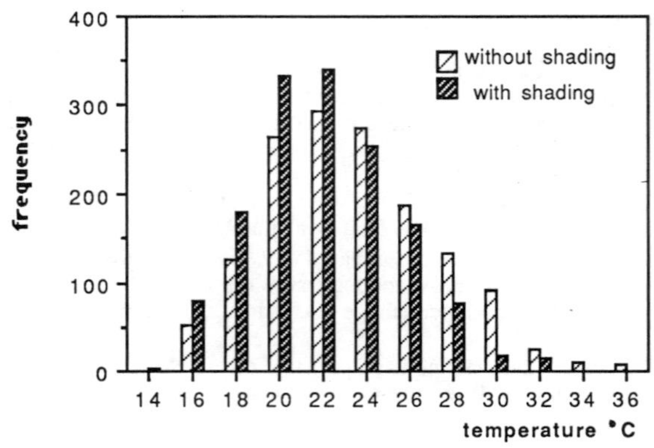

Figure 14 : Hourly frequency of indoor temperature of the reference test cell with and without shading for the two monthes July and August.

Conclusion

This paper gives a brief presentation of what kind of information an architect can get from thermal simulation results in order to analyse the influence of design choices on the indoor comfort to be generated in a building, at the design stage.

Today, manual plotting of simulation results was used in the case of the psychrometric diagramm. Tomorrow final diagramms will be able directly from the computer. It appears fruitfull that architect learn how they will be able to utilize the contained information at the very early design stage to build more comfortable and more energy saving buildings.

References :

- P. Achard, M. Tantot, Characterization of the variation of the solar aperture of a direct gain cell equipped with an automatic on-off controlled shading device, Work shop on passive solar testing in mediterranean climates, Ispra 24-26 June 1987.

- European Passive Solar Handbook, edited by P. Achard and R. Gicquel for CEC, DG XII, Cannes, February 1987.

- P. Achard, A. Ayoob, L. Elégant, Analysis & simulation of the performance of evaporative cooling system in different climates, C I B W 17 conference, Paris Oct. 1988

- B. Givoni, "L'homme, l'architecture et le climat", édition du Moniteur 1978.

- P. Achard, A. Ayoob & L. Elégant "Study of the influence of different passive system configuration on the thermal behaviour of a test cell in mediterranean climate. ISES, Cairo, Egypt 1989.

Implications de la question du confort sur la conception architecturale

ooo

Michel RETBI (MELTM/DAU), Gilles OLIVE (BEGO), Luc ADOLPHE (ENTPE/GISE)

ooo

Dans le secteur du bâtiment, on assiste à un regain d'intérêt pour la question du **confort**, et ceci pour plusieurs raisons qui impliquent la nécessité d'une nouvelle approche de cette question.

1 - La nécessité d'une nouvelle approche de la question du confort :

1.1 - L'intérêt d'une approche globale du confort :

Deux remarques s'imposent?

D'une part, les conforts hygrothermique, acoustique, visuel et olfactif ont des fondements physiologiques mais également psycho-sociologiques.

D'autre part, jusqu'à maintenant, la question du confort a surtout été l'objet d'une approche fragmentée : confort hygrothermique, confort acoustique, confort visuel, confort olfactif, confort spatial, confort fonctionnel (d'activité). Cela s'est d'ailleurs traduit par un intérêt inégal et variable dans le temps pour les différentes composantes du confort.

Or la finalité des études sur le confort est bien d'aider à concevoir des bâtiments dans lesquels on crée des **ambiances intérieures, perçues comme confortables, de tous les points de vue**.

Ainsi, malgré des progrès certains dans la connaissance des différents aspects du confort, de nombreuses incertitudes subsistent sur la **perception d'une ambiance confortable**, intérieure à un bâtiment. La question ultime est donc la suivante : quel **ensemble de conditions**, dites de confort, faut-il réaliser pour **satisfaire** les occupants d'un bâtiment, selon sa finalité d'usage ?

La réponse à cette question passe par une **approche globale du confort.**

1.2 - L'intérêt d'une approche intégrée du confort :

Trois tendances actuelles créent un nouveau contexte à la question du confort, suffisamment fort pour inciter à **une approche intégrée du confort**.

Tout d'abord, on peut remarquer que la demande de confort est intégrée dans une demande plus globale de **qualité des bâtiments**, affirmée ou potentielle, relevant de l'exigence ou du souhait, finalisée sur un souci de **protection** (sécurité, santé), de **confort d'ambiance** (hygrothermique, acoustique, visuel, olfactif, spatial), de **confort d'activité**. Actuellement, le progrès technologique fait progresser sensiblement le rapport "offre/demande" en matière de qualité des bâtiments. Aussi la question du confort devrait être traitée en tenant compte des caractéristiques de son contexte : la qualité des bâtiments.

Ensuite, on peut remarquer que la question de la **sécurité** est actuellement posée avec une acuité certaine. Il est donc difficile de bien traiter la question du confort sans se soucier simultanément des implications que peuvent avoir les exigences de sécurité sur la réalisation des conditions de confort.

Enfin les questions du confort et de la **santé** se trouvent confrontées pour des raisons conjoncturelles.Dans les pays industrialisés et donc fortement urbanisés, à la fin du 19ème siècle, il y avait beaucoup trop de bâtiments insalubres. Dans la première moitié du 20ème siècle, une lutte constante contre l'insalubrité a été menée, avec succès. Elle a été accompagnée de premiers efforts pour créer des conditions de confort thermique dans les bâtiments que l'on construisait. Dans la deuxième moitié du 20ème siècle, la volonté d'avoir des bâtiments confortables, des points de vue hygrothermique, acoustique, visuel, olfactif, spatial, est devenue dominante. Cependant des problèmes de santé sont réapparus sous des formes bénignes ou graves, du syndrome des bâtiments malsains (angoisses, improductivité, maladies) jusqu'à la mortalité (cancers, légionellose). Ainsi, on savait que la **salubrité** des bâtiments n'entraînait pas le confort dans des bâtiments. On sait maintenant que le confort n'implique pas la salubrité. Les questions du confort et de la santé devraient donc être l'objet d'un traitement simultané.

Ainsi plusieurs tendances actuelles poussent à une **approche intégrée** du confort , c'est-à-dire pour un traitement de la question du confort en relation avec les questions de **sécurité** et de **santé**.

Quelles sont les implications conceptuelles, techniques et architecturales, d'une **approche globale et intégrée du confort** ?

2 - La problématique technique d'une approche globale et intégrée du confort :

On a vu précédemment qu'une **approche globale et intégrée** du confort était souhaitable. Ce type d'approche, relativement complexe, suppose une définition aussi rigoureuse que possible de ses objectifs. On peut faire quelques remarques à ce propos.

2.1 - La nécessité d'une approche systémique du confort :

La question du confort correspond au traitement d'un système **"Bâtiment - équipements - occupants"**.

L'**approche structurelle** de ce système n'est pas trop complexe, même en prenant en considération les différents points de vue de :
- **confort d'ambiance** (hygrothermique, acoustique, visuel, olfactif, spatial),
- **confort d'activité**,
- contraintes connexes de **sécurité** des personnes et des biens et de **santé** des personnes.

Seul le repérage des moyens structurels permettant la dynamique du système (par exemple les éléments de stockage), peut présenter une certaine difficulté.

En revanche, l'**approche fonctionnelle** de ces systèmes est d'une complexité certaine. On doit pouvoir identifier :
o les éléments fonctionnels du système :
- les flux (de matière, d'énergie, d'information) traités,
- les décisions,
- les rapports "décisions/flux traités".
o la dynamique du système.

Quelques remarques sur les objectifs et les moyens fonctionnels peuvent aider dans cette approche fonctionnelle des systèmes "Bâtiments - équipements - occupants".

2.2 - L'expression fonctionnelle des besoins :

La norme expérimentale X 50-151 "Guide pour l'élaboration d'un cahier des charges fonctionnel (expression fonctionnelle du besoin)" peut guider dans la formulation de la demande, affirmée ou potentielle.

Elle distingue :
- les fonctions de service et les contraintes,
- leurs critères d'appréciation,
- les niveaux des critères d'appréciation et ce qui les caractérise,
- les flexibilités (ensemble des indications exprimées par le demandeur sur les possibilités de moduler ces niveaux),
- les limites d'acceptation,
- les taux d'échange (rapport déclaré acceptable par le demandeur entre la variation du coût ou du prix et la variation correspondante du niveau d'un critère d'appréciation).

Il est clair que les notions de flexibilité, de limite d'acceptation et de taux d'échange seront très utiles pour une **approche fonctionnelle pratique** du confort et des exigences connexes.

Pour ce qui est des fonctions de service et des contraintes, il faut citer les exigences retenues dans la directive "Produits de la construction", adoptée le 22-06-88 par le Conseil des Ministres de la Communauté Européenne. Elle sont au nombre de six :
- la résistance mécanique,
- la sécurité en cas d'incendie,
- l'hygiène, la santé et l'environnement,
- la sécurité d'utilisation,
- la protection contre le bruit,
- l'économie d'énergie et l'isolation thermique.

2.3 - L'approche micropsychologique du confort :

Dans l'expression fonctionnelle de la demande de confort, il faut enfin ne pas oublier que la perception de certains détails, concernant les conditions de confort ou les moyens de créer ces conditions, a une importance qualitative certaine. On peut citer quelques un de ces "détails" :
- l'hétérogénéité des ambiances,
- la variabilité dans le temps et dans l'espace (d'une ambiance à une autre) des caractéristiques des ambiances,
- les possibilités de gestion des conditions d'ambiance (contrôle et commande).

2.4 - La notion de qualité :

Pour ce qui est de la définition des moyens fonctionnels, la notion de qualité peut être d'une aide précieuse.

La norme NF X 50-120 "Qualité. Vocabulaire" définit la qualité comme "l'ensemble des propriétés et caractéristiques d'un produit ou service qui lui confèrent l'aptitude à satisfaire des besoins exprimés ou implicites".

3 - <u>Les implications d'une nouvelle approche du confort sur la conception architecturale</u> :

A première vue, le fait de faire avancer l'état des connaissances pour chaque aspect de confort et les savoir-faire pour la réalisation des conditions de conforts spécifiques dans chaque domaine pourrait sembler suffisant pour entraîner une amélioration satisfaisante de la qualité architecturale.

Force est de constater que cela n'est pas si simple dans la pratique et que dans les dernières années de nombreuses erreurs ont été commise dans la définition et l'application de règles de conception architecturale spécialisées, et en définitive le plus souvent rejetées à la longue par les concepteurs, bien heureusement. Il en a été ainsi, pour ne citer qu'un exemple parmi d'autres, de la tendance à diminuer les dimensions de fenêtres des bâtiments d'habitations au nom d'une exigence énergétique, négligeant l'interactivité de la fonction énergétique de la fenêtre avec ses autres fonctions (éclairage naturel, vision sur l'extérieur, ouverture des fenêtres pour entrer en relation avec l'ambiance extérieure, ...).

Ainsi l'on peut dire que la seule logique qui consiste à définir des exigences architecturales spécialisées, voire au mieux à les additionner dans la pratique de conception tend à favoriser une conception réductrice, et qu'il apparait nécessaire de réfléchir à des **exigences pour la conception architecturale qui soient les implications d'une approche globale du confort et d'un certain nombre d'exigences connexes**.

On peut d'ores et déjà faire quelques remarques relativement à l'élaboration de ces nouvelles exigences architecturales.

3.1 - Une méthodologie d'élaboration de nouvelles exigences pour la conception architecturale :

L'élaboration de nouvelles exigences pour la conception architecturale est analogue à un problème d'optimisation.

Il y a un certain nombres de paramètres de conception architecturale, concernant le bâtiment projeté et ses équipements, qui participent à la création de l'ambiance intérieure du bâtiment. La satisfaction d'une **exigence de confort spécifique**, en respectant des **contraintes simultanées**, implique des **contraintes résultantes** sur certains **paramètres de conception.** Ce sont ces contraintes résultantes qui déterminent les exigences pour la conception architecturale.

Tout d'abord, il est à remarquer que le repérage des paramètres de conception concernés par un ensemble donné "exigence de confort spécifique - contraintes simultanées" n'est pas du tout évident.

Ensuite, on peut suggérer d'explorer des ensembles **"exigence de confort spécifique - contraintes simultanées"** de types différents :

o les contraintes simultanées ne sont pas des exigences de confort spécifiques : on essaie alors de produire des exigences pour la conception architecturale, réalistes pour la réalisation d'un confort spécifique. Un travail de systématisation est engagé par l'équipe pour les différents types de conforts spécifiques, à partir de travaux exploratoires surtout relatifs au confort hygrothermique.

o les contraintes simultanées sont d'autres exigences de conforts spécifiques : on essaie alors de produire des exigences pour la conception architecturale qui permettent la réalisation d'un confort multiple (la réalisation simultanée de plusieurs conforts spécifiques). Ce travail correspond à la mise en évidence des implications architecturales d'une approche plus ou moins globale du confort, sans pour autant être vraiment réaliste (prise en compte de contraintes qui ne sont pas des exigences de confort spécifique). Quelques éléments existent concernant surtout les conforts hygrothermique et acoustique.

o les contraintes simultanées sont d'autres exigences de conforts spécifiques et des contraintes connexes : on essaie alors de produire des exigences architecturales réalistes pour la réalisation d'un confort multiple (la réalisation simultanée de plusieurs conforts spécifiques). Ce travail correspond à la mise en évidence des implications architecturales réalistes (prise en compte de contraintes qui ne sont pas des exigences de confort spécifique) d'une approche plus ou moins globale du confort. Ce travail ne semble pas encore avoir fait l'objet d'efforts exploratoires significatifs.

3.2 - L'interactivité entre la question du confort et la conception architecturale :

Le renouveau de l'intérêt pour la question du confort ne se réduit pas à la question de l'émergence de nouvelles exigences pour la conception architecturale.

Il y a en effet l'annonce d'une nouvelle interactivité entre la question du confort et la conception architecturale :

o d'une part, l'évolution des idées sur la question du confort peut se traduire par la modification du discours de l'architecte. Il suffit de prendre pour exemple la réalisation de l'Institut du Monde Arabe.

o d'autre part, le souci du confort chez l'architecte peut l'amener à une conception architecturale qui, non seulement aboutit à la création d'ambiances intérieures confortables physiologiquement et psycho-socio-logiquement, mais propose des ambiances qui sont satisfaisantes culturellement et esthétiquement. En ce sens, le travail de conception architecturale ne se réduit pas à faire l'addition d'exigences et de règles. Il est un véritable travail de synthèse, d'invention et de création, où l'utilisation de connaissances ne s'oppose pas au travail sur l'émotion, mais l'enrichit.

ooo

PASSIVE COOLING ANALYSIS TO A HOLIDAY COMPLEX IN IRAKLION CRETE

M. Kopidakis*, E. Athanassakos**, C. Theofylactos***

* 136 Ionias st., 713 05 Iraklion, Crete, GR

** Centre of Renewable energy Sources (CRES)
6 Frati st., Fousia 194 00 Koropi GR

*** 7 Ioustinianou st. 114 73 Athens GR

ABSTRACT. A two-storey holiday building located in the summer resort area of Gournes, Crete, is designed with passive systems to cover cooling and part of its heating requirements. The building contains ten flats which are in use mainly during summer and Christmas holidays. Due to the longer summer use, priority has been given to cooling making use of natural cross ventilation, shading and a hybrid system named ventilation chimney. This system is sized to provide cooling for the extreme summer dry bulb temperature. During the short winter use, the direct gain system covers 26% of the heating load. All flats are provided with an energy efficient fireplace which can amply cover the remaining heating load. Also appropriate levels of thermal insulation and thermal mass contribute to reducing the energy requirements of the building and the sizing of the installed heating and cooling systems.

1. INTRODUCTION

In Greece, the cooling of buildings is as important as the heating especially in the non-residential sector. With the high price of electricity in Greece, the conventional air-conditioning systems are expensive solutions to cooling. On the other hand passive cooling can provide the buildings' occupants with thermal comfort and it can also be an economic solution. The present work deals mainly with the passive and hybrid cooling of a holiday complex and the design of a fireplace - which is again very fashionable in Greece- as an energy efficient system for both heating and cooling.

The study on the design of this building was partially supported by the Building 2000 programme of the DG XII of the European Communities.

2. SITE, BUILDING AND CLIMATE

The building will be located in the summer resort area of Gournes, 15 km east of Iraklion, Crete. The 1000 m² site has an altitude of 15 m above sea level. It is completely flat and has sea views on the N, NW and W sides and a view of open country to the S. The building has two storeys and a basement, each of which has a floor area of 200 m². There are five small (40 m²) flats on each floor. The flats are arranged so that they all have windows on opposite sides, at right angles to the prevailing summer winds - which are in an NW-SE direction. All the entrance doors are on the SE side of the building. Each apartment has a private 9 m² balcony (fig. 1,2,3). The roof can be visited and is covered with a bamboo trellis to provide shading for the tenants and reduce the cooling load. The building's support structure-beams, roof and floor slabs-is of reinforced concrete. The walls consist of two layers of brick with thermal insulation between them.

The overall heat transfer coefficent of the building is lower than the required one by the Thermal Insulation Regulations imposed on this region. Also both the heavy masonry materials used and their appropriate positioning give high thermal inertia to the building.

The climate may be characterized as mild Mediterranean with well defined seasons and rare extremes. The annual glabal solar radiation is 5872 MJ/m² and the averge ambient temperature varies from 12°C to 26.5°C annually. The prevailing winds are cool summer sea breezes of medium intensity (3-4m/s) from the NW. The direction and frequency of the cool summer breezes influenced the layout of the building and its siting on the lot.

3. THE PASSIVE SYSTEMS

The quick turn around of tenant was the main reason for choosing the following passive systems because each flat's cooling and heating system had to be autonomous, simple and straightforward to operate.

3.1 Natural Cross Ventilation

All the apartments have a double exposure to the prevailing cool sea breezes. The openings on both the windward and leeward sides were designed to give comfortable air speeds in the living room.

3.2 Shading

Movable shading devices which can be revolved vertically and horizontally can be used to obstruct direct solar radiation during the warmest part of the day. The shading devices are placed on the parapet of the balcony to improve the natural ventilation. The movable and revolving shading devices perform three functions. They can be used to:

- obstruct the direct solar radiation
- minimize to a certain extent the intensity of the natural illumination;
- deflect the air steams to one direction or another or even to increase their velocity by the manual adjustment of the shading device fins.

As mentioned above, the roof is covered by a sloped double-layered bamboo trellis which:

- creates a shaded 200 m² area which tenants can visit for the beautiful sea view;
- protects the roof from the intense solar radiation, which is over 200 KWh/m² a month during the three summer months.

The shading, the appropriate levels of thermal insulation and the natural movement of fresh air through the apartment as well as the large thermal inertia of the building result in a low cooling load which can be met by the hybrid "ventilation chimney system".

3.3 Ventilation Chimney

The SE side of the basement is partly buried while the NE side is 1.5 m above ground level. Water from a nearby well runs freely over the floor of the basement at a constant temperature of 14°C. Fresh air enters the basement from five openings on the NW side. Each of these openings houses an axial fan sized to deliver the air at an appropriate flow rate to two apartments, one on the first floor and

one on the second floor of the building.(fig.4). The axial fans distribute the air over the wetted surface of the basement and the cooled air is delivered to the two apartments through a diverging pipe system. A plastic cover is spread immediately above the wetted surface to minimize the movement of air throughout the basement. As the dry air flows past the wetted surface, there is a transfer of both senssible and latent heat. The air mass temperature drops and its relative humidity rises. The cooled fresh air is deliverd to each flat separately from the basement. The exhaust (warmer) air leaves each flat via openings near the top of the dynamic fireplace. From ther it is piped to a point some 3 m above the covered roof of the building. The end of the chimney consists of a 1 m black metallic sheet which, when it reaches high temperatures, enables the temperature and density differences needed for thermodynamic operation to be created. Alternatively, an air circulator can be used to achieve proper operation of the system.
If the flats are occupied outside the summer season, the outlet openeings are kept closed with dampers to prevent heat loss.

3.4 Heating

During the short winter use, the heating load of each flat can be met by direct gain and an energy-efficient fireplace. The fireplace is built in the partition wall between the two main rooms and consists of metallic hollow slabs (60X120 mm in cross section and 4 mm thick) in contact with high thermal capacity bricks.
The metallic heat exchangers are arranged in the shape of a parabola. Room air passes naturally through them, gains heat, and is conveyed to the bricks, delivering to them part of the heat gained from the heat exchangers. Part of the air returns to the room from openings 1 m above the floor. In addition, the combustion gases passing through the chimney deliver some of their heat to the brick wall surrounding the chimney(fig.4).

4. ENERGY PERFORMANCE

The energy performance of the building was simulated on a computer using a simplified thermal network method /1/. The ten flats were divided into four zones according to their exposure to climatic conditions. For each zone, hour-by-hour simulation runs were carried out using mean monthly hourly values of solar radiation, ambient temperature and internal gains resulting from the occupants´activities. The internal temperature distribution was calculated with and without the effect of the passive techniques incorporated in the building and the cooling and heating loads computed whenever the temperature was above 26°C or below 20°C respectively. This enables the efficiency of the passive measures to be evaluated under mean monthly conditions. In addition, the ventilation chimney system was sized under extreme summer conditions (36°C and 55% relative humidity) and full occupancy in order to insure the thermal comfort even under these rare conditions.

4.1 Cooling season

With the shading devices for the roof, windows and balcony doors in operation, the total mean annual cooling load of the whole building was calculated to be 60 GJ. Without shading, the annual cooling load was 99 GJ. The peak cooling load to achieve an internal temperature of 28°C under extreme summer conditions was in the range 2.5 kW to 3kW per apartment, depending on the zone (fig.5).
The effect of cross ventilation was investigated under mean monthly conditions. Air flow rates were calculated using ASHRAE recommendations /2/. The average air speeds achieved in the living space were computed with a semi-empirical equation produced from Givoni´s experimental data /3/. Fig (7, 6) show the variation in internal temperature with and without natural ventilation for a typical apartment in zone B. In late morning and afternoon when wind speeds are high the internal temperature nearly equals the ambient temperature. This is because the large ventilation openings give rise to high air flow rates. To reduce the maximum air speeds produced in the living space, the size of the windward and leeward openings were adjusted to make them equal. Thus, in the bedrooms, outlets were incorporated in the loft on the leeward side of the building.
Average air speeds in the living space range from 0.14 m/s to 0.97 m/s. The higher value is encountered in the warmest months (July and August). In these months the internal temperature of the ventilated rooms reaches 29°C with a relative humidity of 55% using the temperature index chart - though in this case, it is conservative because of the assumed clo- it can be concluded that under the forementioned conditions, the temperature can be reduced to just below 26°C with an air speed of a little less than 1m/s. It can also be concluded, for dry bulb temperatures from 26°C to 29°C (the range of internal temperatures found in the ventilated space) and relative humidities varying from 55% to 60%, the desired temperature of 26°C can be produced with air speeds of 0.14 m/s to 1 m/s. This shows that, under mean monthly conditions, the cross ventilation system can provide occupants with thermal comfort.

4.2 Ventilation Chimney

After a careful study of the parameters involved in the design of the chimney the procedure depicted in fig 4. was proposed. It was found that the sized axial fan is suitable for cooling each flat when the dry air passes over a wetted covered surface 3m long. Five openings house the fans which are connected to a flexible piping system to deliver air at the needed flow rates to a couple of apartments. The diameter of the piping system is .4m and .35m for the first and second floor respctively. The calculation of the cooling rate is based on determination of the flow of air, on the local Nusselt Number, and the average convection coefficient. Data analysis showed that, under the extreme summer conditions, the temperature of the air entering the apartment is 25 °C and its relative humidity 76%-values judged to be comfortable for the occupants. Further simulation analysis for a range of different external temperatures (33-36 °C) and relative humidities (55-75%) showed that, if the initial relative humidity was above 60%, the apartment could become uncorfotable because, although the air entering the room would be at an acceptable temperature (ie in the range 23-25 °C), its relative humidity would be intolerably high (ie in the range 85-93%).

4.3 Heating Season

The mean annual heating load of the complex was calculated to be 74.5 GJ, of which the solar contribution was found to be up to 28.5 GJ. For the only winter period when it is expected that the building will be occupied (ie the one month around Christmas), the load is 16 GJ with a solar contribution of 26%. These calculations assumed a thermostat setting of 20 °C (fig. 8,9). For extreme winter conditions (ie an ambient temperature of 3 °C and no solar gains) the capacity of the heating system installed in each flat would need to be in the range 2.2 KW to 2.6 KW, depending on the flat's

exposure to ambient conditions. The heating capacity of the fireplace installed between th two rooms of each flat is 4 KW, which more than meets this requirement.
A detailed analysis was carried out on the internal configuration of the fireplace, the damper location and the height and location of the lintel. The best chimney size was established from the system resistence coefficient and the pressure losss /1/.

fig. 1

fig. 2

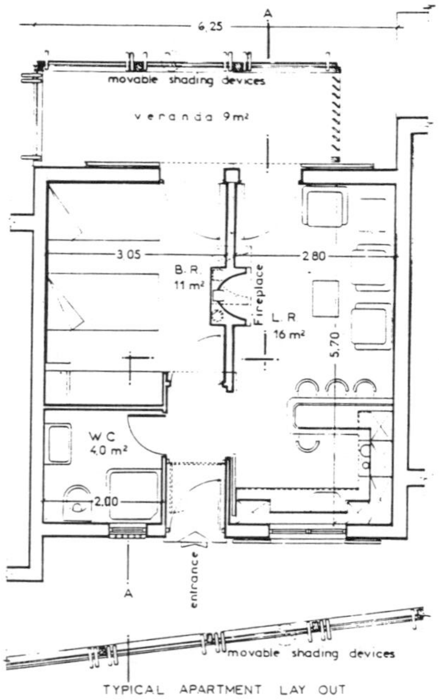

fig. 3

5. References

1. E.K. Athanassakos, C.G. Theofylactos, "Thermal Performance Analysis on the Passive Heating and Cooling Design of Project 734/GR", B2000 Programme, Commission of the European Communities, Contract # EN 3S-0155-NL.
2. ASHRAE Handbook of Fundamentals, pp. 22.1- 22.22 (1981).
3. B. Givoni, "Man, Climate and Architecture", 2nd edition, Applied Science Publishers LTD, London (1976).

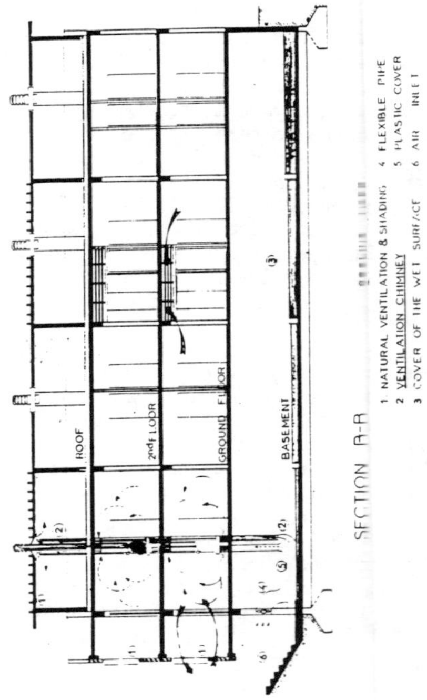

fig. 4

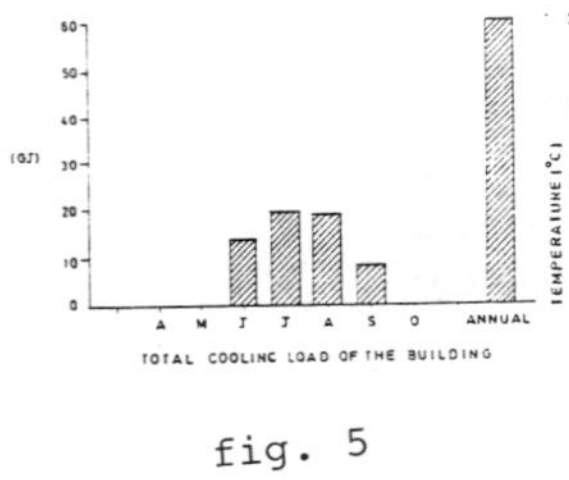

fig. 5

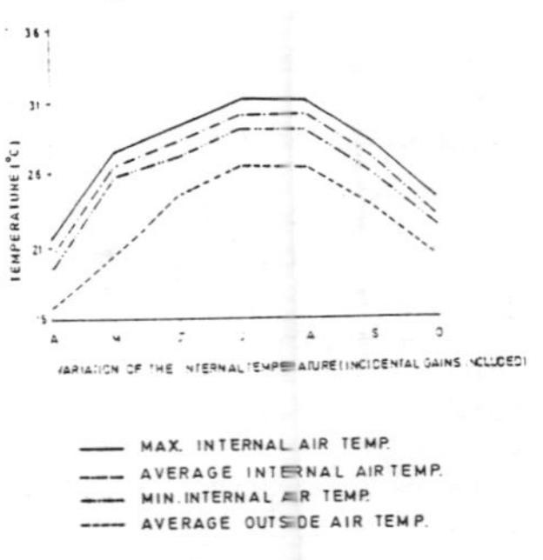

fig. 6

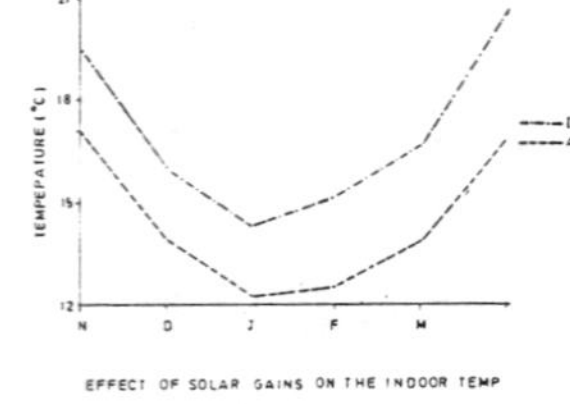

fig. 8

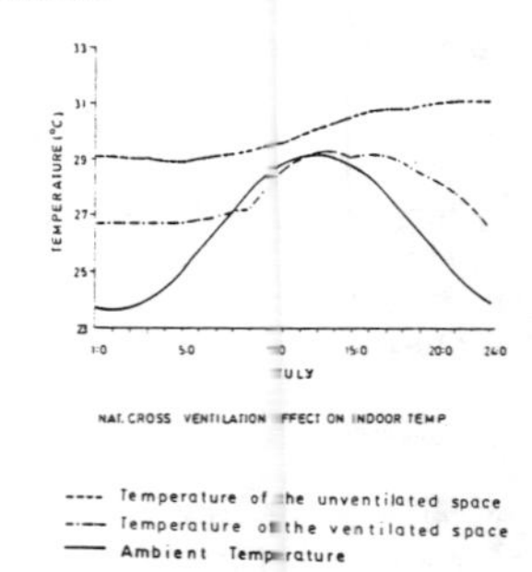

fig. 7

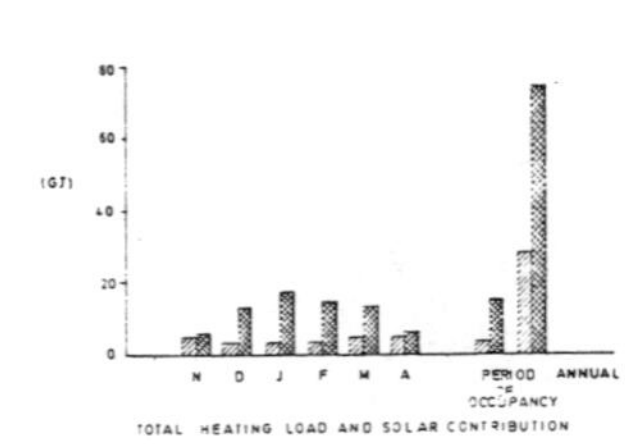

fig. 9

CLIMATIC BUILDING DESIGN MEASURES FOR HUMAN COMFORT AND ENERGY CONSERVATION IN KUWAIT

S.A.A. Al-Awadhi and A.A.M. Sayigh

Department of Engineering, University of Reading, Whiteknights, PO Box 225, Reading RG6 2AY, UK

ABSTRACT: A considerable amount of work has been done by architects, engineers and scientists, in the field of passive designs, so as to achieve human comfort in buildings. This has to take into account the need for energy conservation which has been arising since the last two decades. The aim of this paper is to show the importance and use of climatic data in building design in Kuwait. This approach will result in better human comfort and energy conservation in buildings. An approved statistical methodology has been used to get a model (representative) year of various climatological parameters, namely: temperature, relative humidity, sunshine hours and solar radiation. It also shows how a bioclimatic chart has been established for Kuwait as an initial approach for calculating the cooling and the heating loads requirements. Some architectural examples for buildings in Kuwait are also discussed in the way of applying passive approaches in their design.

1. INTRODUCTION

Knowledge of the climatic data of any place is an essential milestone for designing and constructing buildings in that particular place. Kuwait, which lies between latitudes 28°45'-30°05' North and longitudes 46°30'-48°30' East, is mainly a desert country and its weather is generally hot and dry. It is a typical arid climate. It has been shown in a government report that over 60% of the total electricity consumed in the summer is used to meet the air-conditioning loads of buildings.

An approved statistical methodology has been applied in order to get an accurate picture of Kuwait climatic conditions. The method utilises a period of 22 years of weather data measured in the Kuwait International Airport and has been analysed by using sophisticated computer software. These are temperature, relative humidity and sunshine hours. An empirical relationship between the sunshine hours and solar radiation has been used to get an optimal result for the solar radiation data. In this study a model (representative) year has been carried out for each of the previous parameters.

The model year is made up of those months for which both of the mean and standard deviation are the closest to the long term average mean and the long term standard deviation for that month. A building bioclimatic chart has been established for Kuwait on the basis of the previous climatic analysis. This chart is an important tool to decide where, when and which passive approaches can be applied so as to reduce the energy consumption for the cooling and for heating loads requirements, which can achieve both human comfort and energy conservation at the same time. Also in the study, an architectural point of view has been discussed on some modern buildings in Kuwait. The focal point of discussion is how far these buildings are from using the passive approaches in their design. The paper also shows that the imposition of modern architecture in the Kuwaiti domestic environment has resulted in an improper utilisation of the local climate, materials, people's needs, traditions and aesthetic appeal. This, in fact, has paved the way for Arabic/Islamic architecture to retake its place in the local environment of Kuwait.

2. CLIMATE OF KUWAIT

Kuwait lies between latitudes 28°45'-30°05' North and longitudes 46°30'-48°30' East. Kuwait climate has the characteristic of hot arid maritime regions. It has a long hot summer with frequent dust storms and short mild winter with a relatively small number of cloudy and rainy days /1/.

Figures (1), (2) and (3) illustrate the annual variation of dry-bulb temperature, relative humidity and sunshine hours durations respectively.

3. THE STATISTICAL METHODOLOGY OF THE MODEL YEAR

The importance of the model year is due to the fact that the climatological data of a country varies from year to year. Climate can have a cycle of 30-years before it repeats itself or in some cases 12-years, /2/, therefore a climatic model year for each country is very much needed to predict the performance and sensitivity of energy conservation in buildings in the same country. The model will be more accurate if it utilizes 20-years data or more. A period of twenty two years has been considered in this study /1/. The following steps show the process of developing the model year:

1. Examining the data to make sure that they are reasonably reported and no gap exists among them that is the hourly data.
2. The daily values of the data are then averaged for each month of each year.
3. The standard deviation of the daily values relative to the monthly mean value is calculated for each month of each year.
4. The model year is made up of those months for which both of the mean and the standard deviation are the closest to the long-term average mean and standard deviation for that month.
5. The months chosen to make the model year are real months, i.e. the model year of real months represents the total period for the available data (1962-1983) /2,3/.

In this study the model year has been established for the dry-bulb temperature, the relative humidity and the sunshine hours. From the final results, Fig. (4) illustrates the annual variation of dry-bulb temperature, while Fig. (5) illustrates the annual variation of the relative humidity. Figure (6) illustrates the annual variation of the sunshine hours.

An estimation of the total solar radiation values on horizontal surface, H, (global radiation) has been carried out by using an empirical formula utilizing the data of the model year of the sunshine hours and its respective values of dry-bulb temperature and relative humidity. The empirical formula used for estimating the total solar radiation is expressed as: /4/

$$H = 1.53\ K \exp[\phi(D - \frac{R}{100}^{\frac{1}{3}} - \frac{1}{T_{max}})]$$

where H = total solar radiation in cal./cm^2 day
ϕ = the latitude angle of the place in radians
D = the ratio of the actual mean sunshine hours to 12 hours
R = the mean relative humidity in percentage
T_{max} = the mean of the maximum temperatures in C°
K = the latitude factor
$= (\lambda z + \Psi_{i,j} \cos \phi)100$

where $\lambda = 0.2/(1 + 5.73\ \phi)$
$\Psi_{i,j}$ = the seasonal factor (i = 1,2,3 as inland station, coastal station and hilly station respectively, while j = 1,2,3,... 12 stand for January to December) /4/, i = 1 is considered in this study for Kuwait International Airport, inland.
z = the length of the day in hours and can be given by
$z = (\frac{2}{15}) \cos^{1}(\tan \phi \tan \delta)$

where δ = the solar declination angle for the specific time required.

The annual variation of the total solar radiation is illustrated in Fig. (7).

4. BUILDING BIOCLIMATIC CHART OF KUWAIT

Building bioclimatic chart, /6/, is an essential tool to decide where, when and which passive approaches can be applied to reduce the energy consumption for the cooling and/or heating load requirements, hence achieving both human comfort and energy conservation at the same time /5/. This chart has been established for Kuwait on the basis of the data from the model years of dry-bulb temperature and relative humidity respectively. The extreme seasons have been taken to define the required zones namely the climatic zones of both summer and winter. The hottest months in summer are June, July and August, while the coldest months in winter are December, January and February. In order to define the boundaries of these two seasonal zones, the average of the maximum and the average of the minimum of both dry-bulb temperature and relative humidity are used /6/. The climatic data is plotted on the chart and analysed with reference to the basic zones, Fig. (8). Comfort is assumed in this study to be achieved at dry-bulb temperature 20-26 C° and at a relative humidity of 25-65% /7/. In winter the dry-bulb temperature ranges from 6.2 C° to 21.7 C° and the relative humidity ranges from 28% to 91%. In summer the dry-bulb temperature ranges from 27.8 C° to 44.8 C° and the relative humidity ranges from 6.6% to 39.0%.

5. EXAMPLES OF SOME BUILDINGS IN KUWAIT

The imposition of modern architecture in the Kuwaiti domestic environment has resulted in an improper utilisation of the local climate, materials, people's needs, tradition and aesthetic appeal. This, in fact, has paved the way for Arabic/Islamic architecture to retake its place in the local environment of Kuwait. Two examples of government housing are illustrated in Fig. (9) and Fig. (10) In Fig. (9) the windows shading technique is far more effective than the one in Fig. (10). Fig. (11) shows a massive glazed facades on one of the new private buildings in Kuwait. This building will act as a solar collector, hence increasing the cooling load required to maintain the human comfort conditions. Figure (12) is another example of the government housing in the form of apartments complexes. These stepered buildings give an appropriate shading effect among them. In order to minimize heat conduction flow especially from the outside to the inside of the building (during the summer days) an insulation material is the best to use in the passive design strategy. An example of insulating material (light insulative white blocks) in one of the houses under construction is shown in Fig. (13). Finally Fig. (14) and Fig. (15) show an attempt to utilize the old architectural approaches into the modern buildings. Some approaches have been used efficiently, namely using the white colour and good window shadings, while some of the old building elements were used without clear function, namely the black wood beams.

6. CONCLUSION

In order to conserve energy in buildings and maintain human comfort in these buildings, it is clear that one should promote passive cooling approaches, namely windows and window shading effects, radiative cooling, evaporative cooling, ventilative cooling, coolness storage, building materials, thermal insulation, orientation and landscaping /7/.

7. REFERENCES

1. Meteorological Department of Kuwait Directorate General of Civil Aviation, Climatological Summaries (1962-1983), Kuwait International Airport.
2. A.A.M. Sayigh *et. al.*, Solar mapping of the Arab World, Proceedings of Physical Climatology for Solar and Wind Energy, Miramare, Trieste, Italy, 1986. Edited by Rodolfo Guzzi and Carl Gerald Justus, 972-986.
3. A.A.M. Sayigh *et. al.*, The importance of model year solar radiation data for use in solar applications, Workshop on Physical Climatology and Meteorology for Environmental Applications, ICTP, Trieste, 1988.
4. A.A.M. Sayigh, Solar energy availability prediction from climatological data, Solar Energy Engineering, edited by A.A.M. Sayigh, 1977, 61-82.
5. B. Givoni, Man, Climate and Architecture, 2nd ed., 1976, Applied Science Publishers Ltd., London.
6. K.H.H. Al-Jamal *et. al.* , Design for comfort in Gaza Strip and the West Bank, Solar and Wind Technology, Vol. 6, No. 3, pp 203-211, 1989.
7. M. Al-Zuhair *et. al.*, Designing for comfort in Iraq, Solar and Wind Technology, Vol. 5, No. 4, pp 369-382, 1989.

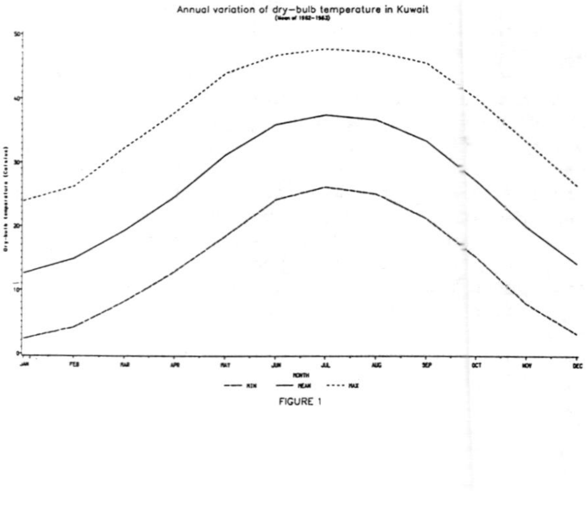

FIGURE 1

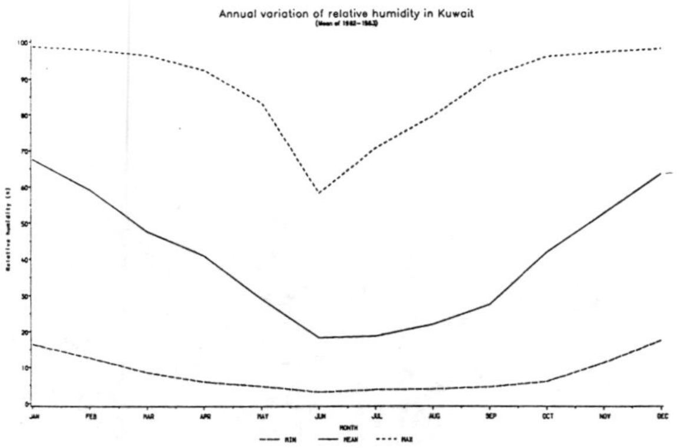

FIGURE 2

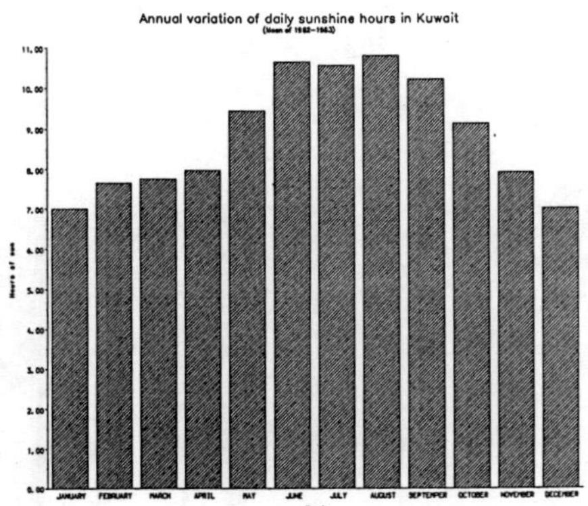

FIGURE 3

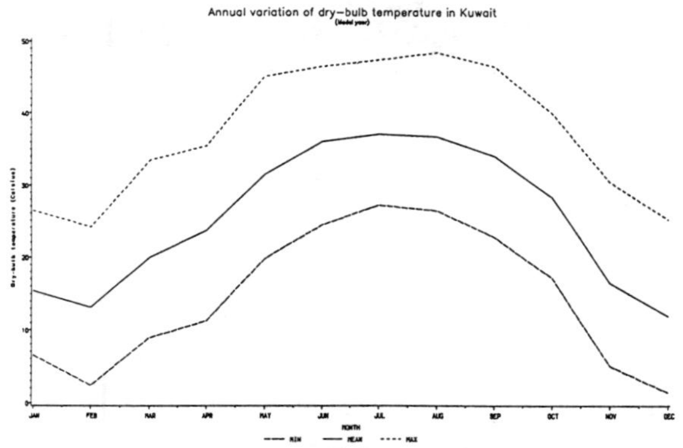

FIGURE 4

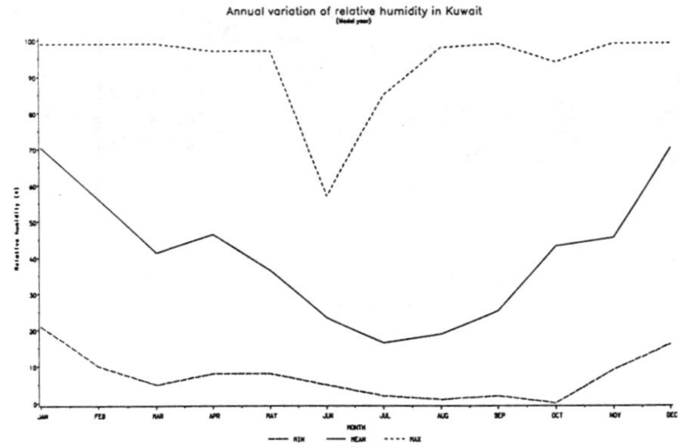

FIGURE 5

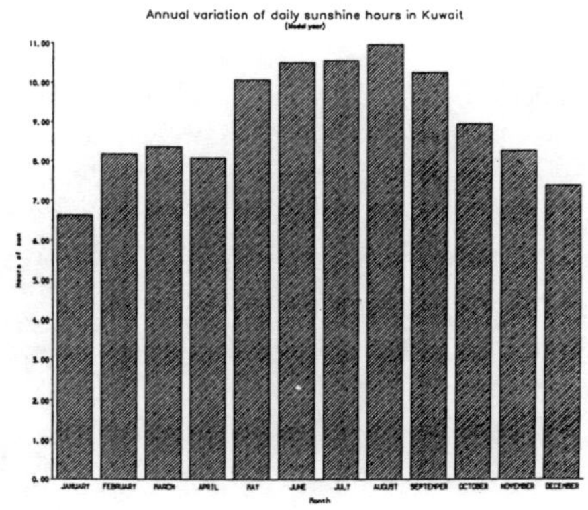

FIGURE 6

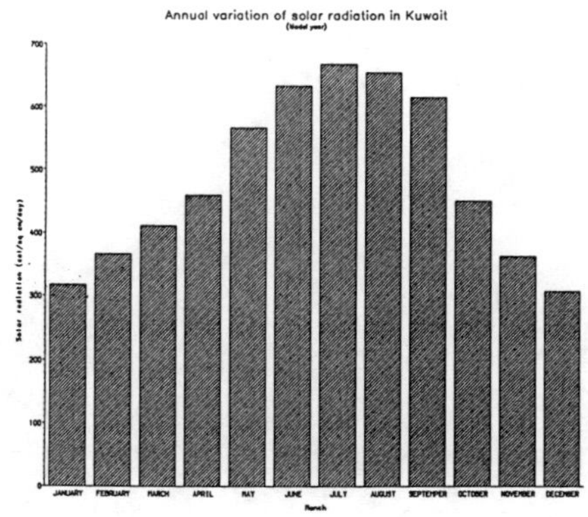

FIGURE 7

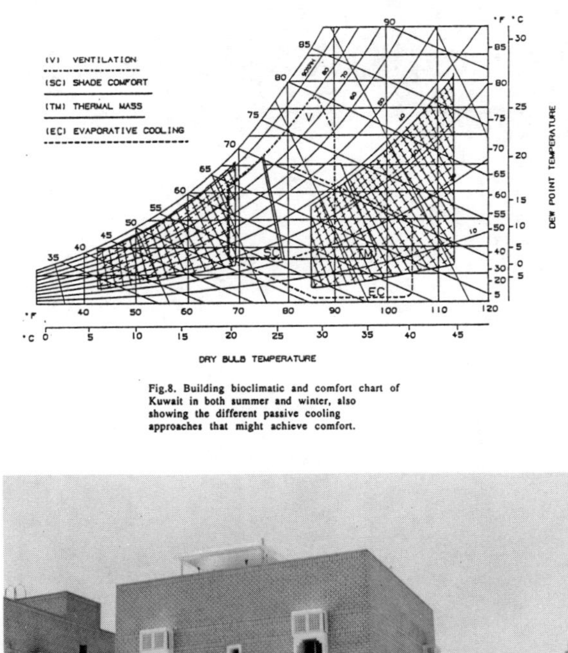

Fig.8. Building bioclimatic and comfort chart of Kuwait in both summer and winter, also showing the different passive cooling approaches that might achieve comfort.

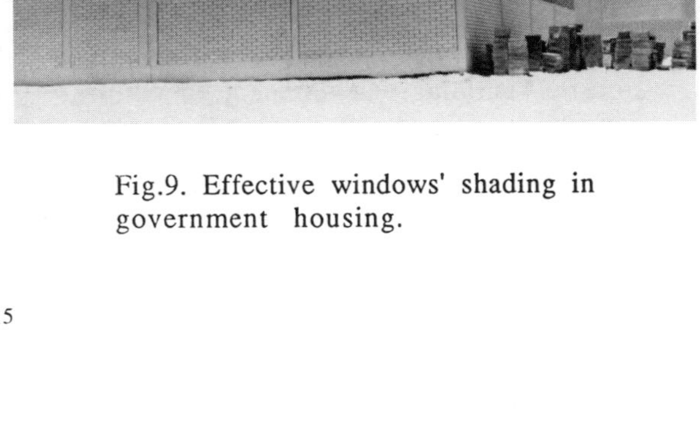

Fig.9. Effective windows' shading in government housing.

Fig.10. Uneffective windows' shading in government housing.

Fig.11. A massive glazed facades on private sector building.

Fig.12. Appropriate shading effect in this stepered building, Government appartment complex.

Fig.13. Light insulative white blocks applied in private sector housing.

Fig.14. An attempt to utilise the old architectural approaches into the modern buildings.

Fig.15. Another example to utilise the old architectural approaches into the modern buildings.

AIR MOTION AND COMFORT AS A DESIGN AID

J.P.B. Nathan, M. Arch
Ma'alot Dafna 134/35, P.O.B. 18367, Jerusalem, Israel
H. Wu, Arch. D.
School of Architecture, College of Architecture and Environmental Design
Arizona State University, Tempe, Arizona 85287—1605, USA

This paper reviews the contribution of air motion to human comfort in hot climates, and describes recent studies which quantify this contribution. Findings from two independent studies performed in the Arizona State University Environmental Testing Chamber are summarized. A general outline for strategy application and design issues is presented as well as suggestions for energy savings.

1.0 INTRODUCTION

"...Human thermal comfort can be defined as the state of mind which expresses satisfaction with the thermal environment."/1/ When conditions are not within the "comfort zone," the choice is either to change the ambient conditions through natural or artificial means or to adjust our sensation of comfort—that is, by extending the comfort limits and ignoring the ambient conditions.

Early work done (most notably by Olgyay) regarded air motion as a key method for extending the comfort zone and quantified this process in various ways.

Air motion acts in two ways: it convects heat away from the body and it encourages evaporation of perspiration, thus cooling. As Milne and Givoni have pointed out, the main limitation of comfort made possible by air flow is when the air flow becomes an irritant. Thus, occupancy type is an important consideration: a factory having quite a different limit than a professional office./2/ In general, an air speed of approximately one m/s (200 fpm) is the maximum suggested air movement according to the new bioclimatic chart (psychrometric format) developed by Arens et al./3/ At this air velocity, the comfort zone should be extendable to 33°C (92°F) with 40% relative humidity.

The combination of air speed, air temperature and moisture, clothing and activity level, all affect the sensation of comfort possible from air motion. Since a decrease in vapor pressure allows for an increase in evaporation of perspiration, the sensation of comfort will be even greater in dry climates. Conversely, when humid conditions prevail, evaporation does not easily occur, and sweating and discomfort prevails.

2.0 QUANTIFYING THE EFFECTS OF AIR MOTION.

2.1 Early studies.

The ASHRAE comfort standard 55—1981 specifies the upper limit of summer comfort at 26°C (79°F). This limit was theoretically extended to 32°C (89°F) with 0.8 m/s (160 fpm) air velocity when the Clo value is a near nil 0.05./4/ Recent studies at Kansas State University by Rohles indicated the potential for using a ceiling fan to extend the summer comfort envelope to 29.4°C (85°F) with a relative humidity of 50% and average clothing levels./5/

More recently, Tanabe and Kimura at Waseda University in Tokyo concluded that the highest comfortable condition for occupancy is 30°C (°F), 50 % relative humidity with 2 m/s air flow./6/

2.2 Studies at Arizona State University.

Two air motion studies performed at Arizona State University confirmed and extended the Rohles' studies, using ambient conditions more closely modelling the climatic conditions of Phoenix, Arizona. One study used a ceiling fan to create air motion and the other, an oscillating fan. The ceiling fan does not intrude on activities and has a stable air plum. The oscillating desk type fan is a "task cooling" device, highly portable and adjustable but with a chaotic air flow. These studies were performed under three different room conditions for both ceiling fan and oscillating fan: 32.2°C (90°F) DB / 21.7°C (71°F) WB, 39% RH; 31.1°C (88°F) DB / 23.3°C (74°F) WB, 50% RH; 33.3°C (92°F) DB / 20.6°C (69°F) WB, 30% RH.

Results indicate that for a lightly dressed individual doing minimal activity, the maximum comfort condition may reach as high as 32°C (90°F), RH 40 %, with an air flow speed of one m/s (200 fpm)./7/ (This research was sponsored by the U.S. Department of Energy under a grant to investigate potential cooling strategies for the Southwestern U.S.)

A total of 189 student subjects were tested, split almost equally among the two fan types. The tests were highly controlled. Subjects experienced 5 different air flow conditions throughout a two hour period, different room locations having different air flow speeds.

For the first 25 minutes subjects experienced a still air condition, allowing for a base level of response and for adjusting to the environmental testing chamber. Only a minimum of activity was allowed, including reading, writing, and casual conversation (assumed MET: 0.8).

Participants had lived in the Phoenix area for at least six months, were young college students, and dressed lightly (Clo 0.4—0.5). Ballot questions were similar to those in the KSU study with comments invited. In the oscillating fan test, a new category of questions was added to investigate the subjects frequency preferences.

Ceiling Fan Results.

TEST	DB	WB	RH	Thermal Votes	Comfort Sensation
1.	32.2°C	21.7°C	39%	87%	83%
2.	31.1°C	23.3°C	50%	85%	90%
3.	33.3°C	20.6°C	30%	87%	66%

In the ceiling fan study, for locations with the ideal maximum air flow of one m/s (200 fpm), 83% of subjects indicated a "comfort sensation" between slightly cool and slightly warm under conditions in test 1 [32.2°C (90°F) DB / 21.7°C (71°F) WB, 39% RH].

Even with a relative humidity as high as 50% as in Test 2 [31.1°C (88°F) DB / 23.3°C (74°F) WB], 90% of the subjects felt comfortable, which exceeded the Rohles studies. However, only 66% of the subjects indicated experiencing a comfort sensation at test condition 3 [33.3°C (92°F) DB / 20.6°C (69°F) WB], reduced humidity appearing not to counteract the temperature rise.

Relation of Comfort Responces

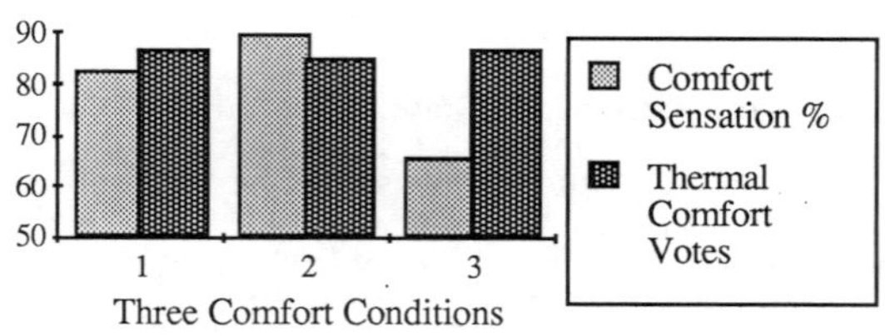

Nevertheless, the thermal comfort vote (based on scores of six adjective pairs), showed higher acceptance of climate conditions and comfort (see data chart above). Test condition #3 resulted in favorable thermal votes this being a strong discrepancy with the comfort sensation results. The favorable response seems to indicate that subjects do after all accept a higher dry bulb temperature when humidity is significantly reduced.

Data gathered from a temperature preference scale seems to support this conclusion. In test condition #1, test subjects actually asked for more temperature reduction than they did in Test #2.

Oscillating Fan Results.

Findings from the oscillating fan study indicated a need for air velocities greater than the one m/s (200 fpm) which had proven acceptable in the ceiling fan study. This appears to have been due to the oscillation and oscillating frequency of the fan which results in a less consistent air flow then with a ceiling fan. Results indicate that 80% of the test subjects felt comfortable if air flows approached 1.5 m/s. Three locations received this higher rate of flow but at different frequencies, and results reflected a strong preference for the more frequent oscillation. All other air flow locations had less than 80% comfort votes.

Other factors being equal, air flow of approximately 1.5 m/s, with a frequency of 6 to 7 seconds seemed to be the most favorable condition. In locations with frequency of oscillation every 14 seconds, results indicate a lower comfort level, even when the air flow in the tested location exceeded 1.5 m/s.

Another important factor for the oscillating fan study is a psychological consideration. Whenever there were perceived obstructions between the subject and the fan even though air flow was not obstructed, the sensation of comfort was far lower. Of students who received an average velocity of 0.5 m/s only 43% felt comfortable when there was an apparent obstruction, whereas 60% were comfortable with only 0.25 m/s air flow when no blockage could be imagined.

With oscillating fans, the air flow should be higher, between 1.25 m/s (250 fpm) and 1.50 m/s (300 fpm). This air flow achieved thermal sensation results more favorable than any of the positions in the ceiling fan study. In oscillating fan tests thermal comfort sensation response was generally good, greater than 80%, and suggests that the subjects were comfortable overall, even though the thermal comfort votes as an individual indicator, tended to be low.

2.3 Drawing conclusions.

Occupants should be comfortable at thermal conditions 1 and 2, and to a limited degree at condition 3, if they have a fan providing air motion at one m/s or approximately 1.5 m/s for oscillating fans. If this is taken as a satisfactory level of comfort, then there are significant implications for building design strategy and energy savings.

For example, it means that ceiling outlet locations should be carefully selected and if possible flexible, in order to provide ceiling fan ventilation. Higher ceiling heights is another obvious design consideration. Locations for "task cooling" desk fans should command attention in much the same task lighting locations are now considered.

Passive, natural ventilation will demand attention to room geometry, designing without "dead" air pockets and with door and window openings oriented to facilitate natural ventilation. Design of proper air velocity and proper "air wash" of the interior space is essential. Wind stacks should be used and may be designed in such a way as to shield occupants from dust and air born debris. Natural ventilation design requires careful understanding and calculation of air flow patterns—the instability of the wind being the only limiting factor.

Reductions in mechanical cooling can be significant by allowing for a warmer thermostatic set point temperature. Accurate calculation of the energy saving effect is complicated requiring consideration of individual contributions of heat load and source. For example, an internally loaded building would have less significant savings of energy than an externally loaded building where the thermostatic set point and outside dry bulb and wet bulb temperatures are more critical.

In order to make a simplified evaluation of the potential energy savings, a building previously designed for mechanical cooling by Diskind Consulting Engineers, Jerusalem was considered, using simple computer calculations for two occupancy load conditions. One calculation considered the building as being of high occupancy, and therefore internally loaded, and the other considered the building as being low occupancy and externally loaded.

A new set of design calculations were performed raising the thermostatic set point temperature from 25°C (77°F) to 30°C (86°F), an increase of five degrees. Occupants should theoretically have no problem accommodating this rise in temperature as long as they are exposed to one m/s air flow from overhead ceiling fans or 1.5 m/s air flow from task cooling devices. A calculation of energy use suggested a reduction of approximately 25% for the externally loaded condition but only about 19% for the internally loaded condition.

A greater energy and cost saving advantage can be achieved through the complete elimination of an air conditioning system. For example, in a hot dry climate such as Phoenix, Arizona, residential evaporative space cooling is effective for about 30 to 50% of the overheated period. /8/ Through the well considered use of fans, occupants should be comfortable without an air conditioning system, based on the assumption of an equivalent temperature (ET) of

88°,the use of high efficiency evaporative coolers, and of course proper location and use of fans.

3.0 PRESENTATION TO ARCHITECTS.

This information should be communicated to architects in order to encourage more systematic use of fans and ventilation and through this, result in energy savings.

The Olgyay comfort diagram is an example of a good communication tool and includes reference to air motion and comfort. /9/

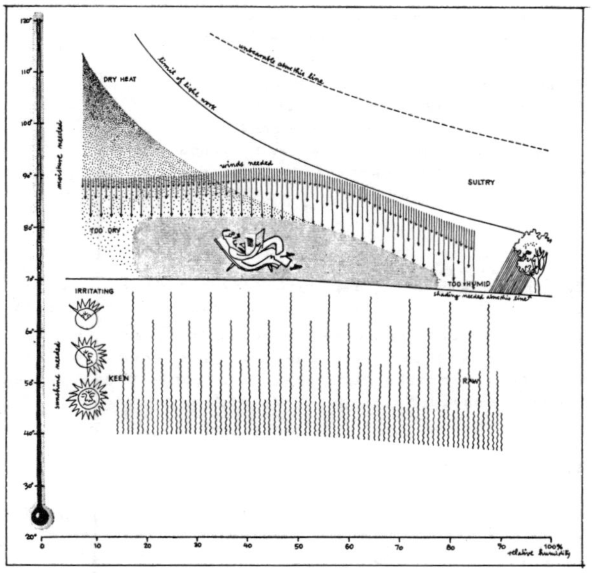

The Olgyay comfort diagram includes reference to air motion.

One example of a descriptive chart developed for communication of the specific comfort research presented in this paper is shown below. It relates temperature and humidity for an air motion of one m/s, subjects at rest, summer clothing. The horizontal line shows where comfort stops. Future development is anticipated in this area.

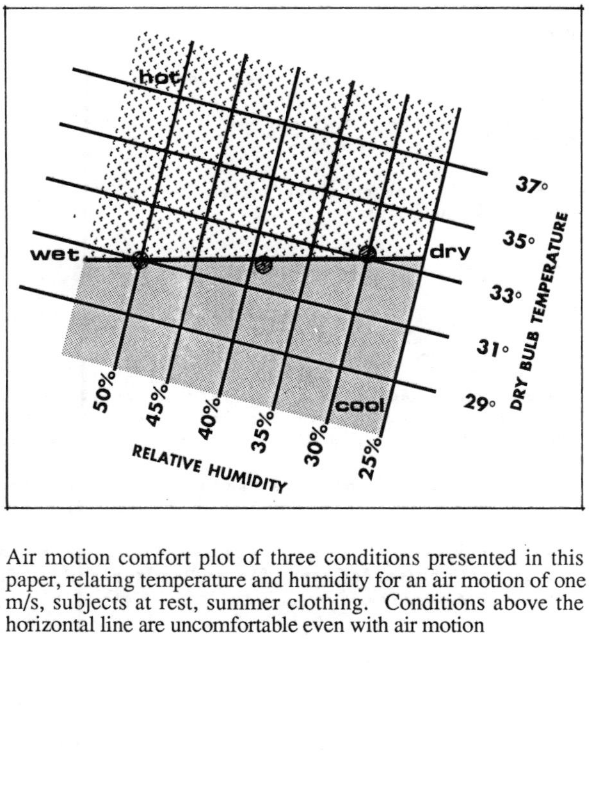

Air motion comfort plot of three conditions presented in this paper, relating temperature and humidity for an air motion of one m/s, subjects at rest, summer clothing. Conditions above the horizontal line are uncomfortable even with air motion

4.0 FUTURE RESEARCH AND CONCLUSION.

4.1 Future research

Although there have been several laboratory studies of air motion, response studies in actual occupancy situations is needed, based on accepted comfort study results. A good starting point would be an air conditioned office where mechanical set point conditions are easily controlled.

Psychological comfort studies are another area of needed research. Aside from the psychological effect recorded when obstructions existed between air flow source and subject (result: uncomfortable responses in the oscillating fan study), it would be useful to determine if those workers employed in more interesting professions are more comfortable than those in more boring positions, and likewise, if engrossing activities in general can produce quantifiable differences in comfort perception.

4.2 Conclusion.

The studied comfort conditions are obtainable using low energy cooling such as evaporative coolers even in the severe Phoenix climate on the warmest summer day. Results indicate that for a lightly dressed individual involved in minimal activity, the maximum comfort condition may reach as high as 32°C (90°F), with a relative humidity of 39 %. This with the assistance of approximately 1.0 m/s (200 fpm) of constant type air flow such as from ceiling fans, or a maximum of 1.5 m/s (300 fpm) air flow from more unstable sources such as oscillating fans.

---

[1]Watson, D., and Labs, K., Climatic Design, McGraw—Hill, New York, 1983, p26.

[2]Milne, M. and Givoni,G.,1979, "Architectural design based on climate," D. Watson, ed., Energy Conservation Through Building Design, Architectural Record Books, New York, pp. 96—113.

[3] Arens, E., Gonzales, R., and Berglund, L. , 1986, "Thermal comfort under an extended range of environmental conditions," ASHRAE Transactions, Vol. 92, Part 1.

[4] ASHRAE Standard 55—1981, "Thermal environmental conditions for human occupancy," Atlanta, American Society of Heating, Refrigerating and Air-conditioning Engineers, 1981.

[5] Rohles, F.H., Konz, S.A., and Jones, B.W., 1983, "Ceiling fans as extenders of the summer comfort envelope," ASHRAE Transactions, Vol. 89, Part 1.

[6] Tanabe, S., and Kimura, K., "Importance of air movement for thermal comfort under hot and humid conditions," ASHRAE Far East Conference, pp. 95—103, 1989.

[7] For a full description of the ceilling and oscillating fan studies see Wu, H., "The use of fans to extend the summer comfort envelope in hot arid climates," CLIMA 2000 1989. The full reports of the individual studies are available through the College of Architecture and Environmental Design, Arizona State University, Tempe, Arizona, USA

[8] Op.Cit. Watson, p. 242.

[9]Olgyay, V., Design With Climate, Princeton University Press, Princeton, New Jersey, 1973, p. 23.

NEW CRITERIA FOR DEFINING COMFORT IN BUILDINGS

V. Corrado, P. Oliaro

Politecnico di Torino
Dipartimento di Energetica
Corso Duca degli Abruzzi, 24
10129 Torino - Italy

ABSRACT. Studies on comfort conditions in buildings are felt necessary, since buildings are "living spaces" for people. In this paper an analysis of parameters affecting comfort has been done. Basic and "traditional" parameters are air temperature, wall surface temperatures, air relative humidity and air velocity, together with occupants parameters such as the kind of cloths and the level of metabolic activity. A further improvement is achieved taking into account air turbulence. Moreover, "new criteria" are presented concerning air quality, i.e. contaminant concentrations: relevant contaminants and global statistic parameters are considered. Target values for all mentioned parameters are supplied.

1. INTRODUCTION

Usually the parameters taken into account for defining comfort in buildings are:

- temperatures (air temperature and wall surface temperatures, or mean radiant temperature);
- air relative humidity;
- air velocity;
- levels of illuminance and sound.

The target values are to be related to the activities performed by people staying in the rooms. Neglecting lighting and acoustic aspects, the attention has to be focussed on air conditions and connected parameters: under this point of view, temperature, humidity and velocity seem to be no more sufficient to define acceptable air quality, because also pollutants concentrations should be taken into account.
In the following paragraphs a short discussion about "traditional" parameters has been reported. Afterwards, air contaminants have been considered.
Starting from the aforementioned parameters, some global indicators of comfort can be defined. Paragraph 5 deals with them, expecially with new ones, such as olf and decipol.

2. AIR TEMPERATURE AND HUMIDITY

A lot of studies have been performed on air temperature and humidity conditions that produce good comfort conditions /1/, /2/, /3/. The same studies took into account also mean wall radiant temperatures. As a result, many indications related to comfort have been given.
For example, Fanger found some "best" values for surface body temperature and for thermal flux due to sweating /1/; in consequence, optimal values for air temperature and humidity can be found as a function of kind of activity, of cloths, etc. Other authors (e.g. Givoni, the Kansas State University /3/) gave minimum values for the same parameters.
Finally a Standard has been set up by ASHRAE /2/, under the following hypothesis: metabolic activity of 1 ÷ 1.2 met, air velocity of 0.15 ÷ 0.20 m/s, mean radiant temperature similar to air temperature, cloths of 0.5 ÷ 0.7 clo. The acceptable comfort conditions are met if at least 88% of people is satisfied.
In figure 1 the results of the aforementioned comfort criteria have been reported (dots reported as "comfort according to Fanger" refer, respectively, to 0.5 clo, 1 clo, 1.5 clo) /3/.

On the basis of temperatures, humidity, air velocity, activity and clothes of occupants, global comfort parameters can be defined. The best known and mos diffused ones are the Predicted Mean Vote (PMV) and the Percentage of People Dissatisfied (PPD) proposed by Fanger /1/, /3/.

3. AIR VELOCITY AND TURBULENCE

Air velocity is often kept into account together with temperatures and humidity; on the contrary, turbulence is normally neglected. Recent studies /4/ proved its importance in generating a sensation of draught. Fanger et al. made a serie of experiments on a sample of people, finding the percentage of dissatisfied subjects as a function of mean air velocity and level of turbulence intensity (figure 2).
The same authors developed a matematical model, which calculates the percentage o dissatisfied persons according to the following formula:

$$PD = 3.143\ (34-t_a)\ (v-0.05)^{0.6223} + 0.3696\ v\ Tu\ (34-t_a)\ (v-0.05)^{0.6223}$$

where:

PD = percentage of dissatisfied subjects $\leq$ 100%
t_a = air temperature
v = air velocity $\geq$ 0.05 m/s
Tu = turbulence intensity

A graphical representation of the above formula is given in figure 3 /4/.

4. CONTAMINANTS

The importance of contaminants as factors affecting comfort increased quite a lot in the latest years, due to the reduction of air change rates (infiltration) produced by energy savings strategies.
A more complex analysis should be done for industrial environments, but it is out of the goals of this paper.
The main contaminants affecting indoor air in residential buildings are listed in table I /5/, /6/, /7/, /8/, /9/.

Air pollutant	Frequency of occurrence	Sources
Asbestos	rarely	insulation or fire retardant lining ventilation ducts
Biological material	occasionally	stagnant water associated with HVAC systems (fan coil units, cooling towers, humidifier pans, damp ducts or plenum), plants, water-damaged furnishing
Carbon dioxide	frequently	people
Carbon monoxide	occasionally	natural gas heating units and stoves, tobacco smoke, incoming air
Formaldehyde	occasionally	unsealed plywood or particle board, UFFI, fabrics, adhesives, glues, carpets, furnishing
Nitrogen diox.	rarely	combustion of natural gas or kerosene used for heating or cooking
Ozone	rarely	photocopies, electrostatic dust filters (electronic air cleaners)
Particulates	occasionally	contaminated air intake, tobacco smoke, paper, insulations, people
Radon	rarely	soil, groundwater, building materials
Sulfur dioxide	only if outdoor air is polluted	no indoor sources
Volatile organic compounds	occasionally	carpets, furnishing, cleaning materials, pesticides, tobacco smoke, glues, adhesives

Table I - Contaminants, frequency of occurrence and sources

Each of them has an influence both on indoor comfort conditions and on health risks for occupants, depending on the concentrations. For defining those values, a study of each air pollutant is to be done.
Asbestos is hazardous only when in contact with ventilation air (inside ducts or plenums) or when released during renovation; in those cases the health risks are high because asbestos is carcinogenic, so those shall be avoided at all.
As far as biological materials is concerned, problems are caused mainly by fungi (mould, mildew, slime, etc) and their spores. Health risks come from bacteria called legionella that could be associated with air-conditioning systems; in that view good maintenance of air conditioning systems shall be done. Other risks of spreading diseases are more occasional.
Carbon dioxide has people as its most important indoor source; high levels are due to the presence of too many people ar to inadequate ventilation. In any case it is not toxic until 50,000 ppm; it is irritating above 10,000 ppm.
On the contrary, carbon monoxide is a by-product of fuel-burning and is highly toxic. The main risks are connected to incoming air, if it comes from a busy street or a parking; tobacco smoke contributes only small quantities; modern appliances (i.e. burners) contribute small quantities too.
Formaldheyde is an important component in the manufacture of many synthetic materials, from which can later be released to the surrounding air; the main problems seem to came from UFFI (urea formaldehyde foam insulation). The risks are more of irritation than of illness, but health damages can overcome if the concentration is too high.
Nitrogen dioxide is produced during fuel-burning. It may enter from outdoors; there could be indoor sources in kitchens and bathrooms (unvented gas stoves), not usually in office buildings.
As far as ozone is concerned, the indoor concentration is usually low because it is a reactive chemical and is easily destroyed by contact with surfaces and other chemicals in the air.
The category of particulates includes fibers, dust, smoke and mists. From the heath or comfort poin of view, the most important particulates are the ones that do not settle on surfaces, but remain in the air and are inhaled.
Radon is a short-lived gas produced during natural decay of uranium and is a health hazard because of the radiation produced during its own decay. All soils and groundwaters contain some radon. It enters buildings either as a gas released directly from the soil or in water that has recently been in contact with soil; entry poins include cracks in foundations, crawlspaces with exposed soil and poorly ventilated shower areas. The effects of radon are not demostrated yet, but it is supposed to be carcinogenic. The scaling of foundations and a good ventilation of lower spaces of the buildings can avoid too high radon concentration in buildings.
Sulphur dioxide is an important air pollutant too and is a constituent of acid rains. The main sources are industrial processes. In any case there are not such sources inside the buildings.
Finally, volatile organic compounds (VOCs) can enter the air by evaporation from solids or liquids. There are many VOCs and some of them are helth hazards; their detection is usually difficult.
As far as acceptable contaminant concentrations are concerned, a new sandard has been issued by ASHRAE this year /10/. The values are reported in tables II, III, IV; other indications are reported in tables V (Canadian exposures guidelines for residential indoor air quality) and VI (WHO working group consensus of concern about indoor air pollutants).

5. GLOBAL PARAMETERS

Indoor comfort conditions can be defined by limiting contaminant concentration (see paragraph 4). One more immediate way is to define one or more "global" parameters: in this way the olf and decipol units have been defined by Fanger /11/, /12/.
The olf unit represents the contaminant capacity of a source; the decipol unit represents contamination of the air. Both are statistically defined and give a global idea of air comfort level.
The olf unit is the air contamination generated by an adult subject performing a sedentary activity under good thermal comfort conditions and taking a medium of 0.7 baths per day.
The decipol unit is a contamination of 1 olf under a ventilation rate of 10 l/s.

In well ventilated buildings the contamination level is less than 1 decipol and less than 15% of dissatisfied people can be found; in a bad ventilated building 10 decipols can be reached with 60% dissatisfied people.
Figure 4 shows the decipol scale. In figure 5 a correlation between perceived air contamination (decipol) and dissatisfied people (%) has been reported /11/.

6. CONCLUSIONS

The performed study stated the following concepts on defining and quantifying indoor comfort level:

a. Traditional approaches are based on air temperature, wall surface temperatures, relative humidity and air velocity on one side, and on metabolic activity and kind of clothes on the other side.
b. A refinement of those approaches can be gained taking into account also air turbulence.
c. On the other hand a completely new approach can be employed on the basis of air quality defined with refence to contaminant concentrations.
d. The new approach has been suggested by low air change rates typical of well insulated buildings, with very low leakages.
e. The new approach shall not be an "alternative" to traditional ones, of course, but it deals with new comfort aspects that should not be neglected any more.
f. The concentrations of single contaminants shall be limited; global parameters (such as olf and decipol) can be defined too.

7. REFERENCES

/1/ P.O. Fanger - "Thermal Comfort" - Robert E. Krieger, Malabar, FL, 1982
/2/ Standard 55/74 - "Thermal Environmental Conditions for Human Occupacy" - ASHRAE, 1974
/3/ CNR/ENEA - "Guida al controllo energetico della progettazione" - PFE, 1985
/4/ P.O. Fanger, A.K. Melikov, H. Hanzawa, J. Ring - "Air Turbulence and Sensation of Draught" - Energy and Buildings, 12/1988
/5/ B. Davidge, G. Kerr, T. Nathanson - "Indoor Air Quality Assessment Strategy" - Public Works Canada, 1989
/6/ AES/SAG 1-4:88-4 - "Indoor Air Quality Test Kit User Manual" - Public Works Canada, 1988
/7/ P.E. Mc Nall - "La qualità dell'aria negli ambienti" - CDA, Novembre 1986
/8/ W. Raatschen - "Demand Controlled Ventilating Systems. State of the Art Review" - IEA Annex 18, 1989
/9/ W. Raatschen - "Market Analysis of Sensors for the Use in Demand Controlled Ventilating Systems" - 9th AIVC Conference, Gent, 12-15/9/1988
/10/ Standard 62/89 - "Ventilation for Acceptable Indoor Air Quality" - ASHRAE, 1989
/11/ P.O. Fanger - "Olf e decipol. Nuove grandezze" - CDA, Aprile, 1989
/12/ P.O. Fanger, J. Lauridsen, P. Bluyssen, G. Clausen - "Air Pollution Sources in Offices and Assembly Halls Quantified by the Olf Unit" - Energy and Buildings, 12/1988

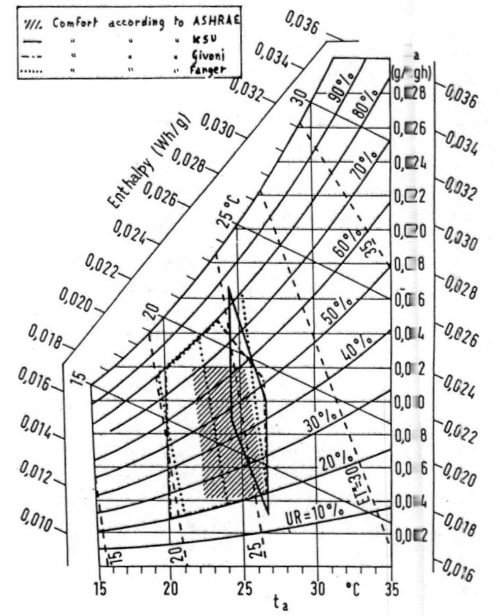

Figure 1 - Different Comfort Criteria

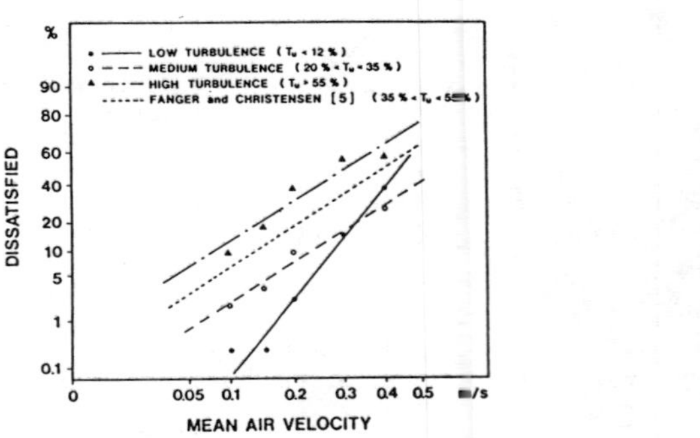

Figure 2 - The percentage of dissatisfied subjects as a function of mean air velocity and level of turbulence intensity /4/

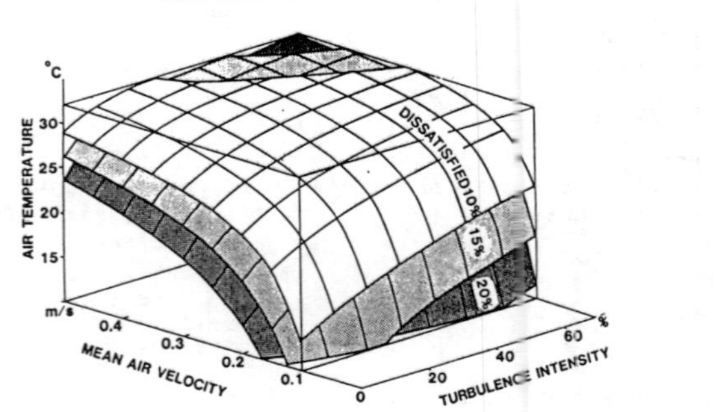

Figure 3 - The percentage of dissatisfied subjects as a function of turbulence intensity, mean air velocity and air temperature /4/

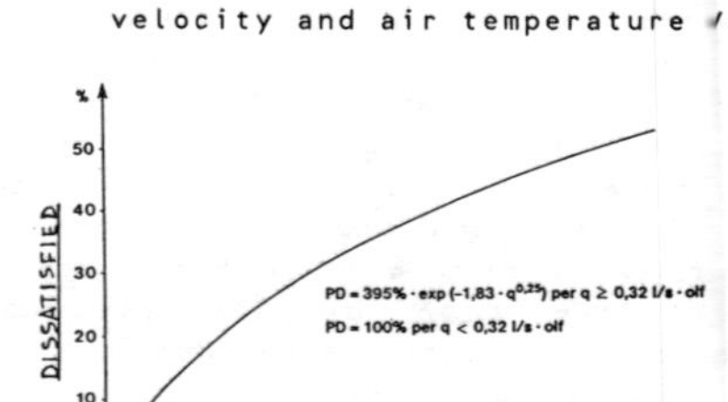

Figure 5 - Correlation between percentage of dissatisfied subject and air contaminants /11/

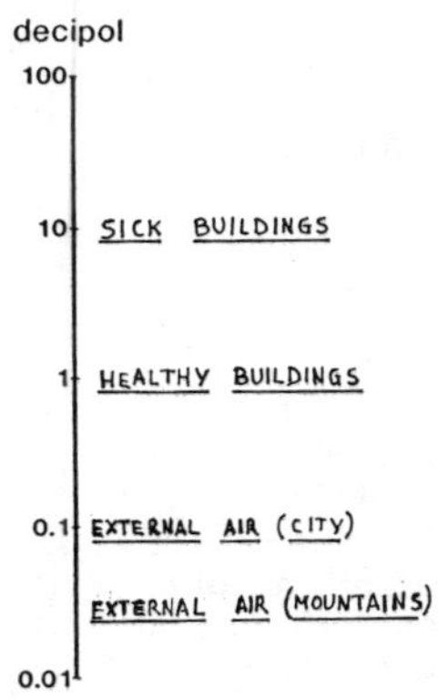

Figure 4 - Decipol scale /11/

Contaminant	Acceptable Exposure Ranges: ASTER[g]	Acceptable Exposure Ranges: ALTER[h]
Aldehydes (total)	$\Sigma c_i/C_i<1$[a]	—
Carbon dioxide	—	× 6 300 mg/m³ (<3 500 ppm)
Carbon monoxide	<11 ppm—8 h[b] <25 ppm—1 h[b]	—
Formaldehyde		[d]Action Level 120 μg/m³ (0.10 ppm) Target Level 60 μg/m³ (0.05 ppm)
Nitrogen dioxide	<480 μg/m³ (<0.25 ppm)—1 h	<100 μg/m³ (<0.05 ppm)
Ozone	<240 μg/m³ (<0.12 ppm)—1 h	—
Particulate matter[c]	<100 μg/m³—1 h	<40 μg/m³
Sulphur dioxide	<1 000 μg/m³(<0.38 ppm)—5 m	<50 μg/m³ (<0.019 ppm)
Water vapor	30-80% R.H.—summer 30-55% R.H.—winter[f]	—
Others See below[i]	Minimize exposure	Minimize exposure

[a] C_i = 120 μg/m³ (formaldehyde); 50 μg/m³ (acrolein); 9000 μg/m³ (acetaldehyde), and C_i are respective concentrations measured over a five-minute period.
[b] Units given only in parts per million so that guidelines are independent of ambient pressure.
[c] See Aldehydes (total).
[d] Although the epidemiological studies conducted to date provide little convincing evidence that formaldehyde is carcinogenic in human populations, because of this potential, indoor levels should be reduced as much as possible.
[e] <2.5 μm mass median aerodynamic diameter—MMMD.
[f] Unless constrained by window condensation.
[g] ASTER—Acceptable short-term exposure range
[h] ALTER—Acceptable long-term exposure range.
[i] Other contaminants of concern: biological agents; chlorinated hydrocarbons; pest control products; product aerosols; fibrous materials; lead; polycyclic aromatic hydrocarbons (PAHs); and tobacco smoke.

Table V - Summary of Canadian exposures guidelines for residential indoor air quality /10/

Contaminant	Long term: Concentration ug/m³	Long term: Concentration ppm	Long term: Averaging	Short term: Concentration ug/m³	Short term: Concentration ppm	Short term: Averaging
Sulfur dioxide	80	0.03	1 year	365	0.14	24 hours
Total Particulate	75[a]	—	1 year	260	—	24 hours
Carbon monoxide				40,000	35	1 hour
Carbon monoxide				10,000	9	8 hours
Oxidants (ozone)				235[b]	0.12[b]	1 hour
Nitrogen dioxide	100	0.055	1 year			
Lead	1.5	—	3 months[c]			

[a] Arithmetic mean
[b] Standard is attained when expected number of days per calendar year with maximal hourly average concentrations above 0.12 ppm (235 ug/m³) is equal to or less than 1, as determined by Appendix H to subchapter C, 40 CFR 50
[c] Three-month period is a calendar quarter.

Table II - U.S. Environmental Protection Agency quality standards for outdoor and indoor air /10/

Contaminant	Concentration	ppm	Exposure Time
Carbon Dioxide	1.8 g/m³	1000*	Continuous
Chlordane	5 ug/m³	0.0003	Continuous
Ozone	100 ug/m³	0.05	Continuous
Radon gas	0.027 WL		Annual Average

*This level is not considered a health risk but is a surrogate for human comfort (odor).

Table III - Guidelines for selected air contaminants of indoor origin /10/

Pollutant	Indoor Standards
Asbestos	Consumer Product Safety Commission has banned use of asbestos in artificial logs for fireplaces, in patching compounds, and in certain garments (16 C.F.R. 1304, 1305, 1500.17(a)(7)); voluntary ban on use in hairdryers (C-10) EPA regulates use in schools (40C.F.R 763.100) and removal projects (40 C.F.R. 763.120); bans installation of friable asbestos for facility insulation (40 C.F.R. 61.150) (C-11) (C-7)
Formaldehyde	Federal: 0.4 ppm target ambient level, HUD standard for manufactured homes, achieved through product emissions standards of .2 and .3 ppm (HUD, 24 C.F.R. 3280.308, 1984) (C-14) State: 0.4 ppm standard for indoor exposure (MN statute 144.495, 1985) (C-15)
Lead (See Table II)	CPSC has banned in paint for consumer use or use on consumer products (16 C.F.R. 1303) (C-10)
Ozone (See Table II and 3)	FDA prohibits devices (e.g., germicides, deodorizers) that result in more than 0.05 ppm in occupied enclosed spaces such as homes, offices, or hospitals, or that result in any releases in places occupied by the ill or infirm (21 C.F.R. 801.415) (C-16)

Table IV - Standards applicable in the U.S. for common indoor air pollutants /10/

Pollutant[a]	Concentrations[b] reported	Concentrations[b] of limited or no concern	Concentraton[b] of concern	Remarks
Tobacco smoke (passive smoking)				
Respirable particulates	0.05-0.7	<0.1	>0.15	Japanese standard 0.15 mg/m
CO	1-1.5	<2	>5	indicator for eye irritation (only from passive smoking)
Nitros-dimethylamine	(1-50) × 10⁻⁶	—	—	Mutagens under investigation for carcinogenicity
NO_2	0.05-1	<0.19	>0.32	
CO		2% COHb	3% COHb	99.9%[c]
	1-100	<11	>30	Continuous exposure
Radon and Progenys	10-3000 Bq/m³	-0	70 Bq/m³	Swedish standard for new houses
Formaldehyde	0.05-2	<0.06	>0.12	Long- and short-term
SO_2	0.02-1	<0.5	>1.35	SO_2 alone short-term
CO_2	600-9000	<1800	>12000	Japanese standard 1800 mg/m³
O_3	0.04-0.4	0.05	0.08	
Asbestos	<10 fibres/m³	-0	10⁵fibre/m	For long-term exposure
Mineral fibres	<10 fibres/m³	—	—	Skin irritation
Organics				
Methylene chloride	0.005-1	—	350 260	TLV[d] NIOSH[e] recommendations
Trichlormethene	0.0001-0.02	—	270 135	TLV NIOSH recmmendations
Tetrachloroethene	0.002-0.05	—	335	TLV
1.4 Dichlorobenzene	0.005-0.1	—	450	TLV
Benzene	0.01-0.04	carcinogen	carcinogen	
Toluene	0.015-0.07	—	375	TLV
m.p.-Xylene	0.01-0.05	—	435	TLV
n-Nonane	0.001-0.03	—	1050	ILO (1980)
n-Decane	0.002-0.04	—	—	
Limonene	0.01-0.1	—	560	TLV turpentine

Table VI - WHO working group consensus of concern about indoor air pollutants at 1984 levels of knowledge /10/

CONCEPTION OF A THIRD GENERATION THERMAL MANIKIN

L. Imre; A. Bitai; C.D. Horváth; P. Niedermayer and L. Bánhidi
Technical University Budapest
1521.Budapest, Hungary, Tel.: 1812-192

ABSTRACT. Thermal manikins of third generation are to be used for comfort measurement in hot environment by simulating physically the metabolic heat generation even as the sweat secretion of the human body. Conception of the presented way is based on the hypothesis that the sensation of the human comfort is in correlation with the body temperatures and with the surface heat and mass flow rate densities of the body. Manikin's body is divided into 15 parts. Devices for serving as heat and mass flux sources are to be inserted under the surface covering of the manikin's body. Sources are to be regulated according to a numerical algorithm based on the thermoregulation of the man. For the regulation the manikin's body surface temperatures determined by sampling measurement are to be used. Hardware means for the operation of the heat and mass flux sources are to be computer actuated.

1. INTRODUCTION

For the analysis of thermal human comfort in closed spaces the Fanger's method is generally used in relation to the whole body /1/. Use of the method is well supported by diagrams elaborated for comfort dimensioning. Standards are often based on the method /2,3/. In cases when the thermal comfort analysis is required for the parts of the human body, computer simulation procedures can be used /4,5,6/.
Thermal manikins are of great importance for thermal comfort analysis in cases when the thermal interactions between the human body and its surroundings can not correctly be characterized mathematically.
The first and the second generations of the thermal manikins are usually micro-processing controlled instruments of 1:1 scale puppet of the human body and are adequate for the simulation of the "dry" heat exchange /7...12/. In "hot environment" simulation of sweating and the heat transfer by evaporation is essential. In the last few years various institutions and laboratories devoted research and development work to design and construct the third generation "sweating manikin".
Bennett, Yount and DeWitt repôrted the development status of the design, construction and testing of an antropomorphous robot /13/. The robot is a manikin composed of several distinct systems jointed with 38 degrees of freedom movement. It has a skin temperature system of flexible heat electrical heaters to form multiple temperature zones of the body. Some active cooling of the skin surface is provided by a perspiration system operating by injection of water at multiple locations on the surface of the skin by an array of capillary tubes. The manikin has a breathing system, skin structure and sampling system controlled by computers.
Meinander reported the development of a "sweating cylinder" for measurements of heat and moisture transmission through clothing materials /14/. The outer surface of the cylinder wall is covered with an electric heating wire covered by a metal layer to spread the heat and a plastic one to give some mechanical protection. Furtherway it is covered by a laminated material the inner side of which spreads the water from tubes to larger areas. Water is supplied through a computer controlled micro-valve system. The amount of water supply can be chosen within certain limits. The work aims at the development and construction of a sweating and movable thermal manikin.
In the Institute of Thermal Energy and Systems Engineering of the Technical University Budapest a project started in 1987 aiming at the development of computational methods of the human thermal comfort analysis and to design and construct a "sweating manikin". The two parts of the project are in close connection because the numerical algorithm of the manikin's control is based on the computer model of the man's thermal behaviour /15/. Herewith, as a status report, the concepcion of the Hungarian way is presented.

2. OBJECTIVES

Conception of the development work is based on the hypothesis that the sensation of thermal comfort is in correlation with the body temperatures and with the surface heat and mass flow rate densities of the body. Body temperatures are resulted by the interaction of the human body with its surroundings. Skin temperatures are of special importance since the body is influenced by the environmental effects through the skin. The thermal sweating manikin can be the substitute of a man in the sense of the environmental interactions only. It is able to simulate physically and to measure the skin temperature-time history supposing that the proper heat generation and sweat mass flow is continuously induced in its surface layer according to the thermoregulatory processes of the man. The actual values of the heat and mass sources should be determined by computer simulation and commands for the regulation and control of the actuating system formed. The work aims at the elaboration of the proper additional appliances and methods to make a second generation "dry" manikin able to produce sweating functions too.

3. BUILD UP OF THE MANIKIN

A functional scheme of the sweating manikin is presented in Fig.1. Under the surface of the manikin's body electric heating devices are embedded for simulating the effects of the metabolic heat generation. Surface of the manikin's body will be covered by an artificial Skin serving for the simulation of sweating. The artificial Skin may be covered by the layers of the Clothing to be tested under the direct influence of the environment.

Heat and sweat generation are operated for each body parts by the Actuating Systems according to the commands of the Regulation and Control System. Commands are formed by the Central Computer using the input data of the person to be tested and, the skin temperatures of the body-parts determined actually by sampling measurements.

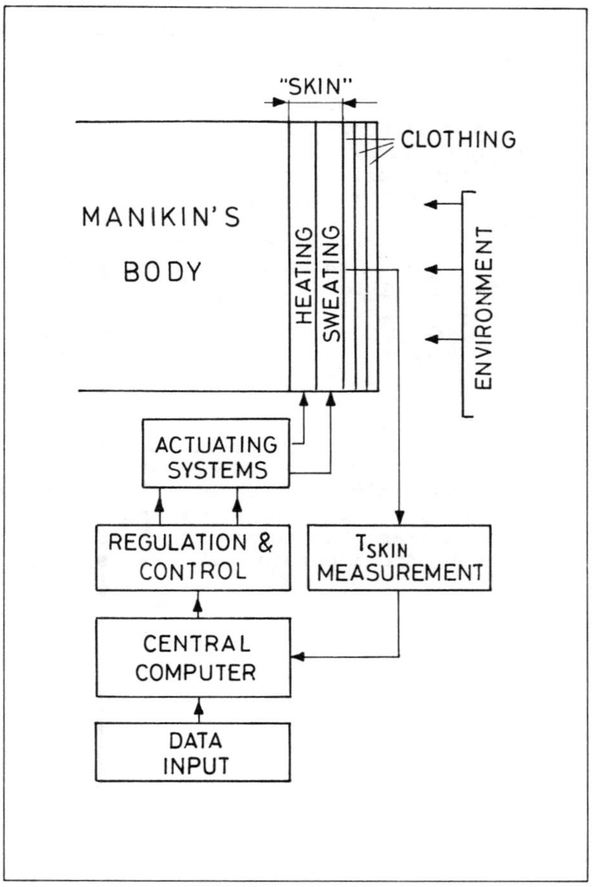

Figure 1: Functional scheme of the sweating manikin

4. CONCEPT OF THE REGULATION AND CONTROL

Skin temperatures /at $\tau = \tau_o + \Delta\tau$ / of the manikin's body-parts determined by direct measurement are used as boundary conditions of the first kind for the solution of the thermal model of the human body /4, 15/. The initial / $\tau = \tau_o$/ thermal state of the body, the position and the activity are also used as input data. The solution is produced by the Central Computer and,blood mass flow rates, core /hypothalamus/ and shell temperatures as well as the surface heat fluxes will be computed for $\tau = \tau_o + \Delta\tau$. By the use of the calculated and measured values, the metabolic heat generation and, the mass flow rate of the sweat secretion will be determined according to the source functions of the thermoregulation of the human body /4,15/. These values will serve as commands for the regulation actuating system in the next sampling time interval $\Delta\tau$. Hardware means for heat and sweat generation are to be actuated quasi-continuously in impulse-operation during the next sampling time interval $\Delta\tau$.

5. CONCEPT FOR THE PRACTICAL SOLUTION OF THE PERSPIRATION

The physical simulation of the perspiration is one of the most difficult problem to be solved when developing a "sweating manikin". Requirements - listed as follows - are very strict and should be satisfied perfectly.

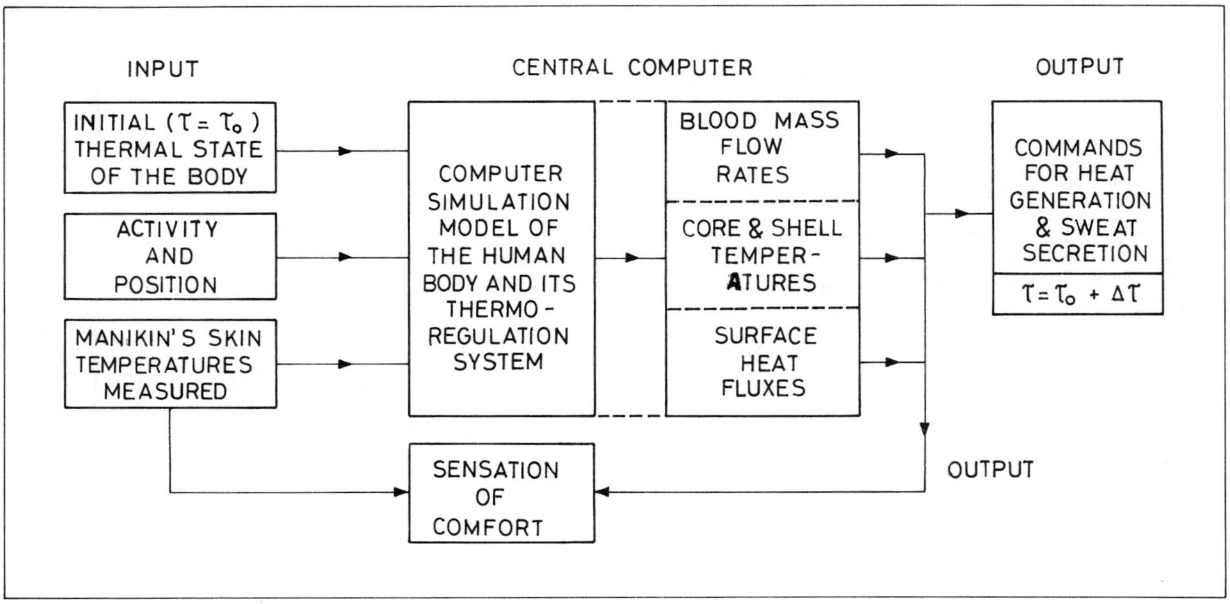

Figure 2: Flow-chart of the regulation

1. The amount of water according to the mass flow rates of sweating should be dis tributed equally to the body-parts.
2. The "secreted" /distributed/ mass of water should be transferred to the surfaces of the manikin's body-parts similarly to that in case of a human body: in vapour or in liquid form actually.
3. The dosed and distributed water should not be "stored" in the artificial skin of the manikin and shouldn't be bounded to that by strong adhesive forces. It means that the equilibrium partial vapour pressure on the artificial skin of the manikin's body should practically be a close value to that of the human body.

6. SUMMARY

The "sweating manikin", according to the concept presented herewith, consist of two main parts. The manikin's body serves for simulation physically the surface heat fluxes and the sweating of a man to be substituted. It serves also for the direct measurement of the actual surface temperatures of the body parts.
The other main part is a central computer with a program simulating the thermal processes of the human body and its thermo--regulation system. The computer forms commands for the actuating system of the manikin by computing all of the parameters needed to identifying the heat and mass sources.
Hardware means of the manikin's heating and perspiration system are to be operated quasy-continuously during the sampling time interval.

ACKNOWLEDGEMENT. The authors wish to thank the Hungarian Ministry of Traffic, Telecommunication and Building for support in this work.

7. REFERENCES

1. Fanger, P.O.: Thermal Comfort. Robert E. Krieger Publishing Company, Malabar, Florida, p.244. 1982.
2. ASHRAE Standard 55-81: Thermal environmental conditions for human occupancy. New York, 1981.
3. ISO 7730: Moderate thermal environments. International Standards Organization, Geneve, 1984.
4. Imre,L.; Bitai,A.; Horváth,C.D.; Bánhidi,L. and Pammer,Z.: Thermal Analysis of Human Body-Clothing-Environment System, Int.Journal for Num.Meth. in Engng, Vol.25, 357-371, 1988.
5. H.Arkin and A.Shitzer: Model of thermoregulation in the human body I-II-III, EEC-148, ISSN 0333-8048 Haifa, 1984.
6. J.Werner and T.Reents: A contribution to the topography of temperature regulation in man, Eur.J.Appl.Physiology 45, 87-94. 1980.
7. T.L.Madsen: Description of thermal mannequin for measuring the thermal insulating capacity of the clothing. Technical University of Danmark. Report No.48. /DTH/
8. D.P.Wyon, T.Lund-Madsen, E.Temmala: Intercalibration of four Nordic thermal manikins. Nordtest project report,1987.
9. D.P.Wyon: The Use of Thermal Manikins in Environmental Ergonomics.3rd Environmental Ergonomics Conf.,Helsinki,1988.
10. D.P.Wyon; C.Tennstodt; I.Lundgren and S.Larsson: A New Method for the Detailed Assessment of Human Heat Balance in Vehicles-Volvo's Thermal Manikin, VOLTMAN International Congress and Exposition Detroit, Michigan, No.850042. 1985.
11. Bánhidi,L.: Hungarian Experiment of Voltman duplicate in heating and ventilation research. New Method in Applied Ergonomics. Taylor and Francis, London. p.159... ...164. 1987.
12. McCullough,E.A. and Wyon,D.P.: Insulation characteristics of Winter and Summer indoor clothing. ASHRAE Transactions, Vol.89. 1983.
13. Bennett,D.W.; Yount,J.A. and DeWitt, C.E.: Design, Construction and Testing of an Antropomorphic Robot: 1988 Development Status Report, Pacific Northwest Laboratory, Richland, Washington 99352.
14. Meinander, H.: Clothing Physiological Measurements on Winter Workwear Materials with a Sweating Cylinder. J.Textil Institute of Finland. No.4. p.620...633. 1988.
15. Imre,L.; Bitai,A.; Horváth C.D. and Bánhidi,L.: Computer model for predicting man's thermal behaviour. Paper presented at the Int.Conference of "Science and Technology at the Service of Architecture, CEC, Paris,1989.

ACTIVE OR PASSIVE ACOUSTICS FOR AUDITORIA? - a preliminary survey.

by G. Dodd,
Laboratoire d'Acoustique Associé au CNRS, Université du Maine, France
and Acoustics Research Centre, Auckland University, New Zealand.

ABSTRACT As the number of "variable acoustics' systems for auditoria continues to grow, hall designers are being tempted by their apparent advantages. For multi- purpose halls they offer selectable acoustics with small hall volumes. This survey looks at the different approaches to variable acoustics that have been tried. It covers the implications of the choice of either of two main categories of electro-acoustical installation, and summarises available data on the comparative performance and experience with both Passive and Active techniques.

1 PASSIVE ACOUSTICS - FIXED OR VARIABLE?.

1.1 Definition

It is the reflections off boundary surfaces which create the acoustics of a hall. No energy input is needed to make the surfaces reflect (or absorb), thus the process is a PASSIVE one and the resulting hall acoustics we describe as PASSIVE. As long as the materials of a hall retain their original physical properties and the hall remains substantially unchanged, then the acoustics will be fixed.

If 1) a hall is intended for a particular purpose e.g theatre, chamber music or symphonic works, and 2) the design proceeds successfully then fixed acoustics is desirable.

If, on the other hand, our need is for a building which provides for a range of performance activities from the theatrical to the symphonic, then this "fixedness" is a frustration. This is the problem facing designers of multi-purpose halls.

1.2 Variable Acoustics

The conclusion appears to be that the ideal multi-purpose auditorium should have the ability to change its acoustics. What, then, should we be looking to make variable? As far as we in New Zealand are concerned the specification and design for a hall would consider, as a minimum, the following criteria:-

1) Loudness
2) Reverberence
3) Clarity
4) Spatial Impression

Given that the suspicion is growing that different listeners have different preferences there is perhaps another criterion by which to judge halls and variable acoustics systems, and that is by their ability to provide a range of acoustical conditions within the same hall at the same time.

1.3 Variable systems in existing halls.

The data published by the Journal of the Acoustical Society of America, J.A.S.A. /1/, on halls built in the twenty years between 1962 and 1982, shows that the predominant acoustical criterion considered (or at least reported on) was the Reverberation Time.

A quick look at Sabine's R.T. formula - RT = 0.061V/A - shows us that in order to vary a hall's R.T. we need to change it's volume, or the total sound absorption inside (or both). The sort of change in R.T. we would be looking for in a multi-purpose hall intended to cover the acoustical spectrum from speech to music would be from roughly 1 to 2s, i.e. a ratio of approximately 2:1.

For the larger halls(e.g. those in excess of 1200 seats) where we could not bring all the audience close enough for theatre use, the range would be somewhat smaller since we would then need only to accomodate the musical repetoire (typically a range from 1.4 to 2.2's-i.e.a ratio of 1.6:1 instead of 2:1).

An analysis the J.A.S.A. compilation,(comprising 51 halls in the USA and 36 from other countries), shows that 1) 49% of halls are intended as multi-purpose, and

2) 39% actually include a system for varying the acoustics (seen primarily as a way of varying the R.T.).

In order of effect the measures have on the reverberation time, they are:

1) a removable orchestra/stage 'shell' - 0.1 to 0.2s

2) curtains, banners and rotatable wall sections for altering the room absorption - 35% average change in RT

3) variable absorption techniques plus volume and seating changes (e.g. hinged ceiling sections, lowerable ceilings (fig), division of the room volume by curtaining, hard-wall division - e.g. separating off sections into separate rooms) -

46% average change in RT
Thus, on average these techniques cannot provide the adjustment required, - even if we consider only musical performances, where the required variation in R.T. would be approximately 60%. Only where drastic changes to room volume and seating are tolerated can such alterations be achieved. Also passive systems have only a very limited ability to change the value of the RT with frequency, as would be needed when changing from speech to music use.
Furthermore, we must note that any philosophy for passive control requires halls to be built with a large enough volume to give the longest required R.T.
It would be foolish to rule out the possibility of new techniques evolving for PASSIVE variable acoustics systems, but it is clear that present interest is centred on research and development for ACTIVE acoustical control.

2 ACTIVE ACOUSTICS

2.1 Definition and Aesthetic Acceptability

A simple definition of an ACTIVE system is one which supplies sound energy to the auditorium over and above that supplied naturally by the source. This means that active systems will use microphones and loudspeakers, and this fact alone is a difficulty for some musical purists.
The idea of having "working" or "active" loudspeakers evokes the sense of sounds being replayed. Hence, conciously or unconciously, connections are made with those features that are seen as undesirable in the process of creating and reproducing recorded works - for example,
1) loss of control to an engineer,
2) limitationsfrom the restricted dynamic range of certain conventional recording media
3) distortions from loudspeakers, etc.
In addition, there may be the feeling that sound supplied to listeners coming from energy emitted by loudspeakers constitutes a perversion, purely because it is not part of the original mechanical energy produced by the source. Further parallels drawn with the spread of P/A systems can also cause real fears about changes in musical sensitivity and listening patterns that we might unknowingly induce.

2.2 IN- LINE and NON-IN-LINE Active Systems

Given the choice between a recording and a live performance, which would people choose? In the case of both recorded and live performances the artists are communicating with their audience. But the live performances contain an all-important element of two-way communication. This may be subtle during the actual performance (though performers frequently comment on how strongly they feel it) and only become overt during applause etc., but any components we include in a hall which inhibit or imbalance this two-way communication are clearly detrimental. Thus the designer of a concert hall must attempt to preserve all the subtleties of audible two-way communication.
IN-LINE active systems, which place between performers and listeners an electro-acoustical link which requires the performer to play into a microphone and the listeners to listen to a loudspeaker at the other end, clearly contribute a one-way element to the communication, and thus by imbalancing the potential two-way communication process become destructive, at least to some small degree, of the live performance experience.
The more desirable alternative of what we might term NON-IN-LINE systems, while still using microphones and loudspeakers, keep them remote from both performers and listeners even hidden from sight if this is desirable. But most importantly these systems may be seen not so much as processors of sound signals but more as mechanisms for adjusting the acoustical properties of the boundary surfaces of the auditorium.
One fundamental and very important feature of all the ACTIVE systems (at least those published to date!) is that they <u>add</u> reverberation to a given space. Thus their planned use in buildings offers freedom to architects to design room volumes considerably smaller than those dictated by the requirements for passive reverberation.
Add to this 1) the convenience of being able to change the acoustics at the flick of a switch,
2) the potential for easy adjustment of the frequency
dependence of the reverberation, and
3) in principle, the provision of control over the other features of the acoustics as outlined in the list above,
then ACTIVE systems look almost irresistable.

2.3 ASSISTED RESONANCE and MULTI-CHANNEL REVERBERATION.

Parkin/2/, observing that "ringing" in P/A systems constitutes a long reverberation at the ringing frequences, harnessed the effect and developed a system of multiple channels of controlled feedback into the ASSISTED RESONANCE (AR) system - one of the two commercial NON-IN-LINE systems that have been around for some years.
Each channel of an AR system is individually tuned to a frequency in the room response. Hence one drawback of AR is its sensitivity to movement of the response as room occupancy and conditions change. Another drawback is the large number of channels (each containing a minimum of one microphone, amplifier, and loudspeaker) that is required to cover a significant frequency range, e.g. of the order of 100 channels are used to cover the bottom 1 kHz range of the audio frequency spectrum.
An alternative approach taken by Philips in their MULTI-CHANNEL REVERBERATION /3/, instead of

supplying the required energy in the very narrow bandwidths at high gain (as in AR),is to use a comparable number of wide-bandwidth channels working at very low gains. Individually the channels would be inaudible. A difficulty arises because the basic principle of the technique means the spatial requirements are very high (around 40m for each loudspeaker and each microphone). Also it has been shown /4/, that colouration imposes a strict limit on added reverberation, - the increase in RT must be limited to about 50% to be acceptable. In consequence the scope for changing the RT and shaping its variation with frequency is rather restricted - as can be seen from the results from a recent installation of MCR in Norway /4/.

Assisted Resonance does not have this same limitation. However, its gain in RT is predominantly at low and mid frequencies, for two reasons:

1) this is the range where the major difference in RT is required in multi-purpose halls, and

2) the need to restrict the channels to an economic number.

3 NEW DEVELOPMENTS.

Two systems were announced at the Concert Hall satellite symposium of the 12th I.C.A. in Canada (1986), and the firms of Roland and Yamaha are also marketing new devices. Regrettably all of these are essentially of the IN-LINE variety, and for them feedback is a large problem.

3.1 Acoustical Control System (ACS)

One of these recent systems which merits special mention, is the Dutch ACS system/5/. This system also uses multiple channels of mics and loudspeakers but by means of changing the interconnections between the mics and loudspeakers as time proceeds, controls feedback instability to the point where large changes in RT can be produced. These are sufficient to satisfy our requirement for a factor of two variation referred to at the beginning, and its developers point out the potential of the system for varying clarity and the lateral energy fraction.

Recent results/6/,/7/, confirm the system's ability to produce large changes in RT but show little change for other parameters, and are inconclusive on the subjective acceptability of perceived changes, as indeed for all the variable systems tested. It is, however, essentially an IN-LINE system, and its performance cannot be fully assessed against others until we have adequate means of assessing colouration.

3.2 The Future

Guicking et al's experiments at Gottingen with new types of electret loudspeakers/8/, may lead to the most aesthetically satisfying of NON-IN-LINE systems. Their performance, albeit relating to impractical, experimental conditions, varies effective absorption coefficients between 1 and -0.5 (i.e produce reflections with a controllable amplification from zero to a maximum of 1.5). Perhaps these are the embryonic "active wall linings" with which we could produce our multi-purpose halls of the future.

4 CONCLUSION.

A) Passive Acoustical Design offers

1) Maintenance - free acoustics for single-purpose auditoria

2) "Natural" - uncoloured sound

3) Inherently stable acoustics from adjustable systems.

but has these disadvantages:

1) Room volume is fixed by the required maximum RT

2) For variable acoustics, major changes to room volume and seating capacity are needed if the RT is to be varied more than about 40%

3) Designs with variable acoustics are likely to be a) time consuming for altering the room conditions b) inflexible if adjustment of frequency dependence is required.

B) Active Acoustical Design offers:

1) Smaller room volumes

2) Freedom from constraints on the positioning of reflectors and room boundary surfaces when designing for "lateral" sound

3) Economical correction of acoustical errors

4) For variable acoustics-a) rapid changes, and b) flexibility of control over both the degree and frequency dependence of changes.

but has these disadvantages:

1) It may incur aesthetic offence

2) Without protection devices it is inherently unstable and may cause coloration of sound

3) It requires maintenance for the electro-acoustic equipment

4) The present generation of active systems have one or more of the following limitations a) a restricted range of RT variation b) large spatial requirements for loudspeakers and microphones c) sensitivity to changes of room configuration d) an insufficient number of installations to establish the dependability and subjective acceptability of the technique.

REFERENCES

[1] "Halls for Music Performance: Two deacades of experience", R.H.Talaske, B.S.Wetherill &W.J.Cavanaugh (eds), (1982), American Institute of Physics for ASA;
[2] P.H.Parkin, JASA,48, (1970), p1025 ; [3] S.H.de Koning, Philips Technical Review, Vol 41, (1983/84), p12; [4] A.Krokstad, Trondheim NTH, Report No 408507, (1985); [5] D. de Vries & A.J.Berkhout,13 ICA 1989 Vol2 p261; [6] J.S.Bradley & R.E.Halliwell, 117th Meeting ASA Syracuse; [7] Mendel Kleiner, Chalmers University, Sweden,p199; [8] D.Guicking and M.Albrecht, Journal of Vibration, Acoustics, Stress and Reliability in Design, Vol 106, (July 1984), p397

ACOUSTICAL POETRY IN ARCHITECTURE

F. Daumal
Dr. Arch., Titular Professor of Dept. Constr. Arquit. I, Escola Tècnica Superior d'Arquitectura de Barcelona, Av. Diagonal 649, 08028 Barcelona, SPAIN.

ABSTRACT. In the architectural acoustic discipline there is a hole to fill: "The Acoustical Poetry". The architecture makes poetics for the "homo videns" but not yet for the "homo audiens", in order to be more complete. Exist the acoustical poetry?. We know some examples around the world, but they are incompletes, only represents the top of a great iceberg. The knowledge of the state of the art in acoustical poetics stills scarce. However, my teaching of the Architectural School of Barcelona demonstrates the interest of the students for those unknown subjects. It's necessary to find this examples to study its poetic language and acoustical foundation. With this state of the art, we can make a complete study of new examples. This research claims to synthesize this knowledges for an intelligent application of space designs.

1. PURPOSE

Close your eyes and take a walk to your last designed building (or garden).
Can you perceive the sounds in correspondence to the acoustical personality that you wanted for the spaces?.
Can you make the diference of sound between each space and identify it?.
If all the spaces sound without definition, so we make anodyne architectural in acoustics.

1.1 What is the cause?

Actually, the architectonic design can't find yet enough acoustical basis to express its "poetry" with the sounds.
The architecture makes poetics for the "homo videns" but not yet for the "homo audiens", in order to be more complete.

1.2 Exist the acoustical poetry?

We know some examples around the world, but they are incompletes, only represents the top of a great iceberg. It's necessary to find this examples to study its poetic language and acoustical foundation . With this state of the art, we can make a complete study of new examples.

2. METODOLOGY OF APPROACH

Rasmussen /1/ suggested a way; diferenciated spaces , Mariétan /2/ searched for simfonic spaces, Moles proposed surprise stimulus, etc, but now in acoustics we are as late as the beginning.
In other works /3/4/5/6/, I stablish the principles poetic language, with a concrete aplication, in accordance to the acoustical personality desired for the designer.

2.1 Acoustical Characters

A dual contrast system is employed like a preliminar methode to make attributive those characters. The acoustic space can be in accordance to;

a) Spatial Effect
Symetrical-Asymetrical
Centripetal-Centrifugal
Central-Perimetral
Directional-Omnidirectional
Concentical-Radial
Global-Local
Lineal-Punctual
Dynamic-Static
Connector/Filter-Barrier
Etc.

b) Activity
Work-Game
Exterior-Interior
Constant-Changeable
Etc.

c) Accent
Reverberant-Dead
Tonal-Atonal
Noiseless-Noisy
Exposed-Isolated
Faithful-Fictitious
Neutral-Suggesting
Aerial-Corporeal
Monotonical-Emphatic
Attenuant-Amplifying
Concentrating-Diffusive
Masking-Loose
Proper-Foreign
Etc.

2.2 Acoustical Personality of Spaces

Those characters can be blended, and the mixture define the specific acoustical personality of each space.

2.3 Acoustical Itinerary

In other work /7/ I propose the Acoustical Itinerary Design Method to integrate several acoustical personalities in an unity. This method was applied in two buildings.

3. SIGNIFICANT RESULTS

The knowledge of the state of the art in acoustical poetics stills scarce. However, my teaching of the Architectural School of Barcelona demonstrates the interest of the students for those unknown subjects.
The preliminar aplications in two buildings were wellcomed by its architects an users.
Gardens and Urban spaces were considered in others projects, because several aspects of the sound can be used like a therapy, especially against the noise of the city.
In this moment, the research of the Acoustical Poetry is proposed to be sponsored. The Blind Spanish Organisation (ONCE) are interesed. Why not the Clairvoyants ?.
This research claims to synthesize this knowledges for an intelligent application of space designs.

4. BIBLIOGRAPHY

/1/ Steen Eiler Rasmussen (1973), "Experiencing Architecture", MIT Press.

/2/ Pierre Mariétan (1979), "Vers une conception paysagère de la musique", Urbanisme num. 168-169.

/3/ Francesc Daumal (1988), "Parc acústic", Pap. Tècn. ETSAB.

/4/ Francesc Daumal (1988), "Proposta d'exposició efímera d'introducció al parc acústic", Pap. Tècn. ETSAB.

/5/ Francesc Daumal (1989), "Acoustical park & City of the sound", Proceedings 8th FASE Zaragoza, Sociedad Española de Acústica, Madrid.

/6/ Francesc Daumal (1989), "Proyecto de exposición Sonidos para la Cultura", Jornadas Nacionales de Acústica, Zaragoza, Sociedad Española de Acústica, Madrid.

/7/ Francesc Daumal (1989), "Acoustical planning & Itinerary in buildings", Proceedings 8th FASE Zaragoza, Sociedad Española de Acústica, Madrid.

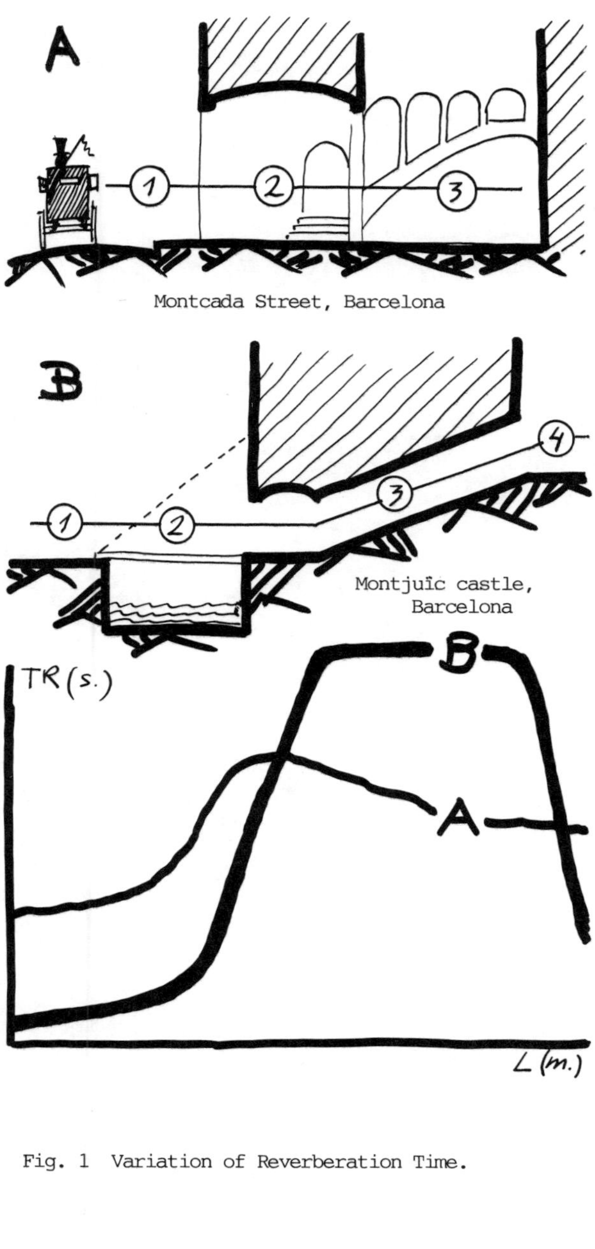

Fig. 1 Variation of Reverberation Time.

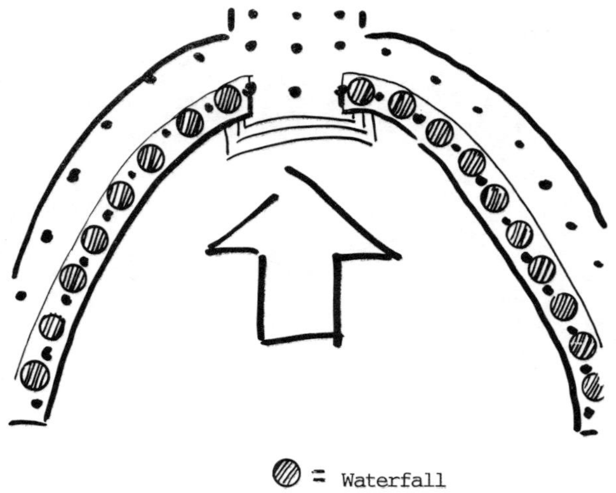

Fig. 2 Acoustical door, Sevilla.

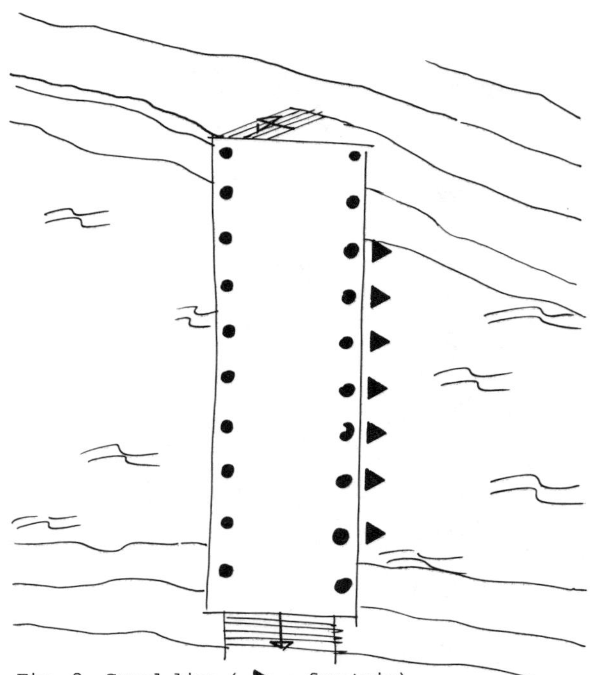

Fig. 3 Sound line (▶ = fountain).

Fig. 4 Acoustical Park.

Ministry of Housing, Physical Planning and Environment in The Hague - A Transparant Ministry

Hoogstad, Prof. Jan
Hoogstad Van Tilburg architecten
P.O. Box 1047 -3000 BA ROTTERDAM
THE NETHERLANDS

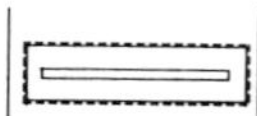

Double facade in connection with insulation from noise and temperature differences standard office blocks since ± 1920

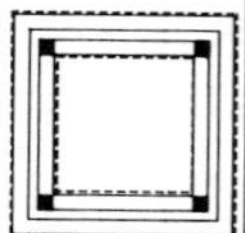

Standard office building in sqaure formation (atrium building) As a spacial system improved with a central inner space

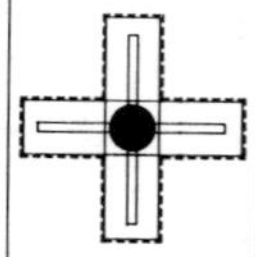

Standard office building in crucific shape

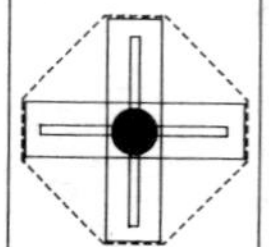

Crucifix shape +4 glass houses

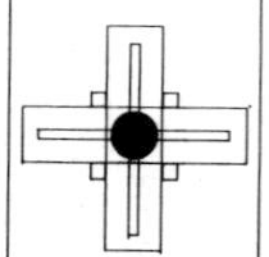

Crucifix shape +4 glass houses + terraces

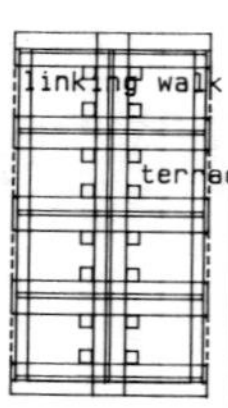

Single glass + sun blinds + windows that can be opened

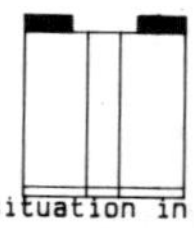

situation in summer

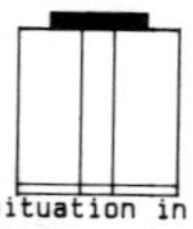

situation in winter

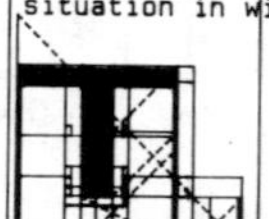

mathematical plan of the elevation

The new building for the Ministry of Housing, Physical Planning and Environment is an area marked already by large scale buildings. In the near future that character will be strengthened further. An attempt has been made to locate a large building on a small site, which, despite its size, does not give the impression of a colossus. Although it is not a dominating building, it is nevertheless, clearly a ministry.

1. If science and technology are to support architecture, it will be necessary to identify exactly what it is that needs supporting.

2. There is a substantial difference in the level of theoretical developments between architecture, science and technology.

3. Three centuries of scientific developments have by-passed architecture - architecture is still on the level of chirurgeons and alchemists.

4. Succesful support is only possible when discussion takes place on equal level.

5. Development of an architectural science is therefore a prerequisite.

6. The core of architectural science is in my opinion the theory of spacial systems.

7. Since the turn of this century office blocks have been developed by one of the most pathetic spacial systems produced during the history of architecture. They consist of no more than a central corridor with small poky rooms along either side. These buildings are without a "soul".

8. Similar working conditions have generated spacial systems without variety, and moreover with a constant climate. This together makes people sick.

9. This Ministry building is an attempt to find answers to this by way of:

9.1. Variety of working activities that are to be accomodated under one roof, i.e.
- research
- issuing of instructions
- execution
- control
- recreation

all under one roof.

9.2. A spacial system which offers:
- a wide variety
- different climates
- individual adjustment of the interior climate and windows that can be opened
- a minimum inconvenience caused by either wind, noise or smell.

10. The scientific and technical support of this project took place by means of computer simulation, monitoring these computer simulations by means of model analysis and checking the practical examples with the same computer programmes.

TRANSPARENT INSULATION - A NEW SOLAR ENERGY COMPONENT

A. Goetzberger
Fraunhofer-Institut für Solare Energiesysteme
Oltmannsstraße 22, 7800 Freiburg, FRG

ABSTRACT. A survey of the recent history, present status and future outlook of Transparent Insulation Materials (TIM) is given. A number of different generic types of transparent insulation materials are either in the early stages of production or under development. The most important one today is based on the honeycomb principle. Physical parameters presently achieved and expected improvements are outlined. Application of these materials in practically all thermal solar energy conversion systems is possible and leads to significant improvements. Most advanced is use in buildings for passive solar energy gain. Several pilot projects exist today demonstrating that good architectural solutions are possible. One advantage of the concept is that it can be applied to old as well as new buildings. Therefore, the potential for energy savings is very large. Other applications of TIM are related to solar collectors. Highly efficient collectors have been realized. A new domestic hot water system based on TIM application will be on the market very soon. It combines the collector and the storage tank and can be fitted directly into the domestic hot water system. Other applications are under development and will be described in the paper.

1. INTRODUCTION

1.1 Materials

In this introduction the young but exciting history of transparent insulation materials is outlined.

The topic "Transparent Insulation" consists of two equally important parts, namely materials and applications. One has to start with good material with specific physical properties and then one has to find and test the appropriate technical application. Transparent Insulation Materials (TIM) and solar energy conversion are ideally matched. In fact one can say that the conversion efficiency of solar energy is as good as the transparent insulation. Ideally, its light transmission should be as high and the thermal conductivity as low as possible.

In practice there are many constraints including economic ones leading to a large diversity of materials being preferred for different purposes. Good thermal insulation is more important if a high temperature difference is desired and if storage of thermal energy for instance through the night has to be provided.

There are a number of generic types of TIM which were described previously. /1/. The most important one today is based on the honeycomb principle. This structure provides convection suppression while at the same time light is transmitted with very little loss, albeit with a high degree of scattering (Fig. 1).

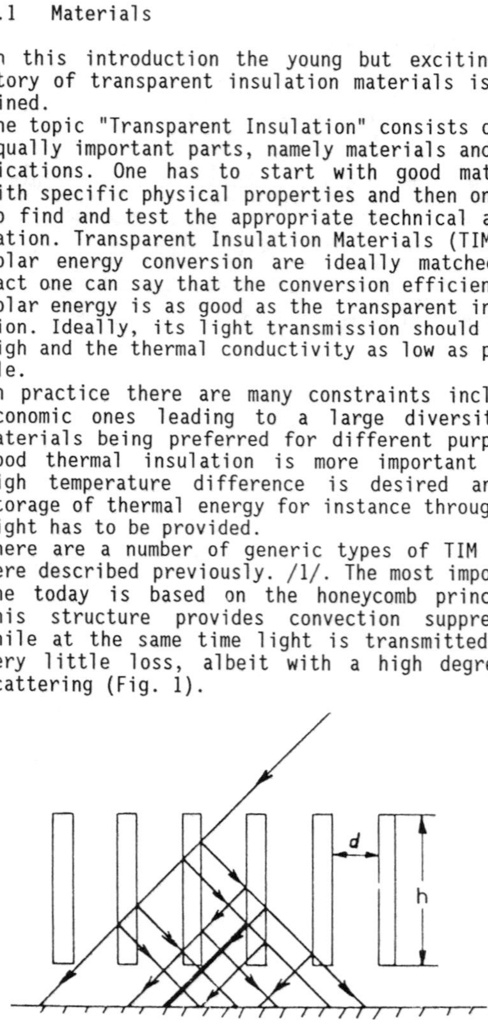

Fig. 1 Cross section of honeycomb structures with light transmission

In our institute we became interested in this principle in 1981. At that time it was a collaboration of Dr. Schmid, Dr. Wittwer and myself. Honeycombs were then already an old concept and the general attitude was that it had been tried and did not work. What were the points that were missed by the other researchers? First they concentrated on convection suppression and did not pay enough attention to infrared radiation (Fig. 2).

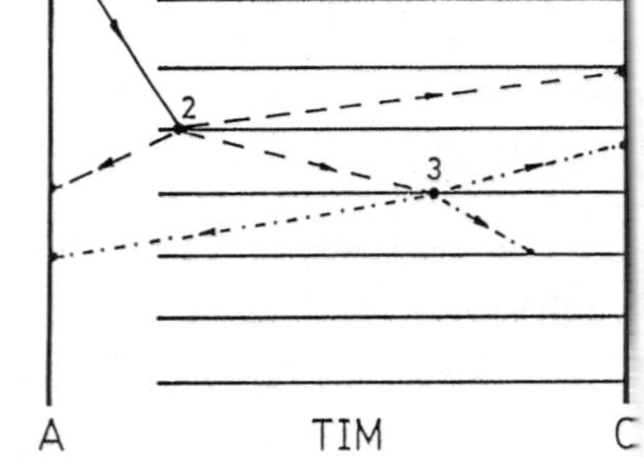

A absorber plate with temperature T_A

C glass cover with temperature $T_C < T_A$

1 directly transmitted IR ————

2 re-emitted IR (first stage) – – – – –

3 re-emitted IR (second stage) –·–·–·–

Fig. 2 IR-radiation in a TIM-system

It is necessary to use honeycomb material with good infrared absorption, and the aspect ratio should be large enough to permit only a narrow cone of infrared light emitted from the absorber to escape directly. This condition is more restrictive than convection suppression.

Another point is that previously the honeycomb was placed directly on the absorber. Thus the material was heated and the radiative transport was enhanced. By simply leaving an airgap, this effect could be greatly reduced. Only then a selective absorber had any benefit.
We were helped in our early work by the availability of PMMA capillary material by Okalux which had just the required properties, although it had been developed for different applications. Very soon the Arel material became available which not only had excellent properties, but was also available in large quantities.
Our collaboration with both companies resulted in a continued improvement of the materials. We also experimented with acrylic foam, a material that was available on the market, but this product could not compete with the honeycomb structures. It is, however, useful in special applications.
In parallel, work started at BASF on aerogel granules. Solid aerogel plates had already been made by Airglass in Sweden and at Berkeley Laboratories in USA. This material is based on a different principle and holds great promise for the future, but will not be treated further here.

2. APPLICATIONS

2.1 Space heating by transparently insulated walls

I now want to turn to applications. TIM can not only improve existing solar conversion systems, but also make entirely new ones feasible.
Historically, our first application was in buildings. The concept simply consists in replacing a conventional, opaque insulation with TIM (Fig. 3). /2/. It is related to the well-known Trombe Wall principle but simpler.

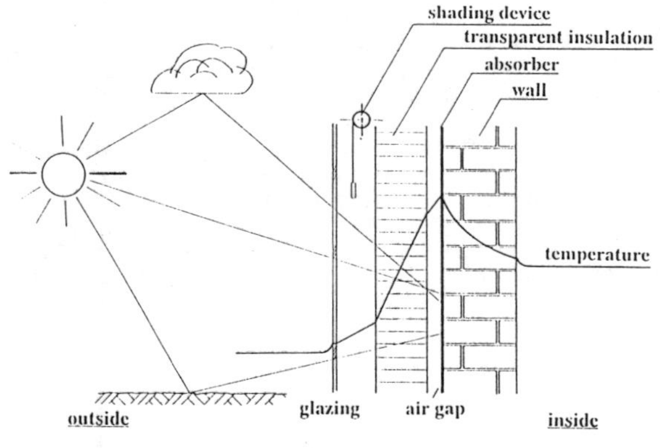

Fig. 3 Principle of passive heating with transparent insulation

The dark-coloured wall surface acts like the absorber of a collector by converting the solar radiation transmitted through the transparent insulation material into useful heat. Because of this transmission heat losses of the wall can be dramatically reduced and parts of the solar gains can be stored in the wall itself. The contribution of these gains for heating purposes occurs with a time shift, dependent on the wall material. The inner wall surface acts like a large low temperature wall heater. The advantage of such a system is that it behaves like a typical passive solar component without the use of an active heat distribution device.

Because of the time delay of heat flow through the wall a very good match with direct gain through windows is achieved. Direct gain and TIM on massive walls together can give a very uniform delivery of heat to the inside of a house.
The great advantage of the concept is that it can be applied to old as well as new buildings. Therefore, the energy savings potential is very large, as will be seen later.
For our first experiments we had to find a building with brick walls without insulation. This description happened to fit my own house in Merzhausen. So this building became the first transparently insulated house. Fig. 4 shows how it looked in its first version in 1982. Insulation material was acrylic foam. (I have to mention here that our colleagues at the IBP in Stuttgart started independently with their own experiments at about the same time).

Fig. 4 First experiment on an occupied building

Many improvements were installed in this house as time went on. In particular, the material was changed to the capillary type with much improved results. We also encountered the effect of overheating in summer very drastically. In the first summer we had some very unpleasant weeks before we managed to put a temporary cover on the outside of the insulation elements. Since that time I have been very convinced of the need for radiation control.
For this purpose an automatically controlled roller blind system was developed (Fig. 5). The present state of the house with roller blinds in operation is seen in Fig. 6.

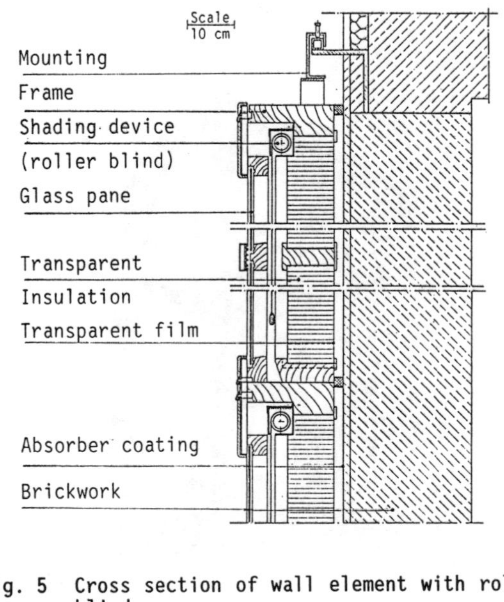

Fig. 5 Cross section of wall element with roller blind

Fig. 6 Test building with roller blind partly opened

In the meantime, application of TIM for passive solar gain in buildings has made great progress. A number of new buildings have been constructed. We have reached the demonstration stage. Examples of these buildings are seen in Figs. 7 - 10.

Figs. 7 - 10: Examples of recent demonstration projects.

Fig. 7 One-storey building in Munich (ISE Freiburg and Municipality of Munich)

Fig. 8 Bollin house, Freiburg

Fig. 9 House in Ardon, Switzerland, realized by ISE

Fig. 10 Terraced houses in Düsseldorf, Kaiser Bautechnik and ISE

Nevertheless, this field is only at the very beginning. The potential of TIM in buildings was recently investigated by Lohr and co-workers for the FRG. /3/. Their results indicate a very large potential dependent on the boundary conditions, in particular the physical properties of the materials, the cost of the systems and the competing energy cost. The technically realizable saving of heating energy nationwide by opaque insulation is 50 %. This goes up to 80 % when TIM is used wherever feasible. TIM can not only be used for heating purposes but also for daylighting. This application is described at this conference in the paper by J. Schmid.

2.2 Collectors

2.2.1 Flat plate collector for process heat (90 to 150 [°C])

Flat plate collectors can be tremendously improved by using highly effective transparent insulation materials as the front cover With a 10 cm honeycomb structure made of polycarbonate, the front losses can be reduced to values of 0.9 [W/m^{2}K] at ΔT = 100 [°C]. Such a collector was recently developed at our institute. /4/. A collector field of 60 [m^2] area has been tested in combination with a transmembrane water distillation unit on the Canary Islands since May 1988.
A reasonable criterion for the comparison of process heat collectors is the efficiency at an irradiation of 800 [W/m^2] and a temperature difference ΔT = 100 [°C]. Under these conditions, a conventional flat plate collector with a selective absorber and single glazing has an efficiency of approximately 20 % only, whereas vacuum tube collectors reach about 50 %. Using a 10 cm TIM-cover in the flat plate type the same efficiency value as for the vacuum tube type is measured (Fig. 11).

2.3. Thermal energy storages with radiatively heated walls

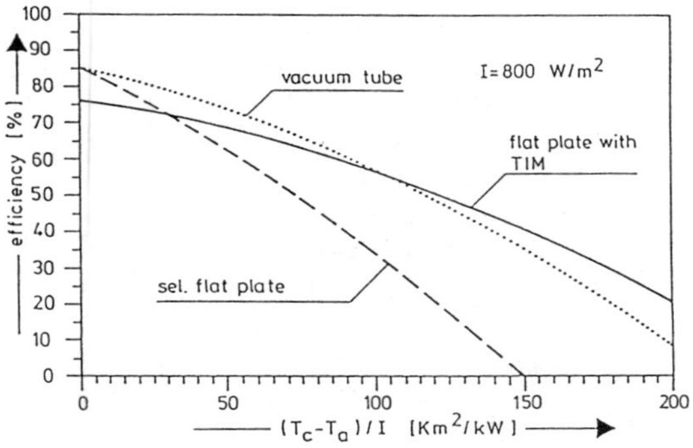

Fig. 11 Efficiency characteristics of different collectors

The production of these collectors is as simple as the production of flat plate collectors without honeycomb structures. The price of such a collector should be only slightly higher than the price of a flat plate collector, but considerably lower than that of a vacuum tube collector.
According to calculations, still further improvements of optimized flat plate collectors with TIM are possible so that they will even outperform vacuum tube collectors in the future.

2.2.2 Integrated storage collector (ISC) for domestic hot water

An integrated storage collector consists of a storage tank whose dark-coloured surface acts as an absorber for the solar radiation (Fig. 12). The tank is connected to the mains water under standard pressure conditions. The aperture of the collector is covered with a layer of transparent insulation material to reduce thermal losses and prevent freezing. /5,6,7/. The advantages over a conventional (solar) DHW-system are

- compact construction, only one single unit
- reduced aperture, because of high efficiency
- no heat exchanger, no pump, no control system
- no anti-freeze fluid.

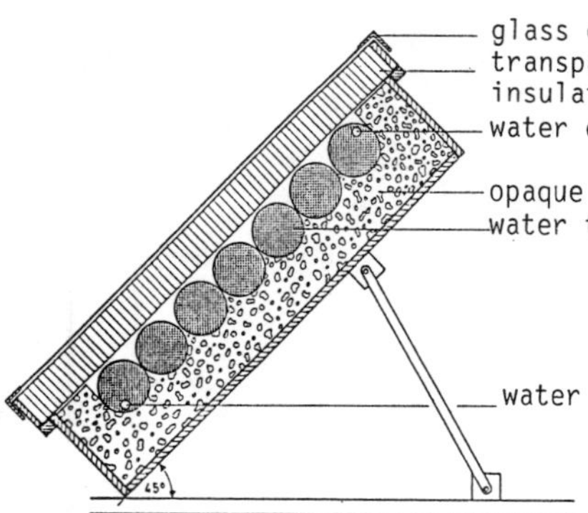

Fig. 12 Cross section of ICS prototype

These advantages will help to bring down the cost of solar hot water systems.
Up to now, integrated storage collectors, having simple single or double glazings, could be used successfully in warm climates only. Under Central European weather conditions freezing occurs if simple glazings are used. These difficulties can be overcome, however, by using transparent insulation materials as front covers of integrated storage collectors. Extensive computer simulations have been made and compared with experimental results to investigate these questions. It has been shown that there is no danger of freezing even under Central European weather conditions.

With TIM completely new concepts of thermal energy storage can be thought of. Conventional storages with opaque insulation materials always have energy losses. But if a storage tank is considered that has its walls insulated with TIM, it is possible to have no losses at all, because the losses which occur can be compensated by the radiative energy gain during daytime. /8/. This simple but very effective principle offers very promising prospects.
A small cubic water tank (0,5 m edge length, approximately 120 l water volume), insulated on four sides and the top with 14 cm honeycomb in combination with 1 cm airgap and single glazing, was tested during 1987 (Fig 13).

Fig. 13 Model of seasonal thermal storage tank

An overall U-value of 1 [W/m^{2}K] and a $\tau\alpha$-product of 0.6 was determined. Boiling occurs on 58 days up to the 20th August.
In transferring the measurements of this cube to a real storage system it can be expected that heat at a temperature of approximately 70 [°C] can be stored from March to September under central European weather conditions. This is a very important conclusion for example with respect to short term storage like weekend storages of waste heat of factories.

3. CONCLUSION AND OUTLOOK

The experimental, theoretical and practical work during the last few years has led to an essential understanding of transparent insulation materials. Materials now available reach heat loss coefficients of 0.8 [W/m^{2}K] with diffuse transmission values of 0.7. The theoretical models and the experimental set-ups have shown their validation and will be used for further improvements of materials.
The initial applications have shown that well-known thermal solar systems can be significantly improved by the use of TIM (e.g. process heat collectors), and completely new concepts are made possible (e.g. house heating with TIM-walls). In order to develop more economic systems, cost reductions are now required for the materials as well as for the whole systems.

A concept for a completely self-sufficient solar house has been devised in our Institute combining TIM-systems with photovoltaic and hydrogen/oxygen units. Transparently insulated walls will provide energy for space heating, and the integrated storage collector with TIM-cover will be used for domestic hot water. Photovoltaic cells drive a closed loop hydrogen/oxygen system for high exergetic energy demands, and pressurized gas storages are planned for compensating differences in solar supply and energy demand (Fig. 14).

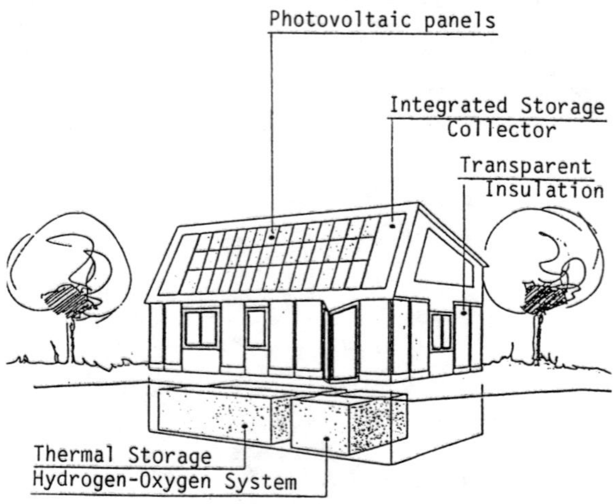

Fig. 14 Sketch of planned self-sufficient solar house

The realization of this project within the next few years will demonstrate the possibility of meeting **all** the residential energy needs by the solar radiation incident on the surface of a house without any fossil back-up or connection to the electric grid. This is an outlook into the distant future, because it will take a long time before such a house will be economical, but it points the direction in which solar energy applications can be developed.

4. ACKNOWLEDGEMENTS

Most of the work reported here was supported by the German Federal Ministry for Research and Technology. I want to thank the many members of my institute who contributed to the work reported here and helped in preparing this manuscript.

5. REFERENCES

/1/ For a survey of the properties of materials and of applications see Proc. 1st, 2nd and 3rd Workshop on Transparent Insulation Materials 1986, 1988 and 1989 published by the Franklin Company Consultants Ltd., 192 Franklin Road, Birmingham B30 2HE, UK.

/2/ A. Goetzberger, J. Schmid and V. Wittwer "Transparent Insulation Systems for Passive Solar Energy Utilization in Buildings", Int. Journal of Solar Energy 2, p. 289, (1984).

/3/ A. Lohr, B. Weidlich "Systemstudie - Lichtdurchlässige Wärmedämmung", Köln 1989.

/4/ M. Rommel and V. Wittwer "Flat Plate Collector for Process Heat with Honeycomb Cover - an Alternative to Vacuum Tube Collectors", Proc. of the Biennial Congress of the Int. Solar Energy Society, Hamburg 1987, Pergamon Press, Vol. 1, p. 641 f.

/5/ A. Goetzberger und M. Rommel "Prospects for Integrated Storage Collector Systems in Central Europe", Solar Energy, 39, p. 211, (1984).

/6/ Ch. Schmidt, A. Goetzberger and J. Schmid "Test Results and Evaluation of Integrated Collector Storage Systems with Transparent Insulation", Solar Energy, 4 , p. 487, (1988).

/7/ Ch. Schmidt and A. Goetzberger "Single-Tube Integrated Collector Storage Systems with Transparent Insulation and Involute Reflector", Solar Energy, in press.

/8/ A. Goetzberger "Seasonal Storage of Thermal Energy with Radiatively Heated Storage Walls", Int. Journal of Solar Energy, 2, p. 521, (1984 .

IMPROVEMENTS ON SOLAR APERTURE

Dr.-Ing. F. W. Grimme
Project Management for Biology, Ecology, Energy
of the Nuclear Research Centre Jülich (KFA Jülich),
Postbox 1913, D-5170 Jülich 1
Tel.: 02461/61-4873

ABSTRACT. Due to the two energy crises, energy saving and rational energy use have become the order of the day. Furthermore, the forecast climatic catastrophe can only be combated by a drastic reduction in energy use. Energy saving with opaque insulation has been used for a long time but the use of transparent insulation provides an opportunity to save energy and also to gain solar energy with the same material. In combination with other components, various transparent or translucent insulation materials (TIM), now being investigated or ready for the market, supply a powerful new impetus to the wide field of solar energy utilization. In this field, the German Concerted Research and Development Programme "Energetic Optimization of the Solar Aperture" includes extensive R&D work on materials, components and active and passive systems, e.g. daylighting of buildings, low temperature heating for houses and DHW as well as process heat e.g. for cooking, sterilization, drying, absorption cooling and thermosolar water pumping.

A wide variety of transparent insulation materials (TIM) are now being investigated or are on the market as new products. Sponsored work on the use of TIM began in 1984 /1/, leading in 1986 to a German concerted research and development programme "Energetic Optimization of the Solar Aperture" /2/. Within this framework, research started in a wide field. Table 1 shows the connections and the working context.

1. Research and Developments on TIM itself

Large and small companies are developing the material itself /2-C,2-I/, e.g. for higher temperatures on fireproof or nonflammable materials (Teflon, glass, aerogel), (physical properties are given in Tables 2, 3) and also for special uses, such as TIM with very high transmissivity for maximum solar gain, very important for process heat collectors with high stagnation temperatures.

To fulfil the German building regulations for higher buildings (> 3 storeys) and to maximize solar energy gain by highly transparent covered solid walls TIM must be fire-resistant. Newly developed glass capillaries or aerogels (quasi-homogeneous silica gel) can fulfil this obligation due to the fireproof starting

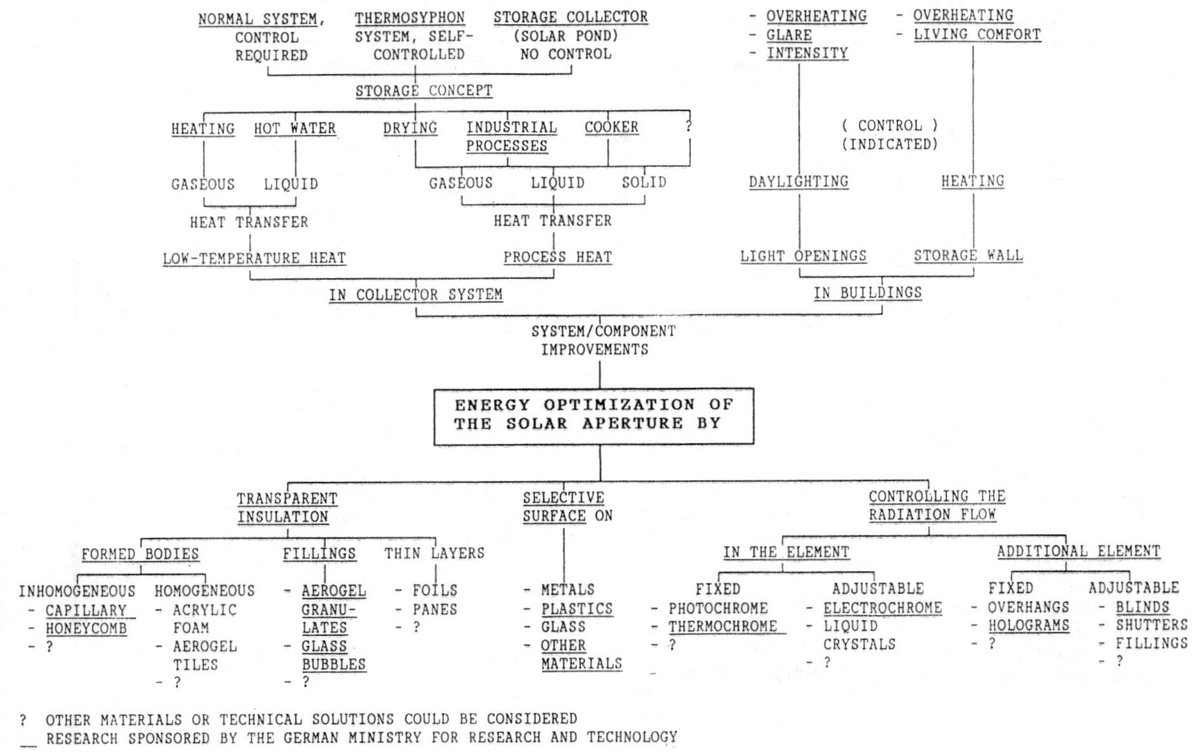

Table 1: Connections and Working Context on the German Concerted Research and Development Programme "Energetic Optimization of the on Solar Aperture"

material. Special certification fire tests on wall-connected glass-covered high-temperature plastics and glass (honeycombs or capillaries) and aerogels still have to be carried out. For overheating protection an effective regulation system is also required.

Producer	Material	Λ	$\tau_{dif.}$	$T_{max.}$	UV
AREL	Polycarbonate	1.07	0.78	≤ 125	+
OKALUX	Polycarbonate	0.98	0.73	≤ 125	+
KAISER	Polycarbonate	0.83	0.75	≤ 125	+
OKALUX	Polyacrylic	0.83	0.73	≤ 95	++
OKALUX	Polyacrylic HW	0.67	0.77	≤ 115	++
OKALUX	Polyestercarbonate	0.76	0.65	≤ 170	O
OKALUX	Polymethylpentane	1.34	0,63	≤ 180	+
OKALUX	Hostaflon ET	0.78	0.68	≤ 170	++

Λ(W/m²K) by T_{mean} = 10 K
$\tau_{dif.}$ = diffuse/diffuse transmission
$T_{max.}$ = max. utility temperature °C
UV = UV stability after 1500 h Xenotest

Table 2. Characteristic Data of Honeycomb and Capillary TIM (10 cm) /2-A,2-K/

	$\tau_{dif.}$	U value (W/m²K)	air gap	
Granulate Filling: air	0.55	1.38	12 mm	#
"	0.43	0.90	20 mm	#
"	0.27	0.50	40 mm	#
SF_6	0.55	1.10	12 mm	#
Ar	0.55	1.03	12 mm	#
reduced to 0.03 mbar air	0.59	0.70	12 mm	#
Filling: air				
Tiles	0.57	0.81	20 mm	##
Granulate				
6 - 8 mm ø	0.43	0.96	20 mm	##
4 - 6 mm ø	0.42	0.95	20 mm	##
3 - 4 mm ø	0.40	0.88	20 mm	##
≤ 2 mm ø	0.22	0.84	20 mm	##

$\tau_{dif.}$ = diffuse/diffuse transmission
measurement results VEGLA, Aachen
measurement results FhG-ISE, Freiburg

Table 3. Characteristic Data of AEROGEL in Double Glazed Openings

TIM with a lowered transmissivity can be used to avoid the need for a shading device for overheating protection and to put TIM on facades as insulation material with a transparent plaster. The development results achieved to date show /2-I/ that a partly TIM wall covering can solve the overheating problem and the solar gain by assisting to save energy. Both systems, opaque (e.g. styrene foam) and transparent, can be used as additional outer insulation systems together on the building facades without architectural problems caused by outer material adaptation.

From the scientific point of view a number of properties are still unknown, e.g. complete understanding and clear description of the total radiation transmission/absorbtion parameters in capillary systems. In /2-A, 2-S/ this scientific research has been started, not only working on TIM but also on micro- and macrostructures of thermochrome or electrochrome materials and other fields important for the work in the field of "Energetic Optimization of the Solar Aperture".

2. Work on Components

For a suitable use of TIM, additional special components may be very important. To protect the material itself or to prevent overheating of walls and buildings or collectors, fixed/movable shutters such as blinds or other components are now being used. Special developments to tighten the fabric or film roller blinds by means of aerostatics /2-A/ gave a good additional impetus in automatic temperature regulation and energy saving. Closed blinds with IR-reflective coating can increase the thermal insulation property of walls covered by TIM by up to 40 %, using such systems in normal compound windows the reduction factor is up to 2.5.

Research on thermochromic shading devices has produced an automatic system without movable elements /2-L/. At the moment the physical properties should be adjusted to fulfil different conditions, such as to prevent overheating or glare, coloured chameleon effects may possibly be used for TIM covered facades or light windows.

Initial work was done on a simple actively regulated electrochromic absorber system for transparent insulated walls /2-A/. Using such electrochromic switching, the absorber of a TIM-covered wall can be black for solar absorption and as a consequence for solar room heating or reflective to obviate overheating by cutting off the solar gain. The electrochrome window on a WO_3 basis has warming-up problems and thus gives additional cooling load by good shading performance.

Later on, although work has not started yet, an improved electrochromic system, with the extremes of transparent or reflective, integrated in light openings or windows may be used to protect against overheating and glare.

To fulfil the criteria for a convenient solar heating by TIM-covered walls, such as overheating prevention, optimal solar gain and total performance, a special regulation system was developed /2-A/, and has been used with success in prototype buildings to open and close the solar aperture for gaining or reflecting solar energy.

For daylight purposes, holograms are being investigated /2-R/. Using such holograms (diffractive grids on transparent substrates) a redirection or reflection of radiation can be generated for a certain spectrum or incident angle. Deep rooms can be lightened by such light-guiding elements for diffuse or direct solar radiation. To save cooling load, caused by solar gain through windows and light openings, holograms can be used as reflective spectral filters, e.g. for solar IR. The goal is to achieve mass-produced low-cost window-integrable holograms for improving daylight, visual and thermal comfort.

For improvements to transparent surfaces such as foils, plastics or glass, additional IR-reflective foils are under development. The first aim is to achieve maximum solar transmission by high visual transparency and maximal long-wave infrared reflection (the present stage is $\varepsilon = 0.12$; $\tau_{vis.} = 84$ %; $\tau_{tot.} = 60$ %). The second aim to produce a coating on polymer foils in a continuous process.

3. TIM Use on Active Systems

Good progress on collectors can be made with TIM, e.g. a standard flat plate collector can be improved for process heat so that the same performance as a Corning vacuum collector may be reached /2-A/.

Not only can the collector itself be improved by TIM, but the whole DHW collector system can also be modified, or rather simplified with TIM /2-A/. Fig. 1 explains the differences between the two standard (pure active and thermosyphon) and the new collector system. Using direct water grid coupled pressurized stores covered with TIM under German climatic conditions provides an efficiency of up to 43 % and 59 % of the annual solar fraction (integrated in a vertical south-facing wall). Integration in roofs or walls is easy if multi-pipe storage is used for installation reasons (saving storage diameter). With a newly designed electronic thermostatic back-up heating system near the DHW-user maximal savings can now be achieved by reducing losses.

An active system for DHW or low temperature heating can be built by putting a pipe system in a TIM-covered solid wall. Especially in the period when room heating is not required the TIM-covered solid absorber can be used as an active system. The familiar uncovered solid absorber for heat pump use and modifications with TIM covers were tested /2-H/. First results show that the TIM-covered element, open to the air at the rear, can be used as a combined hp-absorber/collector system.

In the field of simple technology (not only for the Third World) the very simple collector has been reborn, namely the pressureless foil-storage collector with or without heat exchanger for water grid coupling /2-F, 2-M/. Tests have shown that under Mediterranean conditions with simple film covering from early spring to late autumn throughout the day enough DHW can be produced; using additional TIM covering over 24 hours the total DHW can be provided in the same period by solar energy.

Not only are storage collectors /2-A/ and saltfree solar ponds /2-A/ for process heat under investigation, but also air collectors with undercurrent absorber /2-E/, now improved by a selective absorber, remaining dustfree for extended periods due to a tight covering. The collector performance is drastically increased. The collector can be adapted for special uses by adjusting the air gap height behind the absorber.

Using open glass capillaries as the air collector covering to recover the thermal losses of the capillars with the incoming air stream can be a very interesting new process collector development. Developments are under way /2-O/.

For conditions in developing countries, solar cookers with /2-D/ and - a new type - without /2-F/ heat transfer fluid are now under investigation, and also solar sterilizers /2-F/. First results for the solid TIM-covered solar cooker and for the sterilizer have given a new impetus in this field.

4. Improvements to Passive (Building) Systems

After initial research /1/ and prototype developments /2-A/, attempts are now being made to realize and measure actual buildings with different TIM facades (partly sponsored in /2-A, 2-N, 2-U/). The physical, structural and architectural design currently involves considerable difficulties, many parameters or circumstances are not fixed by measured results or are still unknown. In particular those in the architectural planning field are suspicious not only of the chameleon effects on the regulated facades but also about the large outer facade glazings, above all with respect to TIM facades on buildings to be renovated.

Impressive studies on the aspects of the architecture and construction physics of old and new buildings have been carried out, including a material and building analysis sample catalogue. A scenario on the potential of present and future use of TIM in wall coverings in commercial and non-commercial buildings of standard types and ages gave very emphatic answers for the construction industry and in particular for the facade industry/2-A, 4/.

In particular the problems caused by inhomogeneous walls are being investigated /2-A, 2-D/. Thermal stress due to solar heating on strongly insolated inhomogeneous walls, well insulated by a good TIM insulation will be examined to find error-free conditions.

Better room comfort due to warmer walls and big savings in primary energy will be achieved in solid buildings improved by a TIM facade. Under wet climatic conditions, near the coast or e.g. in Scotland or Ireland, TIM-covered walls can be "solar dried" by heating the walls from outside. The solar surplus, not useful due to good room temperatures, can be employed under all climatic conditions for more ventilation.

The better the transparency and the heat loss coefficient of the TIM covered walls is, the better the regulation system must work. Four elements are necessary for this:

- outer glazing
- TIM
- regulation system by high transparent TIM and a good insulation value for the wall and outer elements
- solid wall as a solar store, given by the normal outer wall

If the regulation system inside the outer glazing is mechanical, then the facade must be built as a normal window. At the moment the system price for such a TIM-covered, regulated wall in the prototype projects is very high (>500 < 800 DM/m² without the solid wall). The price will be reduced by a wider market with standard solutions. After simulations under good weather and building conditions no standard central heating is required, later on after good research results this additional investment saving can be used to improve economic efficiency.

After research and demonstration results have been made known in courses and studies, designers and architects will be able to build and reconstruct not only walls or windows but also entire buildings in an appropriate way, if the construction market is prepared.

5. Future Aspects

With the new class of transparent insulation material not only can energy savings be made but the use of solar energy can be improved in active, hybrid and passive systems for room heating or daylight purpose or for process heat using solar energy gain for air conditioning of buildings, cooling/freezing of food or for industrial process heat, e.g. for drying, sterilization etc. Considerable efforts will be made in this direction in the near future.

Considering the aspect of economics and ecology in the heating domain in buildings, it is clear that under present conditions the decision to make energy savings not only by improving insulation but by gaining solar energy via passive and active means will help us and the whole world. Using the TIM developed in the German Concerted Programme, components and systems with a high degree of comfort for the occupants will be the result. The development and improvement of components and systems based on a twin evaluation of energy and comfort should be followed by system adaptations for different building types and their specifications and perhaps by new standards for building improvements.

The total concerted research programme has been financed up to now by approx. DM 24 million from public funds by the German Ministry for Research and Technology, Bonn and approx. DM 5 million from industry (as of 10.89).

/1/ 03E8411A "Solar Energy Gained Insulation System" "LEGIS", sponsored by the German Ministry for Research and Technology; Contractor: Fraunhofer-Gesellschaft (FhG);

/2/ 0335003A to U "Energetic Optimization of the Solar Aperture" Concerted Research and Development Programme sponsored by the German Ministry for Research and Technology.
Status report 1989 (only English abstracts) distributed by Fachinformationszentrum Karlsruhe, Vertrieb, D-7154 Eggenstein-Leopoldshafen; (Tel: 07247/808-330/301; Telefax: 07247/808-666)
- A "System Developments"; Contractor: FhG; Institute for Solar Energy Systems (ISE), Freiburg
- C "Transparent Insulation Aerogel"; Contractor: BASF, Ludwigshafen
- D "Solar Thermosyphon Cooker"; Contractor: Cambra, Calw
- E "Solar Air Collector"; Contractor: Grammer, Amberg
- F "Uncomplicated Solar Systems for Heating of Liquids"; Contractor: FhG-ISE, Freiburg
- H "Solid Solar Absorber"; Contractor: FhG; Institute for Construction Physics (IBP); Stuttgart
- I "Compound Insulation System"; Contractor: STO AG, Stühlingen
- K "Transparent Insulation for Higher Temperatures"; Contractor: OKALUX, Marktheidenfeld
- L "Thermochromic Shading System"; Contractor: FhG-IPB, Stuttgart
- M "Plastic Storage Collectors"; Contractor: University of Hohenheim, Stuttgart
- N "Pilot Project Transparent Insulation China"; Contractor: FhG-ISE, Freiburg
- O "Glass Capillary Air Collector"; Contractor: University of Stuttgart
- P "Transparent Insulations on Inhomogenic Walls of Old Buildings"; Contractor: University of Stuttgart
- R "Holograms for Daylight Purposes"; Contractor: Fachhochschule Cologne
- S "Optical Investigations on Microstructure of Transparent Insulations"; Contractor: University of Würzburg
- U "Crew Training Centre; Experimental Halls"; Contractor: DLR, Porz, near Cologne

/3/ 03E8640A "Transparent IR-reflective Foils for Windows"; Contractor: Renker, Düren/Freiburg; report see /2/

/4/ Study on Transparent Insulation - Transformation into Practice, Criteria and Consequences for the Application of Transparent Insulation Systems in Buildings. (only summary in English)
Contact address: Büro für Energiegerechtes Bauen, Oldenburger Straße 68, D-5000 Cologne 60 - FRG

NEW CONCEPTS FOR DAYLIGHTING AND HEATING OF BUILDINGS

J. Schmid, W.-S. Wilke, F. Sick

Fraunhofer-Institut für Solare Energiesysteme
Oltmannsstraße 22
D - 7800 Freiburg

Abstract:

It is expected that the use of transparent insulation material (TIM) for heating and daylighting of buildings has a high energy saving potential. Proper design will not only reduce energy but will at the same time increase thermal and lighting comfort for the inhabitants however.
One of the most important aspects of TIM is the fact that it can be applied not only to new buildings but also to existing ones. This new technology will change the appearence of buildings dynamically: the facade reacts to the external climate in an intelligent way. During times in which heating demand occurs, the facades absorb solar radiation and look therefore dark whereas during summer season the sunrays are reflected turning the facade into brightness. This changing appearence is a critical item for architectural acceptance. On the other hand it is a challenge for architects to deal with these new elements and to create flexible building skins.

1. Introduction:

For the conversion of solar energy into thermal energy, transparent insulation material (TIM) properties have a great influence on conversion efficiency /1/. In general high transparency and good thermal insulation properties are leading to good performances. For solar collector application, in which thermal energy is to be produced, the optimum ratio between transparency and thermal insulation depends on the desired temperature difference between product heat and ambient temperature as can be seen in equation 1,

$$\eta_{koll.} = \alpha \cdot \tau - \frac{k \cdot \Delta\vartheta}{G} \qquad (1)$$

in which η is the collector efficiency, τ the transparency of TIM, α the absorbtion coefficient, k the heat transfer coefficient, $\Delta\vartheta$ the temperature difference between the absorber temperature and ambient and G the global irradiation.

For passive solar energy utilization by means of large areas of TIM, the mean energy production is being reduced towards a level at which more or less the "collectors" loss will be covered. In the sense of equation 1, the collectors' efficiency will become zero and thus

$$\alpha \cdot \tau = \frac{k \cdot \Delta\vartheta}{G} \qquad (2)$$

$$\text{or} \quad \frac{\tau}{k} = \frac{\Delta\vartheta}{\alpha \cdot G} \qquad (3)$$

In this special case, the relation $\frac{\tau}{k}$ determines the qualitiy of the TIM-element. For α and G being constant, the relation $\frac{\tau}{k}$ is then directly proportional to the temperature elevation capability of the TIM. In table 1 the figures of τ, k and $\frac{\tau}{k}$ are presented for different TIM configurations. The high quality of the optimized TIM structures becomes evident. These new features do not only increase solar energy utilization in general, but allow for new concepts, which have not been possible in the past.

Table 1: Physical properties for different TIM-structures

Glass-thickness 4mm, ε-Glass = 0,87, low ε-coating: 0,1
g= energy transmission coefficient

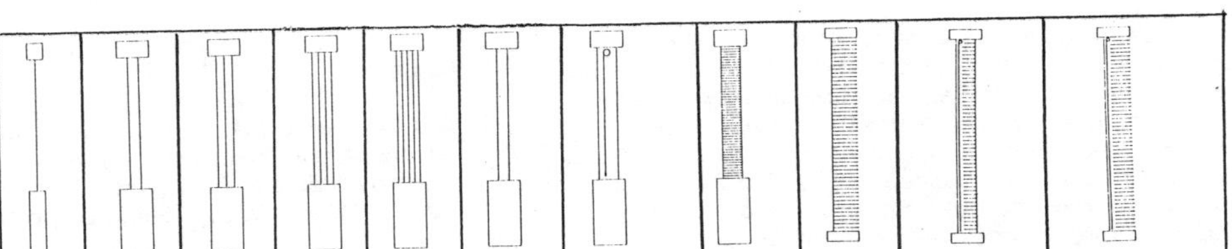

						low ε	blind open/closed			open/closed	open/closed
τ	0,78	0,62	0,50	0,41	0,33	0,62	0,62/0,00	0,45	0,57	0,45/0,00	0,57/0,00
g	0,80	0,67	0,58	0,51	0,45	0,68	0,67/0,17	0,61	0,67	0,62/0,02	0,69/0,02
k	5,80	2,89	2,18	1,91	1,76	1,80	1,78	0,93	0,92	0,58	0,58
τ/k	0,13	0,21	0,23	0,21	0,19	0,34	0,35	0,48	0,62	0,78	0,99
g/k	0,14	0,23	0,27	0,27	0,26	0,38	0,38	0,66	0,73	1,07	1,20

2. The importance of solar energy for daylighting

The use of solar energy for daylighting is one of the classical tasks performed by windows. By increasing glazing areas towards the south thermal solar energy gain becomes possible. On the other hand serious drawbacks have to be considered this way namely: Risk of overheating, of glare and bleaching outs and of reduced comfort during nighttimes due to the relatively low thermal resistance of conventional glazings. By using TIM to heat up house walls, thermal energy can now be produced more efficiently and window areas could therefore be reduced.
The question, which is the optimum relation between measures for daylighting and heating can only be answered when comparing the "value" of both light and heat. For light the expression of lumenhours (lmh) will be used in a similar way as Watthours for the energy since for the conversion efficiency of lamps the expression lumen per Watt [lm/W] is being used.

Table 2:

	lm/W
incandescent lamp	14 - 20
fluorescent lamp	50 - 80
solar light	90 -120
filtered solar light	150 - 200

According to Table 2, the "value" of one kWh solar light, which can fully be used for lighting purposes ranges between 1 DM/kWh and 2 DM/kWh if a price of 0,2 DM/kWh for electricity and an incandescent light bulb is considered.
The importance of the high solar luminous efficiency is even increased when considering the summer case in which heat associated with light is not welcome and sometimes has to be removed by air conditioners needing additional energy. In this case solar energy utilization using filtered light becomes even more attractive as shown in Table 2.

3. Examples of new daylighting concepts

3.1. TIM-lighting elements

Advanced TIM structures allow for the design of elements having the same thermal insulation properties as well insulated conventional walls. Replacing the wall by a TIM-element, as shown in Fig. 1a, leads to the following features:

The light is of diffuse nature which prevents glare. Since the relation between the TIM-element area and the room area is maximized, full daylight illumination can be obtained during longer period of times. Through the large apertures' area, illumination at a higher level is obtained as shown in Fig. 2.

The integrated roller blind allows to control the incoming light flux in such a way that the maximum tolerable values are not passed. During summer season, the influx of solar energy can be adjusted to the minimum requirements for lighting purposes. Undesired room heating is reduced to a level which could not be obtained by artificial lighting methods. A further reduction of heat influx is obtained by using a second roller-shade, which reflects the near infrared part of the solar spectrum leading to a luminous efficacy of 150 - 200 lm/W.
During winter season, the conventional way to reduce glare is to use external shading elements. In parallel to the desired reduction of light, the incoming solar heating energy is reduced the same way however. The use of a partially absorbing interior shade as shown in Fig. 1b separates light and heat influx in such a way that the undesired amount of light is converted into heat on the absorbing shade.
The large translucent TIM area is of multiple benefit:

a) The large area produces more light and its distribution is more uniform than that from a window. The transparency of TIM-structure elements for sunlight is further improved by the elimination of absorption on the wall surface as shown in Fig. 3, which has to be considered in case of glazings.
b) Since its good thermal insulation properties (0,3 - 0,5 W/m^2K), the inside surface temperature lies nearly at room temperature.
As a matter of this fact no draft or other thermal uncomfort is existing and the working area can now be extended directly up to the TIM wall which further improves the daylight utilization.

3.2. TIM lighting and heat storage elements

In the past, glass brick walls have been used for daylighting purposes already. Their high thermal conductivity, has restricted their use to applications in which thermal comfort is uncritical such as storage rooms or staircases. By combining glass bricks with TIM as shown in Fig. 4, translucent walls can now be realized providing high thermal comfort /2/. By tuning the absorption factor for solar irradiation of the glass bricks, storage of solar heat becomes also possible this way.
As can be seen in Fig. 5, the thermal efficiency can be made even higher than using TIM in combination with an opaque wall.

This figure applies for homogenous absorption in the glass. By discrete absorption, as shown in Fig. 6, the thermal properties of the system can be adjusted. If the absorption layer is placed towards the TIM-side (Fig. 6a), properties similar to those of an opaque wall are obtained resulting in maximum storage times. By applying the absorption layer inside as shown in Fig. 6c, immediate use of thermal energy is possible whereas a location in the centre (Fig. 6b) leads to results similar to the configuration shown in Fig. 5.
According to the actual need, these layers can be introduced statically or dynamically by means of electrochromic or thermochromic layers.

4. Impact on future architecture

By introducing such new elements into architecture, the following requirements can now be fulfilled:
For daylighting purposes first priority to translucent or transparent elements will be given. The relation between elements for direct gain and elements with storage capabilities may be selected according to the desired dynamic response of the building. For visual contact of the inhabitants conventional windows will be selected but their size can now be restricted to fulfill this requirement. In room zones, where furnitures are to be placed, conventional walls in combination with TIM can be used.

References

/1/ A. Goetzberger, J. Schmid, V. Wittwer
Transparent Insulation System for Passive Solar Energy Utilization in Buildings
Int. J. Solar Energy, 1984, Volz, p. 289-308

/2/ W.S. Wilke, J. Schmid
Modelling and Simulation of Elements for Solar Heating and Daylighting
to be published

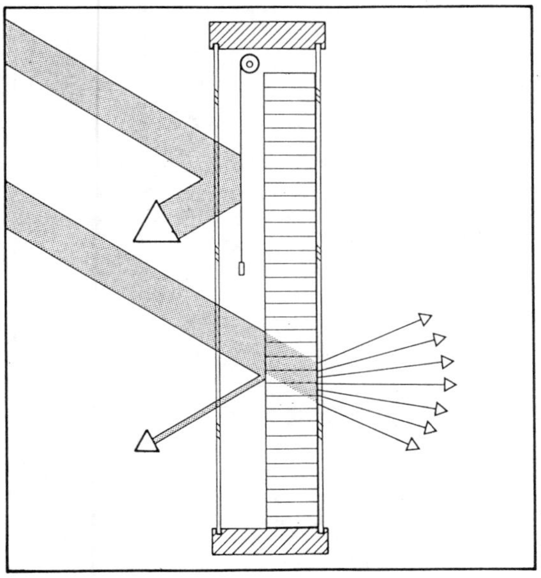

Fig. 1a (case a):
Daylight element with good insulation properties leading to high thermal comfort. Case a shows a translucent TIM element with integrated roller blind for the control of solar energy influx.

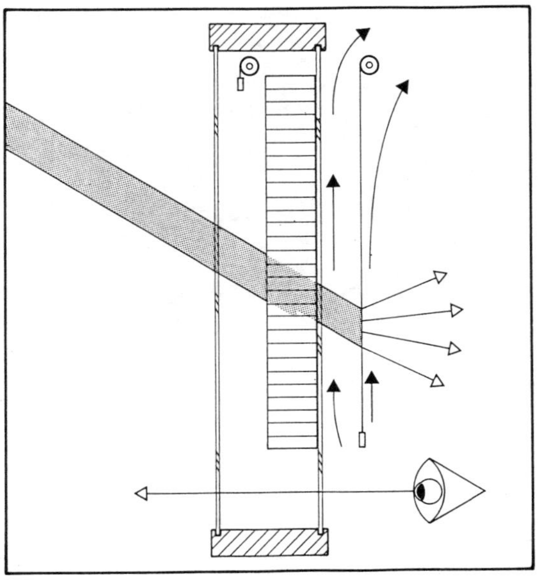

Fig. 1b (case b):
Case b shows an element in which a transparent area for visual contact is left. For the full use of solar energy during winter season <u>and</u> shading to prevent glare, a partially absorbing interior roller blind is used.

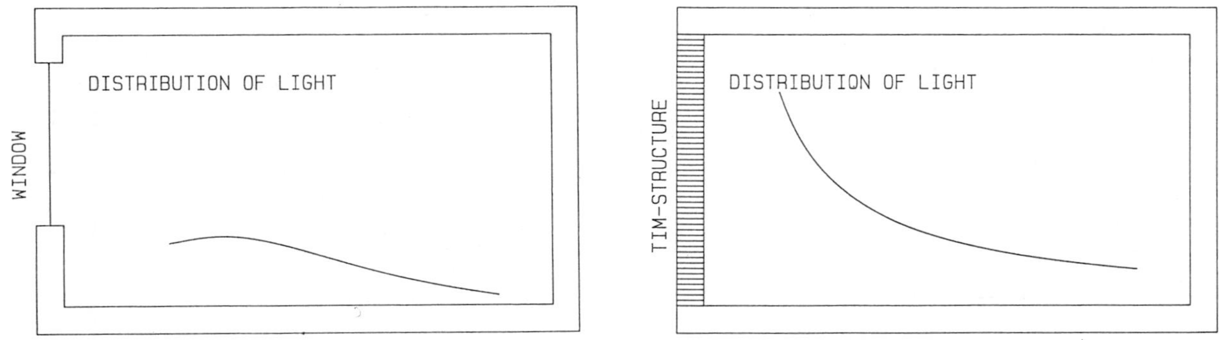

Fig. 2: Enlargement of the aperture allows for more uniform illumination of rooms

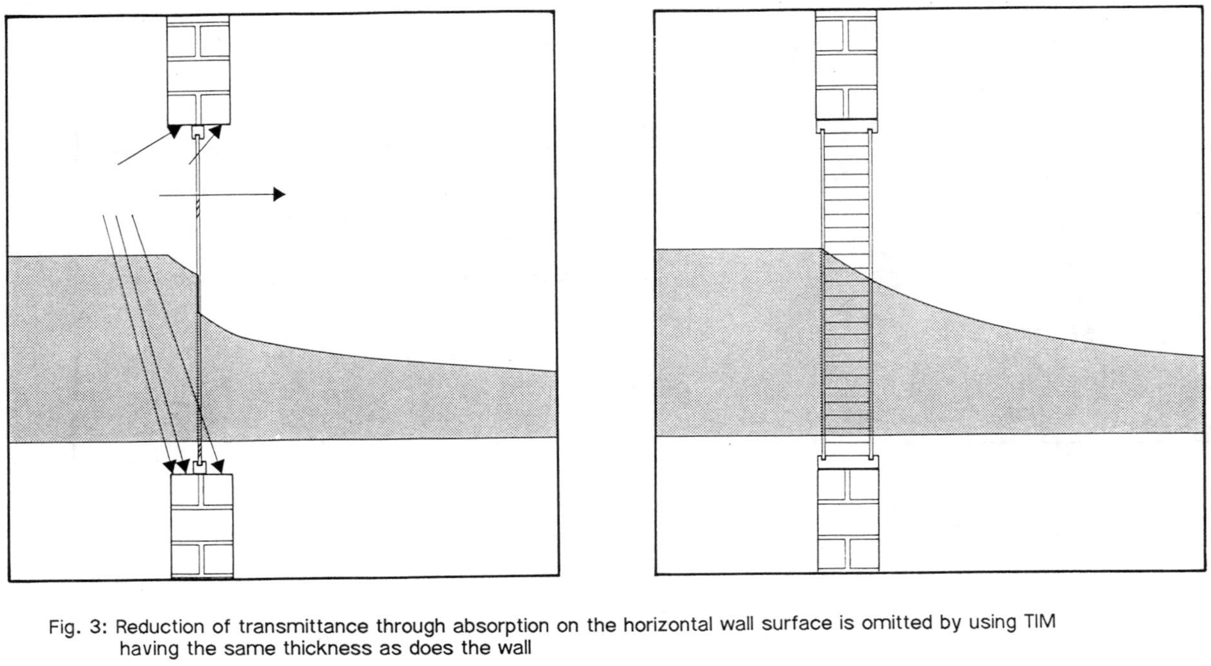

Fig. 3: Reduction of transmittance through absorption on the horizontal wall surface is omitted by using TIM having the same thickness as does the wall

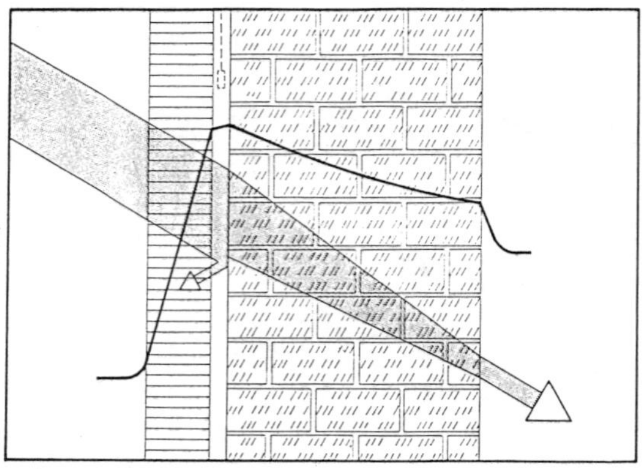

Fig. 4:
The combination of TIM with partially absorbing glass bricks leads to highly efficient heat gains <u>and</u> to daylighting capabilities

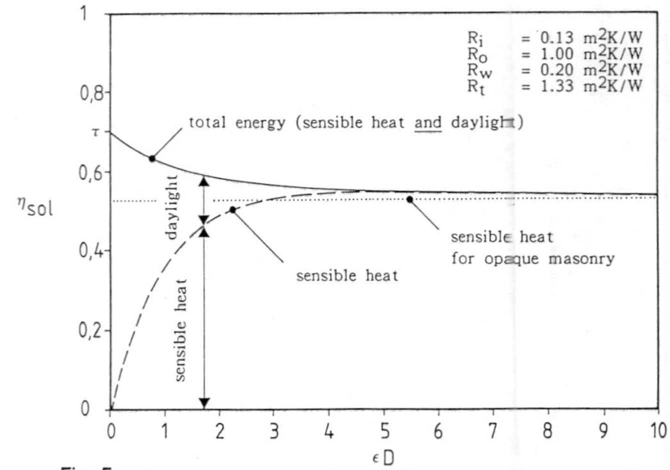

Fig. 5:
Comparison of energy conversion efficiencies. By applying TIM on opaque masonry, a constant efficiency is being obtained. For semi-transparent bricks, heat and daylight vary as a function of relative brick-absorption.

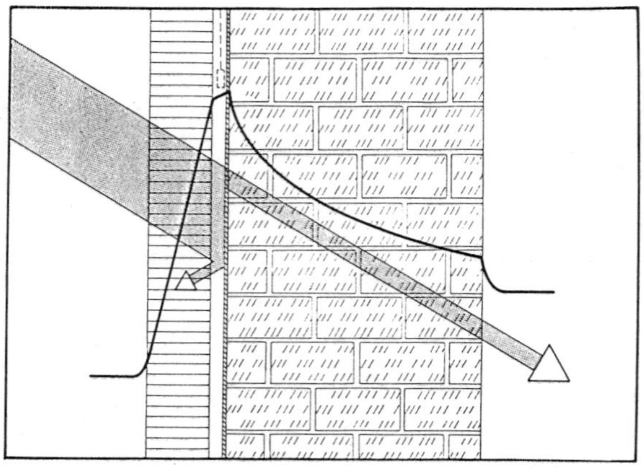

Fig. 6a

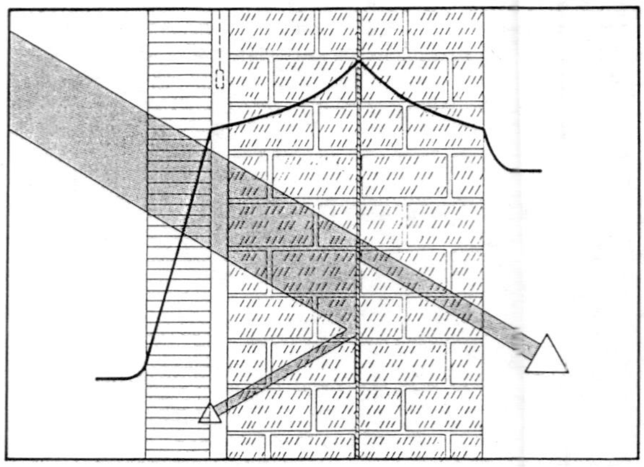

Fig. 6b

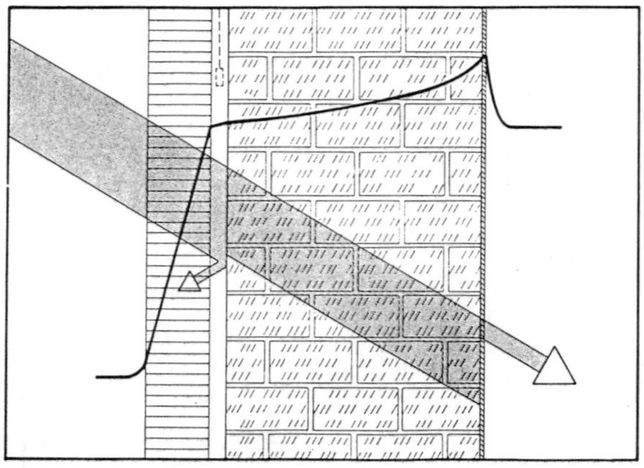

Fig. 6c

Fig. 6:
Control of temperature distribution inside the glass-wall by using discrete absorption layers. Three cases are shown: in Fig. a the absorption layer is located at the outer wall surface producing similar temperature profiles as an opaque wall system.
In Fig. b the layer is in the wall centre whereas in Fig. c the layer at the inner wall surface (seen from inside) leads to high surface temperatures.

WALL HEATING WITH TRANSPARENT INSULATION - RESULTS FROM REALIZED DEMONSTRATION PROJECTS

W. Stahl, E. Bollin, J. Schmid, J. Vahldiek, K. Voss, A. Wagner

Fraunhofer-Institut für Solare Energiesysteme
Oltmannsstraße 22
D-7800 Freiburg
Federal Republic of Germany

ABSTRACT

The development of Transparent Insulation Materials has led to a significant number of new solar thermal systems. An application with a high energy saving potential in the european climate is the Wall Heating with Transparent Insulation. Together with companies a number of projects with these TI Space Heating Systems are now executed.
Besides important practical experiences the measurements are leading to informations about physical parameters of the TI Elements and the contributins to the heating demand. Further aspects of these projects are the new architectural possibilities.

Introduction

After years of intensive research TI Technology is coming closer to the target of market penetration. A large variety of applications of Transparent Insulation Materials are possible in solar thermal energy systems. Space heating with TIM is an outstanding new approach due to the huge potential for saving conventional energy in the low temperature range.
The main problem of solar space heating is the mismatch between solar supply and demand. Heat losses and low energy density are disadvantages of solutions with seasonal storage. On the other hand the mean solar radiation during heating season in our climates should be sufficient for direct solar heating. With horizontal global radiation in the range of 250 kWh/m2 the solar input is comparable to the average heating demand of conventional houses in the FRG.
The increased efficiency of TI Systems is the key to utilize even low radiation levels for direct solar space heating.
The passive design of transparently insulated building walls is the decisive part of the new solar houses described in this paper.
Including the first experiments in 1982 the institute is involved in 13 projects with TI Space Heating applications. Besides constructions with more scientific approach these are:

- single storey bungalow, 40 m^2 TI, retrofit, Munich
- semi-detached houses, 70 m^2 TI, Ardon
- semi-detached houses, 61 m^2 TI each, Freiburg Tiengen
- 8-family terrace housing, 150 m^2 TI, retrofit, Sonnenäckerweg Freiburg
- 9 terrace houses, 22 m^2 TI each, Düsseldorf Hellerhof
- student residential college, 1200 m^2 TI, Glasgow.

The following projects are in a planning stage:

- single family house, totally heated by a TI System, Freiburg
- apartment with only TI Heating, Shanghai
- two training halls, app. 1000 m^2 TI, Köln.

Of the large variety of TIMs the PC honeycomb and the PC or PMMA capillary structure are used most frequently. Aerogel materials are still subject to research. Its superior thermal properties do not yet outweigh its low solar transmission. Being an anorganic material is a significant advantage of aerogel.
To prevent overheating in summer, the roller blind technology has been developed to operate reliably. Nevertheless there are still some problems occurring with new constructions. Other shading devices are under developement.

Measurements

Basic parameters for evaluation of TI Space Heating Systems are:

- room air temperature
- indoor wall surface temperature
- heat flux through the wall
- outdoor wall surface temperature (=absorber temperature)
- ambient air temperature
- global solar radiation on the TI Element area.

The data are the basis for determining

- U-values of the elements
- U-values of the TI Walls
- g-values
- efficiencies
- U_{eff}-values of the TI Space Heating System

and for several other evaluations. With measurements of all other heat inputs, the heating balance of the considered building can be determined. The solar fraction of the TI System for the heating demand is important for economic considerations.

Simulation

Parallel to the measurements, simulation programs have been developed and measured data are used for the validation of these programs. The multi-zone building programs fulfill the following requirements:

- TIM modelling with temperature-dependent U-values
- switching parameters corresponding to the TI Space Heating Control Unit
- consideration of the comfort level.

TI Walls

The realized TI Space Heating Systems are totally passive. The TI Elements are mounted onto the wall of the building. The thermal conductivity of the wall ensures that solar heat is conducted into the buildings.

A basic design for TI Elements includes:

- glass cover (low iron content, impact resistant)
- element control and shading with roller blinds
- TIM (honeycomb, capillary structure or aerogel)
- transparent film or glass
- absorber (non-selective, black, water soluble dispersion paint)

U-values of such elements should be between 0.7 and 0.8 W/m^2K with the roller blind open, 0.4 - 0.6 W/m^2K with the roller blind closed.

Wooden frames are best with respect to heat insulation. The frame area should be as small as possible.

In mounting the TI Elements, it is important to achieve an airtight connection between the frame and the wall. This may be difficult especially in retrofit applications. An uncontrolled air flow can reduce system efficiency significantly.

Walls for TI Systems must have a high thermal mass. Thus energy collected during the day is available indoors with the appropriate time delay. System efficiency increases with increasing thermal mass. Good results are achieved with brick or concrete. Also clay may be a good wall material, but to our knowledge, no experiments have been reported up to now.

Control and shading of the elements is done with roller blinds in the projects discussed here. The roller blinds are driven with 12 V DC or 220 V AC current. For maintenance the drive should be easily accessible. A specially developed textile with a low emissivity IR-coating on both sides is used. A special frame edge construction provides an airtight seal of the drawn roller blind. Such measures are leading to the decrease of element U-value for a drawn roller blind which has already been mentioned.

For high efficiency TI Elements, complete light blocking or another form of effective heat control is indispensable to prevent overheating. Among various new developments, "Switchable Windows" for example are very interesting.

Control of the roller blinds is done with a new commercially available TI Space Heating Control Unit. Up to three different TI Walls can be controlled in response to

- continuous 24 h-average of ambient temperature
- minimum irradiation
- maximum absorber temperature

A second generation control unit is under development. The main improvement is the indoor comfort control and the interaction of the TI Wall and a conventional heating system.

One energy saving effect had not been adequately considered up to now. Due to the increased inside wall surface temperature, excellent room comfort temperatures are achieved. Therefore the indoor air temperatures can be decreased. Typical savings in heating energy of 10 % result from this effect.

Energy Savings

An important aspect of TI Space Heating Systems is the reduction of the heating demand. In our first experiments, we could demonstrate that the TI Walls have no heat losses. Of course heat losses are also due to ventilation, windows, ceilings and floors. While an opaque insulation cannot compensate for these losses at all, the question of the extend to which TI Walls can decrease the absolute heating demand has been discussed at length.

The considerations resulted in the effective U-value, U_{eff}, as a new parameter for TI Systems. While for opaque materials the U-value is always a positive number, i.e. describing heat losses, TI Systems should have negative U_{eff}-values due to the solar energy utilization. The smaller the U_{eff} value is, the better the TI System performs. The dependence of U_{eff} on the ratio of monthly radiation and monthly mean air temperature difference between indoors and outdoors is shown in Fig. 1 for an opaque wall and a TI Wall with and without a roller blind.

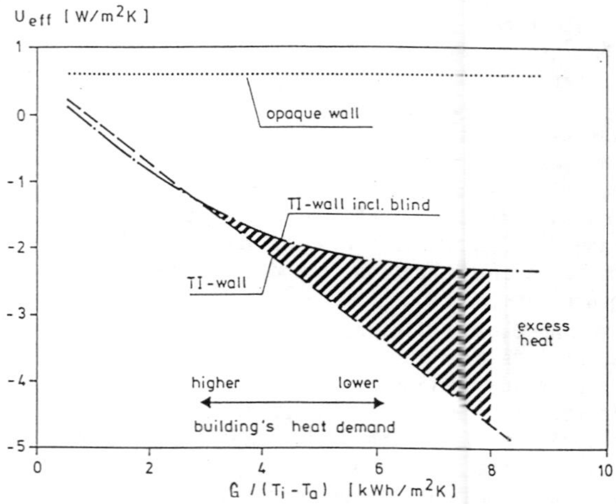

Fig. 1: Measured U_{eff} as a function of the ratio of monthly radiation and mean air temperature difference between indoor and outdoor for a test building: Conventional opaque insulation, TI System with a roller blind, TI System without a roller blind.

Example:

G = 60 kWh/m^2 monthly mean solar radiation on the TI Wall

ΔT = 10 K mean difference between outdoor and indoor air temperatures

opaque wall U_{eff} = 0.6 W/m^2K

regulated TI Wall U_{eff} = -2.2 W/m^2K

unregulated TI Wall U_{eff} = -3.3 W/m^2K

The energy Q is calculated by multiplying U_{eff} with the corresponding number of degree-days. In this example with 31 days per month:

opaque wall $Q = 4.5$ kWh/m^2 month
regulated TI Wall $Q = -16.4$ kWh/m^2 month
unregulated TI Wall $Q = -24.6$ kWh/m^2 month

The unregulated TI Wall has the lowest U_{eff}-values. This corresponds to the maximum amount of collectable energy. Depending on the specific parameters of the application the amount of usable energy is smaller. This is expressed in the larger U_{eff}-value for the regulated TI Wall. Parameters such as thermal mass, orientation of TI Wall, area of TI Wall and window area influence the amount of usable energy and therefore general statements are difficult.
Summarizing our experience from measurements and simulation calculations, we expect a contribution to the heating demand of 100 to 200 kWh per m^2 TI Wall of usable energy per heating season. In calculating the total energy savings, these contributions have to be divided by the efficiency of the conventional heating system.

Examples
The above mentioned list of TI applications expands from first pure research projects to almost commercially realized buildings. After the foundations for TIM and TI Space Heating /1,2,3/ have been elaborated by the scientists it is gratifying that companies take over these new developements. In commercializing the systems the future role of the Institute has to shift from main contractor to subcontractor for scientific advice, simulation calculation and monitoring. Semi-detached houses in Freiburg-Tiengen were built on a private basis by E. Bollin, a colleague in the department for System-Technology, H. Gessler and D. Novakov. Architecture and management was done by CEREC, Ardon CH, Planerwerkstatt Vörstetten and E. Bollin. The role of the Institue is to compare TI Walls with honeycomb and aerogel material. TI Elements habe been made by Glaserei und Fensterbau SCHMIDT, Freiburg.

Fig.2: Semi-detached house in Freiburg-Tiengen.

In cooperation with the Siedlungsgesellschaft Freiburg and the Freiburger Energie- und Wasserversorgungs AG a 8-family house in Freiburg was retrofitted with TI. A similar house at the same location has been renovated with conventional opaque insulation according to the swedish standard for low energy housing. Simulation calculations predict a further decrease in heating demand of 50% for the TI house compared to the swedish standard house. Measurements will be done by the Institute over a period of 2 years to validate the predictions. The retrofit application of TI is the special importance of this project. The TI Elements have been built by Fensterbau HELD KG, Freiburg. The architect is R. Disch, Freiburg.

Fig.3: Sonnenäckerweg Freiburg, retrofit of a 8-family terrace housing.

KAISER Company has managed a 9 single family project in Düsseldorf-Hellerhof and a large student residential house in Glasgow. The architects were Kaiser Company for the houses in Düsseldorf-Hellerhof, in Glasgow KENNEDY&PARTNERS.

Fig.4: Terrace houses in Düsseldorf-Hellerhof.

Fig.5: Student residential college in Glasgow.

Conclusion

With the basic scientific work in the laboratories and the now running demonstration projects TI Space Heating Systems are going to be well understood. A comprehensiv report was recently published by the Institute /4/. Several companies have shown that practical solutions are possible. The energy saving potential is verified. Of course some improvements still have to be made. The new technology is moving from the institute into a small market. This is also the merit of architects, manufacturers of TI Elements and building companies. They have drawn attention to aspects such as

- new architectural possibilities
- change of building appearance
- adaptation of the building to changing climatic conditions
- increase of thermal comfort
- new daylighting possibilities
- environmental considerations.

The ongoing projects show that these aspects are not less important than the economic point of view.

This has to be taken into account if one judges the system only by the still relatively high cost of approximately 500 - 1000 DM/m^2. The recently published TI Study /5/ is an outstanding example of how the scientific work spreads to other disciplines. Although all these successful approaches have been made, the number of applications is still small. Further projects are necessary. The institute's knowledge can increase the success of these new TI undertakings.

Fig.6.: Sonnenäckerweg Freiburg, south-west fassade

Literature

/1/ Goetzberger, A., Schmid, J., Wittwer, V., Transparent Insulation Systems for Passive Solar Energy Utilization in Buildings, Int. J. Solar Energy, 1984, Vol. 2, 289-308

/2/ Goetzberger, A., Schmid, J., Wittwer, V., Transparente Wärmedämmung zur passiven Solarenergiegewinnung an Gebäudefassaden, ARCUS, 1, 32, 1984

/3/ Platzer, W., Pflüger, A., Stahl, W., Wittwer, V., Charakterisierung lichtdurchlässiger Isolationsmaterialien, Tagungsbericht 5. Int. Sonnenforum Berlin, 123, 1984, DGS Sonnenenergie-Verlags GmbH, München 1984

/4/ Transparent Insulation Technology, Report published by the Fraunhofer-Insitut für Solare Energiesysteme, Freiburg, 1989.

/5/ System Study "Transparent Insulation", Weidlich Ingenieurgesellschaft, Berlin, Büro für energiegerechtes Bauen, Köln 1989.

Fig.7.: Honeycomb Transparent Insulation Material

Acknowledgement

The presented results have been mainly obtained in the System Technology Department. It is a pleasure to express gratitude and appreciation to all colleagues for their good work.

Furthermore the continuous support of the Ministry of Research and Technology in these new TI Technologies has made significant contribution to the described state of development.

Of great importance are the companies and public organizations which are involved in the mentioned projects. Special thanks are due to their efforts.

TRANSPARENT INSULATION: DESIGN RECOMMENDATIONS, SENSITIVITY STUDIES, SHADING AND MATERIALS AGEING

K. Wellinger, B. Schläpfer, P. Schneiter
E. Schweizer AG Metallbau
CH 8908 Hedingen (Switzerland)

ABSTRACT: This comprehensive Swiss research project on Transparent insulation lasts from march 1987 to march 1990. It involves a number of Swiss and one German organization and is managed by E. Schweizer AG, a major façade and window building company, with financial support from the Swiss Federal Office of Energy. The main objective is to give the practitioner a general overview on this new technology, and design guidelines. The activities include * laboratory measurements of optical and thermal properties of TI materials by Fraunhofer Institute for Solar Energy Systems, Freiburg FRG (ISE), * an outdoor test stand for thermal and optical monitoring of systems (wall + TI + shading) with Oekozentrum Langenbruck (OZL), * a torturing stand for forced ageing of TI materials with Rapperswil School of Engineering (ITR), * developing a particular line of TI materials with Dr. R.W.Peter, Zürich (RWP) , * developing a particular shading system with fixed louvres with RWP, * writing a TI subroutine for a building simulation model by ISE, * validating the model with the measurements of the open air test stand by Swiss Federal Laboratory for Materials Testing and Research (EMPA), and * conducting parameter studies to gain design recommendations for Swiss climates and building types by Schweizer. To avoid overlapping with other papers of this section this paper concentrates on ageing, shading and a new TI material.

1. AGEING TESTS

The Rapperswil School of Engineering (ITR) having several years of experience in the testing of Solar materials, a collaboration was natural. Their method of testing plastic collector covers consists of stagnating metal sheet boxes with selective black inner coating and different materials as covers in open-air sunlight, and periodically measuring the spectral transmission of the covers.
Vapour diffusion estimations for the system wall + TI soon made clear that humidity as an important ageing factor had to be taken into consideration. Therefore a special test stand for forced ageing was built with the following features (See fig. 1):

- An Aluminum sheet behind the TI material reflects the solar radiation, forcing it to pass through the material a second time
- A radiator replaces the absorbing wall and follows a sine wave temperature curve between the ambient and 100°C even on cloudy days
- A humidifier yealds a constant vapour flow into the air gap between the radiator and the TI. Its level is adjusted to produce condensation during the night, which dissolves during the day.

Four elements have been built, which allow to test two materials with and without humidifier. The cover being one part glass, one part acrylic (UV transparent) and one part opaque, we will be able to separate the influences of radiation, humidity and temperature on the ageing process.
So far no ageing has been observed, therefore the test will be carried on for one more year.

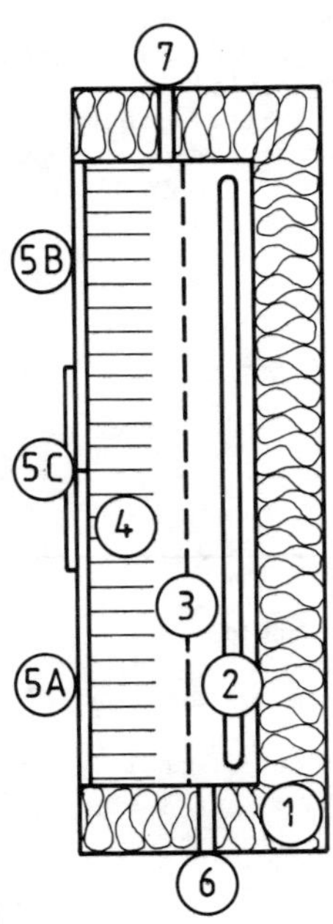

1 Back and frame insulation

2 Radiator (sine wave temperature curve between ambient and 100°C every day)

3 Aluminum sheet (perforated to let humidity pass) reflects the solar radiation so that it passes a second time through the TI

4 TI material

5 Covers
A Glass (UV-opaque)
B PMMA (UV-transparent)
C Al Sheet (opaque)

6 Vapour injector(in 2 of 4 elements)

7 Ventilation holes

Figure 1

Test stand for forced ageing of plastic TI materials, consisting of 4 elements (for 2 materials with and without humidifier)

2. "GEOMETRIC" SHADING

Most of the German projects use reflecting roller blinds with an electronic control to prevent overheating during the summer season. They have the additional advantage to enhance the insulation properties of a TI if lowered during nighttime, but as a mechanical device they somehow do not fit into the philosophy of "passive" solar energy.
Omitting the shading completely is only feasible if the TI has very poor thermal properties, which would mean that such a TI is not a heat gain system any more, and that as a heat loss reduction system it may be worse than an opaque insulation.
So we decided to reconsider the classical overhang roof which keeps the high standing summer sunrays off the house, while it lets the low angle winter sun enter. The idealized form is a set of fixed louvres. Fig. 2 gives the geometric description of such a system. Practical realizations include Venetian blinds, stretched or stampeed metal sheet, as well as sophisticated reflecting bars like the ones shown in fig. 3.
The main disadvantage of this sort of shading is its poor discrimination: if the summer sun has to be disclosed completely, the shading still lets pass part of the diffuse light, while on the other hand it already reduces the spring sun. For winter conditions the complementary effect happens: While the direct radiation enters, much of the diffuse light, which is the most important part of the winter energy balance, is kept away. Moreover, the structure thickness of the shading element may absorb as much as 40% of the direct radiation.
A first attempt was made with structures that emphasize the overheating prevention, and the poor transmittance is shown in fig. 4. The thermal measurements indicate that overheating is not such a severe problem, so that further tests and calculations will investigate more "open" structures that emphasize on winter energy gain.

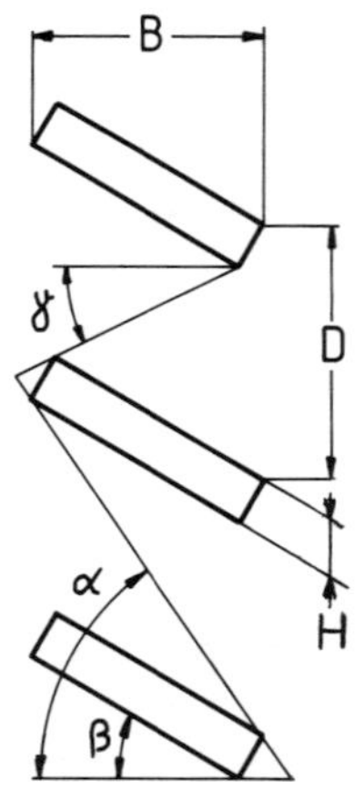

α Cutoff angle

β Max transmittance angle

γ Albedo cutoff angle

B Width of louvres

D Distance of louvres

H Thickness of louvres

(1 - H/D) = max. transmittance. If the louvres are held by vertical bars, these bars further reduce the transmittance!

Figure 2

General description of a geometric shading device

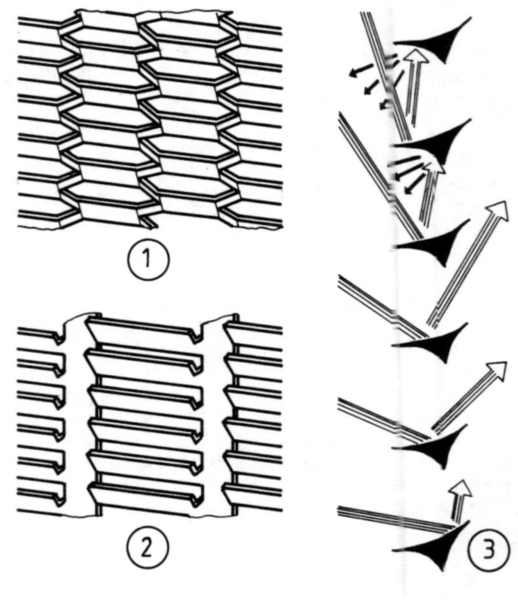

Figure 3

Practical realisations of geometric shading

1 Stretched metal sheet

2 Stamped metal sheet

3 Reflecting bars (OKALUX)

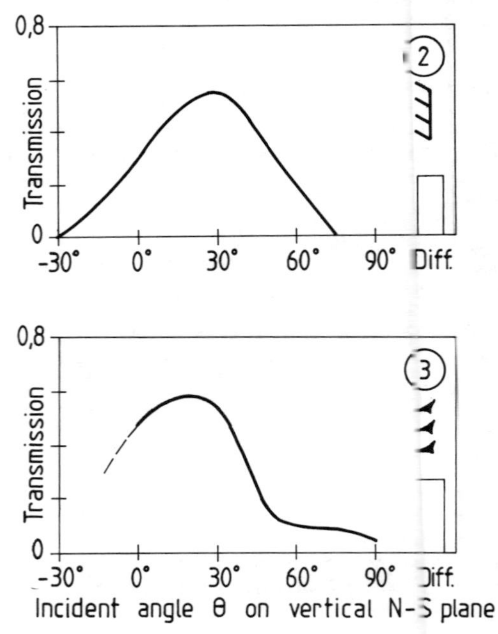

Figure 4

Transmittance of shading devices No.2 and 3 from fig. 3

"Diff": the diffuse transmittance is an integration over all angles

3. A NEW TI MATERIAL

Starting from the facts that stuctures parallel to the absorber have too much reflectance, that honeycombs have too much infrared transmission, and that V-corrugated structures are too floppy, Dr. R.W. Peter invented the "P-layer insulation" (P for pyramid), which is a sort of 3-dimensional V-corrugation: a surface of closely adjoining pyramids or cones. The structured films can be stacked, and the stack is self-supporting (fig. 5).
Measurements with 55° cone angle and 7mm film distance revealed good thermal properties thanks to good convection suppression, but the light was rather reflected outwards than guided towards the absorber, resulting in poor transmittance values. Larger cones (12 instead of 7 mm film distance) reduced the number of films for a given element thickness, and steeper cone angles (38°) enhanced the light guiding effect, resulting in better, but still not satisfying transmittance. The convection suppression stayed at an equally high level.
At this point, further improvement of the geometry stagnated because of unexpected difficulties in the technology of plastic forming. Polystyrene, a well mouldable plastic, which is often used in food packing, is too low in its melting point. Polycarbonate has a melting point of 120°C, a good UV stability and a reasonable price, but even heated positive and negative moulds failed to give steeper cones. So this problem has to be postponed to a subsequent project.

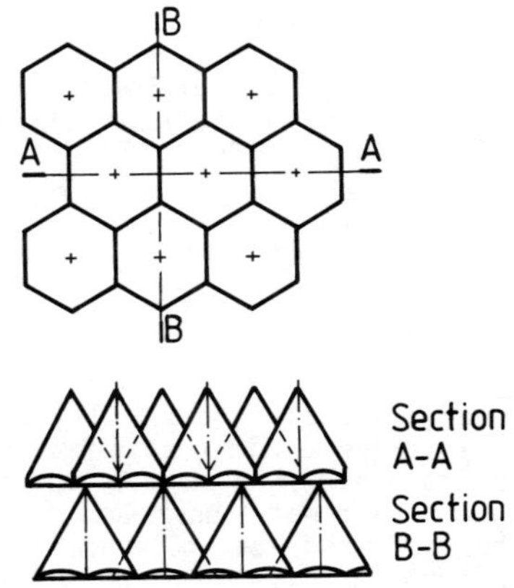

Figure 5

"P-Layer Insulation", stuctured transparent films that give a 3-dimensional convection suppressing structure when stacked and alternatingly turned by 90°

4 REFERENCES

/1/ K. Wellinger, B. Schläpfer:
Transparente Wärmedämmung
in: 5. Schweiz. Statusseminar Energieforschung im Hochbau, Zürich 8.+9. Sept 1988
EMPA-KWH, CH-8600 Dübendorf

/2/ R. W. Peter:
Beschattungselemente für Transparente Isolation
in: 5. Schweiz. Statusseminar Energieforschung im Hochbau, Zürich 8.+9. Sept 1988
EMPA-KWH, CH-8600 Dübendorf

/3/ R. W. Peter:
P-Layer Insulation, A New Transparent Thermal Insulation Material
in: Proc. of 2nd International Workshop on TI
Freiburg (FRG), 22 - 25th March 1988
The Franklin Co. Consultants, 192 Franklin Rd.
Birmingham B30 2HE, UK

2ND EUROPEAN CONFERENCE ON ARCHITECTURE 4–8 December 1989 PARIS, FRANCE Sponsored by Commission of the European Communities

THE THERMAL ENVELOPE - AN INTEGRATED PART OF THE HEATING SYSTEM?

Bjarne Saxhof
Thermal Insulation Laboratory, Technical University of Denmark
Building 118, DK-2800 Lyngby - Telefax +45 45 93 17 55

ABSTRACT. As indicated in the title, this paper focuses on the possible integration of heating systems and the thermal envelope - with one or two digressions on systems integrated in internal building components. The paper discusses a number of ways to integrate solar collectors - active as well as passive - in the roof construction and in the wall constructions. In the first case, both integration in the roof itself and utilization of the roof space as collector in combination with storage in the building mass are treated. Danish experiments with "diffuse solar walls" and so-called "Multi-Function elements" are described. The paper also briefly discusses the topics of dynamic insulation, preheating of fresh air in special window units and finally the somewhat futuristic topic of "smart" windows/walls, components for which it is possible to change the transparency for light and heat according to the demand. The paper attemps to evaluate the different systems, to describe the advantages and possible drawbacks.

1. INTRODUCTION

The low heating demands in energy efficient houses achieved through superinsulation and/or increased utilization of passive solar energy introduce new possibilities for placing and designing heating systems. The systems may be much smaller than in traditional houses and/or run at lower operating temperatures, and it is possible to obtain excellent thermal comfort conditions, practically no matter where the heaters are located in the room, ie it is no longer necessary to have heaters placed directly under the windows.
Actually, it will soon be possible to build comfortable houses completely without traditional heating systems, even in fairly cold climates. An important step forward towards this goal is the application of systems integrated in the building constructions, eg one or more of the systems described in this paper.

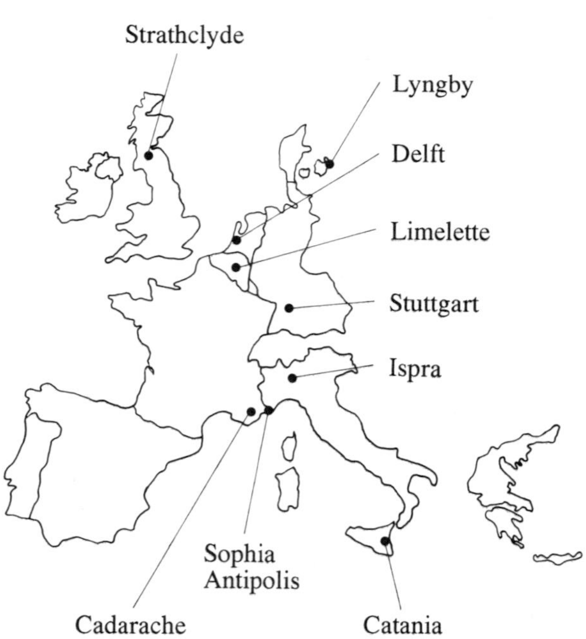

Figure 1: Map of EC member countries, the northernmost (climatically) being Denmark. The marked locations indicate the present PASSYS test cell sites.

In most of the selected designs, air is the heat transport fluid - very few hydronic systems have been included. There are two major reasons for this.
Firstly, preheating of fresh air for ventilation purposes is an efficient way of using the thermal envelope for heating. Secondly, it should be emphasized that the experience relayed here has chiefly been gained in the nordic countries, ie in cold climates (the Danish being the mildest) and freezing is thus a problem in connection with outdoor hydronic systems, eg solar collectors.
The Danish climate is the coldest in the EC (Figure 1), the average temperature in the heating season being 4 C, the design temperature -12 C and the number of heating degree days and sunshine hours respectively 2900 (17 C base) and 1500 (of which only 150 from November to February), /1/.
In connection with some of the systems described, annual energy gains are quoted - these figures relate to the Danish climate.

2. SYSTEMS INTEGRATED IN INTERIOR BUILDING COMPONENTS

To start off with the digressions from the main theme, a couple of integrated systems within the thermal envelope are briefly discussed.

2.1 Ventilation through hollow-core concrete slabs

Prefabricated concrete floor slabs are usually partly hollow to reduce the weight - in Denmark a 1.20 m wide element has typically 5 ducts (Figure 2).
These ducts may be used as ventilation ducts or air ducts in a heating system, utilizing the thermal mass of the concrete slabs. Systems of this type have been widely promoted in Sweden and have also been used in the other nordic countries, chiefly in large office buildings. In these cases, the main objective has been to remove and in some periods store excess heat (often from large internal gains) during office hours - in the summertime even precool the concrete slabs by night ventilation.
It must be said that although the systems can meet the objective, they are not very flexible, and it is extremely important at the design stage to know the internal loads in the different rooms or sectors of the building and then zone the system so it is possible to control zones with different loads individually (the typical example being the executive office next to the computer room).

In houses and appartments the systems are best suited for preheating the fresh air before letting it into the rooms, and for drawing exhaust air from the rooms. If alternating ducts are used for supply and exhaust air, some heat recovery can be obtained, however not very efficiently.

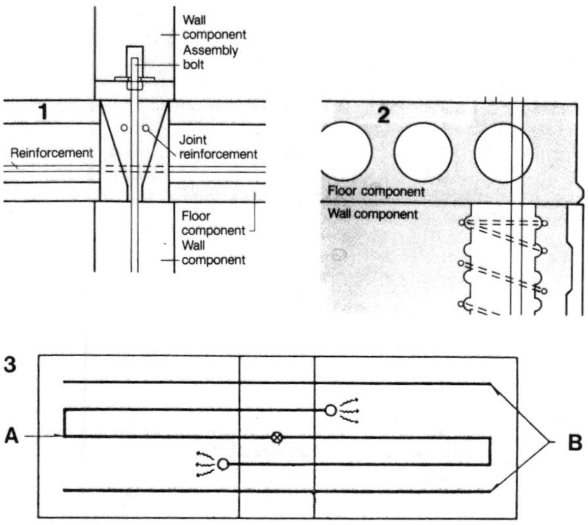

Figure 2: Danish prefab concrete floor slab. (1) and (2) vertical sections, longitudinal and cross respectively. (3) Element plan in smaller scale with indication of 5 ducts, (A) 3 ducts connected to ventilation system, and (B) 2 passive ducts.

2.2 Underfloor or Ceiling Heating (Radiant)

Flor heating is probably the earliest example of a heating system integrated in building structures, and it has had a revival as a low-temperature system for well insulated houses, often supplied from active solar systems or heat pumps, benefitting from the low operating temperatures, eg design surface temperatures in the Danish climate of 29-30 C.
However, there are some drawbacks to it.
The thermal response is very slow, which is unacceptable in most well insulated houses, but may be acceptable in houses with large thermal mass - the problem is of course temperature control, difficulties in changing the room temperature and, in thermally light houses, overheating problems caused by slow response to casual and solar gains.
Also, people tend to like carpets on floors - and floor heating systems do not.
This problem does not occur with ceiling heating, but as warm air rises, ceiling heating is generally less comfortable than floor heating - the "radiant shield" formed by eg a table literally gives you cold feet.
Technically, a good thermal effect could be obtained from heated walls, but the practical problems (obstructions by eg well insulating book cases, the risk of damage from all drilling and hammering activities in a normal home, etc) discourage the use.
Getting back to the thermal envelope, it should be added that in a slab-on-grade construction in a house it takes 200 mm of high quality mineral wool to counteract the extra heat loss through a heated floor, /2/.

3. ROOF INTEGRATED SYSTEMS

The roof of a building presents a large surface area well suited for some type of energy collector - especially in one- and two-storey buildings where the ratio between roof area and living area is very large.
As a result, roofs have had to host a large number of strange boxes, solar collectors "designed" and built by HVAC companies, attached to the buildings like alien substances, appendices. There is a need for "house-friendly" collector design, and one possible way forward is to integrate the collectors in the roof construction, either in the main roof or in the roof of an attached sunspace.

3.1 Active Solar Collectors

An example of an integrated solar collector built on site is shown in Figure 3. A corrugated metal sheet closes the "normal" insulated roof (left side of photo). A framing system is screwed onto the roof (seen between corrugated sheet and the finished collector), and between the profiles sheets of insulation, absorber and - in this case - a corrugated acrylic as transparent cover, are installed. A description of the system is given in /3/.
The systems built so far have been hydronic systems that have to be frost protected (application of anti-freeze additives in a closed loop or design of a drain-back or -down system), but an air collector could easily be built according to the same principles.

Figure 3: Photo of roof integrated solar collector built on site (under construction). Completed collector seen to the right, corrugated metal sheet topping the "normal" roof construction seen to the left.

Another example of an integrated active system is a complete solar thermosiphon water heater (Figure 4) with a flat 150 l water tank integrated in a sloping roof construction, /3/. In this case, the collector was part of the sloping roof/ceiling in a bathroom, but the system could also be used as roof for an attached sunspace when a sloping glass roof should be avoided to limit overheating.

3.2 Roof space collectors

Another way of integrating an air collector in the roof construction is to replace a normal opaque roofing material with a transparent material and then use the roof space or a sectioned part of it (painted black) as an air collector, /4/. An example is shown in Figure 5 - in this case the heat is stored in a rock bed, but the system is well suited for connection to hollow-core parts of the building, eg the concrete floor slabs described earlier. Full scale experiments in a house in combination with parametric studies, /4/, have indicated that the system can deliver approx 50% of the heat and hot water demand in a new house (built according to the rather strict Danish code), and approx 75% of the corresponding demand in a low-energy house.

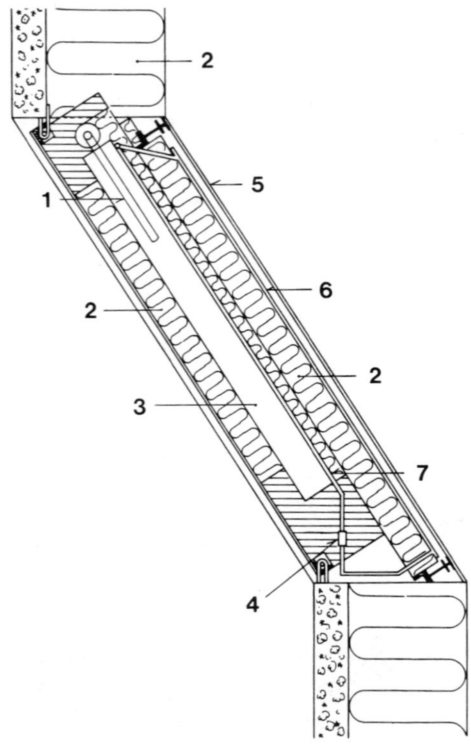

Figure 4: Solar water heater integrated in sloping roof. (1) Back-up heater, (2) Insulation, (3) 150 l water tank, (4) One-way valve, (5) Glass pane, (6) Absorber, (7) Heat exchanger.

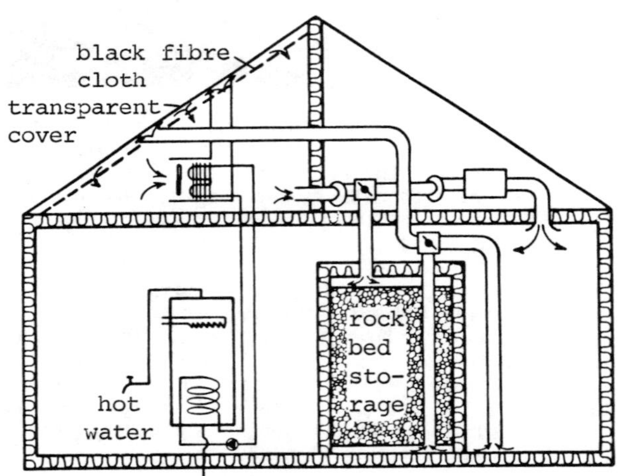

Figure 5: Section of experimental house with roof space collector connected to rock bed storage.

3.3 Photo-Voltaic Systems

The solar cells marketed so far have been fairly small panels often with 40 rather distinct "shooting target" cells. However, a company in USA is currently introducing a new type integrated in a roofing material that would be applicable in most buildings, /5/. The usefulness of photo-voltaic systems is discussed later in the paper (section 7).

3.4 Dynamic Insulation

The principle of dynamic insulation is to counteract the heat flow through the thermal envelope by establishing a uniform air flow through the insulation in the opposite direction of the heat flow.
The attempts to use dynamic insulation in the walls as well as in the ceiling/roof have not been very successful - it is very difficult to secure a wind-independent uniformly distributed air flow through the walls. However, for the roof alone it is possible, and recently, a Swedish company has marketed a system of this type, /6/, where the air supply for the house is preheated in the ceiling, the normal vapour barrier having been replaced by a penetrated plactic sheet or a textile sheet (Figure 6). The illustration also shows a heat pump for recovery on exhaust air.

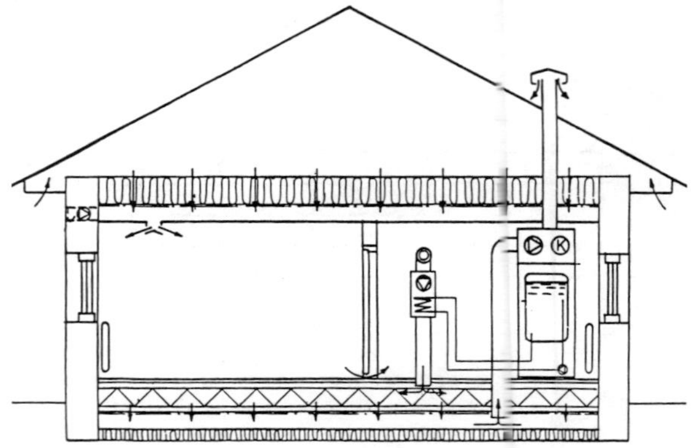

Figure 6: The Swedish Optima house, featuring dynamic ceiling insulation, /5/.

4. WALL INTEGRATED SYSTEMS

Most wall integrated systems can be termed "solar walls" and be described as "normal" walls (insulated or not) with an absorbing exterior surface (simply painted black or treated with a selective coating) with a transparent cover. The cover reduces the heat loss from the black surface. The absorbed heat may be transferred to the house by conduction through the wall or by ventilation of an air gap behind or in front of the absorber.
The Danish climate is not very favourable for solar walls because of the low solar irradiation during the four coldest months, but still simple south facing solar walls with a selective coating and a glass cover yield up to 125 kWh per sq m per year.
The use of new transparent insulation materials improves this performance considerably (eg Honeycombs and one layer of glass about 170 kWh, and two layers of glass with evacuated monolithic silica aerogel about 210 kWh, per sq m per year, /7/).
The main drawback of solar walls especially unvented walls, is the lack of temperature control possibilities. The walls transfer the heat to the rooms, time-delayed, whether there is an actual demand or not. Careful design studies are essential.

4.1 Unvented Solar Walls

The operating modes of a simple mass wall are shown in Figure 7. With one layer of glass, one layer of low emissivity glass and two layers of glass respectively one sq m of south facing wall gains 65, 85 and 100 kWh per year, /7/.

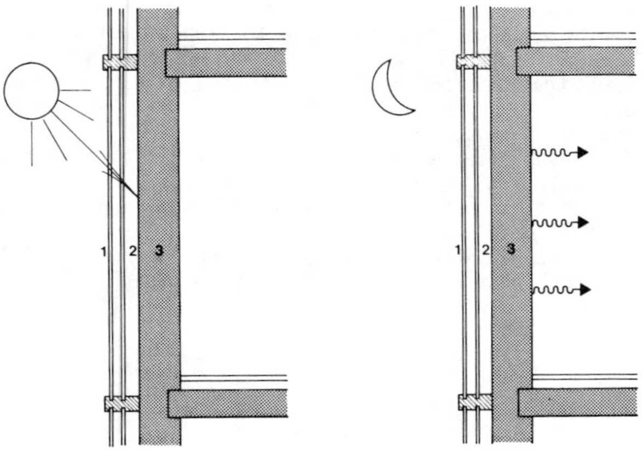

Figure 7: Operation of unvented solar wall during a solar day (left) and during the night (right). (1) Transparent cover, (2) Absorber, (3) Uninsulated thermally heavy wall (concrete or brick).

4.2 Vented Solar Walls

The operation modes of a Trombe wall with a simple flap type one-way damper are shown in Figure 8.

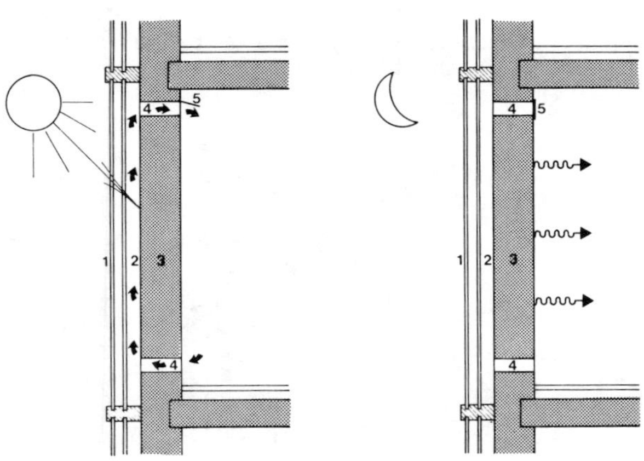

Figure 8: Operation of vented solar wall during a solar day (left) and during the night (right). (1) Transparent cover, (2) Absorber, (3) Uninsulated thermally heavy wall (concrete or brick), (4) Opening in wall, (5) Flap type damper.

As this is an open loop system, dirtying of the absorber and especially the transparent cover is a problem. The problem can be solved in two ways, either by closing the open loop with a sealed air gap at the interior wall surface (which is not very practical - as mentioned in section 2.2) or by using an absorber plate separated from the wall by an air gap (Figure 9).
Finally, an example of a solar wall designed mainly for preheating fresh supply air is given (Figure 10). This wall should be insulated (in cold climates), and there should be some thermal mass between the absorber and the insulation to stabilize the supply air temperature. The room/house is depressurized with an exhaust fan.
The two main problems are dirtying and cold draught on successive cloudy days.

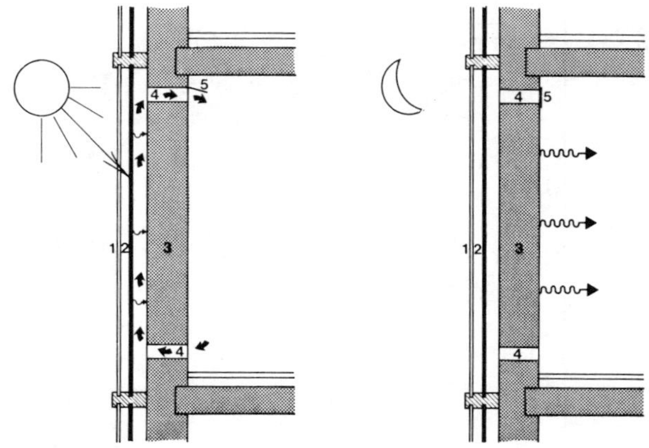

Figure 9: Operation of vented solar wall with separate absorber plate, during a solar day (left) and during the night (right). (1) Transparent cover, (2) Absorber, (3) Uninsulated thermally heavy wall (concrete or brick), (4) Opening in wall, (5) Flap type damper.

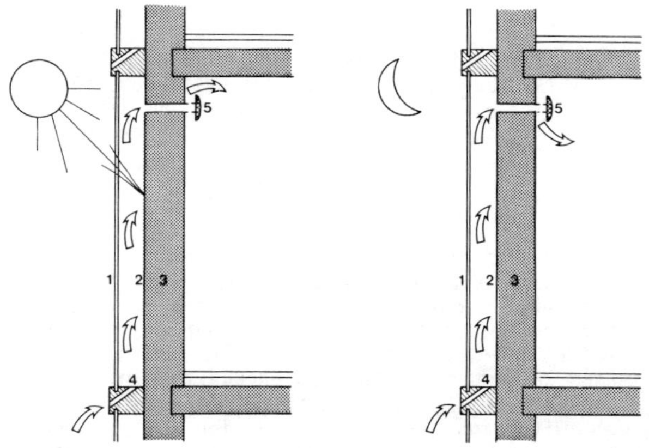

Figure 10: Operation of vented solar wall for preheating fresh air. (1) Transparent cover, (2) Absorber, (3) Insulated wall with some thermal mass at the absorber, (4) Opening, (5) Damper and diffusor.

4.3 "Multi-Function" Wall Element

Figure 11 shows a ventilation air heat recovery wall based on a concept by the Danish architect Arne Meldgaard. The idea is to have an insulated wall with a trapezoid corrugated metal sheet absorber and a transparent cover (eg a double ribbed plastic plate). The ducts created between the absorber and the wall are connected to an exhaust air ventilation system, and the gap between the transparent cover and the absorber is connected to a supply air duct. An alternative version of the wall has an opaque cover, a black metal sheet.
Laboratory experiments and parametric studies show that the heat exchanger efficiency is approx 30% excluding solar gains and the possible total gain 225-1350 kWh per sq m per year, depending on a number of conditions regarding orientation, connected systems etc, /8/. A full scale demonstration project is being co-funded by CEC.
A number of drawbacks are anticipated, one being dirtying (in the transparent cover version) another being short-circuiting of exhaust and supply air at the bottom of the element. An advantage is that the marginal cost is reasonable.

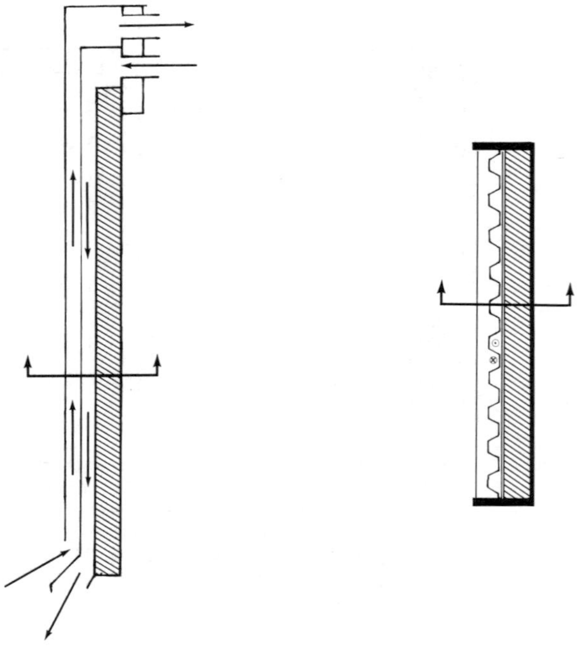

Figure 11: Vertical (left) and horizontal (right) schematic sections of Multi-Function wall element for heat recovery on exhaust air and solar preheating of supply air.

4.4 "Diffuse Solar" Walls

Some parametric studies and a few experiments with glass-covered walls with different orientations have been carried out at the Thermal Insulation Laboratory, eg one pane of glass covering a north facing black-painted insulated wall. The thermal performance was improved, but not enough to turn the heat loss into heat gain. Studies of a west facing mass wall with high efficiency (and high cost) transparent insulation - an evacuated silical aerogel window - indicate a heat gain of approx 30% of the value for the same wall facing south.

5. WINDOWS

Windows are included in this overview for two reasons. Firstly, direct gain is still the dominant passive solar factor, and the superinsulated highly transparent windows that are currently being developed promise to make windows energy gainers, regardless of the orientation. Secondly, also integrated window systems exist, in the form of ventilation windows.

5.1 Future Superinsulated Windows

At present, researchers aim at U-values of 0.4 W per sq m K and high transparency, pursuing two different principles. In USA (at the Solar Energy Research Institute in Colorado) prototypes of evacuated coated laser beam sealed Pyrex glass thermopanes have been produced. The space between the panes is only 0.3-0.5 mm, a network of almost invisible glass balls being used as spacers. The panes are laser welded in a vacuum chamber.
Similar work is being carried out in Australia, with less advanced, but also less costly assembly methods.
In Denmark, research and development work is done on the application of the transparent insulation material monolithic silica aerogel, /7/. The material is produced (in limited quantities, for research and development only) by the Swedish company Airglass. The advantages of aerogel are that only a slight depressurization is necessary to obtain the low U-value, that the material is almost 100% solar transparent and that it by itself provides the spacer for the thermopane.
The disadvantages are that it is extremely fragile, visually not perfectly transparent, at present extremely expensive, and that it disintegrates in contact with water.
However, thermally efficient translucent "windows" and skylights could be produced with today's technology.

5.2 Ventilation Windows

In Sweden, research and development of ventilation windows have been carried out for a number of years, and at least one product has been made commercially available.
The principle is shown in Figure 12 for a supply air window. The window is triple glazed and fresh air is drawn along the glass surfaces (by fan depressurization of the house) to collect (part of) the heat loss through the window and eventually bring in some of the solar gain as warm air that may be distributed to other rooms. An alternative version of the window has an electrically heated inner pane to ensure against cold draught from the supply air - not necessarily an energy efficient solution.
According to Canadian experiments, /9/, 50% of the energy required to heat ventilation air can be supplied this way. The corresponding overall reductions in purchased energy compared to application of normal double and triple glazed windows are 25 and 20% respectively.

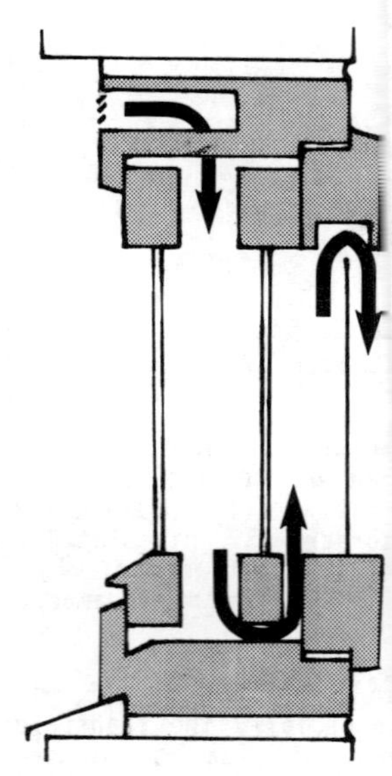

Figure 12: Supply air ventilation window.

Also extract air windows have been produced, the concept being that exhaust air was drawn up through the interior air gap in a triple glazed window and then through a heat exchanger. The aim is to raise the exhaust air temperature through solar gain, but the result may often be increased heat loss (when cloudy or dark) less solar gain for the room and poorer utilization of the gain through losses in the heat exchanger (might be utilized in a heat pump, however). Actually, the extract air window works best as a temperature control unit for extreme direct gain rooms.

It may be considered a problem that a ventilation window (supply or extract air) gets dirty sooner than a normal window - and it may not. In a way, it is better to be able to see the dirt in the ventilation system and thus be reminded to clean it - how many ventilation systems in private homes (or anywhere) will be cleaned regularly?

6. "THE SMART WALL" - WALL OR WINDOW, YOUR CHOICE

One of the problems of using large apertures with the supergaining superinsulating windows described in the previous section is overheating. Improved storage systems may decrease this problem, but still a need for outside shading of the apertures must be anticipated.

The switchable glazings that have been (and still are being) developed can provide this shading within the window unit. The three glass types, photochromic, thermochromic and electrochromic, can all be used to change a building component from wall to window and back, but only the electrochromic glass is user-controlled. The first two types change transparency depending on light and temperature level respectively, while the transparency of the electrochromic glass is changed by running an electric current through the pane. A future energy efficient house may be built mainly from such units, sectioned into small areas that can be individually controlled by the inhabitants, probably from one of the personal computers in the house. At least one prototype of a house with "smart walls" (although not with the extremely low U-values) has already been built, /5/.

A further possible integration into the same smart element is a photo-voltaic system - it is technically feasible to apply amorphous silicon junctions to glazing materials almost like selective coatings - and sell the product as the 3-in-1 (or 5-in-1 or even more, depending on how many functions you like to list).

7. HYBRID VS PASSIVE SYSTEMS - ELECTRICITY CONSUMPTION

The integrated systems described in this paper have mainly been passive systems, the main exception being fan-driven ventilation systems. One reason for this is of course that the paper should focus on the thermal envelope and not on the further distribution of heat within the house or the storage in the house. If these aspects had been considered, more hybrid systems would have appeared. This section has been included as a reminder that in connection with semi-passive (hybrid) measures a major problem is the risk of introducing a serious increase in electricity consumption. In one of the first Danish attempts at a rock bed storage the air flows were beautifully balanced by oversized fans, consuming about 2000 kWh in a year. Better fan sizing and control could reduce that figure by at least 50%, but it would still be a considerable consumption, and a typical cross-flow heat exchanger consumes 800-1000 kWh in a year.

In hybrid systems there is often a simultaneity between solar input and the power demand for fans or pumps that may make small photo-voltaic systems interesting, systems that are not connected to the utility grid and thus can do without expensive frequency transformers etc. Even the need for batteries may be limited, depending on the hybrid system.

8. CONCLUSIONS

As the main purpose of this paper has been to give an overview of possible energy saving systems integrated in the building envelope, only a few conclusions should be drawn.

A large number of solutions has been presented and evaluated - it must be emphasized that the evaluations are subjective, but based on research results, generally from more than one source.

Quite a few of the concepts have performed well, and promising developments are expected in the near future. From the single components and systems presented here, numerous combinations can be made.

One of the major challenges for the architects is to use some of these new energy saving possibilities and create the architecture for them, eg by cooperating with conservation engineers already in the early phases of the design and development of "technical" building components for the envelope.

9. REFERENCES

/1/ CEC Project Monitor, issues 4 & 41, June 1987 & Feb 1989.

/2/ Bjarne Saxhof, Kirsten Engelund Poulsen. Foundations for Energy Conservation Houses. Thermal Insulation Laboratory, Report 130, Nov 1982.

/3/ Peder Vejsig Pedersen. Innovative Solar Heating Systems for New Building Projects in Denmark. ISES Solar World Congress, Hamburg 1987, Proceedings Vol 1, pp 604-09.

/4/ Søren Østergaard Jensen, Vagn Korsgaard. Roof Space Collector. ISES Solar World Congress, Hamburg 1987, Proceedings Vol 2, pp 1135-39.

/5/ Buildings of the 21st Century - Developing Innovative Research Agendas. Workshop Proceedings, IEA, May 1989.

/6/ K. Allan Andersson, Tor Wadmark. Use of Ventilation and Insulation in Single Family Houses...The Optima Concept. Proceedings from Healthy Buildings ´88, Stockholm 1988, Vol 3, pp 49-54.

/7/ Svend Svendsen. Solar Walls and Transparent Insulation - The Use of Airglass in Passive Solar Heating. Invited paper for the 2nd Feasibility Study Workshop of IEA SHCP Task XIII, Advanced Solar Low Energy Buildings. Copenhagen, May 1989.

/8/ Søren Østergaard Jensen. Results from Tests on a Multi-Function Solar Energy Panel. Thermal Insulation Laboratory, Sept 1989.

/9/ Sharif A. Barrakat. Thermal Performance of a Supply-Air Window. Proceedings of the 12th Passive Solar Conference, Portland, Oregon, July 1987, pp 152-58.

INNOVATIVE SOLAR COMPONENTS FOR OFFICE BUILDINGS

G. Scudo
DPPPE, Politecnico di Milano
Via Bonardi, 3, 20133 Milano, Italy

S. Fumagalli, G. Rizzi
ENEA - FARE
ISPRA (Va), Italy

ABSTRACT. A new office building incorporating some innovative solar active and passive technologies built for ENEA at Ispra, Italy is presented. The most interesting feature of the building is the design and production of an active solar panel and a passive solar chimney integrated into a curtain wall. Solar chimneys have different storge-distribution systems. ENEA is carrying out the monitoring campaign.

1. INTRODUCTION

Innovative processes in the Italian building industry have generally a very low diffusion because of many economical, cultural and technical reasons. Low-cost components and systems with integrated building and thermical functions, appropriate to the variety of the local contexts are rare. A number of innovative climate responsive cladding elements have been recently produced in Italy; few of them, including the solar chimney light wall applied to the ENEA office building at Ispra, are described in ref. /1/.

2. THE ENEA OFFICE BUILDING

The building is located in Ispra, which has a continental-temperated climate (DD 2400, design temp. - 5°C, mean temp. in January and July: 13 and 20°C, total average solar radiation on horizontal surface 8400 kJ/sq.m day in the heating season). It incorporates a mix of innovative active and solar technologies in order to achieve a low energy use and to test the efficiency and reliability of different solutions.
The building is East-West (1) oriented and opens the extensive south façade to solar radiation. It

Fig. 1: External views of the ENEA building at Ispra: (a) South façade (b) solar chimney curtain wall of the block B; (c) solar active air panels integrated to the south façade; (d) East front of the block A with the shed on the roof supporting the active water panels.

has a total heated volume of 9.300 sq.m and it is articulated into four building blocks: administrative office and direction (A), research office and laboratories (B), atrium (C) and technological hall (D) (Figs. 1,2). The south façade of the block A (2500 c.m) is designed as a huge solar air panel integrated into the building structure. The active system is composed by air collector, a rockbed storage and a heat distribution network. The system behaves as a free heat source integrated with the HVAC plant.
The south façade of the block B (4700 c.m) is designed as an integrated solar chimney-window enclosure. The solar chimney thermical behaviour is well known /2/. The innovative features are:
- differentiated air distribution and storage systems: metal false ceiling at ground floor, ducts in the reinforced concrete bearing structure at the 1st floor, and metal pipes in the ceiling of the 2nd floor (Fig. 3);
- air change system designed to fit the passive system with its air loops.

3. INNOVATIVE SOLAR COMPONENTS

The main design objective was to produce reliable, low cost, high efficient industrialised solar components which can be integrated in a wide range of office typologies.
The design prototypes and the experimental production of the two solar components (solar chimney and air panel integrated into pre-fabricated building components) were specifically done for the ENEA office at Ispra. Both components are built in GRC (Fibreglass Reinforced Concrete).
The innovative feature of the GRC solar components are linked to GRC characteristics:
- very low weight: 60 kg/sq.m (20% compared with a traditional reinforced concrete panel);
- a soft productive process which allows to produce panels with complex form at a cost competitive with traditional reinforced concrete prefabricated elements.
Solar chimney curtain-wall can be used with many configurations in passive and hybrid solar technologies. Solar panel curtain-wall can be used as component in traditional active system.

4. ENERGY PERFORMANCE

Metereological data, monitoring system, instrumentation, testing conditions of solar chimney curtain-wall are described in ref. /3/, /4/.
The daily average efficiency of solar chimney connected with short ducts in sunny days is around 30% which is quite similar to the performance monitored in the Marostica demonstration project /5/, but with different constraints due to the particular design case of Marostica (ducts in the ceiling 6.40 m long, return air opening with section reduced about 50% in relation to the chimney section). Chimney efficiency at Ispra was expected to be higher than in Marostica, but drawbacks found in the installations cannot allow definitive evaluation.

A more detailed monitoring campaign in the 1989-1990 heating season will measure the performance of the different distribution systems.
Measures confirmed the efficiency of the solution with solar chimney ceiling distribution also for office buildings; furthermore they suggest to evaluate and measure cooling performance in Summer when the chimney acts as a ventilated wall during the day and as a free cooling system night-day.

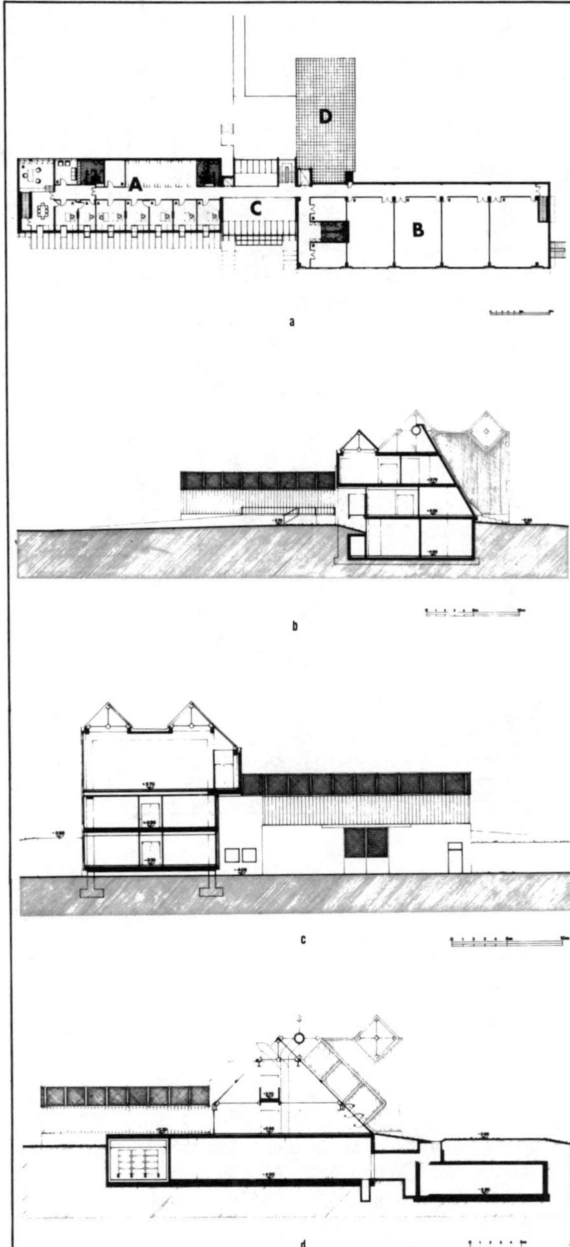

Fig. 2: plan and sections of the building
(a) plan of the first floor with the indication of the four blocks (A,B,C,D)
(b) cross section of the administrtive office (block A)
(c) cross section of the research area (block B)
(d) cross section of the atrium (block C)

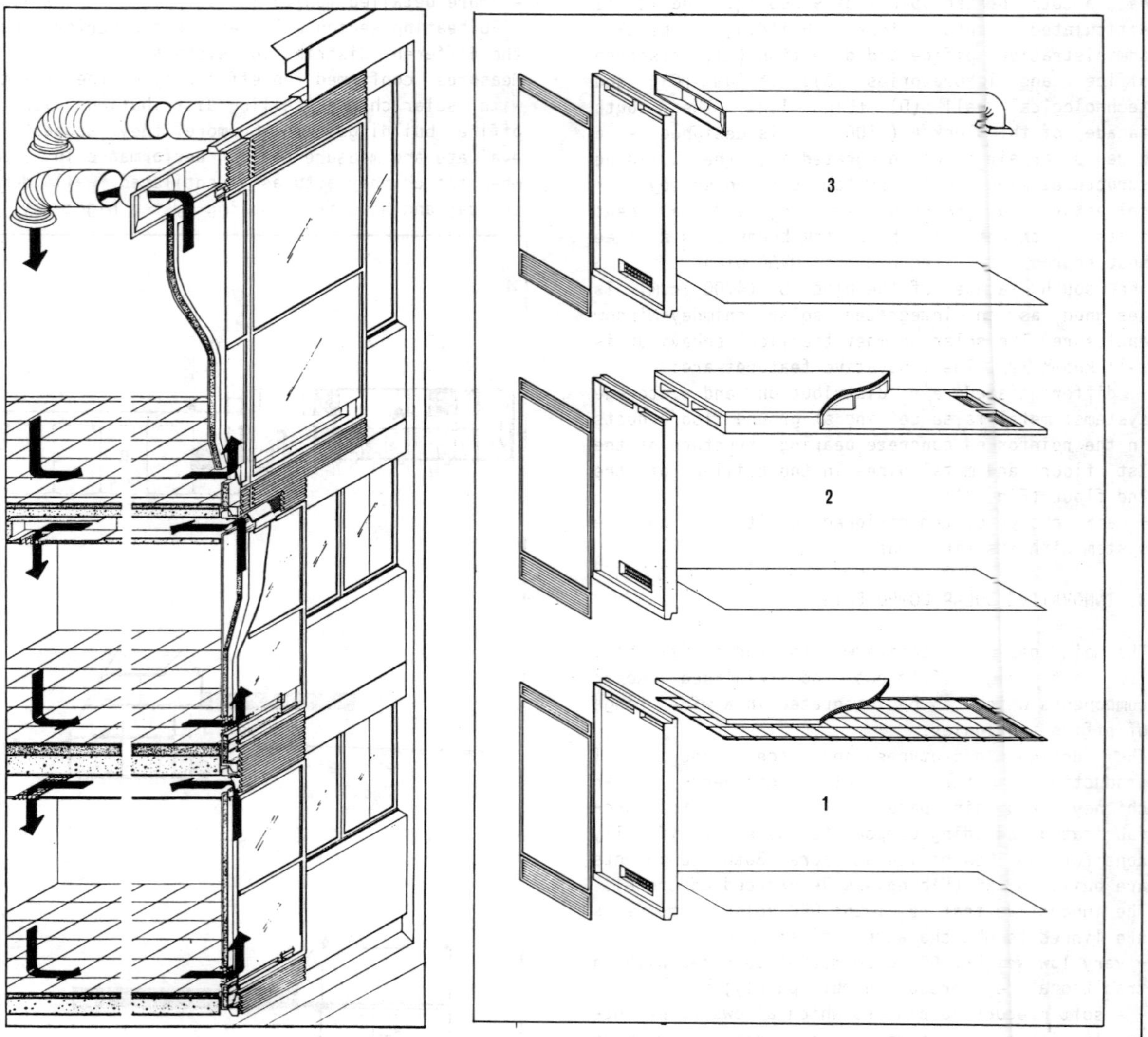

Fig. 3: Axiometric view of the solar chimney curtain wall with three different distribution systems: (1) metal pipe in the ceiling; (2) ducts in the reinforced concrete ceiling structure; (3) metal false ceiling.

NOTES

(1) Building and solar components were designed by Architects G. Scudo and A. Seassaro, Sin/Arch. s.n.c., Corso Garibaldi 50, Milano, Italy; Services Engineer and Energy Consultant was Ing. G.C. Rossi, Via Palladio, 24, 30175 Marghera VE, Italy.

REFERENCES

/1/ P. De Pascali, G. Scudo, Climate responsive cladding components, Science and Technology at the service of architecture, 2nd European Conference on Architecture, Paris, 4-8 Dec., 1989 (to be published).

/2/ A.O. Barra, La conversione fototermica dell'energia solare, Etas Libri, Milano 1981.

/3/ S. Fumagalli et al., I camini solari dell'edificio bioclimatico dell'ENEA di Ispra, primi risultati sperimentali, RTI-FAE (89) 9, ENEA, Ispra (VA) Italy.

/4/G. Contadini et al., Monitoring of a solar chimney and different ceiling storage/distribution systems in an office building located in northern Italy, Proc. of ASES Annual Conference - 14th National Passive Solar Conference, Denver, Colorado (USA), 19-23 June, 1989.

/5/ Camino solare B-C Complesso Panica - Marostica, Research Contract N. 22180, Final Report, Venice, Gen. 1977.

GLASS COATING: ADVANCED TECHNOLOGY FOR HIGH PERFORMANCES.

ANNE MINNE, ir. Architecte
GLACERIES DE SAINT-ROCH
Bvd.E.Jacqmain,162 bte 48, 1210 Bruxelles, Belgium.
Phone: (32) 2/211.61.11

Abstract: This paper describes the properties of coatings on glass. The intent is to define how thermal insulation, solar control and daylighting performances of glass can be improved by means of coatings. It provides some ideas on the possibilities and the flexibility offered by coated glass in the search of comfort and of maximal energy efficiency in commercial and residential buildings.

1. APPROPRIATE WINDOWS.

1.1. Solar gains and thermal insulation.

In temperate and cold regions, climatic conditions range very often beyond the limits of normal human comfort, requiring the use of a heating system.
The greenhouse effect can greatly help heating the building. Windows and sunspaces properly designed allow to make use of solar radiation as much and as long as possible.
But that heat must be contained inside.
The building insulation plays an important role, and insulating windows are very efficiently minimizing heat losses.
In this first case, the requirements for windows are: increasing solar gains and providing efficient insulation.

1.2. Solar control and daylighting.

In warm climates or during some periods of the year, solar and internal gains can create discomfort by raising the interior temperature above the comfort zone.
This problem is greater in commercial buildings, due to the high level of occupancy.
It is then necessary to limit solar gains through glazing elements but care must be taken to never dark the spaces.
Additional energy can be saved by making use of daylighting, as artificial lighting systems release some undesirable heat and of course use energy.
In this case, requirements for glazings are: keeping the sun's rays from entering the building while providing sufficient daylight.

1.3. A sole material for several applications.

To meet those contradictory requirements, the glass properties has been modified.
This can be done by pigmenting the glass, which increases its absorption, or by lining it with a thin metal or metal oxide layer. Combination of both methods is also often used.
In this paper, we will discussed only coated glass, on clear or tinted substrate.
Coatings technology allows to produce high level insulating glazings and efficient solar control glazings, depending on the nature of the coating.
Several deposition techniques and different metal and metal oxides can be used, leading to a wide range of products.

2. COATINGS ON GLASS.

2.1. What is a coating on glass?

Coating is one or more thin solid layers of metal or metal oxide applied onto the surface of glass by various deposition methods.
Coatings are selective with regard to radiation. This means that they do not reflect, absorb or transmit in the same proportions at every wavelenght.
A wide range of "selective filters" can be created by acting on the thickness and on the chemical composition of materials applied on glass, and by the superimposition of different layers.
Coated glass is characterized by its colour and its spectral curve of reflectivity, transmissivity or absorptivity, over a range of wavelenghts.

2.2. Methods of deposition.

Several processes have been developed, for applying thin coatings on glass. They can be divided into two main groups:

2.2.1. Depositing by pyrolysis on the hot glass:

In this process, a solution of metal salt and a reducing compound is sprayed onto the hot glass surface (500 to 600°C). A reduction reaction takes place and an oxide coating then forms on the glass surface by the pyrolysis of the original compounds.
Compounds in a vapour phase or a chemical powder can also be applyed on hot glass and form a coating by pyrolysis.
These coatings are very hard and chemically resistant, and can be placed on the external side of the window.

2.2.2. Vacuum process:

Vacuum coating is the application of thin films on glass surface by evaporation or sputtering.
In the evaporation process, a material is evaporated in high vacuum by heating it. This material condenses on the glass surface.
In the sputtering process, ions bombard a target in a gas discharge and sputter material which condenses on the glass surface.
This process is very flexible and allows the production of a wide range of coatings, of various colours and light and energy transmissions.

2.3. Changing the properties of glass.

Depending on the effect requested, the modification of transmissivity, reflectivity and absorptivity will not be the same at every wavelenghts.
The wavelenghts of solar radiation range from 0,25 to 4 microns (figure 1, curve a).
Solar radiation is composed of ultraviolet, infrared and visible radiation. The visible radiation covers the range from 0,38 to 0,78 microns.
The superimposition of the spectral curve of human eye sensibility on figure 1 (curve b) indicates that all wavelenghts are not perceived the same way.

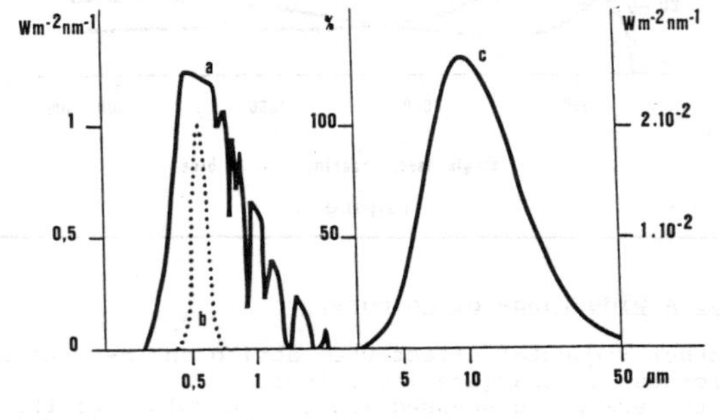

a. solar emission spectrum (earth level)
b. human eye sensibility curve
c. black body emissivity spectrum, 10°C

figure 1

This diagram shows clearly the interesting wavelenghts for an efficient daylight.
Figure 1, curve c gives the emission spectrum of a black body at 10°C. The working zone for low emissivity coatings is thus defined.
Coated glasses are classified in two families of products, in accordance with their properties.

3. SOLAR CONTROL COATINGS.

3.1. Solar control glass.

Transmissivity of clear glass is high in the visible range, but also in the short wavelenght infrared range (figure 2).
Light transmittance of a 6 mm clear glass is 90%, and solar factor is 0.86.
Those characterisitics will be changed by the coatings.
Coatings are increasing the reflectivity and the absorptivity in proportions and at wavelenghts depending on the material applied. They allow the production of solar control glass with light transmittance lying in the range from about 80 to 10%, and solar factor from 0.8 to 0.15.

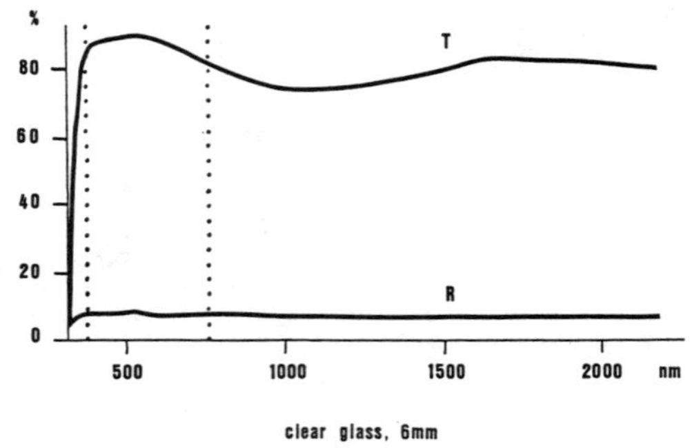

figure 2

3.2. Different products.

Figure 3 shows a typical example of transmission and reflection curves of coated glass in the range of solar spectrum wavelenghts.
We can see that the requirements for solar control glass are achieved: the transmission is reduced drastically in the near infrared range and remains higher in the visible range.
Figures 4 and 5 show the curves of the same coating, applied in greater thicknesses.
The aspect of the transmission curves is similar to figure 3 but with lower values while the reflection of near infrared is increasing more rapidly.

figure 3

3.3. A wide range of colours.

Another requested effect when acting on the coating thickness is changing its colour.
In the example discussed above, the colour of the coating varies from blue (fig.3) to ocre (fig.4) and gold (fig.5), when its thickness is increasing.
This is one of the possible ways to obtain various colours and shades associated with different transmission charateristics.
Other ways are to modify the nature of the product deposed on glass, and to combine different coatings.

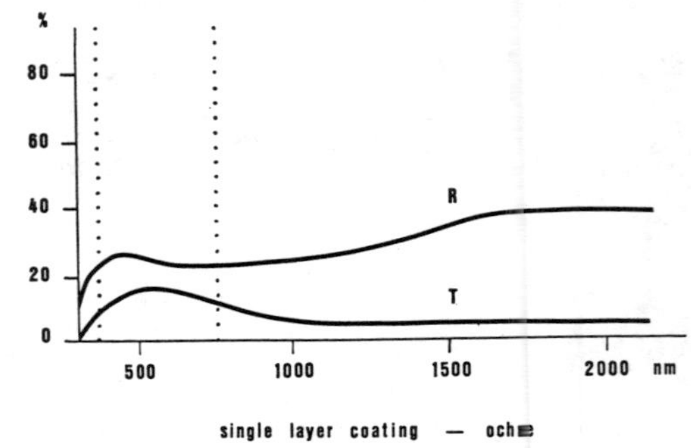

figure 4

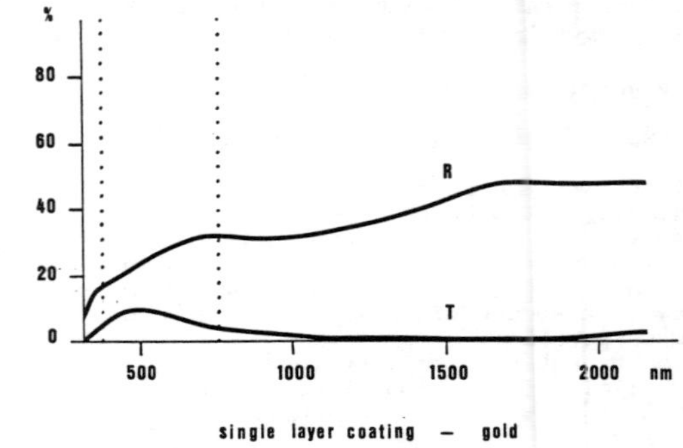

figure 5

Figure 6 shows the transmission and reflection curves of a bi-layer silver-grey coating.
One can obtain some interesting effect in the point of vue of aspect and performances by applying a coating on tinted glass. Figure 7 shows the curves of a tri-layer coating on green glass substrate. As the initial colour of the coating is blue, this coated glass is aquamarine.

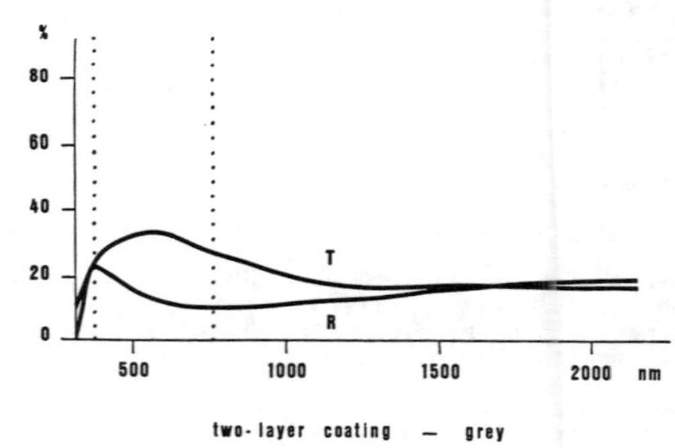

figure 6

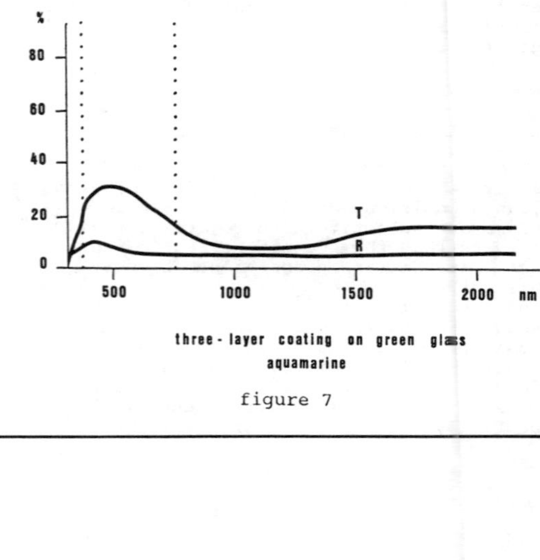

figure 7

4. LOW EMISSIVITY COATINGS.

4.1. Clear glass.

Greenhouse effect is due to the absence of transmission of clear glass in the range of 5 to 50 microns.
This longwave infrared radiation is partially reflected and partially absorbed by the glass. The absorbed fraction is re-emitted on either side of the glazing.
Figure 8 shows that the reflectivity of clear glass is relatively low for longwave infrared radiation.
At the opposite, some coatings have high refectivity in the medium and distant infrared.
Low-emissivity glazing takes benefits of this property.

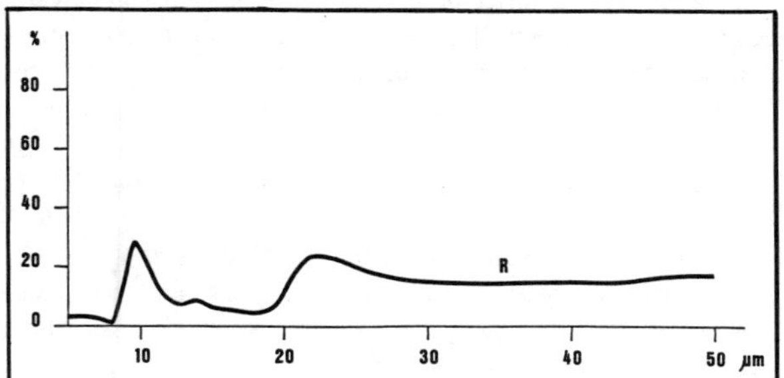

clear glass

figure 8

4.2. Low emissivity glass.

Low emissivity means the same as high infrared reflectivity. The aim is to reduce the radiation exchange between the internal and the external panes of an insulating glass unit.
The U-value of ordinary insulating glass with 12 mm air space is 3.0 W/m K.
Thanks to low emissivity coating, this value can be improved to 2.2 - 1.8 W/m K. If the air in the interspace is replaced by an insulating gas, the U-value can be about 1.4 W/m K.
In addition, such a glass gives a better comfort in the proximity of the window.
Figure 9 shows the infrared reflectivity curve of a low emissivity glass. The values are above 80% for every wavelenghts between 5 an 50 microns.
The transmissivity in the visible range is still remaining very high, as shown on figure 10.

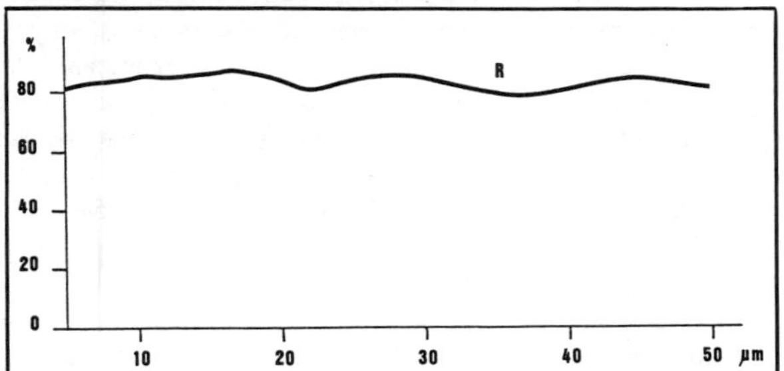

low emissivity glass

figure 9

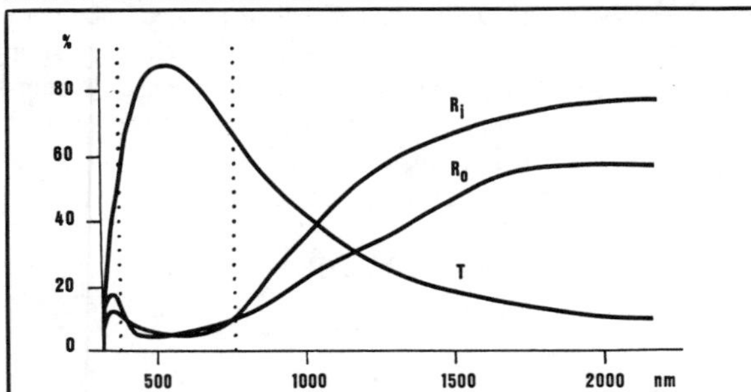

low emissivity glass

figure 10

Those two curves explain why low emissivity glazing are so efficient in passive solar design: the high solar transmission allows sun to penetrate the building and the high longwave infrared reflectivity helps keeping the heat inside.

4.3. Emissivity of solar control glass.

Some coatings used for solar control are also low emissive.
Figure 11 shows the reflectivity of the gold coating already mentioned on figure 5. The values are above 70%, for every wavelenghts.
For that reason, U-value of insulating glass unit with a solar control coating on one pane is generally better than the U-value of ordinary I.G. unit without any coating.

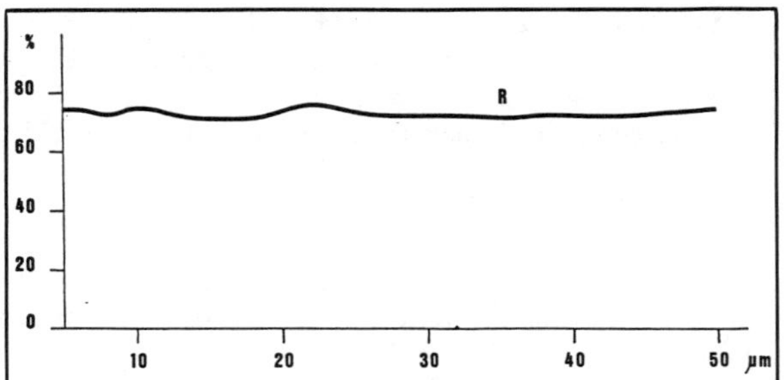

solar control glass

figure 11

5. CONCLUSIONS.

By changing the glass surface properties, coatings are introducing flexibility in the light and energy characteristics of windows.
They allow the optimal combination of daylight and solar control, of thermal insulation and solar gains.
The wide range of aesthetic aspects and the shades of colours obtained with coatings open new perspectives to the architects creativity.
Glass industry is still doing research to improve processes and products. The aim is to provide you with the most efficient and aesthetic products.
The field of coating on glass is not completely explored yet, and future appears to be very promising.

THE PERFORMANCE OF EXTERNAL SHADING DEVICES
Integrating Shading and Daylighting

Koen Steemers
The Martin Centre for Architectural and Urban Studies
University of Cambridge Department of Architecture
6 Chaucer Road, Cambridge, UK. Tel.: 0223 332981

ABSTRACT. Lighting is known to be a significant energy consideration, particularly for office buildings. With the increasing awareness of energy efficiency, the adoption of external shading devices to reduce solar gains, as well as for architectural effect, is becoming more widespread. However, it is important not to overlook the consequences of such systems on the availability of daylight. This paper presents a simple approach to, and initial results of, an integrated analysis for the performance of generic shading device types.

1 INTRODUCTION

One of the most important functional roles of the building envelope is to temper adverse external environmental impacts in order to improve internal building conditions; to act as a selective filter. Another important aspect of facades is the architectural expression and the relationship to the functions. The adoption of external shading devices is a common strategy to intercept excessive solar radiation before it enters the building; however these devices also reduce useful daylighting. A blank wall, having a shading efficiency of 100%, could be said to be the most effective form of shading, which obviously contradicts the object of having a window. This highlights the importance of combining several parameters to be able to assess the overall performance. This paper briefly summarizes methods of performance assessment, and discusses an integrated performance analysis where these other factors are taken into account. Of particular interest is the effect of a sun-breaker on the availability of daylight, as this relates to the overall performance of a building in terms of its energy use. A computer program is used to analyse various simple generic shading types and to produce initial performance results relating to the UK climate. These indicate a means of optimizing the configuration in relation to the requirements and location, and are discussed in terms of architectural implications.

2 AESTHETICS AND PRAGMATISM

The earliest preoccupations with architectural aesthetics, and more particularly with the elevational treatment of buildings, were often directly associated with the interplay of light and shadow. The moulding of forms to heighten the effects of sunlight were based on personal aesthetic decisions, governed by a superimposed order. In a similar manner, the sun-breaker, which is now an accepted architectural element, is the result of personal expression controlled by rationalized performance requirements. It is this pragmatic factor related to shading devices that is of interest to us, as it provides the framework for individual creative expression.

2.1 Historical Background

Solar shading techniques provide a strong regional character and great scope for individual or cultural aesthetic expression. The emergence of the solar shading device in Western architecture is the result of several developments in the more recent history of building technology. Traditionally the masonry wall performed a multitude of environmental roles. The structural implications of the load bearing wall meant that openings tended to be limited in size and number. As a result, the impact of solar gain was limited, being tempered further by the thermal mass of the construction. With the development of new materials, and an architectural movement dedicated to a functionalist expression of these materials, the simple wall element started to be broken down. The structural possibilities offered by concrete and steel led to the use of the glass pane as a main element separating the internal environment from adverse weather conditions. This vulnerable glass curtain, however, does not control effectively the impact of solar radiation, one of the most important tasks of the building envelope. A further element, the 'brise-soleil', was introduced to perform this role. Thus an elemental deconstruction of the traditional load bearing wall had occurred, with frame, skin and 'brise-soleil' as the main layers, each specialized to fulfill specific functions.

2.2 The Modern Office Building in the UK

If one looks at the contemporary office building there is an increasing amount of communications equipment producing significant internal heat gains. Given this situation, it becomes important to shield the internal environment from any further gains to prevent excessive energy penalties in the form of additional cooling loads in the summer, and in some cases, all year round. The main area of concern is to prevent significant solar gains whilst still allowing windows to afford views and provide natural light. An effective daylighting strategy will further reduce cooling and electrical loads by displacing the need for artificial lighting. However, with deep plan speculative office buildings this may not be a prime consideration, although the admittance of daylight is psychologically beneficial. The use of atria in such situations can improve the potential for natural lighting and ventilation (1) by reducing the depth of the plan and affording greater control over the reflectances of surfaces outside the windows.

By adopting external shading devices rationally, solar gains can be significantly reduced whilst also contributing to the architectural appearance of the building envelope by giving the facade detail, depth and delicacy, and a sense of human scale. However, although the aesthetic contribution may be significant, the functional qualities need to be analysed to judge the success of the overall design.

3 SOLAR SHADE AND NATURAL LIGHT: Direct and Diffuse Radiation

It is important to take account of the relative amounts of solar radiation that each orientation receives, to be able to judge the requirement for solar shading. From a pragmatic point of view, north facing shading devices are functionally unjustifi-

the south, but as the orientation becomes more northerly (Figure 1c), fins become very effective and overhangs of little use.

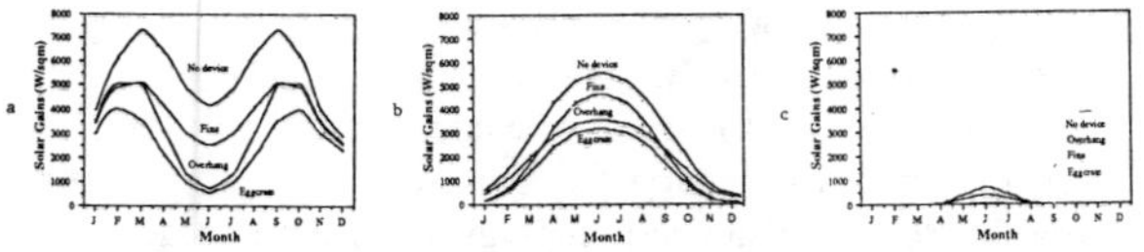

Figure 1: Maximum daily solar gains for the three generic device types for south (a), west/east (b) and north (c) orientations.

It is important to note that these results are for a latitude of about 52° N. If a more northerly location were to be taken then the performance of fins on west and east facades would improve. Conversely, a more southerly latitude would mean that overhangs become more effective at intercepting high altitude sun on west and east elevations.

5.3 Views

The quality of the views afforded from within the building, or perhaps more specifically the quality of visual contact with the outside, is difficult to assess. Two obvious parameters which affect this quality are the degree of obstruction caused by the shading device, and the possible unsightly weathering of it. An analysis of the qualitative parameters is complex, however a view index has been devised for the computer program which, like the daylight efficiency equations, measures the degree of obstruction caused by a device, not only of the sky, but also of the view below the horizon line, which usually contains the more important components that make up an interesting view.

Visual discomfort due to glare is usually minimized if direct radiation is effectively eliminated. Most external devices will reduce the contrast glare by virtue of presenting a surface of intermediate brightness, between the sky and the interior, to the occupier in the room. The effect of a device on the distribution of daylight within a building may also cause discomfort if it excessively reduces the penetration of natural light. This is a significant consideration which is receiving attention amongst specialists, though it is too involved and dependent on room characteristics for the purpose of a simple performance indicator of an external shading device. One example, the light-shelf, deserves a mention here as it a shading device specifically designed to minimize glare and to improve daylight distribution.

5.4 Comfort

The degree of control a device has over solar gains is important in determining not only effects on the energy consumption, but also in assessing comfort conditions in a building. The thermal comfort of a human being depends approximately equally on three main factors: mean radiant temperature, air temperature and air movement. If there is an obvious asymmetry in these factors, this may result in discomfort. Some compensation for one factor by another is tolerable, but if, for example, an air velocity of more than 2 m/s is required to compensate for a high air temperature then the result is an annoyingly draughty condition. In this particular context the aim is to prevent discomfort from direct solar radiation. If, for example, a comfortable environment is maintained for a building in general, particular consideration has to be taken of the conditions near the building envelope. Solar radiation should be prevented from impinging on people at work spaces. Conversely, the potentially lower mean radiant temperatures due to the glazing, caused by conductive losses in cold weather, need to be balanced by other radiant sources at that point, to avoid discomfort. Shading devices can prevent excessive solar radiation, and radiators can improve the mean radiant temperatures near the perimeter of the building. The effectiveness of a sun-breaker at preventing discomfort depends on the amount and intensity of solar radiation allowed to enter and so is directly related to the shading efficiency described earlier, so that no further specific quantification for comfort is necessary to define the performance of a device.

6 THE PERFORMANCE OF SHADING DEVICES

To assess the performance of a shading device one could argue that maximum shade from intense direct solar radiation should be balanced with a minimum reduction of diffuse radiation, whilst still providing a reasonable view. It may occasionally be the case that daylighting has not been an important design consideration. In such a situation it would be biased to judge the performance of the sun-breaker in relation to its effects on natural lighting. The main purpose of openings is then to provide views, and the measurement of performance should adopt the view index as the main factor to compare with the shading efficiency, instead of the daylight efficiency. However, as shading devices are usually employed to reduce energy consumption for cooling, and as artificial lighting, particularly in offices, accounts for both a large fraction of energy use, as well as significant internal gains, it is sensible to judge performance as a function of both shading and daylight efficiencies.

To represent results for a specific type of shading device, a simple set of curves can be plotted showing shading efficiency, daylight efficiency and view index as functions of the aspect ratio of depth of device to dimension of opening. The examples shown here are standardized to the following conditions:

- Simple solid generic device types used as examples are horizontal over hangs, vertical fins and egg-crates directly adjacent to a unit square window.
- Site conditions are urban, at a latitude of 52° N, assuming a 20° 'equivalent obstruction line' from the centre of the window.
- Occupancy is standardized at 8 a.m. to 7 p.m.
- Orientations are either due south, west, east or north.

As can be seen from the graph (Figure 2) showing typical results and as would be expected, an increase in aspect ratio, i.e. enlarging the shading device, results in more shading but also reduces potential for natural lighting and unobstructed views.

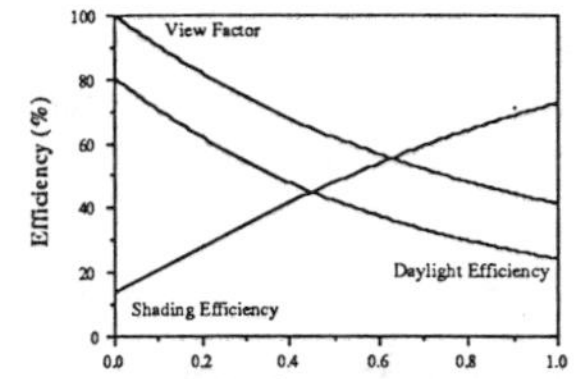

Figure 2: Shading, daylight and view factors for a south facing window with a solid overhang plotted as a function of aspect ratios.

To find the optimum condition, the product of shading and daylight efficiencies has to reach a maximum. Here, this product has been called the yield of a device. The yield can be plotted either against the daylight efficiency, or a range of

able and not cost effective. Similarly, adopting identical devices on all the facades, as is often the case, indicates that orientation has not been seriously taken into account.

Just as orientation is important, so is the height of a window on a facade in relation to obstructions. In an urban situation, for example, windows at lower levels will require no shading from low angle sun and maximum access to natural lighting, possibly by some means of reflective devices such as light scoops, mirrored louvres or prismatic glass. Higher up the facade it becomes more important to shade windows from solar radiation, but daylighting is easier to achieve; some form of light shelf may be an suitable option. The implication is that, where obstructions are significant, there should be a transformation in the shading strategy from the base of a building to the top.

4 AVAILABLE METHODS OF ASSESSMENT

The conventional method for assessing the shading characteristics of different devices is to establish the shading mask and superimpose it on a sun-path diagram.

This popular and relatively simple graphical tool unfortunately does not provide accurate analytical information to assess the performance of a shading device. It will only show when a particular window will be 100 % or 50 % in shade, depending on which criterion is used. Using the 100% mask a simple formula can be adopted to estimate a shading factor. This has a value of zero for total shading, and 1 for total exposure and is defined as follows (2):

$$\text{Shading factor} = 1 - (\text{no. of hours of 100\% shading} \,/\, \text{total no. of hours available})$$

This formula is inaccurate for two reasons. First, it simplifies the situation to represent a point on the facade, it is not integrated over the total area of glazing. Second, it assumes that shading at one hour is as important as providing shade at another. This is obviously not the case, as for example, preventing solar penetration during a summer afternoon is much more significant than shading from morning winter sun.

Any shading device carries a penalty in that it reduces the amount of daylight entering a building by obstructing parts of the sky and reflected components of diffuse light. The benefit of a sun-breaker in reducing solar gains needs to be judged against performance indicators such as the penalty incurred due to loss of natural light, or the obstruction of views.

5 PERFORMANCE INDICATORS

The overall success of a shading device depends upon all those factors which may be affected by it and include daylight, solar shading, views and comfort. Each of these performance indicators contributes to arriving at an integrated performance analysis, and are discussed below.

5.1 Daylight

Daylighting is an important concern both in terms of energy and psychological benefits. Natural lighting can potentially reduce the need for artificial lighting, and thus reduce the lighting load and energy consumption. It also provides some degree of contact with the outside world, and relief from an artificially controlled environment. The effect of a device on daylight can be estimated and is a useful parameter to evaluate the performance of shading devices.

Any external obstruction in the form of an overhang or fin, will cut out parts of the diffuse sky and reflected ground components of daylight. Thus a device may reduce the internal illumination excessively, adversely affecting the overall performance.

The amount of light entering an opening can be approximated and divided by the total amount available, thus a daylight efficiency is defined as in the equations below. It is, however, important to mention the assumption that no reflections from the device or the surroundings are taken into account at this stage. This is an acceptable approximation only if the sky component of daylight is high, but not in a situation where the dependence on the externally reflected component is significant. However, for the purpose of assessing the performance of generic device types on a comparative basis this definition of daylight efficiency can be acceptable.

Tilted horizontal overhang

$$\text{Efficiency} = [\sqrt{1-2(c/h)\sin\partial+(c/h)^2} - (c/h)\cos\partial].100\%$$

Tilted vertical fins

$$\text{Efficiency} = \{0.5[\sqrt{1-2(c/w)\sin\partial+(c/w)^2} + \sqrt{1+2(c/w)\sin\partial+(c/w)^2}] - (c/w)\}.100\%$$

c = depth of device
h = height of opening
∂ = angle of tilt
w = width of opening (3)

Moving on from existing sources, for this study, the above mentioned equations have been adapted for a range of shading device types and incorporated into a computer program. Furthermore, the C.I.E. overcast sky used in the program, is assumed as the limiting daylight condition for most locations under consideration and is an approximation of polluted city environments. It is, however, not applicable in situations where clear skies are the prevalent weather conditions.

5.2 Solar Shading

With the use of a simple computer program the solar intensity can be calculated for any hour of the day, for each month. This enables a weighting of the effect of a shading device according to the quantity of solar radiation allowed to pass into the building in relation to the total available. A monthly shading efficiency thus gives an accurate indication of the seasonal effectiveness of a device. If the efficiency is 100%, this indicates total exclusion of solar gain, whereas a factor of zero implies no shading. However, the result is that a blank wall, having a shading efficiency of 100%, could be said to be the most effective form of shading, which obviously contradicts the object of having a window. It is thus necessary to evaluate the effects on other parameters involved, which determine the need for the openings. Natural light, which has already been mentioned, and the provision of views are the major concerns. Apart from the obvious requirements of providing shade from unwanted sun, the sun-breaker affects each of these to a greater or lesser extent.

If one examines the calculated maximum daily solar gains for the three generic shading devices, some important conclusions can be made. A simple overhang is most effective at cutting out summer sun on a south facade (Figure 1a), whilst having little effect on winter solar gains. An egg-crate device also performs well overall, but it reduces potentially useful winter gains, obstructs more natural light unnecessarily, and requires a greater amount of material to manufacture. Vertical fins perform least well for southerly orientations For west and east orientations (Figure 1b) the conclusions are similar as for

aspect ratios of a particular device. The former will give an indication of the performance of a shading device in relation to its effect on daylighting, so that this can be compensated for by increasing the amount of glazing in relation to the optimum configuration. The latter has more significance in terms of the appearance of the building envelope, and is therefore of particular interest to the architect. It reveals which devices are most effective given a specific aspect ratio, and also whether there are significant differences between shading device types.

The annual yield curves produced for daylight efficiency indicate that for southerly orientations (Figure 3a) the overhang is clearly the most effective at minimizing solar gain whilst maximizing potential for natural lighting. Egg-crates perform only marginally better than fins due to the increased obstruction they present to daylight. For west and east facing windows (Figure 3b) overhangs again perform better, although the difference is less marked, and fins are marginally better than egg-crates. As would be expected, fins are the most suitable for north facades (Figure 3c).

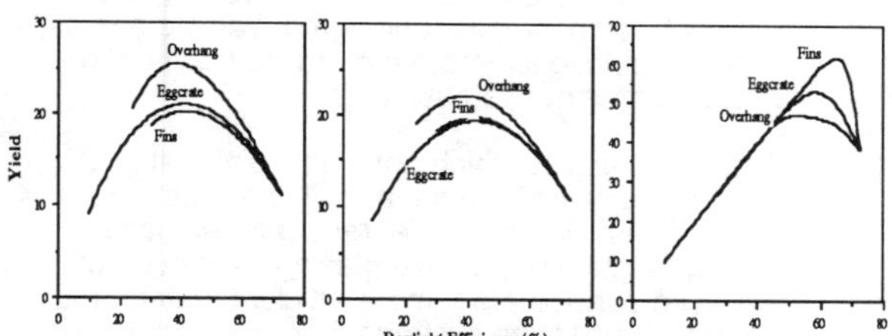

Figure 3: Yield or performance curves for generic devices as a function of daylight efficiency, for south (a), west/east (b) and north (c) orientations.

If one examines the yield in relation to the depth of the device for each orientation, some interesting subtleties are highlighted. When very small aspect ratios are required, of less than about 0.2, for architectural or cost reasons, then south facades (Figure 4a), and west and east orientations (Figure 4b) are best provided with egg-crates. Otherwise overhangs with aspect ratios of approximately 0.5 offer the maximum performance characteristics for these orientations. For north facades (Figure 4c) fins are the most suitable, having the highest yield, or performance, for very small depths of devices, the optimum being about 0.1 of the width of the opening.

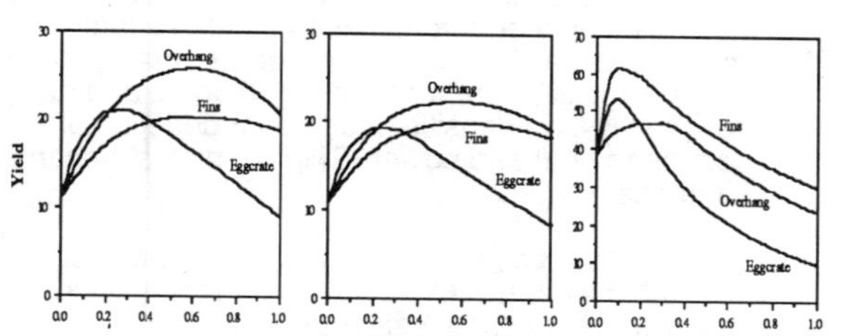

Figure 4: Yield curves as a function of shading device aspect ratio for south (a), west/east (b) and north (c) orientations.

Similar curves can be drawn for any orientation or latitude where the daylight assumptions are applicable. External obstructions and atmospheric turbidity can be taken into account, so that for particular site conditions the optimum generic external shading configuration can be defined.

REFERENCES

(1) BAKER, N., *The Atrium Environment*, Building Technical File, April, 1988.
(2) ACHARD, P., and GICQUEL, R., (ed.), *European Passive Solar Handbook*, prliminary edition, CEC, 1986.
(3) OLGYAY, A, and OLGYAY, V., *Solar Control and Shading Devices*, Princeton University Press, 1957.

7 CONCLUSIONS

In general terms the conclusions reached are very much as would be expected and are summed up below:

- The performance of shading devices cannot depend on the effectiveness of controlling solar radiation alone. The effects of the device on the various functions of a particular window must be taken into consideration. These may include quality of view, daylighting, and thermal and visual comfort.
- Each latitude and orientation will require a different response to shading, and thus a different strategy, for optimum performance characteristics.
- In situations where obstructions are significant, such as in the urban context, there should be a transformation in the shading strategy from the base of a building to the top.
- In the northern hemisphere, appropriate shading to south, west and east facades is the most effective and proves to be worthwhile.
- A simple overhang proves to be the appropriate generic solution for a southerly orientation. Although a louvered overhang may improve natural lighting, any high angle intense summer sun that may penetrate reduces the overall performance. Louvres should thus be spaced to cut out the high sun angles.
- On east and west facades, overhangs perform slightly better than vertical fins of the same depth, by distributing gains more evenly throughout the year. Fins interfere less with natural lighting and views than egg-crates, but are also less effective at reducing solar gains. The extra cost of egg-crates may be a determining factor.
- North facing shading devices are largely redundant, having only a limited effect in the summer months. Simple fins spaced relatively broadly, may be the only appropriate solution that is functionally justifiable.
- The performance, in relation to daylighting, of shading devices can be improved by using light finishes on the device or by adopting louvered devices, possibly moveable, to cut out direct sun but limit interference with diffuse light. The latter may entail complexity and cost that may make a simple fixed system preferable.
- For latitudes significantly greater or less than 52° N the performance characteristics will vary: in higher latitudes fins become more effective on west, north and east facades, whereas for lower latitudes overhangs are the most suitable option for north-west, south and through to north-east orientations.
- Once the generic type of shading device has been chosen, fine tuning in the of form moveable louvres providing increased control, may be the next step. In this way useful solar gains can be admitted, limiting the potentially detrimental effects on daylighting and views.

The principles that govern the need for solar shading are often overlooked by adopting the same form of device for all orientations. From an aesthetic point of view, it should not prove difficult to produce a pleasing architectural solution whilst optimizing performance. The initial attractiveness of the continuity provided by adopting the same shading strategy on different elevations is, to a certain extent, lost upon further consideration.

An architecture that does not adopt an 'order' defined by a pragmatic rationale, within which creativity can produce an individual expression, loses a degree of intellectual conviction. This is particularly the case when that order is used largely as a decorative element, abandoning the principles upon which it was based.

CONTROLLING ENERGY USE THROUGH URBAN PLANNING

P. Steadman
Centre for Configurational Studies, The Open University, Milton Keynes MK7 6AA, England

ABSTRACT. Problems of global warming bring new urgency to the need for energy conservation through reduced use of fossil fuels. This paper reviews the possibilities for controlling energy use by means of transport and planning policies at the urban and regional scales. A series of experiments is reported in which incremental changes are made to the typical land use patterns of English provincial towns and their surrounding regions. These options are evaluated using a land use and transport model which has been specially adapted for calculating energy use in housing and in passenger vehicles. New directions for policy and research are proposed for the 1990s.

1. ENERGY CONSERVATION AND THE GREENHOUSE EFFECT

The recognition by scientists and politicians of the seriousness of the greenhouse effect and the likelihood of global warming brings renewed urgency to the need for energy conservation. Issues which were on the research and policy agendas in the '70s after the OPEC shock of 1973/74, and then seemed to recede again during the '80s as oil prices stabilised, are now back with a vengeance.

When governments met at Noordwijk in Holland last month to discuss atmospheric problems, many were prepared to enact policies to stabilise CO_2 emissions from coal-fired power stations and motor vehicles at 1988 levels by the year 2000. (They were only stopped from making a firm commitment to this date by the prevarications of Britain, the United States, the Soviet Union and Japan.) The Toronto conference on The Changing Atmosphere, held in 1988, and the Inter-Governmental Panel on Climate Change, have both called for global CO_2 emissions to be cut by 20% by the year 2005, just fifteen years from now. It has been argued that, to achieve this global goal, much larger percentage cuts will be required from the developed nations.

These targets have startling implications for the use of fossil fuels, and hence for urban planning and the built environment. In Britain for example, if present trends are simply extrapolated, then emissions of CO_2 are predicted to increase by nearly 40% in the next fifteen years. In those countries, like Britain, which rely heavily on coal and oil for power generation and space heating, there are serious implications for the use of energy in buildings. And in all industrial countries there are major implications for energy use in transport, above all in the use of petrol in private cars. In Britain the use of energy in transport has been growing especially fast, exceeding energy use in industry for the first time in 1986, and now representing close on 30% of delivered energy nationally. Virtually all of the energy used in transport in the UK is derived from petroleum, about 75% of this in road transport.

2. WAYS OF CONSERVING ENERGY IN BUILDINGS AND IN TRANSPORT

Numerous ways of conserving energy in buildings and in transport have been suggested, studied and in some cases implemented over the last fifteen years (cf. Steadman 1977, 1980; Owens 1986). We can, for convenience of discussion, class these under three headings:

1) Those measures which apply to vehicles in themselves. For example: improvements in fuel consumption (increased km/litre) through better engine design, smaller and lighter vehicle bodies, speed restrictions, improved driving habits and perhaps changes to new fuels.

2) Those measures which apply to individual buildings or small groups of buildings within themselves. For example: reduced heat loss through insulation, double glazing, reduction of exposed surface area, more efficient boilers and controls on heating systems; improved solar gain through orientation and glazing design; use of natural lighting and ventilation in preference to artificial; heat recovery from refrigeration and lights.

3) Those measures which have to do with spatial relationships on a larger scale between the activities housed in buildings, and so have consequences for energy use in transport. In this case, energy uses in buildings and in transport may interact, in potentially complex ways. Some examples are the clustering of activities together - residences closer to work-places, or residences closer to shops and other services - such that average journey lengths are reduced, or journeys can be made on foot or by bicycle instead of by motorised transport; the arrangement of land uses and transport networks so as to encourage the transfer of trips from energy-inefficient modes such as private cars or aeroplanes to more efficient modes such as buses or trains; or the arrangement of buildings in appropriate configurations, and at sufficiently high densities, for the introduction of combined heat and power (CHP) or district heating schemes.

Some conservation measures which might at first sight look as though they apply just to vehicles in themselves, can be seen on closer examination to have consequences for, or to be dependent on patterns of land use. The efficiency of fuel use in passenger vehicles for instance, when measured on a per seat basis, can clearly be improved by increasing the load factor - that is, by filling empty seats. In private cars this would mean encouraging car pools, perhaps with financial incentives or by allocating special traffic lanes as in California and Japan. Greater use of public transport might be encouraged through fare subsidies, special bus lanes, or discouragements to car use such as road pricing, higher parking charges or other fiscal measures. But whether vehicles - either private or public - can and will in practice be filled nearer to capacity, is a function among other things of the demand for trips along the routes in question, and whether travellers share origins and destinations, which in turn will be affected by patterns of land use.

It is possible that potential ways of saving energy could be <u>in conflict</u> with each other. For example the reduction of average travel distances and the introduction of CHP might seem to call for increased densities of development; whereas the effective exploitation of solar energy in building design might seem to require the opposite. (Although it has been argued by Owens (1986) that compromise densities are possible which can reconcile these goals.)

Different measures would apply at different <u>scales</u>: CHP or district heating at the scale of estates or urban 'quarters'; the transfer of commuting trips from car to public transport at the scale of the urban or metropolitan region; substitution of plane by rail travel at the national and continental scales (Figure 1).

Several measures of types 1) and 2), internal to vehicle and building design respectively, offer larger and more immediate energy savings than those in group 3), and should therefore be higher on the policy agenda. They are technologically straightforward, well tried and understood, politically uncontentious, and can be introduced on a piecemeal basis. Turnover in the national vehicle stock is fast, and design improvements can therefore be spread quickly. Turnover in the national stock of buildings is much slower - less than 1% per annum in the housing stock of Britain at present- but even here there is great scope for retrofitting.

By contrast, road and rail networks and larger patterns of land use represent huge investments in infrastructure, and are only changed significantly over decades. What is more, the actual consequences and effectiveness of type 3) energy conservation measures are generally less well understood and less easy to predict. In some cases the assumption is made that people will change their behaviour, to travel less far, less often, or by different modes. In practice people may resist such changes, finding them difficult or unattractive.

On the other hand, the ways in which transport networks and the built environment are being shaped presently through planning policies will clearly have important constraining effects on patterns of energy use into the next century. Precisely because the rate of change is slow, if action is to be taken, it must be initiated now. It is even imaginable that the energy efficiency of buildings and vehicles might be improved in the future up to the limits of practical or economic possibility, at which point further savings would then have to be sought through planning and transport system changes.

3. ENERGY USE AND ALTERNATIVE PATTERNS OF REGIONAL SETTLEMENT

Over the last ten years my colleague Peter Rickaby and I have been studying some of these relationships between energy efficiency, land use patterns and the structure of transport networks. We have chosen to look first at the regional scale; more recently at the urban scale. Because of the complexity of the systems involved, and the interactions between various conservation options, we have been using a land use and transport simulation model which has been specially adapted to calculate energy uses, both in buildings and in vehicles. This is the TRANUS model developed by our collaborator Tomas de la Barra and colleagues at the Central University of Caracas in Venezuela (de la Barra and Rickaby, 1987).

Because of the inertia and slow rate of change in settlement patterns, we thought it best to start from the existing distribution of land uses and the existing transport network in a representative region of Britain; and then to make modifications which could reasonably be achieved by the application of planning policies over an assumed twenty-five year period (Rickaby 1981, 1985, 1987; de la Barra and Rickaby, 1982). We began therefore with an empirical analysis of a large area of central and eastern England (Figure 2). Land use maps were used to determine the distribution of parcels of land devoted to different activities. Road maps were used to measure the geometrical and topological characteristics of the road network. On the basis of these empirical measurements an idealised or archetypal city-regional configuration of land uses and roads was developed, which shares many of the observed characteristics of the actual pattern, but from which the effects of accidents of topography and local history have been removed (Figure 3). The average lengths of road links, numbers of roads meeting at junctions and numbers of road links bounding each enclosed area of land are all representative of the real network.

There are eight main roads meeting at the central town,

Figure 2: Study region of central eastern England

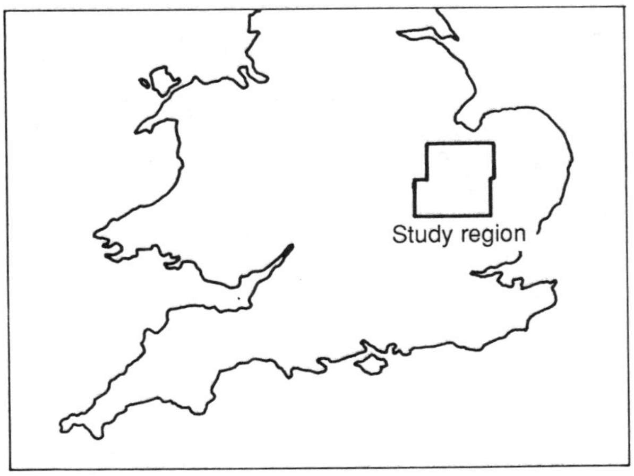

Figure 1: Structural variables significant for energy use in buildings and transport at different scales, from Owens (1986) p. 5 Table 1.2

Structural variable	Scale
Settlement pattern (for example, size and spacing of settlements, etc)	↕ regional
Communication network between settlements	subregional ↑
Size of settlement (area)	
Shape of settlement (circular, linear, etc)	
Communications network within settlement (radial, grid, etc)	individual settlement ↓ ↑
Density	
Interspersion of land use	
Degree of centralisation of facilities	
Layout (estates, etc)	neighbourhood ↓
Orientation (of buildings or groups of buildings)	
Siting (in relation to microclimate)	building ↕
Design	

whose population is 72,000. The boundary of the built-up area of the town resembles a starfish in shape, whose arms extend out somewhat along the main roads. The town has a 'rural hinterland' with a population in villages and hamlets of 43,300. Between the hinterlands of neighbouring towns is the 'rural background' which for each town has a population of 8,800.

The changes to this pattern which were studied for their impact on energy use, all involved relocating 25,000 people from the rural hinterland, along with some or all of their associated places of employment and services, within five alternative configurations of land uses (Figure 4). The road network was left unchanged throughout.

In option 1 the relocated population was all concentrated in the central city, reflecting the proposition that higher densities, and the closer integration of residences with work-places and services, would result in shorter average trip lengths.

There is however no necessary correlation between density of development and the concentration or dispersal of that development in different spatial patterns. As March (1967) has pointed out in an important series of geometrical studies of land use, 'It is as possible to have a high density dispersed pattern as a low density concentrated pattern'. Options 2 and 4 were designed to test what seemed to be promising features, from an energy conservation standpoint, of March's proposals for high-density linear developments, or going back to earlier precedents, the linear cities of Soria y Mata and the Russian 'disurbanists'. In option 2 these high density corridors follow the main arterial roads. In option 4 the main roads are left free for through-traffic, and the ribbon development is run along the secondary roads.

It was thought that linear developments of these types might be specially conducive to an increased use of public transport, since the concentration of population onto a small number of routes might allow fast, frequent services, so encouraging a transfer from private cars.

Option 3, by contrast, concentrates the displaced population into a pattern of eight small new satellite towns at junctions of the major road system. It has some affinities with Ebenezer Howard's (1898) 'federation of garden cities' - although Howard's satellites were linked by rail not road. Option 5 involves dispersing the displaced population into twenty-four villages, either new or expanded.

These different patterns were then evaluated using the TRANUS model. Given a transport network and a distribution of basic employment, the model allocates housing and services to zones within the constraints of available land. It predicts resulting numbers of passenger trips (journeys to work and journeys to services) disaggregated by mode (principally private car and bus, although other modes may be represented). These trips are then assigned to routes, taking into account journey times and the effects of congestion.

The model incorporates routines for calculating fuel consumption in two sectors: domestic buildings, and passenger transport. (Non-domestic buildings and goods traffic are not yet included.) The average specific fabric heat loss of domestic buildings is calculated for each zone taking account of average internal/external temperature difference and floor area per dwelling. Larger dwellings are assumed, on the basis of data from Leach et al. (1979), to have higher rates of heat loss per unit floor area than smaller houses and flats. A domestic insulation parameter can be changed in value to represent projected future improvements in insulation standards in housing. Energy use can then be computed from total heat loss, allowing for the average efficiency of central heating boilers or other heating systems.

Fuel consumption in transport is calculated separately for public transport vehicles and private cars, on the basis of average km/litre figures for typical British cars and buses. These values too can be changed to allow for future improvements in vehicle and engine design. The fuel consumption rates vary with average speed, as calculated by the model for each link in the road network, and depending on road capacity and congestion. Allowance is also made for the different energy content of petrol and diesel fuels in estimating total energy use.

These six regional settlement patterns - one existing and five hypothetical - were compared against three national scenarios of future economic and energy-supply conditions in Britain: a high growth 'business as usual' scenario, a medium growth 'technical fix' scenario, and a

Figure 3: Archetypal city region. Heavy lines show main roads. Broken lines show limits of urban area, 'rural hinterland' and 'rural background'.

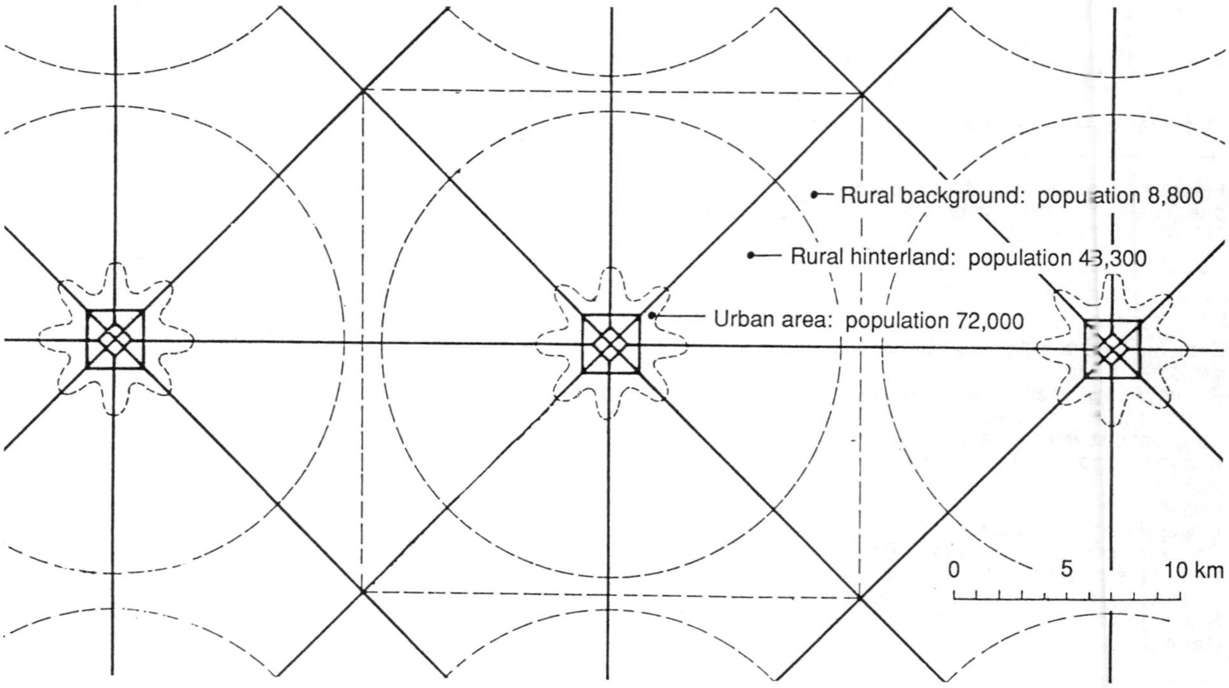

Figure 4: Five alternative city-regional configurations, patterns 1-5, for comparison with the existing situation, Figure 3. Heavy lines show main roads, thin lines show secondary roads, shading shows urban areas.

Pattern 1: concentrated development in the central city

Pattern 2: high-density linear development along main roads

Pattern 3: new satellite towns

Pattern 4: high-density linear development along secondary roads

Pattern 5: development dispersed to villages

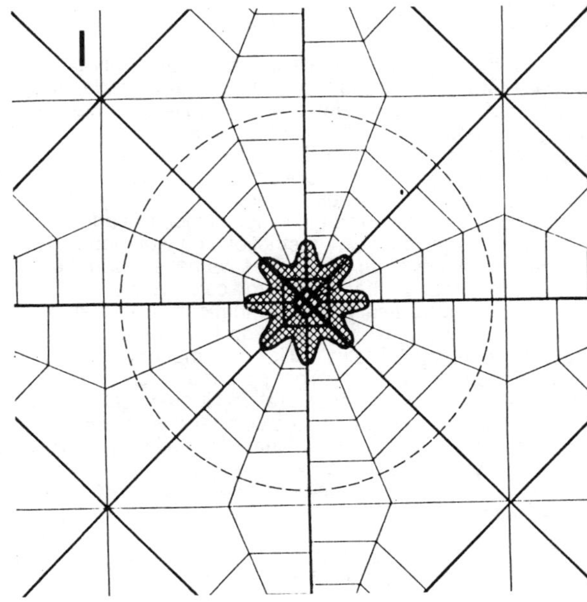

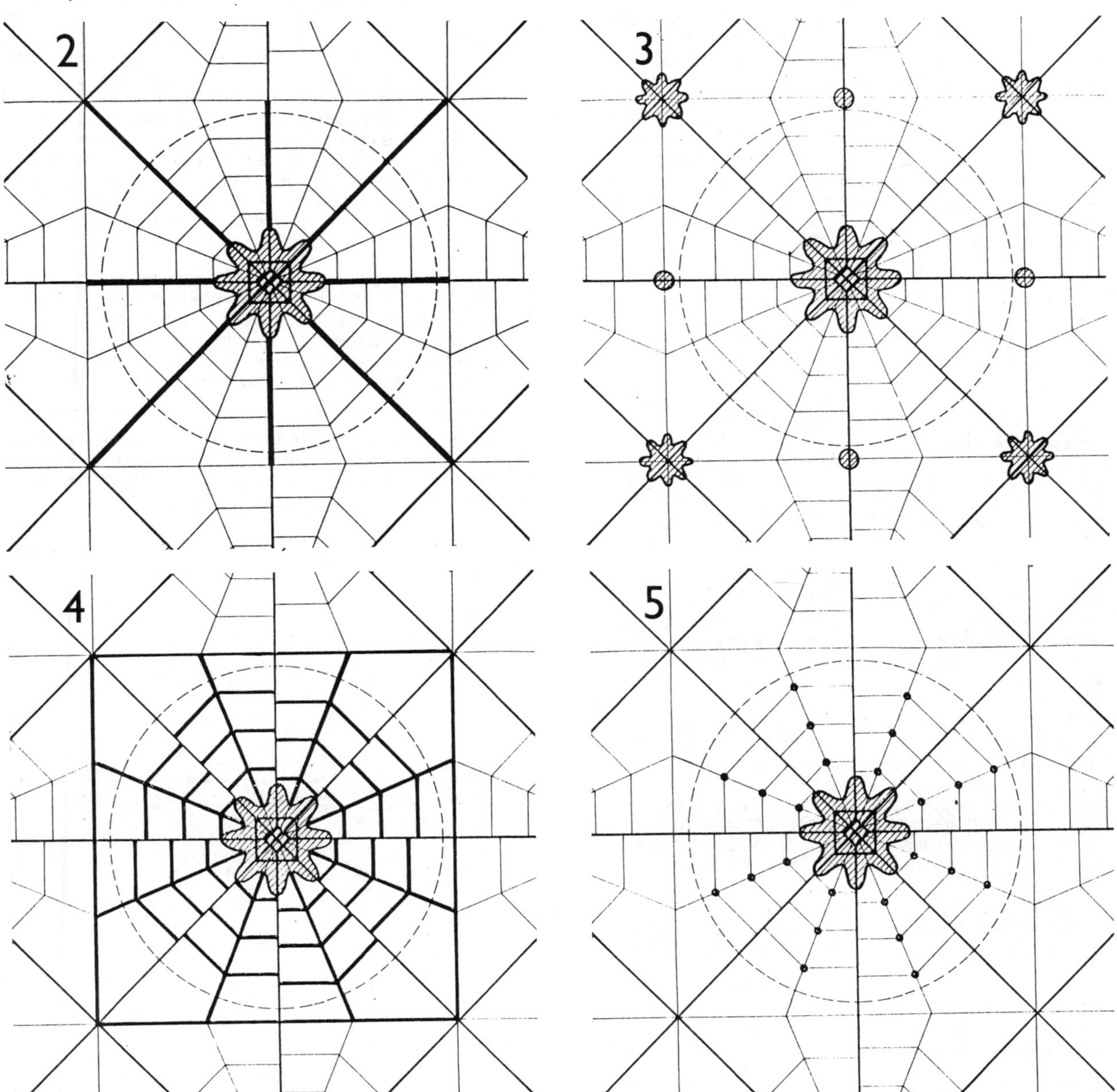

low or no growth 'conservation' scenario.

It is clear that energy use in transport can in theory be reduced simply by limiting the number of trips made, or by reducing trips in length. However, it is quite possible that energy savings of this kind might be bought at the expense of restricting people's choice of housing locations, job opportunities, or access to shops, entertainments and services. There might equally be economic penalties for the efficiency of industry and commerce, associated with restrictions on work journeys and freight transport. In our studies we have made the effective assumption that, for energy conservation measures to be politically and economically acceptable, they would have to involve little or no such losses of social or economic benefits. (This assumption, for the circumstances of the '80s, might perhaps be challenged for the new climate - in both literal and ideological senses - of the '90s and beyond.) In the comparative evaluations made with the TRANUS model then, the estimated levels of energy use were set in each case against the levels of accessibility obtained by the population to places of work and services.

Comparing the existing pattern with the five new arrangements, in all cases overall energy use varied substantially, as would be expected, between the three scenarios, 'high', 'medium' and 'low'. It fell by 29% going from the 'high' to the 'medium' scenario, and by 48% going from the 'high' to the 'low' scenario. In all scenarios less energy was used overall in all the hypothetical configurations than in the existing pattern, and the rank order of the new patterns in this respect did not change between scenarios. However these fuel savings were achieved in every case at the expense of reduced accessibilities or increased costs, relative to the existing situation; suggesting that the evolved historical pattern of land uses and transport routes is actually rather 'efficient' in balancing accessibility and travel cost.

On the other hand two new patterns - option 1, the concentrated central city, and option 5, the village dispersal scheme - showed significant savings in transport fuel use ranging between 9% and 14% for option 1, depending on the scenario, with rather smaller savings in option 5. In both cases it appears that, by concentrating the previously scattered population of the rural hinterland, average journey lengths both to work and services have been shortened. This result is in broad agreement with the findings of a number of other studies, which have found reductions in energy use in transport produced by concentrating new development in small to medium-sized clusters (cf. Owens 1986, chapter 3).

Option 3, the pattern of satellite towns, is not however successful in reducing average journey lengths. This is because the rural population has been moved outwards to the periphery of the region; and although some jobs and services are relocated to the small new towns, there is still extensive commuting from these towns to the central city.

The advantages which it was imagined that the two patterns of high-density linear development, options 2 and 4, might have for encouraging public transport, were not born out by the analysis. The public transport in question was assumed to consist of bus services, but without special bus lanes. The concentration of car journeys along the main routes in option 2 resulted in traffic congestion, which in turn slowed the buses.

It seems at least possible that these results might be different for other public transport technologies, such as light rail or tram, running at higher speeds on separate tracks. Light rail might be suitable too for linking central and satellite towns in option 3 - as in Howard's 'garden city federation'. The TRANUS model is capable of representing the characteristics of these types of system; and we plan new experiments to evaluate such options.

In all five new patterns substantial savings in domestic energy use were predicted over the existing situation, as a direct consequence of higher housing densities for which it was assumed that dwelling sizes would be smaller on average, and rates of heat loss per unit floor area would also be lower.

4. ENERGY AND ALTERNATIVE URBAN FORMS

Meanwhile we have been carrying out other work, as mentioned, at the urban scale, using the same analytical tools and a similar methodology (Rickaby in press). As with the regional work, the research started from empirical analysis, in this case of twenty provincial English towns with populations between 50,000 and 150,000. An archetypal town form has been derived whose overall shape, as in the regional study, is an eight-armed 'starfish' following the major radial roads, with inner and outer ring roads and a pattern of streets and blocks whose topology and dimensions correspond to characteristics of the twenty actual towns (Figure 5).

Five policies for planning and control of new development over a twenty-year period have been compared (Figure 6). These do not involve any net change in population, but assume a decrease in average household size, and consequent increases in the number of households, number of jobs and total area of floorspace. In three 'containment' options, this new development is kept within the same town boundary, either by building up existing sub-centres (option 1), channelling new building into high-density corridors along main roads (option 2), or a combination of both policies (option 3). The remaining two patterns assume that expansion takes place on the periphery of the town, either at low densities between the arms of the star but inside the outer ring road (option 4), as is indeed happening now in many such towns in south-east England; or else into smaller areas of higher density (option 5), which might be suitable for the introduction of CHP or district heating.

The results were essentially negative ones. No significant variations in fuel use were found between the five patterns, either in transport or in domestic heating. (The possible independent savings for CHP or district heating schemes in option 5 were not taken into account.) There was very little shift in travel mode in any of the

Figure 5: Archetypal town form, based on 20 real English towns of around 100,000 population. Heavy lines show main roads, thin lines show secondary roads, broken line shows 'starfish' shape of urban boundary.

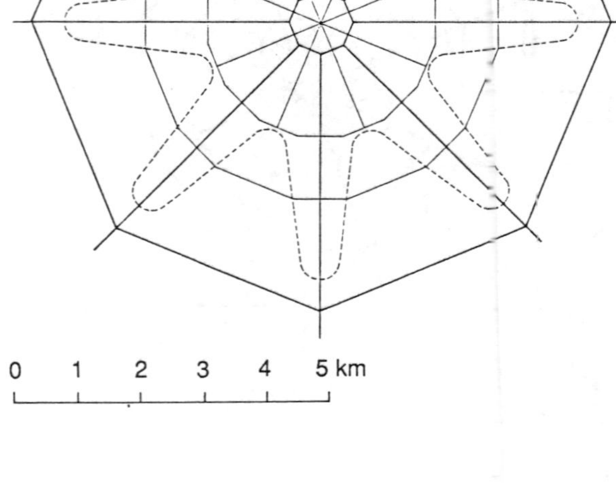

Figure 6: Five alternative urban forms, patterns 1-5 for comparison with the existing situation, Figure 5.

Pattern 1: development of existing sub-centres, with improved radial and orbital road links

Pattern 2: development of high-density corridors along main roads with improved capacity and better public transport services

Pattern 3: combination of Patterns 1 and 2

Pattern 4: low-density peripheral development outside urban boundary with new local centres and improved radial and orbital road links

Pattern 5: high-density peripheral development suitable for CHP/district heating, with improved radial and orbital road links

1

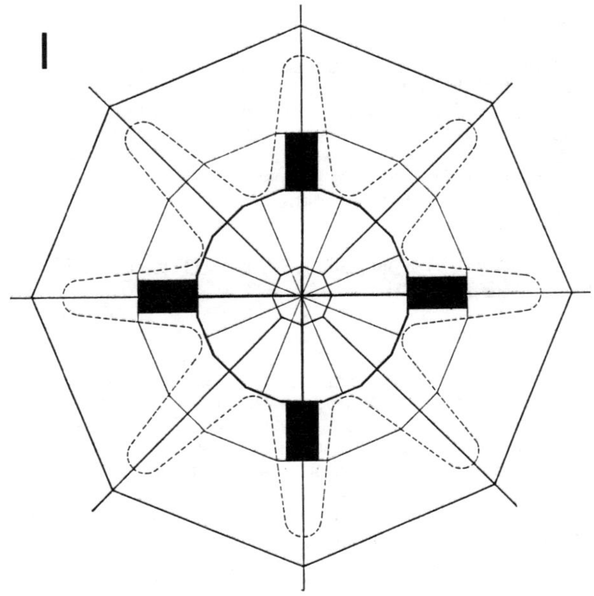

2

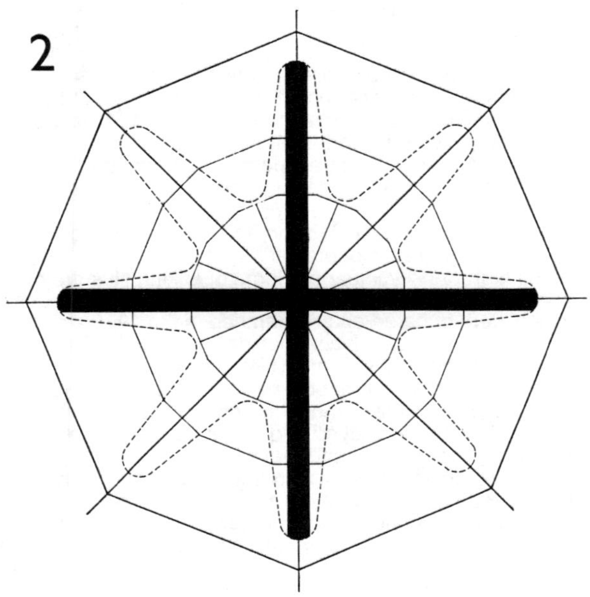

3

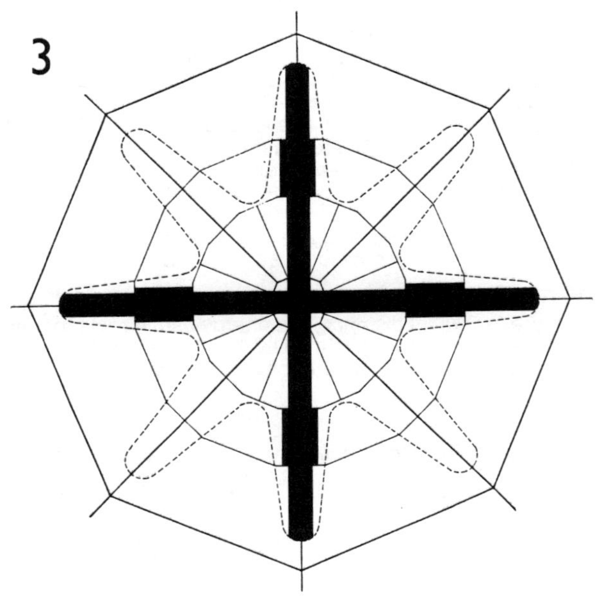

4

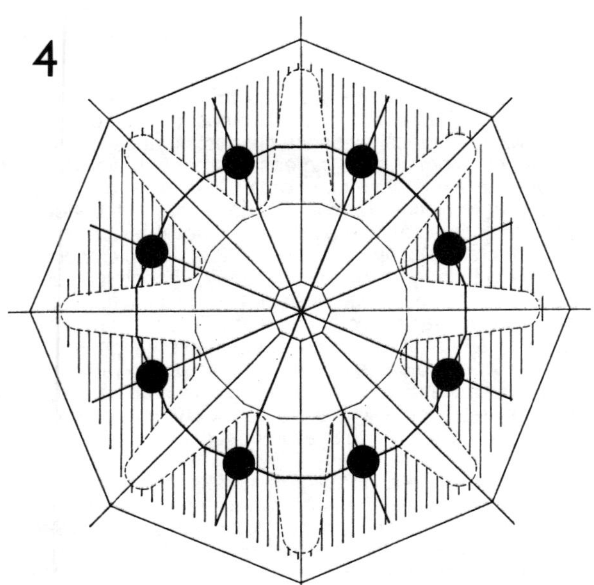

5

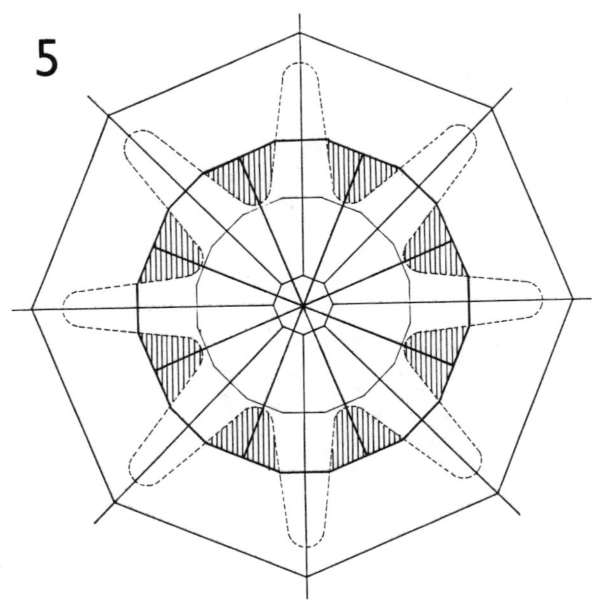

patterns, from private car to bus.

It should be remembered that these are medium-sized cities (100,000 population), not metropolitan regions. They are quite compact, with relatively short journeys to work, and yet do not experience serious problems of traffic congestion. In these circumstances it seems that improved bus services to the newly developed areas fail to tempt drivers away from their cars.

5. DISCUSSION AND NEW DIRECTIONS FOR RESEARCH

It is instructive to compare our results generally with some empirical data on transport energy use and average distances travelled in actual rural areas and towns of different size in Britain (Figure 7). These gross data obscure many of the complicating factors of course, but there is still a clear trend visible, in which both distance travelled and energy use decrease as the size of town increases, up to 'provincial conurbations'. The towns studied in our theoretical exercises would fall somewhere between these and 'large urban areas'. In this light it is perhaps not surprising that rearrangements of land uses within or at the edges of the urban boundary of towns of this size do not result in reduced fuel use in transport.

The trend is reversed for the one case of Greater London, where energy use in transport is higher even than in rural areas. There are several possible explanations for this, including higher levels of car ownership in London, and a greater range of opportunities for shopping, entertainment and other services. But another major factor is likely to be levels of traffic congestion and slow average speeds in the metropolis.

It would be unsafe to imagine that a policy of concentrating new development back into the centres of

---

Figure 7: Travel distance and energy use in transport by different types of area in the UK, 1972; from Owens (1986) p. 29 Fig 3.2, taken from data in Maltby et al. (1978)

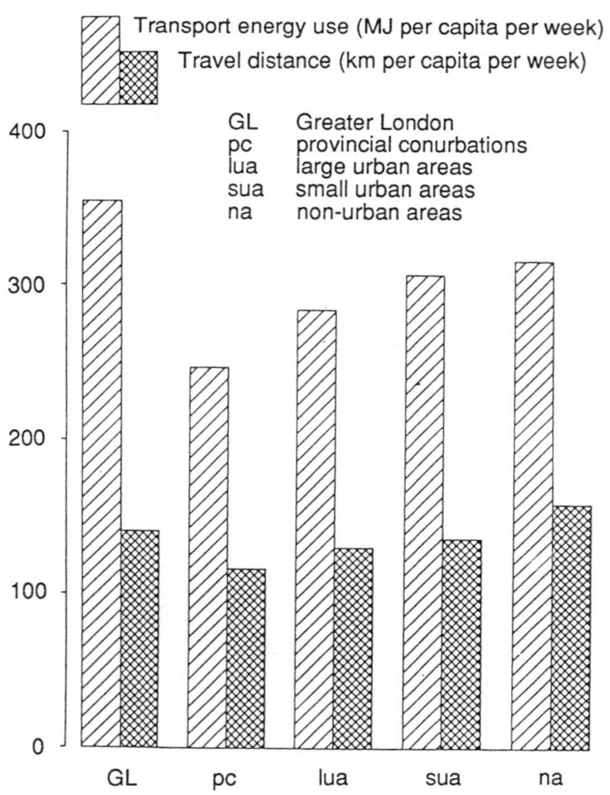

large metropolitan areas - even if such a thing were feasible in terms of available land - would result in any energy savings in transport. This is not just because existing problems of traffic congestion could be made worse; but because housing is already dispersed into low-density suburban areas, and much of the shopping and industrial floor-space has followed, into urban sub-centres or peripheral sites. The goal of creating 'medium-sized clusters' might thus be achieved rather by directing new development to these multiple sub-centres throughout the metropolitan area. It may be easier to pull <u>more</u> shopping and employment to the existing residential areas, than vice-versa. All this is however speculation, untested by experiment.

We have shown, then, how quite radical changes in patterns of new development within towns of around 100,000 population have little consequences for fuel use in transport; and how at the regional scale although conservation is possible, nevertheless there is little modal shift from private car to bus encouraged by new land-use patterns which might seem to be particulary favourable to public transport.

But our assumptions have been perhaps conservative ones, and we have not investigated scenarios in which current levels of accessibility are curtailed, large changes in life style are presumed, or the real prices of fuels are raised very sharply and abruptly. If the cost of petrol were to be increased substantially, say by the application of a 'CO_2 tax', or if petrol was to be rationed, presumably by governments acting together internationally, and if at the same time heavy investments were made to improve the quality, frequency and general economic attractiveness of public transport, then quite other possibilities might be on the agenda.

We should certainly be looking again, as suggested earlier, at light rail, monorail and tram systems running along high-density linear corridors or linking clusters of integrated land uses. We might reopen some questions which were debated and researched in the 1970s, as to the extent to which telecommunications could substitute for travel (Reid 1973, Elton 1974) - for example through 'tele-shopping' where goods are ordered by television link and delivered to the home; through increased 'home working' by self-employed and professional people, especially those in the information technology and financial industries; or through education in the home via broadcast media, as in my own institution, the Open University. (Historical evidence however suggests that the introduction of telecommunications in the past has, on the contrary, acted to stimulate <u>more</u> travelling - much as the computer revolution, so far from creating the 'paperless office', seems to have generated ever more paper. On the other hand, growth in telephone use has of course coincided with the era of low oil prices.)

But one should certainly not think that people will be easily prised away from their dependence on the motor car. Most empirical evidence suggests that people's response to petrol price increases is not to travel less, but to pay more and drive more carefully in the short term, and over the longer term to purchase cars with better fuel performance so as to maintain their previous journey patterns (see Dix and Goodwin, 1982). People become locked in to long-distance commuting, shopping by car, and so on, by major 'life decisions' such as taking a job, buying a house and entering into family commitments. It is possible that, if travel costs became <u>very</u> high, then people would relocate or change their jobs to reduce travel distances. This is psychologically plausible; but it is still only really guessing about behaviour which there have been few opportunities to observe in practice.

Current land use and transport models, such as the TRANUS model used in our own experiments, are in effect 'tuned' to high-mobility, motorised, suburbanised societies, and assume corresponding patterns of trip-making. They offer little behavioural explanation of the underlying human motives for making journeys, and thus are likely to have rather weak predictive power under radically changed conditions of fuel availability and transport choice. This does not mean that we should give up trying to make predictions. Meanwhile the new imperatives for conservation demand that action be taken now, to put in place policies based on our partial knowledge, but whose detailed consequences it will be frankly difficult to foresee. In the 1990s it seems that researchers in energy and planning may have trouble keeping up with events in the world around them.

REFERENCES

de la Barra, T. and Rickaby, P. (1982) 'Modelling regional energy use: a land use, transport and energy evaluation model', Environment and Planning B, 9, pp. 429-443

de la Barra, T. and Rickaby, P. (1987) 'A hierarchical land use and transport model for energy evaluation' in Hawkes, D., Owens, J., Rickaby, P. and Steadman, P., eds. Energy and Urban Built Form, Butterworths, London, pp. 5-28

Dix, M.C. and Goodwin, P.B. (1982) 'Petrol prices and car use: a synthesis of conflicting evidence', Transport Policy Decision Making 2, pp. 179-195

Elton, M. (1974) 'Developments in telecommunications: implications for planners' in Earp. J., Macdonald, M. and Williams, T.E.H., eds., Transportation and Environment: Policies, Plans and Practice, Southampton University Press

Howard, E. (1898) Tomorrow: A Peaceful Path to Real Reform, republished in 1945 as Garden Cities of Tomorrow, Faber and Faber, London

Leach, G., Lewis, C., Romig, F., van Buren, A. and Foley, G. (1979) A Low Energy Strategy for the United Kingdom, International Institute for Environment and Development/Science Reviews, London

March, L. (1967) 'Homes beyond the fringe', Journal of the Royal Institute of British Architects, August

Owens, S. (1986), Energy, Planning and Urban Form, Pion, London

Reid, A. (1973) Telecommunications - Transportation Substitution, Report DOT-TST-73-2, Department of Transportation, Office of Telecommunications, Washington DC 20590

Rickaby, P. (1981) 'Six regional settlement patterns: alternative configurations for energy-efficient settlements', Environment and Planning B, 8, pp. 191-212

Rickaby, P. (1985) Towards a Spatial Energy Model: A Theoretical Comparison of Accessibility and Energy Use in Regional Settlement Patterns, Centre for Configurational Studies, Open University, Milton Keynes

Rickaby, P. (1987) 'Six settlement patterns compared', Planning and Design: Environment and Planning B, 14, pp. 193-223

Rickaby, P. (in press) 'Energy and urban development in an archetypal English town'

Steadman, P. (1977) 'Energy and patterns of land use', Journal of Architectural Education 30, 3, pp. 62-67; reprinted in Watson, D., ed. (1979) Energy Conservation through Building Design, McGraw-Hill, New York

Steadman, P. (1980) 'Configurations of land uses, transport networks, and their relation to energy use', Foro Internacional de Asentiamentos Humanos, Mexico City, May 6-15; published in Vivienda 5, pp. 456-481

UN SYSTEME INTELLIGENT POUR LA GESTION DES REGLEMENTS D'URBANISME

Prof. A. DUPAGNE
LEMA-Université de Liège
15, av. des Tilleuls, B-4000 Liège Belgique

ABSTRACT The paper presents the main findings and conclusions of a research work who came to an end in september 1989. This research work was developed at the LEMA on behalf of the Wallonian Ministry of Urban Planning. It is mainly aimed at testing the possibility of applying new computer technologies on urban regulations in order to improve their efficiency and "desirability". It consists in producing a prototype of a knowledge based system that will be used in a test city for its Urban Regulation management. According to the scanty resources allocated to the research work, its scope has been restricted to a peculiar aspect of the regulation i.e. the specification of the maximum authorized built volume on a plot (outline profile). The main interest of this aspect is its strategic position between the architect, responsible for the creation of an "individual" building and the urban designer,responsible for the urban built form production. This point, crucial for the best transfer of information from both parts is peculiarly well adapted to the application of artificial intelligence technologies. And its complexity let us think that any help will be welcomed. What is more, the building form control is closely related to the architect power to produce energy conscious designs. Even if it is not taken into account in the scope of the present reseach work it will surely facilitate further developments in this direction.

1. INTRODUCTION

Il est maintenant admis que la création ou le maintien d'un paysage urbain cohérent est une impérieuse nécessité sociale, puisqu'il contribue largement au développement harmonieux des activités humaines qui y trouvent leur cadre naturel, particulièrement en Europe. Si l'on ajoute à ce fait que l'apparence d'une ville est une part déterminante de son image de marque qui, par ailleurs, conditionne son attractivité pour d'éventuels investisseurs, on comprendra l'importance d'une gestion ordonnée de l'espace urbain.

Si cette exigence n'est plus guère contestée aujourd'hui, elle n'est pas toujours recherchée avec la dernière énergie. Sa mise en oeuvre effective reste encore assez marginale, malgré quelques belles réalisations, faute de moyens appropriés et largement diffusables. Elle est cependant cruciale pour la Communauté Européenne qui concentre dans un nombre considérable de villes, une large part de sa population et de ses fonctions sociales les plus fondamentales.

Il est dès lors opportun de s'interroger sur la nécessité de mettre au point des outils efficaces permettant à la collectivité d'assurer le contrôle des transformations de son environnement bâti. La présente recherche se situe dans ce courant de développement et son objectif est, dans une modeste part de ce vaste domaine, de montrer aux autorités responsables comment l'emploi des technologies nouvelles, et plus particulièrement les découvertes récentes de l'informatique en intelligence artificielle, fournit une aide précieuse pour aborder efficacement certains aspects de ce difficile problème.

1.1 La maîtrise de l'environnement urbain

On peut penser qu'il y aurait quelque présomption à prétendre créer, dans un contexte démocratique, un paysage urbain cohérent qui ne suscite pas les objections habituelles de subjectivisme et d'autoritarisme administratif.

Il est vrai que, dans le passé, ce furent souvent les "régimes forts" qui se signalèrent à l'attention par de remarquables interventions sur l'environnement urbain, dont on doit bien admettre la qualité même si l'on en critique la finalité. Mais qu'on ne s'y trompe pas, il n'y a aucune fatalité qui lierait la qualité de l'environnement aux régimes autoritaires et les démocraties ne conduisent pas nécessairement à l'anarchie de l'espace urbain. La cohérence y est simplement plus complexe à organiser par la diversité des acteurs et les conflits de leurs intérets, la différentiation de leur rôles ainsi que des moyens dont ils disposent. Cette situation contraste évidemment avec la simplicité du "fait du prince". Quant à la ville, objet de ces transformations, quel que soit le régime politique, elle n'en est ni plus ni moins complexe.

Le paysage urbain est en fait le produit de l'imbrication ou plutôt du rapport dual existant entre les espaces urbains et les volumes construits. Chacun gardant son indépendance et sa logique propre, ils n'en sont pas moins liés par une réciprocité complète qui transfert à l'un toute action exercée sur l'autre.

L'ensemble des volumes construits, appelé parfois "forme urbaine", est le produit de toutes les interventions réalisées par les particuliers sur les bâtiments pris isolement et qui réalisent, tant bien que mal, les intentions des urbanistes. Ceci souligne évidemment l'importance du moyen de communication dont devrait disposer les urbanistes pour réaliser cette tâche.

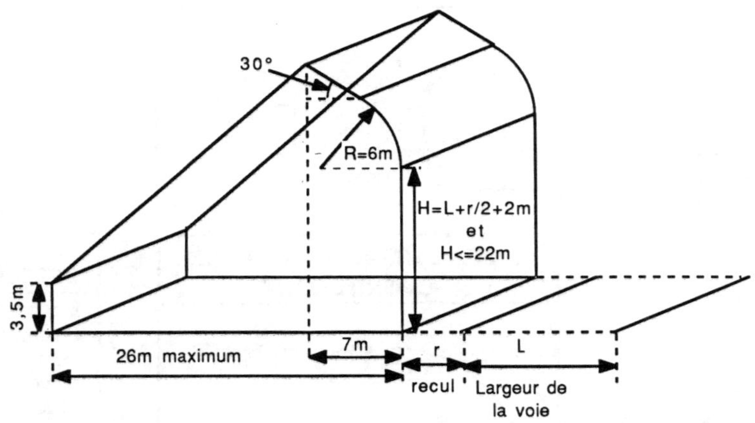

fig 1: éléments de connaisssance permettant la détermination du volume maximum sur une parcelle située à Liège

Parmi les outils réglementaires dont disposent les urbanistes pour tenter d'harmoniser les actions des divers intervenants, citons les plans communaux, généraux et particuliers, et les règlements de construction et d'urbanisme. Les deux premiers qui sont des instruments très puissants mais coûteux et lents à établir n'entrent pas dans notre propos. Nous nous limiterons volontairement à l'examen du troisième groupe c'est-à-dire l'aménagement urbain réglementaire.

1.2 L'aménagement urbain réglementaire.

Un règlement de construction ou d'urbanisme est un ensemble de règles qui s'appliquent à tout ou partie du territoire d'une commune. Ces règles sont regroupées en chapitres qui se rapportent à des domaines nombreux et variés, qui vont des prescriptions sanitaires à la réglementation des obstructions dans les rues, en passant par la détermination du volume maximum bâti autorisé en tout point de la ville. Pour la recherche décrite ici, nous considéreront le règlement de la ville de Liège comme exemple, et dans celui-ci, nous ne retiendrons que la partie relative à ce dernier point de vue. Nous chercherons simplement à fournir aux constructeurs un moyen de connaître sans ambiguité le gabarit applicable à la parcelle sur laquelle ils veulent construire.

1.3 Les éléments de la connaissance pris en considération par le prototype.

Les éléments de connaissance contenus dans le règlement de construction de la ville de Liège que nous avons retenus pour la réalisation de cette application, sont ceux qui permettent la détermination du volume maximum autorisé sur la parcelle considérée.
Cette information a été synthétisée dans le croquis présenté à la figure 1.

2. DESCRIPTION DU PROTOTYPE

Le système est représenté schématiquement à la figure 2. Il permet de localiser le noyau réellement développé dans le prototype parmi l'ensemble des modules du système complet qui seront développés ultérieurement.

2.1 Le module de gestion de la base de données

Ce module de gestion est principalement constitué d'un puissant outil de représentation et de gestion cartographique d'une ville. Pour des raisons évidentes de facilité nous utilisons à cet effet le logiciel CA.MUR [1] (Computer Aided Management for Urban Renewal) développé au LEMA.

2.2 L'interface vers le système :

Ce logiciel d'interface intervient lorsque une nouvelle session est lancée.
On peut le décomposer logiquement en 2 sous modules :

• un module d'accès à la base de données proprement dite,
• un module de génération des objets de la classe correspondante dans l'environnement symbolique.

2.3 Le système en programmation orientée objet

Une fois la session entamée, les données se trouvent alors en mémoire et peuvent être expoitées par le système. Dans l'état actuel du prototype, il n'est pas possible de modifier dynamiquement des données à la fois sur la base de données et sur le système. Ce problème est résolu lors de la réalisation des tests en relançant une session sur la base des nouvelles données.

2.4 Le prototype de consultation

Le développement complet de cette interface nécessiterait un investissement considérable en temps qui ne se justifie pas dans le cadre d'un prototype. Soulignons toutefois que la version exploitable de SEREBA comprendra une telle interface qui devra permettre aux personnes non spécialisées en informatique de se servir du logiciel complet. Nous avons préféré, dans un premier temps, donner la priorité au développement de l'interface en sortie, indispensable aux tests.

2.5 Le prototype de gestion des résultats

Trois sous modules interviennent lors de l'affichage des résultats du système :

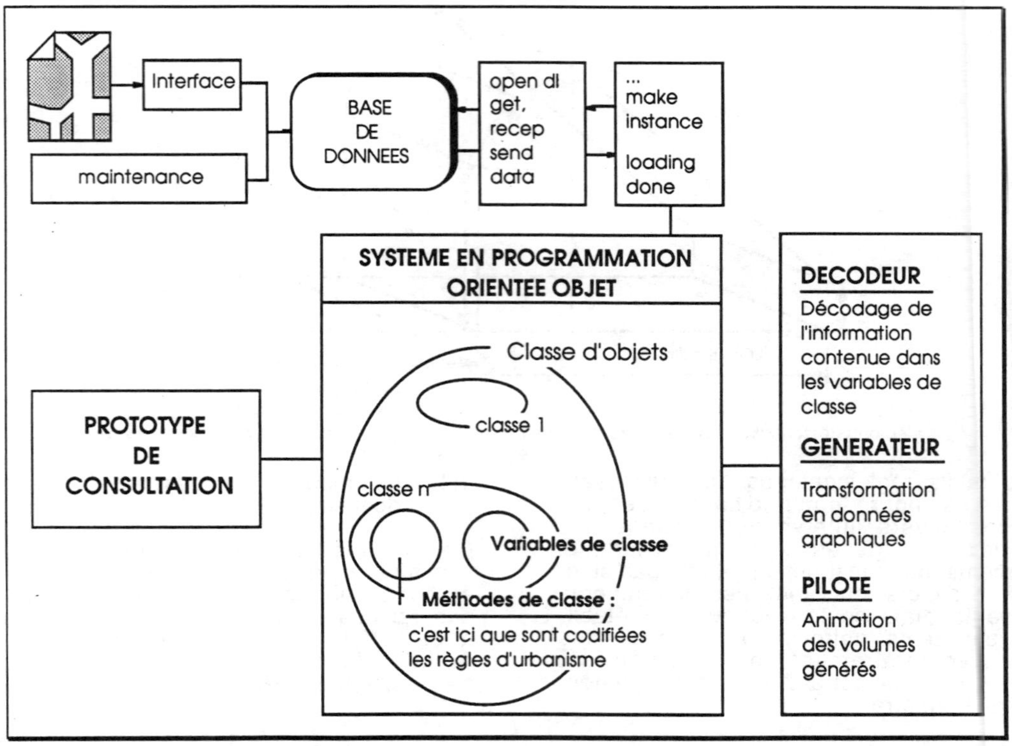

fig 2 : le prototype SEREBA

• le décodeur : Ce module permet le décodage des informations internes manipulées par le système. Cette phase préalable à l'affichage des résultats s'exécute au sein même du système. A ce stade, on récolte des informations contenues dans les variables de classe. Il s'agit des données 2D, concernant le plan de la ville (parcelles, clusters et cluster-limits) et 3D, concernant les divers gabarits formants le volume maximum). Les fonctions gérant ces tâches sont implémentées en LISP.

• le générateur : Il assure la transformation des données du monde symbolique vers le monde graphique, avec production d'un fichier ASCII correspondant au format exigé par le module pilote. C'est à dire contenant les informations nécessaires à l'affichage des résultats. Ils sont alors lisibles par les fonctions d'aquisitions de données de l'algorithme du peintre. Ce module est également implémenté en LISP.

• le module pilote : Ce module réalise la dernière étape du travail. Elle comprend d'abord le transfert des données à partir des fichiers ASCII générés, vers la structure en mémoire centrale de l'algorithme du peintre. Elle permet ensuite la présentation des résultats proprement dite, via une interface - utilisateur "active". Les implémentations du transfert comme de la présentation ont été effectués en C , avec l'utilisation du gestionnaire de fenêtres X-Windows pour la gestion de l'interface-utilisateur.

3 L'INTERFACE-UTILISATEUR

La figure 3 présente une première version simplifiée de l'interface-utilisateur construite pour les besoins de tests du prototype. Elle ne comprend évidemment qu'une partie seulement des possibilités envisageables pour le futur système.

La fenêtre principale de l'interface représentée à la figure 3 se divise en cinq parties. La première ne comporte que des boutons se rapportant aux relations avec le système d'exploitation (exemple: quit). Les quatres autres sont dédiées à l'affichage et aux manipulations de scènes 2D ou 3D.

Ces quatres scènes sont contrôlées par l'algorithme du peintre, ce qui sous-entend que l'interface-utilisateur gère quatre structures de données distinctes en mémoire centrale. On qualifie d'actif cette interface parce l'utilisateur a la possibilité d'effectuer - grâce à l'algorithme du peintre - des rotations selon les trois axes, un choix entre le mode filaire ou le mode "vus et cachés", des mises à échelle, des simulations de variations de "distances focales" ou des combinaisons de ces manipulations sur les scènes affichées.

La sous-fenêtre supérieure gauche montre le plan de la ville ou du quartier considéré avec ou sans les parcelles. Celle de droite permet de voir (en "tout vu" sur l'exemple) un gabarit généré par le système. Ces deux panneaux sont effectivement en connection avec les résultats du système, tandis que les deux panneaux inférieurs simulent ce qui sera possible d'obtenir avec, à droite, le volume maximal intégré dans son environement direct et, à gauche, une coupe (ou un développement) du gabarit représenté juste au-dessus.

Tous les boutons n'ont pas encore reçu leurs fonctions associées : ceci vient de ce que le prototype sera, dans le futur, amené à évoluer et, il fallait donc permettre l'ajout de nouvelles spécifications fonctionnelles sans modifications du design de l'interface vers l'utilisateur.

4. CONCLUSIONS

Le développement du prototype décrit dans ce papier voulait vérifier la possibilité de gérer une règlementation d'urbanisme à l'aide de techniques qui relèvent de l'intelligence artificielle. A l'évidence cette "faisabilité" est démontrée puisque la plupart des modules produisent les résultats attendus dans les cas de figure les plus fréquents. La gestion des cas particuliers ou dégénérés dont la présence n'est pas négligeable dans le contexte des villes existantes devra bien entendu être assurée avant toute application pratique d'un tel système. Il faut toutefois rappeler que la plupart des cas non couverts ne le sont pas non plus par la règlementation sous sa forme actuelle qui demande aussi l'intervention directe de l'expertise humaine.

A la fin de cette recherche, ce qui frappe le plus l'observateur c'est le contraste existant entre une réglementation formulée en termes rigides et la grande diversité des situations réelles exprimées dans la base de données urbaines. Ceci conduit bien évidemment à la gestion d'un très grand nombre de cas particuliers demandant l'intervention directe du responsable administratif ou la non-application du règlement. Il est impossible, lors d'une appication manuelle du règlement, de prendre en compte l'extrême richesse de la réalité urbaine et il est normal que ce règlement reste formulé en termes très simples. Il est toutefois certain que fixer un plafond maximum des constructions ou lier une hauteur de façade à la seule largeur de la voirie adjacente est trop simple, puisque cela ne fait pas intervenir les multiples relations que ce volume autorisé a nécessairement avec les nombreux objets qui constituent son environnement. Par exemple sa liaison visuelle avec des places publiques qui ne lui sont pas nécessairement immédiatement adjacentes, ou encore la projection d'ombres d'un immeuble tour sur son environnement, sont difficilement maîtrisables par un expert humain, même experimenté.

Soulignons que traiter le problème de la réglementation à l'aide des techniques de l'intelligence artificielle a précisément comme intéret de fournir une représentation très riche et bien structurée des objets manipulés. De plus, dans cet environnement de développement, la définition des objets se fait pratiquement de manière interactive, ce qui permet une évolution du système très rapide et aussi fréquente que nécessaire. Dès lors, le développement d'une nouvelle règlementation plus riche et surtout plus réaliste est possible, mais elle demandera une implication plus opérationnelle des experts urbanistes. Elle trouvera dans les techniques de l'intelligence artificielle un cadre approprié et stimulant.

5. BIBLIOGRAPHIE

1. A. DUPAGNE; A. LORQUET
   CEC-CAM.UR; Task Nb. 5 : Data Base Specifications; final report; june 1989.

2. D. PIROTTE
   Realisation d'un système basé sur de la connaissance appliqué aux règlements des constructions de la ville de Liège.
   Rapport final; 28 septembre 1989.

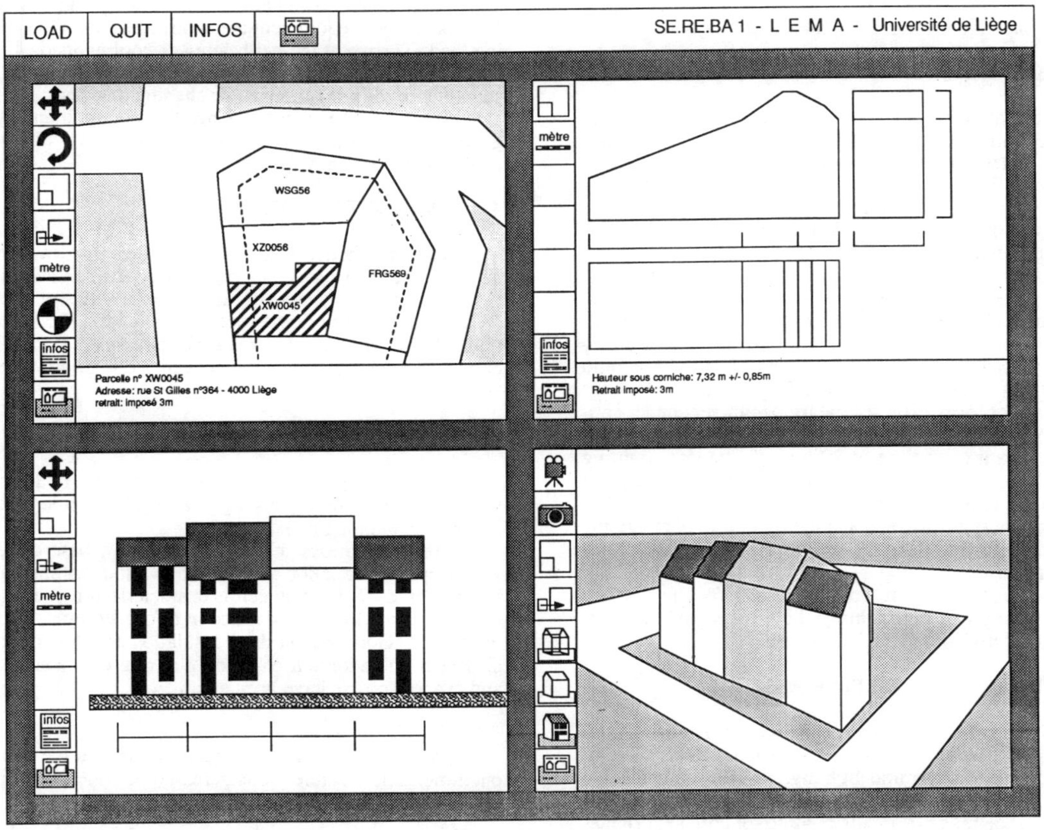

fig 3 : l'interface utilisateur

PARAMETRIC ENERGY MODELLING FOR URBAN SITUATIONS
Detailed Analysis

Koen Steemers, Nick Baker and Dean Hawkes
The Martin Centre for Architectural and Urban Studies
University of Cambridge Department of Architecture
6 Chaucer Road, Cambridge, UK. Tel.: 0223 332981

1 INTRODUCTION

The overall framework of this study is to assess and compare energy implications of alternative approaches to the redevelopment of typical central urban areas in the UK. The specific purpose of this paper is to present an approach to computer analysis which looks at the wide spread of parameters presented by the urban context, in order to assess their significance and to inform urban design proposals. This paper should be read as an introduction to the paper entitled *Energy Use and Urban Development: Urban Block Analysis* by Steemers, Baker and Hawkes (1) in these proceedings, which looks specifically at generic urban forms and their energy uses.

Energy modelling needs to integrate all relevant parameters which may influence the total energy balance, so that the effect of one is shown to have repercussions on others. For example, in the non-domestic sector, and for office buildings in particular, the potential energy advantages of solar shading devices in reducing cooling loads, will be offset by reduced availability of natural light, which in turn incurs an energy penalty. In the urban context there are many such interrelated factors which need to be assessed. In the initial series of parametric studies the effects on energy demand of *urban* variables (horizon angle, orientation, street width and external surface reflectance) have been compared with those of conventional *building* variables (glazing ratio, room depth, internal reflectances, insulation standards and plant efficiency). This is followed by a more in-depth analysis of the interrelationship between the main architectural variables and energy use.

2 METHODOLOGY

An integrated energy model, based on accepted and proven algorithms, has been developed to test the various parameters. Attention has been concentrated on those factors which are specifically related to the urban situation, such as: obstruction to daylight (street width, building heights), obstruction to solar radiation (orientation, overshading), building geometry (plan depth, configuration) and facade design (glazing ratio).

Other parameters which relate to the building fabric and the environmental systems are studied to evaluate their potential influence on urban form. For example, a well insulated building may perform better when the surface to volume ratio is high, whereas a badly insulated building could benefit from having a compact form. The secondary parameters under consideration include the following: insulation standards, systems efficiencies (heating, cooling and lighting), internal lighting datum, ventilation rates, internal and external reflectances, and casual gains (other than lighting or solar).

Primary energy is used as a common unit to compare the effects of these factors in terms of either heating, cooling or lighting energy use.

This paper presents preliminary results for the UK climate, categorising and highlighting the most significant factors and their relative degrees of influence.

3 DESCRIPTION OF THE PROGRAM

The program used, referred to as the Lighting and Thermal (LT) Program, integrates the thermal effects of solar gains and the value of natural lighting. It is a mathematical model used to predict annual primary energy consumption as a function of a wide range of inputs relating to: climate, orientation, building form, urban situation, glazing ratio and type, building fabric and systems efficiencies.

The model evaluates the energy use of individual 'cells', or rooms, allowing the representation of all the variations which make up a complete building in its urban context. The original program, before the inclusion of the urban parameters as inputs, was developed by Baker and Hoch (2).

3.1 Heat Losses:

For each cell, monthly heat losses conducted through the fabric and by ventilation are calculated using mean monthly temperatures, a thermal reference point and a correction factor for intermittent heating, similar to the degree-day method.

3.2 Solar Gains:

Solar gains are assessed based on orientation, shading coefficients, glazing area and a utilization factor for non-domestic buildings (van Dijk and Arkesteijn, 1987) (3).

3.3 Daylighting and Heat Gains:

The daylight factor is calculated using established equations (Hopkinson *et al*, 1966) (4) with an empirically derived formula for the externally reflected component to take account of urban geometry and reflectances. Using the daylight factor, an internal lighting datum and hourly diffuse sky illuminance on the facade (Littlefair, 1984) (5), the monthly electrical consumption and the associated heat gain is calculated.

3.4 Primary Energy Consumption:

Solar gains, heat gains from lighting and casual gains from people and equipment are subtracted from the gross heating load to arrive at the net heating or cooling load. If there is a cooling load, direct solar irradiation is eliminated, corresponding to the use of moveable shading devices.

To reduce the energy balance to a common unit, all energy use is converted to primary energy use by applying the appropriate efficiency factors.

4 PARAMETRIC STUDIES

The parametric analysis is divided into two sections. The first, termed *elemental parametric assessment*, is to test individually the relevant significance of a series of parameters on a standardised notional room or test case. The second section, termed *integrated parametric assessment*, looks specifically at the interrelationship between the main urban parameters.

5 ELEMENTAL PARAMETRIC ASSESSMENT

A series of parametric studies were carried out, using LT, for a theoretical room, varying a range of characteristics relating to both building and context.

For the purpose of standardisation, a base or test case was adopted, in relation to which the various parameters could be compared. This basic cell is characterised by being a south facing second floor office space in a four storey urban context in the UK. Good standards of insulation and systems efficiencies, with the adoption of shading during overheated periods, and office type use have been assumed.

5.1 The Variables:

One characteristic was altered relative to the test case at one time. The variables considered include the following:

1 glazing ratio (% glazing of facade)
2 orientation (S, SW, W, NW, N, NE, E and SE)
3 obstruction height (as described by the angle from the centre of a window to the *equivalent obstruction line*)
4 room depth
5 external reflectances
6 internal reflectances
7 insulation standards (poor, standard or improved)
8 datum internal lighting level

The first four of these characteristics relate directly to the built form, and are of particular relevance to the architect. The others could be labelled as secondary parameters related to detailing or engineering questions. It is important to point out that there is a strong relationship between the two sets of inputs. For example, if a particular room requires a greater amount of light, the increase in datum light level will inevitably increase energy use, however, increasing the glazing ratio may reduce this. Similarly, insulation and glazing strategies may determine the most appropriate building form and orientation.

5.2 Glazing Ratio:

Important to the architectural implications on the design of facades, and of particular interest to energy conscious designers, is the optimum glazing ratio to balance heat losses, lighting requirements and cooling loads. The graph below indicates that, for the test case, total primary energy varies by about 20% from the standard, when changing the glazing ratio. The curve does however indicate a optimum range for the glazing ratio between 40 and 60%. For reduced obstructions, the effect of altering the glazing will inevitably be greater, and the minimum will lie at a different point.

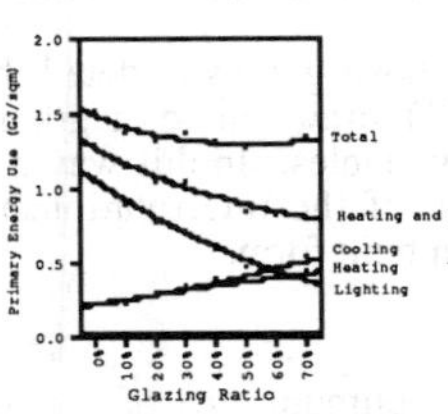

Figure 1: Energy use as a function of glazing ratio.

5.2.2 Orientation:

The graph indicating the effect of orientation on energy use shows that orientation is of little significance, with about 10% variation.

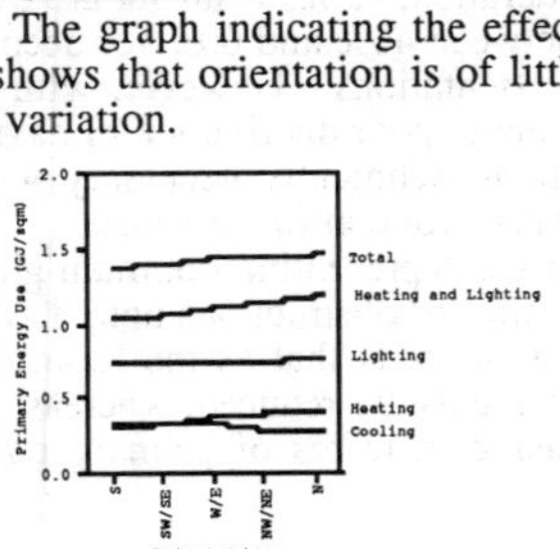

Figure 2: Energy use as a function of orientation.

5.2.3 Obstructions:

The degree of obstruction, as expressed by the urban horizon angle (UHA), indicates the effect of the level of a room under consideration in relation to the surrounding urban morphology. This could be an important design parameter when determining the proportions of buildings, streets and squares.

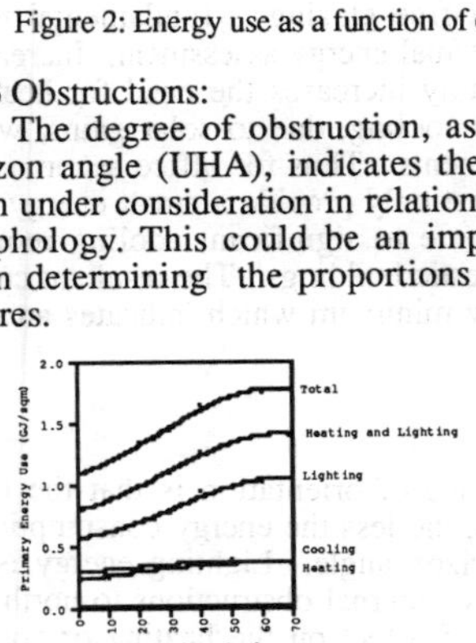

Figure 3: Energy use as a function of obstruction angle.

Results are as expected, in that the lighting load increases the greater the obstruction. The curve is particularly steep for the most common range of UHA's (10° to 40°). Heating and cooling loads increase only slightly, whereas lighting costs may increase by as much as 30%. It is important to mention that, although there is a rise in the energy use as the UHA increases, on average, when all the floors of a building are averaged, results will be less dramatic. This does however highlight that when considering individual floors, the UHA will be important.

5.2.4 Room Depth:

This parameter is of interest with respect to built form as it is related to building depth. Generally the results indicate an increase in energy use for deeper plan depths. Lighting loads increase by 40% in this particular case, but heating demand is reduced. Cooling also generally reduces until gains from light sources begin to play a important role, resulting in a minimum cooling load for a 7.5 m deep plan. Overall the optimum room depth for the test case is about 5 m, with the total energy increasing by about 25% for deep plans.

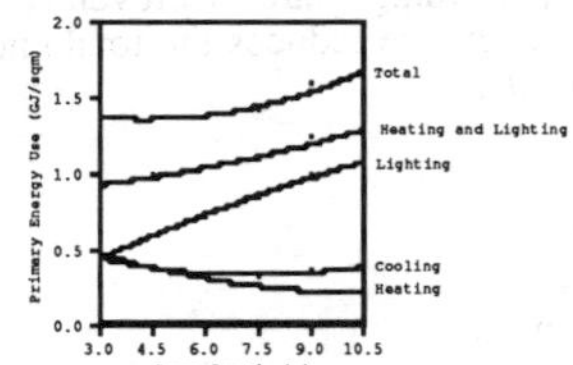

Figure 4: Energy use as a function of room depth.

5.2.5 External Reflectances:

If the external reflectances are under the architect's control, as in courts and atria, then by increasing the reflectances from an average of 0.2 (brickwork) to 0.7 (whitewash), energy use for the test case can be reduced by about 20%. This is mainly as a result of the contribution to natural lighting of the externally reflected component. In highly obstructed situations the savings can be significant.

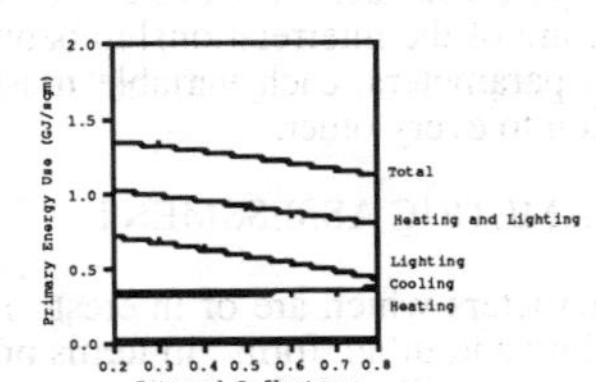

Figure 5: Energy use as a function of external reflectances.

5.2.6 Internal Reflectances:

Generally more under the designer's control, the average internal reflectance reduces lighting requirements, being particularly significant for values over 0.5 in the test case.

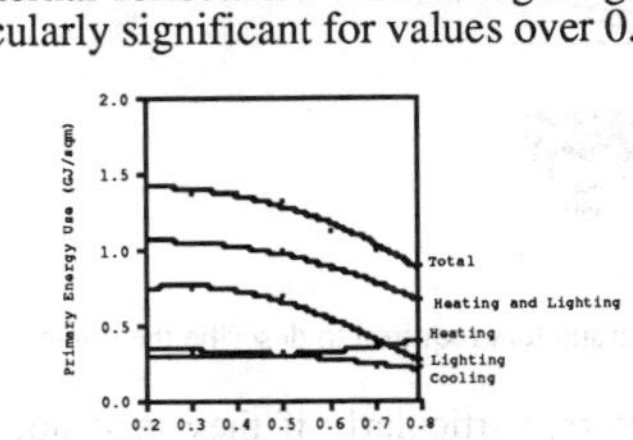

Figure 6: Energy use as a function of internal reflectances.

5.2.7 Insulation Standards:

Three standards of insulation were examined: The first assumes current UK building regulation U-values (0.6 walls, 5.7 single glazing), the second is to the new regulations (0.45 walls, 2.85 double glazing) and the third adopts high standards (0.3 walls, 1.7 triple glazing). The results for the test case cell indicate that although heating requirements drop as insulation increases, lighting energy increases due to the reduced transmittance of glazing. The result is that energy use is reduced by approximately 15% due to improved insulation standards.

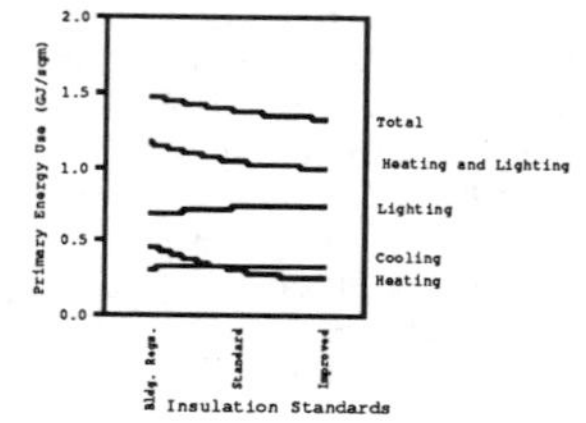

Figure 7: Energy use as a function of insulation standards.

5.2.8 Internal Datum Light Level:

As seen from the results for thermal insulation, the lighting requirements for office type accommodation is at least as important as the heating energy use, as is reinforced by the graph below.

The required lux level in a building depends on the activities and accepted standards. The recommended level for offices has been 500 lux, but a minimum of 300 lux is now often considered adequate (assuming glare is prevented and distribution is even). This difference reduces the total energy use for the test case by over 30%.

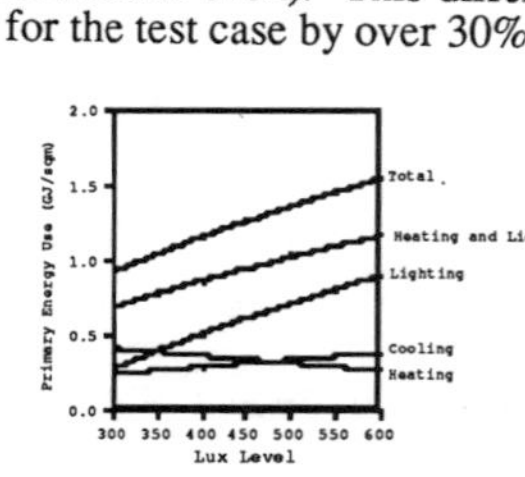

Figure 8: Energy use as a function of datum light level.

5.2.9 Interim Conclusion:

The results seem to indicate that factors which have a direct bearing on urban and built form may be as significant in terms of primary energy consumption as parameters related to building fabric or systems performance. However, to be able to gain a more realistic picture of the interrelationship between specifically the built form parameters, each variable must be allowed to change in relation to every other.

6 INTEGRATED PARAMETRIC ASSESSMENT

The main design parameters which are of interest to the architect relate to the building and urban form. In terms of the inputs for the LT Program, these will include building or room depth, glazing ratio, orientation and angle of obstruction from the centre of the window. The obstruction is simplified to a horizontal line, or the 'equivalent obstruction line', and the angle has been termed the urban horizon angle or UHA.

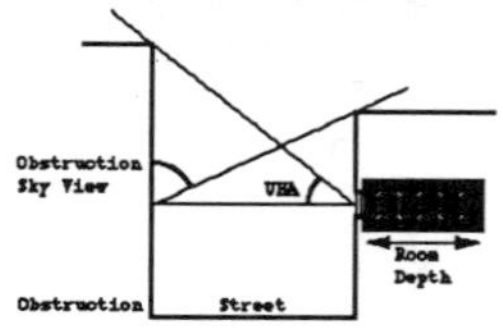

Figure 9: Basic urban parameters required to describe the context.

Other urban parameters, particularly if they tend not to be within the architect's control, such as the reflectances of the various external surfaces, are standardised. Secondary and more detailed variables are also standardised and represent a typical UK office building type designed to the new Building Regulations.

Each of the main design parameters was varied in relation to the others within reasonable ranges defined as follows:

- room depth:3.0 m, 4.5 m, 6.0 m, 7.5 m and 9.0 m.
- glazing ratio: 10, 20, 30, 40, 50, 60 and 70 % of free glass.
- orientation: due south, north, west or east.
- UHA: 10, 20, 30, 40, 50, 60 and 70 degrees.

6.1 RESULTS

The results, for the UK climate (52 degrees N), are expressed in terms of the primary energy consumption for heating, lighting, sub-total, cooling and total. For each orientation there are two sets of graphs, one for heating and lighting energy use and the other for cooling and the total energy use. Every set of five graphs shows results for the five variations in room depth, and each graph plots the glazing ratio against the primary energy use for the range of urban horizon angles.

For example the two graphs below indicate results for a south facing room 6 m deep.

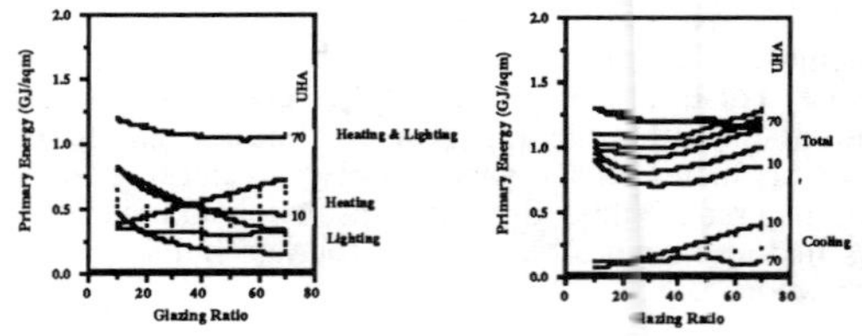

Figure 10: Typical results for a south facing room 6 m deep.

The area between the curves indicates the range of urban horizon angles from 10 degrees to 70 degrees. What is immediately noticeable from these graphs is the significant effect that obstructions can have on the energy balance. An increase in the UHA from 10 degrees to 70 degrees can more than double the primary energy use in buildings which use no cooling, although cooling load does decrease, making the total difference about 75%. It is also interesting to notice the presence of slight minima in the curves, indicating the potential optimisation of the glazing ratio.

Having looked at the two above graphs in detail, below are a full set of results (Figure 11), drawn more diagrammatically but representing the same variables. In this way a complete picture can be established of the interrelationship of those parameters relative to urban built form.

6.1.1 Room Depth:

This factor determines the optimum area of window to reduce primary energy consumption. Generally, and as might be expected, an increase in the room depth requires an increase in glazing ratio. The optimum amount of free glass for the particular case under consideration, varies from about 20% for shallow plans (3.0 m), to between 40% and 60% for deep plan buildings (9.0 m), for all orientations. However, with such deep plans it is likely that due to poor distribution of daylight, higher levels of lighting than technically necessary will be required at the back of the space for comfort reasons.

The minima in the curves representing optimum glazing ratios are particularly emphatic for obstructions upto about 35 degrees. The mechanism at work is that as the room depth increases the heating requirements are reduced, whereas lighting demands, more expensive in terms of primary energy, increase.

6.1.2 Glazing Ratio:

The relative area of free glazing is fundamental to an integrated lighting and thermal energy assessment. Increasing the glazing ratio predictably increases the need for heating, due to heat losses, and cooling, due to solar gains, whilst reducing lighting requirements. Thus for office accommodation, solar gains can at best only provide a small energy saving, and at worst contribute to significant cooling requirements, in contrast to domestic buildings. The resultant curves generally dip to a shallow minimum which indicates an optimum glazing ratio.

6.1.3 Orientation:

The main consequence of orientation is that the more northerly the facade faces, the less the energy consumption is affected by the urban horizon angle. Lighting energy is the exception. In other words, external obstructions to north elevations have only a limited affect on the heating or cooling loads, but do increase lighting loads. The more southerly the orientation, the wider the spread of results due to obstructions.

6.1.4 Urban Horizon Angle:

This is a basic factor defining the urban context, combining street or space width and relative obstructing building height. It also proves to be an important parameter in terms of assessing the primary energy use of urban buildings. External obstructions have the effect of increasing lighting and heating requirements, whilst reducing cooling loads. The effects tend to be greater for large glazing ratios and shallow plans, but are less significant the more northerly the orientation. This suggests that the uses accommodated and the associated lighting datum should relate to floor level and orientation (e.g. drawing offices should be on north facing top floors). The total of heating and lighting loads can vary as much as from 0.4 GJ/sqm (for 10 degrees UHA) to 1.2 GJ/sqm (for 70 degrees UHA) for south orientations. This factor of three difference represents a dramatic change in the angle of obstruction, although this not unreal when comparing ground floor conditions in a city centre with a top floor, but indicates the degree of importance that the urban horizon angle plays. It is important to note that when considering the effect of obstructions on the primary energy use for a specific floor of a building the UHA is of greater significance than when considering a complete urban block.

7 CONCLUSIONS

These preliminary results indicate the importance of urban parameters, and that for the estimation of energy consumption they must be taken into consideration. The factors dealt with in this paper are the most significant ones in terms of both primary energy use and for their potential implication on urban design.

In general terms, orientation seems to be a less important consideration than glazing strategies due to the effect of the urban context presenting obstructions. The graphs show that for the best results, glazing ratios should vary according to relative obstruction heights. In architectural terms this implies that facade design in terms of window distribution should progressively change from floor to floor, optimising the amount of glazing at each level.

Figure 11: Full set of results from the integrated parametric analysis.

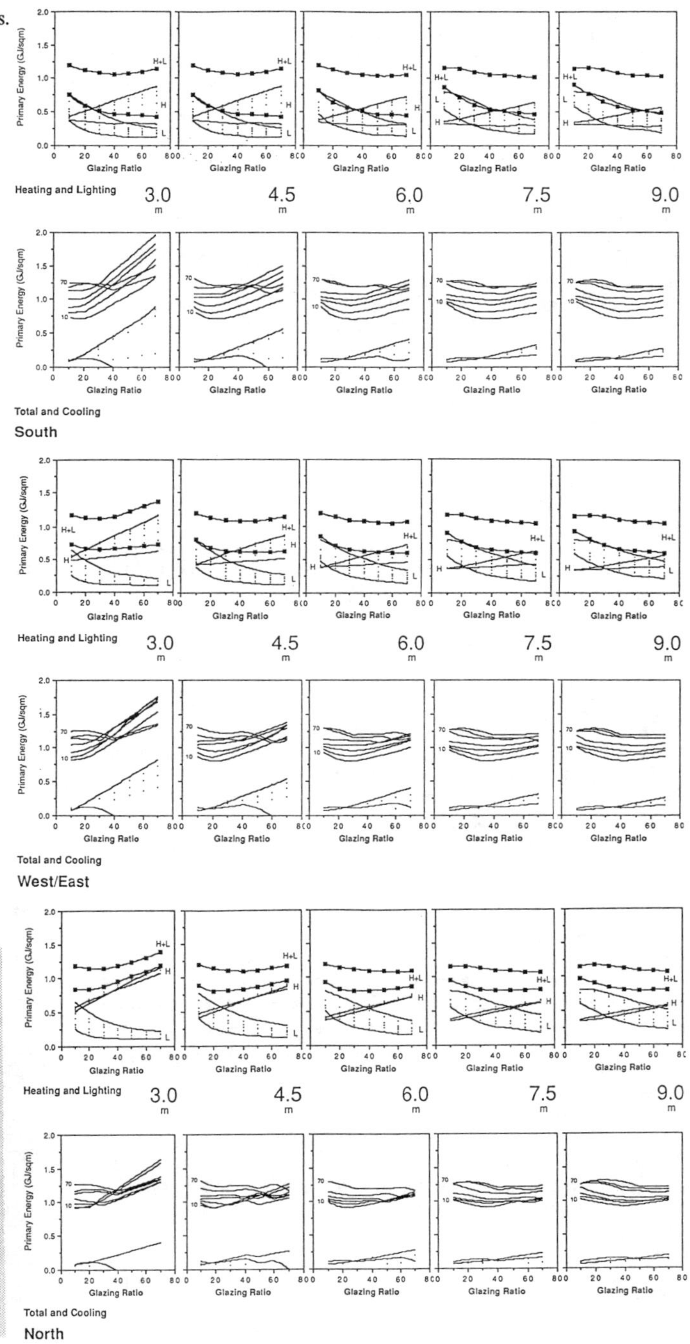

REFERENCES

(1) Steemers, K., Baker, N. and Hawkes, D., *Energy Use and Urban Development: Urban Block Analysis*, The Proceedings of the Second European Conference on Architecture, Paris, 1989.

(2) Baker, N. and Hoch, D., *Experience with a design tool which combines the energy value of daylight and the hermal value of solar gain*. Daylighting Buildings Conf. Proc., UK-ISES, London, 1989.

(3) van Dijk, H.A.L. and Arkesteijn, C.A.M., *Windows and Space Heating Requirements; parameter studies leading to simplified calculation method*. The Netherlands National Report on activities within Step 5, Delft, TNO Inst. of Applied Physics, 1987.

(4) Hopkinson, R.G., Petherbridge, P. and Longmore, J., *Daylighting*. Heinemann, London, 1966.

(5) Littlefair, P.J., *A new method for predicting energy saving from on/off controls*. BRE Research Information Paper 14/84. BRE, Garston, UK, 1984.

SOLAR POTENTIAL OF REGULAR GRIDIRON URBAN ENVIRONMENTS IN WESTERN ARGENTINA

M.A. Atencio, C. de Rosa, A. Esteves, M. Basso and J.L. Cortegoso
Laboratorio de Ambiente Humano y Vivienda (LAHV)
Centro Regional de Investigaciones Científicas y Tecnológicas (CRICYT)
Casilla de Correo 131 5500 Mendoza Argentina South America
Phone: 061 - 241797 FAX 5461 - 241020 Telex 55438 CYTME AR

ABSTRACT. The gridiron plan is the most distinctive feature of Latin American cities of Spanish origin. In regions with temperate-cold winters and a generous solar resource, such as the Andean provinces of Argentina; a significant contribution to future energy budgets can be achieved by the massive implementation of greater conservation measures and passive solar utilization in urban buildings. The assessment of this potential, the main goal of the study, would provide valuable data for urban and energy planning. The city of Mendoza was chosen as base case. Conservation and solar access computational programs were specifically written. 37 sample city blocks were surveyed, processed and analyzed. The results obtained were correlated with several parameters descriptive of the urban form. Solar Saving Fractions (SSF) ranging from 34.52 to 51.56% are obtainable at the upper and lower construction densities by the massive retrofitting of the existing building stock. Preliminary conclusions indicate a high potential of total energy savings, with mean values of 43.5 KWh/m^3 year. The study is being continued in order to extend the validity of correlations by considering a much larger sample.

1. INTRODUCTION

The "Laws of Indies" dictated under Philip II of Spain in 1573 were perhaps the first urban code of modern times. Most Spanish founded colonial cities in Latin America were laid following their rigid prescriptions. In many cases, new settlements established prior to their coming into force, anticipated the "Laws" simple, straighforward approach to city planning. The "gridiron" plan was their most distinctive feature and possibly the only one which has prevailed into the present. A regular matrix of square city blocks was determined by a street system, running generally towards the cardinal points at a distance of approximately 130 m on centers. With time, the foundational cores extended following the same pattern and constitute nowadays their predominant typical urban tissue. After four centuries of urban development these city environments present different stages of densification from extremely dense city centers to low rise residential neighbourhoods.

In regions with temperate and cold winters, (1300 to 2600 heating DD°C) and a generous solar resource (16 to 20 MJ/m^2 day) such as those of the Andean provinces of Argentina, the potential utilization of passive solar in conjunction with added conservation measures in urban buildings, could make a significant contribution to regional and national energy needs in the near future. On a national basis, the domestic sector consumes a 28.8% of all primary energy; besides, the population distribution in the country is predominantly urban, 75% and growing.

Any serious attempt to develop this potential involves the implementation of land use policies which result from urban and energy planning processes at the regional and national levels. An accurate assessment of the potential, the primary goal of this study, must take into account the possibilities of solar access as conditioned by several parameters descriptive of the present urban form and its predictable development.

2. LOCATION OF THE STUDY

The city of Mendoza, founded in 1561 on the temperate arid lands of central-western Argentina has been chosen as a base-case for the study. Its location and relevant data is as follows:

. Population: 600000
. Geographical coordinates: latitude 32.88 S, longitude 68.85 N, altitude 827 m a.s.l.
. Climate: Dry mesothermal with mild sunny winters and warm moderately dry summers.
. Degree days (C) at airport, regardless of the urban heat island effect: heating (base 18°C) 1384, cooling (base 23°C) 163.
. Hours of the year (%): in confort 21.53, heating needed 70.14, cooling needed 8.33.
. Solar global radiation (MJ/m^2 day): 18.06 (yearly mean).

The city features a gridiron layout growing around an extremely regular core, relaid after the 1861 earthquake, with square city blocks of 100 m to each side and 20 m wide streets, with clockwise deviations of 12.5° from the cardinal points (Figure 1). The building massing varies greatly in shape and size according to density. In denser downtown areas tall buildings up to twelve stories in height predominate. Taller structures are avoided due to the high seismicity of the region (Figure 2a). Low density residential neighborhoods feature constructions one or two stories high with generous open spaces at the blocks cores (Figure 2b)

3. METHODOLOGY

3.1 MODELLING URBAN ENVIRONMENTS

The process aimed at determining descriptive models

Figure 1: Typical regular gridiron urban tissue in the downtown area of the city of Mendoza.

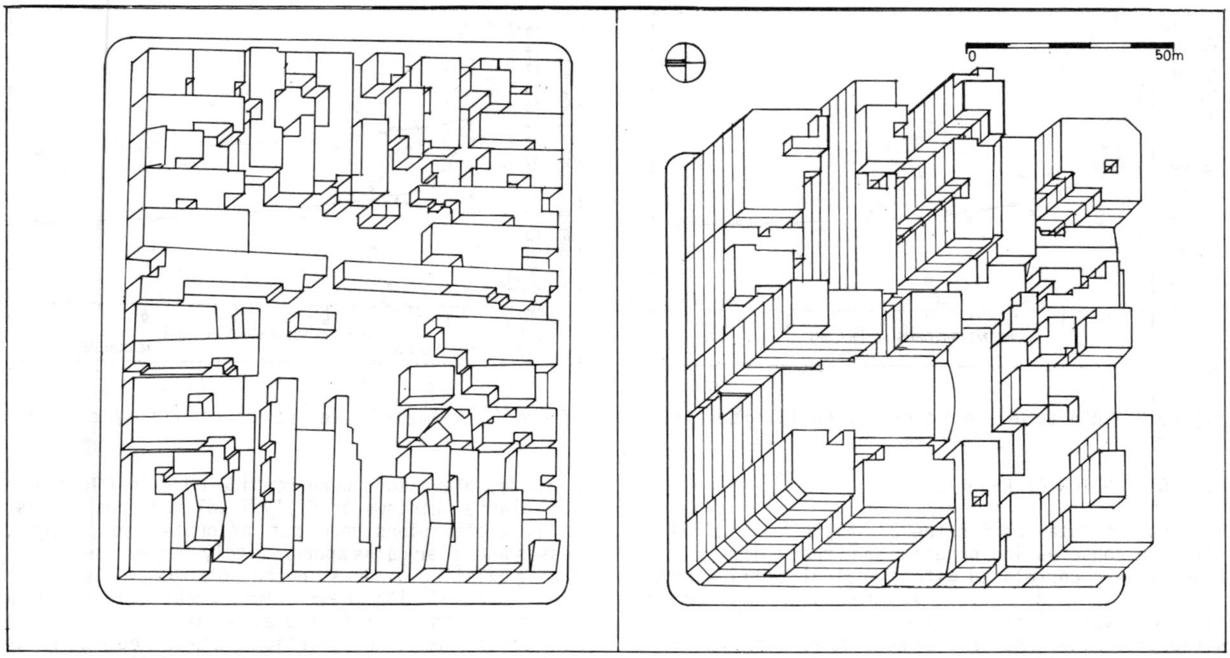

Figures 2a and 2b: Typical low (2.08 m^3/m^2), and high (11.57 m^3/m^2) density city blocks of Mendoza.

of the existing urban environments was accomplished in several steps:

. Determination of zones with maximum, mean and minimum building densities over the urban area.

. Selection of a first batch of 50 sample city blocks, within the zones previously determined, according to the following criteria: homogeneous land use within each block, features of geometric regularity : shape and dimensions of blocks, street width, orientation.

. After a second selection, a group of 37 sample blocks was defined: 9 blocks of maximum density, 11 of mean density and 17 of minimum density.

. Survey of buildings, 1240 in total, for the selected sample blocks, aimed at the determination of tipology, shape, dimensions and technology of components of each building. Thermal conductance (U) values were calculated for every component. In cases of uncertainty alternative values of U were calculated, as well as for infiltration rates, resulting in maximum and minimum alternatives of heat losses per building and city block.

Two computer programs were written along the previous stage, with the following purposes:

. The first program allows the input of the information on buildings shapes and dimensions entering the data of coordinates (x, y and z), of each building corner from scale drawings by means of a digitizer.

. The second program calculates several parameters: areas of thermal exchange, areas of potentially collecting facades: (due North $\pm$ 45°), volumes, skin to volume ratios, heat losses per component, global (Q) and volumetric losses (G) per building and city block. This program calculates also the values of a series of urban indicators for each sample block:

. Volumetric Density (VD): Σ of volumes / Σ of parcels area.

. Weighted Mean Height of buildings. (WMH)

. Orientation Index (OI): weighted mean deviation of potentially collecting facades.

. Typological Homogeneity Index (THI): Σ of volumes per building tipology / Σ of all volumes x 100.

. Spatial Distribution Index (SDI) : describes the greater density of building volumes North or South from the center of a given block.

Several correlations of results are possible after this stage, mainly those descriptive of the energy conservation features of each sample block, with the corresponding urban indexes.

3.2 MODELLING SOLAR ACCESS

An analitical computer program was specifically written for calculating on an hourly basis, the impinging solar radiation on potentially collecting facades. The program, which became a key instrument for the study, constitutes by itself a powerful tool of general utilization in any given situation of building masses and solar position. A detailed description of the model has already been presented /1/; an abridged one of its main operations is as follows:

. Imput of volume data from graphic information by means of a digitizer. As a difference with the conservation program previously described, the building volumes have to be broken-down into elemental, rectangular based, prismatic bodies.

. Determination of the sun's position by calculating the azimuth and altitude angle values for the chosen hour of the day and day of the year.

. Generation of the volume of shadows cast by the building system and subsequent calculation of the position of intersections between building volumes and shadow volumes.

. Computation of areas of potentially collecting facades of buildings, "seen" by the sun at the given hour.

. Calculation of direct solar radiation available on collecting facades, from meteorological data already processed and filed in the program. The difusse and reflected components of radiation are subsequently determined by simplified procedures and added-on to the previous calculations.

. The shadows cast on the buildings of a city block by structures of neighboring blocks can be determined by two alternative procedures: 1- Full use of the main program. 2- By using a simplified program which estimates the shaded areas of collecting facades as a function of the mean elevation of buildings across streets and the streets widths.

The amount of seasonal global solar radiation available for space heating, by means of passive solar systems incorporated to potentially collecting facades, is finally calculated by the program by adding-up the results of 35 hourly runs: seven hours per day (from 8.30 a.m. to 3.30 p.m.) for mean monthly days of a winter period of five months, in which the extraterrestrial radiation is closest to the month's average (from April 15 to September 15).

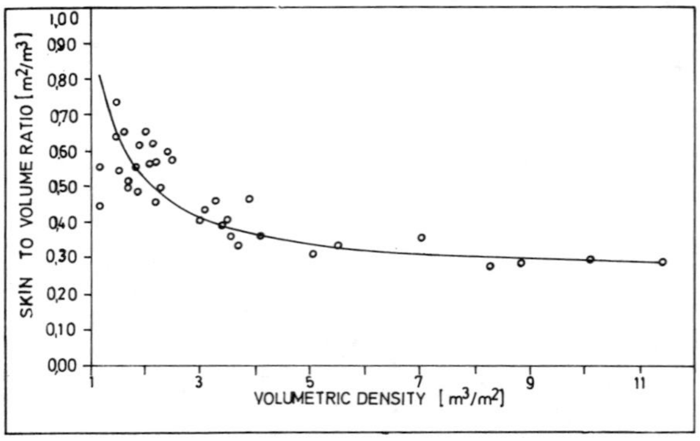

Figure 3: Skin to volume ratios vs. building density.

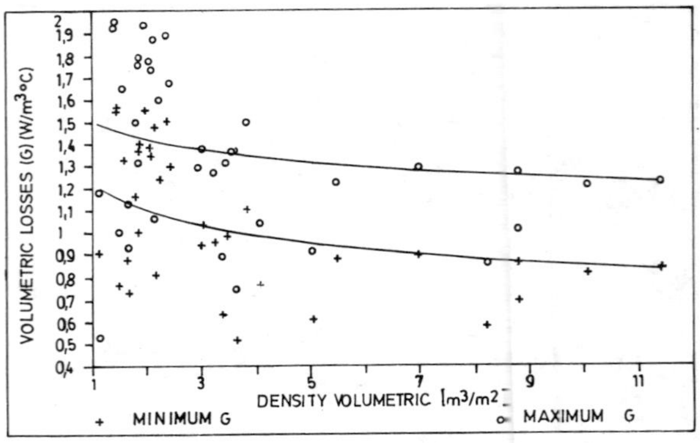

Figure 4: Volumetric losses vs. building density.

4. DISCUSSION OF RESULTS

The results obtained for the 37 sample city blocks, with the conservation program and the solar access program were correlated with the calculated five urban indexes described in 3.1., yielding a great volume of results: over 135 correlations.

Although further analysis of results is still pending, it is so far evident that the most significant urban index is the Volumetric Density (VD) i.e.: the ratio between the sum of building volumes and the sum of corresponding land parcels per city block. Therefore the results presented in this paper will be limited only to correlations with the VD.

In all cases the calculations of dimensional parameters, energy features and solar access were made for the city blocks in their present state (Solar 0.0) When determining the potential energy savings obtainable by retrofitting the whole building stock, (Solar 1.0) the following strategies were assumed to be implemented throughout:

Conservation: 1- Additional insulation on roofs. 2- Exterior insulation of walls. 3- Double glazing and night protection on windows. 4- Improved air-tightness.

Solar: 1- 85% of potentially collecting (non-party) exterior walls turned into direct gain, single glazed with night insulating rolling shutters ($R=0.53m^2°C/W$) Adecuate storage mass is generally provided by the common local heavy masonry construction. 2- 85% of potentially collecting (party) exterior walls turned into unvented Trombe walls, single glazed with night insulating rolling shutters. In both cases 15% of walls is allowed for horizontal and vertical structural members and edge conditions.

From the conservation stand point in the present conditions (Solar 0.0) of the building stock the correlations of skin to volume ratios and volumetric heat losses (G) to Volumetric Density are of particular interest (Figures 3 and 4).

In both cases, it is clearly shown that conservation features improve with increasing density, for two reasons: reduction of exposed walls and foundations by a higher proportion of common walls and the growing presence of voluminous multi-storeyed buildings at higher densities.

The availability of solar radiation on potentially collecting facades (± 45° from due North) per unit of building volume, a highly significant correlation, is presented (Figure 5). It is expressive of the relationship between the potential utilization of the solar resource and the heating need in any climate. A consistent decline of the values of solar energy is observed with the increase of the VD, with greater sensitivities along the exponential fit at the lower end of its scale. The potential of solar space heating by passive systems was calculated through the use of the Solar Load Ratio method of the LANL /2/. In order to determine values of Net Loss Coefficients (NLC), it was necessary to take into account the results of the solar access program. In this respect, North, East and West windows on shaded elevations were considered as net losers in the balance. Windows on fully insolated areas of East and West elevations were not accounted for, since, losses and solar gains approximately compensate. North facing existing windows in the Solar 0 0 alternative are considered with their actual areas as direct gain systems. In the retrofitted alternative (Solar 1.0), most of the insolated elevations were turned into solar systems as previously explained. The differentiation between insolated and shaded areas of potentially collecting facades were computed on a percentual basis over a whole heating season, by the adding of the hourly product of insolated areas and radiation intensities. It is obvious that finer analysis is necessary on the treatment of the "grey areas" of North facing elevations, i.e. those which are insolated only part of the time.

Solar Savings Fractions (SSF) were subsequently determined for minimun and maximun alternatives of the N.L.C. for the existing building stock (Solar 0.0) and their retrofitted counterpart (Solar 1.0)(Figures 6 and 7) It can be observed that for the existing conditions of buildings in city blocks, the solar contribution is almost negligible from the energy view point: SSF values ranging from 0.0 to 4.3% with mean minimums of 1.3% and 2.3% for the mean maximums. The linear best fits show little sensitivity to the VD's increase. The fact that slightly higher values of FAS are obtained at the upper extreme of the DV scale can be explained by two reasons: greater ability of larger buildings to conserve energy and the relatively lesser role of the solar access an contribution in this case.

For the retrofitted buildings, on the other hand, the values of SSF range from a minimum of 34.52 to a maximum of 51.56% with greater concentrations between 44 and 50% of the scale. The linear fits show a noticeable decline with the increase of the VD, a clear sign of the greater significance of the solar contribution over the whole energy balance, due to higher solar access at lesser densities.

Potential annual energy savings have been determined for maximum and minimum alternatives of heat losses. In both situations the consistent linear fits show significant increments of the savings with the increase of the volumetric density. (Figure 8). The mean values range from 0.74 to 3.91 MWh x 10^3/year for the maximum alternative and 0.46 to 2.21 for the minimun. As it is obvious, the potential savings are more significant for larger buildings and denser environments, considering global values. For the lowest densities however, the potential savings per unit volume: 0.0?7 to 0.046 MWh/year, are 2.5 times higher than the values obtained for the upper extreme of the existing VD scale: 0.032 to 0.018 MWh/year.

Over the total savings, the percentual contributions of added energy conservation and passive solar systems show significant variations with the increment of the VD: at the lowest extreme of its scale ($VD=1.0\ m^3/m^2$) conservation and solar stand for 78.0 and 22.0% respectively; at the upper extreme ($VD=12.00\ m^3/m^2$) the pro

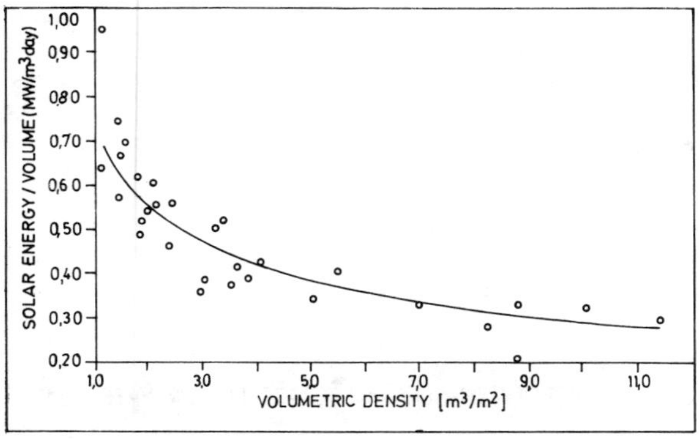

Figure 5: Available solar energy vs. building density.

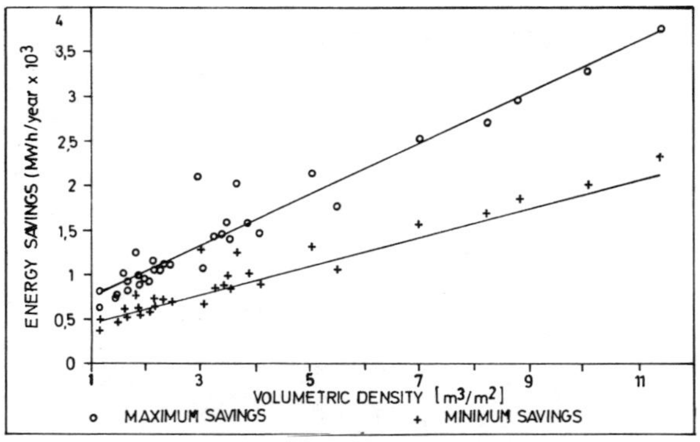

Figure 8: Total energy savings vs. building density.

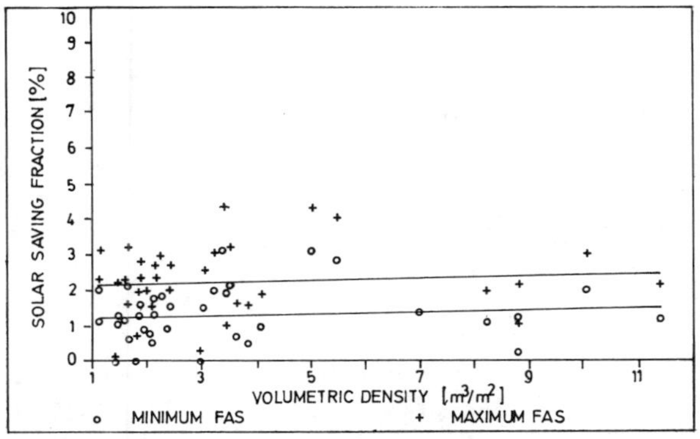

Figure 6: Solar Savings Fraction in existing building stock vs. building density.

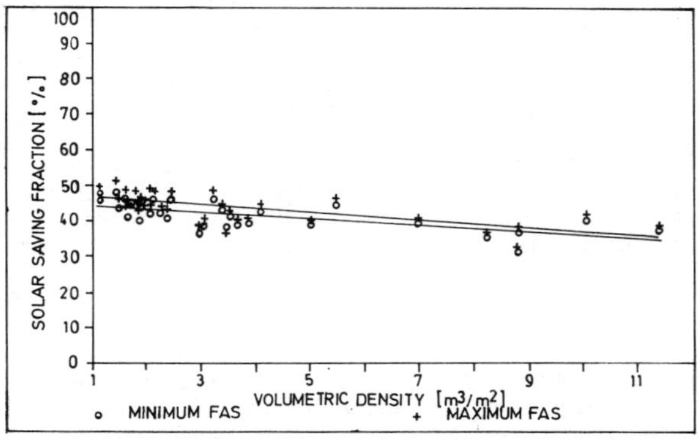

Figure 7: Solar Savings Fraction in retrofitted building stock vs. building density.

portion shifts to 86.5 and 13.5%. As predictable, the denser the environment, the more significant the conservation benefits and the lesser the solar contribution due to a more limited insolation on potentially collecting facades.

5. PRELIMINARY CONCLUSIONS AND FUTURE RESEARCH

The results so far obtained indicate a high potential of energy benefits: mean values of 43.5 KWh/year per unit of building volume, and 17.300 KWh/year for an average housing unit are possible by the retrofitting of the whole existing building stock. Added conservation measures will reduce volumetric heat losses (G) to an approximate mean of 45% of their present values. Solar Savings Fractions (SSF) are significantly higher than what was predictable, with means between 38 and 46%, which represent a 55% of the optimums obtainable in the region: 75 to 80%, for fully insolated solar buildings.

However, it has to be taken into consideration that these figures are highly theoritical. In practice it is hardly conceivable that the retrofitting of the whole building stock is likely to materialize under the present economic conditions. In the midterm, under the pressure of ever growing energy prices, the foreseeable shortages of natural gas and the enforcement of a more demanding energy legislation, a high percentage of the savings potential will most probably be recovered.

Other considerations should also be made. A much higher number of sample cases should be analized in order to achieve statistical consistency, and a reliable generalization of correlations, particularly in reference to the solar potential.

In this respect, the study is going to be expanded by considering a sample of approximately 200 cases modeled with the use of a random program. The effects of the predictable densification of the urban tissue on solar access, will also be considered in this stage. It is unlikely that for the densest city blocks, volumetric densities would ever grow beyond 18.0 to 20.0 m^3/m^2, due to limitations of building heights to a maximum of twelve stories impossed by the high seismicity of the region. Presently the highest value of VD is 11.57 m^3/m^2

Finally, many cities in the region and Mendoza in particular, are dominated by the massive presence of tall trees of deciduous species along every street and within block's cores in residential neighborhoods. A study of the year-round solar permeability of tree crowns of the most usual species: mulberry, China tree, plane tree and European ash, is being performed by measurements /3/. Its results will be incorporated in the solar access computational model.

6. REFERENCES

1. Esteves, A., de Rosa, C. and Cortegoso, J. 1988. "Modelo analítico-computacional para determinación del enmascaramiento solar en zonas urbanas" (Analitic-computational model to determine solar masking in urban zones). Proceedings of the ENCIT 88 - Second National Meeting of Thermal Sciences. Aquas de Lindoia-Brazil p.p. 375-378

2. Balcomb, J.D., Jones, R.W., Mc Farland, R.D. and Wray, W.O. 1982. "Performance Analysis of Passively Heated Buildings- Expanding the SLR Method". Passive Solar Journal. Volume 1, Number 2. p.p. 67-90

3. Cantón, A., de Rosa, C. and Cortegoso, J. 1989 "Bloqueo de la radiación solar a través de la copa de especies caducifolias" (Blockade of solar radiation through the crowns of deciduous tree species). Proceedings of the International Congress: "Energía, Ambiente e Innovación Tecnológica". Caracas, Venezuela. In print

EXPERIENCE DE STOCKAGE INTERSAISONNIER SOLAIRE EN MAISONS INDIVIDUELLES

AUTEUR : MOUETTE HENRI ARCHITECTE D.P.L.G.
ATELIER D'ARCHITECTURE SOLAIRE TEL 16 1 69 01 09 38
17 ROUTE D'ORSAY 91460 MARCOUSSIS

PRACTICE OF BETWEEN-SEASON STORAGE FOR PRIVATE HOUSE

ABSTRACT :After three realisations of this type of storage in solar houses,I am able to communicate the first results of the working of two of them.
The storage takes place in the ordinary ground below the house.Hot air supplied by a roof captor equipped with fans circulates in tubes buried at a depth of two metres.
In winter, the hot air is sent into the living quarters through insulated tubes.
In summer the shutters are closed and the hot air circulates in the underground tubes.
The heat stored is released in two ways :
1/ By conduction through the ground up to the ground floor, with a time lag of three months.
2/ By preheating of fresh air drawn in by the MVC (Mechnically Controlled Ventilation) through the underground tubes.
The economy of heating energy has resulted in progress from 51% to 58% in four years, for the first house.
For the second house (lived in since 1986) the progress has been from 53% to 64%.
As expected, the contribution of between-season storage began after the third year, the first two years taken up with the preliminary heating of the ground below and around the house.

1 ORIGINES

1-1 Ayant eu connaissance des premières expériences réalisées par un architecte français (J.P CORDIER) dans la région du sud-ouest de la France, j'ai eu envie aussi d'adapter ce principe à mes maisons solaires de la région parisienne.

1-2 Dans la recherche et la mise au point de cette technologie, j'ai été aidé par l'Institut Universitaire de Valenciennes en 1983 dont les étudiants avaient travaillé sur ce sujet, en réalisant pour un de leurs ateliers un système de stockage associé à un capteur à air, proche de celui que je cherchais à réaliser. J'ai conçu mon premier projet avec l'aide de leurs calculs et ai adapté leur technologie à la maison individuelle.

Fig 1 : 1ère maison Argenteuil 95

Fig 2 : 2 ème maison Lardy 91

2 FONCTIONNEMENT

2-1 Description générale.

Le stockage est constitué par le terrain situé sous la maison, dans lequel sont enterrés à 2 mètres de profondeur, des canalisations de drainage en P.V.C. de 16 cms de diamètre. Chaque tuyau est raccordé par son extrémité nord à une gaine isolante verticale de même diamètre, qui rejoint le long de la paroi nord le haut du toit capteur. A cet endroit, sont placés les ventilateurs (un pour chaque gaine) d'un débit de 300 m³/h. En hiver, l'air chaud du toit capteur est soufflé dans le volume habitable, par des bouches placées le long des gaines verticales isolées. En été, ces bouches sont fermées, l'air circule dans le stockage. La marche des ventilateurs est commandée par une régulation différentielle et 3 sondes : capteur, ambiance, stockage.
La restitution de la chaleur stockée se fait selon 2 modes :
a/ par conduction naturelle à travers la couche du terrain entre canalisations et plancher du rez-de-chaussée avec un déphasage de 3 à 4 mois.
b/ par préchauffage de l'air neuf de la VMC aspiré dans les canalisations enterrées, par les bouches de soufflage.
De plus, dans une des maisons étudiées, une canalisation d'alimentation du ballon ECS en eau froide a été placée dans les tranchées à côté des canalisations d'air pour le préchauffage de l'eau chaude.

2-2 Particularités.
Les 2 maisons qui fonctionnent effectivement sur ce schéma sont très différentes de par leur volume et structure.
La 1ère a un volume de 233 m³ sur terre plein accumulateur de 55 m² avec un capteur de 45 m².
Mise à part la dalle du rez-de-chaussée en béton,sa construction est entièrement en bois sur les deux niveaux habitables.
La 2ème maison a un volume de 510 m³ sur un terre plein accumulateur de 85 m² avec un capteur de 85 m².
Les 2 planchers habitables sont en béton ainsi que les murs intérieurs et extérieurs avec isolation extérieure.

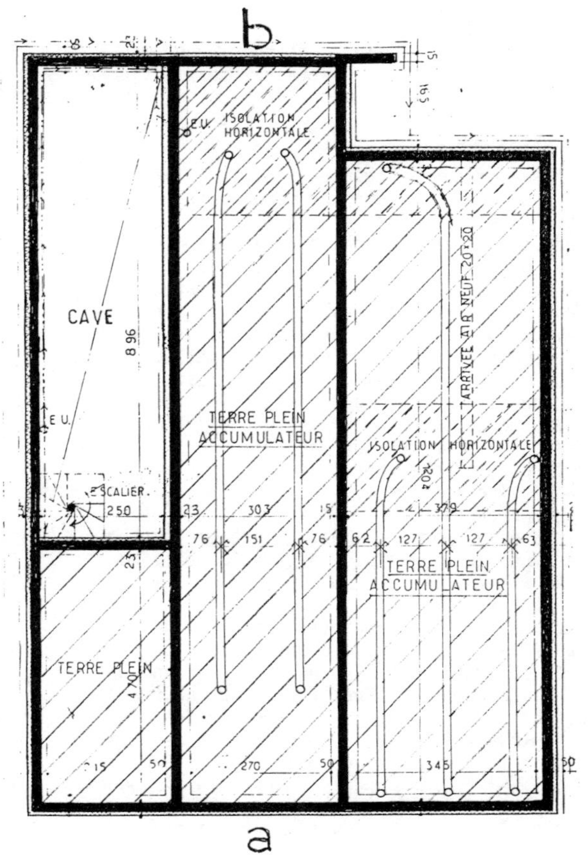

Fig 3 : Plan du sous-sol et coupe

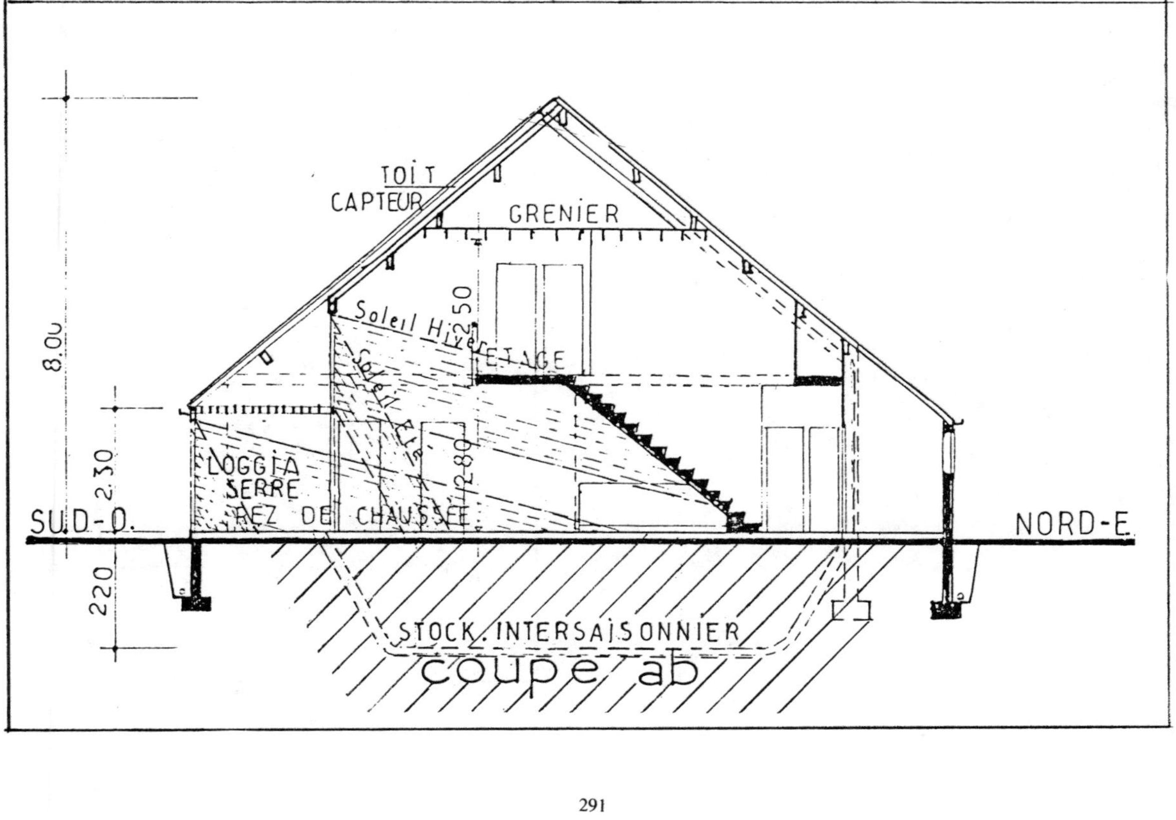

La première maison n'a pas utilisé, jusqu'à présent, sa VMC et n'a donc pas bénéficié du préchauffage de l'air neuf. De plus, les orifices de retour de l'air intérieur vers le capteur ayant été obturés par une erreur du propriétaire, l'air soufflé dans la maison est pris en totalité à l'extérieur, ce qui occasionne un renouvellement d'air 6 fois plus important que nécessaire.
Ces erreurs d'exploitation sont dues à une mauvaise interprétation par le propriétaire, du fonctionnement du système, à l'insu de l'architecte.

3 RESULTATS

3-1 : 1 ère maison - Argenteuil (Val d'Oise)

Elle a comme unique chauffage complémentaire des convecteurs électriques et un ballon électrique pour la production d'eau chaude, sans apport par capteur solaire. Cette maison a été équipée en compteurs divisionnaires qui mesurent les consommations dues au chauffage et celles dues à la production d'eau chaude. En outre un compteur mesure le volume d'eau préchauffée consommée. Cette maison est habitée depuis 1985 et les les relevés ont été faits à peu près chaque mois par le propriétaire. Quatre bilans de chauffage sont basés sur les consommations enregistrées et les degrés-jour fournis par la station météo la plus proche (Pontoise). Ils font apparaître successivement des économies de 53%, 51%, 56% et 58% le dernier hiver.
Pour le préchauffage de l'eau chaude, la consommation est passée progressivement de 48 watts par litre à 46 w/l.

3-2 : 2 ème maison - Lardy (Essonne)

Le chauffage d'appoint est constitué par des convecteurs électriques et par un poële à charbon. Malgré l'absence de compteurs spécifiques au chauffage, 2 bilans ont pu être établis en fonction des consommations globales sur les 2 dernières années (en convertissant la consommation de charbon en équivalent Kwh. En 1987-88 (2ème année d'occupation) l'économie a été de 53% ; elle montée à 64% en 1988-89.
C'est à dire que la consommation du chauffage d'appoint a baissé de 8800 Kwh à 6450 Kwh sans différence notable ni dans les températures de chacun des 2 hivers (2400 et 2335 degrés jours) ni dans les heures d'ensoleillement correspondantes.(641 et 659)

3-3 : Enseignements

Comme les calculs l'avaient prévus, les apports du stockagage intersaisonnier ont commencés à apparaître dans les bilans de chauffage et d'eau chaude, à partir de la 3 ème année de fonctionnement : les 2 premières années ayant servi à réchauffer progressivement la masse du terrain située en dessous et autour de la maison.
Le phénomène a été moins rapide pour la 1 ère maison du fait des erreurs d'exploitation mentionnées plus haut, qui ont limité les apports du stockage intersaisonnier à la seule conduction à travers le terrain sous les plancher du rez-de-chaussée.
Par contre, cette maison a bénéficié du stockage pour le préchauffage de l'eau chaude, mais dans une proportion encore assez faible. La température de sortie de cette eau préchauffée ne pépasse pas 21° pour le moment.

4 PERSPECTIVE D'AVENIR

4-1 Fonctionnement

L'auteur se propose d'améliorer le fonctionnement du système par la mise en place de bouches de soufflage à ouverture automatique en fonction de la température intérieure de chaque pièce habitable. Ce dispositif associé à une régulation spécifique à chaque conduit et à chaque ventilation permettra une automatisation complète de l'installation en toutes saisons, ce qui devrait augmenter le rendement du système.

4-2 Contrôle des résultats

Il serait douhaitable qu'un organisme français ou européen prenne en charge à l'occasion d'une prochaine réalisation, une campagne de mesures diverses des performances du système sur plusieurs années.

4-3 Domaines d'application

Ce système peut s'appliquer à des constructions de formes très diverses ne comportant pas obligatoirement un toit capteur. l'air chaud peut, en effet, être prélevé en partie haute d'une serre habitable.
La figure 4 montre l'exemple d'une maison construite entièrement à l'intérieur d'une demi-coupole vitrée
L'air chaud est aspiré en dessous du sommet par un seul ventilateur et envoyé dans le stockage enterré par 3 gaines rayonnantes, à partir du centre de l'escalier où passe le conduit isolé.
Ce champ d'application ne se limite pas à l'habitation individuelle,mais peut s'étendre à toutes sortes de bâtiment à rez-de-chaussée sans sous-sol construit (sauf partiellement)

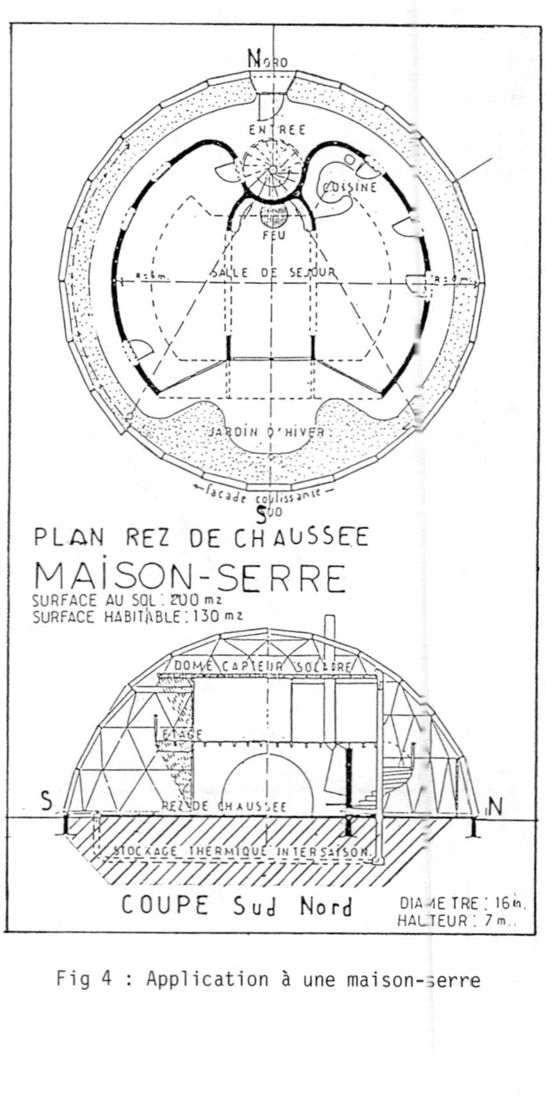

Fig 4 : Application à une maison-serre

NATURAL VENTILATION IN COURTYARD AND ATRIUM BUILDINGS

R. Bensalem and S. Sharples,
Building Science Unit,
School of Architectural Studies,
University of Sheffield,
Sheffield S10 2TN

ABSTRACT. This paper reports on an experimental investigation into the ventilative performances of courtyard and atrium buildings for cooling purposes. Several models of these structures were tested in isolation in a wind tunnel. The ventilation was assessed from actual airflow rates measured by the means of orifices plates inserted in the models. The effect of the courtyard geometry and its orientation to the wind were examined as well as the possibilities of inducing higher ventilation rates by roofing the courtyard and exploiting the pressure field over the roof. The study highlighted the prime importance of the orientation of the building to the oncoming wind and the potential and limitations of some ventilation strategies that could be used in the atrium structures.

1. INTRODUCTION

Open and covered courtyards are common architectural features which provide light, ventilation and other functions to buildings. The potential that these structures can offer to reduce the energy loads of buildings, and in particular cooling, have recently been given closer attention /1/. In warm humid climates the main exigence to achieve thermal comfort without resorting to mechanical refrigeration is to provide strong cross ventilation. In this respect is is not well known whether courtyard or atrium structures will meet these requirements and what are the most influential parameters in their ventilative performances. For example, covering the courtyard with a roof might induce higher air movement by either exploiting the large suction over an adequately designed roof or by deflecting the air into the atrium. Again little is known about the efficiency of any of these ventilation strategies. The scope of this study is:

- To assess the ventilative performances of some courtyard and atrium buildings.
- To examine the effect of the courtyard geometry and the orientation to winds.
- To test the efficiency of several possible ventilation strategies that an atrium structure can use.

2. EXPERIMENTAL APPARATUS AND TECHNIQUES

To have the facilities to test systematically the effect of several courtyard geometries and atrium roof types called for the use of models in a boundary layer wind tunnel. Of the ventilation measuring techniques available, it was chosen to measure directly the ventilation rates through the model by the means of small orifice plate devices. The use of pressure distributions to predict airflow rates was considered inadequate when roof vents are used /2/.

2.1 The Wind Tunnel

The tests were carried out in the Sheffield University's 1.2 x 1.2m boundary layer wind tunnel, which was exhaustively described by Lee /3/. A suburban velocity profile was generated with a power-law exponent of 0.24.

2.2 Modelling Criteria

The most significant parameter to ensure similarities in the flow pattern between the model and the prototype are the geometry similarity, the Reynolds number and the blockage ratio of the model in the wind tunnel. To match the full scale Reynolds number with that of the model is practically impossible. Fortunately, once the flow is turbulent, the flow characteristics becomes Reynolds number independent for sharp edged bluff bodies. The Reynolds number independence of the internal flow was assessed from the pressure loss across the model. This parameter remained fairly constant for Reynolds numbers between 3.5×10^3 and 1.3×10^4. Although these values were lower than the value suggested for internal flow Reynolds number independence (2×10^4) /3/, it was believed that dynamic similarity was achieved for almost all the working range. The blockage in the wind tunnel was kept less than 4%.

2.3 Model Buildings

2.3.1 The Courtyard Models

The models that represented four storey courtyard buildings at 1:100 scale were assembled in Lego-like model elements. Eleven pieces were constructed to cover the range of courtyard sizes tested. The models were made of perspex and had perforated facades, with 10mm diameter circular holes totalling a facade porosity of 11.4%. Four of the model elements had one orifice place device mounted, each in a different floor height (see Fig. 1). The device was a square edge plate of 17mm diameter inserted between two short brass pipes of 25mm diameter, and fitted with two corner pressure tappings. The device was calibrated against a precision commercially available flowmeter.

Figure 1: Orifice Plate Model.

2.3.2 The Atrium Models

Aluminium models fitted with pressure tappings were used to gain knowledge of the pressure field around atrium models with several roof shapes. For the airflow measurement, the orifice plate model elements were used. Five types of roofs were tested (flat, pyramidal, pitched and monopitched roof). Only the monopitched roof was selected for the flow tests since it had a large side under high suction (Cp = -0.95), as well as a side under positive pressure (Cp = +0.26). Orientated differently the same roof had a large area under a slightly negative pressure (Cp = -0.09). Cp is the pressure coefficient related to the dynamic pressure on roof height.

2.4 Test Procedure

The courtyard depth (D) and width (W) were changed from 0.5 to 1.5 times the model height (H). The measurement of the pressure drop across the orifice plate and the dynamic pressure at the gradient height were taken at wind approach angles from 0° to 360° in 30° and 45° increments, and for the atrium test from 60° to -60°. The roof effects were tested on one building size (H/W = 1,H/D = 1). The porosity of the roof was altered from 0% to 1.9%, 3.8%, 5.7% and 23% when testing the roof under suction, but kept to 5.7% and 23% for respectively the positive and near atmospheric pressure strategies. The results were presented in the form of a non dimensional flow coefficient CQ:

$$CQ = Q/(A\ Vr)$$

where Q is the flow rate measured, A the area of the openings in one storey and in one facade and Vr the velocity at the roof height. CQ is in other words the ratio of the velocity at the opening to that at the roof height.

3. EXPERIMENTAL RESULTS AND DISCUSSION

3.1 Courtyard Models

3.1.1 Effect of the wind incidence

The orientation to the wind had the most significant effect on the ventilation. Orientating obliquely the building to the wind not only increased the overall flow but resulted also in a more even distribution within the building than when it was perpendicular to the wind. At normal wind incidence, the air velocity at the openings in the windward sides were between 0.38 and 0.50 time that at the roof height, whilst on the lateral sides CQ was 0.05 to 0.18 and only 0.03 to 0.13 on the leeward side. At a wind angle of 30 degrees the average air velocity passed from 0.18 to 0.27 (a 50% increase) and a more uniform distribution within the building existed. CQ was increased by a factor of 2 in the leeward side and by a factor of 2 to 4 in the lateral sides. At 45 degree wind incidence, the highest averaged values of CQ for the whole building were obtained (0.28), and a more even distribution. Yet, the improvement of the downwind side were accompanied by a decrease in the air velocity in the windward sides (CQ was only 0.24 to 0.36). At a normal wind incidence, a large vertical rotating vortex dominated the flow in the courtyard. The negative pressures encountered in the inner walls were close to that in the downwind external walls. When the wind striked the building at an angle, not only the external walls should have been under a higher suction but the inner walls probably experienced a rise in pressure due to a downwind movement of air caused by vorticies emanating from the upwind roof edge. As a result, the pressure drop rised significantly.

3.1.2 Effect of the Geometry of the Courtyard

In contrast with the dramatic changes in CQ caused by the change in the orientation to winds, the geometry of the courtyard had fewer effects. The changes in the overall flow coefficient CQ was at most 18% (and even lower, 5%, at oblique wind angles). The windward side was particularly insensitive to any alteration of the courtyard configuration. On the leeward side, the changes of the depth or the width affected the CQ magnitude and the flow direction at discrete points rather than the average values. The changes on the vertical distribution of CQ with the alteration of the depth are shown in Fig. 2. At the biggest depth (1.5H), the air was moving outwards (from the courtyard to the outside) in the two extreme floors and inwards in the two mid-height floors. This particular flow distribution resulted probably from changes in flow regime in the courtyard, where first the flow was skimming the roof crests, generating a slow rotating vortex. At bigger depths the flow starts to wash over the courtyard face without a stagnation point. Similar observations were made on the pressure distribution in rows of buildings when the spacing was changed /5/.

3.2 Atrium Buildings

Depending on the ventilation strategy, the atrium building had lower or higher air velocities than the open courtyard. Among the atrium configurations tested, the atria exploiting the low pressure on the roof had generally the lowest average air velocities. The deficit of pressure that existed in the atrium and resulting from the roof suction affected the pressure drop on the downwind sides of the models where the external walls experienced a relatively low pressure. On the other hand, the air velocity on the windward side increased. As a result, the distribution of the flow within the model was very uneven. The more openings there were in the roof the less even was the distribution of the flow and the lower the CQ values on the downwind sides. Locating the roof openings in a positive pressure region resulted in the highest CQ averaged value (0.28) at normal incidence. The pressure drop on the downwind sides benefitted from the built up pressure in the atrium. At an oblique angle, however, the outside upwind walls experienced lower pressure than at normal wind incidence and consequently the pressure drop at this location was affected. The atrium with a closed roof and with the roof having openings located near the atmospheric pressure had at any wind angle one of the fourth highest CQ values. The reason was that the resulting pressure in the atrium building was half way between the pressures on the upwind and the downwind external walls. Orientating obliqely the buildings to the wind had a very beneficial effect on the ventilation except for the atrium using the roof suction strategy.
The atrium models showed all higher average CQ values than the open courtyard, (up to 40% higher) at normal wind incidence. At 30 degrees, the highest increase was 12% and reduced to 5% at 45 degrees. To appreciate the ventilative performances of these buildings the best indice was the ratio of the CQ values averaged over the whole building to its standard deviation. Fig. 3 summarises the results.

4. CONCLUSION

In conclusion, orientating the courtyard with an angle (30° or 45°) to the wind had a very beneficial effect on the ventilation, resulting in higher air velocity on the downwind sides. On the

other hand the geometry of the courtyard had marginal effects. Exploiting the suction effect over a covered courtyard was not rewarding, the pressure forces not being large enough to overcome the suction experienced on the downwind walls. The best results were achieved by closing the roof or locating the openings in a near atmospheric pressure zone. These ventilation strategies resulted in higher air velocities than for the open courtyard.
The experiment was undertaken on isolated models. It is possible that these concluding remarks might not apply for courtyards or atria in an urban settlement, where the flow field could be affected by neighbouring buildings. A study of shielding effects is currently in progress.

REFERENCES

/1/ Leung S.K. & Al (1981), 'Thermally induced ventilation applications in atria", State of the art report, California, Eureka labs, Sacramento.
/2/ Vickery B.J., Baddour R.E., Karakatsanis C.A. (1983), "A study of the external wind pressure distributions and induced internal ventilation flow in low-rise industrial and domestic structures'. Boundary layer wind tunnel laboratory, University of Western Ontario, Ontario, Canada, BLWT-SS2-1983.
/3/ Lee B.E., (1977), 'The simulation of atmospheric boundary layers in the Sheffield University 1.2 x 1.2m boundary layer wind tunnel', Department of Building Science, Sheffield University, Internal Report No. BS38.
/4/ Cermak J.E., Poreh M., Peterka J.A. and Ayad S.S. (1982), 'Wind tunnel investigations of natural ventilation', Preprint 82-519 ASCE Convention.
/5/ Soliman B.F., (1976), 'Effect of group geometry on wind pressure and properties of flow', Department of Building Science, Sheffield University, Internal Report No. BS29, April 1976.

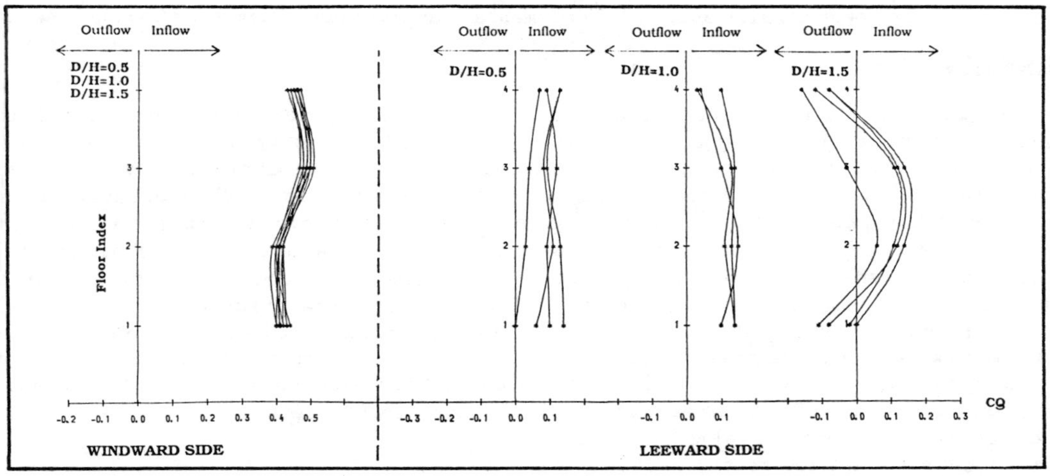

Figure 2: Effect of changing the courtyard depth on the vertical distribution of the flow coefficient on the leeward and windward sides. (The relative depth (D) was changed from 0.5 to 1 and 1.5 the model height (H)).

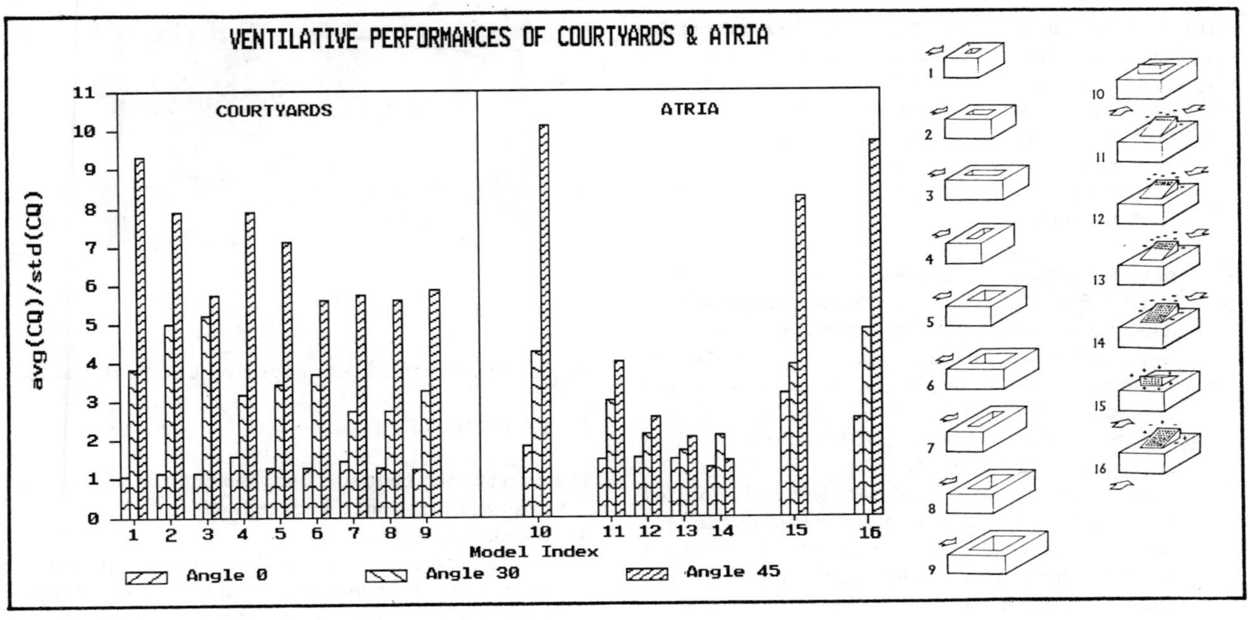

Figure 3: Ventilative performances of courtyard and atrium buildings for 0, 30 and 45 degrees wind incidence. Avg (CQ) and Std (CQ) are respectively the average and standard deviation of the flow coefficient.

EXPERIMENTAL ANALYSIS OF THE ENERGY PERFORMANCE OF A PASSIVE ATTACHED SUNSPACE

I.Meroni, F.Scamoni, P.Tirloni, C.Pollastro, R.Lacci
National Council of Research
I.C.I.T.E. - Institute of Building Technology
Via Lombardia, 49 - 20098 - San Giuliano - Milano - Italy

ABSTRACT. During the studies carried out at ICITE, Institute for Building Technology of the National Research Council, a method has been developed to analyse the energy performance of passive solar components in the building industry by using test cells that simulate a conventional dwelling environment. The paper describes the tested components, the experimental analysis of energy performance which has been carried out, the complementary analysis performed and the obtained results.

1. INTRODUCTION

The development of studies concerning the applications of solar energy suggests that it would be interesting to use passive solar components in the building industry. In this paper the authors compare the experimentally analysed performance of two components: a window and an attached sunspace that was capable of operating either as a completely passive system or as a hybrid system. During the execution of the performed tests, both the possibilities of the sunspace were evaluated. The results are discussed, with an analysis of both the explanations of the results and of the recommended modifications and improvements to be made to the systems in question.

2. DESCRIPTION OF THE EXPERIMENTAL SET-UP

The test cells used for the study were optimized to compare the thermokinetic behaviour of systems designed for handling solar energy.
The main features of the experimental set-up are schematically summarized as follows:

- the cells are structurally designed to simulate a conventional type of living environment;
- each cell consists of a three-dimensional prefabricated components measuring 2.80 x 2.80 x 5 m, made of light concrete with a density of 1800 Kg/m^3 reinforced with wire netting, with a wall thickness of 10 cm;
- the cells were provided with light external thermal insulation, designed to offer a thermal transmittance of the outer wall of approximately 0.3 W/m^2 °C;
- a heat barrier realized between the floor of the cells and the ground simultaneously ensured the thermal insulation of the floor itself and the impermeability of the systems;
- the thermal mass in the cells is distributed more or less in the same way as in a conventional living environment.

The tested systems were positioned on the southward facing walls of the test cells: a window with double glazing and a system consisting of a similar window with an attached sunspace. The performance of the systems was also compared with that of a building technique which is traditional in Italy consisting of a double brick and hollow-brick wall with an insulated air gap, positioned facing the same way as the first two systems and fitted onto a third test cell. The arrangement of the systems tested can be seen in figure 1.

Fig.1:The tested components on the test cells

3. PERFORMANCE OF TEST

The monitoring that was carried out for each cell covered measurements regarding:

- weather conditions;
- radiation incident on the vertical plane;
- average air temperature in the test cells;
- average air temperature in the sunspace;
- energy absorbed during the test period.

The data were acquired and recorded with a scanning of 15 minutes. The comparison between the performances of the systems was based on an evaluation of the amount of energy needed to keep the cell on which they were fitted in thermostatic conditions with a temperature level as similar as possible. This quantity was measured over periods of time ranging from 1 to 5 days.
The quantity of energy considered for the calculation included the electric power supplied for heating the cells, as well as the power supplied to the instruments located in the test cells. The size of the energy contribution is the same as the overall volumetric coefficient of thermal dispersion, and is defined as:

$$Cg^* = q/(h \cdot v \cdot \Delta T) \qquad (W/m^3\,°C)$$

where:
- Cg^* = volumetric coefficient of energy contribution ($W/m^3\,°C$);
- q = energy absorbed during the measurements (Wh);
- h = duration of the measurements (hours);
- v = volume of the cell (m^3);
- ΔT = average difference between internal and external temperature during the measuring period in question (°C).

4. DISCUSSION OF THE OBTAINED RESULTS

4.1 The attached sunspace as a hybrid component.

With this mode of operation, the circulation of the air between the sunspace and the environment was actively ensured by a fan with a shutter located on the upper part of the transparent element separating the sunspace from the environment; another shutter was located on the lower part to make the air start circulating. The presence of the shutters guaranteed protection from stray currents of air capable of dissipating the heat in the internal environment during the night. The shutters were opened and the fan switched on automatically whenever the temperature difference between the air in the sunspace and the air in the environment exceeded 4°C. The temperature of the environment air in the cell was set to a thermostatically controlled temperature of 16°C. The data relating to the test periods in fig. 2 show that:

- under conditions of limited solar radiation the cell fitted with the sunspace dispersed approximately 30% more than the reference cell. This was easily explainable in view of the difference in thermal transmittance between an opaque and a transparent component;
- under conditions of not particularly abundant solar radiation (which unfortunately characterized the measuring period), the cell fitted with the sunspace saved around 15% with respect to the reference cell, with considerably higher peaks on very sunny days.

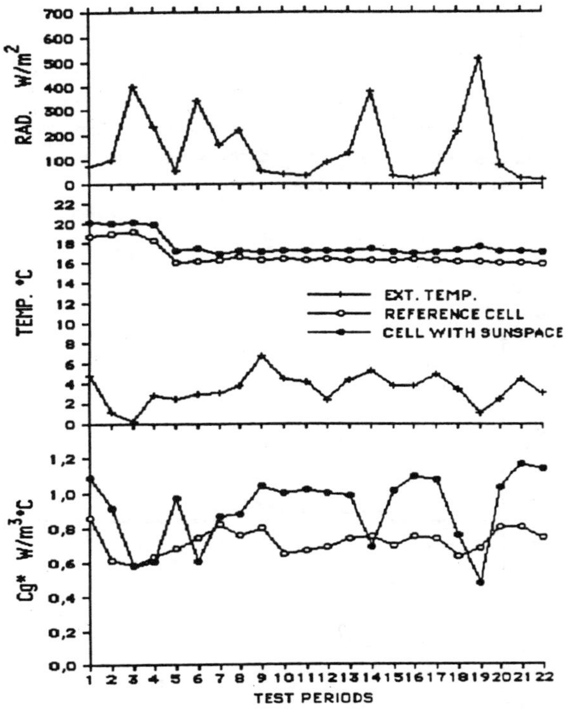

Fig. 2: Data relating to the test periods.

Fig. 3, shows a daily trend of the recorded parameters during the test periods.

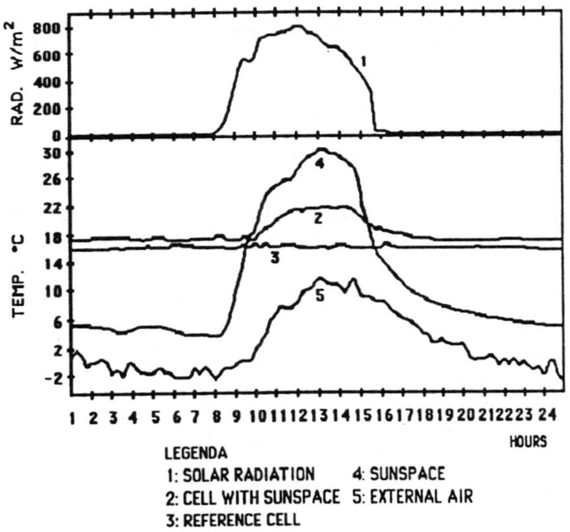

Fig. 3: Daily trend of the recorded parameters.

4.2 The attached sunspace as a passive component.

With this mode of operation, the circulation of the air between the sunspace and the internal environment occurred by natural convection. Once the fan eliminated, the two shutters were opened under the same conditions described for the hybrid system.

The data relating to the test periods in fig. 4, show that:

- under conditions of low intensity solar radiation, as foreseen, the obtained results were the same as those obtained for the sunspace used as a hybrid component operating under the same weather conditions;
- under very sunny conditions, the sunspace saved 40% of the energy.

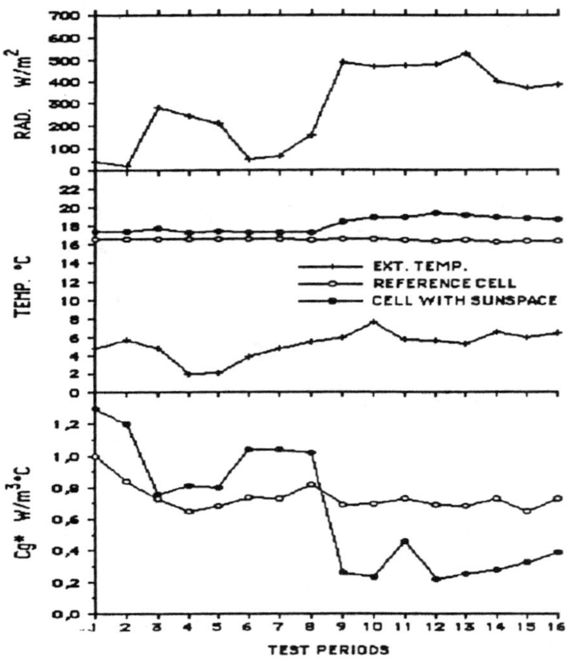

Fig.4: Data relating to the test periods.

Fig. 5, shows a daily trend of the recorded parameters during the test periods.

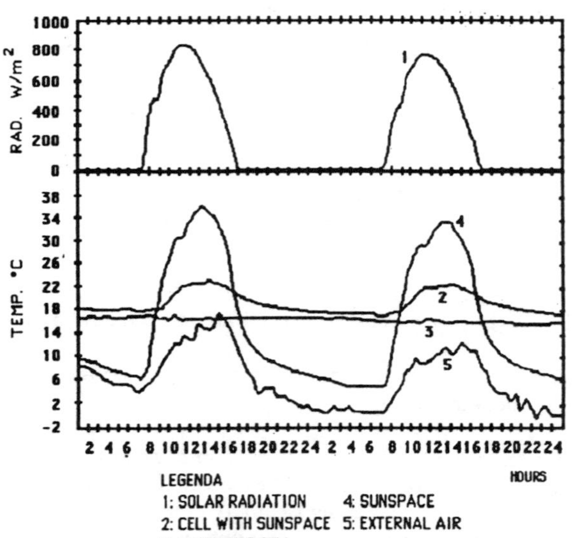

Fig. 5: Daily trend of the recorded parameters.

4.3. Comparison between attached sunspace and window.

The temperature inside each cell was kept at 20 °C by means of hot air heaters while the rolling shutters of the two transparent systems were raised at 9 a.m. and lowered at 3 p.m.. Whenever the temperature of the air in the sunspace exceeded that of the test cell, the window with the attached sunspace was opened in order to admit the heat accumulated in the sunspace into the cell. In these conditions, measurements were taken of consumptions, internal temperatures and climatic conditions, and the energy contribution Cg* of the systems was calculated. The data relating to the test periods are shown in fig.6. Fig. 7 shows a daily trend of the recorded parameters during the test periods.

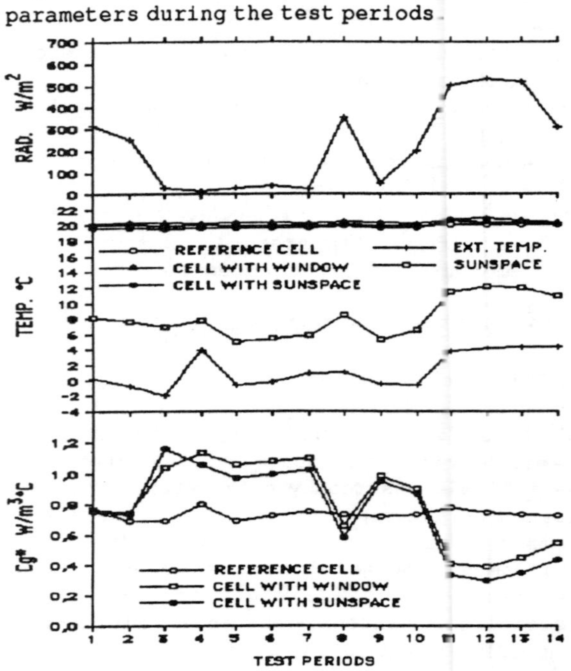

Fig. 6: Data relating to the test periods.

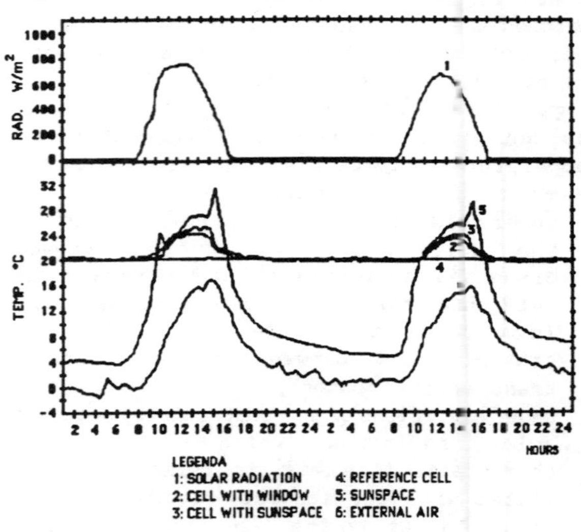

Fig. 7: Daily trend of the recorded parameters.

On the basis of these data, the following considerations can be made:

- with an overcast sky and low solar radiation, the measurement of the energy contribution coefficient is highly repeatable: 1.1 W/m³ °C approx. for the window and 1 W/m³ °C approx. for the cell bearing the sunspace element respectively, and it is always higher than the opaque reference;
- with solar radiation, even when not particularly intense, the cell with the sunspace saves around 10% more than the cell with the window;
- under the same conditions, the cell fitted with the sunspace element saves 20% compared to the opaque reference;
- with intense solar radiation, the amount of energy saved in the cell fitted with the sunspace is in the range of 20% more than that fitted with the window.

Fig. 8 shows the trend of the energy contribution related to the average daily radiation.

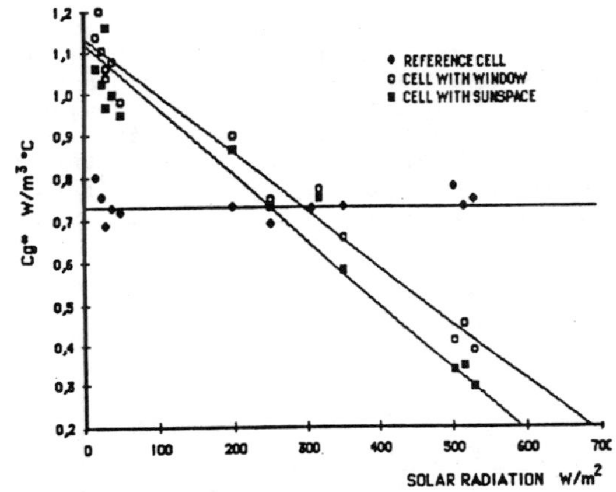

Fig. 8: Trend of the energy contribution.

4.4 The system in natural conditions of heating

The auxiliary heating was subsequently turned off and measurements were taken of the temperature trends inside the cells and the climatic conditions.

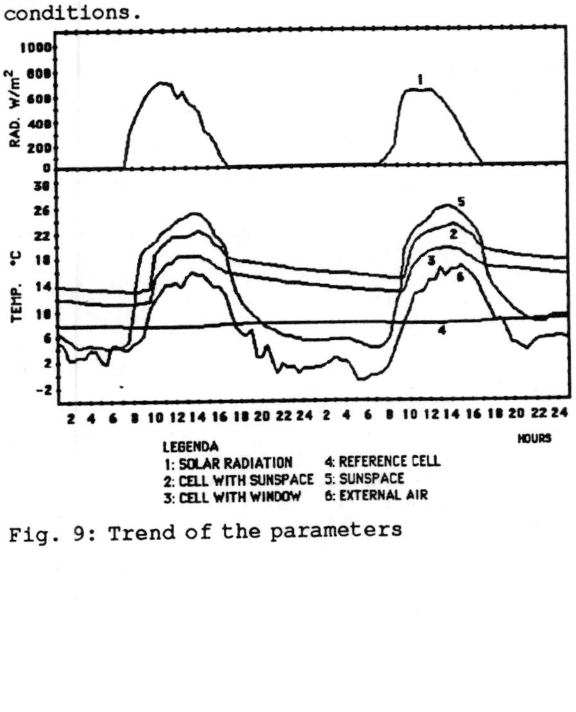

Fig. 9: Trend of the parameters

As in the previous situation, the rolling shutters of the two transparent systems were raised and lowered at pre-established hours and the window with the attached sunspace was opened each time the average temperature of the air in the sunspace exceeded that measured in the cell. The trends of the parameters referring to this period are shown in fig. 9. They show a trend consistent with the ratio between climatic stress and thermal behaviour of the two test cells. As was foreseeable, the sunspace proved to be superior to the other systems.

5. CONCLUSIONS

The results show very clearly that the passive version of the tested components offers advantages with respect to the hybrid version. A comparison between the thermoelectric power data of the first case (passive) and the second (hybrid) shows that the influence of the air movement needed in the latter, is such as to considerably increase consumption (energy supply). There is therefore no reason to turn the passive configuration into a hybrid configuration since this would increase the cost (due to the fans) and decrease the reliability of the system.
Comparing the energy consumption of the three cells during the overall period we can see that the sunspace behave more or less in the same way as the insulated wall assumed as a reference. This can be considered a positive result bearing in mind that the climatic conditions during the whole measuring period have been rather bad, very foggy and misty and have been only four periods with clear sky and hight solar radiations. This is very promising for the application of the sunspace solutions in countries where usually good weather and solar radiation conditions can be found. In conclusion, the test that have been carried out, showed that the energy saving of the transparent systems is greatly conditioned by the climatic parameters and by the location. In fact, the advantages of the systems can only be appreciated under the most favourable temperature and radiation conditions. It should be pointed out, howerer, that even though the integration of a sunspace involves an investment that cannot be underestimated (in the region of 50 % more than the integration of a normal window) it can be balanced by the energy saving (in long term only) and by the capability of the sunspace to offer a living space with fine environmental characteristics.

6. BIBLIOGRAPHY

1) W.Esposti, I.Meroni, C.Pollastro, P.Tirloni "Performance measurement and evaluation of two hybrid solar components" Third International Congress on Building Energy Management, Losanna, 1987

2) I.Meroni, F.Scamoni, P.Tirloni, C.Pollastro, W.Esposti "Experimental analysis of the energy performance of industrialized passive solar components", ISES Congress, Kobe, Japan, 1989

PLANCHER SOLAIRE DIRECT
UN SUCCES THERMIQUE ET ARCHITECTURAL

P.PAPILLON

ASDER
299 rue du Granier
73230 SAINT ALBAN-LEYSSE
FRANCE

LABORATOIRE GENIE CIVIL ET HABITAT
Université de Savoie
BP 1104
73011 CHAMBERY CEDEX
FRANCE

Between 1984 and 1986, the Association Savoyarde pour le Développement des Energies Renouvelables has developped a demonstration program for the Direct Solar Floor in conjunction with architects and thermal engineers and with the help of the Region Rhône-Alpes and the Agence Française pour la Maîtrise de l'Energie. The Direct Solar Floor (DSF) technic is a result of various improvements to equipment and simplified arrangements. The antifreeze liquid, heated in solar panels, circulates directly through hoses cast into the concrete slab which acts as a low-temperature heating floor.
More than 50 detached houses are currently fitted with DSF in this region. Futhemore, about 15 community buildings have thus been equipped.
Technical follow up has demonstrated that DSF is a simple and reliable technic (no trouble has been experienced with existing installations), and is efficient (a correctly dimensioned project will allow an economy of 40 % to 50 %).

INTRODUCTION

Depuis 1984, la technique du PSD s'est développée en Rhône-Alpes et en particulier en Savoie, grâce à l'action de l'ASDER. Elle a permis de réaliser un programme de démonstration de 14 maisons entre 1984 et 1986 avec le concours financier de la Région Rhône-Alpes et de l'Agence Française pour la Maîtrise de l'Energie, en relation avec les architectes et les thermiciens locaux.
Depuis 1986, on assiste à une diffusion assez importante de cette technique en maisons individuelles mais aussi en équipements collectifs.

1. LA TECHNIQUE DU PLANCHER SOLAIRE DIRECT

1.1 Généralités

A partir des constats effectués sur les premières installations de chauffage et d'eau chaude solaires, l'Ecole Supérieure d'Ingénieurs de Marseille a travaillé à la simplification des installations et à la réduction des coûts d'investissements.
Les deux caractéristiques essentielles du PSD sont l'absence de cuve de stockage et l'utilisation de planchers chauffants hydrauliques.

1.2 Principe du Plancher Solaire Direct

Le fluide antigel, réchauffé dans les capteurs solaire, circule directement dans les tuyaux noyés dans la dalle de béton (voir schéma 1).
Cependant, afin de réaliser un stockage et donc un déphasage important entre l'apport d'énergie dans la dalle et la restitution à l'ambiance, les dalles chauffantes solaires ont une épaisseur plus importante que les dalles chauffantes classiques.
Les épaisseurs le plus souvent préconisées sont :
- 0,30 m en dalle de rez de chaussée (n'émettant que par la face supérieure).
- 0,20 m en dalle d'étage (émission sur les deux faces).

Comme le système PSD a de meilleures performances que les systèmes présentés en 1.1, les surfaces de capteurs ont pu ainsi être réduites (de l'ordre de 10 à 15 m² de capteurs pour 100 m² de dalles chauffantes).
Le fonctionnement d'un Plancher Solaire Direct est très simple. Un régulateur différentiel ordonne le fonctionnement du circulateur lorsque la température du fluide est supérieure à la température de la dalle chauffante.
Le Plancher Solaire Direct n'assure pas la totalité des besoins de chauffage. Il constitue une base à laquelle il convient d'ajouter un appoint séparé et à réponse rapide (convecteurs électriques, cheminée, chauffage central avec distribution par radiateurs).

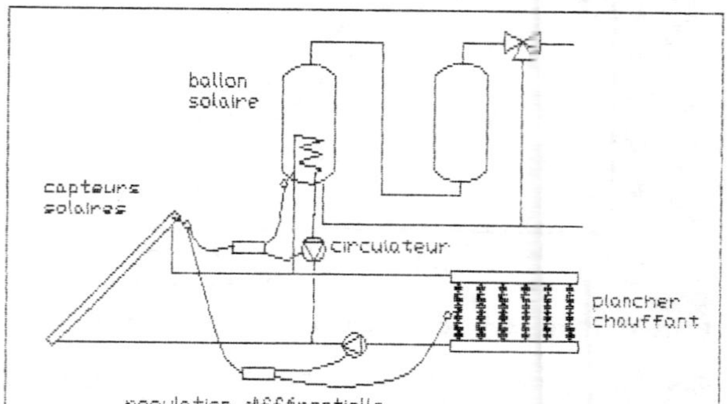

Schéma 1 : PSD en maisons individuelles

1.3 Plancher Solaire Direct en maisons individuelles

Afin de valoriser au mieux l'énergie solaire, les capteurs solaires sont aussi reliés à un ballon d'eau chaude sanitaire (voir schéma 1). Chaque circulateur est alors commandé par son propre régulateur différentiel.
Le fonctionnement au cours de l'année est alors le suivant :
.hiver:base du chauffage + préchauffage de l'eau chaude sanitaire (ECS)
.mi-saison:totalité du chauffage + couverture de 50 à 100 % des besoins en ECS
.été:totalité des besoins d'ECS

1.4 Plancher Solaire Direct en équipements collectifs

Toujours afin de mieux valoriser l'énergie solaire, diverses utilisations sont couplées :
. chauffage par Plancher Solaire Direct
. préparation de l'eau chaude sanitaire
. préchauffage de l'air neuf
. chauffage de piscines

Afin de tirer le plus d'énergie possible du fluide caloporteur, un circuit en cascade a été mis au

point par les thermiciens des cabinets ITF et Cena Ingénierie de Chambéry.
Par exemple, pour le foyer de ski de fond de La Féclaz située à 1 300 m d'altitude (photo 1), le fluide antigel réchauffé dans les capteurs arrive dans une bouteille casse-pression qui réduit sa vitesse (schéma 2).

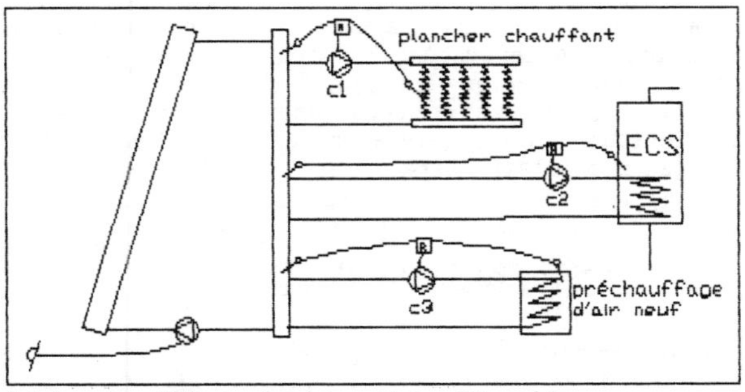

Schéma 2 : PSD en équipements collectifs.

Chaque circulateur C1, C2 et C3 est mis en fonctionnement dès que la température dans la bouteille casse pression au niveau du départ est supérieure à celle de l'utilisation (régulateurs différentiels).
En période de chauffage, le plancher chauffant nécessite la température la plus élevée ; le circuit de distribution chauffage est donc placé en tête.
Le préchauffage de l'eau chaude sanitaire est assuré par le second circuit qui récupèrera une partie de l'énergie non dissipée dans le plancher chauffant.
Le troisième circuit, assurant le préchauffage de l'air neuf, finira d'épuiser l'énergie recueillie par les capteurs solaires. Ce circuit, qui permet une, deux ou trois utilisations simultanées, favorise le rendement des capteurs car la température de retour est la plus faible possible. Ceci se traduit par une productivité très élevée de l'ordre de 2 700 MJ/m²/an.

2. DEVELOPPEMENT DU PLANCHER SOLAIRE DIRECT EN RHONE-ALPES

Grâce à une politique régionale favorable et à la présence d'un industriel fabriquant des kits PSD, une diffusion assez importante de cette technique a pu voir le jour dans la région.
Les premières maisons, instrumentées de façon à connaître les performances du PSD en zone montagneuse (voir paragraphe 4), ont été les fers de lance de cette diffusion.
Depuis, environ 80 maisons individuelles ont vu le jour dont 50 en Savoie (jusqu'à 1300 m d'altitude).
On peut noter une grande diversité dans l'architecture (voir photos 2 et 3) et le coût global de construction (de 500 à 1000 kFF).
Le coût spécifique du système solaire est de l'ordre de 40 à 60 kF pour les 10 à 15 m² de capteurs, le ballon de préparation d'eau chaude solaire, le tuyau pour planchers chauffants (500 à 700 m), le coffret hydraulique de transfert.
Parallèlement, une quinzaine d'équipements collectifs (voir photo 1 et 4) ont été réalisés (gîtes de groupes, auberges communales, complexes touristiques, centres sportifs, foyers de ski de fond). 1989 a vu la construction du complexe hospitalier de Montmélian où les 400 m² de capteurs assureront la fourniture d'environ 34 % des besoins de chauffage (par 5000 m² de plancher chauffant) et d'eau chaude sanitaire de la maison de retraite et d'un camping contigü en été avec une productivité de 1960 MJ/m²/an.
Pour 1990, de nombreuses maisons, dont les études thermiques sont en cours de réalisation, seront construites.

Photo 1 : Foyer de ski de fond de la Féclaz

3. DIMENSIONNEMENT DES INSTALLATIONS. CALCUL DES PERFORMANCES

3.1 Méthode numérique de l'ESIM (PSD)

Parallèlement au développement de la technique du Plancher Solaire Direct, l'ESIM a réalisé une méthode de calcul /1/ permettant de dimensionner l'installation solaire. Cette méthode numérique, d'un usage relativement simple, est basée sur un code lourd de simulation du système. Elle permet d'exprimer le rendement de récupération de l'énergie solaire à partir des conditions météorologiques (ensoleillement, température extérieure) et des caractéristiques de l'installation (capteurs solaires, dalles chauffantes). Cette méthode permet ainsi de connaître mois par mois la production d'énergie solaire utile dans l'habitation et, par comparaison avec les besoins de chauffage, le taux de couverture solaire.

3.2 Méthode numérique du CSTB (eau chaude sanitaire)

Pour le calcul des performances d'une installation de production d'eau chaude solaire, le Centre Scientifique et technique du Bâtiment a réalisé une méthode de calcul /2/ /3/ permettant de connaître mensuellement le taux de couverture solaire des besoins d'eau chaude sanitaire. Les données nécessaires au calcul de ce taux de couverture sont les données climatiques du site (ensoleillement, température extérieure et d'eau froide) et les caractéristiques du système (capteurs solaire, ballon, systèmes d'échanges).

3.3 Logiciel de calcul ASDER/T2I

En 1988, l'ASDER et la société T2I (fabricant de matériel solaire) ont élaboré un logiciel de calcul utilisant les deux méthodes de calcul présentées ci-dessus. Ce logiciel permet ainsi de dimensionner

Photo 2 : maison individuelle à St Jean d'Arvey

les installations de Plancher Solaire Direct et eau chaude solaire en calculant mois par mois, l'énergie solaire fournie mensuellement à chaque utilisation, le taux de couverture solaire ainsi que la productivité annuelle. Ce logiciel fonctionnant sur micro ordinateur compatible IBM est disponible auprès de T2I.

Photo 3 : Auberge communale d'Entremont

3.4 Performances en maisons individuelles

Comme il a été indiqué précédemment, un système Plancher Solaire Direct et eau chaude sanitaire solaire permet d'obtenir des performances thermiques élevées.

On constate ainsi qu'un système correctement dimensionné permet d'obtenir une couverture solaire des besoins de 35 à 60 % suivant les sites (voir tableau 1) tout en conservant des productivités de capteurs importantes comprises entre 1400 et 2000 MJ par m² de capteur et par an.

D'autre part, le coût d'investissement spécifique est relativement faible (de 2 à 2,5 FRF par MJ économisé annuellement).

Photo 4 : Maison individuelle à la Compote

4. SUIVIS TECHNIQUES - /4/ /5/

L'ASDER a réalisé, entre 1986 et 1989, un suivi technique de 14 maisons individuelles afin de vérifier le fonctionnement et les performances thermiques des installations solaires.

L'instrumentation comprenait, outre les compteurs d'énergie et d'eau du système solaire, divers comptages sur l'appoint chauffage et ECS. D'autre part, un solarimètre et un compteur de degrés-jour complétaient cette instrumentation.

Les principaux résultats obtenus sont :

- les productivités et fractions solaires réelles sont supérieures aux valeurs théoriques en plein hiver (novembre à février). Ceci résulte du fait que la méthode de l'ESIM est pessimiste pour les faibles valeurs d'ensoleillement
- les productivités et fractions solaires réelles sont inférieures aux valeurs théoriques en mi-saison (octobre et avril, mai). Ceci est dû au logiciel de calcul qui suppose une régulation parfaite du système (basculement chauffage/ECS) alors que cette dernière ne l'est pas.

Cependant, au niveau annuel, on constate un accord théorie-mesures excellent puisque la différence reste inférieure à 10 %. Ce phénomène traduit donc la validité globale des méthodes de calcul utilisées, et garantit ainsi les futurs propriétaires de maisons équipées en PSD des résultats.

D'autre part, une enquête auprès des occupants nous a indiqué que le système est tout à fait fiable et que la satisfaction des propriétaires est totale. D'ailleurs, en cas de construction d'une nouvelle habitation, ils installeraient à nouveau un système de Plancher Solaire Direct.

	Chauffage		Eau chaude		Productivité	Fraction solaire
	Besoins (MJ)	Apports solaire	Besoins (MJ)	Apports solaire	(MJ/m² an)	(%)
LILLE	44 400	15 550	13 600	4 600	1 440	35
BREST	38 000	18 450	13 050	4 450	1 640	45
ORLEANS	41 800	16 250	12 700	4 600	1 490	38
CHAMBERY	44 700	16 650	13 050	4 800	1 530	37
BORDEAUX	32 550	16 700	12 700	5 850	1 610	50
PERPIGNAN	23 050	13 950	11 550	5 750	1 970	57

BV = 174 W/K Surface habitable = 110 m2
14 m2 de capteurs (B=0.68 K=4.2 W/m2°K) sauf Perpignan 10 m2.

Tableau 1 : Performances en maisons individuelles

5. INTEGRATION ARCHITECTURALE

Une des caractéristiques du PSD est la faible surface de capteurs mise en oeuvre. Il en résulte ainsi une intégration architecturale facilitée. Le chauffage solaire n'impose plus d'architecture spécifique (voir photo 1 à 4) et s'adapte parfaitement à tous les projets de construction.

D'autre part, l'existence de couvertures héliothermiques (capteurs totalement intégrés en toiture) facilitent grandement ce succès architectural.

CONCLUSION

Les suivis techniques ont permis de montrer que le PSD :

- est une technique simple et fiable (aucun problème rencontré sur les installations existantes)
- est performant (une installation correctement dimensionné permet d'atteindre une économie de 40 à 50 %).

De plus, cette technique utilise une énergie gratuite, non polluante et inépuisable ; gage de sécurité pour les générations futures.

Il convient maintenant de diffuser le Plancher Solaire Direct à d'autres régions françaises mais aussi dans les pays de la Communauté Economique Européenne.

/1/ D.Roux/D.Mandineau/M.Chateauminois. "Calcul des Planchers Solaires Directs". Editions EDISUD 1983.
/2/ CSTB - AFME "Extension de la méthode de calcul des chauffe-eau solaires" 1984.
/3/ CSTB - AFME "Extension de la méthode de calcul des chauffe-eau solaires à d'autres climats que ceux de la France Métropolitaine" 1985.
/4/ S.Bedel - ASDER "Campagne de suivi des performances de 14 maisons individuelles à Plancher Solaire Direct" Ingéniorat Physique de l'Habitat. Université Paul Sabatier. Toulouse. 1987.
/5/ ASDER "Campagne de suivi des performances de 14 maisons individuelles à PSD. Bilan au 31 mai 1988". ASDER - AFME - Rhônalpénergie. 1988.

THE USE OF THERMAL STORAGE TO CONSERVE ENERGY FOR COOLING BIULDINGS

Sabbagh, Jafar A., Ph.D.
Mechanical Engineering Department
King Abdulaziz University
P.O. Box 9027; Jeddah 21413
Saudi Arabia

Abstract: Energy consumption by air conditioning in hot climate is very large. Several measures can be taken to reduce this consumption, e.g. building insulation and shading. In this work the author investigates the use of energy storage to reduce the energy consumption. He studied the use of thermal mass of the buildings and use of chilled water storage for this purpose. Energy storage is used for load shifting and load sharing. This can be manipulated to reduce the peak and total energy demand and to shift part of the required energy to more appropriate time. In hot arid areas air cooling is widely used for the refrigeration plant and gas turbines are commonly used for power generation. Both air cooled refrigeration machines and gas turbines are sensitive to ambient temperature. Shifting the load to the cooler period of the day improves performance of the machines and reduces energy consumption. This work shows the different ways of energy saving using thermal energy storage. This will include description of two projects designed under the supervision of the author and using energy storage as a method of cost and energy saving.

1. Introduction

Studying the electric load in the summer of a hot climate city such as Jeddah, Saudi Arabia, shows that air conditioning represents over 60%/1/ of the total load. This can be seen from Figure (1). Using of insulation and shading can reduce energy consumption by as much as 50%/2/. In this paper a study for the thermal energy storage as a means of improving energy consumption by air conditioning will be presented. The utilization of energy storage is more effective in hot arid region where the difference in temperature between night and day can be as high as 20°C . In a sea side city such as Jeddah, with high humidity during the early morning, the temperature difference is usually equal or less than 10°C. Therefore the situation in Jeddah where the study was carried, represents a less favorable situation for the optimum utilization of energy storage than inland cities such as Riyadh, the capital of Saudi Arabia.
The different ways of energy saving using energy storage can be summarized in the following:

a. Store the cool of the cold breath of the night in the thermal mass of the building (in arid areas), for partial cooling during the night and morning.

b. Store the excess in the cooling capacity of the refrigeration plant during the reduced load period, for cooling during the peak periods.

c. Reduce the peak air conditioning load by appropriate load shifting and sharing. This will reduce the capacity of the plant, and will also help to shift the required electricity load to the cooler period of the day, when the performance of the power plant and air cooled refrigeration plant are better.

2. Energy Storage Using Building Mass

Ambient heat is transmitted by convection to the external surfaces of the building and then transmitted to the interior by conduction. Similarly solar and thermal radiation heat the outside surfaces, and then transmitted by conduction to the inside of the building. Increasing the thermal resistance and the thermal capacity (weight x specific heat) of the building material, reduces the heat flow to the inside and causes time lag for heat penetration. The application of this phenomena has been studied by many investigators and air conditioning engineers, for utilization to the benefit of heating and cooling of buildings /3/, /4/, /5/, /6/.

Numerical simulation of model houses response to intermittent cooling, using mechanical air conditioning, and using cold air of the night are shown in Figures (2) and (3). The simulation was carried as part of joint study with SOM of Chicago for the KAU Campus design, under the supervision of the author, /7/, /8/. Figure (4) shows a theoretical simulation and experimental verification in a prototype model, versus soil-air temperature carried under the same project. The figures show the possibility of utilizing the thermal capacity of buildings, to reduce the energy consumed by mechanical cooling, and also the possibility to eliminate mechanical cooling in dry arid areas, where the difference in ambient temperature between day and night is very large. Based on the above studies the early Master Plan of King Abdulaziz University branch at Makkah, was planned in such a way that during the day most of the chilled water from the refrigeration plant goes to the Academic buildings. Student housing are cooled during the night, and have enough thermal mass to keep them cool during the day, and ventilation is provided by ceiling fan. This causes peak load reduction from 22800 ton to 17205 ton, future reduction to 16000 ton can be gained by using chilled water storage of about 4000 m^3 capacity. Both housing and academic buildings were designed to use masonry walls, concrete slabs and insulation from outside.

3. Additional Thermal Storage

Using additional thermal storage such as chilled water storage for load sharing in large projects, is helpful in reducing the installed capacity, improve maintenance and operation of the refrigeration machines. Examples of such projects are the ones mentioned above. The addition of 4000 m^3 of chilled water storage reduced the peak load by 1000 tons. Another example is the KAU Jeddah Campus, currently being designed under the supervision of the author. The Medical Campus of KAU is currently under construction, and has a peak load of 18000 T.R. It will be supplied by 3 x 6000 air cooled chillers, with a fourth additional standby chiller. During the Master Plan phase of the Main Campus, it was decided to interconnect the two central utility plants, along with other services in the two campuses. In the early phases the Medical Campus CUP will be used to supply both campuses with chilled water. This requires the construction of 30000 m^3 chilled water storage and the operation of four chillers, with one additional (fifth) chiller as standby. In the final phase of construction estimated total peak is about 70000 T.R. The addition of chilled water storage of 30000 m^3 will reduce the peak load by 17%, adding storage capacity of 70000 m^3 will reduce the peak load by 37%.

4. The Cooling Load in the city of Jeddah

The estimated electric load for cooling at the city of Jeddah during a typical July month is shown in Figure (5) together with a typical weather data /1/, /2/, /10/. Most of the houses in Jeddah are relatively heavily constructed from 200 mm hollowed or solid masonry walls and concrete slabs, this includes 200 mm masonry partitions. Interior walls are plastered, and exterior walls are plastered from inside and outside, or claded with stones. Additional external insulation is rarely used. The effect of the thermal capacity of this relatively heavy construction can be detected in the cooling load, shown in Figure (5). During early morning period the high relative humidity is responsible for the high load. Because of the thermal capacity of the buildings, the cooling load starts to drop at 4 a.m. The load rises again at 6 a.m. due to the solar radiation and the warming of the atmosphere. The load drops again due to the thermal capacity, and drop of the relative humidity. The peak at 4 a.m. is due to the delay caused by the building mass, plus the entering of many window type units into operation, for the afternoon local habit of taking short rest at that time of the day. Using other measures, such as insulation and shades, together with thermal capacity of buildings, will reduce further the total peak load by more than 50%, and produces a smoother load curve. The effect of the combination of added insulation to the existing relatively large thermal capacity of the existing buildings, is much larger in inland hot arid climate such as Riyadh, due to the large drop in temperature during the night, and the absence of extreme humidity during early morning hours.

5. The Effect of Thermal Storage in Total Energy Consumption

Reducing the peak load of air conditioning in a house unit, in a major project or in the total city, will produce saving in the cost of the operation and maintenance, as well as the cost of the installed capacity of the air conditioning and refrigeration equipments. In a city similar to Jeddah, it will be reflected also in the total cost of power generation, because the reduction in the required installed capacity, and because smoother load curve for the power plant, allows better performance and maintenance.

As shown in the paper, using insulation and construction with large thermal storage capacity, will reduce electric energy consumption and produce smoother load curve. Additional energy saving due to the fact that air conditioning peak load occurs at the warmest period of the day. This has a major impact in increasing the actual energy consumption. In arid land air cooling is widely used for refrigeration machines. Ambient temperature in the afternoon of a typical summer day reaches 45°C and it is possible to reach 55°C in some days in inland city. Using a moderate initial temperature difference of 20°C, will put the condenser temperature of the refrigeration plant at 65°C, and can reach 75°C in some days. At this extreme conditions the coefficient of performance of the plant drops. The energy consumption of the plant increases as much as 30%, for an increase in the condenser temperature from 50 to 60°C. Further drop occurs in C.O.P. if the refrigeration plant is large and has to operate at partial load during the peak period. In the power plant side gas turbine is widely used for power generation in arid regions, because of the scarcity of water. The efficiency of gas turbines depends on the ambient temperature as shown in Figure (6). Therefore operating gas turbines at the cooler period of the day requires less fuel. Additionally the peaks of air conditioning load may require the operation of gas turbines at part load, and to increase the number of start and stoppage of the gas turbine units. This will add to the maintenance cost, reduce the performance and increase the fuel consumption.

6. Conclusion

In conclusion the use of load sharing and shifting using thermal storage will help beside other measures such as thermal insulation and shading to conserve air conditioning. This is of great importance in hot tropical climates where electric consumption by air conditioning is considerable.

7. References

1. Publication of EWR
2. Saggagh, J., "Effect of Passive Solar Design of Buildings on Electric Energy Consumption", European Conference on Architecture, April 1987.
3. Markus and Morris, "Buildings, Climate and Energy", PITMAN, London, 1980.
4. Bollington, N., "Building Physics & Heat", Pregamos Press, Oxford, 1968.
5. Sabbagh, J., "Passive Solar Cooling" Proceedings of Solar Cooling Workshop, Dahran, 1980 MRI/SOL-0601.
6. ASHARE Task Group on energy, "Algorithms of Building Heat Transfer Subroutines", N.Y., 1976.
7. S.O.M. Technical Library on the Master Plan for King Abdulaziz University, Makkah Campus, 1979.
8. Communications with King Abdulaziz University Projects Deparment, 1989.
9. Publications of the Department of Meteorology and Conservation, Jeddah, Saudi Arabia, 1989.

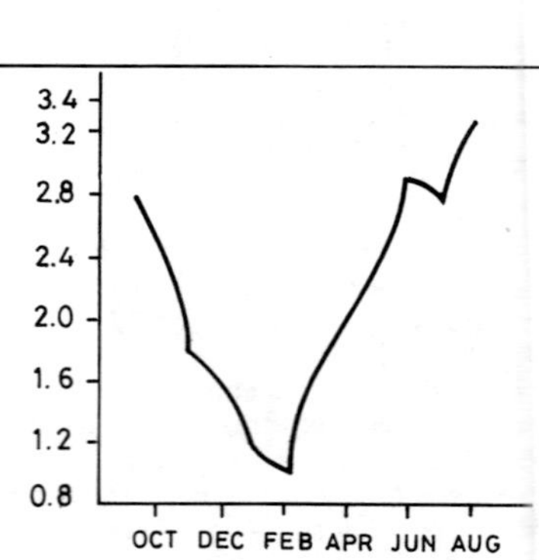

FIG.(1) RATIO OF A-C LOAD TO TOTAL LOAD IN JEDDA

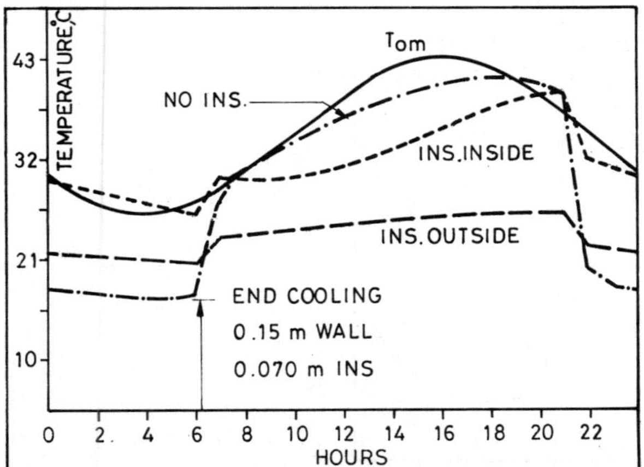

FIG.(2) BUILDING RESPONS TO INTERMITTENT COOLING

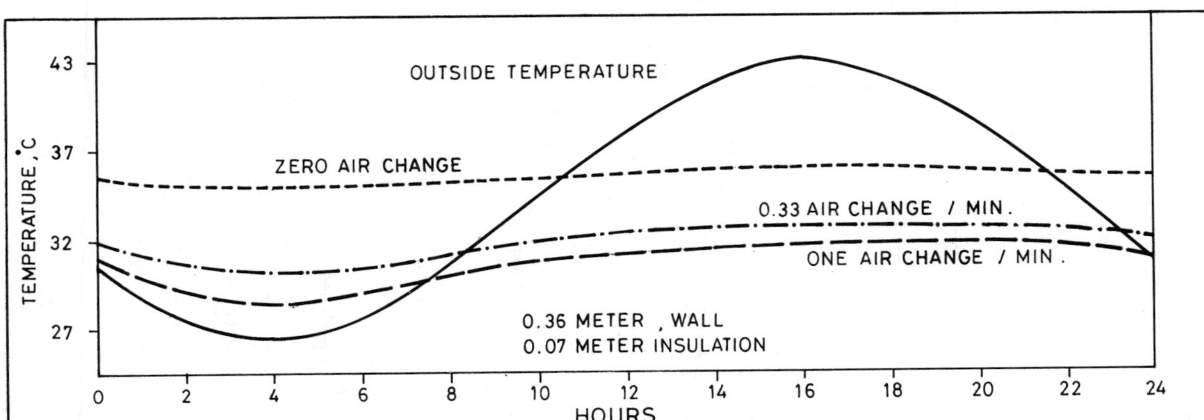

FIG.(3) COOLING BY NIGHT VENTILATION

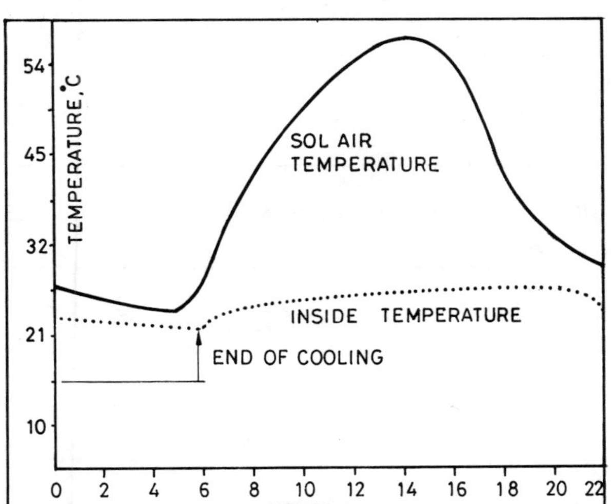

FIG.(4) BUILDING RESPONS/SOL AIR TEMPERATURE

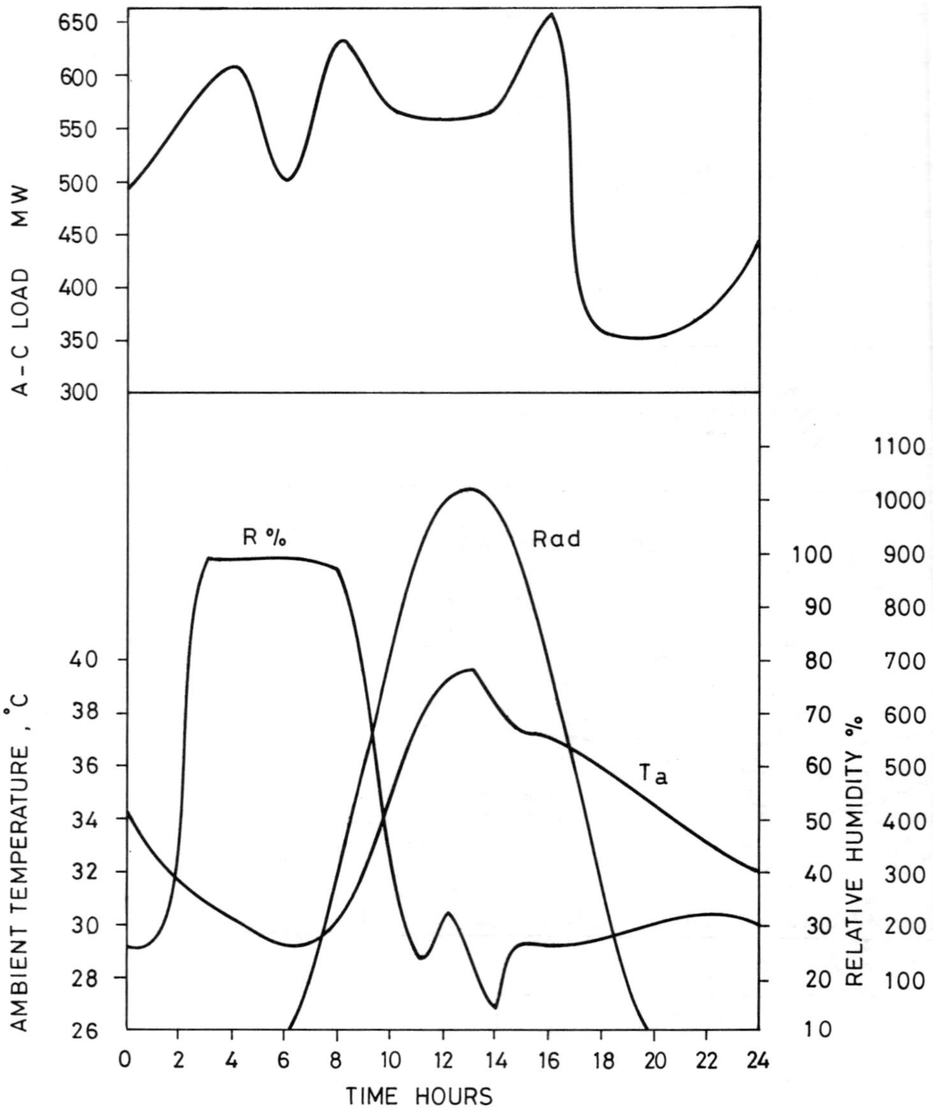

FIG (5) ESTIMATED AIR CONDITIONING LOAD FOR TYPICAL SUMMER AND WEATHER DATA AT JEDDAH

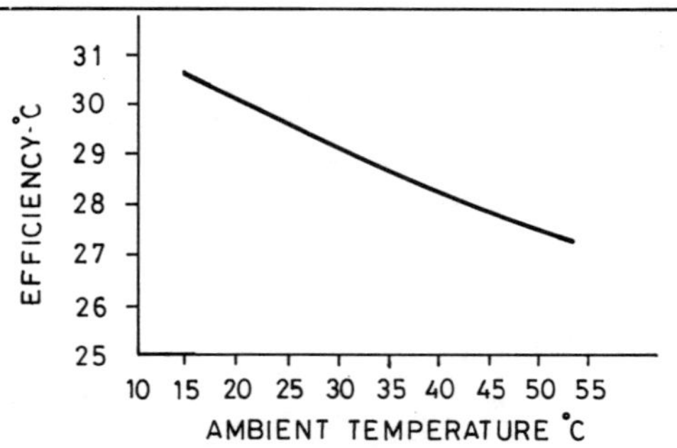

FIG(6) EFFECT OF AMBIENT TEPERATURE IN THE EFFICIENCY OF GAS TURBINE

CALIBRATION OF THE SPANISH SOLAR TEST CELLS

Heras M.R., Tellez F.M., Perez M., *De Ignacio G.
Solar Division/IER-CIEMAT
Avda. Complutense, 22/28040-MADRID (SPAIN)
*E.U. de Arquitectura Técnica de Madrid
Avda. Juan de Herrera, 6/28040-MADRID (SPAIN)

ABSTRACT

This paper aims to develop testing methodologies for Passive Solar Components (Heating, Cooling or daylighting), as well as to obtain characterization parameter of components to be tested. Two identical Test Cells have been built in Tabernas (Almeria), located in the South-East of Spain. Description of these test cells is the subject of another paper presented in the conference. Each cell have two rooms, one designed for instrumentation and the other, highly insulated, for testing purposes. Test room have the possibility of changing a wall and the insulated ceiling. Starting from the basic knowledgement of all materials used in the construction of these cells and using a simulation model, the global heat losses coefficient was determined. At present a simple comparation with the experimental measurements is being carried out.
A comparation of both test cells without excitations was carried out. At present a comparative test is being developed to test both the submitted test cells to heating sequences.
This paper describes the present work and testing plan to obtain a final calibration.

1. INTRODUCTION

In the test site two equal test cells was built. Detailed description of this cells is presented in another paper in this conference. Test site is located at the Plataforma Solar de Almeria (Spain). Test cells are strongly isolated (with 0,40m of styropor of 0,037W/m°C conductivity) and have the possibilities of change a wall, the ceiling of the test room and to turn over a concrete platform. To test different passive solar components and systems with different inclinations can be instaled by desmounting the wall, the ceiling or both.

Before starting test of passive components a "calibration" period is required. This period have, besides to detect contruction defaults, the following objectives:

- To obtain a overall loss coeficient value and to compare it with the theorical value obtained whith the S3PAS simulation model (developed by E.T.S. de Ingenieros Industriales de Sevilla and supported by IER-CIEMAT).
- To check the performance of the Data Adquisition System and to elaborate a "user-friendly" Data Adquisition Program for feature testing and choice a valid calculation procedure. This calculation procedure will be in consecuence with the test methodology chosen.

1.2.- THEORETICAL OVERALL LOSS COEFFICIENT.

Final desing of insulation frame (type of material and dimensions) was determined evaluating differents options with S3PAS program. Until now this program have not been fully validated, but partial comparation with other simulation models shows,/1/, that results are trustworthies. An stimation of the overall loss coefficient of test room in the cells was calculated in a double way: by simulation using S3PAS and by manual calculus appling recomendations of the Spanish Standard NBE-CT-79.

For the calculus with the S3PAS program the following data was introduced:

CONSTRUCTION:

- Dimensions of each component of the test room (adiabatic movable wall, adiabatic ceiling and fixed walls)
- For each layer of each component the values of conductivity (W/m°C), density (Kg/m^3) and specific heat was required.
- Absortivity, emisivity and exchange convective coefficient (W/m2°C) of exterior surfaces.

OPERATING CONDITIONS:

- The program was request to calculate energy needs to maintain the interior temeperature with a constant value of 22°C.
- Air change rate of 1/4 volume/hour.

EXTERIOR CONDITIONS:

- non real values was introduced to simulate laboratory conditions:
  - Constant exterior temperature
  - Solar radiation void.

With the S3PAS model heat loss coefficients of components of the test room was obtained. this are shown in Table 1.

TABLE 1. EXCHANGE LOSS COEFFICIENTS OF COMPONENTS CALCULATED

Component	U
FLOOR :	U = 0.105 W/m2.K
MOVABLE WALL:	U = 0.077 W/m2.K
FIXED WALL:	U = 0.076 W/m2.K
MOVABLE CEILING:	U = 0.077 W/m2.K
DOOR:	U = 0.170 W/m2.K

The power consumption to maintain a constant difference of temperature (17°C) was calculated using S3PAS. The rate between this value and the difference of temperatures, for stationary conditions gives the overall loss coefficient. Table 2 shows the results.

TABLE 2. OVERALL LOSS COEFF. CALCULATED

LOSS COEFF. (W/K)	CALCULUS METHOD	CONVECTIVE COEFF. (W/m2.K)	AIR EXCH. RATE (Vol/hour)	SURF.
6.55	S3PAS	hi=3.5;he=10	0.25	EXT.
10.41	S3PAS	hi=9;he=17	0.25	EXT.
6.37	NBE-CT-79	hi=9;he=17	0.25	INT.
8.20	NBE-CT-79	hi=9;he=17	0.25	EXT.

To calculate overall loss coeficient using NBE-CT-79 only mentioned construction data was required. The resultant value obtained for two possible surface dimensions (interior or exterior) of the components are shown in table 2, also.

2.- EXPERIMENTAL CHARATERIZATION.

2.1.- INTRODUCTION

With the initial configuration of both test cells with all components insulated (meterials are descripted in other document in this conference), the following test have been programmed:

- to compare free performance of both test cell submited to heating sequences,
- to measure air infiltration as a function of presure diferences (interior-exterior), and
- to make thermographies during heating periods to detect thermal bridges.

Because of delay in final installations the calibration period is not finished and only preliminary results can be shown.

2.2.- COMPARATIVE TEST.

Firstly, the two test cells were oriented toward the south (this can be changed as desired) and a first visual analysis of the final arrangement was checked "in situ" to detect possible sources of infiltrations.

The object of the first test was to obtain the approximate value of the overall loss coeficient.

Test configuration was composed of: a heating source of costant power (100 W), measurement of interior ambient temperature in the central point of the test room with thermocouples (radiation protected), and measurement of exterior temperature (from meteorological station data). With this configuration and the door closed, conditions was held. Same conditions was stablished in both cells. The data adquired was hourly averaged, and Figure 1 shows this measurements.

To obtain the first approximation to global overall loss coefficiente a single equation of balance was stablished

$$Qaux + UA(Tint-Text) + Qsto = 0 \qquad (1)$$

where Qaux = auxiliary power
UA = global heat loss coefficient
Qsto = energy storaged into the components per second

A simplification was: to take only values obtained during night (solar radiation void) and to applie equation (1) for values of diference of temperature averaged in night period. Also the term Qsto was void. This asumption is justified in stationary conditions. Values calculated are:

UA = 12.37 (test room of cell 1)
UA = 12.99 (test room of cell 2)

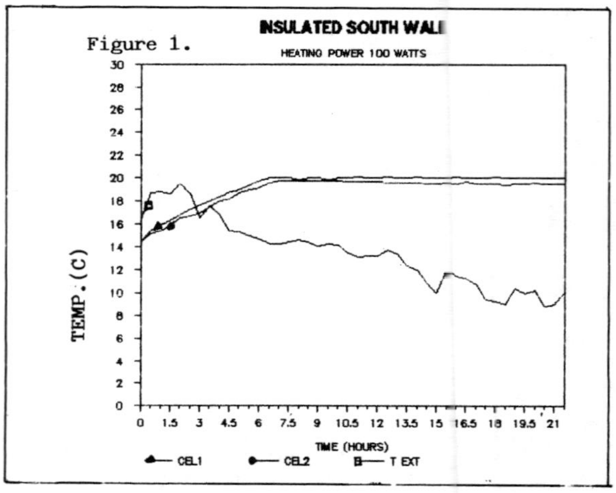

Figure 1.

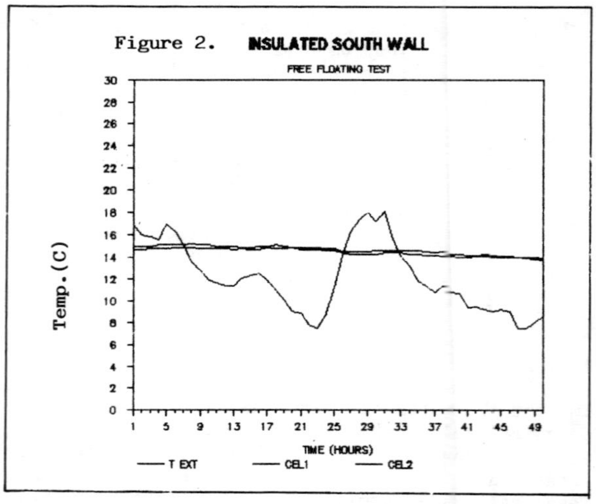

Figure 2.

To compare both cells in free floating conditions was also developed a secuence of measurements, these are drawed in Figure 2.

2.3. RESULTS ANALYSIS AND CONCLUSIONS

Supposing that the test cell's dynamic can be represented in terms of a unic node with time constant

$$\zeta = MC/UA$$

where mc = overal heat capacity
UA = overal heat loss coeficient
ζ = response time

we have calculated for mc values of 3500000 Jules/K (previusly estimated) and for UA approx. 12W/°K the response time is approximately ≈ 82 hours.

This time stimated is in accordance with the empirical results achieved. That is, in both cases, power supply and free performance experiences, it is observed that the test cells are apparently insensible to their contour conditions. But it can be observed in Figures 1 and 2, the indoor temperature have a slow tendence in both cells and both experiments. Indoor temperature is continuosly disminishing in order to achieve an stationary régime. Time periods greather than [illegible] days are needed to this.

In consecuence, the previous UA values stimates (13 W/K) is only a firts approximation to the real one. At present we are trying to determine it appliing dynamical identification methods to the test cell's performance.

however, we can conclude:
1. Both test cells show a very equal behaviour and then they could be used in comparative test methodologies.
2. It can be expected that the UA real value were around the before estimated value. Then both test cells could be used in absolute methods for passive components characterization.

REFERENCES

/1/. S.Alvarez and others "Metodo de Simulación y Calculo de Sistemas Solares Pasivos en relacción a su comportamiento térmico". E.T.S.I.I. Sevilla (1987).

/2/. C.Cools, R.Gicquel,F.P.Neirac:"Identification of building reduced models. Application to the characterization of passive solar components". PASSYS doc. (1988).

/3/. N.Milbank, L.Shorrock, J.Uyttenbroeck, K. de Cuyper:"Recomendation for calculating the energy needs of buildings with particular reference to dwellings". C.E.C. doc.(1987).

/4/. M.R.Heras, F.Téllez,M.Pérez, G.de Ignacio: "Descripcion de las células para ensayo de componentes Solares Pasivos-Almeria (CESPA)". 2nd. Draft. IER-CIEMAT.

SPECIFICATION, CONSTRUCTION AND INSTRUMENTATION OF THE SPANISH SOLAR TEST CELLS

Heras M.R., Tellez F.M., Perez M., *De Ignacio G.
Solar Division/IER-CIEMAT
Avda. Complutense, 22/28040-MADRID (SPAIN)
*E.U. de Arquitectura Técnica de Madrid
Avda. Juan de Herrera, 6/28040-MADRID (SPAIN)

ABSTRACT

This paper describes two passive solar test cells built in Almeria (spain). This cells are focused to develop realistics tests and constitute an intermediate facility for the analysis of physics mechanisms in real buildings, among the simplest ones, like "hot-boxes" with minimal realism but maximal control of exciting variables, and the monitorized buildings, occupied or not with maximal realism and minimal control of exciting variables.
Testing objectives in these cells are to obtain a short number of parameters sufficiently separate from testing conditions to characterize the passive solar system or component and being able to be introducted in a simplified model to extrapolate the behaviour of components in a more complex building. This behaviour is refered to mechanisms like thermal transmision, ventilation, evaporation, ilumination, etc.
The final design and constructive specifications adopted, have taken into account recomendations and previous experiences and the possibility to test non vertical components and different orientations.
Equipments are focused to simulate real interior conditions and to control interior climate and meteorological data in order to provide all necessary data for different test methodologies. Data acquisition system is composed of distributed modules and has connected a sufficient diversity and quantity of sensors.

1. SPECIFICATION

1.1. Introduction

As a project belonging to the IER-CIEMAT "Energía Solar Pasiva" program, whose final objective is to extend utilization of passive solar technics, offering the architects and designers the knowledge and tools for energy conscious design of building, two test cells have been built. Other projects connected with these test cells, within the program are: Development of a software, a simulation model for passive solar systems, "S3PAS", (1988) and a simplified calculation method for evaluation of energy saving opportunities for a given design (foreseen 1990).
Passive systems appropriate to Mediterranean Climates are less developed that those with only heating purposes. Besides economical reasons it can be mentioned that the complexity is higher in passive systems for cooling like roof components, shading devices, components aimed at inducing ventilation, evaporation systems, etc... where the physical behaviour is strongly dependent on microclimate conditions.
This paper presents two spanish solar test cells that have been built in Almeria (shout).

1.2. Situation and climate

Two test cells have been built in the "Plataforma Solar" of Almería. This is a R&D center devoted to solar energy directed by the Instituto de Energias Renovables (IER) of the Centro de Investigaciones Energéticas, Medioambientales y Tecnológicas (CIEMAT).
This center is located in Tabernas (Almeria), town with latitude 37°N and 500m height over the see.
Both cells are shadow free in south direction.
Almeria is located in A zone: less than 400 annual degrees/day with base 15-15 (according to the spanish standard UNE-24.046). Other meteorological data have been measured on site (fig. 1) or take from references, (fig. 2).
With this location, climatic conditions with cooling needs are expected during more than 6 months.

1.3. Other Facilities

Before developing the spanish test cell design, a short review of existing test facilities was realized (1987):
"In More and others (1982), /7/, a revision of existing North American" test cells can be found. Baker (1983), /2/, describes existing test facilities in North America and Europe, establishing the design principles for the actual PASSYS test cells (described for example in Wouters and others 1987), /8/. A brief design for a Spanish Solar test cell was drawed by Cejudo and Heras (1987), /4/ considering posibilities of testing passive solar components for cooling. With a definition of the characteristics and of the areas belonging to the "Mediterranean climate" proposed by Canha da Piedade (1987), in Antinucci (1987), /1/, a Feasibility Study for Mediterranean test cells can be found. Last reference shows a short review of Mediterranean test cells experiences" that includes the design of the spanish test cells.

1.4. Objectives

As we have mentioned, two test cells have been built in Almeria with the main interest in providing design tools that increase utilization of passive system in Spain. Other objectives are:
- to define test methodologies for passive solar

components used in Spain (for cooling, heating and Daylighting when were possible) to obtain dynamic models for component characterization,
- partial validation of the spanish simulation program S3PAS,
- to carry out comparative studies between different commercial passive components,
- to develop and to study the behaviour of new passive components,
- characterization of components (using a reliable methodolgy) to obtain specific parameters for the passive solar component that could be introducted in simplified models in order to extrapolate the behaviour in complex buildings.

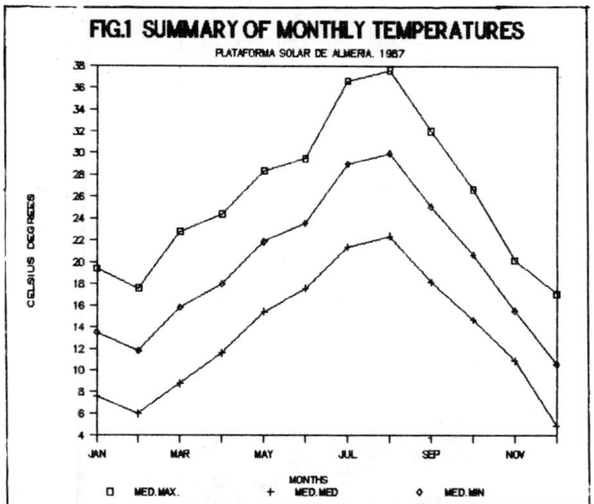

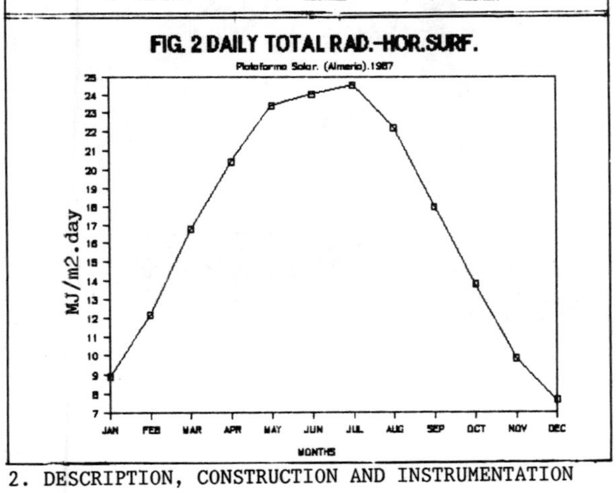

2. DESCRIPTION, CONSTRUCTION AND INSTRUMENTATION

2.1. Design Priciples

The design principles are according with the objectives defined in the last paragraph. From PASSYS design some principles have been accepted:
- construction of 2 cells in order to allow comparative test methodologies,
- possibility to change the "south" wall,
- especial care has been taken to avoid uncertainties like thermal bridges, floor losser or infiltration, ventilation has to be controlled by force to eliminate uncertainties, other design principles to extend test possibilities have been considered,
- the possibility to turn the test cells has been added to investigate the behaviour and sensibility of non horizontal components in different orientations,
- possibility to change the ceiling to allow testing of passive components located in the roof,
- the inertial mass could be distributed like realistic configuration or concentrated in the interior and will be placed after a "good calibration" of test cells.

2.2. Description of the cells

The cells have two rooms, the services room, to place equipments and the test room where the components are mounted (fig. 3) a general draw of the cell.
Constructive criteria:
roof load = 4.9kN/m2
floor load = 4.9kN/m2
wall load = 0.69kN/m2
concret ground load = 1.72kN/m2
Outside dimensions of the test cell are 9,6x4m2 surface for floor structure and 3.7m high plus 0.3m wheels plus 0.5m of the "V" roof.
A brief description of the characteristics of test room is the following:
- Floor structure surface = 22 m2
- Floor interior surface = 16 m2
- Floor thickness ≈ 0.45 m
- Floor insulation = 0.30 m (styropor: density 40 kg/m3; conductivity=0,023 W/m.K)
- Movable wall surface = 4.0x3.8 m2
- " " insulation = 0.4 m (styropor)
- Movable ceiling surface = 6.4x4.0 m2
- " " insulation = 0.4 m (styropor)
- Fixed walls insulation = 0.4 m. (styropor)

A wider description of the cells can be founded in Heras and others (1989), /6/.

3. INSTRUMENTATION

The instrumentation is composed of equipments for data acquisition and monitorization, and equipments for controlled heating, ventilation and simulation of casual gains.
The data acquisition system has to cover three zones separate 50 meters approximately. It is based on the conexion of multiplexors located near each measurement site to an interface lodged into a personal computer (fig. 3) by a 3 wire cable.
At present each test cell disposes of fourty measurement chanels and the meteorological station disposes of a multiplexor module of twenty chanels. With this system and the acquisition software, it is possible to obtain one scan of all the chanels per second showing physical values of each chanel on the screen. Average and stadistic results are storaged in hard disk.
Sensors installed are shown in table 1.

TABLE 1. LOCATION AND TYPE OF SENSOR

Meteorological station (fig. 3)
. Piranometer on horizontal plane
. Pirheliometer for direct radiation
. Pirheliometer for diffuse radiation
. A combined wind velocity and direction meter
. Dry temperature
. Wet temperature
Test cells
. Piranometer on vertical plane
. A combined humidity and ambient temperature meter,radiation protected
. Termocouples "T"
. Termorresistences "PT100"
. Watimeter for electric power comsumption

4. CONCLUSIONS

The test cells that we have presented, have been built and partially equiped. At present these installations are starting on measurements with a calibration with all the adiabatic movable components. These cells are focused to component characterization and we hope that they will be a powerful tool to increase research in the field of passive systems in Spain. Otherwise, we expect these cells will contribute to extend utilization of passive systems applied in Mediterranean climates.

REFERENCES

/1/ Antinucci, M. (Nov. 1988). "Feasibility Study for Mediterranean Test Cells". Final report PASSYS Program. CONPHOEBUS. S.C.R.L.

/2/ Baker, N.V. (1983). "Passive Solar Test Facility". A report preposed for the Commission of the European Communities DG XII.

/3/ Bedoya, C., Carril, A., Cembranos, L., Macía, M. and Neila, F.J. "Las energía alternativas en la Arquitectura". COAM. Madrid, 1982.

/4/ Cejudo, J.M., Heras, M.R. (1988). "Spanish Solar Test Cells". PLEA 88: Energy and building for temperate climates. Porto.

/5/ Dijk V. D. (1985). "A test facility for Passive Solar Energy Systems in the Netherlands". Rep. of TPD-TNO

/6/ Heras, M.R., Téllez, F.M., Pérez, M. and De Ignacio, G. "Descripción de las células para Ensayos Solares Pasivos-Almería (CESPA)". 2nd Draft. Report. IER-CIEMAT.

/7/ Moore, E.F. and McFarland, R.D (1982). "Passive Solar Test Modules". Rep. Los Alamos National Laboratory.

/8/ Wouters, P. and Ph. Voordecker (1987). "Passys test cells: ground plan, section, furcades and details". Rep. of PASSYS program. BRI. Brussels.

Figure 4. SECTION OF TEST CELL AND MOVABLE ADIABATIC COMPONENT

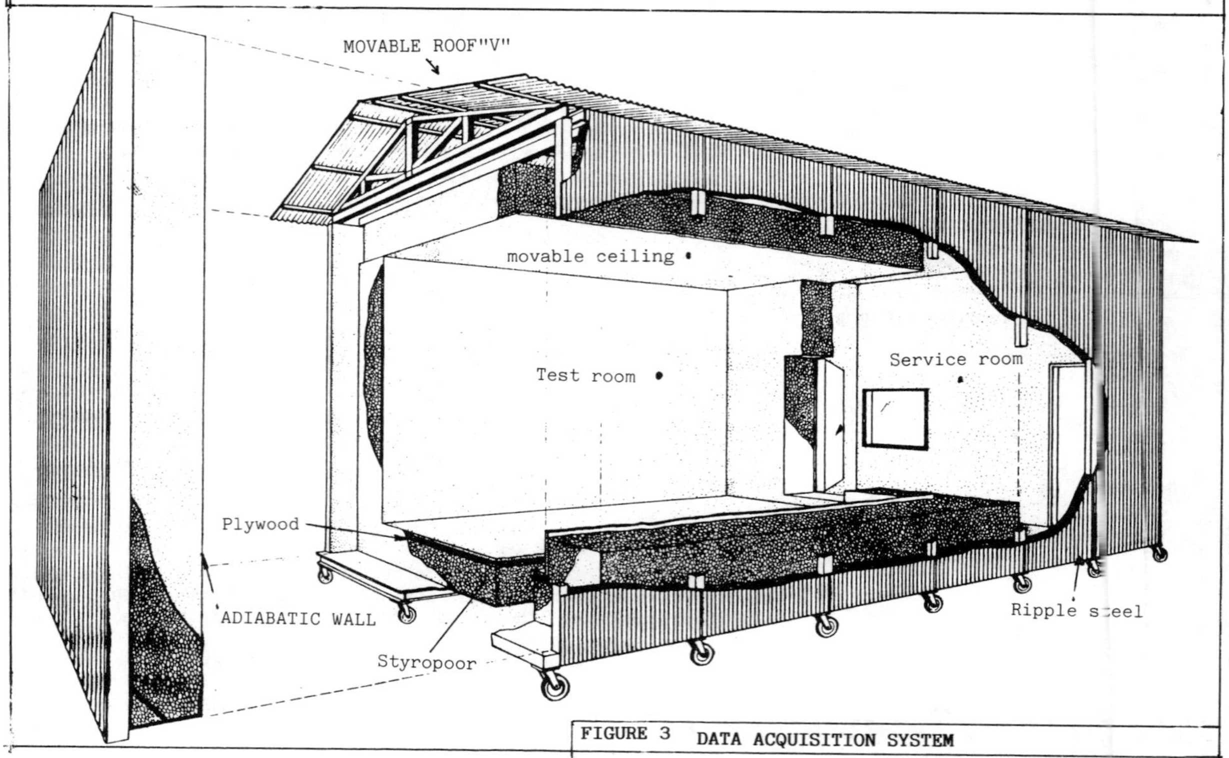

FIGURE 3 DATA ACQUISITION SYSTEM

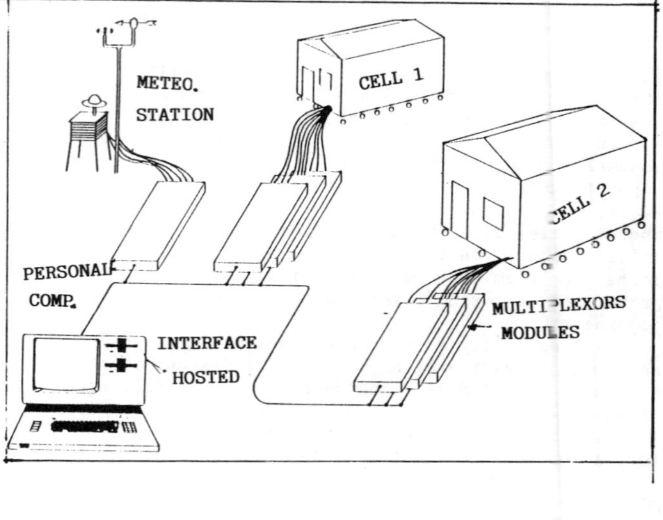

CLIMATE RESPONSIVE CLADDING ELEMENTS

G. Scudo
DPPPE, Politecnico di Milano
Via Bonardi, 3, 20133 Milano, Italy

P. De Pascali
ICIE - Istituto Cooperativo per l'Innovazione
Via Nomentana, 133, 00151 Roma, Italy

ABSTRACT. Four climate responsive cladding components are presented: a ventilated cavity wall, a double window, a solar chimney light wall and a solar façade. Basic configuration operating modes technological features, costs and energy saving are briefly described.

1. INTRODUCTION

Climate responsive cladding/elements components are used as building technologies to control daily and seasonal flux exchanges (light, heat, energy) and to maintain comfortable indoor conditions reducing non-renewable energy use.

The basic configurations of these technologies are:

a1 - heat transfer (converted from solar radiation), without air exchange: close-loop Inside-Inside (I-I) in passive systems;

a2 - heat dissipation: close-loop Outside-Outside (O-O) in ventilated walls;

a3 - heat distribution inside the space: close-loop Inside-Inside (I-I), inside ventilated walls;

b1 - heat transfer (converted from solar radiation), with air exchange: open-loop Outside-Inside (O-I);

b2 - heat transfer with heat recovery and air change: open-loop Inside-Outside (I-O).

The main variables to classify climate responsive components are the type of fluid (air, water), the air circulation (natural, forced), the type of circulation (closed, open), the operation modes (manual, mechanical), the regulation systems (at component level, or integrated with HVAC systems).

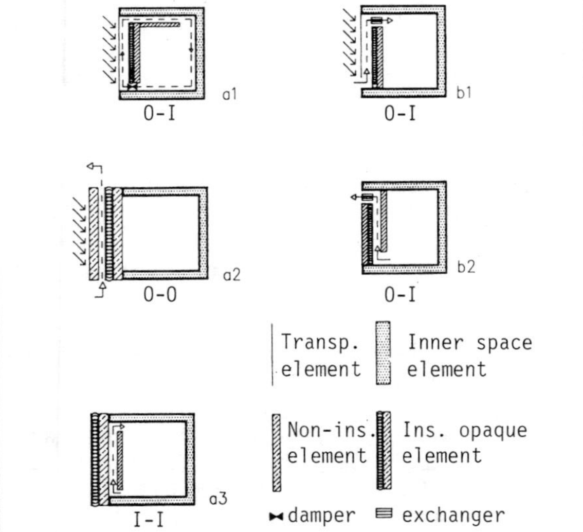

Fig. 1 Dynamic configurations of the climate responsive cladding components.

2. VENTILATED CAVITY WALL

The component (1) is a multilayer wall which operates as traditional ventilated cavity wall in Summer (a2, closed-loop O-O), and as glassless solar panel in Winter (b1, open-loop O-I). When air in the cavity is heated over 18°C in Winter the damper allows air to flow inside the building.

The main elements of the component are:

- external layer: a 0.01 m thick glassfiber reinforced concrete panel;
- insulation: variable with climate; in this case 0.07 m polystyrene panel;
- air cavity: a 0.035 m air-space;
- opening: a 0.03 x 2.50 m opening at the bottom;
- damper: a 0.05 x 2.50 m insulated dumper electrically powered and controlled by a local regulation system;
- structural layer: a 0.1 m thick reinforced concrete layer.

The main technical and economical features are:

- dimension: height 2.80 m, width 2.80 m, thickness 0.22 m;
- weight: 260 kg/sq.m;
- conductance: U = 0.5 w/sq.m°C;
- air tight (only damper): 12 c.m/h sq.m.(100 Pa);
- energy saving: 10% solar heating fraction in Winter in mild climate (yearly DD 1000) and solar collector/space volume ratio 1:7 (sq.m/c.m).

The extra-cost is 60.000 lire/sq.m (1987 price) if related to the cost of the traditional cavity wall which is about 160.000 lire/sq.m.

The component is used as a traditional ventilated wall in East-West façades and as glassless solar collector in South façades. In Mediterranean climates it can satisfy the demand of pre-heating air change at a rate of 0.5-1 cubic meter hour per cubic meter.

3. DOUBLE WINDOW

The component (2) is a double window with a space inside equipped with two Venetian blinds (one black and one white), dampers and fans which allow different operating modes in natural or forced convection.

In forced convection fans and dumpers are regulated by a simple microprocessor which controls the

different configurations.
Heat and mass exchanges between the cavity and the internal-external environments are:
- in Winter: - a1, close-loop I-I, to heat inside space; b1, open-loop O-I, to pre-heat incoming air;
- in Summer: - a2, close-loop O-O, to dissipate heat; b2, open-loop I-O, to cool inner space.
The prefabricated component is made by following elements:
- aluminium frame with doubled glazed pane;
- air cavity 0.18 m thick;
- two venitian blinds in the middle of the cavity;
- aluminium frame with single glazed;
- return air opening operated on/off by electromagnete;
- climate responsive microprocessor regulating seasonal and daily commutations of dampers and switching on/off fans.
The main technical and economical features are:
- dimension: height 2.80 m, width 1.20 m, thickness 0.24 m;
- weight: 55 kg/sq.m;
- conductance: 2 w/sq.m°C;
- air tight (average glazed panel and external damper): 8 c.m/h sq.m (press.diff. 100 Pa)
- energy saving: 15% solar heating fraction in mild climate (1000 yearly DD), with window/space ratio 1:7 (sq.m/c.m).
Prototype was designed to be used as external transparent cladding component in traditional office building technology (floor to floor) or to be integrated into a curtain wall system.
The extra cost is about 100.000 lire/sq.m (1987 price), if compared to a traditional double window or glazed curtain wall.
When the Venitian blinds are used to reduce light flux, the heat converted from incident solar radiation if not dissipated causes overheating. In winter this component acts as a solar collector for space heating; in summer it operates as a solar chimney for cooling.

4. SOLAR CHIMNEY LIGHT WALL

The component is a solar chimney integrated into a prefabricated light cladding element, performing both building and energy functions.
Solar chimney is a thermosiphon air panel (configuration a1, close-loop I-I), connected distribution and storage systems which allow heat transfer in space (South-North) and time (day-night).
The main configurations are:
- in Winter: - a1, close-loop I-I during the day which transfers heat converted by the panel;
- a1 modified, with a plastic membrane which closes the return opening to prevent reverse air circulation;
- in Summer: - a2, close-loop O-O, to ventilate the South wall; b2, open-loop O-I, to ventilate inside space, structure and people.
The prefabricated component is made by the following elements:
- metal frame with single glazed;
- air space: 0.05 m;
- absorber: black 1 mm thick aluminium sheet;
- air space: 0.05 m;
- seasonal dumper with top and bottom openings;
- the containing structure: fibreglass reinforced concrete 0.01 m thick with insulation;
- daily damper and return air opening;
- local control and regulation system.
The main technical-economical features are:
- dimension: height 4.50 m, width 2.00 m, thickness 0.20 m;
- weight: 60 kg/sq.m;
- conductance: U = 0.5 w/sq.m°C;
- air tight (average on glazed panel and external damper): 10 c.m/h sq.m (press.diff. 100 Pa);
- energy saving: 30% year solar heating fraction in Northern Italian climate with a chimney/space ratio 1:10 (sq.m/c.m).
The extra-cost of the component is about 100.000 lire/sq.m (1987 price) if compared to a traditional cladding system.
The most appropriate application of a solar chimney component is in the residential sector because the conversion-storage system fits well the residential pattern of use. Nevertheless it can be used also in non-residential buildings, provided an alternative distribution-storage system is designed to fit a discontinuous pattern of use and an adequate sizing of air change system appropriate to the air loops of the chimney system.

5. SOLAR FACADE

The solar façade is composed by solar panels integrated into prefabricated reinforced concrete panels.
The solar façade was designed to be used in office building with different storage and distribution schemes, according to the specific use pattern.
The façade operates in configuration a1, close-loop I-I and was produced for a sport centre built in Modena, Italy (Polisportiva Modena Sud).
In Winter insulated ducts in the ceiling distribute heated air to the exchange unit (a concrete cavity wall) which stores and distributes by radiation heat to the inside space with different modes (intensity and time lags) depending on specific heating demand.
In Summer heat is discharged outside through a damper and solar panels work as a traditional ventilated wall.
The prefabricated component is made by the following elements:
- double glazed panes, supported by steel bars;
- metal steel absorber: 1 mm thick;
- reinforced concrete panel: 0.25 m thick;
- insulation: 0.05 m thick;
- damper for Summer operation mode;
- control system through the fans regulated by HVAC microprocessor.
The principal technical and economical features are:
- dimension: height 8 m, width 2 m, thickness 0.46 m;
- conductance: U = 0.4 w/sq.m C;
- air tight: 8 c.m/h sq.m (press.diff. 100 Pa);
- energy saving: 17% year solar heating fraction (average Northern Italy climate), solar panel/space ratio 1:10 (sq.m/c.m), storage 0.05 cubic m of concrete/sq.m of solar façade.
The extra cost is about 160.000 lire/sq.m (1987 price) if compared to a similar prefabricated

ventilated wall.
One of the limiting factors to the diffusion of active solar technology in building is the high cost due to the small dimension of solar components and to the cost of distribution and storage systems.
The huge spaces of multipurpose centres (shopping, sport, free-time, etc.) allow large solar panels whose extra-cost is very low; horizontal and vertical bearing structures (ceiling and walls) can be designed to integrate thermal functions (heat distribution and storage) reducing furthermore the global extra-cost of active systems down to acceptable levels.

CLOSING REMARKS
The high level of building integration of the above described components determined low extra-costs compatible with their life cycle (about 5 years for ventilated wall and double window, 10 years for chimney wall and 15 years for solar façade).
Energy performance for heating season was carried out with a simulation code validated with experimental data (ref. /1/, /2/, /3/, /4/). The cooling performance for central and southern italian climates is yet to be done and we expect a reduction of energy consumption for heating and cooling of about 50%.

NOTES
(1) Produced by CMC, via Emilia, 37038 Sant'Arcangelo di Romagna (FO), Italy, in the frame of a research contract between ENEA (Natl. Committee for R&D of Nuclear Energy and Renewable Energy Sources) and ICIE (Istituto Cooperativo per l'Innovazione, Roma); scientific responsibility of W. Mebane, ENEA, P. De Pascali, ICIE, G. Scudo, Politecnico di Milano.
(2) Produced by Coopsette Infissi and CIR, via Volta 8, 42049 Sant'Ilario d'Enza (RE), Italy, in the frame of the a.m. research.
(3) Produced by OCM-VALMA, via Lombardia, 99, 22050 Rovagnate (CO), Italy, and designed for ENEA Solar Laboratory Building at Ispra, designed by G. Scudo and A. Seassaro, Corso Garibaldi, 50, 20100 Milano, Italy.
(4) Produced by CMB, via Karl Marx, 101, 41012 Carpi (MO), Italy, in the frame of the EEC demonstrative project.

REFERENCES
/1/ G. Rossi, G. Solaini: "Contribution to Building Climatization given by an Experimental Ventilated Cavity Wall"; Proc. of the 6th Int., Conf. on Passive and Low Energy Architecture (PLEA 88), p. 363, Porto, Pergamon Press, 1988.
/2/ B. Merello, G. Rossi, G. Solaini: "Thermal Behaviour of a Fenestration Component with Controlled Air Flows", Proc. of the 2nd World Congress on Heating, Ventilating, Refrigerating and Air-Conditioning Clima 2000, Sarajevo, 1989.
/3/ G. Contadini et al. "Monitoring of a solar chimney and different ceiling storage/distribution systems in an office bilding located in northern Italy", Proc. of ASES Annual Conference - 14th National Passive Solar Conference, Denver, Colorado (USA), 19-23 June, 1989.
/4/ G. Rossi, G. Solaini, Solar façade integrated with a H.V. A.C. system, Sicence and Technology at the Service of Architecture, 2nd European Conference on Architecture , Paris, 4-8 Dec. 1989 (to be published).

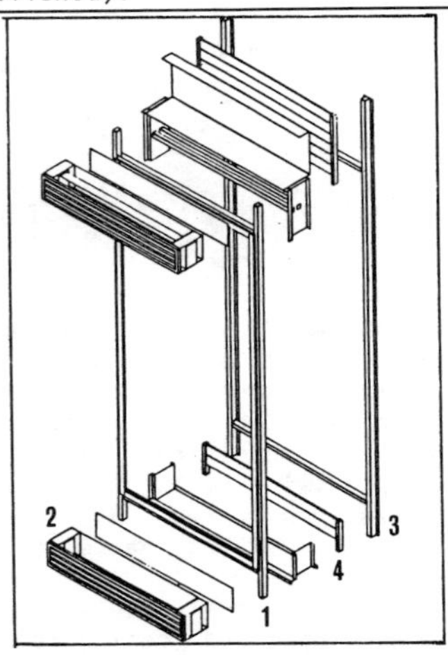

Fig. 2: Dynamic double window:
1-outside double glazed pane; 2-seasonal dampers; 3-inside single glazed pane; 4-return air opening.

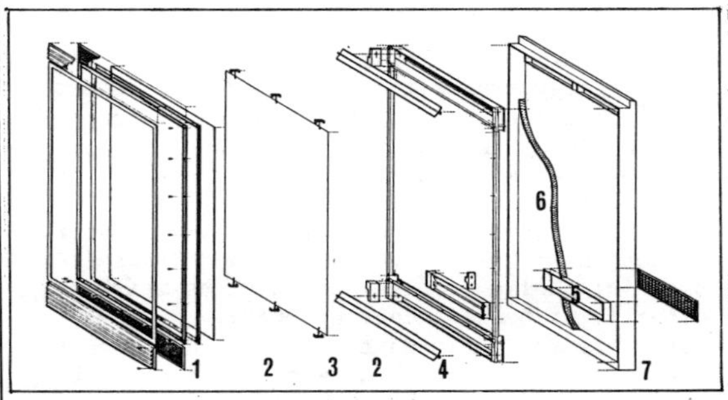

Fig. 3: Solar Chimney curtain wall:
1-single glazed pane; 2-air gap; 3-absorber; 4-seasonal dampers; 5-return air opening; 6-insulation; 7-G.R.C. containing structure

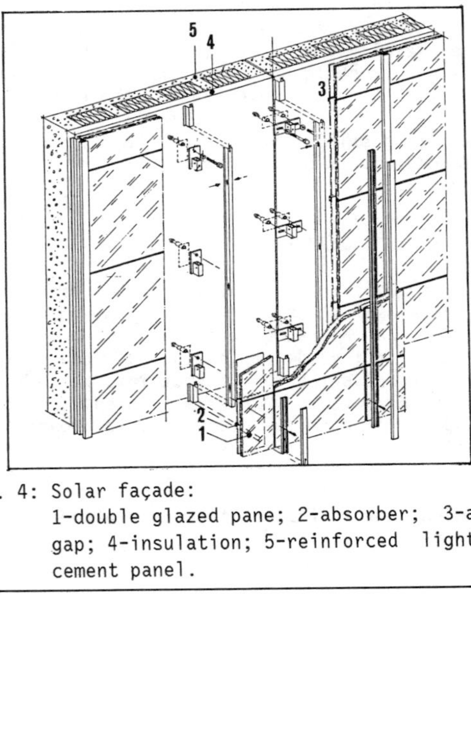

Fig. 4: Solar façade:
1-double glazed pane; 2-absorber; 3-air gap; 4-insulation; 5-reinforced light cement panel.

PASSYS SOLAR TESTING AND PERFORMANCE PREDICTION

H.A.L. van Dijk
TNO Institute of Applied Physics (TPD)
P.O. Box 155
2600 AD Delft
The Netherlands

ABSTRACT: Within the context of the CEC research PASSYS a passive solar test facility is under development in Delft. The PASSYS project is introduced in another session of the conference. A special addition to the TPD testcells is a movable cold box for steady state, dark, hot box tests in addition to the tests under real climate conditions. Parameter identification is one of the promising techniques under development within PASSYS if combined with a carefully selected test control strategy.
At the TPD a specially adapted parameter identification method, MRQT has been developed.
The main test-results can be used as product information and as input data in a simplified design tool, for example the multizone correlation method TCM-heat, developed at TPD. Some typical results are presented.
In series of calculations the dynamic behaviour of a translucently insulated wall is investigated and energy savings are predicted.

1. INTRODUCTION

A Passive Solar Component (PSC) is mainly characterised by the thermal transmission coefficient or U-value (heat loss) and the solar energy transmittance or g-value (heat gain). The measure in which the PSC is capable to store heat and/or the air permeability can be other main characteristics. Unfortunately it is not an easy task to determine these properties: the main heat transfer mechanisms like free convection and transmission of solar radiation cannot be measured directly with sufficient accuracy.
Moreover, the performance of a PSC depends on the size, orientation and position, the building mass and the presence of competing heat sources, etc. The main task in an experiment is to distinguish these factors and to reveal both the performance of the PSC itself as its interaction with the building and the occupants.
A series of experiments, in which each test is aimed to find the effect of one of the influencing factors would result in an excessive overall test duration. Therefore, statistical techniques are being developed and employed in combination with the available physical knowledge on testcell and PSC, to make it possible to extract a maximum of information from a relatively short test sequence.

2. RESEARCH OBJECTIVES

In the research at the TPD with respect to passive solar energy systems the following objects can be recognized:
- quantitative determination of the system characteristics for product information;
- testing of new system concepts before application in practice, to decrease the risk of costly mistakes;
- system optimization;
- system modelling;
- development of simplified design tools.

3. INTRODUCTION PASSYS

The TPD is the Dutch participant in the international research project PASSYS which was initiated in 1986 by the Commission of European Communities.
The Dutch participation in PASSYS is financially supported by NOVEM, the Dutch agency for the management of energy and environmental research and by TNO.

At nine locations, spread over the CEC countries as PASSYS testfacility has been set up. This facility consists of four identical outdoor testcells. The open fronts of these full scale cells are faced Southwards; here the passive solar components can be mounted for testing (figure 1).
Within the testcell a realistic indoor environment is created. The highly insulated envelope can be provided with high or low mass internal lining to approximate various situations which can be found in practice.
More extensive information on the PASSYS project can be found in another section of the Conference.

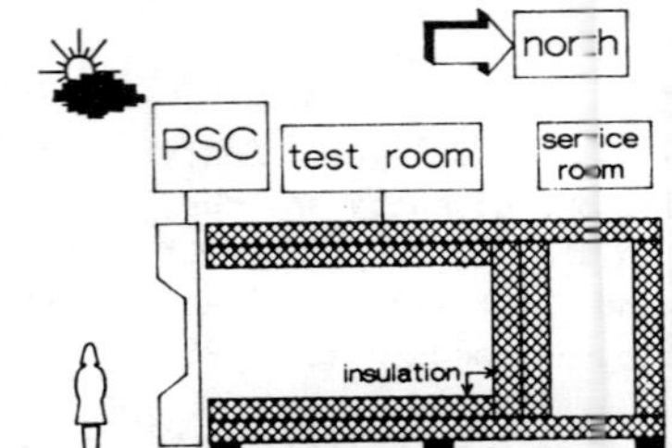

figure 1. The standard PASSYS testcell, with full scale testobjects (PSC).

4. DEVELOPMENT TESTMETHOD

One of the major tasks within PASSYS is the development of an optimal teststrategy and appropriate test evaluation techniques which enable to obtain sufficient information from a relatively short test period.

Promising test evaluation techniques are the socalled parameter identification methods.
Parameter identification is a technique which starts by assuming a mathematical model for the testcell and the PSC, both for the steady state (heat loss and gain) and the transient (accumulation!) thermal conduct; the values of the model parameters, for instance heat resistances and - capacitances are unknown before the test. With the aid of a computer-program the parametervalues are found (identified) as those values which provide the best agreement between the output of the test and the result of a model calculation using the same input data as during the test, see figure 2.
Usually the testroom temperature is appointed as output variable; input variables are outdoor temperature, solar radiation, amount of heating, a.s..
Within PASSYS different methods are being compared at the moment.

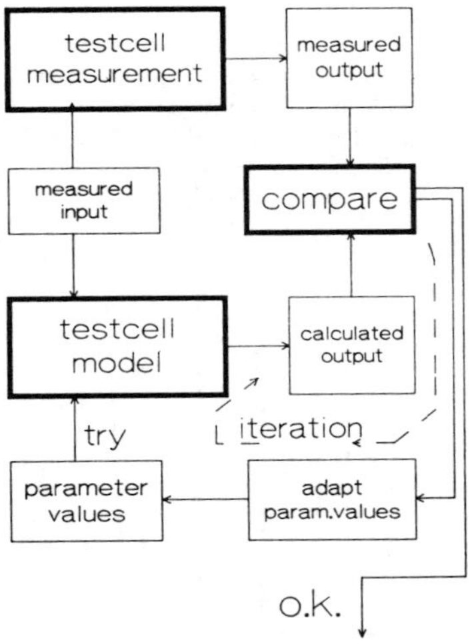

figure 2. The principle of parameter identification to derive the properties of the PSC from the test-cell measurements.

The TPD has developed the parameter identification method MRQT, which is specificly adapted for the PASSYS type of experiments: it can cope with non-linear effects (like temperature dependent heat transfer coefficients), discontinuities (blinds up/down), variable timesteps (successive periods with high and low frequency variations), it can select specific parts of a test sequence for identification, a.s.. Moreover, since in MRQT the model is just a black box with input- and output dataset and a set of parameters, it is very easy to exchange the model for another one for whatever reason. At the moment one of the main topics of investigation is to select a model which requires only a small number of parameters, but still gives sufficiently accurate results for a wide range of conditions. Evidently, one of the basic requirements is that the main PSC and testcell characteristics can be derived from the identified parameter values.

5. THE TPD MOVABLE COLD BOX

Outside the scope of PASSYS, the TNO Institute of Applied Physics developed the movable cold box, see figure 3. This cold box can be placed over the outdoor facing side of the Passive Solar Component to obtain controlled, dark outdoor conditions (low temperature, constant air speed) to measure the heat transmission properties of the component, like the UA-value. This addition to the PASSYS test cells thus converts the cell into a 'hot box' arrangement.
But also dynamic tests can be carried out with this combination e.g., to examine the response on changes in ambient temperatures.

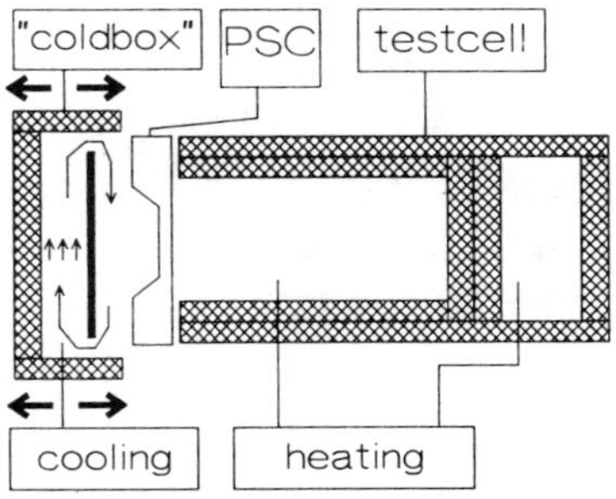

figure 3. Illustration of the TPD movable cold box in combination with a PASSYS testcell.

6. TESTRESULTS

During the period 1988/1989 first testseries were carried out.
These tests include:
- a test on a windowsystem with the testcell-coldbox combination to derive the U-value;
- a comparative test between two identical walls, one of which was provided with internal blinds on the window;
- a test on a prototype wall with exterior translucent insulation material (figure 4).

Figure 5 shows one of the results from the tests on the translucently insulated wall; the characteristics of the wall were identified with the measured heat flux entering the testroom through the wall as reference output variable. The agreement between measured and calculated heat flux is very good.
However, the identified thermal transmission coefficient and solar transmittance were lower than predicted by theoretical studies. Part of this is probably due to minor deficiencies in the testcell and the prototype wall; this will be further investigated.

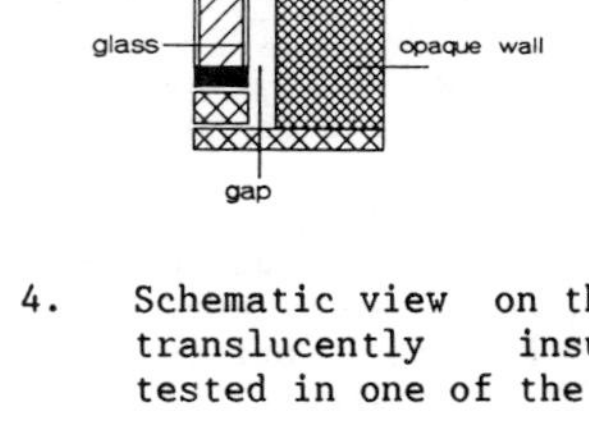

figure 4. Schematic view on the prototype translucently insulated wall tested in one of the PASSYS cells

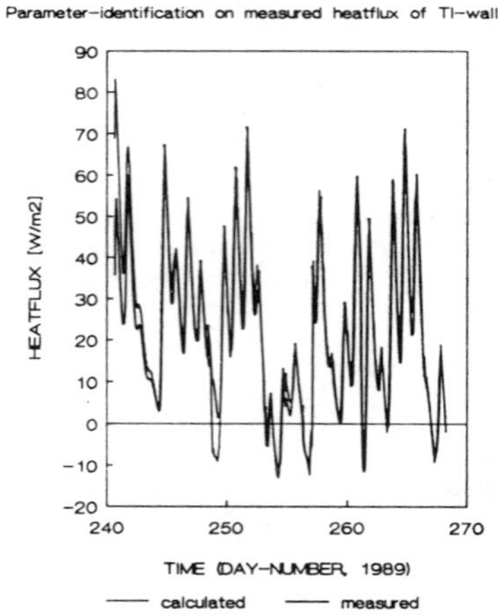

figure 5. One of the results on the translucently insulated wall; measured: heat flux measured in the cell; calculated: result of parameter identification

7. THEORETICAL STUDIES

Computer simulations involving the dynamic thermal behaviour of passive solar components and their interaction with the indoor environment are aimed at the optimization of a specific concept or at the development of more general guidelines.
An example is the development of the simplified method 'TCM-heat'.
TCM-heat is a multi-zone correlation model to calculated the monthly heating requirements in a building. TCM-heat uses the concept 'utilization factor' for the monthly mean solar and internal heat gain. The utilization factor is mainly a function of the room's capability to absorb excessive heat, the type of occupation (dwelling, office) and the ratio between the total amount of gain and total thermal load by transmission and ventilation heat loss, the socalled gain load ratio GLR, see figure 6.

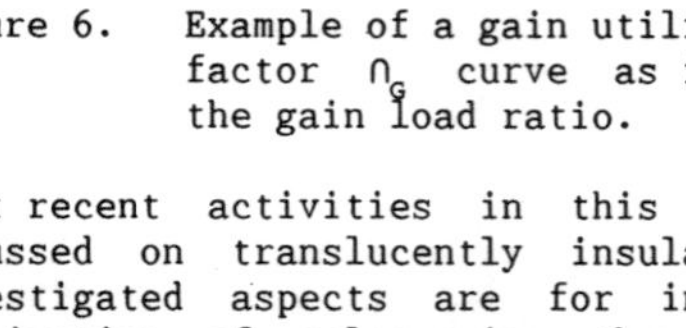

figure 6. Example of a gain utilization factor η_G curve as function of the gain load ratio.

Most recent activities in this field were focussed on translucently insulated walls; investigated aspects are for instance the utilization of solar gains, thermal comfort and local peak thermal loads on the materials.
Figure 7 shows a typical result: the mean heat flux through a 10 cm. wall provided with PC honeycomb material as a function of material thickness, for the period 1 november-31 march, in the Netherlands.

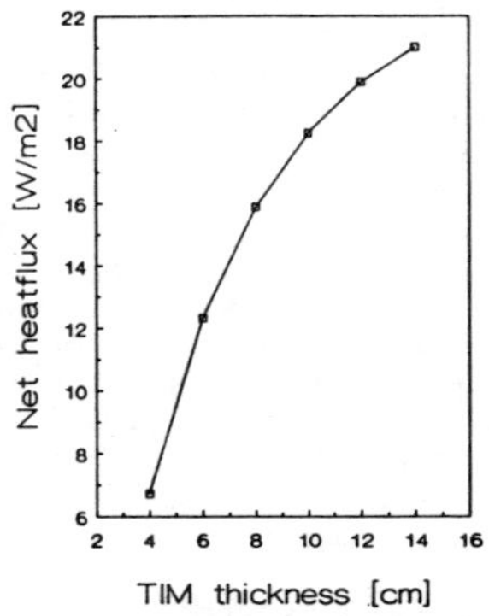

figure 7. Influence of honeycomb thickness on the mean net heat flux of a translucently insulated wall; period november - march; South vertical.

Figure 8 shows the effect of the thermal inertia of a 20 cm. wall with exterior translucent insulation: even on an East facade the heat is supplied to the room at an hour which makes it possible to complement the direct gain through a South window.

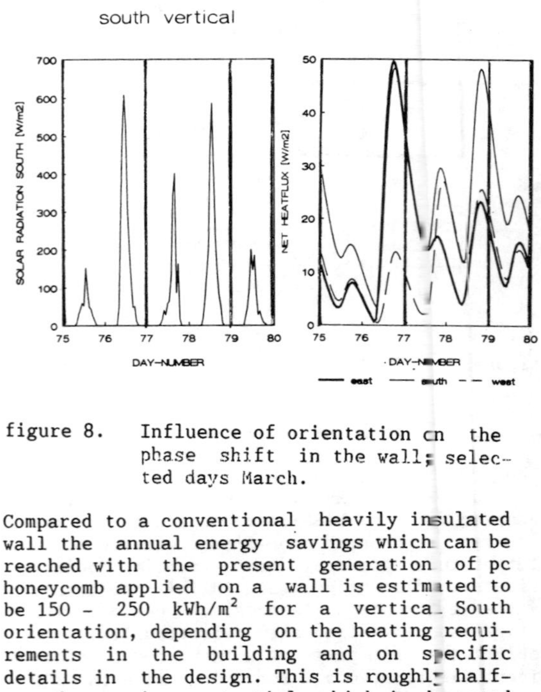

figure 8. Influence of orientation on the phase shift in the wall; selected days March.

Compared to a conventional heavily insulated wall the annual energy savings which can be reached with the present generation of pc honeycomb applied on a wall is estimated to be 150 - 250 kWh/m^2 for a vertical South orientation, depending on the heating requirements in the building and on specific details in the design. This is roughly halfway the maximum potential which is dictated by the limited supply of solar energy during midwinter and by the limited heating requirements in spring and autumn.

AIR FLOW WINDOWS

F.Bonvehí
Consultant Engineer
C/. Pintor Pahissa nº 27, 4º E-08028 Barcelona
X.Traver and M.Miró
Ecotècnia
C/.Demóstenes, nº 6 E-08028 Barcelona
A.Trias
Architect
C/. St. Joan, nº 8 E-08629 Torrelles de Llobregat (Barcelona)

ABSTRACT. Air flow between two panes of a building's skin may provide optimal solutions to effectif control of energy fluxes through walls and by ventilation. The results reported in this paper refer to glazed walls only. The air flow window consists of two glazed panes with venetian blinds in between and air circulation in this space. This device is able to control Daylight Transmitance, Solar Protection, Solar Thermal Collection and Dynamic Insulation. This technology can be applied to windows, courtain walls and greenhouses. Our work has evolved from simulation by computer models and experimental analysis of prototypes on test-cells to several case application studies, single house equipment and retrofitting of a primary School under a EEC Energy Demonstration Project.

1. INTRODUCTION

Air flow windows associate ventilation inside the building and control of energy fluxes through the glazed walls. The basic equipment consists of two glass panes with a venetian store in between, the air flows in this space. This system can be applied to classic windows, double windows, courtain walls and greenhouses.
Design of air flow windows require a through analysis of kind and use of the building as well as of the local climate. Applications may enhance residential, office or leisure buildings, hospitals, schools, etc.
Among the institutions and companies revolved with the project at different stages are: Center National de la Recherche Scientifique, Agence Française pour la Matrisse de l'Energie, Commission of the European Communities, Generalitat de Catalunya, Asolba, Pechineg-Btiment, Folcrà, CICSA, ICIE and Ecotècnia S.Coop.

2. DESIGN CRITERIA AND ISSUES

In designing an air flow window for a particular application we have several elements to play with: glass panes, store and air circulation. Monolithic clear float panes are usually adequate. For settings with very cold winter seasons low emissivity glass may be considered or even a traditional double glazing for the external wall. To reinforce solar protection tinted or reflective glass for the external wall can be of interest. Material, color and dimensions of the venetian blinds must be adjusted to the actual requirements of visual comfort and esthetic appearance.
Concerning air flow several patterns are to be considered. The ventilated wall may function as an intake air system, in this case fresh air introduced into the building is being preheated in its passage through the air space in the wall.
The air flow window may operate as an exhaust air system as well. In this case used air is impelled off the building through the window, the effect is to level temperatures of room air and internal surface of the windows. As a consequence comfort nearby the glass walls is improved otner in winter or in summer.
The air flows itself can be driven either by forced or by natural convection. Air flow windows may operate as an autonomous device or, if requested, the air flow can be integrated with a central ventilation system. On many days air temperature rise at the outlet of air flow windows can be 15 to 30% depending on flow conditions and blinds.

3. BASIC PRINCIPLES

The energy fluxes in an air flow window are described in figure nº1. Solar radiation impiping the external glass pane is transmitted, reflected or absorbed, the last fraction is remitted at infrared wavelengths. When the transmitted fraction strikes the slats of the venetian blinds a fraction will be absorbed depending on incident angle. This angle determines the function that eventually reads the internal glass pane which in turn undergoes the same process as in the external wall. The surface temperature of the internal glass wall determines infrared radiation exchanged with its surroundings. The air flowing in the cavity cools the venetian blinds an the glass walls as it increases its temperature. By evacuating heat from the window the equilibrium temperatures of the different parts is diminished.
The main properties of air flow windows related to energetical phenomena are:
- Natural Light Transmittance.
- Solar Protection.
- Solar Heat Collection.
- Dynamic Insulation.

To express the these properties in a quantitative way the following parameters are used:
- Daylight factor.
- Solar factor.
- Collector efficiency.
- Heat transmission coefficient.

For many window arrangements these parameters have a single value. Whereas for air flow windows these parameters offer a wide range of variation. An important part of our work has been dedicated to establish measured for these basic parameters.

4. METHODOLOGY ON RESEARCH AND DEVELOPMENT

4.1 Simulation

A mathematical model running on computers has been developed. The model is based on a set of equations describing dynamic thermal equilibrium. Temperature distributions are calculated through an interactive process, then thermal fluxes are evaluated. This model allows the determination of values of the basic parameters under variable environmental conditions and different system designs.

4.2 Experimental testing

Prototypes of air flow windows about 90 cm. width and 200 cm. high have been extensively tested. This

experimental work has been performed on test-cells under natural climatic conditions and on laboratory with artificial radiation. Locations of such tests include Toulousse-France, Barcelona-Spain and Bangkok-Thailand.
Experimental data obtained have enabled the validation of the mathematical model and the determination of the characteristic curves of solar radiation, solar heat collection and dynamic insulation as a function of air flow rate.

4.3 Case studies and projects

In parallel with the basic work described, several architectural projects for real buildings have been performed and some instructed. Windows, and attached greenhouses for single houses have been studied, also fenestration for hotels, hospitals, schools and office buildings.

5. NATURAL LIGHT TRANSMITTANCE

The ratio of daylight in a particular point of a room to the external available daylight, both measured on a horizontal plane, is a constant trough daytime and known as daylight factor.
In air flows windows the daylight factor may be adjusted to the needs by properly setting the inclination of the venetian blinds. This setting can be manually operated by the occupants or automatically controlled by sensors, timers or computers. With venetian blinds set at 45% inclination in south exposure we can have about 500-600 lux at 2 meters from the window in central day hours.

6. SOLAR PROTECTION

The efficiency of a shading device is expressed by the solar factor. It is defined as the ratio of the total energy flux into the building to the total solar radiation impiping the shading device. The air flow window has a variable solar factor.
When the store is retracted and without air circulation, we have the solar of our double glazing which is 78 for two 4 mm panes of clear float. This solar factor is suitable for direct heat gain. With white venetian blinds and slats inclined at 45º the solar factor is 26, it can be reduced to 20 by ventilating with exhaust air at 25 m3/hm2 rate flow. When the window's frame is taken into account the solar factor becomes 14. This results are shown in figure nº2. Typical values of solar factors for external shading devices with double glazing are about 10. The reduction of temperature on the surface of the internal glass has two effects: reduction of the cooling load and greater comfort nearby tnat surface.

7. SOLAR HEAT COLLECTION

Air flow windows under sunshine function as solar collectors. This characteristic is expressed by the solar collection efficiency, which is the ratio of the thermal energy evacuated by ventilation to the total solar energy striking the window. Experimental data are plotted in figure nº 3. With an air flow rate of 20 m3/hm2 efficiency is about 20%, with white/black venetian blinds and 15% with white only. Air temperature increase when air flows along the window may vary from 15 to 25% for a 2 meter high window.

8. DYNAMIC INSULATION

Ventilating air flow windows with exhaust air has the effect reducing the temperature difference between the room ambient air and the glazed wall. This differences may be about 1,5% at a flow rate of 20 kg/hm2 and outside/inside delta T of 18C. Heat loss through the glass wall are dismissed and discomfort nearby the window almost disappear.
The dynamic insulation effect is shown in figure nº4. The K value of the window varies with air flow. The K value of an air flow window with 20 m3/hm2 is 1 W/C m2. The K value of a window without ventilation is 2 W/or m2.

9. BUILDING APPLICATIONS

The air flow window system has been successfully applied to verandas in single houses and apartments. A greenhouse attached to a single house has been built using the technology of air flow windows.
The glazed pyramidal dome of an aquatic park in Le Tonquet France has ventilated double glass panes. The inclined walls have no venetian blinds.
Figure nº 5 shows the basic layout of the application. Renewal external air after preheating on an air exchanger flows in the dome's skin. The aim is to avoid condensation on the glass and therefore to suppress water drops falling over the users.
The retrofitting of a primary School in Agramunt/Lleida is under way. The ancient fenestration is being replaced by air flow windows working with air of classrooms. This project has been related as a Demonstration Energy Project by the E.E.C.

10. AGRAMUNT'S SCHOOL PROJECT

The primary school of Agramunt was built in the thirties. In consists of two floors of 750 m2, the ceiling height is 4,5 m. The fenestration surface is about 500 m2, which is a third of floor surface and with good exposure.
Retrofitting is under way and will be implemented in two stages. In 1989 the ground floor's window are being replaced and ventilation of the classrooms installed. In 1990 the works will be completed by replacing windows in the first floor and installing ventilation as well, the roof will be insulated and space redistributed.

10.1 Monitoring

Measurement of local climatic conditions, comsumption, thermal fluxes and internal ambient conditions will be under way for two years. Two classrooms will be intensively instrumented in order to have complete information of the performance of the system.

10.2 Working principles

An outstanding feature of this technology is its capacity to adapt to different weather conditions and school's occupational patterns.
In winter the system will function on the basis of heating the air renewal. The forced ventilation extracts air from the classrooms and fresh air is introduced by the air flow window. Intake air temperature will regulate blower's operation. Figures nº 6 7 shown the layout.
In summer the air circuit will be inversed and exhaust air will pass along the air flow window to the exterior by blower intaking air from the north façade. Figures nº 8 9 show this set-up.

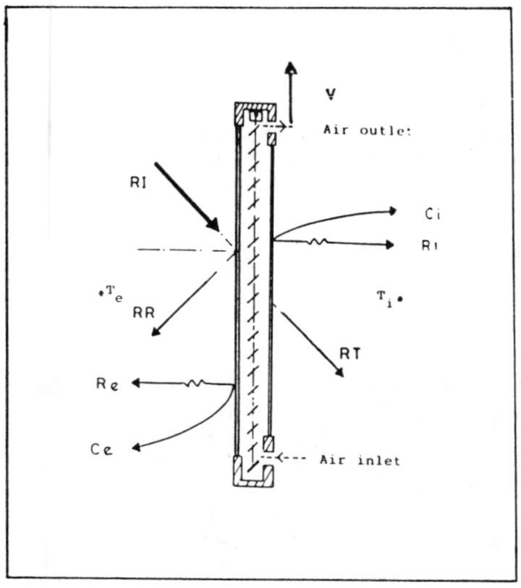

Fig. 1: Energy fluxes upon an Air Flow Window on sunshine:
RI: Incident solar radiation.
RR: Reflected " "
RT: Transmitted " "
V : Thermal energy avacuated by ventilation.
Re: Outwards infrared radiant flux.
Ri: Inwards infrared radiant flux.
Ci: " convectif flux.
Ce: Outwards " "
Te: External temperature.
Ti: Internal temperature.

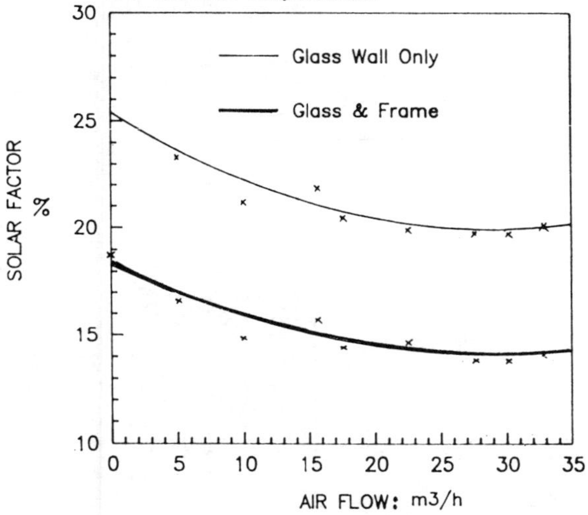

Fig. 2: Experimental data of solar factor vs. air flow.

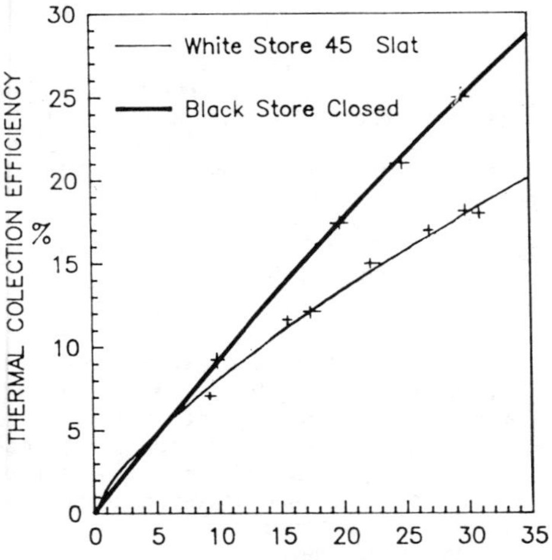

Fig. 3: Efficiency as a thermal solar collector.

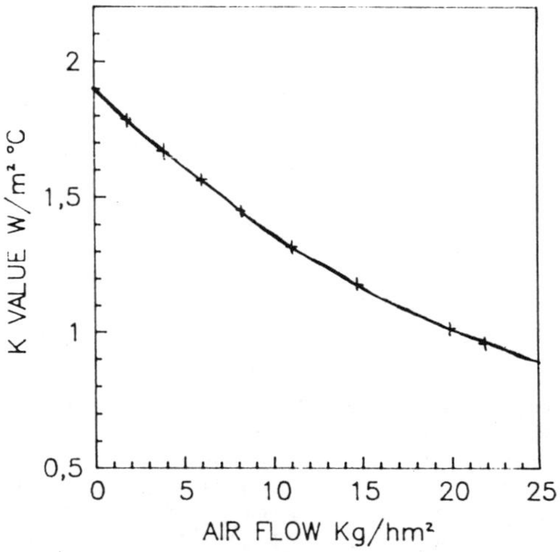

Fig. 4: Experimental data of the thermal transmission K as a function of the air mass flow.

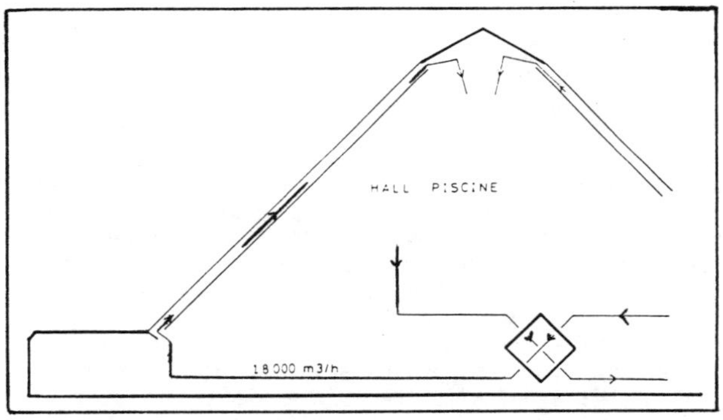

Fig. 5: Working principle of the ventilated skin of Le Tonquet's acuatic park (H. Farmellier I.C.)

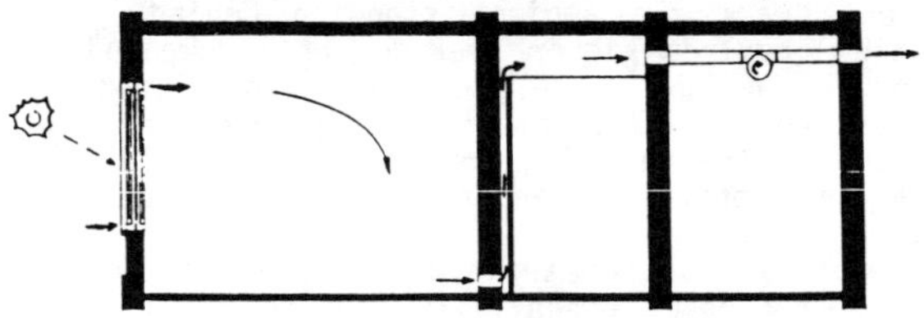

Fig. 6: Winter sunny day with forced ventilation (slats 15-25°).

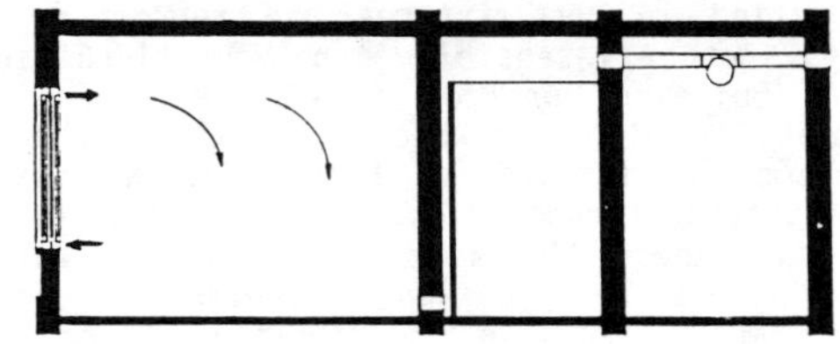

Fig 7: Winter sunny or cluded day without exhaust air extraction.

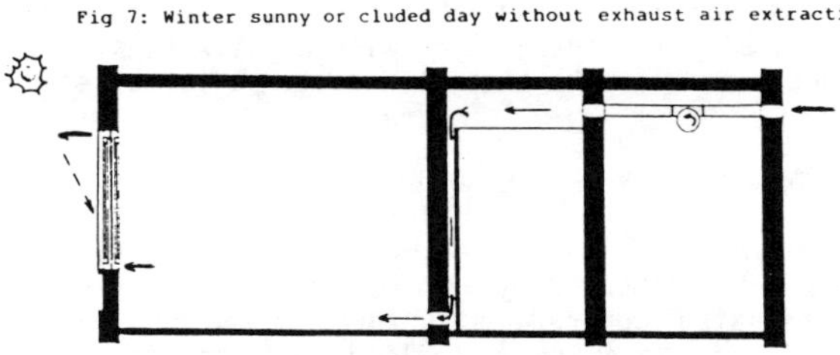

Fig. 8: Summer day with forced ventilation.

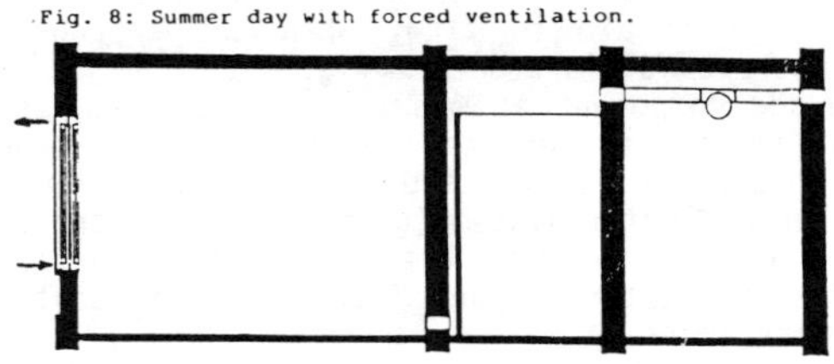

Fig.9: Summer day without exhaust air extraction.

SOLAR FACADE INTEGRATED WITH A H.V.A.C. SYSTEM

G. Rossi* and G. Solaini**
*Istituto Universitario di Architettura, Santa Croce 197, Venezia, Italy
**DPPPE, Politecnico di Milano, Via Bonardi 3, 20133 Milano, Italy

ABSTRACT. The scarce diffusion of solar technologies for building climatization is mainly due to the following factors: 1) solar components are often insufficiently integrated both with building structures and with traditional plants: 2) the presence of solar systems is a restraint with respect to building design and to furniture distribution. In this paper a solar system is proposed, with the aim of overtaking these limits. The paper deals mainly with the integration of the component with a HVAC system: the problem of control of energy and of safety functions is analyzed, and a solution concerning the achievement of a high efficiency is described. Finally, a short analysis concerning the energy contribution of the component to building climatization, as well as an economic evaluation, are presented.

1. INTRODUCTION

Active and passive solar technologies for heating developed in Italy as systems and components are almost always separate from the building system and rarely integrated with the plant engineering system. So the average cost of the interventions could not get so low as hoped, causing a very spare diffusion of the solar technologies, even in the presence of public incentives.
For instance, residential buildings using passive solar technologies often fall in the "passer-by" living unit type, with the main ambients oriented to south, and north-oriented utilities, in order to allow the maximum employment of the collecting components (windows, conservatories,Trombe walls) which can only warm the ambient standing behind. This solution involves higher average costs compared to deeper living units (e.g. double body buildings with one-front living units). Such problems are emphasized in the case of non residential buildings, which are generally deeper than the residential ones.
The second reason for the scarse spread of passive solar technologies essentially regards the production of components: they are usually realized by firms acting in the fixtures and tehrmohydraulic branches, not very sensible to polyfunctional integration (building function + technical functions) or flexibility in use, which - on the contrary - are determining factors for the diffusion of innovative building technologies.
To overtake these limits, a polyfunctional collecting wall has been designed - within a former research /1/ - featuring flexibility and integration at moderate costs. In this paper we consider specifically its integration with a H.V.A.C. system, in order to evaluate also its energy performances.

2. DESCRIPTION OF THE COMPONENT

The solar facade is formed by vertical panels, each panel being composed by a solar collector joined to an insulated concrete structure; moreover, panels are placed alongside a ceiling with canalizations allowing warm air to flow, when the appropriate temperature conditions occur (fig. 1).
The facade component is modular, and can be assembled in two phases: erection of the concrete panel, which ensures insulation and closure; assembling of the solar collector system (glass, absorber and dampers), which provides the energetic functions. The solar collector is formed by three main parts:
a) a transparent part;

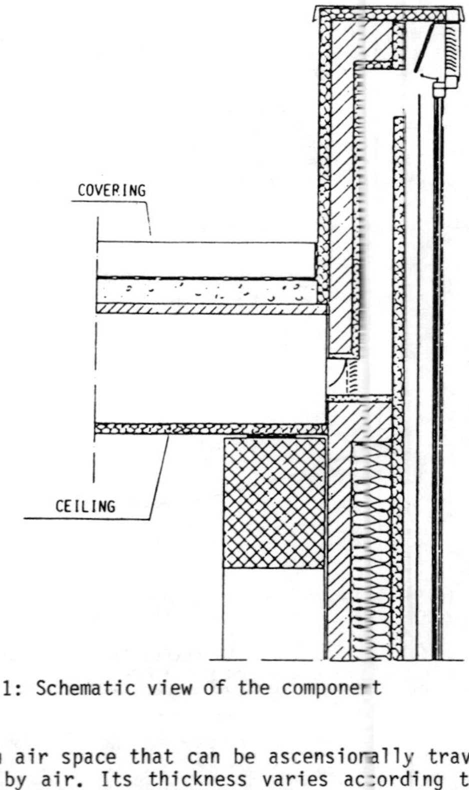

Fig. 1: Schematic view of the component

b) an air space that can be ascensionally travelled over by air. Its thickness varies according to the energy system adopted (active or passive);
c) an absorber placed in median position in the airspace, to turn solar radiations into thermal energy.

3. PLANT ENGINEERING CHOICES

We consider a non-residential building, with a south-oriented facade formed by the above-mentioned pannels,and we assumeto integrate it with a heating (and air-conditioning, if necessary) system. We chose the plant engineering system selecting the most suitable alternatives for the technological system and for the innovative component proposed Features of this choice are:
active typology: the most suitable compared to solar chimney, since the courses of canalization are partly horizontal and the employment of dampers considerably hampers a natural convection system;

- utilizer plant: air-heating type in order to make the integration of the solar plant with the traditional one easy and efficient. This choice allows the heat generation to be both centralized and self-sufficient; futhermore the coil can be fed by cooled water as to obtain the summer cooling, when required;
- collecting: indirect-gain type with forced air collectors. The building technologies demands and the typological flexibility needs are made compatible, keeping the efficiency of the collectors high;
- circulation: closed circuit type. It takes from outside the air flow needed for the renewal in the living unit;
- heat storage: daily. It utilizes the solid bulks of the ceiling where the warm air coming out from the solar collectors makes contact;
- final heat transfer: we largely chose the radiative exchange (warm floor) and the ambient inlet of the air that has already warmed the ceiling bulk. A specific study reagarding comfort conditions and energy saving achievable by means of this system is reported in /2/;
- control facts: an automatic system composed by air temperature sensors placed in the ambient and at the collectors exit and entrance; a regulation valve for the water feeding of the air-heating coil; motorized dampers placed in the plenum; dampers for the collectors ventilation during summer with on/off driving gear; a control unit with seasonal selector.

4. OPERATION OF THE INTEGRATED HEATING PLANT

Factors (fig. 2 and 3):

t_1 : average collectors-bottom temperature
t_2 : average collectors-top temperature
t_i : internal air temperature
t_s : set point for t_i
t_{max} : collectors safety temperature (setting range 50 - 70 °C)
t_{min} : collectors threshold temperature (setting range 20 - 40 °C)
D_1 : damper for the inlet of the collector air in the air-heating engine
D_2 : damper for the recycling of the internal air
D_3,D_4 : dampers for the cooling of the collector air
D_{in},D_{out} : dampers for ventilation
WH : water heater
HC : heating coil
F : fan
DF : diffuser
VA : three-way valve to regulate the water flow in HC
CU : control unit

The manual selector allows to choose between two regulations: first one suitable for cool seasons; second one in the other cases.
When the electrical feeding goes out the state of the devices is: WH/off; fan/off; D_1 closed; D_2, D_3, D_4 open. These conditions form the starting state for the electrical feeding of the system by switching on the general switch.
The control process starts examining the state of the seasonal selector. From position 1 (cool season) it goes on as follows:
a) the water heater WH is on; the three-way valve VA is positioned to obtain complete water flow toward the WH;
b) the collectors-top temperature (t_2) is checked out to be lower than the lowest t_{min} limit; if the check out turns out to be positive, D_2 and D_4 are closed, otherwise they are opened;
c) t_1 is compared to t_2: if $t_1>t_2$ collectors can be used; D_1 is then opened and D_2 closed; on the contrary D_1 is closed and D_2 opened. In both cases the ambient temperature value is compared with the t_s

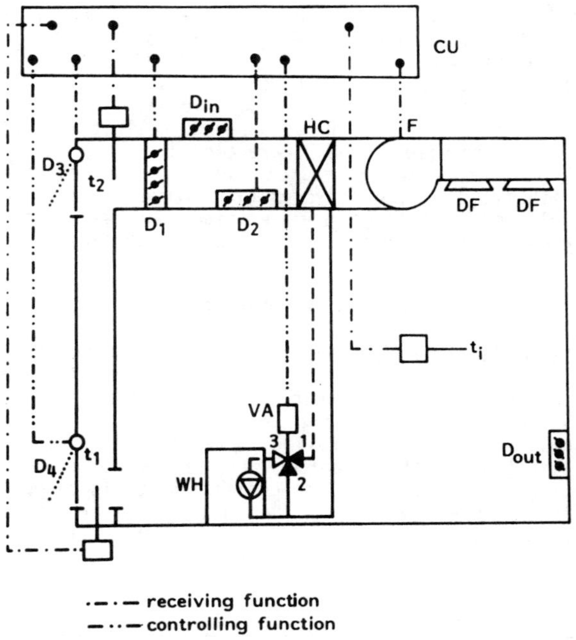

Fig. 2: Plant scheme

value scheduled in the control unit;
d) if $t_i > t_s$ the fan stays off; starting again from point b); on the contrary the fan is open and the VA valve is regulated as to let the water flow in the HC coil;
e) the check out of the points c), d) is carried out until the VA is completely open; so done, the state

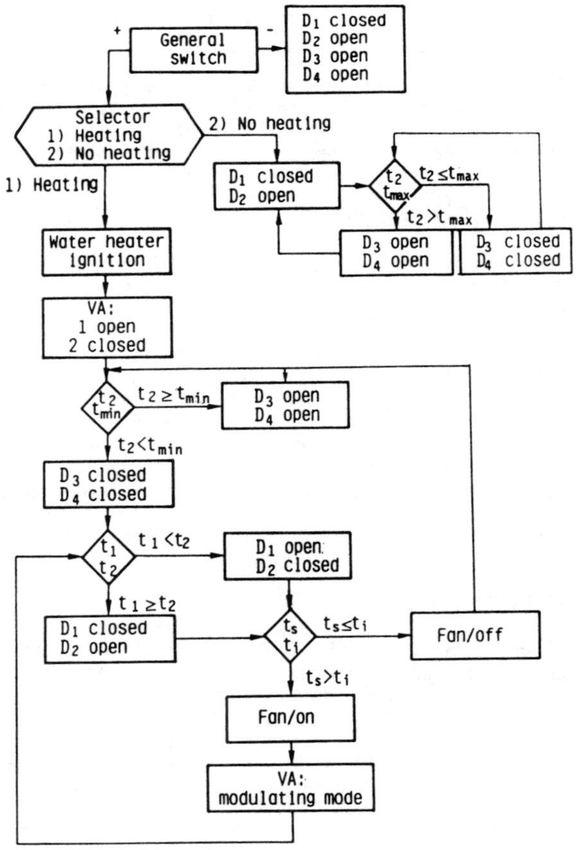

Fig. 3: Control logic

of the seasonal selector is examined again; if nothing has changed the process is carried on, otherwise see point f);
f) D_1 is closed, D_2 and D_4 open;
g) t_1 is compared with the t_{max} highest limit value; if $t_1 > t_{max}$, D_3 and D_4 are opened (collectors cooling); otherwise D_3 and D_4 are closed;
h) in both cases go back to point g) (t_2 compared with t_{max}). After n cycles of g) type the seasonal selector conditions are verified and the process goes on according to the two possible choices.

5. CALCULATIONS AND SEASONAL SAVINGS

We utilized the F-CHART calculation program /3/ to evaluate the performances of the component integrated in the plant-engineering system above-said.
We assume the component to be used as perimetral wall of a gymnasium having volume V = 7,200 m^3 and situated in the Venice climate; moreover, the gymnasium total heat loss coefficient C_g is supposed to be 0.29 W m^{-3} °C^{-1}. The wall is assumed to be south-oriented, vertical position, and collecting surface = 260 m^2. Results are reported in table I.

TABLE I: Energy Balances

Month	T_e (°C)	I_H (kWh/m^2d)	K_t	DDM	Thermal load (kWh/month)	Energy supply (kWh/month)
Jan.	2.8	1.10	0.34	471	23,613	4,567
Feb.	4.6	1.70	0.36	375	18,802	5,217
Mar.	8.3	2.83	0.42	301	15,069	7,331
Apr.	13.0	3.43	0.47	150	7,517	5,020
Oct.	14.4	2.39	0.45	112	5,592	5,919
Nov.	8.5	1.15	0.32	285	14,282	4,005
Dec.	4.5	0.94	0.33	419	20,972	4,038
			Total	2,113	105,847	36,097

(T_e = monthly average outside temperature; DDM = degree days/month; I_H = monthly average daily global radiation on horizontal surface; K_t = monthly average clearness index).

6. ECONOMIC EVALUATION

The cost of the component is 210 \$/m^2 in case of industrialized production. To estimate the additional cost we consider, as a reference system, a traditional cavity wall with the same thermal, acustical and structural characteristics of the innovative one. The specific cost of the traditional wall is estimated to be 110 \$/m^2. Therefore, the additional cost turns out to be 100 \$/m^2. On the basis of this figure, considering that the gross surface of the solar facade is about 300 m^2, and assuming a cost of energy of 0.09 \$/kWh (based on fuel price and on average heating plants efficiency in Italy), we find a payback time of 9.2 years in the Venice climate.

REFERENCES

1) CEE contract n. SE-230-87: "Pannello di chiusura portato prefabbricato per la climatizzazione di edifici con l'energia solare" (1988).
2) G. Rossi and G. Solaini, Contribution to Building Comfort Given by Experimental Ventilated Cavity Floors. Proc. of the 2nd World Congress on Heating, Ventilating, Refrigerating and Air-Conditioning Clima 2000, Sarajevo, 1989.
3) S. Klein, W. Beckman, J. Duffie, A Design Procedure for Solar Air Heating Systems. Proc. of the Int. Solar Energy Society Conference, Winnipeg, Manitoba, 1976.

"INTEGRATED BIOCLIMATIC DESIGN OF A RURAL DAYCARE CENTER IN GREECE"

E.S.Triantis* M.Santamouris** T.Argyriou** and M.Vallindras**
* University of Ioannina, Ioannina Greece
** Pro-techna Co.39 Mavromichali St., Athens, Greece

ABSTRACT. This project uses the principles of bioclimatic design in conjunction with the utilisation of both indoor and outdoor spaces by young children in a daycare center located close to the town of Ioannina, in N.W. Greece. Various passive solar systems are used for space heating and natural lighting (direct gain, clerestory and solar greenhouse systems), always closely connected with special functions for each space and particular moments of the childrens'day. As the center will stay open throughout summer, special provisions will be made for cooling both indoor and outdoor spaces, in order to provide comfortable and varied play areas in the hot season.

1. INTRODUCTION

It is well known that Mediterranean climate is quite different from this of Northern Europe, characterized as it is of considerable heating requirements in the winter as well as important cooling requirements in the summer. It thus seems to require specific solutions for passive solar systems as well, which are not necessarily applicable for Northern climatic conditions. Main passive solar components which constitute the innovative aspects of this project, are therefore adressed to this specificity of the climate and ways of life in this part of the world.
The purpose of this project is to use the principles of bioclimatic design in order to help create comfortable indoor spaces in terms of energy and daylighting for this daycare center and, at the same time, integrate indoor and outdoor spaces to be used by the children in all seasons and thus actively involve them in a continuous give and take with their environment.

2. DESCRIPTION OF THE PROJECT

The building complex is located on a beautiful 12-acre site 10kms South of the town of Ioannina in N.W. Greece, and it is composed of two main units connected with pedestrian pathwalks and open and covered play areas of different character. (Fig.1)
The total indoor area is 750 m2 approximately and it is divided in various levels on the ground floor, following the slope of the site (Fig. 2,3). The whole complex is oriented toward the South, with an open pedestrian space at its center, while each building is organised around a high central space covered by a specially designed solar roof. (Fig. 4)
Direct gain solar systems are used in all classrooms and they are combined with separate solar greenhouse systems attached to the South of some units, depending on use, so that children can enjoy cultivating their own plants in all seasons. (Fig. 5,6)
A system of open evaporative cooling will be used in the summer for both indoor and outdoor spaces, bringing the children into contact with running water, while planting and shading of outdoor play areas will provide comfortable surroundings throughout the year. (Fig. 7)

3. DESIGN PROCEDURE AND SYSTEM DESCRIPTION

3.1 Passive solar strategies used

Since energy conservation and passive solar design requirements have been included in the program, this project has opted from the beginning for the best possible combination of novel techniques and conventional heating and cooling systems in order to achieve maximum comfort conditions for its users'requirements while keeping energy consumption for heating, cooling and lighting at considerably low levels.
As far as passive solar design is concerned, the main strategies used are the following:
a) Orientation of each building block toward the South. This strategy has played a very important role for the overall architectural composition, as well, since it creates a pronounced E-W axis along which main indoor and outdoor activities are located.
b) Minimisation of Northern exposure by the use of berming, roof slopes and planting, as combined with reduced Northern openings and pronounced insulation levels.
c) Use of South-facing wall surfaces for passive solar systems in order to reach comfort levels for heating, cooling and daylighting of main activities spaces. It is planned for a combination of direct gain and simple air collector systems to be used, while a series of attached solar greenhouse systems will be constructed, to function in the winter not only as passive solar design features but also as an extention of the "main activities" spaces into individual gardening facilities for the children.
d) A special clerestory-type solar roof is designed over the central multi-purpose space of the daycare center conferring the character of semi-open space to it by increasing its daylighting and thermal comfort levels, both for heating and cooling. In fact this space will be mostly shaded during the summer, while cooling will be achieved not only by natural ventilation of the space itself, but also by evaporative cooling created by a specially designed water system integrated to the solar roof, which will let small guided water streams fall freely into a fountain, located in the main reception area. This will help cool the building on hot summer days, while creating a strong visual element for children.
e) Besides the above bioclimatic principles used for interior spaces, a very important element of this project is the use of planting, landscaping, shading etc. devices in order to create thermal and visual comfort in outdoor areas as well, to be used by children in all seasons, so as to be in contact with nature throughout the year.

3.2 Description of passive solar components

Main passive solar components which consitute the

innovative aspects of this project can have considerable interest for Mediterranean countries, in general. They are mainly the following:

a) Passive solar greenhouses or sun-spaces, to be used by the children for direct contact with growing plants during cold season, as part of their nature-awareness learning experience. Construction of greenhouse spaces will be light, their roof sections will be opaque to avoid overheating while their sides will be double-glazed and can be totally removable in the summer. Ventilation openings will exist on two sides. Two possibilities (totally demountable light greenhouse, or permanent roof structure to be used as open shelter in the summer) will be constructed.

In everyday operating conditions, and depending on outside temperature, doors to the greenhouse will be opened when additional solar coverage is needed, and lessons arranged accordingly. It is estimated that an average of 7-8% of additional heating coverage is offered by the greenhouse during the coldest winter months and even higher percentages in the fall and spring, while greenhouse glazings should be removed from May through October.

Such greenhouses or glazed verandas can have extensive use in Mediterranean climates if succesfully designed both for winter and summer use - which this project will actually help demostrate.

b) Passive solar roof components, used in the place of skylights or as continuous self- supporting solar roof structures, are also quite interesting both from the point of view of thermal and lighting comfort, as well as livability in large interior spaces. Usual solutions for glazed roofs cause overheating in mediterranean climates for most of the year -so that an inexpensive and efficient passive solar solution that could combine winter solar heating and summer cooling, through shading and ventilation, of central gathering spaces or covered courts, will certainly have considerable utility as well.

It is estimated that such a passive solar roof design (Fig. 8) may cover 20 to 30% of a buildings heating and cooling load, depending on the specific building configuration. Operation and control of such a system could by either manual or automatic (fex. using a micro-processor), depending on its size and complexity.

c) Evaporative cooling components

Such components will be developed as part of the cooling strategy of each building and will be specially adapted to the easy access and use of water, both still and in motion, by the children. They will be conceived as free-standing sculptures, offering a permanent decorative aspect when not in operation and becoming a source of movement and sound when in actual use for cooling. Their presence will help the children realize different posibilities of water used as a dynamic medium and will create an agreeable and comfortable assembly and recreation space for them to use in all seasons.

3.3 The back-up system

The back-up system is composed of a flexible warm air heating system connected to an array of solar (low temperature) heat recuperators as shown in the attached diagram. Air circulation in the whole building will be facilitated by the design of the spaces (level and height differences, combination of low and high walls and openigns)both for winter and summer use, when cross-ventilation and chimney effects are used for natural cooling. The building will also be provided with electic fans at selected places, however, (f.e.x. on top of two-level spaces) which can be used at extreme conditions.

Annual savings of this system are estimated at 15-20% for heating, if compared with those of a standard well insulated building using a conventional oil water-heating system.

4. THERMAL PERFORMANCE OF THE SYSTEM

The results of energy calculations in this preliminary design phase can show us that a total of 52% of the thermal load is covered by the combination of direct gain, solar greenhouse and solar roof systems chosen, which also contribute to a considerable improvement of thermal and lighting comfort levels throughout the buildings. These as well as the cooling contribution of the system, will be calculated in detail in the study phase of the project.

5. DESCRIPTION OF THE MEASUREMENT PROGRAM

The project has been selected as an EEC solar demonstration program and it will be monitored for a two year period after construction.

During the measurement phase a data-logger will be used to collect data on air temperatures of indoor spaces at selected positions, ambient air temperature, and greenhouse air temperature, while relative humidity of the air at selected spaces of greenhouses, central assembly spaces and classrooms will also be recorded and compared to relative humidity of outside air.

Ventilation rates, solar radiation and windspeed will also be recorded. Finally heating and cooling contribution of each passive solar subsystem will be computed and the overall efficiency of the energy system will be calculated and compared to a conventional system using a simulation process.

6. CONCLUSIONS

Alhough there is high interest both from parents and educators for the use of passive solar systems and especially for integrated bioclimatic design, in schools and daycare centers, very few such schools have been actually constructed in Greece. One of the reasons for this is that costs seem too high as such buildings will have to deviate from standard construction practices. We hope this project will help demonstrate that such an approach will create successful environments for children, both from the thermal comfort and educational points of view, and can also show new possibilities for passive solar components to be used in Mediterranean buildings without excessive economic or construction deviations from usual practice.

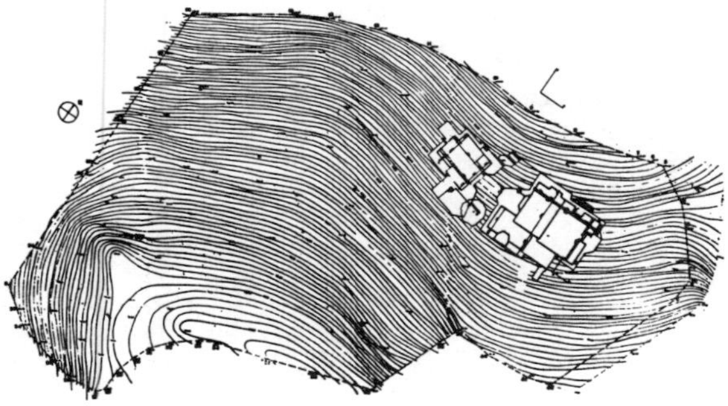

Figure 1: Site plan

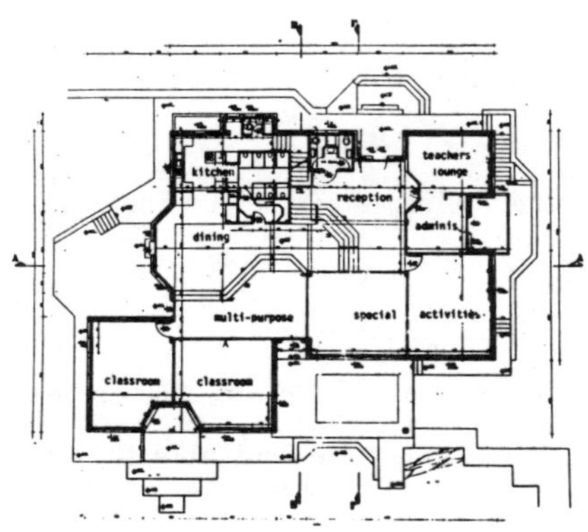

Figure 2: Daycare-ground floor plan

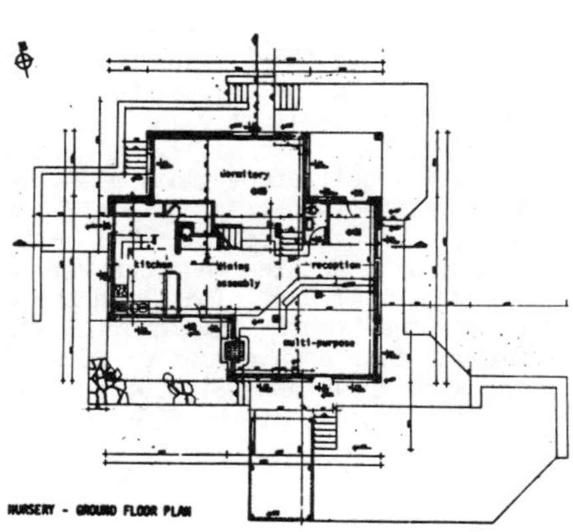

Figure 3: Nursery-ground floor plan

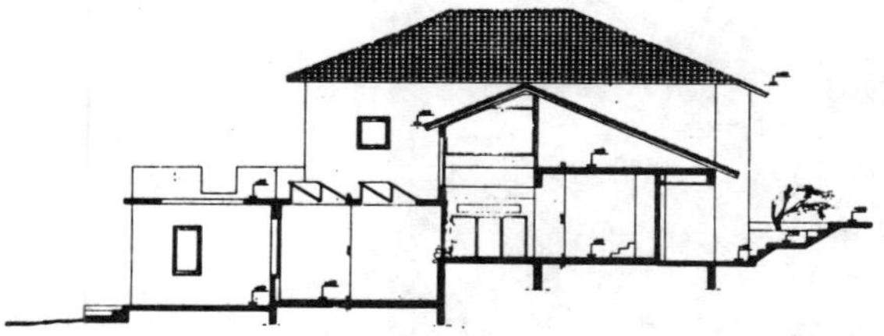

Figure 4: Daycare-N-S section

Figure 5: Daycare-south elevation

Figure 6: Nursery-south elevation

Figure 7: Nursery-east elevation

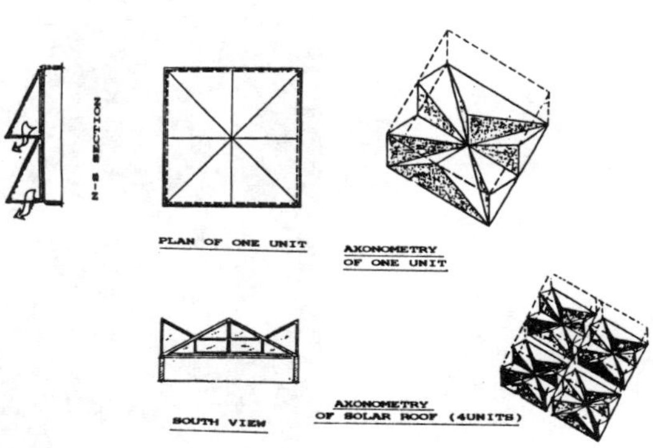

Figure 8: Passive solar roof components

THERMAL PERFORMANCE OF PASSIVE SOLAR SYSTEMS IN MASSIVE TEST CELLS

Helder Gonçalves*; E. de Oliveira Fernandes+
*LNETI-DER, Estrada do Paço do Lumiar, 1699 Lisboa Codex
+FEUP-UP, Rua dos Bragas, 4099 Porto Codex

ABSTRACT. This paper describes the performance of several Passive Solar Systems (PSS), obtained in massive test cells built in Portugal (1988). These cells were tested during the 1988/89 heating season under thermostatic control mode. The results were compared with those obtained by numerical simulation using the ESP code [1] for the same situations. This comparison included not only the overall heat balance but also the temperatures in meaningfully places of the cells.
A parametric study was carried out, which focus on: the influence of the type of glazing in Direct Gain (DG) and Indirect Gain (IG); the influence of the Trombe Wall (TW) thickness in massive structures and the comparison between the light and massive structures for DG and IG.

1. INTRODUCTION

The study of the PSS has been carried out all over the world, mainly since the 70's, particularly in the USA and EUROPE. In the latter case, a large and concerted research program started in 1986: The Passys Project [2] . In all these cases the experimental facilities (building and test cells) used to evaluate the "Performance" of the passive systems were light weight and very insulated structures. The effect of the mass was studied afterwards, adding mass to the previous construction with a different distribution compared with a real situation.
In the case of some systems such as with, the massive storage walls for instance, the thermal mass of the system was greater than the envolving structure. In the Southern Europe regions the construction is traditionally very massive. As a consequences of that, buildings use to have "thermal inertia" greater than the passive components itself. This is one of the reasons, for a concerted project, envolving the University of Porto (UP) and a National Laboratory (LNETI) in this field. It started in 1986 with connections to the Passys Project.
One of the objectives of the above mentionned project was to evaluate the performance of the systems in massive construction, and to try to compare this results with those obtained in light weight construction. In this case, the brick was used, which becames in the last years one of the most typical construction materials. The other objective was to study, the influence of the combination of systems (DG with IG).

2. TEST CELLS DESCRIPTION

The Portuguese test cells are shown schematically in fig.1
Three of the cells have basically the same construction, with 20 cm thick concrete block external walls and 20 cm thick concrete slab. The external envelope is insulated on the outside with 5 cm of polystyrene, and there is a thermal break between the cells and the ground. All the cells have an attic with a typical Portuguese roof; made with a structure of wood with red tiles.
The dimensions of the cells from inside are; 4*3**2.55 m (with*lenght*hight) which correspond to a typical surface area of a bedroom.
Cell number 1 is a direct gain (DG) test cell, with different window areas (1.2 m^2 each) and it can operate with different glazing to floor area ratios: 10%, 20% or 30%.
Cell number 2 is a Trombe Wall (TW) test cell with 20 cm thick, with an area that can vary between 20% to 60% of the floor area. The TW in the cell can operate with the "vents" opened or closed.
Cell number 3 is a combination of DG with IG. The first one with an area of 5%, 10% or 15% and the IG with 10%, 20% and 30% of the floor area.

Fig.1 - General view of Portuguese test cells.

Cell number 4 in terms of PSS is exactly the same than cell number 3. The fundamental difference is the construction, which is a double brick wall with an inside air gap (15*4*7 cm), and it has no insulation. This cell can be compared directly with cell 3, with respect to the thermal inertia upon the performance of PSS.
All the four cells were instrumentated during the construction with the objective of placing sensors inside the mass of the structure. The sensors were located in the walls, slab, Trombe Walls and surface of some walls. All these sensors measured mass, surface, air and resultant temperatures.
The climate date were recorded in site: ambient temperature, global radiation on the horizontal, vertical south, east west and north together with the direct normal radiation. The wind speed and direction are recorded at 10 m height.
The airtightness of the test cells was measured by pressurization with the blower door. The infiltration rate measurements were performed by using the tracer gas technique (CH_4), simoultaneously in the four test cells.
Special measurements with the objective of check in thermal bridges and surface temperatures in the cells were performed using infrared techniques.

3. TEST CELLS EXPERIMENTS

The experiments with the cells started in November 1988, with the calibration procedures, in order to measure the air leakage and the UA coefficients of all the cells, and also the thermophysical properties of the materials such as the emissivity, absorptivity and transmissivity of the glazings.
The tests within the cells started in December 1988 in thermostatic conditions, with a reference temperature of 20°C (Tset), in the four test cells.
In the following months until March several areas of the components were studied. - see Table 1.

	DEC.	JAN.	FEB./MAR.
CELL1-DG	10	20	30
CELL2-IG	20	40	60
CELL3/4-DG/IG	5/10	10/20	15/30

Table 1; Ratios of Glazing to floor area (%)

The heat consumption in each cell was recorded in the D.A.S. and the inside temperature above Tset was free to increase: no heat removal system was used.
To analise and compare the results, we used the Solar Fraction (SF) and the Solar Load Ratio (SLR) as parameters defined as;

$$SLR = Qsol/Qload \qquad SF = 1-Qaux/Qload$$

The experimental results are ploted in fig.2 and 4, together with the fitted curves obtained by numerical simulation.
For the same conditions, obtained in the experimental tests (climate and indoor conditions), the modelisation of the cells was performed using the ESP code.
Simulations were done for all the periods and test cells. The results were compared with some detail (heat consumption, air, mass and surface temperature). Once a "good" agreement was found, extensive simulations were performed with the objective of characterizing other parameters (sensitivity studies).
In DG systems, two main parameters were studied: the mass construction and the type of glazing. In the first one, two different constructions were considered; the concrete block and the brick construction.
Two types of glazing were used; double (DG) and single (SG).
In the IG systems the same parameters were studied, nevertheless the influence of the tickness of the Trombe Wall was carefully studied.

4. RESULTS

4.1 - Direct Gain

In fig.2 three groups of curves are plotted (SF vs SLR) for the DG case. The first group (1) corresponds to the massive construction for double and single glazing. It is possible to see that for the same SLR we found maximum differences of around 10% in the SF (SLR = 1).
The experimental results obtained in cell 1 are represented in those curves.
The second group (2) presents the same results for the brick construction, and the refered differences are lower, around 4%. This means that there isn't a significant benefit in using the double glazing in the brick construction.
The last group (3) compares the two different constructions (concrete and brick).
The differences between the two are very notorious, with values that can achieve 20% (SLR = 1.25), for the single glazing case those values are lower: around 12%.
One important aspect in this analysis, is concerned with the comfort: the internal daily temperature swing according with the type of construction. In fig.3, we plot the maximum temperatures and SF vs the ratio of glazing to floor area, for all the cases. The following conclusions can be drawn;
- In a massive construction we can use ratios of glazing to floor area ratios until 40%, without the risk to obtain a temperature swing larger than 6°C. And with those values we can arrive to SF between 60 and 70%.

- In the brick construction we see that for relations bigger than 35% the system reaches temperature swings greater than 6°C; in this case the SF is smaller than 40%.

4.2 - Indirect Gain

In fig.4 we have the same kind of representation but in this case for the IG system.
The first group of curves (1) represents the influence of the mass in the performance of the Trombe Wall. The following conclusions can be drawn:
- For the same SLR, the differences in SF, when using a concrete construction is very small (1-2%) compared with the brick construction for the double glazing. When using a single glazing the difference is so negligeable that the curves are coincidents.
The second group represents the influence of using double or single glazing in a massive construction. The differences vary until values of 12% in SF for a SLR = 1.5 which corresponds more or less, to a wall to floor ratio of 60%. For ratios between 20 and 30% those differences are lower: 7 to 10%.
The last group represents the influence of the thickness of the TW in the massive construction. The best performance was obtained for thickness of 10 cm.

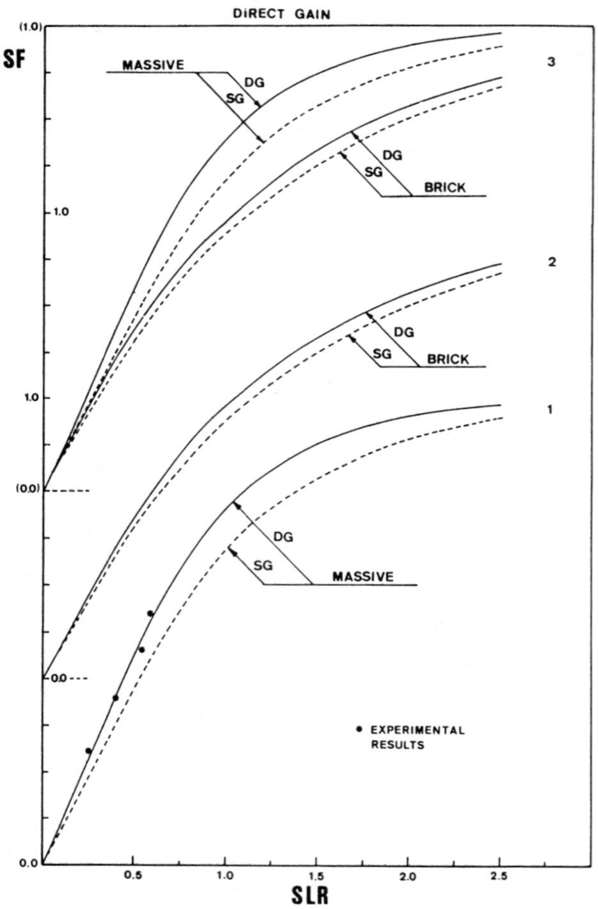

Fig.2 - Direct Gain Results.

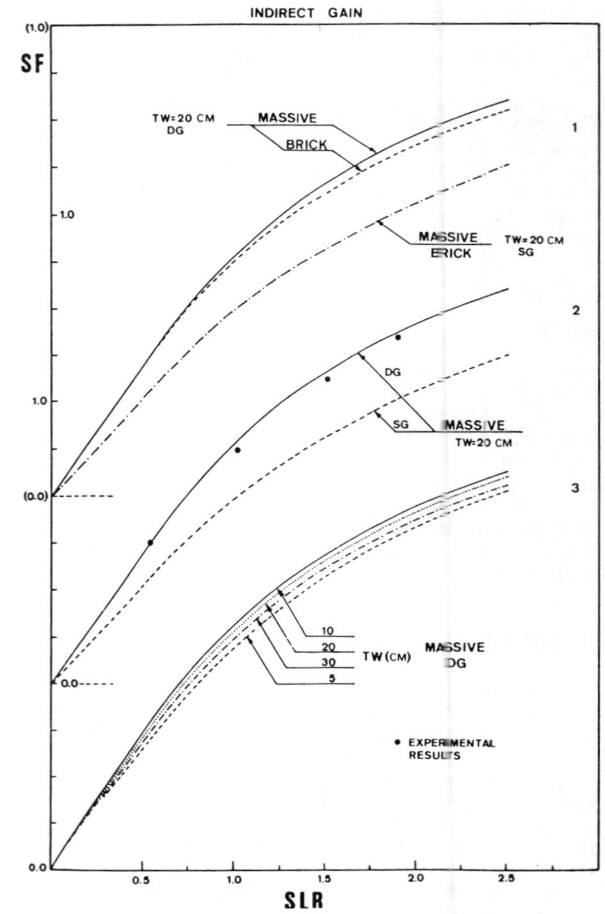

Fig. 4 - Indirect Gain Results.

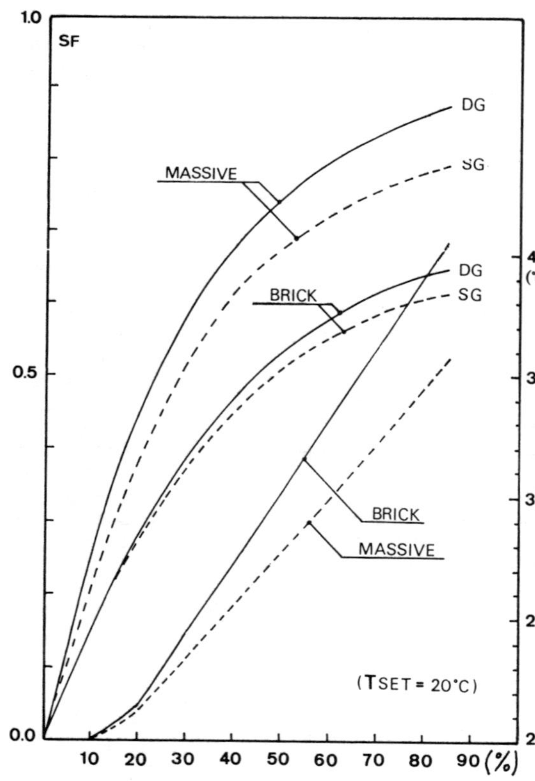

Fig.3 - T_{max}, SF-Direct Gain.

5. CONCLUSIONS

The experimental and numerical results obtained with the test cells, lead us to the following final comments:

- For direct gain the solar fraction curves for massive construction denote a significant difference (up to 20%) compared with the brick construction.

For glazing to floor area ratios, used in practice (up to 40%), the SF values still are in the linear segment of the curve and can reach SF values of 60 to 70% in the massive case, and 40 to 50% in the brick construction.

- For indirect gain the mass of the envelope does not denote a particular influence in the performance of the system.

6. REFERENCES

1 - J.Clark, D.McLean - "ESP Manual", University of Strathclyde, 1987.

2 - R.Gicquel - "The Project PASSYS", Proc. EUROFORUM' New Energies, Saarbrucken, 1988.

OPTIMUM THICKNESS OF NON-HOMOGENEOUS ROTATING WALL ELEMENTS FOR ENVIRONMENTAL CONTROL

Adrian C. Pitts and John J. McDonald
Building Science Unit,
School of Architectural Studies
University of Sheffield
Sheffield
S10 2TN
UK

ABSTRACT. Most passive solar design features are, by their nature, static in response to varying levels of solar radiation. Elements of building mass with high thermal capacity are used to reduce temperature fluctuations which may be produced. This is only a partial solution to the problems of creating comfort in passive solar buildings. Novel building design features with dynamic response attributes are now available, an example being the rotating prism wall. Many studies have investigated heat storage in static components but few in dynamic components. This paper describes a laboratory based experimental investigation which was devised to analyse optimum thicknesses for rotating heat storage elements. The results indicate that thicknesses of about 100mm, using high density concrete, provide a good solution of minimising size and optimising heat storage and discharge.

1. INTRODUCTION

A feature of many passive solar buildings is the utilisation of large areas of glazing to admit solar energy. A particular problem which must be addressed by such buildings is how to accommodate fluctuations in radiation intensity. The heat gained through passive solar design features may be much greater than, or much less than, that necessary to produce a thermally comfortable environment. The situation is further complicated by the fact that a mismatch can occur between the times when a suitable temperature arises and when it is required by the occupants.

The usual means employed to attempt to remove these problems is the provision of a considerable mass of a material with a high thermal capacity. The effect of this "thermal mass" is to reduce temperature fluctuations with the retention of some of the daytime heat gain to provide warmth in the evening. The drawbacks with such a "static" option are that in temperature climates, the solar heat gain may be on some days low, and on other days high. Therefore the amount of "thermal mass" to be provided cannot be optimised and this results in either excessively long warm-up times, or alternatively, periods of overheating. In essence, the "static thermal mass" approach will only work well on certain types of day, with unsatisfactory results, (particularly in conservatory sunspace environments), on other days.

2. ROTATING PRISM WALLS

In recent years novel building design features have begun to appear, some of which offer a means of creating a "dynamic" thermal mass. One form of device is the "rotating prism wall" as described by Faiman and Feuermann /1/. Such walls normally consists of a series of vertical heat absorbing blocks, sometimes of triangular cross-section, to which a layer of insulating or heat reflecting material has been added on one face. These prisms are placed between the glazing and the room interior, though alternative positions have been suggested (Pitts and Patronis /2/). By rotating the prisms, different surfaces may be presented towards the incoming solar radiation. In normal operation, the absorbing surface is presented to the sun during the day and is turned inwards at night, though other operating schedules are possible. Other means of creating a similar effect have been suggested, by Achard and Cools amongst others, using a phase-change material /3/.

These moveable devices do permit a degree of dynamic response (based upon prevailing climatic conditions) which static thermal mass does not. However it should be noted that in the case of the absorbing prism wall, that the devices are often large and heavy and that they can produce increased structural loads and difficulties in construction and installation, by virtue of their size.

3. HEAT STORAGE

The thermal response of building materials to heat flows, particularly those resulting from solar energy, has been studied for a number of years. The emphasis of many studies has been on the performance of building elements subjected to typical daily cyclic variations in solar radiation and temperature. An embodiment of the results of such work is the Passive Solar Design Handbook /4/. The methods described therein, being used to analyse passive design performance.

Several researchers have studied in detail the performance of heat absorbing materials, particularly in Direct Gain types of design (see for example Givoni /5/, Wray and Best /6/ and Gruber and Toedtli /7/). The aim of many studies has been to derive the optimum thickness of the storage element in terms of either an Effective Heat Capacity (EHC) or a Diurnal Heat Capacity (DHC). Very thick elements can have a detrimental effect on performance since they absorb heat at depth which cannot be returned to the surface within a normal daily cycle. Indeed they might be considered as "heat sinks" acting to remove useful heat. This effect may be exacerbated if the internal absorbing elements also have exterior facades.

The focus of studies of heat storage has been on constructions static in both position and heat absorbing effect. It would appear from the results of even such static studies, that the optimum thickness of materials for heat absorption will be less than 140mm (see Table I). This indicates scope for minimizing size and weight of dynamic rotating absorbing elements - especially since their mode of operation is directed towards a daily cycle at the longest.

TABLE I: Optimum Thickness of Building Mass (From Ref. /6/)

MATERIAL	OPTIMUM THICKNESS RANGE (mm)
GRANITE	58 - 135
LIMESTONE	43 - 97
CONCRETE	61 - 137
PAVER BRICK	51 - 117
COMMON BRICK	43 - 97
STABILIZED ADOBE	41 - 89
PINE	23 - 53

4. EXPERIMENTAL STUDY

In order to improve the information relating to the use and optimum thickness of rotating prism walls, a laboratory based investigation was instigated. In this the thermal performance of slabs of material of a non-homogeneous nature (i.e. slabs consisting of both an absorbing and insulating element) were investigated.
Various samples of concrete were produced; to each was added a layer of insulation material (expanded polystyrene) 25mm thick along one face. Temperature measurement points, using calibrated thermocouples, were installed throughout the thickness of each test sample. In order to simulate the action of solar heat gain, an array of solar lamps was positioned above the test sample. The output from these lamps was closely monitored and controlled.
Below the test sample was positioned an aluminium container as a base plate. Water at a constant temperature was circulated through this base plate so as to mimic the interior of a building which might exist behind a rotating prism wall. The whole of the test rig (sample and base plate) was insulated and placed within a rigid container. Figure 1 illustrates the basic layout.
High density concrete (2500kg/m^3) was chosen for the tests so as to maximize storage at minimum volume - samples of various thicknesses were produced under standard conditions for use in the test rig. The aim of the testing was to produce practical data which might at a later stage be used to verify a theoretical model.
Unlike conventional passive heating and cooling schedules which tend to follow sinusoidal type variations; on/off heat flows were used to simulate the dynamic rotation envisaged for use of rotating prisms. Unequal temperature differences between the facades of the test samples were also used. Rotation was simulated in the tests by simple turning of the samples.

5. RESULTS

Cycles of various length with maximum radiation intensities of between 800 and 900 W/m^2 were analysed. For use in a "dynamic" role a cycle of rather less than 24 hours was considered appropriate. A four hour warming or charging period followed by a four hour discharge period was chosen for the main test sequence. Figure 2 illustrates the results for such a cycle in terms of the Heat Storage per square metre of absorber surface. The results would seem to indicate an optimum thickness for storage and discharge of about 100mm.
The Heat Storage Capacity per square metre of absorber for the 100mm thick block was studied for various time cycles, the results are shown in Figure 3.

6. DISCUSSION

On the basis of the results so far obtained, the thickness of rotating absorbing elements need not exceed 100mm for heat storage and discharge during normal cycling.
More extensive testing is still to be carried out with the aim of providing data on dynamic building features and how best to control them.

REFERENCES

/1/ Faiman D. and Feuermann D., "Performance Details of a Rotating Prism Solar Wall", Energy and Buildings, Vol. 7, P.301, 1984.

/2/ Pitts A.C. and Patronis J., "Improved Sunspace Environments Using Climate Prediction and Intelligent Controls", Energy and Buildings for Temperate Climates, Proceedings PLEA '88, P.435-439, Porto, 1988.

/3/ Achard P. and Cools C. "Passive Solar Component using Rotating Blades including a Phase Change Material", Proceedings 1st European Conference on Architecture, p.589-593, Munich, 1987.

/4/ Balcomb J.D., Barley D., McFarland R., Perry Jr. J., Wray W., and Noll S., "Passive Solar Design Handbook (Volume 2)", D.O.E., Washington, 1980.

/5/ Givoni B., "The Effect of Heat Capacity in Direct-Gain Buildings", Passive Solar Journal, Vol. 4, No. 1, P.25-40, 1987.

/6/ Wray W.O. and Best E.D., "Thermal Storage in Direct-Gain Buildings: The Effective Heat Capacity and a Generalized Solar Load Ratio Correlation", Passive Solar Journal, Vol. 4, No. 1, P.41-61, 1987.

/7/ Gruber P. and Toedtli J., "On the Optimal Thermal Storage Capability of a Homogeneous Wall Under Sinusoidal Excitations", Energy and Buildings, Vol. 13, P.177-186, 1989.

FIGURE 1 : TEST ARRANGEMENT

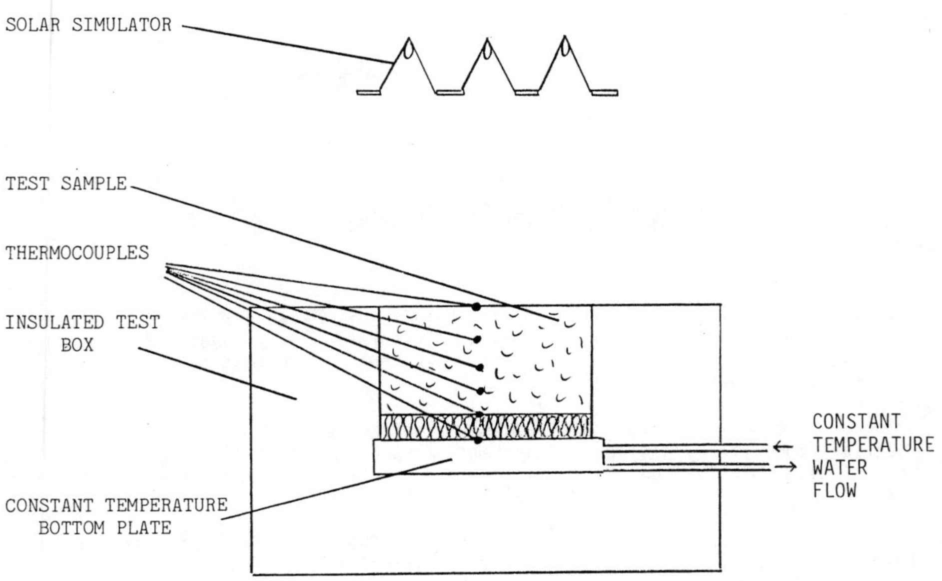

FIGURE 2: VARIATION OF HEAT STORAGE/DISCHARGE CAPACITY WITH THICKNESS

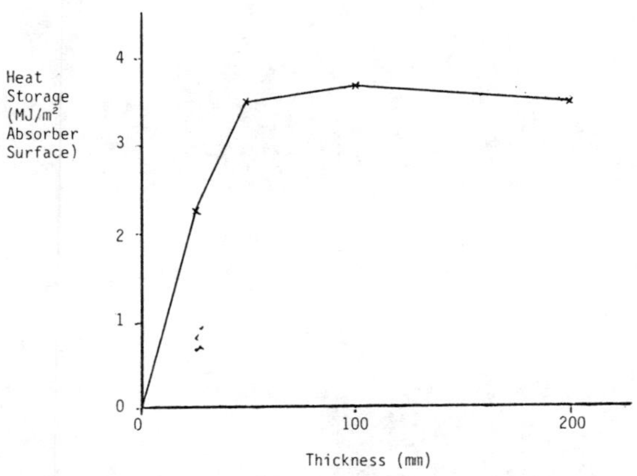

FIGURE 3: HEAT STORAGE/DISCHARGE CAPACITY 100mm THICK BLOCK

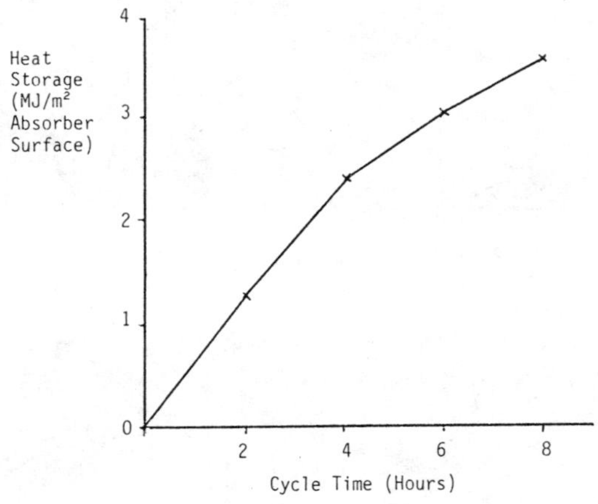

CONSTRUCTIONAL AND ARCHITECTURAL ASPECTS OF FACADES USING TRANSPARENT INSULATION MATERIALS

Architect Dipl.-Ing. Alex Lohr
Dipl.-Phys. Bertram C. Alexa
Büro für energiegerechtes Bauen
Oldenburger Strasse 68
D-5000 KÖLN 60, FRG

Dipl.-Ing. Bodo Weidlich
Dipl.-Ing. Alfred Kerschberger
Assmann Ingenieurges. mbH
Baroper Strasse 237
D-4600 DORTMUND 50, FRG

As part of the system-study TRANSPARENT INSULATION /1/ three main topics have been investigated:

- o Presentation of interrelations between Transparent Insulation Materials (TIM), solar heat gain systems and the overall system of the building as far as functional, constructional and aesthetic matters are concerned;
- o Estimation of the potential for energy savings and market penetration of TIM within new and existing buildings;
- o Fixation of requirements for further application oriented research, development and demonstration in order to find satisfying TIM products as early as possible. Those products should fullfill architectural and constructional claims and should live up to proper building physics, energy efficiency and economic standards.

INTEGRATION OF TIM

Within this study, also new areas for the application and integration of transparent insulation materials have been investigated, both for old and new buildings /2,3/.

Depending on the selected material and passive solar system, a number of different facade constructions arise, each of them creating a distinctive architectural form (figures 1-5).

Forms must be found which offer technical/constructional solutions for these new materials and at the same time provide for their aesthetic integration into the building /4,5/. These criteria must be met to satisfy designers, architects, engineers and users.

Figure 1: Rowhouse with solar walls in Boston/USA. Architect S. Hale (1980).

Figure 3: House Bollin with TIM facade in Freiburg/ FRG. Architect H. Novakov (1988).

Figure 2: House with solar walls in Ludlow, Vermont /USA. Architect R. Weinschenk (1977).

Figure 4: Detached house with TIM facade in Ardon/ Switzerland. Architect H. Novakov (1988).

Figure 5: Rowhouses in Düsseldorf - Hellerhof. Construction by Kaiser Bautechnik (1989).

LEVELS OF INFLUENCE

Different levels within the transparent insulation system can be influenced by the designer (figures 6-13):

- the structure in front of the TIM-system (like protection-bars)
- the structure and design of shading devices (outside/inside)
- the surface structure of the transparent cover
- the width, depth and profile of the frames
- different forms of mounting (screws, metall sheets)
- the material and texture of the frames

Figure 8: House with solar roof and solar wall near Boston/USA. Architect N. Saunders (1978).

- the arrangement of mullions within the transparent component
- the spacial order of the transparent insulation material
- the colour of all components of the TIM system including:
    * the shading device
    * the frame/mullions
    * the transparent cover (may be tinted, printed over)
    * the transparent insulation material (in whole or in part)
    * glue (like stripes, dots)
    * the absorber
    * coloured foils inserted between TIM layers

Figures 6,7: Outside and inside appearances of the Wellesley Science Center (USA). Architects Perry, Dean, Stewart (1978).

Figure 9: Gymnasium with a translucent facade (Kalwall) in Pittsburgh/USA.

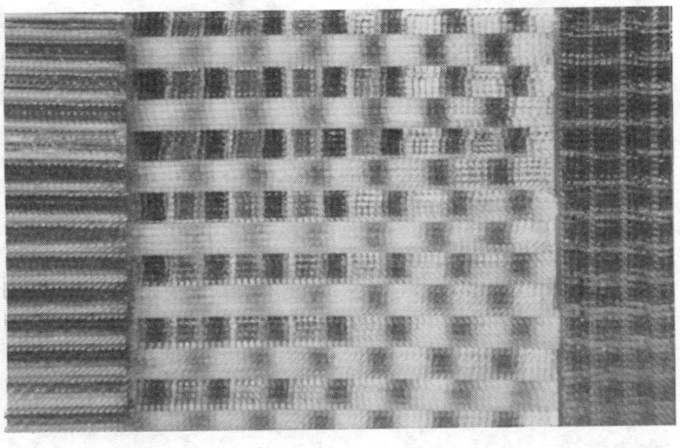

Figure 10: TIM structure of an office building in Düsseldorf/FRG. Architect G. Böhm (1989).

Figure 11: Tawaraya Inn in Kyoto/Japan with transparent sliding doors and windows.

The use of translucent components in Japanese architecture and elsewhere in the past reveals further aspects for the wide range of possible applications of TIM (figure 11 and 12).

Figure 12: Drafts room with translucent roof at Taliesin West near Scottsdale/USA. Architect Frank Lloyd Wright (1937 ff).

Increased importance also must be given to transparent facades at night, since TIM will have a distinct impact on the appearance of those buildings.

Figure 13: Appearance of TIM building with translucent facades at night.

A designer must be thoroughly acquainted with materials and components of TIM systems as well as their functions before using them in real buildings. This also applies for the numerous possible design modifications /6/.

Examples of recent TIM facades as well as facades of similar appearance are shown to illustrate their constructional and formal implications of transparent insulation systems.

As can be seen, numerous chances for new and enhanced ways of designing have been opened up by means of TIM systems, leading towards implications on the appearance of energy saving architecture which cannot yet be foreseen completely.

ACKNOWLEDGEMENT

The study on Transparent Insulation has been performed within the frame of a BMFT (Bundesministerium für Forschung und Technologie) concerned R&D action, named "Energetische Optimierung der Solarapertur" (Energetical Optimization of the Solar Aperture) as a subcontract by order of the FhG Institut für Solare Energiesysteme (ISE), Freiburg.

Copies of the study (German language only) can be ordered from Architect Dipl.-Ing. Alex Lohr.

ADDITIONAL INFORMATION IN:

/1/ A.Lohr, B.Weidlich, A.Kerschberger, B.C.Alexa
Systemstudie Lichtdurchlässige Wärmedämmung
Köln/Berlin, 1989

/2/ A.Lohr, B.C.Alexa, A.Kerschberger B.Weidlich
Realisierte und geplante Gebäude mit TWD-Anwendung
Proceedings of 6. Internationales Sonnenforum, Berlin 8/1988

/3/ A.Kerschberger, B.Weidlich, A.Lohr, B.C.Alexa
Solarer Modernisierungs-Entwicklungsschub durch transparente Wärmedämmung
Bausubstanz, issue No. 2 and 3, 1989

/4/ A.Lohr, B.C.Alexa, A.Kerschberger, B.Weidlich
Konstruktion und Gestaltung von TWD-Systemen
Proceedings of 6. Internationales Sonnenforum, Berlin 8/1988

/5/ A.Lohr, B.C.Alexa, A.Kerschberger, B.Weidlich
Transparente Wärmedämmung -
Passive Solarnutzung mit neuen Baustoffen
Deutsche Bauzeitung (DBZ), No. 3/1989

/6/ B.Weidlich, A.Kerschberger, A.Lohr B.C.Alexa
Planning of transparent insulation systems shown in a case study
Proc. 3rd Int. Workshop "Transparent Insulation Technology for Solar Energy Conversion"
Titisee, Freiburg, 9/1989

RENOVATION OF A ONE-FAMILY HOUSE WITH TRANSPARENT INSULATION

A. Wagner
Fraunhofer-Institut für Solare Energiesysteme
Oltmannsstr. 22, D-7800 Freiburg, FRG

ABSTRACT

The application of transparent insulation material for space heating was tested as part of a building renovation programme in Munich. For this purpose, transparently insulated facade elements - developed at the Fraunhofer-Institut für Solare Energiesysteme (ISE) in Freiburg - were mounted on two walls of a single-storey house, covering a total area of 40 m^2. The project, a co-operation between ISE and the City of Munich, is accompanied by a monitoring programme for two heating periods. The results indicate good performance of the transparently insulated walls so far: the wall temperatures were raised significantly and the dampness in the walls has been eliminated, resulting in higher residential comfort.

1. INTRODUCTION

The use of more than 30% of the primary energy in Germany for space heating shows the high potential savings which would result from reducing the heating energy demand of buildings. This reduction can be achieved by the passive use of solar energy with architectural elements such as south-facing windows etc., as well as by cutting down heat losses with improved thermal insulation. While opaque insulation materials can only diminish the thermal losses of a wall, new transparent insulation materials (TIM) make it possible to compensate heat losses and to use solar gains for space heating. As shown in figure 1, the solar radiation penetrating the TIM is absorbed on the black exterior wall surface. Due to the good insulating properties of the material, the major portion of the transformed heat flows into the wall, which acts as a heat storage and a large low temperature heater at the same time. A highly reflective shading device prevents the wall from overheating.

For practical applications of TIM, new facade elements have been developed and tested at the Fraunhofer-Institut für Solare Energiesysteme (ISE) in Freiburg. With the aims of gaining more experience with the handling of these facade elements as well as getting information on their long-term characteristics, two facades of a building were renovated with transparent insulation. The project was carried out as part of a building renovation programme in co-operation with the City of Munich in 1988.

2. THE RENOVATION

The flat-roofed single-storey house, which is attached to the machinery building of a large sports facility, has a living area of 76 m^2. The exterior walls are built of solid sandstone (36.5 cm). Due to its inadequate thermal insulation, which resulted in cold and damp exterior walls, the house had a very high energy consumption and a low level of residential comfort. Beside the remodelling of the facades, the renovation included the improvement of the building's opaque insulation as well as the installation of new windows.

The facade elements, a composite construction of wood and aluminium, were mounted on the SW and the NW walls of the building. They include the TIM, a polycarbonate honeycomb (PCHC) structure of 10 cm thickness, and a highly reflective roller blind for shading (see figure 2). A thin frame guarantees that the solar aperture area is as large as possible. Due to the complete prefabrication of the elements, the time and cost for mounting were reduced significantly and no dust could degrade the optical properties of the TIM at the building site.

As part of the remodelling of the facade, new windows with double glazing and integrated heat-reflecting roller blinds were installed. The windows were integrated into the elements together with the TIM to create a homogeneous appearance of the facades. A total area of 40 m^2 is covered with transparent insulation. A view of the SW facade with the roller blinds open is shown in figure 3.

The roller blinds in the facade elements are controlled by a personal computer, according to the following conditions:

- exterior wall temperature > 60°C,
- ambient temperature > 15°C (mean value for 24 hours),
- insolation < 20 W/m^2.

The blinds are closed automatically if one of the conditions applies, so that the wall is prevented from overheating, and heat losses are reduced at night.

3. RESULTS

In order to evaluate the effectiveness of the transparent insulation and to check the reliability of the TI-wall system, the project is accompanied by a monitoring programme for two heating periods. The temperatures, heat flows and the insolation are measured at two positions on the facades. The good performance of the transparently insulated SW wall is shown in figure 4. The absorber temperature rises simultaneously with the insolation, whereas the increase of the interior wall temperature follows with a time lag which depends on the wall's physical properties. As long as the interior wall temperature exceeds the room temperature, a heat flux into the room occurs. The measurements show that solar gains could be used for heating on the SW facade, whereas the losses on the NW facade could not be compensated completely because of unfavourable insolation conditions during the winter. However, the level of the mean wall temperatures could be raised on both facades resulting in higher residential comfort.

Additionally the dampness in the walls could be eliminated. The U-value of the whole wall system is 0.7 W/m^2K with the roller blinds open and a transmissivity of 52% was calculated for the facade elements.

4. ACKNOWLEDGEMENT

The renovation is part of the project "Energetische Optimierung der Solarapertur", which is sponsored by the German Federal Ministry for Research and technology.

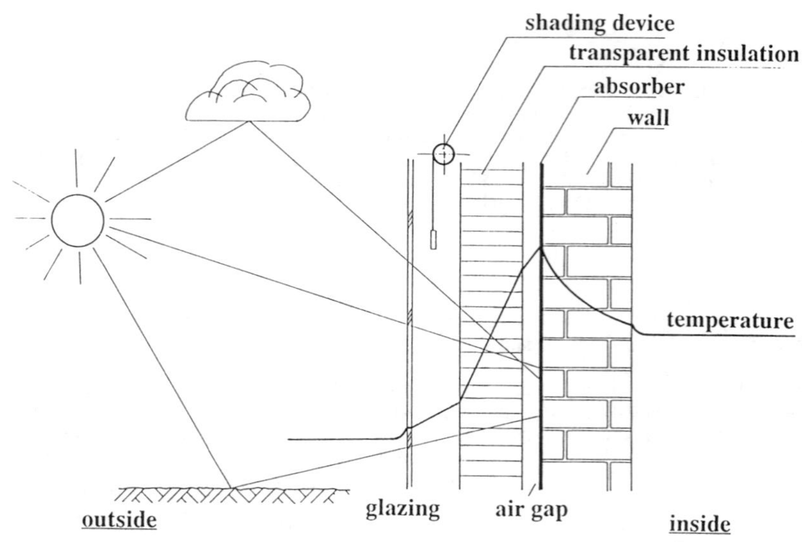

Figure 1:

Transparent insulation for space heating

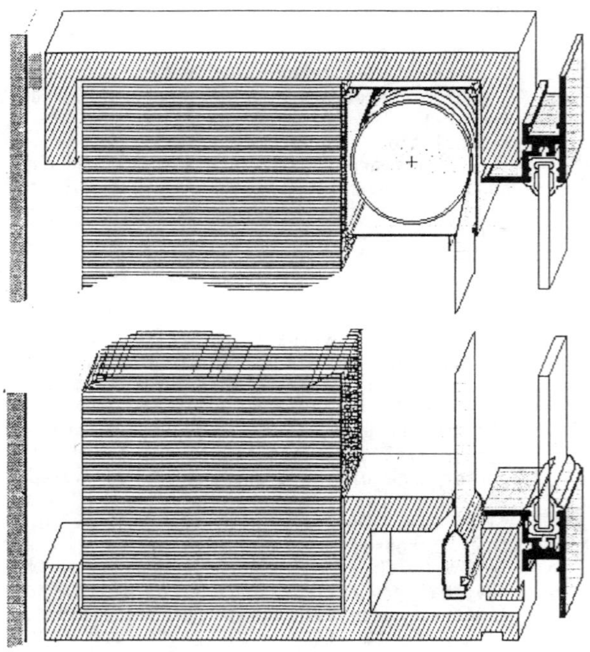

Figure 2:

Transparently insulated facade element with integrated roller blinc

Figure 3: View of the SW facade with the roller blinds open

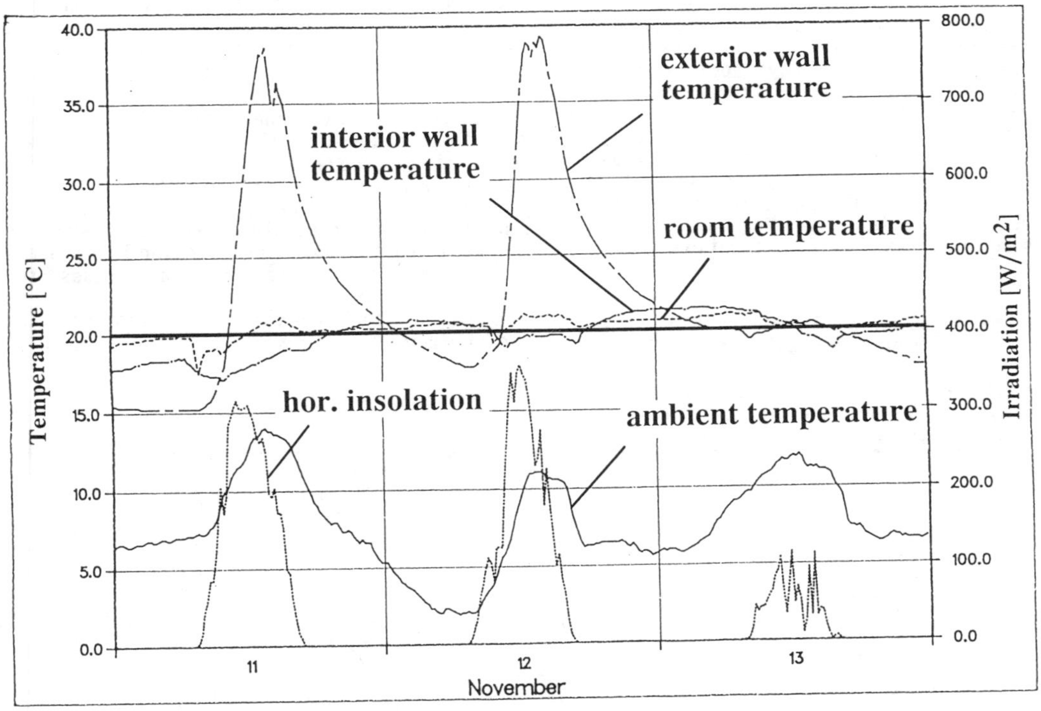

Figure 4: Temperatures and insolation on the SW facade

SIMULATION TOOLS FOR PREDICTION OF THE THERMAL BEHAVIOUR OF TRANSPARENTLY INSULATED BUILDINGS

F. Sick, W.-S. Wilke, J. P. Kummer
Fraunhofer-Institut für Solare Energiesysteme
Oltmannsstr. 22, D-7800 Freiburg, FRG

Abstract:

Correct design and reliable performance estimates of passive solar systems incorporating transparent insulation (TI) materials require computer simulation programs capable to handel the special TI characteristics. There are several programs available at the Fraunhofer-Institute for Solar Energy Systems (ISE). WANDSIM does dynamical calculations of transparently insulated walls using finite differences and is the base module for HAUSSIM, a multi-zone model describing the thermal behaviour of buildings. Several TI wall modules were developed for the simulation package TRNSYS, one based on WANDSIM, and one using detailed transfer functions. Based on the latter, a simplified model was developed. With the help of the described TI software, design procedures can be accelerated and improved. Examples for practical applications of the programs are given.

1. Introduction

The simulation of solar energy systems on computers reduces the time required to conduct experiments to a few minutes after the mathematical model is set up and implemented. The simulation allows a quick quantitative estimate of important properties and - more important- a qualitative comparison of different system variations. Sensitivity analyses may be carried out easily and the engineer's and architect's feel for the system behaviour grows.
The increasing interest in transparent insulation materials sets new demands for the simulation of active and passive solar energy applications. Unlike opaque insulation materials which are thermally described by a constant U-value, TI equipped walls and collectors show a strongly irradiation and temperature dependent - and thus transient - behaviour. It is therefore necassary to run dynamic computer simulations in order to provide correct and reliable predictions of the thermal performance of transparent insulation.

2. Software

The detailed investigation of TI material physics lead to a number of new and modified simualtion programs for active and passive solar energy systems incorporating the use of TI materials. Those concerning the passive use of solar energy are described in the following.

The thermal performance of walls consisting of any combination of transparent and opaque layers is dynamically simulated by the WANDSIM program. WANDSIM uses the method of finite differences to solve the differential equations of heat transfer. One unique feature of WANDSIM is the correct description of the physics of TI materials like temperature-dependent U-values and the specific angle-dependence of the transmission. Another one is the capability to handle time dependent shading devices like roller blinds. In fig. 1, a typical transient temperature profile through a TI wall is shown.

The calculation was done with WANDSIM for a fixed room temperature of 20°C. The ambient temperature is slightly above 0°C. Note that due to the temperature profile there is always a heat flux into the room. It can also be observed from the figure how the temperature profile wanders across the wall and how the closed roller blind increases the thermal resistance for heat losses to the ambient at night.
WANDSIM is often used for optimization calculations for TI walls.

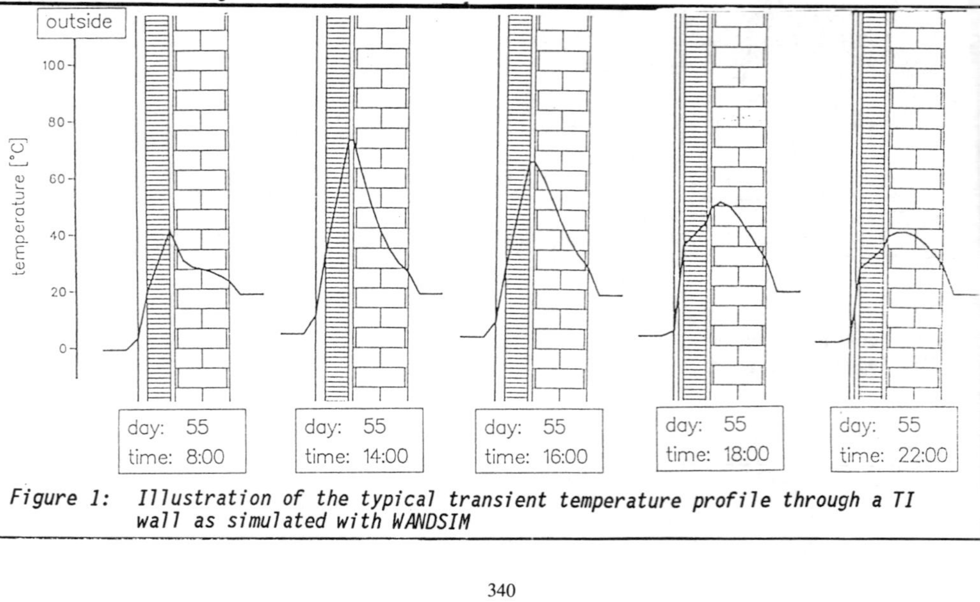

Figure 1: Illustration of the typical transient temperature profile through a TI wall as simulated with WANDSIM

The WANDSIM software module is used as base element in HAUSSIM, a multi-zone model that simulates the thermal behaviour of buildings with or without TI. Like WANDSIM, HAUSSIM allows the use of shading devices and their control. A flexible structure allows virtually any control strategy for the zone climate. It also calculates comfort temperatures. Artificial skylines can be defined for each wall and internal heat gains for each zone. HAUSSIM produces comprehensive graphical output that allows a quick evaluation of the results.
More information on WANDSIM and HAUSSIM can be found in /1/ and /2/.

The WANDSIM module was also included in the TRNSYS program /3/. A model using transfer functions instead of finite differences was included. Based on this, a simplified model could be developed /4/. This new way to simulate TI walls needs a lot less computation time. In one case of a single TI wall, the simplified transfer function model was more than 20 times faster than the finite-difference model. At present, the TRNSYS TI walls are being included to the TRNSYS multi-zone component "TYPE 56".

In fig. 2, the monthly heat input through a TI wall into a zone in Freiburg, FRG, in March and April 1989 is shown. The left bar shows the measured data, the 3 bars to the right are simulation results of WANDSIM, the detailed TRNSYS model and the simplified one, respectively. The good agreement is obvious.

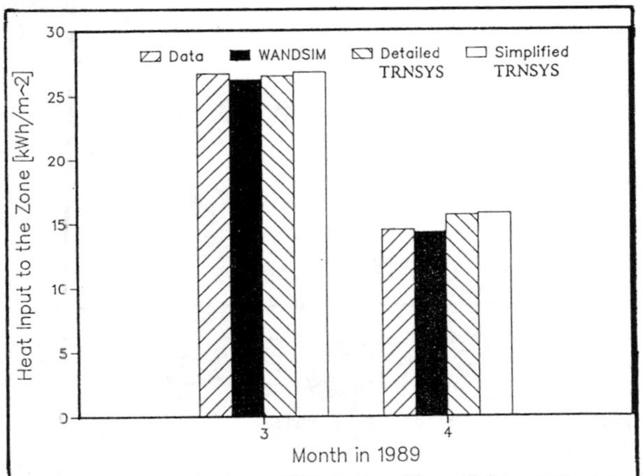

Figure 2: Comparison of 3 TI wall models with data taken in Freiburg,FRG, in March and April 1989

3. Examples

During the process of designing a building with passive solar elements, there are many questions to be answered concerning the thermal behaviour of the projected building. The reason is simple: Any solar system that works well in the winter time works even "better" during the summer, but then causing overheating problems. In order to obtain a comfortable climate in the building throughout the entire year, the solar system must adjust itself to climate changes. The architect as the building designer is on the one hand responsible that the passive solar system is efficient in reducing heating loads and increasing the comfort level and daylight availability. On the other hand, he has to make sure that he does not impose an (additional) cooling load to the building. This is a non-trivial problem and has to be solved carefully. Simulation is the easiest and quickest way to find reliable answers to these questions. In the following a few examples with past experiences are given.

3.1 Optimization of the massive part properties in a TI wall

The maximum incident radiation on a south oriented TI wall occurs at solar noon. Part of the radiation is stored in the massive portion of the wall. The time shift between solar noon and the moment when the inside surface temperature of the wall becomes maximum, is a crucial figure since it determines at what time heating energy is supplied to the room. The example presented in fig. 3 shows how this time shift can be determined.

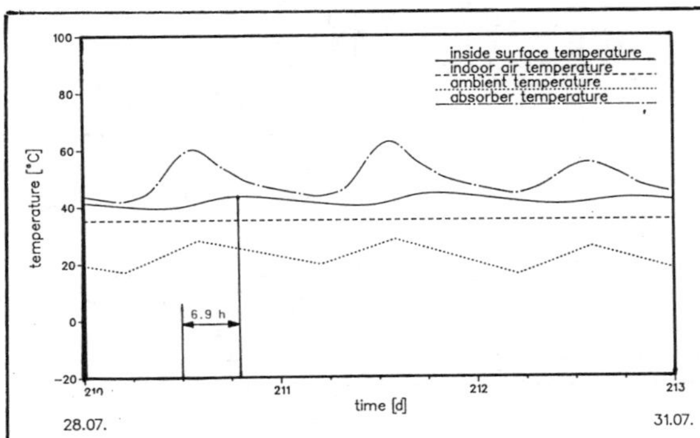

Figure 3: Temperature profiles of the Crew Training Center EVA hall during 3 days in July with a typical Cologne climate. 10cm TI material in front of an air gap and 30cm of concrete were used in this calculation

The application is a new "Crew Training Center" (CTC) for astronauts in Cologne. In the so called EVA hall with a large south oriented windowless wall, an air temperature of 32°C to 35°C shall be maintained. This is an ideal case for the use of transparent insulation. The figure shows the constant indoor air temperature, the ambient and absorber temperature and the inside wall surface temperature as a function of time during 3 days in July of a typical cologne climate. Since the simulation uses solar time, the difference between noon and the time when the maximum inside surface temperature occurs, gives the time shift. In our example this is 6.9 hours. Using different material thicknesses and properties, this time shift can be optimized for any desired application.

3.2 Free floating room temperatures in 2 Shanghai buildings

Central heatings are prohibited south of the Jangtse river in China. The city of Shanghai, located in this area, has cool winters and hot and humid summers. Thus, transparent insulation promises to be a good solution for passive heating purposes. Sun protection by shading windows and TI walls can decrease indoor temperatures in summer.
The Shanghai Institute for Energy is currently planning the construction of a TI house and a conventional one of the same geometry next to each other, in order to compare them and evaluate the benefits of solar architecture. The ISE carried out computer simulations to determine room temperatures in both houses and to optimize the solar house design. The basic building structure in both cases consists of 24cm brick walls with plaster layers. The extreme daily zone temperatures for a room with external walls oriented towards South-East and South-West are shown in fig. 4 for the conventional and in fig. 5 for the optimized TI house equipped with 5cm of honeycomb structure.

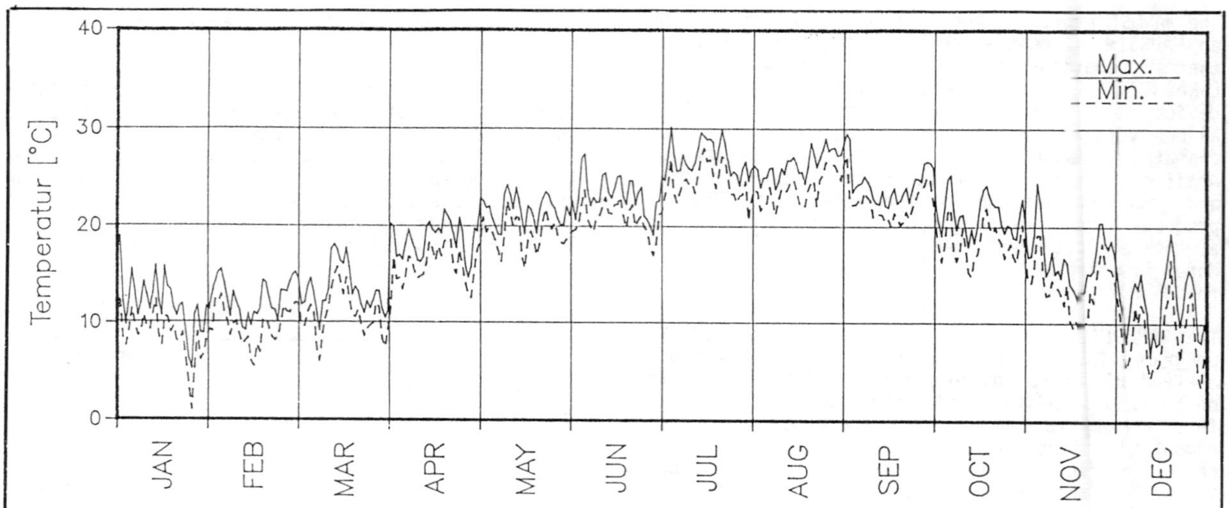

Figure 4: *Maximum (solid line) and minimum (dashed line) daily room temperature of the south-oriented zone of the conventional Shanghai house*

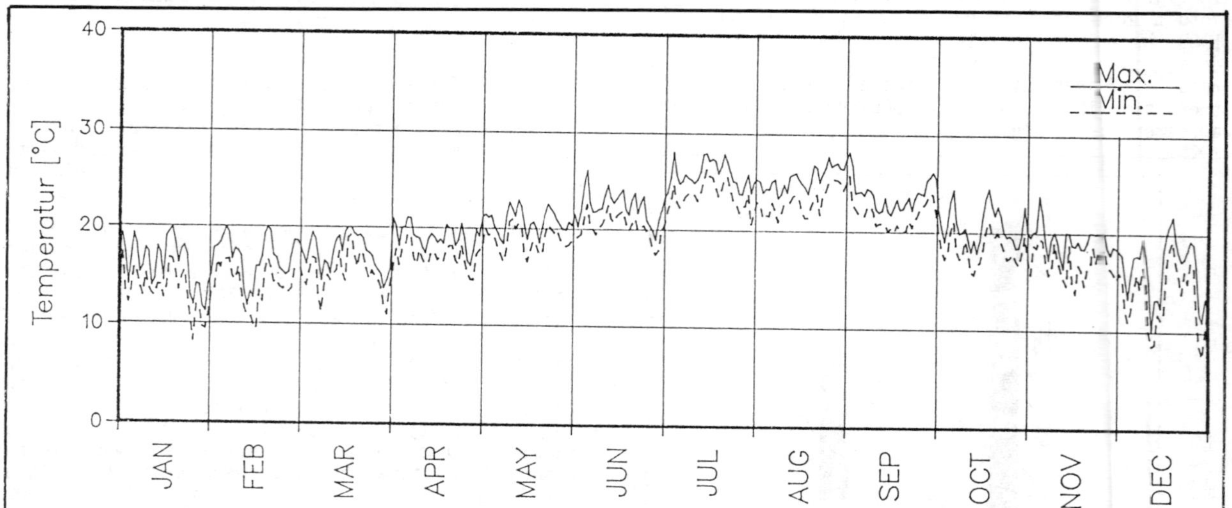

Figure 5: *Maximum (solid line) and minimum (dashed line) daily room temperature of the south-oriented zone of the transparently insulated Shanghai house*

A substantial temperature increase in the TI house in the winter of about 5K can be observed. The temperature decrease of 2 to 3K in the summer months is due to good shading of both windows and TI walls. Several optimization steps were necessary to obtain these results. Without the use of simulation tools during the stage of planning, a considerably worse concept would be used now.

4. Summary

A number of simulation programs for TI applications were presented. The examples presented in section 3 and many others are real projects where architects or engineers asked the ISE for research on the thermal behaviour of their buildings. In all cases changes to the building design were made according to the simulation results that acted as inexpensive and especially fast experiments. The tool of simulation proofed to be necassary and essential for a good building design.

References

/1/ Sick, F.
Transparent Insulation - Simulation Programme Proceedings of the 3rd workshop on transparent insulation technology, Titisee 1989, to be published

/2/ Wilke, W.-S.
Transparent Insulation - Building Simulation Proceedings of the 3rd workshop on transparent insulation technology, Titisee 1989, to be published

/3/ Klein, S.A., et al.,
"TRNSYS, A Transient System Simulation Program", Solar Energy Laboratory, University of Wisconsin-Madison, Engineering Experiment Station Report 38 - 12, March 1988

/4/ Kummer, J.P.
A Simplified Model for Transparently Insulated Walls Proceedings of the 3rd workshop on transparent insulation technology, Titisee 1989,to be published

Atria and Conservatories in Primary Schools; an illustrative analysis of the effect of varying atrium width.

B. Norton
Department of Building
University of Ulster
Newtownabbey BT37 0QB
N. Ireland

D.J. Harris, S.N.G. Lo, P. Waterfield, P. C. Grindley
School of Mechanical Engineering
Cranfield Institute of Technology
Bedford MK43 0AL
England

& R.A. Hobday
Databuild Ltd
Aston Science Park
Birmingham B7 4AP
England

ABSTRACT. A building description file for use with a dynamic thermal building simulation model was developed to describe a particular school building in southern England. To determine that this and the modelling assumptions were valid, a comparison between predicted and measured behaviour was made. Using the validated model, parametric analyses were undertaken. An illustrative example of the effect of varying atrium width is presented.

1. INTRODUCTION

Atria and conservatories are often appropriate in primary school buildings, for reasons of function, scale and construction, occupancy, aesthetics and economics (1). Design challenges arise in preventing summer overheating and glare. An unheated atrium should always improve the thermal performance of adjacent classrooms. It provides a temperature above that of the ambient environment and thus reduces heat losses. Natural daylighting and ventilation may also be enhanced. However the occupancy of an unheated atrium in northern Europe will be restricted by the incidence of low temperatures in winter and will thus be an unsuitable location for timetabled teaching activities in schools. This curtails it's potential amenity value. In contrast, a heated atrium may be used for sedentary activities at all times. However, achieving energy-efficient performance requires more careful analysis: it is essential to ensure that the energy used for heating, during inclement weather, does not overwhelm heat gains in autumn and spring, the year-round displacement of electic lighting by daylight in the atrium and spaces facing into it and the benefits of summer ventilation. This paper addresses the question of the effect of the width of heated atrium on this balance. It is an illustative case study, pertaining directly only to the context of the fabric, form, geometry orientation occupancy of the particular school building with a heated atria examined. This is located in, and subject to the climate of, southern England.

2. HOOK SCHOOL

The considered base-case building, Hook School (1) is located in the small Hampshire town of Hook, about forty miles WSW of London, at a longitude of 1 degree west and a latitude of 51.25 degrees north. The "SCOLA" 1970s system-built built was enlarged and remodelled in 1988. Essentially each "leg" of the H-plan building was extended and the bay between them glazed over and enclosed. Thus two pitched-roof heated atria now run the length of school, interrupted in the middle by a staff room. The atria are carpeted and serve as practical teaching areas, as well as ciculation spaces. They are occupied almost to the same extent as the classrooms, all of which adjoin the atria. The total useable area is 2586 square metres and there are 430 pupils.

Manually-operated opening sections in the glazed roof facilitate the out-flow of buoyancy and wind-driven ventilation drawn from the adjacent classrooms.

The atria are constructed from a steel frame, with reinforced double glazing. The lower sections of the south-facing atria roof incorporated opaque composite sheeting intended to eliminate glare and inhibit overheating.

3. MODEL VALIDATION

The thermal behaviour of Hook School was simulated using the dynamic model SERI-RES (2,3,4). Physical dimensions were derived from plans, sections and elevations. Where necessary these were corroborated and augmented by on-site measurements. Environmental (e.g., heating system and controls) and use (e.g., occupancy) data was obtained by observation. The building description and the modelling assumptions were validated by comparing actual monitored hourly zonal air temperatures with those predicted. The validation simulations were undertaken using appropriate meteorological data measured on-site. Typical examples of such comparisons are shown in figures 1 and 2. Given the prevailing complexities of an occupied building that remained unmodeled, the agreement obtained was viewed as acceptable. Diurnal variations of measured and predicted temperatures also showed adequate concurrence, thus indicating that the dynamic responses modelled were satisfactory.

4. PARAMETRIC ANALYSIS

Four widths of atrium were investigated, namely 2.5m, 5.0m, 7.5m, and 10.0m. Cross-sections of these are shown in figures 3 to 6. This range encompasses the likely feasible dimensions for an atrium in a building of this type and scale. The base-case height of the atrium and inter-zonal walls was maintained. This consequently gave rise to varying strium glazing angles between the four cases considered. Since the overall plan area of the school and all other dimensions remained unchanged, the width of the atrium was directly proportional to its floor area. Thus in the 10.0m wide atrium constituted nearly half the total floor area of the school. This is justified in the context of state-supported educational buildings in the U.K., where a larger area of atrium would not be allowed generally to facilitate a larger overall school area. Local education authorities have strict limits, exercised at the design stage, on the maximum conditioned area per pupil in primary schools.

Ventilation energy required to remove any overheating was simulated via an "imaginary" mechanical system triggered at a particular temperature.

Each atrium was modelled with the roof-plane incorporating the lower opaque composite sheeting orientated either north-south or east-west. Furthermore, in each orientation, the simulatons were repeated without those opaque elements being present. Sixteen separate individual building descriptions were thus simulated.

5. RESULTS

Each simulation was run for a whole year using a test reference year for Kew (5,6). For ease of data processing the output specified was for a representative day of each month. Data files were obtained consisting of the heating and lighting loads for each width of atrium for each of twelve days

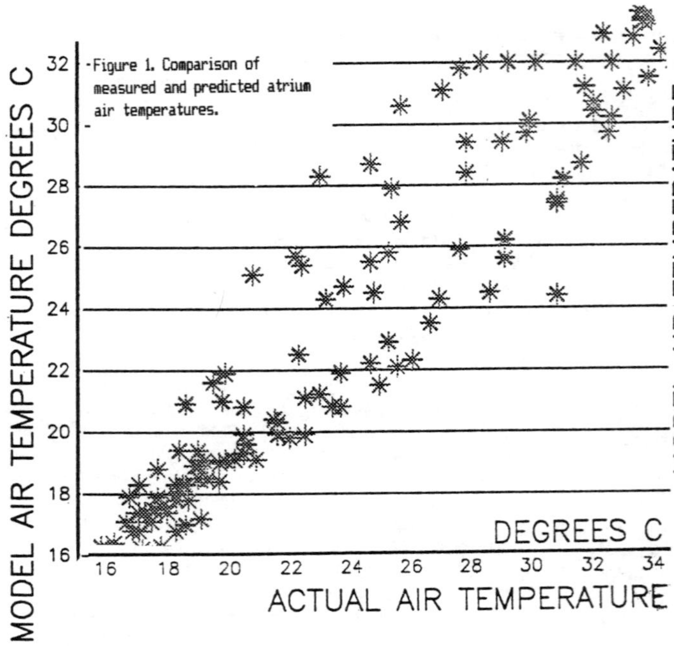

Figure 1. Comparison of measured and predicted atrium air temperatures.

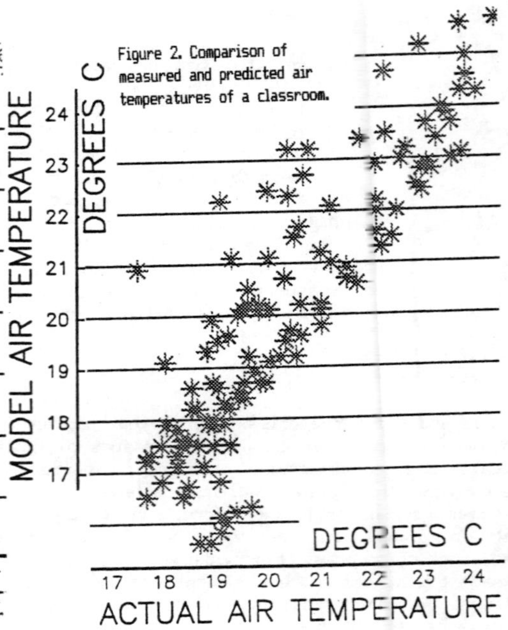

Figure 2. Comparison of measured and predicted air temperatures of a classroom.

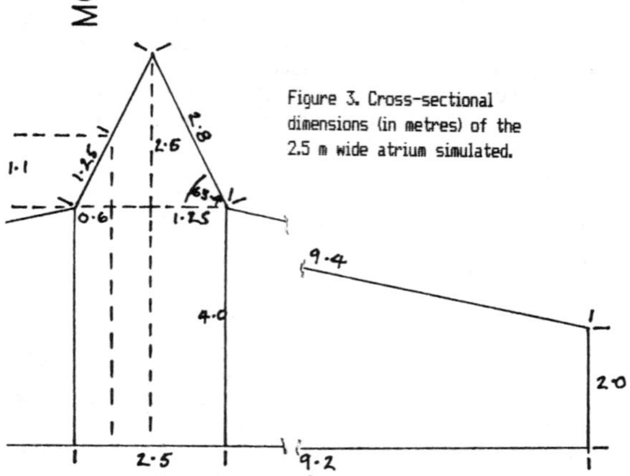

Figure 3. Cross-sectional dimensions (in metres) of the 2.5 m wide atrium simulated.

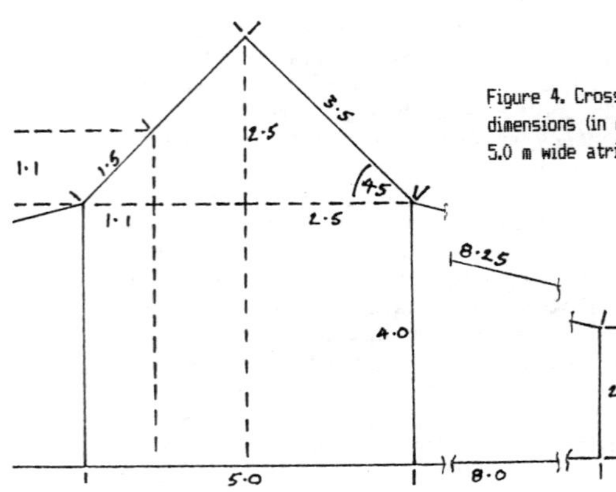

Figure 4. Cross-sectional dimensions (in metres) of the 5.0 m wide atrium simulated.

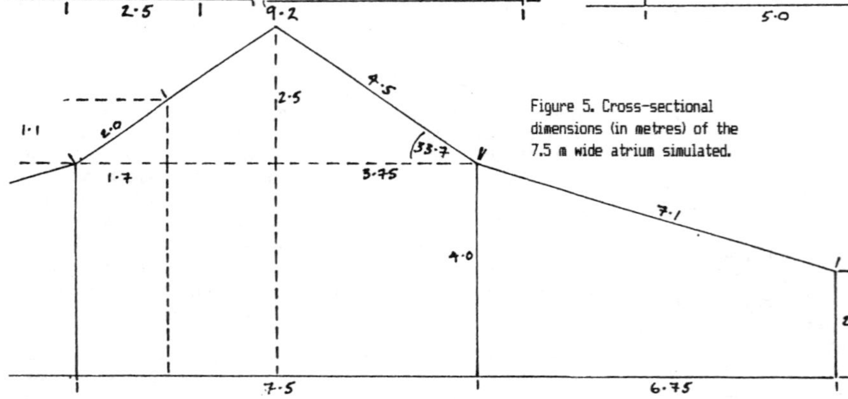

Figure 5. Cross-sectional dimensions (in metres) of the 7.5 m wide atrium simulated.

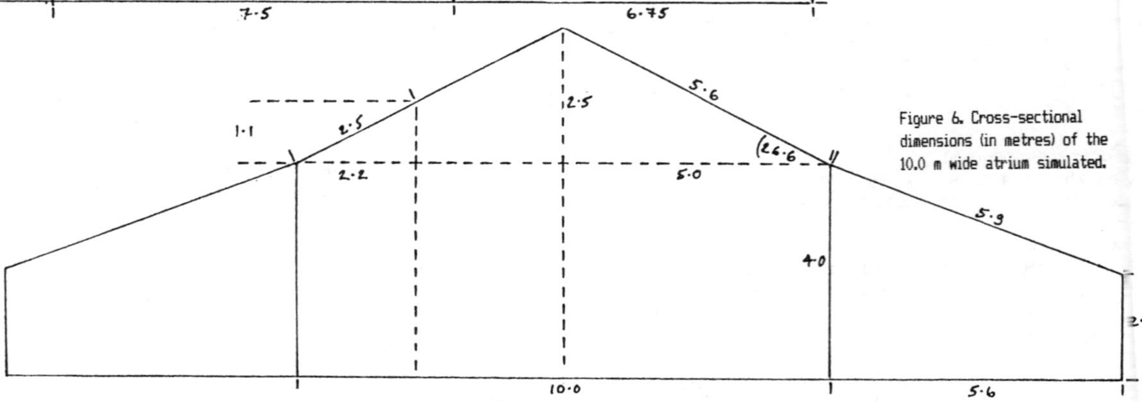

Figure 6. Cross-sectional dimensions (in metres) of the 10.0 m wide atrium simulated.

representative of the year. The values shown on ordinate axis in the graphs shown in figures 7 to 10, refer to the sum of the energy loads calculated for these twelve days. These values should be multiplied by thirty to give predicted full annual energy loads, (e.g., an energy load of 2200 MJ in figures 7 to 10 corresponds to a full annual energy load of 18,333 MJ).

6. CONCLUSIONS

The variation of the heating, lighting and ventilation loads with width for each orientation, and with and without shading, are illustrated in figures 7 to 10.

Overall, a 5.0m wide north-south orientated system with shading was the optimum option amongst those considered. This corresponds to the base-case building, which is fortuitous as it's atrium width was a determined by the dimensions of the bays of the pre-refurbishment H-plan building.

All the widths of the north-south orientated atria with the integral opaque shading in place, as shown in figure 7, gave the lowest overall energy loads in comparison with the other cases examined. For each orientation, the shaded option performed better. Shading has a great an influence as orientation; the east-west shaded atrium out-performed the north-south unshaded.

ACKNOWLEDGEMENTS

This work, though forming part of a study undertaken for the Architects and Buildings Branch of the U.K. Department of Education and Science (DES), does not necessarily represent their views. The results presented illustate briefly one aspect of an on-going research programme and should not be taken to present design recommendations.

The support of the DES and the co-operation of Hampshire County Council and of the staff and pupils of Hook School is gratefully acknowledged.

REFERENCES

1. Norton, B. Hobday, R.A. "Passive Solar Schools" Report to the Architects and Buildings Branch of the Department of Education and Science, London, U.K. February (1989)

2. Palmiter, L., Wheeling, T., "SERI-RES, Residential Energy Simulator", Version 1.0, Ecotope Group, Seattle, Washington, U.S.A.

3. Haves, P. Littler, J., "Refinements to SERI-RES", Report ETSU S-1130. Energy Technology Support Unit, Harwell, Oxfordshire. (1987)

4. Haves, P. Littlefair, P.J., "Daylight in dynamic thermal modelling programs: Case study" Building Serv. Eng. Res. Technol, 9, 183-188, (1988)

5. Loxsom, F.M., "PCLTRY - SERI-RES Format Standardised Kew Weather Data" Report to Energy Technology Support Unit, Harwell, Oxfordshire. (1985)

6. Waide, P.A., Norton, B., Applicability of short reference years for passive solar building simulation" Proc Second European Conference on Architecture, Paris, France, December (1989)

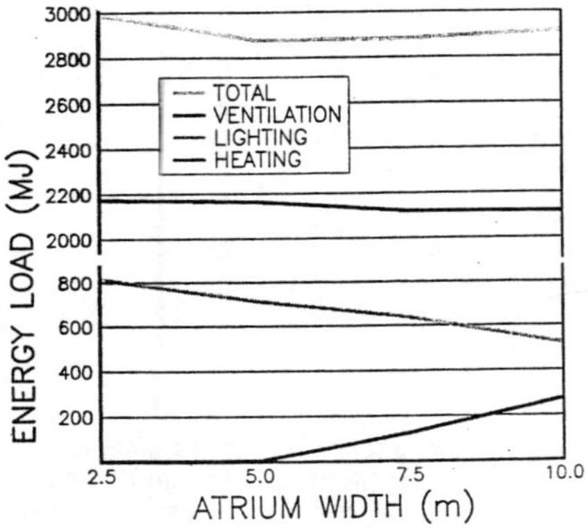

Fig 7. Variation of Energy Loads with Width for a north-south orientated shaded atrium.

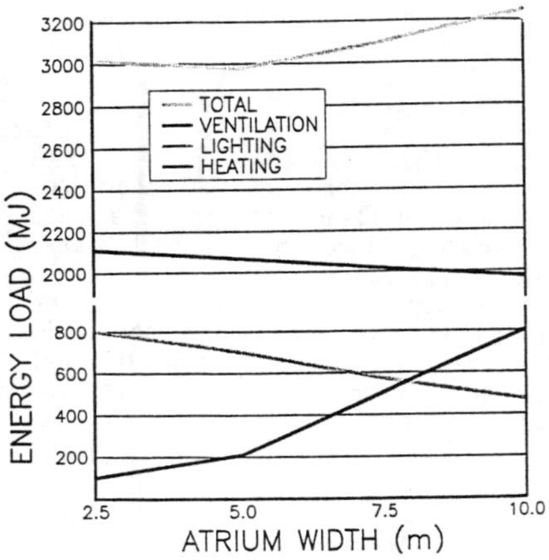

Fig 8. Variation of Energy Loads with Width for a north-south orientated atrium without shading.

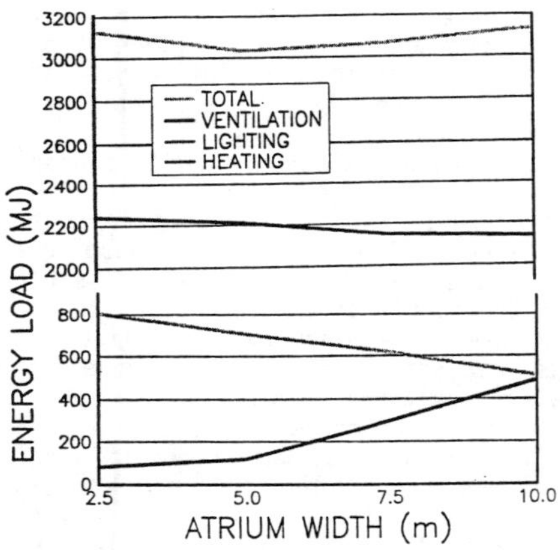

Fig 9. Variation of Energy Loads with Width for a east-west orientated shaded atrium.

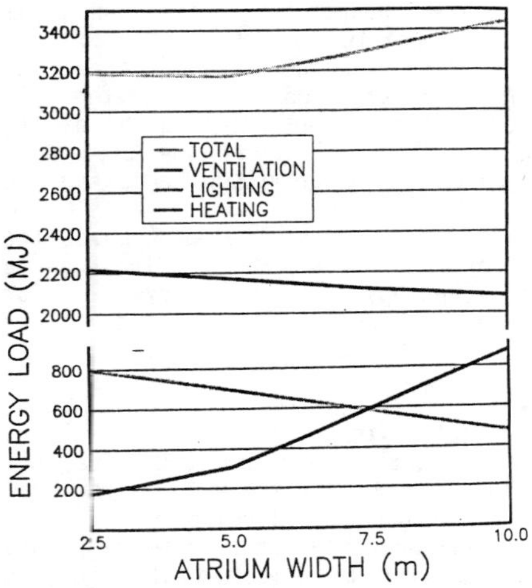

Fig 10. Variation of Energy Loads with Width for a east-west orientated atrium without shading.

CONFLICTS BETWEEN PHYSICS AND ARCHITECTURE IN SOLAR SYSTEMS OPTIMIZATION

Federico M. Butera
Dipartimento di Energetica e Applicazioni di Fisica - Università di Palermo
Viale delle scienze - 90128 Palermo - Italy

ABSTRACT. The lack of diffusion of energy/environment conscious design tools among architects is analyzed in terms of technology transfer. The Design System, i.e. the complex system made of designers and technological devices that performs a design process for producing the project of a building, is described as an open thermodynamic system subject to random perturbations either internal and external, made of neg-entropy packages containing novelty and confirmation. The design process is described as a strange attractor. This unusual "reading" of the architectural design is used for defining the specifications of a computerized design tool based on physics and speaking the language of the architect.

1. INTRODUCTION

Solar energy use for energy saving and/or for improved thermal and visual comfort imposes a new order of boundary conditions on architectural design, by affecting both lay-out and building materials. These new boundary conditions must be embodied into the design process, both in terms of new principles and new tools. New principles and new tools cannot be separated: if the new principles (the ones of solar or bioclimatic architecture) are not included into the design process, tools are useless and, vice-versa, without appropriate tools principles cannot be applied.
Actually, the interaction principles/tools is even more complex than that, otherwise it could not be explained the poor effectiveness of the efforts made in the last 10-15 years in the diffusion of either simple or sophisticated design tools for bioclimatic buildings design: in spite of the tools availability and of the apparent acceptation of the principles, only very few architects have used them.
If -as it is fair- we consider the new available design tools (especially the sophisticated ones, i.e. the computer simulation models for predicting the dynamic behaviour of thermal and/or luminous environment) as new technologies, the specific problem we are dealing with may be considered as a special case of the more general problem of technology transfer. This conceptual approach can reveal itself useful for a correct analysis and for finding solutions to the widespread failure in the diffusion of the energy-related design tool for architects.

2. TECHNOLOGY TRANSFER

The failure in the diffusion process of tools for energy conscious design, may be compared with the failure in the diffusion process of village-level renewable energy technologies in developing countries. In both cases intensive efforts for developing those technologies (or tools) have been made all over the industrialized world, but the success has not arisen as expected, and the systems (the rural village on one hand, and the design team on the other) that should have benefitted from them ignored the innovation or tried first and then rejected it; why?

2.1 SYSTEMS, VALUES AND COGNITIVE TOOLS

In the Sahel, most often, water is some kilometers away from the village and women, bearing children on their backs and a vessel on their head with the washing and some crockery, walk there.
A donor country decides, some time ago, to install a solar pump in one of these villages /1/. It is a demonstration project of the utmost importance because there are many villages sharing the same features. At last, women will no longer have to carry out this cumbersome daily load.
A well is dug in the centre of the village and a pump is fitted; it is also tested and works perfectly well. One year later, when the technician that installed it come back for an inspection visit, the pump is out of use. It works but it is out of use. Women are still carrying out their daily load, walking to the far away pond.
"Women did not want to use the pump", says the head of the village, without other explanation.
Time goes by and the mystery is finally unveiled. The community's social rules prescribe that encounters between girls of marriageable age and their potential intended have to take place out of the village, along the track leading to the water pool, "by chance" and under the watchful and understanding eye of mothers and elder sisters.
In such circumstances there is very little to choose: either the pump or the marriage.
Nobody can deny that the solar pump was an improvement of the system "village", but is also true that the survival itself of the system was possible only by means of the marriage: in order to introduce successfully the new technology in the village it would have been necessary, first, to change the system of social and cultural values.
Many different optimization concepts are conflicting in this story: the physicist's, the economist's and the anthropologist's ones. The problem appears recursive and -therefore- without a stable solution: the new technology is accepted only if social and cultural values are changed, but the change is possible only if the new technology is adopted.
Similarly, in the design team, the new tool/technology can be accepted and integrated into the design process only if the rules/values of the team are prepared to it, i.e. if the system is changed; but it cannot change without adopting the tool.
On the other hand the same can be said about the product of the design process, the building, when a new component/technology (a solar system, for example) has to be introduced and optimized. So, we find the same nature of problem at different scales, when we try to introduce a new technology in a complex system, at territorial (or village) scale, at building scale and at design team scale. The problems appear different in their structure and shape, but the underlying rules are the same: fractal structure? Who deals with technology development usually operates within his own world, made of messages and signs that are familiar to him, and of cognitive tools to interpret them, but he is most of the times

totally unaware of the system that is going to be affected by the technology.

3. THE "DS" AS AN EVOLVING COMPLEX SYSTEM

Let us take as object of our analysis the complex system made up of a physically limited enclosure, interacting with the external environment, containing a design team and the technological devices used; let us call it, for our convenience, DS (Design System). Within the DS all components are mutually interacting with complex non linear relationships, whose parameters are subject to random fluctuations. The design process performed in the DS is like the metabolic process for a living system or the chemical process for an industrial plant.
The degree of complexity of such a system can be found only in natural living systems: biological and ecological, with which some common features are shared:

a. they are open systems, i.e. capable of exchanging energy and matter with their environment;
b. they are made up of sub-systems linked by functional relationships;
c. they are capable of homeostasis, that is of reacting to changes of their environment in order to guarantee their survival;
d. they are capable of evolving, namely of transforming themselves in differently structured and organized systems.

The first two features are also common to "lifeless" systems, whilst the last two have for a long time been considered specific of living systems alone; this has given birth to two sciences for long time incapable of mutual communication.

Things started changing with the development of a branch of the science dealing with inanimate matter: the thermodynamics. Since the beginning of the century authoritative scientists have attempted to use the principles of thermodynamics in order to link the two physical realities: from Shrodinger /2/ on to Prigogine /3/, /4/, /5/ in our days, and Haken /6/, /7/. Other attempts to link thermodynamics with economy /8/, /9/ and thermodynamics with the theory of information were made especially by Brillouin /10/; on the other hand, theory of information is widely used both in biology and in ecology /11/, /12/, /13/; these are sciences that, in their turn, often resort to thermodynamics /14/, /15/.
A strong conceptual support to the analysis of structural changes of complex systems derives also from recent developments in the area of the so-called "deterministic chaos" /16/.
The gap between physics, biology and sociology seems to narrow. Let us see how these advances might be used for our purpose, by recalling some basic notions.

3.1 From order to disorder to order

The word "entropy" is now part of the ordinary language, after having been, for a century or so, restricted to a handful of scientists. The second law of thermodynamics states that, in an isolated system, the entropy cannot decrease; and as entropy provides a measure of the degree of disorder, the general trend of the universe to chaos derives. We are well aware of this: the unavoidability of the deterioration of things unless we repair them is part of our culture and our daily experience.
Only in living things is this principle contradicted. Biological organisms and ecological systems show the capability of moving in the opposite direction: from disorder to order.
A biological system, even the simplest one, in metabolizing organic and inorganic matter creates or at least maintains the pre-existing order.
Shrödinger /2/ asserted that living beings are "decreasing entropy islands", namely systems capable of maintaining their own organization -or even improve it-, to the detriment of the system to which they belong. Hence, there is a flow of neg-entropy (negative entropy = organization, order) feeding the living system and a production of entropy being released in the environment surrounding the living system.
However, there has been some visible evidence, not much to tell the truth, contradicting the statement according to which matter is incapable of organizing itself. Let us examine it more closely.
Convective instability (or instability of Benard): such phenomenon occurs when a fluid, confined in a container of defined shape is heated from the bottom. After a while (the exact moment can only very roughly be predicted a priori), a regular structure of convective motion forms spontaneously.
Ferromagnetic materials: the typical feature of these materials is that when they are heated above a certain temperature they loose their magnetic properties that suddenly reappear as soon as they cool down below this critical temperature. Namely, what happens is that the elementary little magnets that make up the material spontaneously re-arrange themselves; they all direct themselves at the same time in the same direction like soldiers obeying an order.
But in chemistry more than anywhere else do these self-organizing phenomena have been detected (i.e. the Belusov-Zhabotinsky reaction, in which there is a spatial organization, or a temporal one as in the so-called "chemical clocks", in which a periodical oscillation of the components' concentration spontaneously triggers on).
Only very recently, thanks to Prigogine's /3/, /5/ analysis of "dissipative systems", and Haken's, /6/, /7/ "synergetics", a new "evolution paradigm" has been worked out (see also Jantsch /17/ and Lazlo /18/) in which chance plays a direct role in the evolution of systems, and in which the self-organization process becomes a recognized physical phenomenon.

3.2 Neg-entropy, resources and information

One of the most interesting aspect of the system's analysis in terms of entropy and neg-entropy flows is linked to the concept of resource.
For a thermodynamic system (or a system analyzed through the principles of thermodynamics) the resources are neg-entropy flows enabling the system itself to keep its organization or improve it. In this definition it is implicitly admitted that the concept of resource (or neg-entropy flow) is subjective; it is by far evident that anything might be considered as a neg-entropy flow, provided it is useable and used by the system. Hence, oil is a resource in our society structured for its use: in Ancient Rome all Saudi Arabia's oil reserves were not even worth a slave's shoe; grass is a resource for the herbivore but not for the carnivore, and so on.
If we look into rapidly evolving complex systems such as human societies, we note that "progress" is always coupled with the "recognition" of resources by a system itself.
The Industrial Revolution is one of the clearest examples of the transformation process that turned non-resources into resources through the evolution of the system's structure: coal became a resource, then oil and natural gas; through the development of chemistry, also ore deposits previously ignored, and so on.
Naturally the acknowledgement of such a wide range of resources by the social system is coupled also with a transformation of its structure as far as both new components (technologies) and organization are concerned.
The words "recognize" and "metabolize" indicating the

capacity of a system to use a neg-entropy flow have not been chosen by chance. They highlight the important links existing between fields such as thermodynamics, communication systems, biology and ecology.
Also information is "recognized"; any speech is only noise if you don't know the language, i.e. if you are not structured to intellectually "metabolize" the flow of sounds you perceive.
Hence, both physical neg-entropy and information are resources only if the user (the recipient) is structurally prepared to use them.
Apparently this assertion rules out any possibility of the system evolving, be it physical, biological or cognitive: this is obviously false.
As a matter of fact, if we analyze the way in which systems are capable of self-organization, the apparent contradiction fades away.

3.3 Novelty, confirmation and evolution

Systems evolve through accidental fluctuations of internal and/or external parameters that -in special conditions- give rise to processes synergically moving towards a demolition of the pre-existing system and the creation of a new structure.

Any physical system in dynamic equilibrium is always struggling against the surplus neg-entropy it is not capable of metabolizing. The neg-entropy flow, penetrating within a physical system is also a physical flow and, nearly always, never is a pure one. Namely, it is matter containing either elements which can be metabolized by the system and elements which cannot. Hence, the system has to continuously sift out the neg-entropy it recognizes from the rest. The identity of the system, its complexity and its organization exist and are justified on account of the "impurity" of resources. Indeed, the more the impurities have a large fluctuation, the more the system is complex and flexible.
If the same reasoning is translated in terms of information, then the impurities become the so-called "noise"; it can also be a non-recognized message. In other words, a message is always partly decoded, "recognized" and partly not even identified as such. The "recognition" of a part of noise as a message is what happens when we suddenly have clear the solution of a problem, or it happens all the times we re-read (or re-listen to) a masterpiece, and we still enjoy and find new "consonances" between us and the author. We might then generalize and say that both information and neg-entropy reach the system in packages containing novelty and confirmation; by novelty we mean all that is not recognized and by confirmation all that is recognized. Systems tend to discard novelty and retain confirmation: it is an operation which is fundamental in order to survive.
If the neg-entropy (or information) flow was all made up of novelty (that is, if everything was "unrecognizable"), the system would die (for example, closing a lion in a fruit and vegetable shop, or burning wood in a gas boiler or, suddenly and without training, substituting in a DS all the drawing tables and pencils with a CAD system).
If, on the other hand, the flow is "pure", that is exclusively made up of confirmation, the system would tend to simplify itself and reduce its degree of complexity given the "atrophy" of a vast array of its functions. Hence, it would end up by loosing its identity, changing into another simpler and different system (think of a design team in a submarine, without books, journals, newspapers, radio and television).
Hence, all novelty or all confirmation lead the system to death. It goes without saying that impurity is necessary for survival, i.e. the input must contain a bit of novelty and a bit of confirmation. More so, it is exactly the interplay, indeed the continuous fluctuation novelty-confirmation, that has made (and still does make) evolution possible. That is evolution occurs if and when fluctuations within the system's structure are in phase with the novelty-confirmation fluctuations of the neg-entropy input, namely when the system is, accidentally, capable of recognizing a bit of novelty. At that point a very fast reaction may take place, involving all the system's components, re-arranging them according to the new and more advanced structure which is more complex and capable of metabolizing a new resource.
This is also the description of the learning process, which always implies a structural change; we might say that systems, with their internal fluctuations, "explore" the outside environment, searching for new resources (matter, energy, information) to metabolize.

4. DESIGN PROCESS AND STRANGE ATTRACTORS

Most of the thermodynamic systems we are used to deal with may be considered as dissipative systems. "In physics a dissipative system is a system in which high-quality energy is transformed into low-quality energy (heat) through loss mechanisms such as ohmic resistance in electric networks or friction in mechanical systems.... As a consequence of the inherent loss mechanisms, responses of dissipative system to external disturbances will die out, and after an initial transient, the trajectories will approach a permanent regime that is the long-term or stationary behaviour of the system. In phase space these stationary modes of behaviour are represented by geometrical objects called attractors" /19/. An example is given in Figs. 1 to 6 where the Feigenbaum equation $X_{n+1}= rX_n(1-X_n)$ is used /20/, with a graph X vs. n (left) and a time-velocity representation in a radar-type graph (right): for r = 2.98 the trajectory in the phase space converges to a stable point (the center of the radar graph), whilst for r = 3.3 and r = 3.5 the trajectory never converges on a stable point, but it remains always within a well defined portion of space.

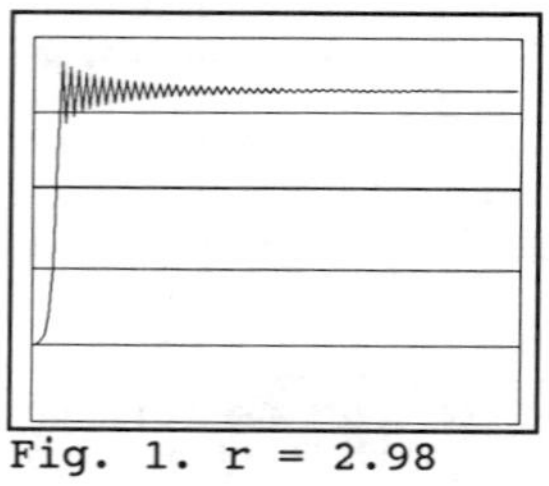
Fig. 1. r = 2.98

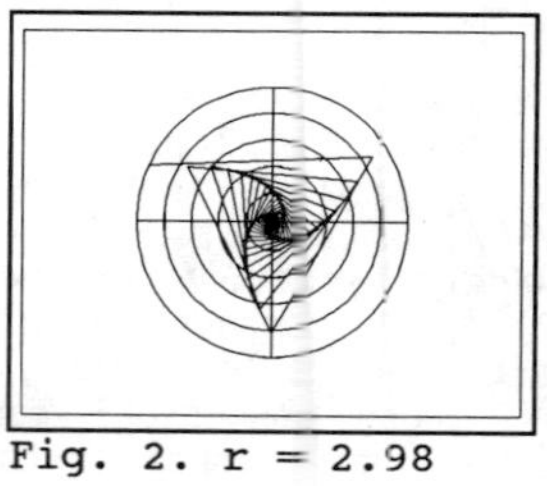
Fig. 2. r = 2.98

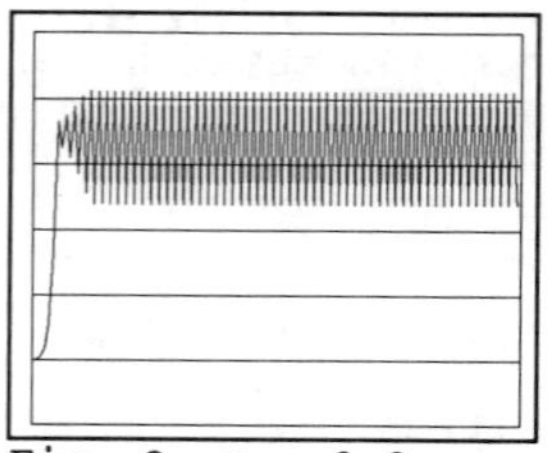
Fig. 3. r = 3.3

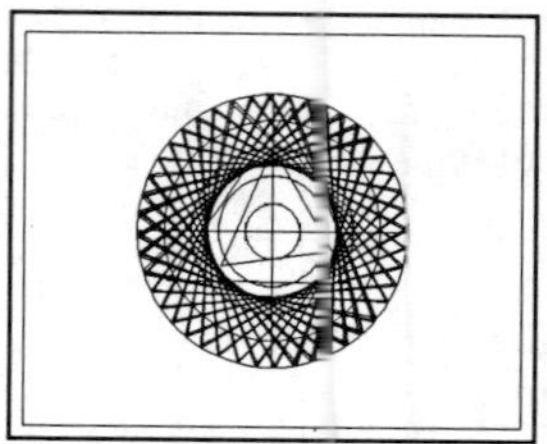
Fig. 4. r = 3.3

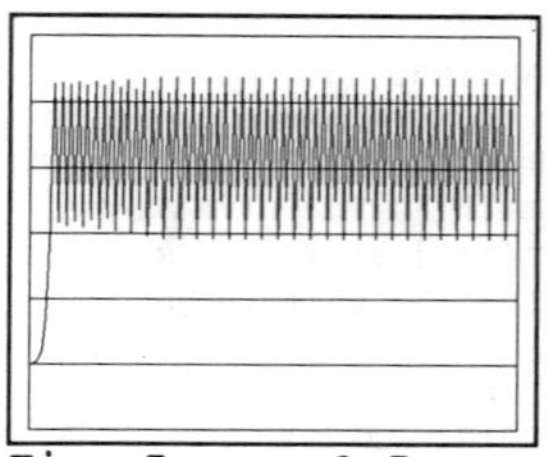
Fig. 5. r = 3.5

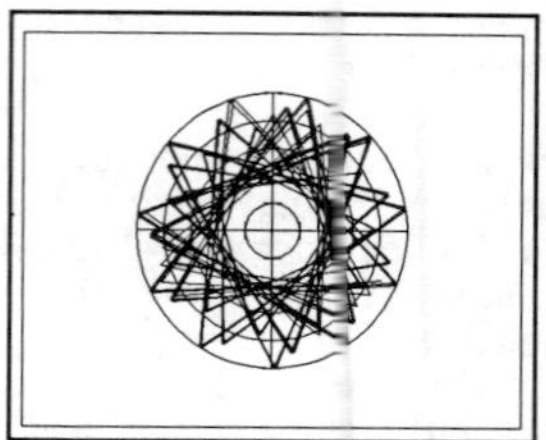
Fig. 6. r = 3.5

The first can be considered as the "physicist approach" to design, whilst the other two better represent the "architect approach".
Another interesting case, in which the number of non-converging trajectories forming the strange attractor is infinite, but always within a confined area, is given by the Lorenz equations, developed for simulating and predicting atmospheric heat and mass transfer phenomena /19/:

$$\frac{dx}{dt} = - sx + ry$$

$$\frac{dy}{dt} = - xz + rx - y$$

$$\frac{dz}{dt} = - bz + xy$$

In Fig. 7 the time variation of X, Y and Z is shown, and in Fig. 8 the projection on the XZ plane of the three dimensional strange attractor "Lorenz' butterfly" that can be derived. Note the switching trajectory from one "wing" to the other.
In Fig. 9 is shown the time variation of X for s = 10, r = 28 and b = 8/3; the only difference among the curves shown is the initial value of X, taken equal to 10, to 10.001 and to 10.0001, respectively. Such small differences at the beginning of the evolution seem to have no effect at all, but suddenly, at an unpredictable moment, the systems' behaviour changes completely. In all three cases the strange attractors have the same shape and occupies the same portion of space.

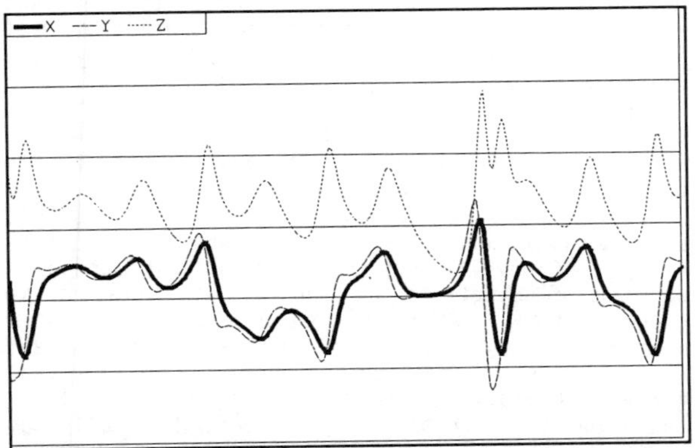

Fig. 7. Time variation of X, Y and Z

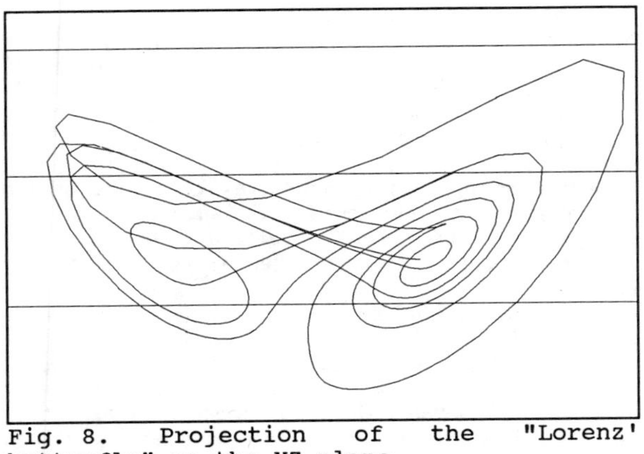
Fig. 8. Projection of the "Lorenz' butterfly" on the XZ plane.

The design process is not much different from all this: even if the same rules are used and the same people and technology is present, because of the impossibility to reproduce the same initial conditions, or because of very small, unpredictable and unavoidable variations during the process, the

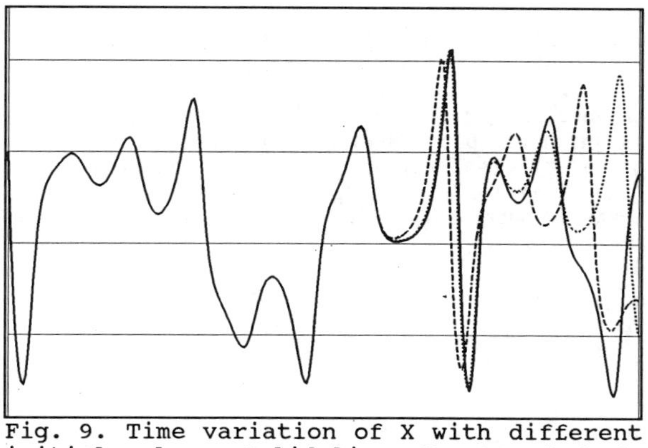
Fig. 9. Time variation of X with different initial values. Solid line, X = 10; dashed, X = 10.001; dotted, X = 10.0001

final product will be different. The different final products, however, are like the points in the trajectories of a strange attractor: all contained within a well defined spatial area, i.e. resulting from boundary conditions that do not give univocal solutions but a "cloud" of possible ones.
This unpredictability of the design process is intrinsic of the DS, and is due to its structure and operation.

5. THE INTRODUCTION OF A NEW TECHNOLOGY IN A DS

We may now try to use all the concepts above recalled for describing a DS in terms that can be useful for designing new kinds of computer tools for promoting energy/environment conscious design, such that will have less probability to be rejected by the architect: a sort of "appropriate" advanced technology.
What will happen if we try to introduce a new component and new relationships among components in our DS, i.e. if try to introduce a new design tool? The characteristics that the tool/technology should have may be condensed into one statement: it must be a blend of novelty and confirmation that matches a wave of fluctuations of the DS, so that it is recognized as a package of neg-entropy/information. By translating the statement into a common language (less rich of information, however), we may say that a new technology is not rejected by a system (any, including the DS) only if it meets the following requisites:

a. hardware and software are a well-proportioned blend of old and new;
b. it is a carrier of messages which can be read by the system as a well-proportioned blend of novelty and confirmation;
c. it can be functionally well-integrated with all the other parts of the system.

Have the design tools for energy conscious design fulfilled these requirements? The answer is: no. The design tools up to now developed have been neg-entropy packages not recognized as resource by the DS.
Actually, in most cases the tool was so designed that appeared to the DS as something that forced the design process to converge tiresomely towards a single possible optimum product: like the attractor in Fig. 2. Moreover, the tool imposed new expertise to be introduced into the system and new relationships among components: why should the DS have accepted such a change? It was too risky (too much novelty), and badly presented.
What should be done, instead, is to try to push the system towards a strange attractor of different shape, a transition like the one from Fig. 4 to Fig. 6. But, in order to do that, a modification of the components and of the relationships among

The first can be considered as the "physicist approach" to design, whilst the other two better represent the "architect approach".
Another interesting case, in which the number of non-converging trajectories forming the strange attractor is infinite, but always within a confined area, is given by the Lorenz equations, developed for simulating and predicting atmospheric heat and mass transfer phenomena /19/:

$$\frac{dx}{dt} = - sx + ry$$

$$\frac{dy}{dt} = - xz + rx - y$$

$$\frac{dz}{dt} = - bz + xy$$

In Fig. 7 the time variation of X, Y and Z is shown, and in Fig. 8 the projection on the XZ plane of the three dimensional strange attractor "Lorenz' butterfly" that can be derived. Note the switching trajectory from one "wing" to the other.
In Fig. 9 is shown the time variation of X for s = 10, r = 28 and b = 8/3; the only difference among the curves shown is the initial value of X, taken equal to 10, to 10.001 and to 10.0001, respectively. Such small differences at the beginning of the evolution seem to have no effect at all, but suddenly, at an unpredictable moment, the systems' behaviour changes completely. In all three cases the strange attractors have the same shape and occupies the same portion of space.

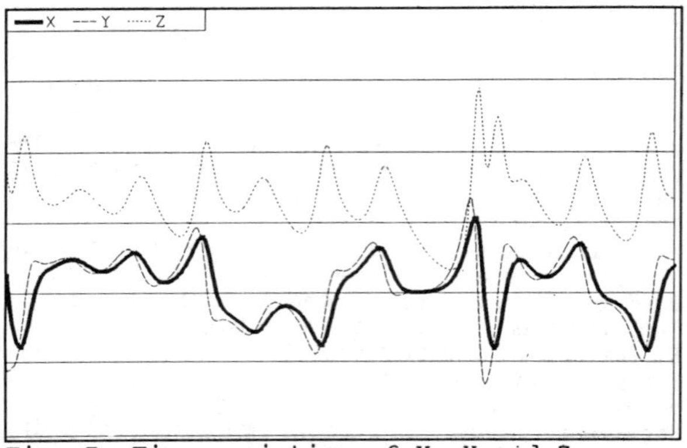

Fig. 7. Time variation of X, Y and Z

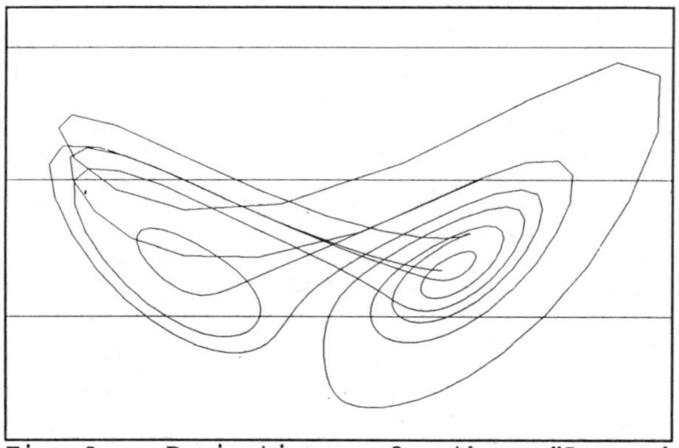
Fig. 8. Projection of the "Lorenz' butterfly" on the XZ plane.

The design process is not much different from all this: even if the same rules are used and the same people and technology is present, because of the impossibility to reproduce the same initial conditions, or because of very small, unpredictable and unavoidable variations during the process, the

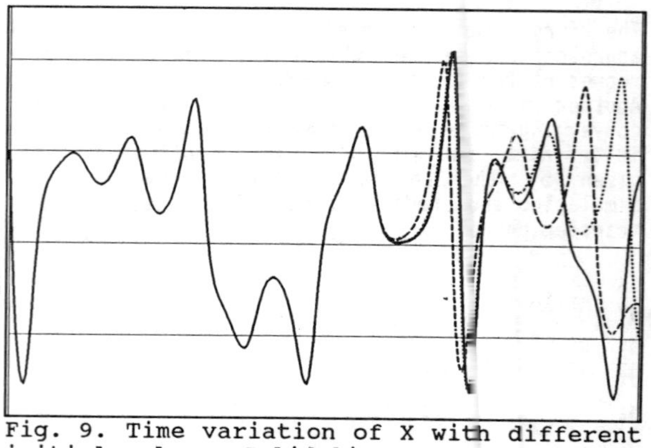
Fig. 9. Time variation of X with different initial values. Solid line, X = 10; dashed, X = 10.001; dotted, X = 10.0001

final product will be different. The different final products, however, are like the points in the trajectories of a strange attractor: all contained within a well defined spatial area, i e. resulting from boundary conditions that do not give univocal solutions but a "cloud" of possible ones.
This unpredictability of the design process is intrinsic of the DS, and is due to its structure and operation.

5. THE INTRODUCTION OF A NEW TECHNOLOGY IN A DS

We may now try to use all the concepts above recalled for describing a DS in terms that can be useful for designing new kinds of computer tools for promoting energy/environment conscious design, such that will have less probability to be rejected by the architect: a sort of "appropriate" advanced technology.
What will happen if we try to introduce a new component and new relationships among components in our DS, i.e. if try to introduce a new design tool? The characteristics that the tool/technology should have may be condensed into one statement it must be a blend of novelty and confirmation that matches a wave of fluctuations of the DS, so that it is recognized as a package of neg-entropy/information.
By translating the statement into a common language (less rich of information, however), we may say that a new technology is not rejected by a system (any, including the DS) only if it meets the following requisites:

a. hardware and software are a well-proportioned blend of old and new;
b. it is a carrier of messages which can be read by the system as a well-proportioned blend of novelty and confirmation;
c. it can be functionally well-integrated with all the other parts of the system.

Have the design tools for energy conscious design fulfilled these requirements? The answer is: no. The design tools up to now developed have been neg-entropy packages not recognized as resource by the DS.
Actually, in most cases the tool was so designed that appeared to the DS as something that forced the design process to converge tiresomely towards a single possible optimum product: like the attractor in Fig. 2. Moreover, the tool imposed new expertise to be introduced into the system and new relationships among components: why should the DS have accepted such a change? It was too risky (too much novelty), and badly presented.
What should be done, instead, is to try to push the system towards a strange attractor of different shape, a transition like the one from Fig. 4 to Fig. 6. But, in order to do that, a modification of the components and of the relationships among

components (i.e. a wave of fluctuations) of the DS is necessary, in order to make the design tool and the "energy consciousness" stable part of the design system.

6. A COMPUTER TOOL AS ASSISTANT FOR ENVIRONMENT CONSCIOUS DESIGN

The problem of the interaction between computer models for environment conscious design and design process has been in depth discussed also elsewhere, mainly with the language of human sciences /21/, /22/, whilst computer tools are developed by other scientists, using the language of physics.
The approach above presented has the underlying ambition to be able to create a bridge, in terms both of language and conceptual framework, between the necessary two components of any research/development group aiming to the realization of an "appropriate" computer tool for environment conscious design.
With this belief, a multisciplinary research group has been set up in Palermo, in summer 1989, for a three years research project, promoted and financed by the italian National Research Council.
The work of the research group (made of mechanical engineers, computer analysts, architects and an expert in computer aided learning) is based on a common understanding and acceptance of the conceptual framework that it has been shortly presented in this paper.
The first stage of the research has been subdivided into two parallel actions:

1. develop an expert system able to navigate in the world of the thermal analysis and lighting, and to evaluate proposed design solutions, in terms of comfort and economic analysis; the aim is that this part of the final tool will be completely hidden to the user, unless explicitly requested to have it in foreground;
2. define the specifications of the interface between the above mentioned expert system and the user, i.e. the Design Sistem.

The first action, that may appear somehow "conventional", is very critical in itself and in its interactions with the second, being the "motor" of the technical/economical evaluation.
The second is actually more open to innovation and, therefore, to errors.
The main question we put to ourselves was: what we expect that the "assistant" should be able to do?
By using our conceptual framework, we answered by defining the following specifications:

A) It should be able to induce modifications into the receiving system, in such a way to make it able to metabolize novelty, i.e. new concepts and new tools.
This, in other words, means that the "assistant" must be not only a powerful design tool, but -before and besides that- a powerful teaching/training tool. For this reason we choose a man-machine interface approach based on heating/cooling strategies classification referred to well known, famous (we call them "archetypal") architectural examples, as lay-out of buildings, shape, materials used, etc.
B) It should embody the results of a detailed backtracking of all the possible implications of each new design concept on the established design process and on the organization of the design team, in order to control the balance between novelty and confirmation.
C) It should be able to avoid, or at least minimize, the "confirmation effect" that could derive from conceptually repetitive optimized design solutions proposed by the tool.
D) It should be able to be continuously adjourned as a rationalized and finalized file of new components, architectural examples, etc., either taken from literature and from the same DS work.

At present a preliminary, simplified, prototype of the expert system has been developed in PROLOG, able to make easier the initial choices of the design process, in terms of walls and glazing characteristics; the ES is designed for an user with a low specific skill, but "obliged" to make energy evaluations and to transmit them to the designer of the building's architecture. An interesting feature embodied in the prototype, that will be retained in the final product, is the possibility to interact with the tool even if the user gives very vague information (may be because at that stage of the design process the requested data are not available).
A data base and some simplified calculation methods for energy analysis and lighting are also being introduced.
As far as the "interface" is concerned, it has been decided:

a. to work for a well defined typology of DS, characterized by a small sized design team (3-5 architects);
b. to concentrate the efforts on natural and mechanical cooling issues, given the importance they have in southern european regions, especially in commercial buildings.

A preliminary set of matrices (one for each building use, i.e., school, hospital, office, residential, etc.) has been prepared, where cooling strategies are classified and analyzed in relation to different climates, taking into account possible side effects on heating and lighting strategies, on building regulations, on common design and construction rules, etc.
In order to make the tool "recognizable" for the user, these strategies are presented as a byproduct of the usual, compulsory, selection process of the envelope insulation. For this reason this part is being implemented first, giving to it an importance that, otherwise, it would not be justified.

7. CONCLUSION

The architectural design process is not ruled only by deterministic laws, therefore also the optimization process in the design of a solar or bioclimatic building cannot be treated as the optimization process of a power station or of a chemical plant, where the main parameters are technical efficiency and cost.
In architectural design the overall optimum is never coincident with the maximum or the minimum of a curve, neither with the stable equilibrium point of a dynamic process. The complexity and the diversity of parameters to be taken into account leave large opportunities to chance to have a role in the identification of the optimum, which is a strange attractor.
Thus, a design tool for energy conscious design, in order to be tailored to the environment in which is supposed to be used, must contain the ability to play with both deterministic optimization processes and with chance; it must be able to interact with the user speaking the same language, or to try to induce the user to learn a new language. Otherwise is like the solar pump in the sahelian village: good for the technicians but not for the intended users.
The approach here described is proving to be a promising starting point for building a necessary but not sufficient prerequisite: a real and positive interaction between human and physical sciences.

REFERENCES

/1/ Butera F.M., Renewable Energy Sources in Developing Countries: Successes and Failures in Technology Transfer and Diffusion, PFE, via Nizza 128, Roma, dec. 1989.

/2/ Schrodinger E., Che cos'è la vita? - La cellula vivente dal punto di vista fisico, Sansoni, 1947

/3/ Nicolis G., Prigogine I., Self-Organization in non Equilibrium Systems, Wiley, 1977

/4/ Prigogine I., From Being to Becoming, Freeman, 1980

/5/ Prigogine I., Stengers I., La Nuova Alleanza, Einaudi, 1981

/6/ Haken H., Synergetics - An Introduction, Springer-Verlag, 1978

/7/ Haken H., Sinergetica, Boringhieri, 1983

/8/ Georgescu-Roegen N., The Entropy Law and The Economic Process, Harvard University Press, 1971

/9/ Daly H. E., Lo stato stazionario, Sansoni, 1981

/10/ Brillouin L., Science and Information Theory, Academic Press, 1962

/11/ Wilson E. O., Sociobiologia, Zanichelli, 1979

/12/ Gatlin L. L., Information Theory and Living Systems, Columbia University Press, 1972

/13/ H. T. Odum, Environment, Power and Society, Wiley, 1971

/14/ Monod J., Il caso e la necessità, Mondadori, 1970

/15/ Eigen M., Self-Organizing of Matter and the Evolution of Biological Macromolecules, Die Naturwissenschaften, okt. 1971

/16/ Mosekilde E., Aracil J., Allen P.M., (eds), Special Issue on Chaos, System Dynamics Review, Vol. 4, 1988.

/17/ Jantsch, The Self-Organizing Universe, Pergamon press, 1980

/18/ Lazlo E., Evoluzione, Feltrinelli, 1986

/19/ Mosekilde E., Aracil J., Allen P.M., Chaos in Nonlinear Dynamic Systems, System Dynamics Review, p. 14-55, Vol. 4, N. 1-2, 1988.

/20/ Ekeland I, Le Calcul, l'Imprevu, le figures du temps de Kepler à Thom, Seuil, 1984

/21/ Yannas S., Physics and architecture. Issues of knowledge transfer and translation to design, Solar and Wind Technology; Special Issue on Interaction between Physics and Architecture in Environment Conscious Design, F.M. Butera, O. Corbella, S. Yannas eds., Vol. 6, N. 4, 1989.

/22/ Los S., Architecture and building science for energy conscious design, ibidem.

INTEGRATION OF SOLAR THERMAL SYSTEMS IN BUILDINGS.

E. Aranovitch, H. Bloem, R. Colombo, D. Gilliaert, P. Tebaldi

Commission of the European Communities
Joint Research Centre - 21020 Ispra, Italy.

ABSTRACT : The paper reports on an activity carried out at the CEC - Joint Research Centre in collaboration with the CEC - DG XVII, concerning the technical evaluation of solar thermal systems. The analysis of these systems requires a systematical approach in order that comparisons can be made with similar systems and that extrapolations to other locations and climates are possible.The JRC activity concentrates on the development of format sheets for the technical description, guidelines for monitoring and technical evaluation of the system.

1. INTRODUCTION

An analysis of the primary energy consumption in the western industialized countries, learns that more than 30% of it is used for heat production at low (< 100 C) or medium (< 200 C) temperature levels. Moreover, these energy needs are largely decentralized. This major use of energy compared to the availability of solar radiation as a low density heat source, and makes the use of solar thermal systems as heat source an attractive solution.
The efforts needed to realize technically good systems and furthermore to enlarge the field of applications and to make them economical, are multiple and are situated on different levels and different disciplines.
One should start from the specific components used in solar systems, of which solar collectors is the main one. It can be stated that the european recommendations for performance and durability test procedures for solar collectors and water heating systems will prevent bad products of being commercialized /1/.
The major problem which should dealt with now, regards the design, realization and operation of the systems. Even the use of qualified components (i.e. solar collector, storage tank, heat exchanger, regulation unit, ...) could lead to unexpected interactions between them and does not give any confidence on the functioning of the system as a whole.

2. SOLAR THERMAL DEMONSTRATION PROJECTS

2.1 Definition.

The projects carried out in the framework of a demonstration programme should fulfill certain criteria before being accepted. Some of the major requirements are that the projects are first and innovative realizations, having reached the level of a commercial potential (research is finished), but for which there is still a major gap for a large scale commercialization due to financial and/or technical risks.
The demonstration projects, especially in the field of solar thermal applications, cover a very vast area. There are not only differences in the type of applications under demonstration, also the size of the systems varies enormously, from some m² to more than 1000 m² of solar collector surfaces. Experience has shown that the limit between 'small' and 'large' systems, is represented by a solar collector surface of about 50 m², or an equivalent of about 20 000 kWh energy production (about the needs of a single family house). This limit is justified by the fact the impact of the performance parameter or performance optimization during the design phase of the project is completely different depending on the size.
For 'small' systems, emphasis is given to simplicity, costs, reliability, while for 'large' systems, performance optimization has a significant impact on the overall cost. Stated in simple terms, this means that a 10 % performance optimization for a 30 m^2 collectors system represents only 3m^2 of collector surface and will have no impact on the overall system cost, while for a 1000 m^2 system this could represent 100 m^2 of collector surface or could justify a sophisticated regulation or operating system.

2.3 Evaluation methodology

Even with the large variety in applications, there is some analogy in design and construction of the systems. This requires a systematical approach for the analysis and evaluation of each of the systems, in order to acquire the maximum of information out from the invested effort. A lack of any systematical approach will hamper to draw up general conclusions on the technology to be demonstrated, and even to intercompare similar systems or parts of it.
The general conclusions to be envisaged through the analysis and the evaluation of the level of success of each of the projects are :

- the performance, i.e. the solar fraction of systems
- the functioning and reliability of components
- the effectiveness of different control strategies
- proposals for improvements in the design, dimensioning and control strategies of systems
- the impact on employment and on the environment of the introduction of solar systems

Furthermore, it would be interesting to develop the necessary tools to assess the replicability of the techniques applied, even extended to other climates. The monitoring enables through the analysis of specific parts of the systems to identify the critical parts in the system.
This whole task could be largely facilitated by the introduction of some level of uniformity at each of the stages in this evaluation process through format sheets and/or guidelines for each of the following items :

- description of the project
- set up of the monitoring of the system

This clear distinction in approach depending on the system size, should be considered during the evaluation of the system performance, reliability, ...

2.2 Scope

All projects and systems being part of a 'Solar thermal systems' demonstration programme have a common goal of reducing the dependency on primary fossil energies. This can be obtained by substituting 'conventional' systems (such as fuel burners, air conditioners, ...) by solar systems, or by improved concepts and/or designs (daylighting, passive systems). This means that these solar thermal systems are in direct competition with so-called 'conventional' heat or cold producing systems.
Major differences exist however, if regarded inside the systems. The 'conventional' systems are factory - packaged and modular systems (consequently they have a rather high reliability), and are site mounted. Solar thermal systems are location dependent designs based on factory-made components. These components have to be connected and dimensioned in order to optimize the energy output in relation to the financial investments, the overall system reliability and durability, and maintenance. In this specific field of solar thermal applications however, the experience and design tools are still lacking.

- presentation of the results

Each of this parts will be discussed further.

3. EUROPEAN GUIDELINES FOR MONITORING

As an integrative part of a collaboration between different services of the Commission of the European Communities, the CEC - Joint Research Centre (JRC) carries out an activity in support to the CEC - Directorate General XVII demonstration programme for solar thermal systems. A major part of the activity consists of the development of common european methodologies and guidelines for the evaluation and monitoring of the projects. It is obvious that the redaction of such documents requires a huge practical experience and a consensus between european experts in this field, before someone can expect that these documents will reach the expected goal and could be made compulsory for the DG XVII demonstration projects.

In support for this activity, the CEC - JRC has set up a concerted action consisting of a working group of european experts called ' Monitoring Group for Solar Thermal Systems'. Meetings are organized on a 8 months interval basis, to which each time about 20 - 25 experts are invited. At each of the meetings, the discussions (which have an informal character) concentrate on one topic, selected from the following priority list :

. format sheets for the technical description of a project
. monitoring methodologies and shemes
. general information on physical parameters to be measured
. guidance on the selection and installation of sensors
. inspection procedures and format sheets
. technical evaluation of a project

In order to structure the discussions during the meetings and make the maximum possible progress, the JRC prepares for each meeting a draft document on one of the selected subjects. From the incoming information, an updated version is made and discussed for the second and last time at the following meeting. At that meeting, a new topic is introduced for which a first draft is put together and which is to be finalized at the next meeting. This approach will furnish reference documents in a rather short time interval. Experience will be gained and these documents will be validated through their use over a period of about 3 years, the time needed to set up all documents. At that time, a revision and updating will be envisaged.

At each meeting, a short time (2 - 3 hours) is dedicated to presentations regarding technical aspects on monitoring (data acq. systems, sensors, new computer software, ...). At present, a document for the technical description of a project (including format sheets) is at the stage of revision (a first discussion was held at the meeting of last september), and a draft document on the evaluation and monitoring of projects is under preparation for discussion at the next meeting, sheduled for may 1990.

These meetings are are well appreciated by the participants, and confirms the lack of collaboration in this field in Europe.

4. PROJECT DESCRIPTION

A project should be presented and described in a readable, functional and understandable way and give all tecnical information to make a technical evaluation. Already at this stage, the level of details will differ according to the system size. Even, if it is not possible to set up uniform rules, due to the large variety of systems involved, one should use the following criteria as far as possible :

- The main subsystems have to be identified. Each subsystem is defined as a well distinguished part of the project and which represent a main energy flow through the system. Four main subsystems are identified (fig. 1) : the solar collector loop(s), the energy storage system, the distibution and/or energy conversion, and the application(s).

- Each of the subsystems should be described with a level of technical detail which will enable to understand and to recalculate the energy flows (production or usage). In general, one could use the same technical information as used in the design phase of the project.

- Each of the different operating modes of the system should be presented and described seperately.

A combined format sheet for 'small' and 'large' systems is under preparation, with the scope of simplifying the task and to make this information available in a complete and uniformized way.

5. PROJECT EVALUATION

The technical evaluation of a demonstration project includes the inspection of the system and the monitoring of its performance, in order to evaluate the operating level of the installed systems and to localize major design errors.

The inspection of the solar installations will give following results :

- statistics about failures in single components
- identification of installation failures
- qualitative statements about the long term system performance and reliability

Quantitative information is obtained through the monitoring of the solar system. It should enable to distinguish each mode of operation and to determine the main energy flows through the system (for 'large' systems, the energy flows per operating mode). The monitoring includes furthermore :

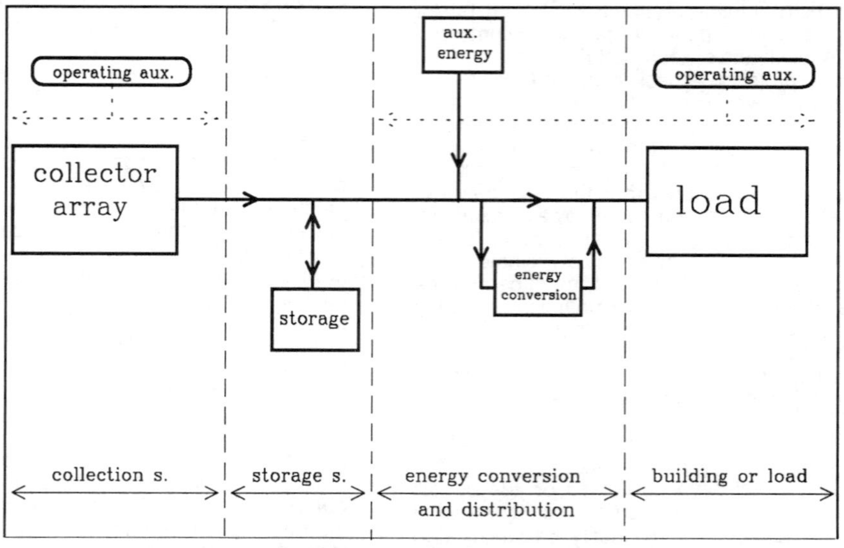

Fig. 1 : Schematical presentation of a project ; general concept

- follow-up of the systems during operation (faults analysis and statistics)
- recommendations for system improvements
- energy- and system operation calculations

These validated measurement data from the solar energy demonstration projects, together with the qualitative information from the technical visits, will enable to make overall and general conclusions.

6. PROJECT MONITORING

In general terms, the monitoring of a project consists of the measurement of energy balances and the determination of the various system operating modes through the measurement of metereological data, mass flow rates, energy flows, electrical energy, status signals,etc. ... .
The number of sensors to be installed is selected, depending on the concept and the size of system. As a rule, a system can be splitted in four subsystems (collectors, storage, distribution and/or energy conversion, load) and all energy flows between these subsystems should be measured, and if feasable in common nodes.
The monitoring could be carried out at two different levels of detail, or a *standard* or a *detailed* monitoring program.

The ***standard monitoring*** program consists of the measurement of the main weekly integrated energy flows through each of the subsystems, globally for 'small' systems, but subdivided or regrouped per operating mode for 'large' systems.

The ***detailed monitoring*** program consists of a registration of the data integrated or averaged over rather short periods, and still representing the dynamics in the system. This should enable to investigate the operation and control strategy of the system.
For 'large' systems, the time interval required for the detailed monitoring corresponds during daytime (from 8 hours before solar noon to 8 hours after solar noon) to hourly values and during nighttime to 4 hours averages, unless there is a change in the operating mode within this time interval. Each time there is a change of operation mode, then an extra average should be made with the indication of the exact time.
For 'small' systems, hourly averaged measurements (during night time, 4 hour averages) are sufficient.

The ***approach*** to be followed for the monitoring of demonstration projects should be a combination of both monitoring procedures. One should start with the detailed monitoring for a short period (one to two weeks). Such period should cover a wide range of weather- and energy consumption conditions. (f.e. the summer should be excluded for house heating systems). This monitoring will permit to understand and analyse the dynamics of the various energy flows in the system.
This period, should be followed by a long period (1 or 2 years) of standard monitoring, These measured data are needed to analyse and evaluate the long term energy performance and operating time of the various system operation modes and the usefullness of each of the system components (f.e. storage tank, heat exchanger ...). The long term standard monitoring period should be interrupted for a short period of detailed monitoring, similar to the first period (preferably 1 week per 3 months).

7. DATA ANALYSIS

The analysis aims to give information on :
- the performance of the overall system
- the functioning of single system components

The evaluation of the productivity or the performance of the overall system consits of the determination of the energy produced by the system, in substitution of the energy which should be produced otherwise by conventional systems with a given efficiency, while the control of single components serve to check the utility of the various control strategies and the performance of the single components.
For the characterization of the performance of a single component or a subsystem, one can introduce the concept of efficiency, which is defined as the ratio of the energy output to the energy input in constant or quasi- steady state operating conditions.

$$\eta = \frac{Q_{out}}{Q_{in}} \qquad \text{parameters : constant}$$

More important for the evaluation of a system in practical operating conditions is the integrated value over a selected time interval of the ratio of the energy output over the input. This integrated time period could include largely varying operating conditions, during its operation. This ratio is defined as the usefullness (g)

$$g = \frac{Q_{out} \, . \, dt}{Q_{in} \, . \, dt} \qquad \text{parameters NOT constant}$$

(*dt* is integrated over a specified period)

Neither these two values are sufficient to analyse the operation and performance of a solar system. A third value which is necessary to characterise a solar system is the solar fraction, which is defined as :

$$s.f. = \frac{Q_{solar} \, . \, dt}{Q_{conv} \, . \, dt} \qquad \text{parameters NOT constant}$$

(*dt* is integrated over a specified period)

Q_{conv} is the total energy required to cover the needs, if NO solar system were available, and Q_{solar} is the energy which substitutes a part of these needs.
Only a combination of the various values can descibe the

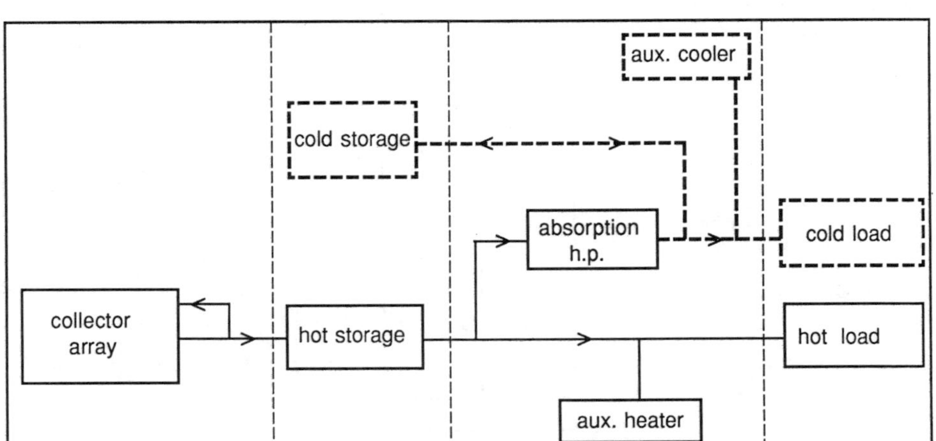

Fig. 2 : Schematical presentation of a project ; example of a CEC - DG XVII project

system in a complete form. The use of energy flow charts and histograms are very useful for the presentation of the results.

8. EXAMPLE.

The methodology described before, can be illustrated through an example of a CEC-DG XVII demonstration project. The project is a solar heating and cooling installation with 400 m^2 of evacuated tubular collectors, two huge storage tanks (90 m^3 cold, 40 m^3 hot storage), an absorption cooling machine and a rather complicated control strategy with many operating modes.
Although its complexity, the system can be schematically presented following the principle of four subsystems (fig. 2). The performance results are presented in fig.3 and table 1. The regroupment of the data per operating mode (instead of integrations over fixed periods) gives a lot more information about the system operation. Besides the general knowledge of the thermal performance of the system, one gets a rapid overview of the funtioning of the solar system. The table shows for example that operating mode 2 is not occuring in practice and that the operating mode 6, is also little useful. A further detailed analysis of these data could lead to proposals for an improved design (shorter pay back time), through f.e. the elimination of the 90 m^3 cold storage tank, compensated by a larger or a better insulated hot storage.

9. CONCLUSIONS.

The goal of a demonstration project is to realize a system on real scale and in real operating conditions, and to learn the maximum from this experience. This requires the analysis and evaluation of the system in such a way that comparisons can be made with similar systems and that extrapolations to other locations and climates are possible.
The existence of common methodologies for the description, monitoring, inspection of the system and the analysis of data would be very useful.
Within the framework of a collaboration between the CEC-Joint Research Centre and the CEC - DG XVII, a major effort is underway to develop common methodologies to fulfill this task. This activity involves the redaction of format sheets for the technical description and inspection of the system, guidelines for monitoring and technical evaluation of the system, the application on a selected set of projects and the updating of the documents.

10. REFERENCES.

1 : E. Aranovitch, D. Gilliaert, W.B. Gilett, J.E. Bates ; "Recommendations and Durability Tests of Solar Collectors and Water Heating Systems' 1988 , CEC - Joint Research Centre - Ispra Italy.

2 : Peuser F.A. , Uhleman R. ; 'Projektbegleitendes Messprogramm zum Zukunft investitions- program 3. Bericht ' ZfS 4010 Hilden (D)

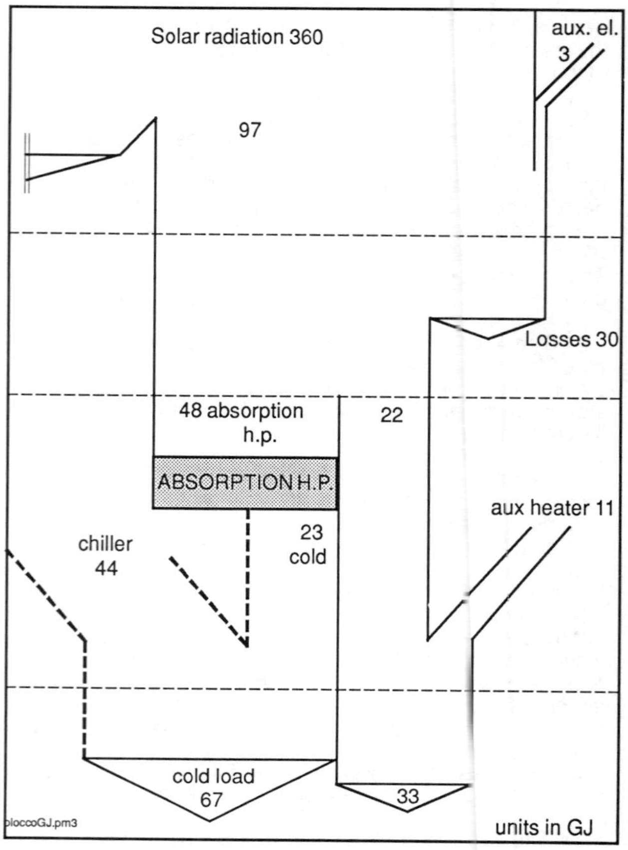

Fig. 3 : Energy flow chart ; example of a CEC - DG XVII project

Description of the operating modes from the high temp. storage tank to cold and hot load	Operating mode	$Q_{coll. out}$	$Q_{load\ hot}$	$Q_{load\ cold}$	Q_{cold} from abs h.p.	$Q_{cold\ store}$ --> load	Operating time (h)
Absorption HP. + electrical chiller —> cold load	1	31	6.3	22	9.8	0	75
Operating mode 1+ cold store —> cold load	2	0.1	0.8	1.6	0.1	0	2
Electrical chiller —> cold load solar collectors —> hot load	3	32.6	15	30	0	0	130
Electrical chiller —> cold load fuel burner —> hot load	4	17	3.4	5.3	0	0	162
Absorption H.P. —> cold storage no load	5	15.3	0	0	13	0	91
Cold storage —> cold load	6	0.3	7.2	8.4	0	5.4	11
	Total	96.3	32.7	67.4	22.9	5.4	

Units in GJ

Table 1 : Performance evaluation per operating mode ; example of a CEC - DG XVII project

SIMPLIFIED METHODS FOR THE SIZING OF ACTIVE SOLAR SYSTEMS (ESM1 and ESM2, two OPSYS products)

J. ADNOT, B. BOURGES(*), L. KADI, B. PEUPORTIER
Centre d'Energétique - EMP
60, bld Saint-Michel - PARIS 6ème

(*) consultant, 6, rue d'Armor, 35760 St Grégoire

ABSTRACT

Within the general objectives of OPSYS (gathering of all theoretical and experimental evidence about the performance of active solar systems), it is particularly important to have well-calibrated simplified methods suitable for engineering of the systems. The method for individual systems (ESM1) elaborated in the previous R-D program (81-83) has been transformed into a convenient software ready for public release (ESM1-4). The method for collective systems, namely medium size water heaters (ESM2) has been established and validated against EMGP2, our european reference. A Handbook has been written, meant as a reference book in Engineering of Active Systems.

1. IMPROVING THE APPLICABILITY OF ESM1

ESM1 is the method, and consequently the piece of software for the computation of performances of individual active solar systems, for space heating and domestic hot water.

A first version of ESM1 was available at the beginning of the working program OPSYS. It was not capable of handling combined Space Heating + DHW systems. A modelling exercise was performed to define the way of combining the two computational procedurec, one for Space Heating, the other one for Water Heating.

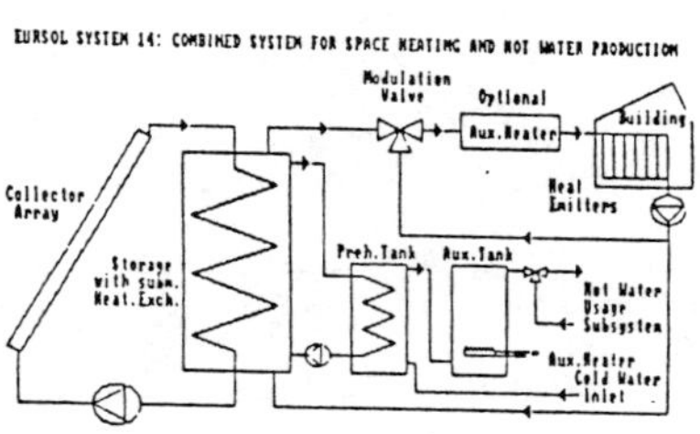

Each possibility was tested with the reference of the agreed European simulation codes, EMGP2 and EURSOL (1,2). The results of the full study (3) shows up a simple computational procedure, valid for a number of system types which are explained to the user (4).

2 - DEVELOPING ESM2 (FOR WATER HEATING SYSTEMS)

At the beginning of the working program the balance of existing methods was made for the treatment of water heating facilities of a medium size, so common in Southern Europe. No specific method existed and we developped completely a new close-form approach to the system (5). It is based on three main concepts :

(a) the repetitive day simulation : the simulation of an average meteorological day (at a monthly level) is repeated until a steady-state is reached (final temperature = initial temperature)

(b) the statistical representation of incident irradiation under the form of Cumulative Frequency Curves (CFC) possible thanks to the definition of a thoreshold radiation (C) a base temperature defined as a time averaged temperature of the storage thanks to a new mathematical treatment of the strati fied store equations.

The method ESM2 was derived using the three assumptions (a)-(c) having in mind the following scheme :

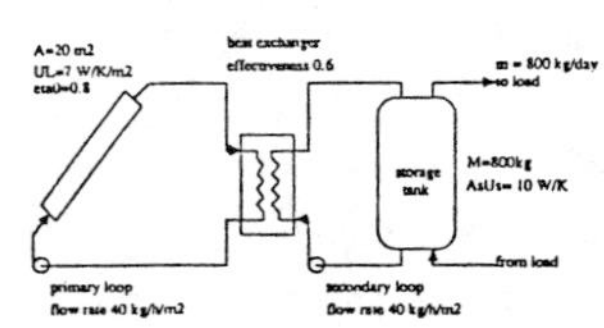

Among the three concepts, the first to be tested was be the full integration of the stratified store equations (C). If we plot the useful energy computed with the integrated formula (on the vertical axis) versus simulation results obtained with the European reference EMGP2 (on the horizontal axis) on a daily basis we obtain results like this one :

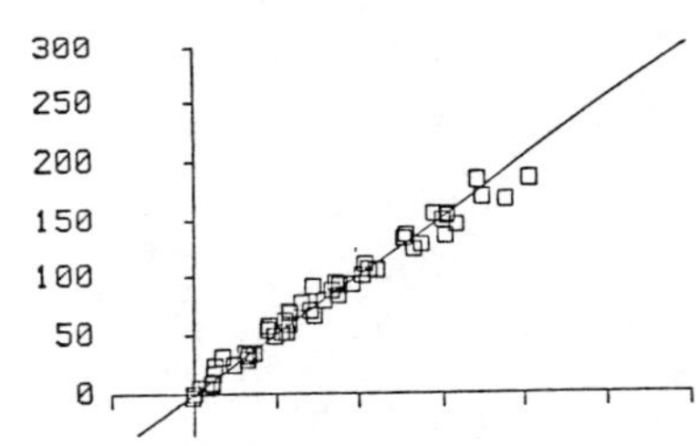

So the integrated solution is accurate on a daily basis. We moved then to the overall questions (a) and (b).

A preliminary version of the full method was issued (ESM2.0) and submitted to testing. As a result a representation of draw-off profiles was selected among three possibilities and the final software package (ESM2.1) was produced.

The most important question that designers have at this stage is the one about sensitivities. Since sizing a system is changing one or various parameters to maximize cost-effectiveness, the validation exercise should produce evidence of the correct reproduction of such a dependancy upon secondary (but costly) parameters. Let's take the case of collector mass flow-rate, ESM2 and EMGP2 showing almost the same optimum :

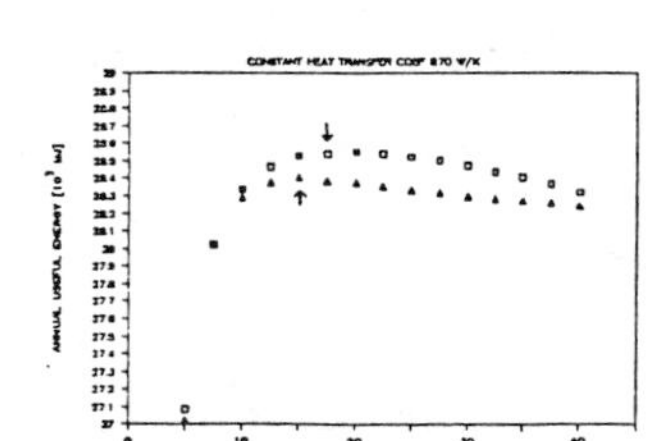

The same for collector area for which significant discrepancies happen only for a sizing which is twice the reasonable and usual choice :

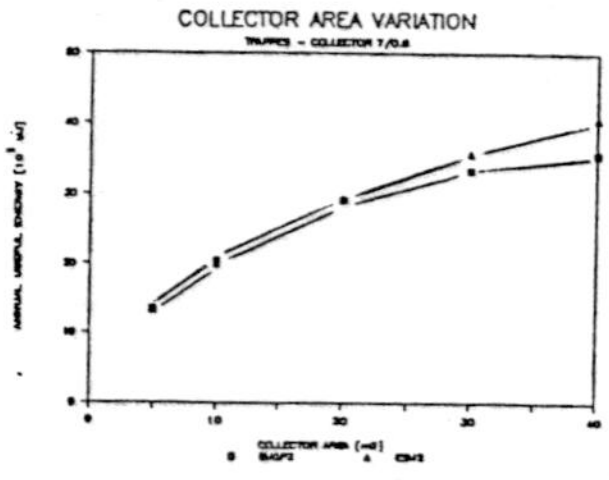

3 - ACCURACY STATEMENT

The concept was defined through a full validation exercise. In fact, both ESM1 and ESM2 were subject to a full parametric study to guarantee an accuracy within ± 10 % in any case, better in general. In the conditions which may create too large a discrepancy, a warning message will be issued automatically to the user. He will then refer to full simulation codes as EMGP2 and EURSOL. As a consequence we give predictive results which are (through EMGP2) consistent with experimental evidence gathered during all the CEC research program on Solar Energy (6).
Both ESM1 and ESM2 use Cumulative Frequency Curves, a meteorological data richer than simple averages but that will be finally as easy to handle as simple averages. They are already available for 14 European locations and will become eventually as common as averages (7).

4 - MAKING THE RESEARCH RESULTS AVAILABLE

The ideal release scheme in the present technical context is made up of three elements :
(i) availability of the computation under the form of a user-friendly software on PCS
(ii) availability of the know-how under the form of an engineering handbook
(iii) availability of a distance-service for preliminary calculations and decision making like in the SOLTEL system by TECSOL (8).
Elements (i) and (ii) are available ; (iii) in under study. So the final products of OPSYS in the field of simplified modelling are :
- the book "Active Solar Systems" : the handbook of European Simplified Methods" by B. BOURGES will be available commercially at the beginning of year'90 from a publisher ;
- the fourth computerized version of ESM1, the method for individual space heating systems with domestic hot water preparation, ESM1.4, is commercialy available from a resource center acting on behalf of the Commission ; there is also a possibility to use ESM1 through worksheets given in the handbook ;
- the first computerized version of ESM2, the method for collective water heating, a developing technique in Southern Europe, ESM2.1, has been prepared with the same quality of programming.
The software is a preliminary engineering tool adequate for a decision process and providing immediate answers to sizing questions.
The handbook is a natural entry point into the ESM world. It gives a description of the systems before starting sizing them and help in using the software. It may be used also to build courses or training sessions.

5 - CONCLUSIONS

The methods ESM1 and ESM2 are based upon different original mathematical derivations of the average behaviour of active systems. They use close form representations of stratification and of the utilisability of the incident solar radiation. This sound basis allows very accurate optimization of design parameters.
A first version of ESM1 was available at the beginning of the working program. It was not capable of handling combined Space Heating + DHW systems. A modelling exercise was performed to define the way of combining the two computational procedures, one for Space Heating, the other one for Water Heating.
ESM2 is a completely brand-new simplified method for DHW systems. It is based on three main concepts :
(a) the repetitive day simulation : the simulation of an average meteorological day (at a monthly level) is repeated until a steady-state is reached (final temperature = initial temperature)
(b) the statistical representation of incident irradiati on under the form of Cumulative Frequency Curves (CFC) possible thanks to the definition of a thoreshold radiation
(c) a base temperature defined as a time averaged temperature of the storage thanks to a new mathematical treatment of the stratified store equations.
The engineering handbook and the two pieces of software (ESM1 and ESM2) are commercially available.

6 - REFERENCES

1 - W. DUTRE : "EMGP2, An European Transient Model for Solar Systems", Reidel, 1985
2 - W. DUTRE : "EURSOL, An interactive program for the simulation of water based thermal solar systems"
3 - EMP-ARMINES, final report of subcontract to KUL
4 - B. BOURGES : "Active Solar Systems the Handbook of European Simplified Methods"
5 - B. BOURGES, J. ADNOT : "A generalized closeform solution for active solar system", Solar Energy
6 - W. DUTRE : "OPSYS : concerted action for solar system model development and validation", same volume
7 - B. BOURGES : "EUFRAT : utilizability and cumulative frequency curves of solar irradiance", same volume
8 - TECSOL : Serveur SOLTEL (sur Minitel, code 3615).

7 - ACKNOWLEDGMENTS

This work was funded by the Commission of the European Communities (DG12) and the personal involvement of T.C. STEEMERS and expertise of C. DEN OUDEN are acknowledged for.
Our activity was coordinated by Prof. W. DUTRE, Katholieke Universiteit Leuven, and he gave us the right impulse and directions.
Significant contribution were made by other OPSYS participants : A. DEBOSSCHER (KUL,BE), A. DEGEUS and P. BERGMIJER (TNO-TPD-TH,NL), S. ORSTERGAARD-JENSEN (TUD-TIL,DK).

OPTIMISATION OF THE HEAT BALANCE OF AN ATTACHED SUNSPACE

Jean-Bernard Gay
Ecole Polytechnique Fédérale de Lausanne - Bâtiment LESO
CH - 1015 Lausanne (Switzerland)

Abstract

Among the various passive solar systems which may be implemented in buildings, attached sunspaces are widely used. This is due to the visual impression of such an architectural element, to the possibility to occasionally use it as an extended living area, and to the ability of the sunspace to contribute to the heating heat requirements of the building. The question is to determine how efficient such a system is depending on its characteristics and on how it is used. In order to evaluate the relative influence of the main parameters, a new type of sensitivity study has been carried out using a *planned experimental design*. Such an approach allows not only an unambiguous determination of the influence of the main parameters, but also of the interactions between the considered parameters.

1. PURPOSE OF THE WORK

Several factors have a direct influence on the heat balance of an attached sunspace. In order to determine the relative influence of each of them, a new type of parametric study has been carried out : for the first time in this field a planed experimental design /1/ has been used such an approach allows not only an unambiguous determination of the influence of the main parameters but also a determination of the interactions between the considered parameters.

2. METHODOLOGY OF APPROACH

Our approach is based on two mathematical tools:
1) "LESOSAI" /2/, a simplified computer calculation code which has been developed mainly for passive buildings;
2) "NEMROD" /3/, a software package dedicated to the determination and analysis of experimental design matrix.

The influence of four parameters has been studied:
- the sunspace size
- the glazing type of the sunspace
- the control of the auxiliary heating
- the inhabitants behaviour concerning the use of the sunspace.

Mathematically every parameter is characterized by two "states", each one corresponding to an extreme situation. From these 8 states, a 2^4 design matrix has been built (annex 1), and used for the determination of the main effects, as well as of the eleven interactions.

3. MODEL USED FOR THE CALCULATIONS

The sunspace is attached to the south wall of a small residential building (12 x 9 x 6 m^3); its depth is 2 m in one case it encloses 1/3 of the wall area (4 x 6 m^2), in the other case the entire wall area (12 x 6 m^2). The sunspace may be equipped either with a single glazing, or with a double IR coated glazing. The auxiliary heating of the building is controlled by either an indoor or outdoor air temperature sensor. Finally the inhabitants behaviour effect has been investigated. Two types of behaviour have been simulated :

- *The inhabitant is energy conscious.* The door between the living area and the sunspace is left open only when the temperature of the sunspace is above the indoor one.
- *The inhabitant is unconcerned with energy problems* : he uses the sunspace as an extension of the living area and keeps the door between these areas open all day.

Parameter	Lower state	Upper state
1. Sunspace size	1/3 of the south wall	The whole south wall
2. Sunspace glazing	Single	Double coated
3. Heating control	Outdoor sensor	Indoor sensor
4. Inhabitant behaviour	"Energy conscious"	"Energy unconscious"

Table 1 : Summary of the 8 parameters levels which have been considered for the calculations.

The 16 cases, given by the 2^4 design matrix, have been calculated using the "LESOSAI" computer code and local climatic data (Lausanne altitude 400 m, latitude 46.5°N).

In each case the considered result was the specific net heat demand of the building for the whole heating season (from October 1st to May 30th). This specific net heat demand (E_H in MJ/m^2) is defined as the total net heat demand divided by the total heated floor area.

Table 2 gives the results obtained for the 16 cases as well as the level of each parameter.

Case	Sunspace size	Sunspace glazing	Heating control	Inhabitant behaviour	E_H [MJ/m^2]
1	Small	SG	Outdoor	Unconscious	481
2	Large	SG	"	"	567
3	Small	DC	"	"	439
4	Large	DC	"	"	451
5	Small	SG	Indoor	"	387
6	Large	SG	"	"	440
7	Small	DC	"	"	336
8	Large	DC	"	"	304
9	Small	SG	Outdoor	Conscious	425
10	Large	SG	"	"	400
11	Small	DC	"	"	415
12	Large	DC	"	"	368
13	Small	SC	Indoor	"	340
14	Large	SC	"	"	301
15	Small	DC	"	"	324
16	Large	DC	"	"	254

Table 2 : List of the considered cases and specific net heat demand as calculated for each case.

Using the mathematical model as defined in the annex, the main effect of each parameter, as well as those of the eleven interactions, have been calculated. Figure 1 presents the results.

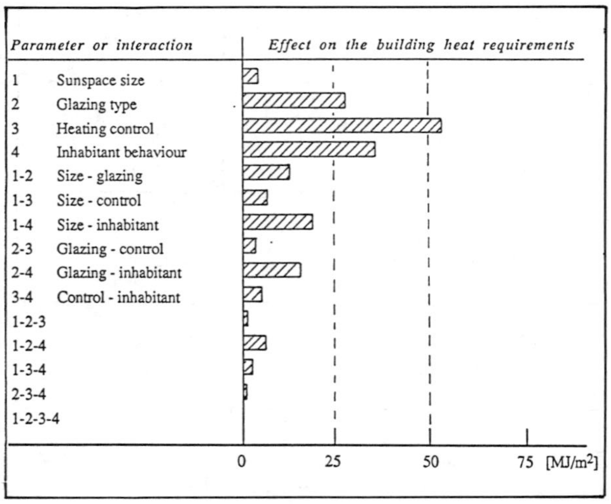

Figure 1 : Effect of each parameter, and of each interaction on the buildings specific net heat requirements. Each value corresponds to half of the variation of the considered parameter, or interaction, when the other parameters are kept at the middle of the experimental domain.

4. SIGNIFICANT RESULTS

It is surprising to note that when optimizing the attached sunspace, the main parameter which affects the building energy requirements is the heating control : only an adapted controller monitored by an indoor sensor, allows efficient use of the solar gains. Two other factors are also important : the inhabitant behaviour and the type of sunspace glazing.

The effect of the sunspace size is strongly coupled with the inhabitants behaviour and with the glazing type. These interactions may be better understood using 3 dimensional plots. In such representations the relative effect of two parameters are considered, the other parameters being fixed at an "average" level.

Figure 2a shows the interaction between the sunspace size and the inhabitants behaviour : with an energy unconscious inhabitant a small sunspace is better than a large one; considering the large sunspace one notes how influential the inhabitants behaviour is.

Figure 2b emphasizes the importance of the glazing type : a high-quality glazing strongly reduces the influence of the inhabitants behaviour.

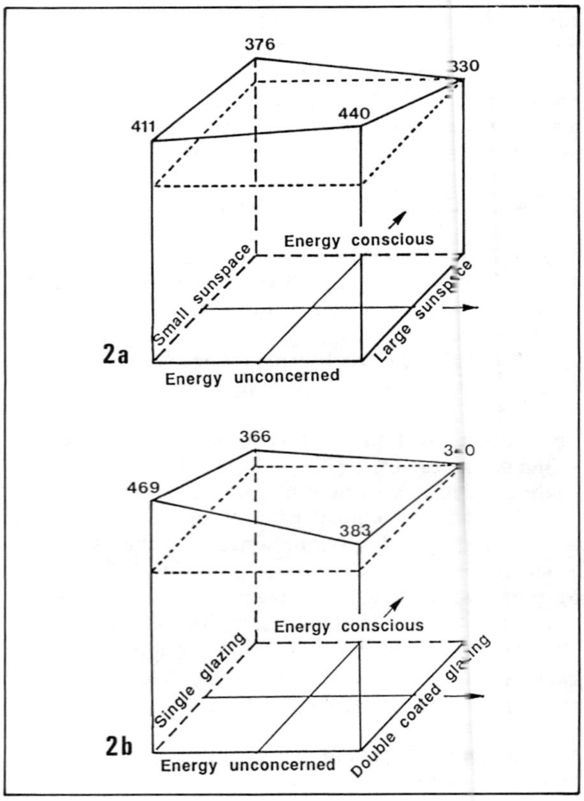

Figure 2 : Interactions between two parameters :
- inhabitants behaviour vs sunspace size (2a) and
- inhabitants behaviour vs glazing type (2b)

5. CONCLUSIONS

- The most important main effect comes from the control of the auxiliary heating : compared to the average specific heat requirement of 390 [MJ/m^2] at the middle of the experimental domain, the use of an indoor sensor instead of an outdoor one permits a 28% reduction of the net heat requirements.
- As a main effect, i.e. without considering the interactions, the inhabitants behaviour effect is less important. However this parameter appears influential as an interaction in connection with the sunspace size and with the glazing type.
- The effect of the sunspace size is negligible as a main factor, while it is strongly determinant with the other ones. The interaction with the inhabitants behaviour is the most influential : it shows clearly that a sunspace may just as well reduce, as increase, the heat requirements of a building depending on the degree of the inhabitants energy consciousness.
- If with an energy-conscious inhabitant the effect of the type of sunspace glazing is small, with an energy-unconscious user the situation is completely different : in such a case the use of a double coated glazing allows a gain of 23% on the building heat requirements compared to a single one.
- With a single-glazed sunspace and an "average" inhabitant, the increase of the sunspace from the lower state (1/3) to the upper one (3/3) induces a 5% increase in the building heat requirements. With a double-coated glazing the effect is overturned : a large sunspace allowing a 10% reduction of these heat requirements.

- Finally, from the best case (a large sunspace, equipped with double-coated glazing, occupied by an energy conscious inhabitant, attached to a house with an indoor-controlled heating system) to the worst case (a large single-glazed sunspace, occupied by an energy unconscious inhabitant, attached to a house with an outdoor-controlled heating system) there is a difference in the specific building heat requirements reaching 313 [MJ/m^2], more than a factor two !

6. RECOMMANDATION

This study shows that a large sunspace attached to the south wall of a building may significantly reduce the heat requirements. However, in order to be efficient the following requirements must be fulfilled :

a) the heating system must be controlled by an indoor sensor in order to allow maximum utilisation of the solar gains,

b) the sunspace must be at least double glazed,

c) the inhabitant has to be informed on how to use the sunspace; he should be motivated :
   - to close the door between the house and the sunspace when the latter is colder,
   - not to heat the sunspace,
   - to accept temperature fluctuations in the sunspace.

7. ANNEXE : MATHEMATICAL TOOLS

a) General factorial design

If we consider k parameters having each two levels, then the general factorial design is made of all the possible combinations between the factors, and the number of experiments is 2^k.

In our case we have four parameters :

S : the sunspace size
G : the glazing type
C : the control of the auxiliary heating
B : the inhabitant behaviour

for each parameter we have two states 0 and 1, the 2^4 design matrix may also be written as follow :

Run	S	G	C	B
0	0	0	0	0
S	1	0	0	0
G	0	1	0	0
SG	1	1	0	0
C	0	0	1	0
SC	1	0	1	0
GC	0	1	1	0
SGC	1	1	1	0
B	0	0	0	1
SB	1	0	0	1
GB	0	1	0	1
SGB	1	1	0	1
CB	0	0	1	1
SCB	1	0	1	1
GCB	0	1	1	1
SGCB	1	1	1	1

b) Relative effect of the parameters

In order to calculate the four main effects as well as the eleven interactions, the following mathematical model may be postulated :

$$Y_R = \beta_0 + \sum_i \beta_i X_{ir} + \sum_{ij} \beta_{ij} X_{iR} X_{jR} + \sum_{ijk} \beta_{ijk} X_{iR} X_{jR} X_{kR} + \beta_{1234} X_{1R} X_{2R} X_{3R} X_{4R}$$

Y_R is the result of the run R (E_H in our case)
X_{iR} is the value of the parameter i in the run R
β_0 is the average value of all the results for the whole set of runs
β_i is the main effect of the parameter i
β_{ij} is the first order interaction between the parameters i and j
β_{ijK} is the second order interaction between three parameters.

Introducing in this model the typical values of the 16 runs we get a system of 16 equations, this system may be solved in order to determine the mains effects (β_i) as well as the interactions (β_{ij}, β_{ijK}, β_{1234}).

REFERENCES

1. G.E.P. Box, W.I.G. Hunter and J.S. Hunter
Statistics for Experimenters, Wiley (1978)

2. LESOSAI - Mode d'emploi
LESO-EPFL (1987)

3. D. Mathieu, R. Phan-Tan-Luu
Logiciel NEMROD, LPRAI - Marseille (1987)

PASSIVE SOLAR ROOFSPACE SYSTEMS: EXTRAPOLATIONS FROM A YEARS MONITORING

A. Tindale and J. Littler
Polytechnic of Central London.
35, Marylebone Rd, London NW1 5LS

ABSTRACT: This paper presents the results of the first measurements to be made in a group of solar roofspace houses. Of particular interest is their application in the urban rehabilitation of the many older, poorly insulated buildings in the UK. The non-availability of control houses for monitoring led to the development of a simulation technique for projection of monitored data results. The performance of the solar roofspace systems over one heating season are presented, as are the projected results for some optimal roofspace systems.

1 NEWHAM ROOFSPACE HOUSES

Solar roofspaces were retrofitted to six $35m^2$ terrace houses during extensive renovation work on an entire estate of local authority dwellings. Standard materials and building techniques were used throughout with only minimal site supervision being required. The other renovation work included re-roofing, insulation and the fitting of new double glazing and wet central heating systems.

The conventional roofspaces were sealed, insulated and glazed on the south slope to admit and retain solar gains. The solar heated air is blown by a fan through a duct into the house and distributed to the lounge and two bedrooms upstairs. House air is recirculated to the roofspace by the fan induced pressure difference. Excess summertime solar gains are dissipated by a thermostatically controlled fan, which is activated when the roofspace temperature exceeds 60°C. The solar roofspace heating system performance was monitored at medium level over one heating season in all six houses.

The fan operates on a simple differential thermostat control, supplying heat to the house when there is a demand and roofspace air is warm. Delivery rates vary in three steps between 250 and 510 m^3/hr as the roofspace varies from 25 to 35°C.

The solar aperture is constituted of 18 m^2 of twin wall polycarbonate sheeting mounted with air tight seals on aluminium glazing bars. The roofspace is insulated around its entire surface with 100 mm fibreglass and sealed with a vapour barrier. The south side is isolated from the north by an insulated and sealed partition of timber and plasterboard construction, hung from the ridgeline.

The project was jointly funded by the CEC as part of DG17.

2 MONITORED RESULTS

2.1 Energy Savings and Control

The average solar heating fraction for the houses, including direct gains, was 38%, with roofspaces contributing on average 1560 kWh over the year. Table 1 shows the annual heat input from all sources for each of the six solar roofspace houses. The results show very wide variations between the houses, with house 1 receiving almost negligible benefits from the system, while house 6 receives 33% of its total controlled heat input from the roofspace system. House heat demand and its relation to solar heat prioritisation control are the primary reasons for the variations. The method used in the Newham houses to ensure solar priority over the conventional wet central heating system was to create an offset in set-points between the two systems. The solar system heats to at least 2-3° above the central heating, thus preventing the latters operation during times of solar heat availability.

In practice, this strategy was found to be reliable only in those houses in which daytime temperatures significantly higher than central heating set-points were acceptable, or where the central heating was never used during the daytime. This was the case at houses 2,3 and 5 all occupied by older people, who enjoy comfort temperatures up to 25°C. The situation at house 1, causing the low solar contribution, was one of continuous daytime boiler usage, combined with high set-point temperatures. Because the

Table 1 Yearly Monitored Heating Contributions kWh

House No.	Roofspace Heat	Central Heating	Direct Gain	% Solar
1	325	12577	861	12%
2	1464	3820	968	34%
3	1648	2360	629	35%
4	1238	4792	781	30%
5	2327	5622	781	39%
6	1122	2234	869	47%

solar heating only has priority by virtue of its higher set point, it must heat the house up to 26-28°C, which on a sunny day, with high radiant temperatures, is uncomfortably hot. The occupant therefore asked for the solar set-point to be reduced. This effectively turns the system off, since during the daytime the set-point is nearly always satisfied.

Figure 1 Monthly Heating Contributions from all Sources

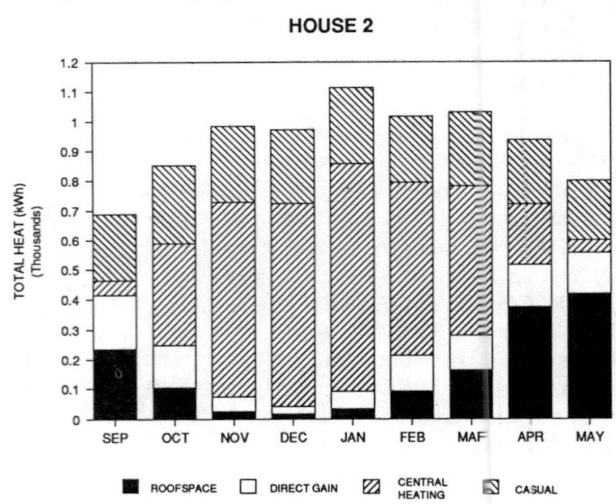

In general; the houses were of low specific heat loss (135 W/K) and being mainly occupied during the day, high incidental gains. This adds up to an overall low daytime heat demand. This especially restricted useful solar contribution in houses 4 and 6, where solar gains were offset by clothes drying and electrical equipment, respectively.

2.2 Comfort

Solar roofspace heating, while categorised as a 'passive' system actually has many characteristics intermediate between direct gain passive and active solar systems. This relates in particular to comfort since the passively collected solar heat is delivered in a very controlled manner. Maximum and minimum delivery temperatures may be easily set to the values that lead to optimal comfort, thus eliminating the possibility of discomfort caused by either overheating or draughts.

It is difficult to isolate the comfort benefits of the solar heating from those due to the other energy saving features fitted at the same time, however most occupants appreciate the extra daytime warmth provided by the solar heating. This is particularly true of the older residents at houses 2,3 and 5 who enjoy higher temperatures but who would not otherwise be prepared to heat the house during the daytime. In this respect *the increased comfort benefits of solar heating systems may be as significant to a householder as the energy savings.*

2.3 Roofspace Design Optimisation

The following comments and recommendations came out of the project:

- Air recirculation between house and roofspaces demands high sealing standards both for roofspace and house. An additional return air duct, control damper and transfer grills are also required. The benefits of this strategy were not proven and so the latter items may be usefully cut from the design.
- In-line centrifugal duct fans perform better under load than axial flow fans. The duct fans had built in speed steps whereas the axial fans required additional speed controller. A single, narrower duct taken down the existing chimney flue was most successful, but is only possible where the occupant does not wish to reserve the use of the flue for a gas appliance.
- Dark finishes to internal surfaces in the roofspace receiving radiation, significantly reduces heat loss due to re-radiation. The absorption was calculated to be 81 and 75% for roofspaces with and without a black back wall respectively.
- The ridge plenum, successfully allowed collection of warm air uniformly along the length of the roof. This may be constructed more cheaply by Additional flexible ducting, with holes cut in the side, hung beneath the ridge and attached to fan inlet.
- The set point offset method of solar heating prioritisation requires a 2-3 K minimum offset between the auxiliary and solar systems. This often involves significant temperature swings over the day, and when this is not desirable, either manual operation or a more complex control strategy is indicated.
- When the temperature set-point offset method is used to achieve solar priority, a relatively fine control of the central heating set point temperature is required for optimal interaction between the two systems. This criterion was only partially fulfilled by the TRV's used at four of the six houses at Newham.
- Great care is required in the design of roofspace systems to minimise ventilation losses to the roofspace due both to night-time leakage past dampers and deliberate recirculation. Both the extra stack height and the exposed nature of the roofspace increase the maximum potential ventilation heat loss due to wind and buoyancy effects. Unless the quality of roofspace sealing is of similar quality to that of the house, it is likely that recirculation strategies will be not only an unnecessary complexity but may actually be detrimental to overall system performance.
- The practical maximum roofspace efficiency for this sort of construction is 30%. This figure relates to a high heat demand, and relatively high solar gain situation.
- The solar savings vary widely depending on house specific loss rate, incidental gains and occupancy patterns. The valuable insights into which factors degrade the system performance imply that future designs could improve significantly on these figures.
- The optimal roofspace glazing elevation increases with decreasing house heat demand. The optimal elevation for the demonstration houses was found to be 55°.

3 EXTRAPOLATION OF RESULTS

3.1 Methodology

The monitoring program allowed calculation of the total solar heat delivered by the roofspace systems in each house, but because this value includes overheat (ie. non-useful heat) it is not necessarily the value of greatest interest. Such monitoring programs often include non-solar control houses to provide data on the energy consumption of a similar house without solar features. In theory subtraction of the total annual energy consumption for the two types of house gives the annual savings due to the energy saving features.

There are problems associated with this approach however, the principle one being one of consistancy. Even when unoccupied, apparently similar houses can vary significantly in physical characteristics /1/. And when occupied, the differences in heating system use and occupancy patterns can render two houses almost incomparable for the purposes of energy assessment. Some large monitoring programs record data for a statistically sufficient number of houses to allow the method to work, but many more struggle with data from just a handful and so meaningful data is less easy to extract from results. The Newham project falls into the latter category and so a different approach was used.

The aforementioned difficulties and the desire to investigate the effect of changes to both design and environment, led to the development of a computer model describing the thermal functioning of the houses and roofspaces in detail. By running the model with and without a solar roofspace the useful solar gains are simply calculated as the difference in house energy consumption between the two runs. Because of the Groups' past experience with the program, and because of its ease of use, SERI-RES was used for the simulations. The technique used is outlined below:

1 A building description is constructed using values from regressed hourly data, to describe the three key heat flow paths: house-roofspace, house-ambient and roofspace-ambient. The incidental gains and central heating schedules were also derived from monitored data. Average values were calculated for each hour of the day over the month.

2 Monitored weather data is transferred into SERI-RES format in such a way as to inject *exact* measured global solar radiation onto the roofspace glazing plane.

3 A series of 'fine-tuning' iterations in which SERI-RES heat and zone temperature output, is compared with the monitored reality, and the building heat conduction values modified in a way appropriate to reducing modelling errors. In practice this meant increasing the house and roofspace thermal mass and making small adjustments to both UA's.

4 Once the model agrees in essence with hourly data and to within 2-5% with measured results on a monthly basis, then it may be considered to be 'calibrated' ie accurately representing the thermal dynamics of the real house.

The main objective was to obtain similar *outputs* using as similar inputs as possible to those obtained in real measurements and subsequent regressions. It is therefore legitimate to adjust the input parameters to ensure agreement between the outputs, provided that in so doing we do not depart from a sensible notion of the mechanics of heat flow occurring in the actual house. The modelling is not intended as a validation of SERI-RES, although if modelling inputs and outputs both agree with monitored data, then it does serve as such. A detailed description of the procedure is given elsewhere /2/.

Although not new in concept, the technique benefits considerably from the ability to inject measured solar radiation directly onto principal glazing surfaces without recourse to more complex building simulation programs.

3.2 The Building Description

The houses were modelled using 3 principal zones: house, solar roofspace and loft. The incidental gains and central heating schedules were derived from monitored data. In both cases, average values were calculated for each hour of the day over the month for each of the 12 months of simulation. The building construction was entered into the building description form, in as much detail as possible, and the model run. The interzone heat losses from roofspace to house, house to roofspace and house to ambient had all previously been obtained by multiple regression analysis of monitored data and where any disagreement with the equivalent SERI-RES calculated values occurred, the model was modified so as to agree with the regressed values. In practice, given accurate construction data entry, this step was found to result in only very small adjustments and is not necessarily time/cost effective.

3.3 A SERI-RES Weather File From Monitored Data

The important requirements for a SERI-RES weather file are hourly values of direct normal solar radiation, global horizontal radiation and ambient temperature. The latter two quantities were monitored directly and so simply required restructuring to the SERI-RES weather file format. However, the direct and diffuse radiation separation was not monitored and so some form of direct normal radiation synthesis was required.

It is possible, to calculate the beam radiation from the known global radiation on any two planes if the position of the sun in the sky is known and if certain assumptions about the radiation distribution are made. SERI-RES inverses this process when distributing beam and diffuse light on the various modelled surface. So if the same Hay algorithms (distribution assumptions) are used when separating the direct and diffuse as those used in the modelled radiation distribution, it is possible to apply monitored radiation values on the modelled surface. By applying the inverse Hay algorithm to the Newham data set, combinations of direct normal and global horizontal radiation, were produced, which, when driving the Newham model, accurately input measured values on the modelled roofspace surface.

At certain solar geometry configurations the inverse Hay algorithm caused singularities and a non-geometrical method was required. A correlation was derived from a Kew Test Reference Weather Year relating global horizontal, extraterrestial, and direct normal radiation. Although this method is not very accurate (typical errors ±20%) it provides a very useful stop-gap.

Because real weather data is frequently required for use in modelling, the above algorithms were written into a user friendly computer program with generalised input capability making it suitable for use for any site and sensor inclination.

The accuracy of the above solar radiation transformation algorithms was investigated by comparing measured and modelled radiation values on the roofspace plane. The hourly RMS error for March 1988 was found to be 5.7 W/m^2 with a total monthly error of +3.5%. Errors for hours of appreciable direct normal radiation were always less than 0.6 W/m^2, while those for overcast conditions were typically overestimated by 0-10%.

3.4 Model Validation

The above procedure was applied to the most representative of the six roofspace houses, house 2. Figure 2 shows the accuracy of the modelling on an hourly basis for a typical March day; and Table 2 summarises the hourly errors over the year. Figure 3 shows the monthly monitored and simulated heat totals over the year. It can be seen that roofspace temperature and solar fan heat delivery are very well modelled on an hourly basis. House temperatures simulations are slightly less so, suffering from 'random' incidental gain and central heating fluctuations from the monthly average values used in the building description. Boiler heat delivery is not accurate on an hourly basis, with significantly different patterns occurring between simulated and measured values. On a daily and monthly basis, however the central heating simulations are accurate to within 5%. The accuracy of the fan heat delivery is heartening in view of the lack of confidence shown in the fan algorithms by SERI-RES users.

Table 2 Hourly Modelling Errors

	R^2 Error	Std error	SERI-RES	Monit
Fan heat (kWh)	0.9463	0.138	162.7	162.8
Boiler heat (kWh)[1]	-0.043	5.694	483.5	490.1
Roofspace temp (°C)	0.9697	1.499	16.01	15.84
House temp (°C)	0.5021	0.990	21.02	21.35
[1] denotes daily totals				

Figure 2 Hourly Simulated and Monitored Results for Typical March Day.

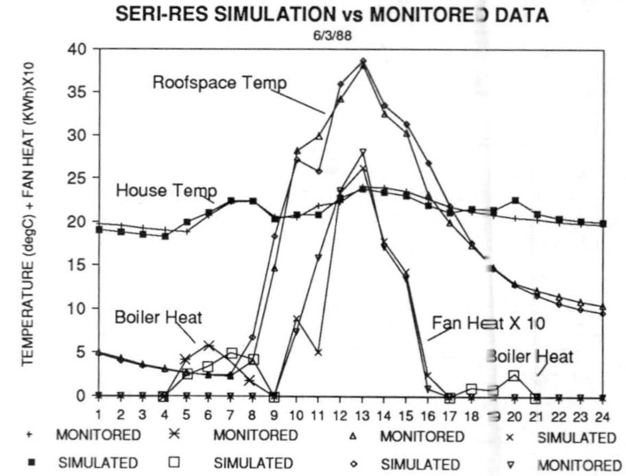

Figure 3 Monthly Simulated and Monitored Totals for 87/88 Heating Season.

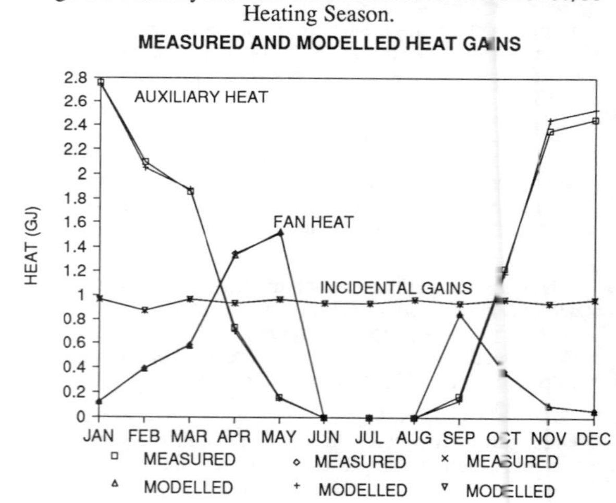

3.5 Displaced Auxiliary

The displaced auxiliary (or true saving) was determined for the *demonstration* systems by replacing the modelled solar roofspace with a conventional one. The increase in house simulated heating load is equivalent to the displaced auxiliary. The true heat saving was found to be 890 kWh per year. This is significantly lower than the 1464 kWh of heat delivered by the roofspace.

The results also show that the roofspace system caused a slight net loss during the months of December and January. This was disappointing since it was hoped that the roofspace would achieve savings by buffering the house from the colder ambient even when the fans are not delivering heat directly. The degraded performance is due to the house-roofspace sealing dampers not sealing as well as expected /1/. Future designs will achieve significant buffer savings.

4 PERFORMANCE PROJECTIONS

The potential for solar roofspaces is investigated by applying the basic roofspace model validated above, to more optimal situations. The following simulations represent three such optimal scenarios.

4.1 Large Scale Renovation

Future roofspace systems incorporating construction optimising measures described in 2.3 would cost approximately £1775. The energy performance of this trimmed down roofspace systems was investigated by simulation. The modelled houses were of solid wall construction and higher heat loss rate (240 W/K) than the Newham houses. The roofspace air change rate was doubled from 2.6 to 5.2 ACH, to take account of reduced sealing standards. The displaced auxiliary heat saving for the house, in this case is 1463 kWh. The equivalent payback period is calculated as 29 years.

4.2 Solar Ventilation Pre-heat (SVP)

When roofspaces are used as a source of solar pre-heated ventilation air, the house ventilation heat loss may be considerably reduced. Warm air heating systems combined with whole house ventilation systems are a rational development to optimise on total installation costs. Solar roofspaces are an ideal way to link solar gains through such systems.

Because SERI-RES does not explicitly model plant, assessment of the efficacy of SVP strategies generally presents rather a problem. However, a simple technique does exist to circumvent the problem /3/. The total house air change rate (modelled as a conductance with ambient in SERI-RES) is replaced by two new conductances: one from house to ambient, representing reduced air exchange with outside air, and one from the house to the roofspace to model pre-heat air flow. The two conductances add to the original equivalent air change rate. In addition a conductance is assigned between roofspace and ambient equal to that between roofspace and house, to account for cooling effect of increased flow of air into the roofspace.

In this way an easily variable proportion of house ventilation comes from the roofspace. The modelling is necessarily rather crude, since in reality heat flow is uni-directional, from roofspace to house and not bidirectional as will occur when the simulated roofspace temperature drops below that in the house. This will tend to cause heat build-up in the roofspace which in reality would not occur. In response to this problem a simple code modification was made to SERI-RES to dissipate this erroneous heat flow.

The modelled house has floor area 120 m^2 and is of timber frame construction, specific heat loss rate 190 W/K, total forced plus natural ventilation rate 0.9 ACH. Pre-heat was taken as two thirds of total ventilation, the rest coming from ambient.

The displaced auxiliary energy saving were found to be 3250 kWh, 42% of which is due to ventilation pre-heat. The payback period for this system is approximately 16 years.

4.3 Replication System (Habitable Roofspace)

The same timber frame house as above was modelled but this time including a well sealed and insulated roofspace with 18m^2 of *vertical* south facing double glazing. The roofspace was thermostatically vented at 40°C to allow some hardy heat loving plants to grow. In the summer, a large window is left open allowing most excess unwanted heat to escape. The construction cost for these measures would cover the extra expense of fitting vertical glass double glazing, medium quality finishing work and allowing for occupant access to the area.

Simulations show that the measures save 2720 kWh of heat yearly with, again, just under half being due to ventilation pre-heat. This roofspace type would have the advantage of not relying primarily on its energy saving attributes to be judged a successful investment. If its recreational value proves significant then, more than the other roofspace systems, the habitable roofspace would add significantly to the house resale value. A payback period for this system has not been given because of the high uncertainty involved in estimating its construction cost.

4.4 Financial Savings

The annual energy and investment performance of the above systems are summarised in Table 3 alongside that of the demonstration system.

Table 3 Summary of Annual Solar Roofspace Cost-Benefits

Roofspace Type (Section No.)	Cost (£)	Energy Saving (kWh)	Adj. Saving (kWh)	Pay-back (Yrs)
Demon. (2.1)	2560	890	1521	71
Replication (4.1)	1775	1463	2508	29
SVP (4.2)	1775	3250	4650	16
Habitable (4.3)	?	2720	-	-

The adjusted savings accounts for the imperfect efficiency of central heating systems which would otherwise have provided the heat, and also for the increased comfort provided by the roofspace systems /1/. The payback period is the simple payback /4/, ie cost divided by annual financial savings.

It should be noted that the above figures do not account for the increased resale value of the house due solar energy features. This factor can drastically improve the economics of any well designed passive solar feature.

4.5 Roofspace System Applicability

The above simulations show that, by applying the findings of this project, future solar roofspace designs can expect to improve significantly on both the energy and financial performance of those at Newham. Both extraction of low grade heat, and cost cutting measures will contribute to improvement of financial viability.

The large daytime solar gains received by passive solar roofspaces make these systems particularly appropriate to large, well sealed buildings, densely occupied during the day-time and thus requiring high ventilation rates. Examples of these include, new schools, old peoples homes and some office buildings. Poorly sealed buildings of high UA can also benefit significantly from solar roofspace technology.

The future for solar roofspace heating systems looks bright as confidence in the technology and performance predictions grows. It seems likely that future installations will make maximum use of both low and high grade roofspace heat by linking the solar and conventional heating systems much more intimately than at Newham.

5 REFERENCES

1 **A. Tindale, C. Hancock, J. Littler,** Hybrid Solar Roofs, Report for the CEC, November 1989.

2 **A. Tindale, C. Hancock, J. Littler,** Hybrid Solar Roofs, Report for the CEC, November 1989 (Annexe)

3 **CAP Scientific,** Green Park Combined School, Building 2000 Thermal Performance Analysis, report for CEC, March 1988.

4 **European Passive Solar Handbook,** CEC publication, Edited by P.Achard and R Gicquel, Brussels 1986

HOUSE DESIGN STUDIES

Howrie A.J., Palmer J.M., Hildon A.
Databuild Limited
4 Venture Way, Aston Science Park, Birmingham B7 4AP

This paper reports on the House Design Studies Project of the U.K. Department of Energy. The aim of the project is to foster the design and construction of passive solar energy efficient houses within the building industry. Lead contractors are appointed to control and record the project. Architects and designers are commissioned by the contractors to produce house designs which both satisfy the needs of a client and address the issue of passive solar design. The designers have the assistance of energy advisors and their designs are analysed for energy and cost performance by simulation techniques. The client and the lead contractor also assess the design for marketability and amenity. The designers are then encouraged to redesign the house with the help of this feedback. It is hoped that some of the designs so produced will progress to being constructed and sold.

The House Design Studies project of the U.K. Department of Energy sets out to advance the building industry's knowledge of passive solar and energy-efficient house design. The project is aimed at understanding the process of passive solar design, providing skills within the architectural profession, and improving the information available to the building industry.

The scientific evaluation of the built environment is an important component of the development of better buildings. The architectural profession has long been served by science, which has endeavoured to provide an objective view of how buildings are performing. For example, information gained from the energy and environmental performance of existing buildings has been used to inform designers of the consequences of their design choices. However, the acquisition of the appropriate information requires that suitable buildings have been constructed, and a scientific investigation can be carried out after they have been built and are in normal service. If all the necessary conditions (not least of which is the financial support for the work) can be met there is still the problem that the building and its occupants, together with the contemporary weather, form a unique and uncontrolled combination of parameters. Such a situation excludes the possibility of a classical scientific experiment, in which controlled conditions are maintained whilst individual parameters are manipulated. For example, the form and fabric of the building itself cannot be modified or changed without invalidating the design. When considering the energy-use of a building the position is further complicated by the variability of the weather.

Despite the problems associated with understanding how buildings perform much has been learned by their continued investigation. Monitoring (measurement of the significant parameters) has benefited the science of building design enormously, but at a rate somewhat slower than is required to influence contemporary building design because of the long time spans of building production and monitoring. However, as the understanding of building performance has developed so it has been increasingly possible to predict the likely outcomes of design options.

This ability to predict has opened up the potential for evaluating buildings at the design stage. With increased confidence one is then able to compare design options by the use of these predictive techniques, at a cost which is lower than examining real buildings and with the facility for controlled comparison.

The House Design Studies project therefore sets out to assess the performance of passive solar building designs by using such predictive methods. This paper reports on the second series of house design studies. The first series of studies were conducted in a time when it was sufficient to concentrate almost exclusively on the energy performance of the houses, this second series has a rather wider view. In this series it is accepted that amenity, marketability, and buildability are important considerations in the take up of passive solar and energy efficient design. Therefore the scope of the study has been increased to address these issues.

Three contractors, from different regions of the country (the south, the north and the midlands and Wales) and with knowledge of the local architectural and building organisations, were appointed to oversee the project on behalf of the Energy Technology Support Unit of the Department of Energy. This paper reports the methods used by Databuild (the midlands and Wales region contractor) in carrying out a series of design studies of passive solar houses.

Since this was the second series of design studies, a basic method already existed and the concept of using architects and designers to design overtly passive solar buildings had been established, together with the predictive techniques for addressing the designs. However, it was possible to identify improvements on the first series, particularly in the approach to obtaining designs which could be produced, evaluated, improved and possibly built.

The major components of the exercise are shown in Figure 1 and their roles and functions will now be described.

The CONTRACTOR manages the process and

facilitates the individual studies, dealing with upto 20 individual designs. This is a major role in the project and must be undertaken by someone experienced in low-energy and passive solar architecture, and who can communicate with both architects and scientists. The role is more than simply a management exercise, the contractor takes an active part in each stage of the process, providing insight and guidance as appropriate.

The contractor engages DESIGNERS (often from a local architectural practice) who are prepared to participate in the programme for a realistic fee. The role of the designer is the central role of the process, and involves taking the BRIEF for a particular scheme and producing passive-solar low-energy designs that can be assessed for their performance. In this second series of design studies it was decided not to enrol experienced passive solar designers but rather designers who had little passive design experience. It was felt that this choice would both increase the understanding of the 'typical' design process, i.e. by the novice solar practitioner, and take the concept of passive solar design to more architectural practices.

The designer and contractor then recruit another participant who is associated with the commissioning and building of houses. This participant, who is referred to as the QUASI-CLIENT, adopts the role of a real client for the scheme. It is preferable if the designer is already engaged in a conventional housing scheme with this quasi-client as the reality of the situation is enhanced. The use of a quasi-client brings to the project the judgement of the market place and attempts a more realistic simulation of the process by which houses are constructed and sold. Each of the schemes which the contractor is managing is therefore set up with a designer and a quasi-client.

These three parties identify a specific site and a house type which complies with the quasi-clients needs and then develop a brief for the designer to reconsider the scheme using passive solar criteria. The brief will require the designer to address the needs of the real site and, typically, the necessity to provide houses which can be used either as north entry or south entry. This means that the designer works in as near to a 'real' context as is compatible with the aims of the project.

Immediately following the formulation of the brief the designers attend an INITIAL GROUP SEMINAR. This seminar introduces the designers to the wider scope of the study and acts as a forum for discussion amongst the participants. The opportunity is also taken to present the fundamentals of passive solar design and to provide them with a compendium of information on energy and house design which they can use for future reference. The compendium, which was compiled by the contractor, contains information on design, building regulations, thermal insulation details, and information on the evaluation procedures to be carried out on their design. The design guidance handbook distributed in this study (Passive Solar Energy Efficient House Design; Simos Yannas) has been especially commissioned for the project and ensures that the designers have a credible starting point for their passive solar design.

At this stage the contractor assigns an ENERGY ADVISOR to the designer. The energy advisor is someone who can give specialist advice on low energy and passive solar design for the period of the study. The energy advisor, who is local to the designers, must assist the designer translate the passive solar concepts into practical building issues. With the assistance of the advisor the designer is then in a position to produce a preliminary scheme design which shows the site layout and initial thoughts on how the brief will be met. These ideas are presented to the quasi-client for comment. Should the sketch design appear acceptable, then the designer proceeds, with the energy advisors assistance, to a more complete INITIAL DESIGN. The initial design is conveyed by the use of a set of architectural drawings and reports showing:-

* site plan and layout
* floor plans
* elevations
* sections
* passive solar design strategy
* constructional details of relevance to energy use

At this stage the design is at a degree of detail amenable for analysis by the PERFORMANCE ANALYSIS SERVICE and COST ANALYSIS SERVICE, and so is passed to the contractor for appraisal of its suitability for the detailed analysis.

The analysis services are provided by organisations who are acknowledged experts in the fields of simulation of building energy performance and costing respectively. These organisations are contracted to ETSU on a continuing basis and support passive solar projects other than the design studies.

The performance analysis service use the mathematical simulation model called SERI-RES to predict the energy and solar performance of the designs produced by the designers. This computer based tool can be used to simulate repeatedly the performance of the house in a controlled and reproduceable fashion. By doing this it is possible to evaluate the design in many forms and compare between design options. These options in design are selected by the contractor, the energy advisor (who has reported on the design to the contractor) and the performance analysis service themselves. This results in a wide reaching investigation of options reflecting the interests of the contractor, the energy advisor and the mathematical simulation experts. Examples of design options which may be investigated in a design include:-

* variations to the glazing areas, particularly to the south and north facades
* double glazing or single glazing options (particularly of conservatories)
* sensitivity of the design to orientation
* evaluation of infiltration air pre-heat
* changes to insulation levels of the fabric.

The cost analysis service performs a similar function to the performance analysis service but this time deals with the cost implications of the design. Likewise they comment on the options which may be adopted to vary the costs of the design.

The absolute values for energy or cost however, are of little value without the ability to compare them with the non-passive designs which they hope to replace. Therefore both of the analysis services must put their comments into context. This aspect of the study introduces the concept of a REFERENCE house.

The performance analysis service uses as a reference a non-passive house matched in size and type to the passive solar house designed. This house will then be simulated in the same fashion as the passive solar designed house (with the appropriate allowances being made for insulation levels and floor areas for example), and comparisons made. Typical parameters used to judge the performance are:-

* space heating energy requirement
* passive solar displaced space heating energy
* window thermal performance losses/gains
* internal temperatures at critical times of the year overheating of occupied rooms
* times when conservatories are warm enough for comfortable occupation.

The cost analysis service does not compare the design costs with an equivalent house but uses two comparisons. In the first case they provide an overall figure for costs of a sample of conventional houses which are considered to be similar to the design. This is referred to as a specific standard reference and is used to reflect broadly the current housing market. In the second case they take the passive solar design and, assuming that the conventional design would not have those features, remove the passive solar elements of the design to produce a 'down graded' reference. This aims to produce a non-passive version of the designer's passive solar house. In both cases the elements of the design are costed by standardised techniques and comments made on the cost of the passive solar measures.

The findings of the performance analysis study of the designers initial design are then reported back to the contractor who, together with the energy advisor, interprets them for the designer. The contractor at this point will interview the designer and quasi-client, take the energy advisor's report and the analysis services reports and then set clear goals for the next stage of the design and analysis.

The designer now has a large amount of information on how his design is predicted to perform, how it could be modified to advantage, and its likely costs. At this point the reaction of the quasi-client may be sought to assess the likelihood of dramatic design changes being acceptable within the brief. The quasi-client introduces into the design process those aspects of the 'real world' which cannot yet be simulated with the precision of a mathematical model. For example, the 'saleability' of houses which may look very different from other houses, or the view down a street in which the north facades are very different and less attractive than the south facades of those across the road. It is these requirements which present a challenge to the designer of mass passive solar housing rather more than does the one-off specialist energy efficient house. It is at times such as this that a not uncommon request from the performance analysis service for less glazing on the north facade is rejected on grounds of 'marketability'.

Should all be agreed the designer goes on to the DETAILED DESIGN of the house armed with the feed-back on performance, cost and marketability, and suggestions as to how to improve the design. The designer is advised to take note of this advice and where feasible incorporate those suggestions into the design. Once again the energy advisor is on hand to assist the designer over some of the more scientific issues.

The detailed design which is now produced is again submitted to the analysis services for comment and appraisal in the same fashion as at the initial design stage. Following the consequent reports from the analysis services and the energy advisor the contractor, the designer and quasi-client have a final debriefing session in which the design and its evolution are discussed.

The outcome of these procedures, which entail one design study is:-

* a passive solar literate designer
* a selection of designs satisfying the brief
* an educated client
* an increased likelihood of passive solar design reaching the market place
* ideally a built scheme.

At the present there are about 50 design studies in the U.K. which represents a very worthwhile exercise in nurturing passive solar design. With the hoped for outcome of several of these designs proceeding to construction, the sale and marketability of passive solar housing will have been shown.

ACKNOWLEDGEMENTS

The authors would like to acknowledge The Department of Energy for funding and the Energy Technology Support Unit for managing the House Design Studies Project. The views expressed are those of the authors.

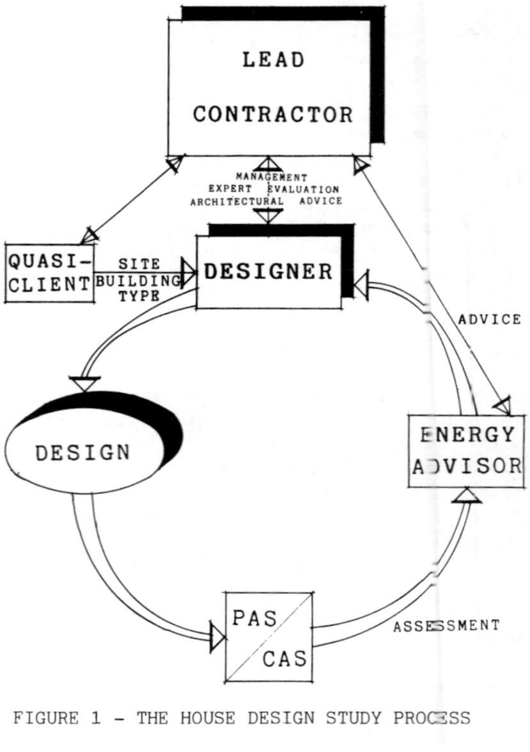

FIGURE 1 - THE HOUSE DESIGN STUDY PROCESS

THE PERFORMANCE AND GENERAL SUCCESS OF PASSIVE SOLAR BUILDINGS THROUGHOUT EUROPE

S Burton, J Doggart and A Crompton
THE ECD PARTNERSHIP
11 - 15 Emerald Street
London, WC1N 3QL, UK

ABSTRACT. The CEC Project Monitor series of 49 high quality, full colour brochures on passive solar buildings throughout Europe is complete. Each brochure describes a building or group of buildings in architectural, structural, energy use and solar terms, with the help of photographs, diagrams, graphs and tables.
The significance of Project Monitor is that it brings together experience in use and monitored energy consumption in more than 1300 solar dwellings and 9 non-domestic solar buildings.
This paper presents an analysis of the results of the 49 projects and draws conclusions which are of interest to all involved in commissioning, designing, constructing and using solar buildings.
The paper concludes with the interim results of the CEC "Passive Solar Assessment Study" which aims to establish the current and potential future contributions that passive solar gain makes and can make to building energy consumption in Europe.

1. BACKGROUND

1.1 From Detail to Europe wide.

This paper looks at passive solar energy in buildings in Europe at two very different scales. Firstly at the individual buildings designed specifically to utilise passive solar energy, and secondly at the Europe wide scale of 130 million dwellings plus non-domestic buildings which all utilise passive solar energy to a greater or lesser extent. It is towards increasing the extent of utilisation of passive solar energy that the CEC's passive solar work is aimed.

1.2 Project Monitor

49 passive solar buildings and groups of buildings were chosen from the thousands of buildings in Europe which incorporate specific passive solar measures. They were chosen on the basis of architectural merit, monitored energy performance and user response. A high quality brochure was produced on each scheme incorporating site, design, solar features, buildability, performance evaluation, user response and costs. The first part of this paper analyses all the projects and draws conclusions of direct relevance to the future of passive solar design.
The significance of Project Monitor is:- that all the buildings have been used for normal purposes and the reactions of users are recorded; that the energy performance has been monitored in use and the analysis is presented is a standard way; and that the whole scheme is described in a consistent and easily accessible form.

1.3 The Assessment Study

The second part of the paper draws on the monitored performance figures from Project Monitor and a host of other information, to assess the current utilisation of passive solar energy in Europe and to forecast the potential and likely use of passive solar energy in the future, based on the years 2000 and 2010.
This part is based on the "Passive Solar Assessment Study" which is currently being carried out by The ECD Partnership under contract to DG XII of the CEC.
The Assessment study does not stop at the energy contribution that the utilisation of passive solar energy can provide, but goes on to assess the pollution reduction that can result from the substitution of passive solar heating (and cooling) for fossil fuel use. Thus the contribution of passive solar design to the amelioration of current concerns of global warming, acid rain etc. is demonstrated and added to the traditional benefits of the conservation of fossil fuels.

1.4 The Conclusions from Project Monitor

The Project Monitor series was a selection of passive solar buildings throughout Europe, and rather than provide a statistical summary of the performances of passive solar buildings in general, the analysis aims to provide answers to five questions of utmost importance to designers and users:-
How much solar gain can be used?
Which passive solar systems work best?
What are the financial and other benefits?
Do users like passive solar buildings?
What are the problems, and the solutions?

2. HOW MUCH SOLAR GAIN CAN BE USED ?

2.1 Analysis

The Project Monitor series contained 10 individual houses, 27 groups of multiple housing (new build), 2 housing refurbishment projects, 1 sheltered housing scheme and 9 non - housing schemes. These have been considered separately for the purposes of analysis. The reasons for the separate analysis of new housing are twofold, firstly the individual houses are generally more expensive with fewer financial constraints on the energy saving and solar measures, secondly the occupants tended to know the background to the solar features and operated them to maximum advantage. This is a very different situation to multiple housing where "standard" passive solar features are applied to a block of dwellings which are occupied by a range of people who do not necessarily know about solar energy and may well not be interested in maximising the energy benefits. Thus whilst the individual houses may show the maximised solar contributions, the multiple housing schemes show more achievable contributions under typical patterns of usage. The latter group also have more replication potential.

2.2 Multiple Housing

In more than a third of the multiple housing schemes, passive solar gains contributed more than 30% to the space heating demand (Diagram 1). Although the highest contributions, around 70%, were in the southern half of Europe: Southern France; Spain and Italy, contributions of up to 40% were obtained in the north. Although many projects achieved less than 30% contribution, the analysis suggests that with good design and good operation 30% solar contribution is obtainable. The percentages of solar contribution are not directly related to latitude. The total amount of solar energy utilised per dwelling was above 3,300 kWh per year in a third of the projects and this applied to both large and small dwellings, in both northern and southern climates (Diagram 2). Standardising for floor area between large and small houses shows that 30kWh per square metre is achievable at all latitudes (Diagram 3).

High solar percentages can only result from successful solar design and even where the heat load of a building has been reduced by careful use of insulation measures, high percentages were still achieved. However financial savings and environmental pollution reduction result from large absolute energy savings. Naturally there is a trend towards smaller savings resulting from solar gain as overall heat demand reduces (Diagram 2). Nevertheless at all levels of heat demand (measured in kWh/m2) there is a wide variation in useful solar gain, with a factor of 3 or more between worst and best (Diagram 4). The conclusion is that even with well insulated buildings, worthwhile passive solar contributions can be achieved.

All buildings benefit from solar gain to some extent even if not specifically designed to do so. On average the solar gain for the existing housing stock is estimated to vary from around 10% to 15.5%, for northern climates. To take one example, the UK average solar energy utilised is 2500 kWh, a figure reflecting the low levels of insulation (and therefore high heating demand) of the existing housing. With better insulated new housing, and further insulation to existing housing, the amount of solar heating would decrease over time. However this can be changed if passive solar measures are incorporated. Thus the 'achievable' solar gain of 3300 kWh in a well insulated, passive solar house, represents a saving, maybe around 1500 kWh, over that which would be achieved without any specific solar design.

2.3 Individual Houses

The larger individual solar houses had solar contributions of 35% and above in 8 out of the 10 examples (Diagram 5) and large energy savings of above 8500 kWh were achieved in nearly half the projects. Even where the heating demand per unit area was small, high percentage and absolute solar contributions were achieved in the best examples (Diagram 4). It is interesting that latitude has less influence on these figures than size and heat loss.

Thus designing to maximise solar gain in individual houses can save significant amounts of fossil fuel energy and this applies to all latitudes. The effect of the occupants in controlling their houses was frequently thought to have been one of the most important influences on the level of solar gains achieved.

2.4 Housing Refurbishment

Adding passive solar features to an existing building or incorporating solar features during refurbishment, has great potential for the wider use of solar energy, due both to the large numbers of houses and the high heating demand of the older housing stock. The two examples included, Lievre d'Or in France and Baggesensgade in Denmark, use increased areas of south facing glazing, Trombe walls and sunspaces. Solar contributions of 20% and 27% were achieved and Baggesensgade recorded the highest overall solar contribution of all the projects, nearly 22,000 kWh per year per flat.

2.5 Sheltered Housing

Christopher Taylor Court, the sheltered housing scheme, although recording a low percentage solar contribution due to the requirement for high temperatures (23°C) over the full 24 hours, demonstrates the significant absolute solar contribution possible in this type of use.

2.6 Non-domestic Buildings

It is impossible to assign general figures to the solar energy usable in non-domestic buildings from those projects in Project Monitor, due to the wide variety of designs and uses. In Spain at Polysportive Esterri sports hall, passive solar energy supplied 94% of the heating load and at Los Molinos college, 80% (the remainder of the heating load being supplied by internal gains). In more northern latitudes, the JEL building in the UK achieved 29% solar contribution.

Large and valuable solar contributions are clearly possible for example schools are well suited to passive solar design in at least two respects, they are occupied primarily in daytime when solar gains are available directly, and they are not occupied in the hottest summer months, so that overheating may not be a problem. However the passive solar design of non-domestic buildings is likely to be tailored more specifically to the building and its use, compared to housing where similar solutions are satisfactory across the range of designs.

3. WHAT ARE THE FINANCIAL AND OTHER BENEFITS ?

3.1 Costs

Assessment of the capital costs and benefits of passive solar design in practice is even more difficult than assessment of the energy savings alone. Whilst additional costs for passive design are quoted as around 10% typically, this varies from zero additional costs over normal design or cost yardsticks (Overbos 8 in the Netherlands), up to a 20% increase in extreme cases. Zero or minimal costs are associated with planning and relocation of glazing, whilst sunspaces, Trombe walls etc. can cause large additional costs. Frequently some solar features, particularly sunspaces and larger areas of double glazing, are included for reasons of amenity and the solar on-costs should then be discounted because of these other benefits.

3.2 Returns

Simple payback times based on the energy savings resulting from the passive solar measures vary up to 20 years in some cases but were less than 10 years in half the housing schemes, where figures were available. These figures were based on fuel costs fixed at the time of monitoring and are thus likely to improve in time as fuel prices increase.
In many projects one major objective was to provide housing for lower income families (Barbaricina, Parklands Estate) and the reduced heating costs resulting from the passive solar design were seen as an essential part of this strategy.

3.3 Other Benefits

In non-financial terms, the passive solar measures generally provided occupants and users with pleasant living conditions and reduced heating energy costs lasting throughout the life of the building, since the measures generally require little or no additional maintenance.
Translating the fossil fuel savings into pollution reduction in terms of carbon dioxide production, acid rain etc, demonstrates the long term benefits of passive solar buildings in environmental terms. This is discussed more fully at the end of this paper.

4. WHICH PASSIVE SOLAR SYSTEMS PERFORM BEST?

4.1 The integrated system

Passive solar buildings work as an integrated system incorporating solar collection, distribution, storage, ventilation and auxiliary heating. Technical monitoring of the schemes produced overall energy consumption figures which usually cannot be desegregated to assess the relative quantitative success of the different solar features. However from the social or qualitative monitoring the successes and failures are more obvious, though not exclusively in energy terms.

4.2 The basic passive solar measures

Most schemes adopted the basic solar techniques:- north/south orientation, site planning to avoid over-shading, location of living rooms on south sides, some thermal mass and redistribution of windows to maximise south facing glazing. Many schemes incorporated no other solar measures and gave good solar performances particularly Overbos 8 and Parklands Estate. There is little doubt that these measures provide good solar gains at low cost with few ill effects.

4.3 Sunspaces

Sunspaces are the next most common features, used in 28 out of the 40 houses or housing schemes. Energy savings from sunspaces are generally considered as threefold, the "buffer" or insulation effect, the supply of pre-heated ventilation air, and the supply of sun warmed air to the house via open windows, door or ventilators. Though energy savings do certainly accrue from the use of sunspaces, their actual energy contribution cannot be separated out from the overall savings in the schemes. Those schemes achieving the highest solar contributions almost invariably used sunspaces, however some of the schemes with poor solar contributions also incorporated sunspaces. It should be remembered that sunspaces are not simply solar features, in fact frequently their function as an energy saving device was secondary to their amenity and additional space aspects. As retrofit measures, sunspaces, produced by glazing in balconies or as additions, are one of the best measures and were used in both refurbishment projects.

4.4 Other measures

Other specific passive solar measures used in some projects include:- air heating panels, placed either on walls (ventilated Trombe walls) or roofs; Barra-Costantini panels (Marostica) and unventilated Trombe walls (Christopher Taylor Court); passive water thermal storage (Casa Termicamente Optimizada and Los Molinos); solar reflectors (Polysportive Esterri and Los Molinos); atria (Scuola Massa Finalese and Lou Souleu); passive solar water heaters (Lawrie Park Road and Smakkebo) and composite collection and storage arrangements (Los Molinos and Scuola Massa Finalese). Some of these worked well and provided useful solar gains, notably the solar reflectors, the passive solar water heaters and the composite systems. Others do not seem to have proved their worth, particularly those systems that required regular user control.

4.5 Use of solar measures

In general those passive systems that were immediately understandable to users, and where control was limited to only the opening of windows and the moving of blinds, appeared to work well. Where the system was more "hidden", as is the case of wall panels, and where users were expected to operate unusual controls, results were not so positive. This does not always apply in non-housing buildings, where more organised control is possible, nor in some of the individual houses where occupants were keen to maximise the effectiveness of the solar features. At the other end of the scale, in dwellings where occupants did not directly pay their own fuel bills (Collegno), even "automatic" passive solar systems do not necessarily provide large 'useful' solar gains.
Some "passive" systems rely on a degree of "active" fan assistance and mechanical control and the more straightforward systems can work well (Boegehusene, Arona and Pedrajas de San Esteban). Les Garennes is an example of a less successful system in this respect. A cautionary note on the use of fan power is worth mentioning, since solar gains can easily be eroded by electricity consumption as at Ecole Primaire de Tournai

in Belgium. Here, despite a significant reduction in overall delivered energy due to solar gain, there was a small cost penalty rather than a financial saving due to gas heating being replaced by electricity used by the fans to distribute the solar gains.

4.6 Thermal Storage

Five of the original projects have used rock and brick stores to spread the use of solar gains over longer periods, but not one was particularly successful due to heat losses from the store and low temperatures of both the incoming and outgoing air. Others which specifically required thermal storage but used in situ mass were more successful, particularly Los Molinos and Scuola Massa Finalese. The use of water storage at Casa Thermicamente Optimizada in Portugal was not seen as effective, but at Los Molinos as part of the "Blanco Wall" system, it provided good results. The use of thermal mass in floors (particularly of sunspaces), walls, fire places and stairs is widespread and is generally thought to enhance solar performance, both by extending the use of solar heating and mitigating overheating from "excess" solar gain. The provision of additional thermal mass over "normal" brick or concrete construction, as in sunspaces at Les Basses Fouassieres in France, has not been shown to improve thermal performance significantly.

4.7 Thermal Loops

Distribution of solar gains around a building by natural convection has been developed in some projects with varying success. At La Salut, the lower convective loop worked well, while the upper loop did not. A clearly defined loop with large openings is required for effective heat transfer. Natural air movements for avoiding overheating are of vital importance and are discussed later.

5. DO PEOPLE LIKE PASSIVE SOLAR BUILDINGS?

5.1 General

The projects demonstrate the great variation in the appearance of passive solar buildings both external and internal, and in their performance. Evidence from the surveys of users shows that the buildings are liked with very few exceptions.

5.2 External appearances

Planning to achieve a south aspect for a complete housing estate can lead to regimented parallel rows but Les Basses Fouassieres, Smakkebo and Willow Park are good examples of more imaginative layouts. Am Lindenwaldle demonstrates one of the possibilities for imaginative terraces. Smaller sites sometimes present difficulties in achieving south aspects and unconventional layouts may meet with initial local opposition from planning authorities as is the case of Am Lindenwaldle.

The clear identification of a building as a distinctive "solar building" is sometimes very important to the designer or user. This may be from a commercial point of view as in the case of the JEL building or from the didactic point of view as in Ecole Tournai and Los Molinos.

The inclusion of sunspaces where houses are to be sold is frequently seen as a selling point, though it is likely that it is more the nature of the additional space provided, rather than the solar gain, that is the attraction. At Lievre d'Or, newly attached sunspaces were one of the features used to improve the overall appearance of the estate and help reverse the serious deterioration of physical and social conditions on the estate.

There are several direct references to the higher value that occupiers put on their "solar" houses, both by tenants at Lievre d'Or, Lou Souleu, and Carrigeen Park, where turnover was lower than average, and by owners at Les Cochevis, Les Basses Fouassieres Willow Park and Parklands. At Les Basses Fouassieres, for example, three quarters of the owners expressed the desire to live in another solar house if they ever moved.

5.3 Internal appearances

Most user response relates to the insides of the buildings, their appearance and comfort. Both are highly rated in the majority of schemes particularly in terms of being light and airy, though poor lighting in northern rooms was mentioned at Parklands Estate and Jeux du Soleil. Lack of privacy from large south windows was sometimes mentioned as a disadvantage (Giffard Park and Parklands Estate).

5.4 Appreciation of the passive solar features themselves.

Sunspaces are universally liked by owners and are well used as additional floor space. The main complaints arose from those which were too small to be used other than for storage. Larger south-facing windows are apparently appreciated for the extra light they give but other passive solar aspects are not normally appreciated in themselves, though may be generally seen as good in helping to reduce heating costs.

6. WHAT ARE THE PROBLEMS AND SOLUTIONS?

6.1 Problems

Passive solar buildings are not without their problems and in addition to those already discussed, there are three main areas where problems have arisen with some frequency. Overheating is by far the most commonly mentioned but poor materials and construction have lead to dissatisfaction, and poor "operation" of the buildings by users has both lead to poor use of available solar gains and to overheating.

6.2 Overheating

Even in northern climates overheating can be a problem and sunspaces particularly should include shading devices and adequate ventilation. On Parklands Estate and at Christopher Taylor Court overheating even without sunspaces was the cause of minor complaints and high temperatures were recorded in several sunspaces where blinds and ventilation were inadequate or not operated to best advantage, for example as at Les Garennes. Overheating in non-domestic buildings is common and at the JEL factory considerable effort was made to overcome this

problem without resorting to air conditioning.
Sunspaces used internal blinds or external awnings and ventilation openings to prevent excess solar heating entering the house, while in the buildings themselves, few precautions against overheating were taken except where unusually large areas of glazing were present, as in Hauser auf der Pick, Haus Waldmohr, Parklands Estate, and Christopher Taylor Court. The specific provision of thermal mass was only considered necessary in northern climates in very lightweight structures such as Willow Park, where a heavy stone fireplace was included.
In southern latitudes passive solar design is as much about the avoidance of summer overheating as about the provision of winter heating. The strategies adopted seem to have been successful as there were few complaints recorded about overheating except in the non-domestic buildings, Polysportive Esterri, Liceo Scientifico and Scuola Massa Finalese all suffering to some extent, particularly in their early days before being modified. No projects included any form of active cooling, such as air conditioning, and only at Los Molinos were specific passive solar cooling measures adopted.
Temperatures above 30°C were recorded inside some dwellings (higher in sunspaces) but these were similar to external temperatures and were thus seen as acceptable.
Most schemes in southern latitudes used similar strategies to avoid overheating. Sunspaces had internal blinds or external awnings and ventilation openings at high level to stop heat being passed into the dwelling. The buildings themselves made use blinds to restrict solar gain, ventilation openings on both sides of the buildings to promote through air movements, and thermal mass inside to modify temperature swings. External awnings were used La Salut, Arona, Scuola Mass Finalese, Pedrajas de San Esteban and Omegna and overhangs at Casa Termicamente Optimizada. Solid roofs to sunspaces, which cut out high summer sun, are a common feature of the apartment blocks in Italy.

6.3 Construction Problems

These were mentioned particularly at Am Steinbruch, Parklands Estate and Am Lindenwaldle, but bearing in mind the extent of innovation involved this has been a surprisingly small problem. The main component failure has been malfunctioning of blinds intended to control overheating.

6.4 Operational Problems

Poor operation of the solar features by users was given as a common reason for poor solar performance. Frequently it was said that house occupants failed to understand the solar systems and so wasted available solar gains, or simply did not control the solar and heating systems to best advantage. The more complex systems understandably suffered particularly in this way. Whether this is a failure on behalf of the designers to accommodate the users adequately, or a failure of the users to adapt to the designs, is open to argument. In two schemes, Jeux du Soleil and Baggesensgade, "education" of the occupants greatly improved the solar performance, demonstrating the advantages of bringing the two sides together. At Baggesensgade while in the first year after refurbishment there were very small energy savings, in the second year after education of the tenants, energy savings of 34% were realised.
Whether the occupant directly pays the individual fuel bill or pays a fixed amount unrelated to the individual use, makes a considerable difference to their attitude towards energy efficiency, as demonstrated at Collegno, Pedrajas de San Esteban and Omegna. Designers should clearly bear in mind the capabilities and interests of their clients and design systems appropriate to these, as well as to the likely maintenance facilities.

7. CONCLUSIONS ON PROJECT MONITOR

7.1 Passive solar design

The wide success of the projects described demonstrates that passive solar design in buildings is here to stay. Passive solar buildings can be designed with confidence and a range of benefits will derive from good design, including more useful solar energy collected, auxiliary heating reduced, attractive and distinctive buildings and satisfied users. Solar design can be applied to most buildings, new and old, well insulated and poorly insulated, houses and non-domestic buildings, big and small, and it will be effective at all latitudes within Europe. In design terms the rule is to "keep it simple", and this will ensure that it costs little or nothing, that it is understood and used correctly and that it performs without problems.

7.2 Spreading the message

The main purpose of Project Monitor has been to publicise the benefits of passive solar design in buildings and thus to increase the use of solar energy. The 49 brochures have been distributed to 3500 designers and other interested parties throughout Europe, they have been used on the basis of articles in journals, they are available from libraries and offices and they have been used as course work in colleges. Two surveys of recipients of the brochures confirm their success as mediums for conveying the information and the overall success of the project.

7.3 The future

In future 5 books in different languages of 30 selected brochures are to be published, and they will be on sale late in 1990. But the work is not complete and there is still much that needs doing to establish passive solar design in all areas. Two areas seem to be lacking in reported experience of solar design, housing refurbishment and non-domestic buildings. This is due to the lack of monitored examples in Europe to date. The enormous quantity of existing housing shows that the scope for increasing the utilisation of passive solar energy here is great, but more examples, promotion and encouragement are needed. In non-domestic buildings the scope for both passive solar heating and cooling is, again, very great and the area largely untapped. Brochures such as those produced in Project Monitor, featuring refurbishment and non-domestic projects, as information becomes available, could have a significant influence.

8. THE ASSESSMENT STUDY

8.1 Passive solar in Europe

From looking at the detail of how much passive solar energy can be utilised in individual buildings and how this is best done, we move to the European-wide picture. What contribution does passive solar energy currently make to the energy supply and what contribution is it likely to make in the future? What effect can this have on environmental pollution?

8.2 Heating and Cooling

Whilst passive solar design is used in northern latitudes to increase solar contribution to space heating, an equal function of passive solar design in southern latitudes is to reduce unwanted solar gains in summer, which would otherwise lead to a cooling demand and consequent use of energy. Thus, in the assessment of passive solar contribution we shall include both the contribution to heating and to the avoidance of active cooling.

9. THE PROCESS OF THE STUDY

9.1 The spread sheets

The Europe wide picture is being built up from 12 computer spread sheets, one for each country in the CEC. Analysis has been split between housing and non-domestic buildings and three dates have been considered 1990, 2000 and 2010.

9.2 The Housing Contribution

Within the housing section the current contribution of passive solar energy to heating has been derived from the total housing stock, its average space heating demand and the average proportion of this supplied by solar energy. To this is added the solar contribution from known passive solar designed buildings, a relatively small amount due to the small numbers of passive solar building so far constructed or refurbished. By the year 2000 some existing buildings will have been demolished, some of the existing stock will have been improved in energy terms, some dwellings will have been refurbished to improve passive solar gain and some new dwellings will have been built on passive solar lines. On the basis of likely refurbishment and new-build rates the "technical potential" passive solar contribution can be assessed, ie: the maximum possible passive solar contribution if every possible new-build and refurbished dwelling incorporated passive solar design.

9.3 The non-domestic contribution

Assessment of the solar contribution to non-domestic building heating has to be assessed on a more crude basis due to the absence of available data. Where possible, similar data to those collected for housing will be used, analysed by category - offices, factories, schools etc.

9.4 Cooling

The assessment of the contribution of passive solar to cooling loads is also to be assessed, based on current estimates of cooling loads in the absence of solar design to reduce overheating.

9.5 Pollution

To these total solar energy contributions in each category, the fuel split is applied together with the amounts of the various "pollutants" produced by each fuel type. Energy and pollution savings are then summed to provide the full picture for Europe.

10. DATA SOURCES

10.1 Data Collection

Collection of data to complete the spread sheets for each country is still under way. Library searches have been carried out to assemble published material. There is very little available. Pro-formas of information required have been produced and have been sent to contractors in each country. Response has varied greatly. Project Monitor is a very useful source of solar and other data, and has been widely used in the study. Much of the necessary data is simply not available. Where this is so, estimates have been requested from the contacts in each country.

10.2 Future improvement

In general the spread sheet system is "transparent", meaning that all figures are visible and their source is recorded. The best estimates are currently included but if new, better data become available, new figures can be inserted and the program re-run. In this way the estimates both for individual countries and for Europe as a whole can be upgraded and improved over time as required.

11. INTERIM RESULTS OF THE ASSESSMENT STUDY

11.1 Progress

So far work has concentrated on heating in the domestic sector and data have been analysed for northern countries totalling two thirds of the European housing stock. Work on housing in southern latitudes, non-domestic heating and cooling is in preparation.

11.2 Domestic passive solar usage

Available results so far cover the northern part of Europe, approximately two thirds of the total dwellings in Europe, and are preliminary. By the year 2000 the technical potential for passive solar usage rises to 15% on average for the whole housing stock and to 15.5% by the year 2010. During the same periods if no effort is made to incorporate specific passive solar measures into new housing and refurbishment, the proportion of useful passive solar will fall to 13% overall for the stock, due to improved insulation leading to shorter heating seasons and thus less solar contribution.

If the technical potential could be realised there would be a 2.5% increase in overall solar contribution to domestic heating, or solar energy would replace an additional 2.5% of conventional fuels. This may not seem dramatic but is a large amount of fuel saving, 4 mtoe per annum, worth 11.4 billion ECU, out of a total heating fuel use of 160 mtoe. The total solar contribution is around 26 mtoe. The reason that the overall proportion of solar energy does not rise dramatically is due to the relatively small amount of new building and refurbishment taking place compared to the total housing stock.

11.3 Increasing overall solar gains

For a more significant impact, the existing housing stock requires solar measures which do not require a major refurbishment.
As an example, the use of "transparent insulation" has been examined in relation to these figures. If 50% of house glazing was replaced by transparent insulation, the solar contribution would rise over the whole housing stock to around 23%. Whether this is considered a passive solar measure is open to some debate, but since it allows solar gain with insulation, it can be seen as such. Other measures, such as sunspaces and Trombe walls can also be used outside major refurbishment, and none of these are included in the previous analysis.

11.4 Pollution

Pollution reduction resulting from the substitution of solar energy for conventional fuels varies with the type of fuel used for heating in each country. The differences are great and the adoption of solar measures in different countries will have a varying effect on the levels of different pollutants. This complex picture is not yet complete.

12. CONCLUSIONS

Passive solar energy is currently a major energy source in Europe and in the domestic sector provides around 14.5% of space heating.
Positive action to increase solar gain in new buildings and in housing refurbishment could increase this to 15.5% by the year 2010, while if no action is taken the proportion could fall to 13%. The introduction of passive solar techniques into the vast existing housing stock could increase the solar contribution to above 20%.
Savings of above 4 mtoe are possible as a result of passive solar design in houses in northern Europe alone, giving a total solar contribution of around 26 mtoe by the year 2010. To this can be added the yet unquantified contribution from housing in the southern third of Europe, from the non-domestic sector and from the avoidance of cooling loads.
Significant reductions in environmental pollution result from this use of solar energy and these are currently being quantified.

DIAGRAM 1

% SOLAR CONTRIBUTION AT DIFFERENT LATITUDES (MULTIPLE HOUSING)

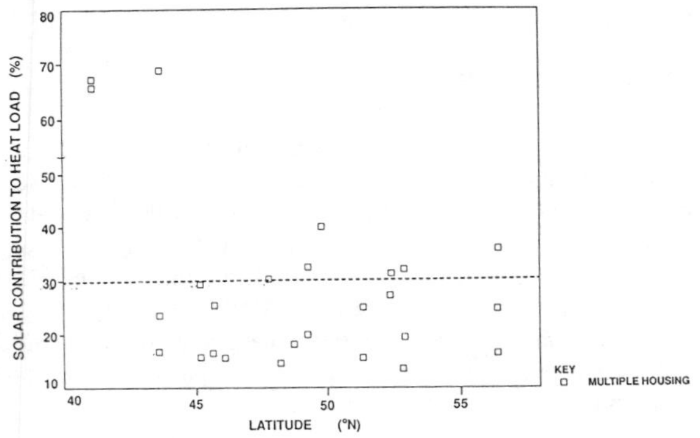

DIAGRAM 2

SOLAR CONTRIBUTION AT DIFFERENT HOUSE HEAT LOADS

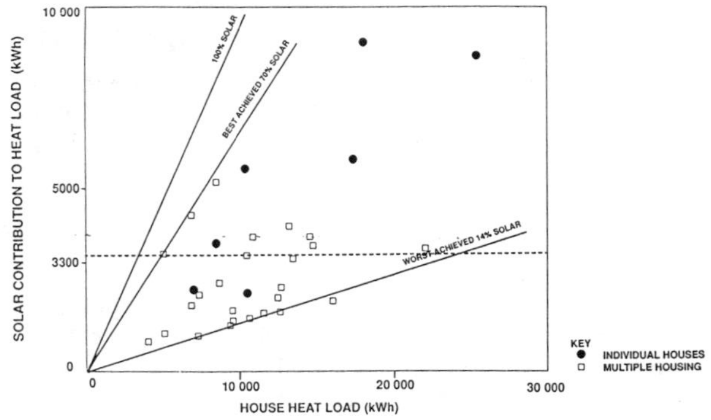

DIAGRAM 3

SOLAR CONTRIBUTION AGAINST HEAT LOAD

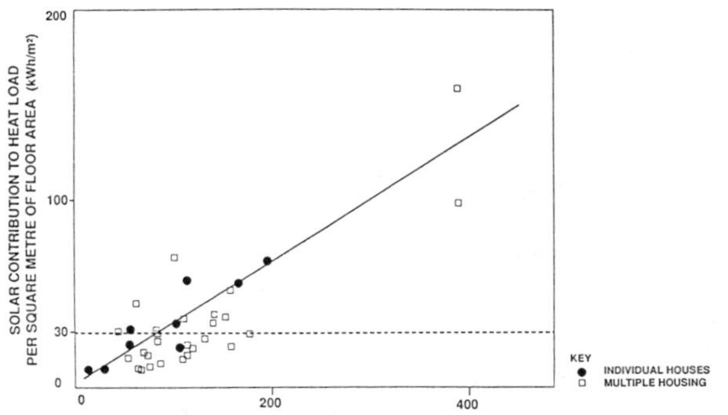

DIAGRAM 4

SOLAR CONTRIBUTION FOR DIFFERENT HEAT LOSS RATES

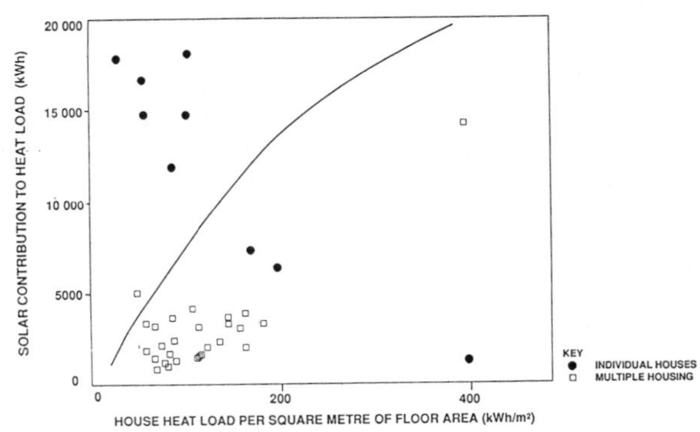

DIAGRAM 5

% SOLAR CONTRIBUTION AT DIFFERENT LATITUDES (INDIVIDUAL HOUSING)

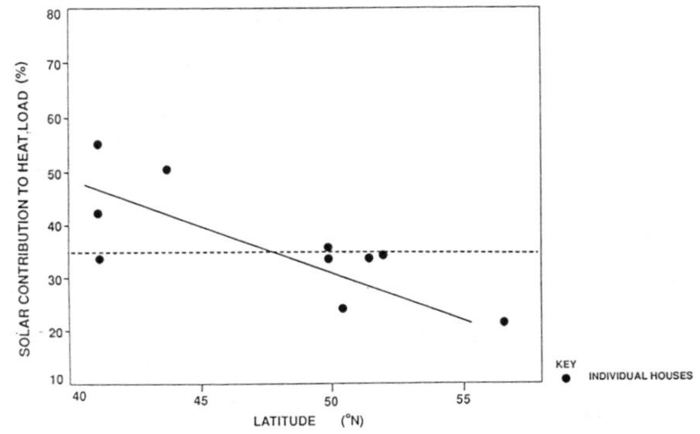

CONTROL OF PASSIVE SOLAR SYSTEM

State of the art

dr.ir. A.H.C. van Paassen,
Faculty of Mechanical Engineering and Marine Technology
Delft University of Technology
Mekelweg 2, 2628 CD Delft
The Netherlands,
Phone: (0) 15 - 78 66 75

ABSTRACT. Without control high solar fractions are difficult to obtain from Trombe walls, direct solar gains, and rock bins. Integration of all these techniques by control is absolutely necessary. A short review is given of the available sensors, actuators, and control systems. The control of individual solar techniques as well as several combinations are discussed; for instance, the control strategy for the combination of direct solar gains and Trombe Wall. Based on literature general conclusions are given about the energy efficiency of these systems in relation with the control strategy. A more detailed analysis is made of a costeffective simple passive solar system with adjustable window devices such as insulating rolling shutters, shading devices and vent windows. The benefits of modern control such as predictive control are compared with respect to simple and straightforward control strategies. Emphasis is put on the application of controlled ventilation for the maintenance of comfort in summer. Natural ventilation is strongly recommended here because of its passive character. In this paper special attention is given to the various digital control systems designed in a concerted action of the project PASTOR of the commission of the European Communities.

1. INTRODUCTION

High solar fractions are difficult to obtain from Trombe walls, direct gains and rock bins. Moreover their contributions have different dynamics, so that maximum values occur at different moments and do not always coincide with the need for heating. A successful application can only be obtained when their complementary contributions are synchronized by a control system. For a conventional control system this is a difficult task, because of the complex nature and interactions of the passive solar techniques. Therefore preference should be given to computer based control systems with more advanced control algorithms. In this paper conventional as well as computer control systems are described.

Collecting solar energy is not always advantageous. Problems encountered in passive solar buildings with poor control include overheating. In summer the solar energy system during a succession of hot days forms a big danger for overheating. This problem can be eliminated by cooling with night air. Of course mechanical ventilation can be used for this purpose, but it is more in harmony with the passive solar philosophy to apply natural ventilation by motorized windows. In this paper special attention is paid to natural ventilation and the way it can be controlled.

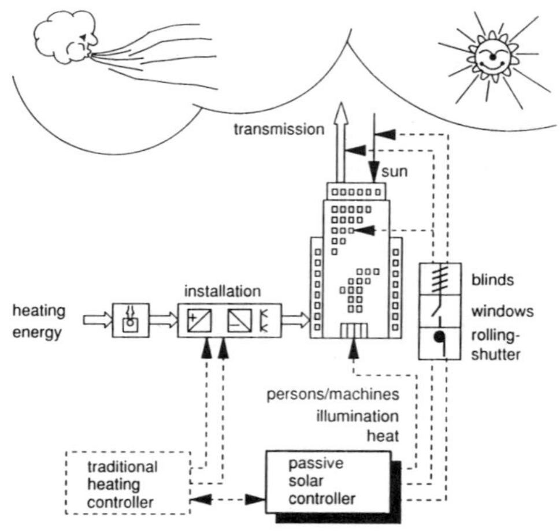

Figure 1. Passive solar and traditional control.

From the above discussion it is obvious that the widespread idea that passive solar systems are self regulating and do not require automatic control is completely wrong. It must be kept in mind that the window is an important factor in the psychological sensation of comfort and that automatic control of window devices is not always accepted by the occupants. Therefore automatic control systems should always have provisions for manual takeover.

In this paper will be discussed: sensors, actuators, control strategies and digital control systems for optimal control. An indication is given of the energy saving and comfort of the various passive solar configurations and their control systems.

2. CONTROL ELEMENTS

2.1 SENSORS

Comfort is the main objective. Consequently comfort should be measured. There are comfort sensors available, but they are too expensive for control applications. An economical solution is the use of the effective temperature in which both the air temperature and the radiant temperature are equally weighed. In passive solar control systems the following sensors are of vital importance:

- temperature sensors for the indoor and outdoor temperature,
- solar flux sensors.

For optimization sensors for the energy consumption and windspeed are very useful. For detailed information the reader is referred to the cataloques of the manufacturers of control equipment for heating, ventilation and airconditioning (HVAC).

2.2 CONTROLLERS

The control system can be realized with components that are standard practice in HVAC. A controller characteristic for solar applications is the temperature difference controller. It compares two temperatures in order to detect when transport of solar energy gains is advantageous, for instance to determine the moment of discharge of a rock bin. Preference should be given to digital control systems, such as Building Managment Systems (BMS) with which the best integration can only be realized without compromises.

Of course, in that case, strategies and algorithms giving this optimal integration should be known. In paragraph 5 several control systems are proposed, which are designed in a concerted action of project PASTOR of the commission of European Communities.

2.3 ACTUATORS

Control of passive solar gains can be carried out by the following essential actions:

- adjusting the position of insulating shutters, Venetian blinds, awnings, etc. by a tube motor,
- adjusting the position of an air damper through an electric or pneumatic motor.

Examples are: control of the convective air flow in Trombe walls, the airflow control through a vent window opening.

3. CONTROL OF INDIVIDUAL PASSIVE SOLAR TECHNIQUES

3.1 CONTROL OF SHADING DEVICES

Shading devices can be used for the following purposes:

- control of direct solar gains,
- control of daylighting
- prevent inconvenient glare.

The shading devices can be controlled in a closed loop configuration or by feedforward. In the first case a temperature, respectively lighting controller adjusts the position of the shading device in order to keep the indoor temperature just below a specific threshold or to keep the illumination at a comfortable level. Feedforward is applied when it is desired to adjust the position only as a function of the sun and the amount of solar radiation, for instance to avoid glare. Both types of control can overrule each other, for instance, in the morning shading is being controlled by feedforward to avoid glare while lateron the temperature controller takes it over by feedback control action in order to avoid overheating.
The window is an important factor for occupants. Therefore automatic window devices always should have provisions to be overruled by the inhabitants by manual control.
Glare generated by direct solar radiation is preceived by people as the strongest factor for sunshading actions. Glare can be controlled automatically by outside or inside Venetian blinds or by awnings. More information about glare is given by Inoue [2]. The best policy is to adjust the slats in such a way that no direct solar radiation can reach the occupants. Proper slat angles can be calculated from the profile angle of the sun.

3.2 CONTROL OF VENTILATION

Ventilation is necessary for fresh air supply and cooling in hot periods to avoid overheating. Mechanical ventilation is a well known technique and it can be controlled by adjusting the speed of the fan motor or by on/off operation. Control of natural ventilation by adjusting vent window openings is much more complicated. A simple control strategy for cooling is the start and stop ventilation by means of the threshold values of the indoor temperature (24°C and 20°C). It will be clear that in this case solar gains can be wasted. Control of the minimum supply of fresh air needs more research. The equations found by Gids and Pfaff [3] may be useful for this purpose.

4. CONTROL OF DIRECT GAINS AND TROMBE WALLS

Control of Trombe walls cannot be discussed without taking the influence of the direct gains into consideration. The same applies for rock bins. The Trombe wall's heat delivery can be controlled by an air damper and/or by switching on the fan in case forced ventilation is applied. The direct gains can be controlled by shading (overhand, awning, outside Venetian blinds).
Control strategies as proposed by Sebold [1]:

1. Night setback 11 PM to 7 AM at 15°C,
   otherwise night insulation on.
2. Trombe wall and direct gain windows.
   Too hot:
   - if $T_{air} > 24°C$, vent to the outside air;
   - if still $T_{air} > 24.5°C$, cancel vent and shade instead.

   Too cold:
   - if $T_{air} < T_{set}$, turn on Trombe wall fan or open air dampers, if the space air temperature of the Trombe wall exceeds T_{air} by 10°C;
   - if still $T_{air} < T_{set}$, use auxiliary energy.

Sebold tested this strategy by computer simulation for various weather zones of the U.S. (Albuquerque, Madison, Santa Maria). From his results the following general conclusions could be drawn:

- The control system reduces auxiliary heat requirements considerably and provides a slightly more comfortable building;
- Night insulation on both the direct gains windows and the Trombe walls is the most useful control (40 - 60% reduction in auxiliary heating).
- The Trombe wall fan control saves only 3 - 8% due to the presence of competing direct gains and is therefore marginal and not cost effective.
- In hot climate zones controls can also be very useful in providing much better comfort.
- Night air cooling was not considered by Sebold.
  However, other research results showed that it is very useful to improve comfort. The number of hours with too high temperatures can be reduced substantially, as will be demonstrated furtheron in this paper.

Sebold's simulation studies show that adding a passive rock bin does not significantly reduce backup requirements [1]. They are of the same magnitude as the auxiliary energy consumed by the charging fan. When a fan is used to increase the air flow through the rock bin the auxiliary heating energy consumption is reduced more efficiently (twice as great) as the passive case, which showes only a small reduction.

5. CONTROL OF A SIMPLE PASSIVE SOLAR SYSTEM

In a concerted action of the program PASTOR of the Commission of European Communities control systems are designed for a simple passive solar system, which can manage the passive solar gains, natural or mechanical ventilation, auxiliary heating and lighting. For the analyses of the control systems no insulation window shutter was considered. From the calculation results of Sebold [1] it may be concluded that Trombe walls and rock bins are not cost effective and therefore omitted in the EC- project.
Three different control systems were designed for this simple passive solar system. Advanced techniques have been applied such as adaptive modelling, optimization, weather predicition, and adaptive control. Participants of this project were: Delft University of Technology, Ecole des Mines de Paris, Centre Scientifique et Technique des Batiments, and a French Company: NAPAC in Paris. In a series of publications these systems are described in detail [4,5,6]. Here, only a global description will be given.

5.1 OPTIMAL CONTROL SYSTEM

The internal structure of the controllers is represented schematically in figure 2. It functions in the following way. The outdoor climate W (horizontal global radiation, outdoor temperature, windspeed and wind direction) and the indoor temperatures (θ_i) are measured. In the preprocessor these values are transformed into relevant parameters for the model; for instance the measured horizontal solar radiation is transferred into solar radiation upon a vertical or inclined surface. These preprocessed data are used to identify the parameters of the model relating the temperatures θ_i to the outdoor climate variables W and the control actions U (position of the blinds, etc.). The parameters are found by a recursive least square parameter estimation method (Iserman, 1977). The weather data W are also used to identify a weather model, which can predict the weather for the next 24 hours. A weather report can be added to improve prediction. This weather predictor W_H and the building model are used to determine the optimal control strategy U (setpoints for room controllers and/or the positions of the actuators.). With feedback control these desired values are realized in each room.

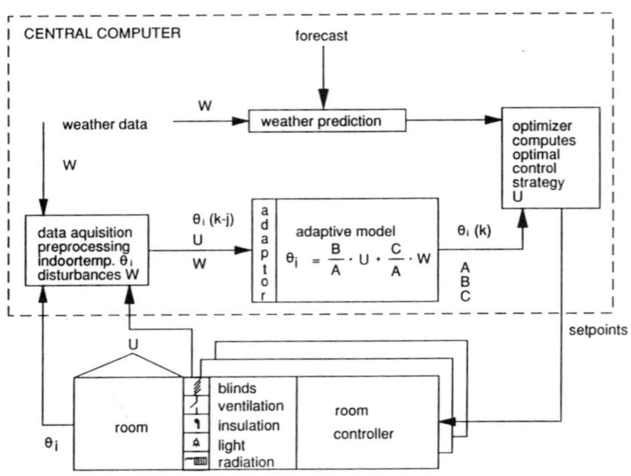

Figure 2. Structure of controllers.

5.2 SIMULATION TESTS.

The control systems have been tested with a simulator that contains a simplified model of the thermal behaviour of the building and its passive components [4]. A one room building is considered with well insulated walls and one window facing south. The calculations are carried out for window areas which are 40 and 80% of the front wall respectively.
Some general conclusions could be drawn:

- The predictive control systems of the Delft University of Technology and Centre Scientifique et Technique des Bâtiments show to be very effective for switching on the heating installation at the best moment in the morning. The number of hours with indoor temperatures lower than 20°C are much lower than the control system, which does not apply weather prediction (NAPAC).
- Although the simulation shows that the control systems are stable, a field test is felt to be necessary. Especially the model identification method, used in each control system may cause problems under real conditions.
- High thermal inertia is the first requirement for getting an acceptable indoor climate in summer.
- Only in colder climates comfort can be realized in summer. In figure 3 hystograms of the indoor temperature in summer are given for Uccle and Carpentras. It showns 69 hours with temperatures exceeding 26°C for Uccle and 607 for Carpentras.
- The heat losses through the windows outweight the advantages of the solar gains, so that night insulation of the windows is essential for passive solar systems. In paragraph 5.4 more details are given.

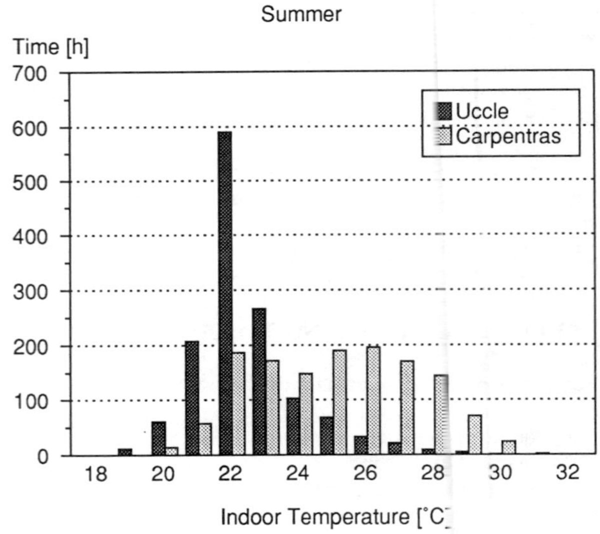

Figure 3. Temperature exceeding for Uccle (Belgium) and Carpentras (South of France). Heavy building and 80% window area.

5.3 COMPARISON WITH A SIMPLE CONTROL STRATEGY

The TU Delft design has been compared with the following simple and straight forward control strategy:

- Starting up and control of the heating installation are carried out in the same way as applied in conventional radiator heating systems;
- Natural ventilation: $T_{inside} > 22$°C, then:
  vent windows open if $T_{outside} > 15$°C
  vent windows closed if $T_{outside} > 24$°C
  vent windows halfway if 5°C $< T_{outside} <$ 15°C
  vent windows at minimum position if $T_{outside} <$ 5°C.
  (The maximum ventilation flow is 600 m^3/h in case mechanical ventilation is applied).
- The minimum ventilation (fresh air control) opening of a high positioned ventwindow is 5 degrees (in case of mechanical ventilation 30 m^3/h is delivered).
- The horizontal awning is put in maximum position if solar radiation on a vertical surface > 400 W and $T_{inside} > 22$°C.

The following conclusions could be drawn:

- For the summer period there is no need for an advanced control system. The rule: "Cool the building by ventilation as soon as the indoor temperature exceeds 22°C" gives results equivalent with the advanced control system.
- In the heating period the advanced control system saves 20% energy with respect to the simple control system.

5.4 INSULATING SHUTTER

As said before an insulating shutter is necessary. With the simulator its effect on the heating requirement is calculated. For example, for heavy buildings with a 80% window surface the insulating shutter gives a reduction in heating requirements of 50%.
Table 1 shows that when a shutter is applied preference should be given to large windows. That does not count when no shutter is applied.

Window surface	40%	80%
Without shutter	1780	2400
With shutter	1576	1200

Table 1. Heating requirement (kWh).

5.5 NATURAL AND MECHANICAL VENTILATION

In the simulator it is assumed that 600 m^3/h outside air can be supplied to the room by the mechanical ventilation system. The natural ventilation Φ is simulated by:

$$\Phi = f\,(\text{windspeed}, T_{out} - T_{in})$$

The coefficients of this equation are such that the maximum ventilation rate equals that of mechanical ventilation. It is emphasized that this rate can only be obtained when a window system with low and high positioned vent openings is applied.
The simulation results showed that natural ventilation can compete with mechanical ventilation. The number of exceedings (> 26°C) for the heavy building in Uccle with 80% window area were 75 and 69 hours respectively.

6. CONCLUSIONS

- Trombe walls are not cost effective.
  The same yields for rock bins.
  The competing direct solar gain overrules other possibilities.
- A minimal system configuration of a passive solar system consists of - a high inertia building with large windows facing south - outside shading device, insulating shutter and a ventilation facility for cooling.
- The optimal control systems designed in a CEC- project meet their requirements. For winter periods they save energy (20% reduction), while for the summer period they can be replaced by a simple and straightforward strategy.
- Only in colder climates temperatures exceeding some comfort threshold can be avoided, provided that the internal load does not exceed 100 W.
  Otherwise, mechanical cooling can not be avoided.
- Natural ventilation with high and low positioned vent windows can compete with mechanical ventilation.

LITERATURE

1. Sebold, A., Clinton, J.R.,
   Impact of controls in passive solar heating and cooling of buildings.
   Report Contractnr. DE AC04 -79AI10891
   US Department of Energy. 1980.
2. Inoue, T. et al.,
   The development of an optimal control system for window shading devices, based on investigations in office buildings.
   ASHRAE Transactions, 1988, Vol. 92, part 2 (OT-88-03- 2).
3. De Gids, W., and J.C. Pfaff.
   Ventilation rates and energy consumption due to open windows. Air infiltration review, nr. 1, November 1982.
4. Paassen, A.H.C. van,
   Digital control of passive solar systems.
   European Conference on Architecture,
   Munich, 6-10 April, 1987.
5. Paassen, A.H.C. van,
   Passive building control.
   CLIMA 2000, August 1989, Serajewo
6. Liem, S.H., Lute, P.J., Paassen, A.H.C. van, Verwaal, M.,
   End report of the concerted action on passive building control,
   Contract nr. EN3S-0036-NL, 1989.
7. Iserman, R.
   Digitale Regelsysteme,
   Springer Verlag, Heidelberg, 1977.

LE PLANCHER SOLAIRE DIRECT (PSD)

Une technique simple et éprouvée pour des bâtiments économiques et confortables

Présenté par :

Thierry CABIROL
Ingénieur Conseil
6, rue de la Mule Noire
13100 AIX EN PROVENCE

André JEAN
T2 I
Trévignin
73100 CHAMBERY

Alain LIEBARD
Comité d'Action pour le Solaire
8, rue de Richelieu
75001 PARIS

Abstract: "The direct solar roof"

Heating buildings using "Direct solar roof" has been asserted as the best performance of solar heating.
By connecting solar collectors and heating floor, without not any tanks and no intermediate heat exchanger, solar energy is directly stockpiled in the concrete body of the heating floor and is slowly restored into the room to heat.
This simple and reliable fitting up makes a basic heating sure with the first rate comfort provided by low temperature radiant heating.
It is highly effective for a fitting up cost rather low because it requires neither tanks nor heat exchanger.
Its very simple setting increases with small equippings that use prefabricated "transference units" gathering necessary component parts together for collectors and roof.
Out of heating season, solar collectors are usually required to produce hot sanitary water or to reheat a swimming pool.
Since 1981, several official sets of measures have confirmed the good results of this technique.

1 LE P.S.D.: UNE TECHNIQUE SIMPLE ET PERFORMANTE

Le plancher solaire direct désigne une technique simple et fiable de chauffage solaire des bâti - ments dont les bons résultats ont été mis en évidence par plusieurs campagnes de mesures officielles récentes (1981 à 1986). Elle consiste à faire circuler directement dans un plancher l'eau chauffée par les capteurs solaires sans utiliser de réservoir de stockage intermédiaire. La fonction de stockage de l'énergie solaire est ici assurée simplement par la masse de béton de la dalle chauffante.
Pendant la saison de chauffage (programme hiver), la chaleur captée par les capteurs solaires à eau est envoyée directement dans le plancher chauffant appelé "plancher solaire direct", par l'intermédiaire d'un ensemble de composants qui assure les fonctions classiques de tout système de chauffage (circulation, régulation, sécurité, etc.).
En dehors de la saison de chauffage(programme été) cette chaleur captée peut être utilisée pour réchauffer l'échangeur d'un réservoir d'eau chaude sanitaire ou d'une piscine (ou les deux dans certains cas).
Le changement de programme hiver-été se fait manuellement au gré de l'utilisateur et des fluctuations saisonnières.
Ce procédé découle de deux observations. Les systèmes classiques de chauffage solaire avec capteurs plans à eau et cuve de stockage sont souvent complexes et coûteux, en particulier du fait de la cuve de stockage. Par ailleurs,on sait que les rendements des capteurs plans sont meilleurs à basse température, doù l'emploi des planchers chauffants qui sont les "radiateurs" émettant aux températures les moins élevées (parfaitement confortables).
L'idée est donc venue de raccorder directement les capteurs solaires à un réseau de tubes noyés dans le béton d'un plancher suffisamment épais pour stocker et redistribuer lentement dans le temps la chaleur solaire captée.

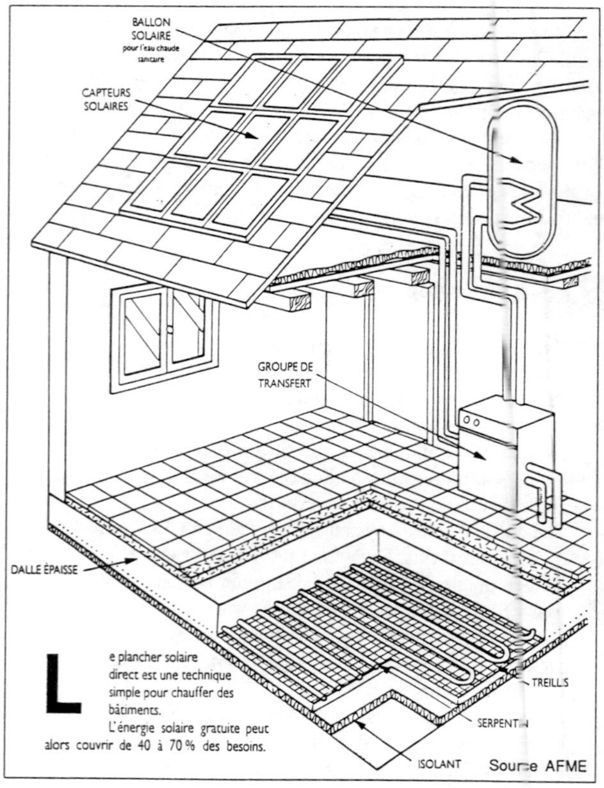

Source AFME

2 CONSTITUTION DU PLANCHER

Le plancher chauffant est constitué par un réseau d'épingles chauffantes réalisées en tubes noyés dans la dalle en béton et raccordés aux "clarinettes" entrée sortie (équipées de robinets, de tés d'équilibrage et de purgeurs automatiques) situées en entrée dans la dalle.
Le tube est soit métallique, soit, plus générale-

ment, en matière plastique(matériaux "spécial plancher chauffant").
On distingue les dalles de rez de chaussée, épaisses, qui émettent par leur seule face haute, et les dalles d'étage, d'épaisseur courante, qui émettent par leurs deux faces. Une même installation peut comporter ces deux types de dalles, moyennant précaution thermostatique particulière sur la cu les dalles d'étage.

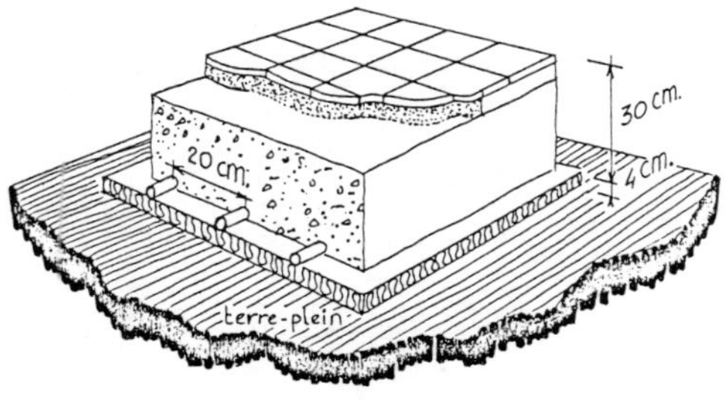

Dalle épaisse sur terre-plein (épaisseur 30 cm)

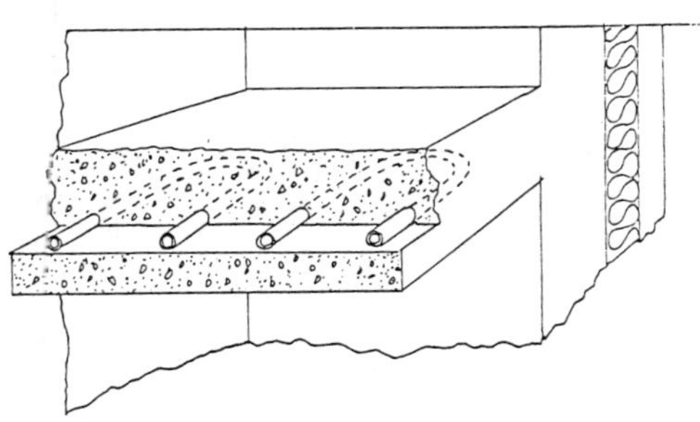

Dalle d'étage (épaisseur 20 cm)

3 PRINCIPALES CARACTERISTIQUES DU PSD

Installation simple et fiable
Bonne productivité des capteurs

Pour un habitat type Hlm	Individuel Isolé	Individuel Groupé	Collectif
Nbre logements	1	6	90
Si solaire couvre 50 % des besoins	350 à 600 kWh/m2.an		450 à 650 kWh/m2.an
	selon utilisation et région		

Nota: Pour les bâtiments qui présentent également une forte consommation d'eau chaude sanitaire en dehors de la saison de chauffage(hôtellerie, internat, hôpital, etc.), on peut atteindre dans les mênes conditions des productivités record qui atteignent 700 kWh/m2.an.

Faible surface de capteurs:
Par logement individuel isolé: 11 m2
Par logement individuel groupé: 10 m2
Par logement collectif: 8 m2.

Surcoût réduit(en F HT):
Habitat individuel isolé: 4 200/m2 = 46 000 F
Habitat individuel groupé: 3 200/m2 = 36 000 F
Habitat collectif: 2 900/m2 = 23 000 F

Autres avantages du PSD:
Le chauffage par le sol basse température apporte une garantie de confort et d'économie d'énergie.
Faible encombrement du fait de l'absence de réservoir de stockage pour le chauffage.
Les surfaces installées garantissent dans la majorité des cas une autonomie eau chaude sanitaire en été qui permet d'éteindre la chaudière d'appoint.

La présence de dalles pleines apporte une amélioration du confort: confort thermique en été (par la forte inertie des dalles solaires); confort acoustique entre logements (dalles d'étage épaisses de 20 cm).
Pour les immeubles de logements collectifs, le système permet une répartition simple des charges:
. Charges chauffage: Du fait que les apports solaires sont automatiquement répartis proportionnellement aux besoins (cette donnée est à prendre en compte par le bureau d'études pour le tracé des épingles du plancher chauffant); du fait que le chauffage d'appoint peut facilement être individuel (gaz ou électrique).
. Charges eau chaude sanitaire: Par la mise en place de compteurs d'eau chaude à l'entrée de chaque logement, ce qui permet de garder l'avantage d'une production collective de l'eau chaude.
Le coût de maintenance de ces installations est réduit du fait de la grande simplicité de ces installations. Nous recommandons toutefois de prendre un contrat d'exploitation; pour les installations moyennes et grandes, il y a intérêt à négocier le contrat en prenant en compte les économies d'énergie attendues (contrat à l'intéressement).

4 METHODE DE CALCUL ET CAMPAGNES DE MESURES

Cette technique de chauffage solaire a été imaginée dès 1977 au Laboratoire de Thermique de l'Ecole Supérieure d'Ingénieurs de Marseille (ESIM). Depuis, les réalisa sont succédées, permettant l'optimisation des schémas et des composants d'une part, et de la méthode de calcul d'autre part, qui fut publiée par l'ESIM en 1983.
Diverses campagnes de mesures financées ou cofinancées par l'AFME et certaines agences régionales de l'énergie ont été organisées:
. Ecole primaire de cinq classes à Eyguières (Bouches du Rhône) par l'Ecole des Mines d'Ales
. Lotissement de maisons solaires (Gironde) par le CETE de Bordeaux
. Maison individuelle à Istres (Bouches du Rhône) par l'ESIM
. Quatre maisons d'éclusiers (Bouches du Rhône) par le CETE d'Aix en Provence
. Immeuble collectif de 90 logements (Alpes Maritimes) par l'ESIM
. Quatorze maisons individuelles en Rhône-Alpes et Savoie par l'ASDER.
Ces campagnes de mesures ont confirmé les bons résultats de la technique et la justesse des prévisions de la méthode de calcul. Elles ont aidé à perfectionner les circuits en évoluant vers la simplicité.
Dans son dernier rapport sur cette technique, le CSTB conclut:"Il apparaît que ce système conduit à des performances supérieures à celles des autres systèmes solaires actifs de chauffage (systèmes avec stockage à eau, ou système à air)". En outre, les performances obtenues en climat froid de montagne ensoleillée sont exceptionnelles.

5 QUELQUES EXEMPLES

5.1 Internat 200 lits du LEP de CAVAILLON (84)

1 800 m2 de PSD en 3 niveaux
150 m2 de capteurs produisent 700 kWh/m2.an avec un taux de couverture global de 40 % pour chauffage et eau chaude sanitaire.
Investissement spécifique: 4,1 F HT/kWh économisé
Ces bons résultats sont obtenus du fait d'une utilisation de l'internat hors période scolaire.

5.2 Ecole Maternelle à TOULON (83)

Les cinq classes de l'école sont traitées en chauffage solaire "passif" par l'intermédiaire de grands vitrages sud est et sud. Les 500 m2 restant sont traités en plancher solaire direct (un seul niveau). Les apports solaires passifs et actifs couvrent 50 % des besoins pour une productivité des capteurs de plus de 500 kWh/m2.an. Ce résultat est d'autant meilleur que l'école est inoccupée pendant les vacances scolaires et que les besoins d'eau chaude de la cantine sont relativement modestes comparés à ceux d'un logement ou à fortiori, à ceux d'un internat (ou d'un hôtel).

5.3 Maison individuelle en SAVOIE

11 m2 de capteurs
Il faut compter de 35 000 F à 45 000 F TTC pour équiper une maison individuelle (capteurs, tuyaux, module de transfert et de régulation, fluide ...). De cette somme, il faut déduire le prix de la toiture(rempla cée par le capteur). En appoint, un poêle ou une cuisinière à bois, une cheminée à récupérateur, des petits convecteurs électriques ou autre système classique.
la consommation ne dépasse pas 3 000 F par an pour le chauffage et l'eau chaude.

5.4 90 logements collectifs HLM et Foyer à MANDELIEU (06)

6 000 m2 de plancher solaire direct. Surcoût moyen 16 000 F HT par logement (valeur Avril 1982). Productivité des capteurs 460 kWh/m2.an; elle augmenterait sensiblement dans des régions à hiver plus rigoureux.

5.5 Immeuble collectif à CHAMBERY (73)

Au centre de Chambéry, un immeuble de seize appartements en accession à la propriété, le Laetitia, est équipé d'un "PSD" et de 70 m2 de capteurs. Les performances devraient permettre une dépense de moins de 2 000 F par an et par appartement.

CONCLUSION

Actuellement, 2 500 m2 de capteurs alimentent près de 20 000 m2 de plancher: logements individuels et collectifs, centres sociaux et culturels, bureaux, gymnase, gîte de montagne.
La technique de chauffage, du type base solaire + appoint est aujourd'hui bien maîtrisée par les différents maillons de la chaîne: fabricants industriels, bureaux d'études et installateurs.
La technique du PSD est facilement accessible aux maîtres d'ouvrage publics et privés et aux architectes, soit pour de petites installations par l'emploi d'ensembles de transfert prémontés en usine, soit par l'emploi de schémas bien maîtrisés par de nombreux bureaux d'étude.
Dix ans de pratique ont conforté les qualités thermiques et économiques du PSD, une technique qui allie la technologie solaire au service de l'architecture.

STRATEGIES THERMIQUES ET USAGES DE L'HABITAT : LA VERANDA

HAUMONT N.
Centre de Recherche sur l'Habitat
Ecole d'Architecture de Paris La Défense
41, Allée Le Corbusier
92023 Nanterre Cedex

In France, the veranda(h) has proliferated in private homes since 1980. It is the product of a new technology both a space to live in and a way of saving energy. Although people are very much concerned with energy saving measures, they consider that the proper part of a veranda(h) is to improve the comfort of their homes, to protect them against the elements and to provide a larger surface. The veranda(h) also allows them to compensate for certain drawbacks in the design of their houses - a small kitchen, the absence of hall - and to enjoy a new way of living, more relaxed, more convivial. Both multifunctional and multivalent, the veranda(h) is a private space in which groups of individuals, different in social position, age and sex, can live according to their specific needs. It can in no way be considered as a mere cabin for saving energy and whose only function would be to let more sunshine into the house.

Depuis le milieu des années 70, on peut observer en France un nouvel intérêt pour les espaces vitrés dans l'habitat : bows-windows et serres se multiplient dans les immeubles collectifs neufs ou réhabilités. A la suite de la crise énergétique, en effet, des recherches qui portaient sur les techniques d'isolation de l'habitat et sur la rentabilisation de l'énergie solaire ont mis en évidence le rôle que pouvaient jouer les espaces vitrés dans le confort thermique du logement et dans les économies d'énergie, à condition que soient respectées certaines règles techniques de contruction : orientation, forme, technologie des matériaux. Cette logique technique ne tient pas toujours compte du fait que les serres font partie du logement et de son système pratico-symbolique. Ainsi des observations ont montré que très souvent les habitants annulent ou réduisent largement les effets escomptés par les spécialistes en matière de gain énergétique, en occultant par exemple les serres lorsqu'elles sont visibles de la rue.
Il existe cependant une demande des habitants, comme en témoigne la prolifération depuis 1980 des vérandas installées par les propriétaires de maisons individuelles. On peut bien sûr se demander s'il ne s'agit pas de simples "gadgets", produits d'une nouvelle mode lancée par les constructeurs de profilés d'aluminium en mal de nouveaux débouchés pour leurs produits. En fait, lorsque les propriétaires de vérandas évoquent les raisons qui les ont guidés dans leur décision et dans leur choix, ils dévoilent une logique d'intervention sur l'espace qui montre bien qu'avant d'être les "simples" clients d'un nouveau produit, ils sont avant tout des habitants.
Afin de mieux comprendre cette "logique de l'habiter", une recherche /1/ dont nous rendons compte ici a été effectuée, dans différentes régions, auprès d'habitants de maisons individuelles qui ont fait installer une véranda sur leur maison, ancienne ou récente, et qui ont donc choisi à la fois son emplacement, sa taille, sa forme et ses matériaux.

1. CONFORT THERMIQUE ET ECONOMIE D'ENERGIE

La grande majorité des habitants considèrent que la véranda joue un rôle toujours positif sur le confort thermique, et qu'elle permet de faire des économies d'énergie quels que soient son emplacement, sa forme ou sa technologie. Toutes ces caractéristiques n'auraient d'effet que sur sa plus ou moins grande efficacité ("elle chauffe plus au sud qu'au nord", "elle isole mieux si elle est bien étanche", etc.) et permettraient d'améliorer une rentabilité énergétique qui serait en fait une propriété intrinsèque de la véranda. L'idée générale est que "de toute façon, avec la véranda, c'est toujours mieux que s'il n'y avait rien".
Cette croyance unanime dans l'efficacité de la véranda repose sur la perception de son double rôle : elle produit de la chaleur et elle protège le logement. Lorsqu'il y a du soleil, la véranda "chauffe", lorsqu'il n'y en a pas, elle protège. Mais si ces deux effets bénéfiques de la véranda sont toujours associés, ils n'ont pas forcément la même valeur. Les habitants savent et constatent que la température s'élève rapidement dans la véranda lorsqu'elle est exposée au soleil, même si la température extérieure reste fraîche ; l'apport de chaleur ainsi réalisée est, comme le dit une habitante, "un cadeau du ciel", précieux mais aléatoire, et dont on ne peut vraiment profiter que si on est là pour s'y tenir ou/et pour ouvrir la communication avec l'intérieur du logement, ce qui permet alors de baisser le chauffage. Par contre, la véranda protège en permanence la maison du froid, du vent, de la pluie. Elle isole la façade ("la véranda c'est une double paroi avec un vide d'air..") et les accès extérieurs. De ce fait, elle est même pour certains plus "efficace" au nord qu'au sud.
C'est à partir de cette analyse que les habitants rationalisent ensuite du point de vue énergétique et économique l'usage qu'ils font de la véranda et, en particulier, qu'ils la chauffent pour pouvoir s'y tenir à leur gré. Pour les habitants en effet, la possibilité d'utiliser la véranda est ce qui jusfifie en grande partie leur investissement : une véranda où on ne peut pas se tenir pendant une période assez longue est "inutile", quelle que soit par ailleurs sa rentabilité énergétique. Dans le nord de la France, les habitants n'hésitent pas à chauffer la véranda, soit temporairement pour l'utiliser en certaines occasions, soit en permanence à un niveau suffisant pour maintenir les plantes "hors gel", ce qui justifie l'utilité de la véranda en hiver. Certains reconnaissent alors avoir consommé plus à cause du chauffage tout en estimant en avoir retiré un bénéfice d'usage justifiant largement la dépense mais la plupart considèrent qu'ils n'ont dépensé en chauffage que ce qu'ils avaient gagné par ailleurs, grâce à l'apport de chaleur et à la protection fournie par la véranda. Tout se résout finalement dans un système de compensation, dans lequel ce que l'on gagne d'un côté compense ce que l'on dépense de l'autre. D'ailleurs le problème pour l'habitant n'est pas tant de dépenser que de dépenser en pure perte, et ce problème leur semble résolu si on ne chauffe que lorsque c'est nécessaire, et surtout uniquement ce qui est

nécessaire ("on ne doit pas chauffer l'extérieur", c'est-à-dire les courants d'air). Ce qui explique que si tous les habitants ont noté que la chaleur accumulée dans la véranda au moindre rayon de soleil disparaissait très rapidement bien que celle-ci soit fermée, ce n'est pas tant ce mode d'évasion de la chaleur, dû au vitrage, qui retient leur attention lorsqu'ils chauffent eux-mêmes la véranda, que celui qui serait dû à sa mauvaise étanchéité à l'air (à cause de joints défectueux par exemple). L'étanchéité à l'air est d'ailleurs presque toujours considérée comme une qualité spécifique de la véranda moderne, par opposition à la véranda ancienne en fer "qui fuit de partout" et qui "ne permet pas vraiment d'économiser le chauffage". La véranda récente, bien calfeutrée, peut être chauffée "sans qu'on chauffe en même temps dehors" et son efficacité énergétique compenserait les dépenses dues à son utilisation. La déperdition de chaleur liée au vitrage est très rarement prise en compte par les habitants. La nuit, ils omettent le plus souvent de fermer les volets si tant est qu'ils ne les ont pas enlevés lorsqu'ils ont fait installer la véranda.
Interrogés sur le montant des économies réalisées grâce à la véranda, les habitants les plus enthousiastes quant à ses propriétés thermiques restent souvent assez évasifs, ou annoncent des chiffres qui dépassent très largement les hypothèses les plus optimistes des thermiciens ("on n'a pas fait les calculs, mais je suis certain d'avoir fait des économies ..." ou "cette année j'ai consommé moitié moins grâce à la véranda.."). En fait le bilan n'est pour eux jamais nul, ni même négatif, puisque le bénéfice d'usage, c'est-à-dire l'agrément, est comme le dit l'un d'eux "incalculable". Le véritable "rapport" de la véranda ne se lit pas sur les facteurs de chauffage, mais dans la vie quotidienne de l'habitant.

2. UN ESPACE HABITE

La véranda constitue un mode d'agrandissement du logement qui n'a rien à voir avec les systèmes d'extension traditionnels (aménagement des combles en particulier). La facilité et la rapidité du montage qui résultent de leur production industrielle font de l'installation de la véranda une "opération éclair" qui n'entraine pas des travaux salissants touchant au gros oeuvre et perturbant la vie quotidienne pendant plusieurs jours ou plusieurs semaines. Il ne s'agit pas d'ailleurs de créer une pièce supplémentaire liée à des pratiques spécifiques mais de constituer un espace polyvalent et plurifonctionnel, qui servira de cadre à de nouvelles pratiques attestant de l'évolution d'éléments du mode de vie dans les domaines de la sociabilité, de la vie familiale et de la relation à la nature et au loisir.
La véranda est le lieu d'un nouvel art de vivre dans lequel les valeurs du loisir et de la détente comme activités privées prennent de plus en plus d'importance. Plus de décontraction dans la vie familiale, plus de convivialité sans contrainte entrent bien dans une évolution de la société vers des normes sociales plus permissives et vers une place plus grande donnée à l'initiative individuelle. Ainsi la véranda est un lieu pour les réunions "décontractées" entre amis ou en famille, où les contraintes du salon disparaissent au profit d'une convivialité moins cérémonieuse. On peut se permettre les verres en carton et les serviettes en papier qui, tout en évoquant l'ambiance des vacances et des piques-niques de la belle saison, allègent les tâches ménagères. Ce qui ne veut pas dire que le modèle d'une sociabilité hiérarchisée et ritualisée a fondamentalement changé et qu'il n'existe plus à l'intérieur du logement d'espace à statut public (propre et rangé) réservé pour la réception. Au contraire, pourrait-on dire, grâce à la véranda, le salon ou le séjour sont à l'abri des risques de dégradation résultant de l'usage quotidien (repas, jeux des enfants, bricolage ou travaux de couture) ou des activités de détente et peuvent pleinement jouer leur rôle d'espace de représentation lors des réceptions plus "officielles" pour lesquelles ils constituent toujours le lieu privilégié.
Espace de liberté, sans contrainte, la véranda n'appartient à personne en particulier et chacun, dans la famille, peut se l'approprier à son heure et apporter avec lui les objets ou petits meubles nécessaires à son activité.
La décontraction vestimentaire qu'elle autorise traduit sans doute aussi, dans le même ordre d'idée, les nouvelles valeurs que l'on accorde au corps, lui même objet d'une liberté nouvelle bien qu'encore soigneusement contrôlée.
La véranda permet également aux habitants d'établir une nouvelle relation à la nature. Sa polyvalence et son statut d'espace habité s'opposent à ce qu'elle puisse constituer autre chose qu'une version très approximative de la serre horticole et les habitants qui s'en servent "comme si" c'était une serre ne manquent pas de signaler par ailleurs que ce qu'ils ont voulu c'est bien une véranda et non une serre. Ce n'est donc pas du côté de cet usage que l'on peut définir le rôle spécifique de la véranda comme nouveau mode de relation à la nature.
Par contre tous les habitants s'accordent dans leurs descriptions du plaisir que l'on peut prendre à tromper l'adversaire en déjouant, au fur et à mesure qu'ils se présentent, les pièges tendus par une nature perçue comme particulièrement capricieuse sous nos climats tempérés. A l'abri de la véranda, on défie le froid, le vent, la pluie et le plus terrifiant des orages n'est plus qu'un "son et lumière" à domicile. Cette dimension spectaculaire de la nature n'est pas cependant une nouveauté et on n'a pas attendu la véranda pour désirer avoir "la vue". Ce qui est nouveau, c'est que cette "vue" est maîtrisée par l'intervention de l'habitant qui devient à la fois metteur en scène et acteur du spectable de la nature qu'il peut alors s'offrir, à l'intérieur comme à l'extérieur de la véranda. En hiver, on y fait pousser les fleurs de l'été ou les plantes exotiques, au printemps on regarde fleurir un jardin réaménagé à cet effet, à l'abri des fraîcheurs de la saison.
La maîtrise de la nature permet de vivre comme on en a envie quelques que soient la saison ou les intempéries : manger dehors sous la pluie, avoir des fleurs en permanence, etc.. La nature ainsi "policée" perd sa rugosité... naturelle : plus d'insectes, d'oiseaux malpropres, ou mêmes bruyants, d'air trop vif, de sols spongieux ou poussiéreux. Avec la véranda, les habitants contrôlent la nature et savourent le plaisir de retrouver, dans cette mise sous tutelle, le goût de la liberté des vacances, sans les inconvénients de la vie au grand air.

3. UN ESPACE PRIVE

Pour les thermiciens et les architectes, l'emplacement de la serre est essentiellement conditionné par l'orientation qui permet de gagner le maxium de calories. Pour l'habitant qui fait installer une véranda, l'emplacement dépend en premier lieu des contraintes et des potentialités de l'habitat existant mais surtout du statut des pièces qui s'ouvrent sur la véranda et de la volonté de privatiser le logement, véranda comprise.
La véranda est le plus souvent installée sur un accès extérieur préexistant de la maison. Si on ajoute à cela la condition évidente de disposer d'un espace assez grand devant l'ouverture pour pouvoir y intaller la véranda, on aura défini les conditions nécessaires à la réalisation de l'opération sans pour autant que celles-ci soient suffisantes.
L'absence de pièce d'entrée dans le logement conditionnera presque systématiquement l'emplacement de

la véranda sur le devant de la maison. Ceci s'explique par le fait que la sociabilité urbaine, dont le modèle s'est largement diffusé sur l'ensemble du territoire, implique une privatisation importante du logement qui se traduit en particulier par l'existence d'une pièce d'entrée. Or dans un certain nombre de maisons anciennes, encore organisées sur un modèle issu d'une société rurale, l'accueil se fait directement dans la cuisine même si celle-ci n'est plus la salle commune et s'il existe par ailleurs des pièces pour la réception comme le salon ou la salle à manger. Dans ce cas, la véranda est installée devant la cuisine pour reconstituer une pièce d'entrée et remédier à ce qui est aujourd'hui considéré comme un "défaut" du logement.
Cependant, sauf dans le cas d'un recul suffisant par rapport à la rue, l'emplacement de la véranda surle devant de la maison s'oppose à ce que les habitants l'utilisent comme pièce à vivre, à moins qu'ils ne la cachent à l'aide de plantation ou de rideaux. En effet, bien que la véranda soit par définition transparente, elle ne constitue pas un mode d'ouverture du logement sur l'extérieur où la vie privée de la famille serait exposée aux yeux de tous. Bien au contraire, les habitants décrivent toujours son installation en terme de "fermeture" du logement ou de "fermeture de la terrasse" et ceci même s'ils désirent instaurer à travers la véranda un nouveau mode de relation à la nature. L'installation de la véranda, quels que soient les objectifs des habitants, entre toujours dans une stratégie de privatisation de l'habitat qui se traduit d'une façon générale par la multiplication des "sas" (porche, jardin de devant, entrée) caractéristique de l'organisation de l'espace de l'habitat dans notre société.
Il n'en reste pas moins que pour certains habitants, relativement peu nombreux, la véranda visible de la rue participe de l'espace de représentation constitué par la façade et le jardin de devant. Les contraintes qu'elle impose aux pratiques quotidiennes sont sans doute d'autant mieux acceptées qu'elle représente pour l'habitant un élément du statut social.
Pour la grande majorité des habitants, cependant, l'emplacement idéal de la véranda est l'arrière de la maison, en relation et de plain pied avec la cuisine ou/et le séjour si celui-ci a déjà un accès sur l'extérieur. Si le jardin derrière la maison n'est pas protégé du regard des passants ou des voisins, on retrouve les mêmes pratiques d'occultation que lorsque la véranda est placée devant la maison. Certains habitants dont la maison ne possède pas d'entrée installeront deux vérandas : une sur l'avant et une sur l'arrière. (cf plan)
L'orientation nord/sud si importante pour les concepteurs de l'habitat solaire ou des logements à serre intégrée, ne joue pas un rôle déterminant pour les habitants lorsqu'il s'agit de définir la place de la véranda. Nous avons vu que les habitants sont plus sensibles au rôle de protection thermique du logement par la véranda qu'à son rôle de capteur solaire. De ce point de vue, toutes les orientations ont leurs avantages et leurs inconvénients. On peut dire cependant qu'ils préfèrent la situer au soleil sans pour autant que ce soit au sud, compte tenu des inconvénients liés à la surchauffe. Le sud-est ou sud-ouest semble l'orientation la plus appréciée.

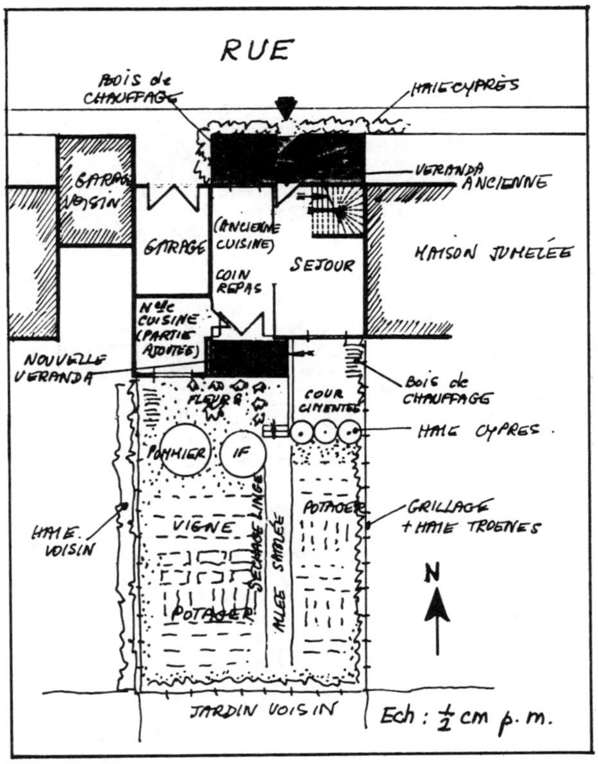

4. UN ESPACE DE REPRESENTATION

Si la véranda n'est pas souvent conçue pour être un espace montré aux passants, elle peut devenir un espace de représentation dans la mesure où elle est utilisée pour recevoir des amis. Elle se distingue cependant des pièces traditionnelles de représentation comme le salon, par une plus grande simplicité. Elle pourra cependant être aménagée avec soin en particulier par certains représentants de la classe moyenne qui à la fois utilisent la véranda pour les réceptions "décontractées" mais aussi qui investissent dans les plantes exotiques et le mobilier en rotin proposé par les revues de décoration.
Toutefois il s'agit moins d'une représentation de signes extérieurs de richesse que de l'accession de ces habitants à un nouveau mode de vie pour lequel la distance marquée vis-à-vis des signes traditionnels de la prospérité (les beaux meubles, les objets de valeur, etc..) devient une nouvelle stratégie de la différenciation. Pouvoir recevoir de façon "décontractée", au milieu des meubles en rotin et des plantes vertes, c'est aussi montrer qu'on a la possibilité de ne pas se référer aux signes les plus évidents de la richesse pour adhérer aux valeurs du loisir, des vacances, de la détente et du bien-être chez soi... à condition que l'on reste "entre soi", c'est-à-dire au sein d'un groupe affinitaire.
L'aménagement du jardin, toujours effectué devant la véranda, même si elle est située sur l'arrière de la maison, peut avoir aussi un caractère ostentatoire. Toutefois cet aménagement n'a pas le même sens que celui du jardin devant la maison. Celui-ci est un espace à caractère public où les habitants ne se tiennent pas mais dont le spectacle, offert aux passants et aux voisins, témoigne du statut de la famille. Il n'en est pas de même pour le jardin situé devant la véranda qui n'est aménagé que pour le plaisir des habitants et de leurs invités et garde un statut privé. D'ailleurs lorsque la véranda est située sur le devant de la maison, les habitants l'aménagent toujours de façon à privatiser par rap-

port à la rue, la partie du jardin proche de la véranda.

CONCLUSION

Le statut très particulier de la véranda dans l'habitat fait qu'elle accueille les stratégies de groupes assez différents tant du point de vue socio-économique que du point de vue de l'âge et du sexe. Les catégories aisées y trouvent de nouvelles possibilités d'affirmer leur statut social en y inscrivant les signes esthétiques de la distinction moderne (mobilier en rotin et plantes exotiques) et en mettant en oeuvre les pratiques qui correspondent à l'évolution de leur mode de vie comme la réception "décontractée". Les catégories plus modestes, souvent propriétaires des logements les plus pauvres ou les plus anciens, y ont recours pour mettre leur maison "à niveau" (création d'une entrée ou / et d'un sas entre cuisine et jardin lié aux activités de jardinage et à l'existence d'un potager).
La polyvalence de la véranda permet à chaque membre de la famille de se l'approprier à son heure, mais aussi au couple et à la famille de se réunir. Les personnes âgées y vivent au soleil, à l'abri des courants d'air et les enfants y jouent en liberté surveillée.
La véranda semble bien éloignée d'un sas dont les propriétés seraient surtout d'ordre énergétique et économique et qui ne trouverait sa place qu'en fonction de l'ensoleillement. Produit de l'innovation technologique et des idées modernes sur la gestion de l'énergie, espace qualifié par la pratique sociale, la véranda peut être considérée comme un nouvel élément typologique de la maison individuelle.

/1/ N. HAUMONT, B. DUSSART, Stratégies thermiques et usages de l'habitat : les vérandas. CRH, Paris, 1987 193 p.
N. HAUMONT, B. DUSSART et al., Concevoir et habiter. L'espace de la véranda. Plan Construction et Architecture. Batiment Econome. Paris, 1988. 144 p.

The CEC Project PASSYS : Achievments in the Period 1986 - 1989

R. Gicquel, C. Cools
Ecole des Mines de Paris
Centre d'Energétique
Sophia-Antipolis
F - 06560 Valbonne

ABSTRACT. The PASSYS project is a European concerted action in the field of Passive Solar Architecture which has been initiated by the Commission of the European Communities as part of its Solar Energy R&D programme and will continue in the forthcoming Joule programme. The CEC's financial contribution is 50% of the overall budget. The PASSYS group is composed of several national consortiums including Belgium, Denmark, France, Germany, Greece, Italy, the Netherlands, Portugal, Spain and the United Kingdom. The project started in spring 1986 and the first phase ends in December 1989. The paper will present the main achievments obtained in the three first years of operation : erection of the PASSYS test sites, validation of the simulation code ESP, principles of a european passive solar simplified calculation method, progress made in the field of test methodology.

1. Introduction

The PASSYS project is a European concerted action in the field of Passive Solar Architecture which has been initiated by the Commission of the European Communities (CEC) as part of its Solar Energy R&D programme. The project started in spring 1986 and the first phase ends in December 1989.The CEC's financial contribution for this phase is of the order of 4 million ECUs which is 50% of the overall budget. For the second phase of the project which will end in 1991, the CEC's contribution will be of about 2,5 million ECUs.

The PASSYS group is composed of several national consortia including Belgium (CSTC-WTCB and Free Univ. Brussels), Denmark (TIL of the Techn.Univ. of Denmark), France (CSTB and Centre d'Energétique of Ecole des Mines de Paris and CEA-CEN Cadarache), Germany (ITW of Univ. of Stuttgart), Greece (Protechna Ltd), Italy (Conphoebus and ENEL), the Netherlands (TNO-TPD Delft and Eindhoven), Portugal (Dept Engenharia Mecanica of Univ. of Porto and LNETI/DER), Spain (IER) and the United Kingdom (ESU, ESRU and ABACUS of Univ.of Strathclyde and BRE and Pilkington R&D Labs and ETSU), and of the CEC's Joint Research Centre of Ispra (I) (see Map below).

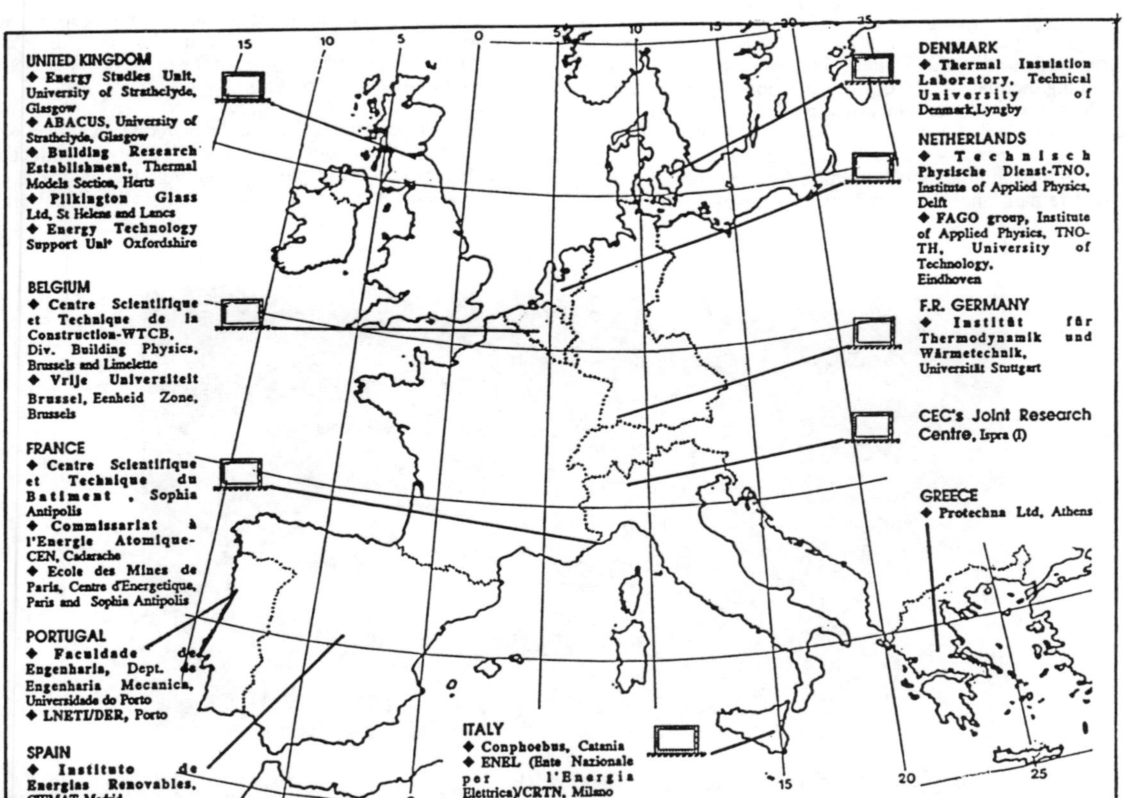

2. Why PASSYS ?

It is generally admitted nowadays that the potential of passive solar systems is high. For example, in both the residential and industrial building sector, current energy consumption in Europe is running at between 30 to 40% of the total primary energy. However, one prerequisite for the more widespread use of passive solar techniques is that designers be able to assess readily the contribution that different passive solar components (PSCs) will make to fuel costs and comfort levels. This requires, in turn, the development of design tools and component characterization techniques from which the parameters of these design tools can be identified. One way in which a PSC can be characterised, is to determine the value of some parameters such as the solar transmission factor STF(the ratio of the component's transmitted solar energy to that which is incident), the heat loss coefficient U_C (W/m^{2}K) and the air permeability P_C (m^3/h), see Figure 1. Unfortunately the determination of these parameters is a non-trivial task. This is in marked contrast to active solar systems which are subject to control action and whose experimental boundary conditions can be strictly maintained. As a matter of fact the complexity of passive solar systems mainly stems from two operational characteristics : they are strongly coupled and the determination of some thermophysical parameters (e.g : convective heat transfer coefficients) is difficult.

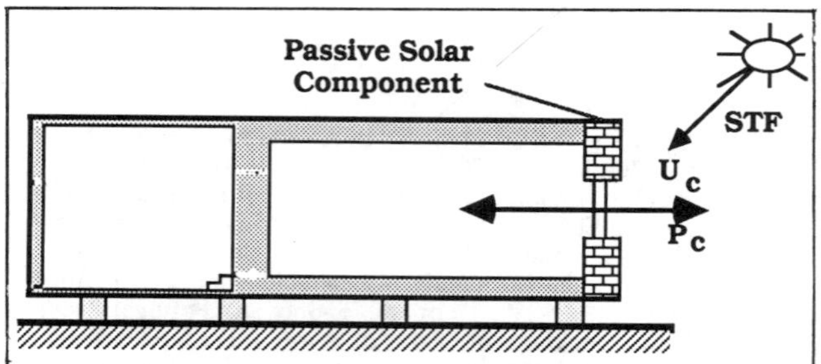

Figure 1 : *Some characteristic parameters of a passive solar component.*

These two points lead to the first general objective of PASSYS : to develop reliable and affordable test procedures for PSC's in a building system.

To overcome these difficulties, there already exists a powerful tool : building thermal simulation. The technique is appropriate for all PSCs, provided sufficient data on their characteristics can be obtained.

While the approach holds much promise for the future, some doubt will remain about the predictive accuracy until such systems have been subjected to rigorous validation. Such doubts stem from many sources, including inadequate representations of some heat transfer processes (for example air flow), through software bugs, ambiguities in the meaning of input data and users themselves who can have a significant influence on the model's predictions.

Therefore the second general objective of PASSYS is to increase confidence in passive solar simulation through validation and further refinement.

Even if the predictive accuracy of the large simulation codes is proven, several problems remain to be overcome. For example, due to their generality and the numerous appraisal possibilities they offer, their use is often time and money intensive. They also require specialist hardware (although recent cost reductions may help to accelerate their rate of uptake). Moreover the CEC's SOLINFO and ARCHISOL projects have shown that, for a number of applications and, in particular, for the design of individual houses, there remains a demand from designers for validated simplified design tools (SDTs) to enable a rapid assessment of the energy and comfort performance of PSCs.

While numerous SDTs exist in relation to building regulation compliance, there remains a need for a better design orientated SDT.

This need gives rise to the third general objective of PASSYS : to increase confidence in passive solar simplified design tools through the activity of development and validation.

The following sections present the structure and activities of PASSYS as devised to address these objectives.

3. The organization of the PASSYS project

3.1. Working structure

During this first phase the PASSYS project consisted of a network of about sixty researchers with the research activity organised into five sub-groups addressing test methodology, simulation model validation and development, simplified design tool development, instrumentation and PSC definition.

The instrumentation sub-group is concerned with such matters as type and position of sensors, while the PSC sub-group is concerned with the identification of the range of components to be examined at the test cell sites.

The sub-groups met regularly and adhered to an agreed work-plan. Issues which concern the PASSYS groups collectively were addressed during plenary meetings held twice a year. Such meetings were attended by, at least, all national project leaders.

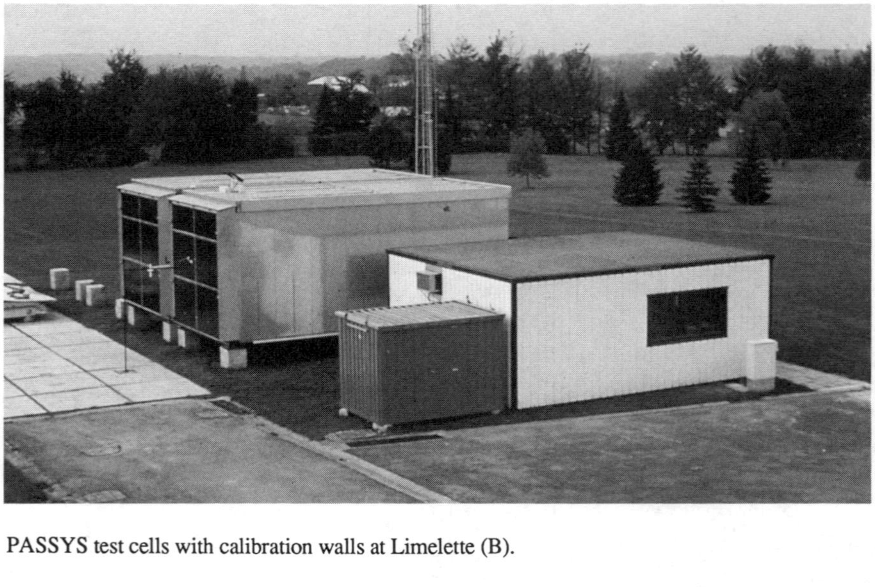

PASSYS test cells with calibration walls at Limelette (B).

Overall, PASSYS was coordinated by a committee comprising Theo Steemers, the CEC Project Manager, Renaud Gicquel of the Ecole des Mines de Paris and several CEC experts and representatives.

For the second phase of the project, the working structure will globally be the same, the coordinator being Peter Wouters from CSTC/BBRI (B).

3.2. The PASSYS Test Cell

3.2.1. Rationale for the Test Cell

It has previously been said that experiments aiming at the determination of PSCs parameters require, at least, the precise knowledge of the experimental boundary conditions. There has already been work on passive solar system testing at several centres throughout Europe and North America.

The results obtained from this work have served to demonstrate that it is very difficult to draw general conclusions from real buildings, with their inherent complexities and occupancy interactions. It would appear preferable to work with unoccupied, simple test rooms or cells. It is also apparent that any inter-comparison of previous results is complicated by the fact that the system configurations and analysis methodologies were different.

These factors resulted in a decision to operate, within PASSYS, with highly standardised test cells to provide a realistic but readily controllable/configurable test environment. These test cells would be duplicated at a number of European sites so that their performances could be analysed in a range of climatic conditions.

3.2.2. Design of the Test Cell

The cell's structure is based on a "hard casing" concept, where the supports and the insulation are separated. The cell is prefabricated with a rugged steel frame shelved with mineral wool. This enables the transportation of a complete test cell while minimising the risk of mechanical distortion. Figure 2 gives some constructional details. The outer surfaces are covered with chipboard and, finally, shielded with stainless steel. The inner surfaces are insulated by styrofoam to give a constructional U-value of < 0.1 W/m^2K.

Internally the cell has two zones, a service room and a test room, divided by a well insulated partition containing a sealed connecting door. The possibility exists for attaching different south wall components contained within a galvanised steel frame which bolts to the cell. The maximum thickness of this component is 380 mm although larger dimensions can be accomodated by the use of special steel clips which attach to the frame. These clips, in turn, contain attachment points for a sun space-type component.

The cells are manufactured in Germany and then shipped to the various test sites.

During the second phase of the project, new cells allowing for the testing of roof components will be delivered to Greece, Portugal and Spain.

3.3. Test Cell equipment

The cells are delivered with an opaque calibration wall and reference wall. The former is used for cell calibration purposes. It can also be used as a base reference component in comparative experiments. Initial experiments have indicated that the loss coefficient of the cell with the calibration wall is of the order of 11 W/K. The reference component is a sandwich of extruded polystyrene between two layers of concrete. It also incorporates a window with 1.2 m^2 of double glazing with the possibility of including a blind.

As the first example of another component, a timber-frame wall with an outer brick leaf has been constructed in the UK. This features an adaptable aperture which can accept windows of varying widths (from 1.2 m to 2.4 m) or a patio door. It is also possible to attach a sun space. It is planned to test Trombe walls at some sites.

A common set of sensors is placed in each cell and each site operates an identical data acquisition system (DAS), Hewlett-Packard A600 real time computer and an HP3852 acquisition box. This is a multi-tasking, multi-user system with a 40 MByte hard disk and tape facility.

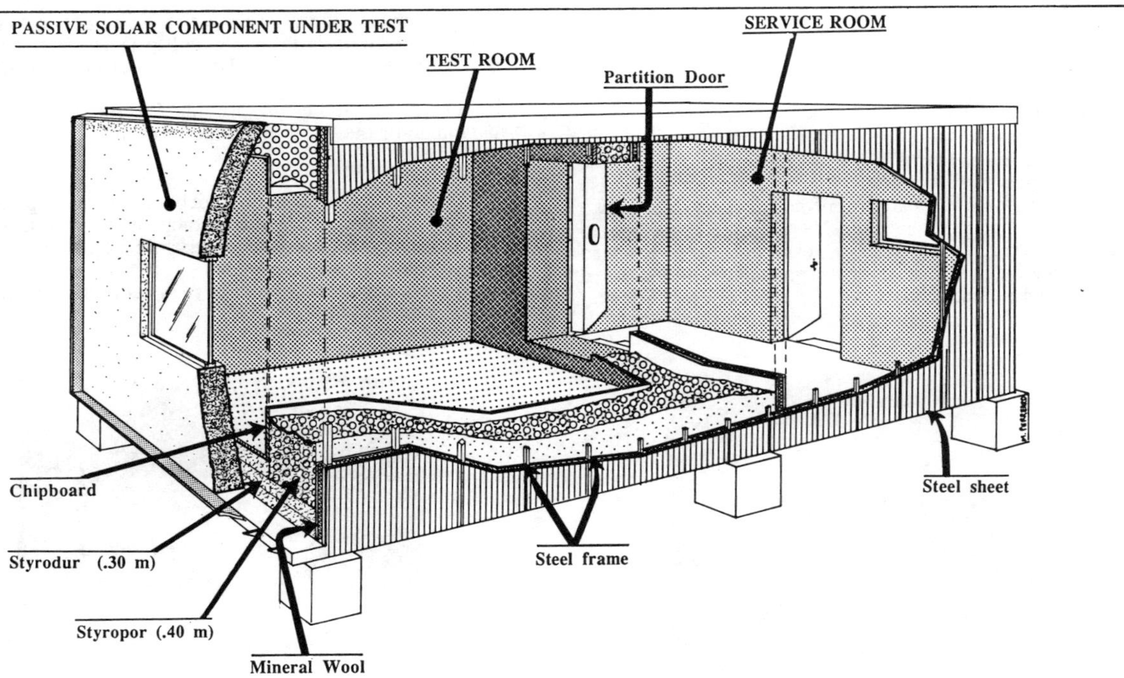

Figure 2 : *Sketch of the PASSYS test cell.*

The DAS software, which was developped by Ecole des Mines de Paris, allows the monitoring of up to 150 channels at a sample period of 30 seconds or greater. Each of ten possible users can define different target channel-sets and their recording frequency. The software then manages the time-sharing of the common devices, ensuring that there is no disturbance between users.

For some sites and/or experiments a means of maximum cell temperature control is required and to this end the cells are equipped with a heat removal system. It is a water-based, frost-protected refrigeration unit providing a high internal air flow rate via a fan-cooled heat exchanger. A sophisticated air distribution system is also incorporated in order to not disturb the natural air velocity distribution within the cell.

3.4. The ESP system

In addition to the test cell and the DAS equipment, each national consortium has been provided with the ESP building energy simulation model developped at the ABACUS Unit at the University of Strathclyde in Scotland.

ESP is equipped to simulate any multi-zone building/plant configuration, taking full account of the dynamic interactions between the various energy transfer mechanisms such as long and shortwave radiation and air flow. The system operates in conjunction with real climatological data and permits the imposition of any control regime on the simulation processing.

To eliminate the problems associated with the central support of a large, complex system such as ESP, the model has been delivered to each site already installed on a low-cost Unix^{TM} workstation. Two compatible machine types have been employed : the SUN3 from Sun Microsystems of the US and the MG-1 from Whitechapel of the UK.

4. Sub-group activities

This section briefly presents the activities of the different subgroups during the first phase of the project. The full description of these activities is to be found in the following sections of this report or in the final reports of the other subgroups.

4.1. Test Methodology

4.1.1. Activities

At the commencement of PASSYS a large number of possible methodologies were considered by this sub-group. These ranged from global comparative methods, where each of two cells are subjected to identical outside and inside conditions, to identification techniques, where the experiment is so arranged as to enable a determination of the parameters of a selected dynamic model of the cell.

At the present time, the group is converging on a consensus methodology which combines the best features of the possible methodologies.

Special attention is being given to the links with other sub-groups, especially the one concerned with SDTs.

4.1.2. Results at the end of the first phase

The methodologies so far explored have been analysed in terms of their assumptions, procedures, applicability and so on, and the results reported in the subgroup final report.

Advice on the hardware/software needs of each method and pragmatic guidelines on method operation are also given. The results from application of the different methodologies to different PSCs, geographical locations and test conditions are also presented.

The report paves the way towards the final methodology and defines the necessary connection between this methodology and the PASSYS SDT .

4.2. Model Development and Validation

4.2.1. Activities

ESP is a simulation model and as such has a large number of features which enables it to model many different passive solar systems. To validate such a model is a non-trivial task demanding a comprehensive validation methodology. The validation methodology developed within PASSYS comprises the following techniques :

1. Theoretical examination of the submodels in order to give insight in the used algorithms and if possible to find more appropriate algorithms.

2. Internal consistency check to detect errors in the used algorithms.

3. Analytical verification in which the model is configured to address a specially established problem for which an analytical (exact) solution exists.

4. Inter-model comparison in which ESP is compared with several other models in an attempt to isolate areas of least agreement and so identify candidates for further investigation.

5. Sensitivity studies to provide information about the magnitude of the uncertainty and to design experiments for empirical validation.

6. Empirical validation in which the model's predictive accuracy is determined by direct comparison with test cell data from an experiment designed to stress some particular feature.

Using these devices it is possible to systematically test ESP in several important respects.

PASSYS test cell equipped with the reference wall at Sophia Antipolis (F).

PASSYS test cells equipped with the calibration walls at the JRC Ispra (I).

In the first phase of PASSYS it has not been possible to cover all aspects of model validation, especially at the whole model level which will remain a development field for some time to come.

4.2.2. Results at the end of the first phase

The principal output is the sub-group report which gives a review of past validation work and details the contribution of PASSYS in the three foregoing areas.

The PASSYS validation methodology is described and the results of the application of the methodology to each of the underlying thermal sub-systems (for conduction, convection, radiation, etc.) reported.

In addition to this an improved version of ESP has emerged.

4.3. Simplified Design Tools

4.3.1. Activities

While, in the longer term, it will be possible to provide simulation-based design tools which owe their ease of use to the elegance of their user interface, the production of correlation-based SDTs remains the method most commensurate with contemporary hardware and designer expectations.

The work of the sub-group is directed to this end with the objective to produce a correlation-based SDT which allows the determination of a building's heat requirements and the performance of its PSCs, together with an appraisal of the prevailing levels of comfort.

In essence the group is seeking to define a robust correlation principle and to improve on the techniques as active in some contemporary SDTs.

4.3.2. Results at the end of the first phase

The principal output of the subgroup is a draft version of a European correlation based SDT, together with the documentation of the process for SDT development in relation to present and future trends.

4.4. Instrumentation

The major aim of this subgroup is to create the necessary conditions for high quality data-sets from the measurements in the PASSYS-cells. Special attention is given to accurate air and surface temperature measurements, solar and infra-red radiation, airtightness and ventilation and thermal bridges.

The principal output of the sub-group is to be found in the high quality of the data sets. Besides this there is a report describing in detail the various aspects of the PASSYS test cells.

5. Conclusions

In its three years of operation, the PASSYS group has become a strongly motivated research team. The group has now obtained its full momentum and has evolved an efficient operational structure with which to address its objectives. With the infrastructure now in place, PASSYS is now poised to make a considerable contribution in the years to come.

As the largest passive solar research project in the world, it is clear that the international community of practitioners and researchers expect much in the form of guidelines and products.

By the end of the first phase of the work the following objectives have been realised :

1. The test cells and associated hardware have been installed and tested throughout Europe. The cells have been calibrated and a base set of experiments performed.

2. The strengths and weaknesses of the different test methodologies have been clearly understood and the basic principles of a common European methodology have emerged for refinement in a second phase.

3. Several valuable contributions have been made to the ESP system at the algorithmic level and the role of empirical validation is better understood.

4. A prototype version of a correlation-based SDT has been developped and the production methodology is more clearly understood and defined.

Thus the project PASSYS is about to provide answers to a number of problems that the building/component industry faces : objective bases for product comparison and optimization, and tools for component integration in different types of buildings and at various locations. It should therefore match in the near future the expectations of european companies , in particular the glass industry. In that respect a new subgroup will be active during the second phase of the project which will be in charge of the relations between PASSYS, the industry and the C.E.N.

THE PASSYS TEST CELLS : A UNIQUE EUROPEAN NETWORK OF HIGH QUALITY OUTDOOR TEST FACILITIES FOR THERMAL BUILDING RESEARCH

P. Wouters, L. Vandaele
Belgian Building Research Institute (WTCB - CSTC)
avenue Pierre Holoffe 21
1342 LIMELETTE - Belgium

1. INTRODUCTION

The characterization of the thermal behaviour of passive solar and other building components is one focal point in the PASSYS Programme. To test such building components, a test environment was chosen where this can be done under real climatic conditions : the PASSYS Test Cells.
These test cells are identical test facilities spread over Europe, equipped with a common set of measurement instruments and control devices.
Thermophysical properties of building components can be determined as well as the resulting effect of outdoor climatic variations on the indoor climate in the space behind that component in a unique and reproduceable way, only differing by the climatic conditions of the various sites. Thus the building industry will obtain an answer to the question how a particular component will intermediate between outdoor and indoor climate. This will provide objective bases for product comparison and optimization as well as for the potential integration of passive solar components in different types of buildings and at various locations throughout Europe.
The development and setting up of this high quality test equipment was the main concern of the Subgroup Instrumentation during the first phase of the PASSYS Programme. The main objective was to provide all the test teams with highly reliable information on the test conditions, resulting in interchangeable and comparable results.

2. THE TEST CELL

2.1. Short history

In between the testing of materials and components in the laboratory (e.g. U-value measurements in hot box - cold box facility) and the overall performance of whole buildings with their complexity and occupants interaction, a wide range of test environments may be suitable for specific thermal building research. The PASSYS group decided on a highly standardized outdoor test cell with two rooms, combined with a standardized test procedure.
This option enables :

. to multiplicate the same test environment at a number of test centers in Europe;
. to take fully into account the real climatic conditions of the site;
. to intercompare the results, obtained with a common test method in a common test environment only differing by the meteorological conditions;
. to preclude the effect of occupancy interactions;
. to perform tests in simplified but realistic size test rooms with full scale building component;
. to perform particular tests where interzonal phenomena have to be taken into account.

2.2. Constructional details

Figure 1 gives some constructional details of the PASSYS test cell. It is based on a hard casing concept, the rigid steel support frame and the insulation being separated. This enables the transportation of a complete test cell from the factory to the test site with minimal risk for mechanical damage. The outer skin of stainless steel lining protects against exposure.
The walls are highly insulated with 400 mm extruded polystyrene foam and mineral wool, resulting in a thermal conductance of about 11 W/K for the test room with calibration wall in its south facade.
The inner steel lining provides a good airtightness of the walls and allows for the adding of various materials onto the floor and the walls.
The cell comprises two rooms : the test room and the service room, separated by a well insulated partition wall and door. The dimensions of the test cell (outside) are : 8.44 m L x 3.80 m W x 3.61 m H or 116 m^3 volume. Those of the test room (inside) are : 5.00 m L x 2.76 m W x 2.75 m H or 38 m^3 volume. The total mass is about 12 tons.
The original design was made by DFVLR/3/ after which the cells were manufactured by CADOLTO, W. Germany. Several physical properties have been investigated by the test site teams, e.g. the optical properties of the external surfaces, the airtightness of the cells, the thermal bridge effects at the partition door as well as at the south wall frames (see figure 2), the stability of the lining, and appropriate actions are being undertaken to recover some problems.

2.3. A new design

The first series of test cells all have a removable south wall enabling the testing of different wall components (see 7). For the testing of new roof components, especially for summer cooling purpose in Mediterranean countries, a new test cell is under design : the so-called 'Mediterranean cell'. This test cell is similar to the original PASSYS Test Cell but has additionally a removable roof.

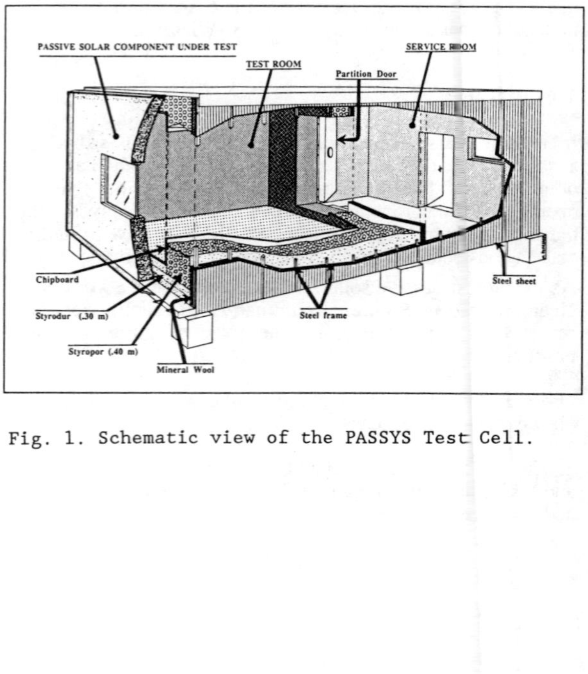

Fig. 1. Schematic view of the PASSYS Test Cell.

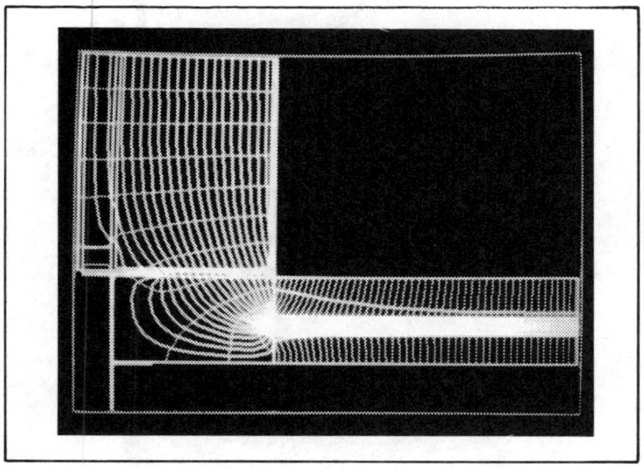

Fig. 2. Thermal bridge of the South frame, as calculated by KOBRU /4/.

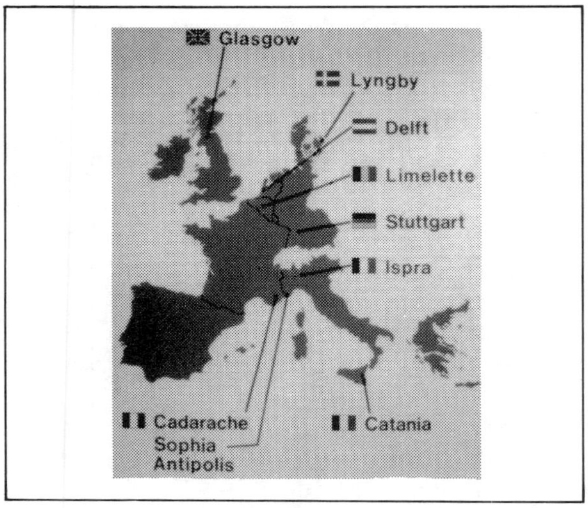

Fig. 3. Location of the PASSYS Test Sites.

3. THE TEST SITES

At present 29 PASSYS Test Cells are located at 9 test sites in 7 EEC Member states as indicated on the map of figure 3. Most of these test sites dispose of 2 or 4 test cells. During the second phase of the PASSYS project, new test sites will be erected in Portugal, Spain and Greece each of them being equipped with one 'classical' and one 'Mediterranean' test cell. Also the Italian test site of Catania will additionally be equipped with one test cell with removable roof. Figure 4 gives a view of the location of the Belgian test site. Figure 5 gives an idea of the wide range in climatic conditions of the various test sites.

4. HEATING AND COOLING SYSTEM

For some experiments the indoor temperature in the test room has to be controlled accurately. For that purpose a powerful heat removal system combined with a heating unit has been developed by several PASSYS teams in cooperation with Weiss Company, the manufacturer of the device. It is a water based, frost protected refrigeration unit providing a high internal air flow rate via a fan cooled heat exchanger. The air is distributed by two textile hoses hanging from the ceiling of the test room, in order not to disturb the natural air velocity distribution within the test room. An

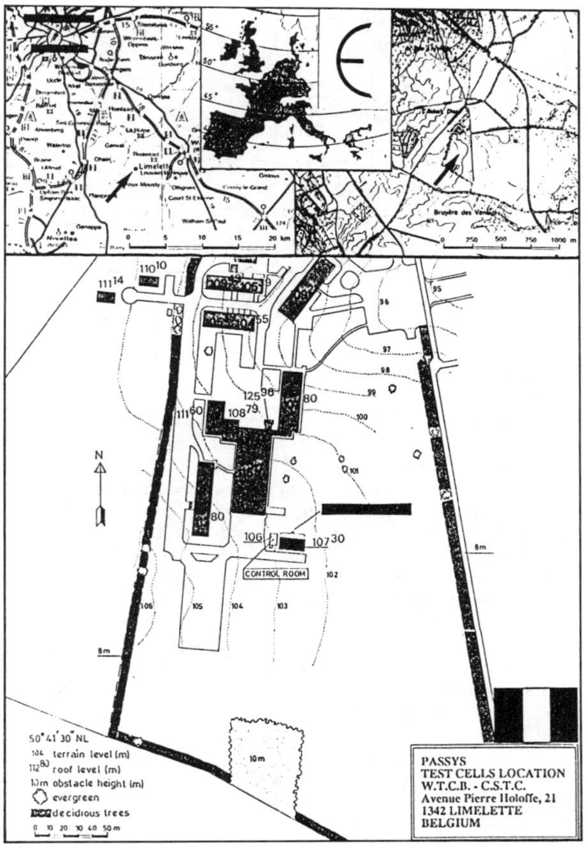

Fig. 4. Location of the Belgian test site at Limelette.

electric resistance heater is incorporated in the air distribution system (See figure 6). Both the heating and the cooling powers can be accurately controlled and measured. The range of operation has been defined for the different climatic conditions at the various test sites as well as for use with different test components. Two cells per site have been equipped with this heating and cooling system.

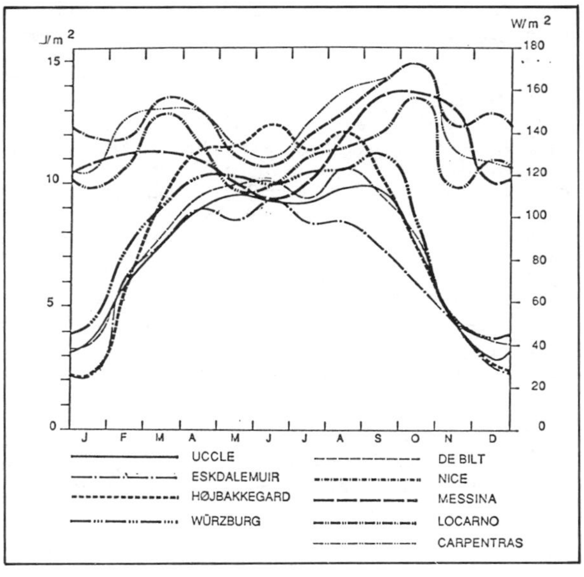

Fig. 5. Monthly mean daily solar irradiation on a vertical south plane for the nearest weatherstations.

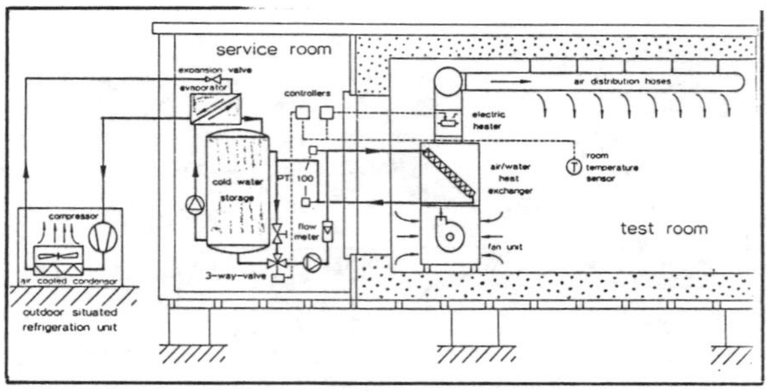

Fig. 6. Schematic view of the heating and cooling system.

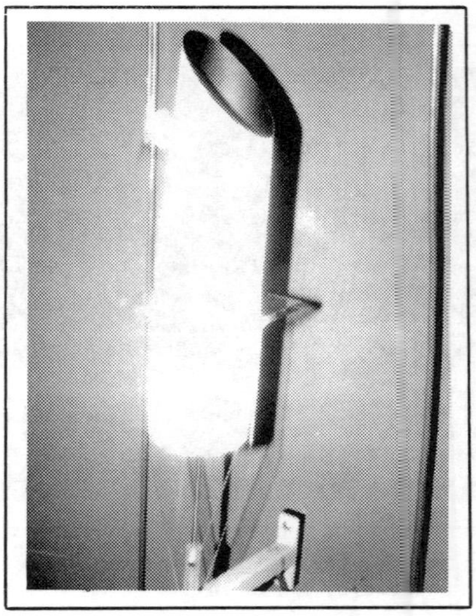

Fig. 7. Shielded air temperature measurement device.

5. INSTRUMENTATION

The choice and the location of the sensors have been based on the requirements from the test method development and model validation activities. This has lead to a common instrumentation for the agreed set of basic measurements, completed by additional sensors for some specific measurements.

The outdoor climatic conditions are characterized by the measurement of the global and diffuse solar irradiances on a horizontal plane and the vertical global solar irradiance on the south facade of the cell (with pyranometers); the wind velocity and direction; the ambient temperature (by a ventilated sensor with thermocouple); the surface temperatures on the roof and walls of the cell (by shielded thermocouples). Several sites measure additionally the longwave radiation (by pyrgeometer) and the outdoor relative humidity. For the latter measurement the recommended measurement system has been developed by TIL-TUD (DK).

The indoor conditions are measured through the air temperatures (with double shielded sensors with Pt100 or thermocouple at seven well defined locations in the test room, see figure 7), the surface temperatures (by Pt100's or thermocouples stuck onto the walls and painted in the same color, at 20 locations). At several sites also the air velocities and other comfort parameters are measured in the test room as well as the heat fluxes through homogeneous walls (by TPD heat flux sensor).

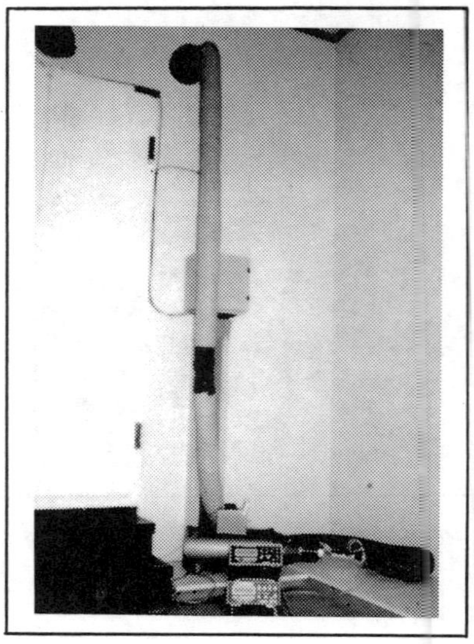

Fig. 8. Fan pressurization equipment.

The measurement of the heating power (electric power of the air distribution fan and the resistance coils) and of the cooling power (water flowrate and inlet-outlet temperature difference of the water-air heat exchanger) is also done by a common equipment.

Since convective heat exchange may be very important in the highly insulated test cells, special attention has been paid to the measurement of airtightness levels and air change rates. The airtightness of the test cell with its test component is measured before and after each experiment by means of the common pressurization equipment, showed in figure 8. The air change rate can be measured continuously using tracing gas techniques. For this purpose a special PASSYS control and measurement system for use in 4 test cells simultaneously has been developed by the BBRI (Figure 9).

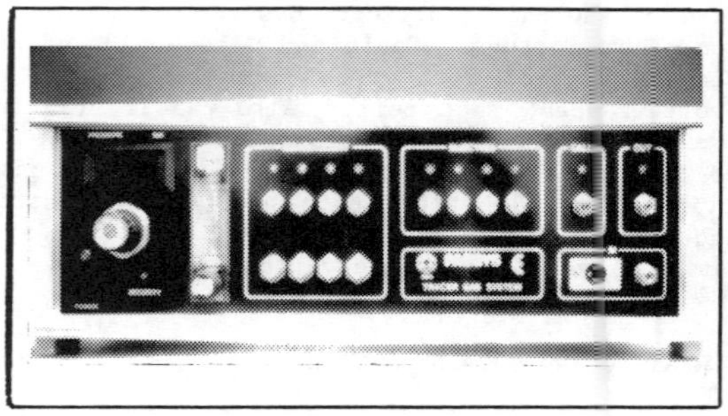

Fig. 9. PASSYS Tracer Gas Equipment.

Specific instrumentation research items dealt with during the first phase of the PASSYS Programme are reported in Part 2 of the Final Report of the Subgroup Instrumentation /2/ : the measurement of surface and air temperatures, heat fluxes, long wave radiation, ventilation rates by tracer gas techniques.

6. DATA ACQUISITION

All test sites are provided with the same data acquisition system. It consists of a HP A600 real time multitask multiuser mini computer with a HP 150 microcomputer as a graphic terminal, a HP 3852A data acquisition and control unit with extender and the necessary accessories.
A specific PASSYS software package has been developed by Ecole des Mines (F).

7. TEST COMPONENTS

7.1. Type of components

The PASSYS Test Cell with its opening of 2.76 m x 2.75 m at its south facing side, is capable to contain full size building facade components (figure 11). The test cells are delivered with 2 components which are common for all test sites :

- The 'calibration wall' is a 40 cm thick insulation layer of expanded polystyrene in between two plywood panels. It is used to define the test cell's basic charateristics : airtightness, UA-value, ...
- The 'reference wall' is a prefabricated concrete sandwich wall with a polystyrene insulation in the cavity and with a double glazed wooden window, capable of containing a shading blind.

New types of test components have been developed by the UK-team :

- a 'timber frame wall' with plasterboard inner face 100 mm glass fiber insulation and covered with a brick cladding, is configurable for different sized windows or patio doors.
- a 'sunspace', a commercially available conservatory with single glazing in a PVC framework and with a timber frame base support, allowing for the adding of thermal mass to the floor. The sunspace can be attached to the 'timber frame wall'.

Other components, such as special glazings, Trombe walls, passive cooling or ventilation elements, overhangs and shading devices, new components under development as well as conventional wall constructions can be mounted into the south frame of the PASSYS Test Cell to determine their thermal performance. Moreover in the new 'Mediterranean test cells the possibility exists to test roof components, e.g. evaporative cooling components.

7.2. Type of tests

Besides the determination of the heat loss coefficient, the solar aperture and the capacitance of passive solar components or building components in general, the PASSYS Test Cells offer the opportunity for the testing of other performances too :

- the dynamic behaviour of the cell with a component (overheating, heating and cooling profiles);
- the gain utilization factor for the component within a given building environment;
- daylighting coefficient of window systems;
- ventilation performances;
- comparison of results in real climatic conditions with those obtained by adding a movable cold box to the test cell;
- the effectiveness of shading devices and control systems;
- condensation and moisture problems by thermal bridging;
- aging and resistance of constructions against exposure to weather;
- ....

8. THE FUTURE OF THE PASSYS TEST CELL

At the end of the PASSYS Project, (December 1991) 35 test cells will be installed at 12 test sites spread over 10 countries of the European Community. Except for four Mediterreanean test cells with removable roof, all these test cells will be identical. They will be equipped with a highly standardised instrumentation set, and common and performant systems for heating and cooling and for data acquisition. The calibration and operation procedures will be standardized and will be recorded in manuals. Together with the common PASSYS test method, now being under development, the PASSYS Test Cells will be fully operational.
A unique network of high quality outdoor test environments will be at the disposal of the building community :

- for designers to develop and optimize new building components;
- for manufacturers to characterize according to a common and perhaps standardized European procedure the thermal (and other) performances of their products;
- for researchers of to study the thermal behaviour of building components coupled with the test room as well as to validate building energy simulation models;
- for companies to obtain a certification for their products on the European market.

Therefore actions will be undertaken towards the European standardization bodies (CEN) and towards the industry in order to bring the test and calculation procedures in accordance with their needs.
Also the other sections of the PASSYS Project will focus on the same European level : the model validation and development activities will partially be executed using the PASSYS test cell as working field, resulting in a well validated version of ESP, the European reference code for building energy simultations. The development of the Simplified Design Tool 'Passport', will also focus on calculation methods supporting the elaboration of heating requirement regulations. The whole PASSYS project, including the PASSYS Test Cells will be of constant influence on the free European building market after 1992.

9. ACKNOWLEDGEMENT

This paper has largely been based on the final reports of the Subgroup Instrumentation for the first phase of the PASSYS Project /1/ /2/.
We wish to thank the members of the Subgroup Instrumentation especially all the authors who have contributed to these reports.

10. REFERENCES

/1/ P. Wouters, L. Vandaele (ed.). 'The Passys Test Cells, Subgroup Instrumentation Final Report Part 1', BBRI, Brussels 1989.

/2/ P. Wouters, L. Vandaele (ed.). 'Subgroup Instrumentation Final Report Part 2', BBRI, Brussels 1989.

/3/ Schreitmüller, et al. 'Design for advanced passive solar test facility, DFVLR, 1984.

/4/ Standaert P., KOBRU 82, Thermal bridge calculation programme, K.U.Leuven, 1982.

PASSYS: TOWARDS A UNIFORM EUROPEAN TESTMETHOD FOR PASSIVE SOLAR SYSTEMS

H.A.L. van Dijk
TNO Institute of Applied Physics (TPD)
P.O. Box 155
2600 AD Delft
The Netherlands

M. Antinucci
Conphoebus
Via G. Leopardi, 60
95127 Catania
Italy

ABSTRACT. The main goal of the Subgroup Testmethodologies in the CEC research project PASSYS is to define test methods to determine thermal and solar properties of building components. During PASSYS I (1986-1989) different testmethods were tried and compared and a preliminary common test strategy was developed. First testresults became available for analysis. During PASSYS II (1990 - 1991) (a) common testmethod(s) will be selected, further developed and verified.

1. INTRODUCTION

The main goal of the subgroup Testmethodologies in the CEC research project PASSYS is to define test methods to determine thermal and solar properties of building components, both for product characterization and optimization and usable as input in design methods.
The priorities are:
- the development of one or a set of reliable and affordable European test procedures for the energetic characterization of components most generally used in the building industry (e.g. direct gain systems); this set of procedures will have the form of a final product immediately usable and to be described in a document suitable as reference for a European standard;
- the evaluation of other performances of building components (e.g. risk of surface condensation, comfort problems, daylighting,..);
- the characterization of more complex test components (e.g. transparantly insulated walls, ventilated Trombe walls).

2. THERMAL CHARACTERISTICS

The main energetic characteristics are:
- the heat loss coefficient by transmission or U-value (per m^2; integrated over the total area: UA-value);
- the heat loss coefficient by ventilation, if applicable;
- the solar heat gain factor or solar transmittance or g-value (per m^2; integrated over the total area: gA-value).

Also important are the heat accumulation in the South wall component and the thermal coupling with the indoor environment: heat is transferred by free convection, thermal radiation and/or directly transmitted insolation. The effectiveness of this coupling and the capability to accumulate potentially excessive heat in the building mass determine whether solar heat entering the room will either be utilized to diminish the heating requirements or lead to overheating and thermal discomfort.
Obviously the identification of the main thermal characteristics is closely linked to the development of a simplified design tool (see other PASSYS presentations).
The highly insulated testcell envelope can be provided with high or low mass internal lining to approximate various situations which can be found in practice.

3. HEAT BALANCE

The main energy flows which determine the heat balance inside the testroom of each test cell are schematicly presented in figure 1.

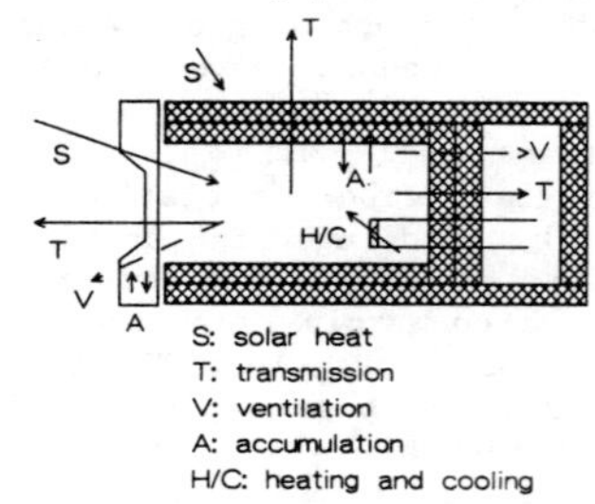

figure 1: main energy flows in the testcell

The heating and cooling power can be used to control the indoortemperature at required levels or to simulate internal heat sources in a room (heating) or additional thermal losses (cooling), to create different gain load ratios in the room.
Different types of measuring and evaluation techniques are being developed and employed, in search for (a) technique(s) which combines the following qualities:
- it provides a sufficient amount of information;
- the results are reliable and usable as product information or in a design tool;
- it requires a relatively short test duration.

4. TEST METHODS

4.1 Introduction

The main quantities measured in and around the testcell are:
- the test- and service room temperature;
- the heating and cooling power;
- the outdoor temperature and solar radiation intensity.

Various types of sensors are applied to measure details, like local temperatures (e.g. thermal bridges, thermal stratification), local air velocity (convection currents) and surface heat fluxes.
The different test evaluation techniques under investigation can roughly be categorized into three groups: integrated absolute approach, transient comparative approach and parameter identification.

4.2 Integrated absolute testmethod

The most straightforward technique which is being tried out is the integrated absolute measurement: the testdata are averaged over a large time interval in order to eliminate the accumulation in test room envelope and South component. The testcell envelope, though, has a very high inertia due to its high thermal resistance and the South component may be heavy or highly insulated as well. Therefore integration periods of several days may be needed to obtain a single reliable value.
The temperature levels of the test- and service rooms are changed in a stepwise manner, e.g. every four days. The thermal characteristics can be deducted by a simple straightforward algebraïc procedure.
The drawback is the low amount of information from a relatively large test duration.

4.3 Transient comparative testmethod

A second approach under investigation is the transient comparative technique. In this approach two identical parallel testcells are used, one with a highly insulated panel as South wall ("Calibration wall"), the other with the component to be tested.
In the first test cell, the reference cell, the indoor temperature is controlled by the heating and cooling equipment in such a way that it continuously follows the indoor temperature in the second cell.
In this way all heat flows through the test-room walls, except for the heat flow through the South component, are the same for both testcells, see figure 2.
The difference in heating and/or cooling power supplied to both testcells is then equal to the heat flow through the South component.

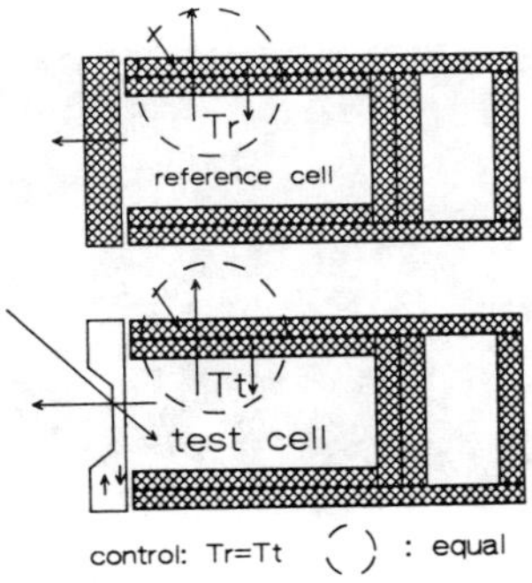

figure 2. Illustration of the comparative test method

This type of method thus yields the transient heat flow through the South component, which in turn can be analysed in different ways. A straightforward form of presentation is the following: the heat flow is averaged over long enough time intervals to minimize accumulation effects in the component. Then the mean heat flow divided by the mean indoor-outdoor temperature difference is plotted against the mean incident solar radiation intensity divided by the same ΔT, see figure 3. A straight line through the points yields immediately the UA- and gA-values of the component: the closer the intercept of the line with the Y-axis is to the origin, the smaller the UA-value; the steeper the line, the higher the gA-value.

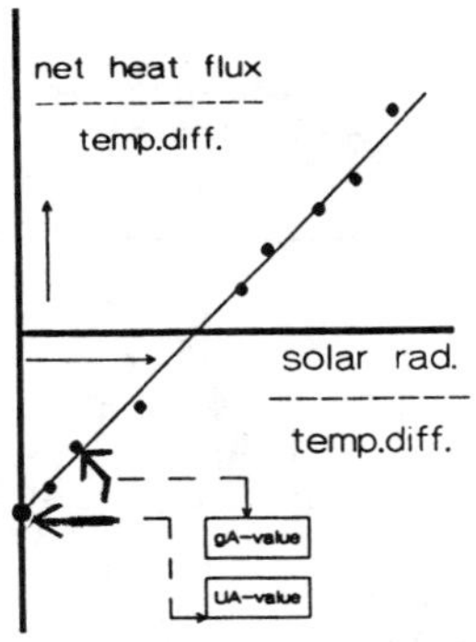

figure 3. Illustrative example of the kind of results which can be derived from the comparative testmethod.

These UA- and gA-values can be applied directly in a simplified design tool.
Notice that the results from the integrated absolute method can be analysed in a similar graphical way.
Another way to analyse the transient heat flow through the component is by applying a parameter identification technique, with this heat flow as output variable and a model of the South component as subject for the identification.

4.4 Parameter identification

Parameter identification is a technique which starts by assuming a mathematical model for the testcell and the South component, both for the steady state (heat loss and gain) and the transient (accumulation!) thermal conduct; the values of the model parameters, for instance heat resistances and - capacitances are unknown before the test. With the aid of a computerprogram the parametervalues are found (identified) as those values which provide the best agreement between the output of the test and the result of a model calculation using the same input data as during the test, see figure 4.

Usually the testroom temperature is appointed as output variable; input variables are outdoor temperature, solar radiation, amount of heating, a.s..
Within PASSYS different models are being compared at the moment.

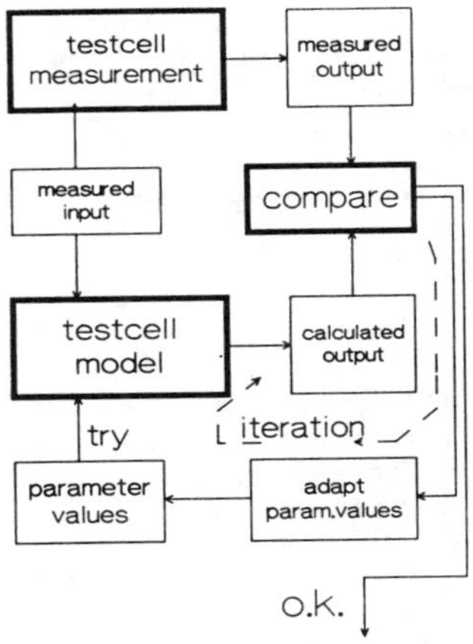

figure 4. The principle of parameter identification to derive the properties of the component from the testcell measurements.

Figure 5 shows an example of a test, giving the measured test room temperature and the best fit calculated curve.

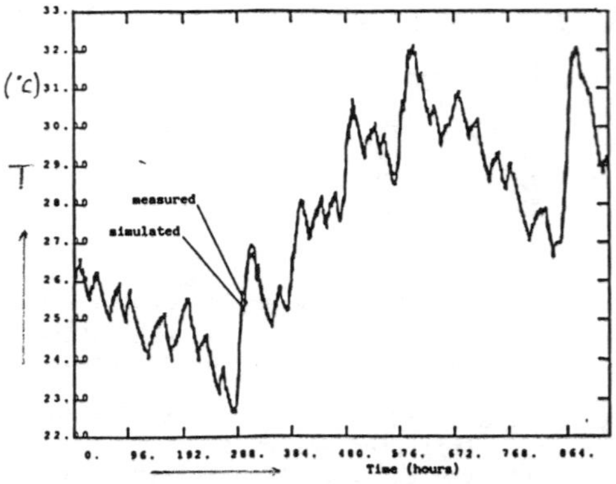

figure 5. Example of a test with parameter identification

The heat transmission coefficient and solar transmittance can directly be deducted from the identified parameters. The transient characteristics can for instance be translated into ingredients for a simplified design tool, like (monthly) gain utilization curves, by carrying out annual simulation runs using the identified model and analysing the computed results.

5. TEST CONTROL STRATEGY

As a consequence of the application of a technique in which different characteristics are derived from the same testsequence special attention should be paid to the correlation between the input variables: if different input variables, like outdoor temperature and solar radiation are correlated, it is not possible to get a clear separation between the parameters which are supposed to describe the thermal conduct of the system with respect to the respective sollicitations.
In order to obtain sufficient and uncorrelated variation in the sollicitations the heating power supply to the testroom is repeatedly switched on and off, with varying lengths of the "on" and "off" time intervals in a pseudo-random manner. The interval lenghts are representative for the kind of frequences which are relevant for passive solar systems; they span the range between 20 minutes and 5 days.
This pseudo-random sequence is preceded and followed by periods with constant but different test room temperatures on which the integrated absolute techniques can be applied, see figure 6.

6. RESULTS

During PASSYS I (1986 - 1989) the testfacilities were built and instrumented and first tests were carried out. At each site tests were carried out on the testcells with calibration wall; moreover, at most of the sites first results were obtained from tests with a common reference wall, a insulated, concrete sandwich wall, with double glazed window. In addition, some of the institutes performed tests on objects like: the reference wall with indoor blinds, timber frame wall with double glazing and low E window and a translucently insulated wall.

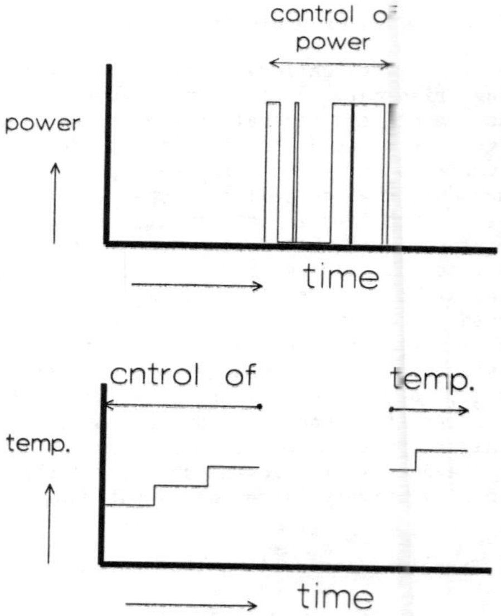

figure 6. Illustration of the common test strategy

These results were however obtained during a phase in which the test equipment and test strategy were not yet fully developed.
Therefore, also simulated tests were used to try and compare the different methods. First results from these comparisons indicate that reliable and informative results can be obtained provided the test strategy is designed with great care. This optimization of the test strategy and the selection of the most appropriate test evaluation method(s) are the main priorities for the next phase in PASSYS.

photo: two testcells with identical walls; parallel test with and without internal blinds.

PERFORMANCE ASSESSMENT OF SUN-SHADING COMPONENTS FOR MEDITERRANEAN BUILDINGS

F. ALEO, M. ANTINUCCI, P. CARBONARO
CONPHOEBUS S.C.R.L.
VIA G. LEOPARDI, 60
95127 CATANIA (ITALY)

ABSTRACT. The Mediterranean climate imposes specific features to bioclimatic building component design. In this study aimed at characterizing some types of solar shading components optimized for the Mediterranean climate, both thermal and daylighting aspects are taken into account. This characterization is based on experimental work performed with Passive Component Test Cells, a useful tool for the assessment of South wall thermal performance. A test-reference experimental method based on integration of data on a daily basis has been adopted. This method was developed in the frame of the PASSYS CEC Programme. Illuminance comparative measurements inside the test cells have been also performed. The thermal performances of the components are presented by a graphical representation. The solar aperture and heat loss coefficient are the parameters characterizing the components. Illuminance comparative measurements inside the test cells are finally shown.

1. INTRODUCTION

1.1 Purpose of the work

The Mediterranean climate is characterized by long hot summers and not too cold winters.
These particular climatic characteristics request specific performances to the Passive Solar Components (P.S.C's).
The moderate but still sensible heating load requires significant solar gains in winter, whereas the relevant summer cooling load suggest:

a) a strong rejection of the solar input in the hot periods;
b) air ventilation as long as outside air temperature keeps lower than inside;
c) possible introduction of passive cooling devices when outside air temperatures becomes too high.

The aim of this work is to characterize two types of solar shading components designed for the Mediterranean climate.
The characterization has been carried out using the PASSIVE TEST CELLS, a useful tool for this kind of problems.
Different methods are available to assess the thermal performances of the P.S.C., as e.g.:

- Absolute methodology based on the use of only one test cell
- Integrated comparative-subtractive method based on the use of two test cells
- Parameter identification techniques.

In the frame of the PASSYS CEC PROGRAMME which has among its purposes the specification of a test method for the P.S.C's the different approaches are still subject of discussion and the choice of a determinate methodology is not yet done.
The experience gained in the P.S.C. testing activity convinced us of the reliability of the Integrated comparative-subtractive method and encouraged us in pursuing its application /1/.
As the components in question shield and therefore reduce the input of solar radiation in the room, not only thermal but also daylighting performances should be evaluated. Trying to respond to this problem illuminance comparative measurements inside the test cells have been also performed during the summer months.

1.2 The components under testing

The first component is a three-side overhang wall supplied with a reflecting double glass window while the second one is a flat wall provided with a solar shield (Koolshade screen) located in front of a standard double glazed window.
Figure 1 shows a view and a vertical section of the three side overhang wall.
The reflecting glass "SOLAR-SIV PALLADIO 10", whose transmittance spectrum in shown in figure 2, has been used.
The Koolshade screen is composed of dark metallic lamellae tilted in such a way as to shut off the direct solar radiation (the selected model has 17 lamellae per inch, each 2 mm large).
In any case both windows have the same transparent area equal to 2.2 m^2.
The opaque part is composed, from inside outward:

- a 12 cm thick concrete layer;
- a 4 cm thick polystyrene layer;
- a 2 cm thick plaster layer.

2. TEST METHODOLOGIES

2.1 The Integrated comparative-subtractive method

The P.S.C. steady-state energy balance leads to the following linear relationship:

$$Q_{PSC} / H = A_S - UA\,(\bar{T}_i - \bar{T}_e)\,Dt / H \qquad (1)$$

In this expression Q_{PSC} is the net thermal energy which passes through the PSC, H is the solar energy density on a vertical south surface, T_i and T_e are respectively inside test cell and outside mean temperatures and finally Dt is the integration time. All these quantities can easily be measured.

Namely the measurement of Q_{PSC} requests the contemporaneous use of two similar test cells, one having the so-called "calibration wall" and the other one having the PSC under testing as South wall.

The "calibration wall" is an opaque and highly insulated component.

It the two test cells are kept at the same temperature, the net thermal energy which passes through the PSC is equal to the difference of the auxiliary energy consumptions corrected by the small amount of energy passing through the "calibration wall", that can be measured by means of heat flux meters.

The aim of the test is the experimental determination of the co-ordinate points:

$$\left[(\bar{T}_i - \bar{T}_e)\, Dt / H \; ; \; Q_{PSC} / H \right]$$

The identification of the performance curve (see equation 1) can be obtained by a linear regression of the experimental data. The y-axis intercept and the slope of the steady state performance straight-line represent the "solar aperture" and the overall heat loss coefficient of the P.S.C.(respectively A_S and UA).

Data are integrated on a daily basis to obtain an experimental point; during this time the calibration and test cells are kept at equal and constant temperatures.

In order to evaluate the PSC performances in different operating conditions three or four different internal temperatures are set inside the test cells and for each internal temperature at least three experimental points are obtained.

The Integrated comparative-subtractive tests have been performed in October 1988 and in March, June and October 1989.

The shading effect of the overhang wall was demonstrated in a set of tests in October 1988, comparing the same window installed on a flat and three-side overhang wall.

The reflecting glass and Koolshade solar screen have been installed starting March 1989.

2.2 Illuminance Comparative Measurements

Illuminance comparative measurements have been performed inside the two test cells equipped with the two different solar shading components in July 1989.

Two illuminance measurements have been performed in both test cells, namely on a plane parallel to the window, very near the glass, and on a horizontal plane located almost at the center of the test cells 80 cm above floor level.

This last test is useful to check if the reduction of luminous flux due to components still guarantees acceptable visibility conditions for a normal office work in the test room center.

The other sensor immediately behind the glass could be used to define the luminance yield of the transparent component.

2.3 Experimental Apparatus

The work has been carried out using three test cells placed at the Conphoebus test site.

The internal volume of the test cells is equal to 59 m^3 (3.2 x 5.8 x 3.2 h).

The loss coefficient of the test cells, obtained by means of calibration measurements, is equal to 17.5 W/°C.

Controlled temperature conditions in the test cells are obtained by a heating and cooling system under computer on-off control.

3. RESULTS

All the comparative-subtractive test results, performed on the components from October 1988 to October 1989, are synthetically showed in table I.

The October 1988 tests are referred to a comparison between three side overhang wall and flat wall both supplied with the same 2-pane window.

The graphic comparison between the two thermal performance straight-lines is shown in figure 3, where the considerable difference between the solar apertures is due to the effect of the overhang.

As the solar aperture values of opaque surfaces are very low, we may admit that the solar aperture of the flat wall is approximately that of the window. The solar factor of the window may be obtained from the experiment as:

$$SF_w = A_S/A_w$$

where A_w is the glass area (here: 2.2 m^2).

Thus the integrated value for SF_w is 0.75 which is very close the value given by the manufacturer (0.78).

The comparison between the two walls supplied with the additional shading devices has been performed in March and June 1989.

It may be noted that the solar aperture of the three side overhang component decreases strongly from March to June because of the change of the solar altitude from spring to summer.

On the contrary, a large increase of the Koolshade screened flat wall component solar aperture from March to June and a corresponding large reduction from June to October (see Table I) is shown.

To explain this result, let's consider that the Koolshade screen is designed to shut off the direct solar radiation, therefore is not effective as regards diffuse radiation.

The diffuse solar radiation reaches about 70 per cent of the global solar radiation in Sicily for a vertical South surface in June.

This percentage drops to 38% in March and 23% in October.

A graphic comparison between the two thermal performances straight-lines obtained in June is shown in figure 4.

The three-side overhang wall with reflecting double glass has a solar aperture sensitively lower than that of the other shading component in

June, and this fact is confirmed by the illuminance comparative measurements performed in the same month inside the two test cells, showing that the illuminance level in the test cell equipped with the Koolshade screen is about three times higher than that measured inside the other test cell (see figure 5). In most part of the industrial nations the recommended illuminance threshold to guarantee a good visibility level is at least 300 lux. The measurement results show that from 8 to 17 this level is satisfied only in the test cell equipped with the flat wall and the Koolshade screened 2.2 m^2 window.
The table I shows that corresponding to each comparative-subtractive test quite similar loss coefficient values were measured on the two components. Loss coefficient differences among double glass, "Solarsiv Palladio 10" double glass or Koolshade screened double glass are quite small. Moreover the glass area represents 20% only of the total wall area.
October and June heat loss measurements are repeatable within 7%, while the sensibly lower March results require further investigations to be explained.

4. CONCLUSIONS

Two different solar shading components designed for the Mediterranean climate have been characterized.
The flat wall supplied with the solar screen is characterized by a very low value of solar aperture during winter.
In order to avoid winter unfavourable situations, as far as solar gains and visibility conditions are concerned, the size of the glass area should be more carefully considered.
The summer shading capability, even if lower than the winter one, is such as to guarantee a good illuminance condition (300 lux), due to the practical absence of direct radiation on a South façade.
The Koolshade solar screen stops the direct solar radiation but it is not very effective as regards diffuse one.
The three side overhang wall supplied with double reflecting glass has a winter solar aperture of 0.49 m^2, that falls down to 0.27 m^2 in summer.
In this season inside the test cell the illuminance conditions were insufficient.
The joint action of the overhang and the reflecting glass in one 2.2 m^2 window in summer is effective to keep low the solar gains, but creates uncomfortable conditions, because artificial light is needed.

REFERENCES

/1/ Van Dijk H.A.L. TNO Inst. of Applied Physics (TPD), Delft, The Netherlands
Antinucci M., Conphoebus, Catania, Italy
Towards a Uniform European Testmethod for Passive Solar Systems.(SECA) Paris 4-8 Dec.89

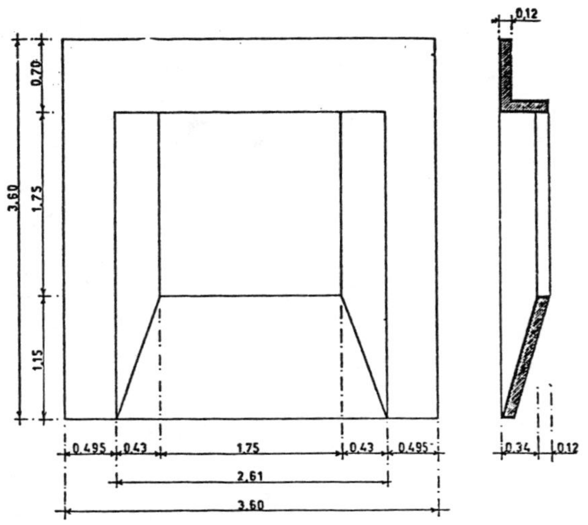

Fig. 1: Three side overhang wall

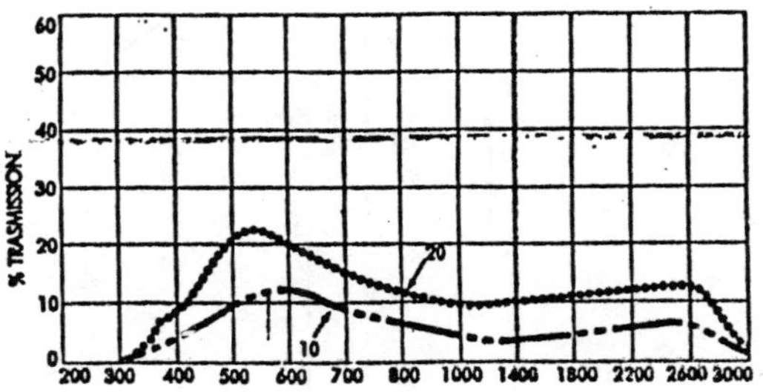

Fig. 2: Solarsiv transmittance spectrum (—··)

COMPONENT	TIME TEST	As sqm	U*A W/°C
THREE SIDE OVERHANG WALL WHIT DOUBLE GLASS	OCT. 1988	1.31±0.04	18.2±0.2
THREE SIDE OVERHANG WALL WITH DOUBLE REFLECTING GLASS	MAR. 1989	0.49±0.03	16.9±0.2
THREE SIDE OVERHANG WALL WITH DOUBLE REFLECTING GLASS	JUN. 1989	0.27±0.01	18.8±0.2
FLAT WALL WITH DOUBLE GLASS	OCT. 1988	1.65±0.04	18.6±0.3
FLAT WALL WITH DOUBLE GLASS AND KOOLSHADE SCREEN	MAR. 1989	0.29±0.03	16.3±0.2
FLAT WALL WITH DOUBLE GLASS AND KOOLSHADE SCREEN	JUN. 1989	0.44±0.02	19.2±0.3
FLAT WALL WITH DOUBLE GLASS AND KOOLSHADE SCREEN	OCT. 1989	0.28±0.01	17.8±0.2

Table I: Comparative-subtractive test results

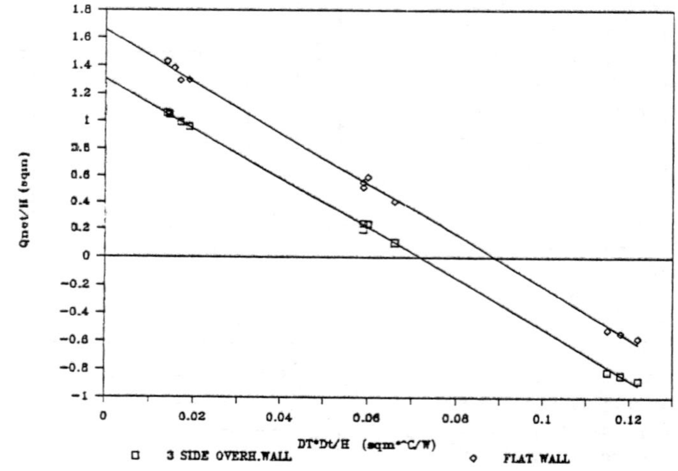

Fig. 3: October 1988: thermal performance curves

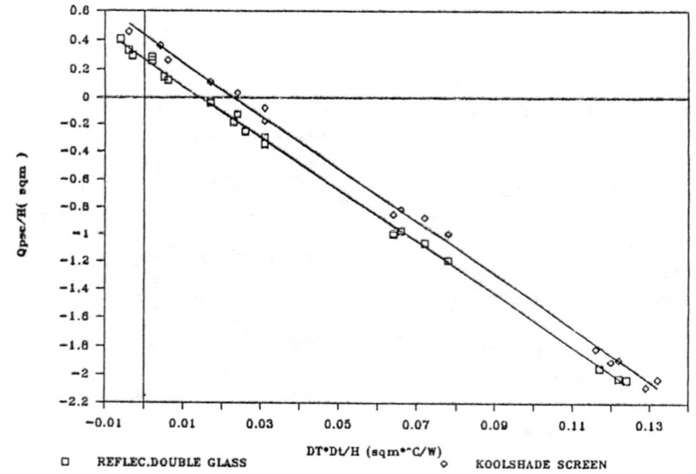

Fig. 4: June 1989: thermal performance curves

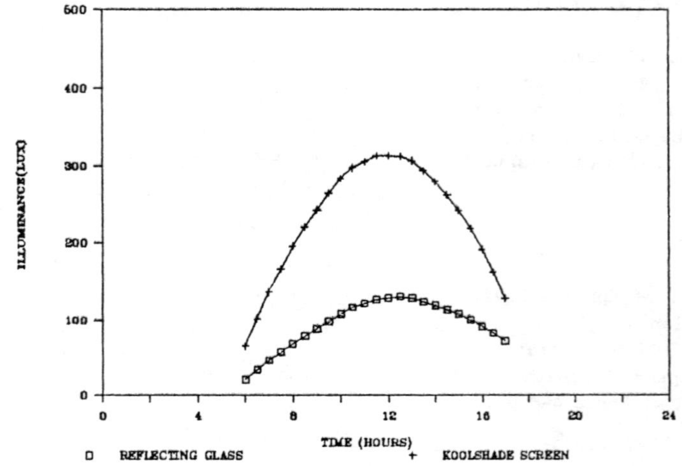

Fig. 5: Illuminance inside test cells

The Model Validation Activities of Project PASSYS

Professor J A Clarke
Energy Simulation Research Unit
University of Strathclyde
Glasgow G4 0NG
Scotland

(On behalf of the participants of the Model Validation and Development Sub-Group)

Abstract

This paper describes the work of the model validation and development sub-group within the PASSYS project. The validation methodology being employed is described and its application is demonstrated. The use of the test cell facility for empirical validation is then discussed. Finally, the paper reports on the conclusions to emerge from phase one of the project and details the objectives as established for a possible second phase.

1. Introduction

In 1986 the European Community's energy R&D directorate established a collaborative venture in the field of Passive Solar Architecture known as Project PASSYS. Involving research consortia from Belgium, Denmark, France, Germany, Greece, Italy, the Netherlands, Portugal, Spain and the United Kingdom, the aim was to increase confidence in passive solar systems through the development of a component testing procedure, the validation and refinement of a reference simulation model and the development of better design tools. The work of PASSYS is undertaken within five specialist sub-groups addressing test methodology, simulation model validation and development, simplified design tool development, instrumentation and passive solar component selection. To enable the collaborative dimension of the project, each national consortia has access to identical test cells, with integral heat removal equipment and data acquisition. In support of the model validation and development work, each team utilises Unix^{TM} workstation technology on which is resident the ESP† energy modelling system [1].

This paper details the work of the model validation and development sub-group (MVDS) in the following respects. Firstly, the validation methodology being employed to investigate algorithmic performance and whole model predictive accuracy is described. Secondly, the use of the test cells for empirical validation is outlined. Thirdly, examples of the application of the methodology are given to demonstrate the process and the result. Fourthly, the changes made to the ESP system as a result of the work to date is summarily described. Finally, the paper reports on the conclusions to emerge from phase one of the project (completed in July 1989) and details the objectives and work plan as established for a possible second phase. For a complete description of the work of this group, the reader is referred to the final report of the MVDS [4].

2. Validation Methodology

The principal objective of the MVDS is to approve/ refine a European validation methodology and to test this by applying it to a common model, ESP. At the commencement of the project two previous studies had been completed which attempted to establish such a methodology for the validation of building thermal models. The first, undertaken by the US Solar Energy Research Institute [5] had resulted in a three part methodology involving analytical testing, inter-model comparisons and the use of empirical data. This methodology was further refined and extended in the second study involving three UK Science and Engineering Research Council funded groups - the University of Nottingham, Leicester Polytechnic and the Rutherford Appleton Laboratory - and the Building Research Establishment [6]. The evolved methodology comprised:

1. An initial examination of a model's theory and a thorough inspection of the corresponding source code.
2. Analytical verification involving a comparison of predictions with analytical solutions which apply to some well defined, usually simplified case.
3. Inter-model comparison involving a comparison of the target model with several other models which are usually better known to the validators and may have been subjected to a greater or lesser degree of previous testing.
4. Empirical validation involving a comparison of predictions with measured data for the same problem domain.
5. A parametric sensitivity analysis to ascertain whether or not the model predictions lie within the uncertainty band associated with the input data.

This was the methodology reviewed and accepted by the MVDS at the commencement of the PASSYS project. In terms of the ESP system, the methodology translates to:

- A critical evaluation of the theory used by ESP to model the different heat transfer processes, including an investigation of possible alternatives.
- A check of the corresponding source code segments severally, and jointly where possible.
- The application of the analytical and inter-model comparison elements of the methodology where possible.
- The identification, design and implementation of empirical validation experiments in terms of the capabilities (and limitations) of the Test Cells.
- The application of sensitivity analysis techniques to determine the acceptability or otherwise of the matched predicted and measured data-sets.

In total some eight ESP sub-systems have been identified as candidates for study in this way (although not all have been, or can be, processed to the same level of detail). They are: climate parameters, external and internal surface convection, external and internal surface longwave radiation, shortwave processes, conduction and air flow. In addition, and principally by the mechanism of empirical validation and sensitivity analysis, an attempt has been made to evaluate ESP's performance at the whole model level.

Because the PASSYS project is concerned with passive solar design, all tests have been devised with this bias. This means that the findings to emerge, while guiding the evolution of ESP, will be applicable to a range of model types operating in this area.

3. Empirical Validation

It has been shown in previous work [7] that there is a paucity of high quality empirical data which can be considered suitable for use in a model validation context. A principal aim of PASSYS is therefore to utilise the test cells, where appropriate, to collect such data. To date it has proved possible to collect only a small number of data-sets. The reasons for this are threefold. Firstly, and most significantly, the establishment and calibration of the test cells and their related data acquisition and heat removal systems, was a time-consuming task, exacerbated by the need to ensure the same high standard throughout Europe. Indeed it was only within the last year of the project that this process was finalised. Secondly, it was deemed essential that each group should gain a technical understanding of ESP before attempting to specify experiments to serve the needs of empirical validation. This process, achieved by the non-empirical techniques previously outlined is now completed. Thirdly, the needs of the MVDS had, necessarily, to compete with the needs of the testing methodology sub-group who required measured data to test the robustness of the different methodologies being proposed.

For these reasons most of the MVDS effort in the area of empirical validation was directed to the identification of those features of ESP

† Selected as the European reference simulation program on the basis of model comparisons by the Passive Solar Working Group [2, 3].

which could be addressed and the approval of a set of related experiments. Nevertheless a few data-sets were captured and applied, experimentally to ESP. But before considering this work, it is useful to consider the strengths and limitations of the cells in the context of empirical validation. On the limitation side, two disadvantages can be identified. Firstly, the cells are not designed to allow the study of fundamental heat transfer processes. Such studies are better undertaken in controlled conditions such as offered by wind tunnels, hot-boxes and the like. Secondly, they cannot be used to represent the real building situation since phenomena such as occupancy and effects of scale cannot be well represented. On the advantage side:

- The cells provide a realistic, intermediate step between the laboratory and real building situations. And being situated in a city centre location, their boundary conditions are realistic and typical of those influencing real buildings. They are also subjected to the same complex set of interacting heat transfer processes.
- The effects of occupancy and infiltration, which give rise to large uncertainties in the monitoring of real buildings, are removed. They can however be reintroduced in a controlled manner if required.
- Their relatively simple structure can be modelled at low computational expense and, more importantly, it is possible to control or predict special effects, such as thermal bridges, to eliminate or quantify modelling uncertainties.
- The calibration tests have shown that the cells are well matched. This, in conjunction with the high standard of the data acquisition equipment, means that the facility is particularly suited to side-by-side experiments in which the relative performance effect of a given design parameter change can be measured and compared with the corresponding model prediction.
- The construction of the Cells, with their removable south wall, provides the possibility of appending validation-specific sensors to a component.

From the experience gained to date, the cells would appear to offer the following possibilities.

Process Studies

In limited cases the cells can be used to study a single heat transfer process. In most cases the experiment in an indirect one in that it is designed to measure the effect of the process, not the process itself (the latter being more readily achieved under laboratory conditions). Because of their design, the cells have very low internal heat fluxes. This gives rise to unacceptably large uncertainties when attempting to measure absolute flux magnitudes or the change resulting from a change to a design parameter. Conversely, these low flux magnitudes mean that the energy processes associated with a south wall component will dominate. This, in turn, decrease component measurement uncertainties.

Inter-Cell Comparison

This technique can be used to isolate the effect on performance of a single design change. In particular it serves to eliminate the uncertainties inherent in the various parameters which underlie the experiment. By producing relative differences, it can provide high quality data sets (of low uncertainty) which can be used to assess model accuracy when the model is applied in parametric mode.

Whole Model Testing

The high degree of controllability of the Cells - in terms of temperature stability and the elimination of complex phenomena such as infiltration - means that they can be used to capture whole system data-sets under well defined conditions. By testing directly the extent to which the model can get close to reality, these data-sets afford the best indication of how the model will perform when used in practice. This technique has been used by the MVDS in conjunction with the differential sensitivity analysis method as proposed by Lomas [8] to establish whether or not ESP predictions will lie within the error bands associated with the inherent uncertainties in the input data.

Assumption Testing

This is a category of experiment that aims to test the appropriateness of the assumptions implicit within a model and to establish the value of the parameters which should be used to represent the system, in this case the test cell.

4. Application of the Methodology

As previously stated, the greater proportion of the work of the MVDS has been concentrated on the theory/ code checking and analytical testing activities. In contrast only a limited amount of inter-model comparisons has been achieved and a modest degree of empirical validation. The activities to date are now characterised by the following brief descriptions.

4.1 *Theory/ Code Checking*

Several ESP algorithms have been subjected to theoretical review and code checking. Of these, the study of external longwave radiation is described here as an exemplar of the process. This commenced with a detailed review of the theory are initially incorporated in ESP and of the possible alternative theories. For example, the following table lists the various theories studied with respect to the estimation of sky radiation.

Formula	Published
Angstrom	1915
Brunt	1932
Bolz	1949
Linquist	1950
Parmelee	1952
Kreitz	1954
Roach	1955
Angstrom	1962
Swinbank	1963
Idso	1969
Unsworth & Monteith	1975
Berdahl & Martin	1984
Czeplak & Kasten	1987
German TRY	

The formula of Unsworth & Monteith was the one implemented in ESP at the start of PASSYS. But because it was shown to possess only regional validity, a replacement was deemed necessary. On the basis of the review, the formula of Swinbank was proposed as the most appropriate substitute. The technique of assessing scene view factors was also critically examined and a representative set of illustrations was devised for inclusion within the system manual.

4.2 *Analytical Testing*

Two ESP sub-systems in particular benefited from the use of analytical testing: conduction and internal longwave radiation exchange. Several analytical formula were derived on the basis of mathematically straightforward boundary conditions such as a constant or sinusoidal temperature regime.

In the case of conduction the conclusion was drawn that ESP's discretisation, in time and space, of the governing differential equations approximates well with the exact solution, even with complex multi-layered constructions. With internal longwave radiation, the conclusion was drawn that ESP's treatment of the process was adequate although in some circumstances it could be important to use view factors calculated by the more rigorous of the two methods offered by the model.

4.3 *Inter-Model Comparisons*

This technique has been employed to investigate ESP's performance with respect to conduction and internal longwave radiation, and to check the processing of the direct and diffuse shortwave flux through windows. In the last case, the ESP predictions were compared with the results obtained from a model based on the fundamental equations and a model which utilised the techniques as expounded by ASHRAE. No appreciable differences were observed. In a similar manner, the results obtained from an ESP's air flow simulation were compared with other model predictions. For the limited cases studied, again no serious discrepancies were found.

4.4 *Empirical Testing*

To be of any real value, empirical tests should have the potential to indicate which parts of a model are most likely to be the cause of a significant departure from the reality. One way in which this can be achieved is to devise a complementary set of experiments which are designed to progress from single process studies to the more complex whole model case. This is a goal of the MVDS. To demonstrate the empirical testing process, an experiment of the latter type is described here.

The experiment, which ran between February 7 and March 10 1989, was carried out on two cells each fitted with the PASSYS calibration wall. One cell was allowed to free-float, to serve as a control, while the other was subjected to the following schedule:

- An initial pre-conditioning period was sustained until the cell temperature matched the control.
- A 4 hour, 2kW radiant heat pulse was applied.
- A free-floating period was maintained until the cell's temperature returned to within 0.5°C of the control.

- A 2 hour, 2KW convective heat pulse was applied.
- The previous free-floating condition was reintroduced.
- A constant temperature of 30°C was maintained for a 5 day period.
- Finally the cell's temperature was permitted to decay naturally.

Throughout the experiment the service rooms of both cells were maintained at 20°C and the cells were sealed to eliminate the effects of infiltration. The captured data-set included climate parameters at locations around each cell, ground temperatures, internal and external surface temperatures, internal air temperatures and heating inputs.

On the basis of best estimates of the various cell parameters, ESP was used to predict the cell response (using the measured climate data but without any prior knowledge of the measured cell performance). Figure 1 shows this base-case prediction against the corresponding measurement.

A test such as this is useful in that it provides an indication of how the model might perform in practice. Indeed such a 'whole model' test is often offered as the final proof of the model's predictive accuracy. In the past, this 'proof' has often been obtained by adjusting the input parameters, within an acceptable range, until the agreement between the measured and predicted data is optimal. The MVDS considered it important to progress further and to attempt to quantify the effect of the various uncertainties which surround model predictions. Such uncertainties may arise from inherent inaccuracies in the measuring equipment associated with the climatic parameters, from poor knowledge of the thermophysical properties of the cell materials or from the need to assume values for unmeasured parameters such as ground reflectivity. The effect of these uncertainties is to produce a band of uncertainty around the predicted variable, in this case the cell's internal air temperature. The differential sensitivity techniques employed by the MVDS is one way to estimate this band.

To give an example of the use of the technique: Assuming that the base case prediction was based on a conductivity of 0.033 $Wm^{-1}°C^{-1}$ for the cell's insulation material. From experience it is known that this value is subject to a small variation, say ±5%. Two further simulations are therefore performed, one assuming the lower conductivity value, the other the higher value. This allows the effect of the uncertainty associated with insulation conductivity to be expressed in terms of a median prediction with upper and lower limits imposed. By repeating this process for each parameter in turn, the total uncertainty band is obtained as the root-mean-squared summation of the upper and lower parameter limits.

Figure 1 shows the result of the application of this procedure in the context of the previously described experiment. As can be seen, the predictions are at all times within the overall uncertainty band although the predicted cooling rate is less than that measured. It was concluded that this was due to the abnormally large difference between the cell's internal and external surface areas causing an additional, but unmodelled heat loss path. The incorporation within ESP of a model of edge effects and thermal bridges has therefore been identified as a desirable future development.

4.5 *Assumptions Testing*

Several experiments were undertaken which fall into this category. One of these involved the measurement of the spatial variation in external air temperature in the vicinity of the test cell. This was done to explore the commonly made assumption that a building's boundary condition can be adequately represented by a single value. From ESP sensitivity studies it had been established that a variation of the order of 2°C would translate to a change of about 0.3°C in the internal air temperature. It was important therefore to establishing the magnitude of the actual variations which could be expected in reality.

To do this, temperature time-series from two sites, in Catania and Glasgow, were studied. These data were obtained from sensors located adjacent to the south and north facades, and beneath the cells (which are not in direct contact with the ground). By comparing these data to data from a reference sensor located at some distance from the Cell, it was shown that in the case of Catania in summer, significant temperature increases, of the order of 5°C could be expected. Whereas, for Glasgow, the 2°C threshold would rarely be crossed. These results suggest that, for hot climates at least, there is a need to take account of micro-climatic effects in future models.

5. Results to Date

In summary the following has been achieved:

- A validation methodology has been approved and selectively applied by the different national groups.

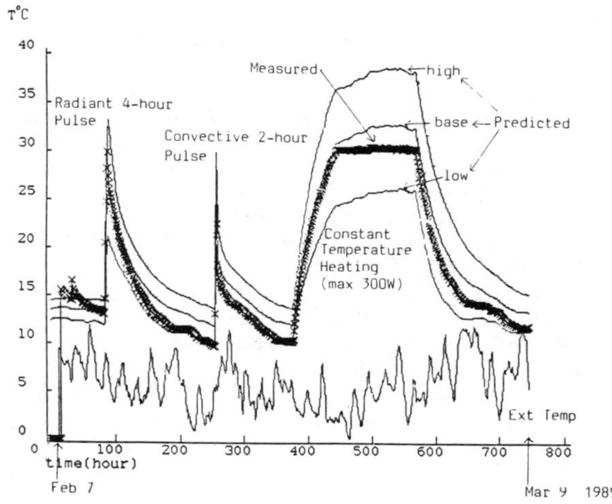

Figure 1: Whole Model Experiment on Calibration Wall

- A strategy for empirical data capture and use has been agreed and a small number of experiments have been initiated to test the process.
- Several programs have been developed to assist in the experimental design and analysis tasks. These include programs for data checking and handling, results analysis by graphical and statistical means, the tracking of principal flowpaths and their causes and the differential sensitivity technique developed by Leicester Polytechnic.

These activities have lead to the refinement of the ESP system in several important respects: several pernicious bugs have been rectified and some algorithms replaced or extended. In addition three new program modules have been developed to permit the installation of monitored data within ESP, to permit a more detailed assessment of human comfort and to permit better results tabulation. These changes to ESP are reported in detail elsewhere [9].

6. Future Work

The stated aims of the MVDS for a possible future phase of PASSYS are:

- To further study the applicability of ESP as a reference code for the PASSYS simplified design tool and for CEN-related activities.
- To continue the validation and refinement of ESP.
- To add new features to ESP to improve its capabilities with respect to passive solar systems.
- To critically assess the input data requirements of detailed simulation models.

With respect to the empirical validation component, the work in the first phase of PASSYS has identified the range of empirical validation experiments [10] which would have meaning in the context of ESP. A more detailed specification was then developed for a subset of these experiments [11] which were commensurate with test cell capabilities. The next stage will be to implement these latter experiments, some running concurrently at two or more sites. Such experiments will focus on testing model assumptions such as the appropriateness of operating with one dimensional conduction theories; on process testing such as inclined surface shortwave radiation under different sky conditions typical of the range experienced throughout Europe; or on passive solar components such as conservatories and advanced glazing systems.

As ESP's predictive accuracy improves so new possibilities are enabled. For example, it is envisaged that the model will be used as a scaling device. Here the intention is to assess the potential for the practical application of the results obtained from the PASSYS test cells. Although the test cells have the advantage that they are room-sized, it is recognised that they are atypical with regard to their constructional materials and heat loss coefficients. There is therefore a need to scale the performance indicators derived from measurements on the test cells to a real building in order to assess the likely performance benefits. It would even be possible to use the model to assess the technologies replication potential. This would require that the model be used to explore the impact on alternative building designs when placed in different geographical locations.

7. Conclusions

The following can be concluded from the work of the MVDS to date.

- Concerning ESP's treatment of climatic data, several routines have been checked and some code errors have been corrected. In particular the routine for the calculation of inclined surface irradiance was found to perform well for some latitudes while under-predicting for others. It has been concluded that there is a case to be made for the use of location-dependent algorithms and to this end the performance of several algorithms have been studied in the context of the different European climates.

  Based on measurements of the spatial distribution of dry bulb temperatures around a Test Cell, some doubt has been cast on the use of a single temperature parameter to define boundary conditions for sites at a southerly latitude.

- Concerning external surface convection, it has been shown that uncertainties in the value of the heat transfer coefficient can lead to significant errors in the computation of the auxiliary heating requirement. It has further been concluded that the present h_c algorithm is too restrictive with respect to the resolution of the free stream wind component on to the building surfaces. The removal of this restriction is especially important in the case of passive solar components such as conservatories.

- Concerning external surface longwave radiation, sensitivity studies have shown that this component has a major impact on the performance of passive solar buildings which incorporate a high level of glazing.

  The original ESP model has been improved in several respects including the treatment of site view factors and the replacement of the equations employed to estimate sky radiation. Other areas, such as time-step estimation of cloud cover, have been the subject of preliminary studies.

- Concerning site shading, some limited empirical tests of ESP's algorithms have been performed which showed good agreement for simple geometries and other, more complex tests have been devised for future application.

- Concerning fabric conduction, the robustness of ESP's algorithms has been demonstrated by the processes of theory and code checking, analytical testing and sensitivity studies.

  However several areas of concern have been identified. These include the need for better user information on the required space- and time-steps; a study of the circumstances in which variable time-stepping should be adopted; and a study of the possibilities for the treatment of 2- and 3-dimensional conduction, thermal bridges and corner effects.

- Concerning internal surface convection, it can be concluded that the ESP data correlations are state-of-the-art although improvements could be made for the case of very low surface-to-air temperature differences.

- Concerning air flow, a limited amount of inter-model comparison has been undertaken and a review of the theory and code checking performed. On the development side, the speed of the numerical solver has been dramatically improved.

- Concerning internal longwave radiation, ESP's algorithms have been shown to compare well with analytical solutions. For those atypical cases where the goodness of fit is unacceptable, improvement opportunities have been identified and in some cases implemented.

- Concerning thermal comfort, new algorithms have been installed according to the theories of Gagge and Fanger (PMV and PPD). The calculation of room mean radiant temperature has also been improved to operate in terms of subtended solid angles and not surface area weightings as previously.

- Concerning controls, sensitivity studies have shown that simulation results are highly dependent on the specified control system (sensor and actuator positions and controller characteristics). As a consequence this topic has been identified for future investigation.

  It has been shown that ESP's controllers perform acceptably although some improvements in the treatment of mixed sensor types in combination with convective/ radiant plant input should be explored.

- Concerning the modelling of passive solar components, work on this subject has been devoted to an investigation of how passive solar components are treated in ESP. While ESP can handle most current passive solar components, it has no inherent knowledge on the best performance assessment methodology, which must be devised by the user. A major task for a possible next phase of PASSYS will be to develop such methodologies.

- Concerning whole model validation, a small number of empirical data-sets have been generated and used, in conjunction with the differential sensitivity technique, to assess ESP predictions. It is envisaged that this activity will increase substantially in the future.

- Concerning ESP itself, new routines and program modules have been developed and several coding errors removed.

In conclusion, over the past three years, a considerable effort has been expended to establish the infra-structure of the MVDS and to approve and apply a validation methodology. Each national consortia now has experience of Unix workstation technology and is familiar, at source code level, with the ESP system. This should mean that the group is well equipped to make a significant, collaborative contribution to the state-of-the-art in the simulation of passive solar systems and to the validation of the model's themselves.

References

1. Clarke J A *Energy Simulation in Building Design* Adam Hilger Ltd, Bristol and Boston, 1985.
2. Dupagne A *Final Report of Passive Solar Modelling Subgroup* Commission of the European Communities, DGXII, 1986.
3. Achard P and Gicquel R (eds) *European Passive Solar Handbook* Commission of the European Communities, DGXII, 1986.
4. Ostergaard Jensen S (Editor), 'Final Report: PASSYS Model Validation and Development Subgroup' Commission of the European Communities, DGXII, 1989.
5. Judkoff R et al 'A Methodology for validating building energy analysis simulations' *Report TR-254-1508* Solar Energy Research Inst. 1983.
6. Bloomfield D et al 'An investigation into analytical and empirical validation techniques for dynamic thermal models of buildings' *Final Grant Report* Science and Engineering Research Council, Polaris House, Swindon, England, 1988.
7. Lomas K J and Bowman N T 'The Evaluation and Use of Existing Data Sets for Validating Dynamic Thermal Models of Buildings', *Proc. 5th CIBSE Int. Symp. Use of Computers for Env. Eng. Related to Bldgs.* Bath, 7-9 July 1986.
8. Lomas K J and Bowman N T 'An investigation into analytical and empirical validation techniques for dynamic thermal models of buildings' *Report for Grant GR/C/62871*, Science and Engineering Research Council, Polaris House, Swindon, England, 1988.
9. Clarke J A 'ESP Upgrades Within the PASSYS Project', ESRU Technical Report, University of Strathclyde, November 1987.
10. Clarke J A 'Some Suggestions for the Empirical Validation of ESP', ESRU Technical Report, University of Strathclyde, November 1987.
11. Baker P, Clarke J A, Strachan P and Pinney A 'Experiments for the Empirical Validation of ESP', ESRU Technical Report University of Strathclyde, May 1988.

DEVELOPMENT OF SIMPLIFIED DESIGN TOOLS FOR PASSIVE SOLAR SYSTEMS WITHIN THE CEC CONCERTED ACTION PASSYS

Luc Bourdeau
Centre Scientifique et Technique du Bâtiment
BP 141 F-06561 Valbonne Cedex

ABSTRACT. One of the goal of the CEC Concerted Action Passys is to provide designers with appropriate Simplified Design Tools (SDT) enabling the assessment of passive solar components and concepts from the points of view of energy and comfort. The SDT working group of Passys is carrying out studies in order to be able to propose in late 1991 a validated correlation based calculation method, called PASSPORT, which will meet the goals described above. Investigations are currently done in order to define a pertinent correlation at the European level and to improve some topics not yet well taken into accounts in the methods available up to now : thermal inertia, outdoor radiation of the building envelop, control on inside conditions, multizoning, intermittent heating, passive solar components, comfort. The main results obtained during the first phase of PASSYS are reported in this paper.

1. INTRODUCTION

The project PASSYS /1/ is a European concerted action in the field of passive solar component and system testing. The project is financially supported by the Commission of the European Communities and by national organisations such as the Agence Française pour la Maitrise de l' Energie. The activities of PASSYS include:
i) to increase confidence in detailed simulation tools /2/,
ii) to develop reliable and affordable test procedures for passive solar components and systems /3/ to be tested on the PASSYS cells,
iii) to increase confidence in simplified design tools, which have to be further developed and validated.
This third important activity of PASSYS is shortly presented in this paper.

2. THE CONCEPT OF SIMPLIFIED EVALUATION TOOLS

One of the concerns of PASSYS is to provide designers with appropriate simplified design tools (SDT) which enable the assessment of passive solar components (PSC) and passive solar concepts from the points of view of energy and comfort. In fact, the architectural design process is a mixture of various preoccupations among which the thermal considerations often appear to the architect as a tedious one and is therefore often neglected or left to the engineers' responsability. As a matter of fact, designing an energy efficient building assumes taking account of the thermal considerations from the first step of the thermal process, when the envelop and the general organization of the buildingbegin to take form.

Several ways can be suggested in which appropriate SDT can be produced. In fact, considering the various approaches and the successive steps of the architectural design process, it turns out that what is needed is a series of SDT, of different complexities, which would allow designers to obtain either rough estimates and design guidelines, or more detailed assessments.
It is clear that the best tool would be a Computer Aided Design environment allowing all the designing participants to work together on the project and to treat all the joined aspects in the same time, instead of taking part one after the other and considering every problem separetely. In this sense, the SDT mentioned above would be a part of all the Building Performance Evaluation tools incorporated in the software environment. This idea is the basis of the new project COMBINE today launched by the CEC.
In this context, the SDT working group of PASSYS decided to concentrate its efforts on the concept of simplified evaluation tools. As a matter of fact, detailed simulation models are not yet suitable for most of possible users (input burden, too large computing time, difficult interpretation of the results,...), even if powerful graphical and intelligent interfaces can be envisaged in order to increase the utilizability. This choice was approved by people from other CEC projects (SOLINFO and ARCHISOL) in which involved professional bodies clearly indicated their agreement for the development of a simplified calculation tool.

3. THE PASSPORT METHOD

3.1 A correlation base evaluation method

Beside simplified simulation tools obtained by mode or data reduction techniques and which may constitute a solution for developing future more sophisticated simplified calculation tools, correlation based tools certainly appeared as a pertinent product to be develop first in the frame of the project PASSYS. As a matter of fact, such tools are currently used in the building world quite widely either for regulatory or performance assessment purposes. However a large need of validation and improvement has been identified for these calculation tools.
It is why the SDT working group of PASSYS is presently carrying out works in order to be able to propose at the end of the project (late 1991) a validated version of a correlation based evaluation tool enabling residential building heat requirement assessment, PSC performance assessment, together with an appraisal of comfort during the heating season. This package will be available in a final report and a software running on a PC will illustrate the using of this tool. The name PASSPORT has been chosen for this package.

3.2 Framework of the method

The simplified model used in the correlation evaluation method PASSPORT is based on the following expression of the heating needs of a building Q_{hn} :

$$Q_{hn} = Q_{load} - \eta . Q_{gains}$$

where Q_{load} represents the total heat losses from the dwelling supposed to have an inside temperature equal to the set-point temperature, Q_{gains} the total gains due to occupancy and solar, η the recovery factor of these gains.
Among the various correlations between η and several parameters which have been so far established by different authors, the one which correlates η with the gains to load ratio (GLR) given by:

$$GLR = Q_{gains} / Q_{load}$$

has been chosen as a starting point and reference in the validation/development work.

3.3 Validation/development methodology

To improve and increase the expert knowledge in the field of passive components and passive buildings, PASSYS has at its disposal two powerful tools: i)one constituted of about 30 identical test cells for studies on components, ii)the other one constituted of a network of about 10 workstations enabling to run the ESP building energy simulation model developed by ABACUS at the University of Strathclyde in Scotland. This simulation model has been chosen as a reference for PASSYS work and is the object of numerous validation work in the Model Validation and Development working group of PASSYS.

By simulating and analyzing through the use of the reference tool ESP the behavior of buildings with different passive solar concepts and designs, by performing sensitivity studies on the influence of the main parameters, information is derived to define appropriate formulations or correlations, pertinent assumptions or simplifications and to precise the domain of application of the methods.

Figure 1 shows the basic methodology used for the development/validation of a correlation based calculation tool. The aim of the work is to be sure that the heating needs HN2 given by the simplified model and the simplified data are coherent with the heating needs HN1 given by the precise reference model which has been chosen and which is supposed to describe correctly the actual behavior of the building, the occupant and the environment.

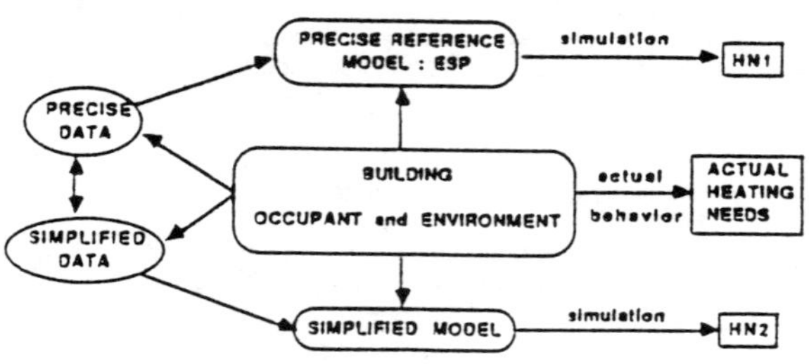

Fig.1. Links between precise and simplified models

4. DEVELOPMENT OF THE METHOD

4.1 Introduction

Several aspects of the correlation based calculation methods have been identified as needing specific studies. Analyses on some of these aspects have been carried out during the first phase of PASSYS. These analyses are briefly presented below.

4.2 Correlation development/validation

The method which is proposed provides monthly heating requirements through a correlation based on the heat losses, the free gains and the thermal inertia of the building.

SImulations performed with ESP have been done and are still under progress in order to validate the correlation on a Europen basis. In a first step, simulations have been performed on monozone buildings with different thermal inertia, glazed surface areas, heat losses and climatic data. These simulations proved the GLR approach to be pertinent.

Additional studies have been launched in order to perform simulation runs on multizone buildings with the view of taking account of such aspects as intermittent heating. This approach is much more complicated since it becomes then complex to derive the value of the correlated parameters from the ESP runs and to link them to the calculation process proposed in the simplified method. A sharp analysis has been carried out on this topic and several procedures have been tested to derive these parameters. These procedures rely on a certain number of simulation runs necessary to define the values of all the parameters for a given point of the correlation curve. Appropriate procedures have then been defined.

4.3 Thermal inertia

Simplified calculation tools have to take explicitely into account the thermal inertia of a building since this parameter quantifies its dynamic behavior. It has been proposed to use a method allowing the designer to approximate the inertia of a building by it first time constant.

Such an approximation is issued from a rigorous definition given to the thermal inertia of a building with the help of the modal method. The modal analysis enables to define the dynamic response of a building by evaluating its inertia as a whole set of eigenmodes. To each eigenmode is associated one time constant. The eigenmode with the highest time constant has usually the dominant effect on the long term building thermal response.

It has been shown that it is possible to approximate the main time constant of a building as the ratio of the energy stored in the building for a 1 K inside/outside temperature difference to its loss coefficient. The energy stored can be easily evaluated from the thermal characteristics of the walls of the building.

4.4 Outdoor radiation of the building envelope

External longwave thermal radiation (ELWR) on both glazed and opaque surfaces is quite important in passive solar applications. As a matter of fact, a surface sensitive to solar radiation is also subject to infrared radiation towards the sky. Sensitivity studies show the importance of this phenomena.

The methods today available generally neglect it or take account of it partially by introducing only heat losses proportional to the air temperature difference between indoors and outdoors. More or less simplified ways can be followed to increase the accuracy of the methods. These ways have been studied and analyzed for various European areas and component types.

Two options have been identified. A first one is to consider that the solar gains through opaque walls are of the same order of magnitude as the heat losses by ELWR not taken into account in the heat losses assessment. A second one is to allow for a more precise evaluation by subtracting the additional ELWR heat losses occuring through all surfaces (both glazed or opaque), from the solar gains occuring through opaque walls. Expressions for this evaluation have been proposed.

4.5 Control on inside conditions

The definition of the indoor temperature plays a role in the control of the heating system, in the determining of the thermal comfort and in heat losses calculation. The resultant temperature appears to be the best foundation for SDT development.

In detailed models different indoor temperatures can be considered as separate parameters. In simplified approaches it is preferred to use one single, representative value for the indoor climatic parameters. It has been demonstrated that there is a strong interaction between this definition and the interpretation of the results from a detailed model, the determination of thermal comfort and the creation and/or use of a simplified correlation method.

The results and consequences of these analyses have been used in the work done to develop the PASSPORT method.

4.6 Multizoning

Most existing design tools treat the building as a single zone, assuming then an internal homogeneous temperature, a unique set temperature and a unique heating control. But passive solar buildings, intermittent heating, as well as buildings with a free-floating heating mode are cases where it is not always possible to treat a building as a unique zone.

For instance, the heat gains from solar contribution during some hours can cause, mainly in sunny regions, overheating in the southern rooms of the building whereas the northern rooms might need auxiliary heating if heat transfer from the southern rooms is inadequate. This heat exchange depends on

both the conductivity of the separating walls and the mutual air movements between the rooms.
Sensitivity studies and analyses have demonstrated two possible ways of dealing with this multizoning effect: an implicit way and an explicit way.
The implicit way does not allow the designer to consider the multizoned configuration as an input to the calculation method. However the effects of multizoning are taken into account when developing the simplified calculation method. In particular, the recovery factor of the solar gains has to be evaluated through detailed simulations of typical multizoned configurations in order to derive correlation curves incorporating south/north transfer effects.
The explicit way generates a more complicated process for the designer but allows him more flexibility. This way is based on the definition of temperature uniform zones in a given multizoned building. The calculation method is then applied to each zone. The only practical difficulty is that the temperatures of the adjacent zones of a given zone are not known beforehand, but have to be evaluated by applying the calculation method to those zones too. So, due to the interaction between the zones, an iterative procedure is needed to solve the heat balances for all zones.

4.7 Intermittent heating and operation

Among the means aiming at saving energy for heating, both user habits and national regulations refer to various possible indoor temperature operation patterns. For instance, the heating plant can be turned on only during selected periods, or the temperature controller can be setted down during nigthtime, or heating can be provided only in specific zones and at different set points. These various patterns will obviously interfere with the recovery of the solar gains.
Preliminary sensitivity analyses have been performed by means of ESP runs on the reference buildings in order to get data on this difficult problem. An intermittent heating pattern used in Italy (a given permitted number of operation hours depending on the location) and three temperature set-back profiles have been particularly studied.
Several existing methods try to take into account the effects of intermittent heating. This is generally done through a correction applied on the inside temperature or the auxiliary heating neeeds, as a function of one or several of the following parameters: the night set-point, the outside temperature, the heat loss rate, the inertia of the building, and the type of the heating system and control.
None of these methods treats separately the two phenomena due to the intermittent heating which are a decrease of the losses due to the lower inside temperature and a decrease of the utilisation factor of the available solar gains due to the fact that these gains may occur when the building is not heated. This separation may appear not pertinent for the night set-back situation where the interactions with the solar gains are very little, but seems more logical in the general cases and particularly in the daytime turn-off or set-back situations.
Therefore the analyses in progress on this topic aim at developing such a way of dealing the intermittent situations in a simplified correlation method.

4.8 Passive solar components

The method which will be proposed must be able to take into account most of the possibly used components of the envelope, for instance through appropriate ways of calculating their thermal characteristics or using data given by tests done on the PASSYS cells. The ways of treating these data by the simplified correlation method have to be validated. Attention is to be beared firstly on specific direct gain components (multiple glazing, movable insulation, blinds and low emissivity glazing), and secondly on indirect components (grennhouses, mass walls,...). The analyses done so far show that single and double glazing can be treated through the same correlation curve, if the appropriate time constant - which depends on the losses and therefore on the type of glazing - is introduced into the calculation.

The method which will be proposed also permits to assess the performance of a component used in a given situation. As a matter of fact results obtained for instance on test cells do not allow to determine directly the performance of components. These results can give only the value of some parameters which can be considered as characteristics of the component such as a U value or a solar aperture. When interested in the actual performance of a component (for instance in term of useful kWh provided by it to the building during the heating season), one has to take account of numerous parameters such as climate, occupants and configuration of the building in which the component is considered to be installed. All these parameters cannot be assessed in a test cell and an appropriate calculation method needs to be used. The proposed method will be able to give this performance.

4.9 Comfort

A review of the most important tools which enable to describe the comfort sensations, to model the excitations or to appreciate the comfort quality of an architectural zone has been performed.
The idea of proposing a method allowing the designer to deduce a unique comfort index from the correlation based calculation of the heating requirements has then been adopted.
From indoor temperature frequency curves and the Fanger's equation giving the variation of the PMV versus the inside temperature, it is possible to derive a PMV frequency curve. A PPD distribution curve can then be built. Probabilistic calculations can be applied to this distribution curve and a mean PPD can be deduced by weighting each PPD by its frequency. As it is more usual to indicate PMV values, the mean PPD value should be transformed in a weighted predicted mean vote (WPMV) which is the comfort index proposed. This index is only valid for the heating season.

5. CONCLUSION

At the end of the first phase of the project PASSYS, the framework of a European correlation based calculation method, called PASSPORT, has been defined. This method will allow designers to evaluate the heating requirements of residential building, together with an assessment of comfort during the heating season. It will also permit the evaluation of the performances of the main passive solar components. A development/validation methodology has been defined and applied. Some aspects of this method have been or will be shortly validated. Studies will be pursued in order to provide a completed version at the end of the program in late 1991. During the second phase of PASSYS, links will be more strongly established with the CEN Technical Committees in charge of standardization in that field.
A report is about to be published which gives the details of the work carried out during the first phase.

6. ACKNOWLEDGMENT

The main contributors to the work of the group SDT of PASSYS in the first phase were I. Blanc Sommereux and G. Lefebvre (EMP-F), R. Van de Perre (VUB-B) "focal point" of the subgroup, S. Sciuto and R. Lo Cicero Vaina (CONPHOEBUS-I), D. Van Dijk (TNO-NL), A. Oliveira (FEUP-P), C. Buscarlet and L. Bourdeau (CSTB-B) leader of the subgroup. This paper could not have been written without their help.

7. REFERENCES

/1/ Gicquel R.,Cools C. The CEC Project PASSYS. 2nd European Conference on Architecture. Paris, December 1989. WIP-Munich.
/2/ Clarke J.A. The Model Validation Activities of Project PASSYS. Idem.
/3/ Van Dijk H.A.L., Antinucci M. Towards a Uniform European Testmethod for Passive Solar Systems. Idem.

PROFESSIONAL ENERGIES: PUTTING SCIENCE INTO PRACTICE

E Shove
Institute of Advanced Architectural Studies
University of York
The King's Manor
York YO1 2EP
England

ABSTRACT. Developments in building technology are of little practical value unless adopted by building professionals. Designers' reluctance to take advantage of both new and established energy saving measures therefore frustrates the application of technological advance.
Work for the UK's Building Research Energy Conservation Support Unit provided an opportunity to examine the practical application of energy related knowledge in three Local Authorities and in the process to explore what at first seemed to be their inexplicable reluctance to improve the energy standards of rehabilitated housing.
This study showed that knowledge of new technology did not ensure its use: it was not enough to provide information and advice about energy saving measures in the hope that designers would automatically take note and amend their ways. More important, failure to adopt such measures could not be explained simply in terms of personal prejudice or blind traditionalism. It was not that designers disregarded the self-evidently sensible message of energy efficiency, but rather that decisions about energy standards reflected the genuinely complex social situations within which they were made. In this, as in other contexts, financial, organisational and professional pressures mediate the practical application of standard scientific knowledge with understandably "illogical" results.

1. INTRODUCTION

Baffled and confused by designers' apparently perverse reluctance to adopt basic energy saving measures, energy experts often conclude that they are ignorant or, at best, wilfully conservative. This paper presents an alternative explanation and suggests that standards of energy efficiency have as much to do with organisational structures and inter-professional relations as with the knowledge, enthusiasm or commitment of individual designers. This general argument is developed with reference to a study of the translation of energy-related knowledge into energy-saving practice in three British Local Authorities.

2. THE LOCAL AUTHORITY PROJECT

Research has long showed that it is easier and cheaper to incorporate energy saving measures when systematically rehabilitating housing than to add draughtstripping, insulation or new heating systems piecemeal. Yet Local Authorities were not taking advantage of these opportunities and it was therefore assumed that their designers were unaware of the benefits of a comprehensive approach or else lacking in the necessary technical expertise. The Institute of Advanced Architectural Studies was asked to develop an educational programme to overcome these presumed deficiencies.

2.1 Knowledge of Energy Efficiency

Interviews with Local Authority staff soon showed that those responsible for rehabilitation work were in fact already well informed - they knew that a combination of draughtstripping, additional insulation and an efficient heating system would help to produce warmer homes with lower fuel bills and less condensation, and they knew that it made sense to include these measures when rehabilitating or upgrading housing. And yet none routinely put this knowledge into practice. Local Authority staff were clearly capable of exploiting the energy-saving opportunities afforded by recent developments in science and technology but failed to make use of this expertise.

2.2 Putting knowledge into practice

It was necessary to understand more about the factors inhibiting the practical application of existing skills if educational efforts were to succeed. We already knew that the energy standards of housing rehabilitation depended on the views and actions of a number of different departments and on a variety of policies and preferences, but needed to find out exactly how these factors affected the translation of energy-related knowledge into energy-saving practice. This led to a focus on the broad political and financial context, on the Authority's rehabilitation programme itself and on its view of the landlord/tenant relationship.

2.3 Policies, programmes and finance

It is easy to argue that programmes of comprehensive rehabilitation provide better value for money than those involving the selective improvement of, say, kitchens, heating systems, or bathrooms. In the event, however, Authorities found it politically difficult to decide whether to concentrate limited resources on only a few houses and only a few tenants or to spread those funds across more properties and more people. Opportunities for energy efficiency obviously depend on the results of these critical policy decisions: designers simply don't get the chance to use their energy expertise if their Authority chooses to make only small improvements to a large number of dwellings.
Decisions about energy standards also reflect very general understandings of the relationship between landlord and tenant. It is the tenant, not the landlord, who benefits from additional energy-related expenditure, and views differ about the degree to which Local Authorities should invest public money in measures which help to reduce tenants' fuel bills. Again there is little that building professionals can do if they work for an organisation which does its best to provide habitable, secure, and weathertight housing but with no particular emphasis on energy efficiency. So, designers will only be in a position to take advantage of technological advances in the field of energy efficiency if their Authority is involved with housing, if it adopts a comprehensive as opposed to a selective approach towards such work,

if it believes that it has a responsibility to provide thermally efficient homes, and if it can adjust financial priorities accordingly. Assuming that all these conditions are met, what then prevents Authorities from incorporating integrated packages of energy saving measures? To answer this question it is necessary to look at departmental structures and their influence on energy-related decision-making.

2.4 Departmental structure

The pilot educational programme was originally intended to inform Local Authorities, to encourage them to take advantage of opportunities for energy efficient housing rehabilitation and to show them how to evaluate effective combinations of energy saving measures. This meant that Authority staff had to take a global view of heating, insulation and ventilation, and review alternative options with reference to this total energy system. In fact designers were already well aware of the value of co-ordinated energy strategies but were again constrained by the organisational context within which they worked.

Detailed decisions about levels of insulation, types of heating system, and forms of tenant advice were made by different departments with reference to quite different criteria. So, for example, the Engineers' department would select and sometimes install heating systems. Providing these were technically efficient that department could do little more to further the cause of energy efficiency. Meanwhile, the Architects' Department was responsible for choices about building fabric. Here, too, opportunities to influence levels of energy efficiency were bounded by the department's function and role. In this case staff were left to make decisions about U values. Maintenance surveyors worked within equally tightly defined limits. Their job was to carry out minor energy-related repairs, not to investigate problems in terms of relationships between buildings, heating systems and users. And on top of all that, questions of energy efficiency emerged amongst the range of financial and social problems confronting Housing Departments. Fuel debt, cold and condensation were all important issues but in this context addressed as domestic rather than technical difficulties. Reasonably enough each department tackled only those problems which it considered to be its proper responsibility. Furthermore, each department's approach towards energy efficiency reflected its specific involvement with only one element of the total energy system. It seemed, then, as if the Local Authority's organisational structure effectively prohibited the kind of conceptual co-ordination required to produce integrated energy efficient designs.

If this was so, then the problem was to by-pass departmental divisions in order to release the energy expertise currently trapped within the bureaucratic structure. The Institute of Advanced Architectural Studies' innocently "educational" programme was therefore designed to change the organisation by developing inter-departmental links and introducing a common policy for energy efficient housing rehabilitation. This was implemented through a series of meetings leading to a one-day workshop and a working group report which in turn formed the basis for an energy policy. The programme involved staff from all relevant departments and, as a result, the three participating organisations now have a shared Authority-wide approach towards energy efficient rehabilitation. To that extent the project was a success.

2.5 Professional practices

But further analysis of the departments involved undermined that happy conclusion. Because the term "energy" had different meaning in each professional language there was in fact little common ground for communication or co-operation. Differences of professional perspective paralleled departmental divisions and, in forming an apparently agreed policy, participants papered over quite significant disciplinary differences. When Architects, Housing Managers and Engineers talked of energy efficiency they still talked of quite different things. Engineers, for example, took the term to refer to the scientifically established performance of a particular heating system; Architects visualised U values; Housing Managers were reminded of debt and fuel poverty; while Maintenance staff thought of condensation and failed appliances.

Perhaps more important, these invisible professional structures ordered everyday decision-making sequences. Common energy policy or no, Architects still expected to pass schemes on to Engineers who would then select appropriate heating equipment, whereupon tenants would move in and cope with the resulting arrangements and maintenance and Housing Management staff would eventually try to resolve any subsequent problems.

Although the "educational" programme circumvented some of the organisational factors obstructing integrated energy efficient design it was clear that traditional beliefs persisted and that established professional practices were altogether more difficult to change. Even if departmental boundaries were demolished, surviving professional divisions would still make it difficult to develop and implement a coherent approach towards energy efficient rehabilitation. Thus, although the project had some effect on energy standards it could not realistically change normal working practices to the degree required to realise the full potential of current energy saving technology.

3. CONTEXTS OF INNOVATION

The Local Authority workshops raise a number of general issues about the problems of putting science into practice. Technical knowledge is clearly a prerequisite for constructive innovation but it is not, in itself, enough to ensure that technological potential is realised. Instead, as we have seen, the chances of putting energy expertise into practice depend on three quite distinct factors.

3.1 Opportunities

First, purposes and policies of employing organisations determine the nature of the workload and available finance and so determine opportunities for energy efficiency. Individuals only have the opportunity to apply skills relevant to their job and it may be that their work situation is such that there simply isn't any scope for energy conservation.

3.2 Organisations

Secondly, even if such opportunities exist, departmental divisions may conspire to prevent co-ordination of the kind required to produce energy efficient combinations of heating system and building fabric.

3.3 Professions

Thirdly, traditional disciplinary distinctions cut across energy-related decision-making. As a result, choices about heating systems and about building fabric are conventionally made by different professions at different times with reference to quite different criteria.

4. A COMPARATIVE REVIEW

These same three limiting factors order the translation of knowledge into action in all sorts of contexts, though the particular formulation of opportunity, organisational constraint and disciplinary division will vary from case to case and from culture to culture. The following comparison of the pressures and interests ordering energy-related decisions in the world of new private sector housebuilding with those impinging on Local Authority housing rehabilitation illustrates the potential of such analysis.

4.1 Opportunities: Private sector housebuilding

In the private sector the volume and character of new housebuilding depends on the market and company policy. Cost limits are similarly determined, as are expectations about appropriate and acceptable building standards. In this world the price of land makes a very real difference to the number and quality of houses built and to expectations about the anticipated financial return needed to justify "extra" energy saving measures. In contrast, opportunities for energy efficiency in the Local Authority context were structured by political pressures and by conceptions of tenant's rights and landlord's obligations which in turn influenced rehabilitation programmes and associated financial priorities.

4.2 Organisations: Private sector housebuilding

Designers control over heating systems and levels of insulation depends on the way in which their housebuilding company is organised, on the relative influence of sales and marketing departments, on relationships between national and regional technical departments and on the role of the sub-contractors. In many cases detailed design of heating systems is left to a sub-contractor who is concerned to minimise risk and at the same time maximise opportunities for profit. Choices about the type and size of central heating systems may therefore depend as much on bulk buying agreements and assumptions about customer preference for colour and appearance as on considerations of energy efficiency. In any event, responsibility for design of the heating system and building fabric is usually divided, if for commercial rather than bureaucratic reasons.

4.3 Professions: Private sector housebuilding

These "non-technical" characteristics influence the application of energy expertise just as departmental divisions ordered that process in the Local Authority setting. And, once more, deep rooted beliefs about the proper division of professional labour underlie many routine working practices.

4.4. Criteria in context

Although methods of energy conservation clearly differ for new and for rehabilitated projects, housebuilders and Local Authority designers face basically the same challenge. Both are involved with housing and both are able to take advantage of opportunities to develop integrated, cost-effective packages of energy saving measures. But, as suggested here, the social, political, and economic characteristics of each situation mean that formally similar energy-related decisions are made with reference to context specific and therefore radically different criteria.

5. THE SOCIAL STRUCTURE OF ENERGY EFFICIENCY

Essentially the argument is that the application of energy saving technology depends on a mixture of practical and financial opportunity, on specific organisational constraints, on taken-for-granted expectations of proper professional responsibility and only partly on the technical knowledge and expertise of individual designers. The future of energy conservation does not therefore depend on the successful conversion of apathetic and prejudiced individuals by a band of crusading "energy evangelists". Such a view is misplaced, for the real problem in the Local Authority world, as in that of private housebuilding, is that the application of designers' existing energy expertise is restricted by entirely comprehensible social and organisational factors. Those concerned to improve energy standards need to redirect their efforts toward understanding the specific combinations of social pressures which constrain and enable the application of technical knowledge in particular settings. To unlock existing knowledge is to find a way through the social obstacles currently preventing its translation into energy-efficient practice. So, perhaps it is social scientists, not technologists, who have the key to energy efficiency.

6. SCIENCE AND ARCHITECTURE

Indeed, the very title of this conference - "Science and Technology at the Service of Architecture" - reflects the central problem. If scientists prove that energy saving measures are more efficient in combination, then that is so whether they are applied to an owner-occupied house, to a house owned by a Local Authority, to a house in France, a house in Spain, or a house in Germany. Hence the standard arguments for energy efficient design and hence, too, the frustration with all those designers who seem to ignore this common-sense advice. The difficulty is that the very real pressures and interests shaping the situations within which energy-related decisions are made have a logic of their own, and the case for energy efficiency may simply crumble in the face of competing priorities. If we recognise that decisions about energy standards reflect situationally specific constraints then we have to conclude that there is no such thing as a single Architecture which Science might serve. Instead, there are architects working within an immense variety of financial, organisational and professional constraints, and consequently only able to utilise scientific and technological knowledge where those social circumstances allow.

ANALYSE COMPARATIVE DE MODELES POUR L'EVALUATION DES PERFORMANCES ENERGETIQUES DES BATIMENTS SUR SITE

V.RICHALET
E.N.T.P.E / C.E.T.E LYON
Rue M.Audin
69518 Vaulx en Velin Cedex

F.FLOUQUET
C.N.R.S / CRTBT
25 Avenue des Martyres. BP 166X
38048 Grenoble

Abstract: Building Energy Signature Models are useful tools to calculate in-situ steady-state building parameters. There are many kinds of them, used for several years over the world. Some results of a larger comparative study of these models, applied to a sample of simulated data set, are presented here.

1 . INTRODUCTION

L'intérêt de l'expérimentation des bâtiments in-situ n'est plus à démontrer mais les moyens dont on dispose pour le faire sont encore peu fiables. Un de ces moyens consiste à analyser le comportement entrées /sorties du bâtiment, par exemple par l'étude de sa consommation en fonction des paramètres climatiques. Pour cela et par souci de simplicité on a d'abord cherché à travailler avec des méthodes statiques, plus connues sous le terme de Signature Energétique, dont il nous semblait primordial de préciser les limites.

2. METHODOLOGIE

Un des objectifs principaux de cette étude étant de connaître l'écart entre les caractéristiques du bâtiment calculées par la Signature Energétique (S.E) et la "réalité" , il était indispensable de connaître avec exactitude le bâtiment étudié. Sur le terrain les données mesurées présentent toujours des incertitudes dont les caractéristiques thermophysiques des matériaux et leur mise en oeuvre . D'où la nécessité de travailler avec des données élaborées, issues de la simulation du bâtiment par un code de calcul thermique détaillé. Notre choix s'est porté sur E.S.P (Environment System Performance /1/), capable de modéliser les flux d'énergie et de masse au sein d'un bâtiment couplé avec son système de chauffage-ventilation, à partir de sa description géométrique et physique et des contraintes liées à l'"occupation" qui lui sont imposées.

Les caractéristiques statiques du bâtiment sont alors obtenues par la simulation d'un régime permanent de température extérieure et intérieure constantes, et de flux solaire nul. Selon l'inertie du bâtiment, ce régime permanent s'établit plus ou moins rapidement comme il apparait sur la figure 1.

L'échantillon des bâtiments étudié est élaboré à partir d'une maison individuelle, type T4, de surface habitable 90 m2, chauffage électrique direct, en faisant varier l'épaisseur d'isolant et sa position (intérieure ou extérieure), la taille des vitrages sud, la nature du plancher (sur terre-plein ou sur vide sanitaire) et les conditions de fonctionnement (température de consigne et taux de renouvellement d'air) et d'occupation. Afin de clarifier les conclusions, l'étude se résumera ici à 6 de ces configurations (description donnée dans le tableau 1).
Les climats utilisés pour la simulation sont ceux de Trappes (froid et peu ensoleillé) et Carpentras (très ensoleillé) .

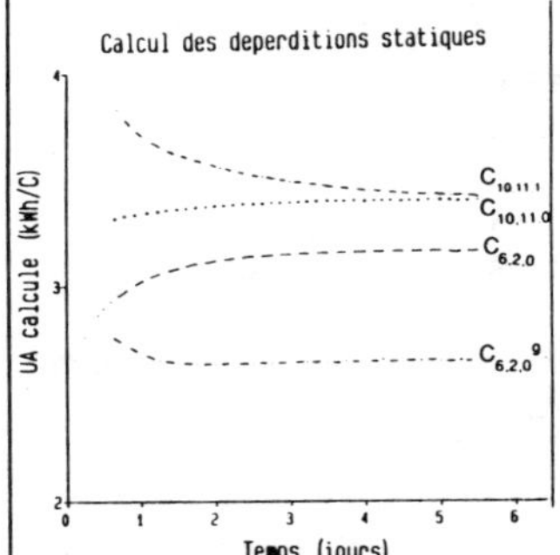

Fig.1 : *Stabilisation de UA en régime permanent (ESP).*

Echantillon réduit

$C_{x,y,z}$	Isolant x cm	Vitrages y m2	Inertie Sol	UA kWh/C
$C_{2,2,0}$	2	2 (S V)	0 : moy V.S	4.40
$C_{6,2,0}$	6	2 (D V)	0 : moy V S	3.66
$C_{10,11,0}$	10	11.2	0 : moy. V.S	3.90
$C_{10,11,1}$	10	11.2	1 : forte V.S	3.89
$C_{6,2,0}^{9}$	6	2	0 : moy T P	3.14 (*)
$C_{10,11,0}^{9}$	10	11.2	0 : moy. T.P	3.84 (*)

(*)écart-type 0.16 kWh/°C à cause du terre plein

Tableau 1

La simulation réalisée sur ESP nous permet de connaître la consommation du sytème de chauffage *H*, la température intérieure et les gains internes et solaires pour différentes fréquences (mensuelle, hebdomadaire ou journalière).

Ces données sont ensuite traitées, comme si elles avaient été recueillies sur site, par une des méthodes de S.E que nous avons recensées dans la littérature.

METHODES TESTEES	
M_T: $H = a - UA. T + \varepsilon$	- C: consommation totale - T : température extérieure
M_{Ti-T}: $H = a' + UA.(Ti-T) + \varepsilon$	- Ti: température intérieure - DJt:degrés jours base t
M_{DJt}: $C = \alpha + UA. DJt + \varepsilon$	- S ensoleillement global vertical sud
$M_{Ti-T,S}$: $H = a'' + UA.(Ti-T) + As.S + \varepsilon$	- As surface équivalente sud
$M_{Ti-T,S,\Delta T}$: pseudo-dynamique	- ΔT,ΔTi,ΔS:écarts entre 2 périodes consécutives de la variable T,Ti ou S
M_{ETP}:dynamique (réseau 2R1C)	- α consommation autres que de chauffage

Tableau 2

L'obtention des coefficients affectés à chacune des variables de la S.E s'effectue par régression linéaire aux moindres carrés, la variable dépendante étant la consommation du système de chauffage pendant la période considérée.

Nous nous sommes intéressés aux coefficients de déperditions (UA) et aux consommations annuelles H_a calculées à partir de la S.E pour un climat moyen du site ou un climat différent et nous les avons comparés aux prévisions de ESP.

Grace à ces observations, nous pouvons situer le domaine de validité des méthodes de S.E en données mensuelles, hebdomadaires et journalières.

3. DONNEES MENSUELLES

Cette fréquence est intéressante car elle ne nécessite pas d'expérimentation lourde et peut même correspondre à une facturation de l'énergie utilisée pour le chauffage. En revanche, le nombre de points obtenu dans une année est faible (8 ou 9 pour la saison de chauffe), ce qui réduit les possibilités de régression multivariable.

La qualité de la régression est généralement bonne, d'autant plus que les apports solaires sont faibles; de même pour les consommations H_a .

$C_{x,y,z}$	UA	Tnc(°c)	R^2	C_a (kWh)	%
$C_{2,2,0}$	5.70	15.7	0.994	12349	1.7
$C_{6,2,0}$	4.57	15.9	0.995	10085	1.3
$C_{10,11,0}$	5.29	14.0	0.991	9366	5.1
$C_{10,11,1}$	5.53	13.6	0.996	9291	5.8
$C_{6,2,0}{}^g$	4.09	15.4	0.998	8581	1.5
$C_{10,11,0}{}^g$	4.81	13.3	0.996	7766	6.9

Tableau 3: *Résultats donnés par la S.E M_T et pourcentages d'écart de H_a par rapport aux valeurs réelles (ESP).*

En revanche, les coefficients UA sont fortement surestimés à cause de la substitution des apports solaires au chauffage en mi-saison alors que la température extérieure augmente. La figure 2 montre les écarts par rapport aux valeurs réelles matérialisés par la distance à la droite UA=UAréel pour le modèle M_T. Ce biais se répercute sur le calcul de la température de non chauffage Tnc également déterminée par la régression.

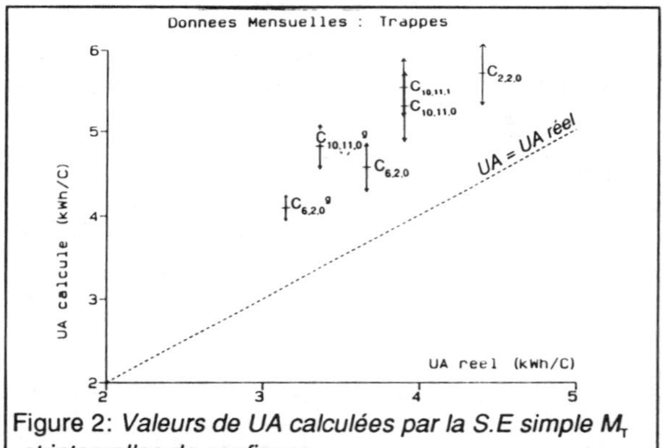

Figure 2: *Valeurs de UA calculées par la S.E simple M_T et intervalles de confiance.*

Les autres modèles de S.E à 2 paramètres conduisent aux mêmes remarques, sauf le modèle utilisant les DJ base 18°C, soumis à un autre type de biais. mis en évidence sur la figure 3 . Sur ce graphe, on a tracé la variation de consommation totale (chauffage et autres usages) du bâtiment en fonction des degrés jours à base 18°C et 14.1°C (température de non chauffage optimum calculée par la méthode PRISM de façon à avoir la meilleure régression possible). Le fait d'avoir des degrés jours base 18°C non nuls alors que la consommation de chauffage est pratiquement nulle (mois de Septembre en particulier), introduit une cassure dans la courbe et fausse la régression linéaire.
On voit donc l'intérêt des méthodes dites à degrés jours variables telles que PRISM.

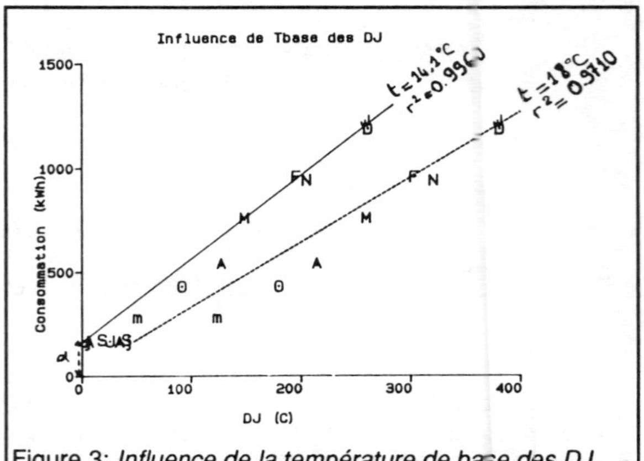

Figure 3: *Influence de la température de base des DJ .*

4. DONNEES HEBDOMADAIRES

Cette fréquence de recueil des données est intéressante car elle permet de disposer d'un plus grand nombre de points sur la saison de chauffe tout en respectant les conditions d'application des méthodes statiques et de stationnarité des paramètres (le cycle d'occupation des locaux est en particulier très souvent hebdomadaire). Elle permet également d'envisager des durées de suivi plus courtes.

La qualité des régressions obtenues avec le modèle M_T est insuffisante dans le cas de batiments fortement solarisés et les valeurs de UA sont biaisées (figure 4). On améliore la détermination des paramètres du modèle en tenant compte de l'ensoleillement global reçu sur site pendant la même période. Les écarts avec les valeurs réelles de UA sont alors inférieurs à 7% même en site ensoleillé, si l'on exclut les configurations solaires passives dont le coefficient UA reste sous-estimé de 12%. A ce niveau on ne distingue pas de différences significatives suivant que la maison est sur terre-plein ou sur vide sanitaire.

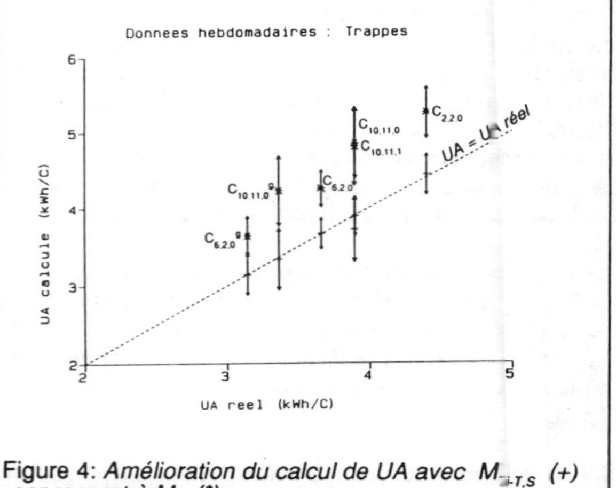

Figure 4: *Amélioration du calcul de UA avec $M_{T,S}$ (+) par rapport à M_T (*).*

En introduisant pour les maisons très solarisées une variable (ΔT) destinée à tenir compte de l'éventuel stockage/ déstockage de la masse thermique constituée par les matériaux (dalle de sol en particulier) on peut constater une amélioration de l'estimation de UA et de la régression. Ce point mérite d'être approfondi et analysé en termes de corrélation possible entre l'inertie du bâtiment et les besoins en chauffage à l'échelle hebdomadaire.

Les possibilités de réduction du suivi sont effectives ainsi qu'il apparait sur la figure 5 dans la mesure où l'on obtient des estimations de UA significatives (test de Fischer) à partir de 8 points seulement. Néanmoins, le large intervalle de confiance associé à la valeur nous fait préférer une durée de 10 semaines minimum. Les séquences démarrant le 1er Janvier sont soumises à de plus grandes incertitudes que les autres, mais nous n'excluons pas que les résultats soient influencés par l'initialisation de la simulation d'ESP. La séquence doit inclure une période ensoleillée et une période froide afin de permettre l'identification de la surface équivalente sud As et une estimation exacte de la consommation annuelle de chauffage. Les meilleures estimations sont obtenues pour la séquence débutant le 22 octobre.

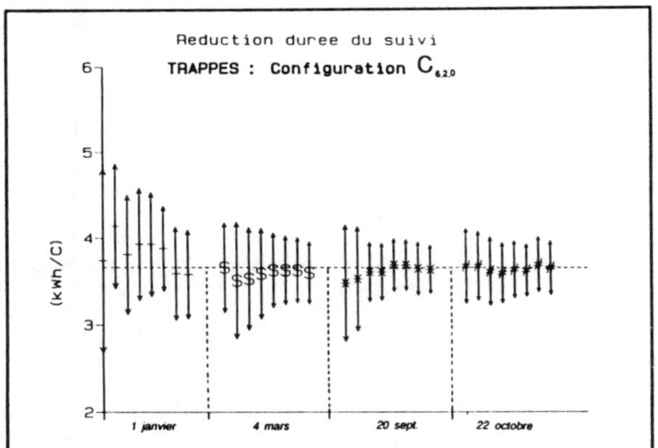

Figure 5: *Valeur de UA déterminée pour des séquences de 8 à 15 semaines débutant à 4 dates différentes*

6. DONNEES JOURNALIERES

Le recueil de données journalières est également très utilisé avec des méthodes statiques, bien que les batiments aient le plus souvent des constantes de temps supérieures à la journée. Un des intérêts réside dans la possibilité de réduire le temps d'intervention sur site à 2 ou 3 semaines.

Comme avec les données hebdomadaires, il est impératif de connaître l'ensoleillement global du site et d'utiliser un modèle $M_{Ti\text{-}T,S}$. Les estimations de UA sont alors très bonnes, quelles que soit l'inertie du bâtiment et la nature du plancher, excepté pour les maisons solaires passives dans un site très ensoleillé comme Carpentras. La figure 6 montre qu'il y a cependant de très grandes disparités de résultats suivant les mois de suivi, en particulier pour une configuration inerte $C_{10,11,1}$. On évitera une conclusion hâtive avec un seul mois de relevés. Les estimations obtenues sur un seul mois sont cependant plus stables, si l'on tient compte de l'inertie du bâtiment par une capacité thermique (modèle M_{ETP}). Ce dernier modèle permet de réduire l'écart de la valeur de UA par rapport à la valeur réelle (9% maximum) pour les maisons solaires passives, indépendament du climat et de la présence ou non d'un terre-plein.

L'utilisation d'un modèle dynamique s'impose donc en données journalières dés lors qu'on ne dispose pas des données relatives à la totalité de la saison de chauffe. L'inertie du bâtiment a des effets non négligeables sur les besoins en chauffage d'une journée à l'autre, qu'il est possible d'appréhender par un modèle dynamique de premier ordre.

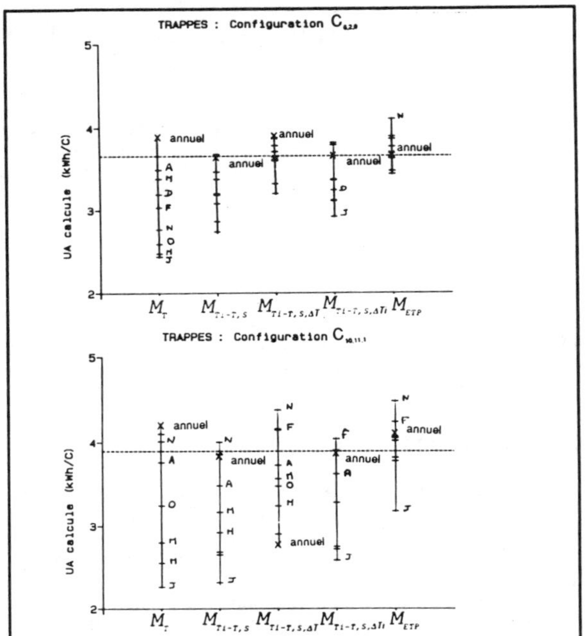

Figure 6: *Estimations de UA par mois (-) et annuelle (*) pour les configurations* $C_{6,2,0}$ et $C_{10,11,1}$ *à Trappes.*

Néanmoins le problème des bâtiments à forts apports solaires passifs n'est pas résolu de façon satisfaisante par les méthodes de Signature Energétique courantes. La non prise en compte de phénomènes rapides telles que l'élévation de température d'air intérieur sous l'effet d'un ensoleillement introduit les limites qu'il faut poser à l'utilisation de ces méthodes.

7. CONCLUSION

Nous avons montré de façon brève à travers cette analyse que l'utilisation d'un modèle de signature énergétique doit s'accompagner de certaines précautions:
- en données mensuelles, seule une S.E à 2 paramètres est possible étant donné le faible nombre de points. Les estimations du coefficient de pertes sont le plus souvent surestimées par rapport à la réalité, mais le calcul de la consommation annuelle est exact pour un climat moyen du site.
- en données hebdomadaires, on obtient de bonnes estimations de UA y compris en réduisant à 10 semaines le temps d'expérimentation, tant que les apports solaires ne participent pas pour plus de 30% au chauffage des locaux., c'est à dire en dehors des bâtiments solaires passifs.
- en données journalières, on atteint la limite d'application des modèles statiques, visible dans l'éventail des valeurs de UA obtenues d'un mois à l'autre. Un modèle dynamique permet de stabiliser les estimations mais ne résout pas le problème des bâtiments solaires passifs.

Remerciements
Nous remercions l'Agence Francaise pour la Maitrise de l'Energie pour le soutien apporté à cette étude.

Références bibliographiques
/1/ "Environment System Performance : Users Manual" JA Clarke
/2/"PRISM" - M.Fels Energy and Building Vol.9 Mai 1986
/3/ A Critical Appraisal of Energy Signature Models - S.Hammarsten Applied Energy 26 (1987)

THERMAL PERFORMANCE OF PASSIVE BUILDINGS MODELLING AND SIMULATION IN THE S3PAS COMPUTER PROGRAM

S. Alvarez, E.A. Rodríguez and R. Velazquez
Dep. Ingenieria Energética y Mecánica de Fluidos
E. S. Ingenieros Industriales
Avda. Reina Mercedes s/n. 41012 Seville SPAIN

ABSTRACT. A new approach to model and simulate the thermal performance of passive buildings is described in this paper. The computer program named S3PAS -one of the results of a national research plan on bioclimatic architecture- addresses this approach and is specially suited to include new developments in conventional and passive architecture (new materials, components and architectural concepts).

1 INTRODUCTION

When compared to conventional buildings, the thermal modelling and simulation of passive buildings require a very different approach for the treatment of many of the issues involved.

In passive buildings, apart from the presence of new elements, a greater emphasis has to be paid to certain heat transfer phenomena (interior re-distribution of solar radiation, sky long wave radiation, etc.) as well as to the coupling between zones (via air movement) and with the ground. None of the existing computer programs was considered suitable enough to provide the infrastructure in which the new models should be integrated. The main reason was that each of those programs allows the use of a unique modelling technique and as it is well known, some of the modelling techniques which are specially well suited to certain building components, are, at the same time poorly suited (or even not suited at all) to others. For example, components including phase-change materials have their own most suitable modelling procedure (explicit finite differences) that is clearly inefficient in the modelling of other elements. Next sections will describe the conceptual modelling and simulation approach carried out in the program S3PAS.

2 TRANSFER PHENOMENA IN PASSIVE BUILDINGS

In general terms, the detailed simulation of the thermal and fluid mechanical behavior of buildings calls for the definition of the transient behaviour of the elements composing the building, with external and internal influences as boundary conditions.

The external influences are:

- Solar radiation
- Outdoor air temperature
- Other external temperatures : sky, ground and surrounding surfaces
- Wind conditions
- Outdoor air humidity
- Outdoor concentration of pollutants

The more common internal inputs are the heat gains delivered from lighting, occupancy and miscellaneous equipment, and the indoor sources of humidity and pollutants, etc.

As far as a passive building is a complex system composed by solid heterogeneous elements (either opaque or semitransparent), liquid elements (water walls, solar ponds, water films, etc.) and gaseous (mainly polluted and humid air), there exist a large number of heat and mass transfer mechanisms between these elements:

- External convection (usually forced) between the external surfaces and the outdoor air.
- Internal convection (usually natural or mixed) between the internal surfaces of the envelope components, occupants, lighting fixtures... and the indoor air.
- Shortwave radiation coming from the sun and in some cases from the internal heat sources.
- External longwave radiation between envelope surfaces and the sky, surrounding buildings and ground.
- Internal longwave radiation between internal surfaces, and between these and the internal gains.
- Fluid flow through cracks and openings between the building and the outdoor, or between zones of the building.

Most of the detailed building simulation programs /1/ /2/ /3/ /4/ /5/ /6/ have retained some basic and well contrasted hypotheses in order to cope with this complex problem. These are mainly:

- One dimensional conduction through walls (except ground-coupled structures and thermal bridges).
- Grey and diffuse radiant behavior of surfaces (shortwave and long-wave).
- Linearized longwave radiant exchange between interior surfaces.
- Uniform room air temperatures.
- Etc.

In spite of these hypotheses the simulation problem of a building is still complex and all the heat and mass transfer modes are coupled. We identify three levels of coupling:

1. Conduction, convection and radiation heat transfers appearing together in each element of the building, and being coupled to each other by the temperature field.
2. Radiation heat transfer in any enclosure providing a second kind of coupling between all the internal surfaces.
3. An interzonal coupling taking place via conduction (through walls separating two rooms), via radiation (through semi-transparent media between two spaces), and/or via air flow between rooms.

Differences between the existing detailed models (/2/ to /6/) dealing with these coupling effects are found partly in the modelling techniques concerned with the transfer phenomena through the building elements and partly in the coupling definition and solving procedures. Nevertheless, thermal and fluid-mechanical coupling can be described in a general formalism.

3 GENERAL THERMAL AND FLUID-MECHANICAL COUPLINGS IN BUILDINGS

In order to describe the general coupling effects in a building simulation problem is useful to establish the modelling procedure for both thermal and fluid mechanical problems first.

Any of the modelling techniques in the thermal problem can lead us to a set of equations coupling the indoor surface temperatures and the room air temperatures.

Each internal surface temperature is defined by the thermodynamic equilibrium of the surface submitted to conduction, convection and radiation heat transfers. This balance equation relates, non-linearly, all the N internal surface temperatures of the building T_{SI} and all the NZ room air temperatures T^J connected to the surface being considered:

$$F_{SI}\left(T_{S1},\ldots,T_{SI},\ldots,T_{SN},T^1,\ldots,T^J,\ldots,T^{NZ}\right)=0 \quad (1)$$

Moreover, the thermal state of each zone of a building is represented by its convective equilibrium, so enabling us to define a representative air temperature as state variable. The writing of an enthalpy balance for the zone under consideration leads us to a first order differential equation relating air temperature to the surface temperatures of this zone (convective effect of walls) and to all the other zone air temperatures (interzone air movement). The discretization of the derivatives in these equations gives a set of algebraic expressions:

$$F_J^T\left(T_{S1},\ldots,T_{SI},\ldots,T_{SN},T^1,\ldots,T^J,\ldots,T^{NZ}\right)=0 \quad (2)$$

The fluid mechanical modelling procedure takes a very similar approach. Typically the dynamic state of each zone is represented by a reference pressure. Flow equations define the different links existing in the pressure network which defines the building's behaviour /2/,/7/. The air mass balance in each zone constitutes then a non-linear system of equations combining these pressures:

$$F_J^F\left(P^1,\ldots,P^J,\ldots,P^{NZ}\right)=0 \quad (3)$$

To be rigorous, Equation 3 must include the air density of each node. These densities are functions of absolute pressure, temperature, humidity and concentration of any pollutant in the zone. Consequently, and in order to close the problem, it is necessary to formulate the balance of the additional species (water vapour and pollutants), for each zone.

In general form, the water vapour balance of each zone relates all the zone specific humidities (by the interzonal air flows) and the temperatures of surfaces and zone air (due to condensation/evaporation phenomena):

$$F_J^W\left(W^1,\ldots,W^J,\ldots,W^{NZ},T_{S1},\ldots,T_{SI},\ldots,T_{SN},T^1,\ldots,T^J,\ldots,T^{NZ}\right)=0 \quad (4)$$

In a general way a similar equation can be written for any pollutant K in zone J:

$$F_J^K\left(C_K^1,\ldots,C_K^J,\ldots,C_K^{NZ},T_{S1},\ldots,T_{SI},\ldots,T_{SN},T^1,\ldots,T^J,\ldots,T^{NZ}\right)=0 \quad (5)$$

where C_K^J is the concentration of pollutant K in the zone J.

These equations are not independent since the mass balance of air defining the pressures depends on all the other state variables. The balance equations require the definition of the interzonal airflows resulting in the pressure network resolution.

In summary, we obtain a set of simultaneous and non-linear equations which, once solved, will give us the complete field of state variables: temperatures (of zones and indoor surfaces), reference pressures, specific humidities and pollutants concentrations.

Though this complex methodology has been formulated in a similar way before /8/, it is not usually solved in this form. Modellers usually look for simpler formulations and solving techniques and therefore have to adopt additional assumptions in order to make both formulation and resolution easier.

The most common simplifications are the decoupling of coupled phenomena (solving thermal and fluid mechanical problems separately by iterative procedures; /3/,/5/), and linearizing systems (longwave radiant exchange between zone surfaces in the thermal problem /6/ or Jacobian system to solve the pressure network system in the fluidmechanical problem /7/).

These simplifications are justified in most cases in that not all couplings existing in a building are equally strong. This allows us to solve the strongest couplings by using values of former iterations in the weakest couplings' variables.

4 CONCEPTUAL MODELLING AND SIMULATION IN S3PAS

4.1 Fundamentals of S3PAS code

S3PAS (/9/, /10/) is a thermal building simulation code on hourly basis. This program distinguishes the three levels of coupling in a building already mentioned and solves them in a hierarchical way from the weaker to the stronger /11/.

S3PAS deals with the first coupling by defining the surface temperatures of each internal wall. This is done by resolving the heat transfer problem in the most appropiate way for each building element.

The conduction heat transfer in each of the elements enclosing a space is expressed in a common formula whatever the method used to model the heat transfer through the selected element.

This equation is:

$$Q_{ci}(t)=B_1+B_2T_{si}(t) \quad (6)$$

where:

$Q_{ci}(t)$: Conduction heat flow from the internal surface of element i at time t

$T_{si}(t)$: Surface temperature of the internal element i at time t

B_1 and B_2: Constants that depend on properties of element i in former time steps

This formulation allows the modelling of the second level of coupling based on the thermal equilibrium of the internal surfaces of a zone.

For surface S_i we have:

$$B_1+B_2T_{si}(t)=h_{cvi}\left(T_{si}(t)-T_i(t)\right)+\sum_{j=1}^{j=N}K_{ij}T_{sj}(t)-\Phi_c(t) \quad (7)$$

In Equation 7, $h_{cvi}(T_{si}(t)-T_i(t))$ represents the convective flow density exchanged between the surface S_i and the zone temperature T_i, $\sum_{j=1}^{j=N}K_{ij}Tsj(t)$ represents the net longwave radiation flow density at the surface S_i,
and $\Phi_c(t)$ represents a shortwave radiation flow density absorbed by the surface S_i.

The conjuction of the N surface equations forms the following system in matrix notation:

$$[K]\{T_s\}-\{h_{cv}\}T_i=\{B\} \quad (8)$$

From this matrix equation, surface temperatures can be expressed as a function of the zone temperature:

$$\{T_s\}=[K]^{-1}\{B\}+[K]^{-1}\{h_{cv}\}T_i \quad (9)$$

In order to close the problem, it is necessary to express the effect of the enclosing surfaces of a zone in the air enthalpy balance of that zone. That effect is the total convective heat flux exchanged along all the surfaces of the enclosure. This flux named $Q_{cv}\mid_{walls}$ is defined as follows:

$$Q_{cv}\mid_{walls}=\sum_{k=1}^{k=N}h_{cvk}S_k\left(T_{sk}-T_i\right) \quad (10)$$

in matrix form:

$$Q_{cv}\mid_{walls}=\{h_{cv}S\}\{T_s\}-\left(\sum_{k=1}^{k=N}h_{cvk}S_k\right)\{T_i\} \quad (11)$$

and by substituting the expression of $\{T_s\}$ given by Equation 9 we obtain:

$$Q_{cv}\mid_{walls}=PIND+COEF\cdot T_i \quad (12)$$

where:

$$PIND = \{h_{cv}S\}[K]^{-1}\{B\}$$

$$COEF = -\left(\textstyle\sum_{k=1}^{k=N}h_{cvk}S_k\right)+\{h_{cv}S\}[K]^{-1}\{h_{cv}\}$$

The interzonal conductive and radiant couplings are implicitly included in the vector $\{P\}$ of the enclosure equations. They have been included in that way because their coupling effect is much weaker than the effect of interzonal air movement.

Finally, the strongest coupling between zones is due to the multizone air flows, resolving the air balances together with the air flow problem.

These balances are expressed by means of the following differential equations written for each zone or controlled volume:

$$V_IC_{pI}\frac{\mathrm{d}\rho_IT_I}{\mathrm{d}t}=Q_{cv}\mid_{walls}+Q_{cv}\mid_{gains}+\textstyle\sum_{J=0}^{J=NZ}\dot{m}_{JI}C_{pJ}T_J-\sum_{J=0}^{J=NZ}\dot{m}_{IJ}C_{pI}T_I \quad (13)$$

where:

V_I : Control volume of zone I

C_{pI} : Specific heat of air in zone I

$Q_{cv}\mid_{gains}$: Convective heat flows from internal sources in zone I

$\dot{m}_{IJ}$: Air mass flow rate from zone I to zone J

The main hypotheses assumed in Equation 13 are:

Uniform temperature in each zone.

Time-derivative of the specific heat of air in each zone neglected.

There is no thermal effect of other pollutants.

According to the dynamic treatment, the mass air balance in each zone is expressed in unsteady state:

$$V_I \frac{\mathrm{d}\rho_I}{\mathrm{d}t} = \sum_{J=0}^{J=NZ} \dot{m}_{JI} - \sum_{J=0}^{J=NZ} \dot{m}_{IJ} \tag{14}$$

Left side of Equation 13 is then written as follows:

$$V_I C_{pI} \frac{\mathrm{d}\rho_I T_I}{\mathrm{d}t} = V_I C_{pI} \rho_I \frac{\mathrm{d}T_I}{\mathrm{d}t} + V_I C_{pI} T_I \frac{\mathrm{d}\rho_I}{\mathrm{d}t} \tag{15}$$

By substituting Equation 15 in Equation 13 and using the mass balance (Equation 14), we obtain:

$$\rho_I V_I C_{pI} \frac{\mathrm{d}T_I}{\mathrm{d}t} = Q_{cv}|_{walls} + Q_{cv}|_{gains} + \sum_{J=0}^{J=NZ} \dot{m}_{JI} (C_{pJ} T_J - C_{pI} T_I) \tag{16}$$

The conjuction of Equation 16 applied to all the zones forms a system of first order differential equations with the air temperatures of each zone as unknowns.

4.2 Internal structure of S3PAS code

In relation to the internal structure, S3PAS is a modular code with a steering controller and different modules. These modules either simulate the thermal behavior of certain elements of the building or resolve the different couplings between them. Other modules do not correspond exactly with real elements and simply calculate things such as solar position, shading effect and boundary conditions.

The various modules are the following:

- **Meteorological**: Calculates the solar position and the direct normal radiation.
- **Shadows I**: Determines the solar obstruction of remote obstacles.
- **Shadows II**: Determines the effect in the solar gains through each building opening due to shading devices.
- **TSA**: Calculates the total external boundary condition of an element in terms of the Sol-air temperature.
- **Envelope Wall**: Simulates the thermal behavior of an opaque multilayer wall of the exterior envelope of the building.
- **Interzone Wall**: Simulates the thermal behavior of an opaque multilayer wall connecting two simulated zones of the building.
- **Envelope Glazing**: Simulates the behavior of an envelope semitransparent wall.
- **Interzone Glazing**: Simulates a semitransparent wall that connects two simulated zones.
- **Floor**: Simulates ground-coupled structures.
- **Trombe Wall**: Simulates a Trombe wall.
- **Colect-RockBed**: Simulates the system formed by an air solar colector and a rock-bed storage system.
- **Zone**: Supports the coupling between the different elements of a zone.
- **Building**: Couples all the simulated zones of the building by calculating the air flows and delivering the thermal balances.

At each time step the steering program calls the different modules in the following order:

1. The modules related to boundary conditions (Meteorological, Shadows I y II and TSA).
2. The simulation of the building itself begins in the element modules which send a pair of values (B_1 and B_2) to the next module (zone).
3. A module "Zone" for each zone is called to perform the longwave radiant coupling between the surfaces of each zone. A new pair of values (*PIND* and *COEF*) is sent to next module (Building).
4. The module Building is called to couple all the zones in terms of air flows and air thermal balances.

5 REFERENCES

1. "DOE-2.1 Reference Manual", Lawrence Berkeley Laboratory, 1981.
2. Walton G.N., "Air Flow and Multiroom Thermal Analysis", ASHRAE Transactions Part 2, 1982.
3. Roldan A., "Etude Thermique et Aeraulique des Enveloppes de Batiment - Influence des Couplages Interieurs et du Multizonage", Ph.D. Thesis, I.N.S.A. Lyon, 1985.
4. Alvarez S., "Análisis Dinámico del Comportamiento Térmico de Edificios", Ph.D. Thesis, E.S.I.I. Seville, 1986.
5. Clarke J. and D. McLean, "ESP, A Building and Plant Energy Simulation System", Version 6, Release 8. ABACUS University of Strathclyde, Glasgow, 1988.
6. "S3PAS 1.0, Simulacion de Sistemas Solares Pasivos - Manual de Usuario". E.S.I.I. Seville, 1986.
7. Walton G.N., "A Computer Algorithm for Predicting Infiltration and Interroom Airflows", ASHRAE Transactions Vol. 90, Part 1, 1984.
8. Axley J., "The Coupled Airflow and Thermal Analysis Problem in Building Airflow System Simulation", ASHRAE Transactions Vol. 95 Part 2, 1989.
9. "Método de Simulación y Cálculo de Sistemas Pasivos en Relación con su Comportamiento Térmico", Informe de Avance $n^o 1$. Cátedra de Termotecnia, E.S.I.I. Seville, 1986.
10. "Método de Simulación y Cálculo de Sistemas Pasivos en Relación con su Comportamiento Térmico", Informe de Avance $n^o 2$. Cátedra de Termotecnia, E.S.I.I. Seville, 1987.
11. Rodríguez E.A., "Acoplamientos Termicos en Edificios", Internal Comunication, Thermal Energy and Fluidmechanics Department, E.S.I.I. Seville, 1988.
12. Rodríguez E. and F. Allard , "Coupling COMIS air flow model with other transfer phenomena " to be published in Energy and Buildings , 1990.

Acknowledgments

This paper is based on a research project supported by the Spanish "Centro de Investigaciones Energéticas MedioAmbientales y Tecnológicas - Instituto de Energías Renovables (CIEMAT-IER)".

THE ENERGY PERFORMANCE ASSESSMENT PROJECT AND SOME EARLY FINDINGS

* Hildon A., Palmer J., Seager A., Brown J.; ** Alexander D., Vaughan N., Jenkins H.; *** O'Sullivan P.
* Databuild Limited, 4 Venture Way, Aston Science Park, Birmingham
** The Welsh School of Architecture, UWCOC
*** The Bartlett School of Architecture, University College, London

The Energy Performance Assessment project (EPA) of the Department of Energy has developed a methodology to assess and report on passive solar buildings. This paper describes the methodology which takes account of the energy, cost and amenity issues of occupied buildings. By 1991 the EPA project will have evaluated over 30 buildings. Some of the initial findings from the first few buildings are reported here.

INTRODUCTION

In attempting to reduce the energy consumption of buildings, designers have evolved a range of design solutions which emphasise high levels of insulation, and maximum use of solar energy. Overall, solutions are required which give the best balance between energy performance, economics, environmental quality, user satisfaction and controllability.

With this in mind, the Department of Energy (D.En.) initiated the development of a rapid, cost effective, multi-disciplinary form of building appraisal. Energy Performance Assessments (EPA), as they are called, seek to determine to what extent individual buildings, representing particular design solutions, save energy, improve amenity, and incur additional costs. Furthermore, they assess the environmental consequences of the energy strategies by examining a number of physical and human parameters.

PASSIVE SOLAR BUILDINGS

In existing buildings the economics are such that improved control of plant is more cost effective than changes to the fabric. These types of projects are well represented in the Energy Efficiency Demonstration Scheme (1). Research and design have also shown that displacing energy supply has a complementary role to reducing energy demand. It is encouraging to see designers utilising available solar energy (heat and light) with the development of a new architectural vocabulary.

Reasons for the popularity of passive solar buildings are difficult to identify with certainty, but the admission of substantial amounts of daylight and sunlight into buildings is well liked by users often accustomed to buildings that reinforce their separation from the outside.

Admitting solar energy into a building can save energy by:-

* solar displaced space heating - SDSH
* daylight displaced electric lighting - DDEL

The design and operation of passive solar buildings is very different for houses and non-domestic cases. In houses the low internal gains, high surface to volume ratio and shallow spaces make SDSH the key factor. In the non-domestic buildings internal gains can be high, buildings may be deeper with a low surface to volume ratio so that the balance of DDEL and SDSH strategies will be different and DDEL may be the greatest opportunity.

Furthermore, passive principles can be employed to introduce natural ventilation and daylight to buildings which otherwise might have to be mechanically conditioned. The high running costs associated with mechanical ventilation and cooling are often overlooked in the mistaken belief that more control will inevitably reduce consumption.

The controlled admission of ambient energy into buildings has been greatly assisted by the advent of new materials, equipment and techniques. For example:-

* low emissivity coatings that reduce heat losses from double glazing to less than that from conventional triple glazing;
* low cost heating and lighting controls which can allow better utilisation of admitted energy;
* techniques such as light shelves to redistribute daylight;
* ventilation systems that can redistribute solar heat;
* atria that can provide existing interior spaces and, with careful design, reduce the need for mechanical ventilation and cooling in deep buildings.

Although these techniques can be shown to be effective in theory and in laboratories, their performance in buildings also needs evaluating by field trials.

ENERGY PERFORMANCE ASSESSMENTS

The Department of Energy's Renewable Energy Research, Development and Demonstration programme is managed by the Energy Technology Support Unit at the Atomic Energy Research Establishment at Harwell (2). The utilisation of passive solar energy (heat and light) in buildings is a component of this programme which includes field trials of real buildings.

Since design, and designers, are informed by experience (3), field trials in occupied buildings have long been an accepted part of the D.En.'s R&D programmes. These trials should be lower cost evaluations of a wide range of buildings with an emphasis on analysis, interpretation and presentation of data to the industry.

Such trials are especially needed in the increasingly popular area of passive solar buildings where developments are taking place rapidly, and where the detailed knowledge of the performance of the fabric and control systems is of paramount importance. Additionally, field trials have an important role in guiding other forms of research.

It was from such consideration that the Energy Performance Assessment project (EPA), and a complementary series of Design Studies (4), was formulated by ETSU. A feasibility study commissioned by ETSU (5) was followed by the appointment of Databuild and UWCOC to develop an EPA methodology. Some of the essential characteristics of EPA's soon emerged from an assessment of traditional field trials methods, for example EPA's:-

* should be credible to the building industry,
* should deal with typical (but 'good') buildings,
* will not be concerned with validation of mathematical models,
* should be 'analysis rich' and not simply 'data rich',
* should be of short duration for valuable feedback,
* must not be solely concerned with energy issues but must also address economic and human factors.

Of particular note was the concern with human factors since passive solar buildings can require greater human intervention:-

* at their inception, when both client and designer need to be aware of the consequences of a passive solar building,
* in use, when both occupants and operators will interact necessarily with the building's environment.

The EPA methodology is regarded as a 'Template of Components' (6, 7). Although the template is quite rigid in the way it defines the EPA method, it also allows for a disciplined flexibility in the response to individual buildings.

In the Preliminaries Component each EPA is carefully defined using information from drawings, initial surveys, etc., and the subsequent components fall into three themes:-

* Measurement of the EPA building
* Modelling of the EPA building
* References

Within the Conclusions Component of each EPA, a holistic appraisal of the building is produced on the analysis of the findings from the individual techniques.

Thus the EPA development has created a simple but robust strategy for dealing with the following issues in both quantitative and qualitative terms:-

* a whole building evaluation,
* the energy performance of the building and its passive features,
* a capital cost analysis of the building,
* comparison of the energy and cost performance indices with a range of credible and appropriate references,
* the human aspects as embodied in the aspirations and experiences of those groups or individuals responsible for creating and using the building.

SOME FINDINGS FROM EARLY EPA'S

At the time of writing this paper (November 1989) the EPA project is analysing the results from the first 10 EPA's with a further 15 in progress. This section reviews some of the buildings.

General Points

The EPA's (and related international studies to which the EPA project contributes (8)) suggest that passive solar buildings typically use about 25 to 50% less heating and lighting energy than similar conventional buildings. Of course, not all the energy benefit is attributable to SDSH and DDEL. EPA's are only able to confidently quantify the solar contribution in houses, where it is of the order of 20 to 35% of the space heating demand. The energy saving is often frustrated by unresponsive heating and lighting systems. There is still considerable room for development of simple, cheap, effective control systems that are specifically designed to enhance SDSH and DDEL without interfering with user requirements.

So what are the cost penalties for such design strategies? Here the evidence is, perhaps, surprising. In houses the additional cost has been only ± 2%. Other studies (4) confirm that an on-cost of about 4% should be able to fund all or most passive solar strategies for houses. In non domestic EPA buildings all had a cost per unit area in the typical range for similar buildings, i.e. they are about average cost.

Therefore, it appears that passive solar design strategies can be incorporated within buildings for little or no extra cost and can, with careful design, achieve significant amounts of SDSH and DDEL, making the strategy highly cost effective. However, the substantial amenity benefits as expressed through the users is perhaps the greatest benefit.

Houses

For domestic buildings the dominant design strategy is to reduce the building's heat loss and to make maximum use of the thermal component of solar radiation to displace space heating (SDSH).

Findings from several houses indicate that the heating season is significantly shortened where good use is made of reduced heat loss and SDSH.

Not all passive solar houses are insulated to much above building regulations, although it is true that most designers seeking to save energy by solar means do attempt to improve other aspects of the house's thermal performance.

Designers have noted that whereas energy efficiency measures such as increased insulation and high efficiency boilers have an identifiable additional cost they do not provide any additional amenity. Conversely attempts to utilise SDSH often cost little extra but can improve amenity.

The simplest and most widely used strategy is to redistribute glazing to give large south and small north windows. EPA's and other trials have observed that this can lead to north rooms with insufficient daylight where electric lighting is almost always required and where the occupants feel the room to be gloomy and unpleasant. Often this room is a kitchen and here the effect is particularly noticeable. In a development of this simple strategy one house employed an integral conservatory with large windows on the south elevation and large windows on the north. It was found that the heat loss from these north windows was more than compensated for by the daylight gains from those windows and the heating gains on the south side.

Much concern is expressed about overheating as a result of large south glazed areas. Generally speaking it appears to be straightforward to protect the house itself from overheating by the use of mass, modest shading and provision for natural ventilation. Monitoring shows that users rarely complain of overheating and that the houses behave well in summer. This is probably related to the UK tradition of high mass (solid floors, blockwork partitions, etc.).

In one house the whole house average internal temperature was lower in summer than winter! This raises another issue. It is often stated and assumed that we do not want our houses to be heated throughout to 21°C or above for all of the day. In passive solar and low energy houses where this is affordable the occupants often choose such a temperature regime. This suggests that assumptions about UK house temperatures, based on experience of conventional houses, may not be applicable to passive solar houses.

A difficulty with large south windows facing public spaces is that the need for privacy may lead to the use of blinds and net curtains which can severely reduce the SDSH. In one EPA the house used large south windows for direct gain and these worked reasonably well facing on to the rear garden. The developer, however, felt that other houses of the same design on the estate but which had their south elevation to the road were much less successful and many occupants used net curtains.

Non Domestic Buildings

Generalisation about non domestic buildings is difficult owing to the vast range in size, function and complexity. Passive solar strategies however are generally limited to displacing space heating (SDSH) and electric lighting (DDEL) in the perimeter zone of a building, and usually these zones are naturally ventilated in such buildings.

Solar displaced space heating: Since these buildings are usually well insulated and have high internal gains, then the potential for SDSH is reduced and the risk of summer overheating is increased. Consequently where experience with domestic design has been transferred directly to non domestic buildings then difficulties have been observed. These occurrences are not numerous nor serious and the buildings are still well regarded by owners and users. The lesson is simply that designers must take full account of the differing thermal inputs. With reasonable standards of insulation (opaque U values of 0.4 $W/m^2/K$, or less, window U values of 3 $W/m^2/K$, or less) then incidental gains can provide the building heat loss at external temperatures close to winter averages. Field trials confirm that in such cases half, or more, of the buildings heat requirement occurs as morning pre-heating, therefore the potential for direct SDSH is limited. Several strategies adopted in an attempt to counteract this disadvantage have been monitored, e.g.:-

Thermal capacity: Mass can be used to carry over thermal gains and limit summertime temperatures. Whilst this works well in sunny periods it also requires large amounts of preheating energy and time. The control of the heating system during preheating and occupied periods is then different and quite complex if optimum use of SDSH is to be made.

Warm air heating: If forced ventilation systems, to provide rapid preheating, are coupled with an east of south orientation then SDSH has a greater potential. In these cases the automatic control of the air system can become quite complicated. For example, if maximum use of recirculated air in unoccupied periods is to be made, when available solar energy is sufficient to warrant running the fans, then those control requirements may conflict with requirements during occupied periods. In one case a shallow plan office with 80% glazed area had an apparently straightforward air heating system and, in theory, constituted the simplest and most direct use of SDSH. However, monitoring failed to detect any significant relationship between the heating system and the solar radiation. Undoubtedly solar gain made a contribution to energy savings in this building but the design and control of the warm air heating system was such that SDSH was not optimised.

Air redistribution: Forced ventilation systems can be used to redistribute solar heat away from a southerly perimeter zone, where it will often exceed demand, to other areas of the building where it can provide SDSH. Attractively simple in concept, the control strategy and equipment for this can become exceedingly complicated if the ventilation system is also used to provide fresh air and heating (which it may have to do to be economically justifiable).

Moveable shading: Moveable shading devices are sometimes installed to allow large solar apertures to displace morning heating but restrict solar gains when internal loads are high. If these devices are manually controlled then it has been observed that they are often left permanently in the worst case condition, thus excluding more useful solar heat at other times. If the shades are automatic then the control strategy and equipment need to be quite sophisticated to avoid inadvertently excluding solar displaced space heating. In one case automatic blinds have been observed in operation when solar radiation and heating demand were simultaneously high.

Concern about overheating in passive solar buildings with a thermal strategy is, rightfully, high. In two cases designers used deep cills at windows to provide some thermal mass, daylight reflection, and to keep desks out of the highly sunlight zone. Understandably, these cills/shelves are used for papers and when windows are open these are disturbed by air movement. An apparently minor point such as this can be very annoying to users and mitigate their enjoyment of their working environment.

Daylight Displaced Electric Lighting: Whilst the benefits of daylight in passive solar buildings have often been noted there have been few examples of buildings specifically designed to admit large amounts of daylight in a way which can displace electric lighting. Recently however this has changed with the realisation that daylight has both energy and amenity benefits that can outweigh any commensurate thermal

disadvantages.
Daylight cannot be redistributed as easily as solar heat, and penetration is limited to a perimeter zone (say 6 - 8m for side lighting, or one level for top lighting). Additionally, and unlike thermal gains, once daylight falls below a certain level and electric lights are switched on the DDEL can fall to zero despite daylight availability. These factors all combine to make design for DDEL a more complex problem than design for SDSH. Hopefully a recent excellent review and design guide will assist (9, 10).
In one case an office design incorporated high ceilings, for good daylight penetration, together with internal and external lightshelves to improve the uniformity of this penetration. The concept functioned well in terms of daylight, and the daylit environment was well liked by users. Several instances of glare were noted, particularly were VDU's were in use.
Although the electric lighting use was low it was not as low as could have been achieved since the control strategy proved to be too coarse and pessimistic, often providing electric lighting when and where it was not needed. Fortunately better advice now exists about controls for electric lighting in daylight buildings (e.g. 11).
An interesting side effect of large glazed areas is their effect on the electric lighting installation. Near to the windows the absence of any reflected electric light from the glazing means that uniformly spaced ceiling luminaires will give a substantially reduced lux level in that zone.
In another study, still in progress, top lighting is used in two sports halls. In one case the adequate daylighting does not displace electric lighting because a user retrofit of an occupant sensor switch, intended to switch lights off after vacation of the hall, switches lights on at the beginning of use, irrespective of available daylight! This is an example of a well intentioned but badly briefed operator frustrating an otherwise successful design intent.
Controls: The most common practical obstacles to successful SDSH and DDEL and to user satisfaction have been related to control.
Often these controls are manual and reflect a design intent "to provide users with control over their own environment". However, manual controls such as shutters, blinds, vents, louvres, etc., are often not used as intended with a consequent reduction in energy savings and comfort.
EPA findings, supported by earlier US research, indicate that occupants are most likely to operate controls if their action:-

* will yield a noticeable benefit
* is connected to consequences by an easily understandable logic
* is required at easily recognisable times and circumstances
* is not counter-intuitive
* does not require an overly complex sequence
* is within their physical capacity

EPA's have observed several cases where one or more of these requirements are not met. For example:-

* an integrated manual and automatic lighting control strategy was not understood by the majority of users, with consequent dissatisfaction and misuse;
* deep light shelves at cill height prevent some users from reaching opening windows;
* a natural ventilation strategy which requires openings to be operated by three separate sets of people.

Where controls are automated then EPA's have observed many instances where they do not function as intended, occasionally with a substantial energy wastage. The rules given above for manual controls also apply to automatic controls. However, we have observed control systems with, for example:-

* little benefit accruing from their action,
* overly complex logic,
* unknown and unidentifiable set-points,
* competing strategies leading to exclusion of potential SDSH and DDEL, and in one case the dumping to outside of collected solar heat whilst another control was calling for space heating.

CONCLUSIONS

EPA's and other field trials in Europe (8) have confirmed that it is possible to build passive solar buildings that make good use of solar heat and daylight to reduce energy use. The evidence is that these buildings:-

* cost little or no extra relative to equivalent buildings of similar standards;
* save about 25% of the energy demand for space heating and lighting;
* produce environments with high amenity value and with few comfort problems.

The major widespread difficulty concerns controls, both manual and automatic, which need particular attention in terms of logic, reliability and ergonomics.

REFERENCES

1. Energy Efficiency Demonstration Schemes. Full details from ETSU, AERE, Harwell, Oxfordshire OX11 ORA and from BRECSU, BRE, Watford WD2 7JR.
2. Details from Renewable Energy Promotions Group (REPG). ETSU see (1) above.
3. H. Marvin. Using Experience and Publications in Building Design. BRE Information Paper IP 13/85. 1985.
4. House Design Studies and Non Domestic Design Studies. Details from ETSU see above.
5. Minor Field Trials, Project Definition Study. J. Littlefair, M. Watson, P. Ruyssevelt, B. Ford. ETSU Report 1983.
6. EPA's - Developing the Methodology. Databuild and the Welsh School of Architecture. ETSU 1987.
7. EPA's - Guide to Procedures Vol. 1 Houses, Vol. 2 Non-Domestic Buildings, Databuild and the Welsh School of Architecture, ETSU 1987.
8. International Energy Agency Annex XI - Passive Solar Commercial Buildings. Details from REPG see (2) above.
9. Daylighting As A Passive Solar Option. V.H.C. Crisp, P.J. Littlefair, I. Cooper, G. McKennan. BRE 1988.
10. Window Design, Applications Manual. CIBSE 1987.
11. Lighting Controls and Daylight Use. BRE Digest 272, 1983.

PROPOSAL FOR THE EXTENSION OF THE NBE-CT-79

M.R. Heras, G. de Ignacio; R. Serra et al*, S. Alvarez et al**
IER-CIEMAT. Av. Complutense, 22. 28040-MADRID
*E.T.S. Arquitectura de Barcelona. Av. Diagonal, 649. 08028-BARCELONA
**E.T.S.I. Industriales de Sevilla. Av. Reina Mercedes, s/n. 41012-SEVILLA

ABSTRACT. One of the most important objectives in Bioclimatic Architecture is the achievement of a design - that incorporating the utilization of Passive Solar Systems gives rise to a rational and conscious use of the energy required to obtain proper comfort conditions in the building. From this point of view, the Spanish Basic Regulations on Building Thermal Conditions (NBE-CT-79), compulsory since 1979 provides the limits in which the global energy losses of the building must be situated. But the NBE-CT-79 is essentially an insulation regulation, that only contemplates thermal losses, and the global thermal losses coefficient values are calculated for the winter season without taking solar gain into consideration and annual period. So the purpose of this research has been to develop a proposal for the extension of the NBE-CT-79 from the Passive Solar Energy point of view. The studies realizated and the consecuences derivated of these allow to have a solid bases where the future Regulation could be spported by the buildings thermal quality control.

1. INTRODUCTION.

The use of solar energy in a passive way, directed to decrease energetics needs in building climate conditioning is an actual fact which has been shown trhoughout time. Internationally, a particular attention has been devoted to this practice in the last years, when new calculation and design methods allowing incorporation of such principles to current building, have been developed.
In our country, one of the main premises for passive solar use on a large scale was the modification of the spanish basic regulations on building thermal conditions (NBE-CT-79), in order to allow not just its use but also the possibility of having a checked tool to perform the suitable designs.
In order to reach that purpose, the "Instituto de Energías Renovables" (I.E.R.), a branch of the "Centro de Investigaciones Energéticas, Medioambientales y Tecnológicas (C.I.E.M.A.T.) - Department of Industry and Energy -, in cooperation with the "Dirección General para la Vivienda y Arquitectura" - Department of Public Works and City Planning -, decided to carry out a research project defined as "Extension of the Building Basic Regulation for Thermal Conditioning of Buildings (NBE-CT-79)".
As it is well known, every regulation directed to control buildings' thermal quality must follow the formal aim of bringing their energy consumption to a minimum, given that this energy consumption depends on both the energy requests and the conditioning system yield. As for the energy request, it is got from a balance in which heat flows take part that can affect in either a favourable or unfavourable way the indoor conditions to be kept inside the building.
The fact that the current NBE-CT-79 is essentially an isolation regulation so for referred to leakage levels, is to be mentioned as an important objection.
It is significant, regarding to this, to analyze the relationship between buildings' energetics request and K_G of the present regulation (Fig.1). Each building, represented by one point, has both a given K_G value and a seasonal energetics request; and each line holds a group of buildings only differing in their isolation level.
As shown, two buildings in the same lacation, having identical K_G, not necessarily should keep the same energetics request.
At the same time, the fixation of allowable values for the coefficient K_G is just to be applied in winter-time, without considering neither the free profits (favourable in winter-time and unfavourable in summer-time) nor the annual cycle. Because of the above mentioned, the outline of this extension proposal has been carried out according to these premises; an evaluation system has been suggested that integrates the current regulation and allows, in addition, to consider the free profits.

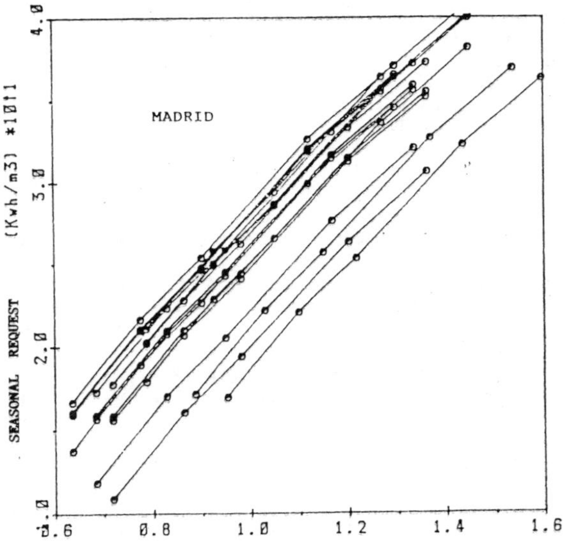

Fig.1 Relationship between K_G and energetic request.

2. THERMAL CHARACTERIZATION OF BUILDINGS.

2.1 Global performance equations.

The thermal situation of a building is the result of a simultaneous action of varying thermal fluxes that occur in time.
Energetic request is the heat quantity that we extract or supply to a room to keep inside fixed temperature conditions.
Supposing that we want to maintain a constant indoor temperature θ_i for a fixed time, we can state the global building energy balance as:

$$Q = Q_T + Q_V + Q_S + Q_{FI}$$

where:

Q Energetic request.

Q_T Heat transfer through the opaque and glazed walls as a result of the difference between indoor and outdoor temperatures.

Q_V Heat transfer because of whished inputs of external air (ventilation), or unwished (infiltration).

Q_S Heat transfer to the ambient because of solar radiation.

Q_{FI} Heat transfer to the ambient because of the internal sources, such as occupiers, lighting and others.

Validity of the global thermal balance for any state is conditioned only by the sign of the different heat flows.

If we suppose winter conditions, and consider the period "t" long enough (longer then 10 days), the former heat flows are:

Q_T: The transmission heat flow may be held in a conservative way suppossing steady state. That:

$$Q_T < \int^t U.A[\theta_i - \theta_e(t)]dt = U.A \int^t [\theta_i - \theta_e(t)]dt$$

where U.A is the global conductance of the building.

The integral of temperature difference may be written as degree-days based on indoor temperature ($GD_{\theta i}$), so:

$$QT \leq 24.U.A.GD_{\theta i}$$

Q_V: $Q_V \approx \int \rho.c_p.n.V[\theta_i - \theta_e(t)]dt \leq 24.0,33.n.V.GD_{\theta i}$

where n is the number of air changes per hour, and V is the volume of enclosed space.

Q_S: The solar flux incident over each wall depends on a set of outputs before becoming converted into heat transfer to the ambient. Anyway, heat transfer to the ambient not always must be completely available, because of accidental existence of overheatings.

Let Φ_S the solar net flux which is transferred to the ambient, and since Q_S is only useful heat, we can put:

$$Q_S = \eta.\Phi_S$$

where η is an output, $0 \leq \eta < 1$.

Q_{FI}: In a similar way to the former instance, we can obtain a net flux to the ambient coming from different internal fonts (Φ_{FI}). According to that:

$$Q_{Fi} = \eta.\Phi_{FI}$$

Obviously, the output is the same as that one of solar net flux, since the overheating must be obtained considering jointly all the solicitations.

The global balance, considering the losses terms as positive, and the gain ones as negative, is then:

$$Q = 24.U.A.GD_{\theta i} + 24.0,33.n.V.GD_{\theta i} - \eta.\Phi_s - \eta.\Phi_{fi}$$

that may rewrite as:

$$Q = 24(U.A + 0,33.n.V).GD_{\theta i} - \eta(\Phi_s + \Phi_{fi})$$

where the forground belongs to the losses, and the second one to the gains.

2.2 Parameter definitions.

As we pointed out at the introduction, we pretend to control the thermal quality of the building from its energetic request. Working over the losses term of the final expression obtained in the last paragraph, we can rewrite that expression as:

$$Q = 24.U_{T+V}.A.GD_{\theta i} - \eta(\Phi_s + \Phi_{fi})$$

where:

$$U_{T+V} = U + 0,33.n.V/A$$

Characterization of thermal request of buildings is based on the concept of Useful Indoor Temperature (θ_{in}), that is defined as the mean indoor temperature that the building would reach during a period of time if it were not conditioned, and we had to consider only the useful gains.

Let Q = 0 and put $\theta_i = \theta_{in}$; then,

$$\theta_{in} = \theta_e + \frac{(\Phi_s + \Phi_{fi})}{24.U_{T+V}.A.N}$$

where N is the number of days along the period.

If, for instance, we suppose that the period is a month, the Q-expression can be written as:

$$Q = 24.U_{T+V}.A.N(\theta_i - \theta_e) - \eta(\Phi_s + \Phi_{fi})$$

θ_e being the mean monthly external temperature.

We can write:

$$Q = 24.U_{T+V}.A.N(\theta_i + \theta_{in} - \theta_{in} - \theta_e) - \eta(\Phi_s + \Phi_{fi})$$

so that:

$$Q = 24.U_{T+V}.A.N[(\theta_i - \theta_{in}) + (\theta_{in} - \theta_e)] - \eta(\Phi_s + \Phi_{fi})$$

and, since:

$$Q_* = 24.U_{T+V}.A.N(\theta_{in} - \theta_e) - \eta(\Phi_s + \Phi_{fi}) = 0$$

we obtain:

$$Q = 24.U_{T+V}.A.N(\theta_i - \theta_{in})$$

If we multiply and divide the last expression by the factor $(\theta_i - \theta_e)$, we have:

$$Q = 24.U_{T+V}.A.N \frac{\theta_i - \theta_{in}}{\theta_i - \theta_e} (\theta_i - \theta_e)$$

that is:

$$Q = 24\ U'_{T+V}\ A\ N(\theta_i - \theta_e)$$

where $U'_{T+V} = U_{T+V} \dfrac{\theta_i - \theta_{in}}{\theta_i - \theta_e}$

Limiting the U'_{T+V} values we can control the thermal request, in the same way that the U ($=K_G$) limitation allows to control of losses at the actual Norm.

θ_{in} parameter is useful for checking the thermal quality of a building, although it does not characterize the energetic request only. In other words, two buildings that have the same compacity and the same value of θ_{in}, not always have the same energetic request per unit of volume. Therefore, the greatest the value of θ_{in}, then the best thermal quality of the building, indeed.

Fig.2 shows the energetic request vs. useful internal temperature in Madrid, Spain. As we can see, a fixed useful internal temperature does not guarentee a specific request, but an enclosed set of energetic request.

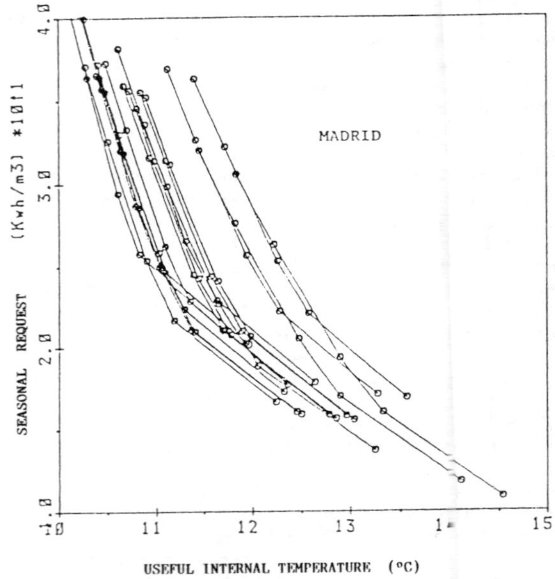

Fig.2 Comparison between used parameter and energetic request.

3. LEVELS OF NORM IMPLEMENTATION.

Thermal quality of buildings is to be regulated in agreement with the three levels or alternative options of use, decreasing in complexity in the same way, then it penalizes the previous case. They are the following:

3.1 Option A: An useful internal estimate temperature.

Control of energetic request of building, starting from the overtaken mean internal temperature, in absence of active conditioning systems. In order to calculate this internal temperature we will take into account not only the losses but also the gains, both solar and internal.

At this level, the criterion is to calculate (in a simplyfied form) all the aspects as possible. The result is an useful internal temperature that must exceed the minimum temperature that is obligatory for any climatic zone.

3.2 Option B: A global coefficient of estimate transmission losses estimate depending on a given shape factor.

Control of thermal exchanges in the building considered as a whole (this is an equivalent version to the actual K_G, with an adjustment). At this option, we can calculate some aspects of the project, less than the former, and the rest of them are prefixed or limited.
We follow the european guidelines of presentation, and we extend and fix the limits of the function:
K_G = Constant * (1/f)
These limits are fixed depending on the shape factor and on the climatic zone, and those ones that must be completed will be the resulting of application of former option, with prefixed values, considered in some aspects as a limit.
For this, we calculate some aspects less than before, such as microclimatic aspects and those whose objective is to introduce the gains into the building, therefore taking into account this aspect through introduction of K_G, but corrected for glazed walls, where also solar gains are included .

3.3 Option C: Limitation of wall types and glazed wall percent.

Control of building thermal losses in view of their components considered in such an individual way.
The objective of this option is to restrict to a great extent the user action, by means of fixing in practice the wall types he can use, as well as limiting the different surfaces percent.
At this level, the indispensable minimum aspects are to be calculated, and the majority of them are prefixed or limited. The limits are taken into account from surfaces (relationships between them and with the total skin surface) and from admissible types of wall elements, following a prefixed typologie.
The calculation to do is more restrictive even than the formers, not calculated yet (or measured) practically no aspects but prefixed, provided that they verify the established limits.

4. GENERAL SETTING OF THE PROPOSAL.

Application of the aforementioned levels includes the following conceptual bases:
a) The three options must be interconsistent, and the Norm's user might choose them according to the energetic and comfort criterions with which the building had been conceived.
b) More detailed options will always allow a better adjustment of the project and a possible inversion economy with regard to the most simplified, where the aspects that will not be calculated, are prefixed with a pessimist weighed value.
c) Formulation of A Option will be made with a general character and, though first it will only be appplicable to winter conditions, later it might be easily extended to control the building thermal quality in summer conditions or annual regime.
d) The formal expression of coefficients or restrictor variables derived from A and B Options has to be homologable with the common regulation which is being prepared by General Directorate III "Inner Market and Industrial Affaires", and XVII "Energy", from the European Communities' Commitee.

5. CONCLUSIONS.

Conclusions on what the Extension of the NBE-CT-79 should take away, in such a way of proposal presented here, are the following:
A set of bound requirements, that buildings must be performed, are marked, from the point of view of the knowledgement of the building thermal quality. These requirements are established in agreement with the actual European Norms guidelines.
For buildings evaluation, we propose three different and alternative options, structured by steps in view of complexity of their use:
A. That is the most complicated option, provided that it is necessary a great afford of information and calculation. Its attraction is based on the fact that it results a more exact evaluation of the building thermal conditions.
B. This option is equivalent to the evaluation method which is included in the actual NBE-CT-79, but with modifications that results of update of knowledge and experience on application of afformentioned existent Norm.
C. This is an option that requires a few of inputs, and the elements that the user can utilize into the design and into the buiding construction, has been prefixed with a pessimist criterion and selecting the most unfavourable performances.
Proposal for the NBE-CT-79 Extension presented here, brings up to date furthermore several aspects to consider them, as well as:
i) Constructive typologies.
ii) Algorithms and calculation methods.
iii) Climatic data.
iv) Solar gains for opaque and half-transparent walls are included, though in a simplified manner.
v) Energetic request of buildings can be controlled.
vi) Value of coefficients and limiting parameters for each different option.
As a general conclusion it may be considered that the studies carried out and their consequences allow to make a solid basis to establish a Norm for the control of buildings thermal quality.
Once the suitable decisions have been taken, an automatic process is established, and no problems are seemed to be found.

BIBLIOGRAPHY.

"Norma Básica de la Edificación NBE-CT-79, sobre Condiciones Térmicas en los Edificios". B.O.E. Real Decreto 2429/79, de 6 de Julio.
"Propuesta para la Extension de la Norma Basica de Edificación de Condiciones Térmicas de los Edificios (NBE-CT-79)". División Solar. IER - CIEMAT. Madrid, Diciembre 1988.

COMPARATIVE STUDY OF PASSIVE SOLAR SYSTEMS PERFORMANCE

Heras M.R., Marco J., Tellez F.M.
Solar Division/IER-CIEMAT
Avda. Complutense, 22/28040-MADRID (SPAIN)

ABSTRACT

This paper is the continuation of the "Comparative study of two solar buildings" also presented in this Conference (SECA) (1), which we have presented the global evaluation results carried out used a macroestatic model to reach the energetic performance evaluation.

In the paper here presented we achieve a comparison of glazed sunspace energetic performance with a macrodynamic models aplication, which is considered the thermal inertia mass effect.

The results comparison is achieve on the same blocks of houses described in (1).

This work is completed with a estimation of the thermal comfort reached in the houses used the Fanger PMV curves.

INTRODUCTION

As is kown in /1/ these houses have been built by Ministry of Public Works and Urbanisme (MOPU) and they are a total of the 100 houses distribuited 40 of them in Alpera (Albacete) and 60 in Pedrajas de San Esteban (Valladolid).

The architectural description of these houses as well as the sensors installations detail and the characteristic of the data adquisition system has been achieved in /1/, although we refer it for anyone aclarations.

The aims of this paper have two different aspects, one to development a macrodynamic model to allow establih a base of comparison between different performances of the each glazed sunspaces of two monitorizated buildings. On the other hand to reach a thermal comfort analisis to allow to evaluate the conditions happens in the houses.

Unfortunatily its imposible to apply the identification models /2/ because for this is neccesary to get ready instantaneous data of the spended power in auxiliar heating, and it has not had this data.

At least to indicate the experimental data where has founded the calculus, have been reached each 30 minutes, and they have been registed the maxima and minima mean and the standard deflection.

MACRODYNAMIC MODELS

The energetic balance of the a system is caracterizated by the thermal out and in flux.
The expresion (1) indicate which is the balance into a buildings separated in each one of the different mechanisms.

To solve this equation instantaneously for each terms supose to know in each moment which is the these value fluxs. Since it supose to have a big number of sensors a calculus models have been developped which can represented the indoor instanteous temperature, only to know a short set of parameters.

One of this models is the electrical analogy, that there is between a electric circuit and a heat transference process.

More recently have been developped other models which represented better of indoor buildings temperature. Amount of them are research to identificatin models /2/.

Following Burch /3/ the monitoritation levels can split of three different steps depended on complexiti of the developped calculus, but all of them follows the monitoritation scheme to development in this works.

The three models level are the following:

MICRODYNAMIC: Are that to solve the all heat transference equations together to happen into a buildings. Generaly it need a big account data, as it repuest to kown all the building characteristics (more that 100 parameters for a buildings)

MACRODYNAMIC: This models are characterized by a notable reduction of parameters number. The buildings split in thermal zones and each of them is characterized for a few parameters which are obtained for regrest of experimental data. This level will be developped in posterior works for the evaluation of the buildings monitoritated.

MACROESTATIC: This is the elemental level and is the described in /1/. It is founded in a means and in a time enough bigger as to eliminate the storage effects (thermal inertia). With only two parameters (loss global coefficient and efective area of energy collection) is defined the building.

II.1.- MACRODYNAMIC MODEL

This model type predict the building thermal performance of a global shape. For this development is neccesary to know the temperature of each thermal zones (rooms), as well as the auxiliar energy consumption and internal gains.

In this models are used a electric analogy which can increase its complexity according of the different number of buildings thermal zones.

In general this electric circuit is represented in the figure 1.

The equations for this model are:

$$C_i \frac{dT_i}{dt} = \sum_j (T_j - T_i) R_{ji}^{-1} + Q_i$$

indicated the j subindex each one zones that has been splited the buildings, and i subindex the concrete zone to study.

Qi is the add of energy solar as well as auxiliar.

Then j equations are written, one for each zone with some coefficients that can be determined from the real data of the buildings performance. Evidently, some restrictions are imposed to this equation. One of them is:

$$R_{ji} = R_{ij} \qquad (2)$$

where R_{ij} is the thermal resistence betweeen the i and j zones.

All this transmition coefficients between different zones and the transmition coefficient between indoor and outdoor are estimated by regretion analysis applied to experimental data.

There are many structures of models thar can be used into the identification process of a linear stationary dynamic system. One of them is the modal basis structure. In our case it is a matrix frame representation of the energy conservation equation as:

$$\begin{aligned} EX &= AX + BU \\ Y &= CX + DU \end{aligned} \qquad (3)$$

where X is the named state vector (temperature), U is the inputs vector (solar radiation, outdoor temperature and auxiliary heating power), Y is the outputs vector (selected variables), and E,A,B,C,D, are the matrices which contains the thermal parameters of the system (heat capacities, transmition coefficients, solar aperture,....).

For the sunspace a mathematical structure as (1) is used into the parameters estimation process. This procedure is not applicable in the description of the indoor temperature because auxiliary heating is unknow.

Then the equation for the sunspace is the following:

$$(Mc)dT/dt=(UA)_0(T-T_0)+(UA)_1(T-T_1)+A_sQ_{sol} \qquad (4)$$

qhere M_c is the heat capacity of the sunspace, UA_0 and UA_1, respectivelly, the conduction coefficients between the sunspace and the indoor and between the sunspace and the outdoor. Outdoor temperature is T_0, indoor temperature is T_1 and air sunspace temperature is T.

The building has been represented by a bizonal structure with the same temperature (see fig 2).

THERMAL COMFORT

Making a survey on different comfot levels, there is a general agrement that Fanger PMV /4/ Curves are the most suitable ones. Therefore that relationship was assumed, with a range from -3 up to 3 according to the following degrees of:

Cold	-3
Cool	-2
Quite cool	-1
Neutral	0
Quite Warm	1
Warm	2
Heat	3

The equation including the above parameters is:

$$\begin{aligned} PMV = & \\ & \{ 0.3521^{-0.042\zeta} + 0.032 \} \times \\ & \times [\zeta(1-\eta) - 0.32 \{R_1\{35.7 - 0.0275\zeta(1-\eta)\} + R_2 - P_a\} - \\ & - 0.42 \{\zeta(1-\eta) - 58\} - 1.4\times10^{-3}\zeta(35-\theta_a) \\ & - 0.0017\eta(44-P_a) \\ & - 0.71\tau F_d \{(\theta_d + 273)^4 - (\theta_{mrt} + 273)^4\} \\ & - F_{dhc}(\theta_{cl} - \theta_a)] \end{aligned}$$

being:

$\zeta = H_M/A_N$=metabolic heat generation rate/body area
$\eta = W/H_M$=Rate of work done/H_M
R_1=4.066 mb/deg
R_2=84.2 mb
P_a=partial vapour pressure of the ambient air
θ_a=ambient air (dry-bulb) temperature (°C)
τ=Stefan-Boltzman constant
F_d=factor of cloting
h_c=convective heat transfer coefficient

This value is assumed as:

$$\begin{aligned} hc &= 2.28\ (\theta d-\theta a)0.25 \quad \text{if } v<0.1 \text{ m/s} \\ &= 12.1\sqrt{v} \quad \text{if } v>0.1 \text{ m/s} \end{aligned}$$

Since we seek to evaluate the comfort level achieved in a house (independent of the each person) we took yhe following parameters as constants:

$$\begin{aligned} PMV = & \ 0.09232 \times \\ & \{48.72 - 0.32\ (56.27-P_a) - 0.0588\ (35-\theta_a) - \\ & 4.72\times10^{-8}\ [(\theta_d+273)^4 - (\theta_{mrt}-273)^4] - \\ & - 0.106\ h_c\ (\theta_d-\theta_a)\} \end{aligned}$$

For a situation in which we hve simplified the equations to relationships with ambient parameters, assuming activity levels and dresses usual in domestic houses.

RESULTS

We proceed to study the house in both ways previously mentioned: thermal comfort and instantaneous response of the system.

1-THERMAL COMFORT

Two typical periods were chosen in the coldest and warmest months in both towns. In Figure 3 and 4 are depicted the results obtained for Alpera and Pedrajas.

As can be seen in no time a discomfort kevel is achieved. Nevertheless some scarce hours of disconfort appears in summer. The performace housesis very well in winter.

ENERGY ANALYSIS

Equations as that represented in $ were used, calculating its derivatyve by means of:

$$dT(i)/dt=(8(T(i-2)-T(i-1)+T(i+1)-8T(i+2))/6 \quad (8)$$

Equations 4 was modified dividing both by the factor AsDt/dt, and result:

$$Qsol/(dT/dt)=[(UA)0(T-T0)+(UA)1(T-T1)]/AsdT/dt -Mc/As \quad (9)$$

This equiation can be fitted with a multiple regression in different times of the year and following a comparison methodology:

1.- It is assumed that the thermal inertia mass (Mc) is constants all the year long.

2.- Loss coeficients ((UA)0, (UA)1), likewise the collection area, will change according to the utilization of the greenhouses.

Taking into account the above mentioned assumptions the collection are is directly comparable from values of the independent term achieved in the adjustment.

From this first comparison, loss coefficients can be compared by means of the relationship.

In the table I are represented the results research for Alpera and Pedrajas de San Esteban in summer and winter as well as an intermediate value (It is in April month).

Since the energy flux from auxiliary heating is unknown only is posible to make a comparison between results for this reason is not posible to use ideentification models, that would allow as to research the absolute values of the parameters.

In the figure 5 are represented grafically of the glazed sunspace thermal experimental evaluation versus the simulation results.

TABLE 1

	ALPERA		PEDRAJAS	
	W	S	W	S
$\frac{UA}{Mc}0$	20.8	5.2	14.2	–
$\frac{UA}{Mc}1$	36.6	27.1	18.8	–
$\frac{As}{Mc}$	34.4	44.9	33.0	–

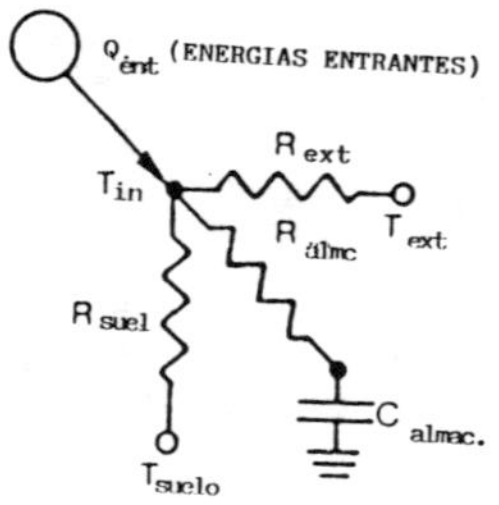

FIGURA 1

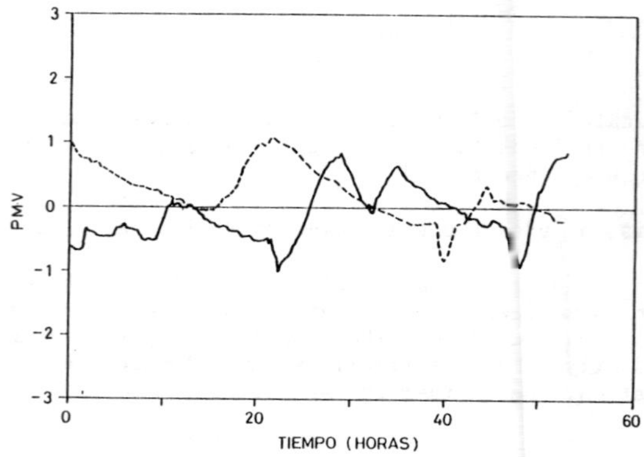

FIGURA 2

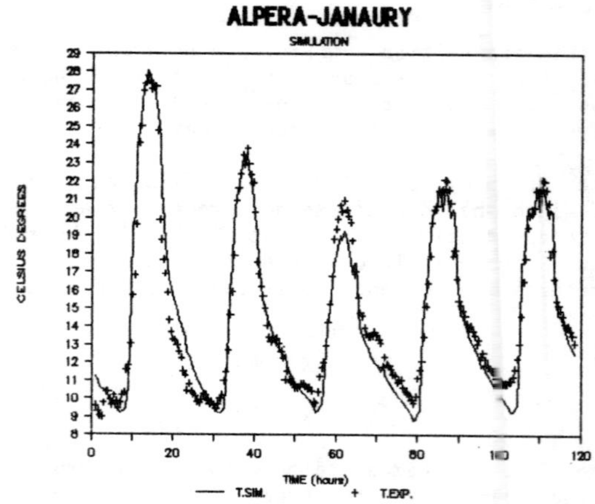

FIGURA 4

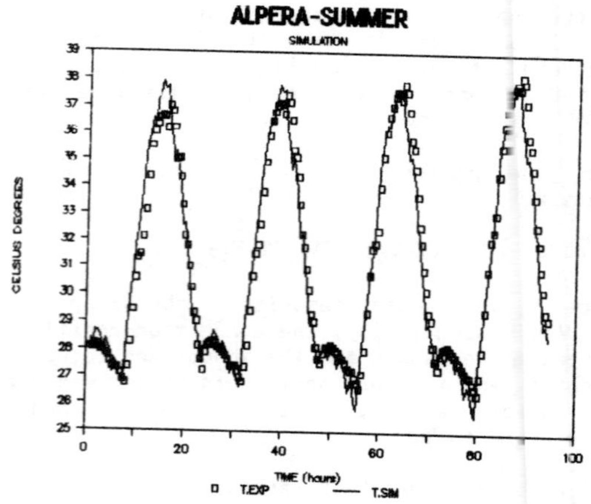

FIGURA 5

/1/ Heras, M.R et al: "Comparative study of two passive solar buildings". SECA 1989.

/2/ Parameter identification techniques for thermal analysis of buildings. ISPRA, October 9-11, 1989.

/3/ Fanger, P.O.L.: "Thermal confort analysis and aplications in environmental engineering". McGRAW HILL. N.Y. 1970

COMPARATIVE STUDY OF TWO PASSIVE SOLAR BUILDINGS

Heras M.R., Marco J., Tellez F.M.
Solar Division/IER-CIEMAT
Avda. Complutense, 22/28040-MADRID (SPAIN)

ABSTRACT

In this work the study of energetic performance of four apartments will be shown. For this study, data acquired during more than one year have been used. Apartaments belong to two projects consisting of blocks of flats located in Alpera (Albacete) (fourty flats) and in Pedrajas de San Esteban (Valladolid) (sixty flats). Bhot blocks of houses have the same design, including passive and active solar components and are within urban areas and continental climates. Building operation, global energy balance using a macrodinamic model and final results will be shown.

1. DESCRIPTION OF BUILDING

These monitorized buildings are part of a public housing, sector promotion of the spanish Ministerio de Obras Públicas y Urbanismo (MOPU).
All the flats (100 in total) have the same design. The two apartments in Alpera and the other two in Pedrajas de San Esteban, chosen for monitorization, have south orientation.
The aim was to build low cost housing, using passive solar design techniques, to give significant energy cost saving for the lowest possible capital investment, in a typical rural area with a continental climate. Despite this, due to the climate, it was necessary to install an auxiliary heating system in each flat.
Each block is a three storey terrace (9m. hight), with two apartments each floor. The access is from a central stairway which is also an axis of simmetry. In the south of each flat, there are the living / dining room and two bedrooms. The bathroom is in the middle and in the north there are the kitchen, a third bedroom and the stairs.
The shouth façade is ocupied by solar elements (air collector and sunspace). Flats in all the floors have the same glassed surface, except ground floor that has less. Separating the sunspace of livingroom there is a massive wall.
North façade is dotted with little windows whith shutters timber frame and solar protection in order to decrease energy losses in winter and make possible cross-ventilation in the summer.
Figure 2 shows transversal section and plant of flats. The principal bedroom can be conditioned with hot air collected in the south façade and forced with a fan controlled by a thermostat.

1.1. Description of construction

Principal façade (south)
- Timber framed, single glazed window
- Massive wall in gallery is constructed of 0.42 m. solid brik provides thermal storage, with a surface of 60% of the façade.
- This massive wall connects the gallery with the interior, with three doors single glazed.

OTHER ENVELOPE ELEMENTS

- Walls are a double skin of 0.12 m. brick, incorporating 0.10 m. of expanded polyestyrene insulation, with an internal plaster end.
- The ground floor slab is isolated with 0.10 m. of expanded polyestyrene.
- Intermediate floors are built from 0.25 m. concrete reinforced covered with ceramic tiles, which provide further thermal storage.
- The main roof is pitched an isolated with 0.15 m. of glassfiber in the roofspace.

With the knowledge of characteristics and properties of constructive elements, first calculation using spanish standard NBE-CT-79, has been developed.

1.2. Description of solar components

To use solar gains and to save energy, the following components have been built:
- sunspace
- massive wall
- air collector controled by thermostatic and conduit to create a convective loop.
- high insulation in the envelope and good frame in exterior windows and doors.

With these components, this building is designed for winter conditions, nevertheless cross ventilation could be created during the summer operation of each component as follows::
GLASSED GALLERY: Short-wave radiation passes through the glass and is partially absorbed in the wall. Green house effect is present between glass and wall. The temperature of massive wall is increased during the day, and there is heat flux from the exterior to the massive wall and directly from exterior to interior through glassed door and window. During the night, insulated protection of glasses are closed and heat flux is from massive wall to interior. The glassed area has 17.1 m^2.
MASSIVE WALL: Its dimensions have been calculated to obtain enough heat capacity and thermal wave delay of about 13 hours. Thus maximun heat flux toward the interior occurs during the night.
AIR-HEATING SOLAR COLLECTOR: it is also located on the south façade of each flat, either to the left or the right of the sunspace as symmetry dictates. By using a thermostatically-controlled fan, the hot air is tranferred through a duct to the hollow wall in the bedroom.
The 4 m^2 full-height collector is buklt from 60 mm. polyisocyanurate foam with a black corrugated surface (to allow maximum solar absortion) a cavity and then fully open glazing. The duct leading from the top of the panel into the flat is made of glassfiber to reduce heat losses. It passes through the flat a high level and into the cavity in the north wall of the rear bedroom. The hot air then follows a serpentine pattern down to the bottom of

the wall, where it enters the bedroom through a grill.

2. OPERATING MODES

In winter, during the day, the sun enters in to sunspace and the south-focing rooms behind. The wall in to the rear of the sunspace collects incident solar radiation and transfers the gains to the adjacent rooms behind. The air in the solar collector heats up until it reaches the thermostat set point, (usually 25-30°C) when the fan operates. Air from the adjacent south-facing room is drawn between the glass and the wall in order that the hot air passes into the insulated duct and then to the north-facing room. The external vent at the top of the panel remains closed.
At night, heat is radiated to the south facing the rooms from the storage wall behind the sunspace, and to the rear bedroom from the hollow wall which has been warmed up by the hot air from the collector. The non-return value behind the fan prevents reverse circulation. In addition, wooden shutters may be closed to insulate the south-facing glazing.
In summer, during the day, the air is drawn from the north side of the flat, by the stack effect of the solar collector which has the external vent open, to promote ventilation. The thermostat is set at 90°C so that the air does not reach the set point temperature and therefore, none is transferred to the north facing bedroom.
Shutters are drawn down over the south facing glaze to reduce solar gains entering the flat. A blind, fitted latter, provides shading to the sunspace in summer.
At night, the sunspace windows, those in the south façade, and the north facing windows can be opened to allow cross ventilation. To avoid condensation problems in these highly insulated flats, positive ventilation is required in winter.

3. ANALYSIS METHODOLOGY.

The calculation model used, considers two kind of heat gains:
a) exteriors:
- solar gains (Q_{sol})
- Heat conduction through the envelope (Q_{con}).
b) interiors:
- auxiliary heating (Q_{aux})
- other intenal gains.
On the other hand, heat losses are summarized in the following terms:
- ventilation and infiltration
- conduction (Q_{con})
- ratiation during night (Q_{rad})
- convection (Q_{conv})
Finally the storage is grouped (Q_{acum}).
An instantaneous energy balance can be expressed as follows:

$$Q_{int}+Q_{aux}+Q_{sol}=Q_{con}+Q_{ven}+Q_{rad}+Q_{conv}+Q_{acum} \quad (1)$$

To evaluate each one of these fluxes, it is necessary to measure a big number of parameters. Then it is justified to do the following assumptions and simplifications [1]:
1.- To eliminate acumulation terms, a long measurement time is required in order to eliminate short time oscilations of the excitations, then:

$$\Sigma Q_{acum} = 0 \quad (2)$$

Samrt and Ballinger [2] consider that it is enough to average for a pericd of one week.
2.- It is possible to group all the terms that depend on the difference between interior and exterior temperatures in an unique term as follows:

$$Q_{con}+Q_{conv}+Q_{vent}+Q_{rad}=UA*(T_{ext}-T_{int}) \quad (3)$$

3.- Solar gains can be expressed as:

$$Q_{sol} = A_s \Sigma H_{svs} \quad (4)$$

where A_s is the effective area for solar gain whose rate to real area exposed to sun gives us a measurement of effectiveness for solar gain of the total surface expounded to sun. Have is the global solar radiation on the south façade.
With these assumptions, equation (1), after integration for weekly periods can be expressed:

$$Q_{INT} + Q_{AUX} + A_S \cdot H_{VS} = U_{A*} (T_{INT}-T_{EXT}) \quad (5)$$

Coefficients A_S and U_A are calculated by regretion analysis using measured values.
Resultant values of A_S and U_A depend on how the building is aparated. For instance, A_S depends on using solar protections and U_A is in function of the air change rate in the building. In order to minimize uncertainties, the balance (5) is calculated in different periods of the year.
With these macrodynamic modes, developed by Subaneo /1/ two important characteristics of building can be obtained: solar aperture (A_S) and global losses coefficient (U_A).

4. SENSORS AND DATA ACQUISITION SYSTEM

Sensors were installed in the four flats, in similar points. Location is showed in Figure 2. Besides these, a meteorological station was used with: sensor combining temperature and humidity a combined wind velocity and direction meter and a pyranometer. For other part in Pedrajas flux sensors in the exit of auxiliar heating and in Alpera special electrical power meters was used.
Data acquisition system to storage data continuously during 11 days.

5. RESULTS

Appling the model descripted, similar results have been obtained with a value of
U_A = 190.00 W/°C
Figures 5a) and 5b) show wind measurements
Figures 6a) and 6b) show medium day for exterior temperature and radiation, averaged for hottest and coolest months.
Table 1 shows resultant values of regretion analysis that correspond to mentioned model. With thiese results Figure 7 shows global energy balance for both cases. In this figure
Besides operating evaluations resulting of individual analysis of sensor measurements that show a non-optical utilization of building can be remarked.
Solar gain is bigger in Alpera than in Pedrajas, the probable reason is that in the microclimate of Valladolid, fog is frecuent and near the monitorized flats in Valladolid there are perennial trees.

RESULTS

	ALPERA		PEDRAJAS	
	WINTER	SUMMER	WINTER	SUMMER
AS(m2)	4.5	2.1	3.25	--
uA(W/°C)	190	240	175	--

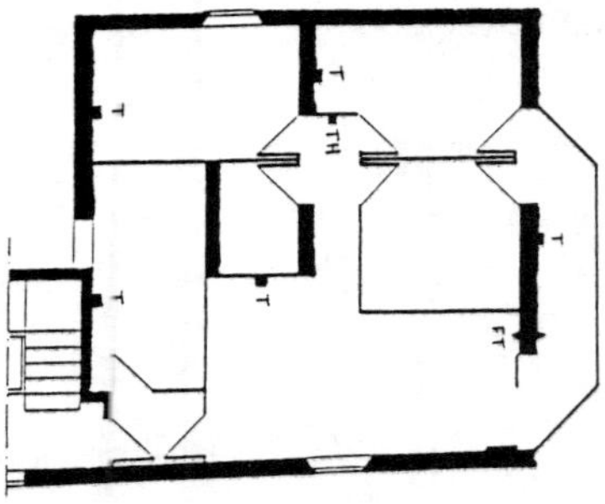

FIGURA 2a

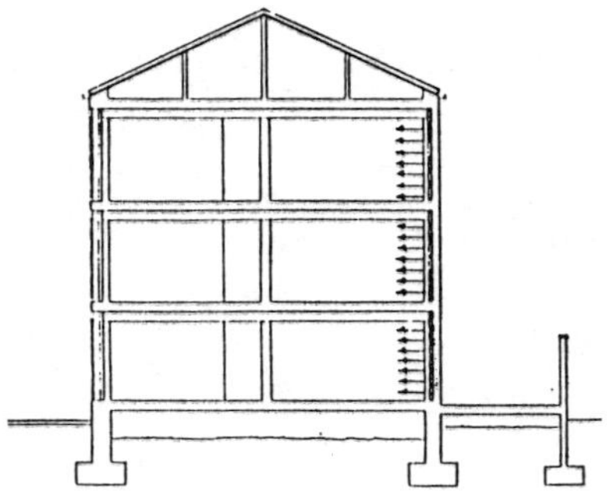

FIGURA 2b

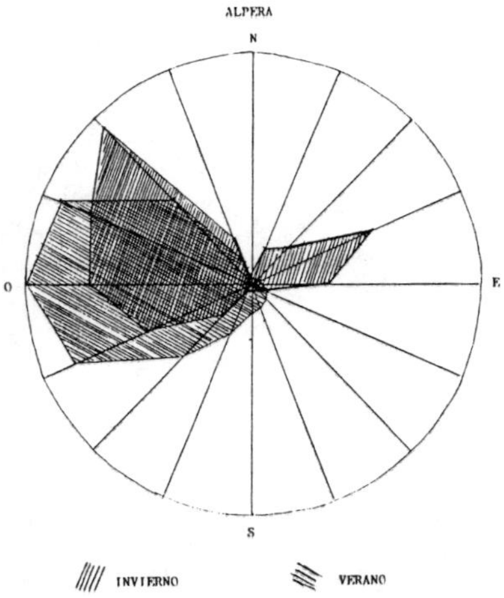

FIGURA 5a

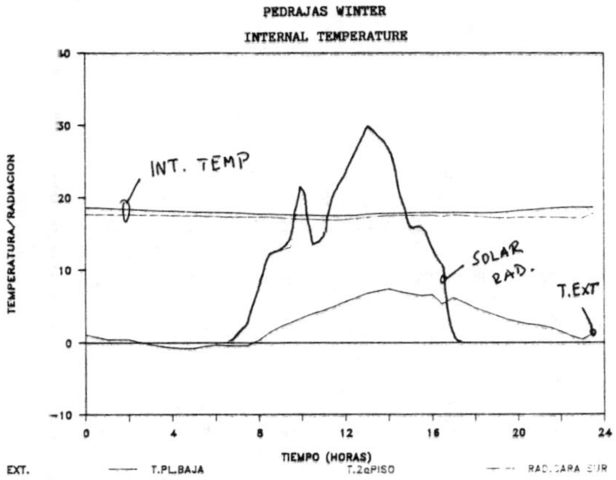

FIGURA 6a

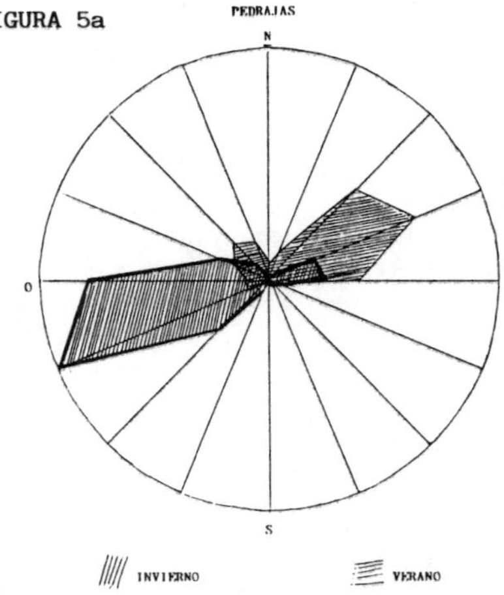

FIGURA 5b

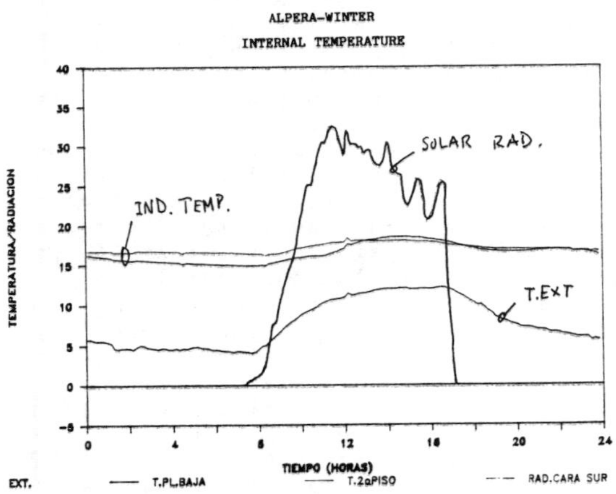

FIGURA 6b

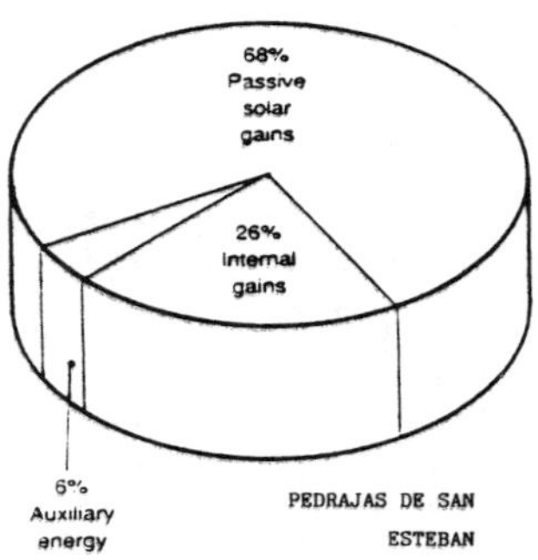

FIGURA 7a

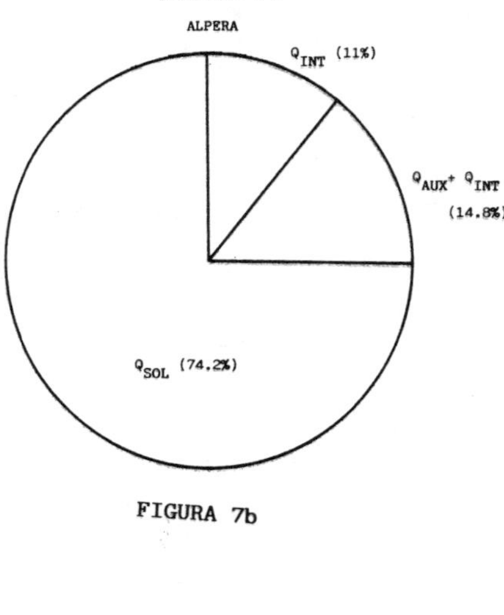

FIGURA 7b

/1/ Subbarao,K. et al: "Short term meausurements for the determination of the envelope retrofit performance." SERI/TP-235-2639. Golden, Colorado. June 1985.

/2/ Smart,M. and Ballinger,G. "Measuring solar and electric Heating contributions in occupied houses." Energy and Building,9 (1986) 213-219.

THE INFLUENCE OF GLASSING SHAPE ON BUILDING THERMAL PERFORMANCE

Juan C. Klainsek
Departamento de Construcción y Tecnología Arquitectónicas
E.T.S. de Arquitectura - Universidad Politécnica de Madrid
Av. Juan de Herrera, 4 - 28040 - Madrid - Spain

Glassing must be especially considered when designing a building since it is the weakest spot of walls and it makes the transmission larger than the one produced through opaque areas.
Buildings with vast glass areas, which is the case of non-residential buildings, are significantly affected by glassing on both their thermal transmission and costs of building and maintenance, Thus, an office building has been chosen to analyse the effect that glassing shape changes have on its energy performance, keeping the carpentry, the type of glass and the size constant.
The Weighting Factor Method has been used in this study because it is possible to determine the energy consumption from the physical description of the building, the environmental conditions and the internal load profiles by using this method. Besides, a computer programme has been developed and applied to the base case building to simulate the building energy requirements and give accurate results about heating and cooling energy consumptions.

1. INTRODUCTION

Thermal transfer through glass areas is more important than that produced through opaque surfaces; considering this and the fact that certain types of buildings, especially non-residential ones, require large glass areas, the energy loss may be so important as to make insulating of the opaque surface practically inefficient. Thus, it is necessary to carry out a realistic simulation of the thermal transmission through windows if an appropriate design and a decrease in heating and cooling energy consumption are to be achieved.
From the several methods which can be used to perform this kind of estimations the Weighting Factor Method has been used. It represents Z-transfer functions and answers for the parametres which have an effect on the building flow. Thus, and from the physical description of the building, the environmental climate conditions and internal load profiles, the heat gains are fixed at a constant temperature. A computer programme has also been developed and applied to a given type of office building.
This paper analyses the effect that different glass shapes have in a base case building (office) situated in Madrid: horizontal continuous windows with a variable height, and vertical discontinuous windows with a fixed height but a variable width, in order they to have the same surface as the previous ones. The frame and glass characteristics are no changed and results about the building energy performance are obtained.

2. BASE CASE BUILDING

It is a 10 floor office building, located in Madrid and with the following characteristics: 40m x 26m floor; 37m height, which corresponds to a gross surface per floor of 1036 m^2, useful conditioned of 795 m^2. Its location is the most disadvantageous one, for the largest façades face towards the east and west.
The months from November to March are included in winter and the rest of months in summer.
In the estimations, the interior temperature in winter is constant: 20 ºC and RH 50%, 25 ºC and RH 55% in summer; the exterior temperature is variable, so the calculations are carried out hourly.
With regard to the exterior conditions, meteorological variables are generated hourly from reliable meteorological data and an auxiliary programme has been used. The daily starting data are: maximum and minimum temperature, total daily irradiance on horizontal surface and day absolute humidity at maximum temperature hour.
The heat transmission coefficient of opaque areas (K-value): -horizontal- roof Kr= 0.43 W/m^2 K, light color (cream); floor levels above open space Kf= 0.66 W/m2 K. -Vertical- traditional wall composed of seven layers, total thickness 23.5 cm, insulating material 4 cm, 244.6 kg/m^2 and Kw= 0.61 W/m^2 K, and four layer curtain-wall, total thickness 5.3 cm, insulating material 5 cm thick, 89.6 kg/m^2 and Kw= 0.42 W/m^2 K, both medium color (green and blue, bright red, light brown, wood and concrete). Metalic windows.
The only shadows on windows are those produced by setback, since glass areas have no shading devices.
A free height in all floors of 2.80m is considered, and the energy consumption calculations are obtained per useful or net surface unit.
The internal loads are fixed by: occupation 55 pers./floor, aproximately equal to 14 pers./m^2; ventilation 2000 m3/h floor, that is 2.5 m^3/m^2; illumination 7.95 kW/floor or rather 10 W/m^2.

3. GLASS AREAS

The glassing used in this paper has the following characteristics: reflectant; insulating 6+6 mm, with an a.s. 12 mm thick and Kg= 3.7 W/m^2 K and 0.48 as corrective coefficient (c.c.) of the solar gain factor.

4. ANALYSIS OF RESULTS

Considering that the base case building windows have different heights, hw=1.00m, 1.10m, 1.20m, 1.30m, and 1.70m, 1.80m, 1.90m and 2.00m all round the building perimeter, the energy consumptions (e.c.) per conditioned surface during summer and winter are calculated. Then, these glass areas are transformed from continuous to discontinuous, with the same façade surface than the former ones. To carry it out, the glassing height is: hw= 2.10m and 2.80m respectively, from floors to lintels. Taking as a variable the external vertical opaque areas (traditional wall and curtain wall) and with this shape change, different e.c. are noticed. Such consumptions are obtained from a type of glassing with its codes into brackets: the first symbol corresponds to the continuous glassing, and the second to the discontinuous glass area, for a 6+6 mm reflectant glass with a 12 mm a.s.: (□) (+); (◇) (△).

A regression line, obtained by the least square method, is adjusted with the e.c. values of each of these shapes of glassing and a line equation is obtained; it represents these consumption variations for glass areas even with a surface smaller or bigger than that of the glassing studied in this paper.

-Fig.1, the values obtained in summer with traditional and curtain-wall, are analysed. The consumptions per conditioned useful unit surface (kWh/y.m^2) are represented on the diagram ordinate axis; the four heights of continuous windows, coincident in the same way with discontinuous shape-transformed windows with hw= 2.10m, and an equal glass area in both cases, are represented on the axis of abscissa.

The aim is to show the difference that can be produced when for an equal glass surface the glassing shape is transformed from rectangular horizontal into vertical.

The e.c. for continuous rectangular horizontal windows and with traditional wall (□), hw= 1.0m, is 21.67 kWh/y.m^2. If the heights are increased to hw= 1.10, 1.20, 1.30 [m], the e.c. increase to 22.03, 22.39, 22.76 [kWh/y.m^2] respectively.

For shape transformed discontinuous windows (+), hw= 2.10m, the e.c. increases to 21.80 kWh/y.m^2; e.c. of shape-transformed windows corresponding to the previous hw= 1.10, 1.20, 1.30 [m] are higher: 22.15, 22.50, 22.85 [kWh/y.m^2].

In the first case (□), e.c. increases as the openings have a higher surface, and it answers for equation:

e.c.= 3.63x+18.04	Correlation Coefficient R^2= 1.00 Estimated statistical error of Y: 0.00

In the second case (+), the e.c. also increases and the equation is:

e.c.= 3.50x+18.30	Correlation Coefficient R^2= 1.00 Estimated statistical error of Y: 0.00

When x= 1 (hw= 1.0m), the e.c. per surface unit would be those shown above, and for the vertical situation they increase to 0.60%.

The following percentages are also calculated for the smallest glass areas.

Two more cases are obtained by using the curtain-wall; the e.c. for hw= 1.0, 1.10, 1.20, 1.30 [m] are (◇): 22.73, 23.07, 23.42, 23.76 [kWh/y.m^2] respectively. For shape-transformed discontinuous windows, (△), as opposed to the previous ones, the e.c. increase to 22.78, 23.12, 23.46, 23.81 [kWh/y.m^2] respectively.

In the third case (◇), the equation is:

e.c.= 3.44x+19.29

In the fourth case (△), the consumption increases and it becomes:

e.c.= 3.43x+19.35

An increase is still stated of 0.22%.

-Fig.2: different e.c. are obtained in winter, with traditional wall and curtain wall.

Energy consumptions calculated with traditional wall (□) are: 16.32, 16.90, 17.48, 18.06 [kWh/y.m^2].

With the same type of shape-transformed glass area (+): 16.34, 16.92, 17.51, 18.10 [kWh/y.m^2].

The equations of the lines are respectively:

e.c.= 5.80x+10.52 (□)

e.c.= 5.87x+10.47 (+)

The increase between one shape and the other is: 0.12%.

In winter, the increase between a shape and its transformed one is noticeably lower than those of the former groups, related to the summer season, so from now on only the equations will be given because the different e.c. can be obtained from them.

For calculations carried out with curtain wall, the equation results in:

e.c.= 6.34x+8.34 (◇)

And for the same kind of transformed glass area:

e.c.= 6.37x+8.31 (△)

There is no practical difference between the e.c. of one shape and the other.

-Fig.3: also for continuous rectangular horizontal windows with traditional wall but with heights different from the former ones and usually used in this building typology, hw= 1.70, 1.80, 1.90, 2.00 [m]; the e.c. during summer are (□): 24.26, 24.64, 25.01, 25.39 [kWh/y.m^2] respectively.

For shape-transformed discontinuous windows, the e.c. increase to (+): 24.34, 24.71, 25.07, 25.44 [kWh/y.m^2] respectively.

The line equations are:

e.c.= 3.76x+17.87 (□)

e.c.= 3.66x+18.12 (+)

The increase between one shape and the other results in 0.33%.

With the curtain-wall, the e.c. are (◇): 25.16, 25.50, 25.84, 26.19 [kWh/y.m^2].

For the same area transformed shape, the e.c. increase to (△): 25.21, 25.55, 25.89, 26.24 [kWh/y.m^2].

The equations of the lines are respectively:

e.c.= 3.43x+19.33 (◇)

e.c.= 3.43x+19.38 (△)

The e.c. between shapes increase in 0.2%.

-Fig.4: with the same windows heights as the previous case, in winter and with traditional wall, the equation is:

e.c.= 5.83x+10.48 (□)

And the equation of the transformed shape

e.c.= 5.87x+10.45 (+)

The e.c. increase between shapes is 0.2%.
With curtain-wall the equation of the line is:

e.c.= 6.4x+8.24 (◇)

For the transformed shape is:

e.c.= 6.4x+8.24 (△)

The e.c. between one shape and its transformed one are the same.

5. CONCLUSIONS

1. Energy consumptions (e.c.) answer for an equation of this type:
e.c.= ax+b
where x represents the height of the different glassing studied and, as width is kept constant, x values may be considered as glass areas.

2. $\frac{d\ e.c.}{dx} = a$; a>0 always ⇒ increasing function

Energy consumption is higher as glass areas become larger.

3. b>0 always

Even if there are no glass areas, there would be thermal transferences through opaque areas.

4. The energy consumption intervals for these shapes and type of glassing studied in summer and with traditional wall, vary from 0.6% to 0.33%. With curtain-wall they vary from 0.22% to 0.2%.

In winter, with traditional wall, between 0.12% and 0.2%; with curtain-wall there is no difference.

To sum it up, energy consumption is higher in summer, when vertical discontinuous windows starting from the floor are used. In winter, the energy consumption may vary if traditional wall is taken into consideration, being the variations of no significance.

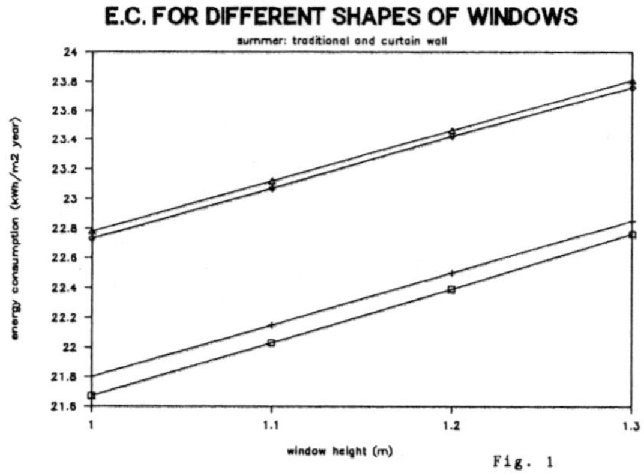

Fig. 1

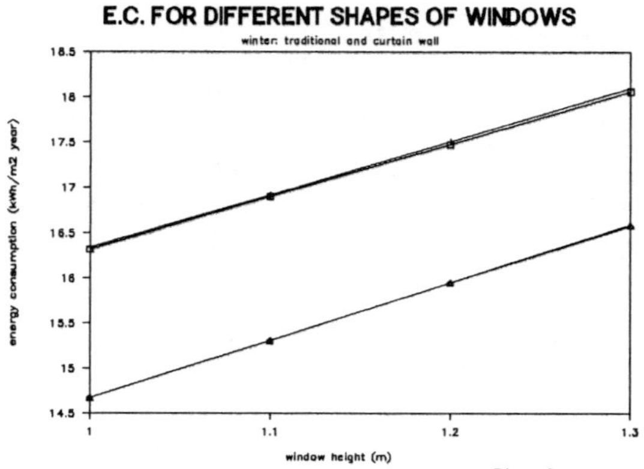

Fig. 2

6. NOMENCLATURE

c.c.: corrective coefficient of solar factor gain. Glassing without blind or shading

e.c.: energy consumption

hw : height of window

y : year

7. REFERENCES

1. ASHRAE Handbook 1989 Fundamentals. SI Edition. American Society, Refrigerating and Air Conditioning Engineers, Atlanta, Georgia.
2. Bennett, J.R. et al. "A Study of the Thermal Characteristics and Potential for Energy Conservation at Four London Office Buildings". Building and Environment. Vol. 20, Nº 2, 1988, pp. 83-94.
3. Löhnert, G. "Thermal Insulation and Shading System-TWS". Energy and Buildings for Temperate Climates. Plea 88. Pergamon Press, 1988, pp. 325-330.
4. McKinley, A.D. and Mitalas, G.P. "Room Thermal Transfer Functions, Computer Programs to Calculate Z-transfer Function Coefficients for Rooms". National Research Council of Canada, 1988, pp. 1-3.
5. Selkowitz, S. "Smart Windows". Glass Magazine. Windows and Daylighting Group, Lawrence Berkeley Laboratory, University of California, August 1986, pp. 86-91.
6. Winkelmann, F. and Lokmanhekim M. "Sun-control Options in a High-rise Office Building". Energy and Buildings 8, 1985, pp. 1-13.

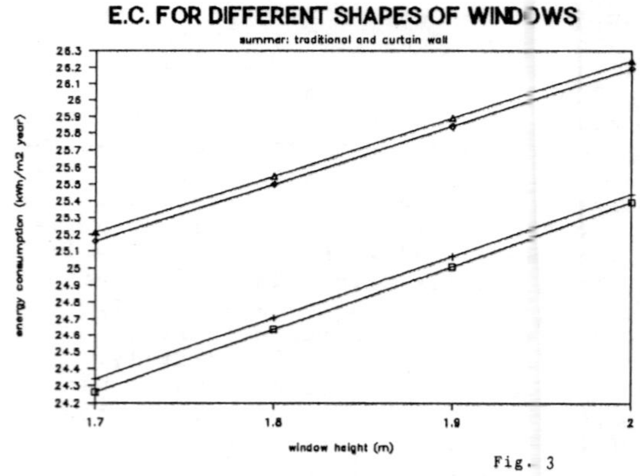

Fig. 3

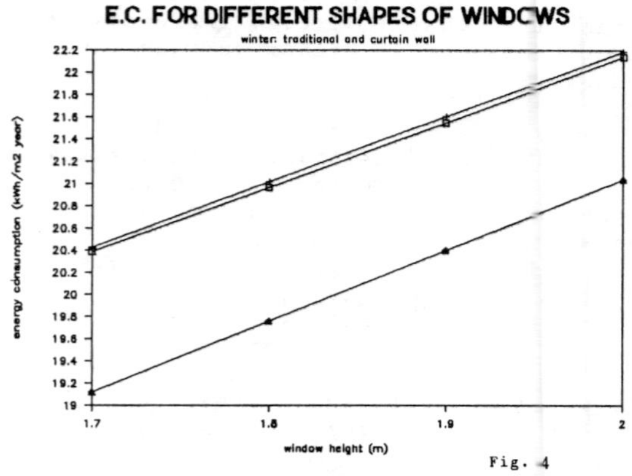

Fig. 4

THE EFFECT OF DIFFERENT TYPES OF GLASSING ON THE THERMAL PERFORMANCE OF NON-RESIDENTIAL BUILDINGS

Juan C. Klainsek
Departamento de Construcción y Tecnología Arquitectónicas
E.T.S. de Arquitectura - Universidad Politécnica de Madrid
Av. Juan de Herrera, 4 - 28040 - Madrid - Spain

Glass areas must be especially considered in the design of most conventional buildings for two reasons: they allow solar energy gains to be stored for thermal conditioning in winter, and they have a lower capacity of thermal insulation if compared with walls. The type of glass used in glass areas has an influence on each factor involved in glassing thermal transmission. On the other hand, radiation energy exchange depends on glassing transmission, and glassing depends on its color. In this paper two groups of glassing are studied: single glass and double or insulating glass (two panes separated by a dehydrated air space). Three types of glass have been taken into account for each group: colorless, colored and reflectant. They have been used in a commercial building (office), in Madrid. Different results have been obtained with regard to the thermal performance of this building in summer and winter.

1. INTRODUCTION

It is necessary the energy analysis to be considered in the development and optimization of the hygrothermal conditioning system from the very early design stage of a building. In doing so, the amount of heat energy which is supplied or extracted from a room is determined and the comfort conditions are achieved.
From the various methods available to carry out this analysis the Weighting Factor Method has been used in this study. With this method it is possible to determine the local gains from a physical description of the building, the environmental climate conditions and the internal load profiles at a constant temperature. Heat removing and air temperature average are calculated by using the local gains, the building characteristics and the cooling systems.
A computer has been used and a computer program has been developed; thus, the energy requirements of an office building and its functions in a climate region are simulated. The energy requirements for heating and cooling can be predicted accurately.
This paper studies the glassing of a given building in order to determine the effect that certain types of glass can have in the energy behaviour of a non-residential building.

2. BASE CASE BUILDING

The base case building is an office building with ten floors, situated in Madrid; its characteristics are: 40m x 26m floor; 37m height (a gross surface per floor of 1036 m^2, useful conditioned of 795 m^2).
This building has the most disadvantageous situation: the largest façades face the east and west.
Winter includes the months from November to March and the rest of months are included in summer.
In the calculations, a constant interior temperature of 20 ºC and RH 50% in winter, 25 ºC and RH 55% in summer is assumed; as the exterior temperature is variable, the estimations are carried out hourly.
Exterior conditions: meteorological variable are generated hourly from reliable meteorological data, and an auxiliary program is used. Daily starting data: maximum and minimum temperatures, global daily irradiance on horizontal surface and day absolute humidity at maximum temperature hour.
The heat transmission coefficients (K-value) of opaque surfaces are: -horizontal- roof Kr=0.43 W/m^2 K, light color(cream); floor level above open space Kf= 0.66 W/m^2 K. -Vertical- traditional wall composed of seven layers, total thickness 23.5 cm, insulating material 4 cm, 244.6 kg/m^2 and Kw= 0.61 W/m^2 K, and four layer curtain-wall, total thickness 5.3 cm, insulating material 5 cm thick, 89.6 kg/m^2 and Kw= 0.42 W/m^2 K, both medium color (green and blue, bright red, light brown, wood and concrete). Metalic windows with different sorts of glass, which are studied in this paper, will be described next.
There are no shading devices, only those produced by setback.
Energy consumption calculation results are obtained per useful or net surface unit; the free height in all floors is 2.80m.
Internal loads are fixed by: occupation 55 pers./floor, equivalent aproximately to 14 pers./m^2; ventilation 2000 m^3/h floor, that is 2.5 m^3/m^2; and illumination 7.95 Kw/floor, or rather 10 W/m^2.

3. GLASS AREAS

The type of glassing used affects the amount of energy transmited indoors, through the rate of incident reflected radiation and it depends on the incidental angle. There is a corrective coefficient of the solar gain factor (c.c.), fixed for each type of glass.
In the present study, the calculations have been carried out taking into consideration different types of glassing, which are the following ones:

Type of glassing	a.s. (mm)	c.c. (-)
single colorless	-	0.94
colorless insulating	6	0.8
colorless insulating	12	0.8
single pink colored	-	0.88
pink colored insul.	6	0.71
pink colored insul.	12	0.71
single grey colored	-	0.74
grey colored insul.	6	0.54
grey colored insul.	12	0.54
single bronze colored	-	0.73
bronze colored insul.	6	0.52
bronze colored insul.	12	0.52
single green colored	-	0.71
green colored insul.	6	0.51
green colored insul.	12	0.51
single reflectant	-	0.55
reflectant insul.	6	0.48
reflectant insul.	12	0.48

Thickness [mm]: single glass= 6; insulating glass= 6+6.
Kg [W/m2 K] : single glass= 5.8; insulating glass, a.s. 6mm=4.0; insul.glass, a.s. 12=3.7.

4. ANALYSIS OF RESULTS

Energy consumption (e.c.) in summer and winter is estimated using the glassing above, considering vertical opaque areas, traditional wall and curtain-wall as variables; with regard to the sorts of windows, they open round the whole building perimeter (continuous windows), with a constant height of hw= 1.30m. In all cases e.c. obtained by using colorless glassing represent reference values.
These consumptions -per conditioned useful surface unit (kWh/year m2)- are represented on the diagram ordinate axis, and the six different types of glassing: colorless colored-filtering (pink, grey, bronze, green) and reflectant, on the abscissas. The energy consumptions of one single and two insulating glass are calculated for each type of glassing, as it was already stated.
In a first group, Fig.1, values obtained during the summer and with traditional wall are analysed. The annual e.c. for colorless glass and 6+6 mm insulating glassing with an a.s. 12 mm thick is 26.57 kWh/y.m2 (100%). If a pink colored glass is considered the consumption decreases in 25.5 kWh/y.m2 (95.97%), that is to say -4.0 %; in the case of grey-bronze-green colored and reflectant glass the results are: 23.51 kWh/y.m2 (88.48%), 23.28 (87.62%), 23.16 (87.17%), 22.76 (85.66%), which represents -11.52%, -12.38%,-12.83%, -14.34% respectively of the reference colorless glass value.
In a similar way, the different consumptions for the other two windows can be studied; it implies a higher consumption due to the increasing of Kg and to the c.c. in the case of a single glass.
In a second group, Fig.2, and still during the summer, a curtain-wall,-less thick vertical opaque area but with a smaller Kw-, is considered. In spite of the latter situation, with the same conditions as for the window with a 12 mm a.s. insulating glass, the e.c. increases, changing from 26.57 kWh/y.m2 (traditional wall), to 27.95 kWh/y.m2 (curtain-wall). In the same way, e.c. increases with the remaining sorts of glass.
Still in this group, on the one hand and taking as a reference the e.c. of the insulating colorless glass with a 12 mm a.s., the result is 27.95 kWh/y.m2 (100%); on the other hand, for reflectant glass 23.76 kWh/y.m2 (85.01%),which represents a difference of -14.99%.
In a third group, Fig.3, e.c. values in winter, with traditional wall, are analysed. Colorless insulating glass with a 12 mm a.s. is again considered, and the annual e.c. is 16.22 kWh/y.m2 (100%); among all the types of glassing studied, the reflectant glass gives the highest increase, 18.06 kWh/y.m2 (111.34%), 11.34% more than the reference colorless glass.
Finally, in a fourth group, Fig.4 and during the winter season as well, the traditional wall is changed into the curtain-wall. The analysis continues for insulating glass with a 12 mm a.s., and it can be stated that the e.c. decreases and changes from 16.22 kWh/y.m2 (traditional wall) to 14.83 kWh/y.m2 (curtain-wall).
There is an obvious trend to decrease in consumption and it is kept in all these sorts of glassing.
The e.c. for each type of glass and for each group are shown in the enclosed figures.

5. CONCLUSIONS

The present conclusions are given for summer and winter separately, since the different types of fuel, conditioning systems and ways of using the building -all of them necessary to carry out a technical economical evaluation- have not been discussed.
Taking into account a base case with hw= 1.30m and from the analysis of results, the following conclusions -which may become general in other cases- have been obtained:
-Summer.
Traditional wall: from all the cases considered the lowest e.c. is achieved with double reflectant glass, with a 12 mm a.s. (22.76 kWh/y.m2); the highest with a colorless glass (26.57 kWh/y.m2). The green colored filtering glass (23.16 kWh/y.m2) shows the lowest consumption when compared with the other three colored ones.
Curtain-wall: it is exactly the same sort of glassing as the former ones (23.76 kWh/y.m2 as contrasted with 27.95 kWh/y.m2; and the green colored-filtering glass with 24.20 kWh/y.m2).
-Winter.
Traditional wall: the least consumption is achieved by using double colorless glass with a 12 mm a.s. (16.22 kWh/y.m2) and the highest with the reflectant glass (18.06 kWh/y.m2). And among the colored glass the pink colored glass (16.72 kWh/y.m2) presents the lowest consumption.
Curtain-wall: the same sort of glassing (14.83 kWh/y.m2 as opposed to 16.58 kW/m2

and the pink colored-filtering glass with 15.29 kWh/y.m2).
Thus, the glassing of a given type of building is an important reference to be considered when calculating the energy consumption in spacial building conditioning.

6. NOMENCLATURE

a.s.: air space
c.c.: corrective coefficient of solar factor gain. Glassing without blind or shading devices
e.c.: energy consumption
hw : height of window
y : year

7. REFERENCES

1. ASHRAE Handbook 1989 Fundamentals. SI Edition. American Society of Heating, Refrigerating and Air Conditioning Engineers, Atlanta, Georgia.
2. Klainsek, J. "Energy Saving in Office Building Design and Retroffiting".Energy and Buildings for Temperate Climates. Plea 88. Pergamon Press, 1988, pp. 173-178.
3. McKinley, A.D. and Mitalas, G.P. "Room Thermal Transfer Functions, Computer Programs to Calculate Z-transfer Function Coefficients for Rooms". National Research Council for Canada, 1983, pp. 1-3.
4. Robinson, P. and Littler, J. "U-Value Assessment of a High Performance Glazing". Energy and Buildings for Temperate Climates. Plea 88. Pergamon Press, 1988, pp. 339-345.
5. Shapiro, M.M. et al. "Transient Three-dimensional Window Thermal Effects". Energy and Buildings, 10, 1987, pp. 89-98.
6. Sullivan, D. "The Influence of Glazing Selection on Commercial Building Energy Performance in Hot and Humid Countries". Lawrence Berkeley Laboratory Report, 1987, BS-247.
7. Turiel, I. et al. "Simplified Energy Analysis Methodology for Commercial Buildings". Energy and Buildings, 6, 1984, p. 70.

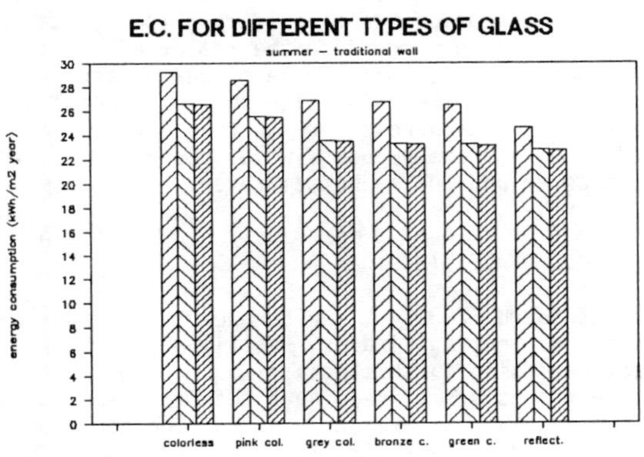

Fig. 1

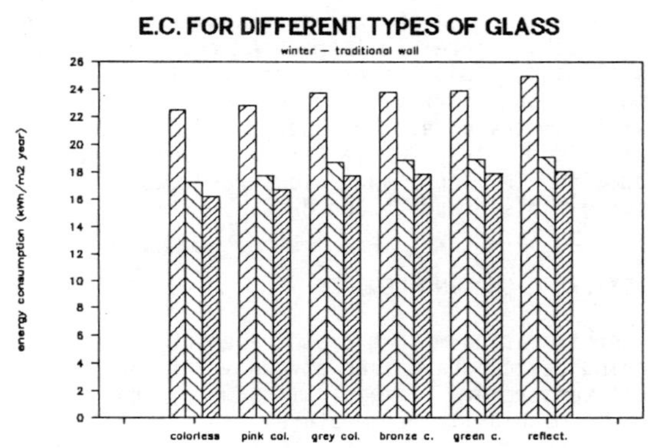

Fig. 3

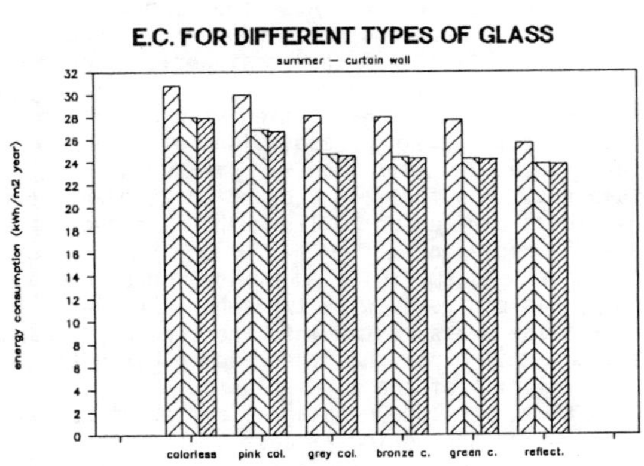

Fig. 2

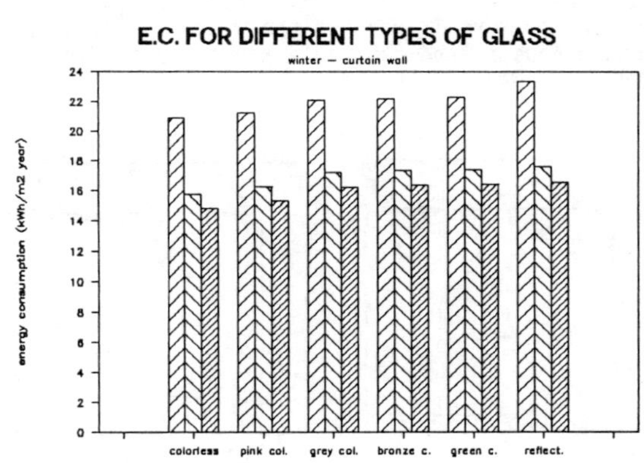

Fig. 4

single glass
insulating glass; a.s.= 6mm
insulating glass; a.s.=12mm

THE AIRTIGHTNESS OF ROLLER BLIND BOXES

Rangel das Neves, A.; Abrantes, V.
Faculdade de Engenharia, Universidade do Porto
Rua dos Bragas, 4099 Porto Codex, Portugal

ABSTRACT. The generality of windows in portuguese housing, are protected from the outdoor by external roller blinds or roller shutters. Pursuing the study of the airtightness of buildings envelope, it is presented the characteristic curves of leakage flow, from some roller blind boxes. Graphs were constructed to permit direct comparison of characteristic curves for the two positions of the roller blind.

1. INTRODUCTION

The airtightness of building envelope is of considerable importance to indoor climate and energy consumption. The building airtightness primarily affects ventilation and thus heat losses. An energy balance for a normal family dwelling shows that the sum of intentional and unintentional ventilation, is an important item in the energy balance. It should be possible therefore to reduce the unintencional ventilation, this means, uncontrolled air leakage flows, through improved airtightness.
Apart from the effects on energy requirements, uncontrolled air leakage also causes discomfort, resulting in loud and widespread complaints of uncomfortable draughts and dust.

2. EXTERNAL SHADING DEVICES

The pressurazing method of the element, specified by Shading devices have been strongly influenced by the local temperature climate. In summer, hot days can occur, mean daily temperatures are within the range where occupant confort is possible, as long as daytime overheating is prevented through shading. The most effective way to reduce the solar load on fenestration is to intercept direct radiation from the sun, before it reaches the window. So, in portuguese housing the habitable rooms are shading with external shading devices.

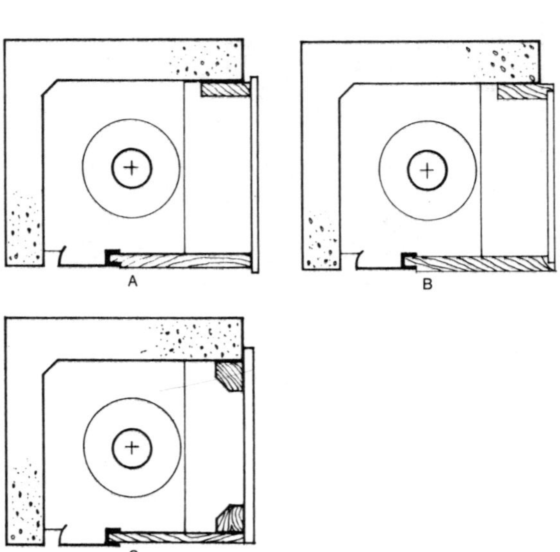

Figure 1: Roller blind boxes

The most representative device is the roller blind, made of rules of PVC and in a small percentage from aluminium. Pursuing the study of the airtightness of building envelope, this study turn up on roller blind boxes illustrates on Figure 1. This set of roller blind boxes is representative of the housing stock.
Figure 1 shows some of the constructional details, where no attention was given during the design and construction in order for the roller boxes to be airtights. Joints between edge beams and roller boxes materials are not design on the principles for achieving airtightness or thermal insulation. A wooden box frame supports wood board or plywood front cover and wooden bottom cover.

3. METHOD OF TESTING

ASTM Standard E283.84, and European Standard EN-42 Methods of Testing Windows, Part 1. Air permeability test, were used in this work.
The test consists of sealing a test specimen against one face of an air chamber, exhausting air from the chamber at different air flow rate and measuring the air flow through the specimen and the correspondent pressure difference.
A metalic enclosure frame was used to make possible to rebuild in laboratory a window with unchangeable dimensions, but allowed the construction under control from roller blind boxes.
Pressure measurements were taken with an electronic micromanometer having resolution of 10E-4 mm of water gauge and air flow was measured with two rotameters, one for flowrates up to 27 m3/h and the other for flowrates up to 175 m3/h. The rotameters had a 2% precision.
The exhaust system, with 700W, designed to provide 70 m3/h was controled with two spheric valves.
The static pressure tap into the chamber pressure was located in order that the reading was unaffected by the velocity of the air exhaust.
The system provided constant air flow at a fixed pressure for the period at last 10 s, required to obtain reading of air flow and pressure difference.
The roller blind boxes were subjected to increasing negative pressure from 40 Pa until 600 Pa or higher, and then the pressure was applied in the reverse order. Each box and for each estreme position of roller blind, was submited to three tests.
For each test, the higher of the two readings at each pressure, increasing as well as decreasing, was noted. The measurements from the three tests were then correlated to an equation, by the least squares method. The caracteristic curve of leakage flow as a function of pressure difference is given by:

$$q = f\ (\Delta p) = c\ (\Delta p)^{n}$$

where

q = air flow (m3/h)
p = pressure difference (Pa)
c = infiltration coefficient, air flow by unit of pressure difference (m3/h Pa)
n = exponent in approximate leakage function (0,5 < n < 1)

In order to compare different roller boxes, the curves were fitted for n = 0.65 and n = 0.67, the best approximation to n = 2/3.

4. RESULTS

Figure 2 shows three record readings corresponding to equal number of tests, for the same roller box and to the same position of the roller blind. This set of air permeability reading enable plott the characteristic curve of leakage flow as a function of pressure difference.

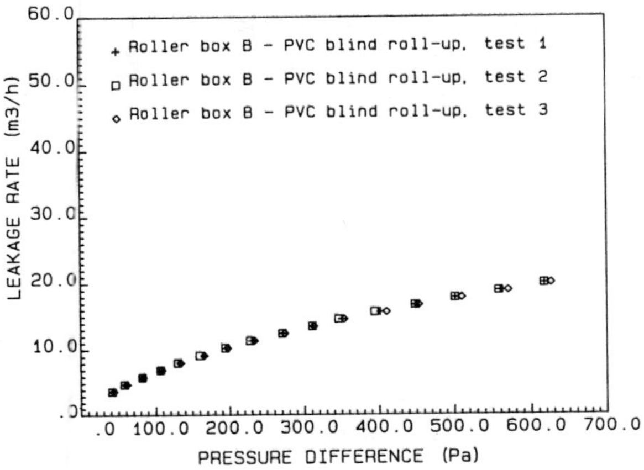

Figure 2: Roller box B readings test record, with PVC blind roll-up

This set of points is obtained by the application of standard EN42. The testing method is reliable. The precision and bias of the test method are quite good.
Figures 3 and 4 shows characteristics curves from tested roller boxes, and for the two extremes positions of the roller blind: roll-up in daytime period and roll-down in nightime period.

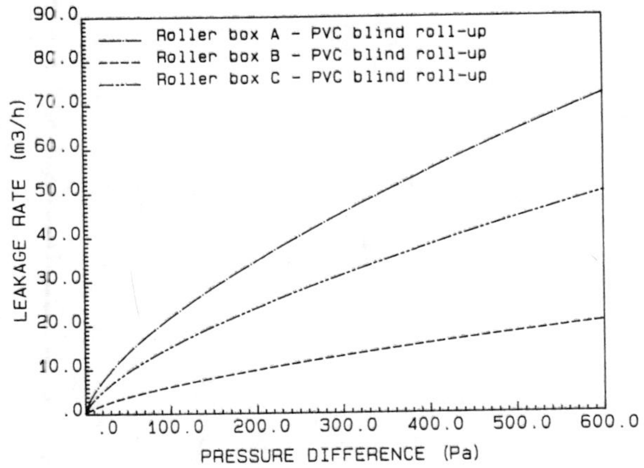

Figure 3: Leakage characteristics with PVC blind roll-up

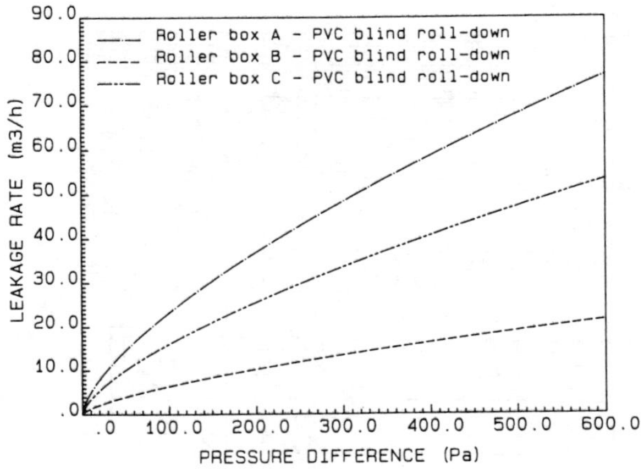

Figure 4: Leakage characteristics with PVC blind roll-down

In Table 1, the values of the flow coefficient c and flow exponent n are shown for all the tests.
The standard deviation and the correlation coefficient shows how the curves are a satisfactory criterion of goodness of fit.

Table I: Experimental values of flow coefficients

Roller box	Flow coefficients c	n	Statistics σ	r
Roller box A				
roll-up	1.106430	0.65	0.902609	0.998556
	0.996955	0.67	0.598897	0.999364
roll-down	1.172166	0.65	1.059389	0.997944
	1.057847	0.67	0.737304	0.999005
Roller box B				
roll-up	0.321034	0.65	0.427048	0.996391
	0.286319	0.67	0.565365	0.993667
roll-down	0.332264	0.65	0.428107	0.996728
	0.296793	0.67	0.573790	0.994114
Roller box C				
roll-up	0.772772	0.65	0.309009	0.999717
	0.690623	0.67	0.268127	0.999787
roll-down	0.821708	0.65	0.139142	0.999958
	0.734138	0.67	0.396956	0.999657

In order to compare these results with the infiltration througt to the corresponding window, it was plotted the Figure 5. For the sake of simplicity, the presentation has been concentrated in one roller box and one type of window. This inswinging casement window is one of the most common in 80's.

5. CONCLUSIONS

Results indicate that infiltration rates through roller boxes are significatives in the contest of the whole infiltration balance of a building.
The set of thermal regulation, to be introduced soon in Portugal, it's very permissive to the envelope airtightness. But the awareness of the need to save energy and money, and improve indoor environment, in conjugation with the presented results, may be a contribution to the design improvement of these components, which could be a challenge to the portuguese architectural concepts of roller blind boxes design.
Depending on materials and design there are a number of available alternatives to ensure airtightness of roller blind boxes.

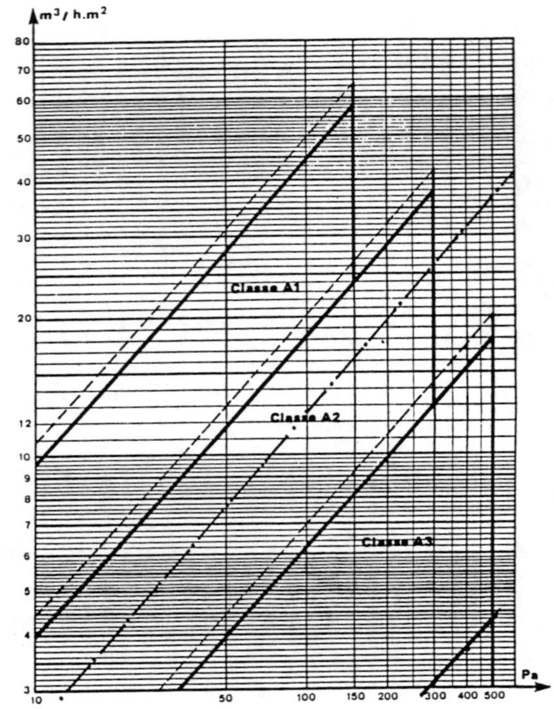

—— Aluminium inswinging casement window
—·— Roller box A with PVC blind roll-down

Figure 5: Inswinging casement window versus roller box, comparative leakages

Airtightness requires very careful sealing of joints with EPDM sealing strips or polyurethane foam, rigid or soft, and joint sealing compounds. For airtightness requirements to be satisfactory during the buildings life, there must be good resistance to aging in the construction materials used on roller boxes, and the design of the joints must be such that they can be maintained easily. Vapour barrier and thermal insulation are of fundamental importance to achieving new performance from this building component.

6. REFERENCES

1. Abrantes, V. and Oliveira Fernandes, E. de (1983) - Experimental Determination of Empirical Flow Coefficients for Air Infiltration Through Pitched Roofs. Proceedings of the Second International PLEA Conference, Crete, Greece, 427-433, Pergamon Press.

2. ASHRAE (1989) - Handbook of Fundamentals. Atalanta, ASHRAE.

3. NYLUND, O. (1980) - Infiltration and Ventilation, Swedish Council for Building Research, Stockholm, Sweden.

THE FIT METHOD FOR ESTIMATING THE HEATING REQUIREMENTS OF BUILDINGS WITH DIRECT GAIN PASSIVE SOLAR SYSTEMS

Armando C. Oliveira, E. de Oliveira Fernandes
Dept. Mec. Eng. - Univ. Porto
R. Bragas, 4099 PORTO CODEX - PORTUGAL

ABSTRACT. A new simplified design tool to estimate the heating requirements of direct gain passive solar buildings is presented. It uses the monthly un-utilizability (FI) and the thermal inertia of the building (IT) as variables. A comparison is made between results of the new method and results from other methods.

1. INTRODUCTION

The existance of simplified design tools to calculate the space heating requirements of buildings is important to building designers, since they allow a quick estimate of the energy consumption. In comparison with detailed simulation (computer) models, they have the advantage of needing much less input data and calculation effort: for instance, monthly averages of the climatic variables are used instead of hourly values.
Most of existant simplified methods are correlation based, /1,2/, obtained through simulations performed for a variety of system parameters. However, their applicability is restricted by the functional form - parameters - used in the correlations. In most of the methods - like SLR - un-utilizability, method 5000 - thermal inertia is not quantified in an adequate way.

2. THEORETICAL BASIS FOR THE FIT METHOD

Let us consider a direct gain building with an auxiliary heating system, which is activated when the air temperature reaches a minimum value: set-point (T_{min}).
Its heat balance for a long period of time (one month) includes the following terms: the solar gains (Q_{sol}) that are equal to the solar energy transmitted through the glazing(s) and absorbed by internal surfaces, the auxiliary energy (Q_{aux}) and the losses to the outside (Q_{loss}). The losses can be expressed as a sum of two terms: the reference heating load (Q_{ref}), which is the energy that the building would require if there are no solar gains, and the excess energy (Q_{exc}) which corresponds to the increase in the temperatures inside the building - due to solar gains. The internal gains (Q_{int}) can be included in Q_{ref}, since they contribute to reduce the heating load. If the reference load is calculated by the degree-days method, the heating degree-days should be calculated for a base temperature (T_b) which accounts for internal gains: $T_b=T_{min}-\dot{Q}_{int}/UA$, where UA is the overall heat loss coefficient of the building (including windows and infiltration).
The contribution of solar energy can be quantified by means of the solar fraction (f), which is referred to the load:

$$f = Q_{sol,us}/Q_{ref} \tag{1}$$

where $Q_{sol,us}$ is the useful solar energy. In a building, at a given instant, the solar gains can be used to supply the losses or stored in the thermal mass for a later use. If the instantaneous solar gains are greater than the instantaneous load and the storage capacity, they will overheat the indoor air (above the set-point temperature), being partially unuseful.

Since $Q_{sol,us} = Q_{sol}-Q_{exc}$, we can also write

$$f = 1-Q_{aux}/Q_{ref} \tag{2}$$

using the heat balance for the building.
For a building with infinite storage capacity, all the solar gains that are not instantaneously useful can be stored and used at a later time. Therefore we have $Q_{exc} = 0$, and

$$f_\infty = Q_{sol}/Q_{ref} = SLR \tag{3}$$

For a building with zero thermal capacity, whenever the incoming solar gains are greater than the load there is excess energy:

$$\dot{Q}_{exc,o} = (\dot{Q}_{sol}-\dot{Q}_{ref})^+ \tag{4}$$

where the symbol + means that only positive values of the difference are considered.
As shown in /2/, the monthly energy excess can be written as

$$Q_{exc,o} = Q_{sol}\ \phi \tag{5}$$

where ϕ is the monthly un-utilizability, that is the fraction of solar gains that are not useful. It is calculated for the critical radiation:

$$\dot{I}_{crit} = UA\ (T_b-\overline{T}_e)/(\overline{\tau\alpha}\ A_{DG}) \tag{6}$$

which is the level of radiation that equals the gains to the load. In Eq.6 $\overline{T}_e$ is the monthly average ambient temperature, $\overline{\tau\alpha}$ is the average transmittance -absorptance product, and A_{DG} is the direct gain area.
Using eq.5 the solar fraction for the zero thermal capacity case can be written as

$$f_o = (1-\phi)\ SLR \tag{7}$$

Equations 7 and 3 represent the two limiting cases for the solar fraction of a building, as already defined in /2/. For a real building with the same characteristics and operating conditions, except for the thermal capacity, the solar fraction shall lie between those two. We can write

$$f = \eta\ SLR \tag{8}$$

where η is the utilization efficiency (or factor) of the solar gains.
Comparing eq.8 to eqs.3 and 7, we have

$$(1-\phi) \leqslant \eta \leqslant 1 \tag{9}$$

The utilization effiency is a function of the overheating if the building had no inertia (ϕ) and of the real thermal inertia. By using dimensional analysis it can be shown, /3/, that the relevant dimensionless inertia parameter is:

$$IT = C_{ef} / (UAt_{ref}) \quad (10)$$

Then, for the utilization efficiency we have

$$\eta = \eta\ (\phi,\ IT) \quad (11)$$

and for the solar fraction

$$f = \eta\ (\phi,\ IT).\ SLR \quad (12)$$

One may find strange that η is not expressed as a function of SLR, since that is a relation used in some simplified methods. However, we must notethat the utilizability already takes the solar gains and the losses into account. We could write

$$\phi = \phi\ (SLR,\ \text{solar radiation distribution}) \quad (13)$$

The function η (ϕ, IT) can be obtained by performing various simulations for a variety of ϕ and IT values, as will be seen in section 4.
Let us remark that the functional relationship now suggested - eq.12 - is more complete than the ones from other methods. In the SLR method, /1/, we have simply f = f(SLR). Since for the same SLR value different values of ϕ can occur, different solar fractions can exist for the same SLR value - which can't be found with SLR method, In the un-utilizability method, /2/, we have f = f(X,Y). X is equal to SLR. However, the same dimensional parameter - Y - includes two different effects: ϕ and IT.

3. CALCULATION OF EFFECTIVE THERMAL CAPACITY

Balcomb, /5/, defined the daily heat capacity, that represents the amount of energy that can be stored in the building structure during one day, and released to the indoor air during the same day, (24 hour period). However, for long-term performance, we shall be interested in quantifying the amount of energy stored in one day and released at a later time - including heat that may be stored for periods longer than 24 hours.
After numerous simulations, Balcomb suggested for the effective thermal capacity a relation of the type:

$$C_{ef} = 0,75\ C_d + 0,25\ C_w \quad (14)$$

where C_d is the daily heat capacity and C_w is the weekly heat capacity - calculated after the building response to a week-period solicitation.
The effective thermal capacity differs from the heat capacity (product mass-specific heat) and differs with the location of the building element. Elements located in rooms with direct sunlight have a higher effective thermal capacity - see Fig.1.
Tables 1 and 2 allow a quick calculation of the effective thermal capacity for common materials. They are valid for elements located in the external envelope of a building, with outside insulation with no less than 2.5 cm thick. They may also be used for internal partition walls, but with a thickness taken equal to one half of the real one. For more details see /5/.

4. CORRELATIONS OF THE FIT METHOD

Several simulations were performed with the ESP computer program, /6/, in order to obtain the η (ϕ, IT) correlations.
A one-zone direct gain building geometry was used, with different construction configurations. Table 3 lists the range of parameters considered.
The operating conditions were typical of residential buildings: for instance, no mechanical ventilation and no control on the maximum temperature (no cooling system).
The hourly climatic data of Porto (Portugal) were used. We must note that the use of only one climatic

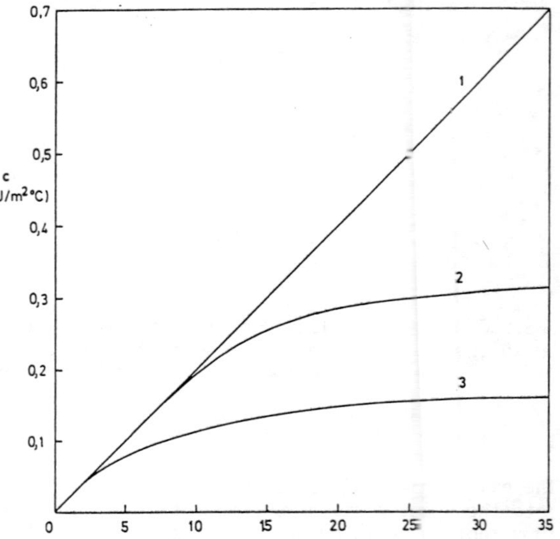

Fig.1 - Thermal capacity of a concrete wall (externally insulated), as a function of thickness: 1-heat capacity; 2-effective thermal capacity for a wall in a direct gain space; 3-effective thermal capacity for a wall in a room with convective gain only.

Table 1 - Effective thermal capacity for single elements, located in direct gain spaces - - values in $Wh/m^2.^oC$.

material / thick(cm)	CONCRETE	BRICK	HARD WOOD	GRANITE
1	5.6	6.0	2.5	6.2
2	11.2	12.1	5.0	12.4
3	16.8	18.1	7.5	18.7
5	27.9	30.0	11.9	31.0
7	38.8	41.4	14.8	43.1
10	53.8	56.1	16.5	59.5
15	71.4	69.8	17.8	78.3
20	79.2	74.7	18.6	86.3
30	85.5	80.5	18.2	93.0

Table 2 - Effective thermal capacity for single elements, located in spaces without sunlight - values in $Wh/m^2.^oC$.

material / thick(cm)	CONCRETE	BRICK	HARD WOOD	GRANITE
1	5.6	6.0	2.5	6.1
2	10.7	11.5	4.9	11.8
3	15.2	16.2	7.2	16.6
5	22.3	23.2	10.6	23.9
7	27.1	27.9	12.5	28.8
10	32.1	32.6	13.8	34.0
15	37.6	37.8	15.0	39.7
20	41.3	41.0	15.6	43.5
30	45.0	43.6	15.1	47.0

Table 3 - Range of parameters used in the simulations

Glazing type :	double glazing, without night insulation
	$\tau_n = 0.72$
	$U = 3 - 4\ W/m^2{}^oC$
Thermal inertia, IT :	5 - 95
T_{min} :	18 - 22 °C
ϕ :	0.1 - 0.9

data-set does not restrict the applicability of the method, since the climate is represent by the ϕ variable.

Fig.2 shows the η (ϕ, IT) curves. For the IT parameter a reference time of 3600s was used. This way, the value of IT is equivalent to the building time constant in hours.

The η (ϕ, IT) curves can be represented by a function of the type

$$\eta = 1 - \phi + a\ IT^{b} \qquad (15)$$

A last squares fit gave for the coefficients $\underline{a}$ and $\underline{b}$:

$a=-.01267+.2028\phi+2.468\phi^2-4.315\phi^3+1.333\phi^4$, if $0.1\leqslant\phi\leqslant0.6$

$a=2.623-8.540\phi+9.585\phi^2+3.293\phi^3$, if $0.6<\phi\leqslant0.9$

$b=-.04346+.7092\phi-3.380\phi^2+6.550\phi^3-3.446\phi^4$ if $0.1\leqslant\phi\leqslant0.6$

$b=-9.102+35.89\phi-45.33\phi^2+18.62\phi^3$ if $0.6<\phi\leqslant0.9$

The mean error of this correlation is less than 2%.

5. EXAMPLE

An example of the application of the method is presented. It corresponds to a direct gain building located in Madison (USA), as defined in /2/. Table 4 resumes its main characteristics.

A low inertia building was considered so that a comparison with results from SLR method could be made. Table 5 presents the results from the FIT method, together with other methods' results. A good agreement was found between all anual solar fractions, which vary between 40% (FIT and TRNSYS) and 43% (SLR).

The use of other thermal mass characteristics - - higher C_{ef}, mass thickness, mass location - can be quantified in an adequate way by the FIT method, which is not the case with other simplified methods.

6. CONCLUSIONS

A new simplified method for calculating the monthly heating requirements of direct gain buildings was presented. A more complete relationship between the parameters involved is used: the ϕ parameter accounts for the climate effect while the IT parameter accounts for the building thermal inertia. Correlations to calculate the utilization factor of solar gains were presented.

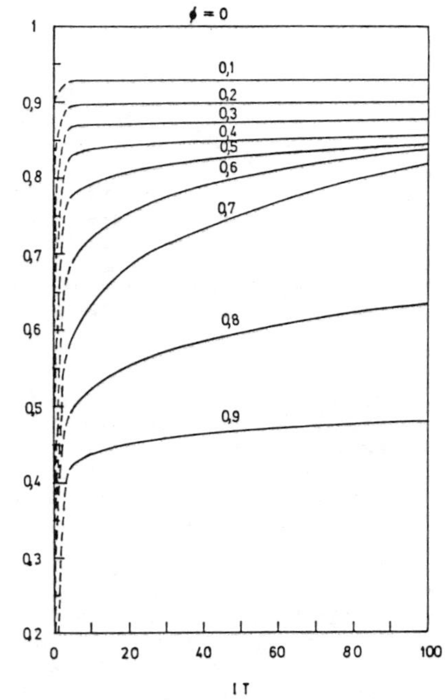

Fig.2 - Utilization efficiency (η) as a function of thermal inertia (IT) and un-utilizability (ϕ).

Table 4 - Parameters for a direct gain system.

UA :	265 W/°C
Windows :	25.6 m² (south)
	double glazing
	$U = 4.17$ W/m²°C
	$\tau_n = 0.72$
T_{min} :	18.3 °C (thermostatic control)
Thermal mass :	concrete, 15 cm thick
	area = 3 x 25.6 m²
Heat capacity :	2.35 x 10^7 J/°C
C_{ef} :	1.98 x 10^7 J/°C
IT :	20.7
Internal gains :	negligeable

Table 5 - Application of the FIT method to the example defined in Table 4. Comparison with results from other methods: SLR (B), un-utilizability (C), TRNSYS (D).

MONTH	DD	$\bar{I}_v$ (MJ/m².d)	SLR	ϕ	η	f	Q_{aux}^A (GJ)	Q_{aux}^B (GJ)	Q_{aux}^C (GJ)	Q_{aux}^D (GJ)
JAN	828	11.4	0.332	0.238	0.888	0.295	13.4	13.4	13.3	13.3
FEB	680	13.3	0.405	0.286	0.875	0.354	10.0	9.73	9.83	9.60
MAR	626	14.1	0.467	0.356	0.856	0.400	8.62	7.74	8.10	7.77
APR	297	10.7	0.647	0.553	0.785	0.508	3.34	2.53	2.96	2.82
MAY	157	10.1	1.148	0.790	0.567	0.651	1.25	0.54	0.87	1.17
SEP	93	13.1	2.680	0.936	0.435	1.00	0.00	0.01	0.00	0.18
OCT	250	12.5	1.110	0.693	0.689	0.765	1.34	1.34	2.06	2.28
NOV	479	9.71	0.464	0.424	0.835	0.387	6.72	6.46	6.83	6.96
DEC	683	8.02	0.282	0.284	0.876	0.247	11.8	11.7	11.5	11.8
PERIOD	4093	-	-	-	-	0.40	56.5	53.4	55.4	55.9
						f_{PER}	0.40	0.43	0.41	0.40

REFERENCES

1-J.D.Balcomb et al.,Passive Solar J.,vol.1,n.2

2-W.Monsen et al.,Solar Energy 27,1981

3-A.C.Oliveira,Ph.D.Thesis,Univ.Porto,1989

4-B.Bourges,F.Lasnier,CEC Report ESF-008-F,1983

5-J.D.Balcomb,LANL Report LA-9694-MS,1983

6-J.Clarke,D.McLean,ESP Manual,Univ.Strathclyde

SPANISH VERNACULAR ARCHITECTURE: ANALYSIS FROM THE BIOCLIMAMTIC POINT OF VIEW

Heras M.R., Montesinos J.L.
Solar Division/IER-CIEMAT
Avda. Complutense, 22/28040-MADRID (SPAIN)

ABSTRACT
In this work, five different climatic areas of the Spanish territory have been selected and once we have analysed their settlements, subsequently a qualitative energy study was made to establish both, the influence of materials used and the building's plan location, orientation and distribution that affect thermal performance of the mentioned typologies.

1. INTRODUCTION

Spanish vernacular architecture offers a great number of different typologies related to its large variety of regions and cultures. This architecture has been of great importance throughout the time, under a historical point of view providing direct solutions to human needs counteracting environmental factors. This work advises from the importance of focusing the architectural design to its final objective, that is, to provide shelter and thermal confort to occupants at low energy costs. Five different climatic areas of the spanish territory have been selected; these climatic zones, as well as their corresponding typologies are the following.

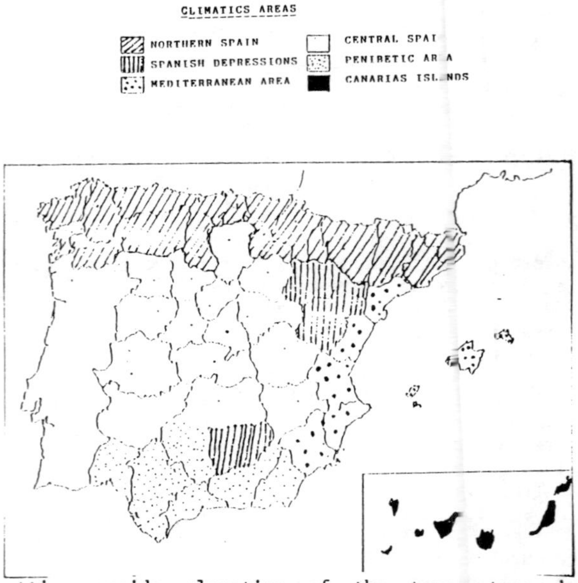

2. TIPOLOGY ANALYSIS

2.1. NORTHERN SPAIN

The climate is warm, with few thermal oscillations and a high indicative of rain. The dominant winds come from the North. In the mountainous area, there is a greater thermal oscillation and low temperatures.
The highest temperatures are produced in July, August and September, with the average of 22°C, the lowest temperatures are produced in January, with an average of 2°C.
Rainfall: from the analysis, we can deduce that raifall maximum values occur throughout the year, July is the driest month with 40mm. and the annual average is between 1000 and 1500mm.

2.1.1. Typology.

A granite store architecture with strong mosoury walls, two sloped roofs, the dadge parallel with the façade, and with another particular aspect, almost without brickcornicer, by windows, solanas or galeries supported by corbels on columns. The facing facade is generally south faced, the exterior starway like a characteristic element of the country man, this starway ends on a platear that sometimes is a terrace, this can be a one or two floors house, in the first can there is and simple separation, one one hand the properly house and on the opposite a space desijnated for domestic animals and working implements. In the second case the first floor is dedicated to the stable and the warehouse, and the second floor is the house with the kitchen like a principal center. The houses have so many interior space, and generally the bedroom is unique, and without a good lighting and ventilation. Also, over the second floor there is one room formed by the attic roof and called sobrado.

2.1.2. Analysis of energetic behaviour.

Compact houses are designed to reduce the exposition to the cool, the windows area are reduced, especially those wich are on the north façade, the solanas receive the solar radiation, the gain of the glazed surfaces an reduced. The attic avoids elevation of the temperature in summer, isolated on winter, constructive materials with big mass capable of acumulating the gain heat and liberate it by night, the roof hoging elevated to deffect the wind, the entrances are considered with blinding wind roof, the point of the roof and the walls in light zones, to reflect the summer sun, and in other areas in dark color to acumulate the heat. They take advantage in party-wall constructions to share the heat and to reduce the cool exposition, they are facing south.

2.2. SPAIN CENTRAL

The climate is continental, although in the Duero basin is more rainy than in the South area. Pluviosity is scarce during the summer. The most rainy season is autumm.
It has large insolation and scanty overcasting winters are cold. Maximum temperature media is 20.6°C and minimum 9.5°C. It reaches its higher average value of 26°C in July. Despite during summer mounths gets to a peak of 32°C and in winter the lowest media is 4°C.
Rainfalls: Yearly average media are measured from soo to 700 mm. highest values correspond to October and November.

2.2.1. Tipology.

Two storey houses; in most cases the upper part projects on the lower; built on large stone columns with capitels and supported by wooden beams. Rubblework in lower floors with adobe and structural timber craft in the upper; wooden eaves and two sloped roofs are constituents to be noticed on façades. Houses are distributed according to two separating spaces at two different levels where agricultural products are stored. Under this floor, stables are located.

2.2.2. Analysis of energetic behaviour.

In order to keep house heating in winter and prevent dwellings from outdoors inclemencies, two storey houses are commonly attached; big wooden beams avoid heat losses; openings on the façades are reduced, using very few on Northern-faced façades and shutters to avoid heat losses. A system of halls and windshielding entrances are provided. There houses are oriented in a way that permits winter sunlighting, and room's principal spacing are opened to the South as to kept as well as in winter, so shutters over the openings are used to benefit the sun thus allowing natural ventilation.

2.3. SPANISH DEPRESSIONS

The climate is the continental character influenced by sea closeness. The Ebro basin is one of the less rainy areas in Spain. It has great insolation, scarce overcasting and it is influenced by dry and cold winds blowing from the Northwest. Maximum temperatures are reached during July-August at average of 22°C, lowest media is reached in January (approx. 2°C).
Rainfalls: From this analysis it can be observed that main rainfalls occur in October-November with a yearly ratio of 400 mm.

2.3.1. Tipology.

The commonest building materials used are stone and rubblework in mountain areas, walling up and adobe in the driest areas, but in both cases the typical "Aragones brick" is used. This monochrome brick is heavier and larger than the Castilian. Covers are made of arabian- like tiles with the same colour as the brick. Floors are higher than those in Castile; openings are also larger and cornice eavens are made of engraned wooden canes. The use of brick as a structural building material is most characteristic in this kind of vernacular architecture.

2.3.2. Analysis of energetic behaviour.

Houses compacts to prevent wind influence and provide a better heating. Vertical glazing South oriented to receive sun input, shadowed to prevent high incidence of sun radiation in summer; isolated with shutters mass and high thermal inertia materials, capable of maintaining to the night the day heat gainings by means of eaves built on South facing windows, it also uses shutters to prevent sun input in rooms located on East and West façades, where sun radiation incide in its lower course: Windows are larger for ventilation purposes, and in order to cool and permit cross ventilation in the hottest periods.

2.4. PENIBETIC AREA

Located at the Andalusian Mediterranean area. Its climate is very dry, with a remarkable difference between coastal and mountainering areas. Temperatures are extremely high and rainfalls are scanty on the coast. Insolation is very high too. Yearly temperature average is 25°C (maximum) and 11.5°C (minimum).
Rainfalls: Pluviometric analysis shows a yealy media of 500 and 600 mm. Maximum values correspond to December, January and Marh with an average to 80 mm.

2.4.1. Typology.

Materials most commonly used are sand, clay and lime, stone is used in a rubblework like way previously limed for plastering. Wood is mainly used in bearning of floors and covers, and sometimes in framework. Houses are of reduced dimensions, arabian-influenced, with two floors in wich the top floor serves as a barn, in other cases as a two sloped roof dwelling. The individual patio is characteristic of this area with a hall that leads as a pathway. This patio is of the utmost importance not only as a pathway but as a living-room.

2.4.2. Analysis of energetic behaviour.

To keep the house isolated from external conditions and increase interior comfort, sets the rooms and its less important facilites (toilets or stairways) between the external walls and the main room spaces, using blinds as heating isolators. Thus preventing night thermal losses, through the openings glazed surfaces. Outdoor spaces as well as the patio contribuite to ventilation, plus cross ventilation provided by narrow floors on hottest seasons and use of eavens projections to protect the interior from sun input. Patios are provided with fountains that help cooling through evaporation in summer, they also count with gardening to increase humidily levels that contribute to general comfort.

2.5. MEDITERRANEAN AREA

For our study we have limited our reseach area between 39° and 42° Northern latitude on a sea level basis from 10 to 300 m. Maximum yearly media 21.6°C; minimus 9.5°C.
Rainfalls: From analysis it can be asessed that rainfall maximum values reach their peaks during the months of September-October, being noticeable the 80 mm. value registered for these two months. Thus indicating that there is a balance in Winter and Spring. Summer is the driest season obviously.

2.5.1. Tipology.

The most significant typology of Spanish vernacular architecture in this area is the "Masia". Eminent agricultural, is provided with a distribution hall for stables and storage rooms, as well as peasants'dwellings and a kitchen that constitues the meeting point and serves as dinning room. The "lar" is located there with a fire-place attached to a smoked wall with benches around; There is sometimes a hall parlour and the bedrooms. Sometimes, the "Masia" has a second floor and in the attics, talking advantage of the space between ceiling and roof spaces for the barn are commonly used. Built on a rectangular plant, it is a covered by a two side sloped roof. They have not central patio and are South orientated.

2.5.2. Analysis of energetic behaviour.

Architecture is basically founded on great mass building materials, as adobe, or brick to soften thermal changes from day to night and to accumulate its subsequent release by night. It also incorporate several buildings opened to the winter sun and are protected from solar input in summer by projecting eaves.

3. CONCLUSIONS

In the above expressed analysis, we state that the climatic environement where housing operates is defined by temperature, humidity levels, winds speed and site's solar incidence.
Achivements provided by vernacular architecture have been marked by a characteristic climatic shelter, considering lack of resources and economic possibilities of the users that have led to their profiting advantages of the sun in winter and rejection of some in summer, using building materials such as adobe, stone, etc. to soften oscillations of indoors temperature and adaptable to the external environement, vegetation etc.
The basis for a conscious energy design to be applied in the described items is an adecquate response to the inconveniences and/or advantages of the climates involved.

4. BIBLIOGRAPHY

-FEDUCHI, Luis. "Itinerarios de la Arquitectura Popular Española." Edit. Blume. Barcelona, 1984.

-FLORES L., Carlos. "Arquitectura Popular Española."
Edit. Aguilar. Madrid, 1973.

-SERRA F., Rafael. "Evaluacion del Aporte Energetico a los Edificios Mediante Sistemas Naturales de Control Climatico en España." Centro de Estudios de la Energia. Barcelona, 1982.

- THE AMERICAN INSTUTE OF ARCHITECTS. "La Casa Pasiva. Clima y Ahorro Energetico." Edit. Herman Blume. Madrid, 1984.

-WRIGHT, David. "Arquitectura Solar Natural. Un texto Pasivo." Edit. G. Gili. Mexico, 1983.

SPANISH PROGRAM ON BIOCLIMATIC ARCHITECTURE: IER-CIEMAT STRATEGIES

Heras M.R.
Solar Division/IER-CIEMAT
Avda. Complutense, 22/28040-MADRID (SPAIN)

ABSTRACT

The aim of this work is to describe the state of art on Bioclimatic Architecture Activities of the IER-CIEMAT after three years and to enumerate the new projects which were started late last year.

During this time, we have finished the first phases of the same IER-CIEMAT projects and we just continue with the second phases of them. The results have been published this year by the CIEMAT.

New projects have been started withinthe national program, which are: to build the Spanish Test Cells in Almeria for caracterization of the passive solar components, to analyze the Spanish vernacular architecture from the energetic performance point of view, to study new monitoring methods. But the most emphasis is in the investigation and demostration of all the aspects related to the Passive Test Cells.

We are going on with the E.C. and I.E.A. Programs participation and the dissemination, coordination of the different Bioclimatic Architecture aspects in Spain.

For these reasons, the most important subjects of interest for the IER-CIEMAT on Bioclimatic Architecture are the use of passive solar components, not only for heating but also for cooling, as refrigeration is imperative. At this stage, our main goals are:-To increase the passive solar components utilization in building.-To facilitate the appropriate Design and Simulation Tools to all Spanish proffesionals: architects as well as scientists and thermal engineers.-To research new passive solar cooling components for Mediterranean climate.

INTRODUCTION

The projects of the Solar Division of the Renewable Energy Institute (IER-CIEMAT) began in 1985, following the guidelines defined in the meetings held throughout the year.

During this time, we have finished some projects, we have outstandings some others and we hope to start some others in the near future, following the guidelines and meetings held with national and international experts as well as Regional Governments, Colegios Profesionales and Universities.

A summary of the present activities of the IER-CIEMAT is shown in table 1, where the following projects are described/1/:

PROJECTS FINISHED

Actually, we have finished the first phase of these projects:

1 - "Studies of architectural design criteria through natural systems for climatic control in Spain" (Bioclimatic design handbook or how to evaluate the building natural energetic performance designed by architects for Spanish geographical areas). The first results of this project is the book titled "Clima, Lugar y Arquitectura, Manual de diseño Bioclimatico" (Climate, Location and Architecture, Bioclimatic Design Handbook) /2/, written by Rafael Serra, at all, from Architecture School of Barcelona and published by CIEMAT in collaboration with the Universidad Politecnica de Cataluña a..d Consejo Superior de los Colegios de Arquitectos de España, figure 1, where the "CLA" Program is included to evaluate the building natural energetic performance designed by architects from the first phases of design.

Figure 1

2 - "Simulation and calculation method for passive systems regarding their thermal performance". The first results of this project is the S3PAS Method, figure 2, which is the accurate simulation method or the simulation as investigation tool. This project has been carried out by the Sevilla Industrial Engineer School group /3/.

Figure 2

3 - "Extension of the Spanish basic construction regulation NBE-CT-79". The first result of this project is the book titled "Proposal for the extensión of the Spanish building thermal performance basic regulations, NBE-CT-79" /4/, figure 3. This projects has been carried out by the Industrial Engineers School of Sevilla and Architecture School of Barcelona and coordinated by IER- CIEMAT. The scientific results are being shown in this Conference (SECA) as other paper submitted.

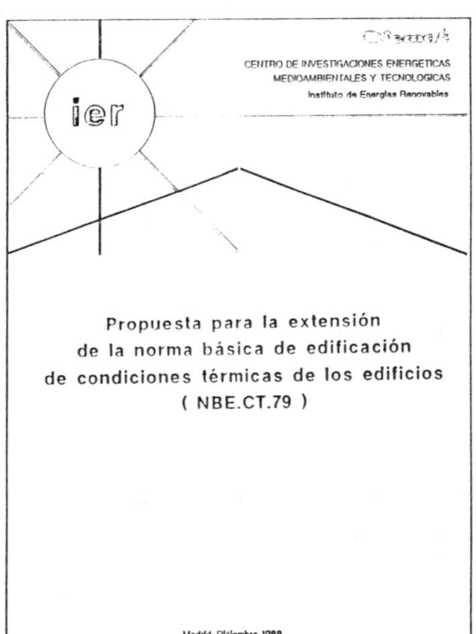

Figure 3

4 - "Monitoring plan of the IER-CIEMAT for the evaluation of passive solar installations". This project has been carried out by IER staff in collaboration with different reaearch teams, who are carried out the monitoring of four buildings construction types and located in different Spanish climatic areas and the results will be published as books which are in the printing. The different books are: one which reviews /5/, and summarizes conclusions of the different buildings evaluations by IER-CIEMAT, other which explains each building with architectural and energetic performance aspects. The energetic performance results of two buildings are being shown in this Conference (SECA) as other paper submitted.

Regarding to the E.C. Spanish Programs participation, we have finished as other countries, the participation in the last R&D projects of the D.G. XII of the Commision of the European Communities which has finished last July.

Regarding to the I.E.A. Solar Heating and Cooling Program, task III "Performance testing of solar collectors" and task VIII "Passive and hybrid solar low energy buildings", have finished.

OUTSTANDING PROJECTS

With regards to the IER-CIEMAT projects, we continue carrying out their second phase, this is:

1 - Applications in the different geographical areas of the "Climate, Location and Architecture, Bioclimatic Design Handbook" with also adaptation of the "CLA" Program to different climate and architectural typologies of the Spanish geographical areas. This project is carrying out by Barcelona Architecture School group and it will be finished in the last month of 1990.

2 - Elaboration of a Simulation Simplified method and a Calculation method from the S3PAS Method, which will be able applicated as design tool and the calculus method to be used in the Proposal of the Extension Spanish Basic Construction Regulation NBE-CT-79. This project is carrying out by Sevilla Industrial Engineers School group and it will be finished in the last month of 1990.

3 - This project has not a second phase from IER-CIEMAT, but we would like that the Proposal for the Extensión of the Spanish Basic Construction Regulation NBE-CT-79 will be used by the Ministry of Public Works and Urbanism (MOPU) in the Committe in charge of the Spanish Regulations.

4 - Monitorization of the different buildings for the evaluation of the energetic performance, we have started to carry out four different monitorizations in Guillena (Sevilla) and Almeria which are two Schools and two houses in Aguilar de Campoo (Palencia) and Torquemada (Palencia). These monitorization are carrying out by IER-CIEMAT staff in collaboration with Regional Government of Andalucia and Castilla-Leon. This project will be finished in 1991.

With regards to the new projects which have been started in this year, 1989, actually we are carrying out the following:

5 - To build the Spanish Test Cells in Almeria (South, Mediterranean Coastal), following the general guidelines given by the PASSYS Program of the E.C., as well as the modifications introduced, considering the Spanish Climate Conditions. These Test Cells have been built, instrumentalized and calibrated. With this project, we want to investigate the different solar passive elements performance with special emphasise in the Mediterranean architectural components, actually the IER-CIEMAT has a group dedicated to this project and the first results are shown in this Conference (SECA) as other papers submitted.

6 - To study new materials and new components for the building. This project will be achieved in the Spanish Test Cells in collaboration with the Spanish manufacturers.

7 - To analyze the Spanish vernacular architecture from the energetic performance point of view, as there are different vernacular construction types in Spain and we would like to know its thermal performance. This project has been started and the first results are shown in this Conference (SECA) as other papers submitted.

8 - To study the new monitoring methods for the evaluation of the buildings energetic performance. This project is being carried out by the Universidad of Valencia in collaboration with the IER-CIEMAT staff.

9 - To help the architects in their design, to make the energetic consultant's office. This project has been started after different requests of the help and it is being carried out by the IER-CIEMAT staff.

Regarding to the E.C. Programs participation, we have been accepted in the JOULE Program, as active member in the PASSYS Program, Second Phase, which will be held in January 1990.

Regarding to the I.E.A. Solar Heating and Cooling Progran, we continue with the Spanish participation in the Executive Commitee, as well as in Task X "Solar materials R&D", Task XI "Passive and hybrid solar commercial buildings", Task XII "Solar Buildings analysis and design tools", Task XIV "Active solar systems"

PROJECTS FOR THE FUTURE

Regarding to the different meetings with passive solar experts: architects, engineers, scientific as well as Regional Government, Colegios Profesionales of Architecs and Engineers and Universities, the IER -CIEMAT will start in the near future, others R&D projects as follows:

10 - To study Thermal Confort for Mediterranean Conditions, as it is very important for us to determine the thermal comfort coefficients and parameters for temperate climates.

11 - To investigate the Natural Cooling aspects for our conditions. This project will be outlined in the meeting to be held in January next month.

REFERENCES

/1/ HERAS,M.R. :"Activities, centers of interest and projects for the future of the Renewable Energy Institute in the area of Bioclimatic Architecture" Proceedings of the EUROFORUM, NEW ENERGIES, Saarbrucken,F.R.Germany, 24-28, Oct, 1988, pp.311-314. P.1.9.

/2/ SERRA,R.: "Clima, Lugar y Arquitectura, Manual de Diseño Bioclimático", Editado por Secretaria General Tecnica del CIEMAT, I.S.B.N.: 84-7834-016-5, Madrid, Nov,1989.

/3/ "S3PAS 1.0, Simulacion de Sistemas Solares Pasivos, Manual del Usuario" Informe Tecnico, Catedra de Termotecnia, E.T.S.I.Industriales, Sevilla, Dic,1988.

/4/ HERAS,M.R.,VELAZQUEZ,R.,SERRA,R.: "Propuesta para la extensión de la Norma Basica de Edificación de condiciones térmicas de los edificios (NBE-CT-79)" Editado por la Secretaria General Tecnica del CIEMAT. I.S.B.N.: 84-7834-009-2, Madrid, Dic,1988.

/5/ HERAS,M.R.,MARCO,J.: "Comportamiento energetico de edificios solares pasivos.Plan de Monitorización del IER" Editado por la Secretaria General tecnica del CIEMAT. (en prensa).

TABLA 1

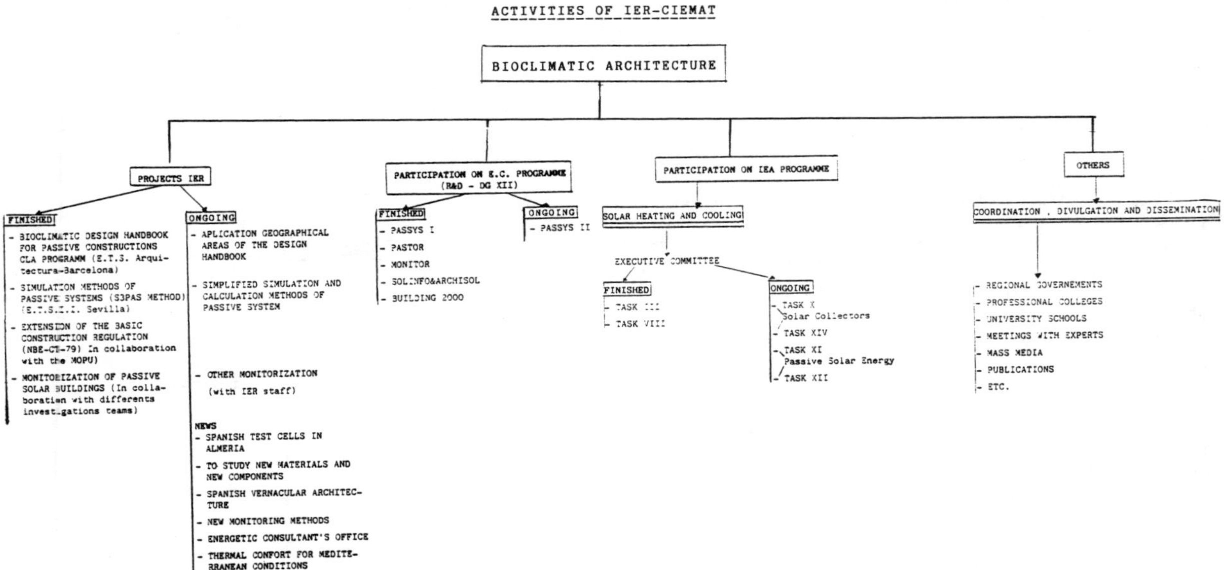

TYPOLOGY ENERGY SIMULATION IN SPANISH VERNACULAR ARCHITECTURE

Heras M.R., Montesinos J.L.
Solar Division/IER-CIEMAT
Avda. Complutense, 22/28040-MADRID (SPAIN)

ABSTRACT

In this work, an outstanding typology on Spanish vernacular architecture has been selected to analyze its energy performance, by simulation methods.

INTRODUCTION

It's unquestionable the influence that a climatic zone has when, adapting a building's typology to its conditions.
The relationship, vernacular architecture location, is mainly developed following two processes:
a) use of autochthonous building materials, and
b) the own's building morphology to be adapted, in order to achieve a better relation between climate and human being.
At this stage, the building becomes a regulating instrument that can preserve stability or indoor environmental conditions through the passage of time and lead us to the study of aspects related to regulation that transform the building into a system that can easily be modified according to variable environmental conditions. For this, a correspondence climatic typology climatic area selected to analyze and evaluate thermal response on a building characteristic of the area, which correspond to the parameters and construction techniques of local architecture, this climatic zone corresponde to central Spain.

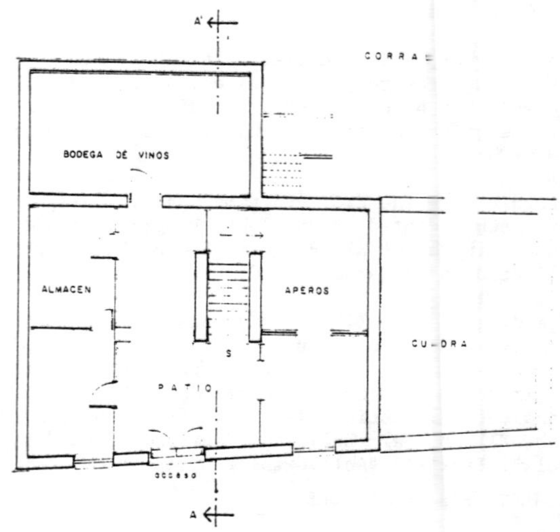

GROUND FLOOR

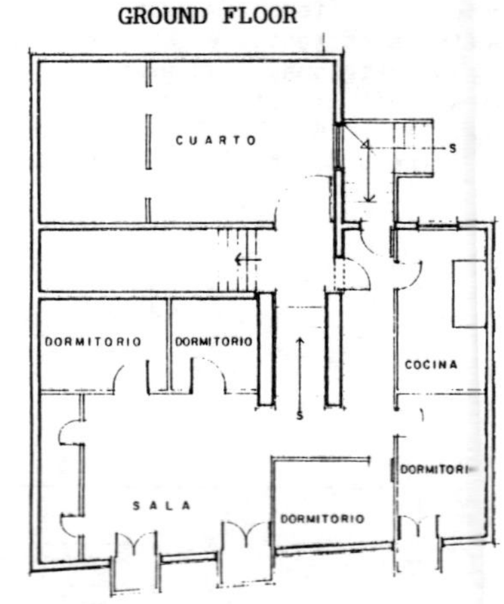

FIRST FLOOR

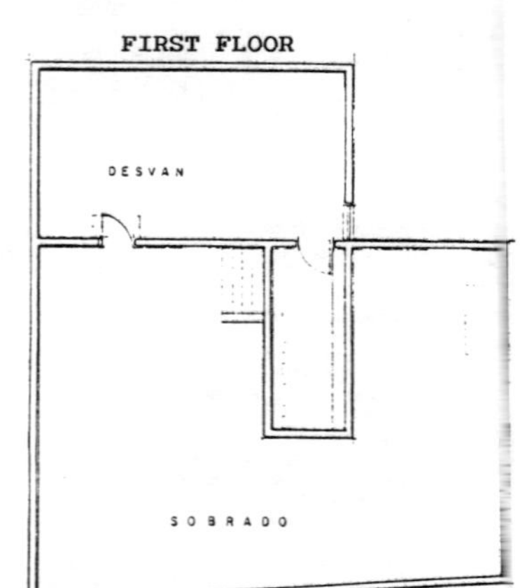

ATTIC

2. TYPOLOGY DESCRIPTION

This typology of the house with truss, is situated in S. Martin de Valdeiglesias, Madrid, 40° 22′ north latitude, 00° 37′ west longitude and 535m. altitude. It is fundamentally composed by the ground floor with a court yard, a stable and stock rooms, the first under floor with the chimney cook, the living-room and the bedrooms, and the second one with an atic and an under roof space. The principal house's face is south oriented.

The most important characteristic of this typology can be summarized in the following terms:

- Masoury walls, usual witewash, used on under floor,
- Tumber frame work on the upper floors with adobe, brick and stone, in the pieces of timber frame work can it be at sight and covered with stucco usually whiteed,
- Play wood with protection of mid walls and promise façade area,
- Bricks of play wood with ingredients of walls to attic,
- Galleries of wood or balconies of this material raised on the façade line,
- Two tiled roof.

The materials used are the tipical ones of the Central Spain: pine wood, walmit, oaks, chestnut and stone. Then we can see that popular architecture is adapted to the medium resources and possible to the climate.

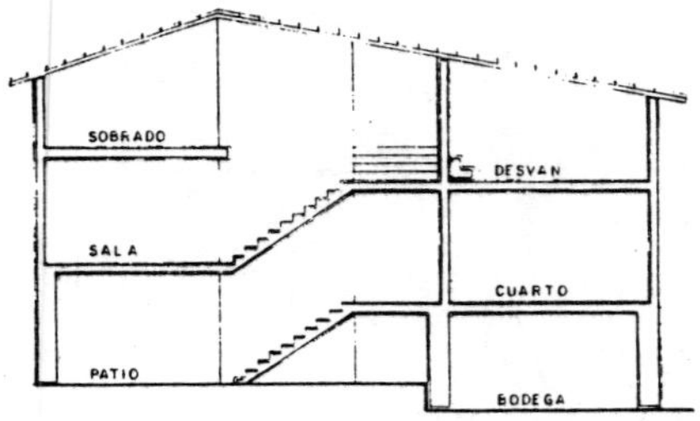

SECTION A-A'

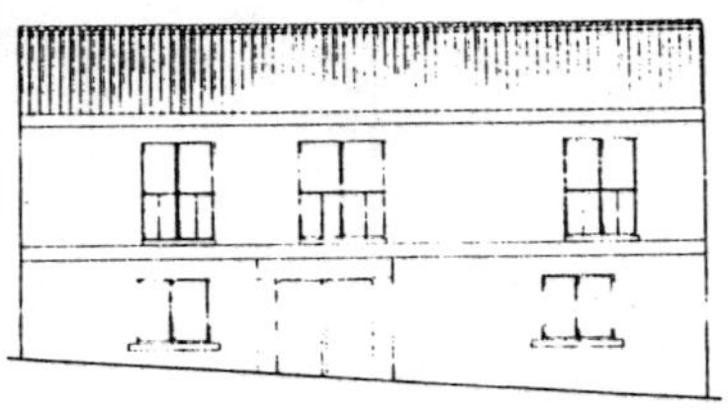

FACADE NORTH

WALL	AREA (m2)	HOLLOW (*) (m2)	THERMAL PROPERTIES
WEST FACE	20	-	M
EAST FACE	20	-	M
NORTH FACE	35	10.71	M
SOUTH FACE	35	3.54	M
SOIL	130	5.25	S
COBERT	130	3.15	C

(*) Window, Door, Balcony or Stairway

THERMAL PROPERTIES

e = thickness rho = density
lamda = thermal conductivity ce = specific heat

	cape number	e (m)	lamda (W/m/C)	rho (Kg/m3)	ce (J/Kg/K)
M	1	0.30	0.87	1800	1380
	2	0.04	0.30	800	920
S	1	0.20	0.14	600	2500
	2	0.05	0.49	1200	920
	3	0.03	0.14	600	2500
C	1	0.20	0.14	800	2500
	2	0.35	0.49	1200	920
	3	0.03	0.14	800	2500

TABLE 1

3. S3PAS RESULTS (1)

Now, we are analysing the results of the S3PAS simulation for typical meteorological conditions on January and July in Madrid.
In table 1 a summary of the most important paramete adopted in the simulation is presented: orientation dimension and thermal properties of the different walls in the specific typology. As we will see late the wall thickness is the more representative one, because it is the principal responsible of the grea thermal inertia in the house.

In the figure 1 the magnitude of the global solar irradiance on horizontal surface, the external ambient temperature and the inertial air temperatur for both months (January and July), are represented against the hour of the day. Nevertheless, January warmer than July, but in both cases, it can be seen that it's more attenuated than the exterior one. Th effect is greater on July than on January.
On January, the house is a few degrees warmer than the external ambient during the day. On July, in th central hours of the day, when solar radiation is greater, the internal air temperature °C is colder than the external one. It can also be seen that, in both months there is a lag of - hours between maxim inertial temperature and maximun solar irradiation.

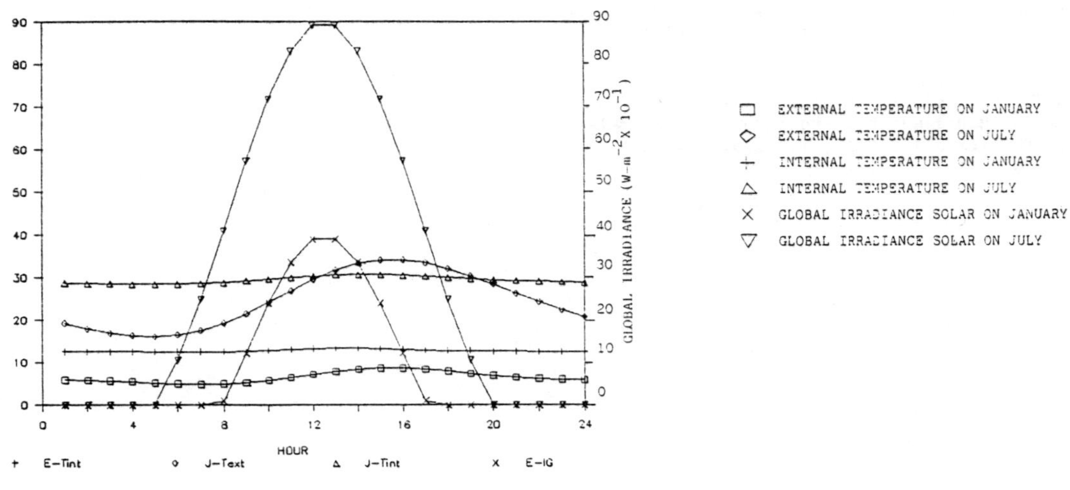

FIG. 1 Results of the simulation with S3PAS in the days January 21 and July 21.
Meteorological variables and thermic evolution of the indoor air.

4. CLA RESULTS (2)

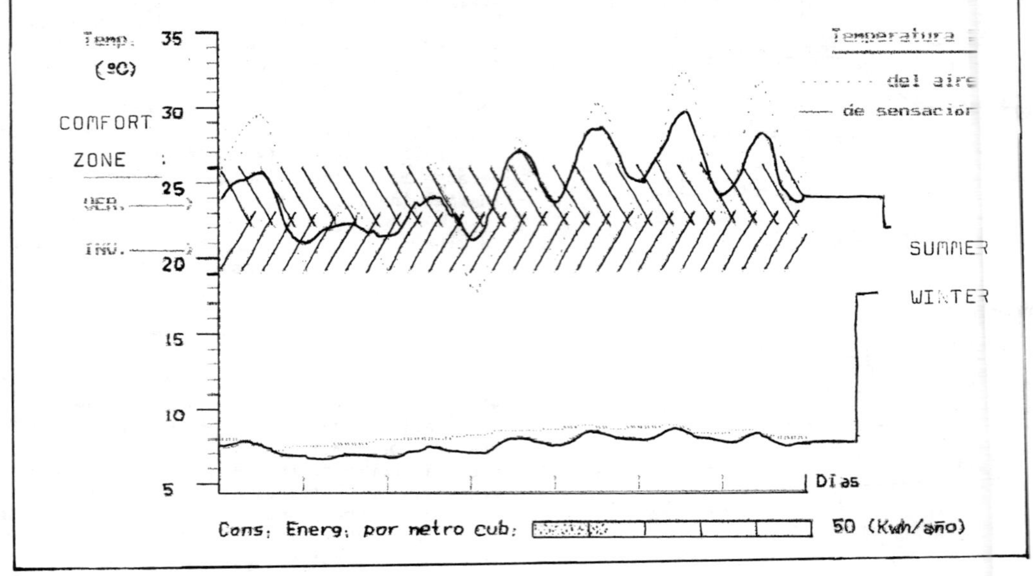

5. CONCLUSIONS

Supposing that the named thermal confort is in the range on 18 to 25°C, we can conclude that the typology analized is enough well adapted to the climate of Madrid.

In summer it is not necessary the refrigeration of the house. In winter the house is capable of maintaining an inertial temperature near the minimu of the thermal confort interval. Then, no much heating power consumption is expected.

Concerned to CLA as tool of predesign we can, say t access to the confort zone in our typology.

6. REFERENCES

(1) "S3PAS 1.0, Simulacion de Sistemas Pasivos, Manual del Usuario" Informe Tecnico, Catedra de Termotecnia, E.T.S.I. Industriales, Sevilla, Dic. 1988.

(2) SERRA, R.: "Clima, Lugar y Arquitectura, Manual de Diseño Bioclimatico, Editado por Secretaria General Tecnica del CIEMAT, I.S.B.N.: 84-7834-016-5 Madrid, Nov. 1989.

COMPARISON OF MEASURED AND REGRESSED PERFORMANCE CHARACTERISTICS AND SIMULATIONS OF A PASSIVE SOLAR HOUSE.

Paul Robinson and John Littler
Research In Building
Polytechnic of Central London
35 Marylebone Road, London NW1 5LS. UK

ABSTRACT. The monitored performance of a passive solar house is compared with the predictions from the SERI-RES building thermal simulation model. The model is driven with real weather data and the building use and components are input in detail. Agreement is very good over the whole heating season and fairly good on a monthly and daily basis. Internal temperatures are underpredicted and solar gains and building heat loss overpredicted during the summer months.

1. INTRODUCTION

Comparison of monitored data from a highly driven direct gain passive solar house has been made with performance predictions from the SERI-RES building simulation model.

The Courtyard "superglazed" design has large areas of south facing advanced glazing, heat recovery ventilation and high levels of insulation. Five superglazed houses of 144m^2 floor area, and south facing windows area of 27m^2 were built together with three double glazed houses with the same floor plan but reduced south facing window area (15m^2) and extract fans instead of heat recovery ventilation. Further details of this demonstration project can be found in the CEC DGXVII Final Report on the Courtyard Houses (SE432/85).

A meaningful comparison was attempted using real weather data, recorded at the site of the Courtyard houses, to drive SERI-RES, and by taking great care to assemble an accurate description file of the building. Measurements on the glazing U-value and shading coefficient, and results of infiltration rate tests were all used to aid an accurate description. Monitored data was used to input space heating and casual gain schedules representative of the actual case.

2. PROBLEMS WITH COMPARING SERI-RES PREDICTIONS WITH MEASURED DATA.

SERI-RES immediately raises the zone temperature to the scheduled temperature, when in reality the rise to the thermostat setting occurs over a period of, possibly, several hours. Hence if the actual zone thermostat setting is used as the SERI-RES zone temperature setting, then the average SERI-RES zone temperature over the period of heating will be an overestimate of the monitored average temperature.

SERI-RES temperatures are environmental temperatures whereas monitored room temperatures are air temperatures.

In the superglazed Courtyard house the heat recovery ventilation system transfers some solar and casual gains in the day zone to the bed zone. Such air flows are not currently modelled accurately by SERI-RES.

3. MONTHLY COMPARISON.

Figure 1 shows the monthly comparison between SERI-RES predicted (space heating + casual gains) and the measured (space heating + estimated casual gains).

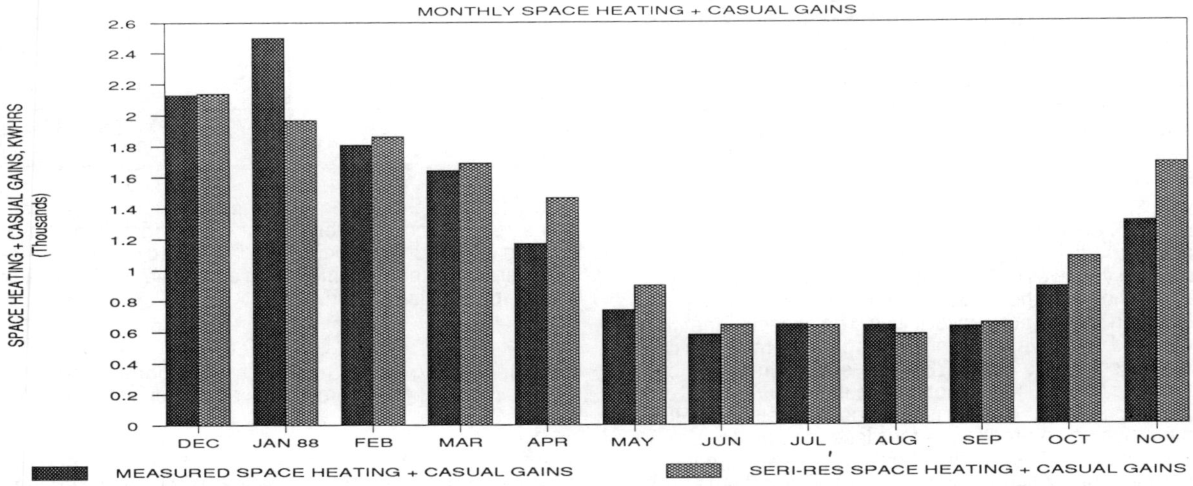

Figure 1. Monthly comparison of space heating + casual gains. Measured and SERI-RES prediction

From inspection of Figure 1. it can be seen that agreement is close for all months except Jan'88, Apr'88 and Nov'88. In Jan, SERI-RES underpredicts by 22%, and in Apr and Nov SERI-RES overpredicts by 25% and 29% respectively.

There seems to be a general trend that SERI-RES overpredicts at the edges of the heating season.

Figure 2 (day zone monthly temperatures) shows that SERI-RES is also underpredicting the zone temperatures during these periods and in the summer months. (The trend is the same for the bed zone temperatures.)

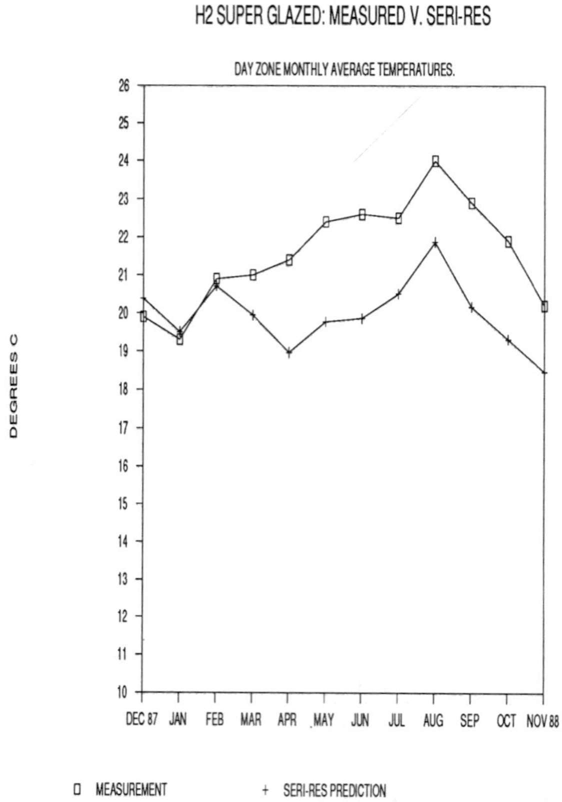

Figure 2. Monthly SERI-RES predicted day zone temperatures v measured.

Figure 3 compares SERI-RES predicted monthly solar gains with those calculated from the effective solar aperture parameter, A_{eff}, resulting from regression analysis of the October to April data. A_{eff} is the equivalent aperture in the south face of the building through which there would be 100% solar transmission and all solar gain would be useful. The A_{eff} determination is strictly only valid for the heating season and is an average for that period; in reality it would vary continuously with solar altitude, and the changing usefulness of solar gains through the year. This method of estimating solar gains gives much improved accuracy over the subtractive or energy balance method.

SERI-RES solar gains are calculated from the user input shading coefficient and "usefulness" considerations. The two agree very well on the whole with some discrepancy in the summer months of May, June, July, where the A_{eff} calculation is not strictly valid; it would be expected that this would lead to an over generous solar gain estimate. As can be seen from Figure 3 the SERI-RES solar gain predictions are even greater than those calculated from A_{eff}. An explanation arises when it is considered that SERI-RES zone temperature predictions for these months are below actual temperatures but the heating schedule imposed does not allow space heating to be used; for this reason the usefulness of solar gains for this summer period will be high according to SERI-RES.

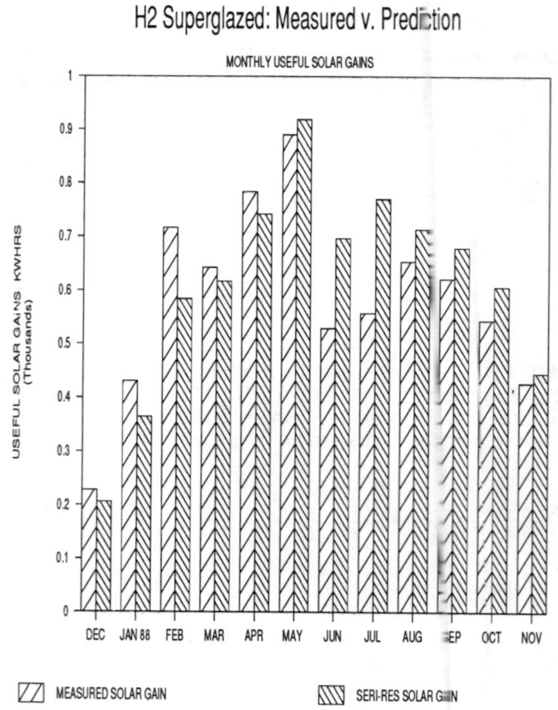

Figure 3. Monthly SERI-RES predicted solar gains compared with those calculated from measured data and regression analysis.

Table I Annual and heating season (Sep-May) totals for heat input to the house: useful solar gains, space heating, (space heating + casual gains), (space heating + casual + solar gains).

ANNUAL TOTALS:	Measured, kWhrs	SERI-RES, kWhrs
Useful solar gains	7026 ± 759	7[illegible]44
Space heating	6466	7457
Space heating+Casual gains	14652 ± 819	15275
Space heating + casual + solar gains	21678 ± 1578	22519
HEATING SEASON TOTALS:	Measured, kWhrs	SERI-RES, kWhrs
Useful solar gains	5285 ± 571	5163
Space heating	6407	73[illegible]3
Space heating + Casual gains	12794 ± 639	13416
Space heating +casual +solar gains	18079 ± 1210	18579

Table I shows the annual total heat inputs to the house. More relevant comparisons are the heating season totals (also given), during which period the predicted temperatures are in better agreement with actual temperatures and the A_{eff} parameter is valid.

For the heating season there is very good agreement between SERI-RES predicted and measured energy inputs.

Figure 4 shows the monthly comparison between building heat loss predicted by SERI-RES and that calculated from the measured house UA (heat loss coefficient) and $\Sigma\delta T$ (the sum of the inside outside temperature differences).

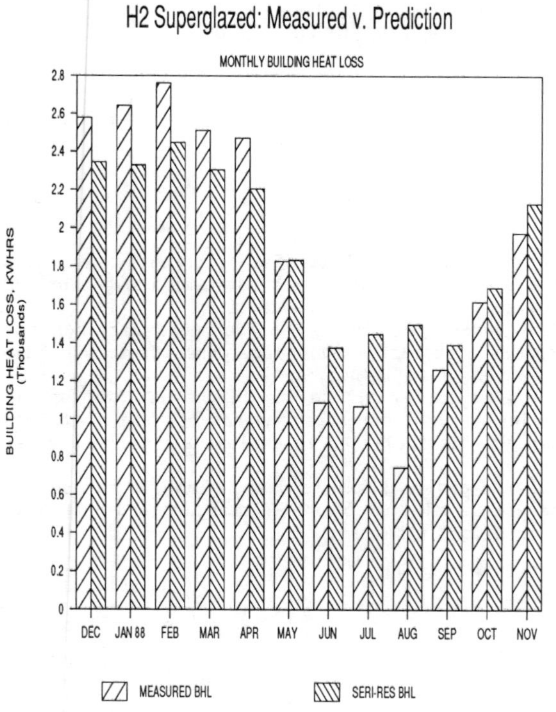

Figure 4. Monthly comparison between building heat loss predicted by SERI-RES and that calculated from the measured house UA (heat loss coefficient) and $\Sigma\delta T$ (the sum of the inside outside temperature differences).

Agreement is fair for all but the summer months. The year total building heat loss is predicted by SERI-RES to be 22888 kWhr, and calculated, from measurement, to be 22548 ± 2438 kWhrs. The more meaningful heating season comparison is:

from SERI-RES:	18873 kWhrs
from measurement:	18848 ± 2038 kWhrs

SERI-RES predicts very accurately for the heating season.

4. HOURLY COMPARISONS OF MEASURED DATA VERSUS SERI-RES PREDICTIONS.

The heating schedule for each day being modelled is taken from the monitored heating periods for that day. The set point temperatures for the SERI-RES simulation are taken as the average zone temperatures monitored during each heating period. This method is effectively the same as for the monthly comparison but on a smaller time scale.

Table II Measurements and Predictions for Feb 3,4 1988.

	space heat + casual gains for 3,4 Feb	house average temperature for 3,4 Feb.
Measured data	143 ± 9 kWhrs	18.3°C
SERI-RES Prediction	132 kWhrs	17.1°C

The heat input profile shown in Figure 5. Over the two days the total heat input from this simulation agrees closely with the measured heat input; 132 and 143 ± 9 kWhrs respectively. The whole house average temperature for this simulation of Feb 3 and 4, is below the actual average house temperature by 1.2°C.

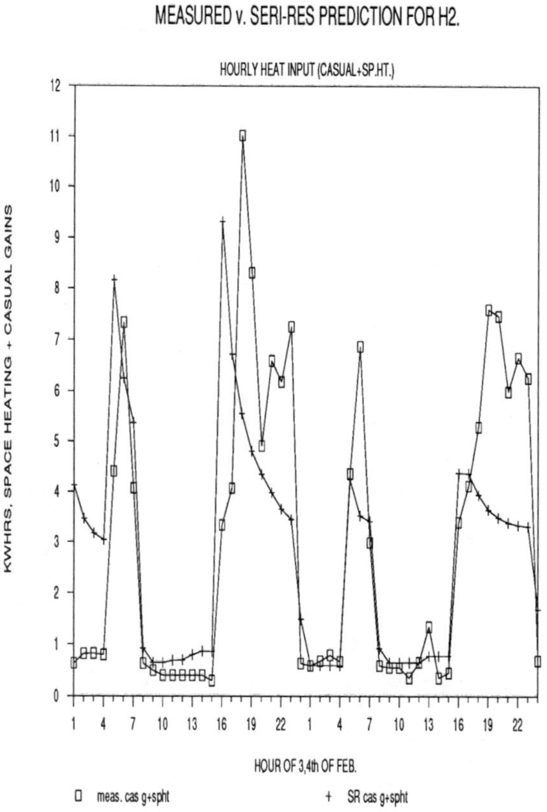

Figure 5. Heat input (space heat + casual gains) profile for each hour of 3,4 Feb: SERI-RES prediction v Measured.

Prediction of useful solar gains is very accurate for all days of February, except for those with high direct insolation when SERI-RES underpredicts.

Over the whole month SERI-RES predicts solar gains of 583 KWhr whereas the estimate from monitored data was 717 ± 74 KWhrs.

TICAD 2: A COMPUTER AIDED DESIGN TOOL FOR ASSESSING THE THERMAL PERFORMANCE OF BUILDINGS EQUIPPED WITH TRANSPARENT INSULATION

L. F. Jesch and S. Gilani

Solar Energy Laboratory
The University of Birmingham
Birmingham B15 2TT, UK
Tel. ++ 44 21 459 4826 Fax. ++ 44 21 459 8206

ABSTRACT. The use of transparent insulation in buildings is on the increase. Some of the currently available simulation programmes have been developed at the time when transparent insulation was not even thought of and consequently they do not provide adequate facilities to cope with the new design concepts.
TICAD 2 has been developed specifically for computer aided design and thermal performance assesment of buildings with transparent insulation. This programme is aimed at architects and energy consultants who have responsibilities for the detailed design and energy flow simulation of buildings.
TICAD 2 is a design tool to calculate the temperatures at any time and at all points in the building shell, including the transparent insulation modules and the outside/inside wall surface temperatures and to obtain room air temperatures resulting from heat gains and losses through the shell due to ambient air temperatures, solar radiation, incidental gains and infiltration.
Thus the purpose of the work reported here is to provide a userfriendly, flexible design tool to be used by architects and energy consultants having access to a machines with a minimum of PC AT standard equipped with a fast coprocessor. Larger computers will do the job faster and slightly better.

1.0 TRANSPARENT INSULATION IN DOMESTIC BUILDINGS

One of the most immediate applications of transparent insulation TI is in domestic buildings. The energy saving and comfort aspects of TI make it an obvious candidate for consideration. The fact that the existing building stock is the largest potential market for using TI in domestic buildings makes the division between retrofit and newbuilt application a natural one. Furthermore the methods of design and construction are also different in the two types of application. /1/

1.1 Retrofit applications.

For both energy conservation and aesthetic reasons the design of retrofit applications of TI require great care and much imagination on the part of the architect. The modules of TI available to date can not always be readily adapted to the architectural style of the existing building. Both shapes and colour of the TI frames have to be selected such that they achieve the energy and comfort objectives without having a detrimental effect on the appearance of the building.
The good aesthetic design solution is the best guarantee for customer acceptance of the new TI technology.

1.2 Newbuilt applications.

The control that the architect and the energy designer have on the formulation of basic design principles and the shaping of the building concepts is best realised in newbuilt applications of TI. Here the aesthetic style and the requirements of the TI technology can be harmoniously married and expressed in architectural language.

2.0 THE SOLAR DEMONSTRATION EXTENSION BUILDING

An extension of the Solar Demonstration House /2/ in Birmingham provides a good example of showing what the computer simulation can do. The floor plan of the building is shown in Fig. 1 and the wall cross section in Fig. 2 gives some information about the construction deatils and the temperature profile in the wall.

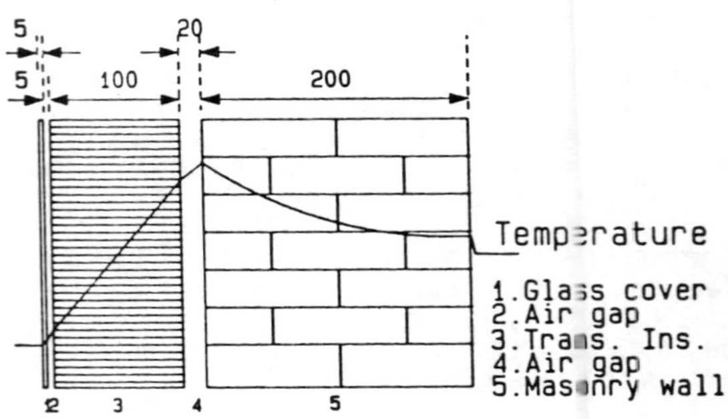

Wall section with Transparent Insulation

Fig 2 **Horizontal Honeycomb cells**

The extension shares a common wall with the Demonstration House on the west side and the South, East and North facing walls are exposed. The North wall is not covered with transparent insulation on the outside but the other two exposed walls are. On all three exposed walls there is a 50 mm opaque insulation also on the inside surface. Both East and South facing windows are double glazed and the floor and ceiling are low mass, highly insulated structures. The windows have movable insulation which can reduce the nighttime heat losses as well as the daytime heat gains. /3/

3.0 WORKED EXAMPLE

The extension has been fully simulated for a whole year on an hourly basis. As a sample the two most critical periods during the year have been selected to show the effect of the transparent insulation. These are the days in January where the minimum ambient air temperatures are experienced and the days in May when the maximum occurs in Birmingham. The first is to test the worst winter condition which also helps in the sizing of the auxiliary heater. The second serves to test the worst summer condition when the overheating occurs. The January period includes dull, cold days without sunshine and also very cold days with high level of beam radiation. The summer period has typical high air

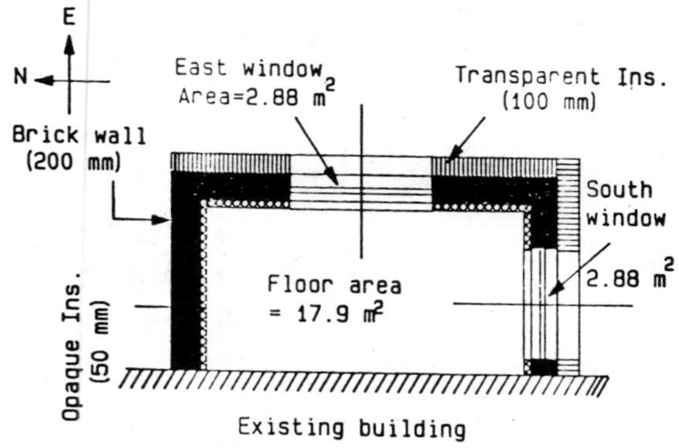

Figure 1 **Floor plan of the extension**

temperatures and high level of beam and diffuse radiation.
These Birmingham climatic conditions for the two extreme periods are caharacterised in Figs. 3, 4, 5 and 6 by solar radiation and ambient air temperatures.

4.0 RESULTS

Although TICAD 2 produces many detailed answers for the purpose of illustration only wall surface temperatures inside and outside two walls shown in Figures7 and 8 and air temperatures in Fig. 9.

4.1 Summer overheating

The simulated results show that although the temperature in the summer goes above the design limit during the worst period for a short time, but it is considered that normal ventilation will be able to reduce it to a tolerable level. It certainly does not warrant special automatic equipment installed to reduce solar gains in a house built in the climatic conditions prevailing in Birmingham UK. /4/

4.2 Winter underheating

Similarly the worst winter days produce temperature in the room which is below the design level. This is the result of previously occuring longer periods of low level solar radiation causing both the cold ambient air temperature and the increased heat loss from the extension. Here too a portable electric fan heater can more than adequately supply the missing heating energy but the extension does not need a permanent auxiliary central heating system installed.

4.3 Transparent insulation bemefits compared

Finally Figure 10 compares the room temperature in the extension before and after the installation of transparent insulation. It shows that the increase in thermal comfort from utilising solar radiation and reducing heat loss at the same time is evident.

Figure 3 **RADIATION DATA IN WINTER**
10 JAN. TO 4 FEB. 1987

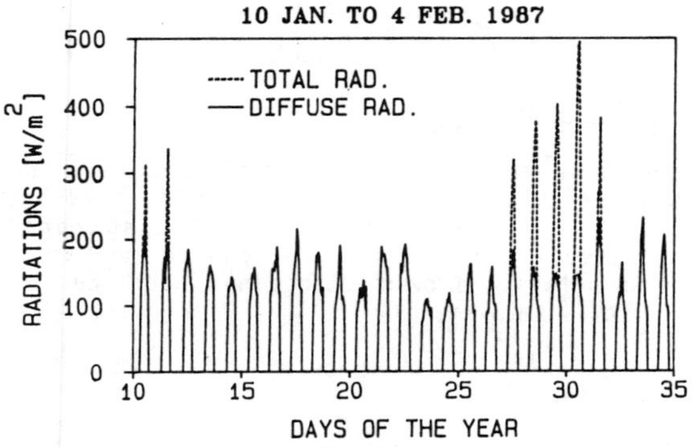

Figure 5 **AMBIENT AIR TEMPERATURE**
10 JAN. TO 4 FEB. 1987

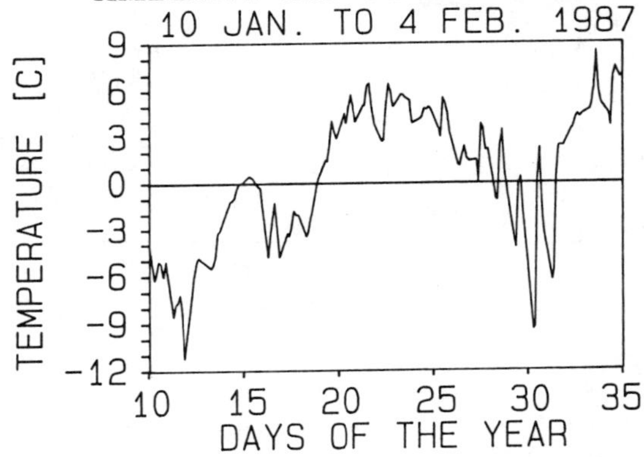

Figure 4 **RADIATION DATA IN SUMMER**
25 APR. TO 20 MAY 1987

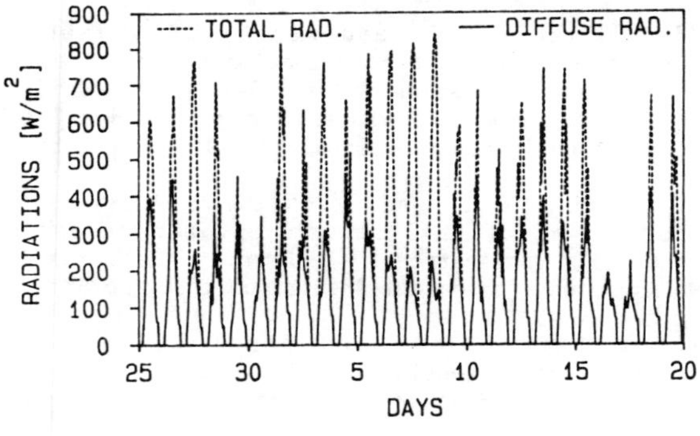

Figure 6 **AMBIENT AIR TEMPERATURE**
25 APR. TO 20 MAY 1987

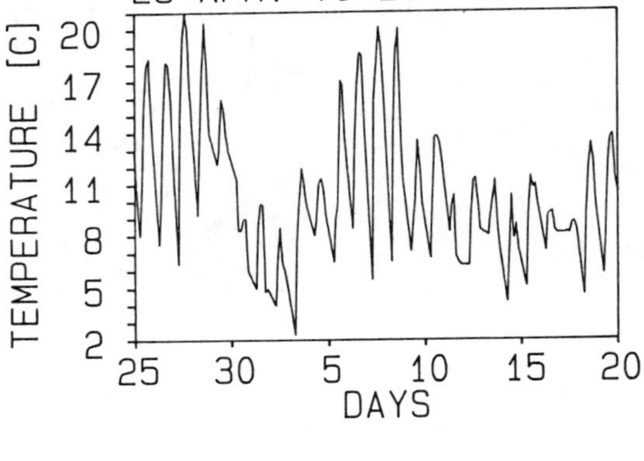

5. THE SIMULATION PROGRAM

TICAD 2 is a modular programme using a special version of TRNSYS purposly adapted to the task. The modifications enable European users to access climatological data and properties of building material commonly used in Europe, all expressed in SI units. The modular structure of the programme makes it easy to change only a single or a few variables for each successive parametric optimisation step. It also allows any portion of the building shell to be fully, partially or not at all transparently insulated. This is important in finding a solution to avoid summer overheating by design rather than by extra mechanisms.
The programme calculates dynamic conditions in the transparent insulation module and in the massive wall where time delays occur due to storage effects.
The wall to room air temperature heat transfer as well as the transparent insulation to ambient air heat transfer are also calculated.
A special feature of TICAD 2 is the ability to consider honeycomb type transparent insulation with the cell structure positioned in a non orthogonal manner. This makes it possible to design every wall separately such that they all have an optimised transparent insulation specifically suited for that wall with all its individual characteristics and its orientation which inevitably means different solar radiation gains. /5/

Figure 7 SOUTH WALL SURFACE TEMPS.
10 JAN. TO 4 FEB. 1987

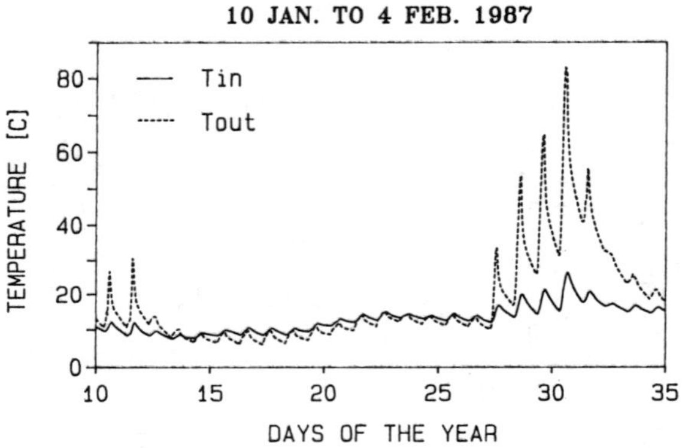

Figure 8 SOUTH WALL SURFACE TEMPS.
25 APR. TO 20 MAY 1987

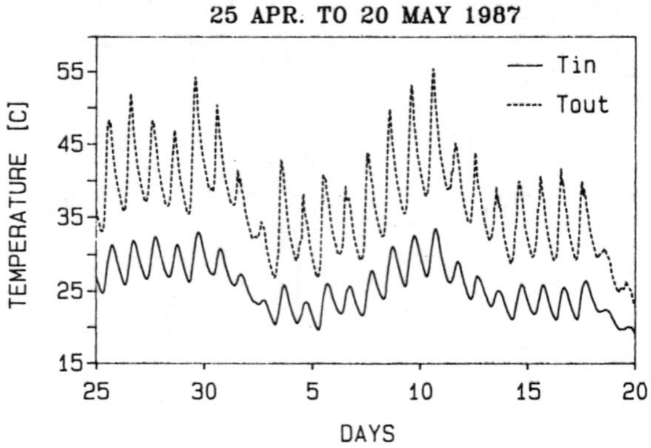

Figure 9 ROOM TEMPERATURES
10 JAN. TO 4 FEB. 1987

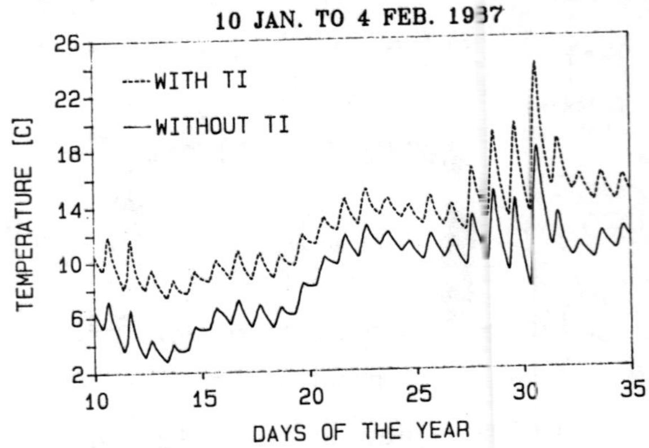

Figure 10 ROOM TEMPERATURES
25 APR. TO 20 MAY 1987

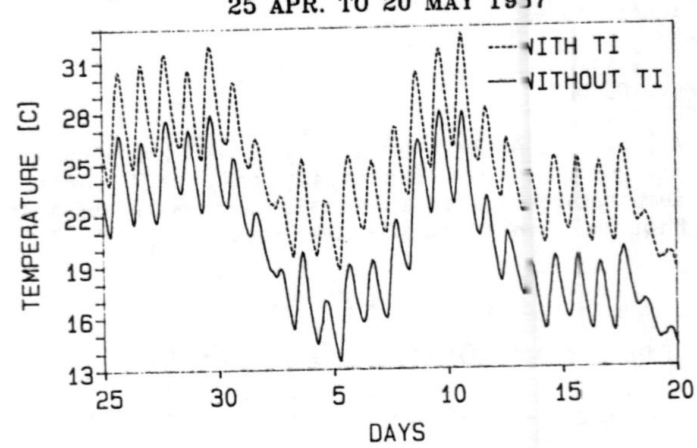

6.0 CONCLUSION

TICAD 2 is able to produce detailed and reliable simulation of energy performance for buildings which are fully or partially transparently insulated and which do not require mechanical shutters to avoid overheating in a North European climate, like Birmingham.

7.0 REFERENCES

/1/ Jesch, L.F.: Energy saving potential of transparent insulation in new houses and in retrofit applications. TI1 Proc. 1st Int'l Workshop on Transparent Insulation, 1986 pp 39 - 40

/2/ Jesch L.F.: The making of a Solar Village and a Demonstration House in Bournville. The Franklin Co. 1985, pp 1 - 24

/3/ Gilani, S.I. and Jesch L.F.: Thermal simulation of a building extension solar heated by transparent insulation and modelled by a modified TRNSYS programme. TI3 Proceedings of the 3rd Int'l Workshop on Transparent Insulation 1989, pp 138-141

/4/ Jankovic L. and Jesch L.F.: Analysis of overheating of wall structures in transparent insulation applications. TI2 Proc.of the 2nd Int'l Workshop on Transparent Insulation 1988, pp 53 - 55

/5/ Chattha J.A. and Jesch L.F.: Solar transmission potential of transparent honeycomb insulation. ISES Solar World Congress, Kobe, 1989

LA PRISE EN COMPTE DES APPORTS SOLAIRES DANS LA REGLEMENTATION THERMIQUE FRANCAISE DES BATIMENTS AUTRES QUE D'HABITATION

JR. MILLET, E. HUTTER

Centre Scientifique et Technique du Bâtiment

4 av. du Recteur Poincaré 75782 PARIS Cédex 16

ABSTRACT : The new French regulation for non residential buildings takes into account the effects of solar heat gains through windows. The regulation is based on the G1 value of the building (W/m^3.°C) which is defined as the heat losses through its opaque and transparent components for a 1°C temperature difference between indoor and outdoor.

According to this regulation, the G1 value must be lower than a reference value. The solar heat gain are taken into account by an increase of the G1 reference value, the new value being a function of windows orientations and areas. This possibility allows to increase the sun oriented windows areas, and leads to a better building design.

1 - INTRODUCTION.

Une nouvelle réglementation des bâtiments autres que d'habitation est entrée en vigueur en 1988 en France. Son objectif est de parvenir par rapport à la réglementation précédente à une économie d'énergie de 25%.

Un des facteurs permettant d'atteindre l'objectif d'économie d'énergie recherché, est la prise en compte des apports solaires, non retenue dans la réglementation précédente.

Après un rappel des principes de la réglementation relatifs à l'isolation thermique, nous préciserons le mode de calcul des apports solaires et en étudierons les conséquences sur les caractéristiques requises pour les parois vitrées.

2 - L'ISOLATION DES PAROIS.

2.1 - Formulation de la réglementation.

Le principe de la réglementation est de fixer une limite au coefficient G1 (W/m^3.°C), égal aux déperditions thermiques à travers les parois d'un bâtiment ou d'une partie de bâtiment pour un degré d'écart de température entre l'intérieur et l'extérieur, divisés par le volume intérieur. Cette valeur de référence est donnée par l'expression :

$$G1\ réf = a\ \frac{A_1}{V} + b\ \frac{A_2}{V} + c\ \frac{P}{V} + d\ \frac{A_3}{V} + e$$

dans laquelle :

A_1 et A_2 sont les surfaces des parois opaques en contact avec l'extérieur, les locaux non chauffés et les locaux non visés par l'arrêté considéré, y compris les parois sous comble et sur vide sanitaire mais non compris celles sur terre plein ou enterrés.

A_1 correspond aux parois verticales ou faisant avec l'horizontale un angle supérieur à 60°,

A_2 correspond aux parois horizontales ou faisant avec le plan horizontal un angle inférieur ou égal à 60°,

P est le pourtour des locaux chauffés sur terre plein ou enterrés,

A_3 est la surface des parois transparentes ou translucides en contact avec l'extérieur ou des locaux non chauffés,

V est le volume intérieur chauffé.

2.2 - Valeurs des coefficients a, b, c, d , e.

Les valeurs des coefficients a, b, c, d, e sont fonction :

- de l'usage du bâtiment (bureau, commerce, hôtellerie, scolaire...),
- du type de chauffage
  . type I:chauffage fonctionnant à l'électricité pour au moins la moitié de la puissance de l'installation,
  . type II:autres modes de chauffage.
- de la zone climatique : la France est divisée en trois zones H1, H2, H3,
- de la catégorie de bâtiment :
  - les bâtiments à occupation continue (catégorie C) qui sont ceux dans lesquels la température doit être maintenue en permanence au niveau normal d'occupation,
  - les bâtiments à occupation discontinue (catégorie D) dans lesquels un abaissement sensible de la température est possible pendant des périodes d'inoccupation suffisamment longues.

Le tableau ci-dessous donne à titre d'exemple les valeurs des différents coefficients pour les bâtiments de bureau et de commerce.

LOCAUX DE CATEGORIE D

Type de chauffage	I			II		
Zone climatique	H_1	H_2	H_3	H_1	H_2	H_3
a	1,00	1,05	1,15	1,15	1.25	1,35
b	0.40	0.45	0.50	0.45	0.50	0.60
c	1.30	1,30	1,40	1,40	1.50	1,50
d	1.8	2,4	3,10	3,10	3.90	4,70
e	0.12	0,12	0,12	0.12	0,12	0,12

LOCAUX DE CATEGORIE C

Type de chauffage	I			II		
Zone climatique	H_1	H_2	H_3	H_1	H_2	H_3
a	0.90	0.95	1,00	0.95	1,00	1,05
b	0.40	0,40	0.45	0,45	0,45	0,50
c	1.30	1.30	1.30	1.30	1,30	1.30
d	1.20	1.40	1.60	1.40	1,60	2,30
e	0.08	0,08	0.08	0.08	0.08	0,08

Les exigences sont plus fortes pour les bâtiments à occupation continue que pour les bâtiments à occupation discontinue. Pour ces derniers, la pratique de l'intermittence du chauffage est en effet un moyen efficace d'économiser l'énergie.

La prise en compte des apports solaires s'effectue par une majoration du coefficient G1 réf que nous décrivons ci-après.

3 - LA PRISE EN COMPTE DES APPORTS SOLAIRES.

3.1 - Majoration de la valeur de $G_{1\ \text{réf}}$

Pour un même coefficient G_1, un bâtiment consommera d'autant moins d'énergie que les apports solaires y sont élevés. Il est donc légitime de moduler la valeur du $G_{1\ \text{réf}}$ en fonction de ces apports.

L'augmentation du $G_{1\ \text{réf}}$ dépend de l'usage et de la catégorie du bâtiment, de son inertie thermique, et de son indice solaire.

Le tableau ci-dessous donne à titre d'exemple l'augmentation de $G_{1\ \text{réf}}$ pour les bâtiments de bureaux et de commerce.

Locaux de catégorie D

Inertie thermique	Indice solaire		
	Inférieur à 0,01	Egal ou supérieur à 0,01 et inférieur à 0,02	Egal ou supérieur à 0,02
Toutes inerties .	0	0,04 W/m³.°C	0,08 W/m³.°C

Locaux de catégorie C

Inertie thermique	Indice solaire		
	Inférieur à 0,01	Egal ou supérieur à 0,01 et inférieur à 0,02	Egal ou supérieur à 0,02
Faible	0	0,03 W/m³.°C	0,06 W/m³.°C
Moyenne	0	0,05 W/m³.°C	0,10 W/m³.°C
Forte	0	0,06 W/m³.°C	0,12 W/m³.°C

augmentation de la valeur $G_{1\ \text{réf}}$ (en $W/m^3{}^\circ C$) pour les bâtiments de bureau et de commerce.

Dans un bâtiment d'occupation continue, la récupération des apports solaires est d'autant meilleure que le bâtiment est solarisé (l'indice solaire caractérisant cette solarisation est défini au § 3.3) et que son inertie est forte.

Dans un bâtiment d'occupation discontinue, une forte inertie thermique favorise la récupération des apports solaires mais se traduit par une moindre efficacité de l'intermittence du chauffage. Cette compensation explique que le paramètre inertie n'ait pas dans ce cas été retenu.

3.2 - Calcul de l'inertie thermique.

La classe d'inertie thermique d'un bâtiment ou d'une partie de bâtiment est fonction de la "*masse utile*" de ses parois en contact avec l'air ; la masse utile est celle qui intervient dans les phénomènes transitoires de période 24 h. Selon la valeur de cette masse, l'inertie est celle indiquée ci-dessous :

Masse utile en kilogramme par m^2 de plancher	Inertie
Inférieure à 150	faible
Egale ou supérieure à 150 et inférieure à 400	moyenne
Egale ou supérieure à 400	forte

3.3 - Calcul de l'indice solaire I_s.

L'indice solaire a pour valeur :

$$I_s = \frac{\sum_i (A_i.S_i.\sigma_i)}{V}$$

où pour chaque paroi transparente ou translucide "*verticale*" i :

A est la surface de la paroi comptée en tableau,

S est son facteur solaire, égal au rapport de l'énergie solaire transmise par cette paroi à l'énergie solaire incidente,

σ est un coefficient dont la valeur est donnée dans le tableau ci-dessous en fonction de l'orientation de la paroi, de la hauteur moyenne des masques vus par celle-ci et de la catégorie des locaux.

Hauteur moyenne pondérée des masques	Orientation de la paroi				
	du SO au SE	du SE au NE	du SO au NO		du NE au NO
			Locaux à occupation continue	Locaux à occupation discontinue	
Jusqu'à 14°	1	0,6	0,6	0,4	0,3
De 15 à 24°	0,6	0,4	0,4	0,3	0,2
25° et plus	0	0	0	0	0

Le coefficient σ traduit la variation d'énergie solaire reçue pendant la saison de chauffage par une paroi. Pour les locaux occupés de façon discontinue, les apports solaires sur une baie orientée à l'Ouest ont lieu en fin de période d'occupation et conduisent donc à une économie d'énergie moindre que si la baie est orientée à l'Est, d'où une valeur plus faible de σ dans ce cas.

La hauteur moyenne des masques est la somme des hauteurs moyennes des quatre tranches d'azimuth représentées ci-dessous :

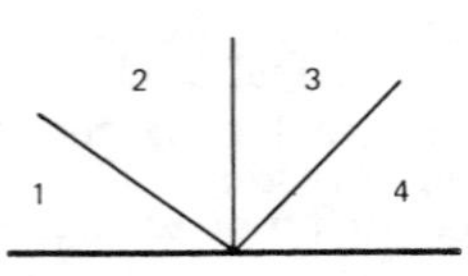

les coefficients de pondération étant les suivants:

Orientation de la paroi vitrée	Tranche d'azimut			
	1	2	3	4
du SO au SE	0,10	0,40	0,40	0,10
du SE au NE	0,05	0,20	0,45	0,30
du SO au NO	0,30	0,45	0,20	0,05
du NE au NO	0,20	0,30	0,30	0,20

D'une façon générale, on effectue le calcul façade par façade en se plaçant au centre géométrique de chacune d'elles. Toutefois, si une partie de façade a un ensoleillement nettement différent du reste de la façade, on la traite séparément.

Considérons la façade A B C D représentée sur la figure ci-dessous, pour laquelle les masques sont constitués par le bâtiment situé vis-à-vis et par le retour du bâtiment.

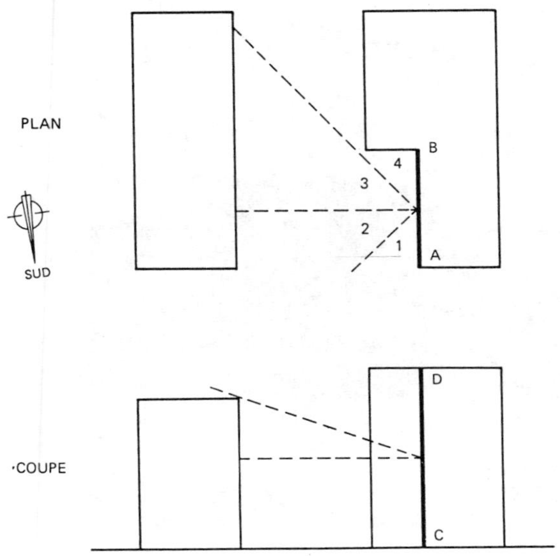

Le profil des masques, vus du centre de la façade, est donné ci-dessous.

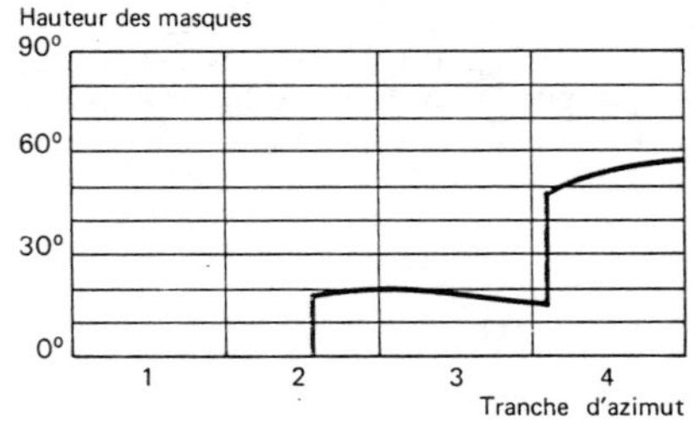

Sachant qu'il s'agit d'un bâtiment à occupation continue, le calcul est le suivant :

	Tranche d'azimut			
	1	2	3	4
Hauteur moyenne	0	8,5	18	52
Coefficient de pondération	0,30	0,45	0,20	0,05
	0	3,8	3,6	2,6
Hauteur moyenne pondérée	10			
Valeur de σ	0,6			

4 - <u>INFLUENCE DE LA PRISE EN COMPTE DES APPORTS SOLAIRES SUR LES CARACTERISTIQUES THERMIQUES REQUISES POUR LES BAIES.</u>

Une façon simple de concevoir un bâtiment respectant la réglementation est d'adopter comme valeur de coefficient de déperditions thermiques pour les parois opaques verticales et horizontales les valeurs de a et b, pour le coefficient de transmission linéique des parois sur terre plein ou enterrés la valeur c.

Le coefficient de transmission thermique des parois vitrées Kv, exprimé en W/m^2.°C doit alors être inférieur à $d + \frac{e.V}{A_3}$.

Cette valeur est d'autant plus élevée que le ratio surface de paroi vitrée/volume du bâtiment est faible et que les surfaces vitrées bénéficient d'apports solaires importants.

Considérons par exemple un bâtiment dont les surfaces vitrées sont uniformément réparties sur les différentes orientations. On peut en fonction du rapport A_3/V calculer le coefficient Kv admissible suivant que l'environnement du bâtiment est dégagé ou non. En retenant un facteur solaire moyen de 0,45, les coefficients de transmission thermique des parois vitrées admissibles sont en zone H1 pour un bâtiment dont le chauffage est de type I :

	environnement dégagé	rapport surface de paroi vitrée/ volume du bâtiment 0,03	0,05	0,07	0,10
bâtiment de catégorie **D**	oui	5,80	4,90	4,40	3,70
	non	5,80	4,75	3,15	2,75
bâtiment de catégorie **C**	oui	3,85	3,50	3,30	3,00
	non	3,85	3,15	2,50	2,10

Coefficient de transmission thermique admissible pour les parois vitrées (zone H1 chauffage de type I).

Quand les baies sont de petites dimensions, l'indice solaire est inférieur à 0,01 et il n'y a donc pas de différence selon que l'environnement est ou on dégagé.

Pour un rapport surface de paroi vitrée/volume du bâtiment supérieur à 0,07, le gain en Kv admissible est de l'ordre de 1 W/m^2.°C et serait plus important si les parois vitrées étaient orientées préférentiellement au Sud.

5 - <u>CONCLUSION</u>.

Par l'augmentation de la valeur limite du coefficient G1 en fonction de l'indice solaire du bâtiment, la nouvelle réglementation thermique des bâtiments autres que d'habitation incite le concepteur à privilégier les parois vitrées sources d'apports solaires. Tout en conduisant à une amélioration de la conception de bâtiment, elle offre une liberté supplémentaire au concepteur en réduisant les contraintes d'isolation et de surface des parois vitrées si cette réduction est compensée par une augmentation des apports solaires.

ENERGY AND USER STUDIES OF A PASSIVE SOLAR SCHOOL IN THE UK

N V Baker
The Martin Centre for Architectural and Urban Studies
University of Cambridge, 6 Chaucer Rd., Cambridge, UK

C Martin
The Energy Monitoring Company
Unit 3, Paggs Court, Newport Pagnal, MK16 OEW, UK

ABSTRACT. This paper describes a passive solar school in the UK. The design incorporates a long SE facing unheated sunspace which also serves as circulation to the teaching areas. Monitoring has shown the school to have a good energy performance and that this is mainle due to the sunspace pre-heating the ventilation air. The passive features were incorporated at no extra cost. The response of occupants is also reported in the paper.

Fig 1 South elevation of Netley School. Source - AJ

1 INTRODUCTION

This scheme originated with a study carried out in 1981 for the Science and Engineering Research Council of the UK for the application of passive solar principles to school design. The study developed the idea of ventilation pre-heating as a passive solar strategy, and as an alternative to direct gain systems which experience had shown were likely to suffer from thermal comfort problems.

The design principles established were then applied to a primary school project for Hampshire County Council. The direction of the design development was consistent with Hampshire's current adoption of shallow plan daylit and naturally ventilated buildings, a passive approach, as distinct from more heavily serviced deep plan buildings, then being pursued by many other educational authorities in Britain.

In the initial study /1/ the the author worked closely with the Architect to produce a design for a school at Locksheath, which due to falling pupil numbers in the area, was never built. However, the principles were then transferred to this infants school at Netley Abbots, fig(1). Since the design incorporated a number of innovative features, it was the intention of the client to carry out fairly detailed monitoring, both of the energy systems and the response of the occupants.

2 PASSIVE SYSTEMS AND BUILDING DESCRIPTION

2.1 Passive Strategy

The most important feature is the use of the sunspace for the pre-heating of ventilation air. The use of this strategy was prompted by the following considerations.

As the standards of thermal insulation have increased, the proportion of heat loss associated with ventilation increases. Schools are subject to a minimum ventilation requirement of 30 cubic metres per child per hour, and although this is rarely met in practice, due to high occupancy density, the actual ventilation demand is still considerable.

Bearing in mind that the utilization of solar energy is a function of the heating load, it seems logical to link the ventilation heat load direct with the solar gains. This carries the further advantage that there is no threshold for solar utilization - any temperature increment to the ambient air by solar gains in the sunspace will reduce ventilation heat demand. This contrasts with other passive systems such as Trombe walls. Simulation studies (Baker/2/) show that the ventilation pre-heating mechanism is the main way that a sunspace reduces energy consumption for a building in the UK climate.

A further consideration was the provision of comfort. Direct gain systems have inherent problems when applied to densely occupied spaces in particular; the overheating of occupants due to direct insolation, and glare. At times of low solar radiation, this problem is replaced with cold radiation and downdraughts caused by the large surface of glass.

Other passive features adopted in the design are concerned with the avoidance of overheating in summer. Automatic vents connect the sunspace with a continuous roof vent, Air enters through grilles in the north gable walls and is drawn through the teaching areas by stack effect or wind-induced suction at the ridge. Internal white rollerblinds are also provided in the sunspace.

2.2 Building Form.

The strategy described above, when applied together with the educational client's requirement and County design policy, lead quite naturally to the building form. The characteristic section, fig(2), has been generated by the relationship between the sunspace and the teaching areas, and the use of the pitched roof. This provides space for plant on the mezzanine level, and increased height for buoyancy-induced ventilation.

The plan, fig(3), is also dominated by the sunspace. This, together with a small top-lit atrium, provides the main circulation for the building. Thermal analysis showed that temperatures in these spaces, although unheated, would be sufficient in winter for this purpose.

Simulation studies also showed that, typical for buildings occupied for a short day, there was a slight advantage to both comfort and energy performance in the sunspace facing a south-easterly direction.

2.3 Building services

The specification of heating and ventilating services were influenced by the passive concept. Each teaching area has an air heating unit located on the mezzanine floor. Warm air is delivered to the teaching area via a rectangular duct within a circular crossection duct, suspended below the ceiling and also carrying low energy fluorescent lighting. Return air is extracted via the concentric duct, and also from the "home base" and the shared "wet areas" located between each pair of teaching areas.

Fresh air can be drawn into the air heater unit from the sunspace. Whether the system is on recirculation or fresh air, is controlled manually from a *green button* in the teaching area, which controls an electric damper in the air-handling unit. This equipment only operates during the heating season. For the remaining part of the year the building adopts natural ventilation.

3 MONITORING AND ENERGY PERFORMANCE

The monitoring /2/ was designed to test the overall performance of the building in terms of energy use, and also to assess the passive performance and comfort by monitoring two teaching areas at a more detailed level. The parameters monitored are shown in Table 1.

The first occupied heating season showed up problems in both building systems and monitoring. By the second season, monitoring problems had been sorted out, but a new crop of system faults showed up. For example, serious problems were encountered in temperature control in the teaching areas - later the cause was found to be the incorrect positioning of the thermostat sensor in the supply air flow rather than the return air. It was found also that the fresh air dampers still delivered about 30% fresh air even when nominally set to "recirculate".

3.1 Global energy consumption

By the third heating season, 1986-87, all of these problems were sorted out and the results of the overall energy performance are shown in Table 2.

The solar performance of the building was assessed by the Siviour /3/ plot shown in fig (4). This analysis assumes the simple building model shown in fig(5). Applied instantaneously, it is likely to show large errors due to heat fluxes into and out of thermal mass. If weekly values are used this introduces other errors but generally shows better correlation. The plot shows that the effective solar aperture of the building was $56m^2$ +/- $5m^2$. This compares with the theoretical maximum of about $90m^2$, calculated from the area of solar glazing and a solar gain factor indicating a relatively high solar utilization of 62%. The intercept gives the total building conductance Ua of 2050 +/- 200 W/K.

3.2 Performance of Sunspace

The monitoring was particularly aimed at evaluating the sunspace, both its energy performance and in terms of its thermal comfort. It was also important to the designers to assess the ventilation pre-heating function, in particular. Two approaches were used to evaluate the energy performance.

3.2.1 Calculating from global energy performance.

The use of the Siviour plot to establish a total heat loss coefficient and an effective solar aperture has already been explained above. If new values of Ua and A are estimated, corresponding to a similar building but without a sunspace, then the energy consumption can be calculated from equation (1) and compared with the measured energy consumption.

The following assumptions were made for the new values of Ua and A. Firstly it was assumed that the insulation value of the separating (south) wall would have been specified at a higher standard if there were no sunspace in the design. We have then assumed that the wall specification would have been of equivalent insulation value to that with the south wall and sunspace together - i.e. no change in Ua. The only change then, is to A. The design has now reverted to a direct gain building and the values for A are obtained from the glazing areas of the south walls of the teaching areas. This comes to $14m^2$, a considerable reduction from the effective aperture of $56m^2$.

The results of the comparison are shown below in table 3.

USEFUL HEAT	kWh
measured, with sunspace	58217
calculated, without sunspace	81245
calculated sunspace contribution	23028
Percentage sunspace contribution	28%

3.2.2 Calculation from monitored teaching areas

Two teaching areas and the adjacent sunspace were monitored for temperature and delivered useful heat to the classrooms.Considering fig (6) we can write

$$Q = U1.(Ti - Te) + U2.(Ti - Ts)$$

for an instantaneous heat balance. By multivariate regression values are found for U1 and U2, Values for the two areas monitored give -

	U1	U2	W/°C
area 1	64 +/- 17	280 +/- 27	
area 2	86 +/- 19	246 +/- 31	

The calculated fabric conductance between the classrooms and the sunspace is about 30 W/°C. It is reasonable to assume that the remainder is ventilation, corresponding to about $600m^3/h$ which is about 2/3 of the recommended value /4/.

Taking a median value of U2 of 260 W/°C, the sunspace contribution can be calculated by the product of this and the mean temperature elevation in the sunspace during occupation, 8°C -

$$Q = 8 \times 260 \times 1620 \text{ (hours) } \times 6 \text{(no.rooms)}$$

$$= 20217 \text{ kWh}$$

This is encouragingly close to the previous derivation of 23028 kWh.

It is also possible to estimate the fraction of this contribution which is due to ventilation pre-heating. The fabric conductance of the separating wall can be calculated giving a value of about 30W/°C, or 11% of the total. This implies that the fraction of energy saving which is due to ventilation pre-heat is 89%.

However, a weakness in this analysis is the implied assumption that the ventilation conductance would be the same if the sunspace were not there. In the real case, the absence of the sunspace would result in the occupants being much more careful to keep the doors closed. However, as shown earlier, the assumption made does not in imply a ventilation rate which is above recommended rates for school buildings.

Monitoring was continued for the heating season 87/88, a milder winter, and it is interesting to note that although the absolute size of the solar contribution was slightly reduced, the effective solar aperture, indicating the proportion of solar to auxiliary, remained the same. Significantly, the intercept of the Siviour plot gives a value for Ua,(taking internal gains into account) reduced from 2260 W/K (86/87) to 1930 W/K, a reduction of 15%. This can be explained by "better behaviour" of the occupants, i.e. less wasteful ventilation. (Probably this unnecessary ventilation the previous season was in response to poor temperature control).,

3.2.3 Temperature conditions in the sunspace

Air temperatures recorded in the sunspace for winter spring and summer, showed the following. In winter, for 94% of the occupied hours (between 8am and 6pm) the sunspace temperature was above 10°C, and 50% of the time above 15°C. Furthermore, these are air temperatures - at times when solar radiation is present, comfort temperatures in the direct sunlight would be about 5°C higher

In spring temperatures are now above 15°C for 90% of the time. However it is interesting to note that they are only above 25°C 12% of the time - this indicates how ineffective a solely circulating convective exchange system would be.

Summer conditions show that the automatic ventilation alone is insufficient to prevent overheating - i.e. temperatures above 30°C. By the following summer moveable shades had been installed which had a beneficial effect, typically reducing temperatures by about 4°C and providing shade and reducing mean radiant temperature. However, overheating is still present to some extent.

4 Human factors

A survey was carried out to assess the attitude of the teaching staff towards the building. There are only about eight fulltime staff, and their responses to most questions varied widely, and so a formal statistical treatment seems inappropriate. Generally responses were slightly on the unfavourable side of neutral,

Most felt that it was too hot in summer, but probably were referring to the sunspace. They also generally felt that it was "slightly stuffy". This is in conflict with what would be expected from the calculated ventilation rates which were quite high. Maybe it could be explained by the fact that people are used to ventilation as a control of overheating - when fresh air enters a room after pre-heating it is no longer perceived as "fresh", because it is warm. This could be an inherent problem with ventilation pre-heat design.

The function of the *green button* was not properly understood, in some cases windows in the north gables were opened in winter, and in some cases the green button was used to provide cooling in summer, drawing in hot air from the sunspace!

There was an even response to the sunspace as a usable space, 1/4 liked it, 1/2 were indifferent and 1/4 did not like it. Most people did not accept the principle of the variable habitability and felt that it should be heated in winter - a real problem for passive design! But we must remember that British schools are designed with modest space standards (about $4m^2$/person) and teachers always want more space.

The teachers felt that they had had poor information about the building - most came by chance meetings with the architect, energy consultant, or monitoring person. In view of the academic origin of this project, the author feels duly ashamed of this; somehow the intended building manual and explanation to the staff never occurred. Certainly the response shows how necessary information is.

In assessing these slightly disappointing responses, we have to remember that the staff really had suffered very poor conditions in the first and second winter. This was due to system control problems, not inherent in the passive design, but perhaps related to the slightly unconventional design of the heating equipment itself.

5 Conclusions

After four years of occupation it is now evident that the principle of ventilation pre-heating by solar gain is sound. The fuel bill for Netley is roughly half of that of comparable schools in Hampshire, and this was obtained at zero extra cost. The use of the sunspace for circulation and certain teaching activities has proved generally satisfactory.

However as the primary energy figures indicate, more attention should have been paid to reducing electrical loads, both by better fan and pump controls and by better daylighting.

Experience showed that system control design and commissioning was initially unsatisfactory. This was partly due to insufficient dialogue between energy consultant, services engineer, and architect.These early difficulties, which proved slow to correct, caused occupants to be generally dissatisfied with the conditions in the first two years. However, both conditions and performance have now shown a marked improvement.

6 REFERENCES

/1/ The Influence of Thermal Comfort and User Controls on the Design of a Passive Solar School, Nick Baker, Energy and Buildings 5 (1982)

/2/ 1st, 2nd and 3rd Monitoring Reports to Hampshire County Council, Architects Dept., Energy Monitoring Company, Unit 3, Paggs Court, Newport Pagnal, MK16 OEW.

/3/ Thermal Design and Performance of the Granada House. Siviour J et al, Electricity Council Research Centre, Capenhurst. ECRC/M1108.

Architect
Dennis Goodwin, Hampshire County Council

Energy Consultant
Nick Baker, The Martin Centre, Univ Cambridge

Monitoring
Chris Martin, Energy Monitoring Company
(with C Edwards)

For further descriptions of the building see -

"Walking on Sunshine - Netley Infants School"
Mark Bowman, Building Services, April 86

"Energetic Design - Netley Infants School"
Dean Hawkes, Architects Journal, 22 June 88

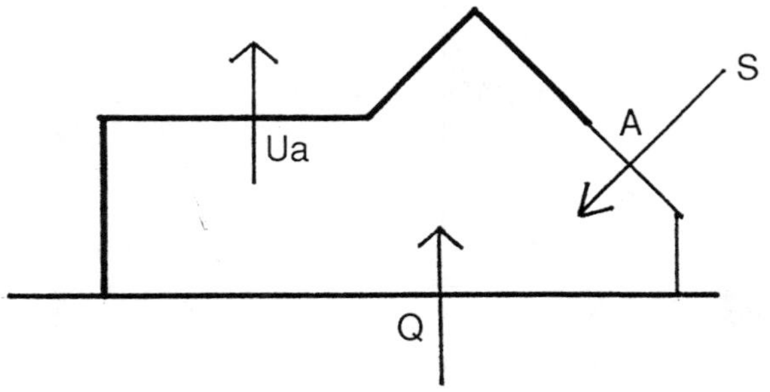

Fig 5 Simple building model on which Siviour plot is based

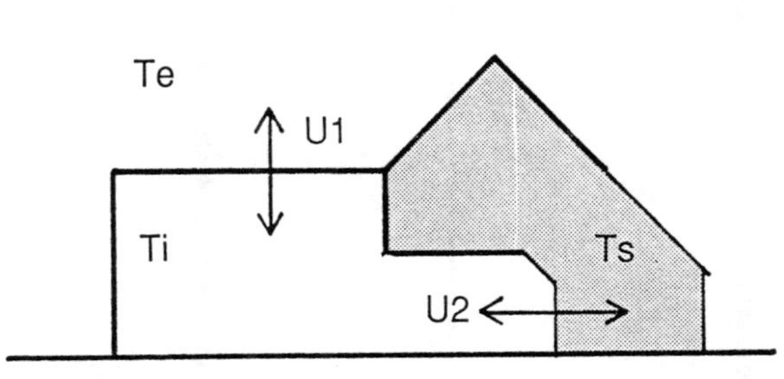

Fig 6 Thermal model for teaching areas

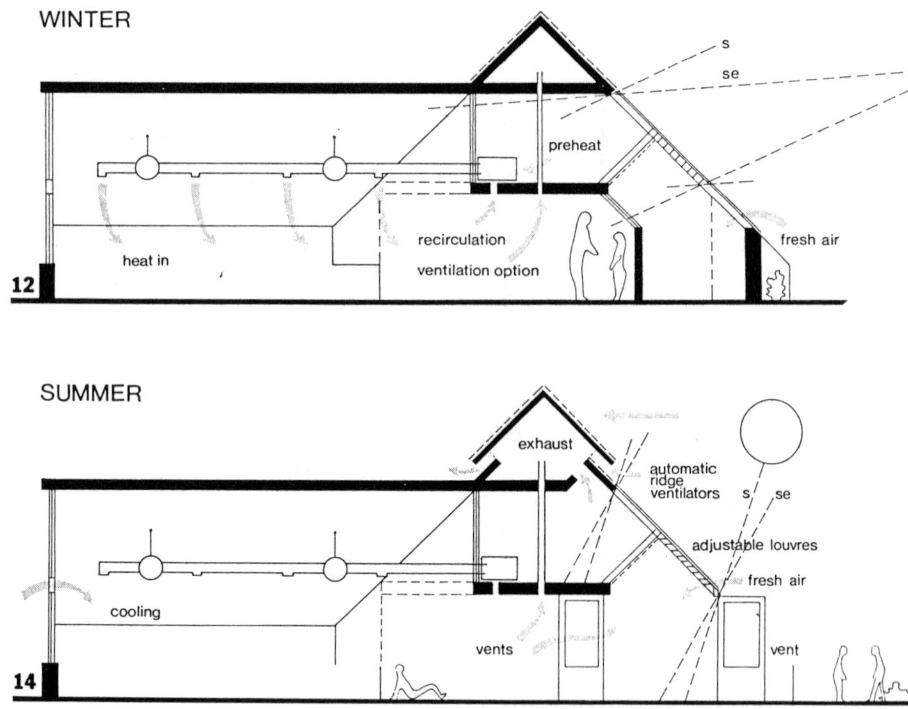

Fig 2 Section showing seasonal operation. Source - AJ

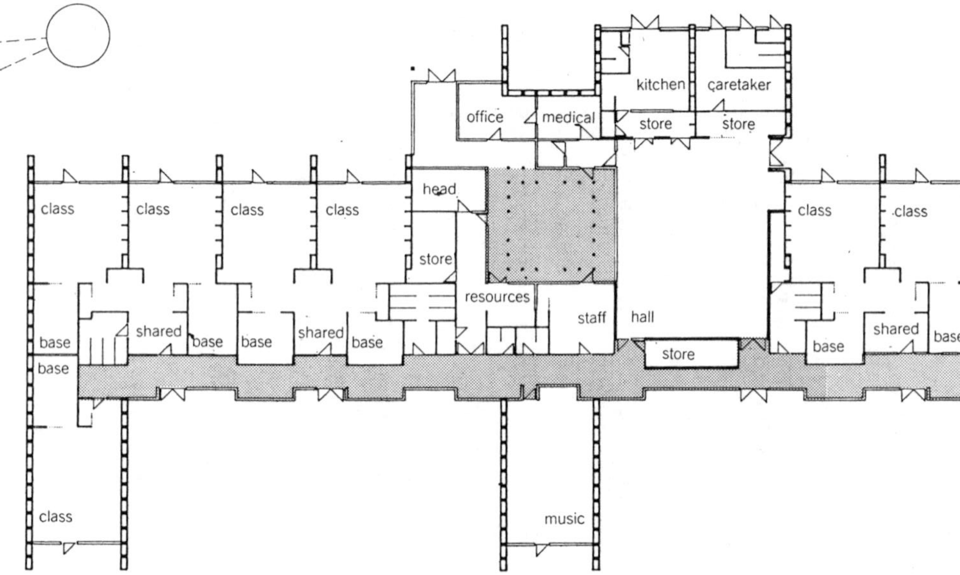

Fig 3 Plan, (sunspace and atrium in tone)

TABLE 2 ENERGY CONSUMPTION kWh/year

FUEL	FUNCTION	DELIVERED FUEL total	DELIVERED FUEL /m^2	PRIMARY total	/m^2
1986-87					
gas	space heatg.	72,771	70.3	77,137	74.5
	water heatg,	9,101	8.8	9,647	9.4
elec.	lighting, pumps,fans	18,785	18.1	69,504	67.2
TOTAL		--	--	156,288	151
1987-88					
gas	space heatg.	65,221	63.0	69,504	66.8

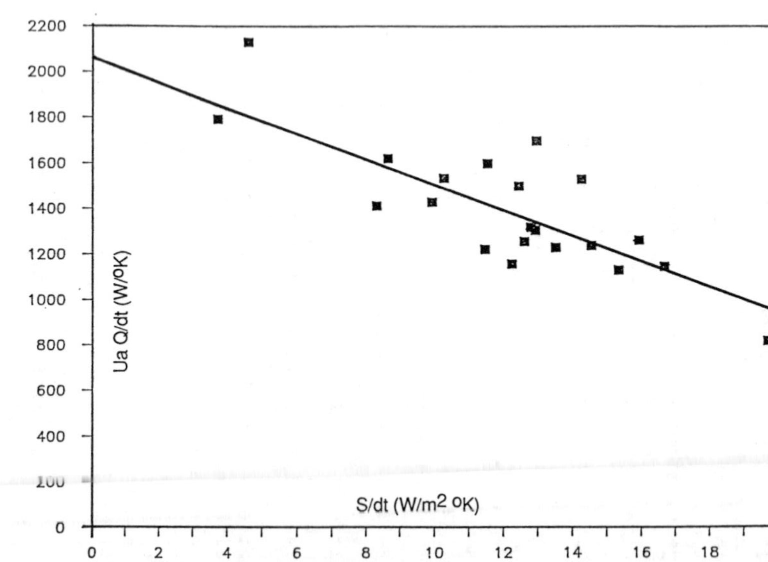

Fig 4 Siviour plot of Q/dT against S/dt
Intercept gives total heat loss Ua
Slope gives effective solar aperture A

Chaînage entre le logiciel de DAO/CAO ARCHITRION et la méthode d'évaluation énergétique HYPERMEGE, dans le cadre du programme PEBI.

L.ADOLPHE (GISE / ENTPE), P.BREJON (EMP)
ENPC / CERMA La Courtine 93167 Noisy le Grand CEDEX, FRANCE

CONTEXTE

La pénétration croissante de l'informatique dans tous les secteurs d'activité a vu naître une nouvelle génération de logiciels, supports privilégiés des formations initiale et permanente. Cette mutation a débuté par le succès de certains logiciels professionnels "détournés "de leur fonction première, qui se sont révélés des outils précieux de transmission du savoir-faire. Cette mise à disposition de logiciels utilisés dans les cabinets d'architectes ou les bureaux d'études, a l'avantage de faire manier des outils professionnels, mais n'offre pas de réelle stratégie pédagogique. Apparraissent ensuite des produits à mi-chemin entre des outils scientifiques et des programmes didactiques: STRATEGY à l'université de Liége, des outils du groupe ABC, SIMULA du CERMA de Nantes, CASAMO-CLIM de l'Ecole des Mines de Paris. Simultanément naissent les didacticiels, logiciels spécifiquement conçus à des fins pédagogiques.

OBJECTIFS

Ce projet vise à évaluer, à partir d'outils professionnels, le coût global d'un projet de bâtiment saisi par l'utilisateur grâce à un logiciel de CAO / DAO, en tenant compte de la composante énergétique, du calcul de structures, des coûts d'investissement et d'exploitation. Le module thermique, fondé sur la méthode HYPERMEGE, permet de calculer les besoins et consommations énergétiques d'un projet pour le chauffage et la ventilation. Autour de ce "package" informatique de modules chaînés entre eux, sont bâtis des didacticiels propres à chacun des champs: conception architecturale, énergétique...

Les objectifs fondamentaux sont donc:

•Analyse de l'offre en logiciels professionnels de conception des bâtiments,

•Amélioration de l'offre en logiciels professionnels de conception des bâtiments, par un chaînage entre un logiciel de CAO / DAO et des modules d'évaluation technique (consommations énergétiques, calcul de structure, coûts globaux, spécifications),

•Transmission du savoir-faire par une mise en situation avec des outils de simulation, accompagnés de tutoriels et d'exercices et familiarisation à ces outils professionnels (Architrion, HyperMege, Accord-Poutre, As-plus...),

•Approche globale de la conception architecturale dans les phases amont du projet (Avant-Projet Sommaire),

•Formation internationale (réflexion sur les pratiques réglementaires propres à chaque pays..)

•Intégration de l'approche énergétique (entre autres) et de ses conséquences économiques dans l'ensemble des préoccupations de l'architecte.

PEBI dépasse donc largement le cadre de l'enseignement, puisque la première phase du travail **(1988)** a consisté à:
-réfléchir sur les pratiques des acteurs de la construction, et sur les besoins d'une approche globale au cours des phases amont d'un projet de bâtiment,
-constater l'absence de configurations logicielles performantes, économiques et largement diffusé de CAO bâtiment associant à un module de représentation graphique du bâtiment, des modules d'évaluation technico-économique (coûts d'investissements, calcul de structure, analyse thermique, coûts d'exploitation...),

ABSTRACT:
This project aims at building performance and cost evaluation by linking a CAD program Architrion and professionnal tools dealing with energy structure, environment and costs. Around this software "package", specific CAL modules are built and offer a unique environment for international building design courses and trainings.The energetic module is based on HYPERMEGE, a powerful correlation to evaluate thermal needs and consumptions of a project for heating and ventilation, and CALMEGE, a program to introduce fundamental concepts of energy conscious design in Architecture.

-choisir un logiciel de CAO / DAO, ARCHITRION, développé et diffusé par la société française GIMEOR, diffusé par ABECON en Hollande, répondant aux critéres précédents et de logiciels d'application performants susceptibles de lui être chaîné,
-élaborer une maquette de chaînage entre ces différents modules. **(voir figure 1)**

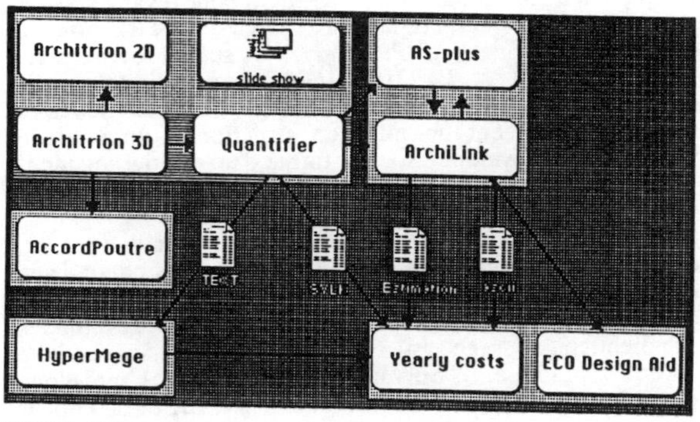

FIGURE 1 : SCHÉMA SYNOPTIQUE DES CHAINAGES ENTRE LOGICIELS.

Ce travail de réflexion a été un préalable au développement du module d'enseignement, assisté par ordinateur qui a débuté en **1989** par :
-le passage des maquettes de chaînage à des chaînages complets, visant à rendre plus explicites les liens entre les différents modules et les échanges de données qu'ils soustendent (interactivité "MacIntosh", valeurs par défaut et tests de cohérence sur les données entrées), dans l'environnement Hypercard,

- le développement de la configuration "MacIntosh" complète, par le développement d'une version d'ACCORD-POUTRE sous ce standard,

-l'amélioration des outils d'évaluation utilisés (HYPERMEGE intègre, par exemple, le calcul automatique de la constante de temps principale; La version 5.0 d'Architrion qui "sort" en ce moment, intègre pratiquement toutes les améliorations sollicitées par le groupe),

-l'élaboration des principes du cours européen d'initiation à la CAO bâtiment:
•structuration des connaissances "bâtiment",
•échange de données entre acteurs d'un projet architectural,
•nécessité d'une approche multi-critère lors de la conception d'un bâtiment,
•nécéssité d'un décloisonnement des actuels modules d'enseignement spécialisé, tant au niveau des écoles d'architecture que des formations d'ingénieurs.

L'année **1990** est centrée sur le développement du cours, en intégrant cet ensemble logiciel dans un module d'enseignement assisté par ordinateur qui sera expérimenté dans le courant de cette année.

Ce didacticiel vise une large cible, puisque les modules d'enseignement peuvent aller de l'initiation pour généralistes à l'approfondissement pour spécialistes:
Etudiants Architectes Elèves Ingénieurs PME du secteurs bâtiment,Cabinets d'architectes,Sociétés d'Ingénierie,Maîtres d'ouvrage ou Investisseurs.

Ce travail a été réalisé grâce au soutien des Communautés Européennes, dans le cadre du programme COMETT, par le Ministère de l'Equipement et du Logement, dans le cadre du programme IN.PRO.BAT. et par son homologue Hollandais, le VROM.

DESCRIPTION DU PROJET

L'interface pédagogique de cette configuration logicielle professionnelle vise à enseigner l'approche du coût global d'un projet de bâtiment lors de sa conception architecturale, en intégrant les diverses composantes que sont la conception et le dessin professionnel, les spécifications et les coûts d'investissement, le calcul des structures, les consommations énergétiques et les coûts d'exploitation. Des variantes de ce projet issu d'une bibliothèque de bâtiments prédéfinis, sont proposées et étudiées au travers des différents outils d'évaluation professionnels disponibles: ARCHITRION pour le dessin et la conception architecturale, HYPERMEGE pour l'évaluation des consommations énergétiques, ACCORD-POUTRE pour le calcul des structures, AS-PLUS pour les spécifications et les coûts globaux, ECODESIGN pour l'évaluation environnementale. Ces études de sensibilité des paramètres de conception servent de mise en situation à un contenu pédagogique, développé sous la forme d'un tutoriel interactif (HyperCard).

Précisons qu'il ne s'agit pas d'un ensemble monolytique (au contenu, à la durée et au public cible fixés), mais plutôt d'une collection de modules dédiés, basée sur une même structuration des données, et proposant une homogénéïté des modes d'apprentissage (usage systèmatique d'HyperCard, interactivité théorie / pratique...). Suivant le public cible, il sera donc possible d'associer certains des modules PEBI et de choisir pour chacun d'entre eux la version "initiation pour généralistes" ou la version "approfondissement pour spécialistes".

Nous développerons dans la suite, d'abord la composante énergétique, son chaînage avec le logiciel de CAO (ARCHITRION) et le contenu du module pédagogique dédié, puis succintement les autres chaînages réalisés dans le cadre du programme PEBI.

Le module Energie

1) Présentation de la méthode HYPERMEGE

La méthode HYPERMEGE développée au Centre d'Energétique de l'Ecole des Mines de Paris est une méthode simplifiée d'évaluation des besoins mensuels pour le chauffage et la ventilation, tenant compte de l'économie apportée par un ralenti de nuit et intégrant l'effet de l'inertie . Elle est issue de corrélations obtenues à partir de modèles détaillés issues de recherches récentes sur l'analyse modale des échanges thermiques dans un bâtiment .
A partir d'une évaluation des déperditions statiques (GV), intégrant une détermination automatique des ponts thermiques, par une description typologique appropriée des parois (configuration, type, place de l'isolant, caractéristiques des parois...), HYPERMEGE reconstitue l'énergie solaire horaire incidente sur chaque paroi extérieure, pour trois journées typiques de la saison de chauffe (demi-saison automne, hiver et demi-saison pritemps) . Le calcul des besoins énergétiques sans intermittence est déduit de cette reconstitution et intègre l'effet des surchauffes via une constante de temps principale du bâtiment et un facteur correspondant au pourcentage des gains non utilisables .
Le calcul des besoins énergétiques avec une intermittence du chauffage la nuit est ensuite calculé à partir de l'économie théorique apportée par cette intermittence. L'étape suivante consiste à évaluer les consommations mensuelles pour le chauffage et la ventilation, pour des configurations classiques d'équipements.

L'intérêt de cette méthode est la simplicité des données descriptives nécéssaires, la rapidité du calcul, l'adaptabilité à tout contexte national et la précision de l'évaluation, compatible avec un outil d'évaluation des coûts d'exploitation en phase d'avant-projet.

2) Chaînage entre Architrion et HYPERMEGE

La grande facilité d'exporter (et d'importer) les données géométriques et topologiques d'un projet de bâtiment saisi sur la dernière version d'Architrion (4.0) a permis de développer un chaînage "transparent", avec le logiciel d'évaluation énergétique HYPERMEGE: pas de boîte noire, mais des écrans interactifs sous HyperCard, permettant à la fois de consulter les données provenant d'Architrion, d'ajouter les données énergétiques manquantes sur l'environnement et le site, les zones thermiques du projet, les composants de chacune de ces zones, et de lancer l'évaluation énergétique
(voir figure 2)

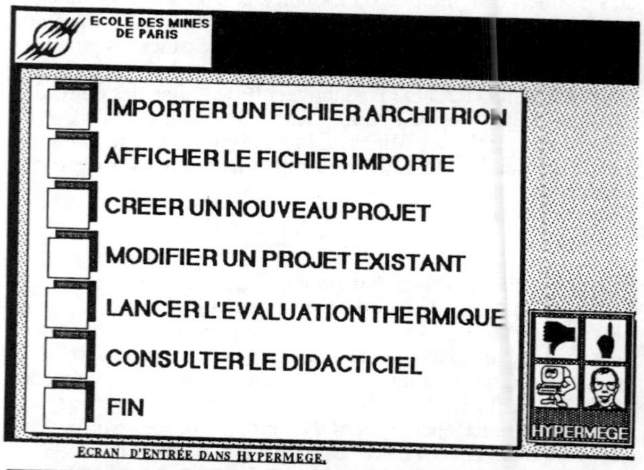

ECRAN D'ENTRÉE DANS HYPERMEGE.

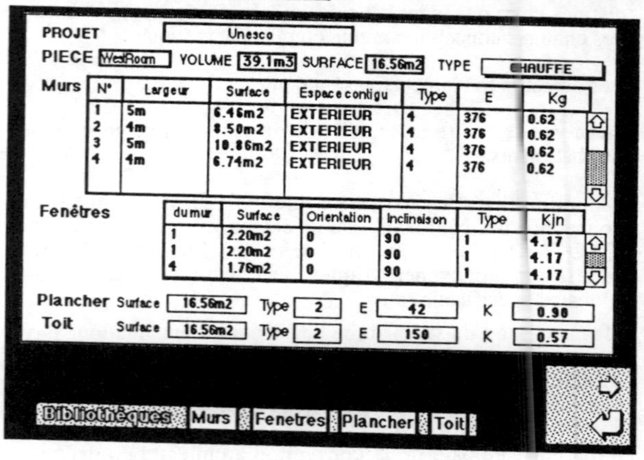

ECRAN CORRESPONDANT À UNE ZONE THERMIQUE D'UN PROJET HYPERMEGE.

FIGURE 2

3) Module didactique CALMEGE

Ce module comporte les parties suivantes:

-une présentation générale (en tronc commun d'une durée de 3 heures) qui met en valeur:
- les exigences énergétiques du bâtiment sous les deux aspects économie et confort,
-la liaison entre la formulation des exigences et la phase de conception du projet architectural,
-l'analyse des données nécessaires pour mettre en œuvre les différents types d'évaluation (données générales du bâtiment-données spécifiques au thermicien).

-une évaluation thermique d'un projet de bâtiment et de ses variantes, saisis sous ARCHITRION, en phase d'avant-projet sommaire, qui peut prendre la forme d'une option simplifiée (en tronc commun de 3 heures), ou d'une option détaillée (en spécialisation de 2 jours supplémentaires), et développe les aspects suivants:
•Déperditions thermiques et dimensionnement,
•Besoins de chauffage et qualification solaire "passive",
•Consommations énergétiques et coûts d'exploitation annuels,
•Approche de la conformité aux réglementations thermiques nationales.

Le module Calcul de Structure - ACCORD POUTRE

Le logiciel ACCORD Poutre est un logiciel de calcul des structures poutres tridimensionnelles .Il permet de déterminer en statique les sollicitations d'une structure et en dynamique les modes propres ou flambement . La structure 3D du projet saisi sous Architrion est directement exportée dans Accord Poutre.

Le module Calcul des Coûts d'investissements - AS PLUS

Le logiciel AS-PLUS permet l'estimation des coûts d'investissements d'un projet à partir de la définition de spécifications techniques et économiques propres à chaque élément, de règles d'estimation propres à chaque élément (de 1 à 10) et de variables économiques générales.

Un module permet la reconnaissance automatique des codes de spécifications (code SFB...) issu du métré d'Architrion.
Le module pédagogique "calcul des coûts d'investissement" met en situation ces notions de spécifications et d'estimation des coûts en les replaçant dans leur contexte économique: tâches de conception et acteurs, matériaux, bâtiment ou équipements.

Le module Calcul des Coûts d'exploitations

Ce logiciel permet l'estimation des coûts d'exploitation d'un projet à partir de la séparation des coûts unitaires en :
-coûts des fonctions nécessaires à l'entretien et à la maintenance du projet et périodicité,
-coûts des éléments du bâtiment (enveloppe et équipements) et évolutions,
-coûts des matériaux et évolutions.
Le module pédagogique "calcul des coûts d'exploitation" permet des exercices sur des exemples de bâtiments résidentiel et tertiaire et compare les coûts obtenus à des coûts réels.

Le module Environnement - ECODESIGN

Ce module pédagogique permet une approche environnementale du projet, mis en situation dans son milieu urbanistique, en cours de conception. A partir d'un modèle de bâtiment issu d'Architrion , il reconstitue toute une parcelle urbaine et y évalue les coûts energétiques (héliodons...) et les conditions de confort thermique dans les bâtiments et l'influence de l'environnement (acoustique, visuel, écologique..).

ETAT D'AVANCEMENT fin 1989

Ce produit sera testé au cours de l'année scolaire 1989-1990 et disponible en milieu d'année 1990.

PROTOTYPAGE

Ce didacticiel est avant tout conçu pour un travail collectif avec un animateur dirigeant l'éxecution d'un ou de plusieurs modules pour un petit groupe de personnes (maximum 10), placées en binôme devant un ordinateur. L'utilisation du manuel d'accompagnement n'est donc pas systématique.

Le matériel informatique nécéssaire est un Mac Intosh SE ou II, équipé d'un disque dur, de la version 4.0 d'ARCHITRION (et de son bouchon!) et d'Hypercard.

ORGANISATION

PEBI est un projet de formation internationale retenu par la CEE dans le cadre du programme COMETT.
PEBI regroupe 4 partenaires universitaires:
- 2 universités hollandaises (T.U.D. et T.U.E.),
- 2 écoles d'ingénieur françaises (ENSTIMD et EMP)

et 3 partenaires industriels associés :
- GIMEOR, auteur et éditeur du logiciel de CAO / DAO Architrion, pour la France,
- ABECON, éditeur pour la Hollande de ce même logiciel et auteur de logiciels d'évaluation économique des projets de bâtiment,
-MACAD, organisateur de cessions de formations à la CAO architecturale.

L'équipe de conception-réalisation est constituée d'architectes, d'experts en thermique, de pédagogues et de sociétés de services informatiques:

Responsables Scientifiques	Equipe (Pays)	Tâche	Logiciel
H.HUBERS-P.DE JONG	VCA /VIBI (NL)	Coordination/Coûts Invest.	
L.ADOLPHE-P.BREJON	GISE-Ecole des Mines de Paris(F)	Energétique	HyperMege
M.AMARA-F.BERLY	Ecole des Mines de Douai (F)	Structure	Accord-Poutre
R.DARU-S.VAN KOL	T.U. Eindhoven (NL)	Conception/Base de Données	Datavisie Bouw
K.DUIJVESTEIN-L.GOMMANS	T.U. Delft (NL)	Environnement	Hypercoen
A.PAVAGEAU	GIMEOR (F)	Conception /DAO	Architrion
G.V.HUISSEDE-K.BAKKER	ABECON (NL)	Coûts d'exploitation	As Plus
A.DEN OTTER-J.BOLDER	MACD (NL)	Dessin et conception	Architrion

CONCLUSION

Outre la grande adaptibilité qui découlera de cette structure modulaire du cours, PEBI est perçu par ses auteurs comme un outil innovant d'E.A.O., renforçant les enseignements traditionnels en proposant une approche réellement multi-disciplinaire du projet de bâtiment et une familiarisation vivante aux outils informatiques professionnels.

BIBLIOGRAPHIE

ARMINES "Participation au programme PEBI", Rapport final MEL-IN.PRO.BAT, Convention n° 8854/05, Février 89

ARMINES/CNEA/DIALOGIC, "Didacticiels en Maîtrise de l'Energie : Etat de l'Art et Analyse des Besoins", Rapport AFME , Conventions AFME n° 5.12.1079 et 6.12.0006, Avril 87

SGAO- Projet de recherche sur le couplage de KEOPS avec des outils d'évaluation thermique et acoustique- Rapport final - Convention N 5.04.11121 -MEL / IN. PRO.BAT - Novembre 1987

G. LEFEBVRE - Analyse et réduction modale d'un modèle de comportement thermique du bâtiment- Mémoire de thèse - PARIS VI- Novembre 1987

THE INFLUENCE OF THE THERMAL MASS ON THE THERMAL PERFORMANCE OF BUILDINGS IN SUMMER AND WINTER.

E. Shaviv
Faculty of Architecture & Town Planning,
Technion-Israel Institute of Technology, Haifa, 32000, Israel.

ABSTRACT. The influence of the thermal mass of buildings on the passive heating and cooling phenomenon is examined. The work discusses the questions of whether there is an optimal total mass and depth for the internal and external thermal mass, and if it is better to have the thermal mass, as a part of the external walls or as an internal partition. The above questions are examined for residential buildings, which operate day and night, and for commercial buildings which operate during day only.

1. INTRODUCTION

This work discusses the influence of the thermal mass on the energy performance of the building during Winter and Summer in the Mediterranean climate. In this climatic zone, traditional architecture dictates heavy structure buildings. We would like to know if the more thermal mass the building has, the better is its thermal performance during Winter and Summer /1/, or is there an optimum amount of thermal mass above which the excessive thermal mass degrades the building performance. The last question might be more important in buildings that operate during day only, like most of the office buildings /2/.

The analysis of the effect of the building's thermal mass on the energy balance of the building (shown in Figure 1) is carried out using a simulation model developed by Shaviv and Shaviv /3,4,5/. This time dependent model solves the heat transfer equation through all exterior walls simultaneously, taking into account the thermal mass in each external wall and in the internal partitions as well. The internal thermal mass is treated in Shaviv and Shaviv as isothermal mass. However, for the particular questions that are raised here, the heat transfer equation is solved for the internal mass as well as for the external walls. This allows a check of the performance of thick heavy internal partitions that are not necessarily isothermal.

The effect of the thermal mass on the behavior of the building is checked during Winter and Summer, and in both cases, the energy consumption of the building is examined (see Figures 2,3). In addition, the model calculates the maximum inhouse Summer temperature, without any mechanical means, or with an operating fan only (see Figures 4,5). The hypothesis is that in many hot places it is possible to achieve good thermal conditions in Summer without operating any air-conditioning unit, provided that the building has a sufficient amount of thermal mass and is ventilated well during the night.

This balance is achieved by the fact that in Summer, heavy external walls delay the transfer of the heat from the outside into the inside spaces. On the other hand, if the building posses a lot of internal mass there will be a slow increase in the air temperature. The implication of these properties is that the building will heat slowly in Summer and will reach its maximal temperature only during the late hours when the outside air temperature is low. With good ventilation in the evening and night, the heat that flows from the heavy walls inside can be removed. We now pose the question, is there a difference in the performance of the building when the thermal mass is located in the external walls, or the internal partitions, during Summer? Clearly, this question has many architectural implication.

The capability to store energy also helps in Winter, since energy can be stored in walls from one sunny Winter day to the next cloudy one. Also, for a building that operates during the day and is heated only during the working hours, the thermal mass can keep the minimum temperature from dropping too low, sometimes even below the dew point. Thus it can prevent the condensation problem. On the other hand, it might store too much energy for the night hours when the building is not in use, which means that the thermal mass can cause the building to consume more energy than light-structure one.

2. THE EXPERIMENT

To check all the above questions a very simple building was taken as a case study. The building examined has the following elements (see Figure 1): The eastern and western external walls (wall number 1 and 2) have external insulation. Namely, the thermal mass added to these walls in different case studies is insulated from the external surrounding. The U value of these walls is 0.682 W / m^2 ^{o}C, which is considered to be very good by Israeli standards. The southern wall (wall number 3) is insulated in the same way as walls numbers 1 and 2 but no thermal mass is added to it. It's U value is almost the same (0.694). Wall number 4 is the non-isothermal partition in which we change the concrete layer while the plaster in both its sides remain the same in all the case studies. We assume not to have any isothermal thermal mass, i.e. we assume that the floor and the ceiling and wall number 5, which face the corridor, have no heat capacity and the temperatures in the adjacent spaces are the same as in the examined space.

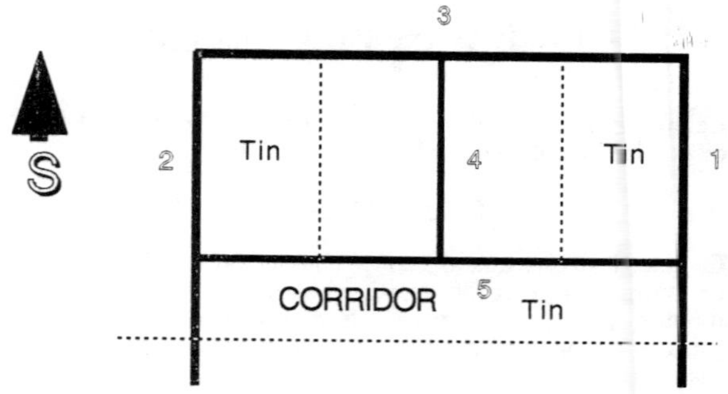

Figure 1. The building scheme.

The following set of simulations were carried:

1. The building was assumed to be in use only from 8 am till 4 pm like most office buildings. These simulations are

denoted by the word DAY (means, day only). These results are summarized in Figures 2 and 3. The air-conditioner and heater can operate only during this period. Also the internal heat gain i.e. light and people, is produced during this time only. As a result, after the heater is turned off in Winter, the temperature inside the building can drop below the thermostat setting point. The amount of the thermal mass and its surface area will determined how low the temperature may get.

2. The building operates day and night, like most residential houses. These simulations are denoted by the letters D&N (means, day and night). The results of these simulations are summarized in Figures 4 and 5. In this case the air-conditioner and heater operate during day and night and the internal heat is produced during this time as well.

For each set of simulation we assumed first that there is no thermal mass in the external walls 1 and 2. The internal partition (wall number 4) consists of a concrete wall with 1.2 cm plaster on both sides. The depth of this concrete layer in the wall was chosen to be 0, 10, 20, 30 and 40 cm. The minimum depth to be examined was 10 cm as it is not easy to build a concrete partition thinner than this. The maximum depth to be examined was 40 cm according to the suggestion of Balcomb /6/, based on the diurnal heat capacity for storing the passive solar energy from day to night. According to Balcomb's recommendations, the depth for half the thickness of a concrete thermal mass should be about 18 cm. Hence the 40 cm is already above this upper limit. The results of this study are denoted by 'Int. mass-Depth'.

Secondly, we assumed that the external walls number 1 and 2 has the thermal mass, and the internal partition (wall number 4) is built from very light material with only 1.2 cm plaster on both sides. Because it is important to insulate the external envelope, the external walls should be made of several layers with different thermophysical properties and the heavy layers should be the innermost. The depth of the concrete layer in this wall was chosen to be 0, 5, 10, 15 and 20 cm. Two such walls exist (walls number 1 and 2), therefore it is the same as having 0, 10, 20, 30 and 40 cm in the internal mass (wall number 4). The results of this study are denoted by 'Ext. mass-Depth'.

The third series of simulations were carried out in order to determine if an increase of the internal mass, while keeping the volume to surface ratio fixed, improves the performance above those in the first set of simulations. In this series of simulations, we increase the number of internal partitions by adding more partitions (each having a concrete layer of 10 cm). We checked for both cases five different alternatives: 0, 1, 2, 3 and 4 partitions (wall number 4 as described above). This is the same as having 0, 10, 20, 30 and 40 cm in the internal mass (wall number 4). The results of this study are denoted by 'Int. mass-Area'.

3. SUMMARY AND CONCLUSIONS

Summarizing the results of all simulations leads us to the following conclusions:

A. In a building that operates only during the day, when the thickness of the internal thermal mass changes from 10 to 40 cm, the energy consumption of the building in Winter and Summer remains practically unchanged. Therefore, the recommended depth for the concrete layer of an internal partition (that exchanges energy with the internal air through its two surface areas) should be 10 cm. Increasing the depth of the internal mass not only does not improve the performance of the building, but it even reduces it very slightly. (See Figures 2,3.).

B. The effect of the internal thermal mass in reducing the energy consumption of the building is largest when the building is in use day and night. The energy consumption and the extreme temperatures in Summer do not chance significantly over the range of 10 to 40 cm depth of concrete layer. Therefore we recommend 10 cm only. (See Figures 4,5.).

C. When the thermal mass is located on an external wall that has only one surface area facing the interior, the recommended depth is only half of the above, i.e. 5 cm. For this case however, the increase in the depth of the external thermal mass up to 15 cm improve the thermal performance of the building that operates day and night (see Figures 4,5). Therefore, for a residential building, the recommend depth for the external thermal mass is 15 cm.

D. In all cases the external thermal mass is preferable to the internal one. However, this improvement is not significant. Therefore, the decision as to where to put the thermal mass should be made according to the building practice, simplicity in building constructions as well as the need to keep the flexibility of the interior spaces (should that be desired).

E. There is no constraint on the number of 10 cm depth of internal thermal mass partitions. In other words, adding more internal mass while keeping the surface to volume ratio fixed, will always improve the performance of both types of buildings during Winter and Summer. We conclude that the amount of thin internal thermal mass (less than 10 cm in depth), is not limited and can be determined according to the architectural requirements only.

F. We would like to emphasis that in both types of buildings the additional number of internal partitions improves dramatically the energy consumption of the building during Summer. In the alternative in which 4 such internal partitions are used, there is really no need to use air-conditioning at all, specially if operating a fan during the evening and night. (See Figure 3,5). The thin thermal mass with the high volume to surface ratio will cool very fast during night and keep the temperature of the building low during the next day.

G. Clearly, one can see from all the graphs, that it is not enough to add just a thermal mass to the building. The additional thermal mass should have adequate surface area that is able to exchange energy with the internal air.

It is important to test the above results for different building materials and climatic zones. The results presented here are only for concrete thermal mass and for the Mediterranean climate. It will be interesting to see if these recommendations change with climate and with different building materials.

4. ACKNOWLEDGMENT

I would like to gratefully acknowledge E & J. Pick Alternative Energy Sources Research Fund, who partially supported this research.

5. REFERENCES

1. Kammerud, R. , E. Ceballos, B. Curtis, W. Place, and B. Anderson, 1984. 'Ventilation Cooling of Residential Buildings', ASHRAE Transactions, Vol. 90 Part 1B, Atlanta.
2. Shitzer, A. , and M. Levy, 1984. 'Dynamic Simulation of the Heating Load of Offices Coupled with Measured Occupancy Distribution', ASHRAE Transactions, Vol. 90 Part 1A, Atlanta.
3. Shaviv, E. , and G. Shaviv, 1977. 'A Model for Predicting the Thermal Performance of Buildings', WP ASDM-8, Faculty of Architecture, Technion.
4. Shaviv, E. , and G. Shaviv, 1978a. 'Designing Buildings for Minimal Energy Consumption', Computer Aided Design J., 10 (4) (July) pp. 239 - 247.
5. Shaviv, E. , and G. Shaviv, 1978b. 'Modelling the Thermal Performance of Buildings', Building and Environment, Vol.13, Pergamon, Oxford, pp.95 - 108.
6. Balcomb, J. D. , 1983. 'Estimating Temperature Swings in Direct Gain Buildings'. In J. D. Balcomb and W. Johns (editor), Passive Solar Design Handbook, Supplement to Vol. III, ASES. Boulder.

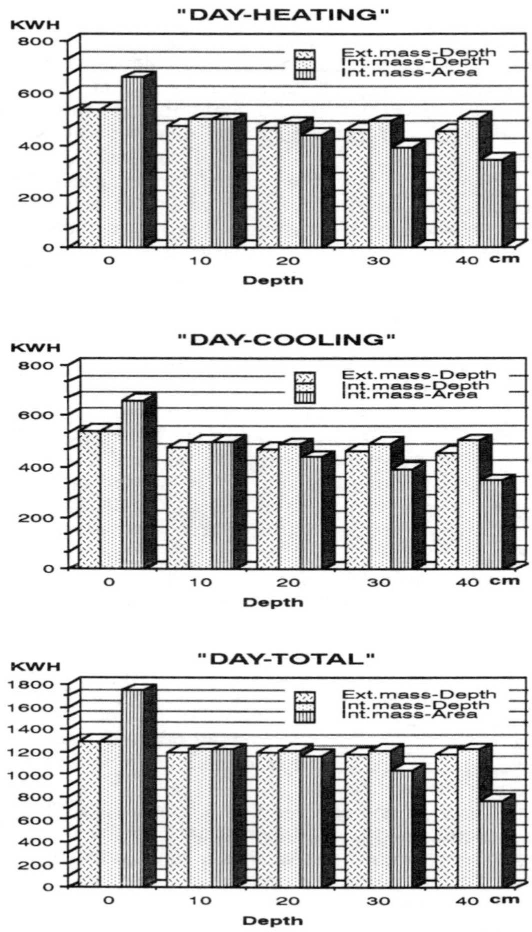

Figure 2: The effect of the thermal mass on the energy consumption of the building that operates only during the day.

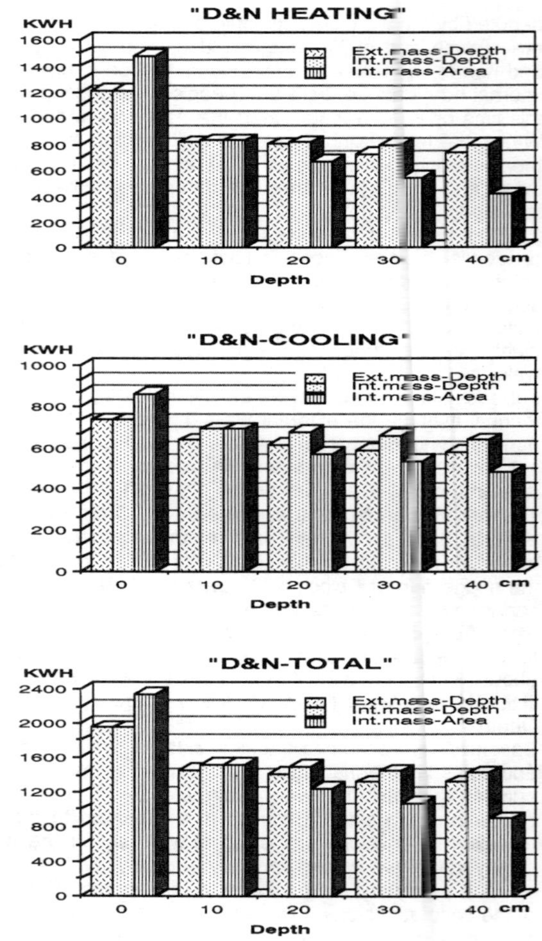

Figure 4: The effect of the thermal mass on the energy consumption of the building that operates during day and night.

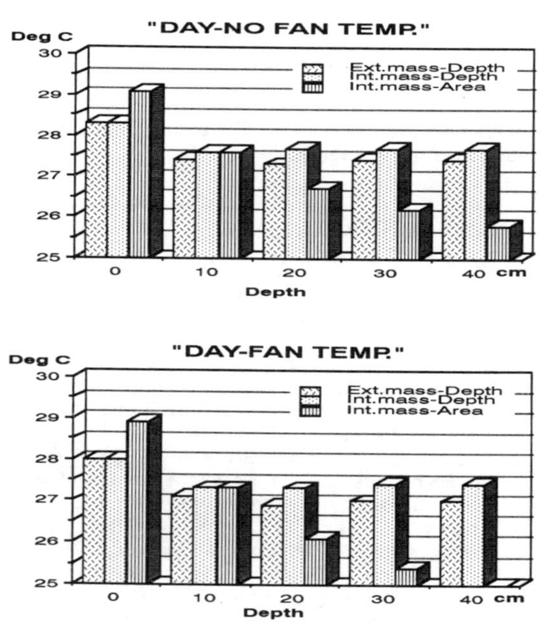

Figure 3: The effect of the thermal mass on the Summer temperatures inside the building that operates only during the day (with and without operating a fan).

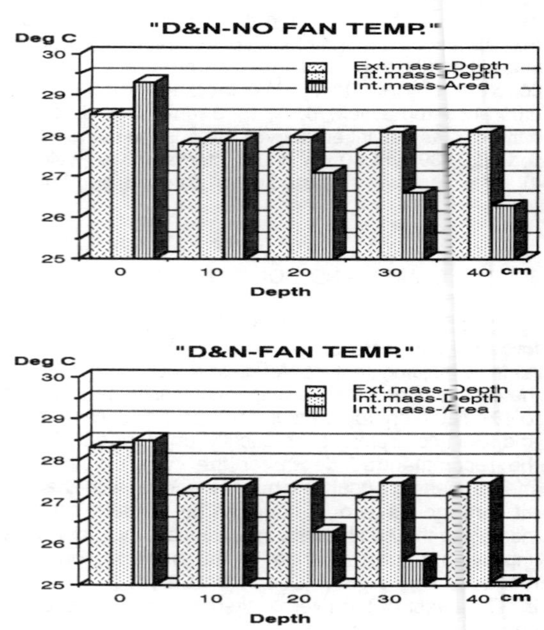

Figure 5: The effect of the thermal mass on the Summer temperatures inside the building that operates during day and night (with and without operating a fan).

THE USE OF SIMULATION MODELS IN THE SCHEMATIC DESIGN STAGE FOR THE DESIGN OF ENERGY CONSCIOUS BUILDINGS.

E. Shaviv
Faculty of Architecture & Town Planning,
Technion-Israel Institute of Technology, Haifa, 32000, Israel.

ABSTRACT. We demonstrate the use of accurate simulation models as design tools during the schematic design phase. Current practice is to apply only approximate methods during this stage and accurate simulation models are reserved for the detailed design one. We claim that approximate methods are anyhow too complicated to be solved manually and might lead to wrong conclusions, while present day computers are strong enough to handle accurate simulation calculations efficiently and at low cost. Hence, the question of what kind of design tools are required for the schematic design phase should be reconsidered again with a new perspective on computing power availability and the knowledge accumulated. The geometrical design parameters are the most important parameters that should be determined during this design phase. Therefore, the design of energy conscious buildings will be presented through the analysis and determination of the geometrical design parameters in the schematic-conceptual design phase

1. INTRODUCTION

Different design tools for passive and low energy buildings were developed mainly since 1976. /1/. Most of these tools were addressed to the different design stages. The comprehensive simulations were aimed for use on mainframes only and specially by researchers in the various research institutes. As at that time personal computers were still a dream, no wonder that simplified methods were developed. Among the large simulations on mainframes developed in those days, we mention here the PASOLE simulation for passive solar design, which was developed and used in Los Alamos National Laboratory (by McFarland). Another large scale simulation is the TRNSYS, developed in the University of Wisconsin (by Beckman, Klein and Duffie), for simulating the dynamic thermal behavior of active solar systems /2/. The cumbersome application of such heavy tools that should be run on mainframes leads to the development of simplified methods, based on correlations of these detailed computer simulations. The "f-charts" was developed from the TRNSYS computer code /3/, and the SLR and LCR methods were developed from the PASOLE simulation design tool /4/.
In the "Design Manual-Passive Solar Heating Analysis" /4/, Balcomb presents the different design sequences. The major stages are: the schematic design, the design development and the construction documentation. For each stage he proposed a manual design tool that is proper for this particular stage. For the schematic stage he developed the LCR method, while for the design development stage the SLR method was suggested. The simulation model PASOLE was used only by the researchers to develop these methods and different design guidelines. As claimed by McFarland, 'PASOLE is not a standard "user oriented" program' and cannot be used 'for other than the specific models for which it is programmed'. He goes on and states that 'the user must have some knowledge of heat transfer principles and FORTRAN programing' in order to use this heavy tool /4/.
Looking back to the years 1976 to 1984, the rational behind the approach suggested by the researchers (that were not architects), is well understood. They could not predict that a time will come and architects will use the computer like using the 6B pencil. However, few architects believed in the use of computers as design tools, even for the early design stages. This were the dreamer architects that enter the field of computer-aided architectural design at the early 70's. In CAD conferences /5/ such ideas were presented by Shaviv, Clark, Greenberg, and few more. In 1977, Shaviv and Shaviv suggested a simulation model, which could be run on a personal computer and could be used by architects /6,7,8/. This particular model was based on interactive mode for introducing the data, and was essentially a questionnaire expressed in daily architectural terminology driven input. /8/. This should be contrasted with the non friendly thermal network physical presentation, like the PASOLE code. This model could be easily used in the schematic design stage. However, as personal computers were rare at that time, this approach seemed more like a science fiction.
The use of accurate design tools on computers seems to day to more people as a realistic solution. The above philosophy was advocated by Klein in the address he carried when he won the Abbot award for 'significant contribution to the field of solar energy' (Ases annual meeting, Denver, CO., 1988). Klein got the award for the development and implementation of the f-chart design method for sizing solar heating systems and his contribution in the development of TRNSYS. In his speech he mentioned the fact that f-chart was an important design tool when it was developed, as designers did not have at that time access to personal computers of the present day power. However, he was not sure that he would suggest this method today.
The purpose of this work is to present the use of an accurate simulation model as a design tool not only through the design development stage, but even during the schematic design stage. Present day practice is to apply only approximate methods during this stage and accurate simulation models are reserved for the detailed design stage. This practice gives rise to few dilemma. First, the approximate methods may lead to a wrong answer, especially in non standard designs and second, the simple formulas that were developed to be manually solved are in most cases quite cumbersome for this purpose. As a result we find that special computer programs were developed to solve the so called manual analysis methods.
We turn to approximate methods because most of the hourly simulation models are heavy consumers of labor and machine resources. It would be very attractive if we could use an accurate simulation method that yields the proper answer in a short time and modest labor and machine resources.
This work presents a way how accurate simulation models can be applied in the early design phase. We investigate what kind of climatological and building data is required to achieve accurate solutions with little effort. Furthermore, we explore the decision making process during the schematic conceptual design stage with the purpose of applying detailed hourly simulation models in this design phase.
The analysis of different design alternatives is carried out using a dynamic simulation model ENERGY developed by Shaviv and Shaviv /6,7,8/. This time dependent model

solves simultaneously the heat transfer equation through all exterior walls, taking into account the thermal mass in each external wall and in the internal partitions as well. The thermal behavior of the building is checked during Winter and Summer. The energy consumption of the building is examined in both cases. In addition, the model calculates the maximum inhouse Summer temperature without operating an air-conditioning unit or with an operating fan only. Most of these results, while of considerable importance to the user of the building, are not available in the approximate methods. The results are summarized in an easy to comprehend and implement graphic format.
Special effort has been invested in the computer code ENERGY, to produce an algorithm that solves the implicit scheme of the time-dependent equation for the heat flow in an efficient way. Thus, to run a full-scale building using this computer code, on a Macintosh II can take about 180 seconds for a short typical year of 90 days. This time can be farther reduced to less than 30 seconds during the schematic design stage, as the representation of the building can be simplified. The running time on personal IRIS workstation, for a comprehensive building, is less than 10 seconds. Our claim therefore is that the common workstations are today sufficiently powerful to revolutionize the design practice discussed above.
The importance of presenting the building in a schematic way has two merits. Not only does it reduce the running time and thus saves computer resources, but it allows to prepare the building data required for the simulation model in a very short time. One should keep in mind that the geometry of the building and the properties of the building materials have not yet been determined in the schematic design stage. In this particular stage the architect is trying to learn what should be the proper geometry of the building. Therefore, the building can be presented by a simple geometry in the schematic design stage. The design tool however, should yield the proper answer in a short time and with modest resources.
The use of an approximate method for this stage is important only if it can save time required to present the building. The amount of work needed to prepare the building's data in the present approach is the same as that required to prepare the building for an approximate method. We ponder therefore: is it really worth the effort to reduce the running time of about 30 seconds needed on a Macintosh II, by using approximate methods?

2. THE CLIMATOLOGICAL DATA

The climatological data can be as accurate as possible, that means that hourly climatological data can be used. This data does not have to change from one design alternative to another. It is best to use for each climatic zone an already prepared data. However, we found that a short year of 6 to 8 days per month gives adequate accuracy. Should computer resources be still a problem (for a while may be) we recommend not to use a full year simulation but resort to few days per month.
The hourly climatological data includes: The ambient temperature, the relative humidity and wind velocity, and the shortwaves and longwaves radiations given by the Israel Meteorological Service /9/. The radiation data includes: the direct (on normal surface), diffuse, downward and net radiations. Based on this data the code ENERGY calculates the total short waves radiation on each surface (according to the azimuth and inclination of the surface) and the longwaves radiations from the ground and the sky /8/.

3. THE BUILDING DATA

The general data of the building needed for the simulation model include the floor area and its volume; the area and heat capacity of the isothermal mass; the initial wall temperature conditions and the internal heat gain according to a given schedule. The building data include the information about mode of operation and size of mechanical equipment (thermostat setting and hours during which the heater or cooler can operate), and the number of air changes during day and night (due to infiltration or natural and forced ventilation). The building data contains also information about the number of walls, including all opaque elements in the external envelope (such as roof and floor). For each external wall, the heat capacity, heat conductivity and width of each layer is provided. However this data may be fixed for all design alternatives during the schematic design stage. For this stage the most important building parameters at our disposal are the geometrical design parameters, which are: the azimuth, inclination and the area of each element of the building envelope, and the area, azimuth, and inclination of each window. The albedo and emissivity of the elements, and the shading factors of the elements and windows can be also fixed at this stage. The building data is stored on a file and can be easily changed by using any full screen editor. (See Figure 1)

```
TESTING THE BUILDING ORIENTATION-SOUTH
180.0                   TOTAL FLOOR AREA
504.0                   VOLUME OF BUILDING
300.0                   AREA OF INTERNAL MASS
8400.0                  HEAT CAPACITY OF INTERNAL MASS
18.3                    INITIAL TEMP. DISTRIBUTION
6                       NO.OF EXTERNAL WALLS
0.0                     AZIMUTH OF REFERENCE WALL

1000.   18.0 24.0      250.0 HEAT GAINS:SCHEDULE;CONST.
10000.  06.0 24.0 17.4 18.6 HEATER:SCHEDULE;TEMP.SET
6000.   07.0 24.0 24.0 27.0 COOLER:SCHEDULE;TEMP.SET
0       00.0 24.0 0.75  4.0 VENT:SCHEDULE;VENT-ACH
0       16.0  7.0 5.0   3.5 U WINDOW:SCHEDULE;DAY;NIGHT

          1             WALL NO 1
          3             NUMBER OF LAYERS

500.0  1.40  0.02   (c,λ,d)'INSIDE PLASTER'
 75.0  0.23  0.20   (c,λ,d)'YTUNG BLOCK 800'
450.0  0.87  0.01   (c,λ,d)'INSIDE PLASTER'

180. 90.  .65 .95   WALL:AZ,INC,ALBEDO,EMISSIVITY
1        37.56      NO OF SOLAR ELEMENTS;AREA OF WALL
90. 90. 90. 90.     SC-SUMMER:DIR,DIF;SC-WINTER:DIR DIF
1        18.00      SOLAR ELEMENT:INDEX;AREA
10. 50. 90. 90.     SC-SUMMER:DIR,DIF;SC-WINTER:DIR DIF
```

Figure 1: Part of the building's data file.

4. THE DETERMINATION OF THE REQUIRED GEOMETRICAL DESIGN PARAMETERS

The results of each simulation run are presented by the total yearly energy consumption of the building that is required to keep predefined thermal conditions inside the building (heating and cooling) (Figures 2a,3a,3b). As in many places in Israel people do not use home air conditioners we calculate and present the August temperatures that are obtained in the house without operating any mechanical means but only with better ventilation at night (4 ach). and with a fan (30 ach) controlled by outside-inside temperature differences. (Fig. 2b).
The use of a simulation model for the determination of the required geometrical design parameters is presented through the analysis of the influence of these parameters on the energy consumption and thermal performance of buildings. This analysis and search for the best fitted design parameters is demonstrated in a case study of designing an energy conscious house in Jerusalem, Israel. We shall include in this presentation only the determination of the building orientation and proportion. The full description of the determination of all geometrical design parameters will be presented elsewhere.
For the determination of the best building orientation (i.e. 50% of windows area is facing this orientation) the building was turned from 60° East to South (E-60) up to 60° West to South (W-60). (Figures 2a,b). The results of the influence of the building orientation on the energy performance of the building show that due south is the best orientation, but 15° East or West to South are acceptable orientations. In Jerusalem the West orientation is slightly better than the

East one. From Figures 2a,b one can see that only the heating requirements determined the best orientation. This is because Summer in Jerusalem is very mild and the windows in this building are well shaded in Summer.
For the determination of the best building proportion the South and North walls were changed from 0.9 up to 1.5 the area of the East and West walls, while keeping the area of the building and windows in each orientation fixed (Figures 3a,b). The results of the influence of the building proportions on the energy performance of the building show that when the building is poorly insulated (C-09 to C-1.5) the squared proportions are the best. (Fig. 3a). These results changed when the building enveloped improved (Fig. 3b). For this case the building proportion has no influence at all (Y-09 to Y-1.5).

5. SUMMARY AND CONCLUSIONS

The preliminary study of the influence of the geometrical design parameters shows that they can be characterized according to three different categories:

a. Design parameters that are not sensitive to other design parameters but have a great influence on the energy performance of the building. These design parameters may, therefore, get their recommended value by rules of thumb. In later design stages they can be examined and determined more rigorously and for the particular case study.

b. Design parameters that are sensitive to other design parameters and also have a great influence on the energy performance of the building. These design parameters should be determined rigorously during the schematic design stage, by series of simulation runs in order to continuously guide the designer as to how he should design the building in the particular case. In this way we overcome the difficulty that the heuristic rules of thumb are good for the average cases only.

c. Design parameter which has a small effect on the final performance of the building. (This means also that it is not sensitive to other design parameters.) The value of these parameters may be set by the user in an arbitrary way. They can be checked and determined explicitly In later design stages.

Using simulation model in the schematic design phase can guide the designer in a short time and modest resources about what he should or should not propose.

6. REFERENCES

1. SERI report, 1980. "Analysis Methods for Solar Heating and Cooling Applications - Passive and Active Systems", SERI/SP-35-232R.
2. Beckman, W.A., S.A. Klein, and J.A. Duffie, 1977. "Solar Heating Design by the F-Chart Method", Wiley-Interscience. NY.
3. Klein, S.A., W.A. Beckman, and J.A. Duffie, 1976. "TRNSYS, A Transient Simulation Program", ASHRAE Transactions, Vol. 82, Part 1.
4. Balcomb, J.D., R.W. Jones, R.D. McFarland, W.O.Wray, 1984. "Passive Solar Heating Analysis-A Design Manual", ASHRAE, ISBN 0-910110-38-7
5. A. Pikes, (ed.),1978. Proceeding of the 3rd Int. Conf. and Exhibition on Computers in Engineering and Building Design, Brighton. IPC Science & Techn Press, , England
6. Shaviv, E., G. Shaviv, 1977. "A Model for Predicting Thermal Performance of Buildings". WP ASDM-8, Faculty of Architecture & TP, Technion. Haifa.
7. Shaviv, E. , and G. Shaviv, 1978a. "Designing Buildings for Minimal Energy Consumption", Computer Aided Design J., 10 (4)
8. Shaviv, E. , and G. Shaviv, 1978b. "Modelling the Thermal Performance of Buildings", Building and Environment, Vol.13, Pergamon Press, Oxford
9. Manes, A., A. Teitelman, I. Fruehling, 1970. "Solar Radiation and Radiation Balance". Israel Meteorological Service, Bet-Dagan.

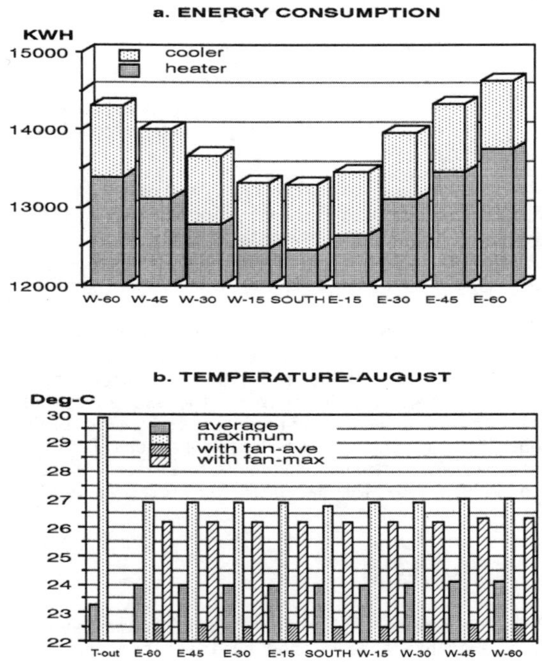

Figure 2: The influence of the building orientation on the energy performance of the building. a. Energy consumption. b. Predicting inside temperature.

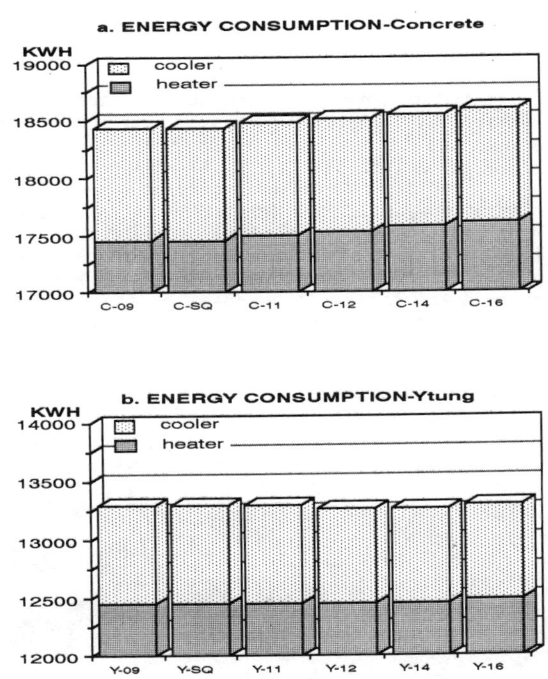

Figure 3: The influence of the building proportions on the energy performance of the building. a. Poorly insulated building. b. Better insulated building.

7. ACKNOWLEDGMENT

I would like to gratefully acknowledge E & J. Pick Alternative Energy Sources Research Fund, who partially supported this research.

CONTRIBUTION TO BUILDING CLIMATIZATION GIVEN BY A GLAZED COMPONENT

G. Rossi* and G. Solaini**
*Istituto Universitario di Architettura, Santa Croce 197, Venezia, Italy
**DPPPE, Politecnico di Milano, Via Bonardi 3, 20133 Milano, Italy

ABSTRACT. In this paper an experimental prefabricated glazed component is presented, whose aim is to shape small balcony-conservatories on the south-oriented buildings facades. In multistoried buildings such components can be set as modules of vertical glazed ribbons. A measurement campaign has been lead in order to test the thermal behaviour of the system composed by the conservatory and an internal space. The experimental data have been utilized for the validation of a model, which has been then used to simulate the thermal behaviour of the system during seasonally varying climatic conditions. Aim of the work is to evaluate the internal and external energy fluxes, and therefore the component's contribution to building climatization. An economic evaluation has been performed as well.

1. INTRODUCTION

This work concerns the energy performance evaluation of a prefabricated glazed building component, which has the function of closing outer spaces such as loggias and balconies. This for the double purpose of retraining the external space of the unit and improving the energy performance of the extent window frame; in other words, the matter is the realization of small "conservatories" with prefabricated glazed components on the facades of buildings.

2. EXPERIMENTAL SET-UP

The tested component is composed by three frontal elements, two lateral elements and an inclined covering, and it has been set on the south-oriented facade of a test-cell (fig. 1); the sunspace and the inside space of the unit have been separated by means of a window frame consisting of an alluminium frame and a partly single and partly double glazed window. Two fans of known capacity have been fitted in this window frame, for the heat transfer from the sunspace to the inner space of the cell every time that the temperature of the former was at least 4°C over the inner space temperature. The latter has been kept at a temperature of 20°C through air-heating.

Figure 1: Setting the component on the test-cell

The experimentation has been conducted starting from the 5th of December 1987 to the 15th of January 1988.

The monitoring has been carried with regard to the following benchmarks:
- incident radiation on the vertical surface of the conservatory;
- filtering radiation in the test cell through the glazed conservatory and the window frame;
- test cell air temperature;
- conservatory air temperature;
- conservatory floor temperature;
- meteorological situation during the experimental period.

Temperatures have been measured by means of thermistors; data have been collected and processed by means of a data logger. The position of the probes is reported in fig. 2.

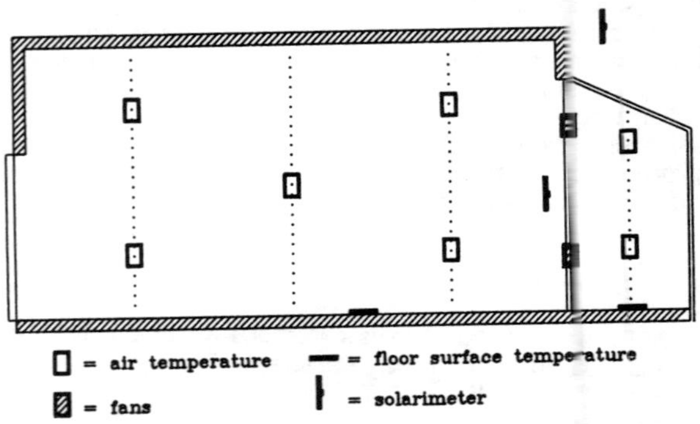

Figure 2: Position of the probes

3. MATHEMATICAL PATTERN

A calculation program, previously developed on the basis of a thermal network model /1/, has been used in order to obtain the thermal behaviour of the system. The radiative heat transfer coefficient has been considered as a linear function of the difference in temperature, while the convective coefficients have been calculated using the relation given by McAdams /2/ (regarding the inside air), and those given by the ASHRAE /3/ (regarding the outside air). A specific subroutine, based on the control logic of the fans, has been developed to evaluate the energy flows from the sunspace to the inner space.

The thermal network corresponding to the experimental set-up is reported in fig. 3. In accordance with this scheme, the experimental values of the air temperature inside and outside the unit have been considered like input data.

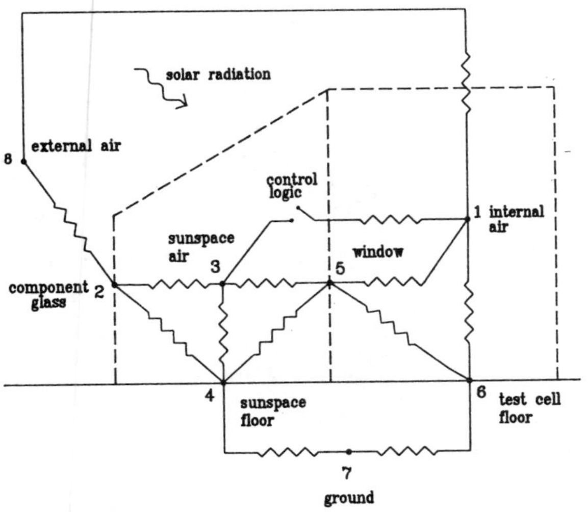

Figure 3: Thermal network corresponding to experimental conditions

3.1 Validation

The experimental data collected during the measurement campaign have been utilized to validate the above mentioned subroutine.
By way of example, the diagram concerning the course of temperature of sunspace air (measured and calculated) is reported in fig. 5. The experimental data were taken on the 8th of January, 1988. External temperature and solar radiation on the vertical surface of the component, referred to the same day, are plotted in fig. 4. Diagrams show that the air temperature measured and calculated values harmonize.

4. SEASONAL SAVINGS ESTIMATE

The calculation code has been then utilized to evaluate the energy savings annually achievable. To this aim it has been considered the case of a 100 m^3 room, facing the exterior and south-oriented, with a balcony closed by a glazed element identical to the tested prototype. The room is supposed to be part of a flat placed in the climates of Venice and Rome, and its total heat loss coefficient per unit

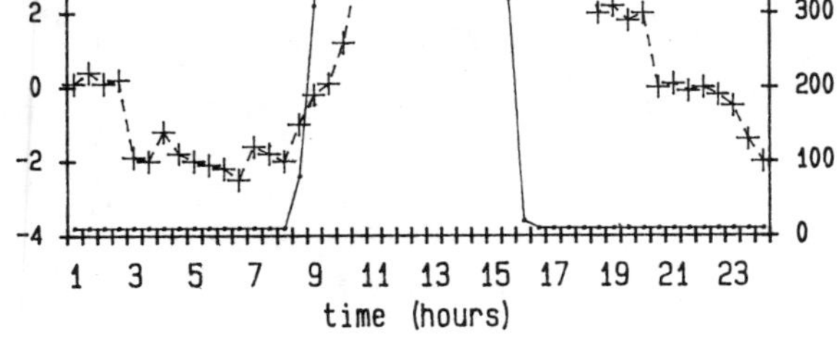

Figure 4: External air temperature and solar radiation on a vertical south facing surface (Milano, January 8, 1988).

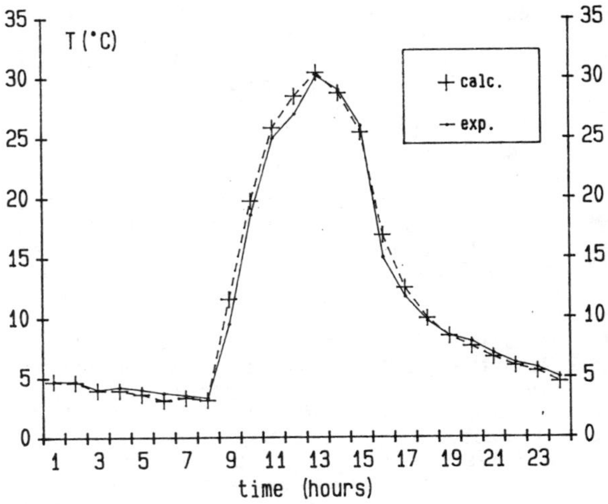

Figure 5: Sunspace air temperature behaviour.

volume, C_g is assumed to be 0.7 $Wm^{-3}{}^\circ C^{-1}$. The energy produced by internal sources is supposed to be 2 kWh/day.
Taking into account the contribution of the south-, east- and west-oriented vertical surfaces, we can assume the realized prototype to be equivalent to a component with a south-oriented working vertical surface measuring 8 m^2. On the basis of this assumption, we could obtain the following mean daily consumption and saving values, monthly and seasonally calculated:

TABLE I: Energy Balances (Venice)

Month	T_e (°C)	I_H (kWh/m^2d)	K_t	DDM	Thermal load (kWh/month)	Energy supply (kWh/month)
Jan.	2.8	1.10	0.34	434	730	71
Feb.	4.6	1.70	0.36	342	574	100
Mar.	8.3	2.83	0.42	264	443	160
Apr.	13.0	3.43	0.47	114	192	123
Oct.	14.4	2.39	0.45	75	125	166
Nov.	8.5	1.15	0.32	249	419	68
Dec.	4.5	0.94	0.33	382	641	61
			Total	1860	3124	749

TABLE II: Energy Balances (Rome)

Month	T_e (°C)	I_H (kWh/m^2d)	K_t	DDM	Thermal load (kWh/month)	Energy supply (kWh/month)
Jan.	6.7	1.66	0.43	313	527	214
Feb.	8.0	2.30	0.43	247	414	195
Mar.	10.6	3.32	0.45	192	323	223
Apr.	13.7	4.49	0.48	93	157	169
Oct.	16.4	2.93	0.49	0	0	0
Nov.	11.7	1.73	0.41	153	258	166
Dec.	7.9	1.32	0.38	276	464	146
			Total	1274	2143	1113

(T_e = monthly average outside temperature; DDM = degree days/month; I_H = monthly average daily global radiation on horizontal surface; K_t = monthly average clearness index).

On the other hand, if we consider, in the same

conditions, a loggia instead of a balcony, with a 6 m^2 south-oriented working vertical surface, we find an energy saving value of 562 kWh/year in Venice and 835 kWh/year in Rome.

5. ECONOMIC EVALUATION

The cost of the component is 175 $\$/m^2$ in case of industrialized production. To estimate the additional cost we need to calculate the cost of a piece of work which guarantees the same enjoyability of an outer space usually exposed to atmospherical agents.
The valuation of the sharp cost of this piece of work is not easy, since it depends on the considered kind of building. With an approximate esteem, we can suppose this cost equal to 50% of the prototype cost.
On the basis of this figure, considering that the gross surface of the window frame to be set is about 7 m^2 in case of loggias, and 11.5 m^2 in case of balcony closure, and assuming a cost of energy of 0.09 \$/kWh (based on fuel price and on average heating plants efficiency in Italy), we find the following figures for the payback times in the different climates:

	Spec. energy supply ($kWh\ m^{-2}\ year^{-1}$)		Payback time (years)	
	balconies	loggias	balconies	loggias
Venice	65.1	80.3	14.9	12.1
Rome	96.8	119.3	10.0	8.1

where the specific annual energy supply figures have been deduced from table 1 and 2.

ACKNOWLEDGMENTS

The component described in this paper has been designed and studied in the frame of a research contract (concerning prefabricated innovative building components) between ENEA (Italian National Committee for research and development of nuclear energy and renewable energy sources) and ICIE (Istituto Cooperativo per l'Innovazione nell'Edilizia). Measurements have been carried out at ICITE (Istituto Centrale per l'Industrializzazione e la Tecnologia Edilizia) a research center instituted by the Italian National Research Council (CNR).

REFERENCES

1) G. Rossi and G. Solaini, Modelli per la simulazione del comportamento termico degli edifici. Condizionamento dell'aria, riscaldamento, refrigerazione, 2-85, 127 (1985).
2) W.H. McAdams, Heat Transmission, McGraw-Hill, New York, 1954.
3) ASHRAE, Procedure for determining heating and cooling loads for computerized energy calculations. ASHRAE, New York, 1975.

APPROPRIATE ENERGY TECHNOLOGY FOR THE DOMESTIC SECTOR

CASE STUDIES IN RURAL SCOTLAND

Dr. H.M. Edge
Robert Gordon's Institute of Technology, School of Surveying,
Garthdee, Aberdeen AB9 2QB, Scotland.

Dr. G.S. Saluja
Department of Building, University of Ulster at Jordanstown,
Shore Road, Newtownabbey, County Antrim BT37 0QB, N. Ireland.

ABSTRACT: A recent study was aimed at assessing the potential for a greater degree of local energy autonomy in fuel supply to rural areas. This paper concentrates on the heating and other energy requirements of rural people, and on how heating installations, building thermal performance, local resource use and hence energy consumption and cost, can be improved in a manner consistent with the needs of a rural populace, which may be very different from those of its urban counterpart. The paper concludes that, given the ageing housing stock, the size of the local biological fuel resource, and the antipathetic attitudes of rural people toward new technologies, low technology solutions are generally the most cost and energy efficient ways of solving the energy problems of the rural domestic sector. The paper considers rural buildings and architecture as an integral part of a rural energy system.

1. Introduction, Rural Buildings in the Context of a Discrete Rural System

A recent study, carried out in the harsh rural environment of the north of Scotland, looked at the heating and other energy requirements of rural people. It considered how heating installations, building thermal performance and hence energy consumption and cost, could be improved in a manner consistent with the needs of a rural populace, which may be very different from those of its urban counterpart./1/

The study considered rural domestic, agricultural and forestry sectors as a discrete system. Data was gathered on building fabric, insulation and fuel-using appliances, local fuel rersource patterns and attitudes to such factors as increasing levels of domestic technology, preferred fuels and heating systems, and the ability and willingness to attain greater levels of local autonomy in energy supply. Most of the data came from 300 questionnaire surveys, aiming at 100% coverage of households in several areas having somewhat different agricultural systems and resource patterns.

Whilst being concerned with technologies for fuel conversion in domestic buildings, the study was much wider ranging than this for a variety of reasons. It was felt that it was inappropriate to consider buildings in isolation, as self-contained systems and as ends in themselves. Rather, domestic and other buildings should be seen as elements in a wider social, economic and resource system, and as tools or machines to be employed by the people existing within that system.

The role of housing as a cog within a system is all the more apparent within an agricultural environment. Here agricultural land acts as an energy capture system, converting solar energy into food. All other elements in the system, namely chemical inputs, farm tractors and machinery, farm buildings for livestock and for food, materials and machinery storage, farm houses and off-farm transportation systems, can be seen as energy using machines, acting as catalysts to enable the energy capture system to operate smoothly.

In this context, any increase in energy autonomy, or 'internalisation' of the energy system, ought to add to the efficiency and robustness of the energy capture process.

This simple model is complicated by various factors: 1. A relatively small proportion of the rural populace is directly in farming or forestry. In the north of Scotland, as elsewhere, this proportion is continually falling. 2. It is over-simplistic to consider rural people, and hence the homes they live in, only as units of production. 3. Even in simple, undeveloped agrarian societies, the economic system is neither a purely agricultural one, nor a discrete and isolated local phenomenon. Therefore the study dealt with the entire rural population within defined areas, considering social and attitudinal factors as well as purely mechanistic ones. The economic and resource system was however restricted to the domestic/agriculture /forestry sectors.

A further reason for aiming to increase the proportion of domestic energy use deriving from local fuel resources, is to increase the flexibility of domestic fuel using systems, allowing more control by building inhabitants. Whilst it would generally be accepted that maximum flexibility of control over heating, lighting and power installations ought to be vested in the hands of the building user, such ideal situations, particularly in the domestic sector, are often influenced by economic constraints on fuel use. In the rural environment, with a widespread, largely biological renewable energy resource, it is possible to use energy conversion technologies which enable many individuals to gain free or reduced price fuels, by substituting their labour for a monetary payment. In areas of high un- or under-employment, such flexibility can act as a useful economic boost to the system. Additionally, the market for local fuels created by those individuals without the ability or inclination to, for example, cut their own firewood, can be exploited by small local businesses, thus providing employment and further boosting the local economy. If increases in local fuel use can be compounded by significant decreases in total fuel used, local energy autonomy may be increased still further.

Treating architecture as an element within a much wider system in this way may lead to very different decisions being made as to appropriate levels of technology to employ in rural, as against urban housing. Rather than talking about 'science and technology at the service of architecture', we begin to consider 'science, technology and architecture at the service of communities. In order to do this we must know something of the nature of the current system, in terms of its resources and the ideas and capabilities of the people who live within it.

2. Energy Resources and the Rural Population

Inthe north-east of Scotland, where the surveys were carried out, in common with many other rural areas, a low percentage of the population are economically active. Though the region's population as a whole has an age profile around the national average, the remoter rural areas have populations skewed considerably towards the old. This is reflected in household size, with 58% of households having only one or two members, whilst only 50% contain any economically active member. 20% of heads of households, and 10.5% of the total population, are economically active in farming, either as farmers or as paid farm workers.

By and large, this rural population has attitudes which are highly independent and fairly anti-pathetic towards new, specifically computer, technologies. Respondents were asked their views on the ability of the community to survive in total isolation from all sources of fuel external to the immediate local area. In more remote, hill farming areas, 38% of the people asked felt that there would be no difficulty in achieving self sufficiency in energy, 17% forseeing that there would be only a few difficulties. The corresponding figures for low-lying, more arable areas were rather lower, though there is a markedly greater confidence in the ability to become self sufficient amongst people living outside the villages. It is worth noting that of those who thought that energy self sufficiency would be difficult or impossible, most felt either that problems would arise for others, rather than for themselves, or that energy for transportation would be the most problematic.

As with many other factors, particularly access to local supplies of fuel, attitudes to self sufficiency are conditioned by the prevailing form of land tenure. Thus in areas dominated by a single, paternalistic landowner, who may also control almost all local employment and most of the housing stock, more people feel that they could become more autonomous, under an almost feudal 'lord of the manor' leadership.

Table I summarises the availability of different energy resources in the hill and arable areas surveyed. Resources are limited to the system under discussion, including only the domestic/agricultural/forestry system. They represent annual figures for maximum sustainable resource use. Wind and solar resources, though theoretically vast, have been restricted to that available through reasonably practicable scenarios. That is, a 5m^2 solar water heating panel on every dwelling, providing half the domestic hot water needs, and a wind turbine on each farm, providing half farm households' electricity needs.

Food crops are included as an agricultural energy resource, though they are obviously not available for heating and other building energy uses. Comparison is also made with agricultural and domestic energy use. Food energy having been deducted from energy resources, reasonable scenarios for the ultimate tapping of these resources suggest that in arable areas around 80% of current domestic fuel use could be achievable from local resources, in hill areas around 7 time current use could be tapped.

Resources vary considerably in the ease with which they can be tapped. The small excess straw resource, available only in arable areas, is the easiest to gather, as harvesting and baling machinery is readily available. Problems arise in providing fuel conversion equipment on such a small scale however, particularly since few agricultural buildings have significant energy needs.

Harder to harvest by purely mechanical means, but much easier to burn on conventional fires and stoves, are the wood and peat resources. Although the timber resource is very large, particularly in the hill areas, the nature of it, largely small thinnings from existing woodland, makes it difficult to harvest in large quantities. Given the cooperation of often intransigent landowners however, the timber resource could be tapped in a sustainable manner, either by the encouragement of orderly casual felling by many individuals, or by small, locally based commercial enterprises. The commercial exploitation of the local peat resource by mechanical cutting is rather more problematic. Peat deposits vary considerably in quality and accessibility, and businesses would need to be highly selective in their cutting activities and more highly capitalised than the equivalent fuelwood organisations. Experiments with small commercial peat cutting operations have thus far failed, largely due to the resistance of landlords of the sporting estates.

These biological resources would be the easiest to utilise with currently available technology in the rural areas. Making reasonable assumptions as to the proportion of the resource which could be tapped in this way, local, sustainable biological fuel resources could quickly make up 26% of total current domestic energy use. This would mainly be an addition to the current fuelwood and peat consumption, much of which is not derived from the immediate local area. Some 20% of households currently gain a significant proportion of their fuel by cutting it themselves, and the willingness to extend these activities is strong, particularly, and perhaps surprisingly, amongst the young.

3. Domestic Fuel Use and Building Fabric

The climate of the north of Scotland dictates a long heating season, and a high proportion of domestic energy use (86.5%) is consumed solely for space heating. This means that the vast bulk of energy demand could be met by relatively low grade fuel sources. Table II quantifies the types of fuel used, showing a preponderance of solid fuels over other types. This is a reversal of the situation for the country as a whole, where gas is the largest heat source in housing.

Climatic factors, together with the size, isolation and construction of rural dwellings, and the inefficiency of many solid fuel conversion systems, contribute towards a domestic energy consumption per household some 1.65 times the national mean. The prevalence of solid fuels is reflected in preferences for types of heating system. Most people expressing dissatisfaction with their heating wished to change either to more efficient coal or wood burning, or to solid fuel central heating. Most of the current solid fuel use is through a single point source open fire.

In terms of building form and fabric, most dwellings are either 1 or 2 storey detached or semi-detached. There is a roughly even split between local authority and owner-occupied housing in most areas, though in some locations private rented dwellings form the bulk of the housing. This last sector is generally of poor quality, though it is difficult to generalise since its condition is entirely dependant on the activities of a few large landowners.

The housing stock is generally old, particularly in the private sector, though its condition depends very much on the income of the owners. In the public sector, 100% of houses have some form of loft insulation, though in other respects construction is poor from the point of view of energy conservation. In the owner-occupied sector 60% of houses have some form of insulation in the roof. In the private rented sector in particular, many improvements, carried out with grant aiding, have not followed a coherent strategy, and ostensibly well insulated dwellings retain, for example, old and draughty windows and doore.

Based on a survey of building structure, U values were calculated for the main elements of each house type. From these figures energy use patterns were compared with building heat loss for a hypothetical heating regime of 20deg.C for 14 hours during the day, and 14deg.C for 10 hours during the night. Since this heating regime was for the whole dwelling, and took no account of times when the building was unoccupied, it probably represents a maximum, rather than a mean level of heating. Figures for energy requirement / +/ energy use (delivered) x 100 gave a mean of 71%. I.e. energy consumption is 40% above that required to heat all of the house continually. The range was from an energy use 7 times the hypothetical requirement, to one providing only 50% of the energy needed. The worst excesses of energy use occurred in dwellings employing open fires as the main or only source of heat. Lowest energy consumption occurred in dwellings using only electricity, reflecting both the high conversion efficiency of delivered energy and the high cost of electricity for low grade end uses.

One of the main causes of this excessive fuel use is the low conversion efficiency of solid fuel on many of the old and poorly designed open fires used. When the figures were recalculated to reflect a probably over-generous open fire conversion efficiency of 30%, useful heat produced was only 61% of that required to fulfil the heating regime. Further calculations, made on the basis of increasing open fire efficiency to the maximum level achieved by commercially available solid fuel stoves, demonstrated a potential saving of 27% of total domestic fuel use obtainable by this strategy. Alternatively, the increased efficiency might be utilised to improve current heating levels.

Currently, energy use per household is 175,000MJ per annum in hill areas, 161,000MJ in arable areas. Of this some 23% comes from the rural fuel resource (though not necessarily from the immediate survey area), and 10.5% is free to the household in monetary terms. In most areas the bulk of this rural energy use is wood. Where peat is available, it tends partly to replace the wood, rather than adding to it.

4. Conclusions

Combining strategies of improving fuel using appliances for wood and peat burning and encouraging greater use of local fuel resources could, by one scenario, easily decrease fuel per household imported into the local rural energy system by some 37.2%. The familiarity of local people with these fuels and their means of conversion, and local resistance to more high technology solutions, would seem to commend these options as easily disseminated to a wide population.

The methods of achieving such strategies have more to do with information dissemination than with technological development. Small, integrated energy businesses in rural areas could be involved in the tapping and sale of locally available fuels, advice on and sale of more efficient fuel using appliances, and advice on means of improving building insulation. Such businesses could provide a much needed economic and employment boost to many areas. There remains however, a role for the technologist in the development and manufacture of efficient, easily installed, easily operated, unobtrusive and, most of all, cheap solid fuel burning stoves, particularly for peat.

Considering rural housing as part of a, albeit simplified, rural system, suggests courses of action very different from those we might wish to employ in a more industrialised, urban setting. A study of the preferences and attitudes of rural people reinforces the need for this locally based, low technology approach.

5. References

1. EDGE, H.M. (1987) 'Present and Future Patterns of Energy Usage in Rural Areas'. Unpublished PhD. Thesis, RGIT, Aberdeen.

Table I Summary of the Rural Energy Resource

	Hill areas (MJ/Annum)	Arable areas (MJ/Annum)
Passive solar housing	1200000	1600000
Human Labour	3600	7200
Sewage	149240	228240
Domestic waste	149240	228240
Food crops	80000000	120000000
Animal products	24543340	24602038
Excess straw	-1710000	7884000
Timber	45000000	6228000
(Minus current use)	-1476300	-640100
Peat (2000yr. Depletion)	5500000	1000000
Peat (100yr. Depletion)	110000000	200000000
Hydro Power	5550336	756864
Active solar	821846	1148061
Wind	424200	983200
Total Energy Resource	**264655000**	**167257000**
Agricultural energy use	6402000	18815000
Domestic energy use	22199616	28364367
Transport energy use	4215023	5175308
Total energy use	**32811088**	**52557912**

Table II Percentage Domestic Fuel Use by Type

Fuel Type	Hill Areas	Arable Areas
Coal	34.5%	31.5%
Electricity	23.6%	31.5%
Smokeless Fuel	12.4%	3.0%
Purchased Wood	11.4%	7.5%
Home Cut Wood	7.6%	4.0%
Oil	5.7%	1.5%
Home Cut Peat	3.3%	6.0%
Purchased Peat	1.0%	5.0%
Gas	0.9%	2.0%
Paraffin	0.1%	0.0%

DEVELOPMENT AND TESTING OF A LOW THERMAL CONDUCTIVITY MORTAR

By Dr. J.A. Tinker,
Department of Civil Engineering,
Leeds University,
Leeds,
England.

Mr. A. O'Rourke,
Department of Building and Environmental Health,
Nottingham Polytechnic,
Nottingham,
England.

ABSTRACT

This paper establishes the need for the development of a lightweight mortar. Five lightweight mortars are introduced and test results briefly discussed.

The current method of moisture factor correction is tested against experimental values. A more appropriate moisture factor is developed and its application outlined.*

NB. The moisture factor is applicable to lightweight mortars in the density range 1080 to 1800 kg/m^3, 0 to 8 % moisture content by volume.

1. INTRODUCTION

Within the United Kingdom the recognised method of assessing the thermal performance of a building element is by calculation of it's thermal transmittance or U-value. The thermal resistance of each elemental component being combined mathematically using a model which approximates to the material's structure. Currently this approximation does not take into account the heat conducting properties of the mortar joint. This wasn't necessary when the method was first proposed as the density and hence the thermal conductivity of the various components used to make up the structure were similar. It was therefore assumed that the effect of the mortar joint, taking up approximately 10% of the wall area, was not worth considering separately.

With the advent of the fuel crisis in 1973 the need to conserve energy became an important issue and this has resulted in strict guidelines being imposed, governing the rate of heat loss through the external walls of a structure. This in turn has led to the development of new 'super-insulating' building blocks to meet the requirements of the current regulations. As these developments on the blockwork have been continuing, mortar has not changed. To utilise the benefits of the new blocks it was decided to develop a more thermally efficient mortar. On the eve of the introduction of even more stringent regulations governing the maximum rate of heat loss through a wall within the UK, we have a situation where building mortars are in extreme cases, six times denser than the building blocks they are bonding. Where, in the past the 10% or so of the total area occupied by the joint could be ignored, it now results in an increase of 26% in the overall thermal conductivity of the wall. This anomaly is currently ignored when the thermal transmittance of a building element is determined. It can only be a matter of time before the regulations are tightened and the inclusion of the mortar joint in the calculation is made mandatory.

Further problems arise when materials have different thermal conductivity values and are used side by side. The process of heat transfer through a porous material such as mortar is complex. With factors such as density and moisture content having a major effect on the process. The denser the material the less air is trapped within it's matrix and as air, in relation to the mortar constituents, is a good insulator it follows that there will be an increase in thermal conductivity. Mortars are very rarely found in the dry state, they often contain moisture within some of the pores. This presence results in a conductive heat bridge across the pore, because liquids are better conductors of heat than air. The presence of moisture also increases the rate of heat transfer in partially saturated pores due to heat being transferred across the air space by an evaporation/condensation process. The cold bridges that are formed increase the heat flow which may lead to condensation and pattern staining.

When determining the thermal transmittance of any building component the thermal conductivity of the various components must be known at an appropriate moisture content so that a fair comparison can be made. Due to the hydroscopic nature of the materials used in buildings, different materials have different equilibrium moisture contents depending on their position within the building element and on the local condition at the time of measurement. Since it is not convenient to take all these factors into account when calculating thermal transmittance, standard moisture contents are assumed. In the United Kingdom values of 3% and 5% moisture content by volume, are assumed for the inner and outer leaves of a cavity wall.

It is very difficult to measure the thermal conductivity of a specimen at one of the standard moisture contents. For the measurement, and in accordance with BS 874 1986 (1), the samples are conditioned to a constant weight in a laboratory environment. As this so called equilibrium air-dry moisture content rarely corresponds to 3 or 5% per cent moisture content, correction factors have to be used to determine the thermal conductivity of the samples at these moisture values.

Over the past few years it has been shown that the present method of correction used in the UK is inaccurate. Several studies (2,3) have been carried out which have shown, for lightweight concretes over specified density ranges, that the current correction factors are inadequate and they have proposed an alternative method.

This paper presents results of an investigation into the thermal conductivity of several new lightweight mortars in the density range 1000-1800 kg/m^3 and presents a relationship suitable for the correction of their thermal conductivity to standard moisture contents.

2.MOISTURE CORRECTION

In the UK the standard factors used to adjust masonry materials to an appropriate moisture content are based on a relationship between thermal conductivity and moisture content derived by Jakob in 1949 (4). These so called Jakob moisture correction factors are tabulated in the CIBSE Guide A3 1980 (5).

Furthermore, the UK method fails to take into account the fact that different materials of similar density can have thermal conductivities varying by approximately 30%. As well as not taking this phenomenon into account there is, as mentioned previously, no recognition of the role that the mortar joint plays in the heat transfer process. Here, UK standards lag behind other European standards, where the effect of the mortar joint is acknowledged by allowing a reduction to be made in the overall thermal conduction of a structure if a convenient mortar is used in preference to a more dense traditional type.

3.EXPERIMENTAL PROCEDURE

Five pairs of specimens were prepared, as a full and partial sand replacement, containing pumice, coal waste, Lytag, Pellite and Pellite plus Filtech. Each with the following densities: 1080, 1245, 1275, 1406, 1806 kg/m^3. In each case the specimens were prepared and tested in accordance with BS 874 1986.

Prior to testing, each sample was conditioned to a constant weight in a laboratory environment maintained at 20°C 65% RH. After completion of the initial thermal conductivity measurement the specimens were then placed in an environmental cabinet and conditioned at 95° RH 20°C for approximately six weeks. The samples were periodically weighed and when two successive weekly weighings were the same the sample was deemed to have reached it's maximum moisture content at that psychometric condition. On average this was around 25% by volume. Their thermal conductivity was then measured. The specimens were then air dried to reduce their moisture content and the thermal conductivity was measured at intervals until sufficient data had been obtained to define a moisture content profile.

At the end of the measurements, the samples were oven dried at 105°C, allowed to cool in a desiccated environment, and then weighed. An attempt to determine the oven dry thermal conductivity of each sample was made. However, because of the hydroscopic nature of the materials under test, the samples always absorbed some moisture from the atmosphere whilst under test conditions and thus no measurements below 0.01% moisture were possible.

4.RESULTS

Table 1 shows the thermal conductivity and associated data for the five mortars tested. Although the mortars were tested over a wide range of moisture contents, in order to check the validity of Jakob's moisture correction factors a comparison at 3% moisture content by volume, was made between his standard curve and the thermal conductivity values determined experimentally. Figure 1 illustrates this comparison, it can be seen that some lightweight aggregates have thermal conductivity values well below those predicted by the standard curve. The Pellite aggregate mortars, in particular, have thermal conductivity values approximately 30% less than the standard curve. A comparison to a traditional mortar is also shown.

Sample	Bulk dry Density Kg/m^3	Moisture Content % by vol	Measured Thermal cond (W/m K)	Sample	Bulk dry Density Kg/m^3	Moisture Content % by vol	Measured Thermal cond (W/m K)	Sample	Bulk dry Density Kg/m^3	Moisture Content % by vol	Measured Thermal cond (W/m K)
Pumice	1080	8.53	0.293	Lytag	1245	8.83	0.362	Pellite	1406	8.58	0.345
	1080	7.00	0.269		1245	7.37	0.360	+ filtec	1406	7.22	0.348
	1080	5.81	0.269		1245	4.94	0.330	+ airent	1406	5.82	0.340
	1080	4.67	0.261		1245	3.46	0.320		1406	4.81	0.366
	1080	3.60	0.260		1245	0.82	0.299		1406	3.62	0.329
	1080	2.18	0.254		1245	0.10	0.296		1406	2.91	0.362
	1080	1.64	0.249						1406	1.31	0.314
	1080	1.29	0.248	Coal	1276	7.50	0.376		1406	0.17	0.304
	1080	0.84	0.244	Waste	1276	7.04	0.378		1406	0.07	0.300
	1080	0.35	0.243		1276	4.18	0.361				
	1080	0.20	0.243		1276	3.92	0.352	Pellite	1806	8.70	0.491
	1080	0.03	0.240		1276	3.20	0.345		1806	4.94	0.470
					1276	2.45	0.341		1806	3.01	0.458
					1276	1.04	0.334		1806	1.78	0.439
					1276	0.10	0.332		1806	0.76	0.438
									1806	0.76	0.439
									1806	0.25	0.437
									1806	0.06	0.438

Table 1 summary of results

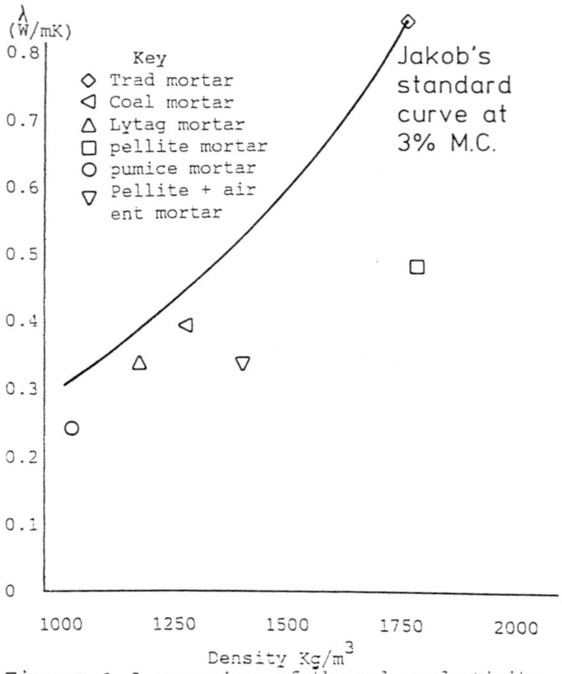

Figure 1 A comparison of thermal conductivity values.

Initial measurements were made with moisture contents in excess of 15% by volume. During test it became clear that the excess moisture within the samples was having an adverse effect on the measurement and therefore the practice was discontinued. The data represented in Table 1 was obtained from samples where no moisture, in the form of condensate, was present on the surface of the sample before or after test. These results are shown graphically in Figure 2 which shows the thermal conductivities of the samples plotted against moisture content. The solid lines shown on represent the best fit equation for each set of data and were obtained using regression analysis.

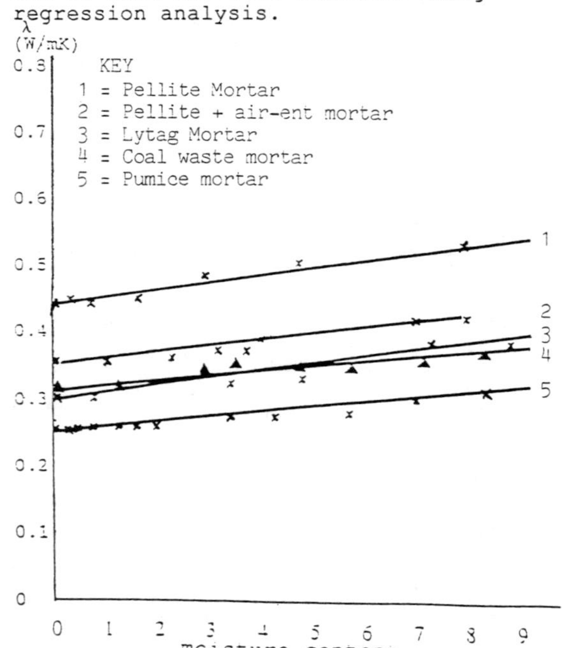

Figure 2 thermal conductivities plotted against moisture content.

In each case the equations were in the form

$$\lambda = \lambda o + AM \quad . . . (1)$$

where A is a constant and M is the moisture content by volume.

5. MOISTURE CORRECTION FACTORS

Using equation (1) the thermal conductivity of each sample at 0% moisture content by volume was obtained by extrapolation. This enabled the thermal conductivity at a finite moisture content to be expressed in terms of a dry value (λo) and this enabled moisture factors (F) to be obtained. Where moisture factors F are given by

$$F = \frac{\lambda}{\lambda o} \quad . . . (2)$$

Using equation (2) moisture factors were obtained at integral values of moisture content for each sample. Figure 3 shows the data plotted as a function of the sample densities at various moisture contents. It can be seen that as with aerated concretes (2) and lightweight aggregate concretes (3) the moisture factors decrease with increasing density. Based on the interpolated points shown in Figure 3 and using regression analysis a general expression for the moisture factor was derived in terms of moisture content (m) and density ρ, of the sample such that

$$F = 1 + 0.04\,M - \rho \times 10^{-5}\,(1.41\,M) \quad . . . (3)$$

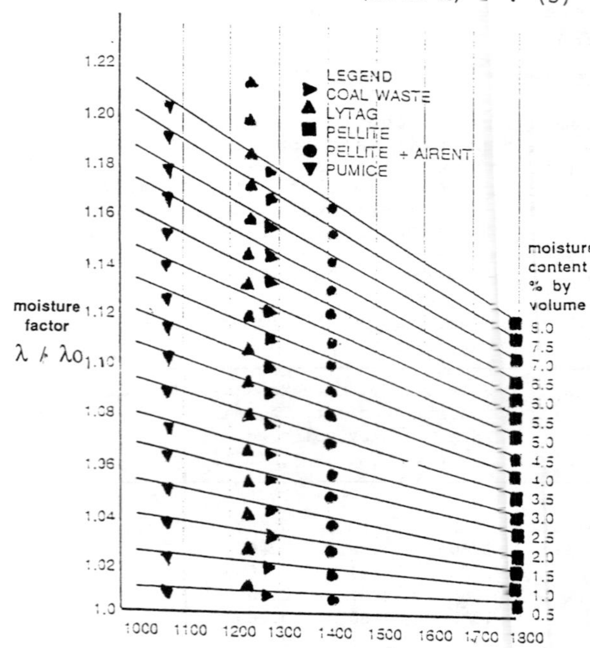

Figure 3 moisture factors plotted as a function of density and moisture content

This expression applies to lightweight mortars in the density range 1080 - 1800 kg/m^3 and from 0-8% moisture content by volume.

The solid lines in Figure 3 were obtained using equation 3 and can be seen to be in close agreement with the experimental data.

6. THERMAL CONDUCTIVITY AND MOISTURE CORRECTION

By using equation (3) to obtain moisture factors Fs and Fm the thermal conductivity s at some specified moisture content Ms can be obtained from the measured thermal conductivity λm, where λs is given by

$$\lambda s = \frac{Fs}{Fm} \lambda m \qquad \ldots (4)$$

CONCLUSION

By using commercially available lightweight aggregates as a replacement for sand in mortar it is possible to reduce thermal conductivity of mortar. It has been shown that this reduction in thermal conductivity is not proportionate to a reduction in density, other factors such as the pore structure and aggregate matrix have an effect on thermal conductivity.

The thermal conductivities of five lightweight mortar samples in the density range 1080-1806 kg/m^3 were measured at various moisture contents up to approximately 8.00% moisture by volume. Moisture factors for each sample were derived on the assumption that the thermal conductivity at zero moisture content could be obtained by extrapolation of the 'best fit' equation for each experimental curve. A general expression was derived relating the moisture factor to both density and moisture content, as opposed to a purely density relationship in the case of Jakob's factors.

A more detailed paper on the development of a lightweight mortar is in preparation in addition to a further paper illustrating the use of the moisture factor.

REFERENCES

1. BS 874:November 1973 and AMD 3006: Aug.1979
   Methods for determining thermal insulating properties with definitions of thermal insulating terms.
2. Stuckes, A.D.and Simpson, A.,The effect of moisture on the thermal conductivity of aerated concrete.BSER&T,6,(2)49-53 (1985).
3. Stuckes,A.D.and Tinker,J.A et al. The effect of moisture on the thermal conductivity of lightweight aggregate concrete BSERT, 7, (1) 27-32 (1986).
4. Jakob, M.Heat transfer, Part 1 (Chapman and Hall, London 1949).
5. CIBSE Guide A3,Thermal properties of building structures (CIBSE London 1980).

PARAMETRIC ANALYSIS OF BUILDING HEATING CONSUMPTION IN ITALY

V. Corrado, P. Oliaro

Politecnico di Torino
Dipartimento di Energetica
Corso Duca degli Abruzzi, 24
10129 Torino - Italy

ABSTRACT. A parametric analysis of space heating energy use in Italian residential buildings has been made using MORE code, a simulation program developed within Italian "Energy Finalized Project". The analysis has been focussed on those parameters that more directly influence the size of solar heat gains (orientation, size of fenestration and kind of glazings). Regression coefficients have been determined in order to correlate thermal loads to the above parameters. Different climatic zones have been taken into account.

1. INTRODUCTION

Building energy use patterns are a complicated interaction of several heat processes characterized by conductive, convective and radiant heat transfer, and mass transfer. Each element of a building includes some of these items, contributing to the thermal load of the whole system. In the past years, a lot of tools and method have been developed to evaluate residential building energy needs for space conditioning. Most of them consist of simulation programs performing complete energy analysis of a building, allowing very accurate modeling of its thermal behaviour.
The goal simulation programs are used for is generally to perform deep studies on a particular building, rather than making parametric analysis of some factors on a large stock of building. The study reported in this document deals with such a second approach, using a simulation code to analyse the effect of varying some parameters on the energy behaviour of the whole building.
The main purpose of such a work is to analyse the way in which the size of fenestration, the kind of glass and the orientation of the building affect the energy need for space heating in a residence building. The reason for this particular interest is that windows have been often considered just as a heat loss source, forgetting to take into account the advantageous use of solar radiation, balancing sometimes completely the heat losses through windows.
The presence of different climatic zones in Italy makes the conclusions of the reported study to be a function of the climate.
A standard single-zone module has been selected for analysis using the MORE energy analysis program, developed within Italian "Energy Finalized Project". Fenestration properties were allowed to vary within a range selected in order to cover the typical building elements utilized in Italian residences.
The weather profiles collected and published by the Italian "National Research Council" were used in the analysis. The utilized data consisted of "mean monthly days" in winter period, referred to three Italian towns (Torino, Ancona, Napoli), placed in different thermal zones. Considering three different sites is not enough to make a complete energy analysis taking into account also climatic factors; however it let us to differentiate conclusions for climatic zones and indicate a direction for further studies.
The correlation between energy consumption and the independent parameters studied was done by means of multiple regression techniques used to analyze the data resulting from parametric runs. It was possible to establish a correlation between the energy need of a buiding and a set of independent parameters using the least-square method. Once this correlation has been established, it is very easy to manipulate the equation for making design optimizations with regard to both energy performance and cost aspects.

2. BUILDING DESCRIPTION

Two building configurations were modeled in the MORE program, the first one corresponding to an intermediate rectangular zone with one external wall, the second one to a corner square zone with two adjacent external walls. The rectangular module is 8 m by 4 m, while the square one is 4 m by 4m. The height of both zones is 2.7 m. Within the corner configuration, three different cases were studied, the first one consisting of two windows on the external walls, the second and third cases consisting of one window on respectively left and right side external wall. All the standard configurations are presented in figure 1.
External walls are made of two layers of hollow bricks with an inserted 4 cm layer of insulation. They have a heat transfer coefficient of 0.58 W/m2.K and a specific mass of 360 kg/m2. Solar absorptance is 0.5. Floors and internal walls have a specific mass of respectively 190 kg/m2 and 168 kg/m2.
A costant infiltration/ventilation rate of 0.5 air changes per hour was assumed, according to the Italian Standard for building design.
A costant schedule (4.7 W/m2) was considered for occupants, lights and appliances, assuming a pure convective heat transfer.
A set point thermostat was used to control the space heating system, supposed to be a fully convective heat supplier. Internal temperature was set to 20°C all over the day.

3. PARAMETRIC ANALYSIS

The simulation program used for the parametric analysis is the MORE code. It is a program for dynamic energy analysis of buildings, based on the heat transfer approach.
The fenestration properties allowed to vary in the parametric analysis were the window size and the kind of glazing. Three sizes of windows were considered, covering respectively 7.5%, 15% and 30% of the glazed wall. Five kinds of glazing were taken into account, including a clear single glass, an absorbing single glass, a reflecting single glass, a

clear double glass, a reflecting double glass. The glazing thermal properties are reported in the following table:

GLASS	SHADING COEFF.	HEAT TRANSFER COEFFICIENT [W/m2.K]
1	0.88	5.1
2	0.65	5.1
3	0.47	5.1
4	0.72	2.9
5	0.32	2.2

Neither shading management, nor external overhangs were taken into account in this study; anyway their influence will be analyzed in future work.
Results were obtained for eight orientations covering a complete 360° rotation in 45° increments and for the four above described building modules.
The parametric analysis was referred to the three towns, placed in different climatic zones (see figure 2). The following table describes the main climatic parameters for the considered sites.

	LATITUDE	DEGREE DAYS [K.days]	MEAN DAILY SOLAR RADIATION ON HORIZ. PLANE [Wh]
TORINO	45°11'	2570.	2260.
ANCONA	43°47'	1590.	1890.
NAPOLI	40°52'	880.	2295.

Within the related study, 1440 MORE runs were completed, each one extended to the whole heating period. The heating energy use in winter period was calculated by the simulation program, at first computing the daily heating load for each "mean monthly day" and then summing such values multiplied by the number of days in each month of the heating period.

4. REGRESSION MODEL

Multiple regression techniques were used in order to generate a simplified expression relating building fenestration parameters (independent variables) to heating energy use (dependent variable).
The general function could be expressed as:

$$E = f\ (S_{fl}, S_f, SC_f, K_f)$$

where:

- E = yearly heating energy use [MJ]
- S_{fl} = flor area [m2]
- S_f = glazing area [m2]
- SC_f = glazing shading coefficient
- K_f = glazing heat transfer coefficient [W/m2.K]

Regression analysis uses the method of leasts squares to characterize the form of the relationship between variables. Energy use for the model can be predicted by means of the simple expression:

$$(E/S_{fl}) = \beta_1 + \beta_2 . SC_f . (S_f/S_{fl}) + \beta 3 . K_f . (S_f/S_{fl})$$

The coefficients β_1, β_2 and β_3 depend on the building site, on the kind of module and on the building orientation. The correlation between these coefficients and orientation is presented in figures 3 through 14 for three different sites (Torino, Ancona, Napoli) and for four different modules.

5. CONCLUSIONS

Some conclusions can be ascertained from the results of the parametric analysis performed in this study:

a. Regression-derived equations seem to be a viabile simple way of performing parametric analysis. They can also become useful tools for solving optimization problems in the building design.
b. The regression coefficients depend on the climatic conditions of the building site. A further study should analyze the effect of climatic factors on the fenestration performance and define proper climatic parameters that could be directly correlated to heating energy use.
c. Sometimes the effect of window exposition becomes very relevant. It should influence, not only a correct window sizing and glazing choice, but the building orientation itself, needing to be taken into account at an early stage of design.
d. The solar gain can be higher in Northern colder sites than in Southern warmer ones, due to the fact that a higher latitude produces a lower sun height during winter season. Besides that, longer heating periods cause the utilization of solar gain also in middle seasons.

6. REFERENCES

/1/ R. Sullivan, S. Selkowitz - "Energy Performance Analysis of Fenestration in a Single-Family Residence" - ASHRAE Technical Data Bulletin, 1985.

/2/ CNR - "Simulazione oraria del comportamento termico energetico degli edifici" - PFE, 1982.

/3/ CNR - "Repertorio delle caratteristiche termofisiche dei componenti edilizi opachi e trasparenti" - PFE, 1982.

/4/ CNR - "Dati climatici per la progettazione edile ed impiantistica" - PFE, 1982.

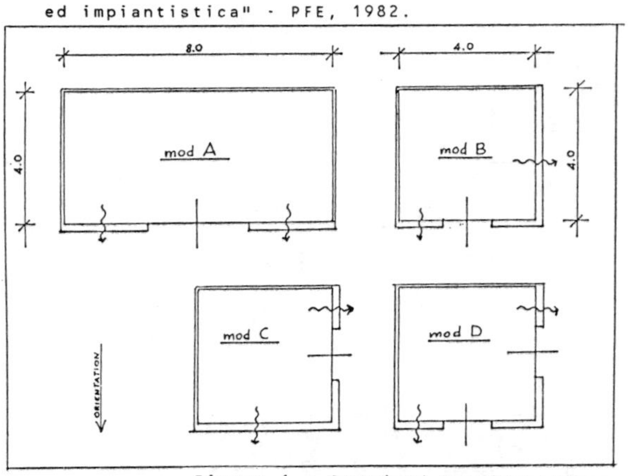

Figure 1 - Standard Configurations

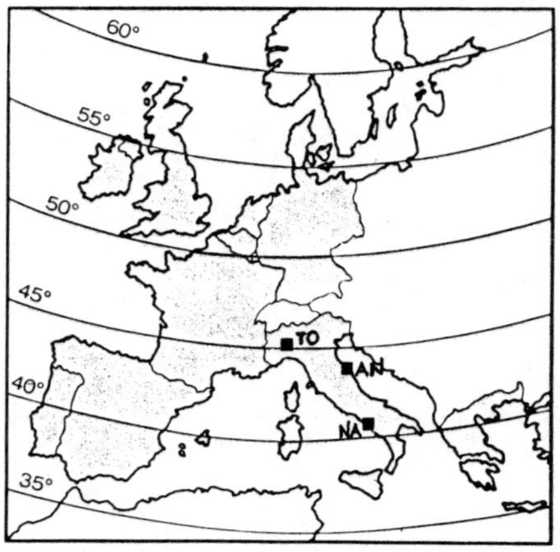

Figure 2 - Reference Sites

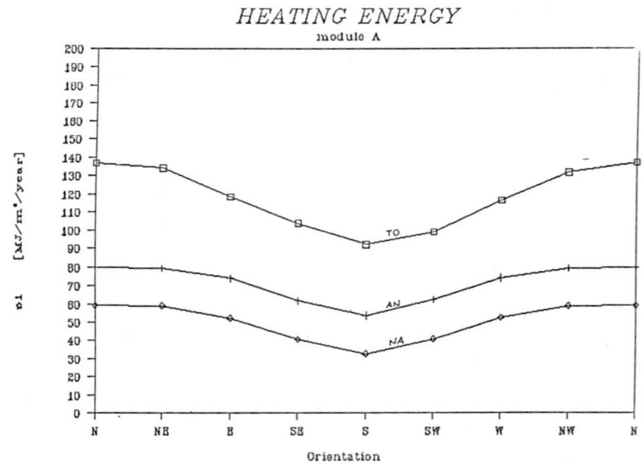

Figure 3

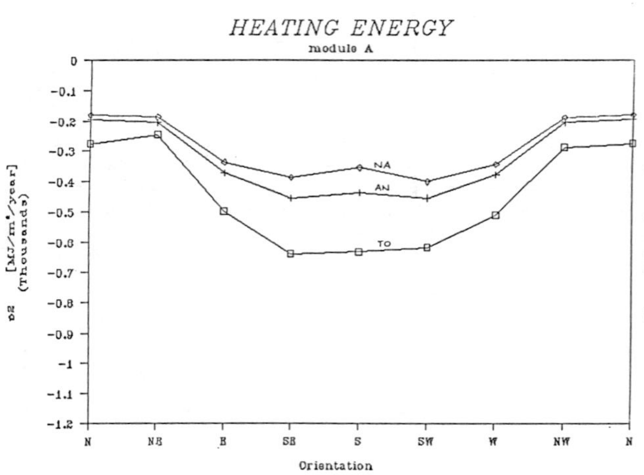

Figure 4

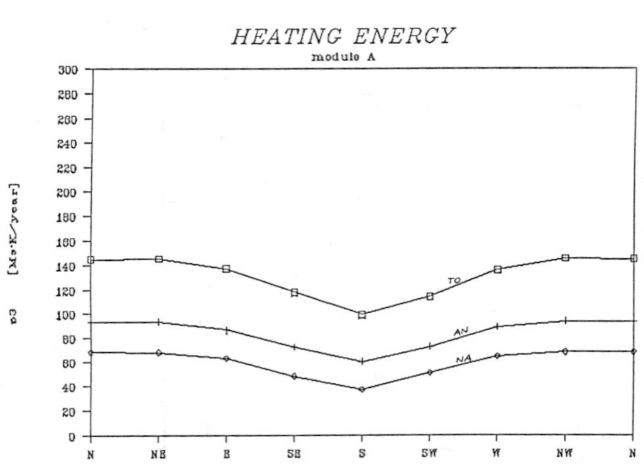

Figure 5

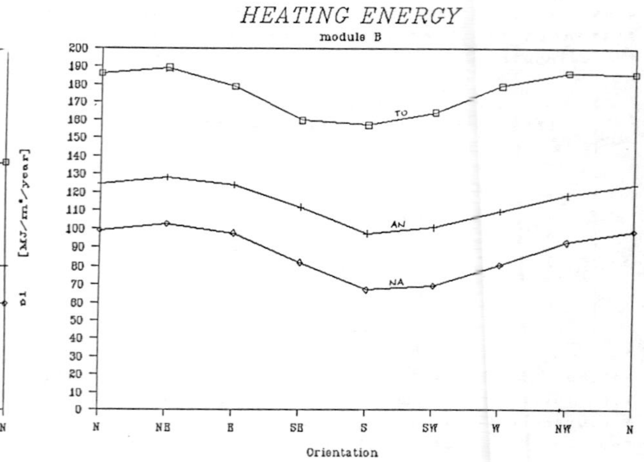

Figure 6

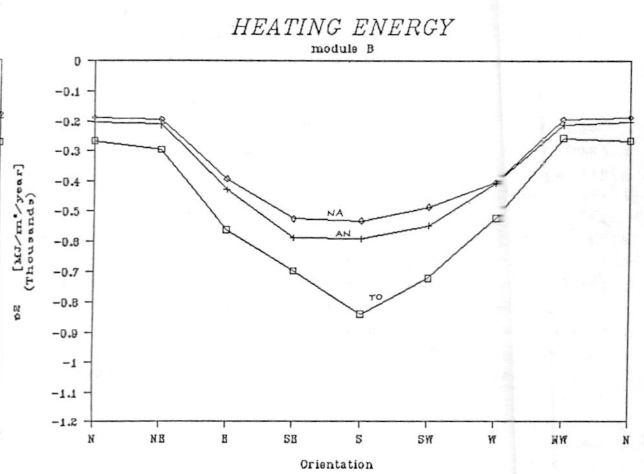

Figure 7

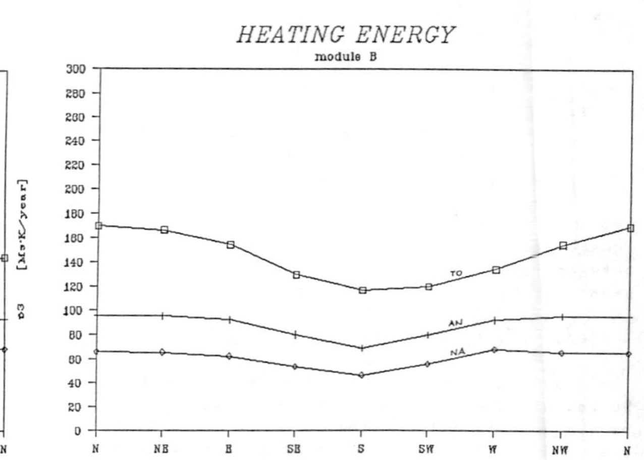

Figure 8

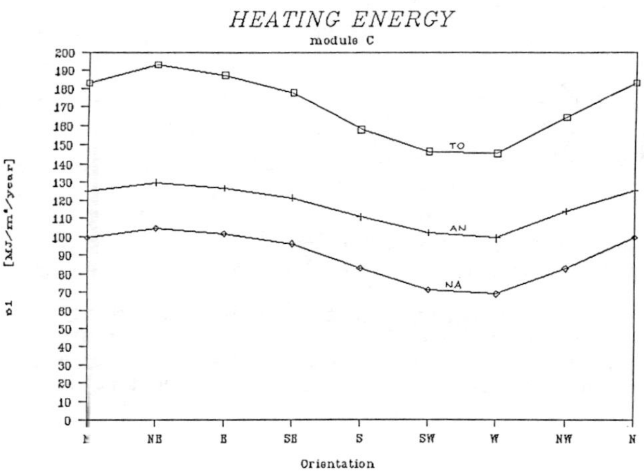

Figure 9

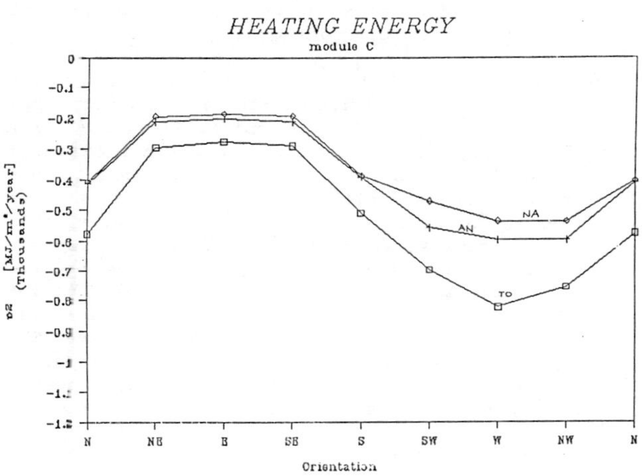

Figure 10

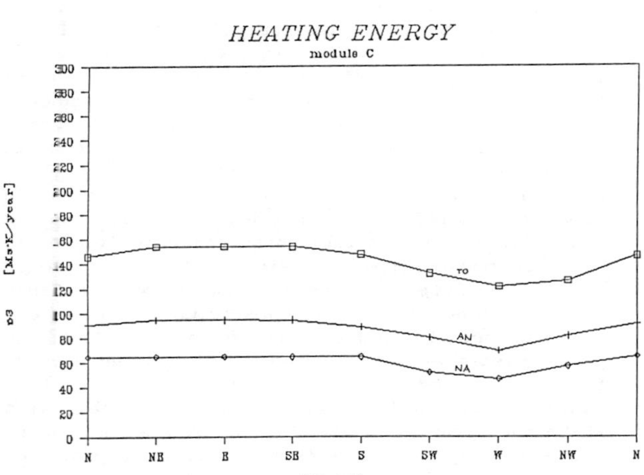

Figure 11

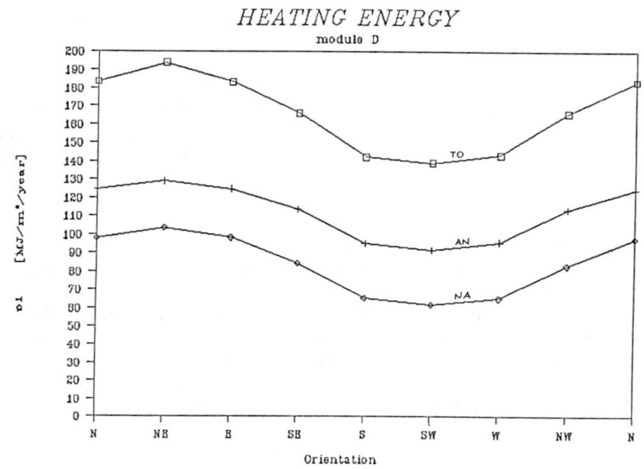

Figure 12

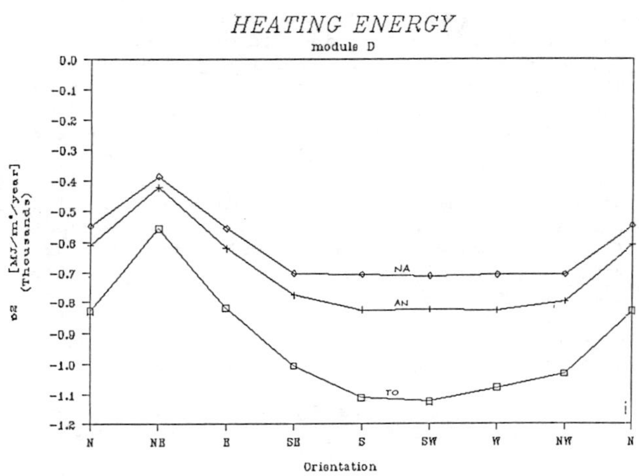

Figure 13

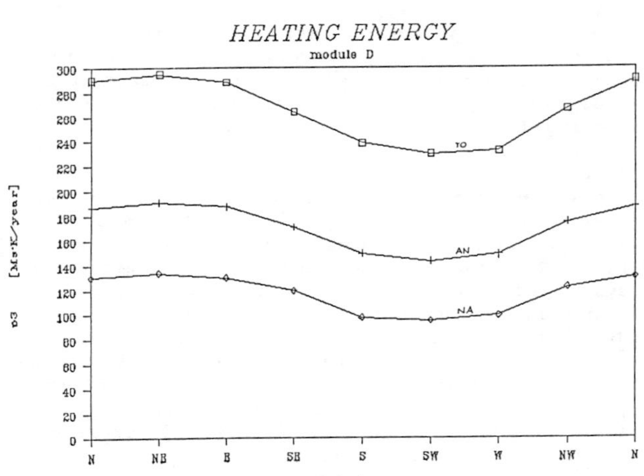

Figure 14

WHY INTEGRAL DESIGN OF BUILDING AND SYSTEMS?

J.L.M. HENSEN
Eindhoven University of Technology
P.O. Box 513 Group FAGO
5600 MB EINDHOVEN The Netherlands

ABSTRACT. The dynamic thermal interaction between building structure and auxiliary heating system is still difficult to predict. Yet it is more than likely that knowledge on this subject will become increasingly important in the near future.
This paper discusses background, set-up, current state and some preliminary results of a research project with the objective to develop or enhance building performance evaluation tools for examining this dynamic thermal interaction. The primary goal is the prediction of fuel consumption and actual temperatures for a given house, heating system and control strategy (as opposed to contemporary prediction of net heating demand based on a presumed indoor temperature profile).

1. INTRODUCTION

The thermal interaction between building, occupant and heating system is still difficult to predict/1, 2/. In practice this may result in non-optimal, malfunctioning or even "wrong" building/system combinations. That this is not an over-statement was already demonstrated/3, 4/. There are however several other examples. It was found, for instance, that the actual seasonal efficiency of high efficiency boilers is markedly lower in practice than in experimental conditions. The reason being that return water temperatures are actually higher then anticipated (i.e. above flue gas condensation temperature). Another example is that of premature boiler aging which is sometimes encountered in multi-family housing. This damage is caused by large temperature stress which occurs on a regular basis during heating-up periods (i.e. when boiler operates at full power and return water temperatures are at its lowest). These are only two examples drawn from a large class of problems which can only be solved by a simultaneous and integral approach of the different system components: environment, building, plant and occupants.

There are several reasons why the problems above and the need for an integral approach have become more important:
1) reduction of space heating demand: for example, the average heating system capacity in newly built Dutch houses more then halved since 1973. The capacity for space heating and for domestic hot water used to be about the same. At present, the ratio between heating capacity for hot water and for space heating is between two and four. Due to these effects, the heating system becomes more sensitive to both external and internal disturbances.
2) energy conscious behaviour: nighttime temperature reduction is only effective/comfortable when heating-up period is short. However, there must be enough additional heating capacity to allow for short heating-up periods. This implies that the system will normally operate at very low loads. Another form of energy conscious behaviour is to shut-off parts of the system (e.g. in unused rooms). In that case the operation of the heating system (e.g. the flow rates) will be strongly different from the design conditions. Then questions are: what will be the effect on system efficiency, fuel consumption and comfort and what are the consequences for the system.
3) higher comfort levels: for example by individual room control, the system may operate quite different from design conditions during a large part of the year. The same questions as before can be asked and again the answer can only be found by an integral approach of building and system.

Up to now the building design process is more or less sequential; first design of the building and then the system. The dynamic thermal interaction is usually left out of consideration; system control is the panacea. Thermal comfort requirements are commonly reduced to required air-temperature, whilst other thermophysiological parameters like radiant temperature, and air velocity are left aside. For system design, usually only extreme external and internal conditions are considered.
This approach is not ideal for neither thermal comfort nor for fuel consumption. However, at present there are no tools for an integral approach.

Previous research in this area mainly focussed on the relation between building design and energy consumption (e.g./5, 6/) or thermal comfort (e.g./7/) with the auxiliary system commonly regarded as a parameter instead of a variable (e.g./8/). Only recently combined building and plant system simulation has received more attention/1, 2, 9/). Usually this research focusses on the commercial building sector. The systems encountered in the commercial sector are usually more specific to a particular type of building than in the case of domestic buildings. The need for generalized tools is therefore not so large. The research outlined in the remainder of this paper on the other hand, primarily focusses on the domestic building sector and on generalized tools.

2. RESEARCH DESCRIPTION

The objective is development and/or enhancement of building performance evaluation tools for examining the dynamic thermal interaction between a building and its indoor climate control system. In this context, thermal comfort and energy consumption are regarded as requirement and optimization variable, respectively. Due to limited resources, the present project is restricted to:
Goal: prediction of fuel consumption and actual temperatures by simultaneous simulation of a house and its heating system subjected to a given control strategy. This in contrast to the prediction of net heating demand as a function of a presumed indoor temperature profile, which is common practice up to now. In that case, the influence of the heating system (i.e. efficiencies, time lag and control inaccuracies) is grossly simplified. Similarly, the building is over-simplified when fuel consumption prediction is based on system simulation with net heating or cooling demand of the building as a known condition. Especially with non-optimal systems (with respect to either design or tuning) this almost certainly leads to large discrepancies between predicted and actual temperatures and fuel consumption.
Tools: 1) generalized simulation (sub-)models with which the dynamic thermal interaction between house and heating system can be examined and 2) knowledge of simplicity (related to necessary computing power) versus accuracy with respect to these models. In view of the hectic information technology developments, it is difficult to predict future possibilities for

building designers and system consultants. There is no doubt however that these will be much larger than at present/10/. In practice this means that it is not clear what kind of form the design tools should take. Because of this unclearness and to link up with recently initiated international research (intelligent knowledge based systems, energy kernel system/11/, the best strategy seems to be to start with enhancing our knowledge concerning the simulation of building and heating system.
Users: researchers and consultants associated with domestic heating systems
Buildings: reference houses for energy research with different levels of energy conservation measures
Systems: 1) wet central heating system, 2) air heating system, and 3) floor heating system

It is assumed that from literature (e.g./12/) and manufacturers sufficient data on system component dynamic behaviour can be obtained. An existing large computer model, ESP (see/13, 14/), will be expanded with system simulation features and sub-models for the various system components and controls. Sensitivity analysis will be used to generate modelling knowledge on for example level of model simplification versus accuracy.

3. PILOT EXPERIMENTS

As a first step in increasing our insight and for generating experimental results, some pilot measurements were carried out in a small apartment with ditto heating system (one-pipe wet central heating system, combi-boiler, room thermostat controlled and exhaust-only ventilation system). An extensive account of these experiments can be found elsewhere/15/; some selected results are reported below.
Temperature measurements near the radiators showed that a radiator not only influences the wall section directly behind it but also a "band" (width approx. 0.5 m) around the radiator. Of course the influence decreases with increasing distance from the radiator.

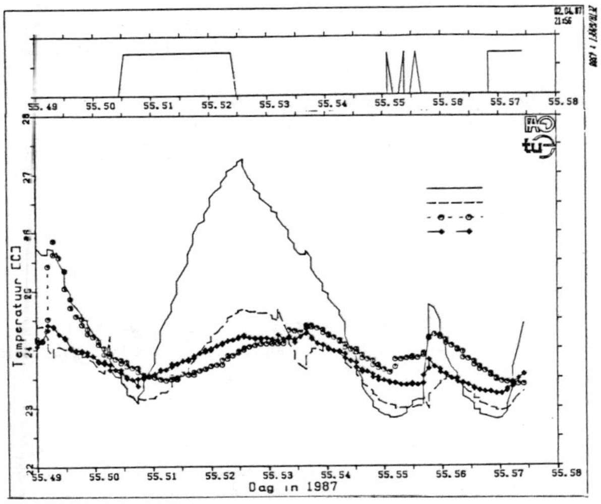

Figure 1 Temperatures in and near the room thermostat and the burner signal for February 24, 1987 (55.49 = 11:45, 55.58 = 14:00). Annotation top down: air near the thermostat, thermostat surface room side, air inside thermostat; adjoining wall surface

From temperature measurements in and nearby the room thermostat it became clear that the switch temperature is not a simple function of ambient temperature and surface temperature near the thermostat. This is illustrated by figure 1. The temperature of the wall section behind the thermostat differs from the wall temperature next to the thermostat, whilst the former is more important with respect to the switch temperature.
The room thermostat proved to be a very important component with respect to the dynamic behaviour of the system. The acceleration element in the room thermostat determines the boiler switch frequency. Without the acceleration element the cycle time was about 90 times longer (180 versus 2 minutes). More importantly, the total burner-on time - over an equal period of time - was approx. 50% shorter which resulted in an appreciable lower fuel consumption. It led however also to considerable increased air temperature fluctuations (up to 3 K difference between highest and lowest value during one cycle). At the time it was not clear whether these fluctuations would be acceptable to occupants. This was investigated in a separate study/16/.

4. SYSTEM SIMULATION IN ESP

The research starts from an existing building energy simulation package: ESP/13, 14/. ESP uses the simultaneous plant modelling technique. In contrast to the sequential modelling technique (used by e.g./17/), this is a rather new development field. This technique involves the representation of plant parts by discrete nodal schemes and by the derivation of energy and mass flow equation sets which represent whole-system, inter-node exchanges over time and space dimensions. The resulting plant matrix is integrated with the multi-zone building matrix, subjected to any control function based on nodal condition statements, and the combined matrix solved simultaneously and repeatedly as the model steps through its finite time increments.

At present the following plant component models are available: mixing box, humidifier, fan, cooling coil, heating coil and duct section for air systems and boiler, radiator, pipe section, convergent and divergent tee-piece and pump for water based systems. These are all single-node (first order) versions. The following plant controllers are presently available: proportional water flow rate throttler to control an air heating/cooling coil and general controllers with basic proportional action on heating/cooling flux or on flow rate as a function of sensed control variable (temperature or enthalpy).

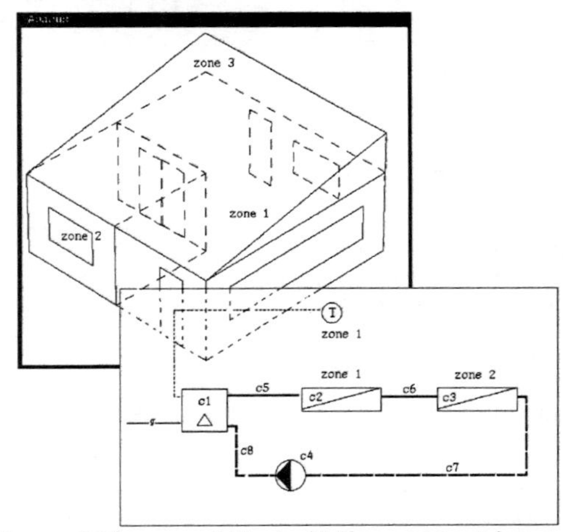

Figure 2 Diagram of the building and heating system

As a brief demonstration of ESP's current heating system simulation possibilities, the 3-zone office of figure 2. was extended with a simple wet central heating system consisting of: boiler (c1), radiators (c2, c3), pump (c4), heating pipe sections (c5 ... c8) and an air temperature sensor located in zone 1. The boiler is switched on/off when the air temperature falls below or rises above 20 °C respectively.

Although the chosen parameter values are more or less realistic, it must be stated that the simulation results must be regarded as an indication rather than in an absolute sense.
Simulations (with building side time steps of 3 minutes and 1 minute plant time steps) were carried out for 1 day. The predicted air temperatures are shown in figure 3. It is immediately clear that the results originate from combined building and plant simulation. It can be seen that the boiler is switched on/off depending on the air temperature in zone 1. However, due to system inertia and the finite simulation time steps the boiler is not immediately switched on/off when the switch temperature is reached. This results in air temperature fluctuations between an upper and lower limit. A similar phenomenon can be observed in reality. Although zone 2 is much smaller than zone 1 they do have identical heating appliances in this particular example. It is because of the system lay-out (one pipe system, so temperature of c2 is always lower than temperature of c3) that the air temperature in zone 2 - although not controlled - more or less follows the air temperature in zone 1. The air temperatures around noon suggest

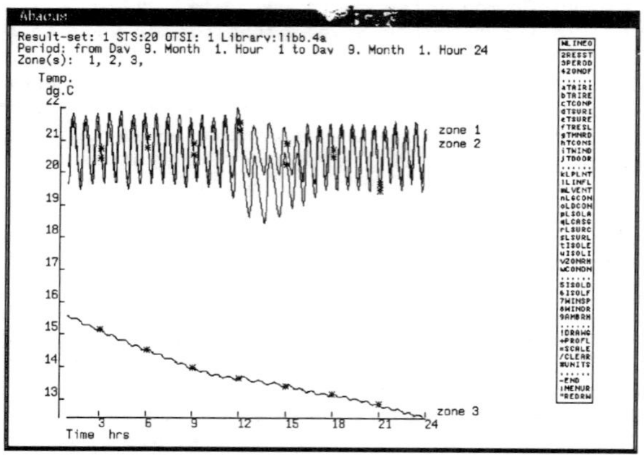

Figure 3 Air temperatures for the three zones

however that control of the heating system on the basis of the air temperature in zone 1, can lead to too low air temperatures in zone 2 during certain periods. The air temperatures in zone 3 are quite different because there is no direct heat input in that zone. The temperature in zone 3 constantly drops, due to too small simulation start-up period (1 instead of 5 days).

5. FUTURE WORK

Future developments within the framework of the present research will be:
- implementation of generalized flow network solver
- addition of parameterized component models
- addition of parameterized plant control features
- interactive plant simulation results recovery
- generation of knowledge on simultaneous simulation of building and plant

As a preliminary conclusion of the research up to now, we feel that we are working in the right direction and are steadily making progress. It will be a long way though before the eventual goal - valid and easy-to-use building performance evaluation tools - is reached.

References

1. CEC, *System Simulation in Buildings. Proceedings of the International Conference held in Liège (Belgium) on 6-8 December 1982,* Commission of the European Communities, Brussels, 1983. J. Lebrun (ed.), CEC report XII/425/83-EN

2. CEC, *System Simulation in Buildings. Proceedings of the International Conference held in Liège (Belgium) on 1-3 December 1986,* Commission of the European Communities, Brussels, 1987. J. Lebrun (ed.), CEC report XII-469-87-EN

3. Hensen, J.L.M. and P.J.J. Hoen, ''Energieproeftuin: a real scale experiment with low-energy dwellings in The Netherlands,'' in *Proc. Int. Climatic Architecture Congress,* Louvain-la-Neuve (Belgium), 1986.

4. Hensen, J.L.M., ''Energieproeftuin: results of an experiment on low-energy housing in The Netherlands,'' in *Proc. European Conference on Architecture,* Munich, 1987.

5. Clarke, J.A., ''Environmental Systems Performance,'' PhD Thesis University of Strathclyde, Glasgow, 1977.

6. Bruggen, R.J.A. van der, ''Energy consumption for heating and cooling in relation to building design,'' PhD Thesis Eindhoven University of Technology (FAGO), 1978.

7. Lammers, J.T.H., ''Human factors, energy conservation and design practice,'' PhD Thesis Eindhoven University of Technology (FAGO), 1978.

8. Hoen, P.J.J., ''Energy consumption and indoor environment in residences,'' PhD Thesis Eindhoven University of Technology (FAGO), 1987.

9. Lebrun, J. and G. Liebecq, ''System Simulation Synthesis Report,'' Energy conservation in buildings & community systems programme; Final report Annex X, International Energy Agency, 1988. Operating agent: University of Liège, J. Lebrun, report AN10 881020-RF

10. Augenbroe, G.L.M. and L. Laret, ''COMBINE Pilot study report,'' Information package for the DG XII programme "JOULE", Commission of the European Communities DG XVII, Brussels, 1989.

11. Clarke, J.A., ''The Energy Kernel System,'' in *Proc. The Future Building Energy Modelling,* Ispra (I), 1987. CEC JRC workshop

12. IEA, ''HVAC component specification booklets,'' Energy conservation in buildings & community systems programme; Final report Annex X, International Energy Agency, 1988. Operating agent: University of Liège, J. Lebrun

13. Clarke, J.A., *Energy simulation in building design,* Adam Hilger Ltd, Bristol, 1985.

14. Clarke, J.A. and D.J. McLean, ''ESP - A building and plant energy simulation system,'' University of Strathclyde ABACUS report, Glasgow, 1988.

15. Hensen, J.L.M., L.C.H. Dings, and W.J.A. van der Ven, ''Meetproject Someren,'' Technische Universiteit Eindhoven rapport FAGO 87.08.K., Eindhoven, 1987.

16. Hensen, J.L.M., ''Literature review on thermal comfort in transient conditions,'' *Building and Environment,* 1990. (To be published)

17. Klein, S.A. and W.A. Beckman et al., ''TRNSYS, a transient system simulation program,'' Engineering Experiment Station Report 38-12, University of Wisconsin-Madison, Solar Energy Laboratory, Madison, WI, 1988. Manual for version 12.2 and later

COMPARISON OF THE INTERNATIONAL THERMAL PROTECTION CLAIMS

Univ.-Prof. Dipl.-Ing. KLAUS W. USEMANN VDI
University of Kaiserslautern
Faculty of Architecture/Civil Engineering
Post-box 3049, D-6750 Kaiserslautern

Chair - Department of Building - Physics
and Building - Service - Engineering

INTRODUCTION

In order to save energy and expenses, the thermal protection for buildings is not only standardized in Germany, but also worldwide; this is an essential factor of the harmonization in building construction, to rationalization, and to the diminution of trade limitations. The thermal protection for building has gained increasingly importance in the last decades in many countries because of the evolution of the energycosts, and will stay a topical problem for many years.
The impuls for an international comparison of the current thermal protection determinations was a report at the occasion of the 3. Intern. Symposium "Energy Conservation in the Built Environment" 1982 in Dublin. In this study, the determinations of the limitation of the specific heat transmission loss got compared after the standing from 1978 to 1980 for 12 european countries. The results of this study are stated summary in Fig. 1.

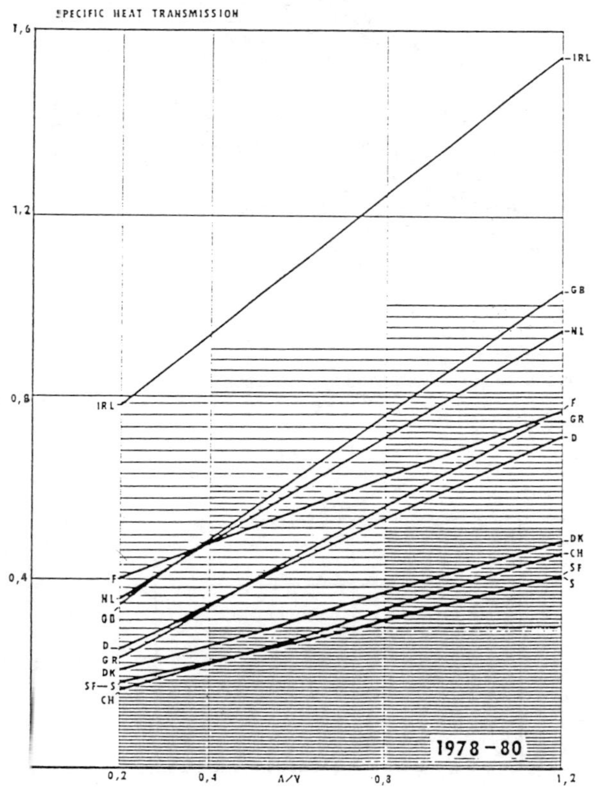

Fig. 1. International european comparison of the demandings to the limitation of the specific heat transmission losses, standing of about 1978-1980

Not-european countries are missing in this study; besides at that time the attemps for standardization were vague and many changements and recommendations of particular national determinations got not considered, respectively got evaluated individually at different times.
In the frame of a study of may chair- and research department about ≈ 35 national determinations respectively recommendations concerning the thermal protection of buildings got analysed after the standing from 1985.

LIST OF COUNTRIES

Australia (AUS), Belgium (B), Bulgaria (BG), Canada (CAN), Republic of China (VRC), Danmark (DK), German Democratic Republic (DDR), Federal Republic of Germany (D) with proof 1 und 2, Finland (SF), France (F), Greece (GR), Japan (Y), Jugoslavia (YU), Irland (IRL), Italy (I), Newzealand (NZ), Netherland (NL), Norway (N), Austria (A), Poland (PL), Portugal (P), Roumania (ROM), Sweden (S), Switzerland (CH), Spain (E), South-Africa (SA), Turkey (TÜ), Tchecheslovakia (CSSR), Hungaria (HU), United States of America (USA), United Kingdom (GB).
For each country, a date table got developed, which includes beneath general geogra ical data, informations about the respective thermal protection demandings (date of publication, publisher), climate conditions, details about the energy costs and a judgement of the thermal protection determinations.
In the interpretation of the protection determinations only those countries got considered in which the thermal protection should or must be proofed. Thereby the most different determinations exist in these countries. In the most thermal protection determinations limits for material data, the coefficient of heat transmission of building elements, thermal diffusivity resistance, the mean coefficient of thermal resistivity of facades or/ and building covering are given, in several countries only for domestic architecture, in other countries for domestic and utilitarian architecture. In many national determinations the window surface percentage is limited at max. 12 or 15 % of the floor surface. Various calculation determinations consider the thermal win because of sunenergy and/ or inner thermal sources. In several countries the thermal protection determinations are related to the kind and of the energyprices. Clear outside-and/respectively room temperatures are not given in all national determinations. Various standards are set for the air change, from the minimal air change rate to the maximal specific ventilation heat demand. 14 countries are setting no standard at all. In some countries only coefficient of joint permeability of doors, windows and glass doors is limited; the demands for the minimal air change are fluctuating between 0,5 times, 0,8 times or 1,0 times, echeloned from 0,5 times to 4 times e.g. after the building dimension or (0,1 to 0,2) times with sufficient building joints.

COMPARING CALCULATION

The international thermal protection determinations are based on different calculation methods. For this examination, an imaginary building, Fig. 2,

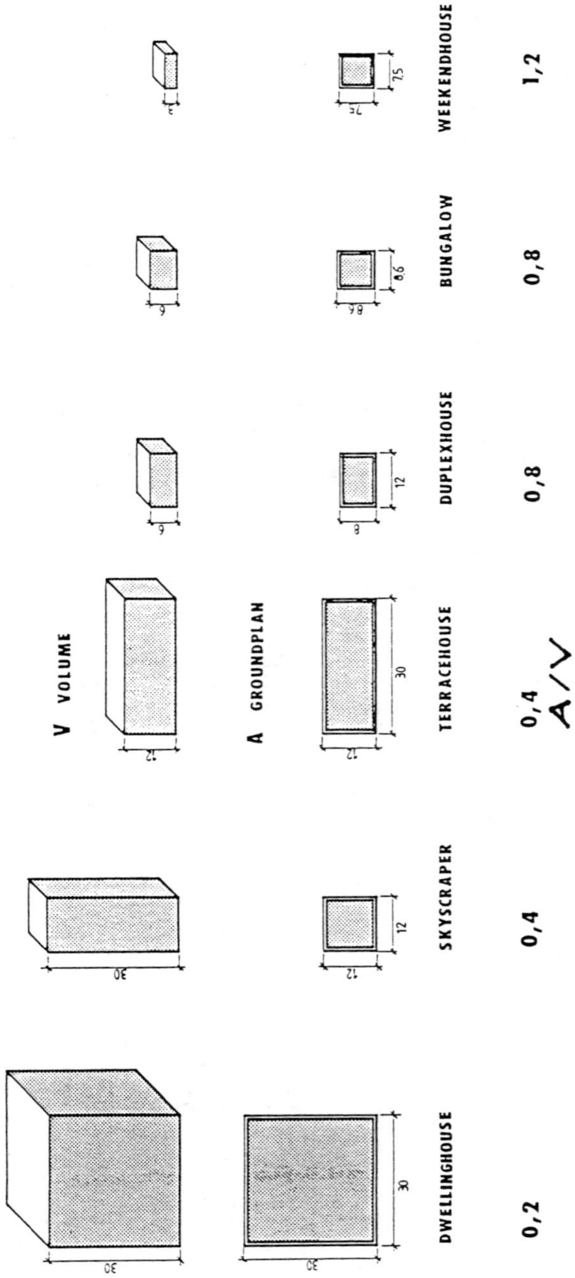

Fig. 2. Building geometry for the calculation of the realation from the heat transferring covering surface A to the thereby covered building volume V for several forms of buildings.

got presumed, which got changed depending to the geometrical conditions. The building geometry (relation of the heat transferring covering surface A to the covered building volume V) results correspondingly to the proof 1 (building cover proof) of the German (D) thermal protection determinations from 1982.

Under the geometrical conditions of Fig. 2 the comparing calculation of heat transmission loss was intended for all countries. The window surface share got presumed with 15 % of the floor surface. So for all countries a mean heat transmission coefficient k_m could be calculated.

From the heat transferring covering surface and the buildingvolume, the volume specific heat transmission loss Q_T/V and with the help of the thermal difference $\Delta\vartheta = \vartheta_i - \vartheta_a$, a volume specific heat transmission loss $Q_T/(V \cdot \Delta)$ could be ascertained. Only this value anables an international comparison of the thermal protection demandings, because it is independent of the outdoor- and indoor climate conditions.

Fig. 3. shows the result. In the international comparison the result pictures a snapshot which corresponds to the standing of 1984/85. For this only for statistical reasons chosen moment intentions to the innovation and revision of the thermal protection demandings are to be registered in the chosen countries.

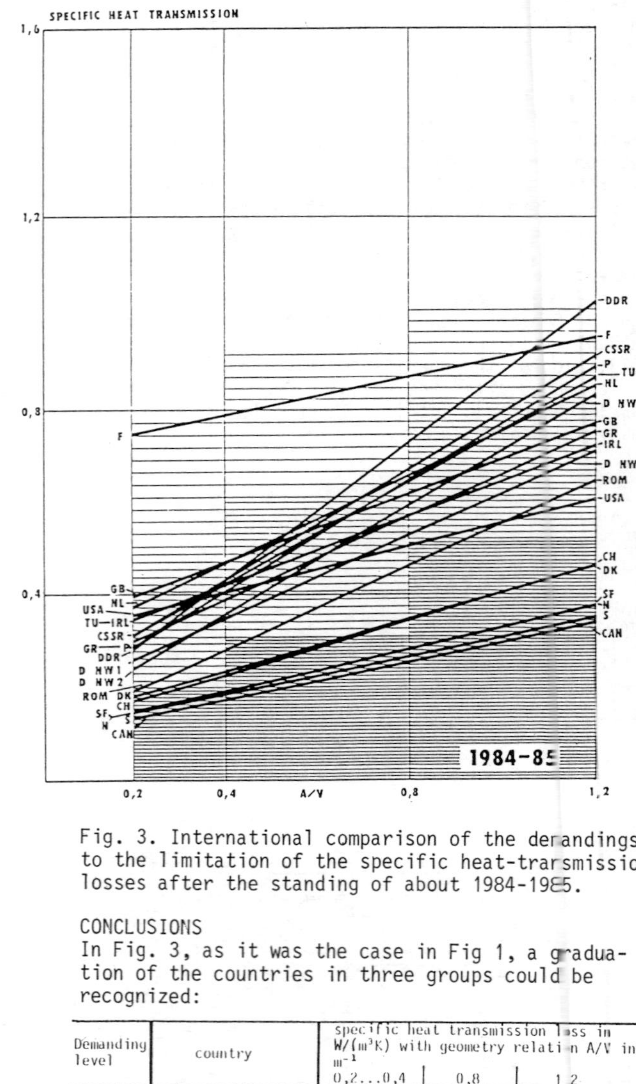

Fig. 3. International comparison of the demandings to the limitation of the specific heat-transmission losses after the standing of about 1984-1985.

CONCLUSIONS

In Fig. 3, as it was the case in Fig 1, a graduation of the countries in three groups could be recognized:

Demanding level	country	specific heat transmission loss in W/(m³K) with geometry relation A/V in m⁻¹ 0,2...0,4	0,8	1,2
I low	DDR,F,TU,NL,D, P,CSSR,(B),(BG), (PL),(HU)	0,2...0,8	0,6...0,9	0,8...1,0
II mean	GB,GR,IRL,D,USA, ROM,(AUS),(Y),(NZ), (SA), (I)	0,2...0,5	0,5...0,6	0,6...0,8
III high	CH,DK,N,SF,S, CAN	0,1...0,2	0,2...0,3	0,3...0,5

A comparison between Fig. 1 and 3 shows a general improvement of the thermal protection measures worldwide. The quick rise of the energy costs in the beginning of the 80's leaded to an important improvement to the thermal protection of buildings. It is easy to recognize, that the demandings improves from 15 % to 30 %, in the average at about 20 %. In Fig. 3 the influence of the shape of the buildings is obvious. The geometrical relation A/V diminishes clear~y with a more compact getting architecture (Fig. 2). At even good thermal construction of the covering surface of a building, a specific heat transmission loss for a free-standing small building (e.g. detached house) results at demanding level

I of about 0,9 $W/(m^3K)$
II of about 0,7 $W/(m^3K)$
III of about 0,4 $W/(m^3K)$.

A comparable large compact building (e.g. skyscraper) has only a specific heat transmission loss, with the same thermal construction of the covering surface. That means with the same mean heat transmission coefficient k_m, at demanding level:

I of about 0,5 $W/(m^3K)$
II of about 0,3 $W/(m^3K)$
III of about 0,2 $W/(m^3K)$

The thermal protection demandings of the countries of level I corresponded approximately to the minimal protection demandings from the hygienic point of view, with a mean heat transmission coefficient $k_m = (1,2...1,4) W/m^2K)$. Level III illustrates a super thermal protection (countries with a very strong and cold winter and/or a high development (industrial country)) in whom the volume specific heat transmission loss is produced to about $(0,2...0,4) W/(m^3K)$, which is realized with a mean heat transmission coefficient $k_m = (0,1...0,5) W/(m^2K)$.
Many countries of level II belong to the favourable climate region with a mild winter. Which requires economically senseful insulation measures only to a certain level, even if the technical standard could demand much more.
Many thermal protection directions of countries of level I refer to foreign (DIN 4108, ASHRAE etc.) or unseasonable standards, which might lead, especially in compact buildings, to wrong conclusions. Partially these directions are going to be revised.
At the transition from level I to II resp. III the volume specific heat transmission losses are diminishing, the ventilation heat losses stay untouched. If we rise the heat protection in level III, the ventilation heat losses represent approximately the whole thermal losses, because no heat transmission losses take place. The question appears, whether a further improvement of the thermal protection of buildings in level III is senseful and economically justifiable, without the reduction of the ventilation heat losses. In this level, several countries seem to have reached the limit of thermal protection of buildings. It only can remain under this level by mechanical ventilation system with heat recovery aggregates, at it is actually the case in many countries of level III.

COMPREHENSIVE ENERGY TARGETING AND COMPLIANCE SYSTEM, BASED ON AFFORDABLE WARMTH, FOR A MAJOR HOUSING ASSOCIATION USING COMPUTER APPLICATIONS.

S Burton and J Doggart
THE ECD PARTNERSHIP
11-15 Emerald Street, London
WC1N 3QL. UK.

ABSTRACT. A major Housing Association wants to improve the energy efficiency of its varied housing stock at appropriate points in its life cycle. The answer is an energy policy, tailored to the particular conditions, which is automatically incorporated into documents and procedures. All opportunities for improving the energy performance of the stock can be thus exploited.
This paper describes the approach taken in developing an appropriate policy including testing it out in use. The development includes:- determination of "affordable fuel bill" levels; use of appropriate computer domestic energy programs to assess fuel bills and condensation risk; assessment of policy for the range of housing types within the stocks, identification of low energy and energy conservation products and techniques acceptable to the Trust; development of architect's briefs and heating specifications; development of a certification scheme to ensure compliance with standards; testing of the system on new-build and improvement schemes to check on capital costs and feasibility.

1. STUDY BACKGROUND

The Notting Hill Housing Trust (NHHT) is a charitable housing association founded in 1964 to provide housing for low income people. It is based in West London and has a varied housing stock of 8000 dwellings including flats converted form 3 - 5 storey 19th century terraced housing, more recently built blocks and newly built housing.
Heating and insulation standards in the older property are frequently poor, leaving tenants with sub - standard temperatures and high fuel bills, and heating with bottle gas (propane) or paraffin may occur. Newer properties have been built with appropriate insulation and heating systems but may still be expensive to heat fully. The Trust decided in 1988 to prioritise the provision of affordable heating to its tenants at every reasonable opportunity. The ECD Partnership was commissioned to propose a complete energy policy, tailored to the particular housing stock, and to suggest how it could be implemented.
There are seven elements to the development, testing and running of the policy :-

- Determination of "affordable fuel bill" levels for tenants;
- Use of appropriate computer domestic energy programs to assess fuel bills and condensation risk;
- Assessment of policy for the range of housing types within stock;
- Identification of low energy conservation products and techniques acceptable to the Trust;
- Development of Architect's Specification and Contractor's Heating and Hot Water Specifications;
- Development of a certification scheme to ensure compliance with standards;
- Testing of the system on new-build and improvement schemes to check on capital costs and feasibility.

2. DETERMINATION OF "AFFORDABLE FUEL LEVEL" FOR TENANTS.

The poor spend a greater proportion of their income on fuel than do the better off, and frequently cannot afford to heat their homes adequately for health and comfort. Research into how much low income families could afford to pay in fuel bills was co-ordinated by the National Federation of Housing Associations (NFHA) and translated into target weekly fuel bills for households and dwellings of different sizes. The NFHA proposed these as "affordable warmth" standards and suggested that all low income housing should be built or refurbished to those standards so that "tenants should have homes they can keep warm enough for health and comfort at a cost they can afford."
Notting Hill Housing Trust accepted these as desirable targets for its housing stock while envisaging that capital finance constraints must be considered.
The target fuel bills were calculated in 1988 on the basis of 7p per square metre of floor area per week for the total fuel bill, but excluding standing charges. Table 1 shows these targets for various sizes of dwellings. Whilst these include all uses of energy in the house it is the heating and hot water costs that can be reduced by the provider of the housing by good design.

There are main opportunities when the Trust could improve the energy performance of its dwelling, first when it is building new dwellings, second when it is carrying out major refurbishment and third when it is carrying out improvements to existing dwellings, sometimes with tenants in place. The capital costs associated with improvements at the three stage is different and was investigated during the project.
As this project was carried out in 1989 for use in subsequent years it was decided to increase the affordable cost per week to 8p to allow for some inflation.

Table 1.	Targets for Affordable Warmth	
Occupancy	Expected Floor Area M^2	Estimated Fuel Target per week
1 person	32.5	£3.50
2 person	47.5	£4.50
3 person	60.0	£5.40
4 person	75.5	£6.50
5 person	86.0	£7.20
6 person	95.0	£7.80
7 person	112.5	£9.00

3. COMPUTER ENERGY ANALYSIS.

Total fuel bill calculations were required for the range of dwellings within the stock, both in their current standards and with energy improvements. Micro computer programs based on the Building Research Establishment's Domestic Energy Model (BREDEM) were used to assess fuel bills under typical conditions of use. These programs calculated all energy used for heating, lighting, cooking, domestic hot water and appliances.

Two programs were used, Energy Targeter for office based analysis of new and existing buildings where plans were available, and Energy Assessor for on site assessment (running on a fully portable computer) where a site visit was necessary and a quick audit was sufficient. Energy Targeter requires a full input of building dimensions including door and window sizes together with construction and heating systems details. The Assessor program includes geometric models of typical houses and so requires less input on dimensions, and an audit can be completed in less than half an hour on the site. The latter was therefore thought to be highly suited to the assessment of existing dwellings where tenants were in occupation and smaller improvement works were envisaged.

As well as assessing energy consumption, Energy Targeter also makes an assessment of the condensation risk arising from combinations of different levels of heating with different insulation and ventilation conditions. Thus the risk of condensation occurring inside a dwelling as a result of the occupant not heating the flat to "full comfort conditions", can be assessed against various packages of insulation and heating. Condensation and subsequent mould growth were of great importance to the Trust and this facility was used in helping to formulate the policy.

Although the programs were specifically used as design tools as an input to the policy making, their subsequent use for certification of designs and in the assessment of proposals both on site and in design offices, was envisaged.

4. ASSESSMENT OF POLICY FOR THE STOCK.

Intervention to improve energy standards was considered separately for the three different construction activities carried out by the Trust.

For new-build schemes previous experience had shown that the 8p per m^2 per week target was easily achievable by improving insulation standards over and above current Building Regulation Standards. Table 2 compares current Building Regulations (1985) insulation standards with those typically required to achieve the "affordable warmth standard".

Additional constructional costs are associated with these measures, but these can be offset by reductions in the cost of the heating system. NHHT was happy to accept the resulting marginal cost increases, subject to testing out on current new-build schemes.

For existing dwellings which were to be refurbished by the Trust, reducing fuel bills to the "affordable" level involves more work in insulating the structure and providing efficient heating systems, than with new construction.

Table 3 shows typical insulation measures required to achieve the 8p per m^2 target.

Additional costs vary greatly with existing construction and existing conditions of windows, walls, floors etc, since,, for example, the costs of providing drylining for energy conservation reasons can be reduced if replastering would otherwise be specified. Being mindful of budgeting constraints, which may vary between schemes, NHHT felt that a lower target might need to be adopted in some cases. Thus the 8p target was increased by 30% to 10.5p with the intention of adopting the 8p target if reasonable or the 10.5p target under difficult conditions.

Table 2. Newbuild standards.

	1985 Building Regs	Affordable Warmth standard
Roof	100mm insulation	150mm insulation
Walls (external)	50mm insulated cavity	100mm insulated cavity
Windows	Single glazed	Double glazed
Floors	No requirement	25mm insulation

When partial works are to be carried out to existing dwellings, such as redecoration between tenancies, providing heating systems or installing sound proofing, it may be appropriate to include some energy measures. Since major insulation works would frequently be necessary to reduce fuel bills to the 8p per m^2 target, the upper limit of 10.5p per m^2 was thought to be more appropriate both to available finance and to working with tenants in occupation.

To investigate works necessary to achieve the 10.5p target audits were undertaken of 15 dwellings, representing a cross section of NHHT housing stock. Both Targeter and Assessor programs were used as appropriate and the fuel bills for various heating systems (existing,electrical,gas central heating) were assessed. Where there were above the 10.5p per m^2 level, fuel bills with appropriate insulation and other conservation measures added, were assessed until the target was reached.

It was concluded that for simplicity the stock should be broken down into three groups with different recommendations for each group . Table 4 shows these recommendations.

Table 3. Major refurbishment standards.

	Affordable Warmth Standard
Roof	150mm insulation
Walls(external)	50mm drylining
Windows	double/secondary glazed
Floors	25mm insulation
Heating system	Condensing gas boiler

Table 4. Existing dwellings improvement standards.

Dwelling Types	Recommendations
A. Mid-Terrace mid floor (inc. high ceiling and large windows).	- Gas Central heating with storage d.h.w or room heaters with instantaneous d.h.w. (small dwellings). - Draughtstripping to all external doors and windows.
B. Mid-Terrace top floor	- Gas central heating with storage d.h.w. or gas room heaters with instantaneous d.h.w (small dwellings). - Draughtstripping to all external doors and windows. - Loft insulation minimum 100mm, 150mm if none existing.
C. End-Terrace (semi detached and dwellings with large external wall areas).	- Assess necessary and most cost effective improvements using Energy Targeter. Measures likely to include:- - Gas central heating with storage d.h.w. and condensing boiler. - Draughtstripping to all external doors and windows. - Loft insulation minimum 100mm. - Drylining to flank wall or all external wall. - Secondary glazing.

5. ASSESSMENT OF INSULATION PRODUCTS AND TECHNIQUES.

The Trust was concerned to use energy products and conservation techniques which had been proved to have good performance and to be maintenance free, as far as possible. The following product areas were investigated and recommendations made:-
Boilers for central heating
Controls for heating systems
Drylining
Draughtstripping
Secondary glazing systems.
Chosen options were incorporated into the documentation.

6. ARCHITECTS ENERGY SPECIFICATION AND CONTRACTORS HEATING AND HOT WATER SPECIFICATION.

In order that energy efficiency was fully incorporated into all schemes, the policy decisions formulated above were converted into an Energy Specification to accompany other documents that the Trust sends to its outside architects and designers. Also a Heating and Hot Water Specification was produced. The Architects Specification includes the requirement to reduce fuel running costs to the 8p or 10.5p target and to provide computer calculations from the certifying agency.(see next section). Insulation levels are proposed for all elements, together with the range of acceptable heating systems, boilers and controls. The Heating and Hot Water Specification is based on a standard "Contractor Design" specification but incorporates all elements of the energy policy as appropriate.

7. SCHEME CERTIFICATION.

Before a new-build or major refurbishment scheme is approved by the Trust, it is proposed that the "energy performance" of the scheme is assessed by an independent certifier. Certification uses the Energy Targeter program and must demonstrate that likely fuel bills meet the appropriate cost standard. With major refurbishment schemes, it may be necessary for two designs to be assessed, so that the Trust can decide whether to adopt the 8p or 10.5p standard for a particular scheme.

8. TESTING THE POLICY OUT IN PRACTICE.

Having put in place all the elements of the policy and its implementation, testing out on actual schemes is now in progress. The purpose is to explore the capabilities of designers to meet the energy targets in the range of schemes, to determine any additional costs and to examine how there costs can be met. This phase includes an element of energy consultancy for designers requiring assistance in bringing schemes up to the energy standards. This is being provided by the ECD Partnership.
It is hoped that in the longer term monitoring of completed schemes in terms of fuel bills, maintenance and tenant satisfaction, will be carried out to confirm the overall effectiveness of the policy.

9. CONCLUSIONS.

This paper, demonstrates the use of computer technology to assess tenants fuel bills levels, to set "affordable" levels and to ensure that these bills are realised in practice in both new housing schemes and refurbished housing. This method of ensuring that tenants can afford to heat their homes adequately is preferable to the alternative method of laying down rigid standards for insulation, heating etc as it accommodates design variations and can also produce least cost schemes.

ENERGY AUDIT ASSESSMENT ON OFFICE BUILDINGS STOCK: Methodology and First Results

M.Brocato (*), F.Parrini (**), S.Sciuto (*), R.Viadana (**)

(*) CONPHOEBUS s.c.r.l.
P.B. 95030 Piano d'Arci
95127 Catania - tel 95/291407

(**) ENEL-CRTN
Via Rubattino 54
20143 Milano - tel 2/88473931

ABSTRACT: Conphoebus and Enel performed the verification of an original methodology for the energy audit assessment on office buildings stock. It is finalized to analyze the stock energy consumptions, find out buildings needing a detailed audit and performe energy rehabilitation designs for buildings having the worst energy performance. The first step was the collection by means of a questionnaire, of general information about geometry, construction typologies, management and energy consumption of office buildings having a volume larger than 5.000 cu.m on which retrofitting works are foreseen. The data were put into a data-base, and statistical analysis, based on energy indicators, were performed. Starting from the ranking through the energy indicators, a methodology was defined for the selection of the buildings on which retrofit is economically convenient. For such a purpose an algorithm for the evaluation of the reference thermal energy consumption was determined. Several interesting considerations are reported on the energy demand distribution, energy saving potentials, payback periods and energy conservation opportunities.

1. FOREWORD

Conphoebus and Enel lined up in 1986-87, by means of experimental verification and comparison analysis with several computer codes, an original approach metodology for the energy assessment on office buildings stock.
In November 1987 Enel decided to verify this approach through several random real cases, which form a test building stock.
An informative questionnaire, requiring:
- geometrical data
- occupational data
- typological data
- system data
- energy consumption data

was sent to the building local responsible, for its filling.
The examined stock consists of about 100 office-buildings, located in the northern-central Italy, having volume larger than 5.000 cu.m and on which retrofitting works are probable in the future.

2. STATISTICAL ANALYSIS OF THE BUILDINGS STOCK AND SELECTION OF THE WORST UNITS

The questionnaire data were verified as congruence and put into a data base, for their statistical analysis. Several attempts of classification through the energy indicators referred to volume, surface, occupancy and degree-days have picked up goods and lacks on each of the same indicators.
Finally, the buildings selection has been performed on the basis of both the total energy demand and the heating consumption. The first utilized energy indicator was the Global Volumetric Yearly Consumption (GC).
It is possible to see that buildings having indicator greater than 400 MJ/cu.m guarantee economically convenient retrofits, with energy saving of about 25-35%. This indicator value, naturally, is not a cut-off threshold: in the 200-400 MJ/cu.m range some buildings have rather convenient single retrofit, that bring to a good global energy saving (about 10-15%), but these are particular cases, to be evaluated time by time.
On the contrary, buildings at GC<200 MJ/cu.m, generally, have low probability that Energy Conservation Opportunity (ECO) has good payback period. For what the other indicators are concerned, it was not possible to pick up similar thresholds, exeption made for the Normalized Volumetric Heating Consumption (HC), because of some peculiar energy uses of certain buildings (such as Computer Center, kitchen,worshops..). In particular, HC indicator was used in setting the 200 MJ/cu.m as the lower GC threshold for leaving out the energy saving retrofits possibility.
It was possible to notice that this lower threshold is not correct: infact, not only units which have GC value in between 200-400 range, but also some buildings at GC<200 MJ:cu.m and HC quite higher than the average, guarantee still good retrofits. At the end, test by test, it was possible to pick up all the buildings on which some energy saving retrofits will be economically convenient.

3. PRELIMINARY CONSIDERATIONS ABOUT THE BUILDINGS STOCK

Up to now, the studied office buildings stock seems rather unhomogeneous, both for construction characteristics and energy consumptions.
Buildings and systems typology, ages of construction and of HVAC systems are very different. In any case a reference energy demand may be defined only for the heating share, since the electric one is strongly influenced by the specific utilization and activities that are present in the building. This reference value may be referred to the same building, as geometry and utilization, but having volumetric heat loss coefficient which matches the Law requirements. Nevertheless the difference between the present heating demand and the reference demand represents neither the upper nor the lower threshold for energy saving retrofits, but it can be considered as a reference value. Infact, while it is always technologically possible to erect building much more insulated than Laws prescribe, it is altough evident that the eventuality of a retrofit is reasonable only for that

buildings whose consumptions are enough higher than the reference one.
The energy demand shares before and after the retrofitting execution is shown in figg. 1-2. It refers to a retrofitting extend up to 44% of the whole stock volume.
The energy demand of such a part passes from the 57% of the total demand (see fig.1) to the 48% (see fig.2), becouse of the retrofitting effect.

Fig. 1 Energy Demand Before Retrofitting

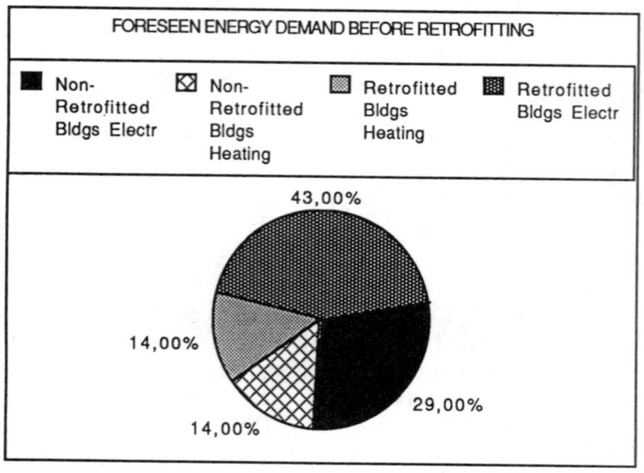

Fig. 2 Energy Demand After Retrofitting

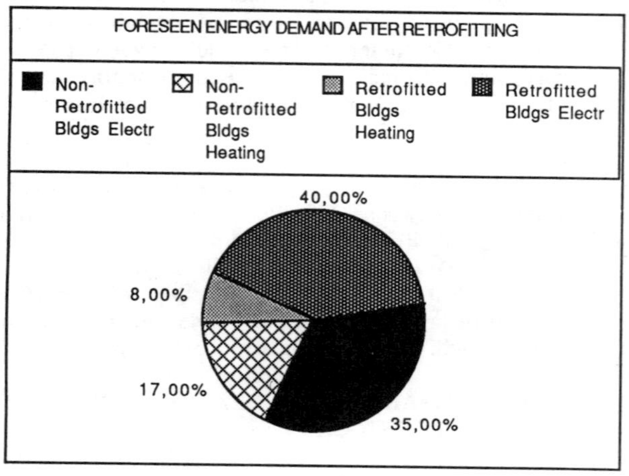

The mean total energy saving potential is equal to 26% shared as in tab.1 and fig. 3.

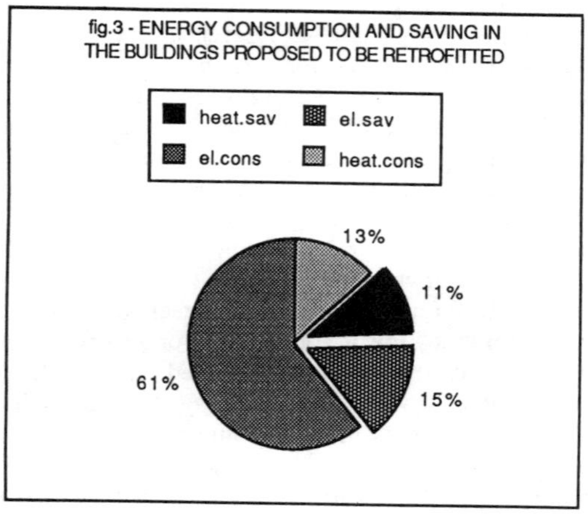

tab 1 - Energy consumption and saving in the retrofitted buildings

Yearly cons.	Total build	Retrof.buid	Saving	Saving%	
[TJ]	(before retrofitting)			whole stock	retrof.build.
Electtrical	333.2	200.1	40.1	12	20
Heating	127.1	61.7	27.7	22	45
Total	460.3	261.8	67.8	15	26

In the retrofitted buildings it is possible to obtain a mean yearly economic saving of about 1.35 ECU/cu.m, with an investment of about 5.25 ECU/cu.m.
The selected ECOs achieve to make more homogeneous the whole buildings stock. In particular, although the electrical consumption would remain rather unhomogeneous, because of the several different activities performed in the buildings (C.C., Services,...) the heating energy demands stabilize to better and less widespread values.

4 RETROFIT HYPOTHESIS IN THE SELECTED BUILDINGS

A retrofit hypothesis was picked up for each selected building, based on a cost-benefit analysis, by means of the computer code "TRAKLOAD". The higher threshold in accepting the payback period, for each ECO, was referred to the mean-life of the same ECO.
A global threshold of about 8-10 years was defined for the whole hypotesis of retrofit. Fig.5 refers percentage distributions of ECOs.

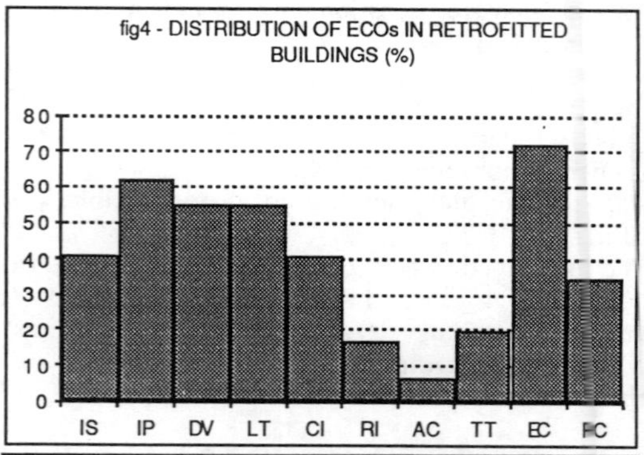

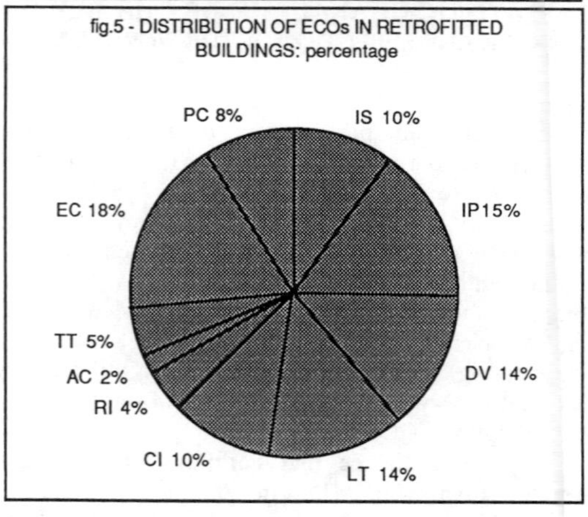

IS	Roof insulation	RI	Descale lighting
IP	Wall insulation	AC	Hot-water system
DV	Double pane	TT	Thermostat reg.
LT	Desk-lighting	EC	Improving heat. eff.
CI	Lighting control	PC	Descale heat. power

5 SOME CONSIDERATION ABOUT HEATING ENERGY INDICATOR

Because of occupational dishomogeneity, the heating energy indicator that better succeeds in ranking the buildings, is the Normalized Volumetric Consumption.
The most evident limits of this indicator are:
- the impossibility to distinguish buildings in order to their size and to their absolute saving potential;
- unaccuracy of the degree-days as normalizing terms to compare buildings located in different climates.
In order to get over these limits, we looked for an algorithm for the calculation of heating energy consumption of a generic building, which matches the present laws. This is the "Reference Energy Consumption", that gets over the evidenced limits:
- it represents a more correct normalizing term than the degree-days;
- it permits the evaluation of a saving potential, like the difference between the actual consumption and that after a retrofitting which brings the building within the Law requirements, come into operation after the building erection.
The study performed by means of the TRAKLOAD code, refers to geometrical, typological and occupational data, peculiar of the examinated buildings stock. The sensitivity analysis allowed to determine an algorithm correlating heating consumptions to degree-days, occupancy and building volume.
The algorithm was explicated by means of an abacus, showed in fig.6, in which, starting from the values of the degree days of the city, it is possible to go vertically to the occupancy and volume curves. From the points so determined, it is possible to read, in the second quadrant, the Reference Energy Consumption, defined in the following range:
- volume : 10.000-30.000 cu.m;
- index of occupancy : 20-200 cu.m/occ;
- degree days : 1.400-2.600

6 RESULTS AND COMMENTS

The main results and comments about the performed test on the 100 office buildings stock, are:
- The quote of electric end-uses, in continental regions, is about 60%, while about 40% is for heating use.
- Considering also the Mediterranean regions, these quotes become respectivelly 70% (electric end-uses) and 30% (heating use).
- When the global volumetric yearly consumption (GC) is greater than 400 MJ/cu.m, the building retrofits is certainly convenient.
- An algorithm was defined for the evaluation of a "Reference Heating Demand", as function of degree-days, occupancy and building volume. It represents an important step for the more correct normalization of the energy demand; furthermore, we can utilize it for the calculation of the energy saving potential.
In the future, this algorithm shall be studied to enlarge the utilization field and improve its accuracy.
- A threshold value for the ratio between the present heating demand and the reference heating demand, may be defined. the buildings which have a ratio value greater than the threshold, present retrofitting opportunity.
Presently such a threshold seems to be about 1.3 but it should be further investigated.

Fig. 6 Simplified tool for the evaluation of the Reference Heating Consumption

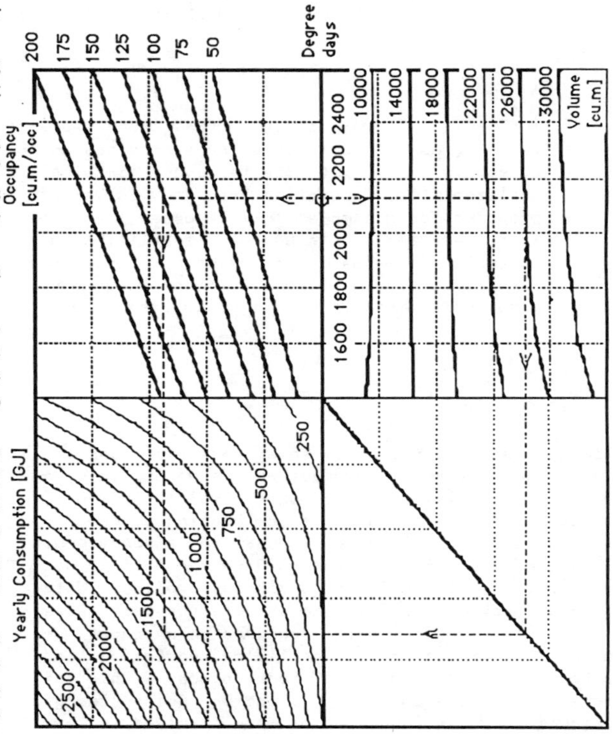

REFERENCES

-IEA XI Energy Conservation - Source Book for EnergyAuditors - M.D. Lyberg , April 1987
-Agence Francaise pour la Maitrise de l'Energie - Guide de Diagnostic Thermique - Eyrolles , 1987
-F. Conti - The IEA work on energy auditing in buildings: dicussion of the main results - ICBEM'87 Lausanne

CLIMATIC DATA FOR THE DESIGN OF SOLAR BUILDINGS: RESULTS AND FINAL PRODUCTS OF THE "EUFRAT" PROJECT

Bernard BOURGES (Consultant, 1, sq. René Cassin 35700 - Rennes France)
M.J.CARVALHO (LNETI, Portugal), M.CHERREY (ARMINES, France)
M.PETRAKIS (Obs. of Athens, Greece), C.ARMENTA (Univ. of Madrid, Spain)
V.RUIZ, F.FRUTOS (ETSII Seville) and J.M.CARIDAD (ETSIA Cordoba, Spain)

ABSTRACT. The aim of the EUFRAT concerted Research Action was to prepare climatological data required as an input for solar system design tools (solar radiation and temperature). The basic objective dealt with the frequency distribution of solar radiation and the derived concept of Utilizability. Cumulative Frequency Curves of solar irradiance and Utilizability data sets were prepared for a sample of 37 European locations, from long-term weather data files (hourly sum). These reference CFC's and Utilizability data sets were used in order to establish simple reconstitution formulas and correlations, suitable for European climates. Related topics were also investigated: e.g., temperature and degree-day values. Final results, - reference data sets and computation methods-, will be published both as a textbook and as a software package.

1. INTRODUCTION

Climatic data (principally solar radiation and ambient temperature) are required as an input by a number of Building Performance Evaluation tools, namely as far as designing space heating systems or using solar energy (active or passive) is concerned. The aim of the EUFRAT concerted Research Action was to prepare, - in addition to already available data sources such as the Solar Radiation European Atlas /1/-, climatological data required as an input for solar system design tools (solar radiation and temperature). This project took place within the framework of the European R. and D. Solar Energy programme and six laboratories were involved in this programme from July 1987 to December 1989.

2. SOLAR SYSTEM DESIGN TOOLS

The long-term performance of solar systems (either thermal active or passive, or photovoltaic) can be assessed by using several types of calculation method. Each of them requires a particular kind of input weather data. Beside detailed simulation programmes requiring files of hourly values (such as Test Reference Years, including solar radiation, temperature, wind, etc.), there are a number of simplified methods which need a more limited climatic information.
In many cases, monthly averages of solar irradiation and ambient temperature are sufficient. For example, Method 5000 recommended in the European Passive Solar Handbook requires such monthly values /10/.

Nevertheless, this data must often be completed by information about the frequency distribution of these parameters at several time scales, either explicitly or implicitly through the derived concepts of Utilizability (for solar radiation) and Degree-days (for temperature).

The efficiency of most solar conversion systems is not constant and varies with the amount of incident solar radiation: many systems present threshold effects or other types of non-linearities. For example, in the case of a flat-plate solar collector /2/ (Optical efficiency: n_0; Heat loss coefficient: U_l) operating at temperature T_b, with ambient temperature T_a, the threshold (or critical) intensity is defined by

$$I_c = U_l \cdot (T_b - T_a) / n_0 \qquad (1)$$

The collector produces heat only if the incident irradiance is higher than I_c (due to the effect of control). Similar phenomena occur in passive solar systems, but in an inverse way: if the solar radiation is low, it is fully used to meet the space heating load; if it is higher than some threshold value, it is either dumped or stored in building inertia (with heat losses). In photovoltaic system, threshold phenomena are not systematic (although they occur in some particular applications) but the efficiency of the system depends on the temperature of the cell (itself increasing with incident radiation).
If the efficiency of a given system depends on the incident solar radiation, the average efficiency for a given period cannot be estimated only from the average irradiation: it requires to know at least the frequency distribution of solar irradiance.
This explains why the frequency distribution of solar irradiance (or the derived concept of Utilizability) is used as an input in many calculation methods. For thermal active solar systems (space heating or domestic hot water), specific methods ESM1 and ESM2 have been developed for Europe /2-3/. For passive solar applications, several methods exist using the concept of "Unutilizability" /4/. For PV systems also, simplified methods based on the same concepts can be used for long-term performance estimate /5/. Most of these applications have been reviewed recently /6/.
Furthermore, the Cumulative Frequency Curve of a given month also enables to define a "Typical Meteorological Day" which reproduces both the frequency distribution and a realistic time profile /7/. This TMD may be used as an input for fast and simplified simulations of any type of solar system.
The aim of the EUFRAT project was to make this information available to the users, from the suitable processing of original extended data files.

3. UTILIZABILITY AND CFC'S: DEFINITIONS

The frequency distribution of solar radiation can be presented as Cumulative Frequency Curves of solar irradiance (CFC). CFC's are defined as curves giving the time n_h (mean daily number of hours, during the month considered), during which the solar irradiance on a given plane has exceeded a value I_c (threshold irradiance or critical intensity) /8/. Examples are given in Fig.1: the time, n_h, is read on the horizontal axis (Unit: hour/day); the critical intensity I_c, is read on the vertical axis (Unit: W/m2).

The CFC intercept with horizontal axis is the mean number of hours between sunrise and sunset, S_0. The intercept with the vertical axis represents the maximal solar irradiance, I_{max}, which can be achieved during the month on the plane considered. The overall area limited by the CFC and both horizontal and vertical axis is simply the mean daily sum of global solar irradiation on the plane considered H_t.
The Utilizability, which is defined as the fraction of incident monthly solar irradiation actually available above a given critical intensity I_c, can be derived from the CFC (Fig. 2):

$$\varnothing\,(I_c) = \int_{I_c}^{I_{max}} n_h(I)\,.\,dI \,/\, H_t \qquad (2)$$

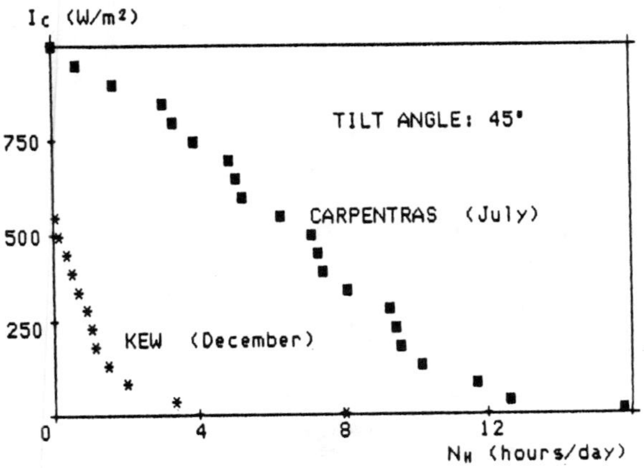

Fig.1: Cumulative Frequency Curves of solar irradiance. The critical intensity I_c has been exceeded during a time period n_h.

4. SIGNIFICANT RESULTS

The basic objective of EUFRAT project dealt with the frequency distribution of solar radiation and the derived concept of Utilizability. Cumulative Frequency Curves (Fig.1) of solar irradiance and Utilizability data sets were prepared for a sample of 37 European locations, from long-term weather data files (hourly sums) /11/.
An accurate computation algorithm for global irradiance on tilted surfaces was used, after additional validation was carried out from measurements made at Seville. Quality control criteria based on time series analysis were also developed. It was also an opportunity to produce high quality values of monthly global irradiation on tilted surfaces on a wide range of inclinations and orientations.
The former reference CFC's and Utilizability data sets were used in order to establish simple reconstitution formulas and correlations, suitable for European climates. A very simple formula has been obtained to estimate the Utilizability curve from standard parameters (Fig.2):

$$\varnothing\,(Xc) = (\,1 - Xc\,/\,Xm\,)^{Xm\,-\,a.Xc} \qquad (3)$$

with

$$Xc = I_c.S_0\,/\,H_t \qquad Xm = I_{max}.S_0\,/\,H_t$$

Eqn (3) depends only on two dimensionless parameters Xm and a, which can be computed using practical correlations.
Some additional related topics were also investigated: for example, temperature and degree-day values (required by many types of computations) were collected and processed. A simple algorithm for the computation of monthly Degree-days at any base temperature has been developed and validated.
The introduction of Incidence Angle Modifier into Utilizability calculation was also investigated. The corresponding method can be used to estimate energy transmitted by glazings in passive solar systems and energy absorbed by solar collectors in active solar systems /12/.

5. FINAL PRODUCTS

The first final product of the EUFRAT project is a document entitled "Climatic data for the Design of Renewable Energy Systems in Europe: Frequency Distribution of Solar Radiation and Temperature" /13/. It contains data tables and practical information to use them in connexion with passive and active solar system design tools (Tables I and II):

- Solar radiation: 37 locations (including data for Concentrators and Compound Parabolic Collectors)

- Temperature and Degree-Days: 230 locations

Simple computation methods which enable to get similar information at any location are also given in this document (e.g. how to calculate monthly solar irradiation on a tilted plane and the corresponding utilizability curve; degree-days at any base temperature).
Furthermore, a set of computer programmes and data bases for personal computers has been prepared: the complete CFC's and Utilizability data set is available on disk with user's routines (pre-processed data for 80 surfaces at 37 locations); the complete temperature and degree-days data base is available under the same form; a specific programme has been developed for the estimate of CFC's and Utilizability at any location, if no measurements are made.

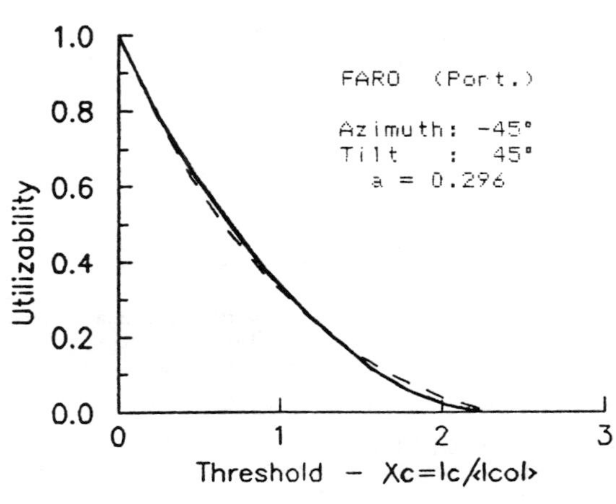

Fig.2: Utilizability curve:

——— Reference curve (EUFRAT data)
– – – Empirical fit (Eqn. 3)

6. CONCLUSION

From the development of reliable summarized solar radiation data sets (simultaneously with the improvement of simplified design methods within the framework of other projects), the present study will result in the improved design of solar systems by engineers and manufacturers. It will produce a useful complement to the actual European Solar Radiation Atlas and the European Passive Solar Handbook. Its applicability could likely be extended to other areas with a large solar potential (Africa, Middle East, etc.).

REFERENCES

/1/ W.PALZ et al. Solar Radiation European Atlas (2 volumes). Verlag TUV Rheinland, Cologne (1984).

/2/ B.BOURGES. Active Solar System Design: Handbook of European Simplified Methods (to be publ.).

/3/ J.ADNOT et al. Workshop P.9, EUROFORUM NewEnergies 88, Sarrebrucken (1988).

/4/ W.A.MONSEN et al. SOLAR ENERGY, Vol.27, pp.143-147 (1981).

/5/ D.R.CLARK et al. SOLAR ENERGY, Vol.33, n.6, pp. 551-555 (1984).

/6/ S.A.KLEIN and W.A.BECKMAN. ASME J. of Solar Energy Eng., Vol. 106, pp.393-402 (1984).

/7/ D.FEUERMANN et al. SOLAR ENERGY, Vol.35, n°1, pp.63-69 (1985).

/8/ B.BOURGES. Int. J. Solar Energy, Vol. 3, pp.241-253 (1985).

/9/ R.PEREZ et al. SOLAR ENERGY, Vol. 39, n°3 , pp. 221-232 (1987).

/10/ P.ACHARD and R.GICQUEL. European Passive Solar Handbook. Prel. Ed. EUR 10683, Brussels 1986.

/11/ B.BOURGES. Workshop P.13, EUROFORUM NewEnergies 88, Sarrebrucken (1988).

/12/ EUFRAT Final Scientific Report (to be publis.). Ecole des Mines, Paris (1990).

/13/ B.BOURGES et al. EUFRAT climatic data handbook (to be published) (1990).

Table I: Example of solar radiation data tables from the "EUFRAT climatic data handbook".

Ht:	Monthly solar irradiation on the considered surface	(Wh/m2)
Imax:	Maximal solar irradiance	(W/m2)
S0:	Number of hours between sunrise and sunset	(h)
Xm:	Imax . S0 / Ht	
K:	Monthly average incidence angle modifier (single glazing)	
A:	Fitting parameter of the utilizability curve (Eqn. 3)	

PORTUGAL — LISBOA
LAT.: 38.70 N LONG.: 9.15 W ALT.: 77.00 M PERIOD : 1982-1987 INPUT:G+D

GLOBAL RADIATION ON A FLAT-PLATE COLLECTOR: ALPHA = 0 BETA = 0

MONTH	JAN.	FEB.	MAR.	APR.	MAY	JUNE	JULY	AUG.	SEP.	OCT.	NOV.	DEC.	YEAR
HT	2130	2788	4487	5311	6412	7215	7138	6703	5082	3643	2198	1850	4579
IMAX	541	689	862	975	1006	1017	995	966	862	734	592	501	1017
XM	2.448	2.612	2.265	2.394	2.212	2.065	2.011	1.950	2.093	2.232	2.680	2.532	2.291
K	.861	.893	.913	.927	.933	.939	.938	.935	.921	.902	.873	.863	.920
A	.25	.24	.10	.18	.13	.15	.14	.11	.12	.13	.30	.25	.19
SO	9.64	10.57	11.79	13.04	14.10	14.65	14.43	13.53	12.34	11.08	9.95	9.35	12.00

GLOBAL RADIATION ON A FLAT-PLATE COLLECTOR: ALPHA = 0 BETA = 30

MONTH	JAN.	FEB.	MAR.	APR.	MAY	JUNE	JULY	AUG.	SEP.	OCT.	NOV.	DEC.	YEAR
HT	3368	3804	5608	5772	6342	6839	6914	7102	6016	4964	3302	2978	5250
IMAX	902	990	1082	1104	1068	1042	1036	1066	1037	995	930	873	1104
XM	2.582	2.751	2.275	2.494	2.374	2.232	2.162	2.031	2.127	2.221	2.802	2.741	2.399
K	.940	.934	.939	.937	.934	.935	.938	.941	.942	.939	.942	.936	.938
A	.37	.36	.19	.26	.24	.27	.26	.20	.20	.20	.45	.41	.25
SO	9.64	10.57	11.79	13.04	14.10	14.65	14.43	13.53	12.34	11.08	9.95	9.35	12.00

Table II: Example of temperature data tables from the "EUFRAT climatic data handbook".

Ta:	Monthly average external temperature	(°C)
S:	Standard-deviation of monthly average temperatures	(°C)
DD(20):	Monthly degree-days at base temperature 20°C	

BELGIUM — Uccle
LAT.: 50.48 N LONG.: 04.21 E ALT.: 104m PERIOD 1931-60

MONTH	JAN.	FEB.	MAR.	APR.	MAY	JUN.	JUL.	AUG.	SEP.	OCT.	NOV.	DEC.
Ta	2.2	2.6	6.0	9.2	13.6	16.0	17.5	17.3	14.7	10.3	6.2	3.3
S	3.0	3.9	2.9	2.0	2.0	2.0	2.0	2.0	2.3	2.3	2.0	3.5
DD(20)	551	487	434	324	199	123	86	91	161	300	414	517
DD(18)	489	431	372	264	136	62	28	32	100	238	354	455
DD(16)	427	375	310	204	74	11	1	1	41	176	294	393
DD(14)	365	319	247	144	15	0	0	0	1	114	234	331

TICAD 3: A PC BASED ARCHITECTURAL DESIGN TOOL USING A LOTUS SPREADSHEET FOR ENERGY ASSESMENT OF BUILDINGS

L. F. Jesch and A.S. Minas

The Franklin Company Consultants Ltd.
192 Franklin Road, Birmingham B30 2HE, UK
Tel. ++ 44 21 459 4826 Fax. ++ 44 21 459 8206

1.0 ABSTRACT

To satisfy to currently increased interest in the use of transparent insulation an easy to learn, easy to use computer aided design tool TICAD3 was developed for architects and designers. It uses the integrated spreadsheet Lotus Symphony which is already familiar to most architects and runs on any IBM compatible PC. The programme enables the user to asses the energy performance of a building in a preliminary simulation mode and make parametric changes during the design process by rapid reruns until the desired effect is achieved. The partial or full use of transparent insulation on the various walls of the building is accomodated by the programme and special modules with transparent insulating qualities can be simulated for daylighting and for thermal comfort design purposes. The limitations and availability of TICAD3 is discussed.

2.0 PURPOSE OF THE PROGRAMME

Available large simulation programs give very comprehensive details for the final design of buildings. These are however very time consuming to learn and to operate. For preliminary design the architect needs a quick answer to the consequences of a small change or a different design idea and does not have the time to run a full programme for an answer which may be abandonned anyway because of an other possible solution. TICAD 3 has been developed specifically for computer aided design and thermal performance assesment of buildings with or without transparent insulation. This programme is aimed at architects and energy consultants who have responsibilities for the preliminary design and energy flow simulation of buildings but who do not have much time in getting into the details of larger and more comprehensive computer simulation

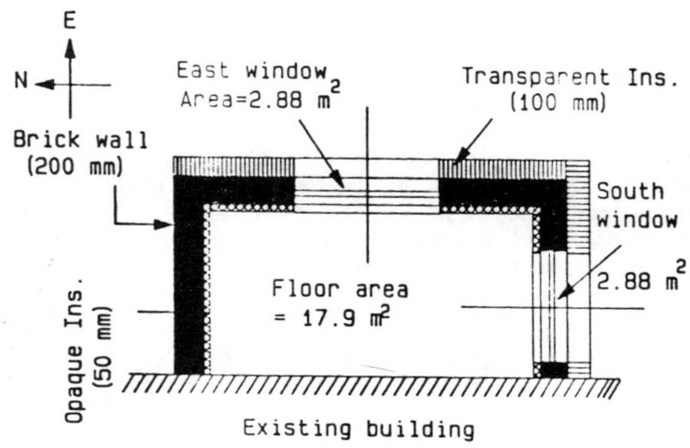

Fig 2 **Floor plan of the extension**

3.0 CHARACTERISTICS OF THE PROGRAMME

TICAD 3 is a design tool to calculate the temperatures at selected points in the building shell, including the transparent insulation modules shown in a sketch in Fig. 1. Transparent insulation allows solar radiation to penetrate it and to warm the wall behind it. At the same time it reduces normal conductive/convective heat losses of the wall. This representsa radiaclly different thermal behaviour from other types of building elements and requires special calculations for estimating the heat gains and losses.

TICAD3 recognises transparent insulation if used, and calculates monthly average heating and cooling energy requirements for a year.

It is not suitable for detailed design because of the limitations associated with the use of large number of input data unacceptable for the spreadsheet method but using condensed information about the climate and some about the building.

Monthly average solar radiation and ambient air temperatures are more readily available and they are quite adequate for the pre-design assesment of ideas before they are evaluated in full detail.

The emphasis here is to provide an easy to use and fast working design tool for architects and energy consultants who need to make quick and repeated preliminary estimates of the energy consequences of design ideas without deatiling. A PC AT standard machine or equivalent, capable of running LOTUS 123 and LOTUS SYMPHONY is all that is needed.

4.0 USING THE PROGRAMME

TICAD 3 is a modular programme using a version of LOTUS 123 specially adapted to the task. The modifications enable European users to access abbreviated climatological data and properties of building material commonly used in Europe, all expressed in SI units. The modular structure of the programme makes it easy to change only a selected few variables for each successive parametric optimisation step. It also allows any portion of the building shell to be fully, partially or not at all transparently insulated, important in finding a solution to avoid summer overheating by design.

The programme calculates average conditions in the building and produces monthly summaries of selected parameters. Graphics display for visual enhancement of intermediate solutions help the design process.

Due to abbreviated input data and monthly average output information the programme takes little space in the memory, not much time to prepare and to run. It is ideal for brainstorming to get feedback from alternate solutions during the preliminary stages of the architectural design process. And graphical output is instantly available.

5.0 SAMPLE CALCULATIONS

The ability of TICAD3 to be used as an effective preliminary design tool has been described elsewhere. /1/
A building extension, shown in a floor plan in Fig.2 was selected for demonstrating TICAD3. The extension was also simulated by TICAD2, a full simulation program. /2/

6.0 PARAMETRIC STUDIES

The objective was to determine three parameters:
a. what percentage of the wall should be covered with transparent insulation,
b. should a 50 mm opaque insulation be used on the inside surface of the wall which has 100 mm transparent insulation on the outside surface and
c. can movable insulation installed on the window control effectively the heat balance of the room.
Twenty one runs were made to try out different cases in which combinations of percentage area covered by transparent insulation were combined with internal opaque insulation and window control. Of these 21 cases four are shown to illustrate the results of heat gain/loss calculation during the process of predesign studies.
All cases have 100 mm honeycomb cell external transparent insulation on the south and east walls applied on 100 %, 60 % and 0 % of the surface area. Case 11 has opaque insulation inside and no window control. Case 13 has no opaque insulation and no window control. Case 16 has no opaque insulation but it has window control. Finally case 18 combines opaque insulation and window control.

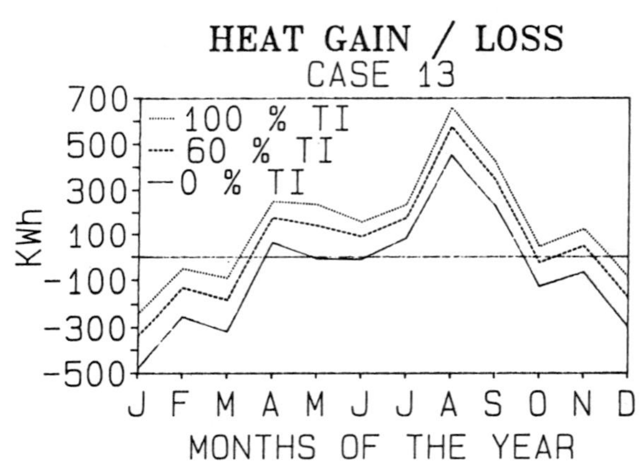

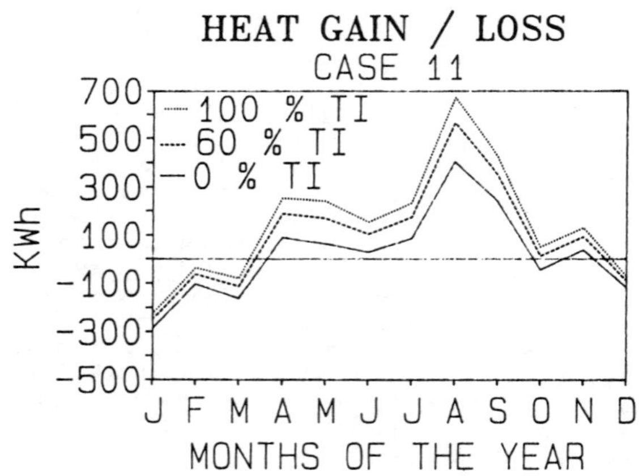

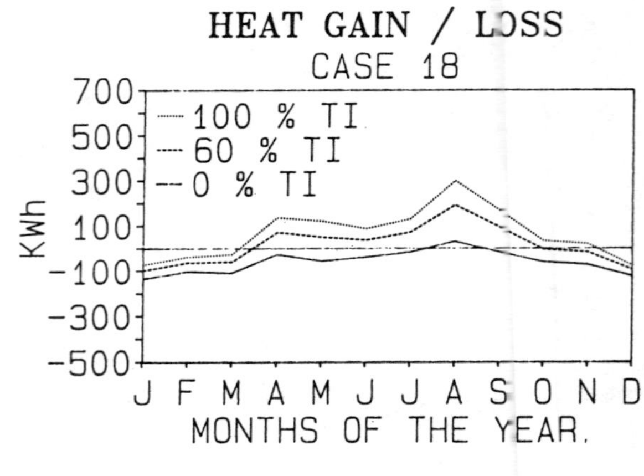

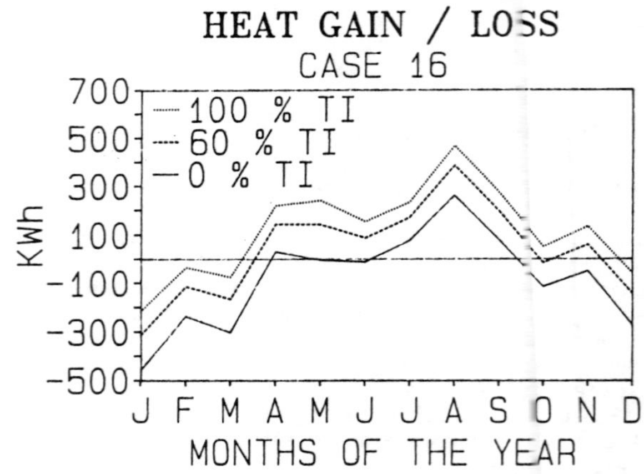

7.0 CONCLUSION

From these selected cases it is evident that Case 18 provides the least heating load and the leas cooling load for a year and therefore it is worth pursuing in a more detailed design study.
This was conclusion was reached in 21 runs which was completed in a day. This demonstrates the usefulness of the programme. Although it is limited in capacity of data handling it is very useful for pre-design parametric studies. Those interested in TICAD3 can contact The Franklin Company Consultants Ltd. at 192 Franklin Road, Birmingham B30 2HE, UK, telephone ++ 44 21 459 4826, fax: ++ 44 21 459 8206

8.0 CREDIT

The authors gratefully acknowledge the work done by Mr. S.U. Chaudry, M.A. Javed, J.A. Chattha and S.I. Gilani in the preparation of this paper.

9.0 REFERENCES

/1/ Minas A., Javed M.A., Chaudry S.U. & Jesch L.F. Spreadsheet methods for preliminary design of transparently insulated buildings. TI3 Proceedings of the 3rd International Workshop on Transparent Insulation, 1989 pp 130 - 133

/2/ Gilani, S.I. and Jesch L.F.: Thermal simulation of a building extension solar heated by transparent insulation and modelled by a modified TRNSYS programme. TI3 Proceedings of the 3rd Int'l Workshop on Transparent Insulation 1989, pp 138-141

PRESENTATION OF CIBAO : A NEW APPROACH FOR INTEGRATED COMPUTER-AIDED DESIGN SYSTEMS FOR THE BUILDING ACTIVITIES

Anne-Marie DUBOIS
CENTRE SCIENTIFIQUE ET TECHNIQUE DU BATIMENT (CSTB)
Sophia Antipolis - BP 141 - 06561 VALBONNE CEDEX (France)

Every building construction is the result of the collaborative effort of a lot of actors, who work at different stages of the design. To handle at the same time the many various problems to be solved would greatly improve the quality and the efficiency of the design itself. It would imply that each point of view should be taken into account. Although CAD systems are more and more in use in the industry fields, they are not really able to cope with the diversity of creative contributions in the construction field. So, a pluri-disciplinary research project, called CIBAO, has been undertaken, with LEMA (University of Liège), FNB, COSTIC and CSTB. It aims at the elaboration of a computer-aided building design environment, that incorporate, from the very beginning of the design process, the contributions of all the domains implied in the construction carrying out. This paper presents some of the ideas issued from discussions between the teams of CIBAO.

1. OVERALL PRESENTATION

1.1 The main objectives of the CIBAO project

This project deals mainly with :

- to emphasize the creativity of each of the actors,
- to facilitate some dialogues between these actors,
- to merge into the design process, the different problems and proposals from the partners, especially from the technical aspects and construction activities,
- not to impose a predefined order, given the fact that the first words can be issued by different actors, depending on the kind of building : an engineer for a factory, a solar specialist for a solar house, an architect for prestigious constructions, etc...

1.2 Classification of actors into domains

Building activities are greatly distributed.
The achievement of a building project and the whole duration of life-cycle, needs so many various expertises and know-how.
However, it does not exist a cooperative process, but a hierarchically structured setting up of the tasks, making it very difficult to solve conflicts and modifications.
In fact, a building is a system, within which every part may ensure or compromise the overall result. The quality of the whole depends on each part quality, and on their coherence.
CIBAO project acknowledges the autonomous creativity of each of the building actors. The design is recognized just like the result of the complex integration of the many activities.
The achievement of this integration process needs some negociation, whose quality is given by partner exchange capabilities, and here may be a major role of architects.
To deal with the great number of actors, A. DUPAGNE proposed to group them together in some sets, that we call "domains" and to study the communications between these sets. Actors of the same domain share some common points of view on the building. Each domain has its own logic, particular knowledge and skill.

- The systems of outer silhouettes : architectural domain, definition of sketch and envelope, layout, urban planning regulations, environment concerns...
- The inner functional spaces : communications and specific uses of the space.
- The technical equipments such as HVAC systems, electrical devices, etc...
- The materials choice, including new technology uses.
- The working aspects, planning, working-site organization, etc...

The integration in the design stage of the constructive knowledge of the contractors may five the opportunity to enhance the creativity of the whole process.

1.3 Domains speak different languages

The domains, as defined, have to exchange proposals and decisions. But each of them has its own logic and conceptual representation. So there exist different points of view, at the same time, on the same project.

The traditional way of designing imposes de facto a hierarchy on the actors' decision ; and the last actor has to manage with the decisions of the first one. In fact, the experienced designer already knows some impossibilities and prepares his proposal not to encounter them. This way of doing cuts down the novelty of overall design, because of the difficulty in getting off the well-known solutions.

A truly comprehensive CAD system must provide some solutions to the exchange problem. These exchanges must be encouraged in both directions between the actors : each one presenting his proposal, and in the reverse, each one being able to access and to evaluate (from his point of view the other actors' proposal.

As a consequence, the sooner a conflict or an inconsistency will be raised, the easier the problem will be solved, and the best the solution will probably be. So the proposals will have to be carefully studied in order to see the new facts, the conflicts of valuation, and the possible consequences of the new facts.

The figure gives an idea of a possible schema of the system.

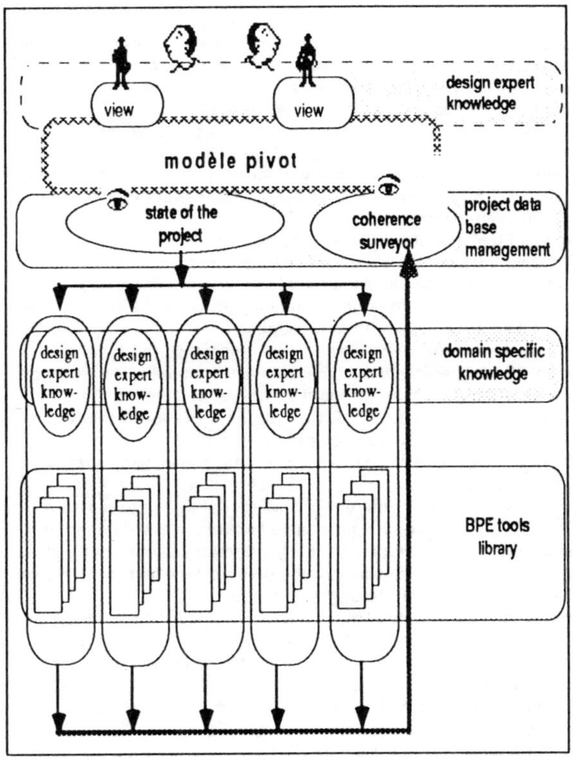

2. MAIN FUNCTIONS OF CIBAO PROJECT

2.1 An exchange tool : pivot model

From all this, it follows that a data model is needed, that allows the definition of views and the expression of constraints given from the briefing stage. The data representation of each domain has to support project description along with first specifications of the building.

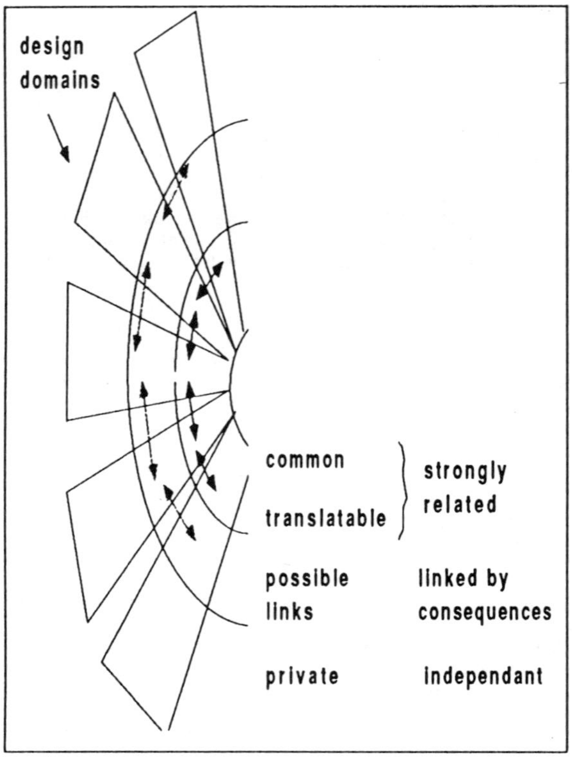

Relational data bases allow views, giving each user a specific conceptual model on the same data. It is coherent with the propagation of the modifications. But in fact, it is not necessary that all the data are accessible by every actor. Due to the enormous quantity of data, this remark could be used in order to build a working common data model. We can think to define a sub-set of data that hold common data and shareable data. So, it could be easier to implement it in a fast working way. Also, this model has to support non-monotonous and dynamic aspect of the project definition in progress.
We call pivot model such a data model.
Some authors think it will be at a low semantic level. Another way of looking at this model is to propose a more ambitious approach, storing more knowledge with the data. In any case, we have to cope with static as well as dynamic coupling.
It is worth stating at this point that the lower the level of the model, the bigger the task of the coherence module. The package including the data model plus the coherence rules is the actual conceptual data model, and it will be as a matter of fact a high semantical model. The prototyping work will show the interest of object oriented approach in this case, that allows easy extension of data definition.
It is not very easy to define this data model. It has to tackle with all data whose valuations may raise conflicts in the design multi-actors process. It concerns mainly geometrical and topological aspects, some data on technical equipments, type and components of works, global data about the environment, etc...
The modelling formalism must support extensions and modifications of the model, and maintain coherence of the data definition. Pivot model should not be confused with the overall project data base. The project definition is composed of very different kinds of information, parts of it being very specific to some profession or some craft.
The schema of the project data base could be physically implemented by different systems ; moreover, some coupling interfaces will be designed in order to re-use existing data bases.

2.2 The coherence module : surveyor and/or advisor

Coherence maintenance needs some functionalities :

- When a new information is send by one domain, to verify if it does'nt raise some conflicts with the existing data. It implies the survey of writing and modifying request to the pivot model.
- To deduce all the consequences of the new facts. This task is more complex than the former, and implies some knowledge about the relations between the entities. Further pivot model definition will point out the difficulty.
- If conflicts are raised, to restore a coherent state : it may imply a simple call to the domains, as well as the call to an arbiter, for exemple the master of the project.
  He may benefit from a specific negotiation knowledge base.

Let us take a heating sheath as an example. The diameter of the duct plus the sheath dimension give the space occupancy of the whole. So a dialogue could be : "I put a sheath into that place", giving some geometrical 3D information. It may be enriched by the qualification as heat transfer.duct These facts may cause some difficulties either by the occupancy of the space, or by the functionalities of the room to be runned across.
Coherence verification will call for some geometrical evaluations (given by the geometrical modeller), and constraints verification (handled by the data base management systems).
Depending on the work setting up the coherence module will work in real time, filtering all the information send by the domains, or it will be waiting for requests, at the end of a dialogue stage between the domains.
A midway solution can be proposed, with two coherence levels :

- First one a shallow one, that only take care of access.
- Secondly, a deeper one, that uses deductive rules and integrity constrainsts, and will be greatly efficient helping to reach a coherent state of the project definition.

Design process is a specification transformation activity, together with dynamical expression of constraints issued by the domains (they are often only restrictions on acceptable velues). In this case, a conflict arises when a proposal for valuation of something is out-of-limit of the domain of acceptable values as restricted by the first choices.

2.3 Geometrical 3D modeller role

Graphical evaluation is the main point of CAD systems- it is often their only capability.
In building field, graphics is the most used mean for expression and communication. It is in fact a powerful tool, either for plans, elevations, sketches, network drawings, etc... It gives in a compact way a big amount of information.
Most of existing CAD software does'nt use volumic representation. They propose in fact very few functions dedicated to architects. A building design mainly uses simple forms of volumes, but it should be useful to handle these volumes as 3D volumes, to identify them, to give them some attributes such as functionalities, and to express some constraints such as links between volumes.
As said before, most conflicts come from geometrical and topological aspects, i.e. the occupancy of the space. Then, the modeller has to propose functions for boolean operations, modification propagations, etc...
An example of such a function really useful is the identification of adjacency, that identify the separator between two volumes. This function is needed by many physical analysis, such as thermal or acoustic analysis. Inclusion identification is needed for the computation of interior spaces, and its dimensions. VOLUME from LEMA has some of these functions. It should be used for the input of data, for the visualisation of the current state of the project ; it should be called by the coherence module, after examination of the pivot model, as far as geometrical entities are concerned.

2.4 General architecture of the system

When thinking about a certain autonomy of the different design actors, and a non-systematic sequencing of the tasks, it suggests the idea of concurrent works, with some means of communication, such as mail box for routing messages, or rendez-vous mechanisms.
Let us see two cases :

- Each actor works from the initial brief, and at each time he holds some new information and he wants to propose it.
  The exchange support is a blackboard, always open for writing and reading, and where all proposals are available, in real time. Then all the partners are supposed to be logged in the same session.

- From the master's proposal (or the client brief), each actor makes his design task alone. Next, there is a meeting for putting together and merging of the proposals, and negotiations when needed in order to reach a coherent state. Each one goes back his office with the new specification (more precisely defined) and continues the job. As many cycles are done as necessary for the full definition of the project.

The second solution gives each partner the ability to work when convenient for him. But, the former prevents the proposals from being incompatible with or too far away from the other's.
At the end of the day, the state must be such that each partner has achieved his design task, and his plan is coherent with the whole of the plans.
Another concept is the necessary arbitration and decision when some conflicts are raised. Who will be the arbiter ? Master or client ? In any case, he should be helped by a negotiation module, with a design knowledge base.
It may be ambitious to say that such a knowledge and expertise could surge from experiments with such a project as CIBAO, however

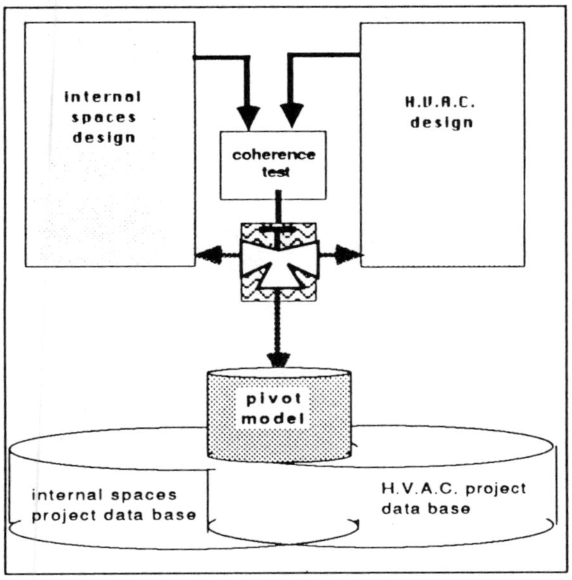

3. BPE TOOLS

It is necessary to include in this environment some evaluators, useable at different stages of the project definition (rough evaluations at the very beginning, stressing the main difficult points, and furthermore giving more precise results when the definition goes along).

In fact, each domain has specific tools that allow for definition of its own solutions and evaluation of proposals.

It should be useful to gather as many of these BPE tools as possible: to build a library where a request translator, and users guide could be proposed. Such a library could help in having a common reference for performance measurements, and making benefit for all the building field of evaluation improvements.

It can be expected easier comparisons between the alternative solutions, and a better fiability. However, to get these BPE tools easy to use, it should be possible that each one expresses its requests using its own language and its own entities. Particularly, due to the distributed aspect of the design knowledge, it will be necessary to handle defaulf values or usual values : these values will be proposed to the user and may help him to define correctly the input of data.

As an example HVAC system default definition is useful for some studies of the envelope.
The definition of translator from the problem request to the BPE procedure call. Some translators will be defined in order to help the user to obtain the answer to his problem. It will use some former works issued from the Modelotheque project, developed at CSTB.

The realization of a common data base will avoid to repeat the data input, by the way of giving access to the common data for all the tools. This data base will facilitate the communication of the proposals between the actors. The task is of course a difficult one, but it is clearly an essential task.

4. PROTOTYPING WORK

4.1 Definition of a sub-problem

In order to give a demonstration of these ideas, it has been decided to show an example on a "maquette", i.e. a working prototype. So a sub-part of the problem has been defined : it takes into account two domains, and two specific aspects, such as they are strongly coupled.
This "maquette" will be restricted to two design domains, and sub-domains where energy requirements are directly concerned:
- inner space design : circulation and volumetric aspects,
- technical equipment design : heating and cooling equipment.

These domains have their specific methods for design tasks. This restricted part of the world should make it easier to define a common data base, the "pivot model". It will be something like a mini product model, using standard concepts. So the method will be tested and if needed, modified. Of course, the definition of the pivot model should be evolutive enough in order to be able to expand it to the five domains in the further phases. The extensibility of this data model should be very useful for the integration and definition of new concepts. The pivot model will be the conceptual node of the data integration.

Some existing BPE tools will be integrated in the system, for two stages of the design process, and for different levels of accuracy. They should be documented with the Proforma documentation formalism (as defined in an another R & D project). The definition of translators will make connection with the project data base, and will give each domain the possibility to express requests to the B.P.E. mini-library.
The 3D modeller is an essential part of the underlying system establishing a strong link with the CAD systems. Of course, it is needed by the first design domain (inner spaces design), but in fact, the use of this kind of tool is definitely necessary in order to cope with the needs of the users. So the data structure has to be closely linked to this tool, in the implementation step.
A coherence surveyor will take in charge the global coherence of the pivot model : it will be responsible for the data updating, and calling to the upper layer whenever some potential conflicts are detected.

Let us note that the architecture of this mini-system should be the same as the CIBAO project one : the mini-system is a first prototype and help to validate the approach.
This restricted first part of the overall project seems to be sufficient in order to show the principles, and to clarify the specific problems that may be raised. This first work is in fact a first step to build on when dealing with the whole of the project: this work should give us the sound bases for the further development.
An executable sketch will be implemented within an object oriented environment, giving knowledge representation capabilities, and coupled with a 3D modeller. This drafting phase aims to refine the proposed concepts of methodology, and to give a testbed for the major ideas.
So, it focuses on three aspects:
- Design transversal approach : data integration by the pivot model.
- Top down approach: black box links with existing BPE tools, by translators processes that may interpret the domain request.
- Bottom up approach will probably be used inside the design domains, and this will open the way for building contractors to participate earlier in the design process itself.

This project is concerned with preliminary design stages. The scheme design is the most critical, in that most all significant decisions are made at this step, and that it is here that coordination and communication are the weakest.
But it has to be made clear that in some parts, the process may require more details than the classical process, because of the different non-sequential organisation. The study should define a better definition of the design process.

A difficult problem is that the dynamic aspect of the definition of the project is an essential part of the task.
Both design domains will have to specify how and when they are able to propose their part of the solution, and what information they will ask for. Process integration is a main point of the research. So the communication requirements of each domain, from its point of view, will be a baseline for the project.
Such questions are raised, as "what minimum information is needed at the preliminary design stage ?", "how to get the building owners' goals expressed ?", etc... A part of the work will need a big effort of knowledge acquisition, with close contact with building professionals.

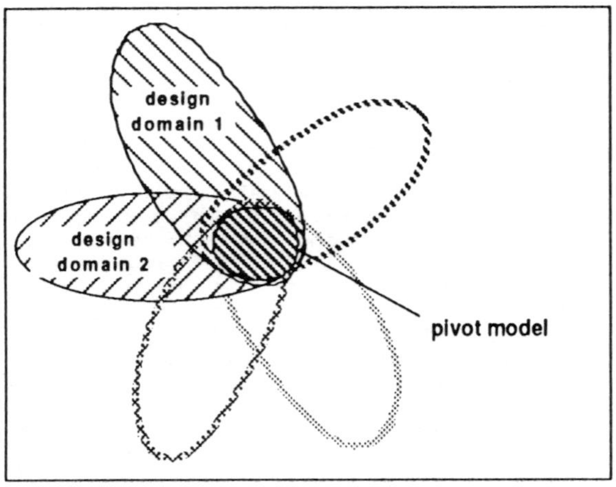

When giving more functions to a CAD system, a question is raised: where will the data base lay ? will it be built by some extension of the CAD internal data base or in an external layer ?
The objects are in fact located in space (and in time axis. However, the functional view upon the building do not need a precise geometrical definition of the parts of it ; in fact, it works at a more generic level.
In fact, first studies do not address space organization : for example, if the brief is "to build a 50 flats block for elderly people", then this fact implies some constraints and consequences on the building design.
Design process implies the ability to handle objects not fully defined, their non monotonous valuation and the saving of the state of the objects. Some attributes may be undefined, or receive a default value, depending on the kind of problems and methods. Our common data model must support such a kind of information.
In this concern, it appears necessary to use an advanced A.I. system, that has together an object-oriented representation of data, and some powerful reasoning control mechanisms : we choose SMECI, developed by INRIA and now available from ILOG Society.

4.2. First results

A first draft of data structure has been defined. It aims at giving the building description needed by a simulation software for energy analysis, CSTBât V.2, that uses an analogic_ representation of thermal transfers. Analogic models are associated to the type of elements, such as walls, air volume, etc... Their parameters are calculated by the system, using materials and geometrical data. The concept of zone has been implemented by the mean of links with the components of this zone, and that allows the automatized selection of the walls of the zone, and of topological data when needed.
Let us note that this zone representation, issued from a thermal analysis, is in fact a structure that is superimposed to the building description, but that does not erase it. It will be the same for any zone decomposition, issued from another kind of analysis of the building, for example fire safety, comfort, acoustic,... The zone concept uses the geometrical definition of the building, and gives a specific point of view upon it. Doing this, it avoids a new input of the data.
The link of SMECI with the 3D modeller VOLUME has been implemented, using ASCII files. In fact, VOLUME proposes a function that scans its data base and writes such files. The data are not only graphical data, but also they give some information about the entities of the project and their attributes and relations : the type of a window, or the adjacency between two rooms for example. Then it is possible to propagate in SMECI these information.

5 CONCLUSIONS

Object oriented representation makes it easy to express different kind of links between the elements, and different points of view. SMECI gives an useful help when modifying the data structure, and is often the case in this kind of research work.

To link a modeller with an A.I. tool is needed to benefit of the graphical interface tool, and to facilitate the communication between the domains. Moreover, it gives the advantage of geometrical and topological functions.

This approach has proved that computer may be a real and powerful tool for design activities, playing the role of a memory, an exchange mean and a visualization tool, that build an efficient dialogue mean for the design actors.

ENERGY SAVING IN BUILDING BY OPTIMIZATION OF CONTROL STRATEGIES. DESIGN USING SIMULATION ANALYSIS

Vinot Benoît
Centre Scientifique et Technique du Bâtiment (CSTB)
Sophia AntipolisBP 141 - 06561 Valbonne Cedex - France

ABSTRACT
Great improvements have still to be achieved for the energy management in building, through the use of innovative and adapted control strategies. Moreover, issued from recent developments in automatic sciences and associated digital control methods, a wide range of control algorithms (adaptative, self-tunning, ...) are today available and could be beneficially applied to control of energy system. Before on-site implementation, these control strategies have to be selected and adapted to the considered energy system. This deals with the class of building (office or dwelling), the energy system (water or electric heating, air conditioning) and the management constraints to respect (intermittent heating, multizone control...); furthermore, the new powerful control strategies are based on advanced theories and usually imply the previous setting of several parameters. For judicious uses, control strategies require preliminary theoretical analysis. The modelling and simulation on computer approach is a general way for these analyses. Provided that adapted models are available, global simulation of system and associated controller can perform reliable results for performance comparison parameter setting of advanced automatic techniques (on-line identification, optimal control).

1. INTRODUCTION

Today, great attention is payed to simulation of global processes. This needs numerical representation of physical phenomena in order to produce similar numerical output, from similar sollicitations as input. Thus, a first necessity was to design models of physical phenomena or system representations, suitable for numerical processing.

Simulation can be used for several purposes : one is to increase physical phenomena understanding, especially when multiplicity of involved phenomena make global understanding or analysis rather difficult. In this way, the entire problem can be split into several local and more simple problems.

Simulation appears also as an helpful tool for systems design and performance evaluation or assessment. In this scope one can either consider single process from a theoretical point of view or total running system in a real sollicitation environment. This second point is a very usual problem in building analysis where macroscopic analyses are performed for a whole construction. Then total approach requires connections between single equipments. The considered systems for building thermal analysis are at first, heating devices (boiler, water or air network, radiators...) and the building characteristics (wall and floor inertia, heat loss, solar gain). Additional essential device is the control system which rules the energy system according to data measured from the building.

2. WHAT ARE CONTROL SYSTEMS ?

"The main feature of good systems is to be self-driving, as much as possible and without human intervention".
- from J.L. PONCELET, 1826.

Most of dynamic human designed systems do not run by themselves. They are neither self-powered nor self-controlled. When necessary, the injected power must be supplied at right times and right levels. Special systems are in charge of these decisions : when and how-much. They are controllers and are able to manage more or less complicated systems. In this way they generally reduce human action but without supressing it. They assist human control assuming only part of global control.

In the field of controllers, several levels can be distinguished. They are all control systems but they take more or less important control actions according to the system they are responsible for. Most of practically implemented controllers are rather simple ones. They are based on a reduced set of parameters and require very little computation. They assume seldom more than one control action and are suitable for driving simple systems. Their failure results from very slight regard to the dynamic behaviour of the system under control.

Today, a large range of control strategies is available for system management, issued from recent development in automatic sciences and advanced data processing on micro-computer.Then, elaborated control strategies can be implemented. Main improvement of control system is resulting from a greater accuracy of the internal dynamic representation, prediction and minimization algorithm (fig. 1).

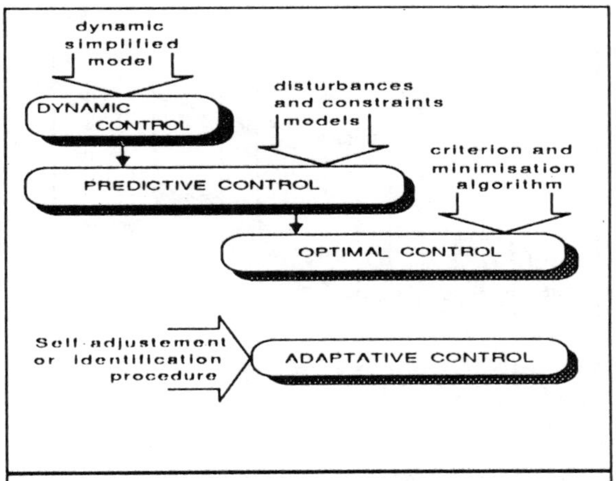

Figure 1 : Improvement levels in dynamic control strategies

2.1 Dynamic representation

This first improvement level deals with the controller internal knowledge about the system it has to control. A very simple system can be linearly approximated through a lone input/output parameter. For greater accuracy and more complex systems, one should consider several simultaneous phenomena. Then it should be used as a numerical representation of the system called "reference model". This requires several parameters for the input/output relation. Such a simplified model, included in all elaborated control strategies may cause much performance improvements. Several control actions can be taken at once, and short term future influence of present disturbances and control action can be known.
For instance, for building energy control, it was shown /1/ that the thermal behaviour of building could be represented with a second of third order model. A general formulation for these model is the state space representation.

$$X = AX + BU + EW$$
$$Y = CX$$

A, B, C, E matrixes coefficients should be initially set up. Such a formulate allows to reach the internal state X of the system (temperature for instance), or the observable output Y, from the knowledge of command action U, disturbances W, and initial state. The parameters of the model represent the different thermal couplings which happen in the building. Previous analyses are necessary for realistic parameters setting, based on physical analysis or theoretical modelling (identification).

2.2 Prediction

The aim of this function is to anticipate any foreseeable input effects on the system. This capability is directly connected to the a priori knowledge of the system and its running surrounding. If one knows now what disturbances will be within several hours, this forecasting can influence the present control decision. The basical knowledge for prediction can be expressed using various

formulations : adapted simplified model or discrete tabulated values. As an example, in building thermal analysis, interesting foreseeable variables are outside temperature and solar irradiation, for which many attempts have been done for global or local predictive models /2/, /3/.
Additional essential information are the possible limits or constraint fields applied to the system. All these information give the controller predictive capability. It can compute not only the control at the current time step, but also determine the set of control actions for every time step over a future temporal horizon.

2.3 Selection

The selection capability allows to choose, among several sets of satisfying predicted control actions, the most suitable one with regard to a given objective function. Therefore, several possible ways are assumed to drive a system from a current state to a target state (fig. 2).

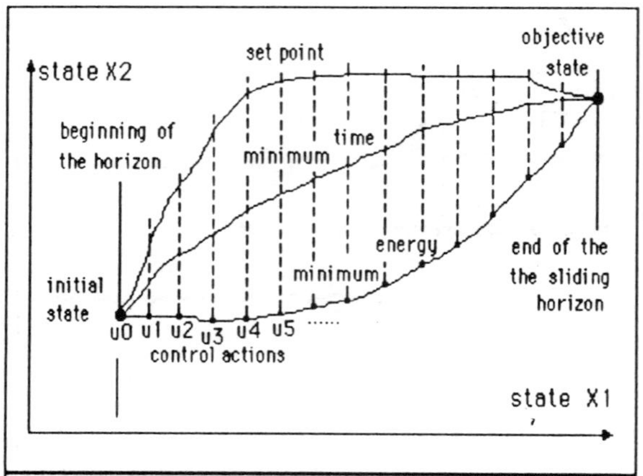

Figure 2 : Optimization problem : possible state trajectories for a different control profile [U0...Un], according to the criterion, in a two dimensional state space.

All the possible control strategies can be evaluated and compared using an adapted function : the criterion. The optimization problem is to detect what the most convenient driving strategy is, to minimize the criterion. According to this, the selected strategy will be considered as the "optimal" one. In a classical optimal control problem, criteria are either minimum time or minimum energy consumption. In building thermal analysis, important aspects such as comfort and heating/cooling cost should be considered, and can be included in a criterion through adapted formulation /4/. Then a minimization algorithm, in charge of the comparison, can reach the optimal strategy . This selection ability requires to formulate additional numerical terms. Some of these have a direct influence on controller performances and sensitivity studies may be necessary to achieve correct settings. For instance, comfort penalty, weighting coefficients or optimal horizon length are crucial points.

3. THE BUILDING THERMAL CONTROL PROBLEM

3.1 General objective

The general problem of building thermal control results from the physical interaction of building envelope, heating/cooling, internal or external sollicitations and close loop controller (fig. 3). The three main physical transitions are : sollicitation toward building, building to controller, controller to equipment. Many analyses have to be performed on building and equipment performances. But the optimization of the global thermal system, in its whole, should take into account the controller behaviour.
Two main goals are in view, in the building thermal control field

- Considering a global problem level, many existing plants are not optimally used and do not operate in an efficient way. Most of the running systems had been empirically set up without any regard to the physical interactions. There is a great lack of global preliminary analysis, due to unsuitable tools.

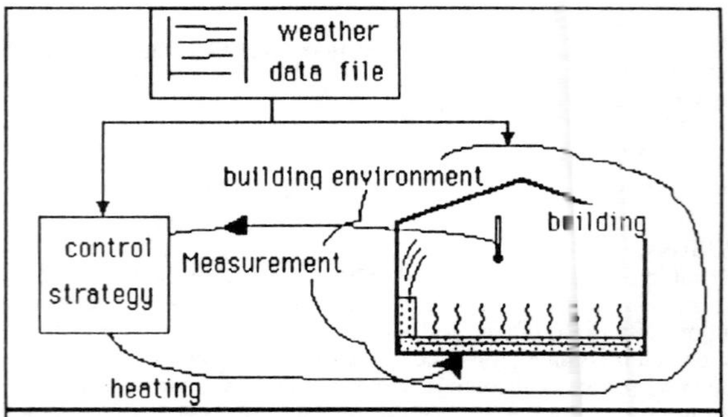

Figure 3. - Heating and control system simulation

There is a wide area of possible improvements for building thermal systems, provided that theoretical means are able to represent the operating system behaviour.

- The other aspect is the design of adapted control strategy for special system. This can be belatedly performed on existing system, but preferably in the system development phasis. In the past, numerous good thermal solutions were not successful due to unappropriate control strategy. For acceptable over-cost, suitable control devices can often provide reliable running conditions.

3.2 System characteristics

Building presents an infinity of thermal characteristics, and a large range of energy systems occur to be installed in it. This multiplicity of possible energy solutions underline the difficulty to choose the good controller which makes the whole system (building + equipment) operating correctly.
A rigourous approach should take care of many parameters which are :

• For the building :

- The inertia of the building. A heavy structure will tend to absorb the oscillation of indoor temperature, and may influence the dead band setting of the controller.
- The building insulation which aims to reduce the effect of outdoor temperature on indoor conditions and the capability to get solar gains. Building thermal behaviour can be very sensitive to solar irradiation, especially in passive building /5/. This depend on orientation, window area. The sensitivity of building indoor conditions to outdoor sollicitation may require high predictive capability of the controller.
- The occupancy, internal gains, directly linked to the office or dwelling function of the building. This aspect concerns the possible programming function of a controller.

• For the heating system :

- Arrangement of the heating system in the building (local or central heating).
- The time response (and delay) as a consequence of the kind of heating system. For instance, air conditioning or floor heating do not have similar effect in term of the resulting thermal state of the building and long term response to a control action.
- Ageing and possibly deterioration of the heating system (fault detection is an additional function of control system).

3.3 Evaluation

As a consequence of these various aspects, some important questions are still under research.

- How should a global methodology be defined, for performance evaluation of controller, energy system adjustement, thermal efficiency. A balance has to be found between purely experimental research work, performed on test-cell or real scale occupied building, and theoretical work using software tools. First ones are limited by physical constraints such as the building configuration, and outside weather conditions, heating system..., but reality is at this price. Second one is a very flexible way to study any class of building and any heating system under various weather conditions. This assumes model availability and the result quality depends on representation accuracy.
- What could be a general criterion for performance objective and comparison of several control strategies among each other?

Energy saving is now a rather conventional way for building thermal evaluation. This considers the primary energy supplied regarding the heating requirements (fig. 4).

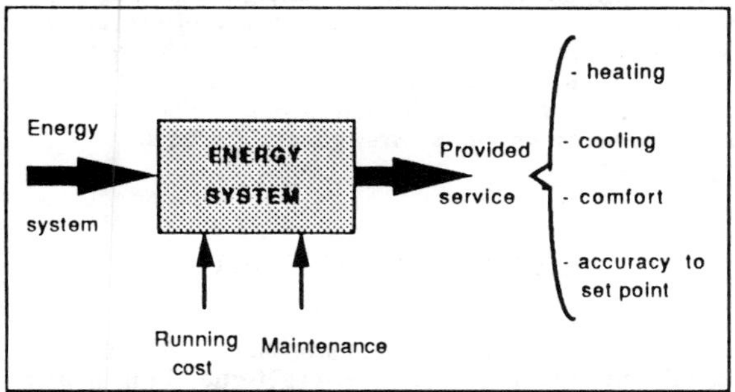

Figure 4. - Energy system simulation scheme.

The quality of the function provided should now be considered, in addition to the service itself. These problems are linked to system efficiency, comfort definition... Either indoor temperature is taken into consideration through the accuracy to a fixed set-point or temporal trajectory for time variable set-points. This is a reducing approach for the comfort which should also include, further to thermal comfort, humidity, air velocity or lighting. Thermal comfort is a function of air and wall temperature among other variables.

4. DEVELOPMENT OF AUTOMATIC MODULES IN A BUILDING THERMAL SIMULATION SOFTWARE

In order to assess the capability to evaluate global energy system under control, through simulation, innovative automatic function has been modelled /2/. This constitutes the feasability study phases for further development in automatic and building control. The aim is to add an automatic component tool box to an existing thermal model library. Then global simulation of building equipped with heating devices and connected to controller can be performed, for quick assessment of systems and control devices performances. General high mark of the work and main results are presented.

4.1 Modelling the optimal control strategy for electric heating

Since 1984, the CSTB has been working within the IRCOSE Research Group which aimed to apply optimal control to energy management in building. The main phasis of the research were :
- development of optimal control strategies,
- test and performance evaluation through simulation,
- on site implementation.

Principally the CSTB contributed to analysis of optimal and classical control strategies using modelling and simulation. Up to now, these results have been partially presented. A model of a deterministic optimal control algorithm was designed. It is based on Pontragyn minimum principle and adjoined system. The objective criterion is obtained using a variable step gradient algorithm /5/, /6/.
First, several parameters were set through simulation. The most important are the relative weighting parameters of the objective function (which rules the conflict between the two contradictory objectives : comfort and energy saving) and the horizon length (future temporal sliding horizon). Sensitivity analyses were performed through simulation, to reach a satisfiying value of the set of tuning parameters.
Then several control strategies were compared (optimal and conventional ones). The global results were analysed in terms of total heat balance and internal comfort provided. It appeared that for the considered heating system, (electric heating floor plus auxiliary), a correct setting has an essential effect on final performance, and sophisticated control cannot make up for unsuitable parameter setting. For instance, well suited suboptimal control reached comparable performances with optimal strategy.

4.2 Thermal control strategies for passive building

In the framework of the concerted action titled "Passive building control" within the PASTOR CEC Research Program, CSTB jointly with EMP, TU DELFT, NAPAC, aimed at design suitable control strategies for indoor climate control /5/. A reference passive solar building was chosen including:
- a wide glazing area facing south (two areas were considered for more or less solar gains),
- a fast response heating device (electric convector or hot water radiator),
- a high or low inertia building envelope depth variable,
- a horizontal overhang used as a controlable solar protection device.

The whole building, equipments and controllers were modelled using the CSTBât simulation software. Many simulations were ran under various weather sollicitations (all year round for Carpentras and Trappes meteorological stations).
Different strategies were designed to drive, in a combined way, the heating, the shading and a fan for air renewal. The goal was to keep internal temperature in comfortable range, in spite of or thanks to external weather conditions : solar irradiation, outdoor temperature.
The three control action to be decided by the designed controller are : shading or no shading, heating or no heating, ventilation or no ventilation.
Through simulation an optimal control strategy was compared with a simple control strategy.
Some general conclusions may be pointed out from the simulation results.
The control strategy performances are fairly linked to the building and environment conditions.
According to the season, advanced control appears powerful during highly dynamic sollicitation period - i.e. mid-season such as spring and autumn - due to its internal reference model. Its advantage is fairly reduced during winter and summer.
The passive design has also a strong influence on controller strategy. High inertia is more likely to increase advanced controller performance with reference to a conventional one.

GENERALISATION - CONCLUSION

The increasing performances of control strategy imply more computation and causes additional parameters. But today, powerful data processors are able to perform rather heavy calculations. The main difficulty for wide use of advanced control systems remains the judicious choice of parameters for effective improved performances. The whole set of parameters can seldom be defined from a-priori knowledge. Some of these may be obtained through physical considerations, while others are linked to purely numerical aspects.
Testing is often the best way to achieve correct settings for later on-site implementation, but real scale experiments are seldom possible. For these reasons, "theroretical experiment" using simulation tools offers an unvaluable help for design and preliminary analysis of control systems.
A significant effort of widening the use of control techniques remains to be done in the field of energy management. Adapted design methods should contribute to this goal. Modelling and simulation methods as well as derived techniques appear there to be essential.

BIBLIOGRAPHY

/1/ LARET (L).-"Contribution au développement de modèles mathématiques du comportement thermique transitoire de structure d'habitation".- Ph. D. Thesis, University of Liege (LPB) Belgium, May 1980.

/2/ Rapports d'activité IRCOSE 1983 (intermed.), 1984-85

/3/ ASTIER (R), DUHAMEL (Ch), BENARD (C),- "Modèle de prévision et de simulation de l'irradiation solaire au pas de temps de l'heure". Revue physique appliquée 18 (1983) 29-38.

/4/ VINOT (B).- "Combined control of direct gains and heating in passive solar buildings".- PLEA 88, Porto (Portugal), July 1988.

/5/ VINOT (B), ACHARD (P), TANTOT (M) .- CEC Concerted Action "Direct gain control" final report CSTB/EMP, june 1989.

/6/ BERNHARD (P).- "Commande optimale, décentralisation et jeux dynamique".- DUNOD automatique 1976.

CONCEPTION ARCHITECTURAL ET BILAN THERMIQUE : UN DIDACTICIEL POUR LA MAÎTRISE DES PARAMETRES ENERGETIQUES LORS DE LA CONCEPTION D'UN BÂTIMENT.

L.ADOLPHE (GISE / EMP)
D.CAMPANA (DIALOGIC)

ABSTRACT:
This CAL program calculates the energetic balance of a simplified building project designed through a 3D drawing module. An expert system offers design alternatives to an improvement strategy choosen by the user.To understand better the rule of each design parameter, a pedagogic module illustrates and comments on them through architectural references.

CONTEXTE

La pénétration croissante de l'informatique dans tous les secteurs d'activité a vu naître une nouvelle génération de logiciels, supports privilégiés des formations initiale et permanente. Cette mutation a débuté par le succès de certains logiciels professionnels "détournés" de leur fonction première, qui se sont révélés des outils précieux de transmission du savoir-faire. Cette mise à disposition de logiciels utilisés dans les cabinets d'architectes ou les bureaux d'études, a l'avantage de faire manier des outils professionnels, mais n'offre pas de réelle stratégie pédagogique. Apparraissent ensuite des produits à mi-chemin entre des outils scientifiques et des programmes didactiques: STRATEGY à l'université de Liége, des outils du groupe ABC, SIMULA du CERMA de Nantes, CASAMO-CLIM de l'Ecole des Mines de Paris. Simultanément naissent les didacticiels, logiciels spécifiquement conçus à des fins pédagogiques.

OBJECTIFS

Ce didacticiel, réalisé grâce au soutien du Service Formation de l'AFME, dirigé par Jean-Gérard SENDER, vise à sensibiliser les étudiants architectes et les architectes, les élèves- ingénieurs et les ingénieurs thermiciens des BET ou des services techniques d'organismes et les maîtres d'ouvrage à la prise en compte de la maîtrise de l'énergie dans la conception d'un projet de bâtiment.
Il fait le lien entre la pratique architecturale et la prise en compte d'évaluations techniques, en abordant le bâtiment de manière globale et en tenant compte des interactions entre les divers composants.
Il permet de calculer le bilan thermique d'un projet simplifié de bâtiment saisi par l'utilisateur.
Il propose des solutions susceptibles de l'améliorer, à partir d'une stratégie définie par l'utilisateur. Pour mieux comprendre le rôle de chacun des paramètres du bilan, une base pédagogique les illustre et les commente au travers de références architecturales.

Ce didacticiel vise à :

•la compréhension des facteurs intervenants dans le calcul du bilan thermique de l'enveloppe d'un bâtiment d'habitation: climat, forme, contiguïté, percements, technologie des parois...

•l'Apprentissage du mode de calcul du bilan thermique, limité aux déperditions et aux calculs réglementaires vus sous l'angle de la conception architecturale, à partir:
-d'une articulation pédagogique autour d'une esquisse de projet proposée par l'étudiant,
-ou bien, d'études de cas préparés à l'avance dans la bibliothèque de projets.

•la maîtrise du rôle des différents facteurs intervenant dans le bilan thermique,et notamment de leur influence sur la maîtrise des formes; aide aux choix de solutions optimales grâce à un système expert,

•l'intégration de l'approche énergétique dans l'ensemble des préoccupations de l'Architecte, dès les phases amont de la conception d'un projet:
-en étudiant la sensibilité de la performance énergétique à différents paramètres,
-en se dégagant d'une approche combinatoire, par une approche typologique (morphologies de référence...),
-en se référant à des architectures remarquables.

CONTENU PÉDAGOGIQUE

Ce didacticiel vise à sensibiliser à l'approche globale du fonctionnement énergétique d'un projet lors de sa conception architecturale. Ce projet est limité à un projet d'habitat individuel ou collectif, défini à partir de zones thermiquement homogènes vis à vis des contiguïtés, du type de toiture et du type de plancher. L'étude d'un bâtiment complet ou de sa décomposition par étages peut répondre à cette notion de zone équivalente. A chaque phase de la saisie est associé un facteur caractéristique permettant d'évaluer l'incidence de la saisie ou de la modification en cours sur le bilan thermique. Dans un objectif de simplicité et de clarté, ce dernier est limité au bilan déperditif en régime permanent: ceci permet une analyse automatique de variantes au projet à partir d'un système expert.

Les différents modules

1-Présentation générale

Ce module introduit les objectifs principaux et la démarche du didacticiel, présente sa structure (les différents modules et les liens entre eux) et les principaux moyens de navigations. **(voir figure 1)**

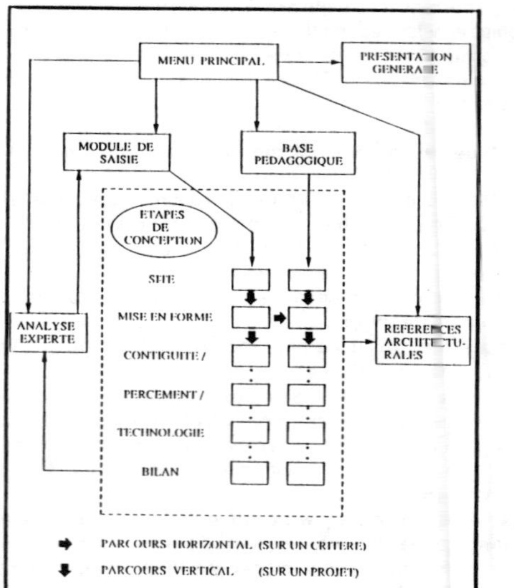

FIGURE 1 : STRUCTURE DES MODULES

2-Saisie du projet

Le module de saisie est le "noyau dur" de ce didacticiel autour duquel s'articule les autres modules. Il permet une saisie simplifiée graphique du projet par l'utilisateur et calcule les facteurs énergétiques caractéristiques. L'utilisateur est libre de construire sa propre démarche, à la fois sur le plan pédagogique et sur le plan architectural.

Pour simplifier le processus de conception, il n'est pas apparu souhaitable dans ce didacticiel de sensibilisation, d'écraser l'apprenant sous toute la richesse de la discipline enseignée, la maîtrise de l'énergie, ainsi que sous toute la compléxité méthodologique de la conception architecturale, démarche de synthèse plutôt que d'analyse, processus plus itératif que linéaire.

A chaque étape de la saisie du projet a donc été associée un "compteur": le paramètre significatif de la phase dans le bilan thermique du projet:

- **la MISE EN SITUATION** du projet dans son environnement; choix de la zone climatique significative du climat du projet; choix du type de situation (urbain, semi-urbain, dégagé); choix du type de programme (individuel, collectif),

- **la MISE EN FORME;** choix du volume générique, limité aux volumes parallélépipédiques, et de l'orientation; mise en place des découpes et extensions,

- **les CONTIGUITES,** des parois horizontales (4 types de planchers, 3 types de toiture) et verticales (donnant sur l'extérieur, un local non chauffé ou un local chauffé), **(voir figure 2)**

FIGURE 2 : ECRANS TRAITANT DE LA CONTIGÜITE DANS LE MODULE SAISIE

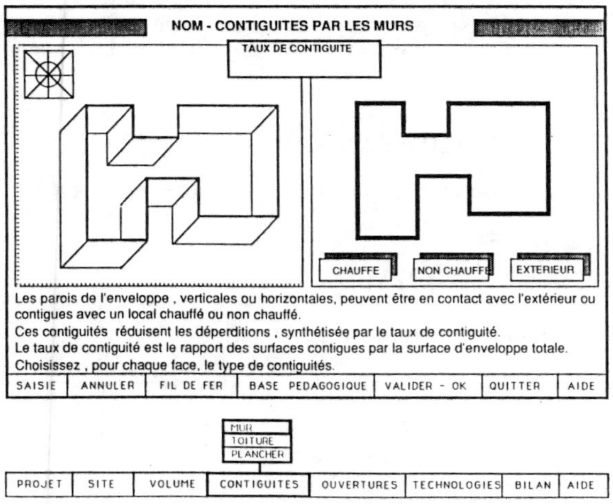

- **les PERCEMENTS,** mise en place des portes et des fenêtres,

- **la TECHNOLOGIE** (des PAROIS et de la VENTILATION), choix de la composition des murs (6 technologies possibles), des toitures (3 technologies), des planchers (4 technologies) -les linéaires de ponts thermiques sont intégrés aux K globaux- et des percements (6 technologies de fenêtres, 6 technologies de portes), choix d'un taux de renouvellement d'air (3 niveaux possibles).
Des aides en ligne liées à la comparaison des paramètres du projet à des seuils réglementaires guident l'usager dans le choix des technologies,

- **le BILAN,** le calcul des déperditions en GV, et le calcul d'indicateurs comme Kmoyen...

Cette structure à base de "compteurs" nécessite l'évaluation, pour chaque projet de bâtiment -ce qui veut dire, en fait, le bâtiment lui-même et sa référence générique, correspondant à un projet de référence de base carrée, la "référence carrée"- d'une centaine de paramètres. Ne restent apparents à l'usager que les plus significatifs, alors que l'analyse du projet se fait sur la totalité des compteurs (voir Traîtement de la connaissance).
Les caractéristiques du projet sont comparées à des valeurs-seuils fixées, sur chacun des paramètres principaux de conception, notamment le coefficient de forme, les taux de percement ou taux de contiguïté.Ces paramètres de conception sont structurés en un arbre d'héritage dont l'exploration par le module d'analyse experte permettra une évaluation de la performance du projet .**(voir figure 3)**

FIGURE 3 : ARBRE D'HERITAGE DES PARAMETRES DE CONCEPTION.

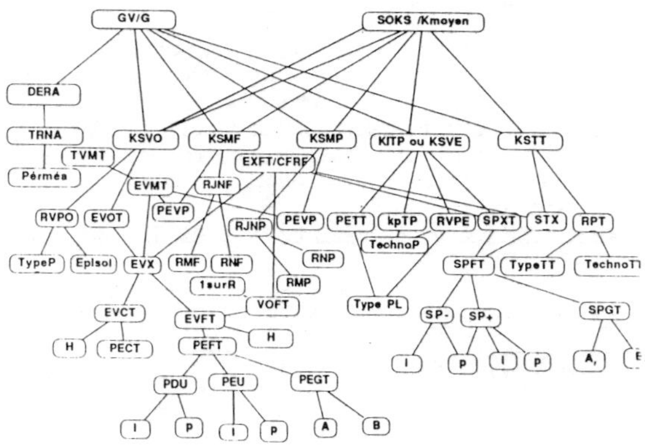

3-Analyse experte du projet

A partir d'un projet initial , soit entré par l'utilisateur grâce au module saisie , soit importé d'une bibliothèque, celui-ci doit définir une stratégie d'amélioration à volume constant de son projet , définie par quatre critères qui sont:
• le paramètre de conception à améliorer , parmi:
GV, Déperditions par les parois, Déperditions par les murs, Déperditions par les vitrages, Déperditions par le plancher, Déperditions par la toiture, K moyen , Coefficient de forme.
• le taux d'amélioration souhaité sur ce paramètre (supposé positif et exprimé en pourcentage): par exemple, amèliorer GV de 10%.
• le type d'amélioration souhaitée parmi :
géomètrie seule, technologie seule, géométrie et technologie .
• le type de recherche faite par le système expert:
-une recherche en largeur d'abord (on améliore un peu le maximum de paramètres modifiables),
-une recherche en profondeur d'abord (on améliore beaucoup le minimum de paramètres modifiables) .**(voir figure 4)**

FIGURE 4 : ECRAN DE DÉFINITION DE LA STRATÉGIE D'AMÉLIORATION DU PROJET.

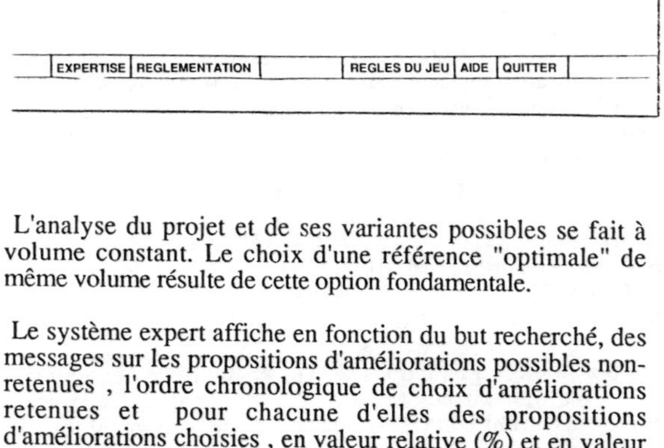

L'analyse du projet et de ses variantes possibles se fait à volume constant. Le choix d'une référence "optimale" de même volume résulte de cette option fondamentale.

Le système expert affiche en fonction du but recherché, des messages sur les propositions d'améliorations possibles non-retenues , l'ordre chronologique de choix d'améliorations retenues et pour chacune d'elles des propositions d'améliorations choisies , en valeur relative (%) et en valeur absolue (m2 ou m...) . Par exemple, "Je propose de réduire les découpes de votre bâtiment de 12% soit une surface horizontale de découpes de 3 m^2."

4-Base pédagogique

Ce module est le plus proche d'un didacticiel thématique classique. Sa structure est la même que celle du module de saisie (voir Structuration de la connaissance). Elle présente messages théoriques et activités pratiques (sous forme d'exercices) liés aux 6 paramètres de conception: site, mise en forme, contiguïtés, percements, technologie, bilan. Elle fait largement appel aux références architecturales desquelles elle extrait des éléments de solutions architecturales qui lui servent d'illustrations ou de supports d'exercices.
(voir figure 5)

FIGURE 5 : ECRANS DE LA BASE PÉDAGOGIQUE LIÉS A LA MISE EN FORME DU PROJET.

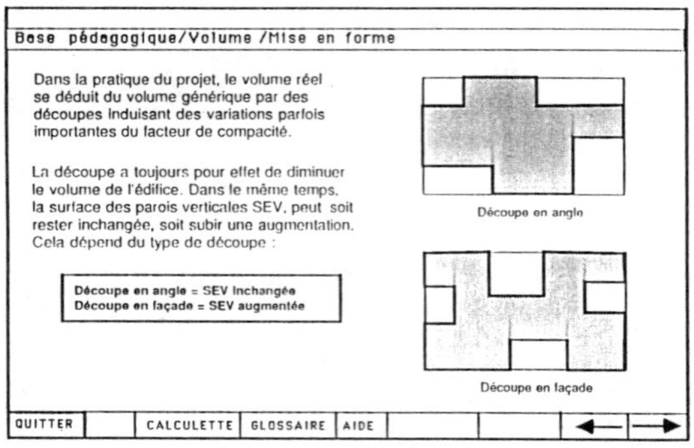

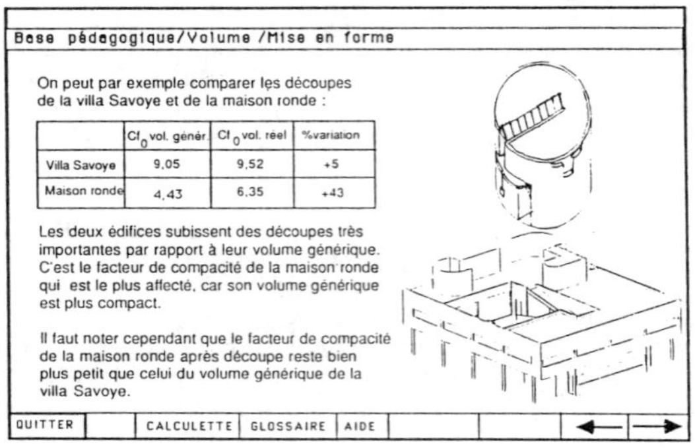

Deux accès sont possibles:
-par paramètre et donc par étape liée à la saisie du projet en cours; à chaque étape, l'accés à la base pédagogique est optionnel,
-directement à partir du sommaire, la présentation est alors linéaire et indépendante du module de saisie.

5-Références architecturales

Ce module a pour objet de montrer les références architecturales utilisées dans la base pédagogique. Ces références sont choisies plus comme référent à l'histoire et à la théorie de l'architecture que pour leur qualité thermique, certains pouvant même être des contre-exemples.
Quatre références architecturales remarquables sont présentées :

la Maison de Verre P.CHAREAU
la Casa Rotonda M. BOTTA
la Glass House JOHNSON
la Villa Savoye LE CORBUSIER

Les possibilités graphiques de l'ordinateur sont exploitées pour présenter en couleur des illustrations détaillées des références architecturales (plans, coupes, pespectives...), articulées autour de leurs points forts . Les textes à l'écran sont limités à une brève description de ces points forts.

ORGANISATION

L'équipe de conception / réalisation est pluridisciplinaire. Elle regroupe des architectes, des experts en thermique, des pédagogues spécialistes de l'EAO, des spécialistes de l'ergonomie, des réalisateurs d'outils informatiques spécialistes de la communication interactive:
Luc ADOLPHE (ENTPE- GISE) • Dominique CAMPANA (DIALOGIC) • Patrick DEPECKER (Ecole d'Architecture de Lyon) • Marie-Annick FRESSE (*Ligne Graphique - Ergonomie*) • Jean-Louis IZARD (Ecole d'Architecture de Marseille-Groupe ABC) • Anne LASSERRE (CNEAO) • Jean-Pierre TRAISNEL (CNRS) • Jean-Pierre PENEAU (Ecole d'architecture de Nantes-CERMA).

PROTOTYPAGE

Le matériel informatique nécéssaire est un IBM PC ou compatible équipé d'un disque dur, d'une carte EGA et d'un écran couleur. Une souris et une imprimante sont recommandés.

ETAT D'AVANCEMENT fin 1989

La conception du produit est achevée; sa rélisation est en cours. Ce module sera donc disponible au cours du second trimestre 1990.

CONCLUSION

Ce travail d'intégration d'un corpus technique dans la méthodologie architecturale relève dune recherche fondamentale sur la confrontation et le transfert de savoirs techniques et architecturaux. Pour être menée pleinement, cette recherche sur les modèles conceptuels des différents acteurs du processus de conception d'un bâtiment mériterait un terme de quelques années incompatible avec les délais impartis au produit. Nous sommes donc conscients que notre modèle puisse paraître un peu caricaturelement techniciste au seul vu du comptage des items architecturaux par rapport aux items technologiques. Une recherche plus fondamentale est donc en cours sur ce sujet dans le cadre du groupe GISE.

BIBLIOGRAPHIE

CNRS, DIALOGIC, CNEAO, Ecole d'Architecture de Marseille "Conception Architecturale et Bilan thermique",Rapport intermédiaire , Marché AFME n° 7.12.0015

ARMINES, Ecoles d'Architecture de Marseille et de Lyon, "Conception Architecturale et Bilan thermique, Extensions du logiciel pédagogique", Rapport final AFME, Convention AFME n°8.12.00444, Juin 1989

ARMINES "Participation au programme PEBI", Rapport final MEL-IN.PRO.BAT, Convention n° 8854/05, Février 89

CERMA, "Etablissement d'un corpus de réalisations architecturales de références pour les didacticiels d'énergétique du bâtiment", Rapport final AFME, Convention AFME n° 7.12.0009

ARMINES/CNEA/DIALOGIC, "Didacticiels en Maîtrise de l'Energie : Etat de l'Art et Analyse des Besoins", Rapport AFME , Conventions AFME n° 5.12.1079 et 6.12.0006, Avril 87

SMART GLAZING AND ITS EFFECTS ON PASSIVE SOLAR DESIGN

J Littler

Research in Building Group
Polytechnic of Central London
35 Marylebone Road, London NW1 5LS

ABSTRACT

Emphasis is laid on the major contribution which energy used in buildings, makes to global warming.

Innovations in glazing are reviewed, and it is suggested that smart glass will make a major impact on the better design of bioclimatic buildings.

Smart glass will allow the use of very large areas of glazing, admitting needed solar gains to houses and daylighting to commercial buildings, without producing overheating, glare or degradation of furniture or works of art.

New windows are discussed which allow space heating energy to be reduced to negligible amounts in housing.

Allusion is made to the place of smart glass in smart walls, and in avoiding indoor pollution.

1 INTRODUCTION

Approximate Division of Primary Energy Use in Developed Countries

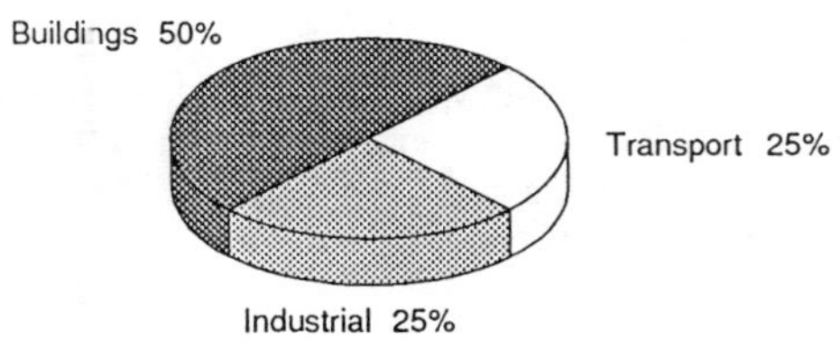

Dramatic improvements have doubled the efficiency of automobiles in the last 15 years. Simple changes, widely adopted, such as aerodynamic shields on truck cabs, have greatly reduced the burden of pollution per person or per tonne carried. Further changes in the pipeline, to engines, controls and aerodynamics, will improve efficiency by an additional 40%, moves to alternative fuels may further cut CO_2 emission. Trains such as the TGV have opened routes to those who would otherwise go by airplanes which cause more pollution per person/km.

In industry, enormous improvements have been made in the efficiency with which many operations are carried out, for example by using targeted microwave heating in drying processes, and by process heat recovery.

Most of the technologies which have caught peoples' imaginations and have allowed these improvements in efficiency, are highly *visible* (eg aerodynamic vehicles) or are highly exciting to the engineers and managers responsible for their introduction (eg the TGV). Measures to save energy in buildings have been the Cindarella partly because they have been invisible and technically unexciting (eg wall insulation with glass fibre).

When technologies lack a glamorous appeal to designers, or obvious economic appeal to the public, changes can only ensue if resort is made to sticks and carrots.

This paper contends that **smart glass** at last provides a technologically glamorous and exciting product, which will also give great freedom to architects in designs for good bioclimatic buildings.

2 THE GREENHOUSE EFFECT OR WHY WE NEED TO REDUCE ENERGY USE

In the buildings sector low fuel prices have not provided the necessary incentive to the consumer to ask for higher levels of energy conservation (the carrot); and in the UK, a combination of Government desire "to lift the burden of controls" and the construction industry's immense resistance to change, conspire to avoid effective legislation (the stick). This laisser faire attitude is typified by the recent UK Government Report which suggests that CO_2 emission in the UK will rise by 73% by 2020 [1]; let us hope that the rest of the world does not share this disastrous head in the sand attitude.

Thus, at least in the UK, rather little progress has been made in reducing the burden of greenhouse gases caused by buildings. The effect is made worse by the reluctance of the electricity industry to add CO_2 scrubbers to its power stations

The following data [2] suggests that as countries develop, the fraction of their total energy devoted to buildings may increase. Since their *total* energy use will increase, the significance of buildings in the production of CO_2 is doubly important.

Energy Used In Buildings as % of Total Energy For Various Stages of "Development"

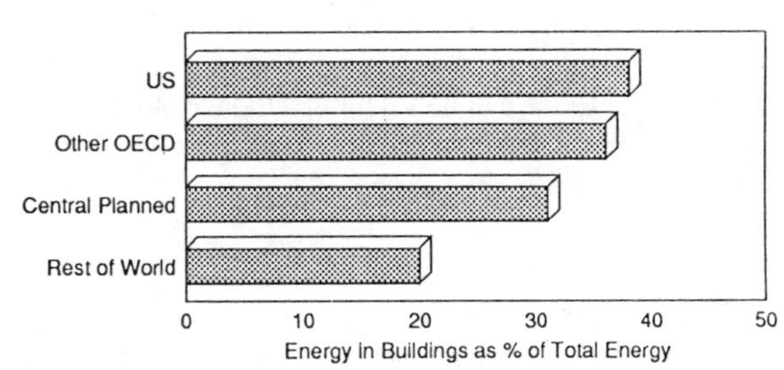

From H Kelly, Buildings of the 21st Century

Disaggregated US data illustrates the origins of CO_2 evolution as a result of energy used in residential and commercial buildings. The dominant sources are space heating in the domestic sector and lighting and space heating in the commercial sector.

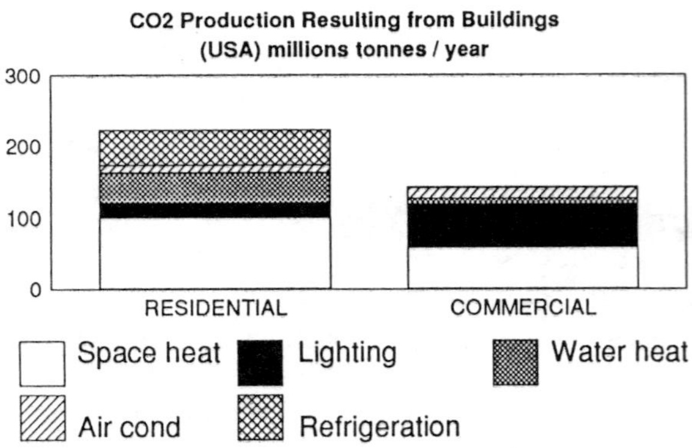

From H Kelly, Buildings 21st Century

The data for CO_2 production may not yet be very reliable, (the UK probably does not approach the US in total CO_2 production as a result of building use) [3]; but the significant observation is that the shape of the two distributions is similar in both countries, namely space heating and lighting plus appliances, are the important areas to attack.

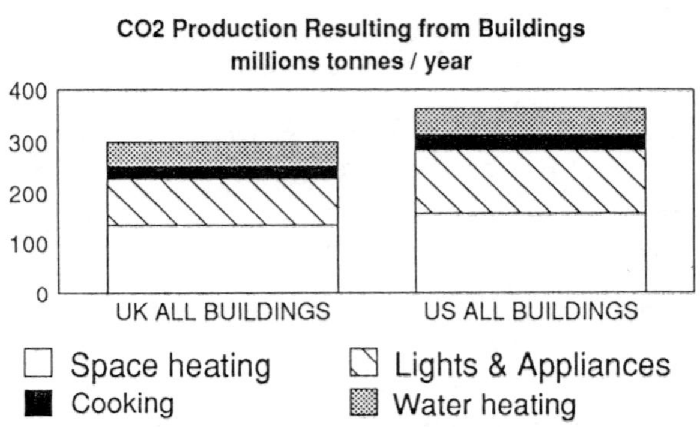

UK data from N. Milbank, Architects J.Aug. 1989
US data from H Kelly, Buildings 21st Century
(IEA Buildings & Community Systems Conference)

Smart glass will reduce space heating by allowing large areas of solar admitting glass, without increasing cooling loads, and will reduce electric lighting by increasing daylighting, also without increasing cooling loads.

However it will always be true that simple insulation, where it can be applied, will be the most cost effective way of reducing space heating.

Even with high levels of insulation and heat recovery, much of the energy cost and thus CO_2 emission, remains unaddressed. The Superinsulated Houses which we demonstrated for DG 17 in the UK [4], are heated to very high comfort levels for the price of a cup of coffee per week, but we were unable to reduce by similar fractions, the electrical energy for appliances and lighting.

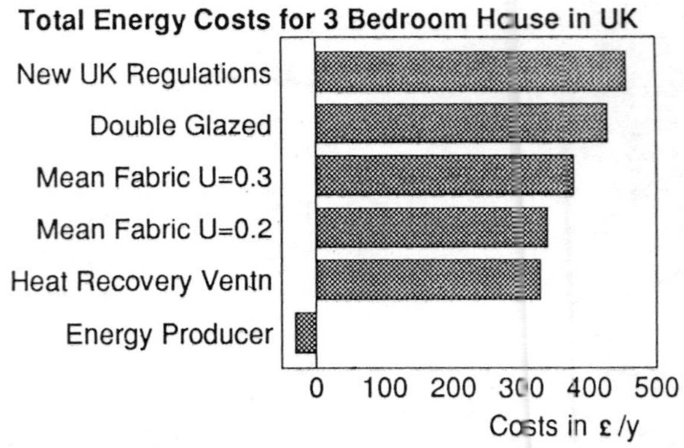

The provision of energy has two sides: the supply side and the demand side. Many developed countries are increasingly choosing to reduce demand by conservation measures, rather than build new power stations. Surely it is this trend which conscientious designers should support?

As the efficiency of appliances in houses rises (and Danish studies suggest up to half the electricity used in homes could be avoided [5]), the casual gains will fall. Frequently, solar gains are not useful because demand for heating has already been satisfied by casual gains - especially in the swing seasons. The opportunities to use more solar gain should thus rise as appliance efficiency rises.

The following paragraphs start with a description of smart glass technology, and go on to suggest where smart glass can help in the reduction of energy use in the domestic sector.

3 SMART GLASS

A wide range of glasses which respond passively or actively to the environment are under development.

3.1 Thermochromic

Cloud gel invented by Chahroudi in his kitchen in 1971, has taken a long time to reach its present day effectiveness. It is a clear film which when heated above room temperature reflects sunlight by turning an opaque white and then turning clear again when cooled. The degree of opacity is also determined by the strength of solar radiation, so that a window in a cool room exposed to high sunlight levels remains clear; but one in a warm room turns cloudy, as illustrated in the Figure [6].

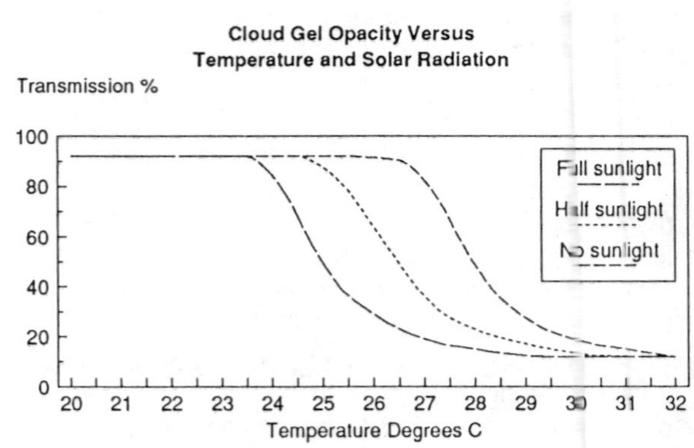

Taken from "The Climatic Envelope", by Roy Chahroudi at Suntek Inc

An analagous product TALD has been developed in

Europe [7].

In general, such organic gels contain water soluble polymers, which alter their chain shape or length dramatically over short temperature spans.

Various oxides of vanadium which can be produced as coatings, also change visible transmission with temperatures near those required [8].

3.2 Photochromic

Large numbers of compounds undergo reversible change on illumination, and photochromic sunglasses provide a familiar example. Silver halides with small amounts of Cu^+ are possibly the most common dopants used in glasses. For plastics, spirooxazines in cellulose acetate butyrate sheet is a strong contender [9].

3.3 Electrochromic

Certain compounds undergo reversible colour changes when a small voltage is applied across a thin layer, as illustrated in this Figure adapted from [10].

Electrochromic Switching Glazing

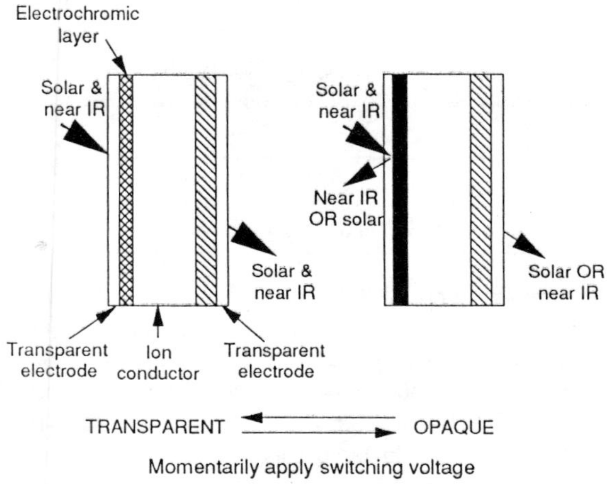

The switching voltage is needed only to change the state. Most investigations have concentrated on transition metal compounds, notable WO_3; but oxidation/reduction reactions in some organics, for example the phthalocyanines, can produce the desired effect.

Asahi Glass has installed large numbers of prototype electrochromic windows in the Seto Bridge Museum and in the Daiwa House [11].

Liquid crystals, commonly seen in calculator displays, are now available in large areas from Taliq as switchable privacy screens in offices. Liquid crystals require continuous application of the switching voltage.

Finally, some suspended particles can be aligned, so that transmission is increased, by the application of an electric field. Battery powered sunglasses using the technology are in the prototype stage [12].

For architectural applications, a great merit of solid state electrochromics over passive switching such as thermo- or photochromics, is that of control. For example, in our Roofspace Collector Systems [13], it is desirable in the winter to allow the roofspace to heat up as much as possible (below 60°C), so that warm air may be pumped into the house; but in summer, it is desirable to keep out the sun at much lower temperatures. Indeed, the cost of extract fans and controls to avoid the risk of roofspace systems overheating the attics, greatly diminishes the cost effectiveness of the systems. Control is even more desirable in airliner cockpits, and PPG recently announced at the Paris Air Show, an electrochromic windscreen for jumbo jets. Unlike the electrochromic glass shown above, which controls *either* the visible *or* near infra red, the PPG unit has multiple layers to control *all* the incoming solar radiation.

3.4 Holographic

Holographic films are currently used to allow drivers to see the road ahead, but yet also to reflect light at a different wavelength, say green, coming from the instruments below, thus providing a "head up" display on the windscreen.

Such holographic films may redirect beam radiation, throwing it to distant parts of a space for daylighting, and may be tailored to reflect or transmit any selected waveband, thus acting as selective filters reflecting longer wavelength infra red (heat) back into rooms but transmitting visible wavelengths.

3.5 Other redirecting coatings

Prismatic glazing has been most effectively applied in the renovation of the Billingsgate Fish Market in London, by the Richard Rogers Partnership [14].

Redirection of daylight may also be effected by obliquely evaporating chromium so that inclined cylinders 100 - 200nm long by 20 - 40nm wide are deposited on the glass [15].

3.6 Selective coatings

It has taken about ten years for selectively coated glazing to become commonplace, following its discussion in the technical press [16]

The same article also pointed out the importance to the energy balance at a window (the energy *gain* arising from incoming solar and near infra red radiation minus the thermal *loss*), of inert gas filling in the space between sealed glass units, and the value of antireflection coatings.

A two pane 3mm unit currently provides a shading coefficient (SC - a measure of the solar heat gain relative to a single 3mm pane of glass), of 0.89. If low iron glass is used the SC rises by about 8% to 0.96, and if the glass is antireflected on all surfaces, by a further 11% to about 1.07. Since the thermal loss remains the same, the window now has the potential to collect an additional 18% of the available solar radiation.

In 1986 PCL designed superglazed houses as an EC DG17 Demonstration, in which large areas of highly insulating glazing (U-value 1.0 W/m^2 K, and shading coefficient 0.6 [17]) were installed in 6 houses. Robinson has reported on the results in the Proceedings of this Conference and elsewhere [18].

In 1989 Lawrence Berkeley Laboratories built and installed in several test homes, sealed triple pane windows with two low emissivity coatings (e=0.08) and gas fill consisting of a mixture of argon and krypton. The U value is about 0.85 W/m^2 K and the shading coefficient about 0.7. LBL has developed units which are less heavy and thinner, by slimming down the centre pane to 2mm and relying on the cushioning by the trapped gas layers - an idea related to Southwall's original suspended centre-film Heat-Mirror product.

SERI has a prototype evacuated double glazed unit with one low e coating. The glass is held apart by tiny glass beads, and the centre pane U value is better than 0.57 W/m^2 K with a shading coefficient of about 0.8.

SERI's Evacuated Window

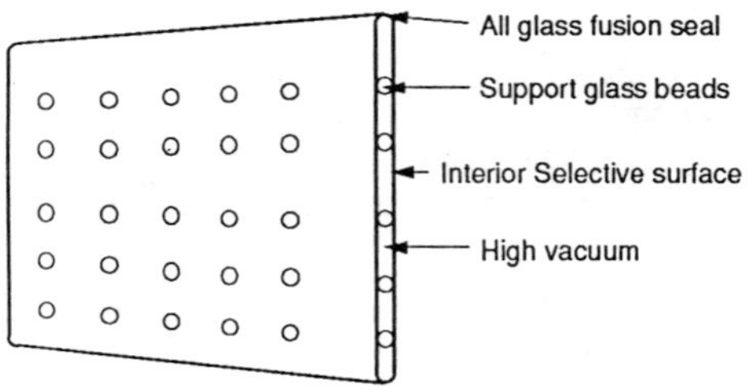

The following Figure illustrates the case of an overcast winter sky in northern climates, with total daily radiation in the glazing plane of 0.9 kWh/m^2. The 24 hour heat balance is adverse for simple double glazing, but positive for several other combinations of U value and shading coefficient. The best technology which could be assembled in a window at the present time from available components, (triple glazing, low iron glass, antireflecting treatment, two low emissivity coatings and Kr gas fill), would provide about 0.4 kWh/day of space heating per m^2 of glazing in the UK.

Heat Balance at Windows in Winter
Outside 0 C, inside 20 Centigrade
Daily incident radiation in vertical 0.9 kWh/m*2

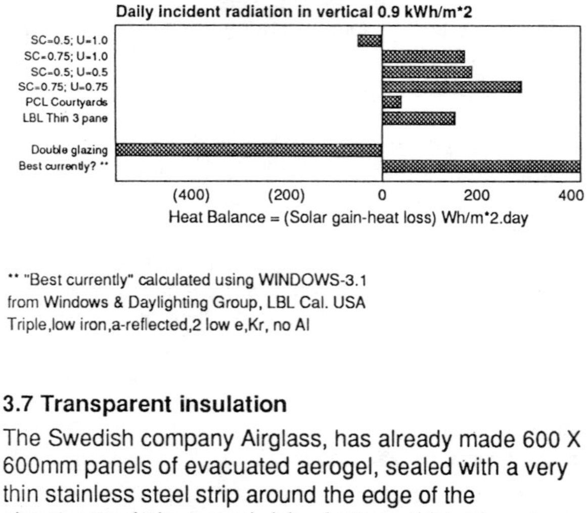

** "Best currently" calculated using WINDOWS-3.1
from Windows & Daylighting Group, LBL Cal. USA
Triple,low iron,a-reflected,2 low e,Kr, no Al

3.7 Transparent insulation

The Swedish company Airglass, has already made 600 X 600mm panels of evacuated aerogel, sealed with a very thin stainless steel strip around the edge of the glass/aerogel/glass sandwich. A 20mm thick slice of aerogel used in this way has a U value of about 0.65 W/m^2 K and the solar transmittance of the aerogel layer is about 70%, whilst the enclosing glass is about 80%.

3.8 Airflow windows

Yuill has reported a 3 pane window [19] in which outside air flows up through a 9mm space. The *laminar* flow ensures low heat transfer between the glazings, by suppressing convection cells yet picks up heat from the inner glass and transfers it into the building in the ventilation air. The U value is about 1.0 W/m^2 K and the air flow about 3.5 l/s.m^2 with a temperature rise of 8 K. (Outside temperature -18 C, roomside temperature 20 C)

4 SPACE HEATING IN DWELLINGS

4.1 Provision of Space Heating

Optimum Window Area for Minimum Space Heating for a Variety of Window Types Facing South: Types Defined by U-Value & Shading Coefficient

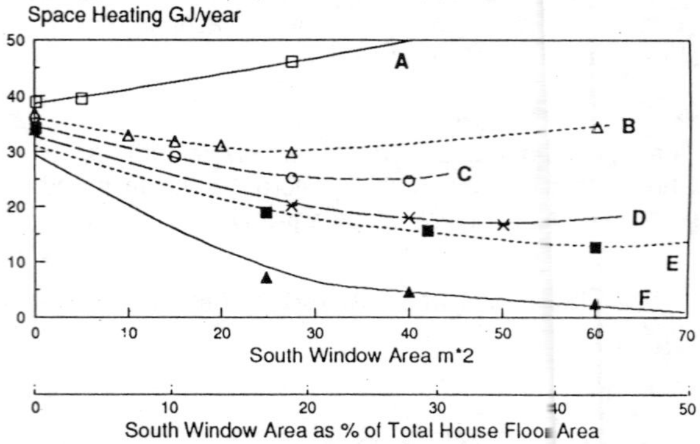

Data is calculated using SERI-RES and the Courtyard Direct Gain Houses

	A	B	C	D	E	F
U value	6.0	3.0	2.0	1.0	0.5	0.0
Shading coefficient	1.0	0.8	0.7	0.6	0.5	1.0

(Information about SERI-RES may be found in [20]).
The illustration shows the value to passive solar heating, of switching from single glazed windows (the top curve A), where the less window, the lower the annual space heating energy, to better windows. The next curve down B, for example, is for double glazing, where the optimum area as a fraction of floor area, is about 17%. Curve E illustrates a window which could readily be built today, and which reduces the space heating in the example house, from 40 to 50 GJ for single glazing, or 32 GJ for double glazing, to about 12 GJ/year.

To reach the dizzy goal of a zero energy house set in the Energy Producer House of an earlier Figure, requires glazing better than we can presently produce. The curve F represents the bounding case of a window with the transparency of single glazing and no heat loss. However it would be quite realistic to assume that glazing between cases E and F *will* appear, allowing houses with large areas of glass, to use direct passive solar gain to reduce the space heating to negligible amounts.

The Courtyard Houses referred to above, do not overheat, even though they expose very large areas of glass to the south. At high summer solar altitudes, the shading coefficient is low and the solar gain modest, due to the high rejection of beam radiation by the multiple surfaces of triple glazing. In very advanced glazing of the sort which approaches case F, the shading coefficient must however be high in the heating season.

The relationship between shading coefficient SC, U-value U (W/m^2 K), south window area A (m^2) and space heating load E (GJ/year), for the Courtyard Houses is:

$$E = 35.2 + 4.27*U - 14.15*SC - 0.275*A$$

The implication is that given a choice between trying to reduce the U value below say 0.7 W/m^2 K, and attaining a high transmittance, it is the *latter* which is more significant, for highly driven passive solar dwellings.

This assertion is even more valid for the design of sunspaces, where the rise in temperature above ambient Δt, (which controls the value of ventilation preheating and the habitablity of the sunspace), is given by a relationship of the form:

$$\Delta t = 3.60 - 0.68*U + 4.25*SC$$

The results for a specific case are illustrated below:

Sunspace Temperatures Excess Above Ambient As Function of Glass U-Value & Shading Coefficient. Sunspace Giving Ventilation Preheating to House

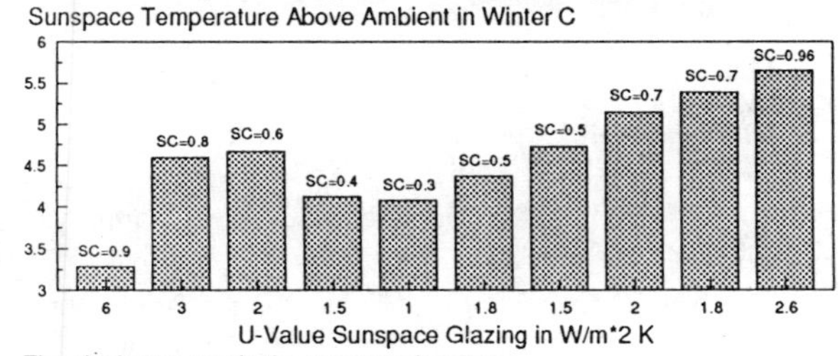

The air change rate in the sunspace has been adjusted to give 0.75 air changes/hour in house

Selkowitz has provided related information for houses in the US, where he shows for a fixed window area of 6.1 m^2 facing east, the relationship between SC and U-value, to achieve a positive annual contribution to space heating. For windows worse than those defined by the heavy line, more heat is lost than gained in the heating season.

Net Annual Useful Energy Flux/m*2 South Window In a House In Madison, Wisconsin, USA

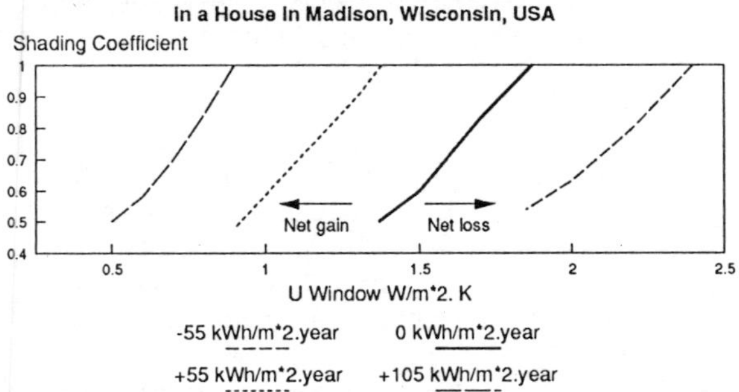

Adapted from Report by the Windows and Daylighting Group, LBL, Cal., USA

4.2 Prevention of Overheating

All these cases suggest then, that highly passively driven solar houses with very low space heating can be achieved even in northern and adverse climates; but the penalty will be overheating unless adjustable means are available to reduce solar gain under controlled conditions.

It is here that **smart glazing** has one major role in dwellings. One product which goes part of the way towards meeting this demand, is shown here:

Climatic Envelope by Suntek Inc.

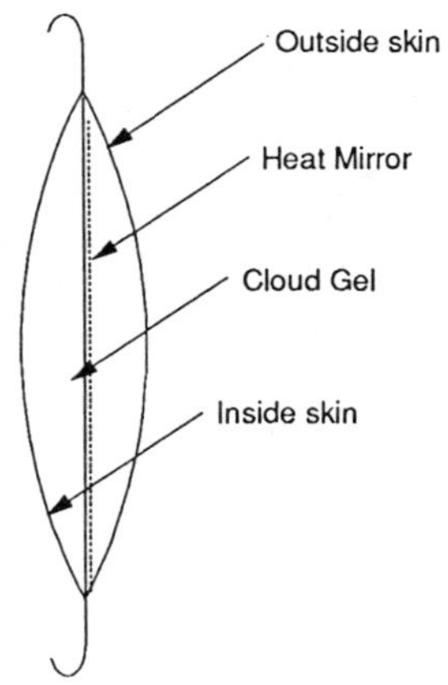

Large areas of the climatic envelope could be used to replace stretched fabric roofs such as that over the atrium of Herron's beautiful Imagination building in London [21], and as the covering for Jeffrey Cook's Solar Oases in hot dry climates.

4.3 Privacy and Preservation

Smart glass allows a particular part of a room to remain in shade as the sun moves round the building. Sections of a window will darken to keep objects of value out of the direct sun. Similarly in the smart homes of the future, it will be possible to program sections of a window facing the street, to become opaque or shaded, for reasons of privacy. Such provision may persuade people not to use net curtains which, in passive solar houses in the UK, have frequently removed much of the benefit of direct gain. Martin has carried out measurements in test cells of the disbenefits occasioned by net curtains [22].

4.4 Daylighting

With very large areas of glass on the south facade, excess daylight may be redirected from the upper sections of the facade by holographic glass and distributed to adjacent rooms by smart walls in which for example upper sections become clear during the day. A smart wall, consisting of active glass panels, may also become the large area video screen, the display area for functions within the smart home and the entry-phone screen. It may be programmed as a wall sized light show.

In cases where sufficient light is not available in north rooms it is possible to envision the collection of diffuse daylight from the roof, with redistribution to the rest of the house. Fluorescent edge collectors have been proposed for concentrating light admitted to photovoltaics. Daylight is absorbed by a dye contained in a thin sandwich. Fluorescence *at a different wavelength* is totally internally reflected at three of the edges of the sandwich and escapes as a line source at the fourth edge. From there it could be ducted by the 3M light pipe system which is already available.

There is also the more distant possibility, that photoluminescent walls may be produced, in which charging occurs during the daylit hours and discharging is triggered at the required time. Already we are seeing photoluminescent materials used in buildings to mark escape routes.

5 INDOOR AIR QUALITY

Time Spent Indoors In Europe

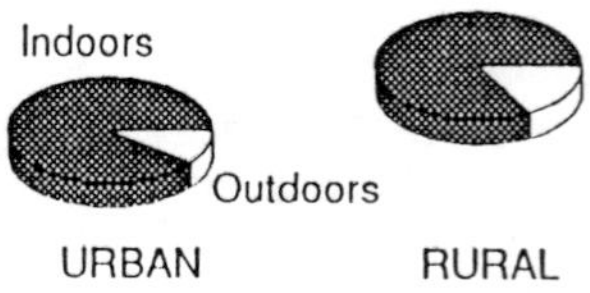

We spend most of our time indoors, where the sick building syndrome is becoming increasingly important. Smart glass has the great advantage over other surface materials of contributing little to indoor pollution.

6 DAYLIGHTING IN COMMERCIAL BUILDINGS

This article has concentrated on smart glass in housing; but clearly after applications in planes and cars, the first use is likely to be in prestige office buildings and art galleries, to control daylighting and reduce glare, particularly adjacent, for example, to computer screens.

7 REFERENCES

[1] Figures from planners at the Department of Energy, reported in The Guardian Newspaper of Nov 17, 1989.

[2] Henry Kelly, US Office of Technology Assessment, in *Buildings of the 21st Century*, International Energy Agency Conference Proceedings of the Buildings and Community Systems Agreement, May 1989.

[3] Neil Milbank, *Response to Climate Change*, Architects Journal, London, August 2, 1989.

[4] J Littler and P Ruyssevelt, *Superinsulation*, Passive Solar Journal, 4.2, 1987.

J Littler and P Ruyssevelt, *Results of Monitoring in Superinsulated Houses at Milton Keynes*, European Conference on Architecture, Munich, April 1987.

[5] Jørgen Nørgård, *Low Electricity Appliances - Options for the Future, from Electricity: Efficient End Use and New Generation Technologies and their Planning Implications,* Editors T Johansson, B Bodlund and R Williams, Lund University Press, April 1989.

[6] Suntek Materials for Buildings Through Molecular Design, Suntek, Albuquerque NM, USA

[7] E Boy and S Meinhardt, *TALD a Temperature Controlled Variable Transparent Glass*, CIB Journal $_4$ 1988

[8] C G Granqvist, *Chromogenic Materials for Transmittance Control of Large Area Windows*, Internal Report from the Physics Dept., Chalmers University of Technology and University of Gothenburg, Sweden, 1989.

[9] Granqvist *ibid*

[10] C Lampert, *Electrochromic Materials and Devices for Energy Efficient Windows*, Solar Energy Materials 11, 1, 1984.

[11] C Lampert *Advances in Optical Switching Technology for Smart Windows*, ISES Solar World Congress, Kobe, Japan, September 1989.

[12] Research Frontiers, Woodbury, New York, USA.

[13] A Tindale and J Littler, *Passive Roofspace Systems: Performance Evaluations,* Second European Conference on Architecture, Paris December 1989.

C Hancock and J Littler, *Retrofit and Monitoring of 6 Local Authority Houses at Newham, with Hybrid Solar Roof Space Systems*, ASES Conference, Boulder, Colorado, June 1986.

C Hancock, J Littler and A Tindale, *Construction and Performance of Retrofit Hybrid Solar Roofs to Low Cost Housing*, European Conference on Architecture, Munich, April 1987.

J Littler, C Hancock and P Ruyssevelt, *Passive Solar Hybrid Roof Space Collector Systems: Results and Simulations at Woodbridge and Newham*, PLEA, Venice, December 1985.

[14] See for example the Architectural Record, January page 121, 1989.

[15] G Mbise, G B Smith, G A Niklasson and C G Granqvist, Proc. Soc. Photo. Opt. Instrum. Engr. $_{049}$ 1989.

[16] J Littler, *Thermal Balance at Heat Reflecting Windows*, J Energy Research, 1979, 3, 173

[17] P Robinson and J Littler, *Methodology of Window Thermal Performance Assessment in the UK and US.* Report to the UK D. Energy/ETSU, 1988.

[18] J Littler and P Robinson, *Superglazing U value 0.9*, Procs Canadian Solar Energy Society, Ottawa, June 1988.

J Littler and P Robinson, *Methodology of Window Thermal Assessment Techniques in the UK and the US*, Report to the Department of Energy, 1988.
J Littler and P Robinson, *100% South Facing Superglazing to Courtyard Houses*, Proceedings of the Third ICBEM Conference, Lausanne, Switzerland, November 1987, p 189.

J Littler and P Robinson, *Monitored Performance of R7-Superglazing at the Courtyard Houses, Energy World Milton Keynes*, European Conference on Architecture, Munich, April 1987.

[19] G K Yuill, *Laminar Air Flow Superwindows*, Proceedings Annual Conference Solar Energy Soc., Canada, Ottawa June 1988.

[20] P Haves and J Littler, *Refinements to SERI-RES, Final Report to the Department of Energy*, ETSU S-1130 1987.

[21] Imagination Building, by Herron Associates, see Architect,Builder, Contractor and Developer, May 1989 p 25.

[22] C J Martin, *Test Cell Studies 1: Measurement of Parameters Describing the Effect of Net Curtains, Curtains and Venetian Blinds*, Report to the UK Dept. Energy via ETSU, 1988.

ARCHITECTURE, RENEWABLE ENERGY AND THE GLOBAL ECOLOGICAL CRISIS.

John Page

Emeritus Professor of Building Science, University of Sheffield, U.K.
Expert Consultant to the Commission of the European Communities' R & D programme "Energy conservation in building".
Address: 15, Brincliffe Gardens, Sheffield, S11 9BG, England.
Telephone (0)724 551570

1. GLOBAL POLLUTION AND CLIMATIC CHANGE.

1.1 Local pollution.

Global atmospheric issues recently have risen rapidly up the world agenda. This emerging crisis in the global atmospheric environment has a chemical and radiative cause. The physical effects of local smoke pollution on the intensity of short wave solar radiation has long been clearly perceived. Smoky polluted cities were well known for their lack of daylight. In an age of heavy smoke, too much atmospheric pollution was all too visible. It literally encased much European architecture in black deposits. Most of the Member States of the Community have made great progress in reducing urban smoke pollution. The widespread availability of natural gas has been an especially important facilitating factor. In spite of these efforts, pollution remains a key factor affecting the local economy in many serious ways, for example sulphur dioxide pollution and oxides of nitrogen pollution leading on to acid rain. Now much of our cultural heritage is being dissolved away in acid water.

1.2 Ozone layer destruction /1/.

Architects have always been interested in the effects of light, but this interest did not normally extend to the influence of light on pollution. This situation now has to change. Our atmosphere is very actively driven photochemically by the sun. A vast range of chemical reactions are proceeding in it all the time. The by products of urban living feed reactants into this photochemical system. The local consequences of these chemical reactions can prove very serious, for example, photochemical smog, caused by the interaction of traffic pollution and sunlight, containing ozone. Such low level ozone is very irritating for the eyes. Ozone is a strong oxident, damaging some building materials. However it is the global atmospheric reactions driven by light, which are now beginning to cause the greatest concern. A few years ago, the numerous interactions of significance between the natural radiative environment and the trace man made chemical contents in the atmosphere were not widely perceived. At the present moment, especially important photochemical reactions involve the widely used CFC and Halon gases. These gases are discharged at the surface. They are very stable in the lower atmosphere and only gradually migrate upwards, until they reach high levels in the stratosphere. There the UV-C radiation from the sun finally breaks down them down, releasing chlorine atoms from the two groups of man made covalent compounds. The breakdown products can enter into chemical chain reactions, consequently destroying some of the ozone. Such CFC gases are very widely used in contemporary air conditioning plant. Halons are very widely used in contemporary fire extinguishing equipment. Designers now have to consider very seriously the alternative technological options in architecture, needed to resolve the problem. CFC's have also been widely used in the production of thermal insulants.

1.3 Global warming /2/.

The second key global pollution problem is how to control excessive global warming. In attempting to develop acceptable future fossil fuel based energy policies, world energy policy makers are now trapped by the adverse effects of the world's ever growing volume of gaseous waste products. The new gaseous threats are invisible. The predicted warming is influenced by the radiative properties and the amounts of the various atmospheric gases which absorb thermal radiation in the thermal infra red region. Most important is carbon dioxide, a steadily accumulating gas in the atmosphere. Other gases like CFC's, methane, and ammonia are important too. Their concentrations are increasing too. The present concentrations of the various CFC's are estimated to contribute nearly one sixth of the predicted warming effect. Thus there are two reasons why there is a need to reduce CFC emissions.

Assessing the architectural impacts of global warming due to the greenhouse effect must now be considered a key factor in future architectural policy. The global climatic warming will not be uniform. The biggest temperature rises are predicted for the winter season in the higher latitudes of the European Community. Such changes will alter the future demands for energy use in buildings.

2. THE ARCHITECTURAL CHALLENGE.

Architecture faces a double challenge. Firstly how should designs be altered to reduce the risks of global warming and of stratospheric ozone destruction? Secondly, accepting the implications of climatic change, how should architectural design change to meet estimated long term conditions? For a start, a warmer world will imply a greater potential demand for air conditioning, and that, in its own right, imposes the risks of the creation of further environmental threats. Research in the European Community might start by making putative architectural bioclimatological analyses for future European climates, examining the seasonal and latitudinal nature of the changed building requirements implicit in the forecasts currently emerging. This will have to involve liaison with bodies like the new centre for forecasting climatic change currently being set up in the United Kingdom.

3. ARCHITECTURE AND POLLUTION.

Architectural designers now need to look at the relationship between pollution and building design at three distinct levels, pollution of the indoor environment, pollution of the local environment, and pollution of the global environment. They must take full account of the recent developments in global environmental control needs.

The role of new technologies, and the use of renewable energy have to be assessed not only in terms of their energy benefits, but also in terms of their benefits in reducing environmental impacts, at all three scales of pollution mentioned. This is a difficult challenge that has yet to be systematically addressed. Architects are involved both in facilitating or controlling, through their design decisions, the internal production of pollutants and, simultaneously, protecting, or failing to protect, their design achievements from the local and global effects of pollution. They currently act with greater or lesser consciousness of what is actually being achieved environmentally, and what adverse effects they have caused, or are about to cause. Approaches tend to be based on specific enthusiasms rather than on balanced knowledge across the whole field of environmental cause and effect in building design and use. Landscaping does not make builings "green". Buildings have to be designed to be "green". They produce pollution through the processes of heating and cooling, and are made from materials like concrete and steel, whose energy intensive manufacture involves the production of local and global pollutants as by-products. Simultaneously buildings are affected by the pollution, which its designers, among others, have themselves helped to create. As about half the energy consumed in certain European countries is consumed to meet building demands, architects are responsible for decisions affecting about half the global warming effect. The pressures on the profession to resolve these problems better are likely to strenghten considerably. The new situation is likely to create pressures for formal environmental impact statements at three levels, the indoor environment, the local environment and the global environment as a requirement for all building designs.

4. THE EFFECTS OF POLLUTION ON BUILDINGS.

4.1 Acid rain.

The effects of pollution on buildings will not be analyzed in detail here. One must however single out the various phenomena associated with "acid rain", which includes mists and fogs, as well as precipitation falling in the field of gravity, as the cause of severe deterioration of much of our European architectural heritage. It is encouraging that Research area 7 of the CEC STEP programme has as its aim the development of the scientific and technical basis for improved protection of the European cultural heritage. This programnme has two sub topics, firstly "Materials, environmental factors and damage", and secondly "Measuring environmental damage". However the fundamental need is to eliminate the basic causes of environmental damage.

4.2 Changes in UV climates.

One must also note that changes in the stratospheric ozone layer could produce adverse radiatively caused deterioration of photochemically sensitive elements of buildings, like paints, plastics and asphalts, in addition to adverse effects for plants and people. Reduced stratospheric ozone levels will lead to increased transmission of UV-B radiation. The ozone layer protects man, plants, animals and the actinically sensitive parts of buildings from the damaging effects of UV-B radiation. At the same time we are seeing increases in tropospheric ozone due to photochemical reactions in the lower atmosphere. The current evidence, is that, leaving aside Antartica, observed UV levels in industrialized countries are currently falling rather than increasing. Some are attributing this fall to the substantial increase in tropospheric ozone, in which traffic pollution plays an important role. Unfortunately much poorly considered and exagerated speculation on the resultant health effects from increases in UV-B from stratospheric ozone changes has recently occurred. I personally have attempted to give a balanced assessment concerning the probable near term impacts of ozone changes on the health of man in my current work for WHO /3/.

5. ENERGY AND MATERIALS USE CONSIDERATIONS.

The energy crisis tended to narrow the approach to overall environmental design. This failure may be found in the work of many energy conscious designers. It is essential that designers now should set their indoor environmental goals on grounds of human health, but adopting means of solution that are compatible with the global ecosystem. This statement has implications both for energy use and for materials use. Materials may be harmful in themselves in the interior, causing sick building syndrome effects and adverse health effects /3/. The production of building materials may cause considerable local environmental damage. The fuel use associated with their production will, through carbon dioxide production impacts, affect the global environment. Just as we distinguish between primary energy use and secondary energy use in the case of electricity, so, in considering the environmental impacts of architecture, we need to distinguish between primary pollution production in the manufacture of the building materials associated with a specific building, and secondary pollution in the use of those building materials in the building itself. The eventual waste cycles need to be considered too. Transport of building materials also requires appreciable energy. In a 1974 report of the UK National Economic Development Office /4/, the fuel cost element as a percentage of total delivered price was 4-8% for aggregates, 3-5% for bricks and 1-3% for cement. The energy cost usage in manufacture of cement was 25-30% of delivered price in 1973, non fletton bricks 10-15%, and aggregates 4-6% The imposition of a carbon tax would produce differential rises in the price of building materials. The materials that needed most carbon burning fuel to make them, would suffer the greatest increases in price.

6. CARBON DIOXIDE BALANCES.

Recent studies in the UK /5/ have shown that buildings are responsible for about half the total carbon dioxide emissions in the UK. By weight, the 1987 building contribution was estimated at about 300 million tonnes of carbon dioxide. This is about 1.5% of the global carbon dioxide production. The breakdown by fuel was 48% electricity, 29% gas, 12% solid fuel, and 11% oil. The high contribution from electricity is due to two factors, the inherent inefficiency of electric generating systems, in the absence of cogeneration, and the very extensive use of coal as the primary fuel. The balance would obviously be very different in France with the high proportion of electricity generated by nuclear energy. Some member states have made considerable progress with cogeneration /6/. Others have hardly started at all. The UK carbon dioxide production breakdown by end use in 1987 was 48% space heating, 16% water heating, 7% cooking, and 29% lighting and appliances. A second study /7/ was concerned solely with the housing stock, which was identified with 30% of the UK carbon dioxide production. While lighting and appliances only accounted for 10% of the total domestic energy usage, the contribution to domestic carbon dioxide production was much greater at 25% of the total. The studies suggest that a 12-17% reduction in current UK carbon dioxide emission levels is achievable solely through the application of energy efficiency measures.

7. CHLORINATED FLUOROCARBONS /1/.

Other environmental issues are issues of chemical policy rather than of energy consumption policy. The use of CFC's in the production of many insulation materials has provided a good example of an environmentally unsatisfactory building materials production process, fortunately being phased out in many situations. The phasing out process will be speeded up if all architects now take the trouble to avoid insulants foamed using CFC's. Specific guidance on CFC free alternative insulation materials has just been issued by the UK Building Research Establishment /8/. The Montreal Protocol on CFC's requires for Group 1 CFC's (R11, R12, R113, R114, and R115) that the emissions from 1st July 1989 be no more than 1986 levels, from July 1993 80% of 1986 levels, and from 1st July, 1998, 50% of 1986 levels. This decision is very significant for future world air conditioning and refrigeration policy. In 1986, the consumption of CFC's in the UK was 66000 tonnes, of which 62% was used in aerosols, 15% as solvents, 18% as foam plastics, of which 4.5% was building insulation. Only 8% was used in refrigeration, of which 4% was commercial refrigeration, 0.5% domestic refrigeration, and 3% air conditioning. BRE note that UK aerosol manufacturers will cut their annual consumption of CFC's by 90% this year. The residual uses then become dominantly located in buildings, and in produce refrigeration. This demands a response from the building sector. As the insulant problem can be simply resolved by change of thermal insulation practice, it comes down to the key issue of what are architects and engineers going to do about air-conditioning usage and air conditioning plant. I will return to this issue later.

8. INDOOR ENVIRONMENTS AND HEALTH /3/.

Energy policy, and both local and national environmental policy, are inextricably interlinked, both one with another and also with issues of human health policy. They are also interlinked with issues of achievable policies for sustaining the future health of the global atmosphere, and of the oceans. In the case of indoor environments it is especially important that the environment be healthy for man /3/. The concept of comfortable environment must not be confused with the concept of a healthy environment. The use of renewable energy does not automatically ensure a healthy indoor situation. For example, indoor air quality may be very adversely prejudiced by a failure to ensure adequate ventilation. In a recent performance assessment of a passive solar heated school at Looe in Cornwell, UK, carbon dioxide levels were found to be in excess of 1200 ppm with a peak of 3000 ppm /9/. The latter figure is three times as high as the recommended level of carbon dioxide acceptable for crew in nuclear submarines. It is hardly surprizing that indoor air quality was consistently rated amongst the most poorly rated attributes of the classroom environment.

Combustion is a significant cause of both outdoor and indoor pollution, and hence of much unnecessary human ill health. It is also a cause of acid rain. A natural product, like wood, burned inappropriately, and much of it is burned inappropriately, is very damaging to health. Considerable respiratory ill health world wide is linked with the use of fossil and, especially biomass fuels inside buildings. Clean needs definition. It is not synonomous with green. Clean has a biological meaning and a chemical meaning. The biological meaning has to consider the multiplicative capacity of living organisms. So, given the right environment and appropriate nourishment, a few organisms can soon produce many. Legionaires disease provides a good example. Current studies are showing "clean" renewable energy, for example, the substitution of passive solar heating for the "dirty" indoor combustion of fossil fuels for heating, simultaneously providing effective ventilation to keep the indoor environment biologically and chemically clean, backed by natural cooling for hot weather, are likely to make important contributions to the respiratory health of non smoking individuals, as well as improving their thermal comfort. Simultaneously passive solar buildings help secure the health of the global ecosystem, in addition to contributing to the energy economy. Renewable energy applications now have to be set in this broader environmental context. For a start, such benefits to human health demand economic evaluation.

9. THE NEED FOR GOOD ALL YEAR ROUND PERFORMANCE.

The need to achieve adequate all year round performance is another indoor environmental factor that is often underrated. There are seasons of overheating and seasons of underheating. The precise balance depends on location in Europe. The building fabric however needs to have season conscious properties. The problem is not how to save heating energy in isolation. For example, in a computer based information technology age, it is becoming increasingly easy to run into problems of overheating through excessive internal gains. In thinking about the indoor thermal environmental regulation of buildings, it is worth pondering the parallel problem of how the human body regulates its internal environment in the face of a wide range of internal and external conditions. It achieves its balance by altering the heat flow characteristics from the heat producing core to the surface in order to balance internal gains with external losses. The range of environments that can be tolerated in comfort is extended through the use of insulating clothing, which is removed when the internal heat production becomes too great for the environment to dissipate the heat, otherwise one becomes too hot. Buildings, like the human body, are internal heat producers, and a balanced rate of heat loss and not zero heat loss, has to be achieved. Building fabrics with seasonally variable properties are can play very useful roles. If the overheating problem is not adequately resolved, one is forced into the excessive use of air conditioning. High insulation is not always a benefit, especially if the internal heat gains per unit floor area are large. One has to explore the year round environmental benefits of different design solutions. Too few designers currently do this. The result is too many buildings are being forced into a position, where air conditioning becomes mandatory.

10 SOLAR ENERGY AND BUILDING.

An important characteristic of buildings is that their surface areas are necessarily large, and so, by appropriate design, solar energy can be captured at the point of demand by elements that are needed for enclosure anyway. No additional separate land for energy collection is needed. The simplicity and low cost of passive heated solar buildings has made them among the most acceptable immediate solar applications in Europe together with flat plate water heaters in the warmer parts of the Community. One must also note the growing interest in natural lighting, as a scientifically based design activity, belonging within the scope of solar energy research /10/. A well designed direct gain building of heavy mass offers good daylighting without summer overheating, as well as significant energy savings in cold weather, and a healthy indoor environment with low pollution impacts. Scientifically passive solar building research and development has been concerned to a large extent with rediscovering what ancient builders knew by pragmatic experience and providing quantitative design methods and manuals to enable designers confidently to produce designs of reliable performance. Improved techniques for energy storage, improved building surface properties and improved fenestration systems are now available. Fortunately we now seem to have moved through the phase when heating alone was considered, in favour of a more balanced all year round environmental approach. This approach places particular emphasis on the appropriate control strategies to adopt for the solar pass through elements at different seasons of the year. Advances in process control are enabling the solar control systems to respond in logical ways to the solar inputs, freeing the solar building from the need for excessive user control. In the future, it should prove possible to link building controls to weather forecasts to establish short term anticipatory control. Such measures will reduce energy demands, and consequently the outputs of greenhouse gases. The new environmental situation substantially reinforces the wisdom of solar energy policies, through their substantial carbon dioxide reduction potential.

11. ECONOMIC APPRAISAL.

The economic appraisal of conventional energy systems has been dominated by the supply side of the conventional energy industries. These industries have often received massive subsidies in terms of loans at special rates, capital write offs and government sponsored R & D at massive levels compared with the renewables /11/. This is particularly true of nuclear energy.

Looking to the environmental side of the balance sheet, the environmental costs of conventional and nuclear energy have been largely transferred to other sectors of the economy. The true economic costs of different conventional energy supply systems have not been evaluated, and imposed on energy users. The polluter has not been paying /12/. As far as the utilities are concerned, the economic costs of waste disposal have ended at the "factory gate or chimney", regardless of the further damage to society and the world environment that resulted. This situation acts very much against the interests of the renewables, because they receive no credit for the environmental benefits they confer. It also encourages excessive consumption of conventional fuels, with consequent increases in the greenhouse gases and discourages investment in improved conservation measures.

The concept of a carbon tax currently under discussion in various parts of the European Community, would significantly shift the economic balance between conventional energy and renewable energy. A carbon tax also would significantly raise the price of nuclear power stations, because of their very large content of materials like steel and concrete, produced using carbon dioxide producing fossil fuels. In view of their carbon consumption reduction role, carbon bearing thermal insulation materials ought to carry a negative tax, i.e. a subsidy, to accelerate their use in conservation and carbon dioxide reduction.

The economic situation will shift even further in favour of renewable energy, if power stations, in addition to extensive desulphurization, are forced towards carbon dioxide removal as well, to offset excessive global warming. European governments have tended to give a low priority to systematic strategic studies of the overall economics of renewables. The environmental benefits have been talked about, but have not been quantified economically in the context of national and international energy policy. I attach considerable importance to getting more environmental economists interested in this field of discussion.

12. POLICIES DIRECTED TOWARDS OVERHOT BUILDINGS.

Precisely why will we need energy for building purposes in a warmer Europe? A hotter Europe means the currently unsatisfied demand for comfort cooling, especially noticable in Greece, Italy and Spain, must grow substantially. The conventional energy intensive way of dealing with overhot buildings and cars is to air condition them, with air conditioning equipments that use CFC gases as refrigerants.

Developing ways of making more effective use of natural passive cooling techniques, becomes especially important, as the only realistic current alternative for promoting human comfort and health indoors under hot conditions, involves using expensive, and often unaffordable air conditioning plant with CFC 's as the current refrigerant gases. In contemporary air-conditioned buildings, human cooling is achieved at the expense of adding to both global warming risks, and to ozone destruction. The issues of natural cooling have not so far received the attention they deserve in the CEC European programme. The relatively recent addition of Greece, Spain and Portugal to the Community, has shifted the balance of attention needed in building energy policy to demand a greater attention to the overheated problem. A forth coming workshop, being organized by the CEC will address this topic in a systematic way. In the meantime, some use could be made of the publications of UNCHS (Habitat), which address this area in detail /13/ /14/.

13. THE REFRIGERANT PROBLEM /8/.

Refrigerants used in compressors can be characterized approximately for their ozone destructiveness and for greenhouse potential by two numbers, the ozone depletion potential which is referred to the refrigerent R11 = 1, and the greenhouse potential, which is referred to carbon dioxide as 1. The group 1 CFC's controlled by the Montreal Protocol have the following properties:-

		Ozone depletion potential R11 = 1	Greenhouse potential CO_2 = 1	Average lifetime in atmosphere years
R11	CCl_3F	1.0	3300	75
R12	CCl_2F_2	1.0	10000	110
R113	CCl_2FCClF_2	0.8	4500	90
R114	$CClF_2CClF_2$	1.0	13000	
R115	$CClF_2CF_3$	0.6	25000	

The hydrochlorofluorocarbons have much lower ozone depletion potentials.

		Ozone depletion potential R11 = 1	Greenhouse potential CO_2 = 1	Average lifetime in atmosphre years
R22	$CHClF_2$	0.05	1100	20
R134a	CF_3CH_2F	0.00	900	Toxicity testing not complete.

The very long atmospheric average lifetimes are very evident. This means changes in existing policies will take a very long time to work through, before significant positive changes are reached. Expert opinion advises that, whenever possible, R11 and R12 equipment should not be used in new plant, and that, until safer alternatives are available, R22 should be used. Great hopes are currently being placed on R134a, but the toxicity testing is not complete. One of the valuable features of CFC's was their low toxicity. This enabled people to remain relatively complacent about equipment losses due to minor leaks and maintenance. It appears about 75% of UK refrigerant sales is used for plant servicing. It is extremely desirable that servicing practices be improved to substantially reduce losses of CFC's from existing plant. This is not a problem with hermetically sealed compressors in use, but the issue of refrigerant recovery at end of useful life remains.

It is also desirable to reduce the overall chemical inventory of refrigerants. The architects contribution comes in the design of buildings that can be effectively cooled with smaller plant. Window orientation, and effective shading are particularly important. The use of efficient artificial lighting systems also helps. Storage strategies for "coolth" can also help even out the daily load, and reduce the peak plant load, so making it possible to operate with a smaller plant. Efficient cooling design is also important for saving electrical energy, and hence for reducing carbon dioxide loadings into the atmosphere.

15. CONCLUSIONS.

The ecological imperatives will put new pressures on architectural designers. Designers can take their own initiatives to make progress, but it will be important they do so in a balanced way, addressing the whole spectrum of environmental issues, rather than personal enthuasms which only address part of the environmental problem. Architecture has long been regulated in the social interest. New national and international policies will lead to greater environmental regulation. The growing interest in green issues, will also mean that well informed protest movements will develop to attack environmentally unacceptable architectural developments. Architects must see the new situation as presenting positive opportunities to help the development of sustainable societies. The new environmental regulatory environment that is likely to develop, must not be seen as a negative factor destroying creative potential. It also follows that renewable energy research and development has to broaden its terms of reference, so the wider environmental interactions are better considered. Multidisciplinary intersectorial working is going to be essential to achieve the sustainable development goals envisaged in the important UN Bruntland Report. This implies a growing collaboration between architecture and environmental science.

It is now clear that a chemical policy for the environment is just as important as a physical policy for the environment. It matters a lot where our chemical inventories reside in the world, and precisely how they might become adversely redistributed by inappropriate developmental policies initiated by man. "Chemical planning" of human settlements is currently almost non existant in contrast with "physical planning"

REFERENCES.

1. UNEP (1987b), The Ozone Layer. UNEP/GEMS Environment Library No 2, UNEP, Nairobi.

2. UNEP (1987a), The Greenhouse Gases. UNEP/GEMS Environment Library No 1, UNEP, Nairobi.

3. World Health Organization (1989). Indoor environment: Health aspects of air quality, thermal environment, light and noise Ed. J.K.Page, WHO, Geneva. (In the press)

4. National Economic Development Office (1974), The increased cost of energy - implications for UK industry, HMSO, London.

5. BRE (1989), The greenhouse effect, Building Services, the CIBSE Journal, July, p 65.

6. Commission of the European Communities. (1988/89). Euroforum New Energies, Proc of International Congress, Saarbrucken, F.R.Germany, Oct 1988. H.S.Stephens and Associates, Agriculture House, 55 Goldington Road, Bedford MK 40 3LS, England, Vol 1 to 3.

7. BRE (1989), Cutting off the gas, Building Services, the CIBSE Journal, October, pp 69-70.

8. Butler, D.J.G. (1989), CFC's and the building industry, BRE Information Paper, IP 23/89, Building Research Establishment, Garston, Watford, WD2 7JR.

9. Anon (1989), Monitoring Looe School, Building Services, The CIBSE Journal, Sept, pp 63-64.

10. Commission of the European Communities. (1990). European Reference Book on Daylighting, in final preparation, CEC, Brussels.

11. Heede, R.E., Morgan, R. and Ridley, S. (1986), The hidden costs of energy, Sunworld, 10, pp 4-8.

12. Hohmeyer, O.H. (1988). The social costs of energy consumption- external effects of electricity generation in the Federal Republic of Germany. Springer Verlag, Berlin, Heidelberg, New York, Paris, Tokio.

13. UNCHS (Habitat). (1984). Energy conservation in the construction and maintenance of buildings, Vol.1, Use of solar energy and natural cooling in the design of buildings in developing countries. Report Expert Group meeting, Nairobi, 1983, Chairman.J.K.Page , UNCHS, Nairobi.

14. Page, J.K. (1989). How to prepare locally based design manuals for passive solar heated and naturally cooled buildings. UNCHS Habitat, Nairobi, In the press.

IDEOLOGY AND INFORMATION IN LOW ENERGY DESIGN

L. Kealy, J.O. Lewis
School of Architecture,
University College Dublin
Richview, Clonskeagh, Dublin 14, Ireland

ABSTRACT: The incorporation of "scientific" information in architectural design has been identified as a stumbling block in the way of a wider adoption of low energy design. Up to now, attempts to overcome this difficulty have paid little attention to the experience of architects who were producing good quality passive and low energy architecture. This paper is concerned with the findings and implications of a study of the work methods of a group of such design leaders in the area mentioned above.
The majority of the architects surveyed have engaged in substantial original work in the integration of "scientific" information in architectural design. Some prefer to deal with scientists (physicists) rather than just with conventional engineering consultants. They agree on the necessity for the architect to understand scientific concepts and have a critical attitude to published information. There is increasing use of computers for calculation and simulation.
Two implications emerge which have a bearing on current efforts in information transfer. The first concerns the role of case studies: the study shows that they must be approached critically and creatively. The second implication concerns the nature of innovation in architecture. The study raises the question whether passive and low energy design can engage architects at the level of insight and theory to a significant degree, or whether its influence will be dependent on percolation of good practice.

1. BACKGROUND TO THE STUDY

The report "Solar Energy Applications in Buildings: Information Requirements for Architects" issued in March 1988, pointed to the need to bring the experience of practising architects to bear on the design of information transfer strategies /1/. It was felt that the experience of architects working in the area of passive and low energy architecture had not been examined in any systematic way, and that this amounted to a neglect of a useful frame of reference /2/. The report recommended a pilot study into the work practices of a representative group, with a particular focus on the integration of the requirements for passive design into their projects, and on the use made of computer-based methods.
The recommendation arose from one of the major issues considered in the report: the integration of information on solar energy applications in buildings into the design process. The necessity for such integration had been widely recognised in the relevant literature, but it was felt that the solutions put forward were based on inadequate models of the process of architectural design /3/. Most solutions appear to transfer a linear sequential model, appropriate for the logical structuring of information, to the design process itself. The important distinction to be made here is that between architectural design as a series of operations leading to the erection of a building on the one hand, and on the other, architectural design as a process of thinking and deciding. The characteristics of architectural design have recently come under scrutiny, notably from those concerned with the compatibility of computers and the cognitive process. Architects, it seems, are wary of this territory, rightly so in our view. But the insights gained in such studies do have implications for those involved in generating information concerning solar energy utilisation in buildings /4/. Taking into account the limitations of time and resources, a modest pilot study was undertaken to further illuminate the parameters of the task of information transfer, and to indicate directions for further study.

2. THE PILOT STUDY

The Pilot Study had two distinct phases. The first consisted of an investigation into the place of technical information in the design process of a number of leading figures in architecture. The second phase was an enquiry into the work practices of design leaders in the area of passive and low energy design. Underlying both stages was a concern to explore the conceptual process and the place of technical information within it.

2.1 PHASE 1: "Idea" and "Technology" in the work practices of leading architects

The intention of the first phase was to explore the relationship between 'idea' and 'technology' in the design process, as exemplified by a number of leading architects, and thereby identify themes to be explored in subsequent study /5/. Six architects were chosen: Marcel Breuer, Louis Kahn, Josef Paul Kleihues, Alvaro Siza, James Stirling and Mies van der Rohe.
The method chosen was to interview architects who had worked in the offices of these leading architects. Although a proforma was devised to assist in these interviews, the intention was to allow each conversation to develop its own momentum, so that differences in approach would not be obscured. The principal themes which emerged from these interviews were as follows:

- the significance of early sketches in exploring architectural ideas in terms of structure and construction;
- the interplay of sketches, models and scale drawings in the design process;
- the importance of experience in making early provision for structural and mechanical service requirements;
- the important role of close collaborators within the office;
- the control of design development exercised by these architects even in matters of detail.

A significant phrase occurs in the account of the work practices of Josef Paul Kleihues, namely "design development work is rarely rushed". Beginning with a sketch by the principal architects, ideas are explored through further sketches, models, and scale drawings. While first sketches are often "poetic interpretations of the brief", "formal diagrams" or "form sketches" which are then developed in the office, nevertheless, they are often highly descriptive in themselves; sometimes of context and brief, sometimes of materials or construction. Ideas are drawn and re-drawn, explored in model form and drawn again, in a way that underlies the centrality of the drawing in the design process. This centrality has been acknowledged in the critical literature: drawing has been described as the language of representation and thought in architecture /6/.
The development of initial ideas was facilitated in

most instances by the work of close collaborators in the office of these leading architects. Thus, while the initial ideas were sketched by the principals, the transition to scale drawings involved the input of other architects capable of interpreting and making more explicit the architectural intentions of the sketch. The key role was carried out by such people as Carles Vallhonrad (Kahn), Mirko Baum (Kleihues), Michael Wilford (Stirling), Gene Summers (van der Rohe), and João Xavier (Siza). At a later stage in the process, many of these collaborators would assume the role of project architect, with responsiblity for co-ordination, detail design and administration of the project. The early process of exposition was always closely monitored by the leading architects.

The development of the design to incorporate the input of other building professionals or specialists shows variations in emphasis, with some architects (for example, Breuer and Kahn) tending to make explicit choices in respect of mechanical and electrical services relatively early, while others (Kleihues, and perhaps Siza) understate these aspects of the building. For the first group, trusted consultants had at times a particularly close collaboration. It would seem that most architects had developed an approach to architecture, a language, within which they had a clear sense of structural and service requirements, based on experience over time.

Following discussions of the interviews, it was decided to conduct the next series of interviews with a focus on a particular building design, to include questions on how these architects obtained their commissions, and on whether they worked regularly with the same consultants.

2.2 PHASE II: Work Practices of Design Leaders in Passive and Low Energy Architecture

In this phase, information was sought from seven architects from six Member States of the EC: France, Germany, the United Kingdom (two), Greece, Italy and Spain. The architects were interviewed, and drafts of these interviews were amended in response to their observations and comments. With a sample of such size, it was of course not possible to reach definitive conclusions which can be applied generally to architects active in the field of passive and low energy design. It was not the intention of the study to reach this kind of conclusion, but rather to shed some light on issues which had arisen during the investigation into information requirements for architects, referred to earlier.

The principal issue to be explored concerned the integration of the procedures required for passive and low energy design with the process of architectural design. In addition it was hoped to achieve an outline of the use made of computer-based methods. The report "Information Requirements for Architects" anticipated that the requirements for energy-conscious design constituted a potentially disruptive element in the traditional design process, at least until such time as experience had ensured that the "technical intuition" so central to architectural design generally, was extended to encompass the necessary areas of knowledge /7/. In addition, it was anticipated that the work practices of these design leaders would be inter-disciplinary to a notable degree, and that this would be most marked in the relationship with heating, lighting and ventilation consultants.

3. STUDY FINDINGS

Allowing for the particular flavour of each interview, it is possible to discern some broad patterns:

- the importance of scientific concepts
- a critical attitude to published information
- an increasing use of computers in the design process
- an emphasis on the primacy of architectural considerations.

The following paragraphs contain brief expositions of the findings of the pilot study, organised under headings which reflect these patterns.

3.1 Scientific Knowledge and Design

It is often maintained that architects' ignorance of principles of physics constitutes a major barrier to the practice of energy-related design. Consistent with this view, software is now being developed with the purpose of assisting architects to manipulate the various energy parameters,without, it seems, having to understand these principles /8/. In this context the design leaders interviewed believe passive and low energy design requires a grasp of scientific concepts. Within their diverse approaches, there are two features which illustrate this point:

- some of those interviewed have found working directly with physicists a creative and liberating experience, because a dialogue developed in which the building could be understood in a dynamic way, thus opening up new possibilities for architectural solutions rather than reducing matters at once to questions of plant and other services provision;
- several architects insisted on the wholistic relationship between their understanding of physical principles and the generation of architectural form. In their accounts of the design process, there is no sense of opposition between the "two cultures" of science and art.

These features relate to the development of the technical intuition which is central to the design process. With regard to the anticipated inter-disciplinary nature of energy-related design, no single picture emerged, except with regard to daylighting requirements. In other respects, the overall profile was one of good practice rather than of exceptional work practices.

3.2 Information and Design

For most of those interviewed, the principal source of information was their own experience in the area. While technical literature was cited as an information source on products, published case studies did not have the prominence anticipated, although they were, in some cases, a source of reference. The European Passive Solar Handbook was not considered an important source of information by this group: the most positive reactions to it were, in relation to its coverage of weather data and its use in teaching /9/. Specialist expertise within the building industry was also an important information source.

At first sight, these findings seem to agree with published profiles of architects as indifferent users of published information /10/. This is not the case: these architects interviewed are a critical audience for published material, with strong and sometimes divergent views on the value of particular material, based on their own experience.

3.3 Integration of Computers

There was considerable difference in personal use of computers in the design process. For some architects,the computer had become fully integrated, while for others it was an adjunct to the central activity of designing. The picture here however, is a fluid one, in which the use of computers appears to be gaining ground, with the emerging potential of CAD, the development of software, and the increasing sophistication in the interaction of programmes with conventional practices. In general, while the larger practices have committed themselves to integrating the computer, the smaller practices vary a good deal in this regard. Where computers are integrated their use by architects extends from CAD to solid and wireframe modelling, to some energy-related calculations. Engineers, whether in-house or outside consultants, generally maintain their role with regard to quantification.

3.4 Energy-Related Design and Architecture

Despite the study's focus on work practices, most design leaders' expressed their intentions and described their methods in terms of architectural concepts such as space, form, light and structure. This was especially evident in two respects:-

- first, the significance of the particular site. It is predictable that for architects involved in passive and low energy design the site is important in terms of energy-related considerations. The site is also a source of inspiration, of historical precedent and of relevant typologies. The coupling of architectural and landscape analysis with scientific data on climate parallels the science/architecture integration noted above;
- second, these architects made intense use of traditional media of architectural design, the sketch, the model and the scale drawing. In addition, however, to their use for studies of form or space, or for p presentation, these architects also used models for daylighting studies.

However, 'comfort' was a concept commonly used even where architectural ideas were being discussed. A more technical tradition was also represented in this small group. In general one could say that the spread of interest was not different to what one might experience in any group of committed architects. The work of the group was a matter of conviction; the underlying dynamic varies, poetic, ethical or pragmatic.

It might be noted that 'design aids' or 'evaluation tools' did not emerge in a way that allows for generalisation. For most, their use was subsumed in the design process and there were some reservations regarding their use. It is clear however, that the definition of "design aid" is elastic, and currently extends from straightforward manual tools to increasingly sophisticated software.

4. SUMMARY AND CONCLUSIONS

A brief study such as this can only be indicative, while the need is for firmer ground. It looked for patterns of agreement, points of congruity in the work practices (and by extension, the attitudes) of the architects involved. This search inevitably relegated the particular architects' approach to a secondary role. The accounts of the individual architects make illuminating reading in themselves, giving life to otherwise abstract accounts of the design process /11/.

In summary it may be useful to list the characteristics of the work practices of these design leaders as they relate to those examined in Phase I of this study.

Points of similarity can be listed thus:

- Salience of architectural values and use of traditional media such as drawings and physical models;
- The design process has a broadly similar pattern;
- The importance of technical intuition in design based on personal experience;
- A clear leadership role in relation to other building professionals.

Points where differences of approach or emphasis emerge can be listed thus:

- increased salience of site climatic conditions in the approach of the Phase II architects;
- the emphasis on comfort in buildings;
- the recognition of a need for a conceptual framework that includes scientific concepts in architectural debate.

The fact that there is so much common ground should not obscure the significance of the points of difference. The integration of concepts of physical science marks a different theoretical base, one which requires further exploration and articulation in mainstream architecture and criticism. This observation has been made before now with respect to energy-conscious design. The converse is also true, and has received little attention. There is an onus on those for whom such design is important to ground their work in the history and theory of architecture to a far greater extent than is presently the case /12/.

The architects interviewed in Phase II of this study are unusual. They are a unique sample not only of architects generally but also of those involved in passive and low energy design in that they are leaders or pioneers in the field. The architects whose work practices were studied in Phase I are also unique, representing aspects of the "avant garde" of twentieth century architecture. A common element is that of original work. The advantage of examining the work of design leaders is that they have had to reflect on what they are doing and have had to explain it to others.

In this context one issue emerges which is central to the question of information transfer. It is the importance of experience and the development of informed technical intuition. For information to be learned, the would-be learner must be able to interact with that information in a creative way. For architects that means they have to be able to use it actively in the design process, and to construct buildings accordingly. One could say that the pool of experience is still limited, and architects still remain to be convinced of the place of climate or energy in architecture. To leap-frog this log-jamb of established attitudes and practices, one must look to the nature of innovation in architecture, as well as to the longer-term impact of more enlightened architectural education.

Two recommendations are made at the conclusion of this study:

First, that the experience of the recent E.C. competition "Working in the City" should be examined in terms both of its take-up by professionals and students, and of the use of the technical literature and design aids provided /13/.

Second, that the role of case studies be examined critically and creatively. There is a recognised need to disseminate good case study material in the area of passive and low energy architecture. But the need to disseminate must be accompanied by a critical analysis of the use of case studies in learning and design /14/.

Innovation in architecture comes in two forms. Generally, innovation occurs gradually, at the margins, through the percolation of good practice. An example here is the development in the servicing of buildings in the last century. Innovation also occurs radically - through the recognition by architects of new insights into enduring values. The question here is whether commitment and passion for climate-sensitive or bio climatic design are motivations for the emergence of an "avant garde" for the twenty-first century. The pre-condition for an affirmative answer is the emergence of an architecture good enough to surpass fashion, not simply an architecture fashionable enough to render its climate-sensitive message digestible-sugar-coating the pill. The quest is for excellence, and true insight into architectural values.

5. NOTES AND REFERENCES

1 Kealy, L. (1988) Information Requirements for Architects. Final Report. Energy Research Group, University College, Dublin. p.46

2 Kealy, L. (1988) p.22

3 Newland, P., Powell, J.A., Creed, C. (1987) "Understanding Architectural Designers' Selective Information Handling". Design Studies, Vol. 8, No 1.

4 Akin, O. (1986) Psychology of Architectural Design. Pion Limited, London. p.170: Developing and Interfacing Design Aids

5 An efficient way of initiating this study was suggested by the fact that several distinguished architects in practice in Dublin had worked in the offices of leading architects, and were available for interview. This avenue was taken with respect to five architects: Marcel Breuer, Louis Kahn, Josef Paul Kleihues, James Stirling and Mies van der Rohe. A chance opportunity led to the inclusion of Alvaro Siza in this group.

6 Crowe, N.A. and Hart, S.W. (1986) "Visual Notes and the Acquistion of Knowledge". Journal of Architectural Education 39/33 p.6 et seq.

Olley, J. (1989), "Drawing: The Language of Representation and Thought in Design" in The Architect and the Drawing Exhibition Catalogue, RIAI Dublin. p.10

7 Kealy, L. (1988) p.17

8 Kealy, L. (1988) p.40

9 Achard, P. and Gicquel, R. (1986) European Passive Solar Handbook. Preliminary edition. CEC Directorate-General XII for Research, Science and Development.

10 Marvin, H. (1985) "Information and Experience in Architectural Design" Research Paper No 23. Institute of Advanced Architectural Studies, York.

11 Accounts of the architects' work-practices are given in Kealy, L., (1989). Pilot Study of the Work-Practices of Design Leaders in the area of Passive and Low Energy Architecture: Final Report. Energy Research Group, University College Dublin. Appendix II.

12 A good example of this endeavour can be found in: Plowler, D. and Fraker, H. (1981) Project Journal: Teaching Passive Design in Architecture. University of Pensylvania. Historical Background.

13 O'Toole, S. and Lewis, J.O. Working in the City. Report of the EC architectural Competition. Gandon Editions, Dublin (in press).

14 Kealy, L. (1989) Case Studies in Passive Solar Design: Bringing the Good News on Comforting the Faithful, Proceedings, Second European Conference on Architecture.

FORMATION PAR ORDINATEUR : POUR UNE ANALYSE EUROPEENNE DES BESOINS

Communication de J. ADNOT, Ecole des Mines de Paris (*)
Coordonnateur CALENER
et de A. DUPORT**, JG SENDER**, Agence Française pour la Maîtrise de l'Energie

----oo----

(*) 60, bld Saint Michel 75006 Paris
(**)27, rue Louis Vicat 75737 Paris Cedex 15

----oo----

ABSTRACT : The computer appears in the Architectural world as a new and promising help for teaching for various reasons :
- because it is already present in the professional world (and so the professional tools should be represented in the teaching)
- because it allows a different teaching (simulation, visualisation) or a teaching of something different (methods and results just arising from Research)
- because it allows learning on the job, at one's own rythm and without Separating Theory and Practice.
We don't speak here of general prospects but of existing products and continuing projects :
- the "GRADIENT" series of five pieces of courseware by AFME on Heat Transfer in buildings'envelopes
- the ARCHIDISC project and its European framework called CALENER.

1 - ARCHITECTURE, THERMIQUE, FORMATION

A partir de la simple bonne volonté d'intégrer dans le savoir de l'Architecte des savoirs techniques parcellaires (p. ex la Thermique) tels qu'ils existent en Sciences de l'Ingénieur, on peut esquisser plusieurs stratégies :
- la stratégie du "savoir de base", où les architectes devraient devenir des physiciens modélisateurs, quitte à perdre leur nature de concepteurs (= hommes des problèmes à données incomplètes, donc non modelisables)
- la stratégie "techniciste" ,où les architectes s'en remettraient aux spécialistes (thermiciens p. ex.) et aux ordinateurs, quitte à ne rien maîtriser
- la stratégie "minimaliste" : le moins de calcul et de notions précises possible mais des exemples, des savoirs faire particuliers, quitte à ne faire que reproduire.
Toutes évitent les difficultés mais ne résolvent pas le problème : la majorité des architectes leur échappent même si chacune génère des marges (de quelques pourcents) qui favorisent le dialogue et produisent des exemples de réalisations progressistes.
Comment être scientifique sans imposer un long apprentissage, comment être opératoire sans être aveugle, comment être qualitatif sans être prisonnier des simples expériences ? On peut dire que l'ordinateur et les nouvelles techniques d'enseignement (et notamment en formation des professionnels en exercice) ouvre de nouvelles possibilités. Avec l'EAO (Enseignement Assisté par Ordinateur, mais en mettant là-dessous toutes les formes d'"assistance" possibles et imaginables) on peut disposer d'un savoir de base et apprendre à l'utiliser avec peu de mathématiques, faire un calcul technique en comprenant le contenu des "boîtes noires" si on le souhaite, apprendre de l'expérience des autres sans être conduit à copier. Nous espérons amener le lecteur à cette conclusion avant la fin de notre propos. Mais il nous faut d'abord passer par quelques exemples.

2 - GRADIENT - THERMIQUE DU BATIMENT

L'ordinateur apparaît en Architecture comme support d'une formation pour différentes raisons :
- parce que l'ordinateur est présent dans l'exercice professionnel (les outils professionnels doivent être représentés dans la formation associée)
- parce qu'il permet d'enseigner autrement (simulation, visualisation ou autre chose (méthodes et résultats récemment issus de la recherche)
- parce qu'il permet d'apprendre sur le lieu de travail, ou à son rythme ou encore sans séparer la théorie de la pratique.
Pour diffuser les nouveaux savoirs ou savoir faire dans le tissu professionnel l'Agence Française pour la Maîtrise de l'Energie (AFME) met à disposition des formateurs et enseignants des outils pédagogiques novateurs. Ces outils sont le fruit de la collaboration du Service Formation avec des enseignants et des chercheurs en Energétique, Architecture et Génie Civil, avec l'appui de spécialistes d'EAO et d'informaticiens.
Avant de présenter la manière dont la décision de créer ces modules a été prise, la stratégie pédagogique retenue, indiquons que la collection de cinq modules est actuellement réalisée et disponible sous le nom de GRADIENT (éditeur : LOGEDIC, réalisateur : DIALOGIC) .

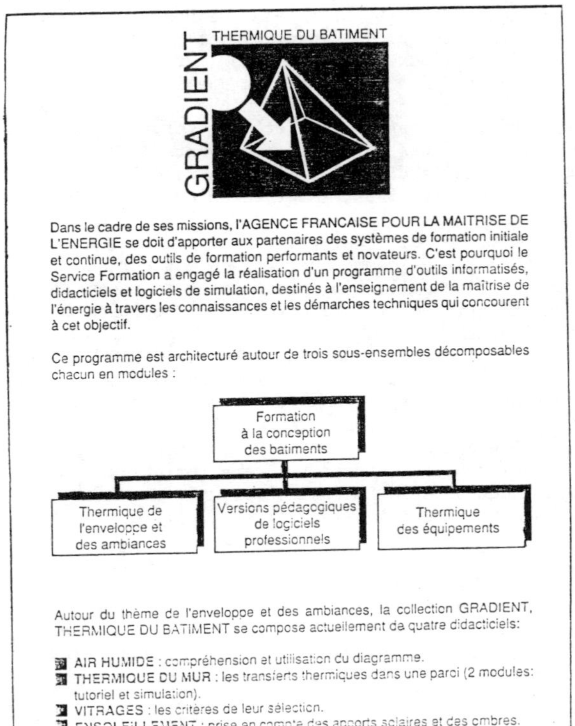
GRADIENT
THERMIQUE DU BATIMENT

Dans le cadre de ses missions, l'AGENCE FRANCAISE POUR LA MAITRISE DE L'ENERGIE se doit d'apporter aux partenaires des systèmes de formation initiale et continue, des outils de formation performants et novateurs. C'est pourquoi le Service Formation a engagé la réalisation d'un programme d'outils informatisés, didacticiels et logiciels de simulation, destinés à l'enseignement de la maîtrise de l'énergie à travers les connaissances et les démarches techniques qui concourent à cet objectif.

Ce programme est architecturé autour de trois sous-ensembles décomposables chacun en modules :

Formation à la conception des batiments
Thermique de l'enveloppe et des ambiances
Versions pédagogiques de logiciels professionnels
Thermique des équipements

Autour du thème de l'enveloppe et des ambiances, la collection GRADIENT, THERMIQUE DU BATIMENT se compose actuellement de quatre didacticiels:

- AIR HUMIDE : compréhension et utilisation du diagramme.
- THERMIQUE DU MUR : les transferts thermiques dans une paroi (2 modules: tutoriel et simulation).
- VITRAGES : les critères de leur sélection.
- ENSOLEILLEMENT : prise en compte des apports solaires et des ombres.

- Thermique du Mur (Régime permanent et régime variable)
- Thermique du Mur (Simulation)
- l'Air Humide
- Vitrages
- Ensoleillement des façades.

Ces produits sont aujourd'hui sur le marché et remportent un grand succès en France.
L'envoi d'un questionnaire à un ensemble d'institutions d'enseignement a permis de prendre contact avec de nombreux organismes de formation et de recueillir ainsi les réactions et les points de vue des formateurs. Ont pu être recensés en particulier des interlocuteurs ayant déjà par eux-mêmes approfondi la question de l'EAO dans leur enseignement. Des rencontres directes avec un échantillon de formateurs ont par ailleurs complété efficacement les résultats de l'enquête : seul un véritable dialogue peut conduire à une définition précise d'un besoin sur un domaine pédagogique aussi nouveau.
Ce dialogue a pris une ampleur plus vaste quand des petits groupes d'auteurs se sont mis à rédiger des listes de contenus, à définir des stratégies pédagogiques, à écrire synopsis et maquettes. Le groupe de travail national, mis en place par l'AFME autour du thème "EAO et Energétique du Bâtiment" a fédéré certaines de ces initiatives. Il a été nécessaire de valider les choix essentiels relatifs aux produits et donc remettre en cause les pratiques usuelles d'enseignement.
On peut dire que le micro-ordinateur n'est pas utilisé au mieux de ses possibilités dans les Etablissements d'Enseignement. Pendant les dix ans qui ont suivi son introduction (1975-1985 environ) on a proposé aux étudiants d'apprendre à se servir de l'ordinateur. Le nombre et la qualité des logiciels professionnels maintenant disponibles et l'informatisation des milieux correspondants a conduit de nombreux enseignants à proposer aux étudiants de se servir de tels progiciels, sans véritables stratégies pédagogiques. Rares sont les cas présentant des supports de formation pensés avec et pour l'ordinateur si l'on excepte les études de sensibilité en Thermique du Bâtiment possibles grâce à "STRATEGY". Le caractère coûteux des réalisations informatiques a tué, et tuera hélas de nombreux projets dans les milieux de l'enseignement.
Dans les milieux professionnels la réalité est toute autre : des modules dont la qualité est graphique et visuelle plus que technique et pédagogique sont réalisés par des sociétés de service sur commande de grandes entreprises souvent à l'occasion d'une campagne d'amélioration des relations et des compétences humaines, internes ou externes. Mais elles ne s'intéressent pas aux problèmes de l'Architecture et de l'Energie. On peut dire que l'AFME s'est comportée comme une "grande entreprise" permettant l'apparition de produits professionnels de qualité qui sont maintenant à la disposition des enseignants.
Le coût d'investissement d'un module d'EAO nous a incités à viser un public large donc à choisir des contenus susceptibles de constituer le tronc commun d'un savoir scientifique utile à l'ingénieur et à l'architecte : calculs thermiques des parois de construction, paramètres caractéristiques de l'air humide et principales valeurs calculables (t° de rosée ensoleillement, etc...).
Le profil des auteurs peut avoir des conséquences directes sur la mise en forme des contenus. Une difficulté classique : éviter de transcrire à l'écran un cours théorique quelle qu'en soit sa qualité.
La composition du groupe de travail - en particulier la présence d'architectes et celle de spécialistes de l'EAO - nous a partiellement protégés de cette erreur.
Les débats furent rudes entre tenants de la démonstration mathématique et partisans d'une pédagogie de la découverte. Evoquons au passage la nécessité de valoriser le fait que l'on utilise un écran. Un bon exemple : doit-on démontrer a priori à partir du comportement thermique impulsionnel d'un mur (équation de la chaleur) l'existence d'un amortissement et d'un déphasage, ou d'abord l'observer pour en dégager les lois et les conséquences. C'est bien entendu cette deuxième voie qui a été choisie dans le module "Thermique du Mur" :

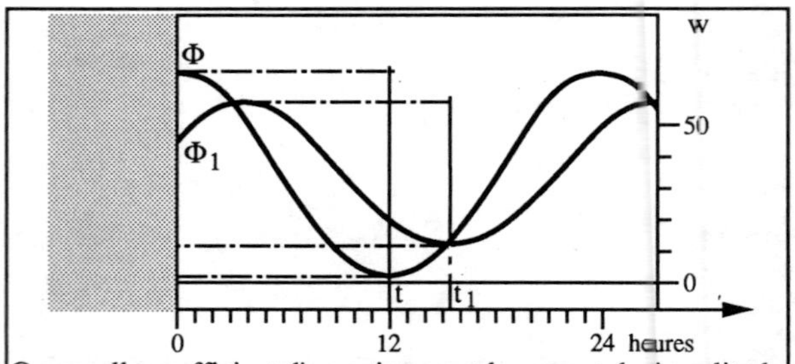

On appelle coefficient d'amortissement le rapport de l'amplitude du flux Φ_1 à l'amplitude de la courbe Φ_f

Les modules traitent d'un domaine ou d'un élément isolé. La nécessité d'une approche globale du fonctionnement énergétique d'un bâtiment au moment de sa conception architecturale, s'est très vite imposée du groupe. Elle conduit à un outil de "Conception Architecturale et Bilan Thermique" présenté par ailleurs dans ce congrès.
Les didacticiels permettent de mieux appréhender le rôle des différents facteurs intervenant dans le bilan thermique et de sensibiliser à une approche énergétique replacée dans l'ensemble des préoccupations de l'architecte. C'est à ce titre que sont menées des études spécifiques sur l'analyse de références architecturales, par J.P. TRAISNEL et J.P. PENEAU.
L'ensemble de ces modules fonctionne sur la configuration suivante :

- IBM PC/XT avec disque dur, 640 K octets
- carte graphique couleur standard IBM
- écran couleur 4 couleurs/écran en mode graphique 320x200
- commande : souris.

On peut bien entendu chercher à utiliser des configurations plus coûteuses et permettant un meilleur graphisme. Compte tenu des capacités financières limitées des établissements d'enseignement on s'est toutefois limité à cette base pour toucher le public le plus large tout en préservant une qualité d'illustration.

3 - EAO, ARCHITECTURE, EUROPE

Pour aller plus loin il convient de donner aux modules d'EAO existants l'occasion de faire leurs preuves : que les enseignants se rendent compte de la manière dont ils complètent leurs cours, qu'ils soient disponibles en auto-formation, et révisions. Nous sommes sûrs que se créeront ainsi des références et des standards de qualité. Déjà plus de 300 modules de la collection GRADIENT sont utilisés dans les établissements de formation.
Pour aller plus loin il faut aussi que les projets qui naissent dans l'esprit de chercheurs, de techniciens, d'enseignants puissent s'intégrer dans des collections plus importantes (avec éventuellement fusion et refonte) pour qu'ils aient la taille critique. Il est important que des moyens matériels et intellectuels (fournis par une société de service, un centre pédagogique, un système-auteur, etc...) soient fournis à de tels projets. Ils peuvent l'être dans le cadre de l'une des collections existantes, ou mieux dans le cadre du groupe CALENER au sein du programme communautaire COMETT.
En effet l'évolution de nos économies et de nos moyens techniques, nos cultures généralistes et techniques, tout rapproche nos douze pays européens dans

le domaine des nouvelles techniques de formation. Non pas pour diffuser un "sabir" technique moyen, mais bien pour que chacun dispose dans sa langue maternelle (l'informatique le peut aisément) d'un produit pensé par une équipe très spécialisée en fonction d'une projection dans l'avenir des évolutions actuelles. L'objectif de CALENER (dont font partie l'Ecole des Mines de Paris, l'AFME, Gaz de France et plus de quarante autres organismes de dix pays de la Communauté) est bien de constituer un observatoire de ce qui change en matière de formation à l'Energétique. En effet ce sont les évolutions de l'enseignement en rapport avec la recherche et les professions, les évolutions dans les entreprises du secteur énergétique en liaison avec la technologie et le marché qui conduisent à promouvoir de nouveaux objectifs, de nouveaux supports. L'EAO est l'un d'eux, riche de possibilités attrayantes dans le contexte de l'informatisation des métiers.
Cette possibilité se manifeste dans une période favorable du fait de l'émergence au niveau de la CCE de deux grands programmes : ERASMUS (formation initiale) et COMETT (formation professionnelle).
Marier les Universités et les entreprises européennes tel est l'objet du programme COMETT, "programme communautaire de coopération entre les Universités et les entreprises dans le domaine des formations aux technologies".
C'est dans le cadre de COMETT qu'a été créé le réseau sur l'utilisation des nouvelles technologies pour la formation en énergétique. Cette AUEF a été baptisée CALENER : Collection d'Aides Logicielles pour l'Enseignement de l'ENERgétique (Computer Aoded Learning of ENERgy techniques, pour nos lecteurs anglophones).

4 - CALENER-ARCHIDISC

Un sous-groupe de CALENER s'est lancé dans la réalisation d'un cours de Thermique du Bâtiment dans l'Architecture, en se basant sur des résultats de recherche thermiques et architecturales de ces dix dernières années. Les exercices et les cours sont mis en oeuvre sur ordinateur mais une image d'excellente qualité était nécessaire pour que l'élève-architecte reconnaisse les objets et les situations auxquels on le forme par ailleurs.
CALENER-ARCHIDISC prend donc la forme d'un vidéodisque interactif sur l'énergie dans le bâtiment. Couplé à des modules EAO existants ou à créer, destiné à former les architectes et ingénieurs en génie civil ou génie climatique à l'énergétique du bâtiment.
Le vidéodisque permet de constituer une banque de d'images soit de documents créés spécifiquement, diapositives, séquences filmées, images numérisées, schémas fixes ou animés, soit de documents existants: par exemple des réalisations architecturales "exemplaires", souvent citées dans les revues, qui sont étudiées, dans ce cas, sous l'aspect performances énergétiques du bâtiment.
Une maquette de démonstration correspondant à 15 minutes d'apprendissage, réalisée par la Société DIALOGIC, existe et "tourne" dans les écoles d'architecture d'Europe pour valider la démarche tandis que l'équipe, coordonnée par l'Ecole d'Architecture de Marseille, accumule les matériaux pour la version finale.
Le produit final (fin 90) intègrera un abaque de sensibilité informatisé qui permet à l'étudiant de maîtriser quantitativement l'influence des paramètres thermiques globaux (compacité, coefficients de forme, K moyen, ...) qui a été mis au point à partir de STRATEGY, produit de l'Université de Liège.

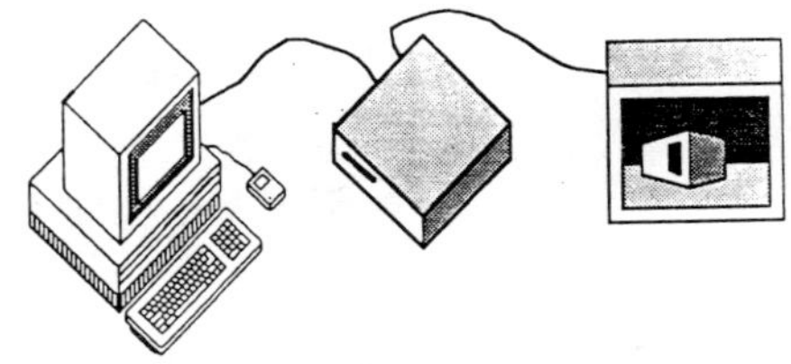

ARCHIDISC : Computer + Laservision + Monitor

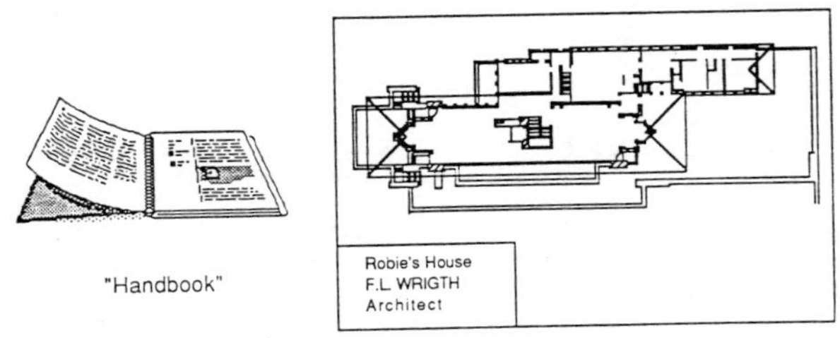

"Remarkable Architectural realization"

ARCHIDISC : CAL Module + Architectural Pictures

5 - QUELQUES POINTS DE REPERE

1 - Les travaux qui sont présentés ici ont été financés par l'AFME (Service Formation) et le programme COMETT (Commission des Communautés Européennes, DG5), les régions Wallone, Pays de Loire et PACA.
Ils ont bénéficié de nombreuses contributions bénévoles, dont celles de l'EMP et de GAZ DE FRANCE.

2 - Adresses

AFME - 27, rue Louis Vicat 75737 Paris Cedex 15
EMP et CALENER - 60, boulevard Saint Michel 75272 Paris Cedex 06
LEMA - Université de Liège, 15, avenue des Tilleuls 4000 Liège
LOGEDIC - 8, rue des Ursulines, 70000 Vesoul.

3 - Lectures complémentaires

Newsletter CALENER, quatre numéros parus
Rapports d'activité CALENER, 87/88 et 88/89.

4 - Liste des participants au groupe de travail national "EAO et Energétique du Bâtiment"

J. ADNOT : Ecole des Mines
L. ADOLPHE : GISE (EMP-ENTPE
I. ARDITI : CNEAO (Paris VI/VII)
J. BRAU : INSA, Lyon
D. CAMPANA : Ecole des Mines
P. DEPECKER : Ec. Architecture Lyon
A. DUPORT : AFME
M.A. FRESSE : Graphiste
J.L. IZARD : Ec. Architecture Marseille
G. KRAUSS : IUT, Lyon
A. LASSERRE : CNEAO (Paris VI/VII)
D. MARCHIO : Ecole des Mines
J.P. PENEAU : Ec. Architecture Nantes
J. SCHMOUCKER: INSA, Rennes
J.G. SENDER : AFME
J.P. TRAISNEL: CNRS
P. VERGUET : LOGEDIC
G. WATREMEZ : DIALOGIC

THE CONTRIBUTION OF APPROPRIATE ENERGY DESIGN STRATEGIES AND APPRAISAL OF A TOURIST RESIDENCE COMPLEX IN CRETE

Koronakis P.
Pr. in the Dept. of Mech. Engng, TEI of Piraeus, 250 Thevon & P. Rali st., 12244 Aegaleo, Athens, GREECE

Kalligeris Y.
Architect, 22 M.Koundourou st., 73100 Hania, GREECE

Santamouris M.
Dr. Physicist, Director of Protechna Ltd., 39 Mavromihali st., 10680 Athens, GREECE

ABSTRACT. The aim of this project is to estimate the contribution and predict the performance of a number of passive and hybrid systems incorporated in the design of an all-year-around tourist residence complex located in a rural area, 4 km outside Hania, a coastal city in the NW part of the Mediterranean island of Crete. The work presents the results of a case study supported by the Commission of the European Communities, under the Programme Building 2000, Project no. 735/GR.

1. INTRODUCTION

There are four basic aspects covered by the assessment, regarding the behavior and thermal performance of the passivily built hotel apartments designed for summer as well as winter time use by regular tenants (students, military men, etc. staying for longer periods), these are :

- Heating and cooling performance
- Daylighting performance
- Thermal comfort status
- Energy saving design improvements

The weather data file for Hania was constructed and presented in the form of Test Reference Year Tables. For this purpose, solar radiation intensities on the vertical were estimated from hourly global solar radiation measurements on the horizontal, taken in Hania, using the Orgill and Holland /1/ model, and the assumption of a uniform sky diffuse solar radiation. Results are shown in Table 1.

2. BUILDING DESCRIPTION

The hotel complex is made of three heavily constructed buildings, partly set in earth, surrounding a central open space plaza containing the swimming pool. Two of them, Buildings A and B, are for residence while the third one, Building C, accomodates the reception and catering facilities. Building A is a three-level construction containing seven double-room apartments covering a total of 300 m^2, while Building B is a four-level construction with twenty double and single room apartment units covering a total of 674 m^2. The site is sloping down the S. side of a small hill with the buildings totally unobscured.

The rooms of all units are faced due south with direct gain glazings, each 4.7 to 6.0 m^2 of double pane glass surface. Light color venetial blinds are placed on the inside to reflect natural light to the white painted ceiling and walls. Curtains are used on all glazed surfaces to cut down on thermal losses and provide privacy. Horizontal awnings consisted of retractable overhangs, made of tent canvas, are to provide shading protection to the whole south facade during late spring, summer, and early autumn months, thus preventing overheating.

3. SUMMER COOLING AND BUILDING PERFORMANCE

Cooling becomes necessary when air temperature goes over the 27°C mark while residents are lightly clad and the relative humidity stays close to 50-55%. In this work cooling of the building spaces is provided naturaly, and electromechanicaly using fans to force fresh air through ducts buried into the ground and then pipe it to the rooms. The following chart shows the modes applied.

COOLING SCHEMES
- Natural process
  - Earth cooling (partly earth set buildings, 260 m^2 of wall and an equal amount of floor surface)
  - Radiative cooling (400 m^2 of exposed roof surface)
  - Cross ventilation
  - Solar chimneys (night breathing only)
- Mechanical (hybrid) system
  - Earth tube air cooling (convective type)

With U-values for the floors and walls in contact with the ground, 1.94 W/m^2K and 0.65 W/m^2K respectively, the average total amount of heat escaping to the earth during a hot summer day is estimated at 130 KWh. In addition, the rate of heat radiated to the sky from the roof during a typical summer night (considering local air temperature and humidity night values) is estimated at 30 W/m^2, on the average. Hence, the radiant heat losses per night are valued at 110 KWh. Figure 1 shows the amount of heat lost to the ground as a function of earth and room temperature. By adding the two cooling effects, and taking the building cell having thermal capacity 140 Wh/m^2K, the resulting temperature drop experienced, mainly, the top and ground floor residents is estimated at 2°C.

Cross ventilation is encouraged by window openings, on opposite walls, covering 15% and 20% of the total window area, respectively. Such an arrangement would creat an air flow rate up to 60 ACH (air changes per hour) depending upon the strength of the suction effect the wind creates on the low pressure side opening.

Solar chimneys with walls made of concrete, thermally insulated on the three sides, painted black, and covered with glass on the fourth side extend verticaly on the roof of each floor facing due S.W. Although, for maximum summer time solar collection they should be faced due West (it is established that for the month of July a vertical surface facing due W receives more solar energy compared to any other orientation. Besides, the late during the day heat charging procedure, minimizes heat losses, thus, the solar chimney starts its night operation at a state of heat charge close to the daily maximum). Figure 2 shows the cross-section of such a chimney, with the thermal analysis and design procedure given in /2/. The air core width is made equal to twice the B.L. (boundary layer) thickness (of the upward air flow created by the buoyancy forces as room air entering

the stack gets heated by the warm concrete walls), taken at the chimney's top end, in order to inhibit the down flow of outside air through the stack's center. As it is shown in /2/, the estimated average air speed through the core of the solar chimney with dimensions, height 2.0 m and length 1.0 m, could be taken as 0.1 m/s, through all night till late next morning. This is translated to an average of 0.5 ACH per apartment unit, for 12 consecutive hours.
The earth tube air cooling system considered was designed to supply up to 3 ACH of conditioned air during the hot hours of the day. Figure 3 shows the outline of the air and water flow streams inside the horizontal buried pipe. The idea is to cool low humidity air by forcing it to travel down a buried into (3m-6m) earth tube partly filled with (running) water. Thus, the combined effect of sensible and evaporative cooling reduces the required pipe length when compared to the case of a tube with air flow alone (sensible cooling only). The water supply system is capable of stopping the water flow inside the tube and drain it, completely. This should be done when the air relative humidity is high (>60%),thus, avoiding the possibility of producing extremely humid air at the pipe's end. Therefore, under such conditions the pipe is operating as a normal earth tube air (only) cooling system. The conditioned air characteristics are then found using the corresponding charts /2/, where the tube's length and diameter are known (their values are found using the created for this purpose algorithm of the combined air-water flow mode) and the exit air temperature and R.H. are the unknown parameters.
The earth tube system takes into consideration three consecutive treatments of the air stream. First,air is cooled (convective cooling only) along its downward movement as it passes through the vertical pipe shaft on its way to the horizontal (main) tube. On the Psychometric Chart, this is a constant moisture content process at the end of which air enters the horizontal pipe a little cooler and having suffered a small increase in its R.H. Next, air enters the horizontal tube, where water flows up to a depth of 1/3 the radius (assuming R.H.<0.60). The length L of this tube is found using an expression given in /3/, for the mass transfer coefficient, H_d, of water to air, and the vapour concentrations at the entrance and at the exit of the tube, taking design inlet values, R.H. 50% and DB temperature 35°C with accepted R.H. at the exit around 75%.

$$H_d = 0.023*D_f*Re^{0.83}*(v/D_f)^{0.44}/D_{hyd}$$

$$L = m_w/(1.5*WW*H_d*(C_w-C_{av})$$

where, D_f is the diffusion coefficient of water vapor to air, Re is the Reynolds number of the air flow inside the pipe, v is the kinematic viscosity of air, D_{hyd} is the hydraulic diameter of the air flow, m_w is the mass rate of water changed to vapour from the water surface that extends down the pipe a length L, WW is the width of water's free surface formed at D/3 away from the pipe axis, C_w is the water vapour concentration evaluated at the mean temperature of the water surface and C_{av} is the average water vapour concentration of the air stream. Finally, the conditioned air is brought to the earth surface through another vertical pipe shaft, losing a little of its temperature and gaining a little on its R.H. The system so designed is estimated to give at the exit end of the return vertical shaft R.H always <86%. This is accomplished using 2X25 m parallel pipes of diameter 0.2 m for the Building A, and 2X60 m parallel pipes of diameter 0.4 m,for the Building B. The estimated temperatures of the delivered air, based on the ambient conditions usually encountered in Hania during the month of July, are shown in Figure 4. It must be pointed out that R.H increases with 4 decrease in temperature, therefore, the higher R.H values for the conditioned air are expected at the lower portion of the graph.

4. HEATING PERIOD BUILDING PERFORMANCE

Estimation of the Building's thermal performance was based on the outcome of simulation runs carried out using the simplified PDT (passive design tool) TCM-heat. Heating is considered continuous and supplied by radiators with the room temperature controlled by room thermostats set at 21°C.The monthly mean daily values of heat losses, solar gains, heat demand, and room temperature can be estimated by running simulations based on monthly average climatic conditions. Results of these runs are shown for both Buildings,in Table 2 of the average monthly thermal energy and room temperature, in Table 3 of the Buildings average thermal characteristics for cold and heavily cloudy days, in Table 4 of the Buildings average thermal characteristics for warm and sunny days, and in Figure 5 where the pipe chart shows the annual contribution to heating of each one of the heat components involved.

5. DAYLIGHTING UNIT PERFORMANCE

Checking the daylighting performance of a hotel room might seem unnecessary unless, of course, the hotel tenants happen to be students, teachers, etc. who need light for their work (the hotel will be used by nontourists during the off season months). In this study the daylighting performance of each unit was investigated considering as a model day the 21st of December, noon time under average cloudy conditions. The building description specifies that all units are with rooms faced due S. with South facing glazed door and window forming a unified area. In addition, some units have an extra bathroom window with a N. or S. orientation, or (and) an all glass door faced due N. or E.
In order to investigate the range and distribution of daylight factors inside each room, at one meter from the floor, along the middle long axis section, simulation runs were carried out using the program MICROLITE version 1.0. Figure 6 shows the results of these runs for each type of room.

6. THERMAL COMFORT CONDITIONS

The Predicted Mean Vote (PMV) and the Predicted Percentages of Dissatisfied (PPD) indices are used to express thermal comfort conditions experienced by occupants of the units, based on climatic conditions expressed by the average monthly values of temperature, R.H, usual air speed, amount of clothing worn, and human activities, expected during the month considered. Values of these indices are calculated using the Bruel & Kjaer Demonstration Program of Comfort Equation by considering appropriate values for the parameters involved. Figure 7 gives the expected average monthly values of the PPM index. Running the Program after feeding it with data expected to occur under extreme conditions (an extremely hot July,and an extremely cold January),the PPM values obtained are 0.62 for July and -0.84 for January, in other words under all circumnstances -1.0<PPM<1.0.

7. IMPROVED DESIGN SCHEMES

The Building thermal performance results reported so far are based on values of construction details and parameters given in the original design. It is possible though to improve thermal performance and energy savings by altering some of these values. So, by varying the south facing glazed area, the number of ACH and the U-value of all floors in contact with the ground, the following table, Table 5, was constructed. The values of energy savings obtained were calculated after carrying out simulation runs on the thermal performance of the Buildings considering the new values of the parameters.

Table 1. Monthly average climatic data for Hania, Latitude 35°30'

MONTH	AIR TEMPER. (°C)	RELAT. HUMID. (o/o)	VAPOR PRESS. (Pa)	DAYLIGHT HOURS (hr)	SUNSHINE HOURS (hr)	DEGREE HOURS (18°C)	PROPORTION EVEN.+NIGHT DEGREEHOURS
JAN	11.9	73	971	10.0	3.5	148.4	0.62
FEB	12.2	70	984	11.0	4.5	157.6	0.61
MAR	13.4	68	1104	12.0	5.8	131.1	0.64
APR	16.4	67	1224	13.1	7.4	53.6	0.86
MAY	20.4	64	1543	14.0	10.3	-	-
JUN	24.7	59	1809	14.4	11.7	-	-
JUL	26.9	57	1969	14.0	12.2	-	-
AUG	26.6	58	1955	13.2	11.4	-	-
SEP	23.6	65	1822	12.0	9.0	-	-
OCT	20.0	70	1610	11.0	5.8	16.6	0.87
NOV	16.8	73	1277	10.0	5.7	47.1	0.78
DEC	13.5	73	1117	9.6	3.6	124.7	0.63

SOLAR RADIATION INTENSITY VALUES IN W/m2

MONTH	HORIZONTAL			SOUTH			EAST/WEST			NORTH		
	INT/TY	A	B	INT/TY	A	B	INT/TY	A	B	INT/TY	A	B
JAN	84	0	.30	122	0	.35	45	0	.10	27	0	0
FEB	117	0	.45	130	0	.40	65	0	.11	37	0	0
MAR	166	.25	.56	134	0	.42	81	.07	.18	50	0	0
APR	234	.46	.68	127	0	.50	108	.13	.17	67	0	0
MAY	276	.50	.71	108	0	.07	124	.16	.26	84	0	0
JUN	313	.62	.76	102	0	0	132	.17	.28	94	0	0
JUL	306	.56	.77	104	0	0	134	.18	.31	88	0	0
AUG	278	.53	.76	126	0	.30	128	.19	.27	71	0	0
SEP	220	.42	.74	155	0	.42	106	.13	.25	54	0	0
OCT	156	.25	.62	166	0	.52	74	0	.14	40	0	0
NOV	106	0	.48	153	0	.50	52	0	.17	30	0	0
DEC	82	0	.30	128	0	.30	44	0	.11	23	0	0

Points to note

(A) stands for "fraction of hours with solar intensity > 500 W"

(B) stands for "fraction of hours with solar intensity > 300 W"

Table 2. Average thermal energy and room temperature values for the A and B Buildings, on a monthly basis.

MON	MEAN TEMPERATURE AND STANDARD DEVIATION (°C)	TOTAL HEAT LOSSES (MJ/mo)	CASUAL GAINS (MJ/mo)	SOLAR GAINS (MJ/mo)	AUXIL. HEATING (MJ/mo)	o/o CONTRIB. TO HEA[T] CASUAL GAINS	SOLAR GAINS	AUXIL HEATI[NG]
OCT	27.3 (1.6)	72984	7058	65926	0	9.7	90.3	0
NOV	23.6 (1.3)	65089	7058	57671	350	10.8	88.6	0.6
DEC	21.5 (0.3)	80151	7058	50279	22814	8.8	62.7	28.5
JAN	21.3 (0.2)	93637	7058	48209	38370	7.5	51.5	41.0
FEB	21.4 (0.3)	82874	7058	46346	29470	8.5	55.9	35.6
MAR	21.6 (0.4)	82235	7058	54813	20364	8.6	66.7	24.7
APR	22.8 (1.0)	60983	7058	51993	1932	11.6	85.3	3.1
		537953	49406	375237	113310	9.2	69.8	21.0

Table 3. Thermal performance of Buildings A and B after a string of cold and cloudy days.

MONTH	ROOM TEMPERATURE (°C)	AMBIENT TEMPERATURE (°C)	HEAT LOSSES (MJ/d)	CASUAL GAINS (MJ/d)	HEAT DEMAND (MJ/d)
OCT	21.0	16.5	1493	280	1213
NOV	21.0	12.8	2661	280	2381
DEC	21.0	9.5	3694	280	3414
JAN	21.0	7.9	4200	280	3920
FEB	21.0	8.2	4103	280	3823
MAR	21.0	10.0	3535	280	3255
APR	21.0	13.0	2579	280	2299

Table 4. Thermal performance of Buildings A and B after a string of warm and sunny days.

MONTH	ROOM TEMPERATURE (°C)	AMBIENT TEMPERATURE (°C)	HEAT LOSSES (MJ/D)	CASUAL GAINS (MJ/D)	SOLAR GAINS (MJ/D)	HEAT DEMAND (MJ/D)
OCT	27.8 (1.8)	22.0	2570	234	2336	0
NOV	24.3 (1.6)	18.8	2471	234	2237	0
DEC	22.3 (1.0)	15.5	2875	234	2213	428
JAN	21.9 (0.7)	13.9	3289	234	2270	785
FEB	22.0 (0.7)	14.2	3185	234	2322	629
MAR	22.1 (0.8)	15.4	2791	234	2157	400
APR	22.9 (1.1)	18.4	2004	234	1763	7

Table 5. Tabulation of sensitivity analysis results after a series of modifications

NO.	RECOMMENDED MODIFICATION	TOTAL HEAT (MJ pa)	CASUAL GAINS (MJ pa)	SOLAR GAINS (MJ pa)	AUXIL. HEATING (MJ pa)	% OF ORIGINAL AUXILIARY HEATING
1	Original Design	537953	49406	375237	113310	100.0
2	As (1) + 20% more S. wind.	583895	49406	441499	92990	82.1
3	As (2)+floors U=0.6 W/m2 C	561599	49406	441495	70694	62.4
4	As (3)+suggested ac/h	527553	49406	441495	36648	32.3

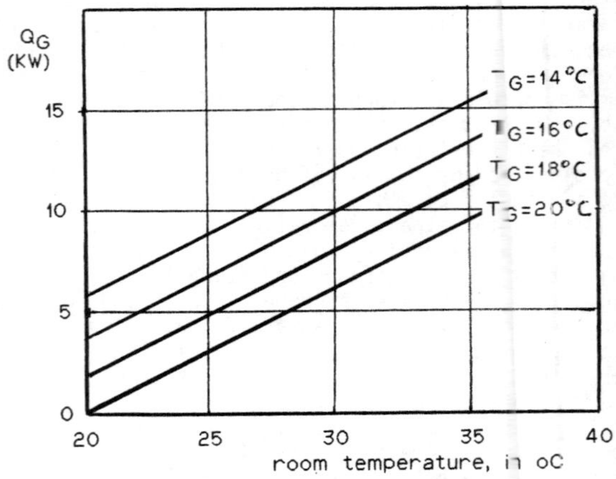

Figure 1. Rate of heat dissipated to the earth as a function of the room and the ground temperature.

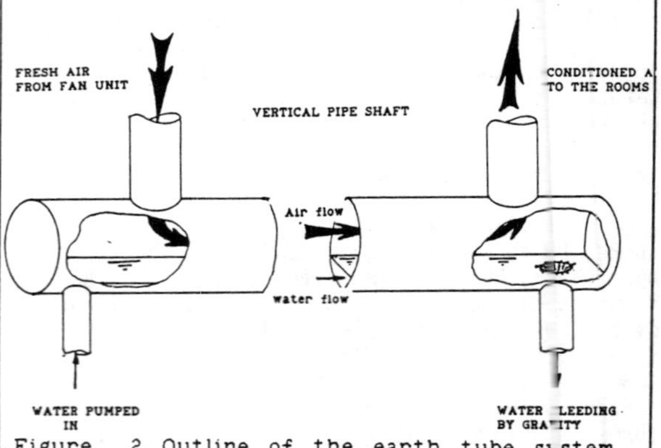

Figure 2. Outline of the earth tube system

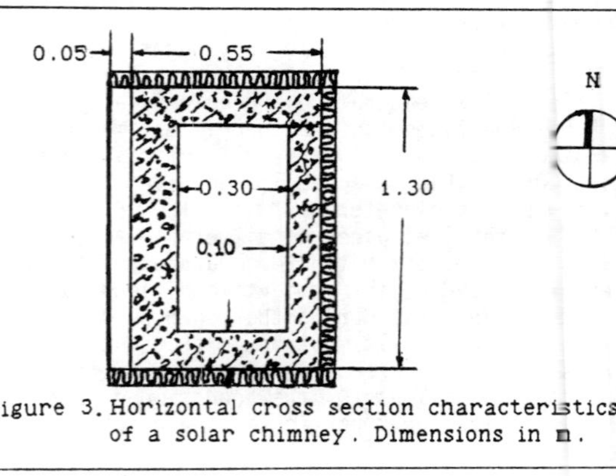

Figure 3. Horizontal cross section characteristics of a solar chimney. Dimensions in m.

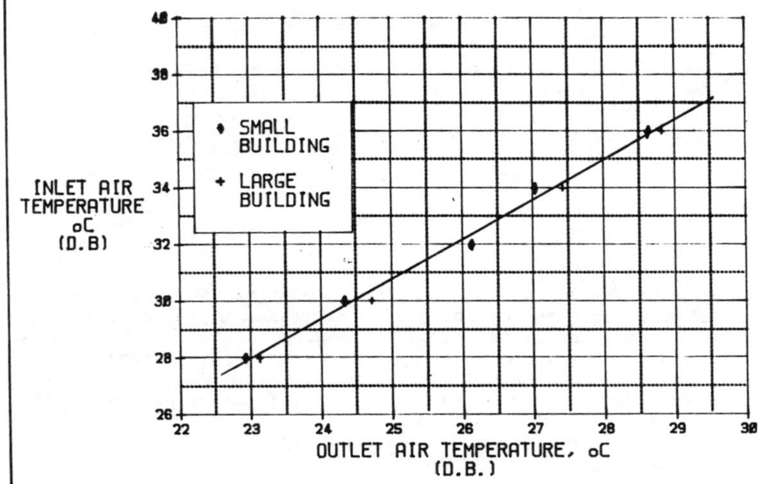

Figure 4. Earthtube System Performance

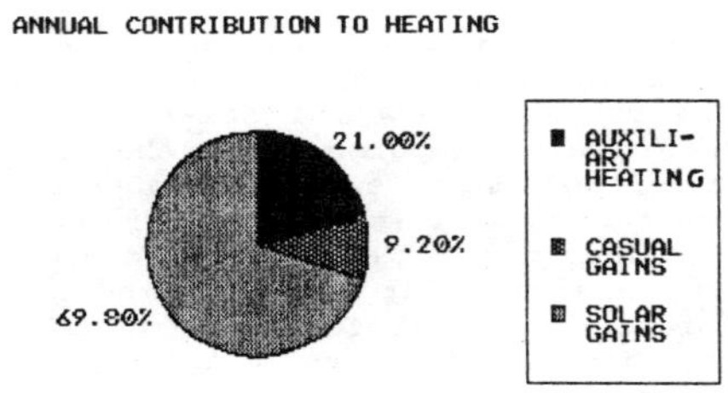

Figure 5. Annual contribution to heating

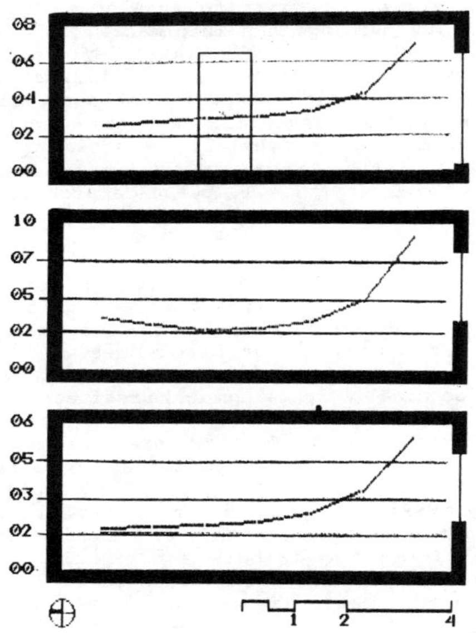

Figure 6. Daylight factor distribution inside each type of room, at 1 meter height corresponding to the 21st of December noon time, a partly cloudy day.

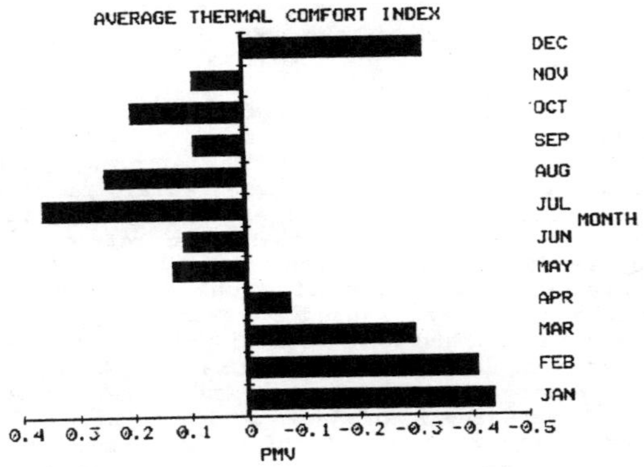

Figure 7. Estimated average monthly PPM index value experienced by occupants inside the units.

REFERENCES

1. Orgill, J.F., and Hollands, K.G.T., Correlation equation for Hourly Diffuse Radiation on a Horizontal Surface,.Solar Energy, vol 19, p.357,1977.

2. Koronakis, P., Alternative Cooling Schemes the Use of Earth Tubes and Solar Chimneys, 2nd European Symposium on Soft Energy Sources and Systems at the Local Level, Hania, Crete, Greece, 1989

3. Gilliland, E.R.,.Diffusion Coefficients in Gaseous Systems, Ind. Eng. Chem. vol.26, p.681, 1934.

ENERGY-EFFICIENT BUILDING DESIGN: A HANDBOOK FOR EUROPEAN ARCHITECTS

John Goulding, J. Owen Lewis
Energy Research Group
School of Architecture, University College Dublin,
Richview, Clonskeagh, Dublin 14 , Ireland.

ABSTRACT. A preliminary edition of The European Passive Solar Handbook, printed in 1986, is being revised to respond to the enlargement of the European Community with increased emphasis on passive solar design for Mediterranean climates and to incorporate other changes. Drawing on material from the preliminary edition and the output of R&D actions within the EC's Third Solar Programme, the revised Handbook addresses topics not previously covered including daylighting and passive solar urban design, with passive solar cooling, atria and climate covered in expanded chapters. The content and forms of presentation of the material are designed very particularly to suit architects and students taking account of varying levels of passive solar knowledge and expertise, particular preferences in learning styles, and temporally -varying information requirements among the target audience. The Handbook will be available early in 1990, initially in English with translations into other EC languages expected to follow.

1. BACKGROUND

1.1 Preliminary Edition 1986

The European Passive Solar Handbook /1/ was first printed in a limited-run preliminary edition in 1986. It reflected the results within The Second Solar Programme of the CEC DGXll for Science, Research and Development and was edited by Patrick Achard and Renaud Giquel. Based on drafts prduced by The Passive Solar Working Group and The Modelling Sub-Group from 1981 to 1983, it provided an overview of the current body of knowledge including passive solar principles, fundamentals, design options, aids to design and case studies in a single European reference book. Copies of this edition have been placed in libraries in Schools of Architecture and other reference libraries used by architects throughout the Community and have been distributed at selected workshops and seminars. Dissemination of the preliminary edition has been undertaken as part of the Commission's project SOLINFO which aims to promote the transfer of information from the solar research community to the European building design and construction professions /2/.

1.2. Potential Audience

There are in excess of 200,000 architects in the European Community as well as other construction industry professionals and students. However a relatively tiny percentage are actively involved in bringing about the construction of passive solar buildings. There is clearly great scope for a handbook which can effectively reach this audience. Only if building designers can be convinced of the advantages and then given the means will passive solar design make a significant impact on mainstream architectural design and the European building industry. The potential construction market is large: the industry employs 8.5 million people with an annual turnover in excess of 350 billion ECUs. Buildings account for close to half of the Community's total energy consumption. Consequently the implementation by increased numbers of architects and engineers of energy-efficient design principles can contribute to the attainment of public energy policies.
To be effective in reaching the potential audience of professionals, to indicate the benefits which passive solar design can offer and to assist them with the necessary design information requires an understanding of their information needs and how such information is sought and used within the design process. It is clear that if this information, which may in content be of great benefit to building design, is not appropriately prepared and presented then it will not be used.

1.3 Audience requirements - the nature of the target audience

Few European architects would claim to be experienced passive solar building designers. Indeed the number of passive solar buildings is still very small compared with the number of 'conventional' buildings erected each year. While some architects will be aware of the broad concepts, others may be reluctant to embark on the design of passive solar buildings because they perceive fundamental and far-reaching differences compared with conventional buildings - which their training and experience have not equipped them to deal with, they may fear. It is important that misconceptions are dispelled before the necessary design guidance is presented.
For the majority of architects and students with little or no passive solar design experience it is necessary to first present an overview of passive solar concepts and fundamentals with strong links to their prior or existing knowledge of building physics. Providing this information in a visually appropriate form, calculated to appeal to architects, will encourage a realisation that passive solar design proposes an essentially more thorough and professionally competent approach to building design in general. Associating energy-efficient buildings with excellence in architectural design will persuade architects that low energy design is worth getting involved in.
For practitioners embarking on the design of passive solar buildings, more detailed design guidance, evaluation tools, data and case studies are needed, relating to European conditions, and presented in an accessible form suited to the design process and its various stages.

1.4. Technical information and architects

To establish what factors are important when providing technical material for use by architects, Loughlin Kealy, (an architect and educationalist), was commissioned to prepare a report entitled 'Information requirements for Architects' /2/. This explores the factors contributing to 'non-technical barriers' to the transfer of information and suggests measures to improve the take-up of such information on solar energy in the short and longer terms. The report discusses the way technical information is used in the design process, giving examples of different 'learning styles' adopted by architects. While in the longer term some of these learning styles may be more easily accommodated if new media, (such as video discs) become more commonplace, in the short term appropriately-designed paper-borne information can

go much of the way to satisfying the needs of architects. In this context an element of redundancy is useful so that building design professionals, teachers and students can find their own 'paths' through the information
The report revealed the need to generate visual 'coat hangers' as devices for encapsulating information, a process which is consistent with the way buildings are presented for study in building journals where space and form are the primary considerations. Kealy indicates that two different kinds of information are required.
The first allows for the translation of concepts of thermal and luminous performance into construction strategies and comprises rules-of-thumb which indicate magnitudes and interrelationships, and methods of calculation which set out simple performance evaluations. Typically these would include graphic tools, such as daylighting protractors, solar diagrams and bioclimatic charts, which permit easy and early quantification.
The second type of information is that derived from experience, principally case studies which can have great value as a key element in the synthesis between visual, verbal and quantification factors.
These and other detailed recommendations in the report have been significant in devising the format for the revised Handbook.

1.5. Analysis of existing material

An extensive analysis of European and international passive solar texts and computer-based information has revealed a wide range of mostly rather technical and topic-specific material. There is little available which provides a coherent, holistic design approach for the various European conditions. Due consideration of energy efficiency and thermal and visual comfort also are important features of a useful source of design information.

1.6. Survey of Preliminary Edition users

A telephone survey was carried out in the spring of 1988 to sample the reaction of a group of users to the preliminary edition of the European Passive Solar Handbook. All of those interviewed had used the Handbook as a source of general information although none had used it as a primary reference for passive solar design projects. Forty percent had used it in education.
In general the Handbook was considered to be pitched at a level of technical complexity which architects, if sufficiently motivated, could follow but there were criticisms of the style of presentation for an architectural readership, particularly in relation to the quality of the graphics and lack of graphical consistency. Other general criticisms were related to the relative emphasis given to different topics, the 'northern European' bias, the rather fragmented structure, a lack of transparency, uncertainty about the accuracy of methods and shortcomings in theory. It was also felt that photographs were, in many cases, not directly related to the adjacent text and that the case studies, although valued, could have been more attractively presented and should have included more background information on the brief and the design decisions taken and processes followed. A section on urban design should also have been included, a number commented in the survey.
Despite these criticisms, the Handbook was valued by nearly all as a landmark in having brought together a diverse range of information on passive solar building design. It was recommended to students by teachers and the revised edition is keenly awaited, judging by the response to this limited survey.

2. OBJECTIVES

2.1 Promotion of passive solar design in the construction professions

To make a significant impact on mainstream architectural design, passive solar design should be accepted as an essential constituent of a more thorough and professionally competent approach to the design of any building and its immediate environment. This process, starting in the schools of architecture, should be seen as a thoroughly justifiable, holistic approach to the design of living and working spaces and not as a technological constraint imposed on the designer by an impending shortage of energy resources. The requirements of improved thermal and visual comfort may be seen as exciting new design opportunities to be integrated in the form and fabric of the building, discouraging the usual practice of adding on heating, cooling and lighting plant and equipment to buildings of fundamentally incomplete design.
A similar change in attitude indicated for practitioners, not only for its own sake but perhaps also prompted by changes in legislation and the demands of building users.
While the intrinsic benefits of a passive solar approach to building design may appeal more immediately to those architects with a technological bias, the other 'school', having a more aesthetic bias, will need to be convinced that there is no conflict between good architecture and passive solar design. For this reason case studies need to show examples as excellent in their architecture as they are in energy efficiency and comfort.

2.2. Address different information requirements

The information requirements of different sectors of the potential audience need to be addressed separately on the one hand to avoid causing confusion by delving into technical detail at too early a stage, before a grasp of principles and fundamentals has been achieved, and on the other hand to allow practitioners, with a knowledge of the principles to find the design information they need easily at any particular stage in the design process. For passive solar 'novices' this can be best achieved (using paper-borne information) as an illustrated text providing a structured overview to be read sequentially allowing the reader to form a general understanding of passive solar design and an insight into the inter-relationships between the requirements for energy efficiency and comfort of heating, cooling and lighting design.
Experienced practitioners embarking on the design of passive solar buildings need design guidelines, recommendations, design tools and data structured and presented in a manner which makes them readily applicable 'at the drawing board'. There are obviously limitations with paper-borne information and with the content of available information suited to European conditions, but while this information may become available through further research it is important that what information is available now reaches the widest audience

2.3. Design for Mediterranean climates

With the entry of Greece, Portugal and Spain into the Community increased emphasis on passive solar design for Mediterranean climates was needed. In addition to providing a chapter on cooling it was felt that these design issues should permeate the whole text.

2.4. To disseminate the results of European research

As the Handbook will play a significant role in disseminating the results of passive solar research and development actions to the construction industry professions. Co-ordinators of actions within the 3rd Solar Programme were asked to provide summaries of their research results together with recommendations and design aids drafted for an architectural readership which would contribute to an increased understanding and would provide readily applied design information. References to large volumes of data such as the European Solar Radiation Atlas (5) and other outputs from R&D actions should be given and only specimen data used in the text.

2.5. Design tools and guidelines

The principal design tool is a revised Methode 5000 including provision for night set back. In addition Methode 5000 will be made available in a computer disk version. It is not expected that an alternative European method will be available until after 1990.

2.6. Architectural emphasis in presentation of technical material

The objective is to present technical material from R&D actions and other sources to a generally non-technically orientated architectural readership. While it is not necessary for architects to know the details of the research and development procedures in generating the final output it is important that an understanding of the appropriate use and the implications of the material are conveyed clearly. The material needs to be presented in an architectural format, closely related to built examples from which inferences can be drawn. The design process itself is a complicated, non-linear one where a large number of dissimilar variables, aesthetic and technical must be reconciled. Precisely how each of these is treated varies from designer to designer and while there are identifiable trends it is certain that technological considerations on their own will not produce good architecture. Therefore the presentation of built examples of excellent passive solar architecture to illustrate the application of passive solar techniques is essential.

2.7. Additional material

As the purpose of The Handbook is to provide coherent, structured design guidance, it is appropriate to include examples of data and their use with a sample of the available design tools. To include large amounts of data and an exhaustive range of design tools would take away from the main purpose of The Handbook.

3 METHODOLOGY

3.1. Working Group

A sub-group was set up in 1988 to advise on the revision. Its members are:-
Alex Lohr, Köln
Patrick Achard, Valbonne
Martin de Wit, Eindhoven - Technical editor
Albert Mitja i Sarvise,
Barcelona - Technical editor
J. Owen Lewis, Dublin - SOLINFO director
John Goulding, Dublin - Co-ordinating editor

Judith Stammers, London, was commissioned to rewrite the English texts to ensure a consistency in style and W. Hastings, Dublin, an architect and graphic designer, was commissioned to visualise the illustrations, page designs and covers. As with the text, material for illustrations has come from a wide range of sources and while consistent presentation throughout the Handbook was a basic requirement, a more fundamental 'visualisation' of the combined text and illustrations was felt to be essential to convey the information in the clearest and most concise manner to the different sectors of the target audience.

3.2. Structure

A structure based on four sections has been adopted to appropriately address the different levels of passive solar knowledge and requirements among the potential readership.
The first section is an introduction for architects with little or no passive solar design experience, in which graphical information predominates, supported by text. This section is organised as a series of 'levels' from The Climate, through The Building to The Occupant with an illustrated glossary, appendices and an index. Material dealing with concepts and fundamentals has been extracted from the Preliminary Edition and revised incorporating new material to achieve a good balance of information on each topic.
The second section, essentially the core of the Handbook, has as its target audience practitioners who want to design passive solar buildings and need easily found and readily applied design information. The structure, devised to allow quick reference, is based on five chapters: The Environment, The Location, The Building, Case Studies and The Detail. It includes a design method, sample data and evaluation tools in appendices together with an index and references. Material is drawn from the Preliminary Edition of The Handbook significantly revised in format and supplemented by results from R&D actions in The 3rd Solar Programme. New chapters on Daylighting, Passive Solar Urban Design and Cooling have been prepared by European experts and are included along with a revised Method 5000, case studies drawing on the Project Monitor /3/ series with additional material supplied by the SOLINFO national sub-contractors. The section on climate has been completely re-written
A Resource Guide in the form of a bibliographic listing of approximately 500 aids to passive solar design, covering publications, design tools, software and audio-visual material, is included and is also available on disk for use with an Apple Macintosh. A catalogue of passive solar components available in Europe, many of which have been used in buildings described in the 'Project Monitor' series of brochures, has been prepared by The ECD Partnership, London, to complement The Handbook.

3.3. PROGRAMME

Due to significant delays in receiving material which in many cases could not be delivered prior to the preparation of the final reports of R & D actions the Handbook programme has been extended by six months from the original July 1989 deadline.

3.4. Publishing

Discussions are intrain with publishers with a view to printing The Handbook early in 1990. A final decision has to be made on publishing The Handbook as a single volume or making the introductory section and the component catalogue available separately. Translations of The Handbook into other EC languages will be a factor in these decisions.

References

(1) P.Achard, R.Giquel, editors, "European Passive Solar Handbook" Preliminary edition, Commission of The European Communities, Directorate General XII for Science, Research and Development, 1986.

(2) Lewis, J.O., O'Toole,S. Energy in Architecture: EC activities in support of the Building Design Process. Proceedings. Science & Technology at the Service of Architecture.

(3) Kealy, L. "Information Requirements for Architects", Energy Research Group, School of Architecture, University College Dublin, for The Commission of The European Communities (Technology Transfer from The Solar R & D Community to the European Building Profession; SOLINFO) 1988

(4) The ECD Partnership, London, for The Commission of The European Communities DGXII under the SOLINFO programme. Project Monitor published as a series of 49 brochures 1988-1989. Copies subject to availability from Energy Research Group, School of Architecture, University College Dublin, Richview, Clonskeagh, Dublin 14.

(5) W. Palz (ed), Commission of The European Communities DGXII, 'European Solar Radiation Atlas': Vol.I, Global radiation on horizontal surfaces, Vol.II Global and diffuse radiation on vertical and inclined surfaces. 1984.

L' IMAGE COMME OUTIL DE DESIGN

Jean ZEITOUN

Centre Informatique et Méthodologie en Architecture
Direction de l'Architecture et de l'Urbanisme
9 rue Barbanègre 75019 Paris France Fax (1) 40 37 07 11

The usual image has taken on a new dimension in design activity since numerical processing tools have been put within reach of designers.
Its use goes beyond mere simulation and creates types of specific and autonomous activities in the design process. Image has become a design tool because it is easier to work on certain sets of architectural concepts than when drawing by hand ; and it is also easier to work on procedures of description, generation and qualification of the architectural space. Thanks to easy decomposition and edition, the technical drawing and plan functionalities are extended by numerical image to architectural representations which can be adapted to the designer's requirements.

Moreover image can no longer be considered as an immediate data; grammatical procedures can be used to generate it and also analytical procedures to attribute it parameters. The distance thus being introduced allows to use image more like a design environment than a product of the design process.

INTRODUCTION

Le design est une pratique aux contours mal définis. Les outils du concepteur ne suffisent pas à définir son contenu théorique, mais ils rendent compte néanmoins de deux grandes classes de problèmes liés à la conception selon qu'ils portent en priorité sur le processus de conception ou sur le produit de la conception.

En architecture, la conduite du projet fait appel à une panoplie d'aides informatiques qui traitent aussi bien de la programmation du projet et de sa gestion, de la simulation visuelle, de la production des documents techniques ou de l'évaluation technique. Ces outils servent tantôt le processus, tantôt le produit. Avec les techniques récentes de l'intelligence artificielle, le concepteur dispose de moyens d'assistance ou de diagnostic attachées davantage au processus, puisque ses connaissances sont mises sous forme de règles établies a priori ou générées par des procédures d'apprentissage. Plus récemment, les méthodes de l'intelligence artificielle ont permis d'aborder des problèmes de modélisation des connaissances mises en jeu par l'architecte concepteur, afin de rechercher comment les mettre en œuvre dans des systèmes à base de connaissances.

Ces outils d'aide à la conception créent autour du concepteur, un environnement qui lui permet de travailler sur une matière et dans un espace virtuels tout en étant confronté aux contraintes bien réelles de l'interaction homme/système[1] . Dans cette panoplie instrumentale, l'image numérique de synthèse, le plus souvent structurée et calculée à partir de modèles géométriques et topologiques, joue une rôle très particulier. Elle constitue un mode d'expression et de communication naturel pour simuler l'apparence de l'objet ou en exprimer certains aspects conceptuels notamment sous la forme de schémas. L'image opère comme un media omniprésent dans le processus de représentation et de communication. La question que nous évoquerons ici concerne la possibilité de faire de cette image numérique, un outil de conception à part entière et non plus seulement un mode de visualisation. Dans cette approche, l'image numérique est à la fois un outil de qualification du produit et un outil de contrôle du processus.

Une seconde question pratique qui se pose dans l'emploi des modeleurs est que l'image ou le schéma induisent des éléments ou des nouveaux que la compréhension visuelle permet d'identifier et qui ne font pas partie de la représentation initiale. Ainsi, en assemblant des formes ou des volumes, on peut induire une nouvelle entité spatiale, liée à la configuration choisie. Cette génération de propriété ou d'entité est une caractéristique du travail graphique de conception. On peut donner l'exemple de création d'un effet de passage par le rassemblement d'éléments architecturaux sous une certaine configuration qui relève essentiellement de la compréhension visuelle, au delà de la création de chaque élément. Dans ce cas, le dessin est une représentation d'un objet spatial plus riche que l'ensemble des représentations élémentaires dont il est constitué.

Une troisième difficulté ou limitation des modeleurs réside dans les aspects de la gestion dynamique des objets et de leur cohérence. Le concepteur peut vouloir résoudre des problèmes de déplacement qui lui imposent de rechercher des trajets répondant à des contraintes, ou d'ajouter des éléments nouveaux dans l'espace qui modifient les relations entre les éléments existants. Dans ce type de situation, on voit clairement que le concepteur manipule une représentation non plus de l'objet mais de sa logique spatiale ou morphologique et que l'image n'est plus qu'une représentation terminale provisoire.

La scénographie intelligente ou assistée constitue une des formes actuelles possibles de réponse au besoin d'étendre et d'adapter les modeleurs des environnements de CAO[4] . Il faut les étendre pour qu'ils puissent traiter plusieurs niveaux de représentation au delà du seul niveau graphique, donc prendre en compte la représentation syntaxique et partiellement, la représentation sémantique du modèle. Il faut ensuite les adapter, car un accroissement des potentialités opérationnelles d'un modeleur nécessite de les rendre cohérentes et efficaces, donc de les réduire à des champs ou des modes d'applications restreints. Cette philosophie correspond assez bien à la tendance actuelle du développement des environnement informatiques adaptatifs et évolutifs en CAO.

Dans ce qui suit nous traiterons de deux exemples de techniques utilisant l'imagerie numérique que sont la scénographie intelligente et la multi-représentation. Les expérimentations et les développements qui sont décrits dans la suite sont les extensions et les applications du logiciel de synthèse d'images IKO[2] développé au CIMA.

1 . LA SCENOGRAPHIE INTELLIGENTE COMME GENERALISATION DU MODELEUR

Nous définirons brièvement la scénographie intelligente ici comme la généralisation du modeleur[3] . On sait qu'en matière de représentation des formes dans un environnement de CAO, la saisie et la modification des données correspond à un compromis entre d'une part les techniques du dessin dans l'espace et d'autre part, les fonctions ou les propriétés spécifiques de l'objet à représenter.

L'une des questions majeures qui se pose dans le développement des modeleurs est sa liaison avec un système de gestion de bases de données capable de représenter les entités et les relations de l'objet et non plus seulement celles du dessin. Par exemple, le dessin d'un escalier, d'un toit ou d'un élément de réseau dans le bâtiment doit pouvoir être modifié ou détaillé en fonction de ses liens avec les autres éléments du bâtiment sans que l'on soit obligé de reprendre le dessin une fois que l'entité a été créée. De même, il est utile de pouvoir exprimer non seulement la forme, mais également certaines propriétés fonctionnelles de l'entité au même niveau que celui du graphique et non pas comme une connaissance extérieure au modeleur qui ne serait disponible qu'à travers l'intervention du concepteur.

Une définition opératoire:

Les objectifs de la scénographie intelligente sont simples. Il s'agit de mettre à disposition du concepteur des outils de création et d'évaluation de scènes spatiales statiques ou dynamiques à partir de descriptions et de procédures dépendantes de sa pratique et de la nature des objets produits. Pour cela le dessin et la représentation graphique ne sont que des réalisations techniques de génération de scène. Le dessin n'est plus la forme d'écriture spatiale mais le résultat d'une description ou d'une prescription. Le modeleur réside dans la description opératoire de la scène plutôt que dans son instanciation graphique.

Les principes opératoires de la scénographie que nous retenons sont au nombre de quatre:

1- Utilisation d'un langage opératif de description de scènes
2- Représentation des entités de la scène en termes d'acteurs
3- Schématisation graphique des structures de la scène
4- Fonctions d'évaluation spatiales et visuelles

Ces principes ont été décrits par ailleurs. Nous indiquerons simplement ici quelques précisions d'ordre fonctionnel sur ces points. La transcription d'un énoncé en une scène tridimensionnelle trouve son utilité dans de nombreux domaines d'application, pourvu que les structures syntaxiques et les objets considérés appartiennent à des domaines bien délimités. Lorsque qu'une scène est donnée, l'introduction d'un élément limité se fait en déterminant sa position relative aux autres éléments.

Les relations entre un élément et son environnement sont réalisées par l'utilisation de messages entre les divers éléments considérés comme des acteurs. Naturellement, il est préférable d'utiliser des mode de résolution hiérarchisés de ces inter-relations pour éviter de produire des des situations de blocage ou des explosions combinatoires. Des fonctions topologiques portant sur les interdistances ou sur les valeurs de paramètres topographiques retournent au système des valeurs numériques et des environnements de l'élément introduit d'un type supposé connu.

La plupart des problèmes de positionnement et d'analyse de scènes se font plus aisément lorsque l'on utilise des structures réduites. En particulier le fait de créer graphiquement des schémas qui représentent en quelque sorte une structure de composition d'éléments de la scène, comme des axes, des zones, des lignes, des points forts, susceptibles de résumer ou de guider des opérations et des actions courantes sur la scène. On réglera ainsi les contournements d'éléments, les alignements le long d'axes, les zones préférentielles d'implantation, des cheminements guides, les zones de hiérarchisation de certains paramètres de la scène, etc. Les fonctions d'évaluation spatiales correspondent à des possibilités de correction de la scène ou à des possibilités d'analyse et de recherche particulières. Ainsi, la recherche d'alignements, la recherche de stéréotypes de types couloirs ou places, la détection de situations[5] de masquage d'un élément par d'autres, sont autant de fonctions de ce type.

Le problème principal de la scénographie réside dans la recherche de cohérence et dans la validation de génération de formes sous contraintes. L'utilisation de systèmes à base de connaissance est rendue nécessaire pour la valider, une fois qu'elle a été définie. Quant à l'utilisation de l'image numérique plutôt que le graphique au trait, se justifie d'autant plus que l'on traite de scènes visuelles et que l'on peut traiter ainsi de paramètres comme la couleur, la matière, les effets optiques et certaines formes de perception, simulables uniquement par cette technique.

1 . 1 Application scénographique en réglementation urbaine

Cette recherche est effectuée au CIMA dans le cadre des développements réalisés autour du système IKOgraph© de synthèse d'images du laboratoire[6] . Le but de cette application est de donner au concepteur un moyen de tenir compte dans des simulations visuelles urbaines de l'effet de certaines réglementations. Ainsi, en produisant des schémas volumétriques et architecturaux, le concepteur sait s'il respecte ou non la réglementation. De plus, il peut lui-même introduire des prescriptions qu'il mettra sous la forme de règles afin de tester certaines règles de conception.

La réalisation de cette application résulte de l'établissement de liens et d'interfaces entre le système IKO et un de générateur de système à base de connaissances. La formalisation de principe est la suivante. Les données de départ sont un plan de parcelles d'une zone urbaine. Sur ces parcelles on peut placer des bâtiments ou figurer ceux qui existent déjà. Un bâtiment est caractérisé par une structure simplifiée telle que:

[Bâtiment] —> [Toit]* [Corps] [Base]*
[Corps] —> [Nombre Etage] [Type Façade Etage]*
[Parcelle] —> [Aire Constructible] [Aire Libre]
[Aire Constructible] —> [Bâtiment] [Eléments d'aménagements]*
[Aire Constructible] —> [Contours] + [Zone Centrale]

Avec la déclinaison des diverses variables:

[Type Général]* / [Règle de Composition] —> [Type 1] + [Type 2] + ...
Le signe "/X" indiquant la contrainte contextuelle "X"

Un énoncé scénique s'écrit alors:

[Arbre(i)] := [Arbre Z] / [Réglementation (u)]
L{*sur* [Trottoir (j)],
devant [Façade (k)],*aligné* (Σ_lArbre (l_m)}
I{*Réflexion* sur [Façade(n)]}

L'arbre nommé (i) est de type Z, localisé par les indications définies par l'opérateur L{ } à savoir, sur un trottoir existant de nom (k), aligné avec la ligne moyenne du groupe d'arbres (l_m) interagissant par les spécifications telles que I{ } à savoir, une réflexion sur une façade donnée (n). La réglementation ou les prescriptions à respecter étant la réglementation (u) sont données d'entrée comme un univers de validation sur le type Z.

La réglementation est donnée par un ensemble de règles portant sur les diverses valeurs et configurations de volumes sur la parcelle. Ces variables sont les valeurs des surfaces bâties, les morphotypes de bâtiments, leurs dimensions et surfaces de planchers les orientations et les distances inter-volumes.

On peut également inclure dans l'application des règles de déclinaisons des règles locales ou globales relatives au style ou à la composition, notamment sur le choix des éléments de façades ou d'ouvertures selon la localisation sur la façade ou selon des critères plastiques de proportions ou d'autres fonctions d'évaluation des paramètres disponibles.

Le lancement de ces règles à partir d'un ensemble de parcelles génère une série d'objets de type IKO, un objet étant essentiellement une structure de liste éventuellement récursive pour spécifier les procédures de générations de la représentation graphique.

Dans une application de ce type, l'image est une réalisation particulière d'un contexte de conception. On peut calculer en retour des fonctions qui à leur tour peuvent redéclencher des règles ou des diagnostics sur des aspects de la scène qu'il est difficile de modéliser dans la base de connaissances.

Ce type d'hybridation des diverses représentations de la scène en:

Structures d'objets (Objets IKOgraph)
Objets acteurs (Macro-procédures de IKOgraph)
Règles d'urbanisme, de composition (Base de règles juridiques)
Fonctions diagnostic ou d'évaluation (Lecture scénographique)

rend compte de la généralisation du modeleur par un ensemble de descriptions et de procédures qui sont orientées ou finalisées par l'utilisateur.

1 . 2 . Application en pré-robotique

Dans ce qui précède, l'image est une ouverture sur une scène, laquelle est construite à partir de règles relatives aussi bien aux objets de la scène qu'à leurs propriétés. Les systèmes à bases de connaissance permettent de traiter des contraintes de composition et d'une certaine cohérence de l'ensemble. Dans l'application en pré-robotique, c'est à dire en aide à la spécification de fonctions robotiques, la scène est dynamique et l'un des sous-système de de celle-ci est en interaction avec son environnement du fait qu'il doit atteindre un but.

Le robot est défini comme un sous-ensemble de la scène considéré comme un acteur, disposant de fonctionnalités établies de comportement, d'interaction et d'évaluation. Il est donc plus général qu'un simple automatisme défini dans un contexte donné.

La planification de trajectoire d'un engin est un exemple de ce type et l'on peut donner à cet engin un objectif et ensemble de spécifications modales. L'engin est un acteur qui doit avancer sur un schéma de routes qui lui est indiquée et atteindre un objectif. Il dispose d'un ensemble de moyens de détection et des règles de comportement mécanique, comme par exemple la gravité qui lui permet de conserver le contact avec le sol, l'équilibre ou le non renversement qui lui permet de déterminer les positions acceptables, la non chute, le non glissement, l'élasticité du sol, etc. Son pilotage s'effectue exclusivement par rotation des roues sans glissement ou avec une loi de glissement. C'est d'ailleurs la seule commande à activer pour lancer le processus de déplacement.

Les messages échangés avec l'environnement immédiat sont ré-évalués à chaque boucle du processus. De proche en proche, l'engin se déplace en respectant un ensemble de lois physiques virtuelles.

Il est clair que ce type de simulation ne peut s'effectuer que parce que l'on a mis en place des mécanismes de propagations. Par exemple, faire tourner une roue est une loi temporelle. A chaque pas élémentaire de celle-ci, les fonctions de contrôle et de positionnement sont testées. Les modifications de l'environnement sont simulées et calculées, comme par exemple la déformation du sol sous l'action de la pression des roues de l'engin. Naturellement cette simulation est d'abord visuelle et compatible avec les propriétés physiques des éléments en présence. Les enchaînements sont décrits par des règles, elles-mêmes instanciées par des messages inter-acteurs.

On dispose ainsi d'une série de règles spécifiques du déplacement dans lesquelles l'engin est d'abord assimilé à une boîte et échange des messages avec les obstacles rencontrés pour établir des autorisations de mouvements ou de contacts.

La caméra est un acteur auquel on donne des consignes de déplacement et qui sont aussi bien des objectifs que des contraintes. Elles permettent de déterminer des trajectoires, les inter-distances et les paramètres de la caméra. Par exemple si la trajectoire est déterminée approximativement par des règles de suivi ou de balisage, les paramètres de point de mire, d'ouverture et d'orientation da la caméra sont des fonctions déduites de l'analyse de la scène. On peut également gérer les vitesses de déplacement en fonction de critères morphologiques de la scène. On peut enfin définir des formats catalogués de déplacements de type voiture, piéton ou autres.

Ces exemples de simulation de mouvement permettent de valider des hypothèses de travail en conception, notamment lorsque l'on cherche à utiliser les règles et contraintes énoncées comme des paramètres du processus lui-même. Ainsi, le fait de rechercher certains points de vue ou de tester la possibilité de prendre en compte certains types de balises revient à préparer des situations de compréhension de certains mouvements.

Le processus de conception étant très souvent qualifié par la boucle:

[Situation] —> [Evaluation] —> [Modification]
—> [Situation]

La question centrale dans l'exploitation de l'image de manière contrôlée est de construire des fonctions d'évaluation typiquement scéniques et visuelles, qui ne sont pas simplement déduites des données numériques et topologiques disponibles.

2 . LA MULTI-REPRESENTATION

Les images numériques font appel à des modèles de calcul, morphologiques ou comportementaux, qui doivent satisfaire aux propriétés que l'on attend d'eux à chaque étape du processus de conception considérée. Une scène d'architecture étant modélisée, son image est à priori la visualisation de son apparence, selon un point de vue. Cette simulation de la vision n'est pas suffisante pour le concepteur qui ne se contente pas de la simple visualisation et qui doit retraiter l'information visuelle à partir d'autres grilles conceptuelles. C'est à l'adaptation des représentations graphiques à ces besoins d'analyse que tente de répondre la multi-représentation.

2 . 1. Principe de la multi-représentation

La multi-représentation est une forme de lecture de l'image à partir d'une lecture multiple de la scène. Elle est en quelque sorte un mode de lecture des éléments de la scène et de mise en relation sémantique entre ces éléments. D'une part il s'agit pour le concepteur de travailler sur un objet en cours de conception donc sur des représentations schématiques et indicatrices compatibles avec cet état d'évolution de l'objet. Il faut donc disposer d'une certaine cohérence dans la représentation d'un élément incomplet ou inachevé. La question de la cohérence est un problème important, mais qui n'est pas abordée pas ici. D'autre part, la modélisation est supposée évolutive ou phasée pour qu'on puisse l'adapter à cette transformation du modèle. Il faut pouvoir travailler avec ces modèles et passer visuellement de l'un à l'autre.
Dans ce cas la représentation par l'image numérique donne la possibilité de traiter les variables spatiales impliquées dans le processus. En particulier le fait de pouvoir se déplacer dans l'espace de l'objet, à l'intérieur même d'une fenêtre pour parcourir l'objet comme un commentaire déroulant

On peut considérer que la représentation visuelle supporte une généralisation à travers la multi-représentation, ce qui sera illustré dans la suite.

La multi-représentation peut se décliner selon plusieurs axes:

Multi-modes de représentations de la même scène.
Multi-scènes différentes dans un même mode.
Multi-scènes dans une même scène.
La combinaison des précédents.

Le mode de représentation est un item dans la liste indicatrice suivante:

Perspective plane ou sphérique
Plusieurs caméras paramétrées
Restitution filaire
Restitution surfacique colorée, texturée, ombrée
Commentaires associés

La multiplicité des scènes peut se traduire de l'une des manières suivantes:

Extraction, regroupement des diverses occurrences d'un type d'élément
Mise en correspondance de scènes liées par un même critère

La multiplicité de scènes dans une même scène consiste à associer des fragments de cette scène et à les ré-introduire dans la scène pour en compléter la lecture:

Utilisation de multi-fenêtres à l'écran
Parcours de groupes non structurés d'images, selon des critères explicites ou non (mode hypertext)

Autrement dit, le fait de pouvoir recomposer les scènes avec des fragments de modèles ou des fragments d'espaces séparés à partir de rapprochements fonctionnels ou sémantiques permet de développer des lectures finalisées ou conditionnelles du projet.

L'emploi de la multi-représentation traduit une forme d'appropriation visuelle du projet ou de la maquette qui vise clairement à se dégager des contraintes de la pratique visuelle qui consiste à rendre compte de la morphologie et de l'apparence.

2 . 2. Exemples d'application

L'image sélective:

Si l'on extrait tous les éléments de connexion d'un réseau dans un espace urbain ou dans un bâtiment, en éliminant les parties courantes, on dispose d'une vue d'ensemble des points d'articulation, donc de la syntaxe de ce réseau. Il s'agit d'une image sélective recomposée avec extraction et occultation tout à la fois des sous ensembles de la scène. Ce qui permet d'une part de s'affranchir de la dimension de l'image en la réduisant aux seuls points d'intérêts et d'autre part, de traiter des points de vue non assujettis à la forme, mais à la fonction ou à l'interprétation.

L'image opérative:

On peut également à partir d'un concept ou un lien sémantique rechercher l'image ou les images qui leur correspondent. Un premier exemple est l'examen quantitatif de l'exposition au soleil des murs et des ouvertures d'un ensemble bâti, ce qui permet une extraction d'information de l'image sur la quelle on peut effectuer ensuite des simulations. On extrait donc des images finalisées de la scène.

Un second exemple est la recherche de points de contrôle de l'espace dans un environnement construit. Il s'agit de déterminer les espaces visibles depuis un ensemble de points de sorte que l'ensemble des zones visibles de l'ensemble des points de contrôle soit connexe. Les positions des points de contrôle sont variables. Dans cet exemple l'ensemble des points de vue liés par la condition de connexité est une fonction de lisibilité de l'objet à travers une multi-représentation.

L'image générique:

La multi-représentation trouve encore son utilité directe dans l'exploitation de bases de données d'images en phase de conception. Ainsi, le concepteur qui travaille sur une scène spatiale peut avoir besoin d'utiliser des éléments schématiques ou réalistes, sans vouloir les traiter spécialement. Il s'agit le plus souvent d'éléments de remplissage dont le détail n'est pas à traiter immédiatement mais dont la présence et les dimensions sont des variables utiles. Dans ce cas il fera appel à une base de données d'objets standard déjà catalogués et dont les paramètres sont compatibles avec le dimensionnement et la qualification utiles à la conception. La base de données objets en multi-représentation fournit alors au concepteur:

Classe et type d'objet dans la classe
Image filaire
Liste des paramètres accessibles
Opérations réalisables sur l'objet
Image colorée texturée et déclinée
Exemple de mise en situation de l'objet
Script de la macro-instruction IKO associée.

Autrement dit, le concepteur travaille avec des images composites qui ne sont pas interprétables simplement mais qui sont disponibles à plusieurs niveaux d'utilité.

De cette manière, l'utilisateur peut développer son travail sans être détourné du traitement quantitatif de l'image que représente le remplissage utilitaire de la scène. Les éléments scéniques étant là comme des concepts ou comme des figures provisoires. Ceci est surtout le cas du traitement des espaces urbains. Dans certains cas, on peut définir des opérations disponibles sous IKO, compatibles avec les énoncés scénographiques tels que:

"poser objet sur ...."
"aligner objet avec ...."
"décrire le trajet ...."

et tous ceux que le concepteur élaborera pour son propre usage en macro-instructions dans le langage IKO. La scénographie assistée nécessite une telle base d'éléments génériques comme arguments du langage opératif.

3 . L'APPROCHE VISUELLE CONSIDEREE COMME UN PROTO-LANGAGE DE CONCEPTION

Dans ce qui précède nous avons indiqué quelques unes des pistes développées en scénographie architecturale et urbaine. Le concept sous-jacent à toute cette approche est celui d'opérateur spatial, qui permet de construire de transformer et d'assembler en une scène statique ou dynamique des objets tridimensionnels de diverses catégories:

Solides et limités en étendue dans la scène
Corpusculaires diffus dans la scène
Non matériels de type champ vectoriels
Filiformes de types schémas de distribution
Virtuels de type "propriété".

Un opérateur spatial est une correspondance triple entre:
une procédure
un objet spatial pris dans les catégories précédentes
une phrase d'un pseudo langage naturel.

Qu'il s'agisse d'un cube, d'un champ gravitaire, d'un mouvement, d'un brouillard, d'une réflectivité, tous ces items sont susceptibles d'une telle description, voire d'une modélisation élémentaire.

On définit également des schémas qui serviront de guides ou de squelette dans le positionnement, l'expansion et la configuration des objets spatiaux. Ce sont des objets intermédiaires de la scène, des formats de syntagmes est utilisables par les procédures de configuration et de placement des objets.

Enfin, pour assurer un contrôle de la scène, il est nécessaire de disposer de fonctions d'évaluation typiquement visuelles et spatiales. Le terme visuel indique que l'on détermine des fragments de scènes qui ont une signification visuelle dans l'espace sans pour autant se réduire à une liste d'objets ou de procédures. Par exemple, une perspective ou un effet d'avenue ou de passage, ne sont pas des objets, mais des effets résultant d'une configuration. Cette recherche de liens entre des fragments d'éléments correspond à la recherche de cohérence spatiale. Celle-ci se traduit par le fait que l'on peut recomposer des liens entre des parties de la scène selon une lecture située à un niveau supérieur de description. Il est fondamental de maintenir dans cette démarche une logique d'emboîtement des niveaux de structuration de la scène. Les emboîtements:

[[[[OBJET] SYNTAXE] SEMANTIQUE] COHERENCE]

sont ainsi pris en compte aux niveaux suivants:

Syntaxe:
Règles de positionnement et de formation des objets. Formalisation des règles d'assemblages ou de génération.
Sémantique (partielle):
Structure des liens fonctionnels entre les objets. Identification de figures spatiales fonctionnelles ou symboliques: sorties d'immeubles, places, rues, alignements visuels, espace périphérie d'un objet, carrefours, etc.
Cohérence (partielle):
Evaluation de figures scéniques par fonctions visuelles. Détection de schémas incomplets ou inacceptables.

Pour se convaincre de l'importance de cette organisation verticale des niveaux de lecture, il suffit de considérer de manière voisine, comment s'effectue la lecture d'un texte, depuis la forme des mots et de leur assemblage, jusqu'à la compréhension des phrases et du discours qu'elle constituent. C'est dans cet esprit, mais à un niveau beaucoup plus modeste que l'on peut élaborer des schémas permettant de construire ou d'identifier des composantes formelles et sémantiques de scènes spécifiques d'architecture.

Nos recherches actuelles nous conduisent à approfondir les modes de description des scènes architecturales ou urbaines dans des termes qui utilisent des multi-représentations. En même temps les procédures statistiques peuvent fournir une première évaluation de configuration spatiales. Sans être des modèles structuraux, les approximations statistiques nous renseignent sur la probabilité de rencontrer tel ou tel type d'objet dans tel ou tel contexte.

La définition de nouvelles entités fait partie intégrante de l'analyse de scène à des fins de synthèse. Ainsi, tous les éléments composant une scène nommée, par exemple un passage, seront susceptibles d'être ré-

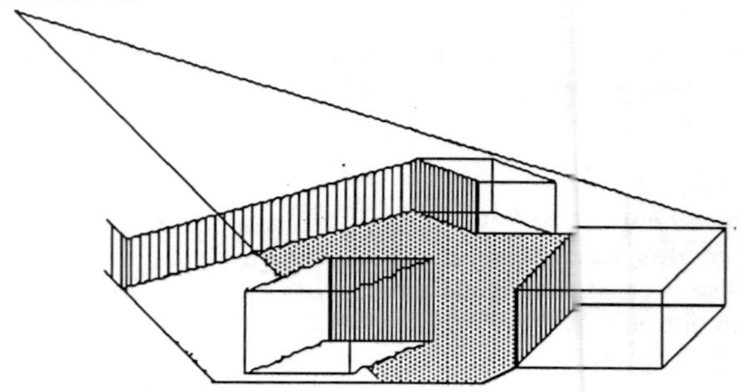

Exemple de fonction d'évaluation scénique: Primitive de détermination des espaces ouverts dans une scène. En entrée la donnée d'un point de vue et d'une direction, en sortie, l'espace ouvert visible et la typologie et la surface des façades correspondantes.

employés dans la génération d'un autre passage.
Afin de conduire cette recherche à un terme opérationnel, deux conditions principales sont requises:

1- La restriction du vocabulaire et des isotopies à quelques catégories de scènes.
2- La réalisation d'une structure de contrôle interactive pour limiter les tâches du générateur de scènes.

Nous procédons actuellement à un travail de repérage et de dénombrement des scènes et figures dans la suite de ce qui a été fait auparavant dans une recherche sur la rhétorique de l'image de synthèse[7]. Ce repérage a pour but de constituer une sorte de dictionnaire des figures spatiales nommées usuellement accompagnées des règles de combinaisons ou d'attachement à leur environnement. Joint à l'ensemble des procédures de génération d'objets, un tel dictionnaire complète le dispositif d'assistance scénographique.

4 . CONCLUSIONS

Les quelques indications données ici, montrent que les diverses approches de l'image tridimensionnelle proposées correspondent à l'établissement d'un système d'énonciation des scènes en vue de leur évaluation ou de leur génération. On obtient des résultats dans des cas simples ou bien structurés, mais il est clair que la génération d'une scène suppose une bonne analyse pour que l'on puisse la faire évoluer de manière cohérente. Le mouvement d'un élément ou la recherche de trajectoires, pour ne citer que ces exemples n'ont de sens que si les contextes dans lesquels ils opèrent sont suffisamment connus.

En quoi, l'approche de l'image est-elle alors un outil de conception? Il est clair qu'il ne s'agit pas dans notre propos d'automatisation mais d'assistance à la génération de scènes à partir d'élaboration de structures para-linguistiques. L'aide à la conception consiste alors essentiellement dans l'implication de registres différents à un même niveau d'intervention du concepteur. Celui-ci peut envisager non plus des modèles de calcul techniques partiels et efficaces, mais également des structures guides qui impliquent concurremment des composantes ou plutôt des lectures différentes des scènes spatiales considérées.

L'image est alors la visualisation, avec évidemment toute sa richesse propre, de composants de la scène qui peut être partiellement intégrée dans la boucle formalisée déjà citée:
[Situation] —> [Evaluation] —> [Modification]

C'est ainsi notamment que les réglementations et les prescriptions constituent en particulier un système de contraintes ou de contrôle traduisibles sous forme d'opérateurs ou de données transférables spatialement.

En considérant l'image comme une représentation compatible avec le processus de conception, elle est un mode de lecture du projet en cours de spatialisation. Les implications fonctionnelles de ce mode de lecture imposent au concepteur d'expliciter des schémas de cohérence, donc à valider spatialement ce qui est énoncé dans d'autres registres sans pour autant perdre la richesse des représentations associées à un même fait spatial.

L'espace construit image devient un objet de reconnaissance et de réidentification des savoirs techniques.

A titre d'illustration de cette forme de connaissance par l'image, on peut précisément inclure la reconnaissance de figures spatiales locales ou globales, de morphologies ou de configurations contraintes dans un espace urbain. Les produits de cette reconnaissance sont regroupables et généralisables en signes complexes, en morphotypes élémentaires diversement configurables et qui constituent une matière de base à la reconnaissance des scènes urbaines. Cette matière scénique qui s'exprime à travers l'image supporte également un script, c'est-à-dire une description structurée et un schéma relationnel ou fonctionnel impliquant d'autres morphotypes selon les contextes rencontrés. On dispose ainsi d'une sorte de formation de l'image de scène par une procédure d'assemblage de figures minimales que l'on pourrait appeler des scénèmes, par analogie avec les morphèmes. De la même manière que le son est une des formes d'acquisition et de transmission du langage parlé opérant sur des assemblages de phonèmes, la scène peut être considérée comme une configuration de scénèmes qui s'intègrent dans une morphologie scénique plus complexe. Le traitement de la scène par l'image constitue ainsi un mode d'analyse avec les catégories les plus proches possibles des opérateurs spatiaux qui ont permis de la générer.

L'image numérique, dans ce qu'elle possède de connexion et de cohérence avec d'autres formes de représentations d'une même scène, fait aujourd'hui l'objet d'une autre forme de simulation du projet. Encore faudra-t-il parvenir à établir les langages d'opération et de manipulation de ces scènes, c'est-à-dire le langage de représentation et de manipulation des connaissances extractibles de l'image à partir de la recomposition des scénèmes. On peut envisager avec une certaine légitimité aujourd'hui la possibilité d'élaborer des langages d'images ou de scènes dans des domaines d'isotopies particuliers, langages qui seront des outils privilégiés dans le transfert des connaissances en matière de génération, d'évaluation ou d'analyse contrôle de scènes. Cela définit sans doute le premier pas vers une assistance à la conception par l'image, mais également une re-définition de l'image de conception.

[1] cf: ZEITOUN, Jean.- Du projet à l'objet: la conception assistée comme production d'espace virtuel.- Cahiers du CCI, numéro spécial, les chemins du virtuel.- Paris,1989.

[2] Le Logiciel IKO est développé au CIMA sous la direction de Michel BRET. Il fait l'objet de versions sur PC "IKOLIGHT©", sur IRIS et SUN " IKOGRAPH"©. Ecrit en C sous UNIX. Il est constitué d'un interpréteur qui sert de langage de programmation d'image et de développement pour l'utilisateur. Sa structure orientée vers la définition de procédures comme objet élémentaire permet de produire de l'image tridimensionnelle sans recourir au geste. Cette condition fondamentale est nécessaire dans toute approche scénographique assistée.

[3] cf: ZEITOUN, Jean.- Considérations sur le développement d'applications architecturales de la scénographie intelligente. - Paris, actes de la journée CIAO du CIMA.- Paris Hermès, 1988.

[4] cf: ZEITOUN, Jean.- Simulation visuelle et conception spatiale en architecture: les apports de l'image de synthèse en conception architecturale.- Paris, actes MICADO 88.- Paris Hermès, 1988.

[5] Nous disposons déjà d'un premier dictionnaire des situations détectables à partir d'un corpus de fonctions d'explorations spatiales données, définies comme des primitives spatiales. La construction d'un pseudo langage de programmation de ces primitives permet notamment de détecter les espaces vides et les caractéristiques principales de leurs formes.

[6] cf: Josef RABIE .-Toward the simulation ofurban morphology. - European Conference on Management and Representation of Urban Change. - Cambridge, England, September, 1989.

[7] ARNOLD, M., BRET, M., PORADA, M., et al.- Langage de création et rhétorique de l'image de synthèse. Rapport de recherche sous la direction de J., ZEITOUN.- Paris, Ministère de la Culture et de la Communication, 1986.

PASSIVE SOLAR HEATING AND ENERGY EFFICIENT HOUSING DESIGN

A Design Guide for Housing in the context of the UK Department of Energy Passive Solar Design Programme

Simos Yannas
Environment and Energy Studies Programme
Graduate School
Architectural Association School of Architecture
Bedford Square, London WC1B 3ES, UK

ABSTRACT. The paper discusses the approach followed in the planning and production of the UK Housing Design Guide, highlighting some of the difficulties that have arisen from lack of sufficient design information on certain key topics, and from the current discrepancy between the output of research projects and the information requirements of designers. The paper focuses on the use of parametric studies as a source of design support and describes the methodology adopted for extracting design guidelines and design aids from the first sets of runs carried out for the project using the SERI-RES simulation model. A further set of simulation studies is now being undertaken, including sensitivity analyses and additional parametric runs.

1 INTRODUCTION

Following the completion of a Definition Study /1/, the Architectural Association School of Architecture was commissioned by the Energy Technology Support Unit of the Department of Energy with the production of a Housing Design Guide in the context of the UK Passive Solar Design Programme. Work on the project was started at the AA School in early 1986 and the effort so far has been of about 2.5 man-years. In addition, contributions were made to various aspects of the project by NBA Tectonics, Yard/Principia Mechanica (previously CAP Scientific), Halcrow Gilbert Associates, Energy Monitoring Company, British Gas R&D and by many other colleagues who have provided technical information and acted as reviewers.

The Design Guide is being produced in two volumes of 100-120 pages each. Book 1 focuses on Principles, Objectives and Design Guidelines and is structured around key stages of building design. A first draft was produced in April 1988. A revised abridged version was completed in September 1988 /2/, and has been in use for over a year now by designers in the context of Design Studies that are being carried out around the country under the UK Passive Solar Design Programme. Both drafts have been reviewed in detail by a group of technical experts. Book 2 focuses on relevant built examples and design studies. Its structure follows a UK housing typology with further groupings of the examples on technical criteria. The book developed as a separate volume out of what was initially one of the chapters of a single-volume Guide. A first draft of Book 2, /3/, was prepared in March 1989 for technical review and is now also distributed to a number of design practices. Over the last and current academic years the books have been used as teaching material and as support to student projects in a number of schools of architecture including the AA School's post-graduate teaching programme in Environment and Energy Studies.

Owing to information gaps in a number of areas and delays with the completion of other research work on which this project had to rely for information or results, the production of the Design Guide has taken longer than originally expected. However, the nature and scope of the work has also changed and the project has taken a more investigative role in pursuing its design support objectives. Following further work that has just begun on the project, it is expected to complete the books for final publication sometime during 1990.

2 DEVELOPMENT OF THE HOUSING DESIGN GUIDE

2. 1 Programme of Work

Our Definition Study had reviewed existing design handbooks in the UK and elsewhere pinpointing areas where design guidance was insufficient or unavailable; it identified the kind of information required, and examined ways in which information should be structured in order to be of use in the course of design. The Definition Study concluded with a preliminary outline of structure and contents for the Design Guide in consultation with an Advisory Group of designers, researchers and educationalists. This was planned around key stages of design, Fig. 1.

Starting from this outline structure the development of the Design Guide proceeded through the following stages of work:

- review of recent field studies and research results as potential sources of design information;
- planning and implementation of parametric studies to identify likely trends in building performance and to help

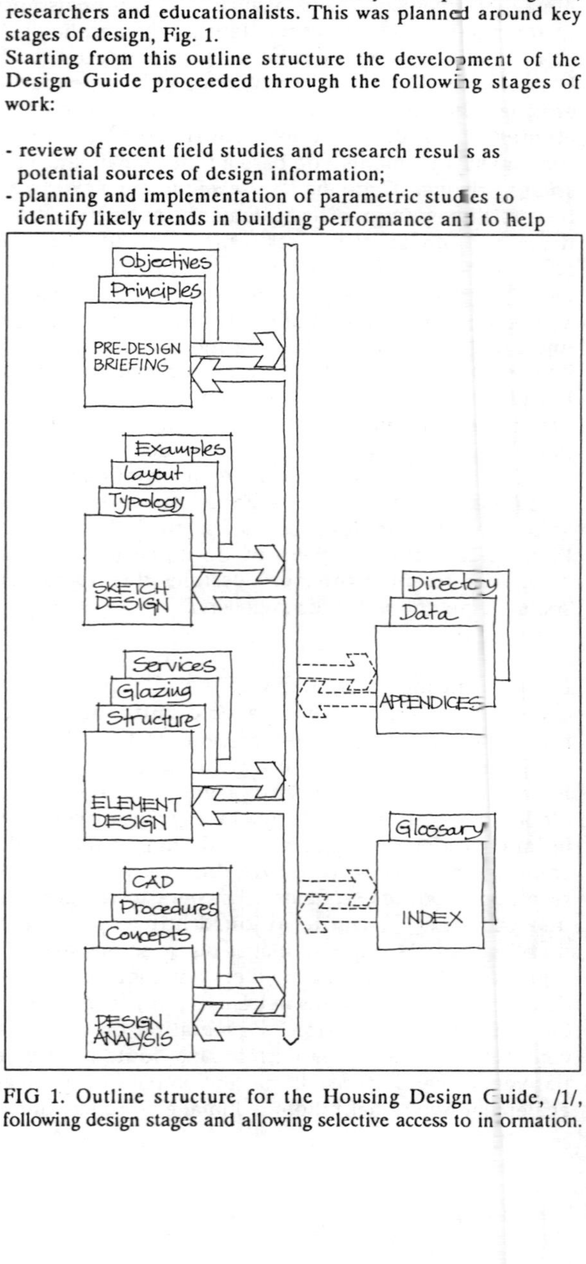

FIG 1. Outline structure for the Housing Design Guide, /1/, following design stages and allowing selective access to information.

develop design guidelines and design aids;
- production and circulation of working drafts of the Design Guide;
- technical review and testing of the Design Guide drafts;
- revisions and amendments, further parametric studies, coverage of additional topics in response to feedback received from designers and technical reviewers; preparation of the books for publication.

The project is presently into its final stage of work.

2.2 Design Information and Issues of Knowledge Transfer

The literature review encompassed most of the work carried out in the UK since the early 1980's including major field studies, results from thermal simulations, and the research projects sponsored by the UK Passive Solar Design Programme. Although a considerable volume of material was available, much of it was either too general or too inconclusive to be helpful to designers in practice. The design variables left unaccounted for in the course of field studies; uncertainty over the effect (or even the nature) of some of these variables; and the focus of many research projects on issues of building physics on the one hand, and on a crude qualification of building performance in terms of energy saving on the other, has a limiting effect on the type and depth of design information that can be extracted out of recent fieldwork or from other research results. The lack of detailed qualification of the processes of passive solar heating in spatial and constructional terms and with respect to weather variations and occupant comfort is a serious problem. Comparisons and generalisations from results obtained at different times by different researchers are commonly marred by differences in techniques used for collecting, treating or interpreting data. Few monitored buildings have been studied in sufficient depth to answer the sort of questions that would be asked by designers at the level of spatial planning or detailed design, and scarcely any of these buildings are of sufficient architectural interest to inspire as exemplars. These issues are by no means unique to the UK, where in fact the Passive Solar Design Programme has responded recently both through reviews of performance assessment methodology and through new fieldwork to help remedy some of the problems. Such issues are likely to become acute at a European level as we move toward a pan-European practice of architecture.

When such discrepancies between research capabilities and design support are set against the background of mainstream architectural practice, influenced as it is by a predominantly formalist architectural discourse and its pull toward global (or universal) forms of expression (as opposed to the regional "problematique" of bioclimatic design), our field emerges as still lacking in credibility and inspiration to influence the course of architectural education and practice, /4/. Designers expect design information and support tools to be precise, specific and reliable, as well as being relevant, interesting and well presented. But this is only one aspect of the problem we need to face, Fig. 2. Still more important is the conceptual gap from the disassociation, for most of this century, of environmental design concerns from architecture. This is at the root of the current predominance of universalised form. A major task must be to re-qualify and re-construct environmental design within the body of architecture, and as part of the design process, Fig. 3. As an essential condition, physical processes such as passive solar heating and passive cooling must no longer be seen as alternative heating and cooling systems (which they still are for many researchers in the field), but as sensations of space and time which designers might wish to create and express, -or avoid and depress-, in the course of giving form and function to a building.

A major stumbling block for designers will be the fact that drawing, - the main tool, and universal language of architects-, abstracts both spatial and time dimensions, leaving the dynamic of environmental variables almost totally hidden from view (and consciousness), Fig. 4. The

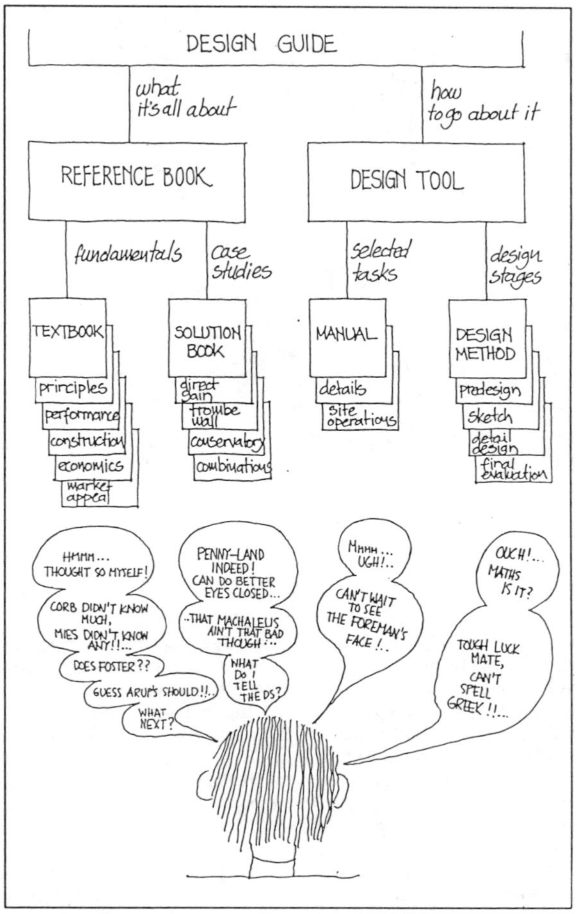

FIG 2. Functions of the Design Guide and need for sensitivity to the users' background and requirements, /1/.

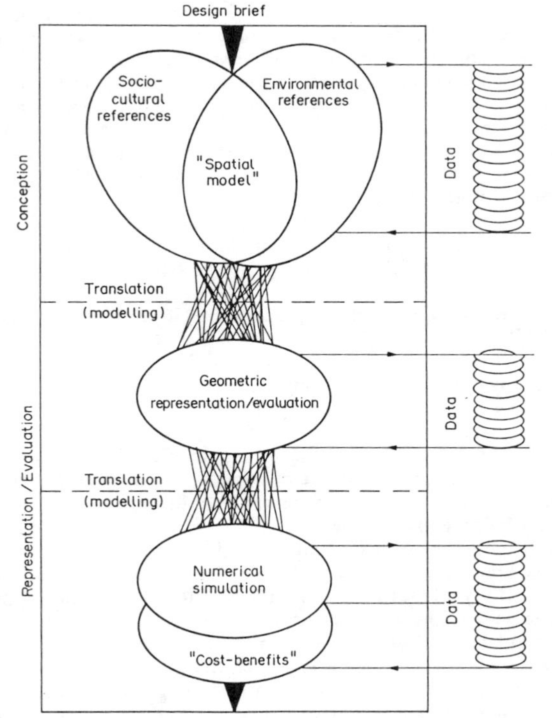

FIG 3. Schematic of design process: there is need to inform the spatial model through environmental references and to provide design support at various stages, /4/.

FIG., 4. The medium is only part of the message , /4/.

transmission of architectural knowledge is proceeding mainly through static images of frozen time. It is neither sufficient nor desirable simply to substitute alternative snapshots; the medium can only carry part of the message. We need to highlight and qualify that which is not readily visible from a snapshot. Architects are well trained in the de-construction and re-construction of the spatial dimensions; this faculty must be now extended to the temporal dimensions of site, climate and building occupancy. Modelling, simulation and visualisation of heat, light and air flows can complement traditional architectural tools, allowing designers access to the temporal dimension of environmental design. The re-adoption of environmental design as an essentially architectural, rather than engineering, concern, and the development of design support tools of varying detail and sophistication for use in the course of design, are the two fundamental conditions for re-establishing a balance between form and function in architecture. Given these conditions, attention to thermal and visual comfort, low use of fossil fuel and improved environmental quality must follow as corollaries rather than being imposed as prescriptive commands. Such concerns have been central to our teaching programme at the AA Graduate School,/5/, and have also influenced the approach taken in this project. In this paper, which focuses on the analytic foundations rather than the didactic, or "reconstructive", content of the UK Housing Design Guide, the central theme is the role and application of modelling and simulation techniques as potential source of design support.

3 PARAMETRIC STUDIES AS SOURCE OF DESIGN SUPPORT

3.1 Review of Previous Simulation Studies and Planning of Parametric Runs for the Design Guide

The gaps, discontinuities and inconsistencies identified in the literature at the early stages of this project indicated the need for an analytic substructure that could help bridge these knowledge gaps and provide a framework for generalising results in the form of simple design guidelines and design aids. Simulation offers a useful tool for investigating the likely sensitivity of building performance to changes in design parameters, and can provide the analytic foundation for the formulation of design guidelines. However, in addition to issues relating to the choice, capabilities and limitations of the simulation model to be used, the modelling and simulation of the environmental performance of buildings also raise important questions of procedure, concerning the approach to be followed in pursuing such investigations. Attempts at comprehensiveness and/or realism quickly lead to impossibly large and tortuously complicated sets of runs, from which only a fraction of the results could ever be of use for design guidance. Oversimplification on the other hand, may compromise accuracy or interpretation. The approach followed in this project is summarised in Figure 5, which also shows parallel developments in, and interactions with, other research projects in the UK Passive Solar Design Programme. Preliminary work for the Design Guide started with a detailed critical review of previous simulation studies. In parallel to this we also looked at the applicability of correlation-based monthly and seasonal calculation methods as design aids and for parametric studies.

The review of previous simulation studies, /6/, focused in particular on the performance assessment methodology and output of an earlier series of Housing Design Studies under the UK Passive Solar Design Programme, /7/, and the extent to which these provided meaningful insights on "passive solar design". The Housing Design Studies had resulted in some twenty designs that spanned a broad range of house types with the aim of exploring the potential of passive solar heating for new and existing housing, /8/. Each of these design schemes had been evaluated, both during its development and after completion, through several full-year hourly simulations and through cost analysis. These evaluations were interpreted by comparison of the designs with a number of variants that had been chosen to represent typical developer's housing or "non-solar" reference houses. A fairly consistent set of conventions and input data had been used for the simulation studies and each design had been analysed under three different occupancy scenarios to identify the likely effects of varying occupancy patterns on heating requirements, useful solar gains and cost-effectiveness. The simulations had been conducted with the SERI-RES model as modified and extended for the purposes of the UK Programme /9/.

In parallel to the performance assessment methodology of the Housing Design Studies we also considered the parametric studies carried out on variants of a typical semi-detached house by Bloomfield, /10/, /11/, and the simulations performed by Everett in the course of interpreting and extrapolating from the monitored results of the Milton Keynes Linford houses /12/. Three important issues were highlighted by these reviews. Firstly, current simulation models present important limitations as tools for qualifying, comparing or directing the development of designs. The modelling of many physical exchanges and spatial configurations is poor or inadequate, and all too often the simulation is little more than a caricature of the spatial and physical entity conceived by the designer. Secondly, certain underlying assumptions, conventions and choices of input data were identified as exerting a strong influence on the simulation results and on their extrapolation into design guidelines. In particular, the way in which these studies attempted to emulate "typical" operational scenarios as the context for the simulations (invariably, on fairly tentative data and assumptions concerning occupancy patterns, activity and behaviour), and the equally tentative (or arbitrary) criteria for defining "usefulness" of solar gains, appeared to obscure the meaning of results, making it difficult or impossible to assess building performance or the effects of individual design parameters. Thirdly, there is the practical problem of the considerable human and machine time required for the undertaking of any comprehensive investigation of parametric variations.

In view of such problems, correlation-based calculation methods cannot be discounted as a possible alternative to simulation, provided their structure and level of abstraction are understood and provided they are open to user inputs on all the parameters of interest. Our review of simplified calculation methods, /13/, assessed and compared a number of monthly correlation techniques and studied their suitability for parametric trend analysis. As a result of this study a monthly calculation method, based on a correlation algorithm similar to that of Method 5000, was implemented on a spreadsheet, calibrated with building, weather and occupancy data from the Milton Keynes Linford house No. 36 and tested

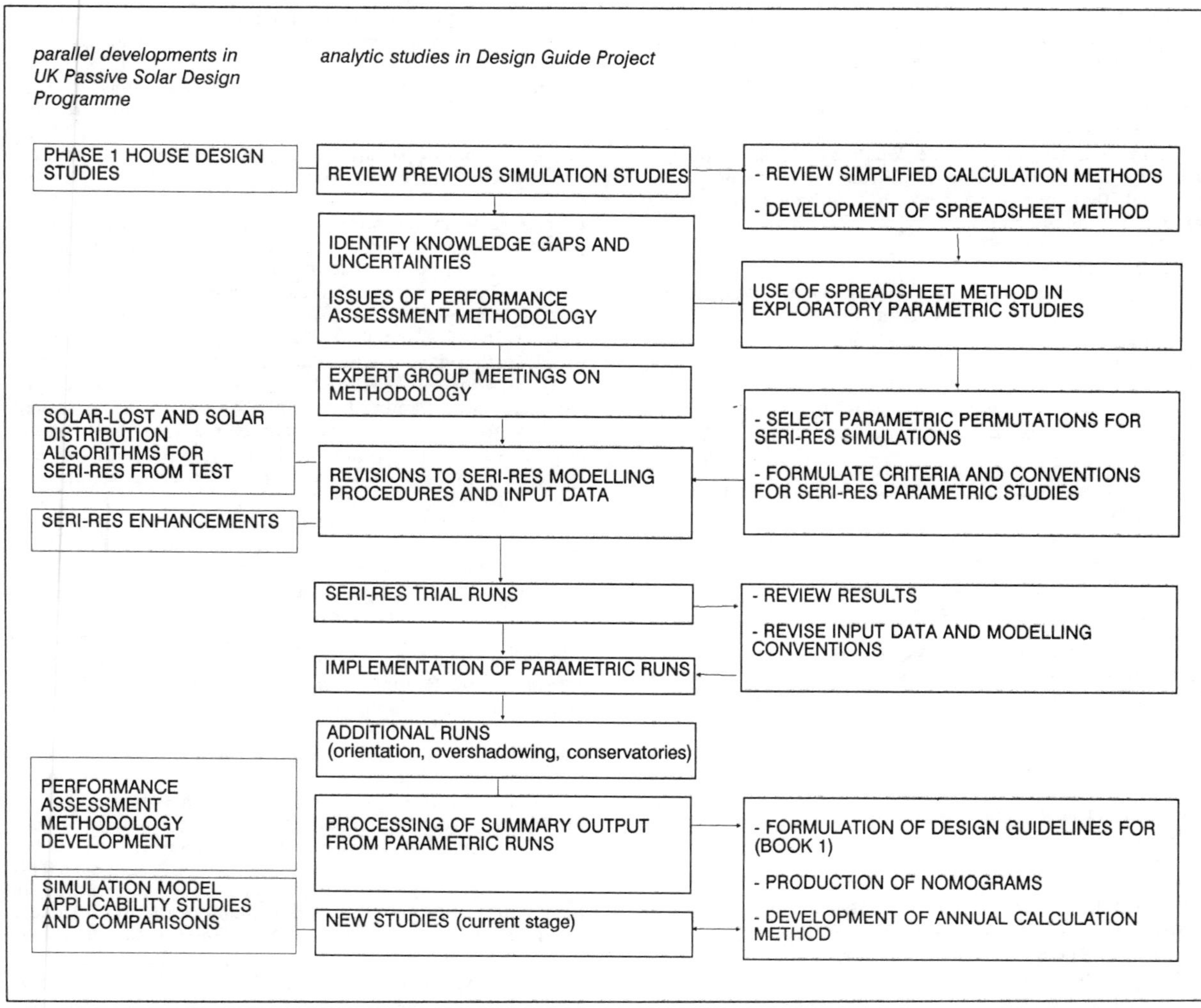

FIG. 5 Planning and implementation of parametric studies for the Housing Design Guide, also showing relationship with parallel work in the UK Passive Solar Design Programme.

against measured values of auxiliary heating energy. The seasonal heating energy results obtained from the monthly method came within 5% of the measured value. Varying the south-facing window area resulted in similar trends to those that had been obtained by Everett, /12/, with dynamic simulation.

On the strength of these results it was felt that the spreadsheet method offered a useful tool for a preliminary exploration of the parametric domain, as such exploration could help reduce the potentially enormous parametric combinatorial to a manageable set of runs in any subsequent use of simulation for the same task. The spreadsheet (a family of interlinked spreadsheets programmed under Ashton-Tate's Framework II) was arranged so as to save an annual summary output at the end of each run and to collect these summaries for trend analysis. As each run took only a few seconds to set, execute and interpret, the spreadsheet provided an interactive and efficient tool for observing the effects of parametric variations. This allowed us to review and compare a considerable number of permutations that encompassed all the likely combinations of insulation values and glazing parameters. Having initially completed a set of runs for London weather data and for a detached house, subsets of parametric permutations were run for several other locations and for other house types (semi-detached, terrace, flats), /14/. The trends resulting from variations in window areas showed the potential contribution from passive solar heating to be fairly sensitive to regional variations of outdoor temperature and solar radiation. On the other hand, comparisons of the trends in heating energy obtained by varying window areas on different house types indicated that these trends were a function of the house heat loss coefficient (for fixed values of internal gains). This suggested that once a fairly wide range of parameter variations had been studied for one house type, the respective trends for other house types could be inferred by interpolation.

In combination the review of previous simulation studies and the simplified parametric runs provided valuable insights on the effect of key design variables and on the objectives of further simulation studies in the context of this project. The exploratory runs using the spreadsheet method had indicated that by careful selection of the parametric permutations a fairly small number of runs would be capable of giving interpretative insights over a much larger domain of design configurations. The review of previous simulation studies on the other hand had highlighted the sensitivity of the "solar contribution" to assumed operational conditions. A trial run conducted for this project together with results from previous simulation studies had already shown that if significant levels of overshadowing and/or heavy use of net curtains were to apply during sunshine hours, or if very high internal gains were assumed, the contribution from solar gains diminished significantly, the scope for designers investing time or capital on special measures for passive solar heating became questionable, and the need for simulation negligible. These observations led to the following guiding principles and objectives for simulation studies for the Design Guide:

(a) Adopt desirable conditions for passive solar heating
It was felt important from a design viewpoint to identify the likely contribution from passive solar heating under conditions conducive to a meaningful operation of its physical processes and in the light of current trends toward higher standards of thermal insulation in the UK. Clearly, if no noticeable benefits were to be found by varying a design parameter under favourable conditions, it could be safely concluded that performance would only worsen under unfavourable conditions.

(b) Bias elimination
Trends and interpretations emerging from parametric runs should be as free as possible from the biasing effects of hidden parameters, underlying assumptions or unmodelled exchanges or variables.

(c) Create an interpretation chain

TABLE 1. PARAMETER VALUES FOR PARAMETRIC STUDIES

INSULATION SCENARIOS (opaque elements)
- 1982 Building Regulations minimum requirements (applicable at the time of the runs)
- Linford base house:
  Wall U=0.34; Roof U=0.26; Floor U=0.9
- 1989/90 revision of Building Regulations:
  Wall U=0.45; Roof U=0.25; Floor U=0.45
- "Superinsulated" :
  Wall U=0.22; Roof U=0.22; Floor U=0.32
  (U-values in W/m2 K)

VENTILATION RATES
- 1.0 ach
- 0.5 ach

WINDOWS
Areas and Distribution by Orientation
- due south (also south-east and south-west):
  6 m^2, 21.6 m^2, 30 m^2
  with northern windows fixed at 6 m^2, 10% of gross floor area;
- due south and due north, 11.5 m^2 on each orientation;
- due east and west, 13.75 m^2 on each orientation.

Glazing Type and Night Protection
- single glazing with night curtains;
- double glazing with night curtains;
- double glazing with 1.0 m^2 K/W shutter at night;
- low-emissivity double glazing with night curtains.

HEATING SCHEDULES
- 2-period (9 hours/day)
- 1-period (16 hours/day)
- continuous (24 hours/day)
- no heating

HEATING SETPOINTS
- Living/dining room zone 21 °C; all other 18 °C
- Living/dining room zone 18 °C; all other 16 °C

WEATHER DATA
1967 weather year for:
- London, Kew (51.5 °N)
- Aberporth (52 °N)
- Eskdalemuir (55 °N)

- Kew Test Reference Year (PCL TRY)

INTERNAL GAINS
Based on measured data from Milton Keynes Linford houses.

SKYLINE OBSTRUCTION
5° (and sample runs for 10° and 30°)

It was thought desirable, where possible, to establish some form of interpretative link with previous work, as well as maintain such links between subsets of our own studies.

(d) Encompass a wide range of design configurations
This was partly because previous studies had only provided fragmented snapshots of the domain, and partly due to the fairly pessimistic results they gave especially for smaller semi-detached and terrace houses.

Following from the simplified runs a set of less than 50 parametric permutations were found to be sufficient, in the first instance, to represent the combinatorial of design configurations for a base detached house. Subsets of those permutations were selected for SERI-RES runs under three different heating schedules for the three UK locations for which weather data were available at the time. In addition two small sets of runs were defined to study the effect of overshadowing and the contribution from attached conservatories (these were complementary runs to be interpreted in conjunction with results from previous studies). In total some 320 full-year runs were chosen out of a combinatorial of around 10,000 for a detached house. However, it is believed that the results obtained allow interpretation over the whole of this domain, as well as providing some meaningful insights on likely trends for other house types. A further 90 runs were performed on a few selected design variants for short reference periods chosen to represent typical cloudy, average and sunny winter days, and average and peak summer days. These were for the purpose of providing hourly zone temperature profiles to be used as illustrations in the Design Guide.

An important consideration in preparing for these parametric runs was that access to the minicomputer version of SERI-RES, which was to be used as the simulation model for the studies, was through the Performance Analysis Service (PAS) provided at the time by CAP Scientific to all the projects under the UK Passive Solar Design Programme. This meant that the runs could not be undertaken interactively in small batches, but had to be defined and run in bulk. Table 1 provides a summary of the main parameters varied in the studies. The base house for the parametric studies was taken as a 2-storey, 4-bedroom detached house with a total floor area of 110 m^2; one of the variants examined matched the characteristics of the Milton Keynes Linford houses for which detailed SERI-RES files already existed. However, the original 13-zone SERI-RES model of the house required 45 minutes for a complete full-year simulation on the PAS minicomputer operated by CAP Scientific and was considered impractical for the parametric runs. After some testing, this was replaced by a three-zone model which reduced simulation time to ten minutes per run, /15/.

3.2 Processing of Results, Design Guidelines and Design Aids

The results obtained from the parametric runs conducted so far present variations in heating energy ranging from close to 14,000 kWh to well below 1,000 kWh. This is as wide a range as to encompass the heating requirements of most house types of recent or near-future construction and for most UK locations and heating regimes. For London weather data (Kew 1967 weather year), a 2-period intermittent heating regime (of a nominal 9 hours/day on duty) and constant internal gains, the range of variation is still of the order of 8.5:1. Figure 6 maps this domain as resulting from the permutations of three thermal insulation scenarios, four glazing technologies and increments in south window areas (for fixed north glazing). The results obtained under a 1-period intermittent heating regime (16 hours/day on duty) and under continuous heating (24 hours/day) show that the trends expressed by the slopes of the curves in Fig. 6 are not significantly affected by change in heating regime. They are, however, sensitive to changes in the gains to loss ratio, and more particularly to the solar gains and heat loss characteristics of windows. This is the main factor underlying the flattening or upward change in the slopes of the curves as a function of increasing insulation for

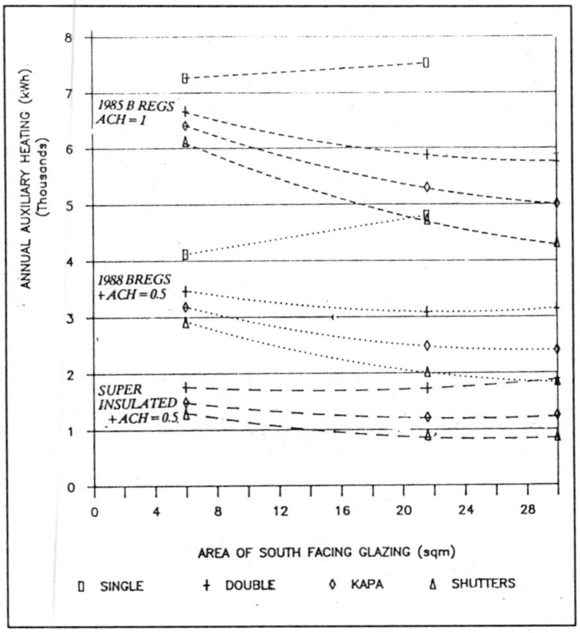

FIG. 6. Simulation results for London, Kew (1967 weather data) and for a 2-period intermittent heating regime showing effect of varying level of insulation and south-facing window area and glazing type.

opaque elements, and the downward trends resulting from improving the thermal resistance of glazing. The flattening of the curves indicates how increasing the insulation standard of the opaque envelope affects the "usefulness" of solar and internal gains in displacing heating fuel. This effect can be assessed from Fig. 7, which shows how a constant input of "raw" gains (same area of glazing, same solar input and same internal gains in all three runs) is transformed into different proportions of "useful" (fuel displacing), and "excess" (setpoint overshoots), gains as a function of increase in the standard of insulation of the opaque envelope (the results are for a 24-hour heating schedule). While increasing insulation yields an increasing proportionate contribution from solar and internal gains (rising from 61% to 83% for the variants of Fig. 7), with every increment in insulation a higher proportion of the gains becomes excess heat. The qualitative value and design significance of this excess heat are still open to argument, as to how such excess may represent unwanted overheating at certain times; a welcome higher temperature

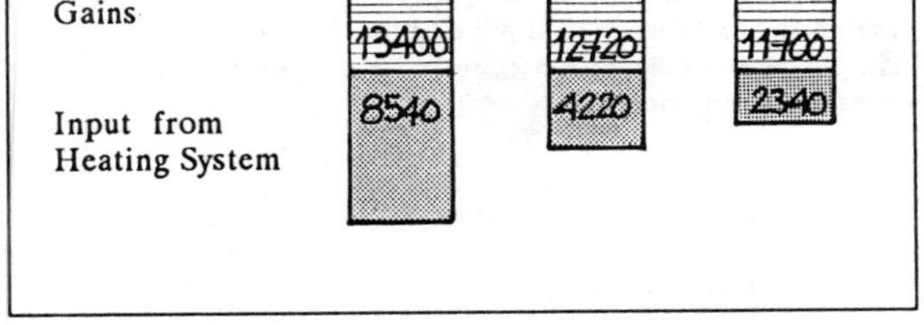

FIG. 7. Effect of increased level of insulation on useful solar and internal gains (values for Kew weather data and continuous heating with fixed glazing parameters and internal gains).

above the setpoint at others; or a source of surplus heat that might be used to meet part of the heat demand in other parts of the house (see /6/ for discussion). It is worth noting that for the superinsulated variant (170 W/K heat loss coefficient; 0.65 W/K m3 volumetric heat loss) the value of "excess" gains is substantially larger than the heat output required from the heating system.

The flattening and change in slope of the curves as a function of increasing insulation is shown across the whole set of heating schedules and locations. The variations in the gains to loss ratio also explain differences in trends found between the three different locations, and also between results for the two weather datasets used for Kew. Results for the Aberporth weather data showed higher energy saving potential from south-facing windows than at Kew, while those for Eskdalemuir were slightly worse; differences of the order of 25-30% in heating energy were found between Kew 1967 and Kew TRY weather data.

While the bulk of these parametric studies was run under favourable conditions of solar access, and for due south orientation, a small subset of selected runs proved sufficient to characterise the effects of overshadowing and of deviations from due south orientation. These were shown to lead to quite considerable increases in heating energy requirements. Severe overshadowing can eliminate the advantages of a southern aspect, and from an energy viewpoint the design implication is to advise for smaller window areas. With heavy overshadowing the choice of orientation becomes independent from energy criteria. On the other hand compared to unobstructed southern orientation, deviations from due south were found to incur significant reductions in useful solar gains, the more so the shorter the duration of the heating season becomes (ie. the higher the insulation standard).

The small set of runs on the contribution from conservatories were conducted with the same base house model, two sizes of conservatory and two glazing variants and were intended to provide a link with the more detailed results obtained for a semi-detached house by Baker, /16/, in his study on the pre-heating of house ventilation air through conservatories. The runs confirmed that ventilation pre-heating is the major contributing mechanism of conservatories with respect to heating energy savings. They also showed that this effect tends to diminish as airtightness of the parent building increases (as is the trend with recent low-energy housing in the UK).

Study and processing of the results of these SERI-RES parametric runs was undertaken in three stages:

- firstly, to help inform and direct the general thrust of the design advice given in Book 1 of the Design Guide;
- secondly, for the purpose of constructing a nomogram which traces the effect of window design parameters and house insulation standards on heating requirements;
- thirdly, to produce a simple manual calculation procedure for estimating seasonal heating requirements.

Design Principles and Guidelines

The numerical results of the parametric studies are not presented in the Design Guide but the key conclusions from subsets of parameter investigations were translated into brief statements as design guidelines, rules-of-thumb, or warnings and spread into the respective sections of the book (siting, internal planning, window design, conservatories, etc.).

Nomogram

The nomogram, Fig. 8b, was produced by linear regression from the simulation results. It is shown here in its preliminary form. The user input to the nomogram is a volumetric heat loss coefficient for a "reference house" (Building Regulations' minimum level of insulation, minimum glazing areas of 10%

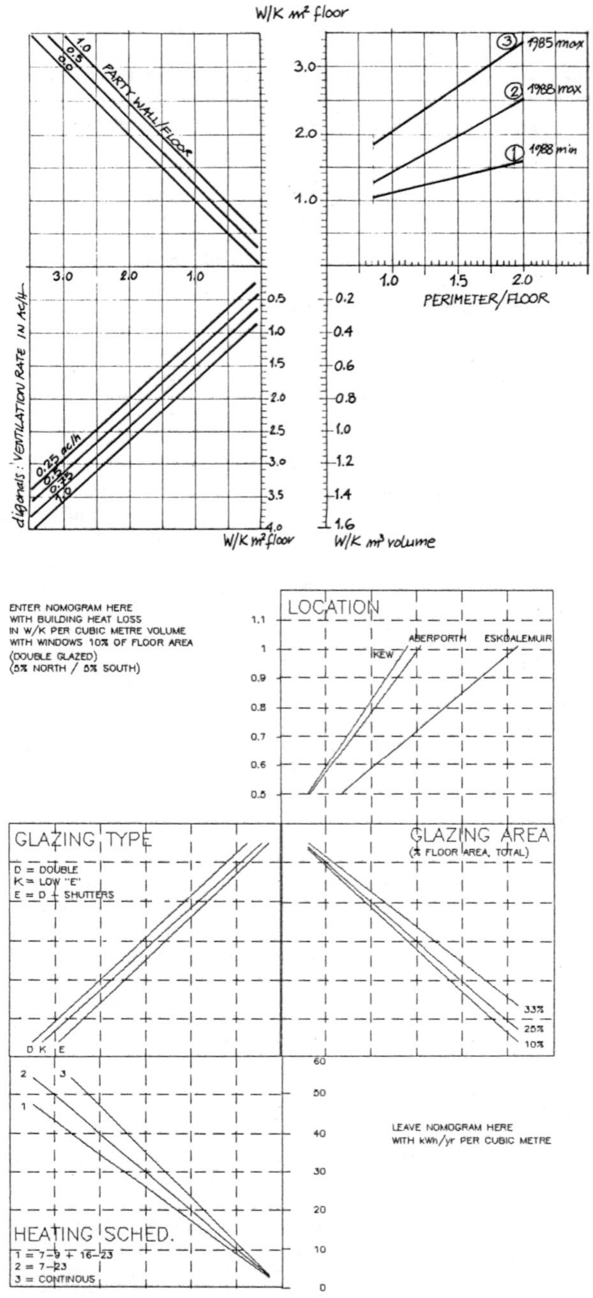

FIG. 8 Nomograms to estimate volumetric heat loss coefficient (top), used as input to obtain heating energy estimates as function of location, glazing parameters and heating regime (bottom).

of floor area for daylighting, and double glazing). The nomogram is intended to be independent of house type and size and to provide indication of whether there is scope (or freedom from energy criteria) to increase window area, or to adopt different glazing technologies. In the current stage of the project the nomogram is being extended to inform on issues of orientation and solar access as well. The reference volumetric heat loss coefficient does not require that a design already exists, or that a calculation is performed by the designer. Such a value can be derived from the information given in the design brief on the type of accommodation that is being sought, the basic requirements of the Building Regulations, and a reference value for air exchange. It was found that values of the reference heat loss coefficient also could be expressed in the form of a nomogram, Fig. 8a, from typical dimensions and relationships characterising the main UK housing types. In this way these nomograms can provide a simple yet powerful exploratory tool which can help inform the discusions with the client at briefing stage, and by providing an early assessment of likely options (this will be configured with basic cost-analysis in the next stage), help clarify priorities and give insights on where emphasis should be given in the course of design development and element design.

Seasonal calculation procedure

The seasonal heating energy calculation was developed around an annual energy balance equation of the form:

AUX = ALR * GHL * IHF (1)

where,

AUX = heat input from mechanical system, in kWh;

ALR = Auxiliary to Loss Ratio, is the fraction of the gross heating load to be met by the mechanical heating system; this is given by the correlation algorithm (see below) as a function of the Gains to Loss Ratio (GLR, the sum of gross annual solar and internal gains to the gross annual heating load);

GHL = gross annual heating load under a continuous heating regime, in kWh;

= 0.024 * 365 * HLC * (Ti-To) (2)

where,

HLC = whole house heat loss coefficient (including ventilation conductance), in W/K;

Ti = mean 24-hour whole house temperature under continuous heating regime, °C;

To = mean annual outdoor temperature, °C;

IHF = a factor for intermittent heating (see below for derivation);

= 0.85 for a 1-period heating regime of 16 hours/day

= 0.72 for a 2-period heating regime of 9 hours/day.

The existence of some meaningful correlation between ALR and GLR was already apparent from study of the simulation results on plots such as those of Figs. 6 and 7, and from the literature, /17/. To establish the correlation, values of GLR were computed by taking the solar and internal gain values from the SERI-RES annual summary output, and calculating the gross annual heating load under continuous heating regime with equation (2). The 24-hour whole house temperature was taken as 18.75 °C (which matches the mean value of setpoint temperatures for the three-zone SERI-RES model). ALR values were obtained by dividing the values of seasonal heating energy calculated with SERI-RES by the gross annual heating load (obtained as above). The resulting plot of ALR against GLR values is shown in Fig. 9. The best fit algorithm found for the correlation was:

ALR = 0.913 * exp (-1.137 * GLR) (3)

A comparison of heating energy values calculated with the correlation algorithm and those obtained from the SERI-RES simulations is shown in Fig. 10. On average the correlation algorithm produces results that are within +/-15% from the SERI-RES predictions. Several other mathematical expressions were tried, including correlation of GLR with a "utilisation factor" for solar and internal gains. However, the correlation with ALR was preferred because it suggests more immediately, through visual inspection of the exponential curve on a graph, how changes in GLR result in higher or lower input from the conventional heating system. It also simplifies the calculation required from the user to a level of mental arithmetic.

The intermittency factors were obtained from SERI-RES results for the two intermittent heating regimes by linear regression against the results for continuous heating and were rounded for manual calculation to the values given above.

The entire procedure (which also includes a step for estimating fuel costs) fits on a single A4 sheet and allows the user to estimate heating energy requirements and fuel costs

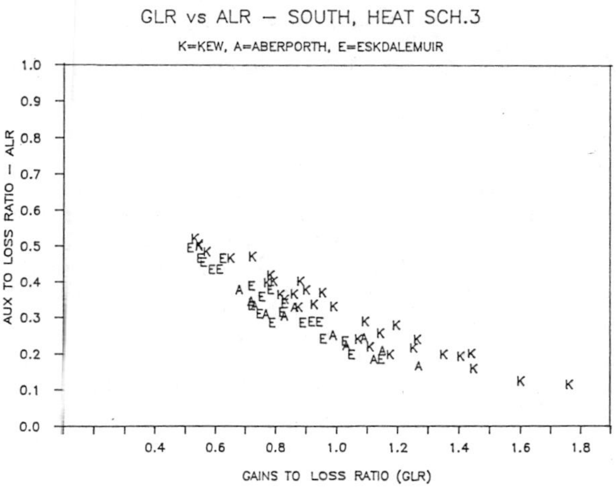

FIG. 9 Plot of ALR versus GLR.

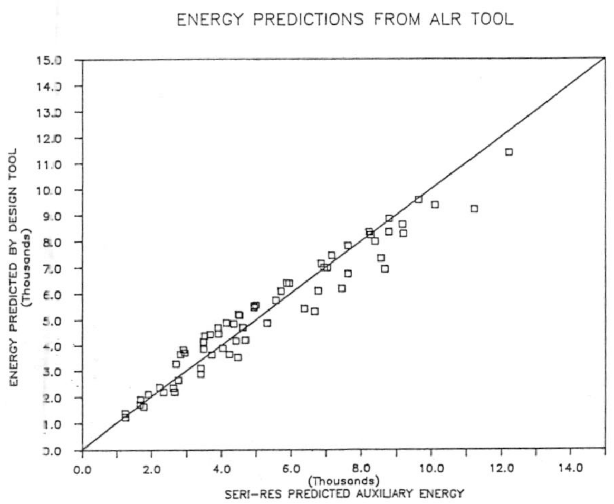

FIG. 10 Predictions from simplified method compared to SERI-RES results.

for the three different heating regimes in seven simple steps. For manual application the value of ALR is read off a graph as a function of the calculated value of GLR. The simplified procedure takes no longer than five minutes to apply manually; in its spreadsheet version, recalculation is instantaneous, and the time required only depends on the speed with which the user can enter the inputs (typically less than one minute). The advantages of this calculation method over the nomogram are that it is open to user inputs of indoor and outdoor temperature, solar radiation, orientation and overshadowing in addition to retaining all the other features of the nomogram. All of the input data necessary for carrying out a calculation are supplied in a 4-page Appendix to Book 1 of the Design Guide. This includes tabulated factors to account for the proportion of incoming solar radiation that is lost from the building as a function of glazing area and the reflectance of internal surfaces. These factors were derived by repeated runs of algorithms developed for calculating the solar-lost and solar distribution factors for input to SERI-RES /18/. Weather data (annual values of mean outdoor temperature, and solar radiation transmitted through single and double glazing for eight orientations) are given for seven UK locations; these were derived from the monthly data produced by Page and Lebens /19/. The effect of overshadowing on incoming radiation is accounted for graphically as a function of the obstruction angle. Further work is now in progress to improve the correlation and other aspects of the calculation method.

4 CONTINUING WORK ON THE HOUSING DESIGN GUIDE

4.1 Book 1

The current draft of Book 1 of the Housing Design Guide is structured into the following sections:

- Briefing
  principles of bioclimatic design for UK housing, design objectives, advice for pre-design briefing with client;
- Design Development
  siting, building form, internal planning and layout;
- Element Design
  window design, conservatories, opaque elements, heating systems and controls.

In addition, there are sections on design analysis which provide an introduction to performance assessment techniques, the design aids presented above and their application.

The objectives of the current (final) stage of work on Book 1 of the Design Guide were established in conjunction with the project's Advisory Group and the feedback from designers and students:

- to extend and consolidate the design guidelines and prescriptive information given in the book.
- to improve the credibility and usability of the nomogram and calculation procedure;
- to provide additional coverage of selected topics including: the effects of overshadowing on typical site layouts; ventilation; conservatories; roofspace collectors; thermal bridges; heating systems; cost analysis.

Currrent work on the project includes sensitivity analyses and some further parametric runs; a series of expert group meetings for resolving issues of guidance in areas of uncertainty; revision of text of the books; and further development of the graphical presentation and illustrations of the books.

4.2 Book 2

Since the beginning of this project our database on low energy housing examples has expanded to some one hundred schemes in the UK and from other European countries with similar climates. Unfortunately, from a designer's viewpoint, few of these passive/low energy buildings stand out as exemplars in the sense of combining outstanding design quality, satisfactory technical performance and adequate documentation. This is not too surprising, since were it to be otherwise there would be little point in spending money on further fieldwork or on producing design guides. The current objective for Book 2 is to provide a critical but fairly comprehensive review of the range of design possibilities across house types, acting essentially as an illustration of the principles presented in Book 1. Due to lack of detailed data on many of the schemes and difficulties with providing comparative performance indicators, firm generalisations are not possible but certain general conclusions are emerging by house type and with respect to particular technical features.

ACKNOWLEDGMENTS

The work presented in this paper is being carried out under contract to the Energy Technology Support Unit (ETSU), Harwell in the context of the UK Department of Energy Passive Solar Design Programme. The views expressed in this paper do not necessarily reflect those of ETSU or the UK Department of Energy. I would like to thank the project officer at ETSU, Dr Jim Riddle, for his continuous support. Also to acknowledge the contribution by Dr David Newton,

Martin Johnson and Dr David Bartholomew of Yard Limited (previously CAP Scientific) in the preparation and running of the first series of parametric studies, and the support of Dr Brian Stringer, then responsible for the Performance Analysis Service at ETSU; Dr Paul Ruyssevelt, Halcrow Gilbert Associates, for his help with the preparation of the nomogram and the second set of parametric studies, and for many useful discussions; Chris Martin, Energy Monitoring Company, has helped with the correlations for the seasonal calculation procedure.

REFERENCES

[1] Yannas, S. and K. McCartney (1985). Definition Study: Passive Solar Design Guide. Final Report to the Energy Technology Support Unit, UK Department of Energy Passive Solar Design Programme. Environment and Energy Studies Programme, AA Graduate School, London.

[2] Yannas, S. (1988). Passive Solar Energy Efficient House Design: Principles, Objectives, Guidelines. UK Department of Energy Passive Solar Design Programme. Environment and Energy Studies Programme, AA Graduate School, London.

[3] Yannas, S. (1989) Housing Design Guide Book 2: Examples. UK Department of Energy Passive Solar Design Programme. Environment and Energy Studies Programme, AA Graduate School, London.

[4] Yannas, S. (1989). Science in a Regional Approach to Architecture. In Los, S. (Ed) L'Architettura del Regionalismo. Franco Muzzio, Venice.
See also: Yannas, S. (1989). Physics and Architecture: Issues of Knowledge Transfer and Translation to Design. In "Interaction between Physics and Architecture in Environment-Conscious Design". Special Issue of Solar & Wind Technology, Pergamon Press, pp 301-308.

[5] See Annual Project Reviews of the Environment and Energy Studies Programme, AA Graduate School, London.

[6] Yannas, S. (1986) Design, Thermal Modelling and Performance Assessment Criteria for Passive Solar Houses. Working Paper 1, Housing Design Guide Project, UK Department of Energy. Environment and Energy Studies Programme, AA Graduate School, London.

[7] The methodology of the Phase 1 House Design Studies is described in reports by Davis Belfield & Everest and by CAP Scientific produced between 1983-86.

[8] A5 Architects (1984-86) Phase 1 House Design Studies. Unpublished compilation of summary reports on individual design schemes. UK Department of Energy Passive Solar Programme.

[9] Haves, P. and J. Littler (1987). Refinements to SERI-RES. ETSU-S-1130. Research in Building Group, Polytechnic of Central London.

[10] Bloomfield, D.P. (1986) The Use of Thermal Models for the Production of Design Guidelines. Proceedings of the International Climatic Architecture Congress, Ciaco Editeur, pp 215-236.

[11] Bloomfield, D.P. (1986). The influence of the user on the results obtained from thermal simulation programs. Proc. of CIB Fifth International Symposium, pp 161-180.

[12] Everett, R. et al (1985). Linford Low Energy Houses. Final Report. Open University ERG 50, ETSU-S-1025.

[13] Yannas, S. (1986). Simplified Thermal Analysis Methods in Passive Solar Design and Evaluation. Working Paper 2, Housing Design Guide Project, UK Department of Energy Passive Solar Programme. Environment and Energy Studies Programme, AA Graduate School, London.

[14] Yannas, S. (1987). Simplified Parametric Studies. Housing Design Guide Project, UK Department of Energy Passive Solar Programme.

[15] Newton, D.H. and M.P. Johnson (1988). Linford Passive Solar Houses: Credibility Review of SERI-RES Prediction by Comparison to Readings from Monitored Buildings. CAP Scientific Limited, London.

[16] Baker, N.V. (1985). The Use of Passive Solar Gains for the Pre-Heating of Ventilation Air. Research Report, ETSU UK Department of Energy Passive Solar Programme.

[17] van Dijk, H.A.L. and C.A.M Arkesteijn (1987). Windows and Space Heating Requirements: Parameter studies leading to a simplified calculation method. IEA Annex XII. TNO Institute of Applied Physics TPD.

[18] Martin, C. (1986). SOLDIST Program. Prediction of solar-lost and solar distribution values for SERI-RES based on results of test cell experiments. UK Department of Energy Passive Solar Programme. Energy Monitoring Company.

[19] Page, J. and R. Lebens (Eds. 1986). Climate in the UK. HMSO.

DEVELOPMENT OF PILOT EXPERT SYSTEM AS AN INTEGRATED DESIGN TOOL FOR THE DESIGN OF PASSIVE SOLAR COMMERCIAL BUILDINGS

O.C. Mørck
CENERGIA ENERGY CONSULTANTS
Stationsvej 3, DK 2760 Måløv
Denmark

ABSTRACT. The paper describes the development of a pilot expert system developed to illustrate and prove the concept of an "integrated passive solar design tool" to be used by architects and engineers. The concept includes an existing CAD system, existing programs for daylight calculation and thermal simulation, and integrates these with an expert system module for assistance to the building designer in choosing the right components for a particular building design. The developed system further includes a module for building cost estimation. The development hardware platform is a standard PC-AT computer using the DOS operating system. The paper summarises the conclusions and ends up with recommendations for future work.

1. INTRODUCTION

1.1 Design evaluation

The design of buildings involves both qualitative evaluation and quantitative evaluation. The visual evaluation of a perspective of a proposed design is an example of a qualitative evaluation. Quantitative evaluations are traditionally performed by engineers calculating the acoustics or ventilation needs. Another, in this context less thought of quantitative valuation is the quality check against building standards and regulations. Both the qualitative and quantitative evaluations are important stages of the design process.

1.2 Background

Computer aided draughting and visualisation systems are becoming more and more common. Often they are given the designation "computer aided design systems". This designation should, however, rather be reserved to systems which includes more than just draughting and visualisation.
Various research communities have developed many advanced (quantitative) design tools. However, they are rarely used by the practitioners, because of cumbersome user interfaces,e.g. many, detailed data needed, difficult results interpretation schemes. Their use are simply too timeconsuming to allow their in the design of normal building projects.
A truly useful computer aided design tool will integrate the the qualitative and quantitative design evaluation capabilities into one, simple to use, effective design tool.
Within the Artificial Intelligence research and development community recent developments have resulted in highly user-friendly and extremely efficient expert system development tools and environments.
At the same time the computer industry have taken "quantum leaps" with respect to the performance/price relationship of personal computers. The personal computer have become a powerful work station within the reach of even small architectural companies.
The dream of an "ideal" truly useful computer aided design system combined with rapid development within the area of personal computing formed the background for the development of a pilot system of an integrated design tool for the design of passive commercial buildings. The system was given the name "EKSPRO", which is the Danish acronym for "expert system building designer".

1.3 Project organisation

A project team was organised consisting of:
-an architectural company
-an engineering company
-a research professor in the field of expert systems
-a company producing HVAC equipment
-The Danish Technological Institute, department of heating and ventilation
-an energy consultance company (project leader)
The development project was partly funded under a special programme of the Danish Industrial Development Board called the TUP-programme. The TUP-programme (Technological Development Programme) was initiated to support the introduction of new information technologies in Danish companies. The total project cost was 100 000 ECU. The project was initiated in May 1987 and finalised in February 1989.

2. OBJECTIVES OF THE EKSPRO PROJECT

2.1 Objectives

The objectives of the project was to develop a prototype expert system to assist an architect/engineer in the design of commercial buildings with respect to thermal comfort, daylight quality and energy consumption, paying special attention to passive solar energy gains.
The aim is to present an expert system which can be used as an integrated part of the design process.
The primary purposes of the project was to demonstrate and prove the concept and to uncover the problem areas to prepare for the development of a complete system. Besides the development of the actual system the emphasis was on the process of development itself, to document the problems met and the chosen solutions.

2.2 Relation to other projects.

Other workers have developed expert systems to be used for particular purposes in the building design process. In brief these are: Intelligent user interface to building simulations(1), design of HVAC systems (2), the generating of building designs based on a simple module (3), and representation of generic design standards (4). in contrast to these systems the basic idea of the EKSPRO project was to investigate the concept of an "integrated tool-box" for building design.

3. THE STRUCTURE OF EKSPRO

The idea of the project was to combine an existing CAD system (SCRIBE MODELLER), an existing daylight calculation module, and an existing thermal simulation programme (SUNCODE PC) with a new expert system. This combined system appears to the user as one integrated system.

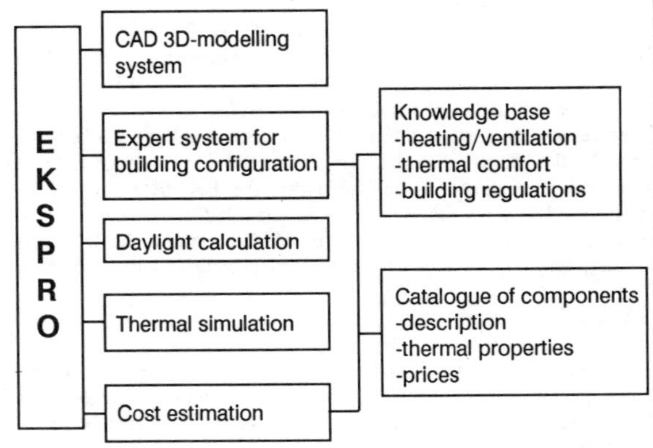

Fig. 1 Ekspro system architecture.

Because of the limited budget available and because the main purpose of the TUP-programme was to gain experience with the expert system technology, the main emphasis of the project was planned to be placed on the development of the expert system module with minimal effort devoted to the necessary interfaces.
The core of the system from this viewpoint is the expert system which besides a number of rules for good thermal design contains building regulations and access to libraries of building components, including thermal physical and cost data. This part of the system is programmed in prolog. On figure 1 above this part is shown as the "expert system for building configuration".
The system is developed under the DOS operating system and thus suffers under the limitations imposed by DOS (640 kB RAM, not multi-tasking). Each application, CAD system, expert system, daylight calculation, thermal calculation and cost estimation has to be executed sequentially. A menuprogram was developed to assist the user in picking the applications. The expert system was developed in Prolog (II) and can be characterized as an open frame structure holding extensive expansion capabilities. As implemented in the prototype the system holds rules for good thermal design, building regulations and links to libraries of building components containing thermal physical properties.

4. THE USE OF THE EKSPRO SYSTEM

Each session of use begins with a sketch of the building design using the CAD system. After that the user has the choice entering into one of the other modules. Upon entering into the expert system the user has the choice of using it as a thermal design advisor for a particular design and choice of building components, or letting the system undertake a complete automated configuration of a selected subset of building components. What the system actually does is, that it calculates development of the room air temperature over a given day. If the temperature raise is too high the evaluated configuration is discarded.
If the system is performing the automated configuration a search for a new combination is made and the next configuration is evaluated. Besides the thermal evaluation the system holds possibilities for establishing relations between certain components, which e.g. prevents one type of wall component to be used along with a given floor component. After the thermal evaluation the user is advised how to change the design. This could for example be to use more efficient artificial lighting systems or lowering the solar gains.
The system also checks the design for conforming to the building regulations with respect to room height, volume per person, and size of windows for fire escape.
Also, some general advise on choosing a heating and ventilation system, is provided.

5. CONCLUSIONS

5.1 Conclusions

The main conclusion from the project was that the established concept proved viable. This was confirmed, not only by the building designers participating in the project, but also by designers to whom the system was presented at different seminars.
An important conclusion is, that even for knowledge-based systems, calculations are fundamental. Very little concerning building thermal performance can be expressed in simple rules on a high level. The development of a rating system for strategy selection might be a solution. At a lower level (for the details) simple rules can still be adequate.
In this light the use of expert systems as intelligent front ends to detailed simulation models to assist in the generation of fail-proof input data files and in the evaluation of results are very useful.
With respect to the user interface it is essential to devote a large effort to make the programs straight-forward to use and to integrate the conventional programs. It is envisaged that the EKSPRO system can be transported to a multitasking environment (OS/2) and thus allow for a more dynamic user interaction. This will also remove the DOS RAM-barrier, which again allows for a better integration of the programs.

5.2 Lessons learned

It is generally said that knowledge acquisition is a bottleneck for the development of expert systems. This was also the case for this project. Part of the problem was due to communication problems between the building designers on one side and the computer scientist on the other. The knowledge for simple rules were not a problem, but for complex strategies and constraints among components it was.
The project also suffered from limited financial resources. A more powerful computer hardware platform with a more flexible operating system would have eased the development considerably. Likewise more resources could have been spent on investigations of different development possibilities and especially on a refinement of the user interface through better integration of the programs.
Finally, the 20 % - 80 % rule was once again confirmed. In this case it turned out that 80 % of the work was better solved with conventional procedural programming and 20 % using the prolog software. This is, however, to a high degree dependent on the choice of expert system development software.

5.3 Future development

The EKSPRO system needs as described above a better user interface with a better integration of modules. As correct cost estimation is of major importance to builders this area will be emphasised in future development. This covers maintenance and energy cost as well construction cost. The areas of correct choice of HVAC equipment and artificial lighting systems are areas which traditionally are timeconsuming task of the design process. Therefore, it is intended to develop EKSPRO to better cover these areas, again with respect to design, maintenance and energy cost.

6. RECOMMENDATIONS

The concept of "integrated design tools" is now under development by several research groups all over the world. The idea is still new, and no complete system has been developed and marketed. This means that there is a very good opportunity for establishing an international forum for exchange of ideas and plans for development of tools following this concept. One way of doing this could be to initiate semiannually work shops either separately or connected to other conferences to meet and discuss this subject at the level of working details, demonstration of prototypes, etc.
As work is going on so many different places, there is an urgent need for a standard way of representing buildings in these systems. An internationally agreed to standard (perhaps on an object-oriented database) would make it possible for the different research groups to exchange programs, mutually test these and even produce integrated systems based on program modules developed by the different research groups from all over the world.

7. REFERENCES

(1) Clarke, J.A., "An intelligent Approach to Building Energy Simulation", Expertsystems for Construction and Engineering Services, London 1986, p. 73-88.

(2) Doheny, J.G. & P.F. Monaghan, "IDABES: An expert system for the preliminary stages of conceptual design of building energy systems", Artificial Intelligence in Eng., 1987, vol. 2, No. 2

(3) Jackson, B., "Development of a Knowledge-based designing system", First national conf. on computer applications, Lagos 1985.

(4) Rasdorf, W.J. & T.E. Wang, "Generic Design Standards Processing in an Expert System", Journal of Computing in Civil Engineering, 1988, vol. 2, No. 1.

APPLICABILITY STUDY 1: A UK INITIATIVE TO DETERMINE THE ERROR CHARACTERISTICS OF DETAILED THERMAL SIMULATION MODELS

K.J. Lomas[1], D.P. Bloomfield[2], F. Parand[1] and A. Pinney[2]
[1]Leicester Polytechnic, P.O. Box 143, Leicester, LE1 9BH, U.K.
[2]Building Research Establishment, Garston, Watford, WD2 7JR, U.K.

ABSTRACT. Applicability Study 1 is a two year UK initiative aimed at enhancing the useability and credibility of Detailed Simulation Models (DSMs) for the design of passive solar dwellings. The joint Leicester Polytechnic, Building Research Establishment project will work with ESP, HTB2 and SERIRES. It will highlight the areas of agreement and disagreement in the design trends predicted by DSMs, provide guidance on the optimum modelling technique, indicate the inherent uncertainty attributable to DSM predictions, identify the attributes of DSMs necessary in order to obtain reliable results and indicate where additional theoretical or experimental research is needed. Early results showed improvements over those obtained in earlier studies by ETSU but there are still underlying differences in the predicted design trends and energy saving predictions. These were attributed to the different algorithms employed by the DSMs.

1. INTRODUCTION

Detailed thermal simulation models (DSMs) play a key role in the UK Department of Energy's (DEn) Passive Solar Design Programme (PSP) as they provide the most cost-effective way of resolving the key issues encountered in the assessment of building designs. The Energy Technology Support Unit (ETSU) acting on behalf of the DEn, is currently evaluating three DSMs: ESP, HTB2 and SERIRES.

In a recent study, expert users of each of these models were requested to predict the annual energy use of a passive solar house, the Linford house /1/. Unfortunately, widely differing predictions were obtained for the absolute annual energy use and, more significantly, the trends in energy use as the window area and type were varied (Fig. 1). Such results seriously undermine the credibility of DSMs and pose a barrier to their widespread application.

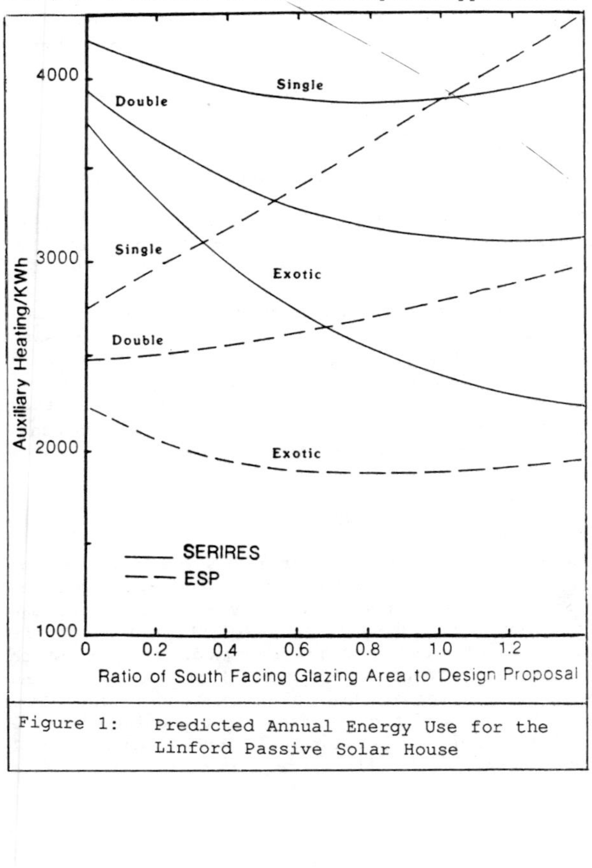

Figure 1: Predicted Annual Energy Use for the Linford Passive Solar House

A six man/year research initiative, Applicability Study 1 (AS1), has therefore been funded by the DEn. This research is being undertaken by Leicester Polytechnic (LP) with the Building Research Establishment (BRE) acting as a major sub-contractor. The aim is to enhance the useability and credibility of DSMs by mapping out their error characteristics. Specific objectives in order of priority, include:

(a) identifying the design problems for which the DSMs can be used with reliability and those for which they cannot;

(b) identifying the attributes which a DSM must possess in order to resolve each design problem;

(c) estimating the inherent uncertainty in the predictions of DSMs;

(d) providing guidance on the optimum method of using DSMs;

(e) identifying the need for further theoretical or experimental research; and

(f) identifying the features of DSMs which need further development and validation.

Given the ambitious goals of the project, it was necessary to limit the scope of the study to domestic buildings with direct gain passive solar features. Issues relating to daylighting and detailed plant component modelling were not addressed. The research programme has five distinct stages. This paper describes the overall research strategy, the analysis techniques and some of the results from Stage 1.

2. SELECTION OF MODEL EVALUATION TECHNIQUES

Five main techniques are available for assessing DSMs and from previous work the authors have a clear understanding of the strengths and weaknesses of these /2/. It is evident that analytical verification is inappropriate as the method cannot be applied to whole buildings and empirical validation is impractical because it is too time consuming. The remaining techniques, sensitivity analysis, parametric studies and intermodel comparisons have played a key role in AS1, but because none of them provide an exact truth model, it is not possible to identify whether predicted trends are correct or not. However, if all three DSMs produce similar results, despite their use of different algorithms and solution techniques, then confidence in the DSMs will have been increased. Where model predictions diverge, experiments or theoretical studies will be defined in order to permit empirical validation or analytical verification to be undertaken.

3. MODEL VISUALISATION

To rationalise the objectives of AS1, it is useful to visualise a DSM as 'black box' containing algorithms (Fig. 2). Some algorithms accept inputs, some produce outputs and some only interact with, and pass calculated values to, other algorithms.

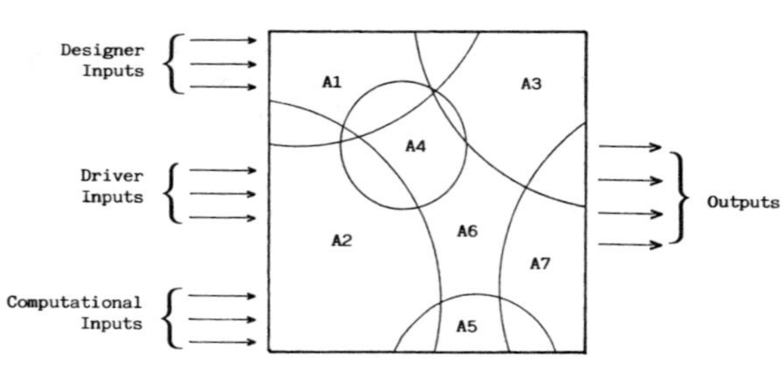

A1, A2, etc. are individual algorithms

Figure 2: Visualisation of DSMs

The designer inputs describe the building and are manipulated consciously by the model user to reflect the form of the building under investigation; in this sense they are determinate. The driver inputs describe the external and internal conditions to which the finished building will be exposed. These are indeterminate at the design stage and the robustness of a proposed design must be tested by subjecting it to a wide range of plausible weather and occupancy patterns.

The computational inputs are values which have to be given simply because a DSM is being used, they bear no physical relationship to, and have no influence on, the designer and driver inputs. They relate primarily to the conduction algorithms used in the models (e.g. node placement and time step).

The outputs describe the predicted performance of the building (e.g. internal temperatures and energy use). All DSMs provide a wide choice of possible outputs.

4. SELECTION OF DESIGNER INPUTS

Early work has focused on simulating a single zone within a house, the living room, with the intention that additional zones will be added later to study inter-zonal heat transfer phenomena. To conduct the parametric studies, it must be possible to vary individual aspects of the building independently of each other. Therefore, the designer inputs were divided into four 'Sections' and within each Section, a number of 'Aspects' which could be varied. By examining 'standard building descriptions' generated for DSMs by others /1,3,4/ it was possible to identify a number of 'Variants' for each Aspect (Table I).

A full set of designer inputs was produced by combining one Variant from each Aspect. Over 13,000 living rooms designs, which cover the likely range of current and future U.K. housing stock, can be investigated.

SECTION	ASPECT	VARIANT	DESCRIPTION
GEOMETRY	TYPE	GT1	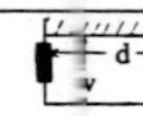
		GT2	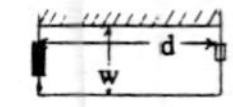
	Shape	GS1 GS2 GS3 GS4	0.5 1.0 } d/w values 1.5 2.0
	Area	GA1 GA2 GA3	$12\ m^2$ $16\ m^2$ } Area of floor $20\ m^2$
CONSTRUCTION[1]	Mode	CMA	Traditional heavyweight U = 1.4 Y = 3.8
		CMB	Modern heavyweight U = 0.5 Y = 2.3
		CMC	Modern lightweight U = 0.5 Y = 0.9
		CMD	Advanced heavyweight U = 0.3 Y = 2.4
WINDOWS[2]	Area	WA1 WA2 WA3 WA4	20% 33% } As percentage 45% } of floor area 55%
	Type	WT1 WT2 WT3	Single glazing, U= 5.6 Double glazing, U= 2.9 Low-emissivity, U= 1.6
	Orientation (Azimuth)	WO1 WO2 WO3	180° 90° 270°
PLANT & CONTROLS	Combination[3]	PCA PCB PCC PCD	mix/mix mix/convective air/mix air/convective
	Capacity[4]	PC1 PC2	Sized Infinite

[1] External wall, internal wall and floor constructions vary with method. Thermal transmissivity (U, $Wm^{-2}{}^{\circ}C^{-1}$) and admittance (Y, $Wm^{-2}{}^{\circ}C^{-1}$) quoted for walls only.

[2] Details relate to shaded window (see GT1, GT2)

[3] Variants relate to control/plant. Mixed plant produces 60% radiant and 40% convective heat, air is 100% convective. A mixed control senses 60% mean radiant temperature and 40% air temperature, air control senses 100% air temperature.

[4] Sized means plant capacity sized by steady state method, infinite means plant has unlimited capacity.

Table I: Designer Inputs

5. SELECTION OF DRIVER INPUTS

Single design day weather data /5/ was used for the early AS1 work to keep the simulation times short. This permitted numerous intermodel comparisons to be made so it was possible to determine the meteorological conditions for which the DSMs produced good agreement and the conditions for which they did not.

The inputs describing the internal conditions were, as for the designer inputs, devised by reviewing standard building descriptions /1, 3, 4/. They span the range of conditions likely to be encountered in the current and future UK housing stock and they permit over 250 different combinations of internal and external conditions to be investigated (Table II).

6. SELECTION OF OUTPUTS

The principal objectives of a passive solar design is to maintain comfortable conditions with minimal auxiliary energy consumption, by maximising the use of adventitious (particularly solar) gains. Of the numerous physical and psychological factors which affect our perception of comfort, only two, air and mean radiant temperature, are generated by the DSMs. In the UK, these two are frequently averaged to give dry resultant temperature. Hourly values of 11 parameters were therefore produced for each

SECTION	ASPECT	VARIANT	DESCRIPTION
WEATHER[1]	Type	MT1	Cold, cloudy winter day, WISZ
		MT2	Cold, sunny winter day, WICL
		MT3	Typical winter day, WIAV
		MT4	Cool, sunny spring day, SPCL
		MT5	Typical spring day, SPAV
		MT6	Hot, sunny summer day, SMSZ
		MT7	Typical summer day, SMAV
OCCUPANCY	Schedule[2]	OSA	No occupants, no heating
		OSB	No occupants, continuous heating
		OSC	Two occupants, heated 07-09 & 16-23 hrs
		OSD	Four occupants, heated 07-23 hrs
THERMOSTAT	Setpoint	TS1	21°C
		TS2	18°C
		TS3	24°C
VENTILATION	Rate[3]	VR1	0.35 ach
		VR2	1.00 ach
		VR3	1.5/10.0 ach

[1] Single days representing complete range of UK conditions, DSM data files devised by Loxsom, WISZ etc. are the Loxsom code letters.

[2] Hourly profiles of curtain opening and closing and radiative, convective and latent gains defined for OSC and OSD.

[3] Fixed air change rates for VR1 and VR2, VR3 rate is 1.5 unless temperature exceeds 25.0°C when occupants open windows and rate increases to 10.

Table II: Driver Inputs

simulation and eleven daily performance metrics (DPMs) were generated from these to succinctly describe overall daily performance (Table III).

7. THE RESEARCH PROGRAMME

7.1 Stage 1: Basic Simulations

In Stage 1 the predictions of the three DSMs have been compared for a wide range of designer and driver inputs to identify conditions for which the DSMs produce concurrent predictions and the conditions for which they do not. It was also possible to gain some insight into the cause of divergent trends and hence to clarify the issues of primary importance for Stages 3, 4 and 5. During Stage 1, the modelling assumptions and the algorithms employed by the models were fixed. A brief description of the Stage 1 research procedure and some results is given below.

7.2 Stage 2: Sensitivity Analysis

The sensitivity of the outputs to the inherent uncertainty in the individual designer inputs is being quantified for a few carefully selected situations. The individual designer input to which the DSMs are particularly sensitive are being identified, so it should be possible to give guidance on the designer inputs have to be chosen with care and those which have little influence on the predictions.

The total uncertainty in model predictions, due to the inherent uncertainty in all the designer inputs, will also be determined. This will give an estimate of the minimum uncertainty which must be attributed to predictions and hence an indication of the maximum resolution capable from DSMs. This minimum uncertainty will provide a benchmark against which to compare the results from Stages 3, 4 and 5.

7.3 Stage 3: Computational Inputs

The aim of this stage is to give guidance on the selection of the computational inputs. Emphasis will be placed on optimising the choice to achieve accurate predictions with minimum computational effort.

7.4 Stage 4: Modelling Assumptions

When modelling any building the model user makes, with or without realising it, numerous assumptions. Typically, window reveals, furnishings, window frames and internal doors, are ignored. The influence of such assumptions will be examined by repeating some of the basic simulations with features that were previously ignored being modelled explicitly. Guidance in the level of modelling detail necessary in order to achieve optimum predictive accuracy will be produced.

7.5 Stage 5: Algorithmic Substitution

This is probably the most complex stage of the project. It involves substituting the algorithms employed in a DSM with alternatives, and repeating some of the basic simulations to assess the effects. Guidance on the algorithms which a DSM must employ in order to resolve specific design issues will be produced. It should also be possible to identify the need for new algorithms or further theoretical and experimental research.

8. STAGE 1 RESEARCH

8.1 Experimental Procedure

To identify the conditions under which the DSMs predict similar design trends, and the conditions for which they do not, a research method based on 'single parameter excursions' was adopted. A particular set of designer and driver inputs were selected for the first 'Base Case' simulation. Then, a particular Aspect (e.g. window area) was selected and a number of simulations are undertaken using different Variants of this Aspect (e.g. 20%, 33%, 45% and 55% glazing). All other input values were held constant, so only the one Aspect was varied at a time. Single parameter excursions of this type were made for all three DSMs and for each Aspect of interest until a complete 'Round' of simulations was complete.

This method produced direct information about the trends predicted by each DSM and an early indication of the Aspects for which agreement was good and the Aspects for which agreement was poor. No previous investigation has placed as much emphasis on comparing the trends predicted by DSMs.

Preliminary Stage 3 work indicated that 19 days pre-conditioning was necessary to obtain reliable results so the Stage 1 simulations ran for 20 days each with output being produced only for the final day.

8.2 Preliminary Studies

Modelling assumptions will be studied in detail in Stage 4, however, for the purposes of conducting the Stage 1 simulations, it was assumed that there were no site or facade features to cause shadows and that there were no internal doors or furnishings in the room. The thermal affects of the adjacent zones were mimicked by placing fictitious materials behind the partitioning walls and ceiling.

The models were used without any modifications to the algorithms and in the mode which a typical user would adopt. Difficulties were encountered when

Hourly	Daily
Auxiliary power input	Total energy use
Zone air Temperature[1]	Peak power output
Glass internal surface temperature	Mean, Max, Min Air Temp.[1]
Six opaque Surface Temperatures	Mean, Max, Min Mean Radiant Temp.
Mean radiant temperature	Mean, Max, Min Dry Resultant Temp.
Dry resultant temperature	(Time of occurance of Max & Min)

[1] For SERIRES replaced by enclosure temperature.

Table III: Outputs

trying to model the ground floor since none of the DSMs have algorithms dedicated to ground conduction. However, a realistic modelling approach was devised which could be implemented in an identical manner in all three programs and which did not necessitate an increase in the pre-conditioning period. Ground modelling algorithms are currently being investigated in a separate project at Leicester Polytechnic.

8.3 Minimising Experimental Error

It is difficult to create accurate and compatible input files for different DSMs and even with simple buildings there is plenty of scope for human error. Therefore, rigorous 'quality control' procedures had to be adopted. Central to these procedures was the completion of a data 'input proforma' for the base case buildings. In the proforma, values are given for every single input parameter required by each DSM and the pages are formatted so that the inputs required by different DSMs, for the same building feature, appear side by side. Omissions, discrepancies or incompatibilities in input data are therefore immediately obvious.

Each researcher manually transferred the relevant values from the proforma to one DSM and checked the input files created. These were then checked by the other two researchers in the team. The process of undertaking a base case simulation, the modification of the input files to conduct the other simulations in the Round, and the comparison and plotting of the outputs was fully automated to avoid human error.

All the model input files, the output files, and the graphs produced for each DSM, were stored both on computer tape and in hard copy form. This permitted any erronious simulations to be repeated and it provides an archive for any future model evaluation work.

9. SOME STAGE 1 RESULTS

The variation in daily energy use as window area and type are varied is discussed, for the three Rounds in which the buildings were continuously heated.

The base case designer inputs used in Rounds 2B and 5 were deliberately chosen to mimic the living room of the Linford house /1/ for which the very poor results were originally obtained (Fig. 1). In Round 2B the building was double glazed whereas in Round 5 it was single glazed. The base case building in Round 4 was of traditional construction with single glazing and a lower set point temperature and higher ventilation rate (Table IV). The results were produced by ESP version 6.8a, SERIRES version 1.2 and HTB2 version 1.0.

In all three Rounds, the cold sunny winter design day (MT2) was used for the base case in order to test the accuracy with which the DSMs predicted the interaction between solar gain and heating system power output. In all three Rounds, the DSMs predicted comparable diurnal variations of plant power output (e.g. Fig. 3) which indicated that the solar gain during the day was sufficient to maintain the set point without any contribution from the heating system.

Round	Designer Input		Driver Input	
R2B	CMD WA3 WT2		TS1 VR2	
R4	CMA WA4 WT1	GT1 GS1 WO1 PCA PC1	TS2 VR1	MT2 OSB
R5	CMD WA3 WT1		TS1 VR2	

Table IV: Base Case Input Rounds 2B, 4 and 5

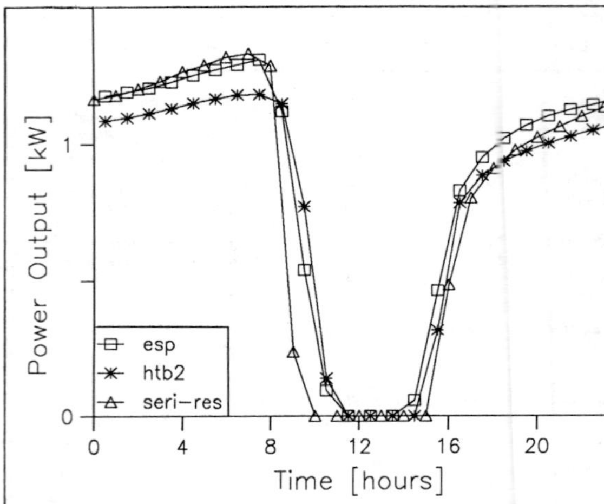

Figure 3: Predicted Hourly Power Output for Round 4.

Compared to the previous results (Fig. 1) the predicted trends are now in much closer agreement as are the absolute predicted values (Fig. 4). This improvement is attributed primarily to the rigorous quality control exercised over the simulations and the care taken to ensure compatible input data for all three DSMs. It seems therefore, that most of the discrepancies in the original Linford house results can be attributed to either poor problem specification, errors in data interpretation, or model user errors. There is though, a need to demonstrate that this closer agreement can also be obtained for complete multi-zoned dwellings.

Despite the improved agreement, there are still underlying differences between the predictions. For the heavyweight buildings (Round 2B and Round 5) the DSMs predicted generally similar trends in energy use as the window area varied (Fig. 4). For the double glazed building (Round 2B) all three DSMs predicting an energy saving of between 14% and 15% when the window area was increased from 20% of the floor area to 55%. For the single glazed heavyweight building (Round 5) the DSMs predicted the reverse trend and the energy savings predicted as window area decreased were more variable (between 8% for HTB2 and 17% for ESP). Combining single

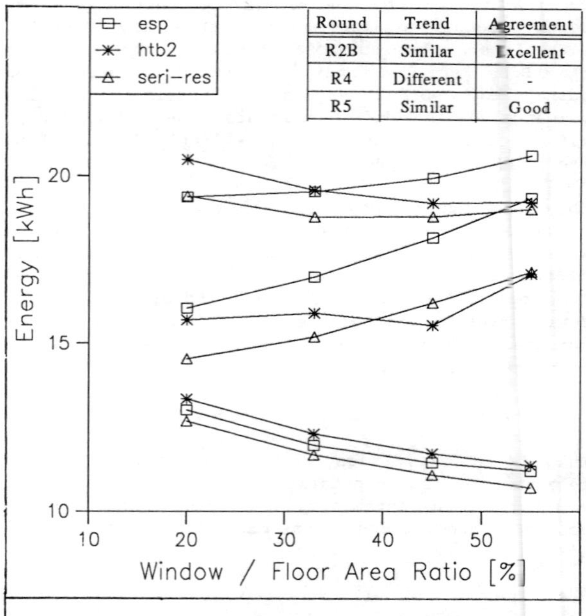

Figure 4: Predicted Influence of Window Area on Daily Energy Use.

glazing with a poorly insulated structure (Round 4) caused the three DSMs to show completely different trends. ESP predicted a gradual increase in energy use as window area increased, whilst SERIRES predicted a minimum energy consumption at 33% window area and HTB2 a minimum at 45%. Such results provide a model user with conflicting advice making it impossible to determine the optimum window area.

In all three Rounds, the models predicted the highest daily energy consumption for single glazing (WT1) and the lowest for low emissivity glazing (WT3). However, the predicted energy savings due to changing window type differed considerably depending on the model being used. For example, in Round 5 (Fig. 5), the energy savings predicted by the three models when changing from single to double glazing were 43%, 23% and 36% for ESP, HTB2 and SERIRES respectively. When changing from double glazing to low-emissivity glazing the corresponding values were 6%, 23% and 13%. Thus, a designer wishing to minimise energy consumption by altering the glazing type, would arrive at the same solution irrespective of the model used, however, the wide discrepancy in the predicted energy savings could result in a completely different estimate of the cost effectiveness of the solution.

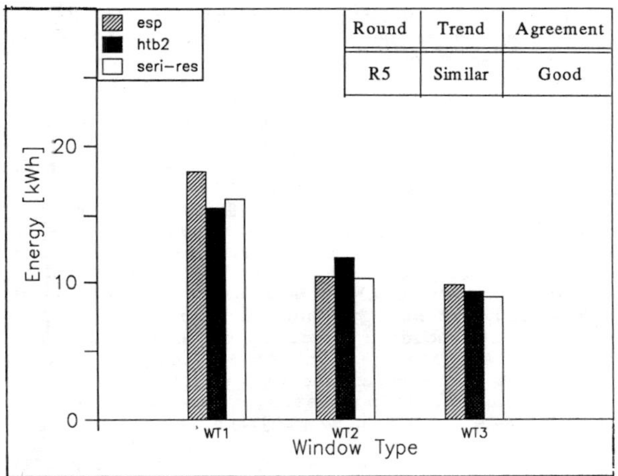

Figure 5: Predicted Influence of Window Type on Daily Energy Use.

Preliminary investigation of the divergent predictions suggests that they are likely to be due to differences in the way the DSMs model the thermal properties of windows.

The windows algorithms in the models will be studied further in Stage 5 of the work, however, ETSU have already commissioned field trials, and other modelling studies, to examine these algorithms in greater depth. Test cells with different window areas, window types (and plant/control strategies) have been monitored by the Energy Monitoring Company and these data are being compared with the predictions of DSMs made by Leicester Polytechnic.

10. CONCLUSIONS

1. The UK initiative Applicability Study 1 has succeeded in showing that the level of disagreement between the predictions of DSM can be minimised by careful problem specification and rigorous quality control procedures. The result obtained in AS1 are markedly better than those obtained previously.

2. Even for a very simple, single zoned building, the three DSMs, ESP, HTB2 and SERIRES, can produce conflicting design advice for some situations (e.g. the change in energy use as window area and window type are varied).

3. The divergent predictions are likely to be due to the differences in the algorithms dealing with the thermal properties of windows.

REFERENCES

1. Everett R. et al., Linford low energy houses. Final Report. Open University ERG 50, ETSU-S-1025 (1985).

2. Bowman N.T. and Lomas K.J., Empirical validation of dynamic thermal computer models of buildings. Building Services Engineering Research & Technology, Vol. 6, No. 4, pp 153-162 (1985).

3. BRE, Private Communication (1983).

4. BRE, Private Communication (1985).

5. Loxsom F., Meteorological data for the passive solar programme, Polytechnic of Central London, Research in Building Unit, Collection of reports for ETSU Model Refinement Study (1985-86).

ENERGY IN ARCHITECTURE:EC ACTIVITIES IN SUPPORT OF THE BUILDING DESIGN PROCESS

J Owen Lewis, Shane O'Toole
Energy Research Group, University College Dublin
Richview, Clonskeagh, Dublin 14, Ireland

ABSTRACT. During the period 1986-89 the SOLINFO and ARCHISOL projects have put in place mechanisms which have improved the efficiency of information flows from energy researchers to the European construction industry. The information requirements of the architectural profession have been clarified and various forms of carefully-structured material have been made available to practitioners and students including several innovative design tools. At a time of reduced general interest in energy issues, the attention of many European architects has been drawn to the opportunities which are available to design buildings which use a fraction of the energy of 'conventional' buildings, that provide more comfortable and better quality indoor environments, and which cost little or nothing more to construct. In the process, contributions have been made to an improved understanding of contemporary building design processes and research requirements. Thus the purpose of the research has been firstly to change the structure of passive solar energy research information to make it more readily communicable to particular groups of users, and secondly to enlarge the audience interested in energy conscious design by eliciting new responses in terms of changed perceptions and criteria for building design.

1. INTRODUCTION

The Commission of the European Communities has conducted research and development programmes on solar energy applications in buildings and the rational use of energy since 1975. Recent programmes have emphasised passive solar energy technologies. Since these are characteristically architectural design approaches, particular emphasis was placed in the now-concluding Third Programme on addressing European building design professionals and students, on improving access to and use of research results in education and practice, and in stimulating awareness of the potential of bioclimatic design.
The EC building industry had a turnover in 1987 of 364,000 MECU, or about 11% of the Community's GDP. Over 8,000,000 people are engaged in construction. There are over 1,100,000 building firms in the 12 Member States, 90% of which employ 10 or less people. Of particular importance in the present context, the architectural profession numbers over 200,000 in the EC.
Energy-efficiency and low energy design has yet to make a significant impact on mainstream architectural design. There is currently some revival of interest in energy as a design issue as part of the new interest in environmentally-benign technologies. Energy conservation techniques are, of course, of primary importance in energy conscious design, but usually have relatively low impact on the design itself. The rich design potential of bioclimatic approaches, coupled with their economic attractiveness has meant that this theme has figured largely in the undercurrent of interest in a more energy-efficient architecture and climate-sensitive design.

1.1 Third Non-Nuclear Energy R&D Programme

During much of the 1970s, EC research and development concerning the use of solar thermal energy in buildings emphasised active space- and water-heating systems employing sensible heat storage. In the early '80s active solar R&D activities were rapidly scaled down in most northern Member States. The R&D programmes of the Commission have reflected these changing priorities. In 1979 the CEC commissioned an investigation of passive solar activities in Europe, and subsequently two architectural ideas competitions were organised for the Commission, in 1980 and 1982. In 1981 the Commission placed a number of contracts as a result of a call for proposals aimed at the development and production of new low cost components to allow passive solar buildings to make more use of solar energy. Most work related to architectural components and ranged from new window and glazing systems to movable insulations to shutters and heat storage devices as well as novel controls.
The European Passive Solar Working Group and the European Solar Modelling Sub-group were established in 1981. Their work led to a series of workshops in the Member States during 1983 and 1984 and a draft European Passive Solar Handbook /1/.
In the Third Programme of 1986-89 some 85% of the solar heating R&D budget was allocated to passive research and technology transfer, compared with 20% in the previous programme. This programme has been described elsewhere /2/. Because of the nature of this energy technology and of the construction industry, it included several projects which used innovative measures to introduce research results, researchers and specialised professionals expertise to European building desingers. SOLINFO and ARCHISOL were two of these projects.

2. TECHNOLOGY TRANSFER FROM THE SOLAR R&D COMMUNITY TO THE EUROPEAN BUILDING PROFESSIONS: SOLINFO

SOLINFO aimed to promote the transfer of information from the solar research community to the building design and construction professions. A very great amount of information is generated in research and development on solar energy applications in buildings; yet this knowledge was not being implemented in the industry quickly or extensively enough. The overall objective of the SOLINFO project was to improve the understanding of the particular information requirements of the different building designers and constructors at the various stages of the building process, and to make available this information in forms convenient for use by practitioners and students.
A group of experts from the Member States advised on the specific actions undertaken through SOLINFO (see Appendix I). They also acted as a two-way channel for transmission of information coming out of the programme to researchers/professionals in their countries and for provision of feedback and other information from practice to the SOLINFO co-ordinators, the Energy Research Group at University College Dublin.
Specific tasks undertaken include:

- studies on information needs and working methods of practising architects;
- preparation of resource/information packs to aid teaching of passive solar design in architectural schools;
- holding workshops on daylighting techniques for mid-career professionals;
- distribution of the preliminary edition of the European Passive Solar Handbook and preparation

of a new revised edition;
- production and distribution of Research Digests;
- preparation of databases of EC contacts for use in dissemination of information from the CEC programme.

2.1 Architects' information needs and working methods

A number of analytical studies were carried out on the information needs and working methods of practising architects to establish criteria for the design of information for ready acceptability.
In the first study, material was reviewed included commentaries on the design process and design methodology, and studies of architects' use of technical information. From the threads which emerged, a set of overlapping criteria were developed and published in March 1988, under the title "Information requirements for architects" /3/. Another study, a pilot study in two parts, examined the work practices of some design leaders in passive and low energy architecture. The first part investigated the place of technical information in the design process of a number of leading international architects. The second was a series of structured interviews (based on themes identified in the first stage) with European designers of passive solar and low energy buildings. The results underline the importance of experience and the development of informed technical intuition, and the study is the subject of another paper at this meeting /4/.

2.2 Teaching packs for architectural schools

The lessons learned in the first study described above were put to use in the design of prototype teaching packs for schools of architecture. The aim is to make teachers aware of the possibilities of passive solar design and provide them with teaching materials where information is presented in a form easily understood by architecture students. The prototypes are being produced in English, French, German and Italian in the form of "valisettes" containing a range of thematically-linked materials in different media. Design of the prototypes was preceded by a workshop in Paris in October 1987 at which leading educators in passive and low energy design compared experiences.

2.3 Workshops

Workshops on passive solar design for mid-career professional education were developed and held in many Member States during the Second CEC Programme. The lessons learned were used in the Third Programme to develop a series of daylighting workshops held in Portugal, the Netherlands and Ireland. Support was also given to workshops in Spain and Britain.

2.4 European Passive Solar Handbook

Reference was made above to a draft European passive solar handbook outlining the basic principles and concepts for passive solar architecture which was prepared during the Second Programme. A limited number of issues of a preliminary edition (prepared by Ecoles des Mines de Paris) was printed in 1986 and distributed to schools of architecture and other reference libraries used by architects, delegates at selected passive solar workshops and meetings, and participants in the "Working in the City" architectural competition.
The handbook is now being revised with increased emphasis on passive solar design in Mediterranean climates, to meet the needs of the enlarged EC. The new edition will also include results from several sectors of the Third Programme. It includes:
- a design primer for architects and students with little or no experience of passive solar design;
- a solar handbook aimed at providing practitioners and students with information directly applicable to the design process;
- a passive solar components catalogue which gives technical specifications of a range of passive solar components and products available from European manufacturers, together with selected construction details;
- a resource guide containing annotated lists of useful publications and computer software.

2.5 Research Digests

A pilot series of four Research Digests has been prepared and disseminated widely throughout the EC to explain the results of the Third Programme to design and construction professionals, those interested in the development of renewable energy, and others responsible for solar research programmes.

2.6 Databases

Extensive databases of researchers, interested professionals, educational institutions and technical journals and journalists who can contribute to the exchange and dissemination of information on passive solar energy have been established for use in SOLINFO, ARCHISOL, Project MONITOR, and the other parts of the CEC programme. These contribute significantly to the efficient transfer of information.

3. ARCHITECTURE AND SOLAR ENERGY: ARCHISOL

Architects have a major influence on the efficiency with which energy is used in the servicing of buildings. Technological feasibility does not in itself ensure the widespread implementation of new energy systems, and the elements of passive solar design cannot be considered only in their technical dimensions as, of their nature, passive systems have profound architectural implications. The ARCHISOL project had the goal of increasing architects' perception of passive solar design, improving their attitude to it and stimulating them to bring such techniques into mainstream architecture.
Specific activities carried out through ARCHISOL include:
- preparation and distribution of a poster exhibition on energy conscious design;
- organisation of architectural competitions for the design of buildings with passive solar features;
- preparation and distribution of a video on solar architecture in Europe;
- development of design and evaluation tools.

3.1 Energy Conscious Tradition Exhibition

The Energy Conscious Tradition Exhibition is in poster form designed for easy dissemination and mounting at architectural schools, institutions, conferences and workshops. It consists of sixteen highly visual posters illustrating traditional architectural devices and building forms which modify the impact of the immediate environment. Text is minimal and passive solar concepts are shown using graphic symbols. An accompanying catalogue contains all the illustrations and a brief text in each of the nine EC languages. The exhibition will be made available to all European architectural schools and professional institutes.

3.2 Competitions

The EC "Working in the City" architectural ideas competition was the vehicle for an extensive education/awareness exercise which reached architects in all parts of the EC. Press coverage was good: publicity was achieved in over 60 journals in nine languages in eleven EC Member States.
Competitors were asked to design a non-domestic building in which passive solar features, particularly daylighting, were of fundamental importance. Each of the 1500 registered competitors received an information pack in their own language. The development of this material was a crucial element in the project, given the underlying educational objectives of the Competition. It included simplified manual and graphic evaluation techniques (specially developed for this competition) for use in the preliminary stages of a project when concepts are fluid and results can be fed back into the initial strategy.
The entries were submitted first to technical

assessment and then to an architectural jury. All the entries were exhibited publicly in Dublin and a selection are being shown during this meeting. A book of the competition results is in preparation /5/.
ARCHISOL also provided some assistance to the PLEA (Passive and Low Energy Architecture) '88 competition for the design of multi-family residential buildings for Vila do Conde in Northern Portugal.

3.3 'Solar Architecture in Europe' Video.

To inform the lay public and those involved in building and funding building or renovation projects, a video entitled "Solar Architecture in Europe" was produced by David Clarke Associates, London. Aimed primarily at an audience with no previous technical knowledge of the subject, it is also being bought by architecture and engineering institutions for undergraduate and mid-career education. The original version (which has an English-language commentary) is available from University College Dublin; copies are also being distributed by a commercial agency with access to an EC-wide network. Danish, German and Spanish-language versions have been prepared by national authorities.

3.4 Design and Evaluation Tools

A series of calculation procedures and performance indices were developed for use in the first place in the "Working in the City" competition by Dr. Marc Fontoynont and Dr. Nick Baker /6/. A daylighting design procedure in wallchart form together with advice on the use of architectural study models also includes guidance on evaluating sunlight in the proposed buildings. The LT Method is a simplified graphical method for assessing thermal and lighting performance. A special device for measuring Daylight Factor in architectural models was developed and distributed in kit form /7/.

4. CONCLUSIONS

A variety of actions aimed at advancing the twin objectives of 'tailoring' the presentation of research results to the needs of particular groups of users and of enlarging the audience interested in energy conscious design has been undertaken. The methodology has relied upon the establishment of a network of well-placed individuals in each Member State, and has sought to cooperate with relevant existing educational and professional organisations wherever possible. A substantial impact has been achieved in the technical and professional media throughout the EC, and many 'products' have already or will shortly be published.
Although the results to date are encouraging, an enormous task remains. In the new JOULE programme several themes will be continued in the Rational Use of Energy in Buildings sector, but with some change in focus and within a more restricted budget. However further work in the area will also be undertaken in the VALUE programme of Directorate-General XIII for Telecommunications, Information Industries and Innovation.

5. REFERENCES

1. ACHARD, P. and GICQUEL, R., editors. 1986. European Passive Solar Handbook. Preliminary edition. CEC Directorate-General XII for Research, Science and Development.
2. LEWIS, J.O. and STEEMERS, T.C. 1988. European Programmes in Solar Thermal Research and Technology Transfer Proceedings, in Soft Energy Sources at the Local Level. Naxos.
3. KEALY, L. 1988. Information Requirements for Architects. Final Report. Energy Research Group, University College Dublin.
4. KEALY, L. and LEWIS, J.O. 1989. Ideology and Information in Low Energy Design. Proceedings, Science and Technology at the Service of Architecture. Paris.
5. LEWIS, J.O. and O'TOOLE, S. "Working in the City" - Results of the Third European Passive Solar Competition. Gandon Editions, Dublin, (in press).
6. BAKER, N.V. and FONTOYNONT, M. 198[illegible]. Experiences with simplified design tools for daylighting and thermal design. Proceedings, Science and Technology at the Service of Architecture. Paris.
7. BAKER, N.V. A Low Cost Instrument for Measuring Daylight Factor in Models. Proceedings, Science and Technology at the Service of Architecture. Paris.

APPENDIX I

NATIONAL SUB-CONTRACTORS

Prof. André DE HERDE, Centre de Recherches en Architecture, Université Catholique de Louvain.

T. Vest HANSEN and Mikael GRIMMIG, Solar Energy Testing Laboratory, Technological Institute, Tåstrup.

Dr. Jerome ADNOT and Dominique CAMPANA, ARMINES, Paris.

Alex LOHR, Büro für energiegerechtes Bauen, Köln.

A. N. TOMBAZIS, Alexandros N. Tombazis & Associates, Athens.

Dr. Paolo OLIARO, Dipartimento de Energetica, Politecnico di Torino.

Joep HABETS, BNA/Royal Institute of Dutch Architects, Amsterdam.

Prof. Eduardo MALDONADO, Departamento de Engenharia Mecânica, Universidade do Porto.

Dr. Mª del Rosario HERAS, Instituto de Energias Renovables, CIEMAT, Madrid.

Dr Dean HAWKES and Dr. Nick BAKER, Martin Centre for Architectural and Urban Studies, Department of Architecture, University of Cambridge.

CASE STUDIES IN PASSIVE SOLAR DESIGN: BRINGING THE GOOD NEWS OR COMFORTING THE FAITHFUL

KEALY L.
School of Architecture, University College Dublin, Richview, Clonskeagh, Dublin 14, Ireland.

ABSTRACT. This paper addresses the role of Case Studies in information transfer in the area of passive and low energy design. It is in two parts: the first summarises the findings of a recent study of user response to case studies in this area; the second places these in the wider context of the issues of information transfer to architects.
The paper concludes by outlining the challenge for case study material in the area of passive and low energy architecture.

1. INTRODUCTION

This paper starts from the position that Case Studies are an acknowledged means of communicating the fruits of experience. Traditionally the study of existing buildings has been seen by architects as an essential component in architectural theory and history, as well as by providing information on constructional and technical matters.
The paper has two main parts: the first part is based on the findings of a survey carried out in 1988 within the SOLINFO project. This project is part of the research and development programme of the Commission of the European Communities on solar energy applications in buildings. The key task underlying the SOLINFO programme was that of improving the utilisation of information on solar energy in the design of buildings. It was understood that this is a complex task. It was difficult to define its parameters, and inevitably there were many assumptions built into any initiatives undertaken. Such projects require the identification of significant or potentially significant landmarks - points of reference through which patterns might emerge which could illuminate the theoretical and practical complexities of the problem. Thus the need for "benchmark surveys" was identified as a means of identifying such patterns, and of providing essential feedback /1/. The survey cited in this paper was one such: a survey of user response to case study brochures published by Project MONITOR; the second part of the paper brings the findings and thus implications into the wider context of information dissemination for architects, and posits the need for critical and creative evaluation of the role of case studies in this regard. It does this with a mixture of urgency and caution: caution because the survey was modest in scale and the issues complex, and urgency because the issues raised need to be faced in the continuing effort to promote an intelligent approach to energy in buildings.

2. SURVEY OF USER RESPONSE

The dissemination of case studies by Project MONITOR provided an opportunity for the type of benchmark survey mentioned in the introduction. These case studies were published in brochure form, each brochure being devoted to a single example of passive solar architecture. The format adopted for these brochures reflected "market research" enquiries and were published in English only /2/. The brochures were disseminated by the Energy Research Group at University College, Dublin, to over three thousand users in all countries of the E.C., the network being established through the SOLINFO project. The survey had a dual purpose: to clarify some aspects of the information problem for SOLINFO, and to provide feedback for the authors of the brochures, which might be useful to them. The survey was conducted by means of a questionnaire, circulated with the brochures issued in January 1988. It was aimed at responses to the first fourteen brochures - the Monitor set would comprise fifty when complete. The report on the survey was based on analysis of returns received by June 1988, which amounted to approximately 8% of questionnaires issued - an acceptable return rate for enquiries of this type. Of these, about 50% were from architects. Completed questionnaires have continued to find their way back to Dublin - perhaps in its own way a commentary on human communication systems. At the latest count they comprise over 9%, with architects maintaining their proportion of those responding.

TABLE I.
Respondents according to Occupation /3/ (June 1988)

Occupation	Number
Architects	124
Architecture/Design Students	13
Engineers/Consultants	35
Construction/Management	16
Reachers/Researchers	23
Town Planners	2
Sundry Occupations*	14
Unstated Occupations	22
TOTAL	249

*Sundry Occupations: Journalists, Librarians, Public Relations, etc.

I do not propose to itemise in detail the responses received, but rather to give an overall picture, and then to focus in on architects as a group. The questionnaire sought to answer six questions:

1. Who uses the information?
2. What use is made of the information?
3. How useful is the information?
4. How useful are the illustrations?
5. How can the format be improved?
6. How can the material be improved?

The first two questions are straightforward and factual. The second two introduce an element of evaluation, and respondents were asked to grade the content of the brochures, and the illustrations. As regards the content, gradings were sought in each information category contained in the brochures: Project Background (Design Objectives, Energy Saving Features, Site and Climate); Design Details (Planning, Construction, Passive Solar Design, Auxiliary Systems, Controls, Passive Operating Modes) and Performance Evaluation (Monitoring, User Response, Performance of Passive Solar Design, Cost Effectiveness, Project Data). Each category of illustration (photographs, general drawings, construction drawings, energy pie diagrams, energy-related graphs) was also rated. Comments were invited regarding improvements in format and material.

Responses to the questions can be summarised as follows:

- The largest user group was that of Architects (50%), followed by Engineer/Consultants (14%) and Teachers/Researchers (9%).
- The material was used more at the level of ideas than as a source of technical information.
- The informational content had a generally high rate of approval. However, a clear need was expressed for more extensive treatment of construction. Less definitive but still considerable were requests for more information on "Cost Effectiveness", "Auxiliary Systems" and "Controls". Architects in particular focused on "Construction" and "Passive Solar

Design" and "Controls".
- Illustrations generally were rated well. The principal exception was in relation to construction drawings, where more detail was demanded. This was true of architects and of other user groups also.

Approximately one third (32.9%) of respondents made suggestions for improvement in the format of the brochures /4/. The predominant response was favourable, even as suggestions were made. In addition to comments relating to the content, which some respondents included under this heading, the principal format-related sugggestions concerned improvements for the sake of those readers whose English was imperfect, requests for references to relevant publications, for a standard data survey on the front page of each issue, for classification under the C1/SfB system, and for colour-coding of text in the interests of clarity. There were several adverse comments on the typographic style adopted for the brochures.

The comments made regarding the material content of the brochures are particularly relevant in the context of this paper and will be dealt with in greater detail /5/. In addition to their rating of the various categories of information and of the illustrations, about 60% of respondents also commented on how they thought the material could be improved. These comments range from specific queries about individual issues to general remarks and detailed suggestions. There were several requests for translation into all EC languages, and for a more restrained graphic style. Several simply noted their appreciation. However several clear sentiments emerge from the comments:

- The need for a great deal more information on construction, and/or the details of the passive solar elements. Almost a quarter of all those who returned questionnaires asked for additional information on construction, many specifically seeking construction details.
- The need to facilitate comparative studies in publications of this kind. This is evidenced in requests for all drawings to be dimensioned or scaled, for the same units for calculation to be used in all instances, that the evaluation methods be described, that the buildings be dated, to give just some examples.
- The question of costs is frequently raised: requests for elaboration on maintenance cost, or savings in energy consumption over standard constructions, how cost effectiveness is calculated and how improvements could be made on the situations reported.

In summary, the survey demonstrated that, while there was general welcome for the case studies, there were important areas which the studies did not, and in the limitations of format probably could not, address. This was particularly marked in the requests for information on construction.

Before going on to consider how these case studies fit into the general field of information transfer, it might be helpful to focus briefly on the comments of architects in particular. These comments reflect also the suggestions of those architects whose completed questionnaires arrived too late to be included in the Final Report of the survey, and which have therefore not been included in the statistics cited above. The proportion of architects to other groups did increase very slightly in this extended population of respondents, from just below to just over 50%.

Of this group, about 30% made no suggestions for improvement and simply rated the information given. Of those who did comment, almost 55% expressed the need for more comprehensive treatment of construction. Specific aspects identified include: positions and types of insulation; wall and floor finishes; positions of vapour barriers; draft sealing; construction details relating to technical installations. More general points were also raised: "Monitoring should rely on standard methods for all projects to make comparisons possible. The dependence of performance upon design should be worked upon and demonstrated. The material is satisfactory considering the limited space provided. Practically all items indispensible. I would rather have more construction drawings and detail photos. I believe Project Monitor should emphasise, by demonstrating how can Passive Solar Design improve energy conservation, the efficiency of design on this matter. It would be worth examining in formal terms the dependence of performance upon design."

These thoughtful comments (Greek respondent) identify issues which bring the function of case study material - and not only those under discussion here - sharply into focus. For architects, case studies do more than illustrate successfully completed projects - they have a role in assisting the architect in the process of designing. This point will be developed below. A further quote from a UK respondent reinforces the point: "There should be a little more information on the general objectives and how these were affected by the decision to apply extensive passive solar techniques. A small section on local vernacular styles and their energy performance where appropriate would be interesting: typical details - showing, through photographs and descriptions, how the designer overcame a few of the unique problems that innovative forms/designs tend to generate......Project data, I appreciate, is difficult to keep consistently categorised, but if possible direct comparisons between issues really is very useful;....Ideally, a small educational element pointing out the more relevant aspects of the region's climate, such as excessive precipitation, seasonal winds etc. would make the series an excellent tool for students and practitioners."

TABLE II.
Respondents (288); Architects classified by Country (145)

	Architects
Belgium	[illegible]
Denmark	3
Federal Republic of Germany	30
France	9
Greece	16
Iceland	9
Italy	19
Luxembourg	-
Netherlands	16
Portugal	10
Spain	8
United Kingdom	21
TOTAL	145

3. ARCHITECTURE: THE CHALLENGE FOR CASE STUDIES IN PASSIVE AND LOW ENERGY DESIGN

The most critical challenge facing architects in the area of passive and low energy architecture is that of linking form and performance. In conventional practice the integration of design concept, the materials to be used, the nature and quality of the constructional strategies comes about partly by training, but vitally through repeated experience of a process which extends from concept to construction of the building. In addition, many architects are accustomed to making study visits to completed buildings and to listening to other architects describe their work. In this framework "constructional strategies" are not a different species of decision-making - it is not unusual for the early "form" or "idea" of sketches of architects to contain implicit and sometimes explicit attitudes to structure, materials, construction or services. Whether such sketches contain evidence of such attitudes or not is a function of the particular architects own process of thinking and deciding - the sketch being an active ingredient in this process. There are two aspects here which require elaboration: the nature of the design process itself and the fact that architects construct their own paths through that process.

The design process has been described as linear process, problem/design/solution, as though it applied to well-structured problems. It is of course well known that this picture is false that architectural design problems are "ill-structured", requiring the designer to have command of a range of mental skills in order to surmount them. Solutions are "satisfying" rather than "optimising", and are the result of a limited exploration of alternatives, with, for the most part, a single option being explored in depth /6/. This facet is sometimes ascribed to time constraints - it is also a characteristic response to complexity of the problem. Thus architects make "soft" or "enabling" decisions which in themselves are capable of elaboration or transformation as projects progress. Typically the architect begins with little information about the specific problem, building up information as the project develops. However the architect also begins with a great deal of information as to how to translate complex requirements into buildings. The "how to" here is highly dependant on the capacity of images to carry multiple messages, and on the degree to which the individual architect can interact with that capacity. Experience is a vital part of this interaction, and results in the development of a "technical intuition" inseparable from the activity of designing buildings. Architects' requirement for visual structuring of information is not simply a need for lots of pictures - it is for a way of presenting information which links the statement of the parameters of the problem (traditionally through sketches) to the solution, as drawn or built. In passive and low energy design the challenge is to link form and performance as stated above, and the "technical intuition" needs to encompass new areas, of which there is still limited experience available.

The second aspect of the design process which impinges on the role of case studies is the fact that architects construct their own paths through that process. It has been suggested that the individual architect's need for and use of information is related to an individual style of learning, developed over time and partly a matter of personality. Broad patterns have been identified, reflecting a balance struck between elements of the learning cycle: concrete experience, observation and reflection, abstract concepts formation and the testing of inplications /7/. The individual develops a strategy that deals with taking risks in situations where there is too much information. The strategy allows the individual architect to ignore a great deal of information and to work with the kind of information that supports the way that architect likes to operate.

Taking these considerations together one can see that the truism that architects respond best to visual material has to be intelligently interpreted. There is a world of difference between material presented visually on the one hand, and on the other visual and illustrative material prepared as a generous appendage to data. The distinction strikes at the heart of much excellent material prepared for architects. One of the respondents in the MONITOR survey put the point succinctly: "What is said in the text could often be said much more efficiently by legends on drawings". The importance of this kind of comment can easily be missed because of its statistical insignificance. The point is not that architects are unwilling or unable to read and assimilate data unless it is visually presented. It is to stress the importance of having graphic material which makes the point. There is an extension to this statement - it is to suggest the importance of having buildings which can be seen to make the point that excellence in architectural design is to be found in buildings which function well in energy terms. This consideration has to be placed alongside the need for the information supplied to cater for a spectrum of approach on the part of its audience.

3.1 The Challenge for Case Studies

It is possible to set out the challenge for case studies in passive and low energy architecture based on the considerations outlined above and on the extensive literature on the criteria for publications aimed at architects /8/.

1. Case studies need to show how a building was generated. In particular they must explore the relationship between the form of the building and the context, natural and architectural in which it is set. In addition, the design should be related clearly to the functioning of the energy strategies followed.
2. The studies must show how the building was constructed in sufficient detail for architects to see and understand the connection between energy performance and decisions taken with respect to fabric, assembly and finishes. The technology associated with its energy performance should be explained, and its constructional implications clear.
3. The studies need to show how the building has performed in a way that allows for comparison between strategies. Costs and user response are important elements here.
4. Studies need to include references to back-up sources of information so that particular aspects can be pursued further if required.
5. There is need for a clear hierarchy in the way information is presented. Thus the emphasis needs to be on graphic material, and text written in architectural vocabulary. The quality of presentation in graphic design terms is important.

In summary, case studies need to deal with their subject at several levels: at a down-to-earth level which will allow architects the knowledge that low energy design is worth getting involved in; at a rational level that shows the connection between theory and realisation; at a theoretical level which establishes the support provided by research and experiment; and finally at a level which asserts the excitement of discovery of a genuine new architectural dynamic. This last is surely the key, and the note on which this paper concludes.

The title of the paper asks whether case studies are a means of spreading a message or of providing support for those already interested. Perhaps it is a false dichotemy, but I think not. To refer back briefly to the brochures published by Project MONITOR, one can see how they confirm the significance of the role of solar energy in buildings designed to make constructive use of it. This is a useful contribution - it is essential for architects to know that there has been validation done in this area. The buildings are not in general architecturally distinguished - one looks hard for an image to put before a sceptical colleague and say "there! I told you, architecture and energy - it is possible." It is the quality of inspiration which will do most to bring the good news.

4. NOTES AND REFERENCES

1. Kealy, L. 1986. "Hidden Agendas". Unpublished Background Paper in relation to the Action Statements for SOLINFO and ARCHISOL. Energy Research Group, University College, Dublin.
2. Project MONITOR was produced by the ECD Partnership London, and distributed by the Energy Research Group, UCD, Dublin.
3. Some respondents gave more than one occupation, for example, Architect and Lecturer. In such cases it was necessary to make a judgment based on internal evidence of the completed questionnaire.
4. Kealy, L. 1988. Survey of User Response to Project Monitor Brochures Nos. 1-14: Final Report. Unpublished. Energy Research Group, UCD, Dublin. Appendix 2a.
5. Kealy, L. 1988. op.cit. Appendix 2b.
6. Akin, Ömer. 1986. Psychology of Architectural Design. Pion Limited London. P.95-108.
7. Newland, P.M., Powell, J.A. and Creed, C. (1987). "Understanding Architectural Designers selective information handling". Design Studies Vol. 8 No. 1.
8. "Architects and Information". Research Paper No. 1. Institute of Advanced Architectural Studies, York. 1971. P.3 and 4.

NEW TOOLS, NEW RULES: Potentials and Constraints

Dr. Reza Beheshti

Department of Design, Planning and Organization, Delft University of Technology
Stevinweg 1, 2628 CN Delft, The Netherlands
Telephone: (015)784893 FAX: (015)785263 Telex: 38151 bhtud nl

ABSTRACT. This paper discusses contributions of the information technology and computer-based design techniques to architectural and urban design in the last decade, concentrating on their potentials and constraints. It argues that since architects and urban designers have met the challenge of new these technologies, they have to investigate new design methods based on an appropriate design philosophy. The gaol of these investigations will be to improve design methods in architecture and urban design.

1. INTRODUCTION

This paper tries to reflect on ways of thinking about design methods, a study of the intellectual means and the mental operations employed in the conception of an architectural or urban design product. Method, in this context, means the organization of mental acts, put in certain patterns to generate designs, fulfilling a set of accepted norms and being within the framework of requirements of members of the design coalition team. Members of the design coalition team are defined as all those involved in, affected by or are able to exert direct or indirect influence on the process of designing the built environment. In the light of designing in an information environment, the process of design will become a complete multi-disciplinary process, involving data communication and employment of large scale knowledge-base systems for architects and urban designers. While computers are introduced as new tools, they necessitate new rules concerning the act of designing the built environment.

The design information environment provides a broader scope for the design activity. It facilitates consideration of all factors affecting the design product, ranging from social, economic, political, cultural, psychological, legal and physical domains as well as direct and active participation of members of the design coalition team. The latter refers to concurrent design activity, a great potential of designing in an information environment. This facility provides simultaneous designing and allows the designer to promptly overseeing consequences of changes brought about at different stages of the design process.

This paper introduces a model of architectural and urban design process in an information environment. This model provides a possibility for 'navigational design', referring to a feedback mechanism or a facility enabling architects and urban designers to return to earlier stages of design process to implement changes and returning where they left out the process to carry on with the design activity. It is assumed that changes made by the architect or urban designer will be implemented and carried forward to later stages by the system itself and automatically. Arguably, it would be interesting for a designer to generate, in a shorter span of time, as many design alternatives as possible. Such a possibility requires methods of evaluating design alternatives using multi-criteria scales. Another important potential of a design information environment is its implemented design knowledge-base containing design expert systems. The latter are expert knowledge of various design domains required by the design decision support system. In this respect implications of artificial intelligence have had a great impact on the development of new architectural and urban design tools and rules.

The possibilities offered by such environments have begun to transform the practice of architecture and urban design, beyond expected evolution of design practice. One of these implications will be steps towards a European Building Standard System, providing a data exchange facility at a European level. This also corresponds to the notion of '1992 EEC Open Borders' the year marking ' border crossing data flow'.

2. THE DESIGN ACTIVITY IN TRANSITION

This century has witnessed great changes in the ways of practicing architecture and urban design. These changes are breaking away from an evolutionary process of the design activity. They have in the first place affected ways of thinking about design and means of implementing them. Nevertheless, these changes have given a new image to architectural trends and from it new architectural movements of this century have been born. Despite all these changes, still certain aspects of architectural and urban design remain unchanged. This refers to the main characteristics of these processes and does not emphasize means and ends. From Vitrovius to Le Corbusier, to Sterling, Rogers, Krier and Rosi, the intention of architectural and urban design activities remain the same. This papers tries to emphasize the necessity for architects and urban designers to accept the challenge of the computer and information technologies and adapt their practices. The main argument is that the essentials of architectural thinking and what was coined earlier as intentions shall remain unchallenged. These are new tools offered by these technologies that concern us. Since these technologies are presented to us and intend to develop further and offer more possibilities, it is our duty, as architects and urban designers, to take a direct part in these developments and effectively influence their course of development and turn them into means appropriate to our ways of thinking and our requirements.

2.1 The Traditional Design Tools

Architectural and urban design activities have been concerned with the transformation of the built environment. In this regard all efforts have been made and all avenues of thought and talent have been followed to enrich the underlying ideas and philosophies as well as improving the techniques and means of realization of these ideas. The design process, particularly in the last three decades, has been engaged in transforming architectural design, alongside other design disciplines, into an empirical science. The success of these efforts have yet to be evaluated. Endless discussions of design theories, arguments on order, symbolism, perception, structure, system, etc. have led to believe that there is enough room for procedural changes in the design process as well as a possibility of totally different approach to design. These efforts at least refer to dividing the involved issues into tools and rules which accompany the use of these tools, whether these discussions being just 'hypotheses', thus theories whose probability has yet to be investigated, or 'laws', hence possesing high empirical probability.

In any case, the realm of theories and the substance of architectural thought and intention remains to be independent of tools, while some aspects of it can be influenced and improved by design tools. For instance, the advent of drawing table and drawing materials did not change the essence of architectural design but allowed a different possibility for the designers for presentation of their ideas and mental process to the target groups of such products.

2.2 The Transformation of Tools

We can take the view that the influence of new technologies and development of new tools on changing rules can be regarded marginal if the design environments remain unchanged. It is necessary to consider that it was not the design tools which changed the way of design thinking, but the changing awareness and social factors which influenced new architectural theories as well as design practices. It was also the new modes of production which required a more efficient means of tackling architectural design and brought about a new face to the practice of architecture.

At the present state-of-the-art, the computer and information technologies, independent of the requirements of architectural and urban design, have developed and influenced almost every aspect of life, labour and production. It can be argued that it is about time that architects and urban designer, who have been bewildered by these developments and by these technological possibilities, realize the greater impact which can be anticipated on their profession.

In this respect we should not look to hardware tools offered for the practice of design, which will inevitably take over the outgoing tools technology, but to domains which can influence our thinking and will benefit the existing theories of design. The challenge of this readiness to accept transformation of the practice, will be an effort by the architects and urban designers to gradually understand the nature of these technologies. They should try to initiate a divergence from its evolutionary path in order to create a specific technology based on the possibilities offered, but yet a natural continuation of the design tradition.

3. THE DESIGN ENVIRONMENTS

The practice of architectural and urban design is indirectly in debt to the developments of the computer and information technologies. These developments in many cases were not intended to be used by the architects and urban designers, but were adapted up to certain degrees to the environments of architectural and urban design activities. It is fascinating to observe these efforts. It seems these efforts, up to certain points with the exception of a limited number of cases, are mainly concentrated on improving the design activity from the technical point of view, trying to take away burdens of repetition and helping to win the time factor as well as introducing an improved marketing element into the architectural and urban design processes.

3.1 The Traditional Design Requirements

Evidently any discussion of the architectural and urban design environments contains, up to a certain point, a closer look at the design requirements. In this paper no attempt is made to look for the specific requirements of the design activity, but the reference is made to the required conditions. These requirements, apart from the physical means of design which have been subject to changes during the long history of architectural design, can be divided into two distinct groups. Firstly, those required by the designer in terms of his or her ability to design, creativity, intuition, talents, attitudes, training and expertise. Secondly, one can refer to the building up of experience, enrichment of human knowledge which inevitably has a great influence on the act of designing as well as methods of using the existing design knowledge in a broader sense of the term. Traditionally, this transfer of knowledge occurs through apprenticeship and documentation of experiences. The evidence if these efforts is the rich architectural literature which almost covers every aspect of the profession. Nevertheless, the existing knowledge is constantly influencing the evolutionary process of design.

3.2 Projected Design Environments

An attempt can be made to project some of these arguments to the future and at the same time combining it with a parallel development in the computer and information technologies. In this discussion one has to pay special attention to the contributions which can be made by the developments in artificial intelligence. This is not an effort to replace architects by so-called intelligent design machines but the benefits can be drawn from the employment of these technologies in appropriate ways to improve productivity of this powerful mental process of creativity.
The distinct difference of human actions with one of a highly developed design machine will remain untouched, the ability to make decisions based on uncertainties and factors such as feeling, dislike, association, desire, etc. It is the embedded design ability of an architect which develops during his or her career which cannot be replaced by any machine. This arguement is true regardless of existing and ongoing discussion of technologists with respect to studying the potentials of a machine for allowing rules of logic alongside irratioanlity and fuzzy modes of reasoning.

4. DESIGN RULES

The act of designing requires certain patterns to be followed, some being technical procedures and others being overlaying patterns of thought and the design philosophy. The term 'design rules' is not a new term. Here we have to define design rules with regards to both tool-related and designer-related aspects. The former can arguably totally transform in order to encompass possibilities offered by the new technologies, but the latter will fall into a set pattern of acceptable rules and can only partially change the rules but not the intentions.

4.1 Design Theories, Methods and Techniques

The introduction of these new technologies, taking into account their present status and their projected future developments, will not contribute to make design theories obsolete. Such a redundancy, if occurs, cannot be totally related to these developments. In other words a sound design theory will be redundant on its own merit and at an appropriate time and will be directly related to so many factors that the influence of the new technologies will be almost next to nothing. The purpose of this discussion is to point out that those design theories which reflect intentions of architectural and urban design intentions will remain more or less unchanged. What tends to changes are ways of approaching the design activity and means of tackling design problems. Therefore, the main area of concern is appropriate design methods and their required design techniques. Arguably the latter will suffer a greater degree of transformation than the former, as it comes closer to tools-related aspects of the design activity. Tools are best appropriate for introduction of the new technologies as far as characteristics of these technologies and the nature of design representations are concerned. There is nothing mystical about the fact that form, structure and systems can be described in terms of simple or complex numerical relationships.

4.2 The Design Rule-base Systems

The effort of architects and urban designers should not be to contribute to further developments of the computer and informations technologies. These efforts can be best left to specialists. The greater responsibility would be to define the realms and rules of architectural and urban design processes in the context of possibilites offered by these technologies. In this respect a parallel development of the design rule-base systems will be a first step towards a fully integrated design information environment. In order to achieve this a model of the design activity have to be found with corresponding layers and rules to that of a design information environment. The proposed model of the design activity will contain all required information for the design activity at all stages of its development and decision making models for both the design activity by the architect and urban designer and the participation of the design coalition team.

4.3 The Design Domain Specific

To construct this model, the design activity have to be structured in a manner suitable for our purpose. The model considers design as a problem-solving activity where each state can be obtained by the application of an operation to the previous state. We can refer to each problem-solving process as a chain or network of states (which are data) and operations (which are processes) in such a way that loose-ends are also attached. The task of problem solving corresponds to finding the right sequence of operations (or action path) that transforms the problem description state into the solution state.
The architectural and urban design are hierarchical processes, controlling the order in which a sequence of operations is to be performed. It seems the design process in certain stadia conforms to a relational structure. Thus we are yet left with a system, embracing both hierarchical and relational structures. In order to achieve this, the complexity of the process has to be disaggregated into chains of simpler conditions of input/output, related to the logical action path of the design activity. The complexity of relationships and their subsequent disaggregation can provide a basis for concurrently executed problem-solving activities. Concurrent design and navigational design will allow a flexible approach and will fit in with the notion of combined hierarchical and relational structures.
Simultaneous or concurrent design activity requires an organisational structure, demanding a system of rules and/or patterns/measures to allow a thorough analysis of the problem state and its subsequent synthesis to attain the decision state. The proposed model adopts an structuralist approach as its basis to replace atomistic attitudes or holistic explanations.
The structuralism is posing systems of interactions or transformations as the primary reality and hence sub-ordinating elements from the outset to the relations surrounding them, reciprocally, conceiving the whole as the product of the composition of these formative interactions.
Relating these arguments to the architectural and urban design processes we can take the view that the structuralist view of the

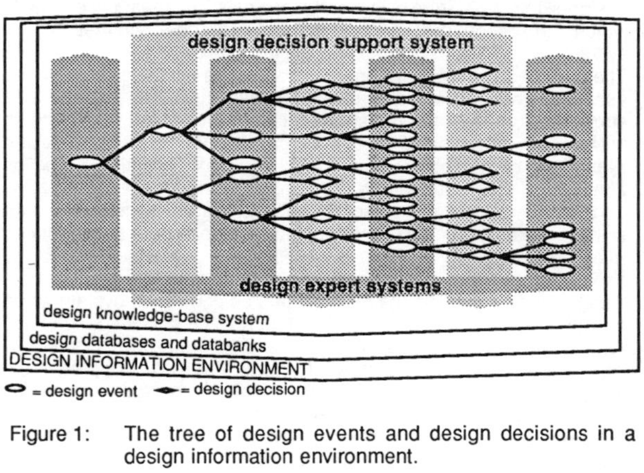

Figure 1: The tree of design events and design decisions in a design information environment.

design process can provide a chart of the design activity wherein all elements and their relationships can be explained in unity. Structures are adequate modes of analysis that can be developed further in the actual design process or synthesis step. They also provide relevant attributes to discuss the possible effects that is likely to be produced by the design activity. The determining factor in this model is change which will take place at a rate that is appropriate to the process and is related to the time intervals of the design process, defined as phases. The actual process of design will be a chain of design events each followed by design decisions, in turn leading to a new design event or coming to an end. The possibilities offered by the new technologies will enable the architect or urban designer to keep track of different courses of action, to foresee consequences of each design event and implementation of required changes to the process, implementing design knowledge and expertise, making appropriate decisions and above all having the freedom of not being bound by these possibilites and be able to by-pass the system as far as his or her intuition and creativity is concerned.

5. THE DESIGN DECISION SUPPORT SYSTEM

These arguments can be further expanded beyond the discussions of the design domain specific and its boundaries in terms of content as well as quality and modes of reasoning. These will be decided by the rules of the design activity which will set the framework for establishing design knowledge-base systems and its specified design expert systems. To complete the picture of the design information environment, it is necessary to establish norms, rules and means of assisting design decisions

5.1 Decisions and the Information Environment

In order to establish the link between the architectural and urban design trends of decision-making and that of a design decision support system, similarities should be traced to define overlapping domains. Such an effort will certainly demands shaping the decision-making process to contain mechanisms favoured by an intelligent decision support system as defined by the rules and boundaries of the artificial intelligence.

The main concern in this activity will still remain adequately determining the consequences of design decisions with regards to different design alternatives. The complexity of the process and the nature of decisions as well as the employed design tools and design rules will increase by the complexity of the employed system. As this complexity of the system (design tools) increases, the necessity of defining and establishing new design rules will also increase accordingly.

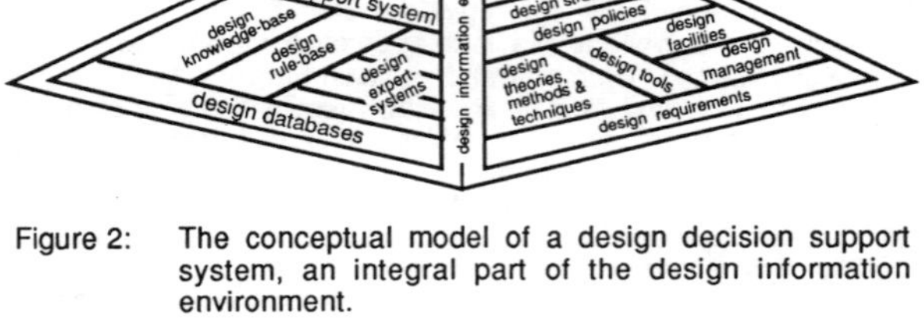

Figure 2: The conceptual model of a design decision support system, an integral part of the design information environment.

5.2 The Design Decisions

One of major differences of architectural and urban design decisions with other forms of decision making process is the element of the design intention which can only be defined by the particular individual designer involved. In this respect, there exist a variety of elements which cannot be programmed and cannot be contained in the system.

The goal of any expert system is to contain as much as possible the essential and crucial parts of expertise of as many experts in a structured form with a well established set of rules and reasoning. In areas such as medicine or commerce, developing such an expert system will be carried out successfully and up to a point will function effectively. The problem with the transfer of the architectural design and urban design expertise to an intelligent integrated design system, would be to find recurrent patterns to establish some general rules, applicable to different situations. As such rules cannot be easily established, it can be argued that only those parts of a designer's expertise can be included in a design expert system which to a great degree is of an objective nature. The subjective rules will remain to be complex as far as translating them into a system is concerned. The present state-of-the-art of the technology does not suggest that subjective rules can be effectively included in a design decision support system. Nevertheless the goal of any design research activity can be to overcome this complexity in order to provide a broader horizon for the designer.

6. CONCLUSIONS

It would not be a wrong assupmtion that the architects and urban designers have accepted the challenge offered by the computer and information technologies. The employment of the new technologies is with great speed affecting these practices and it can be anticipated that the future ways of the practice will change dramatically. This paper pointed out significant factors in this trend and tried to discuss how designers have to meet this challenge effectively and efficiently. The paper proposed a design information environment and tried briefly to outline its contents and characteristics.

Like any other technological revolution, the introduction of the computer and information technologies will transform the designers' working environment. This paper noted out that the most significant of these potentials would be an ability to oversee the consequences of changes brought about in the process of designing The key-word in this system is change and the term navigational design referred to the presence of the designer at all phases and steps of the design process, controlling the quality of the product by implementing new ideas as they come and when they come.

Designing in an information environment allows a designer to benefit from a design decision support system implemented in the environment which is based upon a comprehensive design knowledge- base system, supported by a variety of design expert systems. Also, the design information environment offers its greatest challenge, that of conducting concurrent design and the involvement of members of the design coalition team.

Arguably the designer in this environment will exercise more freedom for the expression of design ideas and arriving at more accurate decisions. The intention of designing in an information environment is designs which can stand up to higher standards and can serve their purpose more effectively.

The introduction of this environment does not necessarily replace the friendly face of a drawing table and sketch-paper with the unfriendly face of a computer screen. The state-of-the-art of the technology shows the possibility of introducing a totally integrated electronic design office, ergonomically designed with the help of designers.

Much can be said about profits that can be gained from designing in an information environment. Also, one can draw attention to what cannot be achieved in this environment and shortcomings of such a system.

This paper argued that a design information environment is not intended to and will not replace the need for designers. The system will not be able to replace human reasoning and judgement. While the system in any possible way supports the design decision making process, it comes short of design judgements and qualitative and subjective decisions.

REFERENCES

Beheshti, M.R. and Monroy, M.R., "Steps towards developing an architectural design information system". in: Maver, T. and Wagter, H. (eds.), CAAD FUTURES'87, Elsevier, 1988.

Beheshti, M.R. and Veer, P. van der, "Decision support environment for civil engineering design". in: CIMA (eds), EuropIA'88 (The application of artificial intelligence to building, architecture and civil engineering), Hermes, Paris, 1988.

Beheshti, M.R., "Design in an information environment", DRS'88 Conference (unpublished paper), 1988.

Beheshti,M.R., "Editorial", Design Studies, special issue Vol.7, No. 3, 1986.

Hillier, B. et al, "Knowledge and Design". In: Cross, N.(ed.), "Developments in design methodology", , John Wiley, 1986.

Lansdown, J. and Roast, C., "The possibilities and problems of knowledge-based systems for design", Planning and Design, Vol. 14, No. 3., 1987.

Moshe, R. and Shaviv, E., "Natural language interface for CAAD systems. in: Maver, T. and Wagter, H. (eds.), CAAD FUTURES'87, Elsevier, 1988.

Norberg-Schulz, C., Intentions in architecture, MIT Press, Cambridge, 1965.

Ohsuga, S., "Toward intelligent CAD systems", CAD, Vol 21, No.5, 1989.

Piaget, J., "General problems of interdisciplinary research and common mechanisms". In: Friedman, S. et al, "Main trends of research in the social and human sciences (Part one: Social sciences)", Mouton/UNESCO, Paris, 1972.

Pugh, S., "Design activity models: worldwide emergence and convergence", Design Studies, special issue, Vol.7, No. 3, 1986.

Zeigler, B.P. and Wael, L. de, "Towards a knowledge-based implementation of multi-faceted modelling methodology". in: Kerckhoofs, E.J.H. et al. (eds.), "AI applied to simulation", SCS, 1986.

ENERGY SAVING IN THE STUDENT'S LIVING QUARTERS STUTTGART-HOHENHEIM

P.O. Braun, J. Schmid, E. Gruber
Fraunhofer-Institut für Solare Energiesysteme
Oltmannsstr. 22, D-7800 Freiburg/Br.
W. Germany

Abstract

In this paper various methods for reducing the heating energy consumption and two types of daylighting systems will be presented. These methods were applied as part of a research project conducted in the student's living quarters in Stuttgart-Hohenheim. The six buildings comprising a complex near Hohenheim University are oriented north-south. In order to improve the thermal insulation the buildings were covered by earth mounds on the north side (U-value 0.35 W/m^2K) and plants were grown on the roof. The large window areas towards the south function as solar collectors; the energy gained can be stored by suitable materials in the floor and in the walls. Four of the buildings (A-D) are four storeys high, two others (E,F) are three storeys high (figure 1 and 2). The buildings were fitted out with different energy saving components, allowing interesting possibilities for comparison. A comprehensive measurement and regulating programme provides information on the efficiency of the measures. This has been assisted by a sociological-empirical programme with questionnaires and interviews carried out by the Fraunhofer-Institut für Systemtechnik und Innovationsforschung.

1. DESCRIPTION OF THE PROJECT

The following energy saving components have been installed:

1.1 Variable heat loss and sun-shielding protection

In building C, variable heat-shielding systems were integrated between the inner and the outer panes of 81 windows, as shown in figure 3, totalling 135 m^2 glass area. The roller blind systems were developed at the Fraunhofer Institute for Solar Energy Systems. The fabric of the roller blinds with aluminium layers on both sides reduces the heat losses through a double glazed window (usually a U-value of 3 W/m^2K) up to 60 % (U-value 1 W/m^2K) during cold winter nights (figure 4). This is achieved by reflection a broad radiation spectrum, the visible and the infrared range. The energy efficiency of the heat-shielding systems was calculated to save 9.500 kWh per year in this building. During summer, the roller blind systems are valuable sun shades and prevent overheating of the rooms, which is very important for passive solar energy architecture such as in the buildings in Hohenheim. Thus the system hinders uncomfortable fluctuations of the room temperature and draughts in the window area. However the effect depends strongly on the students behaviour.
In order to determine the effect of these systems, the heating energy consumption of the buildings was monitored as shown in figure 8. Due to insufficient wall insulation on the north side of building C and D, the average heating energy requirement of these buildings is higher. This and some other differences in the operation of the buildings have more influence on the energy consumption than the roller blinds. On the other hand, the interviews have shown, that the occupants of building C often use the roller blinds and are willing to do so.

1.2 Conscious user-determined increase of the room temperature

In building E a conscious user-determined increase of the room temperature is being tested in the student living rooms. The idea is, that at times when the students are absent, maintaining a constant high level of the room temperature wastes energy. Therefore a special combination of an electronically controlled radiator thermostat valve had been installed, causing two temperature levels. The students press a button, if they would like a higher temperature, for example 20°C. After a period of time, the lower temperature which is 4°C below the upper one, is automatically set; the scheme is shown in figure 5. Estimating that the room temperature is at the lower level for about four hours per day on average, the energy consumption of one student's living room could be reduced from 800 W to 600 W/day (including the night reduction). The desired room temperatures of 20°C, however, could not always be reached fast enough so the students increased the thermostat setting without authorization. Therefore, the heat energy consumption, which had been rather low until March 1987, increased in the following months.

figure 1: view of the student housing from the south

1.3 Radiator heat regulation by the air vents

Underneath the windows in the student's living rooms, air vents are integrated, which the students use instead of opening the windows. Usually, the thermostat valves for the radiators open widely to compensate the slump in the room temperature when a vent is opened, so the radiator heats with full power. To avoid this disadvantageous regulation, magnetic contacts were installed in the vents wings in some rooms (figure 6). If the vents are opened, these reed-contacts cause the radiator thermostat valves to close until the room temperature has fallen to about 8°C below the former level.

1.4 Optimization of the heating control unit

The buildings are heated with gas heating systems, which provide domestic hot water in two heating circuits: the main water circuit for the radiators in the building and the smaller second circuit for the air ventilation heating system, used in the bathrooms, corridoors and lounges in addition to the installed radiators, to maintain a comfortable room climate in those earth-sheltered, windowless rooms. The contribution of the air ventilation system to the heating of the buildings varies considerably between the buildings and influences the energy consumption in different ways. The radiator heating system and its co-ordination with the air heating and ventilation systems has been optimized by micro-computer control during a period starting in March 1989. The optimization operates with both, inside and outside temperatures; thus solar gains are taken into account. As can be seen in figure 8, the heating energy consumption in building A, where the optimization took place, is much lower than in the previous months and in the comparable buildings. However localised low temperatures in some rooms with large exterior walls required the termination of the experiment in early April.

1.5 Daylighting systems

The indoor corridoors get some daylight through the transparent doors of the student's living rooms. In order to bring daylight into the northern lounges, which receive little light from natural sources, two daylighting systems have been installed in Hohenheim. One is a mirror light pipe system, the other a fluorescent planar concentrator system (FPC); the diagramm is shown in figure 7.
The systems, mounted in the centre of the stairwell, collect the light underneath the glass skylight over the stairwells. The mirror light pipe utilizes multiple specular reflection from the walls of the pipe, which is covered with silver-coated plastic film, to guide the light down to the lounge.

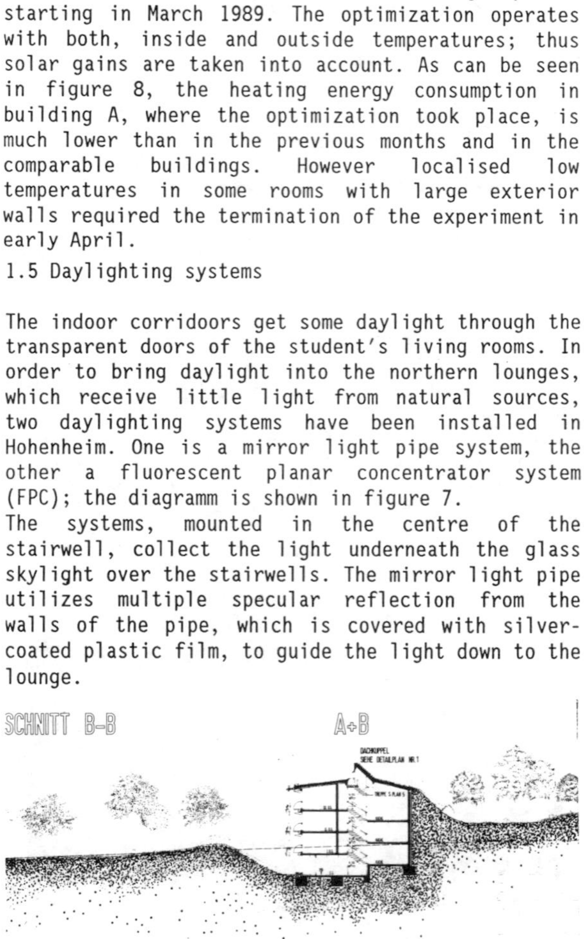

figure 2: cross section plans of the buildings

In the FPC-system, the light is absorbed by the fluorescent dyes incorporated in the coloured acrylic sheet, emitted as fluorescent light and then guided by multiple total internal reflection to the edge of the plate and further within the wall of a transparent acrylic tube.
Whereas the mirror light pipe increases the illuminance level in the lounge by a factor of 2 to 5, the FRC-system, due to various loss mechanisms and inadequate dimensions, does not contribute significantly to the illumination of the room.

figure 3: view of the facade of building C including the roller blinds

2. CONCLUSIONS

Figure 8 shows the heating energy consumption of the houses related to the area per square metre during the period from December 1986 to June 1989. This shows that in spite of the comparable structure of the buildings and the type of users, there is no possibility of gaining values on the influence of the energy saving components directly from the measured heat consumption of the houses. However the total heating energy consumption of 80 kWh/m^2 for the students' living quarters in Hohenheim is much lower than in other comparable student housing.
The climatic conditions inside the earth mounds are very mild; throughout over the year the lounge temperature level is between 18 and 25 °C without any heating.
The installation of the roller blinds led to a better quality of living due to the decreased heat loss at night and sun protection during the day- in this house all of the occupants are satisfied with the level of the room temperature throughout the whole year.
The conscious user- determined increase of the room temperature in house E by a control button did not prove to be successful.
The optimization of the heating control unit showed that there is still a great potential for energy saving.
The daylighting systems still have to be optimized by research and development work to achieve maximum use of the collected natural light.

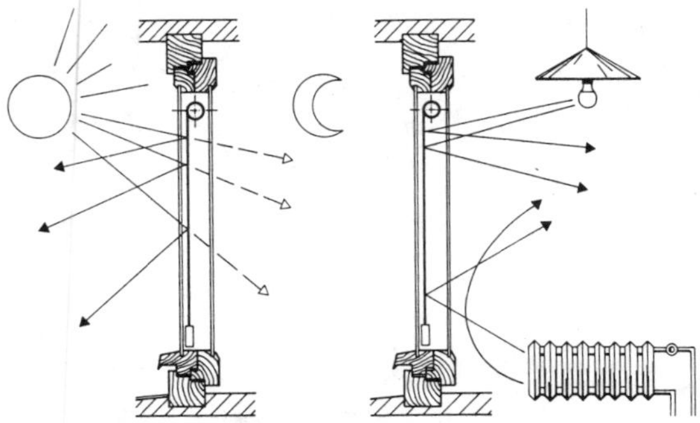

figure 4: diagram of the roller blinds

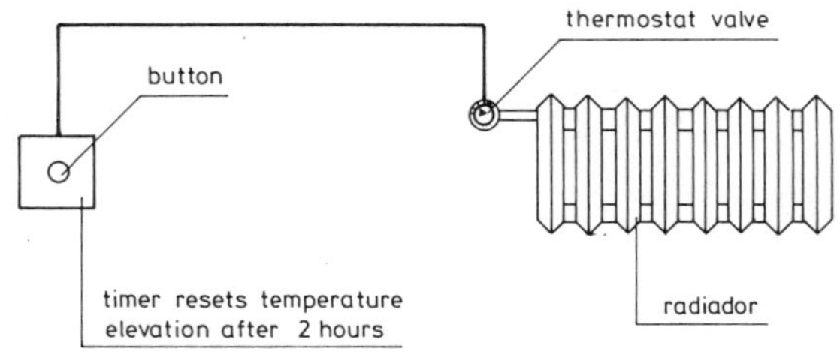

figure 5: diagram of the conscious user-determined increase of the room temperature in building E

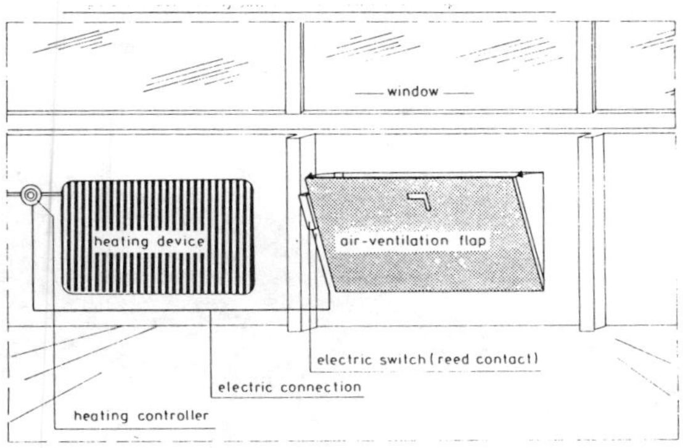

figure 6: diagram of the radiator heat regulation by the vents

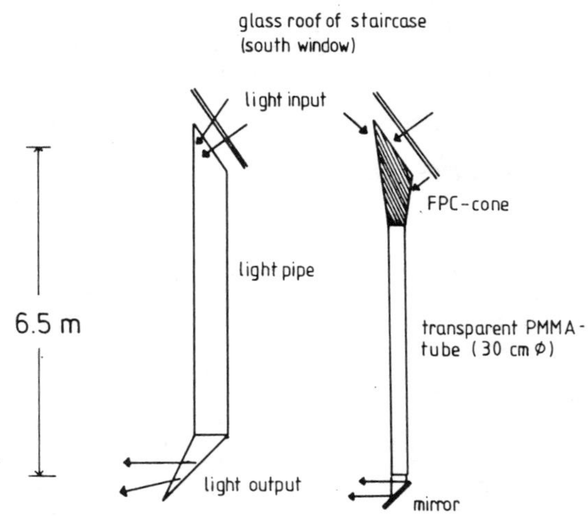

figure 7: daylighting systems

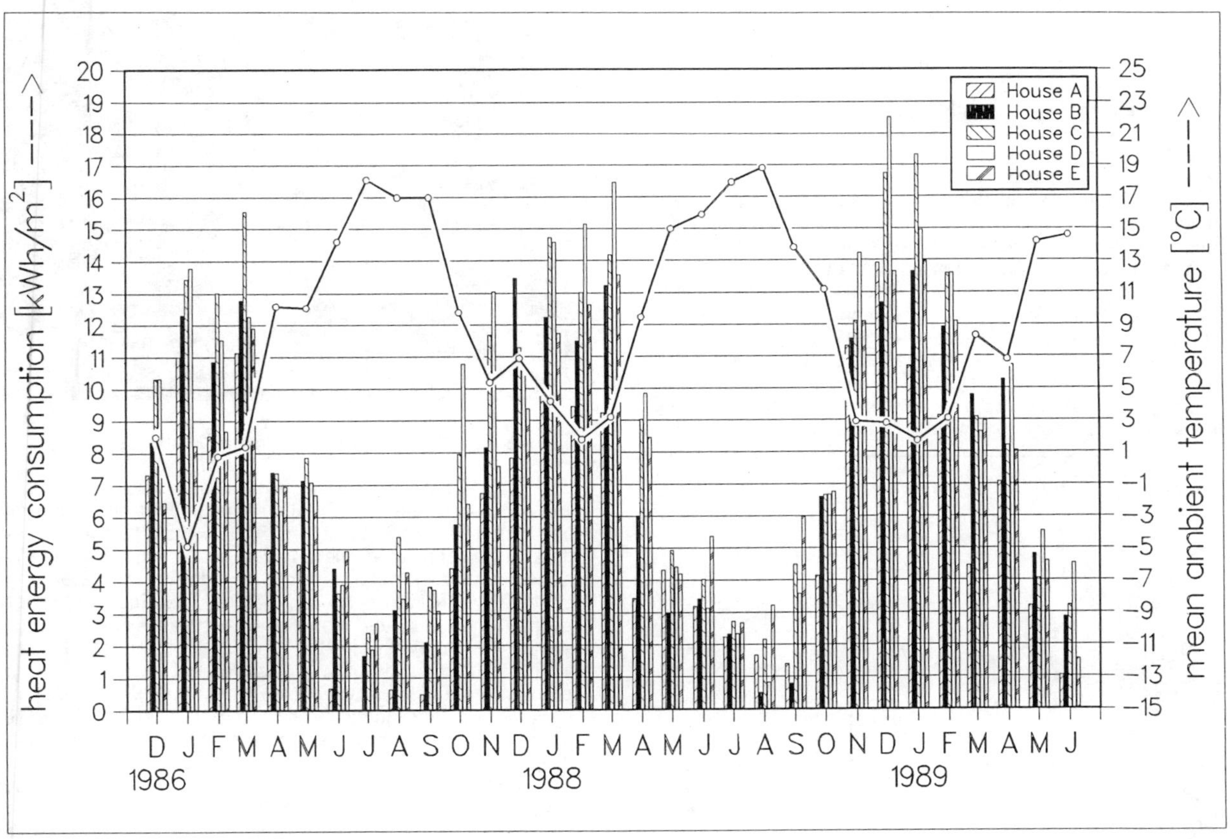

figure 8: heating energy consumption related to the area per square metre of the buildings

MONITORING RESULTS OF PASSIVE SOLAR SHELTERED HOUSING

DAVID CLARKE BSc Dip Arch RIBA
DAVID CLARKE ASSOCIATES
4 TOTTENHAM MEWS
LONDON W1P 9PJ
UK

ABSTRACT. Christopher Taylor Court is a sheltered housing scheme of 42 flats in Birmingham UK. Monitoring since completion in autumn 1986 has shown that energy savings from the conservation and solar features are 51%. Solar contribution is 10.5%. This figure seems low but the building was heated for 24 hours a day and internal temperatures are high. In absolute terms the contribution of solar represents a large financial saving giving a pay back period of 8.5 years for conservation and solar features. There are some moving parts featured in the design, these have proved less important than expected and not all been used as intended. The design for solar access has created controlled open spaces which are well used and liked by occupants.

1. HISTORY OF PROJECT

1.1. Context

The sheltered housing scheme forms one part of Rowheath Solar Village in Bournville, Birmingham, UK. The building was occupied in September 1986. The project was accepted as a Demonstration Project by the Commission of the European Communities who supported a proportion of the extra capital and monitoring costs /1/. Monitoring was completed in 1988.
The scheme is funded by the Housing Corporation who require that the access to all flats must be enclosed and heated to 16°C, the flats must be constantly heated to 21°C (compared to average 18°C for family housing). The brief from the Bournville Village Trust was to achieve significant energy savings through passive solar design and to create a building which complemented the character of Bournville, a much respected example of the English garden city movement.

1.2. Description of scheme

The layout creates two large open ended courtyards as a result of the need for solar access. The construction is conventional masonry walls with insulated cavities, timber roofs with insulating quilt and pre-cast concrete floors. The south walls (Fig 1 and 2) are a combination of direct gain windows and doors and solar mass walls. These consist of an outer layer of diffusing glass, a rendered brickwork wall with a selective surface coating, and plaster internally; there are insulated shutters which are intended to insulate the wall during the summer and during the daytime in the winter when there would be direct gain. There are moveable shades on the exterior

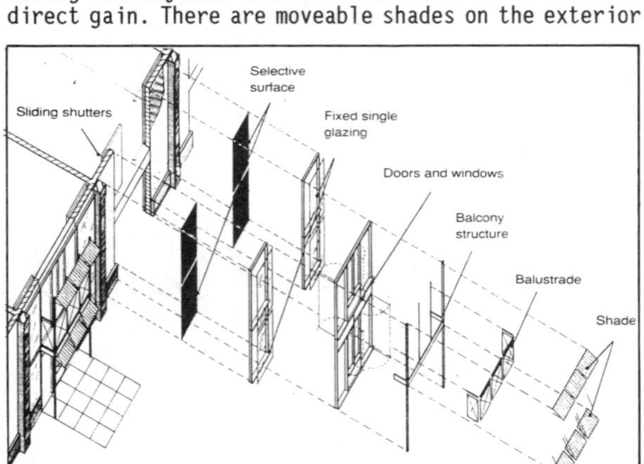

Figure 1 Exploded View of South Wall

Figure 2 View of typical South Facade

of the building to reduce peak temperatures in summer. The selective surface is the main innovative component used on the project /2/. Trials were carried out to determine the best method of adhering the treated nickel foil. The final solution consists of a priming coat of silicon compound which provides a smooth compatible surface for the self-adhesive silicon coating already on the foil. Perforations in the foil allow transmission of any trapped water vapour within the structure.
Heating is provided by a central gas-fired boiler plant; distribution is to the north of the flats Branches to each flat supply low capacity radiators; thermostats control a valve which can quickly shut down each flat. Flow temperatures are modulated in relation to external temperatures and to solar radiation. Each living room and bedroom has a small window for ventilation and a full size door that can provide very high levels of ventilation in summer.

2.MONITORING

2.1.Climate

The climate in Bournville is more influenced by its remote location from the sea compared to most areas of the UK, resulting in higher summer and lower winter time temperatures than average. For the monitoring year in question solar was approximately 10% less than historical average (fig.3.)

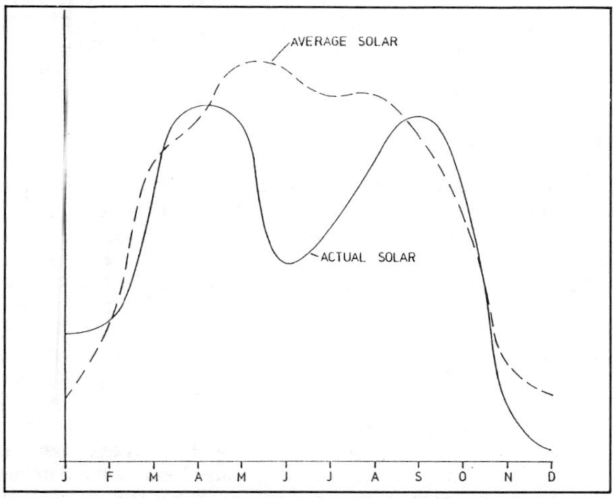

Figure 3 Comparison of Monitored Solar with Average

2.2.Performance Monitoring

The monitoring scheme included 6 flats with many sensors as well as monthly manual readings of meters. Data from sensors were condensed at a central monitoring room in the building. Unfortunately many of the data were corrupted because of insufficient screening. It would have been very difficult within the time scale available for monitoring to correct this. Statistical and mathematical methods have been used to abstract useful information from uncorrupted data./3/

For the monitoring year the overall heat demand for all flats and common parts is 430,640 kWh, of which solar contributes 45,100 kWh (Fig.4). The average internal temperature for the year was 22.8°C for flats and common parts compared to a design temperature of 21° C in the flats and 16° C in the common parts. While the percentage of solar (10.5%) seems at first sight low for a passive solar design at this latitude it does however represent a very considerable financial saving, and with the building heated for 24 hours to such high temperatures it is unlikely that the higher solar contibutions of 20 - 30% found in intermittently heated buildings are possible with simple technology.

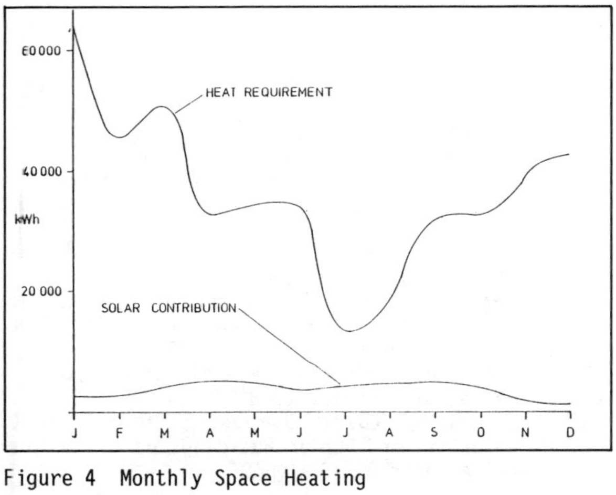

Figure 4 Monthly Space Heating

The overall consumption has been compared on paper to the same building built without solar features and to Building Regulations standards but with the same heating regime. This comparison shows a saving of 51% of annual heating due to the combination of higher insulation standards and passive solar measures. However if the building had been a typical sheltered housing scheme with flats on either side of the corridor the overall building would have been more compact with a lower UA value.

3.COST EFFECTIVENESS

The total project cost £1,003,000 to construct in 1986 including extra costs of £66,690 for passive solar features, insulation measures and controls. The annual saving of 353,650 kWh is equivalent to £7,780, a pay-back period of just over 8.5 years. For solar features alone the pay-back period is about 12 years. For the report to the CEC in connection with the demonstration scheme a different method of assessing cost effectiveness is required, the final result being, comparison of the equivalent cost of the fuel resulting from the energy features and also the pay-back period (see table I). For these calculations a 1% discount rate has been used which is the UK government's recommendation for calculating pay-backs on energy features in buildings./5/

A SOLAR ELEMENTS

Investment	£13444
Energy production	49611kWh
Cost per kWh	1.06p
Payback (against gas)	12.32years

B ALL ENERGY FEATURES -1

average consumption - CTC consumption (gross figures gas and electric)

Investment	£66690
Energy production/ Replacement	600477kWh
Cost per kWh	.43p
Payback (against gas)	5.55years

C ALL ENERGY FEATURES -2

difference in consumption and solar compared to building regs standards

Investment	£66690
Energy production/ Replacement	353651kWh
Cost per kWh	.74p
Payback (against gas)	8.57years

ASSUMPTIONS: 30 year repayment @ 1% real disc rate, annuity factor 0.39
Delivered gas cost 1.4p
Useful gas cost 2.2p
operating cost of energy saving features: nil

TABLE I Cost effectiveness comparisons using CEC economic viability formulae

Part A of the table shows the calculation for solar features alone (cost £13,444); part B shows the calculation considering all the energy features and is based on a comparison of Christopher Taylor Court with the average for elderly persons in the UK (1.79 GJ/m^2)/6/. This is the most favourable comparison resulting in cost per kWh of less than 0.5p and pay-back periods of just over five years. Part C is a comparison against a building built to Building Regulations standard resulting in a pay-back period of 8.5 years.

It can be seen from the above discussion that the cost effectiveness issue is one where different goalposts can be chosen. For this project the comparisons show favourable results against several different criteria.

4.FEEDBACK

4.1.Users

A user survey was carried out on the residents asking about temperatures and use of moveable insulation (see table II). The feedback on temperature shows that 55% are always satisfied with the temperature of their flat, and 40% sometimes found the temperature too hot; this number is almost entirely accounted for by the first floor residents whose flats run at a higher temperature because of the lighter ceiling, however this general perception of sometimes being too hot was not such that it reversed the decision to discontinue the use of the moveable shades (see below).

	Ground floor	1st Floor	Total
Always right	73%	33%	55%
Too hot sometimes	18%	67%	40%
Other(regulation problems)	9%	0%	5%

TABLE II Perception of flat temperature

The use of moveable insulation shows that many residents did not understand its use and/or found it to difficult to use. The operation of the moving shutters is very simple; they have an extremely easy free running track and are no more difficult to operate than a normal door. The original design was to have the shutters linked with internal curtains running on clips that would be used in a different way in different seasons, as it was the intention that the insulation be moved during the winter and left against the wall in summer. The 25% of people who are using insulation every night are using the system incorrectly as are the 35% who never use it. However of those who have insulation 61% do make some use of it and those who mistakenly use the insulation during the summer are not increasing the fuel consumption of the building they are just increasing the night time temperatures in their own flat.(see table III)

Every night	28%
Every night in winter	17%
Some nights	5%
Some nights in winter	11%
Never	39%

TABLE III Use of Moveable Insulation.

The moveable shades were used for the first two summer seasons. For the last two summers they have been left in a withdrawn position because residents are not prepared to trade the reduced daylight that results from shading against what was perceived to be an acceptable level of temperature in summer. This includes the exceptionally warm summer of 1989.

4.2.Construction

As most of the building is of a traditional nature there were no major problems experienced during construction. The standard of workmanship achieved on the application of the selective surface foil was not as good as achieved under a more controlled conditions for the test samples. Despite the use of diffusing glass some of this application is not ideal from a visual point of view. To-date there have been no failures of adhesion. For reasons of economy the glazing on the outer surface of the selective surface is fixed, this causes problems when the void of the wall is disturbed for example by monitoring probes or by other maintenance tasks, in future it would be very desirable to have an opening section for maintenance purposes so that defects can be easily and properly attended to and any deterioration in the fabric can be more closely examined.

4.3.Comparison with performance modelling

The performance modelling carried out using the accepted European simulation model ESP predicted an energy consumption for a typical flat of 2005 kWh per annum. This compares to a measured consumption for such a flat of 3330 kWh. At first sight this appears to be a substantial difference. However one must take into account a) the increased temperature at which the flats are actually operated (nearly 3°C higher than assumed for modelling) and b) solar availability was 10% less than the year used for the simulation. By applying correction factors for these two items the prediction is increased to 2560 kWh per year resulting in a difference of only 28 % A reasonable accuracy for simulation of absolute consumption accounted for by the difference between the generality of the model, and real conditions: for example the common parts of the building are almost 7°C higher than simulated and only one third of the heaters in common parts have their own thermostatic controls resulting in excess heat being dumped by opening windows.

4.4.Designers

For the architect there is a significant time gap between design and performance results being available; building designers will always have to design on the basis of the best simulation available, the results from simular schemes and their own experience. Unlike other sections of industry the prototype is the model that has to be used for the next 60 years. Building designers do not have the luxury of constructing and testing prototypes then refining the end product. The development of energy technology will reduce the inherent risks in this process as tested components and systems become available. Only in this way can less adventurous designers incorporate solar architecture on any significant scale.

At the time of the design, performance simulation suggested that the price being paid for an open layout which would provide good solar access and controlled open spaces for these residents, in energy terms would be compensated by the increased efficiency of the fabric, the increased solar uptake and the benefit of daylit corridors. This fundamental aspect of the design has been vindicated by the monitoring.

5.0.CONCLUSIONS

At the end of a six year period of construction, monitoring and feedback the following conclusions are drawn.

1) For any innovative project use the most advanced simulation tools available but don't expect absolute predictions.
2) For any innovative components carry out full scale trials before construction starts.
3) Simplify monitoring to concentrate on the most useful data rather than most complete.
4) Allow individuals access to control their thermal environment but provide controls to avoid dumping of excess energy.
5) Listen to the people who use your building.

REFERENCES

/1/ D Clarke Passive Solar Sheltered Housing at Rowheath Solar Village. 1987 European Conference on Architecture
/2/ D Clarke Practical Experience of Using Selective Surfaces in Mass Wall Collectors. UK ISES Solar Optical Materials Conference, Oxford 1988.
/3/ L Jankovic Energy Performance Analysis of site 5 flats (to be published)
/4/ Commission of the European Communities Demonstration Programme Guidelines for Final Technical reports. Brussels 1987.
/5/ UK Dept of Environment PSA circular 36/80 amended by circular 25/81
/6/ H. Herring. Energy Use and Efficiency in UK Commercial and Public Buildings up to the year 2000 chapter AI.10 (HMSO)

1er. PRIX DANS LE 1 er. COMPÉTITION NATIONAL DE IDEÉS SUR HABITATION EN BOIS

RICARDO TENDERO & CONCHA GARCIA DE VIEDMA
GLORIETA LOPEZ DE HOYOS, 4 5º DCHA A
28002 MADRID (SPAIN)

ENGLISH ABSTRACT.

PURPOSE OF WORK

TO DESIGN A HOUSE THE SOUTH OF SPAIN, TO BE BUILDED MOSTLY ON WOOD MATERIALS, AND WITH PASIVE CLIMATED SISTEMS, BOTH TO AVOID SUN AND HEAT IN SUMMER, AND TO GET IT IN WINTER.

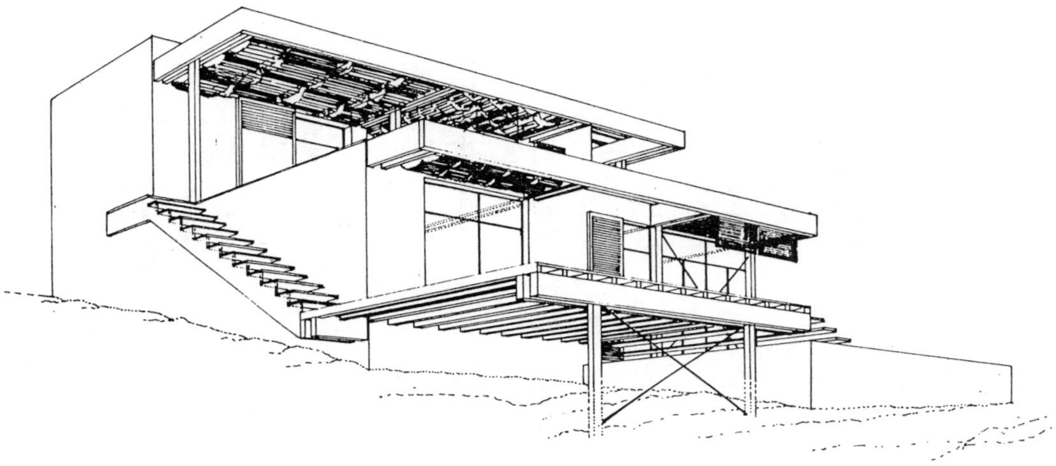

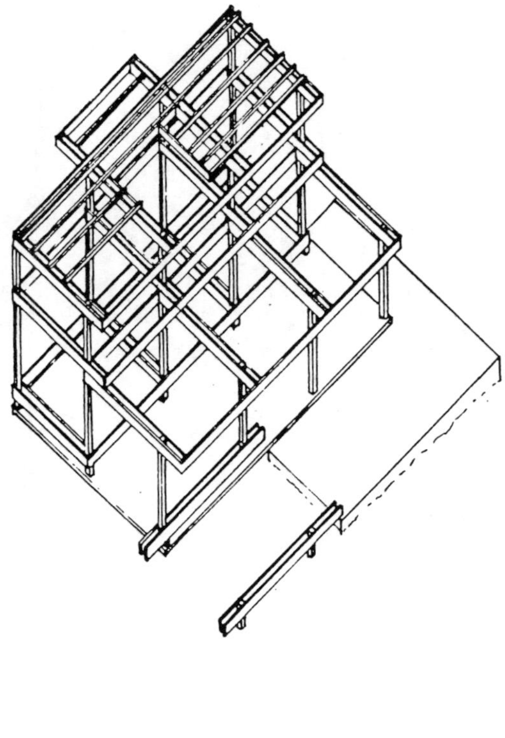

SYSTÈME CONSTRUCTIF

FONDATION. SOCLÉ EN BÉTON ARMÉ.

OSSATURE. POTEAUX EN BOIS APPUYÉ AUX MOYEN DES BOULONGS ET PLATE-BANDS D'ACIER. DOUBLE POUTRE EN BOIS LAMELLE-COLLÉ APPUYE AUX NOYEN DES BOULONGS. DIAGONALES D'ACIER.

PLANCHER. SOLIVES EN BOIS AVEC PANEAUX CONTRE-PLAQUÉS.

DALLE. BÉTON SUR RADIER.

PAROIS. EXTÉRIEURES ET INTÉRIEURES AVEC DOUBLE PANEAUX SUR CHÂSSIS EN BOIS.

ISOLATION THERMIQUE AUX MOYEN DES FIBRES MINERALES.

SYSTÈMES PASIVES DE CAPTATIÓN ET CONSERVATIÓN D'ENERGIE

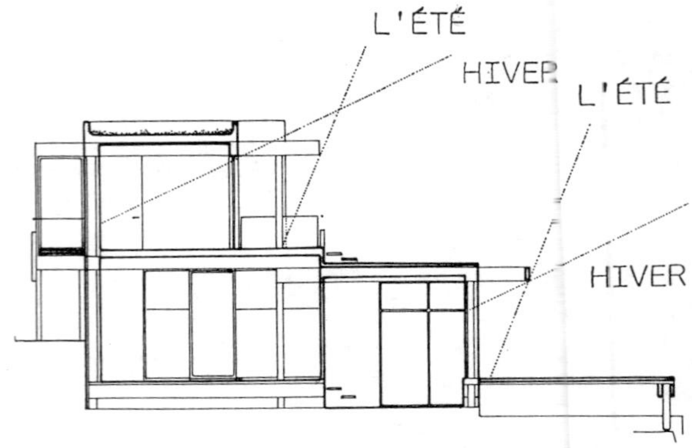

INTÉGRATION DANS L'ENVIRONNEMENT QUAND IL Y A DES CONDITIONS DU CONFORT:

- TERRASE EXTERIEUR A L'OMBRE.
- CLOISON MOVILE ET GRANDS OUVERTURES.

EMPECHÉR L'ENTRÉ DE LA CHALEUR PENDANT L'ETE

- ORIENTATION VERS LES "PATIOS" INTÉRIEURES ET EXTÉRIEURES ZONES A L'OMBRE.
- MURS EXTÉRIEURES A L'OMBRE.
- SYSTÈMES DE PROTECTION SOLAIRE.
- ENFONCANT LE BÂTIMENT DANS LE TERRAIN.
- ARBRES DE FEUILLE CADUQUE AUX ALENTOURS DU BÂTIMENT.

PERMÉABILITÉ AUX BRISES DE MER:

- FENÊTRES AVEC OUVERTURES SUPERIEURES ET INFERIEURES POUR RÉGLER LA VENTILATION.

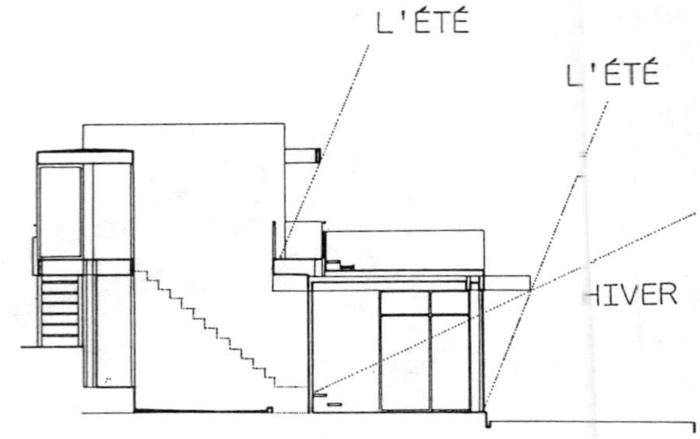

CREATION DE LA HUMIDITÉ ADDITIONELLE:

- "PATIO" INTÉRIEUR AVEC PLANTES ET ÉTANG D'EAU.
- TOITURE PLAT AVEC ÉTANG D'EAU.

CHAUFFER LE BÂTIMENT EN HIVER:

- LES OUVERTURES VITREES SONT ORIENTÉES AU SUD.
- ISOLATION THERMIQUE AVEC VOLET ISOLANT.

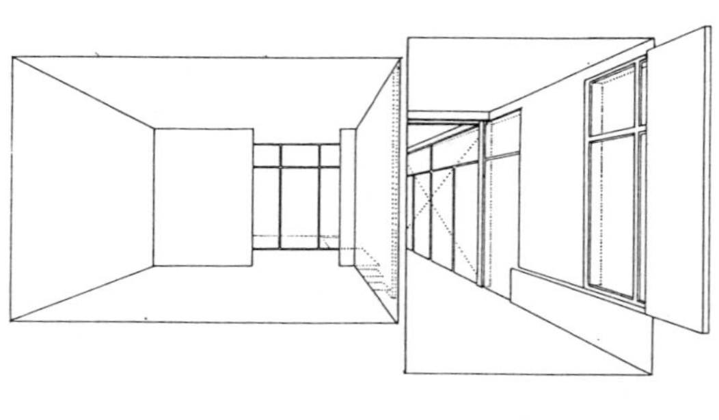

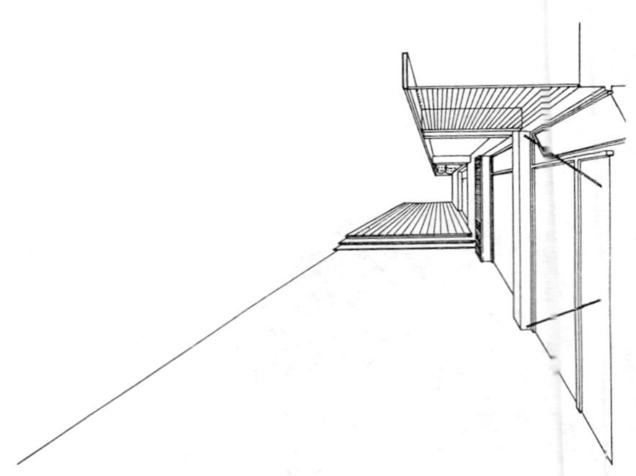

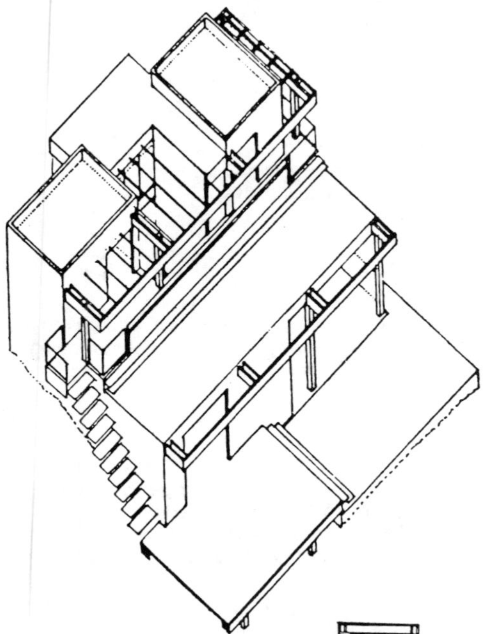

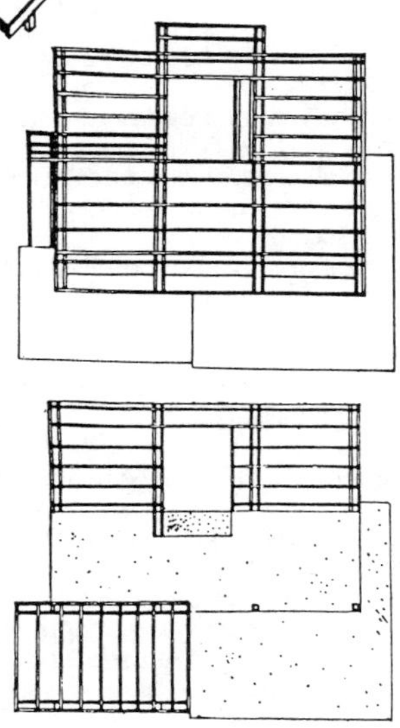

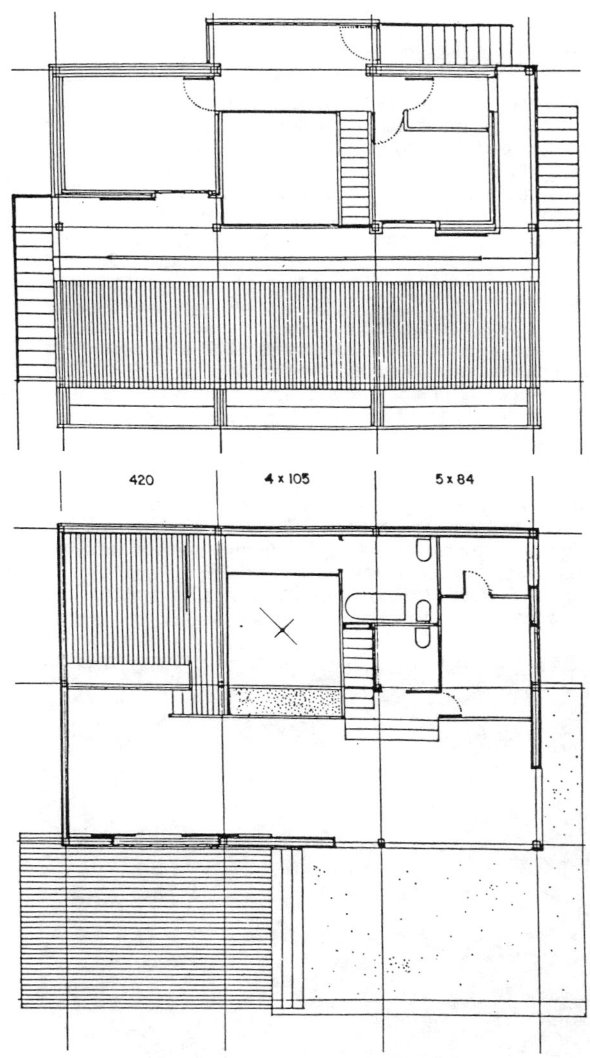
420
4 x 105
5 x 84

SOLAR VILLAGE No 3 - LYKOVRYSSI ATHENS

by: Alexandros N. Tombazis,architect
75, Vasilisis Sofias Ave., GR-115 21 Athens, Greece

architectural design: A.N.Tombazis and Associates
energy consultant: V.Loftness, architect and D.Daskalakis E/M engineer
design and construction of active systems: INTERATOM GMBH, W.Germany
project coordination: Solar Village S.A.

ABSTRACT. The Solar Village no3 built in the suburbs of Athens for 435 families is a result of an intergovernmental agreement between Greece and the F.R. of Germany. The project provides energy savings by appropriate design and increased insulations and the use of different active and passive solar systems that will be compared by monitoring. The construction of the project was completed in 1988.

1. INTRODUCTION

The Solar Village -3 Project is the result of an intergovernmental agreement between the Greek Ministry for Research and Technology and the equivalent W.German Ministry (B.M.F.T.) made in 1978 as a joint Greek-German project in the field of low temperature solar utilization. The coordination of construction of the project was overtaken by the Solar Village Company that was created for this purpose. Design commenced in 1979, construction in August 1984 and was concluded in 1988. The houses were awarded to those eligible by lottery in April 1989 and the village should be occupied by the end of 1989.
The village is intended for 435 families, eligible for subsidized housing from the Workers Housing Association (O.E.K.). The main aim of the project, further to housing 435 families, is to demonstrate the feasibility of energy conservation and the utilization of low temperature solar systems for space heating and heating of domestic hot water (DHW).
A condition put to the designers by the O.E.K. administration of that period was that the project should be as similar as possible to the other current OEK projects as far as the design of buildings, the size of apartments and the materials to be used were concerned. In fact O.E.K. was already planning to erect 47 four storeyed buildings on pilotis on the very same site for a total of 500 dwellings.

Aerial view

2. STATISTICAL DATA

The site is located between the districts of Pefki and Lykovryssi 18Km north of down town Athens. The original plot size was 90.444m2 which was however reduced to 72.100m2 after some areas were designated as wooded areas, the widening of some boundary roads and the allocation of land for the future building of a school and daycare centre. If one assumes 1,4 persons per room (O.E.K. statistic) which means a total population of some 1800 persons the overall density would be 200 p.p.ha for the original site or 250 p.p.ha for the reduced one.
The 435 dwellings are broken down as follows:
74 units with three bedrooms with a net area of approximately 100m2.
191 units with two bedrooms with a net area of approximately 80m2.
25 units with two bedrooms with a net area of approximately 70m2.
145 units with one bedroom with a net area of approximately 60m2.
The total area of apartment buildings is 44.176m2 with a ground coverage of 14.555m2, divided into 25 buildings of two to six storeys.
The total area of common facilities such as shops, a meeting - multipurpose hall, an exhibition hall, a small lecture hall, cafeteria, library and the energy centre is 3353m2. When considering the original plot size of 90.440m2 this gives a plot utilization factor of 0.52 and a total plot coverage factor of approximately 20%.

3. GENERAL ORGANIZATION OF PROJECT

The village is organized into four neighbourhoods each one with its own childrens' play area separated by a herring-bone street pattern (the designers' initial proposal to have no cross vehicle movement was not approved). The remaining streets between buildings are used as pedestrian ways appropriately paved and also for parking with a ratio of one place per dwelling. The main street network does not continue further than the site boundaries in order not to induce through traffic of vehicles.
Common facilities are developed around the centrally located main village square which is sunken in relation to the surrounding ground both for reasons of protection from the north winds and in order to develop movement and its use on two levels. It is thus comprised of the sunken main level and the upper level of the roofs of the buildings. The main square is to be used as an open air multipurpose use space shaded in summer by large deciduous trees (plane trees).

4. CLIMATIC DATA

The psychrometric diagram for Athens shows the basic climatic considerations for the site based on five years of climatic data for Athens.
In general terms the following conclusions can be drawn:
From June through September solar gains are undesirable and shading must be provided. During May and October solar gains are partially necessary. During the remaining six months heating is required and full insolation is desirable.
Ventilation from June through September is required especially during night when temperature swings are the largest for the entire year.
There is a variety of wind directions during winter. Buildings should be sheltered during this period from the N-NE winds. During summer N-NE and S-SW winds can help for cooling.
For solar shading a south-oriented cut off angle of 70°would suffice but would provide inadequate shading during parts of

the year, thus a movable cut off angle of up to 55° would be preferable

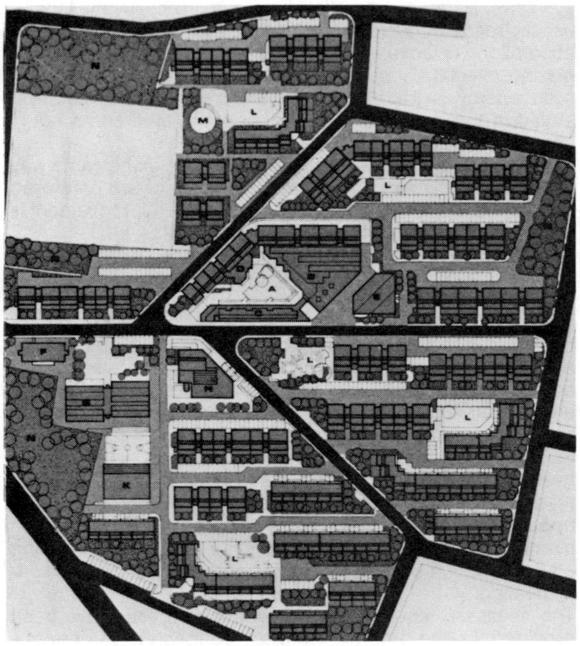

Site plan

5. BASIC STRATEGIES OF ENERGY DESIGN

The basic strategies used in the energy design of the project were:
The reduction of loads by increase of insulations further than those required by Greek regulations. Thus in general, 10cm of insulations were applied. This increase reduced energy requirements to 60% of that of conventional construction. Of this remaining 60%, on average, 70% is covered by solar contributions which reduces the final energy requirements to less than 20% of normal construction. Heating loads are on average 2,2W/m2° C.
The elimination (as far as possible) of thermal bridges and the provision of adequate interior mass in order to reduce temperature swings. For this reason exterior insulations were used over walls and concrete frame. The upper and under surfaces of balcony slabs were also insulated.
Protection from winter winds and full winter insolation. In this respect buildings have a E-W axis, are adequately spaced and staggered in plan to eliminate through north winds. Mostly south-oriented apertures are provided. North apertures are provided as needed only for lighting, ventilation and views. East and west facing apertures hardly exist at all.
Overheating is avoided by full shading of all apertures and the use of light colours for walls.
Cross-ventilation is provided by shallow plans and appropriate openings to south and north.

6. DESCRIPTION OF DWELLINGS

Two dwelling types were designed. The larger units of 100m2 with three bedrooms were designed as two-storeyed row houses with individual north entrances and private south-facing gardens. Living areas are provided on ground floor level facing south while facing north are the kitchen and stairs which lead to the upper floor three bedrooms of which two face south and one north, the latter being the only main room of the entire village to face this direction. The stairwell, further to its main function, serves also to stack-ventilate the entire house by means of the clerestory provided.
The remaining units of 60, 70 and 80m2 are organized in pairs on either side of stairwells and are arranged on three to six floors. Ground floors are drawn out towards the south in order to provide extensive verandahs for the first floor apartments and are provided with private gardens. All bedrooms are south-oriented. Living-rooms are double oriented while kitchens and bathrooms are north-facing. Balconies are provided on both the south and north sides, those of the south being fully shaded by vertical awnings. Metal frame superstructures are provided on the roofs to support collectors (in the active solar buildings) and provide drying space for laundry.

7. DESCRIPTION OF COMMON FACILITIES

The Energy Centre is developed on two levels. All E/M installations are provided on basement level lit by means of a clerestory which also serves as a viewing window for people passing by. On ground floor level administration and measuring apparatus of the solar system have been installed. The triple steel chimney is a characteristic feature of the Energy Centre.
A row of shops occupy the south side of the main square. Their display windows are protected by a continuous overhang while they are lit and passively heated by means of high south- oriented clerestories.
The flat-floored multipurpose hall and exhibition hall are also lit and passively heated by means of skylights.
The library and cafeteria are provided with access, lighting and heating by apertures that link them to the open square.
As already mentioned all common areas surround the main square which functions as a large common open air room for the entire village.

Partial view of central square

8. MAIN MATERIALS USED

Further to the solar systems which in general are simple, especially as far as the passive ones are concerned and the exterior insulation, the remaining materials used and construction methods employed are conventional. This was done further to cost reasons so that the project, where not necessary, would not differ from the other O.E.K. projects and in order to demonstrate that solar houses do not have to differ from conventional ones.

9. ENERGY SYSTEMS

The village includes two large groups of solar systems, the so-called active systems with 401 units and the passive ones with 34 units and the common facilities.
The different systems cover the full range of central to decentralized systems. The central heating systems depend fully on the Energy Centre and can be used in large scale

centrally planned projects. The semicentral systems have common mechanical installations which depend on the Energy Centre within each building. The fully decentralized systems function on the level of individual apartments. Such are mainly the passive systems.

9.1 Active systems

As far as the active systems are concerned five different groups of systems have been provided.
In the first and largest region with 250 units, heating is provided by radiators or underfloor hot water provided by two heat-pumps installed in the Energy Centre which run by power produced by two diesel-generators. The waste heat cogenerated is used to preheat the water thus increasing the system's efficiency. Thus the buildings of this region are provided with roof top solar collectors and are centrally controlled.
In the second region with 78 units, heating is again provided by heat pumps which however in this case are installed in the machine room of each individual building.
In the third region 40 units have flat plate water solar collectors that provide heating and domestic hot water installed on the roofs of the buildings. Back-up heating is provided by oil burners within each individual building.
In the fourth region nine units are provided with individual air collectors, an air distribution system by ducts and individual rock bed storage.
In the fifth region 24 units are heated by water produced in vacuum tube collectors and stored in a 500m3 interseasonal semiunderground storage tank. Here high-temperature water is stored from summer for winter use.

9.2 Passive systems

The 34 passive solar units are distributed in one three-storey building and three two storeyed row house buildings where living quarters are provided on ground floor and sleeping on the first floor. Here back-up heating is provided by electric heaters as quantities are very small.

Two-storey passive solar rowhouses

In general terms six different passive systems have been provided:
1. the direct gains system in which solar energy entering by means of south-facing glazing which is shaded when necessary is primarily stored in the floor mass and secondarily in the walls. In some cases solar radiation is deflected towards the ceiling by means of thin horizontal inverted venetian blinds.
2. unventilated mass walls where a 30cm heavy mass wall, usually made of concrete, (painted black on the outside) for absorption is placed a few centimeters away from the exterior glazing (single or double). In this case, as is well known, solar energy is trapped behind the glazing, absorbed and stored in the mass wall and gradually radiated some hours later to the space behind.
3. the ventilated mass wall which functions in the same way as the previous one with the further addition of two controlled vents in the upper and lower part of the wall. Air movement is thus generated further to the radiation of heat.
4. the water wall which is the same as the second system but in place of the mass wall 30cm metal tanks from floor to ceiling filled with water are provided, forming a complete wall between the solar system and the adjacent room.
5. the different attached sunspace systems where the solar system becomes a whole extra room (greenhouse) which collects solar energy, stores it in the mass of its construction and delivers it either via openings or by radiation of the back wall of the greenhouse to the space behind.
6. the different thermosyphoning systems where air is heated in the space between the exterior glazing and black metal panel behind it which is insulated and provided with upper and lower vents which induce thermosyphoning air-movement to the space behind or into floor or ceiling plenums, much in the way that the Romans used underfloor air heating.
The necessary equipment and central computer have been provided in the Energy Centre, connected to the few thousand measuring points which have been installed throughout the project. Two groups of scientists, one for the active and the other for the passive systems will for at least four years record and analyze weather data and the measurements from the different systems. From these observations and analysis comparative conclusions for the different systems will result.

10 CONCLUSIONS

From the purely logical point of view there would be no reason to install so many different systems. Neither the comparative cost of some of them nor the number of units provided would justify such a decision if the whole project was not one of research and demonstration. They were selected for this reason only, to be tried out under conditions of actual use so that conclusions can be drawn regarding their efficiency, initial and running costs, reliability and final acceptability by the end-users.
It is sure that if there are no special reasons, such as high densities, inadequate insolation or other, many of the active systems cannot be compared from the initial and running costs point of view and from the point of view of simplicity to the passive ones. On the other hand, in contrast to the passive systems, they need no operation from the part of the user. Passive systems however are so simple and with Greek climatic conditions in mind (a mild climate without extremes which however requires both heating and cooling) so effective, that it is really illogical that to a lesser or more extensive degree they do not become part of every project.
This leads possibly to the most basic conclusion:
"Labels" such as passive or non passive systems principle or bioclimatic architecture are of little importance. What counts, even for reasons of principle if you want, but also for reasons of comfort and environment and energy conservation and national economy is that energy-conscious design should always be part of a comprehensive architectural design. There need not be exaggerations, it does not mean that energy-conscious design should push aside other considerations or interests, it simply means and that for many reasons, that it should regain its proper position within the total procedure of architectural design.

THERMAL AND VISUAL STUDY OF A HOTEL COMPLEX USING PASSIVE COOLING COMPONENTS

by A.N.Tombazis*, A.Argyriou**, M.Santamouris**
* Meletitiki Ltd - 75, Vasilisis Sofias Ave., 115 21 Athens Greece
** Protechna Ltd - 39, Mavromichali, 106 80 Athens Greece

ABSTRACT. The Hellas Holiday Hotels tourist complex of approximately 600 beds is to be constructed near Agios Nicolaos, Crete. The project is conceived much in the way of a traditional village with small buildings developed along the natural contours and open squares around which the public buildings are grouped. Landscaping, outdoor living and cooling are of major concern. By appropriate design and shading, cross ventilation, evaporation and the use of air to ground heat exchangers, 52% of cooling loads and 82% of heating loads are covered by passive or hybrid means. Natural lighting has been carefully incorporated into the design. The study is part of the B2000 project organized by the Commission of the European Communities.

1. PROJECT DESCRIPTION

This project of Hellas Holiday Hotels is a luxury tourist complex located in Lasithi county south of the town of Agios Nicolaos-Crete. The site of nearly 180.000m2 adjacent to the bay is inclined 20%-30% to the north. The bay has an east-west axis with a view to the east and to the open sea.
The linear layout of the buildings follows the shape of the bay. The complex is developed much in the way of a traditional village. The different buildings are positioned in such a way so as to receive the late evening sun during the summer. They face mostly to the north and to the view, with their main axis running east west so as to be able to develop passive solar features. Public and auxiliary facilities both enclosed and open are provided in the centre whereas guestrooms are provided laterally to remain quiet and secluded. The layout leads to the maximum possible exploitation of the view and the adaptation of the buildings to the considerable slopes of the ground.

Aerial view of model

Two entrances provide access to the uncovered parking lot and service yard. Vehicular circulation is confined to these areas. One main axis, parallel to the seashore, constitutes the basic circulation path. This together with the central square creates an open air promenade with sitting and recreation areas that promote outdoor living, which is of primary importance.
Adjacent to the main entrance is the main square which functions as the main outdoor living space of the whole development during the day but especially in the evenings. The square is provided with arcades on two sides to be used as outdoor sitting areas. The paving is in a pattern of white marble strips and red brick that will absorb heat during winter and will be shaded in summer by large deciduous trees. A water fountain is to be provided for evaporative cooling. Around the square are developed:

1. a long (E-W axis) two-storey building on the south side housing the reception-lobby and administration on ground floor and administration and top management accommodation on the first floor. The ground floor has been provided with extra height both for architectural reasons and in order to provide extra coolness. The building is cross ventilated and naturally lit from both long sides.
2. a square building providing a dividable multipurpose hall for conferences - exhibitions - games and banquets on ground level and service facilities on basement level. This building is provided with deep recessed door windows on all sides and a central lantern above the wooden roof that will admit daylight into the centre of the plan.
3. a two-storey long (E-W axis) building on the north side that houses the speciality restaurant on the upper level. The restaurant opens onto the square via the arcade on the south side (open air eating under cover) and a terrace with a sea view to the north. The space is provided with south-facing clerestory windows for passive heating, is naturally lit from both sides and is cross ventilated. A main stair off the main entrance covered by a cupola and top lit by means of an oculus, leads to the lower level main restaurant and other service areas (kitchen, laundry etc). The restaurant is provided with an extensive north-facing terrace shaded by a deciduous creeper on trellis. The space being single-sided is provided with overhead fans.

Around a second smaller square or piazetta are developed:
1. an enclosed sitting-room with a large fireplace mainly for winter or shoulder period use. This space too is crosslit and ventilated from the east and west sides being nearly square in plan.
2. a main bar area which directly opens onto the piazetta which will be used as an outdoor sitting area. The back side of this space is dug out of the natural slope.
3. a shopping mall with small individual shops which will spill out into the arcade along which they are arranged.
4. an indoor multipurpose use space arranged as a three sided stepped theatre provided with specially designed top lighting. This space is also dug out of the hillside.

On a lower level are developed a group of swimming pools on different leveled terraces connected by waterfalls, together with a pool snack-bar and a health centre including gym facilities, saunas, an indoor pool and childrens corner. The whole of this area is underground and, apart from the one side that faces the sea, is provided with top lighting in different manners.
Four basic types of 237 double and 28 suite guestrooms are

provided. All are arranged in small two-storey linear buildings (E-W axis) that develop around the bay along the natural contours. All rooms are provided with an entrance and bathroom on the south side and a spacious private verandah on the north seaside. Access to the lower level rooms is gained via an exterior pathway on lower level while to the upper level ones by little wooden bridges in pairs from the higher ground behind. Thus all rooms, even on an extremely sloped site as this, have their back side open and are able to be cross ventilated and lit. Some of the upper rooms are provided with south-oriented clerestories.

View of model

Pathways and walks covered by trellises and climbing plants interlink the buildings and provide access to the bay from different points of the complex. Small sitting arrangements related to secluded corners or special views are provided. Running water is used as a main cooling and environmental feature in the form of gullies, pools, fountains and waterfalls. Along the water-front paths, wooden gangways, jetties and sun decks will provide for seaside activities. Floating decks and gangways will stretch out into the shallow bay providing for activity areas (shaded by tents) for both daytime use and also when lit-up at night.
The morphological treatment of the buildings refers to the architect's intention to express the historic memory which still remains strong in several parts of Crete.

View of central part of model

2. DESCRIPTION OF PASSIVE SOLAR FEATURES/COMPONENTS

The hotel will be occupied mostly during the warm period from April to October and therefore was designed using passive cooling features. Heating in some instances will be provided by mechanical means. High levels of insulation act positively both in summer and winter.

2.1 Passive systems

The most important passive energy-saving systems for buildings are the following:
Nocturnal cross-or fan-forced ventilation reduces cooling loads in periods of large daily temperature deferences between ground and ambient. Small south apertures eliminate solar gains. Whereas large north glazing contributes to heat losses and provides excellent views over the bay.
Shading devices permanent or movable, cut out direct solar radiation during the overheating period. Plantation on pergolas and deciduous trees further contribute to this effect.
The desirable maximisation of daylight results in the reduction in use of artificial lighting. All rooms are mainly north daylit providing uniform natural light in the living areas. Additional rooflights provide natural light into deep and otherwise unlit places.
Temperature reductions due to evaporative cooling are foreseen due to the proximity of the complex to the sea and the existence of the swimming-pool in the centre of the complex. Fountains and streams of water provided in proximity to the buildings will further contribute to this effect.
Air to ground heat exchangers are used for cooling and heating purposes. The system is based on the stable temperature difference that exists between the ground at a certain depth and ambient temperatures. Pipes are buried underground and air is circulated through them. Air will lose or gain heat depending on the season.

2.2 Auxiliary systems

Air-conditioning units and fan coil units for heating and cooling purposes are installed in public areas. This is a requirement foreseen by Greek regulations irrespective of the actual needs due to the provision of passive sola features. Split units with electrical resistances for heating purposes are used in the guestrooms.

3. ENERGY CALCULATION PERFORMED AND DESIGN TOOLS USED

3.1 Heating cooling loads and passive contribution

The main use of the complex during the cooling period gives emphasis on passive cooling systems such as cross-ventilation, ground and evaporative cooling. High thermal insulation of the building fabric and reduction of south oriented openings minimize the possibility of overheating. Permanent and movable shading devices are used wherever needed.
Each building was zoned for reasons of calculation as follows:
In the reception building each room was calculated separately.
The multipurpose building was considered as one zone.
The restaurants were zoned separately (a total of three zones were assumed for this building).
Each room on each floor of the bedroom buildings was taken separately.
Auxiliary and non heated spaces were not taken into consideration
Occupancy parameters depending on the type of building were taken according to the ASHRAE Fundamentals.
The following U values were assumed:

single glazing	5.8	W/m2°C
double glazing	3,7	"
external walls	0.58	"
floors	0.95	"
roofs from	0.38 to 0.44 W/m2°C	

The heating and cooling loads for every building as well as its natural temperature have been calculated for every month

using the admittance method.
Thus the annual passive contribution is 82.1% for heating and 51.8% for cooling.

3.2 Ground cooling

The simulation of the ground to air heat exchangers has been made using the passive cooling simulation components programme PASCOOL.
Calculations show that ground cooling can provide 30% of the cooling load needed for the multipurpose building, for the month of June, 37% for the month of July and 43% for the month of August.

3.3 Night ventilation

Another method chosen for the passive cooling of the project is night ventilation. According to this method a building is cooled by increasing the air-flow during night-time, when the ambient temperature assumes its lowest values.
The maximum coverage of cooling loads by night-ventilation is achieved during the month of June while ground cooling gives for this month its minimum coverage. This shows that night-ventilation and ground cooling can be used very well as two complementary methods.

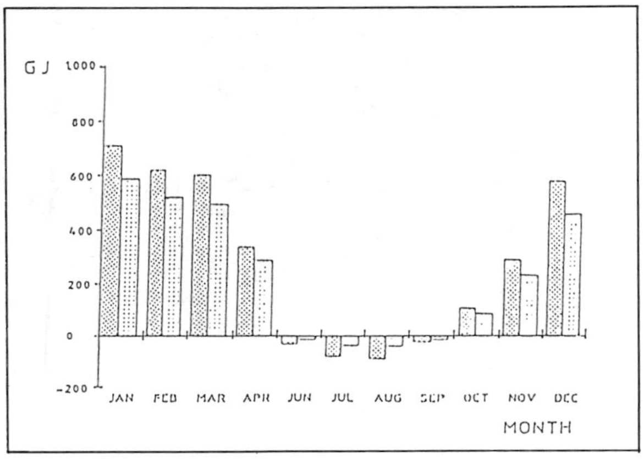

Heating and cooling loads/month

3.4 Daylighting

The daylighting efficiency of the different buildings was estimated using the computer programme Microlight I.
The simulation was run for 12.00 noon on 21 December assuming an overcast sky i.e. the most unfavourable daylighting conditions.
Daylight distribution is uniform and varies between 3 and 9% except for the zone near the glazing where higher values are found. The equivalent range for the office space investigated ranged from 5 to 13% which shows that for both types of spaces daylighting performance is good.

3.5 Thermal comfort

The thermal comfort of the buildings has been estimated both for summer and winter conditions. The PMV (Predicted Mean Vote) method has been used by running the programme COMFORT. The PMV varies slightly below its optimum value during January and slightly above the value in July however still being quite satisfactory.

4. GENERAL DESIGN GUIDELINES/POINTS OF INTEREST RESULTING FROM THIS PROJECT

1. Provide a harmonic relationship between architecture, landscape, human intervention and nature. The dialogue between design and accident.
2. Improve microclimatic conditions around the buildings by using evaporative cooling and shading.
3. Apply simple but innovative concepts in cooling such as by using the ground cooling method.
4. Reduce, by taking the necessary design decisions. auxiliary cooling and artificial lighting in comparison to similar conventional buildings.
5. Passive solar features however important do not have to dominate the architectural design.
6. Promote social contact and outdoor living and activities in a hierarchy of squares, terraces, secluded corners and open spaces.
7. The "historic reference". The use of classical elements in the organisation of the plans and the morphology of the elevations and building volumes states the architect's intention to express the "temporal sequence" in architectural design.

Passive Solar Design of Green Park Combined School

B. Norton
Dept. of Building
Univ. of Ulster
Newtownabbey BT37 0QB
N. Ireland

J.R.R Wagge
Architects Dept
Bucks C.C.
Aylesbury HP20 1UX
England

& D. Newton
YARD Ltd
London WC1N 3LF
England

ABSTRACT. The design of a school building is described which seeks to reconcile passive solar design with the diverse functional constraints on school architecture.

INTRODUCTION.

Green Park Combined School, shown in figure 1, is intended to house 300 pupils and is of 1400 square metres floor area. It is located in Newport Pagnell, Buckinghamshire, England, 67 kilometres north north-west of London. The pertinant meteorlogical statistics are: the prevailing wind is from the south-east, total annual insolation on the horizontal plane is 2,298 MJ per square metre and there are 2468 "U.K. standard degree days". Though loated in a suburban residential area, the arrangement of the site facilitated no overshading of south-facing facades. The building was subject to the usual U.K. planning and building regulations.

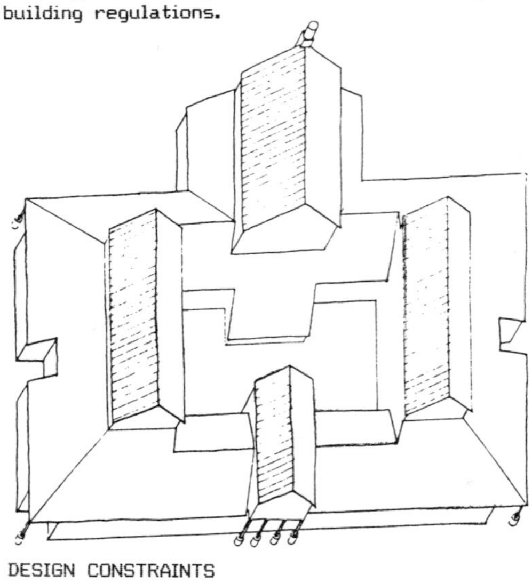

FIG 1

DESIGN CONSTRAINTS

Local education authorities in the U.K. have strict limits, exercised at the design stage, on the maximum conditioned area per pupil in primary schools. Furthermore, they also generally have firm policies on the nature, arrangement of, and inter-relationship between, internal spaces. This is documented usually in the form of a standing design brief for each type of school. This brief reflects both the adopted educational philosophy and a distillation of collective wisdom and experience. Thus innovative manoeuvres in school design are, compared with other buildings, subject to unusually inviolable limits. Substantially, passive solar design in schools must be reconciled successfully with;
* Cost limits
* Educational philosophy
* Occupancy
* Site, and
* Aesthetics

Atria and conservatories are the passive solar feature most often found in school buildings (1,2). Roof-space collectors, in constrast, are an element of only a few, mostly domestic, buildings (3,4,5). They are however increasingly now an option both considered (6), and adopted (7), in the context of school buildings.

At Green Park School, it was decided to include roof-space collectors as the principal "passive" solar feature. This was because;

* the internal layout was unaffected, thus retaining a "design brief" plan
* the solar heat gain system was isolatable from the building and thus more controllable, particularly with respect to overheating
* an existing feature, the roof, is employed
* single-storey school buildings have high ratios of roof area to total floor area, thus there is a large potential area for collection (primary school buildings in the U.K. are single storey to allow direct access to outside)
* the system has a relatively rapid thermal response, which is compatible with the diurnal pattern of occupancy being within the insolation period.

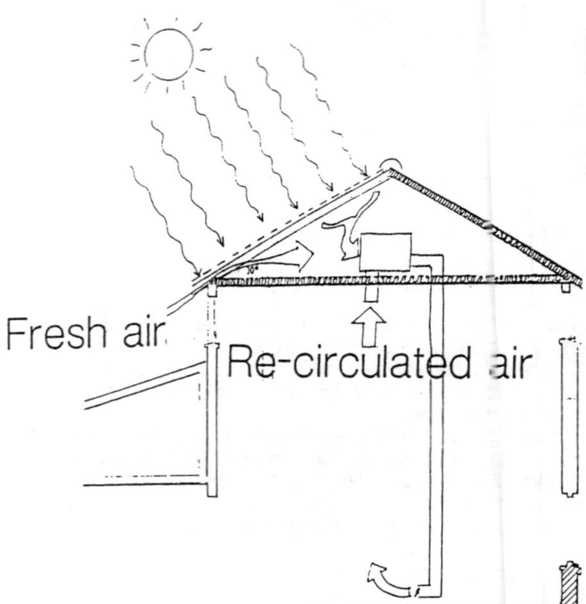

FIG 2.

ROOF-SPACE COLLECTORS IN SCHOOLS

A roof-space solar-energy collector (3) is essentially a pitched-roof which is either fully-glazed 7) or partially-glazed on its southerly aspect. As shown in figure 2, solar-heated air from the roof-space collector is conveyed by an automatically controlled fan via a duct to form a pre-heated supply to a warm-air space-heating system.

Various alternative modes of operation of a roof-space collector can be identified(3). School buildings are occupied primarily during the day. Thus an appropriate mode of operation is for air to be heated to be passed from the outside into the school building via the roof-space collector. If this air forms a large proportion of the total ventilation, then pre-heating of the ventilation air is provided effectively. As with conservatories (8), the roof-space collector does not then have to be at a temperature above that of the building to provide a beneficial effect, since any warming of the air passing through the attic-space will provide some reduction of the ventilation heat load. In newer school buildings in the U.K. it is thought that energy use (in primary units) typically divides into three very-approximately equal portions (1); (i) that lost conductively through the fabric, (ii) that embodied in the air ventilated to maintain human comfort and health, and (iii) the electricity used for artificial lighting. As buildings becomeincreasingly better insulated, so the proportion of energy used to heat ventilation air will become greater. Roof-space collectors are thus compatible with, and compliment, the use of such energy conservation measures in buildings

In summer, ventilative cooling may be obtained if the fan drives air from building into the roof-space collector and thence outside. Replenishing air thus flows into the school from the outdoors.

Roof-space collectors can have a low initial capital cost as the physical construction does not differ greatly from that of a conventional pitched roof. The only additional requirements are the glazing and associated fittings, internal insulation, a fan and associated ductwork and appropriate controls. The fan and controls may only require up-rating or modification of components already associated with the auxiliary air heating system. The ensuent running cost is also small, being only that of the electrical power to run the fan.

ANALYSES

As part of the design process for Green Park Combined School, factors affecting the heating and daylighting aspects of an initial design were subject to sensitivity analyses using the dynamic thermal simulation model SERI-RES (9,10,11). Readily-available annual hourly meteorological data for Kew, London was employed together with realistic diurnal and annual patterns of occupancy. The assumptions regarding heating capacities for each zone (to meet steady-state heat losses at an ambient temperature of minus 1 degrees C plus 10%) and heat losses from the heated rooms are provided in Tables 1 and 2 respectively. It was further assumed that 10 cubic metres per child of the total fresh air intake is taken through the heating system via the roof space collectors. The windows are single-glazed throughout with 4mm thick clear float glass in aluminium frames. The roof-space collector aperture is 16mm thick twin-walled transparent polycarbonate, there is 150mm and 100mm thicknesses of insulation beween the roff-space collector and the ordinary roof repectively and the classrooms beneath. The building fabric is a cavity wall comprised of a blockwork inner leaf, extruded polystyrene insulation, an air gag and a brickwork outer leaf.

RESULTS OF ANALYSES

Annual and diurnal energy load and temperature variations were examined. Illustrative examples of outcomes of the annual sensitivity analyses are summarised in Table 3. The design implications of these results, have to a large extent, been included in the final design for Green Park Combined School; whose construction should be complete in late 1989.

Table 1 Heating capacities for each zone

Zone	Total UA	Setpoint(°C)	Capacity (kW)
Class-N	1420	18	30
Class-E	1845	18	38
Class-S	1845	18	38
Class-W	1520	18	32
Hall	1060	14	17
Total	7690		155

Table 2 Summary of UA values for heated rooms

	U (W/°C/m^2)	A(m^2)	UA(W/°C)
Glazing (rooms)	6.5	290	1880
External walls	0.6	610	370
Ground floor slab	0.75	1430	1060
Ceilings to collectors	0.22	450	100
Ceilings&walls to roofspace	0.78	1210	940
Infiltration (at 2ac/h)			3340
Total			7690

Case		Auxiliary Heating (kWh pa)	Lighting demand (kWh pa)	Change: Auxiliary Heating (kWh pa)	Change: Lighting demand (kWh pa)
1	Original case with no collectors	97,000	19,900	-	-
2	Original case with collectors	91,400	19,900	down 5,600	-
3	Collectors painted black and roof insulated	87,700	19,900	down 9,700	-
4	as 3 + increased pre-heating ($15\ m^3$ per hour per child)	85,700	19,900	down 11,300	-
5	as 3 + 15% less glazing on north facade	87,300	20,160	down 400	up 260
6	as 3 + reduced weight floor slab	84,900	19,900	down 2,400	-

Building orientation

The original orientation specified was 20° east of south. Two cases were run with orientations which are 30° on either side of this, with the following results:

Case		Auxiliary Heating (kWh pa)	Lighting demand (kWh pa)	Change: Auxiliary Heating (kWh pa)	Change: Lighting demand (kWh pa)
1	Original orientation (20° east)	87,300	19,900	-	-
2	10° west	87,450	20,000	up 150	up 100
3	50° east	87,700	19,950	up 400	up 50

TABLE 3.

ACKNOWLEDGEMENTS

The support of the Building-2000 programme of DGXII of the Commission of the European Communities, Brussels, Belgium is gratefully acknowledged. The views expressed herein do not necessarily represent those of Buckinghamshire County Council.

REFERENCES

1. Norton, B. Hobday, R.A. "Passive Solar Schools" Report for the Architects and Buildings Branch of the Department of Education and Science, London, U.K. February (1989)

2. Norton, B., Harris, D.J., Lo, S.N.G., Waterfield, P., Grindley, P. C., & Hobday R.A., "Atria and Conservatories in Primary Schools; an illustrative analysis of the effect of varying atrium width." Proc. Second European Conference on Architecture, Paris, France, December (1989)

3. Norton, B. Crompton, A.L., Lo, S.N.G., and Probert, S.D.,"Hybrid Roof Space Solar-Energy Collectors" Proc. European Conference on Architecture, Munich, F.R. Germany, 423-428, April (1987)

4. Norton, B. Lo, S.N.G., "Roof-space Collectors Integrated into each of Fifteen Newly-built Dwellings of Passive Solar Design in Milton Keynes" Proc. of the Solar World Congress, Kobe, Japan, September 1989)

5. Tindale, A., Littler, J., "Passive Roof Space Systems: Extrapolations from a Years Monitoring", Proc. Second European Conference on Architecture, Paris, France, December (1989)

6. den Ouden, C., (Editor), Proc of the Final Building 2000 Meeting, Chania, Greece, September (1989)

7. Williams, A., Bower, J., Ratcliffe-Springall, C., Thomas, D., Nex, R., "Perronet Thompson School", Building, 45-52, January (1989)

8. Baker, N.V., "The Thermal Design of Conservatories for Solar Ventilation Pre-Heating" Proc. UK-ISES Conf. Greenhouses & Conservatories: Aspects of Thermal Behaviour and Energy Efficiency, (B. Norton, Ed), London, U.K., 1-13, February (1985)

9. Palmiter, L., Wheeling, T., "SERI-RES, Residential Energy Simulator", Version 1.0, Ecotope Group, Seattle, Washington, U.S.A.

10. Haves, P. Littler, J., "Refinements to SERI-RES", Report ETSU S-1130. Energy Technology Support Unit, Harwell, Oxfordshire. (1987)

11. Haves, P. Littlefair, P.J., "Daylight in dynamic thermal modelling programs: Case study" Building Serv. Eng. Res. Technol, 9, 183-188, (1988)

A BIOCLIMATIC SCHOOL IN VENICE

C.CAPPAI - A. VENTURINI
UNIVERSITY OF VENICE (IUAV) - ITALY

Our plan for a bioclimatic school in Venice faces principally two precise problems:1) buildings in the periferic area of Venice,in front of the south lagoon;2) the realization of a passive solar construction which ought to have a particular architectural value because of its environmental situation.We face these problems looking to the realization of solar buildings that we saw and analized.The first problem we want to solve is to create a complexstructure that do not comunicate only its dependance on the sun to the people who examinate it.
A bioclimatic building that can be considered for its architectural value and not only for its tecnical results.

1. ARCHITECTONIC REPORT

1.1 General consideration

It is very stimulating to link the planning of passive solar buildings with the special environment of Venice and its lagoon.The volumetric control and the design of the façades of a plan of passive solar architecture is generally a difficult enterprise.The use of wide glass surfaces can make theese buildings anonimous,making them look like greenhouses rather than architecture works.Our effort is to find solutions which let us build in such a delicate environment as the venetian lagoon,mantaining the utilization of solar energy that can heat our building by itself.We find it usefoul to place side by side a technical description on the utilization of passive solar sistems for the heating and lighting,and a description of the urban and architectural environment to make it clear how the two matters are closely linked like in every architectural work.

1.2 Urban insertion

The area of Sacca S.Biagio is located in the south of Venice,being the last branch of the Giudecca and Sacca Fisola sistem.It is placed behind the latter and can be considered its natural completing (figure 1).Sacca S.Biagio is regarded in the municipal urbanistic programs as an area of sport installation for the urban radious,in spite of its laking connection with the town and its periferic position with rispect to the centre of Venice.In our opinion it can be considered without any doubt as a completation and integral part of the area of Sacca Fisola.Starting with this we project an urban revision of the area

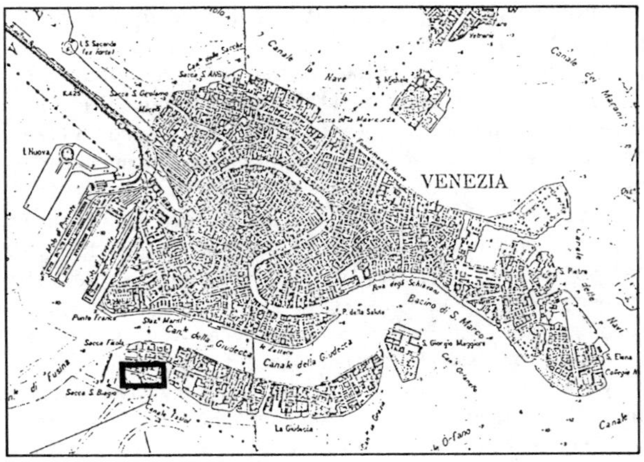

Figure 1 : Localization of Sacca S.Biagio

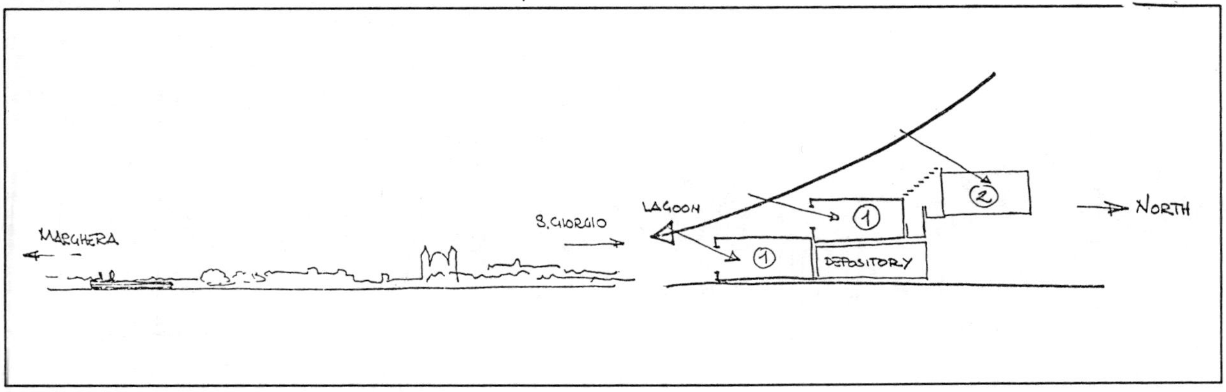

Figure 2 : the south front of Giudecca and section scheme

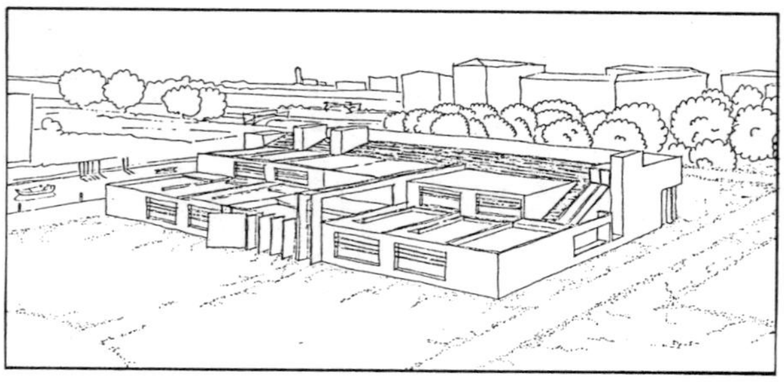

Figure 3 : Exterior view from south side

and making use of tipically venetians elements it has,such as its sistem of islands and canals (absent in Sacca Fisola) and its direct comunication with the southern lagoon,we realize a more complex quarter sistem.

1.3 Architectonic principles

We want to link concretely to Sacca Fisola in order to create a common unifying web and,on the other side,we look at the island of Giudecca in order to understand the way in which we have to interpret this finge area of the southern lagoon.From this consideration springs our plan.The northern side must be compact and hard (like tne norther one of Giudecca)in the southern side the building must scrumble,mantaining a linearity interrupted only by few raised elements (figure 3);the complexity of the volumes must fit in the volumetric sistem of the southern side of Giudecca which is precisely very discontinuous. Therefore,not a massive and regular sign,but a series of small strikes that find in the whole unity and explanation.The school is very compact.It can be included in a rectangle but this is some time contrasted by the elements that break this regularity:the canal goes into the building;the rooms in the south side at the floor level play with the structural septums going in and out the border of the rectangle.The degrading design of the building (figure 2 and 5) allow the classes to have an exterior place (exterior class) useing the roof of the class below.The interior structure is organized by some simple elements:the classes,the corridors,the common hall.We called "the little Venice" the interior composition of those elemnts.

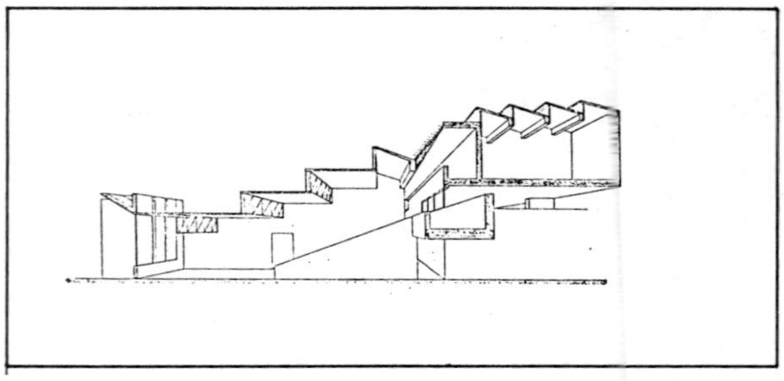

Figure 4 : Interior view of the common hall

The common hall is like the square (campo) in which the streets flow together in various levels.Th superior corridor (calle) forms a bridge with the corridor in the common hall (figure 4)

2 TECHNICAL REPORT

2.1 General directions

The heating system used io our project is a passive solar one.This system has given us a complete and direct gain.The sunlight coming into the big window that faces south,heats the various rooms in a direct way and mostly heats the internal walls that function as an "energy accumulators".

2.2 South classrooms

The windows of the south classrooms (figure 6) are a little bit different from the others,as we use for each of them three "brise-soleil". The function of these "brise-soleil" is to make a shadow in order to avoid dazzling.Their position was studied in order to gain the maximum of light and therefore the maximum heating during the winter,and the minimum of direct light and heat during the summer (theoretically the room should be in complete shadow).Beside that the classroom have a skylight that is used to get a direct lighting of the wall that divided classrooms from corridors,so that they can work as energy accumulators.That wall,having a grat thermic inertia (that is a capacity to realease wery slowly heating to the rooms) avoids a too grat fluctuation of temperature during the day,expecially during cloudy days in which is impossible to

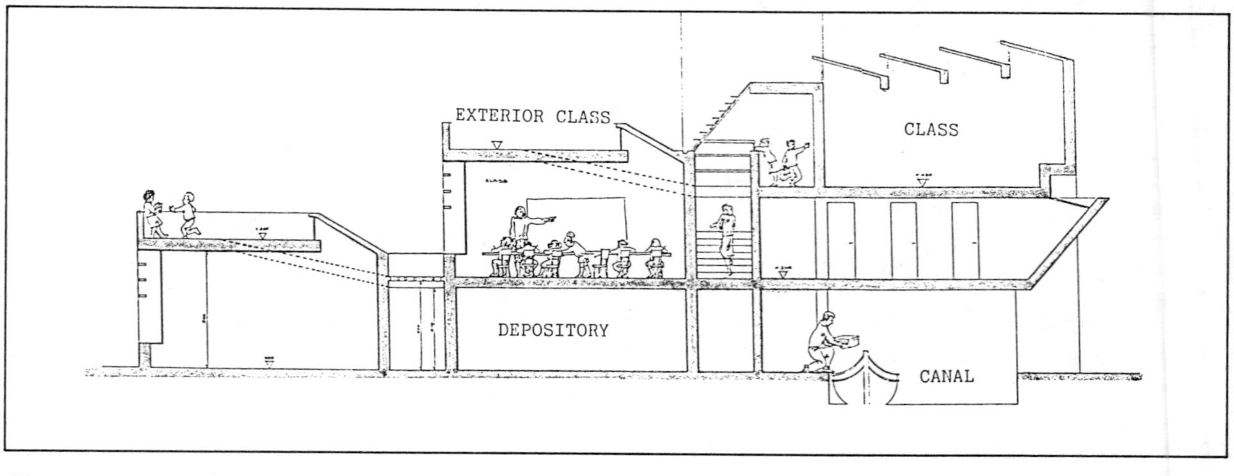

Figure 5 : Section on the classrooms

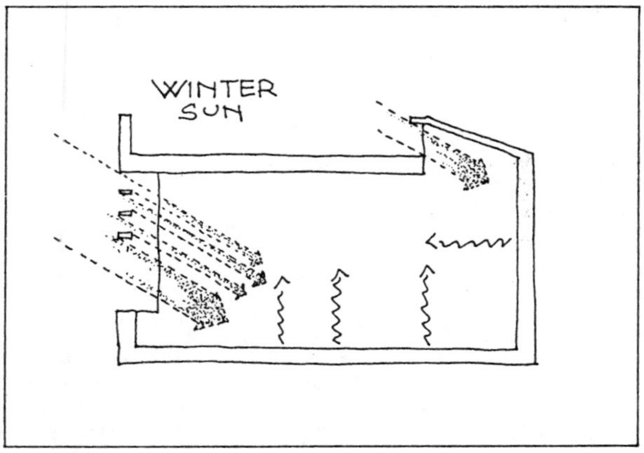

Figure 6 : South classroom

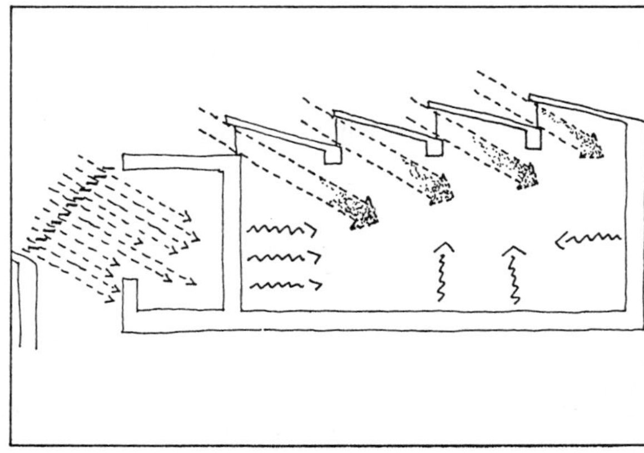

Figure 7 : North classroom

accumulate energy.During the summer the cooling is obtained by a special inclination of the skylight that avoid the coming in of the light and by ventilating the room during the night.The special system that we use,that employs the thermic insulating on the external site of the wall provides an accumulations of heating during the day and gives it back during the night.

2.3 Common hall

The common hall is very big,so the main problem was to avoid the disperction of heat towards other parts of the school.This area might be ruardea as a thermic fly-wheel beetween the different areas of the school.Heating and lighting of this room was provided throught some skylights on the roof.This roof is made in different layers. They are stuggered and this is obtained by a peculiar link of the roof to the reticular girder, that allows the placing of some skylight.To gain even more heat we have other windows facing south, that are very usefoul to light the room and their peculiar shape allow us to have a wide window facing the lagoon without dazzling.

2.4 Second level corridor

One of the most interesting area of this building is the corridor of the second floor (figura 7).This caracterizes the building from an architectural point of view (a linearity which is interrupted only in a few places;figure 3) and from a technical one.This area might be considered as a sorth of greenhouse that heats directly the second floor and indirectly the first floor.The system allows to heat the internal walls;the heat is then given to other parts of this building, expecially by convection.The special screening aaopted for this window,obteined by little alluminium lamellae has the only aim to avoid dazzling and the excessive heating of this area,considering the big dimenction of it.

2.5 North classroom

The façade facing north was meant to be compact and hard.We reduce to the minimum the dimenction of this part of building that is hardly to be heated.We reduced to the minimum the dimension of the apertures of the windows in order to avoid disperction of heat.The heating of these classrooms is obtained directly by the skylights on the roof and indirecly by the wall that divides the corridor from the classroom.

2.6 Daily thermic variation

We calculated the daily thermic variation in a south classroom in a sunny day in december.
Considering the thermic gain given by the whole window area,the thermic dispersion of the room,the external temperature in december,by means of numberbords (cfr. E. Mazdria - "Sistemi solari passivi" - Franco Muzio Editore,Padova 1986) we find the media interior temperature,and its daily variation.In the schedule below we show the school hours.

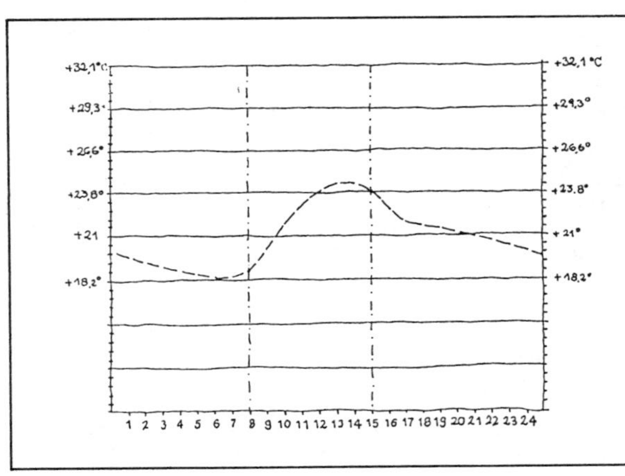

Figure 8 : The variation of the temperature during the day in a classroom in December

THERMAL PERFOMANCE OF VERNACULAR EXCAVATED LIVING SPACE IN THE MEDITERRANEAN ZONE. (VALENCIA. SPAIN). ARCHITECTURAL CRYTERIA DERIVED FROM ITS ANALYSIS

V.SIFRE, F.ARANDA, V.BLANCA
INSTITUTO VALENCIANO DE LA EDIFICACION C/.CAMINO DE VERA 14 (UNIVERSIDAD POLITECNICA). 46022 VALENCIA. SPAIN
ESCUELA TECNICA SUPERIOR DE ARQUITECTURA
UNIVERSIDAD POLITECNICA C/.CAMINO DE VERA S/N 46022 VALENCIA. SPAIN

ABSTRACT. The investigation is centered on the confort perfomance of underground excavated living space which was extensively used in the area of Valencia, Spain. It was been chosen an excavated home, representative of the area, with manu similar living dwellings around it. Those dwellings were excavated on soft sandy stratum, being the roof a water proof hard calcium stratum. It has two "patios" also excavated, one to the front where the entrance is and another one to the back of the house. Those patios favor sun penetration, ventilation and daylighting in the adjoining rooms. Thermal inertia is high on floor and some of the lateral walls, but it is not so great on roof since it is only 60 cm. thick. Its perfomance presents a very uniform thermal environment with fluctuations of 0.7ºC (mean value) during the day. Humidity is high of about 10% more than of the mean outside value. Only when natural ventilation is achieved the inside humidity is the same that the one outside. Mean interior monthly temperature is about 4,5ºC higher than the outside in winter. In summer is only 2,5ºC higher than the outside.

1. THE UNDERGROUND EXCAVATED LIVING SPACE.

The search of people looking for a cheap living space, adecuated to the climatic environment found a simple way, around the city of Valencia, to build it in the underground. There are many examples of these livings units like, Burjasot, Godella, Paterna, Benimamet, Rocafort, etc, which are located to the North-West side of Valencia.
Once it was discovered its possibilities, it extended, shaping whole neighbourhoods in the appropiate areas where it was possible to build them. During the present century, mainly on the 1950's most of them disappeared, united to the concept that it was not a healthy and proper way of living, and also to the especulation of ground property. Only a few of them remain actually under contradictory opinions about its bondage.
From the concepts of Bioclimatic Architecture there is some looking back to the natural environments created by our ancestors, in the believe, that they were not so wrong in dealing with climate. So there is some recognition on the way they were living and this investigation tries to find out the confort conditions achieved by those people that still live in underground spaces, its advantages and inconveniencies.

1.1. Geological description

The example of living dwelling that has been studied is representative of the excavated underground ones that still exist in the town of Paterna. They are known as "LA TORRE" caves. Its spaces has been excavated on sandy therciary deposits, soft but sufficiently cohesive. The stability of the roof is assured by the existence of a uniform thick calcium layer of about 60 cm. made by the cementation of disolved salts from the rained superficial water.

1.2. Constructive description.

The thick water proof uniform layer roof is supported by resistent walls that are the remains of the soil excavated, which are sometimes reinforced by brick walls. The shapes of the rooms are basically ortogonal, giving only a certain cupular shape to the meeting of ceiling and wall. The roof is the proper rock layer with its irregularities. However the walls are covered by tiles to prevent humidity from the ground. The floor is covered with the traditional artificial stones. This way of excavating left wall separations that are not very thick being the mean value of about 30 cm. This will not give much inertia to walls. However, the roof being 60 cm. thick and the floor gives its high thermal inertia that helps to maintain uniformity in the inside climate.

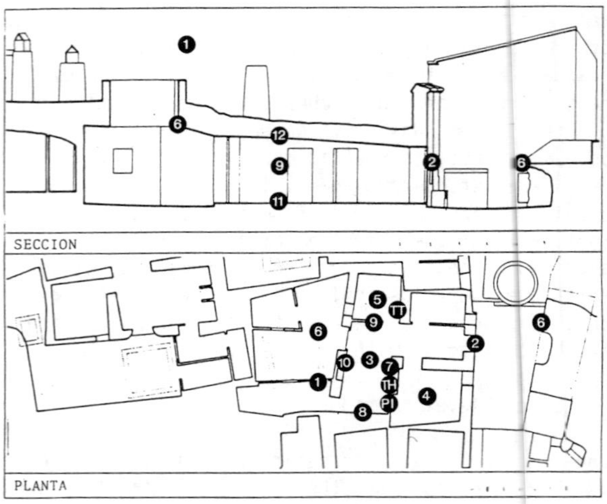

Fig 1. PLAN AND SECTION OF THE UNDERGROUND SPACE. It shows the location of the points were data were taken.

1.3. Architectural description

The living unit imitates those of the outside in the distribution on spaces. It is organized in only one plant, shaping ortogonal rooms, with a central corridor and rooms to both sides of it. The façade is build by escavating a front patio that is shared by several units which gives common access to a certain number of houses. In the back of the house there is a private excavated patio which serves to give natural light and ventilation to the back rooms that are in the kitchen, bathrooms, and other auxiliar spaces.

2. AIMS OF THE INVESTIGATION

The purpose of the investigation is to know the environmental confort conditions inside the living units. The results will be analized and from its conclusions it will be possible to obtain projectual criteria for living units with similar characteristics and also to rehabilitate the existing ones.

2.1 Applied concept of environmental confort.

It has been considered that the higrothermic sensation which depends mainly on the air temperature, relative humidity, the radiant temperature of the near surfaces and air velocity.

2.2. Variables meassured.

One of the variables studied was the heat transmission through the soil and the oscilations of outside temperature to find out how it would affect the inside environment in short periods of time, dayly and weekly. It deals with the capacity of the underground space to maintain an uniform environment.
For that, the variables meassured were dry air temperature, relative humidity, radiant temperature, and air velocity in its instant values, mean dayly values, mean montly values and mean annual values.
The meassures were taken through several sequences of 24 hours and two continuous ones of the annual cicle (2 years).

3. TEMPERATURE ANALYSIS.

The results of the different data meassured were ploted graphs and tables to facilitate the study and sign out the relevant points that are commented on the following paragraphs.

3.1. Annual temperature analysis.

The outside annual temperature oscilation is of 41.2ºC (from -3ºC to 38,2ºC). The inside annual temperature oscilation is of 17,5ºC (from 11,5ºC to 29ºC). The gap between the maximum and minimum outside and inside temperature is of about 4 days. The explanation for not having bigger gaps resides on the fact that these living units are very superficial, with not so great inertia on lateral walls and roof and present a big difference in the behaviour between those of the real underground architecture and the superficial excavated spaces.

3.2. Monthly temperature analysis.

Under the examination of mean monthly temperatures, it can be observed that the mean inside temperature is shifted up all over the year. This rise in temperature is of about 3.7ºC as mean value all over the year. However the increment in winter is of about 4,5ºC and in summer is 2,5ºC. This phenomenon is beneficial in cold weather but presents a problem for the summer months.
The oscilation of temperature during the month is satisfactory since the maximum difference between mean values account only for 4,2ºC giving a very uniform environment.

The inside mean value is always higher than the outside one.

1.3. Dayly temperature analysis

One of the main conclusions that can be deducted from the observation of the graphics is the great stability of the inside thermic oscilation, which in many days accounts for a straight line indicating an uniform temperature. The mean oscilation value of inside temperatures is of 0.7ºC in front of the 10,9ºC in the outside. The stability is furtherly proved by the observation of 5 days cicles in which a drop of 10ºC in the mean outside values, even below 0ºC, produces a drop in the inside temperature of only 1.5ºC. In the temperature climate and with high ventilation rates to provide heat the inside temperature rised 2,1ºC in front of 8,4ºC difference with the exterior one. This delay of 4 days for mean dayly temperature is reduced to 3 days when we refer to instant temperatures.

3.4 Hourly temperature analysis.

The constant experimentation of 24 hour cicles during the different seasons of the year allow to evaluate the radiant temperature of the walls. It was sensibly the same that the ambient temperature, changing only when intense ventilation rates were applied.
Another conclusion was the great homogeneity of temperatures in the inside ot the excavated living unit independently of the orientation of the façades.
It is also remarkable that the patios create a microclimatic environment intermediate between the outside and the inside of the living unit with a tendency to stabilize the inmediate outside temperature.

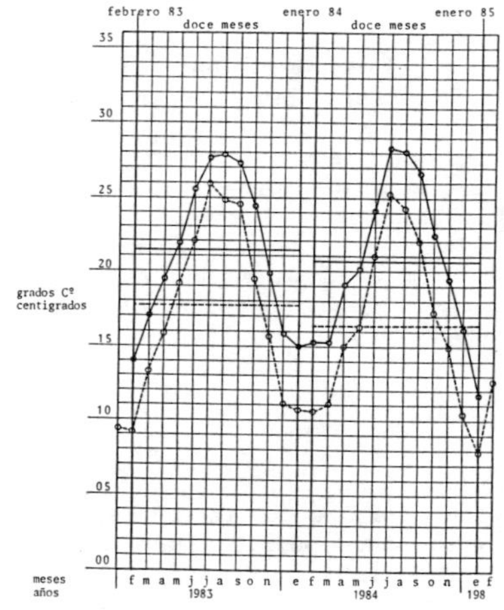

Fig 2. AVERAGE MONTHLY TEMPERATURE DURING THE TWO YEAR TEST.

Fig 3. The table shows the average Effective Corrected Temperature (TEC) during a whole year.

TS=DRY BULB TEMPERATURE

H% RELATIVE HUMIDITY

mes	TS	H%	TBH	V	TEC
febrero	14.0	75.6	11.5	0.10	13.6
marzo	17.1	75.1	14.5	0.10	16.4
abril	19.5	68.5	15.7	0.50	17.1
mayo	21.8	68.3	17.9	1.00	18.4
junio	25.6	71.8	21.7	1.50	21.6
julio	27.7	82.1	25.3	1.50	24.4
agosto	27.8	77.4	24.6	1.50	24.0
septiembre	27.3	74.9	23.7	1.50	23.5
octubre	24.5	69.0	21.8	1.00	21.5
noviembre	19.8	79.7	17.5	0.50	17.9
diciembre	15.8	77.1	13.4	0.10	15.3
enero	15.0	75.5	12.4	0.10	14.5

TBH= WET BULS TEMPERATURE V= AIR VELOCITY

Fig 4.
INSTANT TEMPERATURES ALONG SEVERAL DAYS. It shows

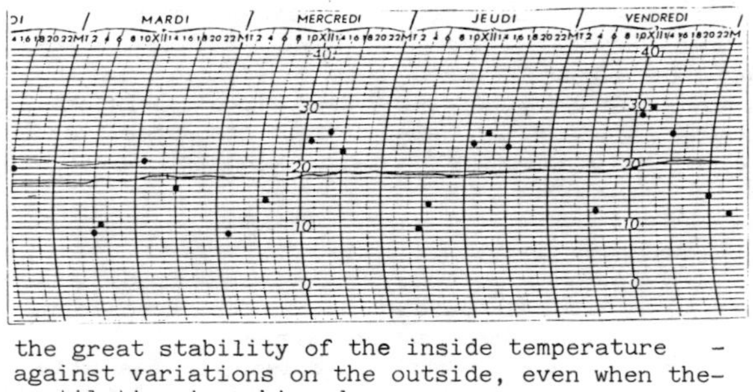

the great stability of the inside temperature against variations on the outside, even when the ventilation is achieved.

4. HUMIDITY ANALYSIS AND AIR VELOCITY.

From the biannual graphs it can be seen that the relative humidity is always higher than the otuside, refering to its mean values. It follows the outside oscilation with some resistan- ce. The inside oscilation is between 56,5% and 90% against and outside one of 24% to 98%. Condensation on the inside surfaces was never observed.
The high values of relative humidity in our Mediterranean climate, which is enhanced in the inside of the living unit is the most crucial problem, more so than the proper tempera- ture. In closed spaces without the human interven- tion the humidity rises and makes that the Effective Corrected Temperature (TEC) is high enough to be uncorfortable.
About 48% of the days during the year, the relative humidity remains above 75%, which shows how critical this variable could be in the environmental confort.
This problem is corrected by the people who live in this underground environment, by ventila- tion. The layout of the house favors cross ventilation with the patios to the front and back of the house. With only two or theree hours of ventilation the relative humidity descends 35% with only a 1.5ºC change in tempera- ture. This drop of humidity, when there is not ventilation rises quickly again to its normal values.

6. CONFORT ANALYSIS

Analizing the different variables that account for the definition of confort: air temperature, radiant temperature, relative humidity and air velocity, it can be deducted that these underground excavated spaces are quite confortable provided that ventilation is achieved by the people living in all year around except in January and February and also July and August, when the TEC remains slightly outside of the confort zone.

7. CONCLUSIONS

7.1. Confort conclusions

The overall result is very good in battering inside conditions with two points that have to be solved. One is the disminution of relative humidity and also the temperature increase in summer. However confort can only be achieved introducing some heat gains in winter and disminu- tin humidity and heat gains in summer, at least on an instant way during the hours that those variables are high. Humidity, as has been said, is corrected by the people throught ventilation and also some gains in winter, since the people who live there are very concious and surprisingly very exact in guessing when to open windows, and close them. This happens in winter when the outside tem- perature is higher than that of the inside environment. The artificial heat used is very simple usually through butano bunners which are the cheapest local source of heat in the area, before there were used local burnners. During the summer no special meassures were taken into account except opening windows and doors.

7.2. Architectural conclusions

The architectural layout being the same that those similar ones built above ground does not present any specific problem. It has some advantages the schnne of two patios, one to the front and one to the back of the house, with the adjoining rooms opened to those. This solves the problem of direct ventilation and direct sunlight inside the rooms. Accoustically the environment is quite good and does not present any specific problems, neither from neighbours nor the outside. The constructive solution could probably be improved with a thicker roof layer.

7.3. Possible design meassures that could improve the inside environment.

In winter passive systems could be used to increase heat gains that could increase inside temperature.
In summer sun protection would be advisable mainly on roof to prevent radiation.
Ventilation could be increased and maintainel through some specific design that will provide a continuous ventilation with some air temperature control.
In summer some passive cooling meassures could be beneficial if they will be active during the time that there is outside high temperatures, mainly cooling air to be introduced as ventilation

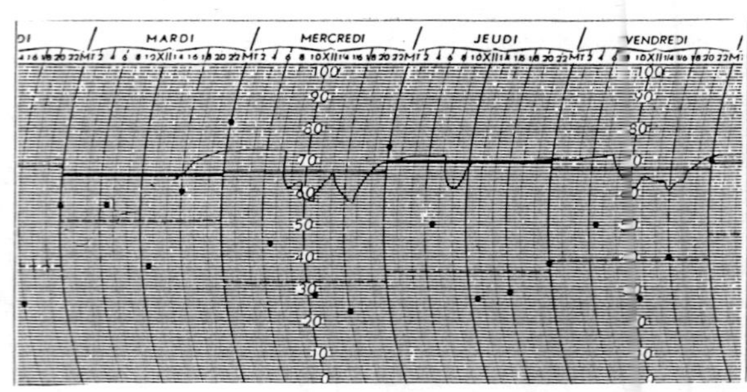

Fig 5. INSTANT VALUES OF RELATIVE HUMIDITY. It shows the tendency to stabilize, or those days, around 70%, only distributed by the drops provoked by the people who increase ventilation by opening to the outside.

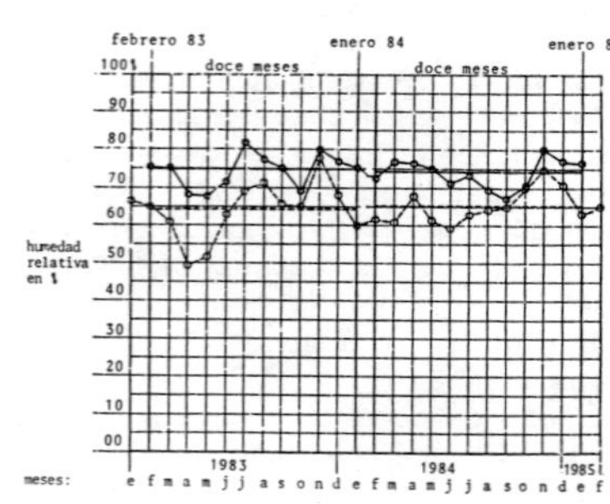

Fig 6. AVERAGE RELATIVE HUMIDITY DURING THE TWO YEAR TEST.

PROJECT FOR A SCHOOL IN VENICE-MESTRE

Arch.: Tassoni Mario — Santa Croce 1 - 30125 - Venezia
Arch.: Dondi Annamaria — Santa Croce 1 - 30125 - Venezia
Arch.: Bonati Nelly — Via Vecchia 2 - 26100 - Cremona

Istituto Universitario Architettura Venezia — Tolentini 191 - 30125 - Venezia

The project examines the possibility of constructing a school in a suburban area located between Venice and Mestre. The neighborhood is a buffer zone between a district built up around 1900 and whose public services and structures are now inadeguate, and a working-class neighborhood zoned for residential development, which is as yet altogether without services. The supporting structure incorporates diverse building materials: primarily, bricks of stabilized earth and, where necessary, reinforced concrete pillars and floors. Earth was chosen as the primary building material on the basis of its excellent insulation capabilities, ideally compatible with passive solar-energy climate control systems (greenhouses, atriums, and trombe-walls).
The proiect proposes the use of greenhouses, atriums, trombe-walls and solar panels, as well as natural illumination suitable for a school enviroment.

The project studies the possibility of building a school on the outskirts of the municipality of Venezia-Mestre in the region of the Veneto (Italy), in an area of particular importance due to its location between the city of Mestre and the historic center of Venice.

The project is to be realized in a buffer zone between a district (San Marco) which was built up around 1900 and whose public services and structures are now inadequate, and a working-class neighborhood zoned for residential development, which is as yet altogether without services.

The initiative, by means of a "projectual geometry" ingrained in the project itself whereby the plan follows and determines the conformation of the area, aims to create a unified organism that will link together the districts described above.
The complex articulates itself along a pedestrian street that connects the two districts.

The school is planned for 160 students and is articulated on two floors: twelve regular and two special classrooms, administration offices and the gymnasium on two levels with its services.
The gymnasium has three entrances: the first is for the use of students only, while the other two are, respectively, for the athletes and the public. In this way the space serves different purposes at different times.
An expanse of greenery is also planned on the south-east side just in front of classroom area.
The entry way, two stories high and on the south-west side, consists of a greenhouse.
Just inside the entrance there is a stairway leading to the administration section, which is connected with the classrooms on both the first and second floor. The groundplan of the project is composed of two square figures which overlap while one of the two rotates around the other.

The supporting structure incorporates diverse building materials: primarily, bricks of stabilized earth and, where necessary, reinforced concrete pillars and floors. The roof is flat and is designed for no secondary uses. The building consists of two stories. The overall thickness of the earthen-brick will be 40 cm; earth was chosen as the primary building material on the basis of its excellent insulation capabilities, ideally compatible with passive solar-energy climate control systems (greenhouses, atriums, and trombe-walls).

Energy-saving devices employed in the project are thus the following: orientation of the building towards the south; use of greenhouses, atriums and trombe-walls; use of solar panels (placed on the roof in conformity with the 45° slope of the sheds) to provide hot water.

In an initial phase of the project, the studies with regard to daytime illumination (with cloudy sky) and direct solar radiation were executed on the basis of a standard-classroom model (scale 1/20). In actually drawing up the project, however we found the statistics furnished by the standard model to be insufficient for our purposes. Therefore we deemed it necessary to modify the relation between the surface area of the classroom and that of the windows, thus guaranteeing a natural illumination suitable for a school enviroment.

1

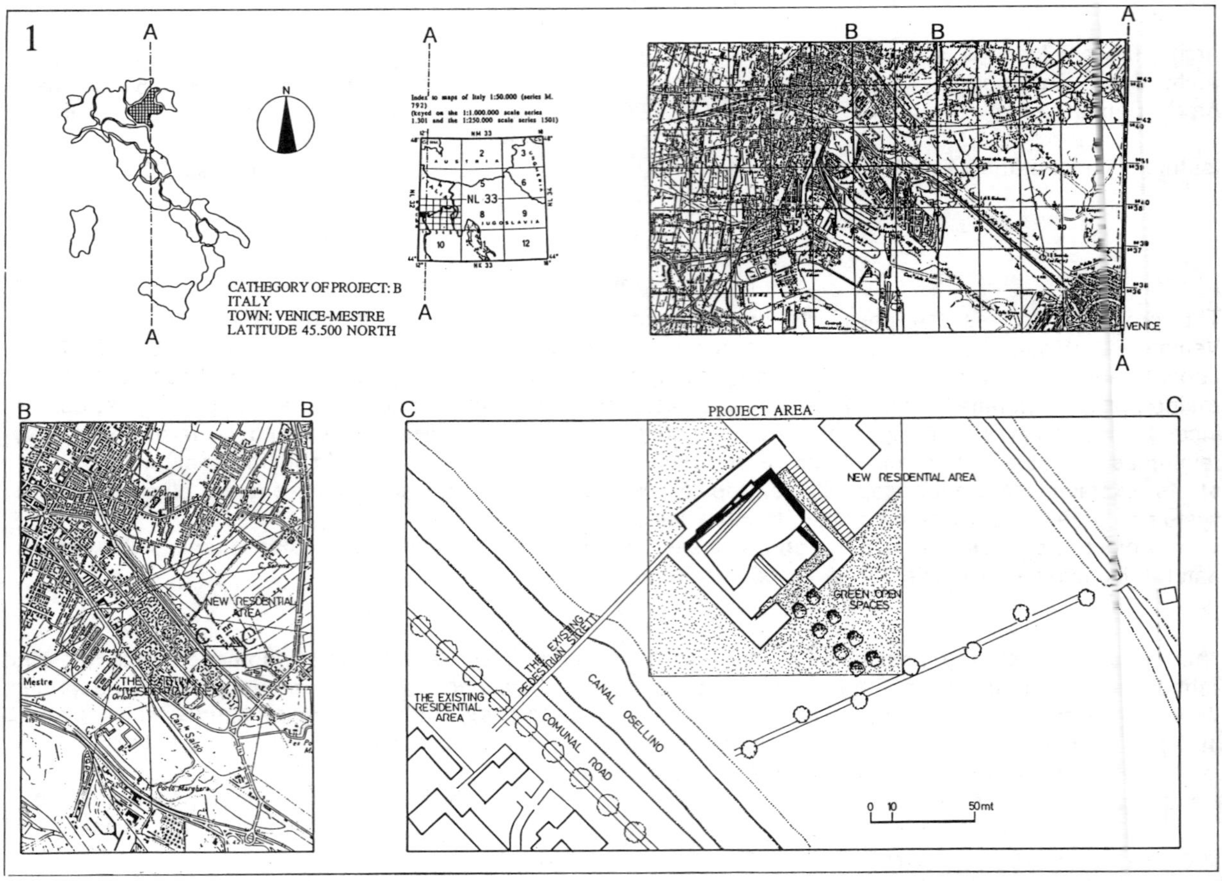

2

SECONDARY SCHOOL

1. ENTRANCES	mq.	245,89
2. CLASSROOM	mq.	720
3. SPECIAL ROOM	mq.	460,80
4. OFFICES	mq.	339,16
5. GYMNASIUM	mq.	842
6. TOILETS	mq.	81,60
7. CORRIDOR	mq.	508,80
8. DRESSING ROOM	mq.	63,92
9. CENTRAL HEATING	mq.	8,25
10. GREENHOUSE	mq.	64,58
	mq.	3344

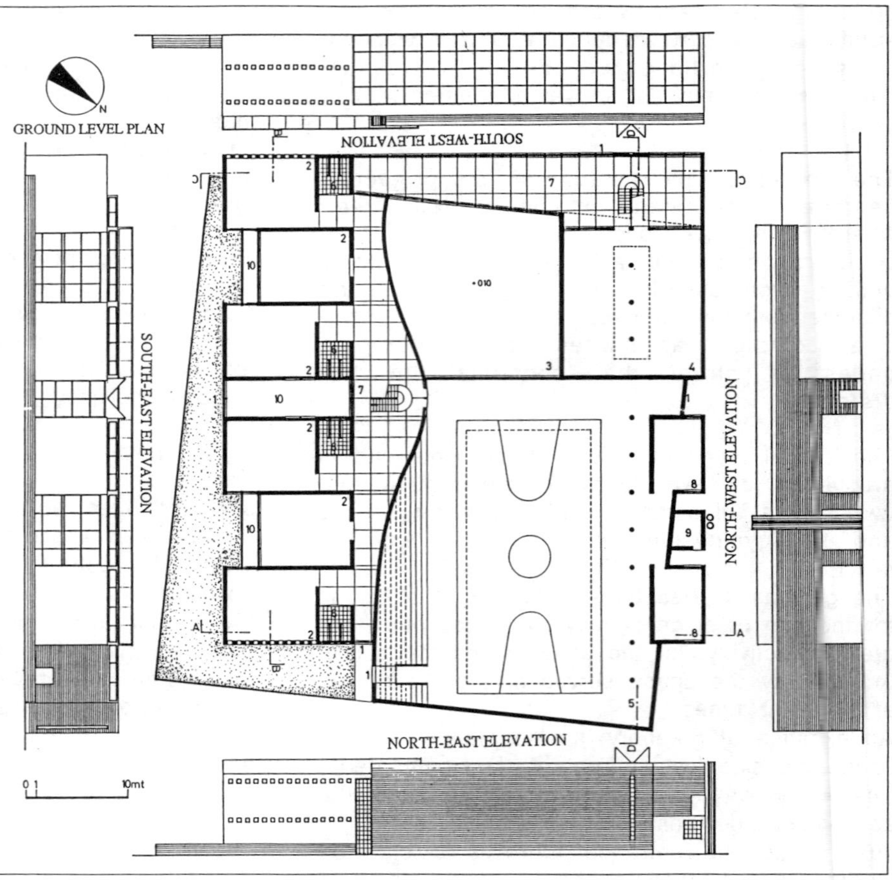

3

SECONDARY SCHOOL

1. ENTRANCES	mq.	245,89
2. CLASSROOM	mq.	720
3. SPECIAL ROOM	mq.	460,80
4. OFFICES	mq.	339,16
5. GYMNASIUM	mq.	842
6. TOILETS	mq.	81,60
7. CORRIDOR	mq.	508,80
8. DRESSING ROOM	mq.	63,92
9. CENTRAL HEATING	mq.	8,25
10. GREENHOUSE	mq.	64,58
	mq.	3344

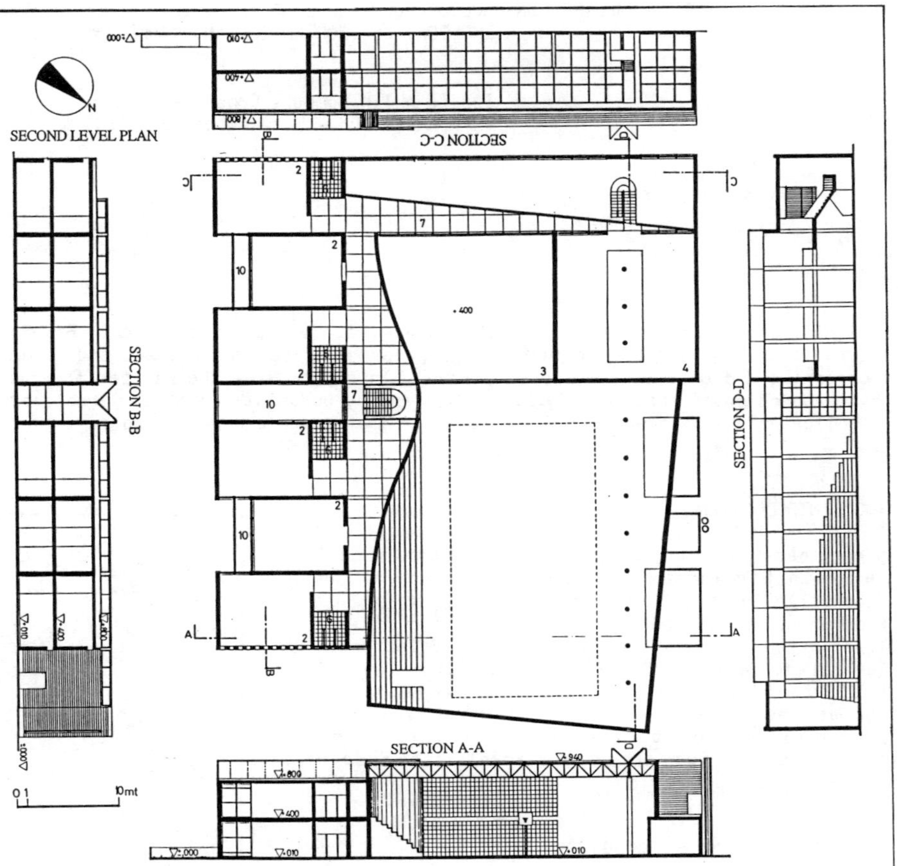

4

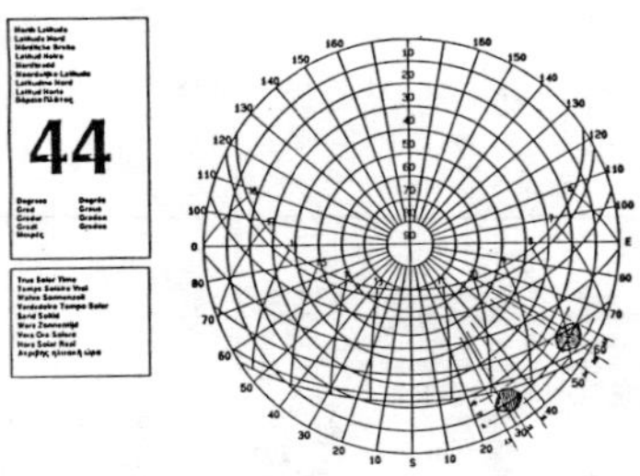

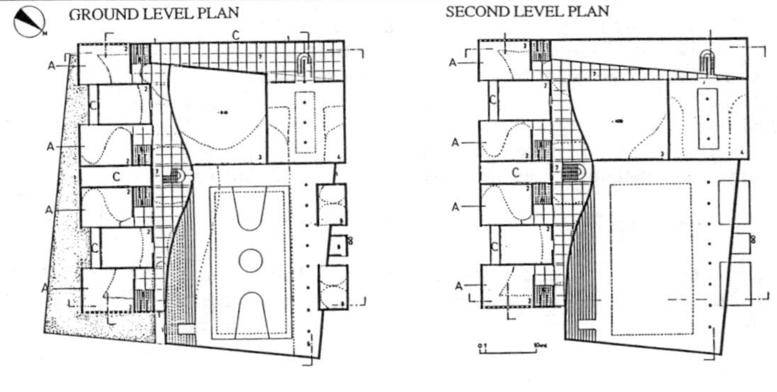

NATURAL LIGHTING PLAN

BRIGHT ZONE

MEDIUM-BRIGHT ZONE

DIM ZONE

THE SHADE IS PROVIDED EXCLUSIVELY BY TREES

ENERGY-SAVING DEVICES

TROMBE'S SYSTEM
(Concrete base min. 0,35 mt. from the ground Anti-streaming upper protective ledge)

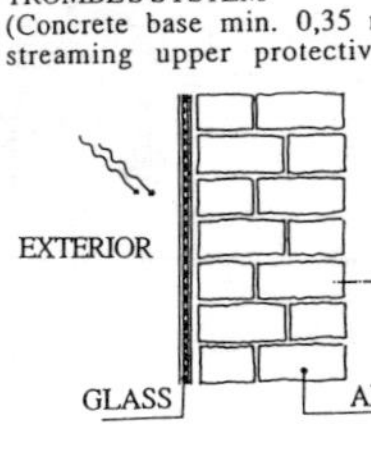

A

WATER SOLAR PANELS

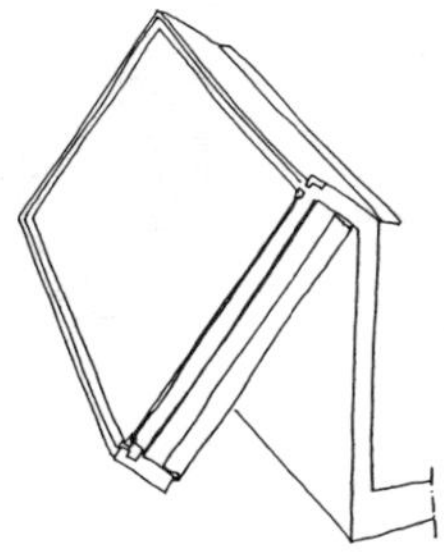

B

GREENHOUSE

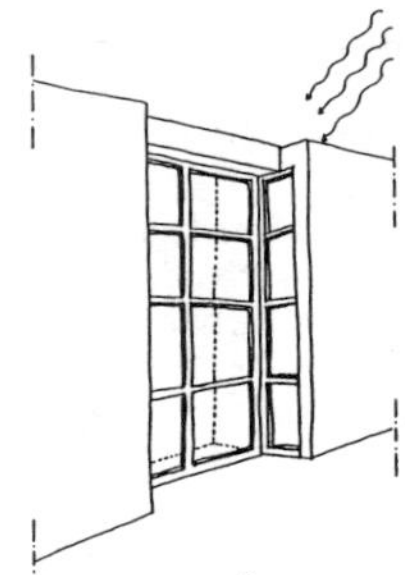

C

THERMAL PERFORMANCE OF 90 PASSIVE SOLAR FAMILY HOUSES IN BIRMINGHAM, UK: IMPLICATIONS FOR DESIGN AND MARKETING

Jesch, L.F.
The Franklin Company Consultants Ltd.
192 Franklin Rd., Birmingham B30 2HE, UK

ABSTRACT. Ninety passive solar family houses were constructed in South Birmingham in 1984/85 as part of the Bournville Solar Village project. This paper describes the project and summarises the results of computer monitoring and of a user response survey. In general the thermal performance was quite good, as indicated by measured solar fractions (0.32 - 0.37) and by occupant response. Problems with components and design are highlighted and recomendations are made for the design and marketing of future projects.

1.0 INTRODUCTION

Ninety passive solar family houses were constructed in South Birmingham in 1984/85 as part of the Bournville Solar Village project consisting of 300 housing units, pioneered by The Bournville Village Trust with the financial support of the Directorate General for Energy (DGXVII) of the Commission of the European Communities /1/. This paper summarizes the main results of three years of performance monitoring and outlines some conclusions which are significant for future projects.

2.0 DESCRIPTION OF THE PROJECT

2.1 Climatic conditions

The site in South Birmingham, in the West Midlands of England, latitude 52.4°N and longitude 2°W.
The weather in Birmingham is cool temperate, with moderately warm summers and mild winters. The normal heating season is from September/October to May. The mean of two years annual insolation on the site was measured as approximately 1080 kWh/m^2

2.2 Design criteria

The aims of the project were to provide low energy solar housing in an extensive building development to meet the needs of first-time buyers and to demonstrate that this type of housing is a realistic and attractive alternative to conventional buildings. Many design features and construction techniques employed in low energy family housing had been proven in the past to be effective in one-off buildings. The challenge of this project was to translate these features and techniques on a large scale into a building development which would at the same time be in keeping with the building styles and traditions of The Bournville Estate. This affected site layout, house design and component specification.

2.2 Thermal design of the houses

SITE LAYOUT: The ninety family houses were designed for a 7.6 acre site. The layout, shown in Figure 1, was constructed to allow optimal solar gain for every house, to maximise privacy and to avoid a dull grid-plan: objectives which seemed mutually incompatible. A major contribution to providing an interesting site layout was in making the main access road a broad curve rather than straight: this counteracted the monotony of the straight row of houses adjacent to the road. In addition, the orientation of houses was allowed to

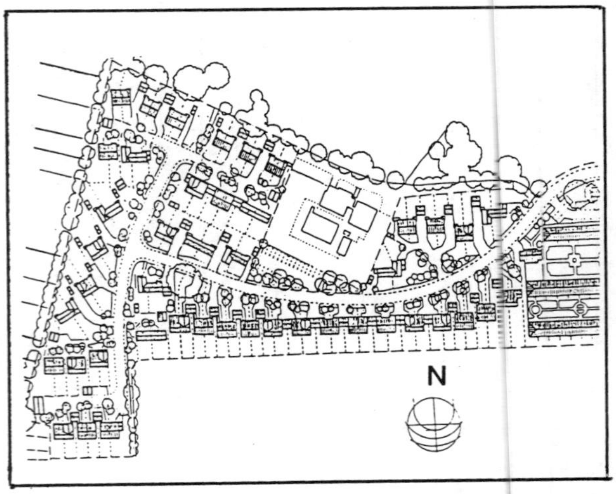

Figure 1 Site plan for the 90 houses

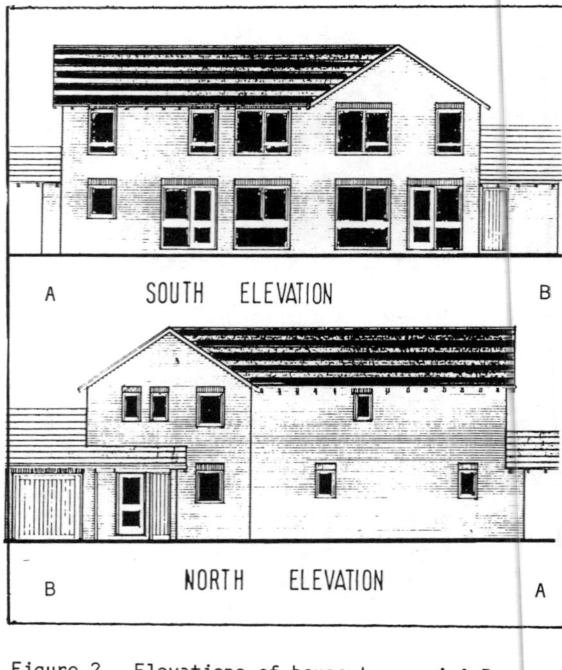

Figure 2 Elevations of house types A & B. In house A all rooms have south-facing windows.

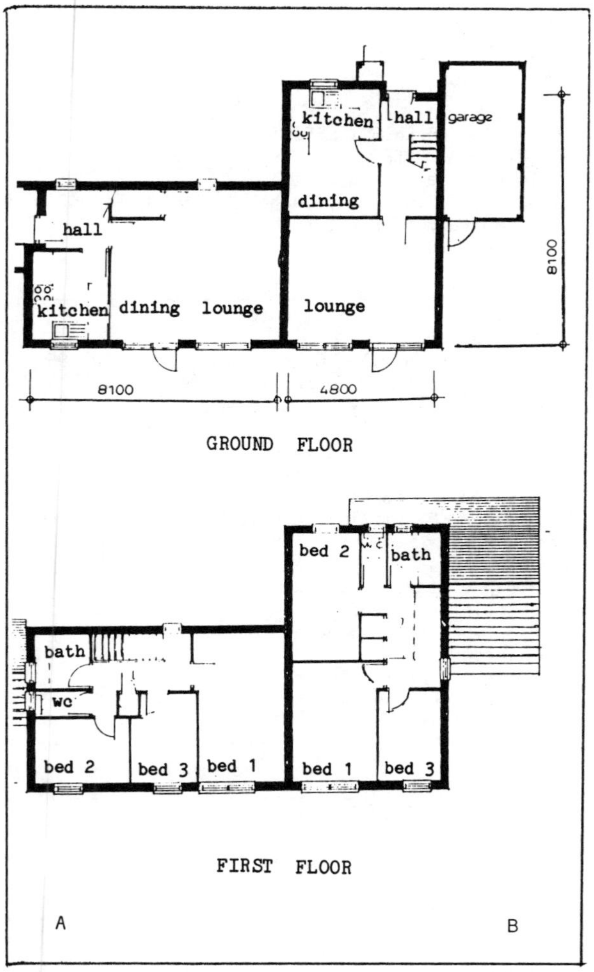

Figure 3 Plans of house types A & B. In house A all rooms have south-facing elevations

addition, the orientation of houses was allowed to deviate by 15° from South. Extensive planting of shrubs and small trees was carried out to provide for the privacy of the occupants as well as to create a pleasant environment.

BUILDING DESIGN: Three types of house were designed, two types with three bedrooms and one with two bedrooms. Elevations and plans of two types of houses are given in Figures 2 and 3 respectively.

In all of the houses large windows face south (usually 7.3 m^2 in ground floor living room) and windows to the north are small (0.7m^2 in kitchen) Most of the houses have a living room and one or two bedrooms on the south of the house with kitchen, bathroom and one bedroom to the north. Some houses were designed so that all rooms faced south.

ENERGY-RELATED BUILDING COMPONENTS: The wall structure consists of 100 mm class B semi-engineering brick on the outside, 100 mm cavity insulation material and 100 mm high density concrete block on the inside. There is 150 mm of glass-fibre roof insulation and double glazing throughout. Windows are all fitted with thermal blinds for additional night insulation or protection from the sun. Trickle ventilators were installed in all window frames to guarantee minimum ventilation. Humidostat fans were installed in the kitchen and bathroom to eliminate condensation. An efficient gas-boiler provides central heating and hot water, controlled by a programmable timer and a thermostat.

2.3 On-costs of the passive solar houses

The additional costs of passive and low-energy features in the 89 passive solar family houses was £ 258,100 /2/. This figure is based on a detailed analysis of the overcost of incorporating the low-energy/passive solar features in the houses. The architect and quantity surveyor's estimate is that approximately £2,900 extra was spent on each of the 89 houses.

3.0 MEASUREMENT AND RESULTS

3.1 Computer monitoring of thermal performance

Energy performance measurement was carried out by detailed instrumental monitoring of six houses, complemented by energy consumption recording in all the other houses.

Instruments and sensors in the six houses and in a weather station were wired to computer equipment in one of the houses where raw data was collected, archived, reduced and analysed /2/.

Figure 4. illustrates the measured degree days in each of the monitored houses for both years, while Figure 5. illustrates the solar fractions in each of the monitored houses for both years.

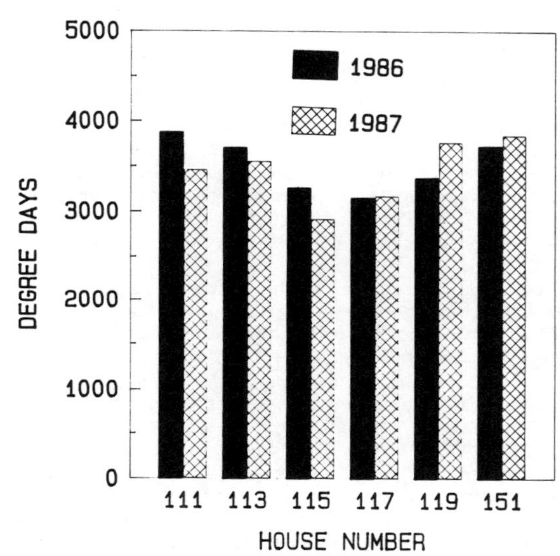

Figure 4 Measured degree days for six houses 1986 and 1987

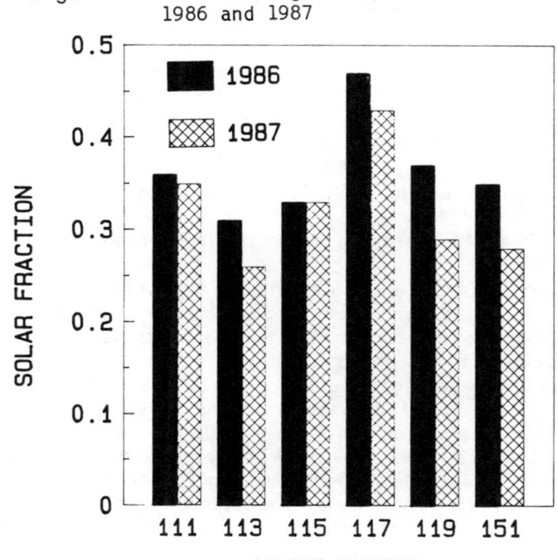

Figure 5 Measured solar fractions for six houses, 1986 and 1987

The annual average solar fraction of heat delivered to all the family houses was measured as 0.37 in 1986 and 0.32 in 1987.

3.2 User response survey

Occupants of fifty of the houses were interviewed, using a standard questionnaire. The results are tabulated in Table 1 /3/.
Users interviewed were pleased with the low level of their energy bills. For ninety-five percent they were lower than in their previous house, and ninety percent felt they were below average. The experience of living in the house had altered their perception of the importance of energy considerations: while only 30% had considered energy use important in making their initial choice, 63% felt it would be important in purchasing a future house.
The main problems encountered by users lay in the impairment of daylighting from small north-facing windows, the leakage of polystyrene granule insulation and the failure of double-glazed window units.
The lessons learned from the project have already been evaluated and the consequences incorporated into current building projects of the Bournville Village Trust.

4.0 CONCLUSIONS: IMPLICATIONS FOR DESIGN AND MARKETING

This project provides a good working model as to how passive solar design can be implemented on a fairly large scale in a suburban development, and highlights potential component and design problems.

4.1 Component standards

It is clear that in the UK it is essential to upgrade the standards of components and their installation to ensure that good passive solar design can be implemented effectively.

4.2 Design problems: the potential of transparent insulation

With regard to design, this project highlighted the sensitivity needed in balancing questions of natural lighting and heat loss on the one hand, and solar gain and privacy on the other hand. Technological developments of the late Eighties may give new answers to these problems. Transparent insulation "windows" could be used in north-facing walls to allow light gain without compromising heat loss /4/. Used in south windows, transparent insulation could facilitate solar gain but would allow greater privacy to householders: like the net curtains so popular in Britian it would prevent outsiders viewing the interior, but unlike net curtains it would not impede solar gain.

4.3 "Passive solar" as a marketing concept

However successful it may prove, "passive solar" is not recognised or understood by the general public. To translate technical success into widespread practice it is clear that new marketing methods need to be developed. The market is clearly ready for this development, as the green consumer movement gains pace. However clever marketing is required to make the link effectively between passive solar design and the environment. It is clear to the confirmed practitioners, but as yet has no recognition by building developers, estate agents or prospective house-purchaser.

QUESTIONS ASKED DURING PERSONAL INTERVIEWS AND THE THREE CHOICES OF ANSWERS

	(percentages) A	B	C
IMPORTANCE OF ENERGY EFFICIENCY IN BUYING A HOUSE			
1 Importance for current house A.very B.some C.none	30	48	22
2 Importance for future house A. very B.some C.none	63	35	2
HOUSE HEATING			
3 Heating costs: compared to previous house A.higher B.similar C.lower	5	0	95
4 Heating costs: compared to average similarly sized houses A.above B.similar C.below	0	10	90
House warmth enough in winter A.always B.mostly C.not	46	42	12
5 Overheating in hot weather A.often B.sometimes C.never	4	56	40
DESIGN			
6 North windows: natural lighting provided A.Above average B.Adequate C.Inadequate	4	22	74
7 South windows: privacy problems A.always B.sometimes C.never	6	26	68
8 Net curtains on south windows A.always B.sometimes C.never	28	4	68
COMPONENT PROBLEMS			
9 Double glazing: failure A.none B.some C. Most	6	38	56
10 Wall insulation granule leakage A.none B.some C. A lot	30	54	16
11 Problems with thermal blinds A.none B.some C.many	38	50	12
12 Roof insulation problems A.None B.some C.many	94	6	0
13 Humidity controlled fan problems A.none B.some C.many	76	22	2
ENERGY MANAGEMENT			
14 Experience of humidity fan effectivenesss A.good B.fair C.bad	39	55	6
15 Net curtains: perceived effect on heat gain A.improves B.no effect C.reduces	6	52	42
16 Use of thermal blinds at night A.Usually all B.Usually some C.Never any	57	43	0

Table 1 Results of the User Response Survey

5. REFERENCES

/1/ Jesch, L.F. "A community of active and passive solar houses at Bournville Village in Birmingham UK" Proc. DGS 5th International Solar Forum, pp 1051-1068, Berlin 1984

/2/ The Franklin Company Consultants Ltd "The Bournville Solar Village: Final Report to the Commission of the European Communities", 1989 (to be published),

/3/ Shrimpton, A.; L.F. Jesch and A. Pollock "The making of a solar village: management view and user response" ISES Solar World Congress, Hamburg, 1987

/4/ Schmid, J, W.S. Wilke, F. Sick "Transparent insulation: new concepts for daylighting and heating" Proceedings, 3rd International Workshop on Transparent Insulation, Titisee/Freiburg, 1989

DEVELOPMENT OF THE CONCEPT FOR SOLAR ENERGY EXPLOITATION ILLUSTRATED BY AN EXAMPLE OF A BUILDING DESIGN

Miomir Dragović, B. Sc. Architecture
University of Split
Faculty of Civil Engineering, Split
YU 58000 Split, V.Maslese bb

ABSTRACT. The concept for passive exploitation of solar energy has been developed in several phases. Continuous observation of the obtained results at the end of each phase can be used to determine the significance of the results achieved at leach level. In Split a multi-storey building with 46 flats was designed so that the obtained results in the reduction of energy consumption were estimated in each phase: urban planning, ground plan zoning, building design, installation of passive solar systems and additional heating and hot water preparation. The evolution of the concept was developed at each level of the project with different effects. The final confirmation of the obtained results can be achieved later when the building is in exploitation under different conditions.

1. CONCEPT DEVELOPMENT

The concept for solar energy exploitation for heating, cooling and hot water preparation in a multi-storey building is developed during the design process generally in three phases: preliminary design, main design and detailed design.
Each phase represents a process of detailed analysis and planning. It is very important when and in which phase of the design process the decisions are made and realized, since in each subsequent of decision making the flexibility is reduced.

In the first phase, while defining the relevant objectives and constraints, the energy problems refer to the main design of the passive solar systems and their general characteristics.
The most significant differences with regard to the evolution of the conventional building concept occur in this phase. Until now the architect has worked mainly alone in the phase of preliminary solution development. The existence of the thermal demands by regulations, programmes or reference energy points, requires that in the earliest phase the architect cooperates with the thermal engineers in developing the new concept of that building. The architect works on several levels simultaneously which means that he works, at the same time, upon the optimization of physical and passive solar properties, although he is not aware of this. It is evident that through the plan of masses,i.e. the factor of the building shape, the architect achieves the first level of the building concept, and, thus, either reduces or increases the real possibilities for the design of a solar energy building.
The further development of the concept is influenced by the selection of either the passive or active system, i.e. by the cooperation between the architect, thermal engineer, structural engineer and installations designer and the investor. It is certain that the design process for such buildings lasts longer than the design of "conventional" buildings, since this process asks for occasional analytical estimations of the obtained and expected results.
Furthermore, efficient energy, i.e. solar energy, exploitation implies an efficient construction either by the optimization of the structural and functional characteristics of the building, or by the optimization of total energy demands. This close interaction is particularly present in the design of residential buildings, especially multi-storey buildings.
The design of a multi-storey building which is the result of cooperation between architects, thermal engineers, installation designers, structural engineers and urban designers /1/ has to be permanently monitored in order to estimate the final energy results in realtion to conventional buildings. The next part of the paper presents, as a general review, the phases of work on the design development.

2. MULTI-STOREY BUILDING DESIGN

2.1 General Review

The investor (organization which organizes the building construction for the future users) order the design for a multi-storey building with solar energy for heating and hot water preparation. Thus, the design solution had to ensure minimum energy consumption and thermal comfort in the winter and summer periods.

2.2. Location and climate

The building is located at a latitude of 43.52° North and longitude of 16.43° East at 38 m a.s.1. It is in the mediterranean climatic zone. The values of the global solar radiation and the mean annual air temperatures are presented in Table 1. The orientation of the southern facade is at an angle of 9.5° towards West.

2.3. Urban Analysis

The urban plans of this housing settlement have already prescribed the outlines of the building, and, thus, the plan of the masses. The value of the building shape factor was not promising. Consequeatly, a simple calculation was performed by the cooperation between the architect and thermal

Figure 1: Site plan

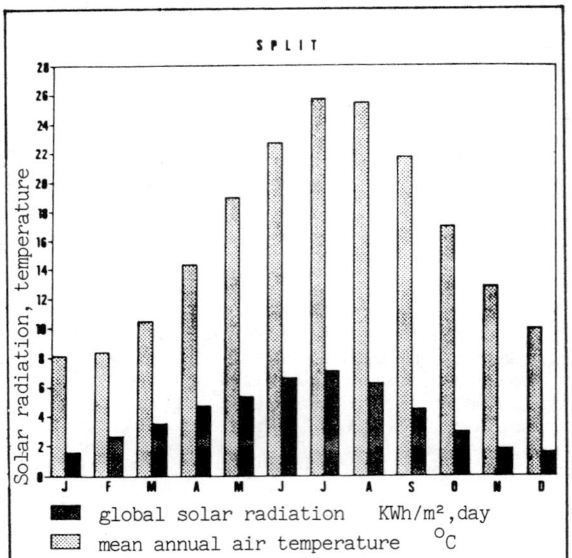

Table 1.

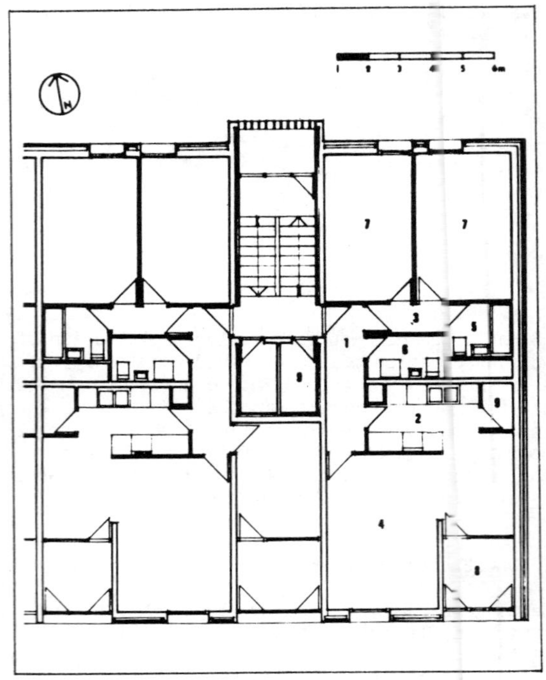

Figure 2: Typical floor plan

engineer, and a change in the plan of the mass was carried out. Thus, the optimal solution with a favourable orientation could be expected to yield satisfactory results. In the end it was proved that a savings of 18% with regard to the initial plan has been realized.

2.4. Zoning of the layout

The building has 5 separate entrances with 8-10 flats, i.e. two flats at each staircas which is characteristic for this type of buildings. The flats are intended for 4 to 5 persons, with thermal zones, so that the living-rooms, dining-rooms and kitchens are southerly oriented. The living-room and the study have exits to the glazed loggia, which is one of the characteristic passive system components and in addition to its energetic properties, represents an alternative area with specific regional tradition.

The proper positioning of doors ensures an appropiate distribution of heat. This thermic zoning of the ground plan results in further energy savings, which could not be effected without close cooperation with a thermal engineer.

2.5. The Building Structure

The lateral building supporting framework is made of reinforced concrete 20 cm thick, whereas the exterior walls, curtain walls and parapets are made of bricks 20 cm thick. The supporting ceiling structure is of reinforced concrete 16 cm thick.

The storage capacity of the building amounts to 23.5 MJ/K. The thermal insulation of the marginal external structure is stronger than required by Yugoslav standards. The external walls (except the walls of the glazed loggia) have been built according to the principle of ventilated marginal structure, with a complete elimination of all thermal bridges on the building.
The concept of the solar energy exploitation, has, in this phase of the design, been most remarkably confirmed. The improved thermal insulation made it possible to achive a 32% savings in the energy consumption in relation to building insulation effected according to existing regulations.

2.6. Passive Solar Systems

Direct gain through the southern window, and the principle of the warm beam in the glazed loggia, were chosen since they are economical and easy to maintain.
The glazing was effected by double glass with an absorption of 0.76 with 12.24 m² facing South. The windows are not with thermal insulation at night, so the sun protection is with PVC roller-blinds with the box externally located. The parpaet of the loggia is made of reinforced glass protected by an iron railing. Heat is stored in the loggia walls

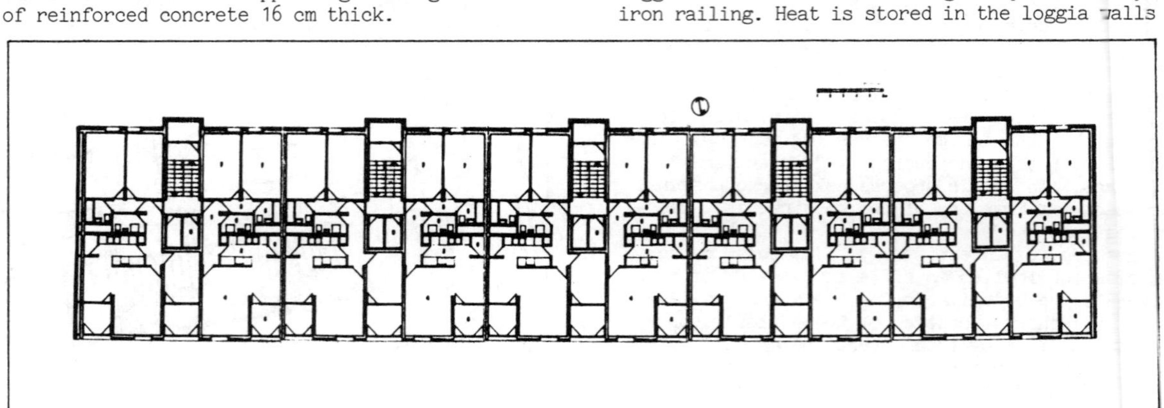

Figure 3: Typical floor plan of multi-storey building in Split

which are made of bricks and concrete without thermal insulation.
The heat is distributed into the living-room by the controlled opening of the loggia door, i.e. by a small fan with a thermostat placed in the upper part of the door. A detailed calculation has proved that in the phase of defining the passive solar system, i.e. by a glazed loggia with the previously provided high thermal insulation, it is possible to reduce the energy consumption for heating by 55-63%.

2.7. Auxiliary Heating and Hot Water Preparation

The contribution of the investors and future users was most significant in this phase. The thermo-accumulation electrical heaters were selected as auxiliary heating sources which are thermostatically regulated. The preparation of hot water was effected by a thermosiphon plant with solar collectors placed on a slanted roof. The system was effected separately for each flat with additional heating of water in the storage area of that flat.

3. THE CONFIRMATION OF THE CONCEPT

Continuous control and analysis of the achieved energy savings were effected in each phase of the concept development for the solar energy exploitation. The economic analysis confirmed the economical justification of the realized concept by taking into account the increased initial investment in the thermal insulation and the passive solar system, as well as other indicator of the efficient construction of the building.
Since the building is still under construction, it can be noted that the thermal optimization of the passive solar system will be studied throughly only when the building is in use under different conditions. The planned permanent recording of the obtained results should confirm the initial assumptions about the level of energy savings.

4. CONCLUSION

The methodology of architectural and urban design integrates the energetic concept in a specific way. The objective of the building design is to create an environment with a pleasant temperature. When the energy concept becomes the concept of the solar energy exploitation as the most natural energy source, it becomes quite evident that the architect must be consulted. The technology, which has been constantly developing in many scientific disciplines, should be applied in architecture as well, not only regarding the uses of materials, but as an awareness, as a concept in the methodological approach to design. Solar energy has always been present in architectural design, but the only question to be answered is to what extent the architectural concept should integrate it.

Reference:

/1/ M. Dragović, L. Kovačević, M. Umberger, J. Grabovac:
Concept for an efficient consumption of energy and passive exploitation of solar energy-preliminary design-S-1, Split, 1989

/2/ M. Dragović, J. Grabovac: Integration of the Concept of Solar Energy Exploitation into the Methodology of Architectural and Urban-Planning Design, Proceedings of PLEA Conference, Porto, 1988.

LE CORBUSIER'S CONTRIBUTION IN CLIMATIC ARCHITECTURE

Jacques J.P. MICHEL
Architecte D.U.H.
B.P. 32 - 14, rue des Poissonniers - 92204 NEUILLY-SUR-SEINE CEDEX
FRANCE T. (1) 47 22 07 94 FAX 47 47 01 71

ABSTRACT. Recreating a basic architecture. The first studies of Le Corbusier in climatic architecture dates as far back as 1928 for a villa at Carthage Algeria and namely on the Chandigarh Capitol designed between 1952 and 1956, where the positive and negative features of the climate, site, daylighting and aspects have produced an architecture of balance, rich in counterpoint.
At Amedabad where the monssoon , the dry hot season had given architectural solutions as roof parasol with hanging gardens to protect the building from the sun, vaults opened to the main direction of dominant winds, where climatic design enabled the conception of a basic architecture in harmony with tradition while respecting the local way of life.
The design of the Brazilian Pavilion at the "Cité Universitaire" on the parisian climate, the vacation houses on the mediterranean coast using natural cross ventilation through the building, movable sun breaker and roof covered with earth to provide a strong insulation.

1. INTRODUCTION

1.1 "The Sun-breaker

For twenty years the attempt has been made to elabore the principles of a rational architecture suited to modern town-planning. The essential elements in this technique were glass panes and pillars. The glass pane is an invaluable acquisition, an achievement of modern technique, but one which has to be perfected because the sun, usually man's friend, becomes his implacable enemy in certain latitudes at the height of summer. Therefore some device enabling the sun to have its full effect in winter and checking it in the dog-days of summer was indispensable. The first studies in 1928, for Carthage, were devoted to solution of this problem.
Since 1950 the same problem has arisen for Algiers where the sun problem must be taken into account in town-planning projects.
Barcelona too was chiefly concerned with the same problem in its researches in 1933.
In this same year, 1933 we find the complete expression of the sun-breaker principle in an arrangement in front of a pane of glass on the South and West façades of a tenement house in Algiers.
In 1934 the same device is used in houses for mass production in workshops in the North of France, intended for Algiers. The 100% glass pane is here again fitted with a sun-breaker in the West and in the South. The sun-breaker appeared again in the vertical section of the large blocks of flats in Ouchaia, where the southern façade is overhanging. In 1958 a sketch of the Law Courts at Algiers illustrates a sun-breaker whose importance is stressed by its being transformed into an open loggia. In 1959 the plans for the business-centre at Algiers (building 140 mètres high with bastion 15) show in its final form the loggia used as a sun-breaker. In the same year the Roscoff oceanographical laboratories were fitted with three kinds of sun-breakers: on the South side cells covering the glass pane; loggias for the flats; finally, on the west side, vertical sun breakers. (Vertical because the sun's maximum heat is at the horizontal level.)
Le Corbusier was called to Rio de Janeiro in 1956 and he placed the offices od the Ministry of National Education and Public Health facing direct North (southern hemisphere), an unusual position in this latitude. He suggested the sun-breaker which was carried out through the good offices of the Brazilian architects committee with Lucio Costa and Oscar Niemeyer at its head. This particular sun-breaker was based on the combined studies of Barcelona and Algiers 1933. The vertical sun-breaker was used in a second skyscraper in Rio, also with an unaccustomed orientation, designed by the Roberto brothers, whilst a third under construction at the present, architects Reidy and Moreira, likewise makes use of the vertical sun-breaker facing the West"/1/.

2. SOLAR PASSIVE ARCHITECTURE

Analysis of several projects using architectural solutions as the sun-breaker, roof parasol with hunging gardens, cross ventilation, glass pane, loggias etc...
Resume of seminars given in 1960 by Jacques MICHEL at the School of Architecture at Yale University, New Haven Con. USA
- Chandigarh Punjab India : Plan of the Capitol. The Palace of Assembly. High Court Building
- Sarabai House At Ahmedabab 1955
- Villa Shodan at Ahmedabab
- Vacation Houses - Cap Martin Mediterranean coast

2.1 Chandigarh Punjab India

The site chosen for the new capital of the Punjab is at the foot of the Himalaya mountains on a vast plateau between two great rivers, the new capital will have in is final phase 500.000 inhabitants. Yhe first stage now in construction will have 150.000 inhabitants.
Chandigarh is a political capital and its focal point is the "Capitol" the place where meeting of parliament is held and where the ministries are, the hights courts and the governor's palace. Following the theory of the 7VS, vertically from bottom to top of the town a great avenue which is 100 mètres wide rises to the Capitol where it will serve the 4 mains buildings. A pedestrian way is coming from the city center to the esplanade of the capitol, the cars drive in a trench (5 mètres) below grade, directly to the foot of the buildings. The complete surface of the Capitol is free from the mecanical circulations "The pedestrian is the master".
The design of the buildings of the Capitol for

which Mr Le Corbusier was responsible, was a great architectural venture using very poor materials and a labour force quite unused to modern techniques, with the tremendous obstacle of the sun and the necessity of satisfying indian ideas and needs.
The problem was accentuated by the ruling factor of the sun, under which this new indian way of life must be created. THe rainy season also had its problems. Le Corbusier had a revelation about the manner of construction to be used. The sun and rain are the two controlling factors in an architecture which must be both parasol and umbrella. The roofs must be treated in "hydraulicien" and the problem of shade can be taken as problem number 1. Here the "brise soleils" take the place of the classical façade but they cover not only the windows but the entire façade and influence the whole structure in contrast to all contemporary problems, the town of Chandigarh and its capitol had a freedom which was in itself a dangerous difficulty. The space devoted to the Capitol is vast, the buildings occupy a very small ground area. How could sufficent visual cohesion be given to such a huge conception, this was one of the most difficult problems.
Plan number 4445 shows the form, ground plan, and exact position of the buildings together with their vehicle and pedestrian traffic circulations (the signs) Le Corbusier expressed by symbols the basis of his philosophy and his understanding of the Art of city design. (Key of the creation of Chandigarh). From this arose the conception of the great esplanade about 400 yards long which joins the parliament Palace to the Hight Court. Along this esplanade are constructed the signs of the Moculor, there harmonic spiral, the daily path of the sun, the play ot the two soltices, the four horizons tower. (Made of concrete on the site or precast, treated with colour, wood frame with iron plating or bronze etc... They will be of great size, 20 mètres hight, 30 mètres long, "la Fosse de la considération") will receive "the open hand" vessel of sheet metal carried by a vertical member supported on ball bearing (about 16 metres height).
Pools are constructed in front of the Hight Court and the Assembly, those of the Government Palace are situated on two different levels, the distance which separate the governor's palace from the principal esplanade of the Capitol is so great, the use of play of reflections and arranging pools at differents levels preserve a precious visual relationships. In this manner the imperative relationships between the objects of the landscape are maintened. Mention must be made of the modulor, which has brought under harmonious rule, all the dimensions which have been used (texturique) (arthmetique) to layout of the Capitol on the site by a "carrure géométrique" two square of 800 mètres, inside square of 400 mètres "obelisques" on the corner.
Landscaping of the capitol : artificial hills have been created with fill from the excavations made for streets and parking strips. In certain parts of the capitol the horizons will be enclosed by wall of green.
No idea taken from folklore or art history could influence the design of these buildings which are build in mass concrete braced occasionally by thin membranes of reinforced concrete, gunnite projected on metal lath etc.. - the whole problem here is one of construction - the aesthetic which will come from this, will be enterely new.

2.2 The Palace of Assembly

The plans show the final version of the Palace of Assembly but contain certain errors rectified at the site by Mr. Le Corbusier in March 1956. For example, the great hall of deputies, with the seating arrangement and the place reserved for the speaker constitued an encroachment on the indian constitutional procedure. All was seating to correspond to indian custom, in which the government and the opposition are situated face to face and discuss issues, instead of adressing their discourse to an assembly and to a gallery filled with an lager public.
One of the interesting points to call attention to is the adoption of a circular form for the hall. The assembly hall is made in a hyperbolic shell with an average thickness of 15 cm (6") constant throughout its surface resulting in a very low cost and a minimum of weight (principle of industrial cooling towers).
This shell does not terminate in a horizontal but in an oblique section which receive a metallic framework (aluminium).
This framework will become a veritable physical laboratory destined to ensure the interplay of natural lighting, artificial lighting, ventilation and acoustic electronic mechanisms. Moreover this framework will lend itself to possible solar festivals, recalling to men, once a year, that they are children of the sun (entirely forgotten in our imfettered civilization crushed by absurdities, particularly in architecture and city planning).
All the acoustic corrections, more or less will be able to be made along the shell according to valid estimates. What is notworthy in the form is that part of the surface of the shell is sound-reflectant, while the other part is sound-absorbent - the air conditionning necessary here, works under the best conditions since the cool air is introduced from above to below, several meters above the deputies and visitors, and will descend by gravity to breathing level while the warm air will rise and be exhausted by mechanical apparatus installed in the framework of the hall.
This project will be striking by its classification into two types of framing
1 - framing of the offices similar to that of Secretariat : brise soleil, portico, concrete etc... 2 - framing of the space called the "forum" in the middle of which the great hall of deputies (lower house) and the hall of senators (upper house) are immersed - this immense promenade surface insures, for the deputies, the possibility of all kinds of favorable encounters - for individual conversations, dealings out of session etc...
The empty space above this forum corresponds to the three floors of superimposed offices. The roof forming the celling of the forum is supported by tall, slender concrete columns.
The run of water from the terraces (during the monsoon) is ensured by what Le Corbusier has called two "rivers" : one in the form of a large concrete canal separating the forum and the grand portico : the other separating the forum and the body of offices where the president and his ministers are located. Yhe rooftop will be accessible to accomodate evening or night time entertaining.

2.3 Hight Court

In exploitation since March 1956. The Hight Court design proceeded from the repetative function of 8 courts of justice and one height court.
- Dominant winds, sun (shadow) have dictated the orientation of the building (as in the all city)
- Architectural solution for the double problem, rain and sun - Roof parasol 102 mètres x 32 mètres, which protect the courts and offices from the direct sunlight, the wind goes freely under the parasol and through the hall. Hydraulic treatment of the roof, which directs the rainwater through a large canal and gargoyles which are connected with the pools outside.
The addition of courts was done following a rythm decided since the early composition of the capitole. Arithmetique first on the dimension of the courts

and the hight court, each room is in itself a plastic body. Height, width, lenght ; that was the base of the decision for the small courts. 8 x 8 x 12 mètres, 12 x 12 x 18 mètres for the height court.
At the begining of the design of the glass pannels and the "brises soleil" (1,40 m deep - 4'7") the use of the Modulor was employed (arithmetique and texturique)
In the transvers section of the building which shows the offices and the courts protected from the sun, the modulor brings "l'unité texturique" everywhere. The three pylons of the great entrance (hall), coated with gunnite cement are painted (one green, another white and the third in red orange), both left and right wall are black.
The circular ramp in the hall is made of concrete rough from formworks. Recessed mouldings appear in the face of the concrete achieving a situation similar to that in which the Egyptians prepared sculptured frescos in their temples 5000 years ago.
650 square metres of tapestry designed by Le Corbusier covered the black wall of the small courts (8 tapestries of 64 sq. m each) and one for the wall of the large court of justice (144 sq. m). These tapestries were made for acoustical reasons.

2.4 House of Mrs Manorama Sarabai - Ahmedabad 1955

The river floods sometimes, the house is build not too far from the river, the first floor is at 4' above the natural grade.
The country is tropical, the monsoon is there for two months of the year. Period of heavy rain and hot humidity. At this time the house must be closed with only the air conditionning in use inside.
During the period of dry hot climate with intense sunshine confort is achieved by currents of air, and shadows. The house is opened to the main direction of the dominant winds in order to be traversed by currents of air.
The half cylinders of the vaults carried by the brick walls are as pipes opened to the direction of the wind, with verandas build in front and also on the south west side, to protect the inside from the sun.
The structure is "catalonian vaults". The first row of flat bricks set in plaster without formwork, coupled with a row of bricks cast roughly in cement as in the Jaoul houses in Paris. These half cylinders are carried to the bricks walls by a rough concrete lintel - the composition serves to create openings in these walls, all parallel, playing solids against voids.
Some walls are white coating (plaster) also present are intense colors. There are pivoting doors in natural wood between the verandas and the house.
The mosquitos are everywhere and windows cannot be left open without special amenagements.
An other research in this house was the roof. The half cylinders of the vaults when the roofing is applied, are covered with earth and the upper part of the house becomes a magnificent garden - in relation to the luxuriant nature around. this solution also provides a strong insulation for the house. Incidentaly we must say that the Indians sleep several months of the year on their roofs during the dry hot season.
A concrete toboggan is build in liaison between the garden roof and the swimming pool. The floors are covered by a stone paving (made of black madrasstone) layed out according to the Modulor.
This system of stone paving is used also at Amed Abad on the Shodan house, Millowners Association building Museum etc... at Chandigarh and lately in Paris at the "Maison du Brésil"
The module in plan of the Catalonian vaults, used at the Sarabai house is 12' with bricks walls of 1' - the module at the Jaoul house in plan is 12' and 7'5" with brick wall of 1'.

2.5 Villa Shodan - Ahmedabad India

The module of this construction is 7'5" or 2.26 m.
The concrete structure is very simple, but there is a great plasticity achieved by the implantation of the differents rooms, in their form, their spaces, their dimensions, in the shadows of the sun breakers, under the ombrella of the roof and with the hanging gardens.
Here is an "Architecture pneumatique" (Résonnace et contre-point)
The roof parasol gives a strong insulation and protects the house from the sun.
The concrete for the façades and sometimes inside is rough, from the forms made of wood planks as they are available.
Only under the parasol and on the ceiling of the rooms is used a regular formswork in sheet steel (3' x 4'7") which relates to the design of the Madras stone paving on the floor.
A strong polychromy covers same of the walls and ceilings and creates an ultimate dimension in the Architecture.
A concrete ramp leads from the first floor to the mezzanine. Above on the space of the hanging garden, on several levels are three apts, separate but in contact with one another.
Originaly this villa was designed for an other site, but the project was so well appropriate with the indian needs and the climatical conditions of this part of India, that the transfer of this house from a site to another was perfectly natural.

2.6 Vacation houses (Unités de vacances) Cap Martin mediterranean coast

These buildings were never build, but the experimentation is interesting for the use of precast cells (room 12' x 12') kitchen elements (type Marseille Unité) etc... which can be used in any location and incorporated in local construction methods. Mr. Le Corbusier decided to build 5 vacation units on a slope, near the shore, on the site of his actual vacation retreat at Cap Martin.
The parallel walls will be made with ruble concrete and stones incorporate (from the excavations) cast on benches.
The roof will be an irregular concrete slab covered with earth and stones, which will look (from the upper roads) as the other terraces cut (since thousand of years) on the slopes of the hills surrounding the mediterranean sea.
The 2 main floors between the walls, for each unit, will be built as a boat deck on wood planks with a slope for the easy washing. After that the rooms 12' x 12' and other elements built at the factory will be placed on floors and will provide the finished and equipped elements of the house.
The empty spaces between the cells and the walls will become the "Atelier" on the first floor. On the second floor the living room with an extra height of 6' over the roof of the cell, where you could sleep in a natural cross ventilation and look from this stage over the large Forum.
The south elevation looking at the sea will be covered with aluminium pannels and wooden framed glass pannels, with vertical ventilation and an movable sun braker (under the roof) will control the sun (winter and summer)
The north elevation (access) will be covered with aluminium pannels, plus windows and vertical ventilation, on the upper part are movable pannels for cross ventilation. Other approachs were studied for the application of the 12' x 12' cell but the only realisation was the prototype unit, where Mr. Le Corbusier spent his vacation each year at Cap Martin.

/1/ Girsberger - Zurich copyright 1946
.Michels'collaboration with Le Corbusier

LE SOLAIRE PASSIF EN FRANCE
Etat de l'art
Axes de recherche

Eric DURAND
AGENCE FRANCAISE POUR LA MAITRISE DE L'ENERGIE
Service Habitat et Tertiaire
Sophia Antipolis - France

* RESUME
Bien que les acquis techniques soient importants, les expériences nombreuses depuis une dizaine d'années, les logiciels de plus en plus performants, il semble que, au vu des enquêtes, campagnes de mesures et de la connaissance du milieu professionnel sur le terrain, il y ait encore de nombreuses actions à mener d'incitation, de conseil et aussi de recherche, tant dans le domaine des produits que des outils de conception. Les analyses sociologiques montrent en outre qu'une composante "humaine" des produits est essentielle (ergonomie, adaptation aux mentalités).

LES AXES DE RECHERCHE - HABITAT

Dans cette réflexion, nous aborderons chaque idée avec le souci de prendre en compte simultanément trois aspects indissociables :

o le chauffage - Maîtrise des déperditions - Valorisation des apports gratuits,
o la "climatisation naturelle" et le moyen de limiter le rayonnement direct en été,
o l'éclairage naturel qu'il faut maintenir à un bon niveau tout au long de l'année.

1. Le gain direct : une paroi "aperditive"

Sa capacité intrinsèque de performance et de bilan annuel positif (gain supérieur aux pertes) est l'atout principal du vitrage Sud.

Une "petite" réponse aux problèmes d'effraction pourrait être le développement de composants fenêtres à effet pariéto-dynamique tels que :

o fenêtres à volets intérieurs formant capteur solaire en position fermée pour le préchauffage de l'air (figure 1),

BLOC BAIE "CLIMATIQUE"
AVEC VOLET ROULANT A 3 POSITIONS

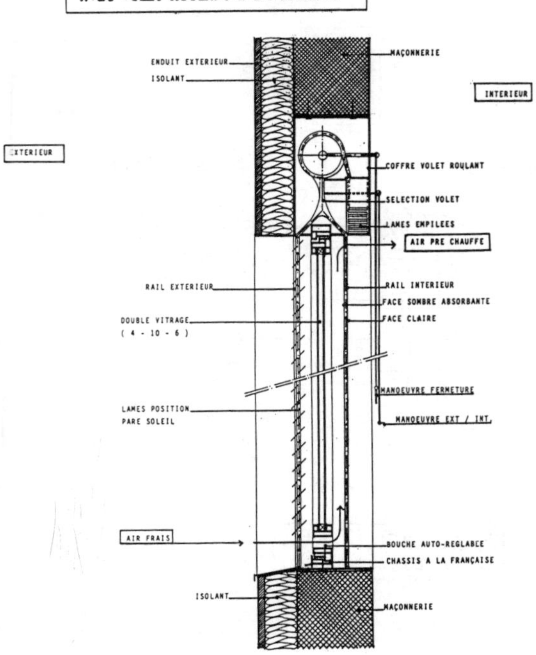

o volet roulant à deux positions (figure 2) :
. Intérieur (hiver)
. Extérieur (été).

BLOCS-BAIES MULTI-FONCTIONS

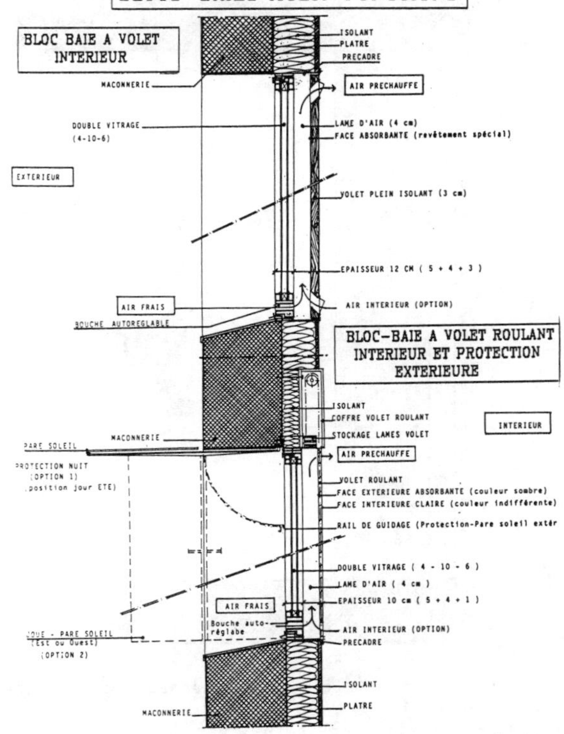

2. La serre

En elle-même, la serre ne me semble pas devoir être un thème de recherche en soi dans la mesure où elle représente un surcoût important non énergétique ; par contre, pour le cas où elle serait un acquis, une valorisation énergétique consiste à l'utiliser pour le préchauffage de l'air neuf.

En ce cas, on pourrait imaginer des kits ventilation permettant, à partir d'une prise d'air préchauffé, de conduire cette chaleur dans les différentes zones de la maison (gaines + ventilateur adapté + clapets anti-retour, etc...).

3. Les murs capteurs

Contrairement au pessimisme général, je pense que l'avenir est plutôt devant nous. Il suffit, si possible, d'oublier le passé et de partir sur des bases nouvelles :

1°) Il ne faut plus parler de noir, ni de couleurs en ruputure avec l'esthétique du bâtiment.

2°) Il serait souhaitable que ces composants soient plus légers et, pourquoi pas, industriels.

3°) Il vaut mieux une performance "moyenne" avec un produit beau qu'un produit très efficace mais inacceptable par un maître d'ouvrage.

4°) L'effet Trombe (thermocirculation) est peu valorisant car le "gain direct" joue ce rôle. L'intérêt du mur capteur est la restitution décalée dans le temps, d'où le qualificatif de gain "indirect".

L'idée que je suggère pour la recherche de composants type mur-capteur repose sur le principe suivant (figure 3) :

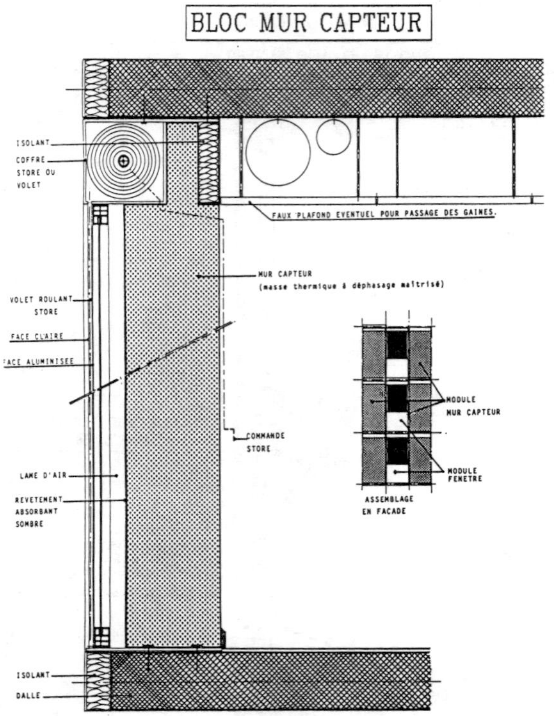

o composant industriel avec "normes européennes" (ou sur mesure),
o dimensions réduites (poids, encombrement),
o esthétique traditionnelle (coloration, aspect),
o fonctions intégrées : . effet serre,
. masse,
. isolation-occultation.

(Les recherches pourraient porter entre autres sur un matériau à forte chaleur massique, à conductivité thermique faible (déphasage) et à émissivité forte).

4. Le préchauffage de l'air - La ventilation

Le préchauffage de l'air est à mon avis un bon moyen de valoriser les apports solaires. Rappelons que le capteur à air en boucle ouverte est le meilleur des systèmes solaires à air. De plus, le préchauffage de l'air améliore notablement le confort.

Une idée de recherche :

Des composants de façade à effet de serre : avec une prise d'air en partie basse et une bouche de soufflage en partie haute (produit assurant l'étanchéité d'une façade): figure 4.

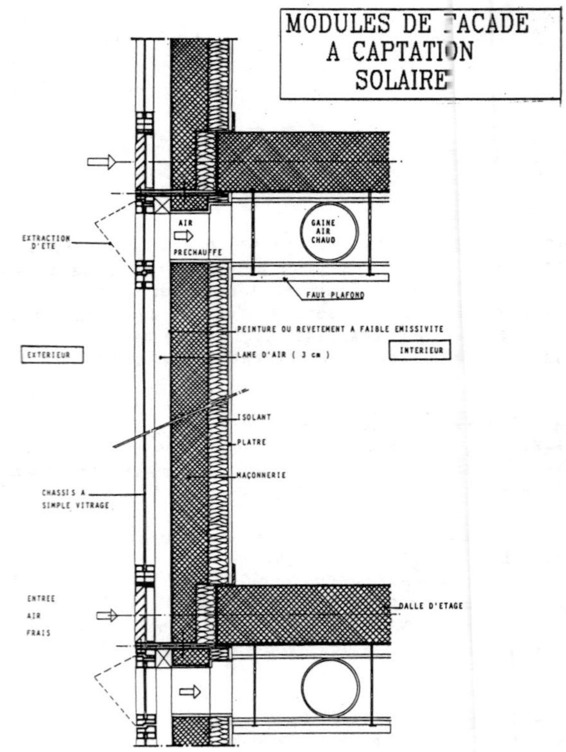

Cette réflexion sur le préchauffage de l'air me conduit aux conclusions suivantes :

o déperditions de plus en plus faibles,
o air neuf très largement chauffé,
o apports internes non négligeables,
o possibilités de récupérer sur l'air extrait.

Dans une telle configuration, un réseau aéraulique ne pourrait-il pas être le seul support à l'intégration de tous les transferts d'air (y compris un chauffage d'appoint pour les besoins résiduels !) (figure 5).

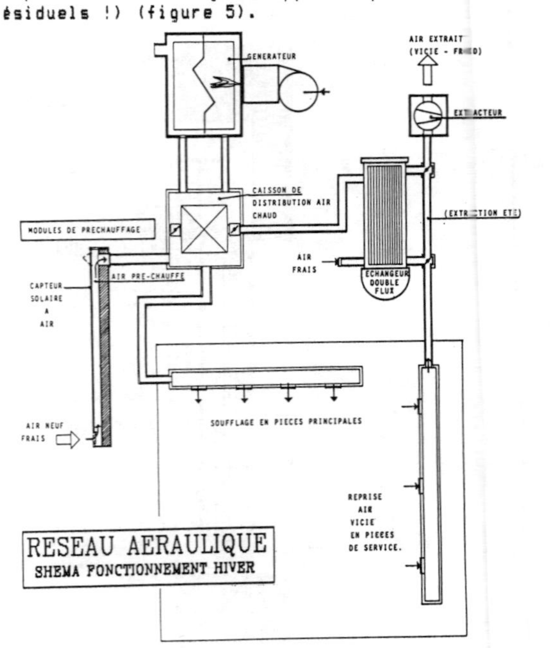

LES BATIMENTS TERTIAIRES

Nous distinguons deux grandes familles de bâtiments tertiaires, ceux à usage intermittent, ceux à occupation permanente. Mais, dans sa diversité, le "Tertiaire" recouvre un nombre important de familles, avec autant de spécificités et donc autant d'applications possibles de techniques solaires passives avec variantes. Le parti technique étant une fois l'inertie lourde, l'autre fois l'inertie légère, les serres, le gain direct, il est aussi délicat de donner une réponse car toutes ces solutions ont leurs avantages. Le choix s'effectue en principe en fonction de l'usage futur.

I. LE TERTIAIRE A USAGE INTERMITTENT

Par nature, ce type de bâtiments nécessite un bon éclairement intérieur, des besoins de chauffage limités (en général concomitants aux apports solaires) et des conditions plus sévères que l'Habitat en ce qui concerne les niveaux de température en confort d'été (26° C maximum pour de bonnes conditions de travail dans un bureau).

o Le gain direct est ici particulièrement intéressant pour les deux fonctions, chauffage et éclairage mais fait apparaître une contrainte liée au confort d'été : nécessité de protection extérieure et/ou des masques fixes.

o Le "gain indirect" type mur capteur est, quant à lui, adapté aux usages nocturnes (gymnases, salles de spectacles).

. Le rôle de l'inertie

o Traditionnellement, les bâtiments à usage intermittent, bureaux, écoles... sont conçus avec une inertie légère. Cette technique est, sur le plan climatique, en contradiction avec une bonne valorisation des apports solaires directs. Une inertie légère conduit à des surchauffes passagères intolérables qui impliquent de baisser les stores, d'ouvrir les fenêtres, annihilant ainsi tous les gains solaires potentiels. Une conception intégrant des masses thermiques assurant une inertie séquentielle de quelques heures est nécessaire pour restituer l'énergie solaire pendant les passages nuageux, en soirée, ou tout simplement pour limiter les surchauffes. Une inertie moyenne, voire lourde dans certaines régions, est indispensable.

o En zone Sud, compte tenu de l'usage intermittent, apports solaires + apports internes assurent pratiquement la totalité des besoins diurnes.

o En zone Nord, il faut compter sur l'efficacité d'un système de chauffage classique. Sa fonction principale sera la mise en température avant l'arrivée des occupants et des apports solaires.

Il faut donc trouver un compromis entre une inertie légère qui permet une mise en température rapide avec un minimum de dépenses d'énergie et une inertie lourde qui permet de conserver la chaleur emmagasinée la journée, mais qui demande un apport beaucoup plus important pour une remise en chauffe. La tendance est a priori de privilégier l'inertie en zone Sud (moyenne à forte) et de se contenter en zone froide d'une inertie moyenne.

. L'effet pariéto-dynamique

o Les ouvertures pariéto-dynamiques qui conjuguent préchauffage de l'air neuf et gain direct semblent être un produit parfaitement adapté à cet usage.

Tous les systèmes à base de préchauffage de l'air neuf ou de récupération des déperditions sont susceptibles de bonnes performances. Par exemple, des composants industriels préfabriqués, hauteur d'étage, intégrant allège et ouvrant, ou parois opaques .

. Cas particulier des écoles et gymnases

o En période d'occupation maximale, les établissements scolaires ou sportifs se caractérisent par des apports internes très importants ; les besoins de chauffage sont alors très faibles ou même négatifs (apports gratuits supérieurs aux besoins même en plein hiver). Dans ce cas, l'énergie solaire a donc une fonction principale :

. Préchauffer ou chauffer le bâtiment en attendant ses occupants : il faut donc pendant la journée stocker l'énergie solaire incidente pour conserver cette chaleur en vue de combattre les déperditions nocturnes, pour assurer le lendemain, à l'ouverture des locaux, une température de confort.

Dans le cas des gymnases, l'utilisation judicieuse des apports gratuits peut être suffisante pour se passer d'un chauffage de base.

Des systèmes type mur capteur à déphasage (12 heures environ) devraient répondre assez bien à cette fonction.

o Dans ce type de bâtiment, la fonction éclairement est capitale ; il convient donc d'imaginer des composants "vitrages" équipés de protections solaires pour les périodes estivales et la mi-saison, protection ne devant pas diminuer le coefficient d'éclairement.

II. - LES BATIMENTS A OCCUPATION PERMANENTE

Ce sont essentiellement des bâtiments d'hébergement tels qu'hôpitaux, maisons de santé et de retraite, hôtels... qui se caractérisent par rapport aux autres bâtiments tertiaires par un usage continu et des besoins importants (chauffage - eau sanitaire - éclairage - froid). Le gain direct, associé à une inertie moyenne à forte, est toujours une technique à mettre en avant, en tenant compte des orientations des différentes façades. En tissu urbain dense, il convient donc de traiter différemment les ouvrants des façades mal exposées (Nord-Est à Nord-Ouest), même si elles abritent des locaux identiques à ceux situés avec une meilleure orientation (Sud-Est à Sud-Ouest).

o Hôtels, hôpitaux, voici beaucoup de bâtiments qui font appel aux composants industrialisés. L'intégration de systèmes de préchauffage de l'air neuf pour les murs orientés au Sud (avec un système de récupération et de distribution de cet air préchauffé) est une solution satisfaisante.

o Le poste chauffage, compte tenu de la possibilité de jouer avec l'inertie, pose des problèmes moins complexes que pour le Tertiaire intermittent. En corollaire, l'aspect protection solaire doit être étudié avec soin dans la mesure où l'impossibilité de rafraîchir (la nuit, entre autres), doit conduire à limiter strictement les apports solaires en saison chaude.

A MICROCLIMATICAL SUNSPACE - UTILIZATION AND THERMAL BENEFIT

Dipl.-Ing. I. Lütkemeyer
Dipl.-Ing. C. Hartmann
Technische Universität Berlin, Sekr. Ro 44
Fachgebiet Klimagerechtes Bauen, Prof. H. Schreck
Straße des 17. Juni 135, D - 1000 Berlin 12
Tel: 030 / 381 71 59

ABSTRACT: In Co-operation between the Technical University of Berlin and the Council of Berlin Charlottenburg the retrofit and extension of an existing Kindergarden was realised. The central issue was to develop an energyconscious concept for the extension without additional energy demand. The extension by a microclimatical sunspace as an unheated space (finished in 1988), the thermal benefit, the utiliaztion and user's acceptance are described here. After one year's use and started monitoring first results can be presented.

INTRODUCTION

The atrium or sunspace as contemporary types of microclimatic spaces may have an impact on energy issues. The case study of a day- nursery in Berlin demonstrates a design strategy of an energyconscious retrofit by the aplication of a microclimatical sunspace. The attached south facing sunspace is an unheated area (225m^2) of intermediate temperatures. The collected solar heat is used for climatisation of the intermediate space itself. The climat is determined primarily by the ambient climat and the heat losses of the adjacent heated areas. The energetical issue is to create a space of high quality without any heat energy demand, and to reduce the transmission and infiltration heat losses of the adjacent spaces.

ARCHITECTURAL CONCEPT

The leading idea was to roof over the south patio of the present building, to create a microclimatical area, a green playing area for eventful experiences in close connection to the groups'rooms of the kindergarden. The reorganisation of the old building is connected with the measures for the new building parts.
The link between in- and outside was improved by an new entrance, attached conservatories on the street side and new interior connections. Climbing plants will improve the elevation and the microclimate on the facades.
The sunspace covers the whole area of the south patio of the kindergarden. The floor area is 225m^2, the volume is approx. 1100m^3. It acts as an weatherprotected 'glas roofed' room outdoors for children's playing activities. As well it is a transition area between inside and outside. According to these two functions two different sheres were created. One of them is a small, introverted and sheltered area. Most of the plants are in this part. Different corners and niches can be used as sitting areas or provide building of caves or other creative activities. The other one is oriented to the playground outside and suitable for activities and games. Elements here are the climbing frame within the sandbox an the large doors to the outdoor space. Most of the grouprooms are orientated to the sunspace with direct access, so that the sunspace becomes a center for playing and moving between different rooms.
The sunspace is a wooden framework construction with steel bars, and purlins which hold the puttyless double low-E-glazing. The floor consists of a reinforced concrete slap with bricks laid in mortar.

PASSIVE SOLAR FEATURES

The south faced sunspace is an unheated area and a bufferzone which works as a sun energy collector. Only the radiation of the sun and the heat losses of the old building are used as heat sources. The climate within the sunspace depends on the climate outside and the character of a microclimatical area is intended. Although the sun radiation is

Figure 1: Interior view of sunspace

Figure 2: View from south-east

primarily used for the conditioning of the sunspace, surplus energy can be directed into the inside rooms by opening the grouprooms' doors.
Main elements of the sunspace solar system are:

- the sunspace itself with its internal thermal masses (bricks on the floor, walls of the old building, spandrels)
- ventilation flaps at ridge and eaves
- interior shading devices for south orientated glazings
- tube system controlled by fans which suck out the warm air from the ridge of the sunspace and store the heat in the ventilated floor areas of the sunspace.

The conditions in the sunspace can be controlled automatically or manually. The automatic control is necessary for a moderate room climate and the growth of plants, especially during weekends and holidays. The manual control allows the individual conditioning according to the users' requirements.

THERMAL PERFORMANCE OF SUNSPACE

As a result of simulated thermal performances, the sunspace was decided to act as a bufferzone in the sense of an independent microclimatical space.
The target is to provide conditions of thermal comfort for the sunspace all year round. During the heating and transition period the air temperature should not exceed 25C. On sunny days, when overheating tendency will occur, the interior shading device acts as an absorber. The heat accumulation of the air space between glazing and shading area will be used by sucking warm air from the top into the ventilated mass construction of the floor and the spandrel walls of the sunspace to be stored. Discharging of the storage occurs by time lag emission of heat. (see Fig. 4 - Calculations by dynamic thermal simulation programme SPIEL - C.Green, Sheffield)
By this way surplus energy can be stored to increase the usable solar gain without overheating the sunspace and to extend the temporal space use.
Overheating in summer is avoided by ventilation flaps at the ridge and the eave. Ventilation and interior shading devices provide thermal conditions within comfortable ranges, only in hot summer periodes, the temperatures are estimated as to warm.
For winter periods, the calculations and first monitoring results show, that the big volume of the sunspace allows only a little usable energy support for the parent spaces. The calculated sunspace temperatures on sunny, cold winterdays remain positiv, although the outdoor temperatures are clearly below zero. The mean sunspace temperature of a measured five days period in November with mostly sunny weather is about 9 Kelvin warmer than the outside temperature (see Fig. 5).

ANNUAL ENERGY PREDICTION

The attached sunspace expands the usable floor area of the building by nearly 50%. The sunspace is unheated as temperature swings are quite acceptable in the play and activity areas. This way the heating demand is reduced to less than 2/3 related to comparable demand per square metre of usable floor. The reduction of heat demand of adjacent heated space is about 15%, if the sunspace-glazings U-value is 3.0 W/mk. The projected "Low-E" glazing (U-value 1.8 W/mk) provides 5% reduction more (see Fig.6). The more important aspect concerning the glass quality is, that inside the unheated sunspace temperatures below zero will be avoided. This particularly means better conditions for the growth of plants.

UTILIZATION

The sunspace is used for playing activities as a weatherprotected room outdoors. The duration of use is depending on weather conditions and season. The times of intensive use are in the early morning, at noon and in the late afternoon,

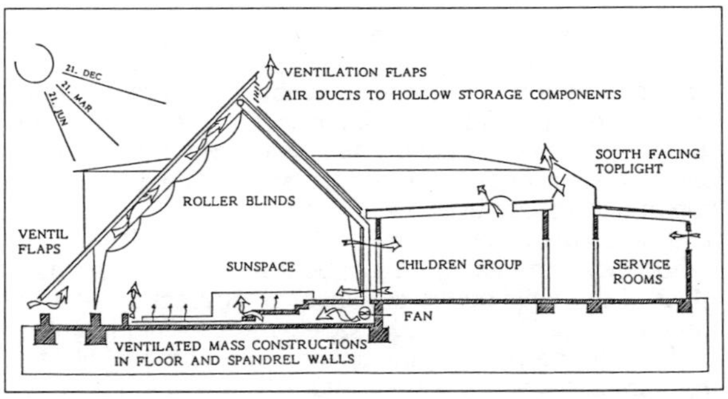

Figure 3: Section showing passive-hybride systems

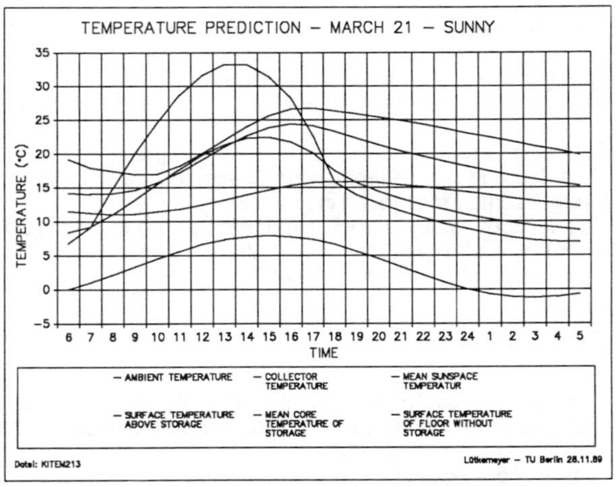

Figure 4: Temperature prediction for a sunny day - temperature distribution in storage comonents

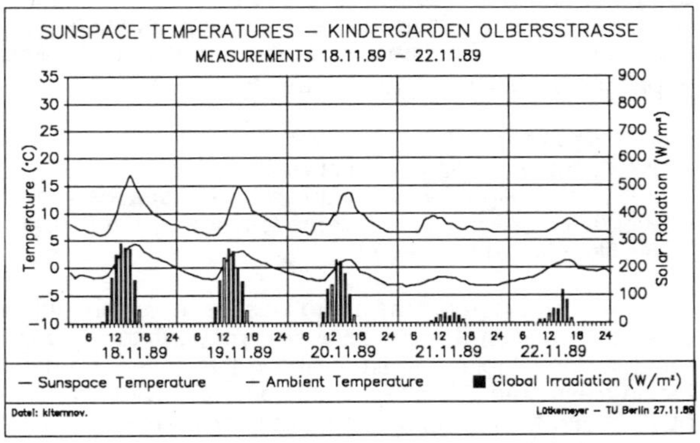

Figure 5: Measured sunspace temperatures

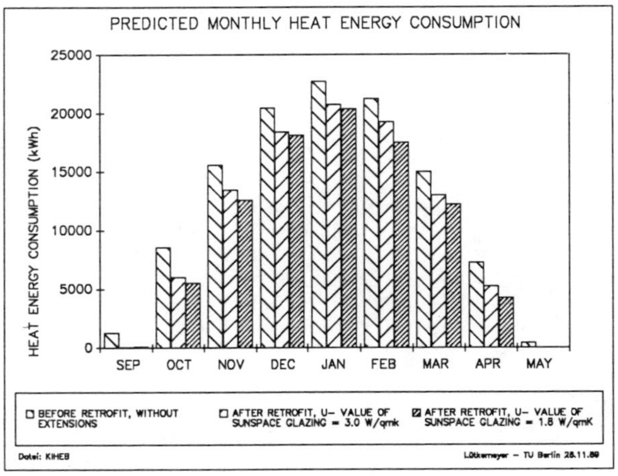

Figure 6: Prediction of monthly energy demand

while there are no observed activities in the groups' rooms (see Fig. 7).
After one year's use of the sunspace the users' acceptance was analysed. To all the educators of the Kindergarden (8 governesses) a questionaire was given to estimate the user's feeling concerning the comfort conditions, the utilization and the charcter of the sunspace.
They should vote between contrary aspects (monotonous-various, exiting - tiring) on a scale from 1 to 7 (see Fig. 8).
Most of the estimations of sunspace character are in the neutral range. The positive aspects are the light ambient and the friendlyness of the space. Further the posibility of frequent use and the various character was identified.
Negative evaluations are the unsafety and the noisy, exacting atmosphere. As well the sandy dust from the sandbox causes some problems.
The estimation of the comfort conditions gives a very heterogenous result (see Fig. 9). While the temperatures in spring and autumn are felt as quite good, the temperatures in summer and winter are perceived as too warm respectivly too cold.
Although the sunspace temperatures for winter periods are much higher than the ambient temperatures (see Fig. 5), the users' estimation is that it is too cold.
The daylight quality in the sunspace is very good but the adjacent goups' rooms do not get enough natural light. The noisy atmosphere causes acoustic problems that must be solved.

CONCLUSION

The results of computer simulation of thermal analysis, the user's response, and first monitoring results indicate that the strategy of a microclimatical space is significant. The useable floor area can be extended without additional heat energy demand. Moreover, the heat energy consumption of the heated space is reduced. A frequent use with several activities is possible and most comfort conditions are acceptable. The estimation of acoustics and the summer and winter temperatures indicate that the design must be developed carefully with notice to some important requirements, to provide conditions of thermal behaviour:

- avoidance of overheating in summer
- avoidance of cooling down in winter
- sufficient daylight for plants and adjacent spaces
- avoidance of acoustic problems (caused by hard surfaces like glas, stone, etc.)

One aspect concerning the energy losses, which should be carefully considered, is the disposition of north facing glass areas for architectural reasons like in this case. The north facing glass area (90 m^2) provides a good daylighting distribution and spatial impression, but effects also enormous transmission heat losses.

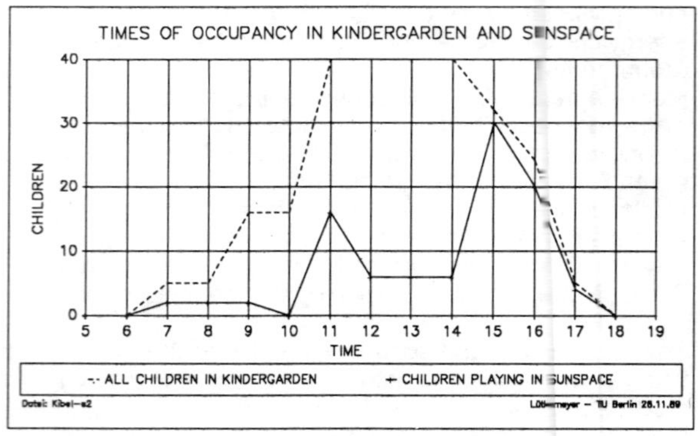

Fig. 7: Times of occupancy in kindergarden and sunspace

ESTIMATION OF USE AND CHARACTER

	extreme 1	2	neutral 3	4	5	extreme 6	7	
unusual			o					ordinary
stimulating				o				tiring
light		o						dark
interactive				o				not interactive
calm						o		noisy
general use				o				specific use
frequent use			o					rarely use
safe						o		unsafe
various				o				monotonous
friendly		o						unfriendly
clean					o			dirty
urban			o					rural
outside				o				inside

Fig. 8: Estimation of use and character of the sunspace by semantic description

COMFORT CONDITIONS

	good 1	2	neutral 3	4	5	bad 6	7
TEMPERATURE CONDITIONS:							
. summer						c	
. winter						c	
. spring			o				
. autumn			o				
AIR QUALITY					o		
HUMIDITY				o			
DAYLIGHT - SUNSPACE	o						
DAYLIGHT - GROUPROOMS						o	
ACOUSTICS							o

Fig. 9: Evaluation of the comfort conditions

THERMAL PERFORMANCES OF TWO LOW-ENERGY MULTISTORY RESIDENTIAL BUILDINGS

J.-P. Eggimann, J.-B. Gay
Ecole Polytechnique Fédérale de Lausanne - Bâtiment LESO
CH - 1015 Lausanne (Switzerland)

ABSTRACT. The energy consumption of two residential solar buildings (in Préverenges, near Lausanne) was measured during the years 88 and 89. These buildings have good insulation, an attached sunspace on each flat and solar collectors for the hot water system.
Compared to an average building, the total energy consumption is reduced by 40 to 50%. This reduction is very important for the heating energy consumption (only 260 MJ/m^2 for a standard year) but it is weak for the electrical energy consumption.
The amount of extra energy required for hot water depends on the building and is quite high. A badly set control and high water consumption reduce the solar part. The performances of these buildings are excellent, especially for the heating energy, but they are very dependant on the inhabitants behaviour.

1. INTRODUCTION

1.1 Purpose of the work

A group of three low-energy residential buildings was erected in 87 in Préverenges, near Lausanne. From the beginning of the project special care was taken in order to reduce energy demand and increase comfort. The following steps were taken to achieve this :

- adequate insulation of the envelope,
- large south-facing windows,
- an attached sunspace on each flat,
- solar collectors for the hot water system,
- an adapted heating and ventilating system.

Detailed measurements were taken in order to determine the effective performances of these buildings and to evaluate the real effect of the inhabitants behaviour.

1.2 Methodology of approach

Two of the three buildings were fully monitored : building C (16 flats, 44 persons) and building A (8 duplex flats, 32 persons) (Figure 1). Temperatures and heat flows were measured in each room. Air change measurements, with a tracer gas method, were taken in five flats. Sensors were also installed to measure the local weather as well as the functioning of the heating and ventilating system.

The third building, B (8 flats, 29 persons), was not monitored; only certain values were noted, recorded by hot water and electricity counters.

1.3 Local climate

The weather, measured from July 88 to June 89, was warmer than that of a standard year, and nearly as sunny :

average outdoor temperature :

- standard year : 9.4°C (Oct-Apr : 4.6°C)
- year 1988-89 : 10.4°C (Oct-Apr : 5.6°C)

total horizontal solar radiation :

- standard year : 4400 MJ/m^2 (Oct-Apr : 1550 MJ/m^2)
- year 1988-89 : 4300 MJ/m^2 (Oct-Apr : 1400 MJ/m^2)

Figure 1 : South façade of building A (8 duplex flats) in Préverenges.

2. MEASURED ENERGY CONSUMPTION

All values were measured from July 88 to June 89. They are given for building C, with building A added in parentheses. They are in the specific form of energy consumption per square meter. The reference area is the floor area of the heated part of the building (flats and part of the stairs) : 1634 m^2 (A : 1245 m^2).

2.1 Heating

The heating energy consumption was 228 MJ/m^2 (A : 226 MJ/m^2). Based on the these measurements, an estimated heating consumption for a standard year would be for both buildings, 260 MJ/m^2, which is approximately 40% of the consumption in an average Swiss residential building.

Heating was supplied by a gas boiler with condensation. Average efficiency was very good : 87% (A : 88.5%).

2.2 Hot water

The hot water collector system has an area of 1.3 m^2/person (A : 1.4 m^2/person) and covered 36.3% (A : 48.7%) of the global hot water demand.

This demand includes :

- the hot water consumption : 58 L/d person (A : 53 L/d person)
- 2 washing dryers (A : 1)
- circulation and tank losses.

The complement, 130 MJ/m^2 (A : 88 MJ/m^2) is assured by the gas boiler. For building C it is as high as the average Swiss hot water consumption (which does not include washing dryers).

The difference between the two buildings is due to the higher consumption of building C and to the control of the water transfer from the solar tank to the hot water tank. (In building C some water is transfered at inadequate moments and this increases the loss).

As the solar radiation for year 88-89 is almost the same as for a standard year, the energy demand for a standard year will be the same as the measured one.

The efficiency of the collectors was 36.6% (A : 40.2%) which is high for this type of collector (single glass, selective absorber).

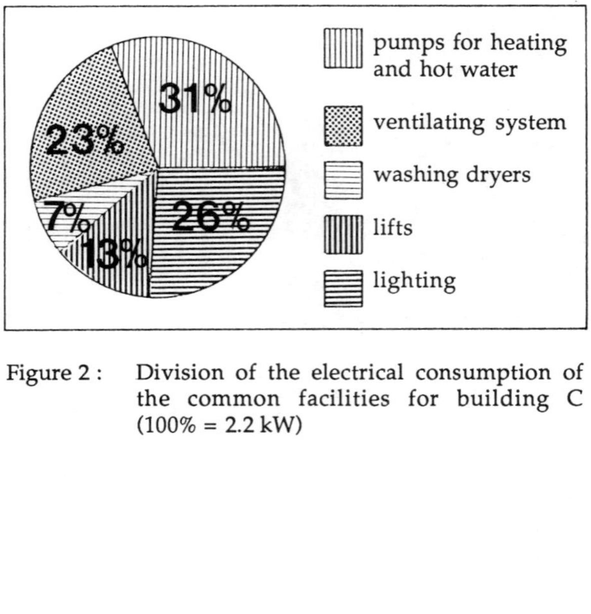

Figure 2 : Division of the electrical consumption of the common facilities for building C (100% = 2.2 kW)

2.3 Electricity

The cooking is done with electricity in building A and with gas in building C. For comparison with other buildings, this gas consumption is added to the electricity consumption.

The electrical energy consumption was 120MJ/m^2 (A : 110 MJ/m^2) and was divided into :

- 24 MJ/m^2 for cooking
- 54 MJ/m^2 or 64 W/person for lighting and appliances in flats (A : 65 MJ/m^2 or 81 W/person but with the cooling added)
- 42 MJ/m^2 or 2.2 kW for common facilities (A : 45 MJ/m^2 or 1.8 kW).

The electrical consumption of the computers used in the experiment is not included in these values.

As the consumption for common facilities is rather high compared to the consumption in the flats, we took more detailed measurements to find the origin. Figure 2 shows the results. The largest part came from the heating and ventilating systems. (This latter simply takes out the moist air from the kitchens and the bathrooms). The lifts are the most powerful appliances in the building but don't account for the largest consumption.

2.4 Total consumption

Figure 3 shows the comparison between the measured buildings and an average Swiss residential building /1/. The reduction of total energy consumption is between 40 and 50%. Heating energy decreased and boiler efficiency increased. The changes in the other consumption values are less important.

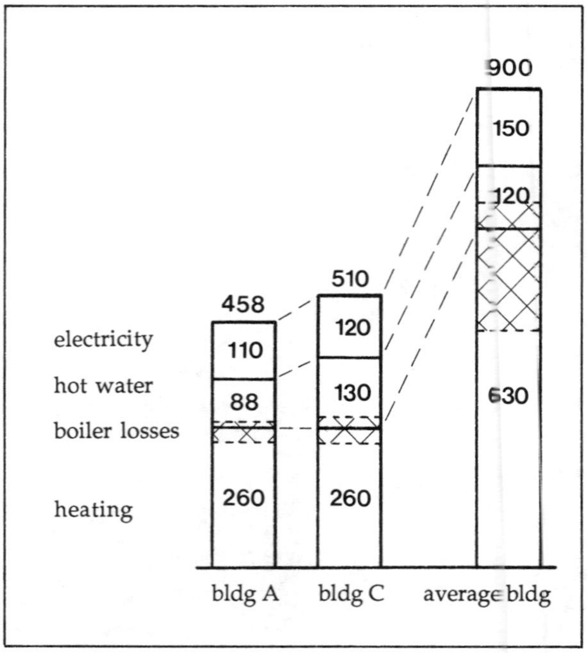

Figure 3 : Comparison of the total specific energy consumption for a standard year between the measured buildings and an average Swiss residential building /1/. All values are in MJ/m^2 of heated floor area.

3. INHABITANTS BEHAVIOUR

3.1 Influence on the heating

Detailed comparison between the flats shows the importance of the inhabitants behaviour.

Due to the indoor temperature differences (extreme flats : min. 18.4°C, max. 22.7°C) the heat transfer between neighbours may reach 23% of the total demand.

The ventilation habits are different from one person to the next. The air change rate was measured in 5 flats with a tracer gas at constant concentration. With doors and windows closed its value was 0.2 h^{-1} : this is the infiltration rate due to the ventilating system and to the outdoor conditions (wind, temperature difference). When the inhabitants behaviour is free, the average air change rate reached 0.4 h^{-1} but depended strongly on the flat. The observed values were between 0.2 and 0.8 h^{-1}. This led to additional variations which may account for 26% of the total heat requirements.

Note that the inhabitants have no influence on a very important parameter for the difference between flats : the location of the flat in the building. Calculations made for building C show that, with identical indoor conditions, the upper-left flat needs 60% more and the center flat on the second floor 39% less heating energy than the average of the 16 flats.

3.2 Electricity and hot water

A comparison between the flats of the three buildings shows that for the electricity (building C : electricity with gas for cooking added) a ratio of 1 to 3 in energy consumption per person is observed (Figure 4).

For the hot water consumption per person the difference is even larger (Figure 5).

4. CONCLUSION

Such a survey shows :

- how a substantial reduction of heat demand may be achieved in a multifamily building when special care is taken, in the conception of the building, to reduce heat loss and increase solar gains,
- that a careful verification of the proper functioning of the technical installations (heating, ventilating and solar collectors systems) is absolutely necessary to ensure good efficiency,
- that the inhabitants behaviour, in regards to the opening of doors and windows, has a very large influence on the consumption of low-energy buildings.

5. REFERENCE

/1/ C.U. Brunner, E.-A. Müller, Strategie zum Energiesparen, Schweizer Ingenieur und Architect, 30/31 1983.

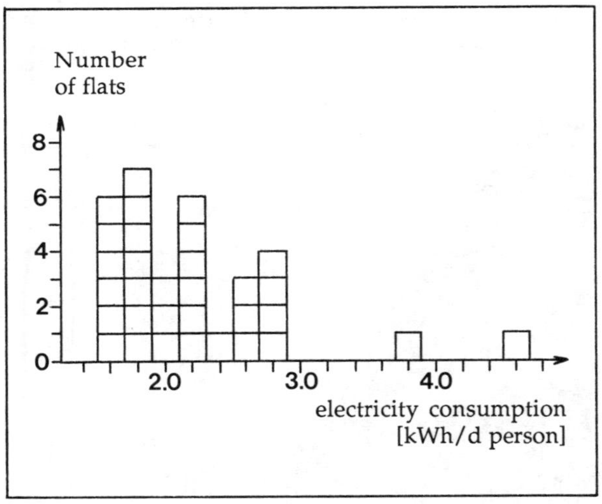

Figure 4 : Histogram of the electrical energy consumption per person for the three buildings.

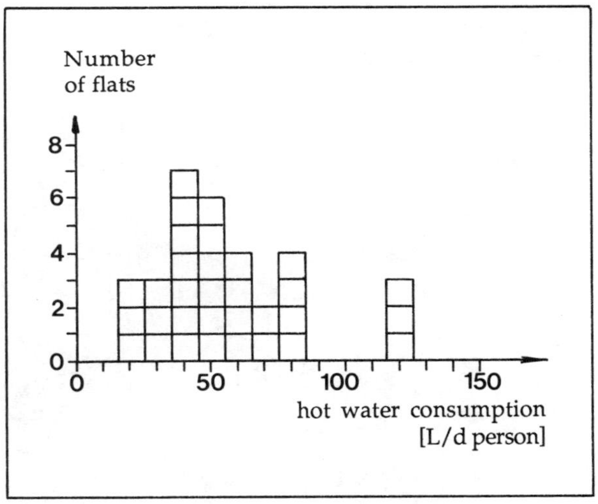

Figure 5 : Histogram of the hot water consumption per person for the three buildings.

ONE YEAR EXPERIENCE WITH A SOLAR HOUSE WITH TRANSPARENT WALL INSULATION (TWI)

Elmar Bollin

Fraunhofer Institut für Solare Energiesysteme
Oltmannstrasse 22
7800 Freiburg, FRG

Abstract:

In 1988 a semi detached 2½ storey house with 65m^2 transparently insulated walls was erected in Freiburg. The building has been occupied since December 1988. This privately financed project demonstrates that a solar house with a large area of TIM facade and additionally a sunspace could provide sufficient heating comfort and could achieve an apealing appearance. The design of the solar components, which is based on research and development at the Institut für Solare Energiesysteme (ISE), is reported. First experience shows the feasibility of transparently insulated buildings.

1. INTRODUCTION

Transparent Wall Insulation (TWI) is an efficient application of Transparent Insulation Material (TIM) which gives new impulses to solar architecture. It creates a new generation of solar houses with a characteristic appearance resulting from a large area of transparent glas facades. Efficient TWI with high energy transmittance could reduce the yearly energy requirement for space heating to one quarter compared with conventional wall insulation. This gives TWI a high potential for energy saving in newly built houses as well as in the retrofitting of old buildings with a low insulation standard. To demonstrate the feasibility of TWI, the author decided to build a privately financed 2½ storey-house with TWI in 1988 which has been occupied since December 1988 by the author's and a friends family.

The concept of the building includes different active and passive solar heating systems such as:

- transparent wall insulation (TWI)
- sunspace and windows with integrated roller blinds as variable heat protection and shading device
- integrated storage collector mounted into the TWI-facade.

The design of these solar components is based on the research and development at the FhG-ISE /1/ /2/ /3/. As a result of appropriate planing, no central heating system was installed and investment costs could be saved. If auxiliary space heating is required, two decentralized tile stoves provide for adequate heating comfort inside the two apartments.

Fig. 1: View to the solar house with transparent wall insulation (TWI) in autumn 1989.

2. THE BUILDING

The building situated to the south-west of Freiburg FRG has two occupied apartments. It is surrounded by conventional newly built houses. The semi-detached building and its roof ridge is orientated from the north-west to south-east.
It has 2.5 storeys, about 190 m^2 living area and a volume of about 1200 m^3. The south-east and the south-west facade are partly transparently insulated. A solar storage collector is integrated into the south-west facade for domestic water heating.The design of the building is the result of an energy concept with different temperature zones. Due to the sunspace, the living space has a large window fraction of 40 %. The exterioir walls are made of massive 24 cm bricks of 1.6 t/m^3 density in the case of TWI and 24 cm bricks of low density (1.2 t/m^3) in the case of opaque wall insulation. Figure 2 shows the plan of the 1st floor with the different types of wall insulation.

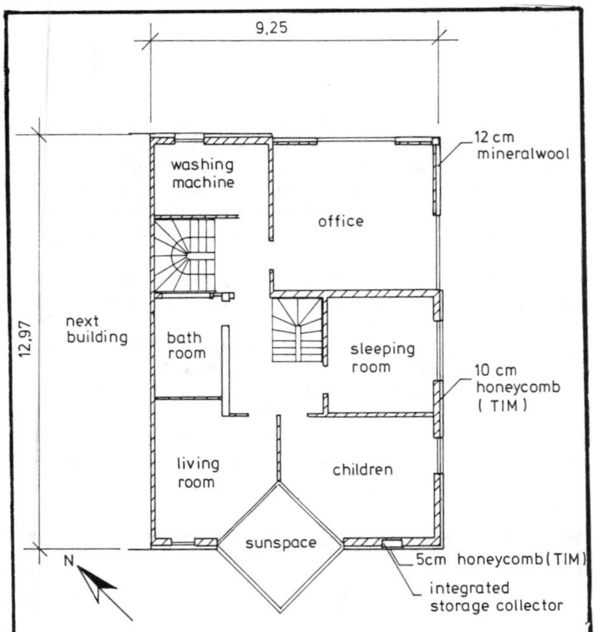

Fig. 2: Plan of the first floor with the different types of wall insulation.

3. TRANSPARENT WALL INSULATION (TWI)

For solar heating purposes, 37 TWI modules with a total area of 61 m^2, cover the south-east and south-west facade of the building. These modules were pre-fabricated by the Schmidt Company in Freiburg with integrated 10 cm honeycomb material and shading devices. Figure 3 shows a cross section of the prefabricated TWI module. The modules are covered by a low iron glass pane. The integrated roller blind protects the building against overheating during summer and against heat losses during winter nights. With a diffuse energy transmittance of 56 % and a U-value of 0,66 W/m^2 with the blinds open, these TWI modules attained a high efficiency.

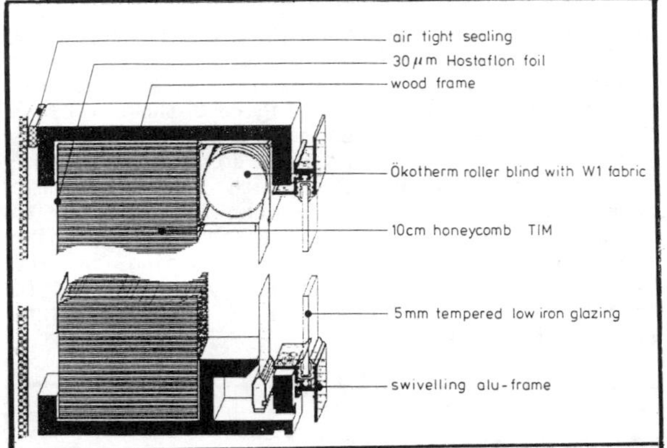

Figure 3: Cross section of the TWI module pre-fabricated by the Schmidt-Company, Freiburg

After the TWI modules were mounted onto the wall, the gaps between the TWI modules and between the TWI modules and the other components of the facade were covered with red aluminium moulding. These red strips match the red wooden strips of the opaque facade and give the Tiengen TWI house its characteristic appearance.
During the course of a year the building changes its appearance due to the operation of the roller blinds within the TWI facade. During summer the roller blinds are closed and the highly reflective coating of the fabric prevents against overheating. During winter these blinds are opened in day time to let solar radiation transmit through the TIM and are closed at night time for better insulation.

Fig. 4: View to the south-east facade with the storage collector at work and the closed blinds of the TWI in summer 1989

4. INTEGRATED STORAGE COLLECTOR

For domestic hot water heating a painted black stainless steel tube of 25 cm diameter, 5.5 m lenght and a volume of about 255 l was mounted in a gap in the south-west wall. Figure 5 shows a cross section of the integrated storage collector. This kind of storage collector with an aperture of 4m² was transparently insulated with 5 cm honeycomb material. The bottom of the tank is connected to the mains water at a pressure of about 6 bar while the top is connected to the plumbing in the bathroom. Auxiliary electric energy can be used to heat the water in a boiler tank if the temperature needs to be raised.

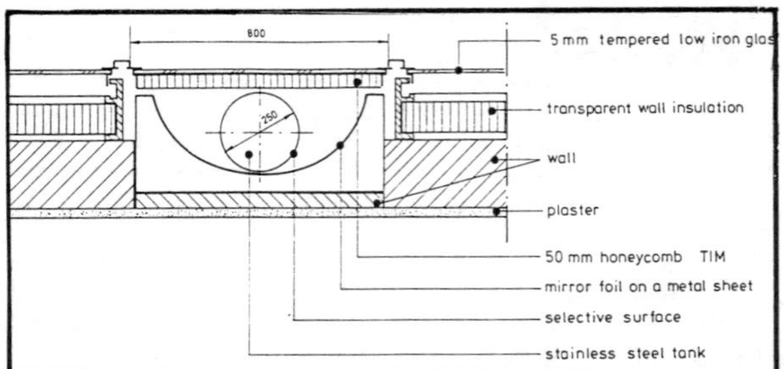

Figure 5: Cross section of the integrated storage collector

Figure 6 shows the hydraulic diagramm of the hot water system. During operation, if a hot water tap is turned on in a bathroom, fresh water from the mains runs through the solar collector tank and hot water is transported to the bathroom. If the water temperature is below the required temperature the auxiliary heating system raises the temperature.

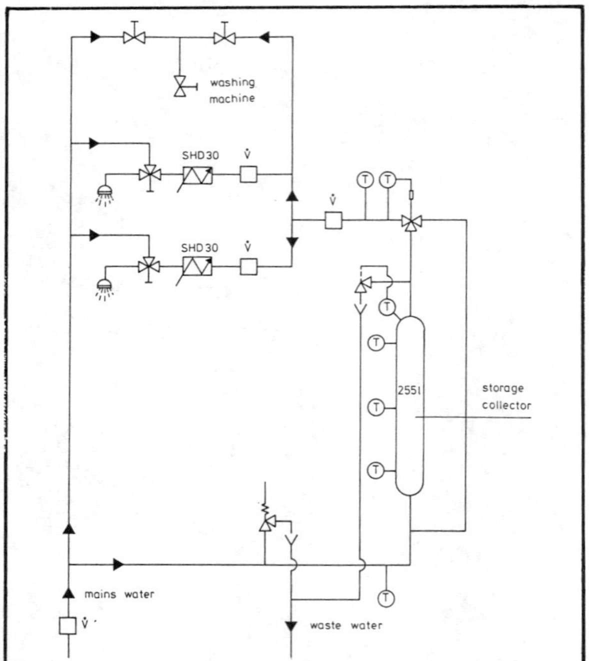

Fig. 6: Hydraulic diagramm of the domestic hot water system

This kind of operation allows even low solar energy input to be transformed into useful thermal energy. Control and regulation instruments are not necessary. The scientific research at the ISE shows that storage collector tanks integrated vertically in the building's facade could achieve a collector efficiency of 40 % and 60 % solar fraction /2/. To protect the occupants against scalding, and to prevent overpressure and boiling various security devices were installed.

5. ONE YEAR'S EXPERIENCE

The architecture and the appearance of this solar house with its large area of TWI has a high acceptance not only among the experts. "Warm" exterior walls during winter and "cool" exterior walls during summer create a comfortable climate inside as a result of the TWI. To a certain extent, this is due to the adequate regulation of the direct solar gains through the windows and the sunspace by the roller blinds integrated inside the glazing construction.
The two tile stoves generate sufficient heating comfort when the solar heating is not adequate alone.
During the winter period from December 1988 to May 1989 6 m³ wood were burned in the two stoves which is equivalent to about 10.000 kWh. In order to assess this amount correctly the high moisture level inside the newly built construction and the incomplete insulation of the building's north-west facade have to be taken into consideration.

The cost balance in Tab. I demonstrates that appropriate planning was able to reduce the additional costs and makes TWI cost efficient.
In order to obtain detailed data about typical values like the wall's time constant, the heating comfort, the energy consumption of the auxiliary heating system, the outside temperature and the solar gains of the building, a monitoring system, provided by the ISE, will be installed in December 1989.

Tab. I.: Cost balance

61 m² TWI	36 600 DM
Credits by the abolishment of	
a central heating system	- 25.000 DM
conventional wall insulation	- 6.100 DM
a space for the heating system	- 3.000 DM
Additional costs	2.500 DM
Yearly energy saving of about 10.000 kWh	1.000 DM

6. REFERENCES

/1/ A. Goetzberger, J. Schmid, V. Wittwer
"Transparent Insulation System for Passive Solar Energy Utilization in Buildings"
Int. J. Solar Energy 2, pp. 289-308, 1984

/2/ C. Schmidt, A. Goetzberger, J. Schmid
"Test results and evaluation of Integrated Collector Storage Systems with transparent insulation"
Solar Energy, Vol. 41, No. 5, pp. 487-494, 1988

/3/ A. Goetzberger, J. Schmid
"Review of Components for Passive Solar Energy Utilization"
Int. J. Solar Energy 3, pp 309-328, 1985

- STOREY AND GLASS ROOF ADDITION TO AN ARCHITECTURAL OFFICE -

Florian + Franz + Wendelin Lichtblau
Dipl. - Ing'e, Freie Architekten
Soeltlstrasse 14, D - 8000 München 90

Static: Dr. - Ing. Pittioni / Munich
Technics: Dipl.-Ing. Struve / Rosenheim
Glass elements: Fa. Okalux / Marktheidenfeld

Sponsored by the Economic Ministery of Bavaria
Translation: M.G. Montry / Munich

WORKING IN A SOLAR COLLECTOR FULL OF LIGHT ' - 60 % OF HEATING AND LIGHTING BY PASSIVE ENERGY GAIN

from the South over Soeltlstrasse

the 'East Office'

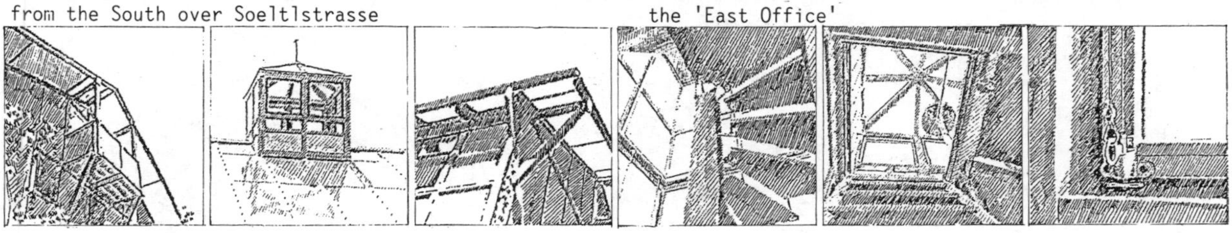

I. PROJECT DESCRIPTION

Building type, extent of construction activity

HARLACHING (MUNICH) In the early sixties we built with the most modest approach a ground level office building of concrete panel / wood frame construction with a flat roof over a two layer brick and concrete basement. To this and to newly layed foundations we added storey additions in 1987 / 88: an upper floor with gallery, of pure wood skeleton construction under a moderately inclined saddle back roof. The rebuilt space added 500 m3 to the existing 600 for a total of about 1100 m3.

Location and considerations

The neighborhood, since the thirties, has been zoned for single family residences with generous greenspace. Building density and design were tightly controlled. Planning considerations are oriented towards general municipal and state building codes.

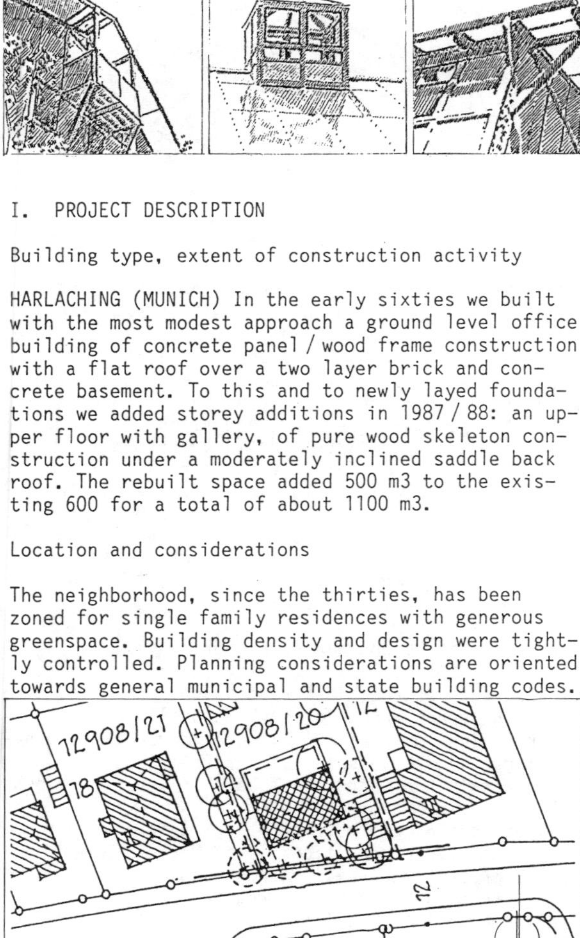

site plan, 0 m

Process, construction, costs

Planning and acquisition of permits for the initial version (façade opening, air collector / gas heating system) began at the end of 1986. X, Y and Z proportions were derived from the existing structural conditions, the column and construction grid, for weight reasons as well, was minimized to a 10 cm cross section.

In the middle of 1987 changes were made to accommodate the newly developed roof glasing. In addition came the related calculations and application for a government grant. After partly adventurous bid solicitation, construction began in September. Quick assembly was a necessity as our office itself was, at this time, without a roof.

Primary construction consists of laminated larch, horizontal members being pegged with zink-coated T profile steel to continuous columns. Further construction consists of mineral insulation and pine plywood with larch trim, screwed to seal strips. To this was added larch flooring, larch stairways with steel details, and window frames of Z profile steel. The installation of the 72 mm thick glass roof elements ensued as dry glasing directly attached to rabbeted wooden rafters, and as visible fastening by means of screws over a watertight profile, to create a second drainage level, with aluminum trim.

After a little less than four months of winter construction, lasting until February of 1988, our storey additions were, with good fortune, complete. We moved in in March. Construction methods and materials remain, to the present, troublefree. Evaluation costs for construction alone increased about a third from 300.000, - DM for conventional construction to 400.000, - DM for the 'pilot version'. We succeeded in getting proportional public funding but on the condition, that after completion we were to carry out a one year program of measurement and analysis. Balance sheets and results confirm all idealistic consumptions.

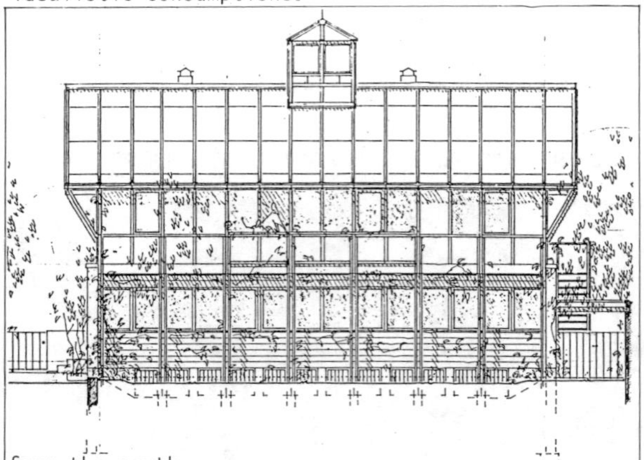

from the south

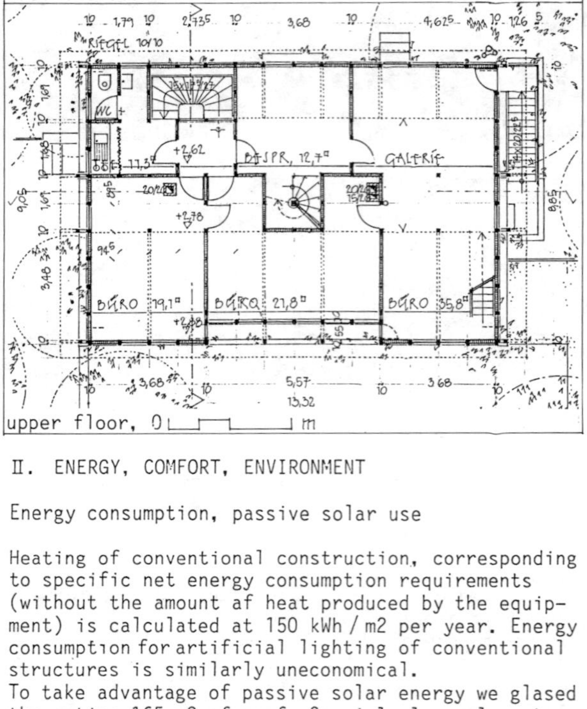

upper floor, 0 m

under the ridge into the basement and should be equipped with reverse blowers.
The glasing of the façade is set back the necessary measure to allow for view and ventilation. In addition to the solar heating we use gliding, thermostat-guided gas single ovens with chimney connection as already existed in the lower floors.

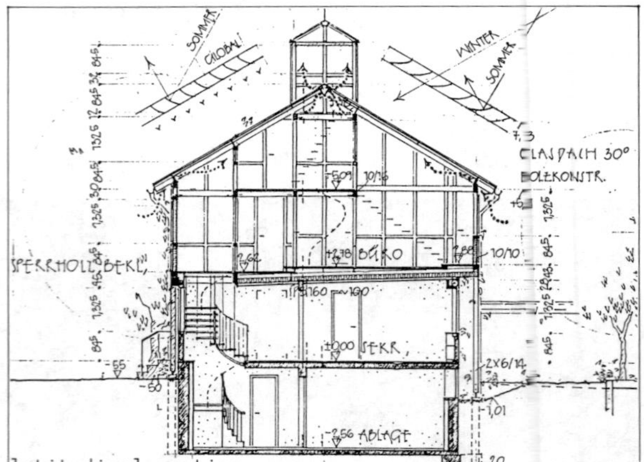

latitudinal section

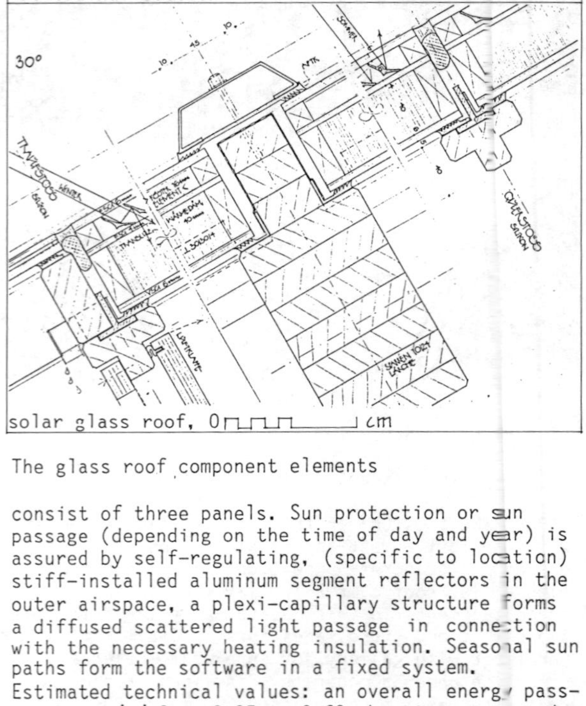

solar glass roof, 0 cm

II. ENERGY, COMFORT, ENVIRONMENT

Energy consumption, passive solar use

Heating of conventional construction, corresponding to specific net energy consumption requirements (without the amount af heat produced by the equipment) is calculated at 150 kWh / m2 per year. Energy consumption for artificial lighting of conventional structures is similarly uneconomical.
To take advantage of passive solar energy we glased the entire 165 m2 of roof. Special glass elements were developed and fitted. These elements combine for the first time all the functions of sun protection, light and energy passage as well as light diffusion and heat insulation into one construction component. And this was accomplished without any technical furnishings. The building itself becomes a solar collector full of light! The specific heat energy consumption decreases to under 40 %. In addition come corresponding energy savings on artificial lighting. Primary energy savings: more than 60 % !

One can also do without technical equipment to aircondition the building. In addition, for heat flow, thermal ventilating flaps were installed. If necessary, waste heat under the ridge can be recycled. For the time being, we use only a temperature regulated blower to support heating of the ground floor, which seems to work extraordinarily effective. Air ducts should extend from two points

The glass roof component elements

consist of three panels. Sun protection or sun passage (depending on the time of day and year) is assured by self-regulating, (specific to location) stiff-installed aluminum segment reflectors in the outer airspace, a plexi-capillary structure forms a diffused scattered light passage in connection with the necessary heating insulation. Seasonal sun paths form the software in a fixed system.
Estimated technical values: an overall energy passage degree 'g' from 0,25 to 0,60, heating passage degree k = 0,8 W / m2K, assessed sound absorption of 52 dB. With such elements were achieved during the current heating period an equivalent 'k' value of less than - 0,65 W / m2K. Solar gain is therefore exceptionally high.

Influence of surroundings, comfort level

The surface of the glass roof is not shaded by surrounding trees or buildings. The quality of the surroundings makes micro climate improvements unnecessary. Nonetheless we cultivated intensively so that within two years the façade will vanish behind climbing vines.
All aspects of comfort, with regard to indoor temperature, air composition, light direction and sound protection, have, through ecologically adapted methods and materials, reached a high level of refinement. The intelligent use of heating and airconditioning equipments sharpens our awareness of

natural processes and correlations and creates a moderate ' bracing climate ' which we find, after an adaption period, phisically and mentally stimulating as well as good for our health.
The self regulating light passage of the 'Köster' segments, in connection with the diffuse scattering effect of the capillary plates and white varnished interior walls create a shade- and dazzle-free daylight of virtually high density, regardless of ambient conditions. The space use is extremely variable. Our mediterranean plants grow in the remotest corners.

III. PERSPECTIVES OF THE PLANNING

The acknowledged design intentions of our project were primarily:

- to continue to establish, with decisive characteristics, a consistent dialogue between 'old' and 'new' until they grow together to form an insoluble whole.
- to ensure quick construction and flexibility of space creation and façade construction through modular order, building principle, and immediateness of materials and construction.
- to do justice to future energy savings requirements by means of reasonable costs and environmental and user friendly construction, and
- to search for, consciously, new methods of realization, after a careful research and planning process, that can also be offered to future clients, and all who are interested, as visible proof of our priorities.

Next and generally, this thesis:
that the resultant problems of the sins of our civilization can at least be dulled through the simple application of ecological knowledge. Inhibition is a function of our state of consciousness and not primarily of technical aspects. There are truly many things that could be changed immediately for the better, if only we wanted to exercise the lost art of anticipation.
And we have ascertained, that modern technology and ecological construction methods are not mutually exclusive. Rather, the appropriate use of that technology, especially in our muddled situation, even seems to be indespensible for the balance and / or completion of natural processes. In other words, to 'domesticize' them.
We believe, with regard to realizible state-of-the-art technology, that our 'Collector House' is an optimal form for utilizing suitably, in accordance with human needs, the gift of sunlight. Both subjectively felt and objectively measured values would speak emphatically in support of our belief.
If the above mentioned glass elements and construction components were made commercially available, then a number of possibilities would open up for effective deployment in new building, renovation, and redevelopment. The amortization factors add up to a considerably decreased expenditure for: construction planning and construction, installation and maintenance of technical equipment, heating, ventilating, air-conditioning and lighting energy.
Therefore, without the additional use of unecological and unnecessary technical apparatus, we arrive at the most efficient use of space design, lighting and energy attainable. And all to the benefit of the human being.
As simple and logical as our end result may seem - it was hard-won. The risks involved at the time of the undertaking piled up uncertainly and problematically. Looking back today, we were compelled to add that in no way would we have forgone the rocky path that led to our result. The absolute value of our journey has been preserved because, for once, we let the others do the talking -
we took the risks. F.L. X'89

former condition from southwest

'night architecture'

working model, m 1 : 20

before the party ...

ENERGY CONSCIOUS MULTI-APARTMENT RESIDENTIAL HOUSES IN FLORENCE

A. Gioli, A. Corsi, P. Puccetti, M. Sala, A. Toti
Dipartimento di Progettazione, Facolta' di Architettura
Via Ricasoli 66 - 50121 FIRENZE (IT)

1 - DESCRIPTION OF THE PROJECT

The aims of this project are to obtain a sample of a multi-apartment energy conscious building, carefully studied both from architectural and structural point of view, not less from solar techniques and energy conservation rules.

The first objective of our design has been to minimize wastes of energy and of heat loss: insulation and compact mass building in order to receive the maximum solar gain in the given area.

The selected layout presents two solar and three normal multi-storey building facing the main road.

The oriention of the buildings follows the surrounding construction, as well the North-South axis, integrating the existing town design.
As regards the project we proposed a four storey construction, with 10 apartments, six of them are two-level flats.
Energy conservation is based on highly insulated Prefab-Panels and on a double range of buffer zones, in the North and in the South facade.

Daily operations of passive solar systems induce the participation of the inhabitants and increase their consciousness of the problems of saving energy.

The choice of large panels technology and of commonly used materials aims to demonstrate that the solar design doesn't need sofisticated equipment or high cost components.

Solar passive systems used in the project are:

1. direct gain through the windows
2. sunspaces in the south terraces
3. vertical air collector on bearing panels
4. direct gain through the roof window

Movable insulation devices:

1. roller shutter on the windows
2. swivel shutter on the roof window
3. sliding windows in buffer spaces

3 - PASSIVE SYSTEMS OPERATING

Main passive system in the building is the complex sunspace-air collector, sited on South front. Sunspace is an ancillary space generally well accepted by the people, and moreover it may act both as solar collector and as buffer space during cold periods.

A vertical double face solar air collector, facing East and West, is attached to the sunspace, in order to supply further thermal energy to the indoor air, both during morning and afternoon.

AXONOMETRIC VIEWS

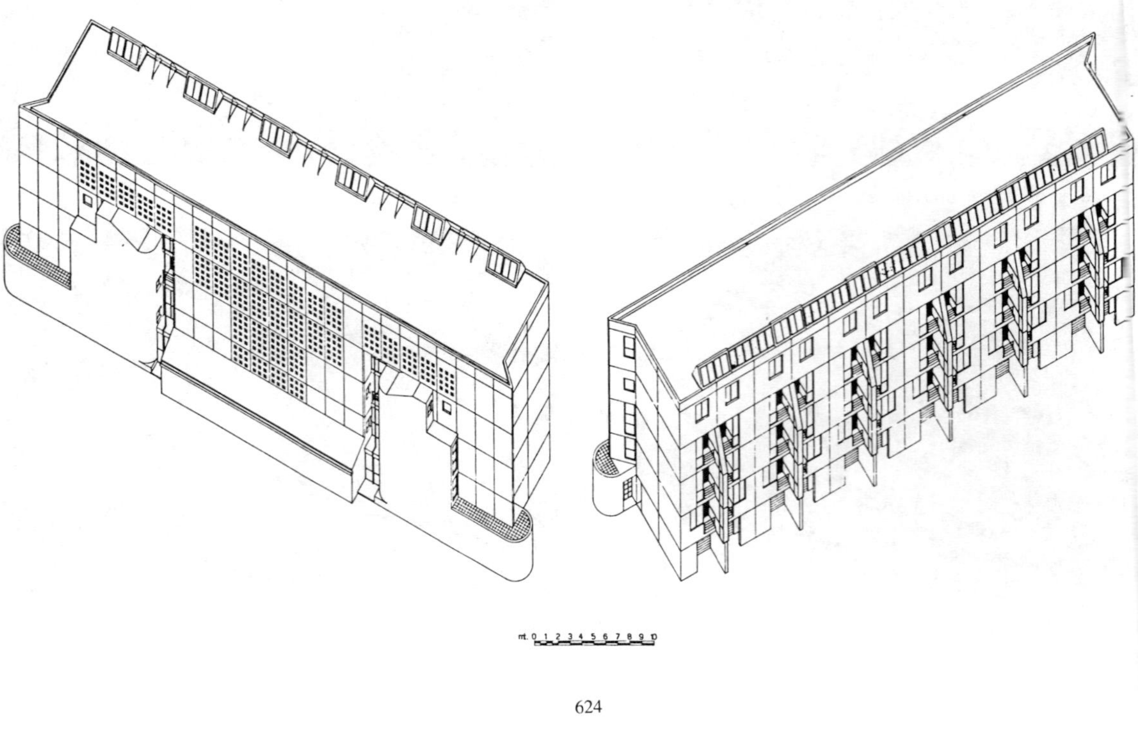

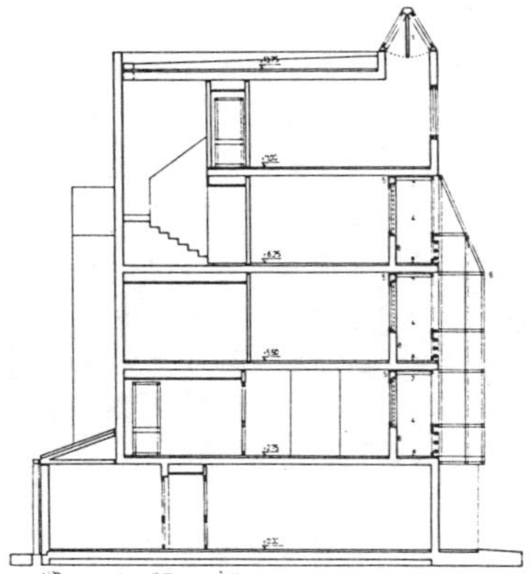

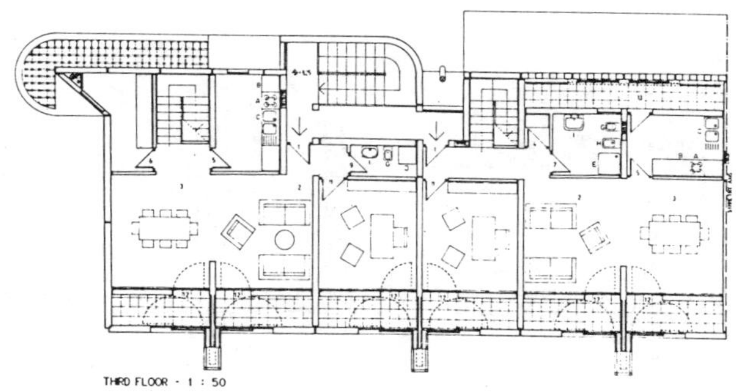

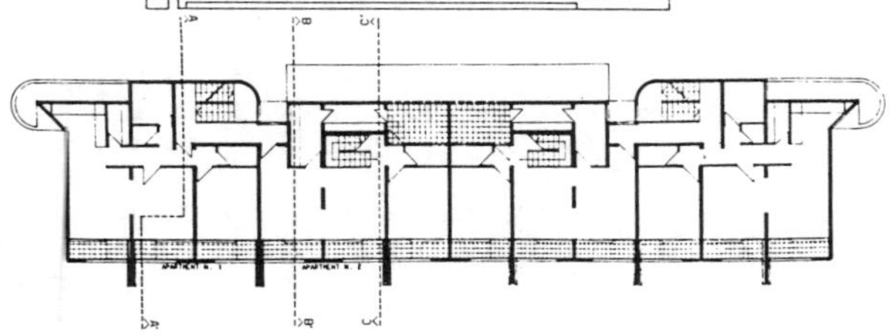

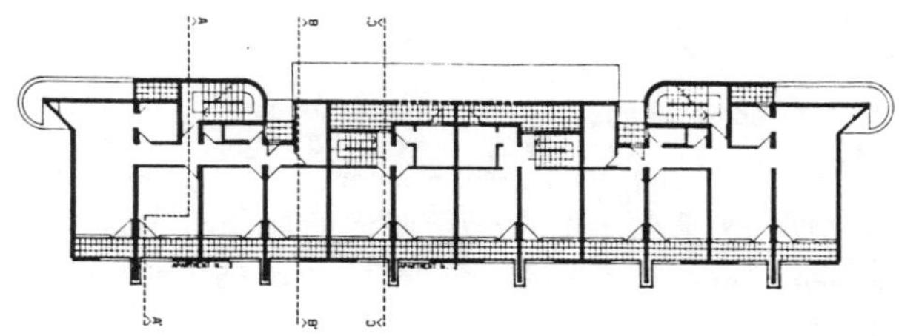

On summertime it provides shading of incident radiation, reducing overheating and ameliorating comfort.

Air convection is forced through the prominent part of the air collector, and slightly forced across the sunspace. Hot air is taken from the higher part of the vertical air collector, where air depressure is provided by the axial fan, controlled by a thermostat. The indoor air flows into the sunspace through the air intake sited at the lower part of the sunspace internal wall.

In the sunspace (during the sunny days) the air is warmed by free convection that arises from the dark pavement (made of black ardesia) and the wall (black painted). Sunspace air flows into the air collector through the lower intake, is warmed in the collector and it is recollected at higher part by the axial fan.

Fan operating is controlled by a thermostat, that switchs ON when air collector temperature is higher than indoor temperature and this one is lower than 19°C.

The four duplex apartments on third-fourth floor are warmed by the front windows, and by the roof windows with reflective swivel insulation. The same device enables also the summer ventilation by free convection of hot air.

5 - HEAT REQUIREMENT CALCULATIONS AND SOLAR GAINS ESTIMATES

CLIMATE ANALYSIS

Owing to the lackness of average hourly values of temperature and of relative humidity, a recalculation process has been made in order to estimate simulated hourly values that enable to draw the representative enthalpy points of local climate on the psychrometric chart.
The hourly temperatures calculator provided by Koenigsberger, Ingersoll, Mayhew, Szokolay in "Manual of Tropical Housing and Building" (p. 62) has been used for the calculations of hourly values. Relative humidity daily ranges for each month have been estimated as comparison with similar climates of European towns sited at the same latitude. Hourly variations have been estimated by approximating a two periods sinusoidal curves to daily extreme values.

Graphical results of comfort analysis on the psychrometric chart are given in the table, where one can see that comfort zone (outlined according to Victor Olgyay) is reached only at some periods of the day from May to September.

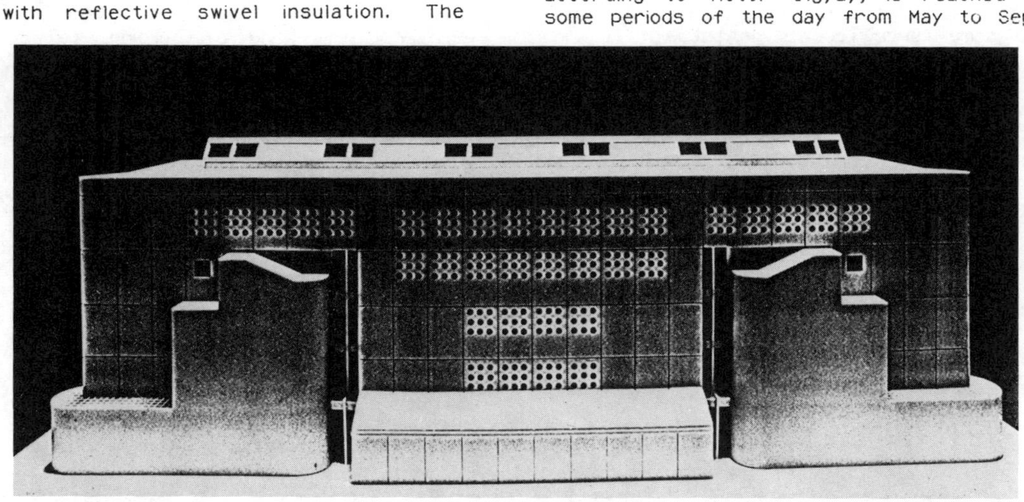

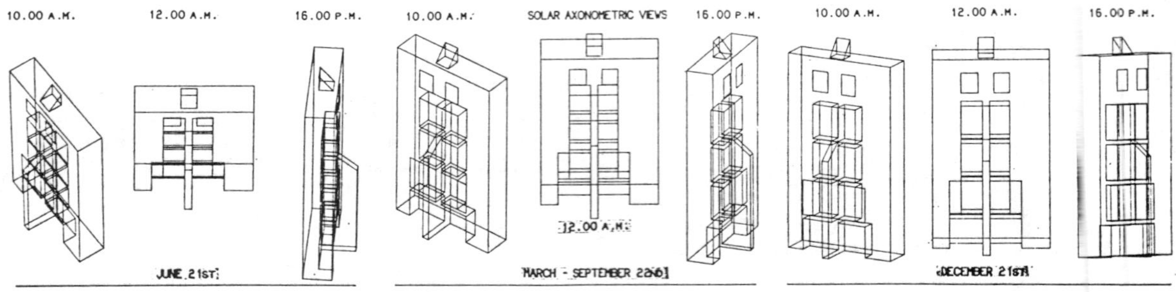

4 - SOLAR COLLECTOR FOR DOMESTIC HOT WATER

Domestic hot water is supplied by a water solar collector system (mq. 24) integrated in the roof structure.
Hot water storage is centralized in two tanks at different temperature, sited in the common cellar at ground floor, from where the hot water is supplied through a distribution pipe ring to each gas boiler at the same temperature, that is achieved by the solar heating. Further desired heating is supplied by the individual gas boiler in each apartment. This solution enables every user to pay only his own extra energy really consumed.

METHOD OF SOLAR GAINS ESTIMATIONS

For the evaluation of the performances of the designed building, one of the 10 multistory block has been chosen, that is sited in the appropriate position to get optimal solar exposure.
Total and specific heat requirements of each apartment have been estimated, (see tables below) with reference to external design temperature = -2·C, given for the project.
Respective heat losses of each surface are estimated, as well the resultant G value of each apartment, calculated including ventilation heat losses.

The duplex apartment n. 2, sited at 1st and 2nd floor level, has been chosen as sample for solar gains estimates.
The apartment has volume = 265.65 m3, and G value = 0.83 W/m3·C.
Calculations have been carried out by using the Solar Load Ratio (SLR) method, as described by O.A. Barra in "Conversione fototermica dell' energia solare", ed. Liguori, IT., with the assumption that sunspace and air collector are assimilated to the 'night insulated direct gain', and same curve for SHF is used.

Calculation process used is summarized below:

1. estimate of monthly degree-days, with 18^C base temperature
2. calculation of monthly gross heat requirement (MGHR), by the equation: MGHR = G * 24 * Vol * ^CDD
3. estimate of internal heat gains (MIHG), as: 0.035 kWh/m3day
4. estimate of monthly net heat requirement (MNHR), as difference of MGHR and MIGH.
5. estimate of monthly solar radiation (MSRI) incident on East, South, West facing collecting surfaces, by integrating the given data referred to correspondent vertical planes and conveniently reduced by appropriate shading factors for the South facing surfaces shaded from East and West respectively.
6. estimate of monthly absorbed radiation by each collecting surface, by the equation: MABR = MSRI * Sup * tau * .9
7. estimate of monthly Solar Load Ratio, as: SLR = MABR / MNHR
8. estimate of correspondent monthly Solar Heating Fraction, from published tables
9. estimate of monthly auxiliary heat energy (MAHE) to be supplied to the building, as: MAHE = (1 - SHF) * MNHR
10. final estimates, as the yearly summatory of saved energy, and the 'merit figure', estimated as:
    MF = (Sum(MIGH) + Sum(MAHE)) / (·CDD * FlatS)

Merit Figure is 0.024 kWh/m2, that is considered fair for this kind of building.

THE ELISA LEMONNIER BIOCLIMATIC SCHOOL

G. GAUDIN
GAUDIN INGENIERIE
11, RUE DE LA BUTTE DE PRAUD
44400 REZE
FRANCE

- Elisa Lemonnier bioclimatic nursery school
- Technology transfer from a research centre to civil engineering
- Measurements campaign
- Development of the SIMULA software program

1. ELISA LEMONNIER BIOCLIMATIC SCHOOL

1.1. Record sheet

- works master : city of St Nazaire - 44 FRANCE
- architect : C. COCHY - St Nazaire - 44 FRANCE
- climatic consultant : G. GAUDIN - 11, rue butte de Praud - 44 REZE - FRANCE
- floor-space : 500 m2 habitable surface + 100 m2 of covered playground
- cost : 3 000 000 FF inclusive of VAT, i.e 425 000 ECU (furniture included)
- capacity : 90 pupils (aged 2 to 4 years)
- further use : activities centre (yoga, dancing) for elderly people on Wednesdays (activities hall)
- inauguration : January 1986

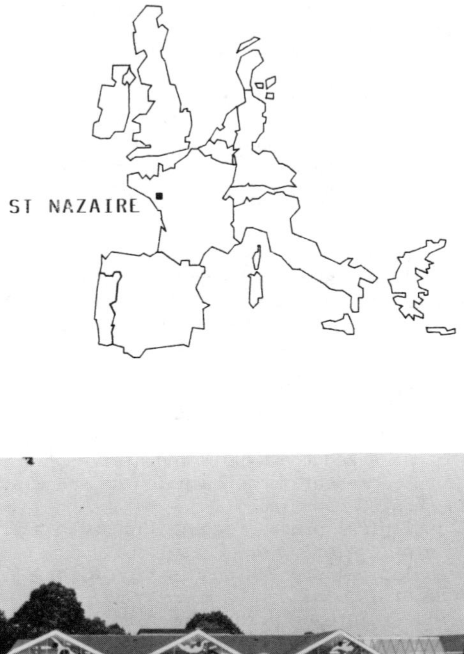

PHOTOGRAPH OF SOUTH FRONTAGE

1.2. School layout

The school is organised around an east/west passageway. From this area, the following are catered for :

* on the southern side, the warm premises :

- 3 classrooms (N° 1 Oldest, N° 2 Medium, N° 3 Youngest) and the youngest children's sanitary facilities

The medium-aged children's classroom is equipped with a wooden mezzanine. The youngest children's classroom is equipped with a heavy mezzanine and a sleeping bay.

* on the northern side, the less heated premises (buffer areas) :

- headmaster's/headmistress's office
- general sanitary facilities
- spare-time activities room

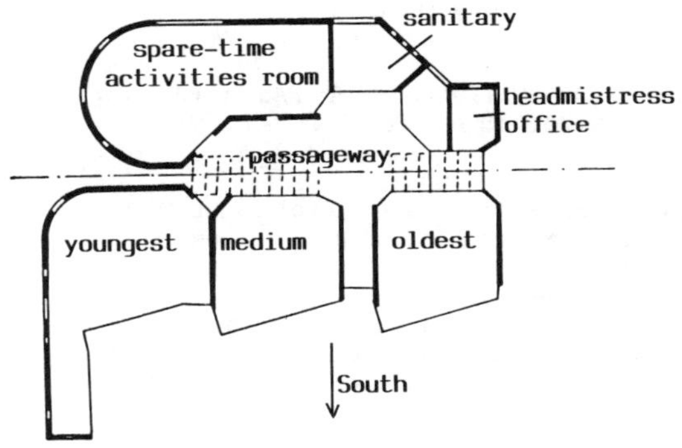

SCHOOL LAYOUT

1.3. Solar techniques (heating)

The solar techniques are of the passive bioclimatic type (direct solar gain)

- the classrooms are equipped with south-facing bay windows
- the passageway is equipped with a (partial) glass roof
- the north-side premises serve as a buffer area. (the activities room also has a south-facing glass roof).

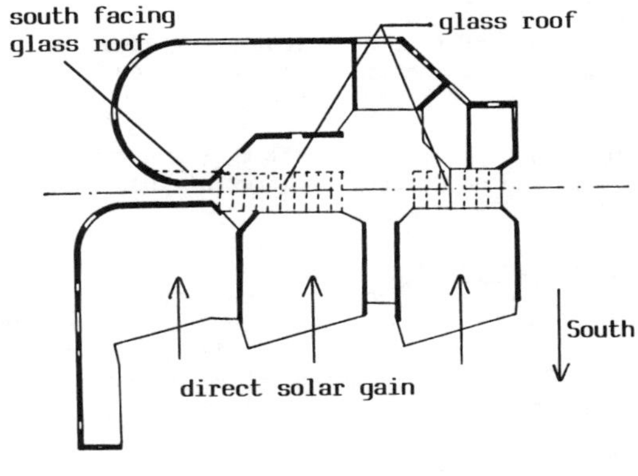

PRINCIPAL SOLAR INPUTS

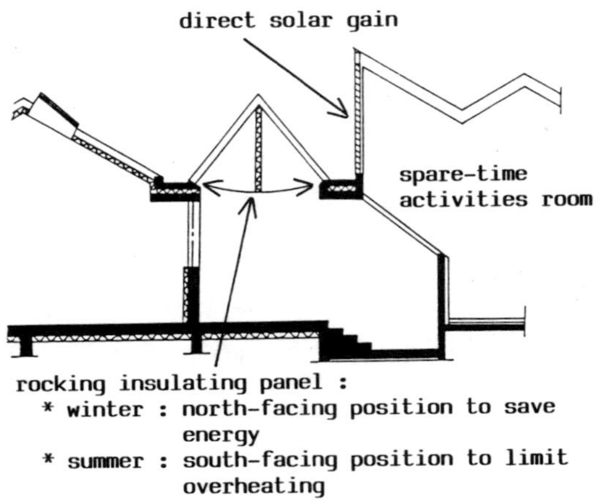

SOLAR GAINS CONTROL AT PASSAGEWAY LEVEL THANKS TO A ROCKING INSULATING PANEL

1.4. Solar techniques (natural lighting)
The architect brought several parameters into play :
- the south-facing lighting
- the north-facing lighting, in broad daylight through the passageway or by VELUX roof windows
- the brightness of the walls or ceilings

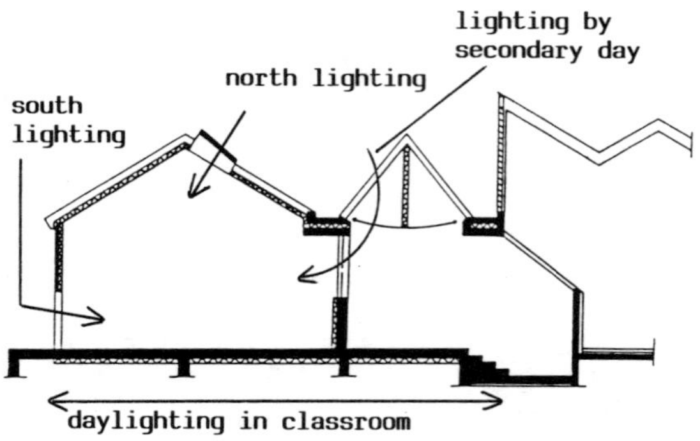

CLASSROOM CROSS-SECTION

1.5. Climatic aims
- a 60 % energy saving
- summertime comfort (in several other kindergartens, summer temperatures are often too high)
- great luminous quality of the various environments
- building a bioclimatic school that would be qualitatively and quantitatively free of cost overruns.

1.6. Design means
In order to achieve such ambitious aims, the architect and the climatic engineer made use of the most top-notch research work in this field. The latter is dealt with in paragraph 2 hereunder, entitled "technology transfer".

2. TECHNOLOGY TRANSFER
2.1. Our partners
CERMA, the Nantes Architecture School's Research Centre developed, starting in 1981, the SIMULA research software program (a thermal software program that is variable over a given period) and a data-processing heliodon.
The GAUDIN Ingenierie office availed itself of the tools developed by CERMA to improve the building's design.
The CERMA/GAUDIN Ingenierie team thereby achieved a technology transfer from the world of research to the field of application.

2.2. Characteristics of SIMULA
SIMULA is a VARIABLE-mode software program (by means of FOURIER series analysis) over any given period (an hour, a day, a week ...).
It is MULTIZONAL, meaning that the zones can be coupled among themselves.
The walls are split up in layers. Each layer is characterised by :
- thickness
- resistance
- heat storage capacity
- density
The solely resistive walls (e.g the glazings) may have resistances that are variable in time (addition of rolling shutters).
The exchange surfaces may be characterized by their absorption coefficient.
The temperature, air renewal, human occupation, heat release (lighting machine) and solar flux stresses are applied at the very moment at which they occur within the period of time considered.
The days taken into consideration are determined by the user ; they can be any kind, average or extreme.
A complete solar modulus (algorithm : Perrin de Brichambault method) makes it possible to calculate the solar fluxes precisely.

2.3. Use of this research product
The Elisa Lemonnier school is the FIRST-EVER OPERATIONAL FRENCH BUILDING IN WHICH ALL PACKAGE PARAMETERS ARE DIMENSIONED BY SIMULATION IN VARIABLE MODE OVER A PERIOD OF TIME (24 hours or 1 week, summer/winter, average or extreme conditions).
List of analysed parameters :
- glazed surface / passive solar input
- fixed insulation, mobile insulation
- inertia, distribution, exchange surface
- active solar system (air pick-up)
- solar protection
- ventilation
- regulation
- thermal load and discharge of building
This analysis was made possible because of the support given by the "Agence Française pour la Maîtrise de l'Energie" (AFME).

3. MEASUREMENTS CAMPAIGN
"There can be no science without measurements"
3.1. Introduction
L'AFME has asked the GAUDIN Ingenierie office to carry out a measurements campaign on schools.
The apparatus comprises :
- 16 analogical probes (acquisition with a time cadence of 2 min.)
- 16 analogical probes (acquisition with a time cadence of 10 min.)
- 16 All-or-Nothing probes (opening and closing of doors, of roofing frame, etc)
- 9 counting tracks (heating, gas consumption)
The files are stored in a central acquisition processor and transferred, each night, via the telephone network onto the computers of the GAUDIN Ingenierie office.
The measurements campaign was begun on 15th July '88 and will be completed on 15th July '90. About 18 million measurements will be carried out.
The analogical probes measure each building's characteristics :
- inside temperature
- outside temperature

- ground temperature at different depth levels
- resultant temperature (black ball measurement)
- wind speed
- relative air humidity
- solar energy transmitted
- luminous energy (lighting campaign)

N.B : We have used more than 50 different analogical probes ; this enables us to alter the measured parameters in accordance with the needs of the experiment.

3.2. The campaign's aims

- impact of solar energy on power needs : quantifying of savings achieved.
- enhancing the knowledge in the bioclimatic architecture field
- highlighting usefulness of variable mode software program for design purposes
- make possible a better manipulation of scientific tools via calculation/measurement comparison
- contribute to deeper scientific thinking about natural lighting (a field often empirically touched upon by architects)
- analysing the interactions between thermality and natural lighting

3.3. Conclusions of measurements campaign

As of december '89, about 40 % of the work relating to the measurements campaign was completed. The natural lighting portion has not yet begun. Nevertheless, several conclusions can already be cited.

3.3.1. Energy savings

The goal of a 60 % energy saving for schools due to the use of bioclimatic techniques and to the setting up of a judicious control system is not only attainable and reasonable but also, can be done WITHOUT INSTALLATION COST OVERRUNS. However, in order to attain this goal, the designers and the engineer must make a major imaginative thinking effort at the design stage.

3.3.2. Summer comfort

If the dimensioning of the glass roofs, solar protections and ventilation is correctly carried out, the bioclimatic schools are more comfortable in May, June, September and October than traditionally-designed schools (an extremely important point for the site under consideration).

3.3.3. Inertia

If a school is to be energy-efficient by virtue of solar techniques (47° latitude), it must have a certain amount of inertia. The latter should be judiciously distributed. Too light a construction would result in :

- overheating
- loss of solar gains (increase in energy consumption)

3.3.4. Comfort/quality of life

The thermal (summer/winter) environments, the quality of the luminous environment, the internal/external space relationships are such that all the occupants are highly satisfied with their working conditions :

Bioclimatic techniques with quality-of-life enhancement.

3.3.5. Utilisation of SIMULA

A building gives satisfaction for at least 3 quality reasons :

- architecture
- power saving factor
- environment (degree of comfort)

The measurements campaign shows decisively and unequivocally that the designers were absolutely right in using this experimental software program for the design.

The energy and environmental aspects owe a great deal to this research work. Thus a popularization of SIMULA was decided upon (see chapter 4).

3.4. Architectural quality

The architect's work carried out by C. COCHY is considered as being highly successful by the many visitors.

IT IS THEREFORE POSSIBLE TO CREATE BUILDINGS THAT ARE TECHNICALLY VERY EFFICIENT AND THAT ARE AT THE SAME TIME ARCHITECTURALLY SUCCESSFUL.

In April '89, the architect C. COCHY was awarded one of the 4 "Architecture, Energy & Environment" prizes for a public building erected less than 10 years ago.

4. DEVELOPMENT OF THE SIMULA SOFTWARE

The use of the SIMULA software program, as witnessed by :

- the Elisa Lemonnier project
- 3 other projects executed by GAUDIN Ingenierie
- the measurements campaign

makes it possible to obtain spectacular results. SIMULA makes it possible to go well beyond French thermal legislation and to offer building works foremen :

- high quantitative efficiency (energy saving)
- qualitative satisfaction (comfort enhancement).

Aware of these advantages, the French Chamber of Consulting Engineers (Technical Equipment Section), a professional body consisting of 300 engineering offices that represents the majority of independant thermal designers in France, has, following upon Mr GAUDIN's initiative, requested CERMA to produce an operational version of said software for all practising engineers.

The operational version incorporates all the attainments gained from the measurements campaign and from the monitoring of buildings built with the help of the operational version.

This software program will be used by the members of the CICF-ET as from 1st January '90. It will run on any IBM AT computer or compatible equipped with a hard disc.

We are dealing here with an exemplary co-operation between a research laboratory and practitioners so as to :

- prepare for the Single Market deadline
- promote technology transfers
- raise the technical level of consulting engineers' offices.

DESIGN FOR ENERGY EFFICIENCY : A CASE STUDY GOVERNMENT OFFICES AT WOLLONGONG, NEW SOUTH WALES

D. M. Rowe, ASTC
Principal Engineer, Public Buildings M&E
Public Works Department, N.S.W.
State Office Block, Philip Street
Sydney, NSW 2000 AUSTRALIA

ABSTRACT. New Government offices at Wollongong, NSW of 14,350 m^2 net rentable area were designed to provide a satisfactory work place at low capital cost and with the least possible annual consumption of energy. Means to that end included building orientation and layout; task/ambient lighting design; and an air conditioning system designed to benefit from the temperate local climate and from internal heat gains. Total energy consumption has been recorded each month from January 1988 and air conditioning components have been recorded since December 1988. Energy consumption has been approximately two thirds of the industry norm.

1. INTRODUCTION

New Government offices were designed in 1984 for Wollongong, a provincial city on the south coast of New South Wales in eastern Australia.
A major aim was to design for the lowest possible energy consumption consistent with a satisfactory workplace and with competitive capital cost. Orientation, envelope, daylight penetration, artificial lighting and the air conditioning system were designed to take advantage of passive energy flows and the temperate climate.

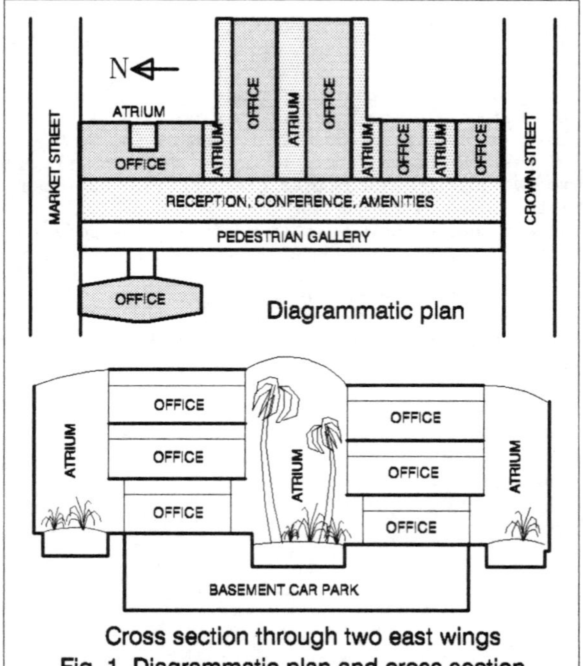

Fig. 1. Diagrammatic plan and cross section

2. ABOUT THE BUILDING

The building of 14,350 m^2 net rentable area stands on a podium over an underground tenant carpark. It has six wings of three storeys (fig. 1). Five of the wings extend east by lengths between 19 and 47 metres. They are connected laterally by glass roofed atria. Each wing is fifteen metres wide. The sixth wing is free standing.
North and south walls of the five eastern wings facing the atria are of clear single glazed curtain construction from floor to ceiling. All walls of the western wing and the north wall facing Market Street are similar but of tinted heat absorbing glass. There are no windows in the eastern walls of the east wings. The pedestrian gallery western wall is of insulated metal clad sandwich construction with narrow windows.
Roofs of the office spaces are of flat reinforced concrete insulated with 75mm. of mineral wool under the soffit and 25mm. of expanded polystyrene below the topping. Atrium roofs are of tinted heat absorbing glass.
Enclosed offices located in the centres of the wings and conference rooms adjacent to the pedestrian gallery result in a high proportion of occupants having access to daylight.

3. LIGHTING

A task/ambient lighting system was designed to reduce visual interference with screen based tasks.
The ambient component is by recessed ceiling mounted luminaires yielding an average design illumination level of 250 lux at desktop from fittings with two 32 watt fluorescent tubes each on a three metre each way grid.
Task lighting is by an 11 watt PL fluorescent lamp mounted on an adjustable arm on a heavy square base which stands on each desk. Artificial light levels up to 1400 lux are available for desk tasks. Each task light is switched independently.
Ambient luminaires are switched off at intervals during the day by a pulse generated by the direct digital control system. Switches located close to work places and each switching two to four luminaires enable people to restore the ambient light in their local area if required. Control is as illustrated in fig. 2.

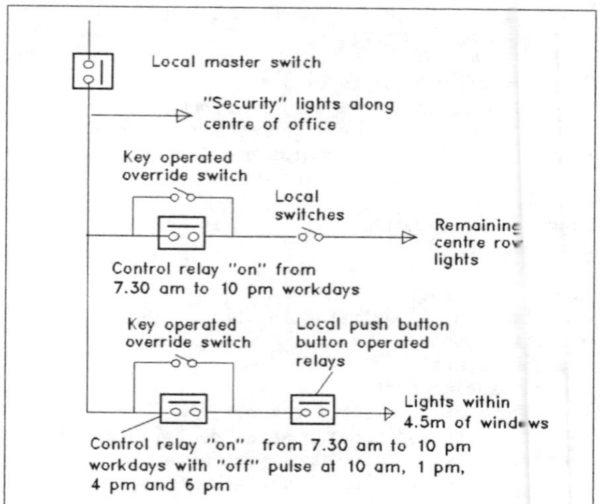

Fig.2 Typical office luminaire control (after Payne)

4. AIR CONDITIONING

Fig. 3 shows the temperate climate at Wollongong [1]. Design conditions are 32°C dry bulb/23.5°C wet bulb (summer) and 7°C (winter). For much of the year the temperature lies within a comfortable range between 20°C and 24°C. Air conditioning is however required to provide the "more rigorous environmental control" noted by de Olivera Fernandez, Maldonado and Cortesao [2] as necessary for the comfort of a large number of people in a given confined space.

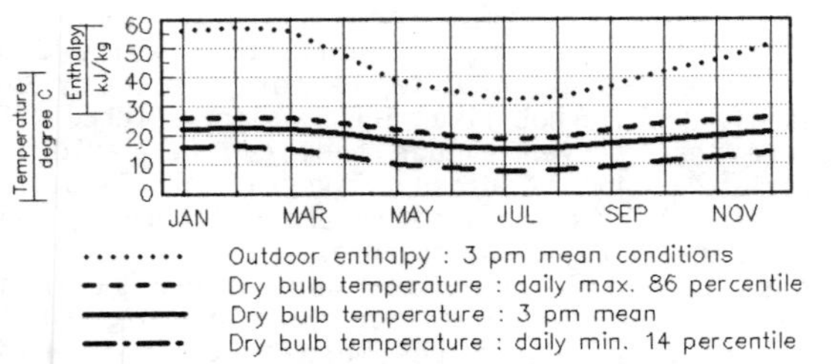

Fig. 3 Climatic averages for Port Kembla, a suburb of Wollongong [1]

The air conditioning system is designed to provide mechanical cooling or supplementary heating only when internal temperature rises above 24°C or falls below 20°C.

Five air handling units supply conditioned air to variable air volume distribution systems in separate wings or pairs of them as shown in fig. 4. Outdoor air is mixed with re-circulated air when outdoor enthalpy is above the enthalpy of return air. When it is lower outdoor air only is circulated until its temperature falls below the supply air set point when it is modulated with return air to satisfy cooling requirements.

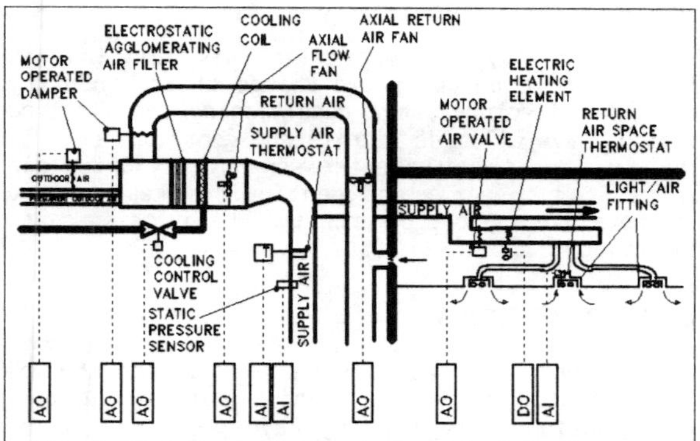

Fig. 4. Diagrammatic arrangement of single zone fan coil unit and typical part air distribution

Supply air temperature from each conditioner is modulated downward under control of the direct digital control (DDC) system until the cooling requirement of the zone of greatest heat gain connected to it is just satisfied. Other air valves modulate to maintain space temperature set point. The cooling medium is chilled water. Mechanical refrigeration is initiated by a zone temperature exceeding 24°C. A reciprocating chiller (400kW refrigeration) cools loads of up to one third of total. A centrifugal machine (800kW) operates between one and two thirds. For greater loads both machines operate in parallel (fig. 5).

"Greenhouse" warming by the glass roofed atria reduces the requirement for supplementary heat. Air is exhausted by fans from the top of each atrium when temperature reaches 24°C. They stop when atrium temperature falls below 24°C and the resulting entrapment of warm air prevents heat loss

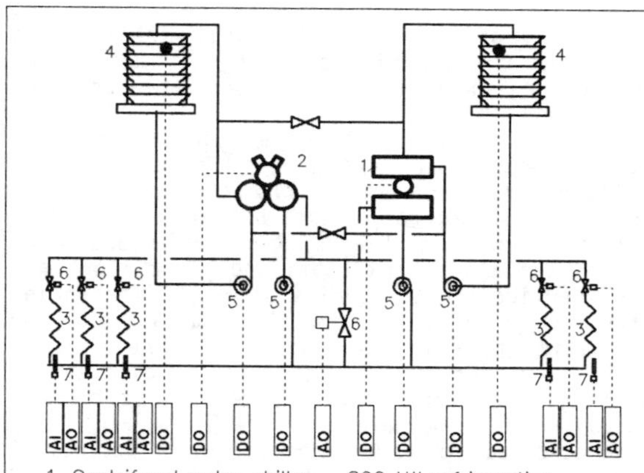

Fig. 5 Diagrammatic arrangement of chilled water refrigeration plant

for much of most winter days.

Investigation of heat losses using the Carrier E20-II heat load and energy flow simulation programme indicated a low requirement for supplementary heating and led to the selection of electric resistance element heaters installed in airstreams after variable volume terminal units. A zone heater is switched on when zone temperature falls below 20°C and off when it rises to 21°C.

Electrostatic agglomerating air filters were selected to ensure clean air in an industrial urban environment and for low pressure loss.

Cool air is circulated on summer nights and warm air is recirculated before occupation on winter mornings; heat evolved in the building in winter is recovered by raising the supply air temperature to the highest acceptable point.

5. CAR PARK VENTILATION

The fully enclosed underground carpark is ventilated by mechanical systems. Supply and exhaust fans start at half speed when carbon monoxide concentration exceeds 10 parts per million and run at full speed when it exceeds 50 ppm.

6. DISCUSSION

Total and component energy for the whole building is shown in Table 1 and figs. 6 and 7.

MONTH	TOTAL	ALL A/C	ALL REF	CHILLERS	C. TWRS	FANS	HEAT
D88	178	83	58	41	17	26	0
JAN	180	74	49	34	15	25	0
FEB	138	79	56	40	16	22	0
MAR	172	74	51	36	15	23	0
APR	134	49	23	34	8	26	0
MAY	144	36	9	5	4	24	2
JUN	130	34	1	1	0	21	12
JUL	140	48	2	1	1	25	19
AUG	140	33	1	1	0	20	12
SEP	141	38	* 7	* 1	7	24	6
OCT	157	33	* 9	* 2	7	23	2

Table 1. Monthly Energy Consumption MW hr.
*** Reciprocating chiller fault**

The Australian Building Owners and Managers Association (BOMA) uses a parameter called Energy Utilisation Index as a means of comparing energy efficiency of office buildings. The index is defined as the quantity of energy in megajoules consumed per square metre of net rentable area per annum (MJ/m^2 PA).

Net rentable area is defined as the sum of areas within the building measured from the internal surfaces of permanent walls including areas occupied by structural columns but excludes stairs, toilets, cleaners cupboards, lift shafts and motor rooms, escalators, tea rooms and plant rooms and other service areas; lobbies between lifts facing other lifts serving the same floor; areas set aside as public space or thoroughfares; and service areas, areas for service vehicles and deliveries, car parking and access thereto where not for the exclusive use of occupiers.

A standard year is 2,500 hours and other periods of occupancy are adjusted pro rata.

The energy utilisation index has varied between 368 and 406 MJ/m^2 PA. This compares favourably with the calculated BOMA target for acceptable energy consumption in a building of this size of 560 MJ/m^2 per annum.

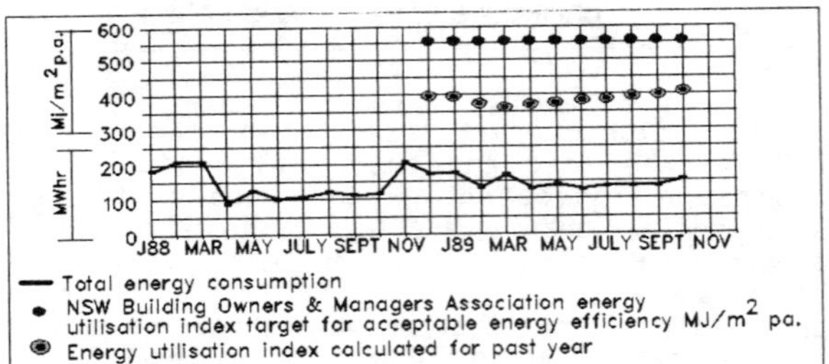

Fig. 6 Total energy consumption

No supplementary heat was used between December 1988 and April 1989. Cooling energy was negligible in June, July and August. Fan energy was approximately constant due to supply air temperature reset as predicted during design.

Energy consumed during the six months January to June considered to represent approximately half the yearly total is shown by components in fig. 7.

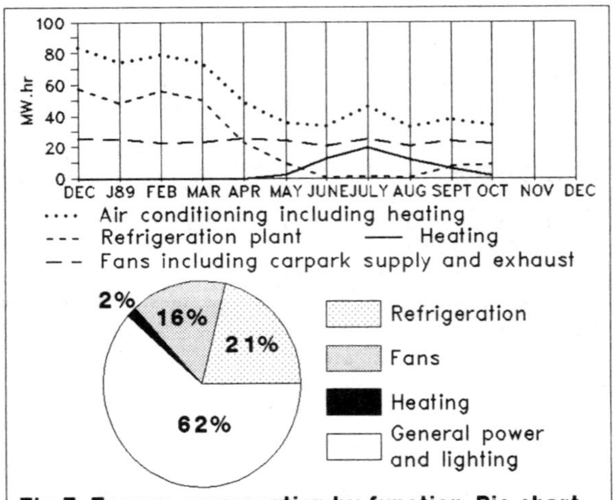

Fig.7 Energy consumption by function. Pie chart based on six months January to June

This clearly indicates the minimal requirement for heating in a deep enclosed office building in a moderate climate provided the heat evolved by lights, office machines and people can be thus "passively" recovered. It also suggests that recovery of this waste heat by similar "passive" means could be effective in conserving energy in quite cold climates.

The maintenance of high ventilation air flow quantities throughout the year has contributed significantly to this reduction in need for supplementary heat. It has also avoided a possible disadvantage of variable air volume air distribution systems resulting from the considerable air quantity reduction in winter months which occurs with some constant temperature systems. It has also contributed to stability of air valve operation by reducing system static pressure variations.

The system has proven highly tolerant of heat load variations.

Payne [3] has reported the results of a survey of occupant acceptance of the lighting systems. Approximately half the occupants of the building responded to his questionnaire. Of these 69% reported the lighting as comfortable while 31% considered it uncomfortable for a variety of reasons. 32% did not like lights being switched off automatically even though they can be restored immediately. And 26% did not like the task lights.

Early complaints resulted in the mid morning and mid afternoon light switching pulses being deleted. The system still provides worthwhile savings however as lights are switched off overnight and are not turned on in the morning until needed locally by workers.

Also many of the occupants work outside the office for part of the time and it is not uncommon for lights switched off in the middle of the day to remain off in local areas because no one is present to require them or because daylight is adequate.

Payne concluded "the system of background lighting and task lighting used in the office areas has proved acceptable to the majority of users although a significant minority have have expressed displeasure. It could be argued that this disaffected minority would be found in a similar survey of a conventional office building although the size of the minority does give the impression that that this particular lighting system does not provide an overall acceptable working environment".

7. CONCLUSION

Measured energy consumption in this building has been found to be about two thirds of the industry norm. Architectural, air conditioning and lighting designs have all made significant contributions.

The results demonstrate the serendipity which can arise from an early and continuing creative and co-operative support of architecture by science and technology.

8. ACKNOWLEDGEMENTS

The Project Architect was Morgan, L. for the NSW Government Architect. Significant engineering contributions were made by Payne, L. (electrical and lighting) and Moore, J. and Yeo, E. (mechanical) of the NSW Public Works Department.

9. REFERENCES

1. Australian Bureau of Meteorology (1975), Climatic Averages for Australia, Australian Government Publishing Service, Canberra.

2. de Olivera Fernandes, E., Maldonado, E. A. and Cortesao, C., The Role of Passive Solar in Nonresidential Buildings in Moderate Climates, Energy and Buildings for Temperate Climates : A Mediterranean Approach : Proceedings of Sixth International Plea Conference, Porto, Portugal, 27-31 July 1988, Ed. de Olivera Fernandes, E. and Yannas, S., Pergamon Press, Oxford, England, pps109-113.

3. Payne, L., Office Area Lighting at the Wollongong Government Office Block, Proceedings of the Electric Energy Conference 1989, Sydney, The Institution of Engineers, Australia, Canberra.

66 SECOND GENERATION SUNSPACE HOUSES IN THE NETHERLANDS

Authors: Ir. J.M. Boonstra (Woon|Energie)
Drs. M. de Langen (Woon|Energie)
Ir. T.H. Reijenga (BEAR architects)
Woon|Energie, Crabethstraat 38 j, 2801 AN Gouda, The Netherlands tel. 01820 - 24233

01. Background

The project presents a second generation design for passive use of solar energy in low cost single family houses, optimized for NW European sea climate conditions.
First generation projects (such as EC Energy Demo SE 389-84 , Goirle NL, Woon|Energie) provide good information on benefits and costs of passive solar design options.
From those experiences it can be concluded that in the NW European sea climate zone, the most attractive passive solar option, in terms of energy conservation, cost benefit ratio and quality of the the indoor thermal zones consists of a combination of

- south facade: second skin sunspace
- ventilation: natural ventilation ducts
- space heating: single plate hot water radiator central heating

In the project mentioned above, user satisfaction with the second skin sunspace design has proven to be very high, while at the same time energy consumption for space heating and ventilation, electricity included has proven to be at the expected level (ca 21 kWh/year.m3 building volume; lower than most other experimental low energy housing designs build in the Netherlands during the same period, including those with integrated hvac heat recovery systems and minimum size windows.

In the present EG demonstration project SE 220-88-NL the main focus has been to achieve a simple low cost second skin design with integrated sunshading/night insulation, suitable for mass production. This is being done in coöperation with a Dutch producer of greenhouses for agriculture.
The stages of the EG demonstration project in Spijkenisse-NL is: preparation in 1989, building in 1990, monitoring and evaluation in 1991/92.

02. Project description

66 single family houses (in five rows), Site: municipality of Spijkenisse-NL
floorspace house/sunspace: 106/9 m2 ; volume 290/50 m3
5 rooms, sunspace, kitchen, bathroom, storage
structure: concrete; N-facade: brick; Inner S-facade: panels; sunspace facade: Al-frame,single glazing
price: Dfl 142,000.-,taxes incl.
Glazing: sunspace facade: nett. 20 m2 single, inner S facade: 10 m2 double, N facade 3,5 m2 double
Ventilation: natural ventilation ducts and cooker hood
Space heating: central heating, single plate, hot water radiators.
Expected yearly energy consumption for space heating: 600 m3 natural gas. Savings compared to actual average house is 400 m3 natural gas and 250 kWh elektricity (no mechanical air system)

Building contractor: Wilma Bouw Rotterdam, Producer sunspace facades: Smiemans Projecten

66 SECOND GENERATION SUNSPACE HOUSES IN THE NETHERLANDS

EC Demonstration project 220-88/NL

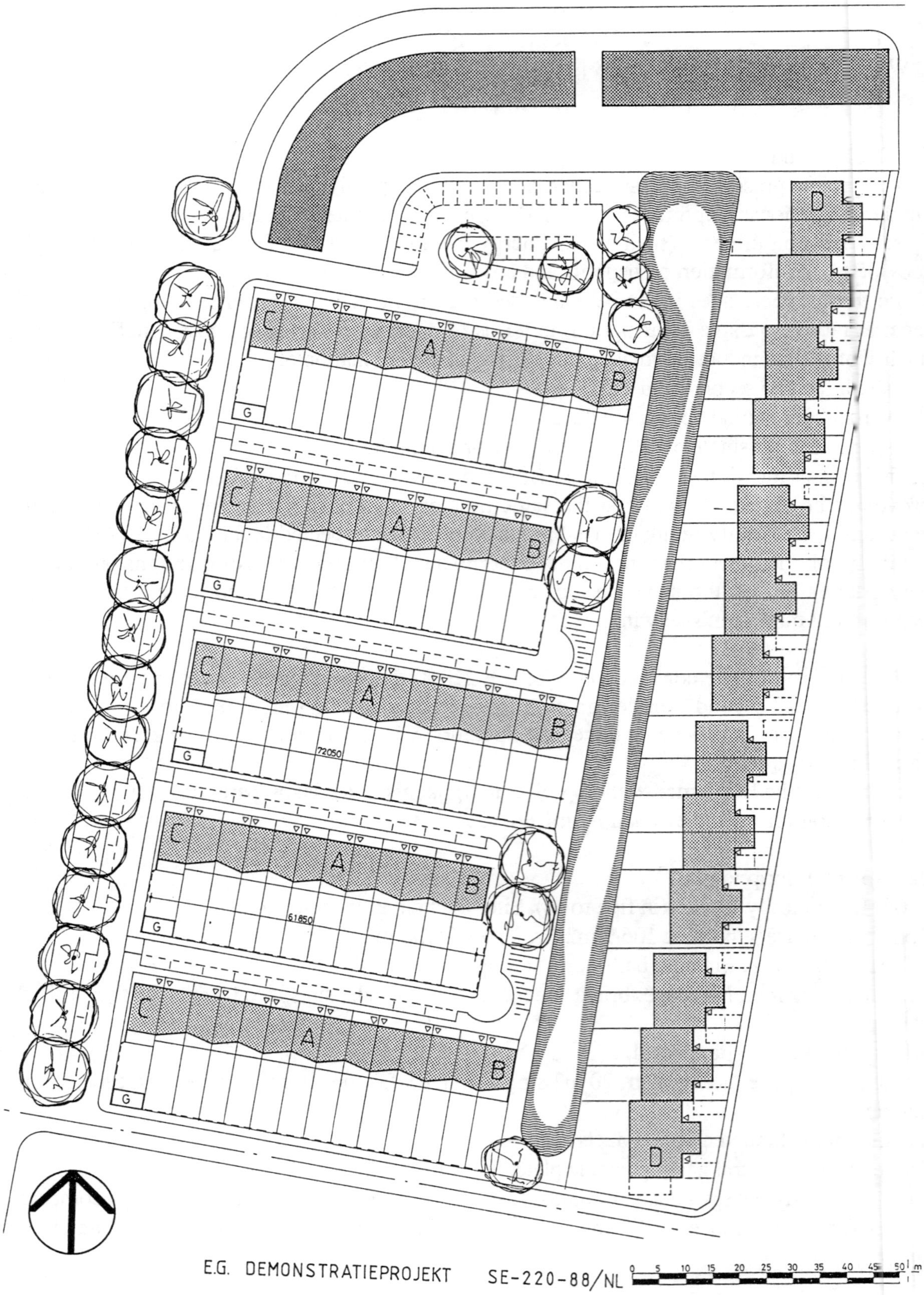

66 SECOND GENERATION SUNSPACE HOUSES IN THE NETHERLANDS
EC Demonstration project 220-88/NL

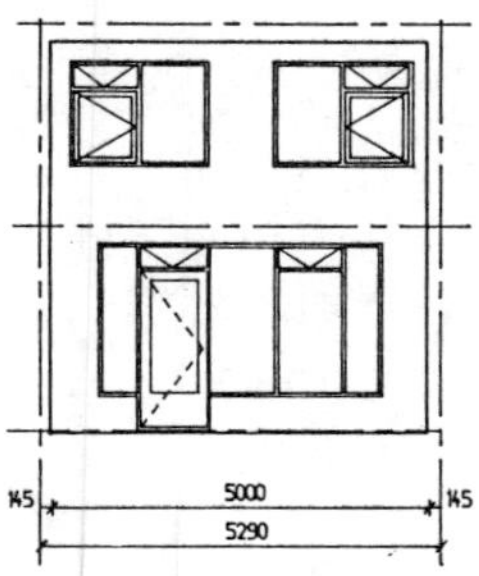

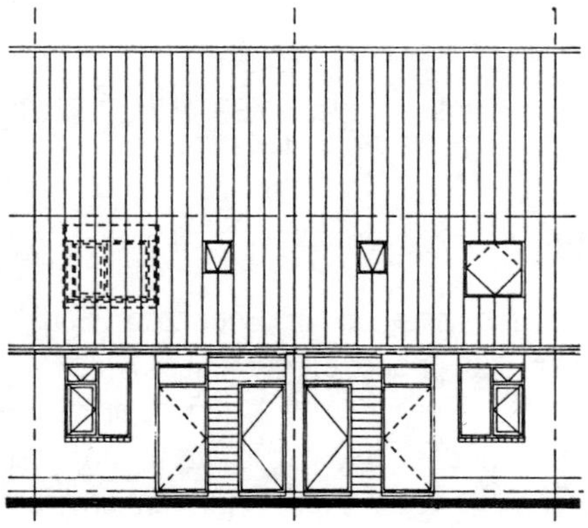

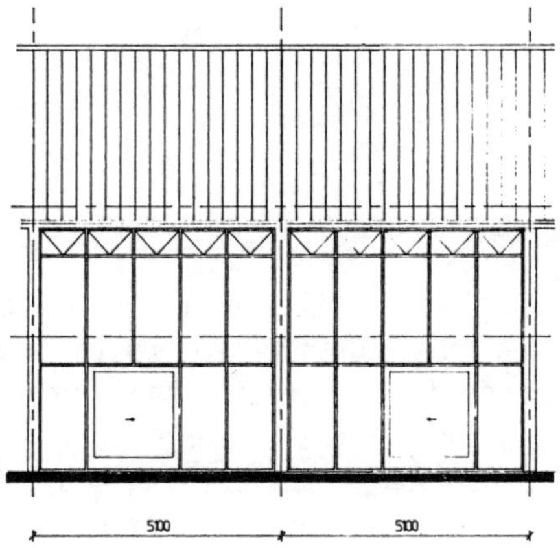

INNER FACADE **NORTH FACADE** **SOUTH FACADE**

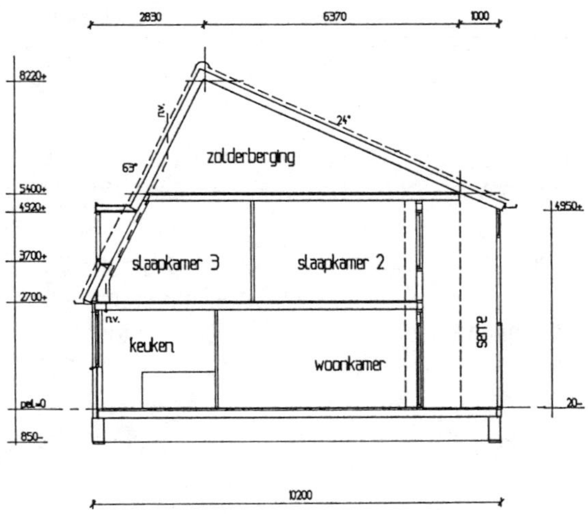

FIRST FLOOR

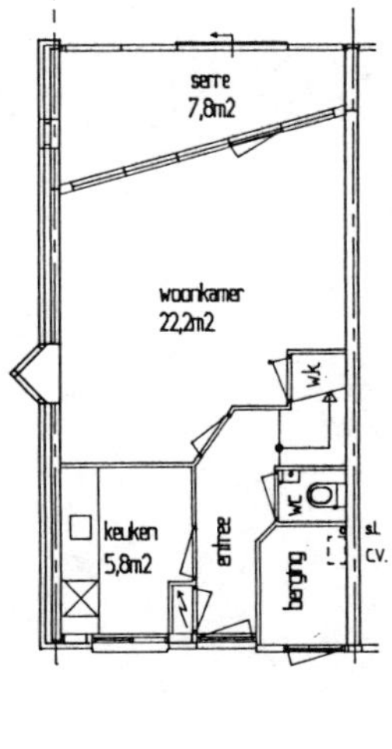

SECOND FLOOR

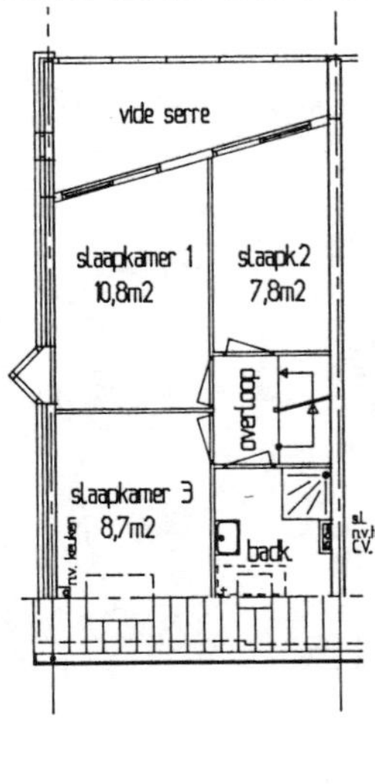

THIRD FLOOR

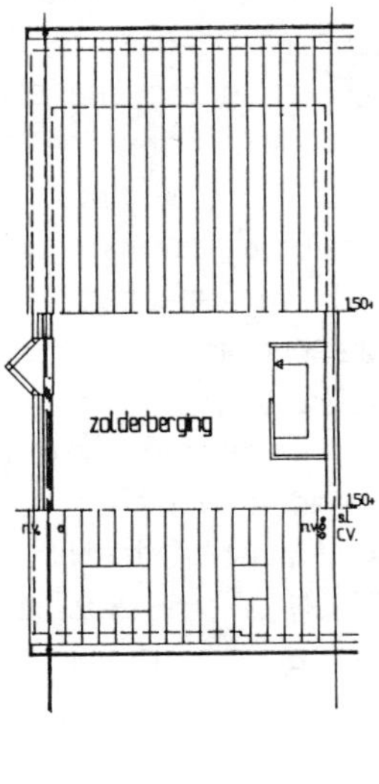

POLYFUNCTIONAL GYMNASIUM IN BRONZOLO (ITALY-SOUTH TYROL)

Romano ZURETTI

Piazza S. Leonardo, 4b 39051 BRONZOLO Italy
telfax 0471-952495

The planning task of the polyfunctional gymnasium of Bronzolo (South Tyrol/North Italy) dates from July 1984 and up to date various proposal projects were formulated, examinated and corrected. A systematic and continuous revision of the basilar principles moved to the settlement of the present model, which includes the exploitation of the synergy in the construction of the building and this thanks to its position on the territory in order to attain the peak of the efficiency as living comfort .
The proposal of new architectural researches, especially for the most important public buildings like the polyfunctional gymnasium of Bronzolo means , here in SouthTyrol, to swimm against the stream.
The solar passive architecture is a cultural provocation. Opening the building in the sun means damaging the tyrolian landscape.

1. THE PROJECT

1.1.BUILDING PROGRAMM FOR THE SPORTS INSTALLATION OF THE GYMNASIUM.

The sports installation of the gymnasium is constituted by the following environmental units :

Section of the sports and recreation activity
- gymnasium 18x33x7 at the ground floor ;
- intered bowling-green 9,50x9x5 ;
- stand 8x4x4 ;

Section of the support services

GYMNASIUM
- hall
- n.3 common dressing-room each for 23 users
- n.1 dressing-room for trainers
- n.6 units for sanitary services, two of which can be used by the public
- n.5 units for the showers for boys
- n.2 warehouses for equipment
- room for the cleaning equipment
- direction room for using the gymnasium as a hall for spectacles
- greenrooms at the side of the stage

BOWLING-GREEN
- n.3 units for sanitary services which can be used also by the public
- warehouse and storehouse for equipment

SECTION OF THE AUXILIARY SERVICE(s)
- bar at the ground floor and at the intered floor

SECTION FOR THE PUBLIC
- hall for the public and admittance entrance at the seats of the associations
- edge for the reception/ticket-office
- linen-room
- hall for the congresses
- tiers for the public at the first floor for the gymnasium, at the intered floor for the bowling-green
- sanitary services for the public at the first floor

SECTION OF THE TECHNOLOGICAL INSTALLATION
- heating room
- air-conditioned room

1.2.GENERAL QUALIFICATION FOR THE RUNNING OF THE GYMNASIUM

The sporting installation for the gymnasium/bowling-green is explained by this scheme

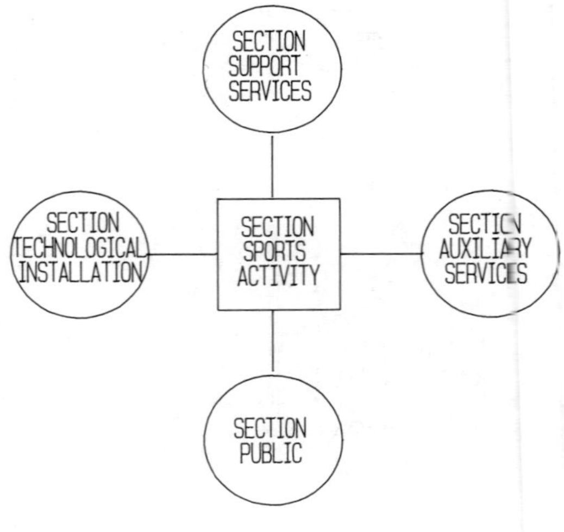

The sections which constitute the sporting system, are characterized by relation of complex kinds, but they must permit the development of the following sports:

- volley ball
- basket ball
- soccer
- gymnastics
- bowls

Thanks to the polifunctional character, the volume of the gymnasium is subdivisible in two parts at one-third of its lenght through a hanging moving wall.This permits to use less space in favour of the administration costs or the contemporary use of more users. Moreover the whole complex is right for sheltering manifestations of sigh under cover.

2. PLANNING PRINCIPLES FOR A PROJECT WITHOUT BARRIERS

The planning principles deal with the constructed elements, wich determine the internal and external environment and the element which separates this two milieux.These elements are therefore distinguishable as external elements, frontierelements and internal elements.

2.1.EXTERNAL ELEMENTS of arrival

- dimension of the parking places

of transit

- width of the side-walks; presence of invitations for the overcoming of the edges between side-walk and street; absence of unforeseeable obstacles (supports for traffic signal, etc); appropriate kinds of surfaces for avoiding skiddings and for signaling changes of level, crossings; stairs; lifts.

2.2. FRONTIERELEMENTS

The whole elements that are able of guarenteeing the services, which the envelope can supply dealing with dimensional organization; services, equipment, so that the environmental requested conditions can be assured as well as the passing of people.

- dimension and structure of passage; doors and main doors; thresholds.

2.3. INTERNAL ELEMENTS

- Dimension and structure of sanitary services ; spaces for services, particularly for what concerns admittances, internal circulation, use of the equipment and the furniture.

3. THE LT METHOD

3.1. The LT METHOD is a tool for energetic planning, which is expressly developed for this use. It refers itself to a mathematical model which can foresee the annual consumption of primary energy referred to 1 m2 of trampling surface as function of:

a) local climat conditions

b) orientation of the facade

c) kind of windows

The calculation method is provided with eight graphics, each of them for the five european climatic zones. Bronzolo makes part of the continental area. First of all, the model values the thermic conduction through the envelope of the building or the losses (or the profits) of heat for ventilation.

Using the medium monthly temperatures, a set point temperature and a correction factor which considers the intermittency of the running (working), you can calculate a total monthly thermic load. Afterwards the model values the solar profits and applies to them an utilization factor. In the meantime, you can calculate the monthly hours of day-luminosity and of the values of internal lighting which are requested.

All this permits the calculation of the monthly consumption of electric energy due to the artificial lighting and the monthly profit of thermic energy.The fractions of hour are considered when the internal lighting level exceeds the 66% of the reference value to consider also the real change in comparison with the monthly average. The thermic profits due to illumination and solar radiation, united to a fixed casual profitvalue due to persons and to installations, are then subtracted from the total load to establish the net load for heating. When the profits exceed the total heat quantity which is necessary it needs a cooling. At the moment when it needs a cooling, the model eliminates the profits due to the direct solar radiation letting only the diffused daylighting. All this,reproduces in a realistic way the mobil screens, which are realized to eliminate all the direct radiations as soon as it needs a cooling. The model considers circa 30 parameters, the majority of which are fixed to provide the LT curve. The values assumed by these parameters are chosen for representing a typical modern schoolbuilding, engaged during the day.

In front of an esteemed thermic load with the application of the low 373 about the contenance of the energetic consumption e-quivalent to 493.200 Kwh/year correspondent to an hour consumption of 274 Kwh per 10 hours/day per 180 days/year we have, by ope-ning the building at the sun, a reduced con-sumption equivalent to 235.110 Kwh/year that means 130 Kwh per 10 hours/day per 180 days/ year, keeping to 52% the theoretical reduc-tion of the energetic consumption foreseen with the low 373.

ENERGY CONSUMPTION

	Heat	Light	Cool(*)	Total	Surf.	Index
	Kw/year	Kw/year	Kw/year	Kw/year	m2	Kw/y m2
Ordinary Project	493.200	227.750	0	720.950	1984	363
Pass.Solar Project	235.110	198.334	-81.200	433.444	1984	218
Differ.absolute	258.090	29.416	81.200	287.506	0	145
Rate Difference	-52%	-13%		-40%		-40%

(*) Produced by the passive solar system during the summer-time and storaged into the thermic storage floor (porphiry stones) and non computed.

COST

Ordinary Project = Lit 2.800.000.000
Pass.Solar Project = Lit 3.086.000.000
Extracost = Lit 286.000.000
Cost of 1 Kw/h = Lit 189
Saving in 1 year = 287.506 Kw/y x 189 L/kw = 54.338.636 Lit
Amortiz. extracost = 286.000.000/54.338.634 = 5,26 years

ENERGY CONSUMPTION

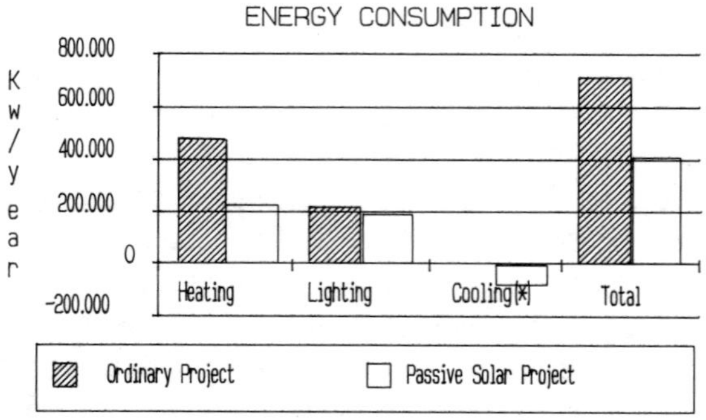

The application of "LT" calculation method proposed by the Energy Research Group of the University of Dublin to the polinfunctional gymnasium of my little town Bronzolo, has full confirmed the previously choices of the project based on well known calculation method(s) but of lengthy elaboration.
The project conside several building parame-ters, in particular :

1) AREAL PARAMETER
Planimetry and orientation of the building lot, microclimate, acustic and environment pollution, access to solar insolation and day-luminosity.

2) BUILDING SHAPE
Dimension of planimetry and section, height of the rooms, dimension of the passive zone, orientation, bordering building.

3) BUILDING TYPOLOGY
Buildings features, floors insulating, walls insulating, roof insulating, type and area of glass windows, thermic mass, vapour boundary.

4) DAY-LUMINOSITY and SOLAR INSOLATION.
Windows area, distribution of glazed sur-faces, shadows, internal and external re-flecting surfaces, projections, dazzling check

5) PASSIVE SOLAR PARAMETERS
Thermic measurement relativeness to the climate and the buildings typology, distri-butive character, solar forheating of air ventilation, effect about the comfort.

6) NATURAL AIR VENTILATION
Predominant wind, chimney effect, opening distribution, slowest ventilation, cooling of the structure, nightly ventilation, coo-ling through air ventilation.

7) PREVENTION FROM OVERHEATING.
Screening, reflecting elements, buildings insulating, casual thermic contributions, standards of comforts.

8) ARTIFICIAL LIGHTING
Lighting source, lighting level and standard lighting, task of artificial lighting rooms lighting, integration with solar light, daz-zling and visual comfort.

9) HEATING
Choice of fuel type, distribution of heatings installation, radiating elements, blowing elements, thermic comfort, distribution and storage of warm water, conditioned air system.

10) MECHANIC SYSTEMS

11) ATRIUM, HOTHOUSE , INTERIOR SPACE

12) MANTENANCE SERVICE AND CLEANING

BUILDING SYSTEM IN CONCRETE BLOCKWORK - DESIGN, CONSTRUCTION AND PERFORMANCE

Josep I. de LLORENS i DURAN & Alfons SOLDEVILA i BARBOSA
Dr. Architects, Professors at the School of Architecture, Barcelona
Castanyers, 11 - 08916 BADALONA - SPAIN

ABSTRACT . A new building system has been developed to improve the constructive technology and thermal performance of low cost housing projects. The main characteristics are: the building system,the erection process, the elementary layout, the richness of the space, the adaptability to the users and to the environment and the thermal performance.

1. THE BUILDING SYSTEM

The building is a double rectangular container made of exposed hollow concrete blocks. The roof is also split into two self-suporting independent layers, the outer one for waterproofing and the inner one for closing and insulation. External doors and windows are sliding and assembled on pre-frames. Reinforced insulated shutters restore the thermic continuity of the wall during the night and supply additional security without bars.

Floors are made of two boards and timber joists. A shock absorber in between improves the acoustics. The joists are supported by the inner leaf of the wall by means of niches distributed in horitzontal rows. Partitions are sliding timber panels hanging from rails fixed to the joists. The same solution is adopted for the cupboards that can be placed between bedrooms if privacity is required. Services are exposed. They don't interfere with masonry walls. Pipes and cables run horizontally above the upper floor (under the roof)

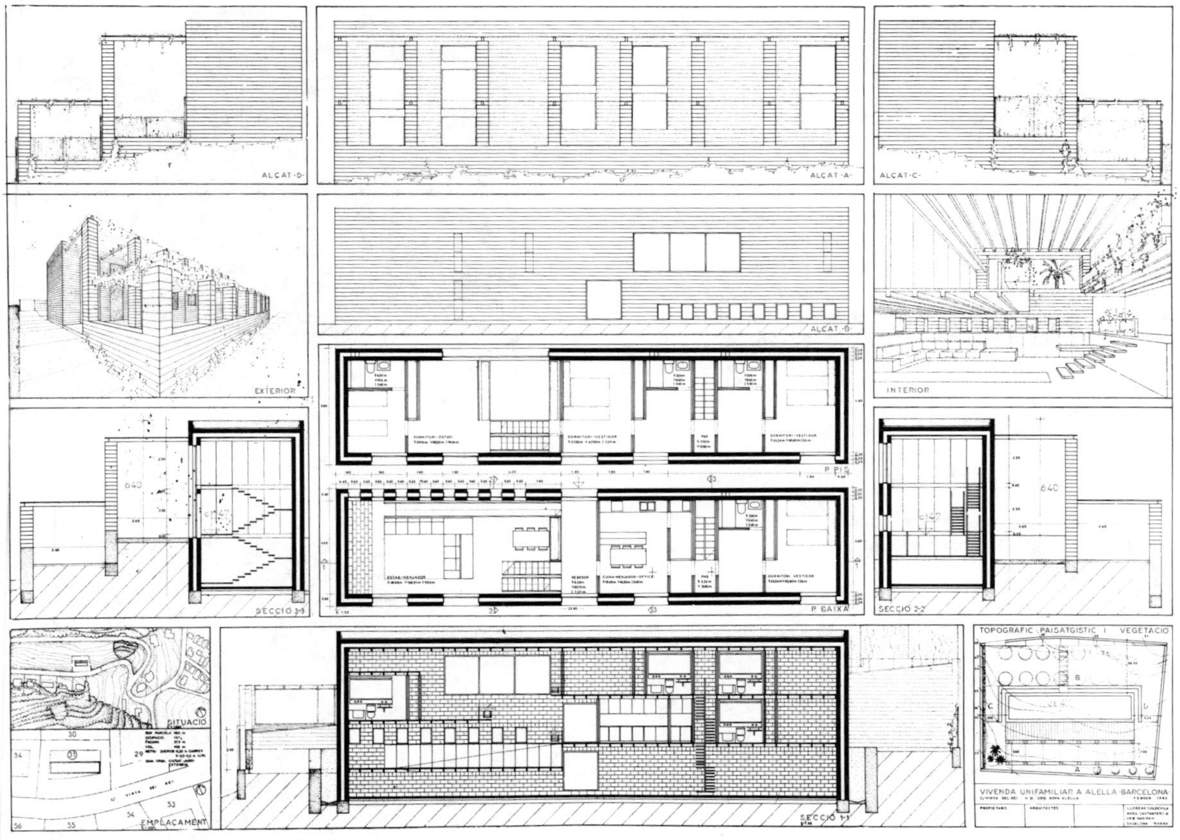

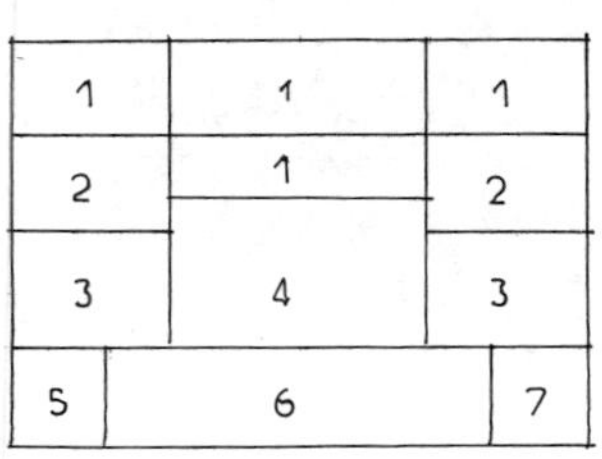

Figure 1: Graphic summary
1. Elevations
2. Perspectives
3. Transversal cross sections
4. Plans
5. Location
6. Longitudinal cross section
7. Site plan

or below the lower floor (in the cavity) and enter vertically into the rooms. Finishes are not used on walls. Nor plaster neither painting are applied on the interior. Exposed concrete blockwork is combined with the wood of floors, shutters and partitions. Rendering of the outer leaf is not necessary since waterproofing is solved by the drainage of the cavity wall.

2. ERECTION PROCESS

The erection process is based on the independence of the building components. The floors, partitions, carpentry and services are placed consecutively after the completion of the walls and the roof (fig. 2). Interferences and simultaneities are avioded, so the works become easier and simplified. To get the independence between building components it has been necessary to study thoroughly all the details and design new solutions unusual in the traditional techniques.

3.- ELEMENTARY LAYOUT OF THE SCHEME

The building is a rectangle 6 m wide. The only one interior span is roofed with simple joists. Nor supports neither bearing walls interfere with the space and all rooms are next to the façade. Blocklayering turns to be very easy.

4.- RICHNESS OF THE SPACE

The interior is not completely subdivided because the floors made of wood horitzontal platforms are interrupted by a big central space and the wood partitions are sliding (fig. 3). Moreover the possibility of laying the floors every 1,20 m connected by a staircase, contributes to the richness of the spatial perception.

5.- ADAPTABILITY TO THE USERS

Partitions and floors can be easily adapted to different users. The joists simply rest on niches. There is a horitzontal row of niches every 1,20 m so platforms can be placed on a wide variety of locations according to the needs of the user. Joists can also be removed and placed in other locations, allowing for changes that usually occur in a family during the life. The same idea is applied to the partitions made of timber wall panels hanging from rails fixed to the joists.

6.- ADAPTABILITY TO THE ENVIRONMENT

The adaptability to the environment is based on the construction of intermediate spaces acting as filters like porches, balconies, verandas or patios (fig. 4).

7. THERMAL PERFORMANCE

The thermal performance is improved by the continous cavity of the walls and roof. It can be insulated according to the climatic environment linning the external face of the inner leaf of the wall and laying the insulation on the lower layer of the roof. The cavity can also be ventilated to avoid condensation in winter and gains of heat during the summer. It has to be pointed out that thermal bridges are totally avoided including corners, jambs, lintels and floors (fig. 5).

CONCLUSIONS

The result is a low cost highly efficient building system suitable for a wide range of users and locations.

Figure 2: Walls, roof and floors are placed consecutively

Figure 3: The interior space is not completely subdivided

Figure 4: The porche acts as a filter

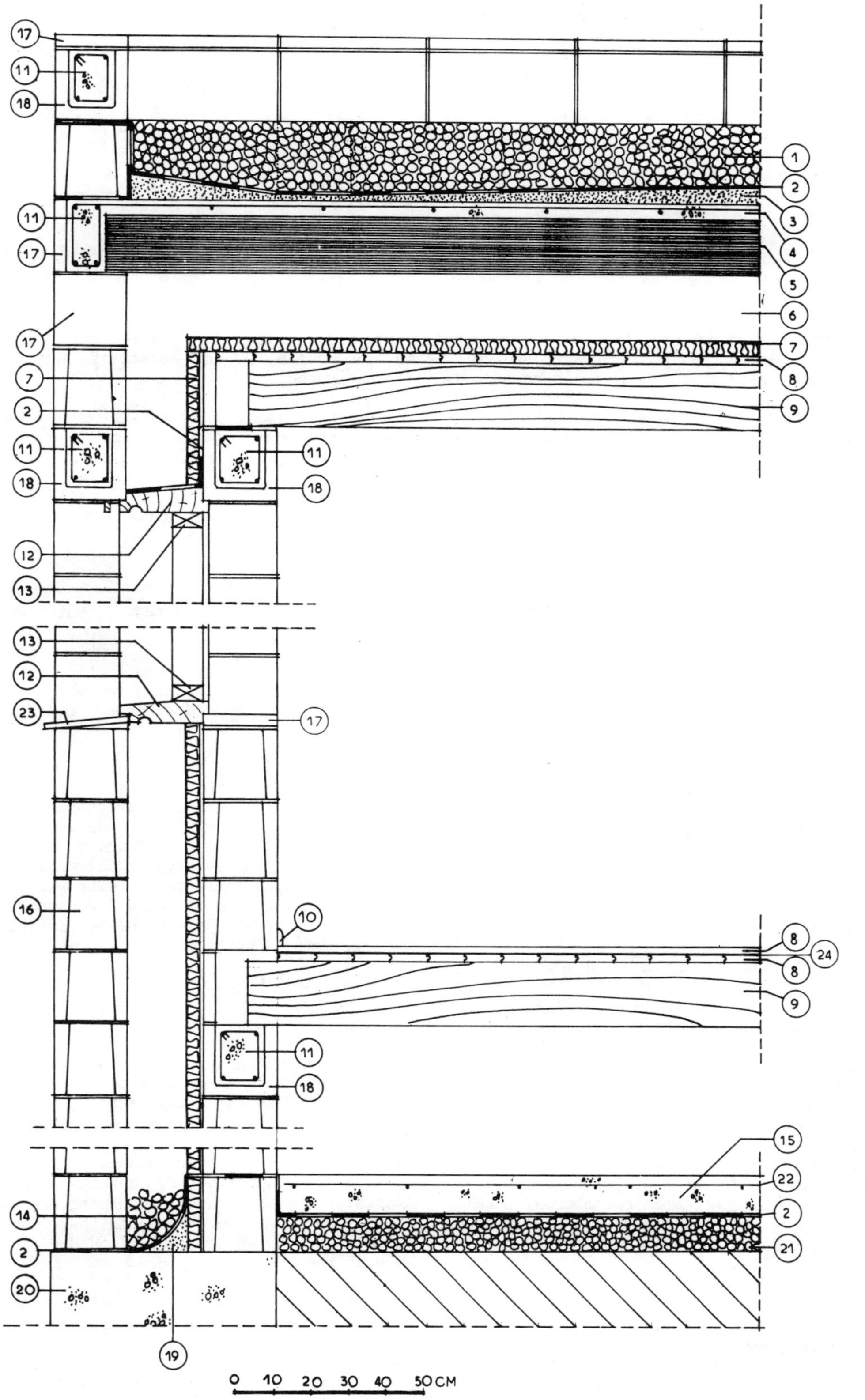

Figure 5: Detail of typical cross section - 1.Gravel - 2.Waterproofing membrane - 3.Lightweight concrete 4.Reinforced concrete - 5.Prestressed concrete joist - 6.Cavity - 7.Insulation - 8.Wood board 9.Wood joist - 10.Socle - 11.Bond beam - 12.Preframe - 13.Frame - 14.Drainage - 15.Concrete 16.Hollow concrete block - 17.Solid concrete block - 18.Bond beam block - 19.Waterproofed mortar 20.Foundation - 21.Crushed stone - 22.Reinforcement - 23.Polished stone - 24.Shock absorber

SOME LOW ENERGY BUILDING DESIGN CONSIDERATIONS FOR MALTA

J.M. CACHIA

UNIVERSITY OF MALTA

ABSTRACT

Buildings in Malta consume over half of all imported energy. Building energy use is not regulated. The existing building laws address public health and well-being. The possibility and impact of introducing basic energy conservation features to local building methods is investigated.

Considering the number of parameters and their complex interaction in space and time, it would be a laborious and time consuming exercise to accurately estimate manually the energy balance of a building. The availability of cheaper and more powerful computing power has made available building energy models which have been improved upon over the last decade or so.

This paper describes a research project which was carried out at Cranfield Institute of Technology in the U.K. using one such building simulation model namely : Tas (Thermal Analysis System).

1.0 CONTEXT

1.1 Energy Overview

Malta depends solely on fossil fuel imports for all its energy needs. 60% of the total energy consumption is to generate electricity with a conversion efficiency of less than 25%.

The building sector in Malta is the largest single consumer of energy and responsible for over half [2] of all energy consumption in Malta.

90% of the delivered energy utilised in buildings for heating, cooling, lighting and appliances is electrical. Nationally, conversion losses account for 45% of the energy bill of Malta.

The pattern of electricity consumption in Malta is temperature related giving rise to peak demand periods above the power station base load in mid-Winter and to a lesser extent in Summer indicating a dependence by and large on electricity for both space heating and cooling, see fig. 1. Peak demand periods which have determined the necessity to invest in a new power station can be reduced by resorting to energy conservation measures. A saving of one unit of electricity results in four units less energy imported to Malta and a four-fold reduction in noxious power station CO_2 emissions the major cause of the climate change phenomenon. Half of these savings can be achieved by energy conservation measures in the building sector.

1.2 Building Construction Techniques

Locally available globigerina limestone and concrete are the two most commonly used building materials. Limestone blocks are utilised extensively in walls which vary in thickness from 150mm for a single skin wall to 500mm for a double skin cavity wall. This material has an average density of 1700kg/m^3, it is a porous material and has a thermal conductivity varying from 0.6 - 1.0 W/mK. The scanty building laws which address public health and well being specify the use of stone headers to tie the separate skins of a double wall together. This gives rise to extensive thermal bridging and damp penetration to the building interior particularly around apertures. The substitution of bond stones in a cavity wall with steel wall ties can reduce heat transmission by over 15%.

Roofs are flat and constructed of suspended reinforced concrete slabs over which are laid a crushed stone bed and screeded concrete laid to falls.

1.3 Climate

The climate in Malta is characterised by hot dry summers and cool winters. Temperature extremes range from 1.7^{o}C to 42.1^{o}C while daily means of 9.4^{o}C and 31°C in January and July respectively are the normal. High relative humidity which can exceed 90% in winter is a main cause for discomfort during the hot summer months when solar radiation as high as 1000 W/m^2 is possible.

2.0 COMPUTER SIMULATION

2.1 Thermal Analysis System

The model used for this research is the Tas software package which runs on an Apollo series 3000 work station. The system utilises the thermal response factor technique to calculate building energy demand, plant sizing and internal environmental conditions on an hourly basis. The 3D drawing process is supported by databases of material thermal properties, occupation patterns and internal energy gains. There are further databases for climate data on an hourly basis and plant which are used in the simulation process.

The simulation takes into account the transient conduction of heat in the building structure as well as air infiltration and internal energy gains. There is the facility to rapidly amend design data and insert building revisions. Although interzone air changes can be specified, temperature stratification in individual zones is not modelled. The simulation of attached conservatories is also restricted.

2.2 The Building Models

Two domestic buildings are simulated. The first is a 200 sq.m. mid-row dual aspect terrace house, representative of the most common housing type in

Malta. The second is a three bedroom detached house having the glazed apertures concentrated on one elevation, see fig. 2. The buildings are assumed to be unfurnished. An air infiltration rate of 0.5 air changes per hour as well as intermittent occupancy is assumed.

Simulation runs were carried out on two variations of the models, one assuming standard local construction and the other assuming the use of insulated cavity wall construction and roof insulation. The simulations were aimed at optimising building orientation and aperture sizing in order to minimise heating and cooling energy demand and to investigate how this is effected by increased insulation levels and changes in building thermal mass.

3.0 SIMULATION RESULTS

3.1 Terrace House Model

Simulations were run on the insulated version of the terrace house model. On day no. 3, a typical Winter day, there is a distinct minimum heating load of 4.9kWh when the elevation having the larger glazed area, $6.2m^2$ in this case, faces South. Considering the dual aspect plan layout, heating demand is a maximum when the building orientation is in an E - W direction. Both these aspects receive 30% less solar radiation than a southerly one. Overshadowing from a parallel two storey block increased the heating load to 6.5 kWh/day. Reverting to standard uninsulated construction increased the energy demand fivefold. This tallied with a similar percentage change in AU value calculations.

Once the heating plant is not in use an average difference of 3^oC in air temperature was registered between the two versions of this model.

The introduction of a 1 sq.m skylight having movable insulating panels in each of the first floor rooms of the uninsulated version, reduced the building energy demand to a value similar to that of the insulated version.

Simulation runs carried out on day no. 205 in mid-Summer indicated that roof insulation drastically reduced the requirement for cooling.

The insertion of thermal insulation above the structural slab and beneath the roof finish effectively producing an isolated thermal mass helps stabilise internal temperatures within the comfort zone. The application of external roof insulation is not a practical solution as the flat accessible roofs are used regularly. Diurnal temperature variations would also induce high thermal stresses. Considering the high thermal capacity of local building materials, flushing the building interior with cool night air would create a heat sink for the following day's summer heat.

3.2 Detached House Model

A cost effective building energy conservation measure is the use of thermal insulation. The application thickness follows the law of diminishing returns whereby the first layer is the most cost effective. The use of thermal insulation is not standard building practice in Malta, its introduction can yield appreciable energy savings both to the individual as well as nationally.
An attempt was made to minimise heating energy demand by optimising the glazed areas. The heating/cooling loads plotted against S. facing glazed areas did not indicate a preferable maximum value. A glazed area equivalent to 30% of wall area though seems to be common design practice.

A number of computer simulations were run for different building orientations on day nos. 3, 97, 205 and 280, respectively. The heating and cooling demands are presented in fig. 3. The compass points relate to the orientation of the elevation in which the glazed area is concentrated.

During the summer (day 205) the load plot is quasi symmetrical on N - S and E - W axes, east and west orientations being associated with a higher cooling load. A southerly orientation is also preferred in Winter.

The concentration of glazing on a south facing elevation provides maximum solar access during the heating season. The addition of external canopies reduce direct solar radiation in summer, however, solar radiation reflected off exposed hard surfaces during the arid summers is not eliminated.

As the sun follows the same path in spring and autumn maximum solar penetration in spring when heating is necessary causes overheating in autumn. The use of movable insulating panels and shading screens such as louvres and bamboo blinds which are traditionally used in Malta help eliminate unwanted solar gains when required.

A further model consisting of a single zone rectangular single aspect building having 30% glazing on one elevation was set up. Internal temperatures were allowed to swing freely in the absence of heating or cooling plant.
Internal day temperature plots on day nos 3 and 205 for different building orientations are presented in figs. 4 and 5 respectively.
Internal air temperature plots for different positions of roof insulation in zone 8 of the detached house model are presented in figs. 6 and 7. The adoption of mid-slab insulation provides the best comprise between thermal performance, ease of application and building use.

4.0 Conclusion

Our fore fathers were aware of the seasonal variation in solar geometry as evidenced in the orientation of Megalithic temples dating to 4000 B.C., the ruins of which still stand.

In more recent examples of rural vernacular buildings, window openings and building access points face south and are protected by loggias which is highly responsive to its surrounding microclimate. It seems that people then understood the environment much better than we do today. Labour intensive agriculture required people to spend more time outdoors and provided the opportunity for gaining first hand experience of the surrounding micro climate. The attraction of the local labour force to the manufacturing industry has led to a more "hi - tech" way of life, less time spent out doors and an increased reliance on mechanical space conditioning. As a consequence, the necessity and the art of designing buildings bioclimatically has been discarded.
The need to make architects aware of the benefits of climate responsive building design is identified. The University of Malta is participating in a UNDP - ECE project : "Solar Applications and Energy Efficiency in Building Design and Town Planning". The main objectives of this project is to make architects aware of the potential of climate responsive building design, in reducing energy demand and increasing thermal comfort and building amenity and to provide the means to and encourage energy efficient building design.
Building energy conservation is a major response strategy to the Climate change phenomenon, its widespread application in Malta will also help reduce energy imports which are a burden on the country's balance of payments and will help curb the 9% per annum increase in electricity demand.

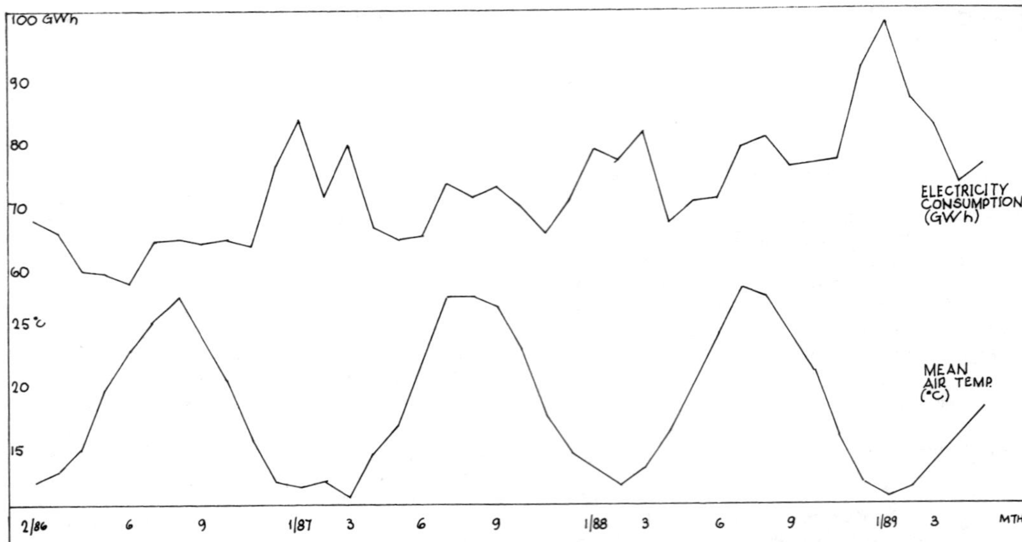

fig 1 electricity consumption, all sectors

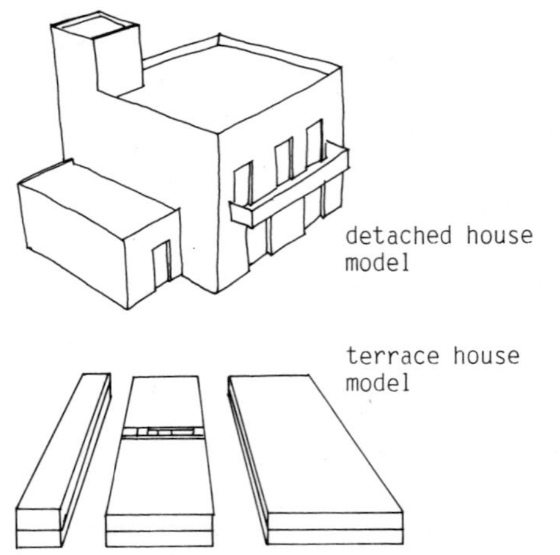

fig 2 the building models

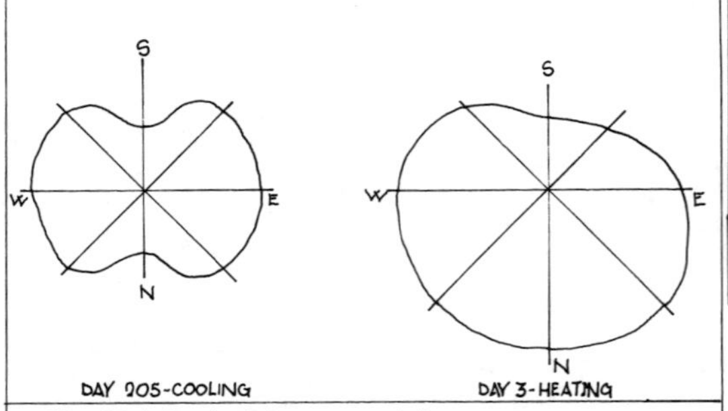

fig 3 detached house model:
orientation and energy demand

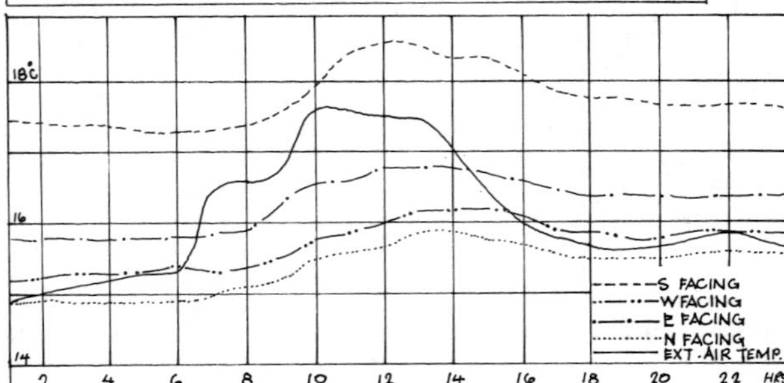

fig 4 single zone model:
air temperature plots for day 3

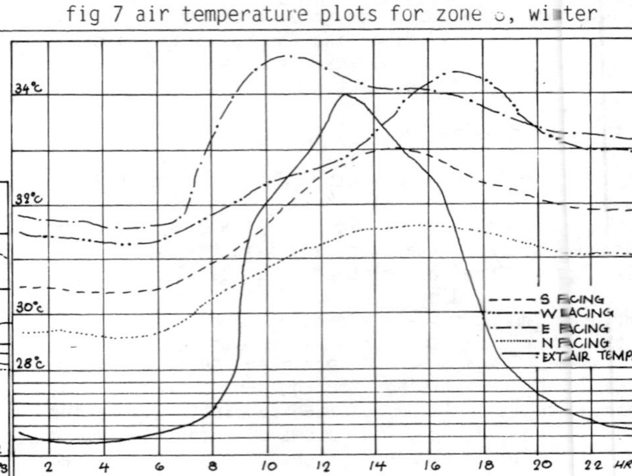

fig 5 single zone model:
air temperature plots for day 205

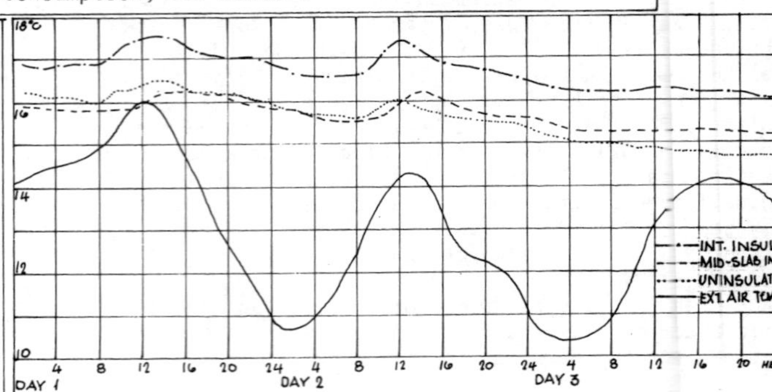

fig 6 air temperature plots for zone 8, summer

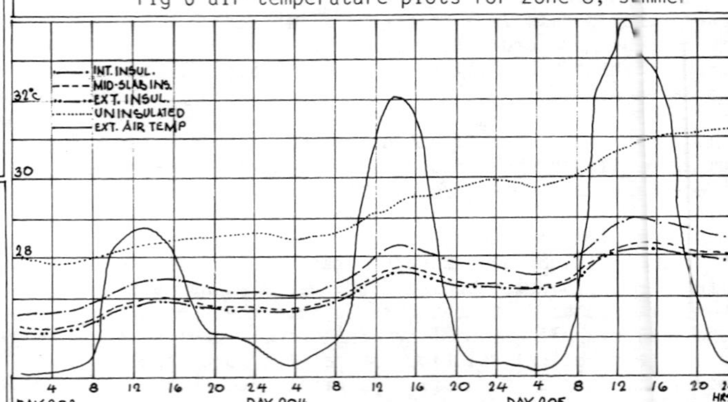

fig 7 air temperature plots for zone 8, winter

ARCHITECTURAL IMPLICATIONS IN INTELLIGENT BUILDINGS

Alain FILLOUX, Michel RUBINSTEIN
Centre Scientifique et Technique du Bâtiment
Sophia Antipolis
B.P. 141
F-06561 VALBONNE CEDEX

1. INTRODUCTION

With the emergence of new technologies in buildings, the design, the realisation and the management of the edifices have to be considered according to the requirements for data processing, telecommunication and automation.
To give satisfaction, a building is first and foremost the result of an intelligent construction or refurbishment to work and manage intelligently.
To face up to the set of problems, designers are developing new solutions concerning the organization and the design of the working places, the quality of ambiances and the integration of cabling systems.
The aim of that paper is to specify the implications of these elements on the frame.

2. DESIGN AND ORGANIZATION OF WORKING PLACES

From a functional point of view, a building can be described by its basic functions allowing the production of a service:

. reception of the human and technical elements of a firm,
. representation,
. production of services of quality,
. preservation of a good state of operation,
. evolution of shapes and functions,
. security and safety,
. communication,
. easy circulation.

The organization and the design of working places implie a preliminary investigation to draw a typology to configurate the space, making distinctions between:

. individual and isolated areas,
. individual and semi- open areas,
. collective and semi- open areas,
. easy circulation,

involving the technical requirements for physical ambiances of the working areas, and for connection to networks, wiring, plugs..., and taking into account functionnal links and paths through the building, characteristics of the working stations and the potentilities of uses of the building.

2.1 FUNCTIONAL LINKS

They concern the different ways and paths that people, products, equipments use to go through the building.
Any path must be identified by its nature, intensity, frequency and length. This evaluation contributes to the definition of the levels of proximity between the different areas by reference to the technical requirements: physical access, load resistance of the floor, sanitary equipment, circulation of low mobility people, safety, ambiances, location...

2.2 CHARACTERISTICS OF THE WORKING STATION

The main typological characteristics of the working stations are:

. specificity of configuration,
. orientation and opening of the area,
. perception of the space,
. processing of the public image.

These characteristics give the designer informations about the ability of interaction between entities concerning:

. circulation between the working stations or groups of stations,
. accessibility, obstructions and ease for circulation,
. mobility of partition walls and furniture in connection with the different networks and their acces units and the hard frame of the building...
. natural and artificial lighting,
. thermal characteristics of the building
. the safety and security requirements.

2.3 THE WORKING STATION

Some specific points have to be investigated to design a coherent and easy to live working station:

. the arrangement, the size and the accessibility of the working equipments and the storage spaces
. the ergonomy (positions, movements...)
. the communication with other stations,
. the reception of people,
. the spatial and technical constraints for setting up and the possibility for sharing furniture.

3. THE QUALITY OF AMBIANCES

Although the quality of the ambiances involves lighting, thermal, acoustic aspects and the orientation of the working stations, the influence of these parameters on the frame is variable. They concern mainly:

. the location and the aperture of the windows,
. the integration of the artificial ligth,
. the protection from the sun (blinds, sunscreens...)
. the location of the HVAC equipements,
. the mass of the frame and the acoustic characteristics of the components,
. the location and the orientation of the working station, in relationship with the windows, the light sources and the access.

4. WIRING CONSIDERATIONS

The building construction is generally the largest factor affecting the cost and the design of the network system.
Before drawing the plans, the network designer should make a detailled site inspection to determine the optimum route for the cable, and he should also ensure that the route chosen complies with all cable routing requirements.

4.1 HORIZONTAL DISTRIBUTION

The criterias for selecting a cabling system are security, cost, exploitation, and design according to the needs of the users, and the type of the equipments to be connected.
Horizontal distribution of cabling can be achieved through ceiling systems, floor systems or wall systems.

4.1.1 CEILING SYSTEMS

Distribution of cabling trough ceiling systems should satisfy specific requirements:

. Regulations and Codes of Practice about ceilings,
. Security specifications about fire and electricity shocks prevention,
. Specifications about installation of telecommunication and low voltage networks,
. Reservation of spaces above the ceilings to distribute HVAC ducts,
. Load resistance of the ceiling, according to the weight of the cabling system.

Zone distribution
The basis of the cabling is the distribution from the wiring closets to the center of zones. The working stations and plugs are then connected by cables to the central point.
This solution allows easy configuration of the area.

Duct distribution
The cables are distributed by ducts from the wiring closet to the plugs of the working station.
Such a system allows predefinite paths for the wiring and is limited and expensive for further evolution.

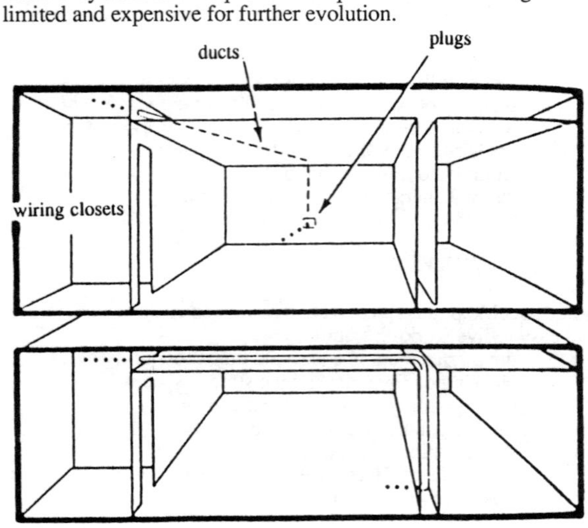

Surface raceways.
They consist of raceways that mount to wall and ceiling. They provide oultlet boxes for access to networks.
The distribution of cables is limited to the wall, then the flexibility of the system is low when it is used as a free standing system.
In opposite, the wire management is easily acccomplished by providing raceways channels.

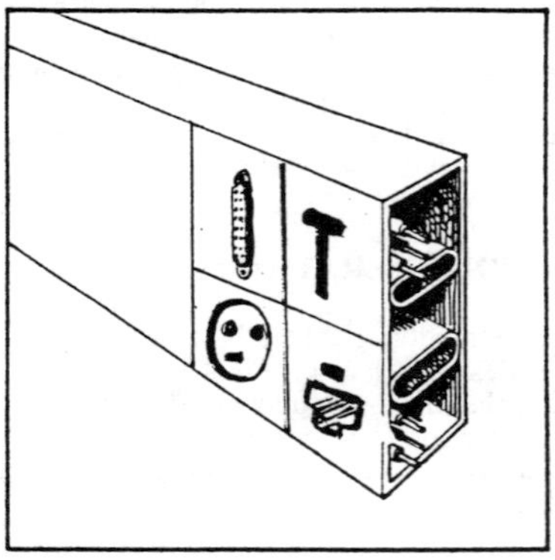

Poke-thru systems.
The distribution is accomplished through the ceiling of the room located under the working station. It is a low cost system but its main disadvantages are to reduce the resistance of the floor and to increase the risk of fire propagation.

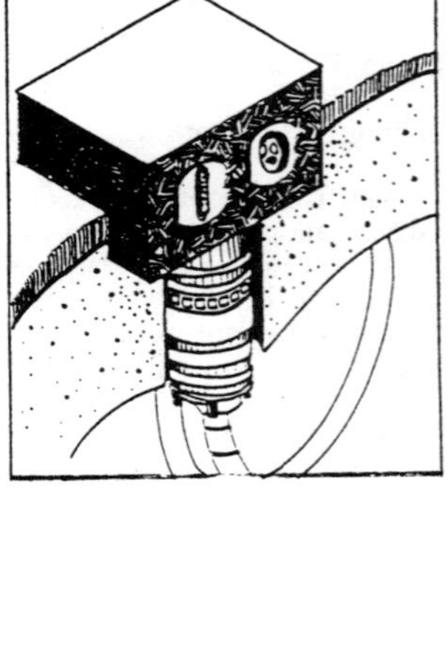

4.1.2 FLOOR SYSTEMS

The most commonly used distibution systems are:
- . in-floor systems,
- . cellular raceways,
- . access floor systems,
- . flat conductor cables.

In-floor systems consist of under floor ducts embedded in the concrete or
included as part of the structural load-carrying floor of the building. The cables in the ducts are accessed by means of boxes.

A cellular raceway is both a distribution system and a floor component. Interconnection of the ducts is accomplished by rigid ducts. The distribution system is covered by floor covering, carpeting...
Access floor systems or *raised floor systems* are metal or composite panels supported by piedestals above the structural floor. Power cables are distributed under the floor through conduits or manufactured wiring systems. Data and telecom cables are encased in conduits or raceways. Connection to cables is achieved through service fittings or panel holes and boxes attached to underside of the panels.

Flat conductor cables are designed to be installed directly on a smooth structural floor and covered by carpet squares. The distribution system is fed by conduit, in-floor systems or raceways.

4.1.3 WALL SYSTEMS

Wall and partition systems contain raceways within the furniture partition. They are prewired for power application. They allow for additionnal cables by means of channels with removable covers. As a general statement, a 3 cm separation is required to avoid interferences and perturbations of data networks by power cables.

4.2 VERTICAL DISTRIBUTION

In multi-storey buildings, distinct sleeves and raceways are required for circulating data (and communication) cables and the power network. They are located in the collective parts of the building and they are open on each storey. They make part of the construction, from the basement to the roof and they are protected from water penetration.
All the ducts are located in continuation. They are free of discontinuity. The distribution through the floors is achieved by opening the bottom of the ducts.
In any case of discontinuity, the distribution of cables must be achieved in accordance with the Regulation and the Technical Rules, by using horizontal raceways or ducts.

5. WIRING CLOSETS

The location of the wiring closets is depending of the shape and the size of the buildings.
In multi-storey buildings, the wiring closets are located close to the vertical ducts and raceways. The number of closets is related to the area of the building in order to satisfy the wiring requirements and to optimize the management of the networks.
In horizontaly-stuctured buildings, the wiring closets are located all along the main passages.
The following characteristics are indicative but they allow acceptable conditions of work for the installation, connection and maintenance of equipments.

Maximal distance from an E.M.I source
Minimal area: 6 m2
Acces by key-locked glazed door: 0.75 m wide
Air ventilation
Power source: 1 kVA at least. Connection to hearth
Lighting: 200 lux min.
Telecommunication equipment: hand free telephone

6. RETROFITTING

When raceways are not available, one can use separate ducts, but specific requirements have to be satisfied:

. collective ducts: applicable to cable distribution in 6-storey buildings maximum. The duct should be non-flammable.
. individual ducts: applicable to cable distribution in buildings where no more than 20 working stations will be connected to networks. A non-flammable duct could be encased in concrete to interconnect the stations.

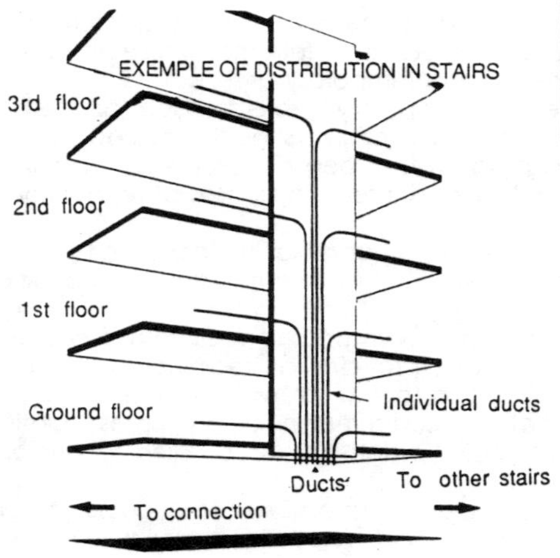

INDIVIDUAL DUCTS

a	less than 10 users	5 cm
	other cases	10 cm
d	minimal distance between cables and power network	20 cm
d'	minimal distance between cables and other tubes and networks	15 cm

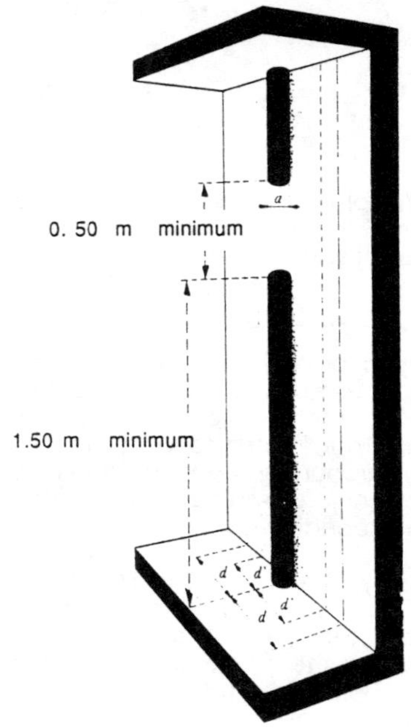

COLLECTIVE DUCTS

7. CONCLUSION

An intelligent building is a building wich is able to provide services corresponding to the actual and future needs of the users within an optimal cost and allowing good conditions for life and work.
The developement of the technologies in the field of telecommunication, data processing and automation, is changing the environment of the working station in buildings.
This evolution makes difficult to predict short-term architectures, configurations and fonctionnalities of the working areas.
The main difficulty is the integration at the earlier stage of a level of flexibility allowing the systems to evoluate without interruption of the services and activities.
From a general point of view, architecture, organization and equipment of a building are directly related to the choice of any system or any strategy. They imply new relations between the technical solutions and the buildings and they lead to consider that buildings are not only a collection of walls, stairs, doors, windows,... but building are systems wich provide services.

HOW ENCOURAGE THE TEACHING OF PASSIVE SOLAR AND LOW ENERGY DESIGN

A. De Herde and the team "Architecture et Climat"
Université Catholique de Louvain
1 Place du Levant - B - 1348 Louvain-la-Neuve

S. Yannas
Architectural Association School of Architecture
34 Bedford Square - London WC1B3ES

ABSTRACT : The present paper is the result of a work which has been executed within the frame of the SOLINFO program of the Commission of the European Communities. At this stage, 2 prototypes have been realized: they should give an answer to the title-question.

1. GOALS

SOLINFO aims to promote the transfer of information from the solar research community to the building design and construction professions. It concentrates on improving our understanding of information requirements, supporting under-graduate and mid-career education, providing aids to professional practices, and speeding-up research dissemination.

More specifically, the aim of the SOLINFO project in the context of undergraduate education is to encourage the teaching of passive solar and low energy design by the production and dissemination of "resource packages" and teaching materials for schools of architecture.

The objectives of this task are:

- to make teachers aware of the possibilities of passive solar design in architecture;
- to provide appropriate pedagogic material and resources for the teaching of passive solar architecture;
- to present information in forms easily understood by students.

2. MEANS

The working group feels that the above issues must be addressed right from the beginning of an architectural course and proposes to relate the introduction of passive and low energy design to the teaching programme of the first two years of the architectural curriculum.

In most schools of architecture the curriculum in these first two years will cover:

- the study of the physical and socio-economic contexts in urban and open sites;
- spacial requirements according to use and occupancy, and some general notions on climate, comfort, lighting, amenity;
- the perception and differentiation of architectural space;
- representation of space and the instruments associated with it (geometry, projections, scale, perspectives, scaled models, photography, video);
- measured drawings and typological studies;
- introduction to structures;
- building materials and construction techniques;
- elementary information on building services.

It is proposed to respond to these objectives by creating an appropriate set of resources packages, or "valisettes". The focus of each valisette will be on a good example of climatic design that is also recognizable as an architectural project of high quality. The Herzet house in Esneux, designed by arch. B. Albert, meets these requirements in a very successful way. It has been chosen to be the subject of the first valisette. The package in the valisette will include a number of items on different aspects of each building, and additional material on relevant issues of passive solar design.

The working group feels that the vehicle of the valisette, with its focus on one strong architectural example and an accompanying booklet on a relevant topic each time, offers a number of advantages. It permits to start disseminating the product of the SOLINFO educational project as soon as the first valisette is completed. It also allows to spread the dissemination process over a longer period of time, thus ensuring there is a better chance for the messages to get thought. The concept is flexible and as feedback is received from schools the package can be made to respond to comments and criticisms. As a container the valisette can be designed to provide an attractive means of presenting the package and to accommodate a number of different forms of material (drawings, booklets, slides, computer diskettes). The working group believes that such a resource package could stand a good chance in meeting the diversity of educational approaches in the Community's schools of architecture.

Regarding the selection of the architectural projects to be presented, it is felt that these must not be confined to contemporary architecture.
Besides contemporary projects such as the Herzet house, the selection should include important buildings from other periods, exemplars of vernacular architecture and interesting cases of renovation.

It is of course impossible to cover all the major building types for the climate and site conditions in the Community.
However, the working group feels that a dozen or so valisettes will provide a reasonable number of examples and cover a wide range of important architectural concerns including issues relating to history, typology, density, and building technology.

Thus the teacher responsible for a course or design studio should find relevant material that will be directly usable in the course; a taste for passive solar architecture; and the stimuli to motivate the students.

3. CONTENTS OF THE RESOURCE PACKAGE

It is proposed that the following criteria should be used for selecting the valisette examples: firstly, the buildings must have an exemplar character in architectural terms; secondly, there should be some clearly recognizable features that mark the building as an example of climatic design. In the case of the Herzet house, the recognizable climatic features are for one part the semi-cylindrical conservatory, and on the other hand the total reorganization of the layout to reach the sun as much as possible. To assess the climatic characteristics of a scheme two complementary approaches are proposed: the statements made by the designers themselves, in this case by Br. Albert, and an objective appraisal of the building.

When talking about their schemes designers will have made reference to climate, the relationship between indoors and outdoors, the use and amenity of internal spaces in relation to orientation and sunlight, the notion of buffer spaces and zoning, and the predicted use of energy or the need to conserve it. Such statements are valuable in helping to understand the intentions and motivations of the designer.

In addition, objective criteria are seen as consisting of technical assessments, through observation and calculations, of how the principles of climatic design have been incorporated and expressed, and may also include measurements.

3.1. The presentation of each example may cover

- plans, sections and elevations to scale 1/50. It is felt that the use of color will be necessary on elevation should also appear on the plans and sections.
- a series of slides and photographs to represent the building; in addition to spatial characteristics these should attempt to reflect the temporal dimension by including shots at different periods of the year and under different weather conditions.
- a demonstration diskette prepared on some well established CAD package to give 3-D representations of the building from different viewpoints illustrating the course of the sun, issues of shading or of overshadowing, the penetration of daylight, etc. This might allow the student a certain amount of interaction.
- presentation of the project by its architect including early conceptual sketches.
- a "guided tour" through the building and architectural critique of the scheme; a video tape would be of use for this purpose.
- the experience of the occupants as perceived through daily activities and throughout the year.
- site conditions and climatic data related to the building.
- climatic design principles presented and expressed by the building.
- data on thermal properties and on energy use providing information on the thermal performance of the building, but also to allow the teacher and students to carry out further calculations of their own if they wish.
- mechanical equipment.
- the structure.
- construction details of particular interest.
- a list of bibliographical references.

It is estimated that the package might contain 4-6 sheets of A1 drawings, 20-30 slides and photographs, 20-30 A4 pages of illustrated text, one diskette, and if the budget permits, a video tape.

3.2. Additional material

Additional material on relevant issues of passive solar design to accompany the selected architectural exemplars:

- Principles of climatic design presented in relation to the different climatic regions of the Community and with reference to different building types (heating, cooling, daylighting).
- Writings by architects on the relationship between architecture, urbanism and climate (Vitruvius, Palladio, Wright, Delorme,...).
- Climatic design at the urban scale; site layout; creating attractive microclimates in open spaces.
- Building form; the building skin as transition between indoors and outdoors; internal layout, daily or seasonal changes in use of space as function of climate, vertical or horizontal zoning; thermal capacity, heat distribution.
- History of glass; the window, loggia, greenhouse, sunspace, conservatory, atrium.
- Shading devices and screens, landscape features near buildings (complements booklet on site layout and open spaces).
- Analytical tools and design aids relating to site layout, shading, daylighting, prediction of heating and cooling loads, prediction of hourly indoor temperature.

These are seen as individual booklets ranging in size from 8-32 pages. The booklets would be self-sufficient and complementary. They are meant to provide compact coverage and design guidance on a whole range of topics that the working group considers to be important for architectural students.

It is thought that there will be some matching between the topics of the booklets and the selected architectural examples so that individual booklets can be inserted as appropriate in each valisette. It was felt that the principles of climatic design had to be included in the first valisette in order to restore correct and complete knowledge at the beginning of the program.

4. QUESTIONS

- Do the basic ideas presented here, seem to you to meet the stated objectives of the project?
- Are the proposed form and contents of the resource package capable of meeting the objectives of the first two years of architectural study ?
- What further requirements do you encounter in your teaching of passive solar architecture ?
- Please make suggestions on architectural projects in your country which might serve as examples for the resource packages.

5. REFERENCES

Solar Energy Applications in Buildings. Information Requirements for Architects Loughlin Kealy B.A. - E.R.G. University College Dublin

Science in a Regional Approach to Architecture - Simos Yannas A.A.S.A. - London

Teaching Passive Design in Architecture - Harrisson Fraker, Donald Prowler - University of Pennsylvania

Record of Solinfo Teachers Workshop in Paris, 5 and 6 October 1987, S. O'Toole, E.R.G., University College Dublin.

SOLAR SYSTEMS

DWELLINGS OF LOW COST OF MAINTENANCE IN CASTILLA-LA MANCHA

SANTIAGO HERNÁNDEZ MARCULETA
CONSEJERÍA DE POLÍTICA TERRITORIAL
C/. Fermín Caballero nº 24
16071 - CUENCA SPAIN Telfno. 966 - 212004

The Junta de Comunidades de Castilla-La Mancha initiated two years ago a scheme in order to building dwellings energetically efficient (or solar passive), constructing 16 family houses in three terraces of 6,6 and 4 houses each ones.

West and south elevation in a 6 houses terraces

DESIGN OBJETIVES.

The aim of the proyect was to warrant that energy efficient housing schemes are both cost effective and, more important, low cost of maintenance.

South elevation. One of the houses keep close the window indoors shutters.

These finality aren´t so primordial for private promoter as for the Goverment who must look for the people with low attainment power. Therefore the research and study of the natural solar systems passive must be develop by the Public Administration. When the users know about adventages of maintenances of natural technics passive they can demand in from the private builders.

The desing and system building solar passive belongs to soft technologies and there application don´t develop by itself the industrial increase and don´t augment the building cost, but it cheapens maintenance cost, along the building life (between 50 and 100 years).

PROMOTION FEATURES.

- 16 family attached houses in Montalbo (Cuenca, Spain).
- Founded by the Consejería de Política Territorial, Junta de Comunidades de Castilla-La Mancha.
- Architect: Guillermo Yañez.
- Delinered up to people in October-89.
- Natural solar passive systems.
- Accumulator walls for cooling.

North elevation; we can see the principal door and the small north windows.

CLIMATE.

The La Mancha plateaw, exceeds 800 metres above sea level and endures the bigger sudden transition between nights and mornings temperatures in Spain as it can see in the next relation.

The large window and the accumulator wall with the south surface black coloured.

Table 0
Arrange Spanish cities with a larger temperature difference.

	ANUALES			SALTO TERMICO	VERANO=MAXIMO MES DE JULIO		
	MEDIA ANUAL	TEMPERATURAS MED. MAX.	TEMPERATURAS MED. MIN.	ANUAL	MEDIA DE LAS MAX.	MEDIA DE LAS MIN.	SALTO TER.
MOLINA DE ARAGON 1	10´2	17´2	3´1	14´1	28´9	10´1	18´8
CUENCA 2 LABORATOR.	11´7	18´7	4´8	13´9	35´2	18´6	18´6
ALBACETE 3	13´4	20´1	6´7	13´4	32´6	15´5	17´1
CIUDAD REAL 4	14´5	21´2	7´8	13´4	34´4	16´5	17´9
TERUEL 5	11´6	18´2	4´9	13´3	29´7	12´9	16´8

These climate that have an abrupt termic variation within a lonely day needs massive materials with hight specific heat in order to buff the extreme temperatures, to storage heat and to delay indoors radiation from walls.

ACCUMULATOR WALL.

They are arranged in the ground floor near the principal window. These brick walls are provide only in 6 dwellings in order to gauge dissimilarities steps, so as in equalice indoors temperatures such as to delay heat, origined by thermal inertia.

The best way to enlarge the effect of delay heat from the brick walls is when it is deal out arround one room. Therefore all the partition wall work also as accumulator and as delayer wall. (Delay 24 cms brick wall = 8, 69 h.).

The accumulator wall is painted in black colour at the south face to improve abssorption.

INDOORS SHUTTERS.

They have high value as insulation and decrease the coefficient thermic transmission of the window from 3,1 w/m^2. C$^{\circ}$ to 0,83 w/m^2. C$^{\circ}$. In winter nights the owner must close the shutters if he wants to avoid cold bridges and gaps of draft.

Indoors accumulator wall view.

DESIGN.

The arrangement is the tradicional in this dwelling tipology. In the south are the living room in the ground floor and two bedrooms in the upper floor.

The kitchen and the baths works as buffer rooms in the north face.

PASSIVE COOLING.

Montalbo (Cuenca) is 890 m above sea level and has sunny weather. The summer originate the bigger thermic jump. You must elude overheating but must also profit by low mornings temperatures. It is possible thanks massive wall and the inner structural walls which improve thermal inertia. The outer layer of the gable walls have insulation with external cladding. This tipology avoid interstitial damp. There is a vent at the top skylight which enable the air circulation in the central staricase and nexts rooms.

CONSTRUCTION.

The houses have got a high level of insulation. The first floor has a sheet of polystyrene under the flag. All the windows have double glass. The space between the last floor and the tilt roof are ventilated so each house can use the staircase and vent which are near the skylight as a chimmeney. The dwellings are buildings with structural brick walls.

AUXILIARY HEATING.

In order to be able evaluate the effective heating cost there are electrical radiators in each room. But as it is very useful in rural customs it is foreseen a fire wood stove in the living room.

In sumer the awning preserve from sun.

The went in the skylight allow natural air movement.

THE OWNERS ACTIV ROLE.

It´s essential the contribution of the people who live in the dwelling. They must every day to manage their own passive systems. In summer they must open the awning and in winter close the window shutters. As the some form that the people dresses diferents times and forms. The clad or envelope of the house must change several times in only one day. In morning summer you must open the windows becouse you must cross ventilation but don´t allow to go in the solar radiation (direct or diffusse), so as you must keep close the shutters. The performance increase 25% so the solar contribution to the total space heating demand reach 65%.

CONCLUSIONS.

The Consejería of Política Territorial have initiated managements with the C.I.E.M.A.T. in Spain in order to gauge the thermic conduct from these houses. The real performance will condition its employment so in Future Urban Planning as in buildings construction that the Junta de Comunidades de Castilla-La Mancha are going to design.

Insulation shutters.

The Public Administratio thinks that if the solar-passive houses are of low cost of maintenance all social dwellings must be solar-passive as well as the families who go into these social houses have very little income.

- In hard climates as in the one of Castilla-La Mancha, the construction buildings must have the bigger and heavier walls as part of their structure.

- When it is possible the insulated layer must locate in the outer part of the exterior wall in order to eliminiate the intersticial damp.

- The owners must colaborate and manipulate the shutters and awning so as they exchange their clothes when the weather transform.

- All the dwellings and new urban plans, that we, the local and Regional Administration buildings must designed with solar passive technics and be subsidized the projects which comtemplate efficiency energetic and low cost of maintenance.

The flag have black colour in order to improve heat accumulation.

APPENDIX

PAPERS IN ASSOCIATION WITH THE EXHIBITIONS

"WORKING IN THE CITY": EUROPEAN COMMUNITIES ARCHITECTURAL IDEAS COMPETITION

S. O'Toole and J. O. Lewis
Energy Research Group, School of Architecture, University College Dublin
Richview, Clonskeagh, Dublin 14, Ireland.

ABSTRACT. An architectural ideas competition for the design of working places and buildings associated with learning, and which emphasised daylight as a design issue affecting energy efficiency and environmental quality was promoted by the Commission of the European Communities during 1988-'89, through the ARCHISOL (Architecture and Solar Energy) Project. The competition was open to EC architects and students of architecture, and a primary aim of the competition was educational. The documentation distributed to all who registered formed an information package containing heating, cooling and daylighting design primers, and new design and analytical tools. The competition was assessed by 17 expert architects and building physicists and a jury of eminent designers. The winning competition entries are being exhibited at this Conference, and a book of the competition results is in preparation.

1. BACKGROUND

The two previous European architectural ideas competitions promoted by the Commission of the European Communities (Directorate-General XII for Science, Research and Development), in 1980 and 1982 /1,2/, were concerned with the design of housing. Then the energy issues were relatively straightforward, being mainly concerned with the use of passive solar gains to reduce auxiliary heating demand.
In "Working in the City", the third CEC architectural ideas competition, the area of interest was non-domestic buildings: either working places (typically commercial offices) or buildings associated with learning (including schools, university buildings and public libraries). Three main differences resulted from this. Firstly, there was a shift of emphasis to the energy value of daylight, and the energy saving of reducing or avoiding air-conditioning. Secondly, a greater concern was shown to the comfort of the working environment, and thirdly, the energy implications of the design were far more complex than for the simpler domestic building types.

2. OBJECTIVES

The competition objectives were to produce good urban architecture which emphasised the quality of working or learning places, to stimulate a wide range of innovative designs appropriate to their climatic and urban contexts which emphasised the thermal and visual comfort of the user while displaying an awareness of economy and value, and to stress the potential of daylighting and other passive solar design strategies for heating, cooling and ventilation to reduce energy consumption.
The competition was organised by the Energy Research Group at University College Dublin, advised by a number of experts (Appendix I), during 1988-'89 as part of the Commission's research and development programme Project ARCHISOL /3/, which set out to introduce design professionals to energy-efficient building design in a stylistically neutral manner. Thus, a primary aim of the competition was educational.

3. INFORMATION FOR COMPETITORS

The competition documents distributed to all 1500 architects and students who registered for the competition formed an information package containing heating, cooling and daylighting design primers and new design and analytical tools. To increase acceptability by architects, this information was presented as a series of wallcharts containing diagrams and text. In addition to a four page (A4) competition announcement brochure and an A2 poster, both in colour, a 40 page (A4) 'Competition Conditions and Procedures' booklet was produced which outlined the rules of the competition and contained:

- A daylighting design procedure,
- A design checklist,
- A 'lighting and thermal glazing' (LT) design tool,
- Worksheets to be filled out by the competitor,
- A polar sundial,
- Diagrams and photographs illustrating the evaluation of daylighting using a simple architectural model,

The competition package also contained:

- Stereo suncharts for different latitudes in Europe,
- A daylight factor meter (card and plastic kit),
- 7 Wallcharts (Daylighting Design Guidelines, Design Parameters and Performance I-IV, The LT Method and Daylight factor meter assembly),
- Letter to competitors, Entry form, Official address label and printed envelope.

The material was devised by Dr. Nick Baker, Cambridge and Dr. Marc Fontoynont, Lyon, /4/, designed by Bill Hastings and edited by Ingrid Gibney, Shane O'Toole and Owen Lewis in Dublin. A team of over 40 individuals and five companies contributed to the successive phases of document production. Translation of all documentation into the nine official languages of the Community and competition promotion was undertaken by collaborators in each Member State (Appendix II).
The adoption of a desktop publishing system, Xerox Ventura, for all typesetting and pagemaking permitted a greater degree of control over both the design and editing of texts than would have been possible using conventional typesetting and resulted in lower print costs. In addition, by carrying out pagemaking once on the English text and subsequently loading translated texts into the same format, much duplication of work was avoided. The original English version of the text sent for translation was accompanied by a set of instructions on how it was to be typed and returned to Dublin on computer disk. This procedure allowed each language to be 'pagemade' accurately, without concerns about proofreading. The finished 'pages' were sent on disk to a bureau which printed them at a high resolution onto bromide paper, from which printing plates were made, after illustration artwork had been added.
While all of the special or accented characters used in Roman script could be accommodated, the Greek version posed special problems. These difficulties were, however, resolved following discussions with the manufacturers of Ventura Desktop Publishing, the Ventura agents in Athens and assistance provided by A.N. Tombazis and his colleague S. Doulgerakis.
Because of the educational emphasis of the overall project, from the beginning special attention was accorded the information requirements and learning styles of architects /5/. The outcome of studies and investigations of these issues directly

influenced the development of various design aids for competitors and their modes of presentation.
In the first place many of the intended competitors would not necessarily be expert in the design of energy-efficient buildings; and available tools are usually more suited to evaluation processes than to design processes. In this ideas competition it was desirable that tools had maximum 'transparency' so that the architect or student was encouraged to consider the energetic implications of design decisions at the earliest stages of the process.

4. DESIGN AIDS AND PROCEDURES

Guidance towards energy-efficient architectural design in the realms of heating, cooling, ventilation and daylighting was prepared and presented in a highly graphic format as a series of five charts designed for wall display. These dealt with various design parameters and related building performance: issues such as comfort, thermal loads in large buildings, the thermal balance of glazing, useful solar gains and the influence of orientation; overheating and 'passive' methods of cooling and night ventilation; daylighting and energy, and ventilation; and atria and sunspaces.
The separate daylighting design guidelines start from the premise that an efficient daylighting design can be characterised by the quality of the indoor luminous environment during daytime, as well as by its ability to offer a good substitute to artificial lighting: consequently, daylighting techniques need to be judged qualitatively and quantitatively. Topics briefly addressed here included:

- Daylighting and visual comfort.
- How can the amount of light entering a building be characterised?
- How often will daylight alone give sufficient illuminance?
- Sunlighting.

A simplified calculation method which combines the energy value of daylight and the thermal value of solar gain, the 'LT method' /6/ was developed to show the implications for energy consumption of early decisions concerning building form and facade design. It is intended to be quick and easy to use, to allow the designer to explore a number of options. It is presented in the form of sets of eight curves, one set for each of five European climatic zones, showing annual primary energy consumption per square metre of floor area for three orientations of facade and for rooflights. Curves are presented for lighting, heating, combined heating and lighting, cooling and total energy, derived from a mathematical model which predicts energy consumption as a function of:

- local climate conditions;
- orientation of facade;
- fraction of glazing area in facade;
- type of glazing.

Competitors were required to indicate the solar availability on their chosen site using supplied stereographic suncharts. They were also asked to make a working model to 1:20 scale of at least one area or group of rooms illustrating the typical design approach to daylighting adopted, or an important daylighting design feature. Guidance was given on evaluating Daylight Factor values using a DF meter specially designed /7/ and supplied to each competitor in kit form. Competitors were asked to submit small-scale plans showing a crude daylight distribution based on these values, and to complete a lighting evaluation worksheet following a step-by-step procedure. Finally, competitors were encouraged to evaluate qualitatively the behaviour of sunlight in their building using the supplied polar sundial and to submit photographs of the interior of the model.

5. RESULTS

479 architects and 1143 students registered for the Competition. The 186 entries were examined by a panel of expert architects and physicists (Appendix III) who prepared reports on each submission. The technical assessment involved 360 man hours, and was structured into four components:

- understanding and communication of environmental strategy;
- thermal quality (comfort and performance);
- Lighting quality (comfort, performance and agreeableness);
- quantitative evaluations (LT Method, daylight and sunlight).

The jury of Henri Ciriani, Arthur Gibney, Alvaro Siza Vieira and chairman Edward Cullinan considered every entry and the reports of the technical assessors and selected schemes to share the total prize fund of 40,000 ECUs.
The Jury commented that "to ask entrants to combine good urban architecture, aware of its context and containing high quality working and learning places, with an innovative approach to daylighting, passive solar design and heating, cooling and ventilation to reduce energy consumption, was to ask a great deal of them. This showed up particularly among the student entries." /8/.

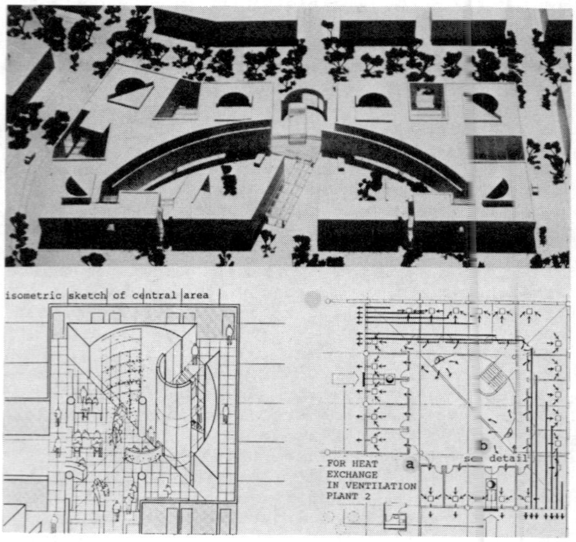

Part of the winning design for Roskilde

The most outstanding scheme, both technically, architecturally and in its urban context, was submitted in the working places category for architects. It was the scheme for an office complex in Roskilde, Denmark, by the multi-disciplinary team of KHR A/S, Esbensen and ISLEF. The other major prizes went to projects for London (both to architects), Madrid (student) and Paris (student).
The results were announced on 17 May 1989 when an exhibition of all entries opened in the Royal Hibernian Academy in Dublin. Philippe Bourdeau, Director of the Commission's DG XII/E, presented the prizes before an audience which included representatives of Europe's architectural profession who were attending a meeting of the Comité de Liaison des Architectes de l'Europe Unie in Dublin. The winning projects have subsequently been displayed in Copenhagen and London, now in Paris at the Second European Conference on Architecture, and will also be exhibited in several other European venues during 1990.
The "Working in the City" competition has proved a very successful vehicle for raising interest in the professional press. Coverage of which we are aware has included notices and editorial comment in over sixty journals in nine languages in eleven EC Member States. A book of the competition results is in preparation.

6. COMMENT

The technical assessors reported that the best schemes were often not the most immediately spectacular; they showed their quality in the integrated and comprehensive way they treated the comfort and well-being of the occupant, against the background of a realistic energy-conserving strategy. Within this framework, they also showed

appropriate innovation. In relation to daylight specifically, the assessors found that the issue of relating room depth to height was well understood, but that the design of the window apertures was not always well developed. There were many examples of devices such as light shelves and reflectors described at a rather diagramatic level, and sometimes great optimism was shown for the performance of light wells and ducts. While it was evident that the majority of competitors had sincerely pursued the goal of a daylit building, comparatively few showed knowledge of detailed daylighting design. It also seems that the communication of environmental concepts is not easy, and is not a commonly held skill amongst architects. It may be a skill more useful than for communication alone, for just as drawing aids the development of the architecture, so the communication of ideas inspires their conception and development.
Nonetheless, as noted previously, the competition and its results have attracted widespread coverage in European architectural periodicals and thus contributed to the renewed awareness and interest among architects of the importance and potential of daylighting as a design issue. Furthermore, there is continuing interest among teachers of architecture in the competition design tools with the result that those are likely to be made available in revised format to European schools of architecture.

7. REFERENCES

1. R.M. Lebens (ed). Passive Solar Architecture in Europe: The Results of the 'First' European Passive Solar Competition - 1980. For the Commission of the European Communities. The Architectural Press Ltd., London, and Halsted Press, a division of John Wiley & Sons Inc., New York. 1981.
2. R.M. Lebens (ed). Passive Solar Architecture in Europe 2. The Results of the 'Second European Passive Solar Competition - 1982'. For the Commission of the European Communities. The Architectural Press Ltd., London. 1983.
3 J.O. Lewis and S. O'Toole. Architecture and Solar Energy: ARCHISOL in Solar Energy Applications to Buildings and Solar Radiation Data. Kluwer Academic Press, Dordrecht. 1988.
4 N.V. Baker and M. Fontoynont. Experiences with Simplified Design Tools for Daylighting and Thermal Design. Proceedings, Science and Technology at the Service of Architecture (Paris, 1989).
5 L. Kealy. Information Requirements for Architects. Unpublished Final Report to Energy Research Group, University College Dublin. 1988.
6 N.V. Baker and D. Hoch. Experience with a design tool which combines the energy value of daylight and the thermal value of solar gain. Proceedings of UK and ISES Conference, Daylighting Buildings. 1989.
7 N.V. Baker. A Low-Cost Instrument for Measuring Daylight Factor in Models. Proceedings, Science and Technology at the Service of Architecture. (Paris, 1989).
8 S. O'Toole and J. O. Lewis (eds). Working in the City. Gandon Editions, Dublin, (in press).

Appendix I

David Clarke, London
Michel Gerber, Toulouse
Alex Lohr, Köln
Professor Cathal O'Neill, Dublin
Shane O'Toole, Dublin
Alexandros N. Tombazis, Athens

Appendix II

Mikael GRIMMIG, Solar Energy Testing Laboratory, Technological Institute, Taastrup.
Dr. Jerome ADNOT and Dominique CAMPANA, ARMINES, Paris
Alex LOHR, Büro für energiegerechtes Bauen, Köln.
A. N. TOMBAZIS, Alexandros N. Tombazis & Associates, Athens.
Dr. Paolo OLIARO, Dipartimento de Energetica, Politecnico di Torino.
Joep HABETS, BNA/Royal Institute of Dutch Architects, Amsterdam.
Prof. Eduardo MALDONADO, Gabinete de Fluidos e Calor, Universidade do Porto.
Dr. M. del Rosario HERAS, Instituto de Energias Renovables, CIEMAT, Madrid.
Dr. Marc FONTOYNONT, LASH/ENTPE, Lyon.

Appendix III

Patrick ACHARD, Ecole des Mines, Valbonne
Nick BAKER, Cambridge Architectural Research Ltd.
Martin DE WIT, Technical University of Eindhoven
Albert DUPAGNE, LEMA, Liege
Aldo FANCHIOTTI, I.C.I.E., Roma
Marc FONTOYNONT, LASH/ENTPE, Lyon
Gian Vincenzo FRACASTORO, Universitario della Bascilicata, Potenza
Michel GERBER, Ecole d'Architecture Toulouse
John GOULDING, ERG, School of Architecture, Dublin
Jaime LOPEZ DE ASIAIN, Seminario de Arquitectura Bioclimatica, Sevilla
Eduardo MALDONADO, Universidade do Porto
Conall O'CATHAIN, Queens University, Belfast
Dermot O'CONNELL, School of Architecture, Dublin
Erwin PETERSEN, Technological Institute, Tåstrup
Mat SANTAMOURIS, Protechna Ltd, Athens
Rafael SERRA, ETSAB, Barcelona
Gabi WILLBOLD-LOHR, TH Aachen

Edward Cullinan, Chairman of the Jury, speaking at the awards ceremony, watched by Owen Lewis

INTEGRATED UTILIZATION OF PASSIVE SOLAR ENERGY AND NATURAL DAYLIGHTING IN AN OFFICE

Kristensen P.E.
ESBENSEN, Consulting Engineers
Havnegade 41
DK-1058 Copenhagen K
Telephone: +45 - 33 11 42 24
Telefax : +45 - 33 11 68 34

ABSTRACT. A 4,500 m² three-storey office block has been designed with special emphasis on utilization of passive solar energy and natural daylight. Furthermore, maximum visual and thermal comfort were prime design objectives.
The project was designed in a team consisting of KHR A|S architects & planners, ISLEF Building and Construction Company, CENERGIA ApS and ESBENSEN Consulting Engineers.
Low energy consumption and a good visual and thermal comfort is achived via an integrated design approach of the building. Glass covered atria, glazed skylights of the stair wells and a large glazed archade act as buffer spaces, and provide natural daylighting and passive solar energy deep into the building.
Ventilation is provided via a displacement ventilation system, which utilizes the chimney effect in the atria as driving force.
The overall primary energy consumption is reduced from 9.000 MWh/year for a traditionel new office building to approximately 4.500 MWh/year for the actual building.
One of the main recommendations of the Brundtland Report "Our Commom Future" is, that the use of fossil fuels in the industrialized world should be reduced by 50% during the coming 30-50 years. This design study illustrates how this goal can be achieved in a commercial building with additional qualities of good visual and thermal comfort.
The design was in May 1989 awarded a first price in the CEC architectural design competition "Working in the City".

1. DESIGN CONCEPT

Commercial buildings are traditionally much dependant upon artificial lighting and large mechanical ventilation and cooling systems.
In such buildings, the glazing area of the facades is a key factor in the process of minimizing the total energy consumption of the building.
The diagram illustrates primary energy use for heating, cooling and lighting, dependent on the size of the windows. The diagram is part of the technical information package, provided by CEC for the Competition.
The diagram illustrates two significant features.

1. Optimum glazing area for heating or cooling or daylighting differs significantly.
2. The benefits from utilization of natural daylight has a very strong impact on the optimum window area, when considering the total fossil energy use of the building.

It can be seen, that optimal utilization of passive solar energy, daylighting and natural ventilation strategies can only be achieved via an integrated design approach of the building.

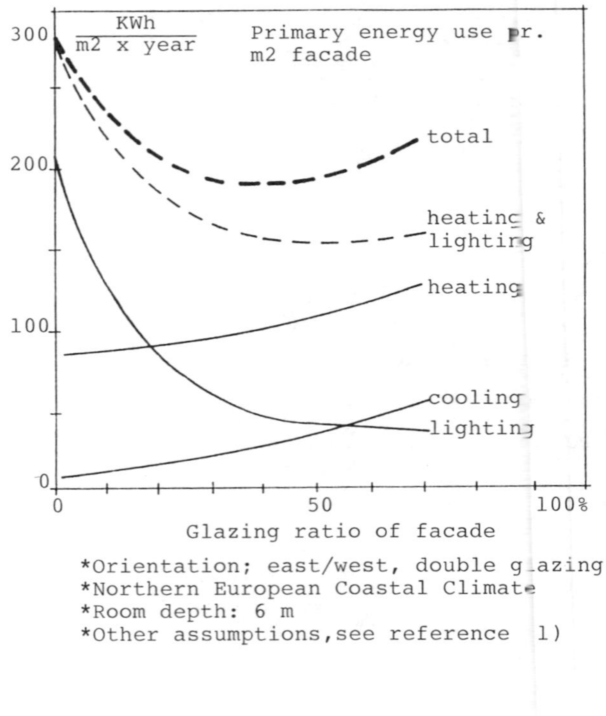

*Orientation; east/west, double glazing
*Northern European Coastal Climate
*Room depth: 6 m
*Other assumptions, see reference 1)

In the actual case the window area cover 30% of the external facades, measured as transparent glass area. Furthermore, the design includes two glass covered atria, three glazed stair wells and a glazed archade, all those elements acting as buffer spaces.

Section, first floor

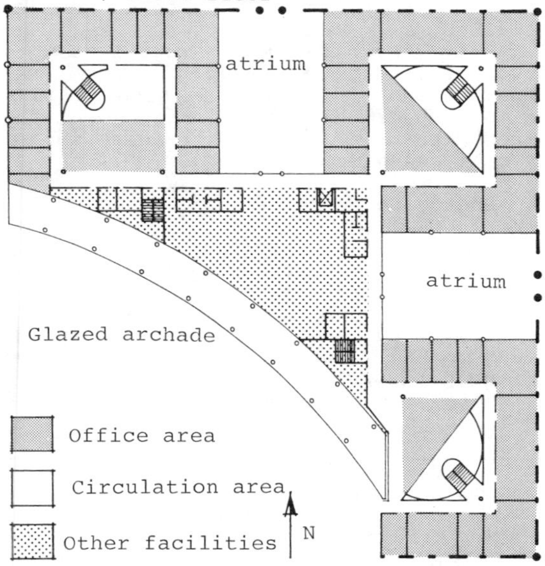

The glazed spaces are open in the three stories, this allowing natural daylight and passive solar energy to penetrate deep into the building.

2. MAIN DESIGN FEATURES

Maximum utilization of natural daylight in the atria means, that the window area at ground floor level is larger, in this case 60% of wall area.

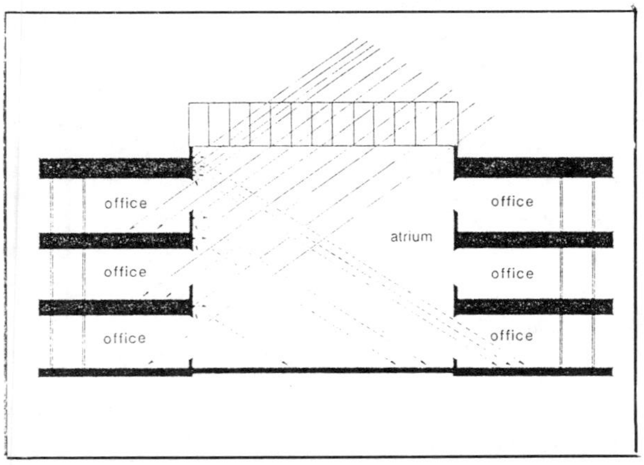

The smaller window area at the upper level, 40% of wall area, leaves a larger opaque wall area, which reflects daylight lower into the atria. The result is a more even distribution of daylight into the offices on each floor.

Ventilation is provided via a displacement ventilation system, which injects preheated fresh air centrally in the building. Exhaust air is drawn out of the building via roof louvres in the three-storey atria and in the glazed archade, utilizing the chimney effect as driving force.

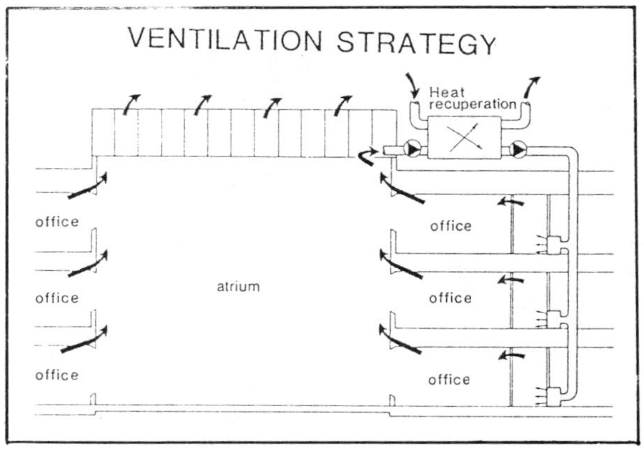

During wintertime, the roof louvres are closed, and the atria acts as buffer spaces. Ventilation air is then drawn from the atria via a fan-driven heat recuperation unit.

3. DESIGN ANALYSIS

3.1 Daylighting

In order to analyse the daylighting qualities of the design, a 1:20 model of part of the building was built.
The daylight efficiency is characterized by the daylight factor = measured light level indoors divided by measured light level outside the building. The figure illustrates the distribution of the daylight dactor on first floor, as measured in the model outside under diffuse sky conditions.

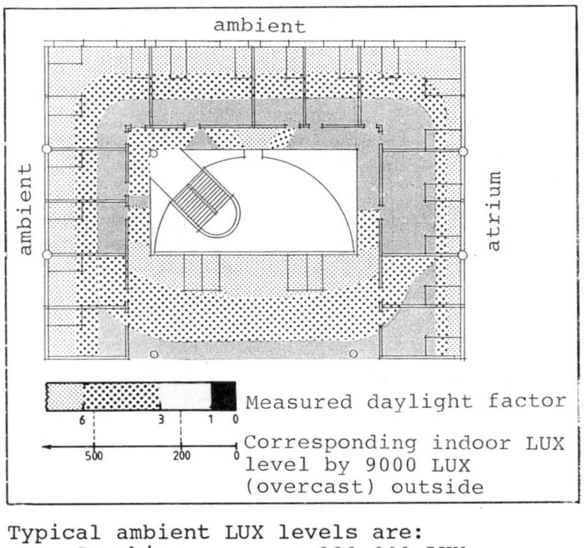

Typical ambient LUX levels are:

Sunshine	:	120.000 LUX
Light overcast	:	18.000 LUX
overcast	:	9.000 LUX
dark overcast	:	3.000 LUX

In offices, a generel light level of 200 LUX is required, whereas 500 LUX is required on the desks.

With overcast skies, 9000 LUX, the 200 LUX treshold is achieved in areas where the daylight factor exceeds 200 LUX/9000 LUX = 2.2%. This is seen to be the case in most parts of the offices.
The 500 LUX level for the desks requires a daylight factor larger than 5.6 to satisfy the requirement on an overcast day. This threshold is exceeded for almost all desk positions illustrated in the diagram.

It can be concluded, that natural daylight will cover the total light requirement in most days, and artificial lighting will only be needed on very dark days and during those hours of the winter days, when the sun is not up.
The detailed calculations shows, that on an annual basis, natural daylight covers 76% of the lighting required during working hours 0900-1900, this period being the standard working day assumed in the Competition.

3.2 Orientation of offices

Utilization of passive solar energy is influenced by orientation, where south orientated windows are normally regarded as optimal.
The actual building has offices towards north and east, atria with horizontal glazing, and the archade has glazing towards south-west. The influence of different orientations for this building was investigated, as shown on the figure. The SUNCODE computer simulation model was used for the purpose.

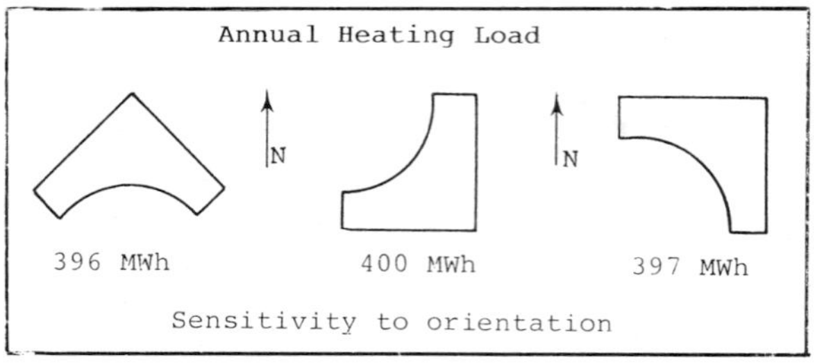

It is seen, that the actual orientation selected, shown to the right, has a heating load almost identical to the two other alternative. This orientation is preferrable, since the offices are then orientated towards east and north, and overheating problems are then easier to control.
Since overcast skies are critical for daylighting, north and east orientated offices are also acceptable when considering daylight utilization.

3.3 Indoor climate

Different measures were investigated to control overheating in the offices during summertime.

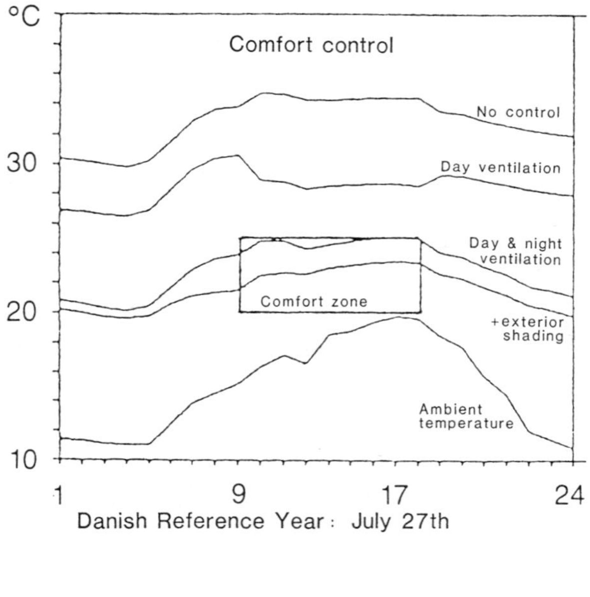

The TSBI computer simulation model from the Danish Building Research Institute was used for that purpose, and on the chart below, calculated room temperatures are shown for an east orientated office on July 27th in the Danish Reference Year.

The acceptable temperature variation during working hours 0900 to 1700, is 20°C-25°C.
It can be seen, that night ventilation is very efficient to bring down the temperature in the room. External blinds are added to further bring down the temperature on this typical Danish summer day with sunshine.

3.4 Energy Consumption

One of the prime objectives was to achieve a substantially lower use of fossile fuels for heating, cooling and lighting, yet maintaining a good indoor climate.
The achievement regarding energy use is shown in the figure below, through the different phases of the building design. The energy consumption is calculated using the LT calculation method (1).

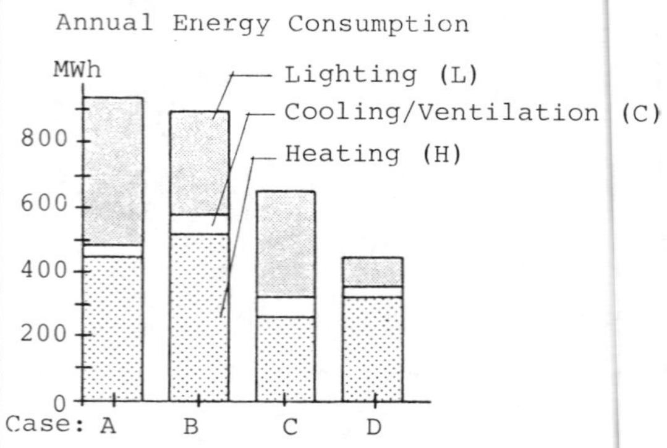

A: Traditionel building with 20% of the facade glazed.
B: Actual building with 30% of the facade glazed, with glazed stairwells etc., but without glazed atria.
C: as B, with glazed atria.
D: as C, with efficient lighting control system and with actual fan assisted natural ventilation system.

Calculation of primary energy use

L: Electricity for lighting = 3,7* x use in building

C: Electricity for cooling =

$$\frac{\text{cooling load}}{2,25} \times 3,7^*$$

where 2,25 = COP of cooling plant

H: Oil for heating =

$$\frac{\text{heating load}}{0,70} \times 1,05^{**}$$

where 0,70 = boiler plant efficiency

* The factor 3,7 reflects an anticipated coal use at the power plant, which rated in thermal value is 3,7 times the electricity use in the building.
**The factor 1,05 reflects, that a 5% waste on oil is anticipated before used in the boiler.

All energy uses in the building are transferred to equivalent use of primary fossil fuel. It is therefore possible to minimize the use of fossil fuel without anticipating any present costs for electricity and oil.
It can be seen that the complex building shape with a large facade area, case B, means an increased heating load, but a decreased electricity load for lighting, compared to the base case, A.
The buffer effect of the glazed atria is very significant, which is illustrated by the substantial reduction in heating load going from case B to C.
The benefits of a good daylight design can only be gained, if an automatic dimming system for artificial lighting is installed. This is illustrated by the large reduction in electricity use for lighting, going from case C to the final design, case D.

It can be seen, that this radical reduction in electricity use for artificial lighting means on increased heating load, since there is now not so much "free heat gain" from lighting. However, the total use of fossile fuels is reduced.

4. ECONOMY

An estimate of the extra cost for glazing the atria shows, that these costs can be matched by savings on the costs of the facades towards the atria. Those facades can be produced cheaper, since they are not exposed to rain, snow or frost.
The extra costs for the system for automatic dimming of artificial lighting is payed back as electricity savings through very few years, and the fan assisted natural ventilation system is estimated to be cheaper than a conventional system, which can maintain the same indoor comfort.

5. APPLICATIONS

This design shows that low consumption of fossil fuels can be achieved in a building with improved thermal and visual comfort, as compared to a conventional office building.
Since the main achievements of the design are the use of natural daylight and natural ventilation techniques, the basic design concepts are applicable under most climatic conditions from Northern Scandinavia to Southern Europe.

6. CONCLUSIONS

The conclusions of the study can be summarized in the following statements:
*All working areas and circulation areas receive natural daylight.
*Good visual comfort and a stimulating working environment is expected.
*Electricity for lighting is reduced to only 20%, compared to a traditional office building.
*Total primary energy consumption is reduced to 50%.

7. REFERENCE

(1) Working in the City
Competition material from CEC DG 12.
Available from School of Architecture, Dublin.

Concours "Working in the city"
1° prix J-P CONAN 11rue Clavel Paris

Ce projet se situe rue des Pyrennées dans le XXème arrondissement de Paris. Il utilise des techniques solaires passives en éclairage,en chauffage et en ventilation.Supportant et profitant de son contexte,ce batiment s'integre dans la ville.La forme qui resoud cette integration dissocie les elements fonctionnels permanents des fonctions caracteristiques du programme.Articulée de l'interieur par un dispositif lumineux,elle donne un sens au batiment tant en plan qu'en élévation.Utilisé principalement de jour,l'institut de langues,orienté plein sud est un capteur energetique autonome

Une fois les potentialités du site etudiées,le projet se situe dans sa conception entre 2 pôles,Le côtexte urbain(sens externe du projet)et le systeme passif(sens interne du projet).Une efficacité conjointe a été recherchée,en utilisant pour cela une solution simple:un batiment lineaire, ce qui pour le tertiaire est approprié.Le parti demandé etait la resolution du probleme urbain et du probleme solaire en une seule et même forme.Ce travail sur la linéarité met en evidence le fondement du plan par rapport au site,mais aussi la dominance de la coupe N/S pour l'efficacité thermique et lumineuse,que des calculs viennent verifier à posteriori. L' implantation urbaine etant resolue autour de la contrainte de l'exposition des classes au sud,au dela du programme fonctionnel,une strategie interne au batiment peut aider à prendre les decisions thermiques et lumineuses.La totalité des facteurs interviennent dans la constitution progressive du projet comme des elements de confort.Il convient donc d'etablir une hierarchie de rapports propres au contexte tendants vers l'unité formelle et la valorisation mutuelle de tous les espaces travaillés.On peut ainsi decider que la lumiere ou la chaleur soit partagée par certains espaces dès le stade fonctionnel, cela donne un organigramme passif.

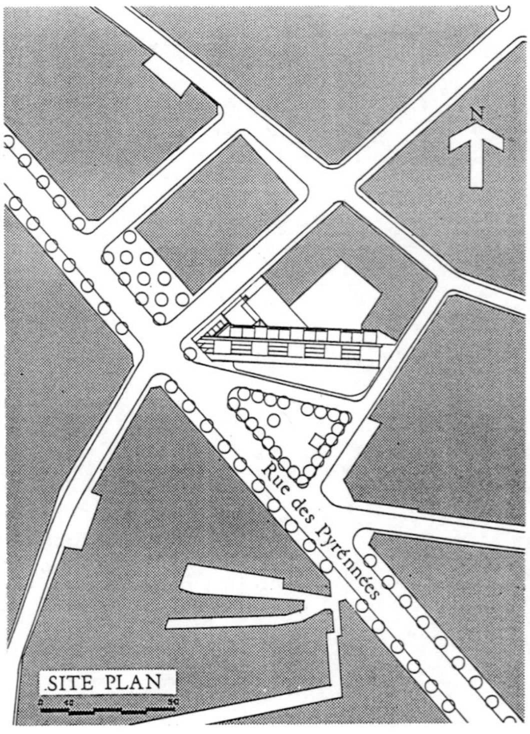

Le corps principal du batiment est composé de 4 serres qui chauffent et ventilent les parties communes,éclairent zenithalement le plan profond de la bibliotheque du rez-de-chaussée ainsi que les salles de classe qui les separent.Ces dernières presentent au sud des murs-Trombe isolés interieurement et reversibles.La grande paroi unifiant verticalement les circulations horizontales est capteur de chaleur tout comme les cages d'escalier situées dans les serres.Dans cette solution les espaces servants sont thermiquement autonome,cela pour des raisons accoustiques.Tres vitré le corps linéaire du projet possède à lui seul 80 % du vitrage total, il est presque entièrement vitré.Une efficacité maximale est viséee ammenant à faire varier trés localement le systeme de circulation d'air en fonction de l'ensoleillement et de la temperature exterieure.Des clapets sont actionnés par un système de controle d'environnement assisté par un programme informa -tique adapté.

Quelques chiffres d'aprés la methode LT:

pour 3965 m2 de programme on a en ECLAIRAGE:200KWh/m2 an
CHAUFFAGE: 90KWh/m2 an
VENTILATION:20KWh/m2 an pour la salle à gradins

On a 80% de zone passive

La consommation nette par m2 est de o,22 kwh/m2 pour une consommation totale annuelle de 793000 KWh/an

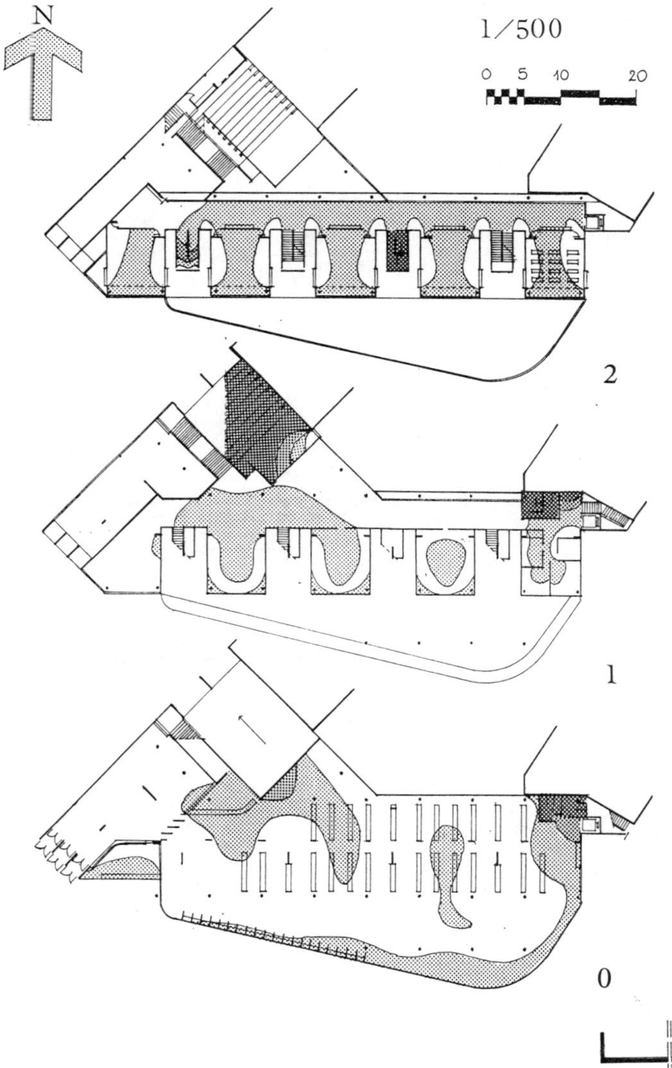

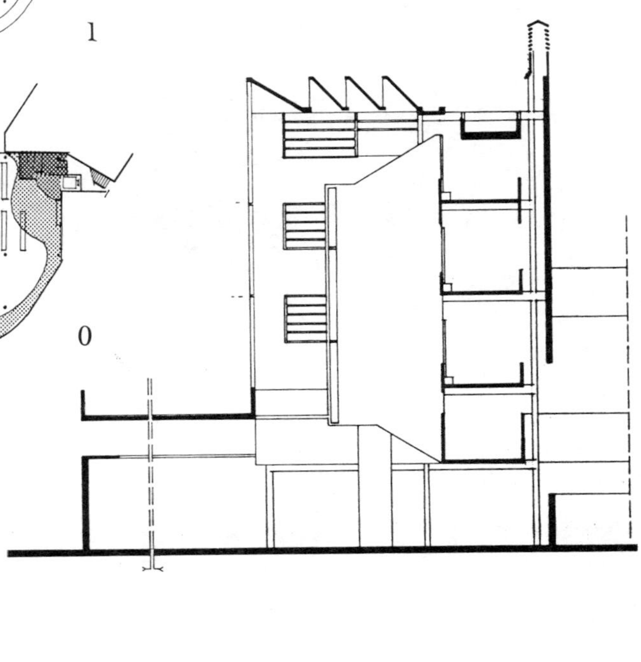

Le batiment,pour s'implanter dans la ville doit etre conçu comme un double espace servi,l'un constitué d'éléments caracteristiques du programme qui peuvent devenir obsolètes,l'autre d'un élément permanent par ses réamenagements possibles et qui est le principal acteur de la qualification du site.Mais l'ensemble site/projet tire ses qualités d'un troisième élément qui les met en relation avec le contextequ'il contient et ou il s'exprime à la fois.C'est l'element relationnel qui est ici principalement d'espaces servants,et determine la spatialité.L'unité du projet ,du batiment au site,reside donc dans 3 niveaux de composition -1-urbain

-2-relationnel

-3-local ou typique

La partie linéaire du batiment donne une mesure au batiment,par sa frontalité et son rythme sur la place triangulaire où il domine les arbres.Il crée aussi un arrière en ouvrant le paté de maisons sur la placette que l'element relationnel,anime et invite à entrer.Le "typique" est representatif du programme dans son rapport au reste du batiment par ses caracteristiques formelles.Les3 niveaux permettent,dans la hierarchie du rapport contextuel,le developpement stable de la spatialité interne au projet.Ils sont tous les trois présents sur la facade principale,et cela montre que par dela le plan;l'unité se fait en elevation.

La structure du projet se constitue sur des critères de/:
TEMPS (permanence ou non de parties de l'edifice
FONCTION (rapport servant/servi)
FORME(composition architectonique et lumineuse)
Le confort thermique est simultanement dependant de ces trois facteurs.

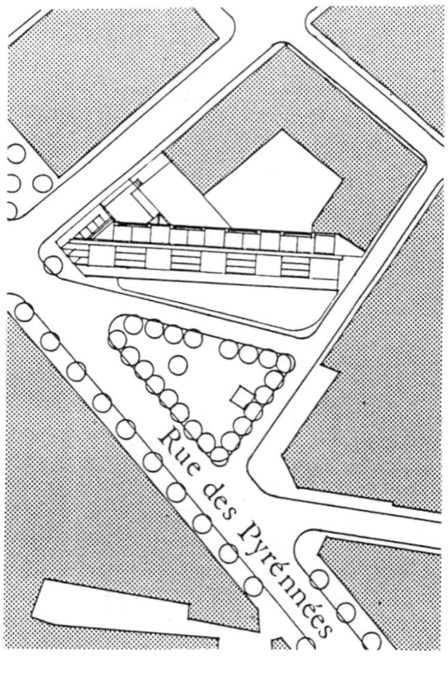

BUILDING 2000, A PILOT PROJECT OF THE COMMISSIONS'S NON-NUCLEAR ENERGY R&D PROGRAMME: SOLAR ENERGY APPLICATIONS TO BUILDINGS

C. den Ouden
EGM engineering bv, P.O. Box 1042, 3300 BA Dordrecht, The Netherlands

ABSTRACT

The main aim of the project "Building 2000" is to stimulate the use of passive solar technologies for the design and construcion of buildings. Within "Building 2000" a call for proposals was published inviting companies, developers and authorities which plan to onstruct a non-domestic building with a high multiplication value e.g. office building, hotel, school, hospital, etc. to participate in this pilot project. In case the plans for such buildings were open for the inclusion of passive solar features, other energy efficient measures such as using innovative controls and modern daylighting features which save energy and can improve human comfort, a possibility for design assistance from the commission is offered to benefit from build-up experience within the sub-programme "solar energy sapplications to buildings". 35 proposals were selected to participate in this pilot project "Building 2000". It is expected that by offering support in the form of the provision of design packages or by free consultations of design support units having close liaison with European experts on the various fields (passive solar, daylighting, etc.), the buildings designed under this pilot programme will become an example for multiplications in the near future.

1. INTRODUCTION

Within the third four-year R&D-programme (1986-1989) of the Commission of the European Communities in the field of "Solar Energy Applications to Buildings" a number of concerted actions are in progress; for details see (1). Within these concerted actions several European R&D-teams collaborate on R&D-topics like::

1. Passive Solar Urban Design,
2. Modelling and Simplified Design Tools,
3. Testing of Passive Solar Components,
4. Control Systems for Passive Solar Buildings/Houses,
5. Comfort/Air Flow Studies,
6. Daylighting

A need for technology transfer from these R&D activities to the design practitioners has already been identified during the second four-year R&D-programme (1981-1985) by initiating the project Monitor, see (2) and (3), and intensified by initiating the projects Solinfo and Archisol; see (4). This sub-programme "Building 2000" aims at stimulating the use of Passive Solar Buildings in Europe by encouraging design teams of larger non-domestic buildings to design their future buildings with the help of built-up experience by the Experts of the European Passive Solar Research Community. It is to be expected that by bringing together the design teams of these buildings and the European Passive Solar Experts the technology transfer will be encouraged.

In this paper a review is given of the category of buildings participating within the 'Building 2000' programme, the possibilities of design support from the R&D-community to the Design teams, the workplan and organisation of Building 2000 and the expected results from Building 2000, final products, brochures, etc.

2. PROJECTS PARTICIPATING IN THE PILOT PROJECT BUILDING 2000

Within Building 2000 in total 35 projects are participating. These are all projects in the design phase of which the probability of realisation is already secured or almost certain. The total cost for the realisation of these future buildings is approximately 140 Million ECU. The variety in building types reflects the built environment of Europe since there are schools (7), office buildings (3), public buildings (6), hotels and holiday complexes (6), laboratory and university buildings (8) and sport and educational centres (5) represented.
It is to be expected that when all these buildings are commisioned, their average yearly energy consumption - compared with traditional buildings without passive solar and energy conscious measures - will be significantly reduced (30 - 50%). In tables 1 & 2 a review of the buildings (types, countries, estimated building costs and floor area) is given. See for more detais 5. and 6.

Table 1. Details per building category

Type of building	no.	floor area (m^2)	cost (ECU)
Schools	7	1,011,800	16,090,000
Office buildings	3	42,500	41,550,000
Public buildings	6	15,500	12,150,000
Hotels & holiday complexes	6	24,000	22,963,000
Laboratory & University	8	81,400	45,311,000
Sport & educational centres	5	7,900	5,046,000

Table 2. Details per participating country

country	participants	category *)					
		1	2	3	4	5	6
B	1					1	
D	6			1		3	2
E	4	1	1		1		1
F	5	1		1	1	1	1
GB	6	3		2			1
GR	6	1			4	1	
IRL	1			1			
NL	2		2				
P	4	1		1		2	
Tot.	35	7	3	6	6	8	5

category *) 1 = Schools
2 = Office buildings
3 = Public buildings
4 = Hotels & holiday complexes
5 = Laboratory & University
6 = Sport & educational centres

3. THE ORGANISATION OF BUILDING 2000

For the realisation of one of its important objectives, viz. the inclusion of passive solar measures and energy consciousness in actual designs during the various design phases of a number (35) of large non-domestic buildings, there is a need for an effective organisational structure. In figure 3.1 a schematic representation is given of Building 2000's organisational structure.

Figure 3.1 Schematic review of the Building 2000 organisational structure

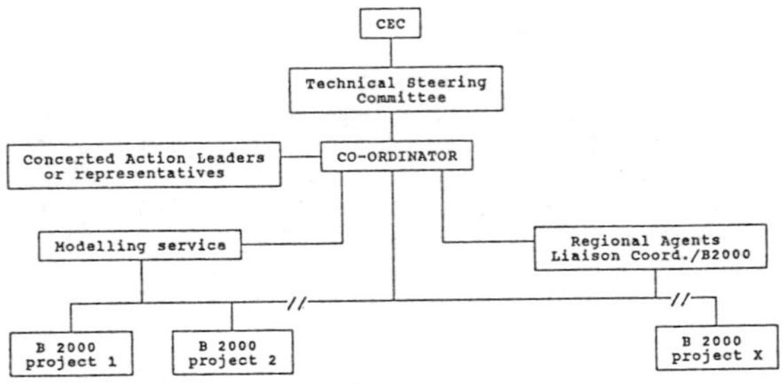

note: X = 35, known at 28.11.1989

From figure 3.1 one can see that - under the guidance of the Commission's Technical Steering Committee - the role of the Coordinator and the Regional Liaison Agents, within the pilot project Building 2000, is essential.
The task of the Regional Liaison agents is to investigate, for a number of Building 2000 projects in their region, the need for design support.

The design (advice) support should be in line with the general objectives of the Building 2000 programme, which means that the support to be given by European Experts or National Design Support Experts has to be with regard to passive solar and related topics. The Liaison Agent will have to define, together with the Building 2000 project teams, the need for design (advice) support.

In principle three different types of design (advice) support activities are possible:

- a. The consultation of an individual European Passive Solar Expert or a National Design Support Expert on Passive Solar and related topics.
Note: Since several of the participating projects are already in the detailed design phase this seems to be the only effective way to improve the design on certain component detailing and/or minor changes in the passive solar and daylighting aspects already integrated in the design concept.
- b. To carry out a small case-study by a recognised Design Support Unit having experience with certain passive solar/daylighting design tools. In this case-study a sensitivity analysis for certain design parameters to be subject to change could be made so as to build up to the selection of the final size of these parameters.
Note: This seems to be of value for those projects in which the preliminary design phase has not yet been concluded.
- c. To conduct a more detailed modelling study. In several Building 2000 projects detailed simulation models might have to be used to provide answers to specific design aspects, such as air movements in atria, approaches to passive cooling, etc.
Notes: When these requirements are identified to the coordinator he will contact the right modelling expert to conduct a more detailed study on a subject preferably of use to a group of similar Building 2000 projects.
- Further detailing of the Building 2000 projects has to proceed and may not or seldom directly profit from the results of such 'horizontal'-studies; however it is to be expected that the results of these studies will be of value for future designs, especially when the results are translated into design guidelines and simplified design tools are developed from these 'horizontal'-studies. After defining the design (advice) support by the Regional Liaison Agent for the Building 2000 projects in his Region he will communicate this need to the Building 2000 coordinator. After approval by the coordinator (who will consult the Technical Steering Committee) the support activities (a, b and/or c) can start. The Liaison Agent and the Coordinator have to keep in touch with the Building 2000 project teams and the involved experts, design support units; and the modelling experts in order to ensure a correct follow-up of the design (advice) support activities as described under a, b and c.

4. TIME SCHEDULE OF BUILDING 2000

This concerted action has officially started on 1.1.1988 and will last 27 months till 1.4.1990. In figure 4.1 the number of working meetings within Building 2000 is given. On top of these meetings many individual meetings between the Regional Liaison Agents and the participating Project Design Teams will be held. Also meetings of the Coordinator and various experts have been and will be planned.

Figure 4.1 Review of various meetings planned in B.2000

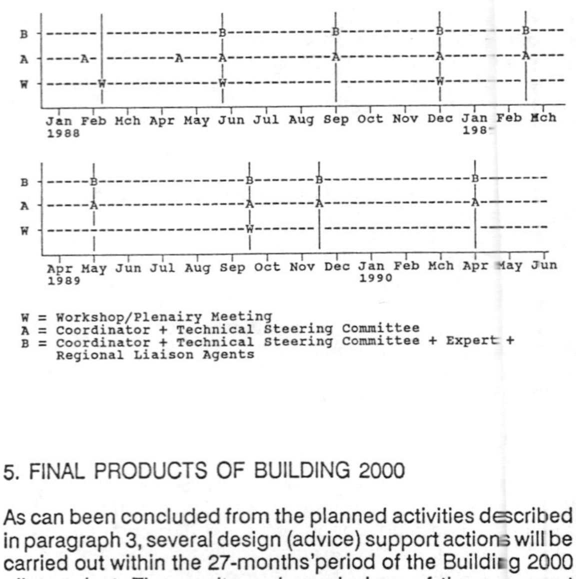

W = Workshop/Plenairy Meeting
A = Coordinator + Technical Steering Committee
B = Coordinator + Technical Steering Committee + Expert + Regional Liaison Agents

5. FINAL PRODUCTS OF BUILDING 2000

As can been concluded from the planned activities described in paragraph 3, several design (advice) support actions will be carried out within the 27-months'period of the Building 2000 pilot project. The results and conclusions of these support actions will have to be reported and summarised in the coordinators' final report to the Commission. It is therefore to be expected that the results of the effectiveness of individual expert consultations shall become known as seen from the point of view of the design-teams within Building 2000 (= conclusions from support activity mentioned in 3 under item a.). Also the results and conclusions from support activity b (= evaluations from the various case studies), and the spin-off in the several Building 2000 projects for which these case-studies have been prepared will be reported both by the Design Support Units who have to carry out these case studies, and the coordinator, when evaluating these results. Furthermore the results and conclusions from the activities described in paragraph 3.c will be reported.

This will be the responsibility of the modelling expert who will have to report his findings to the coordinator. Finally it is foreseen in the Building 2000 pilot project that a coherent documentationset of all the projects participating in "Building 2000" will be prepared and widely distributed in order to ensure the realisation of more and better passive solar buildings in the future. In figure 5.1 a schematic review of the planned final products, resulting from the Building 2000 pilot project, is presented.

Figure 5.1 Schematic review of final products

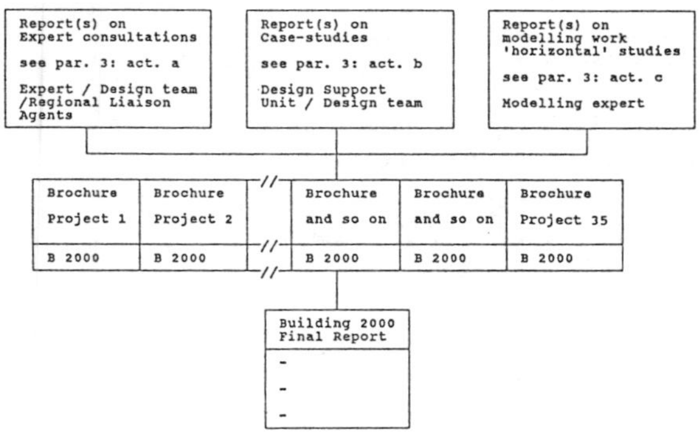

Figure 6.1

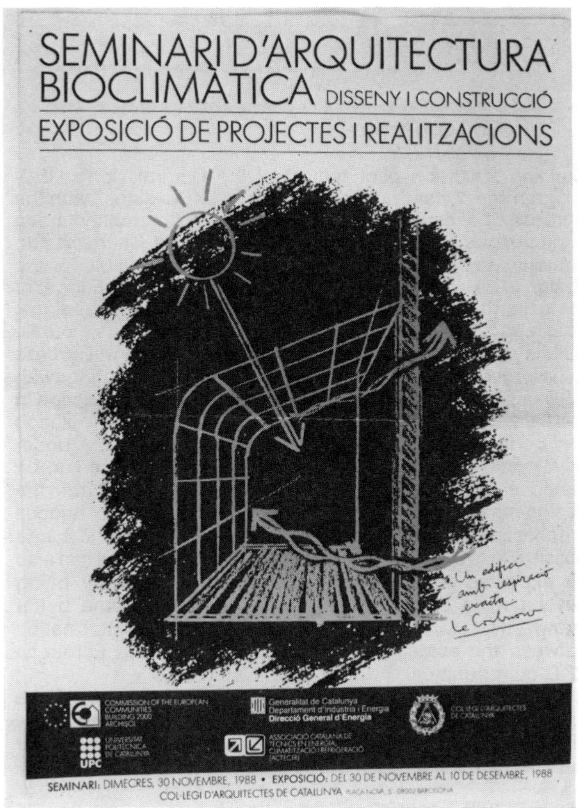

Figure 6.2

6. PROGRESS OF WORK UPTILL NOVEMBER 1989

During the period between the start of Building 2000 (officially 1.1.1988, but effectively since May 1988) and November 1989 a wide range of Design Support activities has been carried out for almost all Building 2000 projects (only a few projects were delayed and, therefore, design support was not useful).

All three indicated possibilities of design support in paragraph 3, namely:

a. individual consultation between design team and selected European and/or national experts;
b. small case studies by recognised Design Support Units, experienced with certain passive solar/daylighting design tools;
c. detailed modelling studies for groups of similar Building 2000 projects;

have been exploited by the design teams. Moreover, several other activities were organised during the last periodo of eighteen months, like:
- the organisation of several workshops, see figurue 6.1;
- the organisation of a few exhibitions, see figure 6.2.

At this moment (November 1989) the situation with respect to the preparation of the final products of Building 2000 as schematically shown in figure 5, can be summarized as presented in figure 6.3. In this figure an overview is presented of the number of reports already finalized and the number of draft texts for the various brochures received so far.

Figure 6.3 Products November 1989

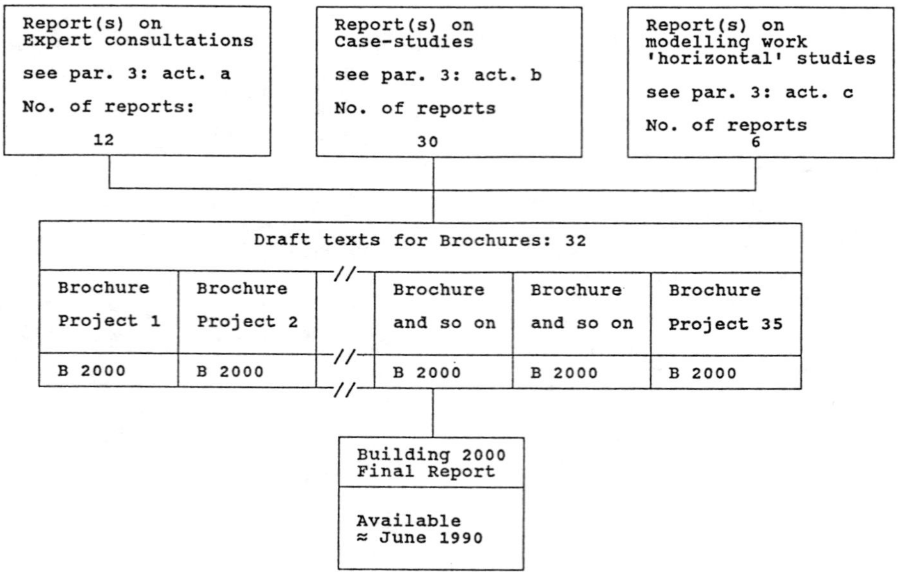

7. CONCLUSIONS

Building 2000 is a pilot project of the Commission's R&D-programme 'Solar Energy Applications to Buildings', with the purpose of encouraging the adoption of solar architecture in large buildings currently in the planning stage. There was an enthousiastic response of project teams, responsible for the design of 35 large buildings with a total construction budget of more than 140 Million ECU, including schools, office buildings, hotels and holiday complexes, university buildings, public buildings and sport and educational centres. The willingness to improve their building concepts by collaborating with R&D-experts was encouraging for the Commission's action in this field. It is to be expected that the design aids and simulation tools, daylighting concepts, comfort criteria, etc. under development in the Commission's R&D-programme can be tested on their practical value. Moreover the feed back from the design practitioners will allow the Commission to refocus emphasis in the R&D-programme on more practical and possibly less theoretical problems and approaches. It can also be expected that alternative design aids currently being developed in other European groups shall become better known. This will result in better exchange of information between the actual design practitioners and the European R&D-cummunity.
A coherent information set (series of brochures) of the participating Building 2000 projects will be prepared and widely distributed in order to ensure the realisation of more and better passive solar buildings in the future.

Within Building 2000 "Science and Technology at the Service of Architecture" became real practice. This was not only realised by the various support activities initiated by Building 2000 but also by the active exchange of ideas by architects and design team members with R&D-workers during the four workshops held between May 1988 and September 1989.

References:

(1) Steemers, T.C. Solar Energy Applications to Buildings and Solar Radiation Data. Proceedings of the EC Contractors' Meeting, 13 and 14 November 1986. Published by D. Reidel Publishing Company for the CEC.
(2) Turrent, D. Godoy, R. and Ferraro, R.
'Performance Monitoring of Solar Heating Systems in Dwellings' Dec. 1981 ECD Londen
(3)Turrent, D. 'Project Monitor: Aims and Objectives' Proceedings of the 1987 European Conference on Architecture, p.483-486 edited by W. Palz, CEC and Published by H.S. Stephens & Associates
(4)Lewis, J. Owen 'SPREADING THE CREED OF ENERGY EFFICIENCY IN EUROPE' Proceedings of the 1987 European Conferences on Architecture, p.622-626, edited by W. Palz, CEC and Published by H.S. Stephens & Associates
(5) den Ouden, C. "Building 2000, Newsletter, published by EGM engineering bv, Dordrecht, The Netherlands
(6) den Ouden, C. "Building 2000, proceedings of the Euroforum New Energies, an international congress held in Saarbrucken, F.R. Germany, 1988, vol. 1, p.143-146, published by H.S. Stephens & Associates.

ALPHABETICAL LIST OF AUTHORS

With the following paper numbers refer to the list of contents at the beginning of the book